T0317475

Trace Quantitative Analysis by Mass Spectrometry

Trace Quantitative Analysis by Mass Spectrometry

Robert K. Boyd
National Research Council, Ottawa, Canada

Cecilia Basic
Basic Mass Spec., Winnipeg, Canada

Robert A. Bethem
Alta Analytical Laboratory, El Dorado Hills, CA, USA

John Wiley & Sons, Ltd

Other Wiley Editorial Offices

John Wiley & Sons Inc., 111 River Street, Hoboken, NJ 07030, USA

Jossey-Bass, 989 Market Street, San Francisco, CA 94103-1741, USA

Wiley-VCH Verlag GmbH, Boschstr. 12, D-69469 Weinheim, Germany

John Wiley & Sons Australia Ltd, 42 McDougall Street, Milton, Queensland 4064, Australia

John Wiley & Sons (Asia) Pte Ltd, 2 Clementi Loop #02-01, Jin Xing Distripark, Singapore 129809

John Wiley & Sons Ltd, 6045 Freemont Blvd, Mississauga, Ontaria, L5R 4J3, Canada

Wiley also publishes its books in a variety of electronic formats. Some content that appears in print may not be available in
electronic books.

Library of Congress Cataloging in Publication Data

Boyd, Bob, 1938-
    Trace quantitative analysis by mass spectrometry / Bob Boyd, Robert Bethem, Cecilia Basic.
        p.   cm.
    Includes bibliographical references and index.
    ISBN 978-0-470-05771-1 (cloth : alk. paper)
    1. Mass spectrometry.   2. Chemistry, Analytic—Quantitative.   I. Bethem, Robert.   II. Basic, Cecilia.   III. Title.
    QD272.S6B69 2008
    543′.65—dc22

                                                                                                        2007046641

British Library Cataloguing in Publication Data

A catalogue record for this book is available from the British Library

ISBN Cloth 978-0-470-05771-1

Typeset in 9/11pt Times by Integra Software Services Pvt. Ltd, Pondicherry, India

The authors dedicate this book to all of our mentors and colleagues,
too many to mention by name, with whom we have had the privilege of working over
many years and who have taught us so much.

Contents

Preface

'When you can measure what you are speaking about
and express it in numbers, you know something about
it; but when you cannot express it in numbers, your
knowledge is of a meagre and unsatisfactory kind . . .'
William Thomson (Lord Kelvin), Lecture to the
Institution of Civil Engineers, 3 May 1883.

The discipline devoted to careful measurement of specific
properties of the universe around us is known as
metrology. The famous statement by William Thomson,
quoted above, summarizes the importance of quantitative measurements for the testing of scientific hypotheses;
indeed, without such quantitative testing, it is fair to say
that hypotheses can not be regarded as scientific at all.
Missing from Thomson's comment, however, is a mention
of the importance of careful evaluation of the uncertainties that are present in any quantitative measurements and
the resulting degree of confidence that can be placed in
them and any conclusions drawn from them. These uncertainties are just as important as the 'best' quantitative
measured value itself.

This book is devoted to the science and art of *chemical
metrology*, taken here to mean the quantitative measurement of amounts of specific (known) chemical compounds

present at trace levels (roughly defined as one part
in 10^6–10^{12}) in complex matrices. Examples are drugs
and their metabolites in body fluids, pesticide residues
in foodstuffs, contaminants in drinking water etc. Such
measurements are extremely demanding, and involve the
use of a wide range of apparatus and of experimental
procedures and methods of data evaluation, all of which
must be used properly if reliable estimates of chemical
concentrations and their associated uncertainties are to
be obtained. While this is true of any chemical analysis,
the modern advances in trace-level analysis are critically
dependent on developments in mass spectrometry.

For several decades before its application to chemical
analysis, mass spectrometry was a major tool in fundamental physics. The invention of mass spectrometry is
usually attributed to Joseph John Thomson, no relative to
William Thomson (Lord Kelvin) whose picture appears
above. In 1897 J.J. Thomson measured the ratio of the
charge of an electron to its mass, thus confirming for the
first time that this then-mysterious entity possessed properties characteristic of a particle. (It is interesting that his
son G.P. Thomson later emulated his father by winning a
Nobel Prize, but for demonstrating that the electron *also*
possesses properties characteristic of a wave!). An account
of the life and work of J.J. Thomson was published (Griffiths 1997) to commemorate the centenary of the first
measurement of mass-to-charge of an elementary particle.

F.W. Aston, a student of Thomson, won a Nobel Prize
for using mass spectrometry to demonstrate the existence of the isotopes of the elements (Aston 1919), and
for developing a higher resolution mass spectrometer
that permitted measurement of atomic masses with sufficient accuracy and precision for the first reliable estimates of so-called mass defects, i.e., deviations of actual
(measured) atomic masses from those predicted from the
sums of the masses of the constituent elementary particles (protons, neutrons and electrons). Later, this work
was extended by K.T. Bainbridge whose measurements of
mass defects (Bainbridge 1932, 1936) were of sufficient
accuracy and precision to confirm for the first time the

famous relationship derived by Albert Einstein concerning the equivalence of mass (in this case the mass defect) and energy (in this case the binding energy of protons and neutrons within an atomic nucleus).

The first analysis of positive ions is attributed to Wien, who used a magnetic field to separate ions of different mass-to-charge ratios (Wien 1898), but the first appreciation of the potential of the new technique in chemical analysis appears to have again resulted from the work of Thomson in his famous book *Rays of Positive Electricity and Their Application to Chemical Analysis* (Thomson 1913). The present book is intended as an introduction to the use of mass spectrometry for quantitative measurements of the amounts of specific (known) chemical compounds (so-called 'target analytes') present at trace levels in complex matrices. This modern day meaning of 'quantitative mass spectrometry' is rather different from its much more specialized historical meaning in the earliest days of application of the technique to chemistry.

In the two decades spanning about 1940–1960, the petroleum industry was the major proponent of mass spectrometry as a tool of analytical chemistry, and indeed the first few issues of *Advances in Mass Spectrometry* (essentially the proceedings of the International Conferences on Mass Spectrometry) were sponsored and published by the Petroleum Institute. Raw petroleum and its distillate fractions are incredibly complex mixtures of chemical compounds, mainly corresponding to chemical compositions $C_cH_hN_nO_oS_s$, and it is impossible to devise a complete chemical analysis of such an extremely large number of components at concentrations covering a dynamic range of many orders of magnitude. However, some knowledge of chemical composition is required by chemical engineers for optimization of the industrial processes required to produce end products with the desired properties. To this end petroleum chemists devised the concepts of compound *class*, i.e., compounds with a specified composition with respect to heteroatoms only ($N_nO_oS_s$), and compound *type*, i.e., compounds with a specified value of Z when the composition is expressed as $C_cH_{2c+Z}N_nO_oS_s$. Clearly the parameter Z is related to the degree of unsaturation. Reviews of this application of mass spectrometry have been published (Grayson 2002; Roussis 1999). The earliest methods yielded information on relative amounts of hydrocarbon compounds in a distillate, i.e. *type* analyses for the compound class with n = o = s = zero. A high resolution adaptation of the original low resolution mass spectral methods was first published in 1967, and permitted determinations of 18 saturated- and aromatic-hydrocarbon types and four aromatic types containing sulfur.

Essentially, the general approach first identified specific mass-to-charge ratio (m/z) values in the electron ionization mass spectra that are characteristic of each compound type, and obtained a calibration based on analysis of mixtures of known composition:

$$S = R.C$$

where S is a vector containing the appropriate sums of signal intensities at the m/z values that are characteristic for each compound type, C is a vector whose elements are the concentrations of these compound types, and R is the (square) matrix of mass spectrometric response factors determined from the calibration experiments (average response coefficients for each type are on the diagonal of the matrix and the off-diagonal elements take into account inter-type contributions to signal intensities at the characteristic m/z values). Quantitative analysis of an unknown thus requires inversion of the response matrix:

$$C = R^{-1}.S$$

An example of such a type analysis for the class $C_cH_{2c+Z}S$, for both a raw petroleum feedstock and one of its products from a refinery process designed to remove the sulfur content, is shown in Figure P.1 (here the carbon number c is replaced by n). Modern developments in quantitative petroleum analysis incorporate new advances in separation science as well as in mass spectrometric technologies, especially ionization techniques and ultra-high resolving power (Marshall 2004) and improved calibration of the response matrix (Fafet 1999).

This early and narrow interpretation of the phrase 'quantitative mass spectrometry' is now badly outdated, although petroleum analysis is still an important branch of analytical chemistry that is still being developed. However, apart from its historical importance it introduced important concepts, including calibration and response factor, that will appear throughout this book. Nowadays, quantitation by mass spectrometry generally refers to determination of target analytes (known and specified chemical species, rather than groups of compounds defined within 'classes' or 'types' as in the petroleum case), present in a complex matrix at trace levels. This book is intended as an introduction to this demanding branch of measurement science, one that is crucial for meaningful studies of a wide range of phenomena including environmental, pharmacological and biomedical studies.

The approach adopted throughout the book is to emphasize the fundamentals underlying the scientific

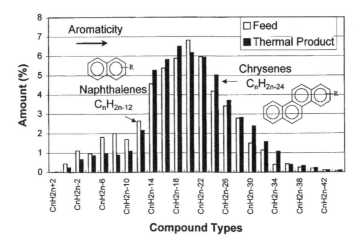

Figure P.1 Type analysis of a petroleum feedstock and its hydrocracked product for compound class $C_nH_{2n+Z}S$. Reproduced with permission from S. Roussis, *Rapid Commun. Mass Spectrom.* **13**, 1031–1051 (1999).

instruments and methodologies, illustrated by historically important developments as well as innovations that were current at the time of completing the manuscript (late summer 2007). Hopefully this will prove to be of more lasting value for the reader. However, a discussion of 'fundamentals' without any description of how 'the fundamental things apply' (see text box) to real-world problems is unlikely to be of much use to anyone, so the final chapter discusses some illustrative examples from the literature.

As mentioned above, although this book is devoted to *quantitative* analysis of specified 'target' analytes, the analyst must have a degree of confidence that the signals being measured do indeed arise from the presence of that target analyte (confirmation of analyte identity), and only from that analyte (signal purity). Therefore, even quantitative analyses inevitably involve some degree of confirmation of analyte structure and identity, and also a check for potential contributions to the recorded instrumental signals from other compounds. The degree to which such checking of analyte identity and of signal purity is necessary will vary from case to case, e.g., analysis of a synthetic pharmaceutical drug in blood plasma following a clinical dose is much less likely to require a high degree of identity confirmation than that of a chlorinated pollutant in an environmental sample. Determination of the appropriate degree of such checks is an example of application of the concept of '*Fitness for Purpose*', another major theme of this book. This principle is discussed more fully in Chapter 9.2, but will appear in several intervening chapters so a very brief introduction to the concept is presented here, based on a discussion (Bethem

2003) of its applicability to mass spectrometric analyses. The following principles are quotations from this work (Bethem 2003):

1. Ultimately it is the responsibility of the analyst to make choices, provide supporting data, and interpret results according to scientific principles and qualified judgment.
2. Analysts should use methods which are Fit for Purpose. Analysts should be able to show that their methods are Fit for Purpose.
3. Fitness for Purpose means that the uncertainty inherent in a given method is tolerable given the needs of the application area.
4. Targets for measurement uncertainty describe how accurate and precise the measurements need to be. Targets for identification confidence describe how certain one needs to be that the correct analyte has been identified.
5. Establishing method fitness consists of showing that the targets for measurement uncertainty and identification confidence have been met.

In its simplest terms, the 'Fitness for Purpose' principle corresponds to the commonsense notion that an analytical method must provide answers with sufficiently low uncertainties that the requirements of the user of the data are fully met within specified constraints of time, cost etc. On the other hand there is no point in developing, validating and using an analytical method with extremely low uncertainties in the data (high precision and accuracy) if a considerably less demanding method (generally less expensive in terms of money, time and effort) will suffice.

'The Fundamental Things Apply —'

with apologies to 'As time goes by' by Herman Hupfeld (1931); Warner Bros. Music Corp.

This famous song was featured in the films *Casablanca* and *Sleepless in Seattle*, as well as the well-known British TV comedy series of the same name. This is a long stretch from the subject of this book, but the above quotation from the lyrics seems appropriate.

A distinction between 'fundamental' and 'applied' science is drawn by some, including (alas) by some funding agencies! One of the themes of this book is best expressed by quoting one of the giants of 19th century science:

'There does not exist a category of science to which one can give the name applied science. There are science and the applications of science, bound together as the fruit of the tree'. Louis Pasteur, 'Revue Scientifique,' Paris, 1871.

Louis Pasteur

It is not essential for an experimental scientist to be familiar with and understand every detail of the theoretical underpinnings of his/her laboratory work. However, to be able to properly plan an experimental investigation so that the results can be meaningfully interpreted, it *is* essential that he/she should understand the background of the relevant theory, its basic assumptions, and the limits of its applicability and the magnitude of the consequences of the approximations involved. This general theme was a guiding principle in writing this book, and hopefully this approach will ensure that the book will have a reasonably long useful lifetime. However, fundamentals without much discussion of how they apply to real-world problems in trace analytical chemistry are not of themselves very useful, and discussions of how the 'fundamental things apply' will appear later in the book.

A specific example of the continuity between 'applied' and 'fundamental' research, of direct relevance to the subject of this book (see Chapter 5), is provided by the direct line of development starting from an electrostatic paint sprayer designed for industrial use, through attempts by materials scientists to prepare single molecules of synthetic polymers in the gas phase to enable fundamental studies, to the eventual award of a Nobel Prize to John B. Fenn for the invention of electrospray ionization mass spectrometry and its application to biochemistry and molecular biology.

Figure P.2 shows a generalized procedure for achieving Fitness for Purpose.

This book is not intended to cover important branches of mass spectrometry that provide accurate and precise quantitative measurements of *relative* concentrations, e.g., of variations in isotopic ratios of an element by isotope ratio mass spectrometry (IRMS) and accelerator mass spectrometry (AMS). Rather, this book is mainly concerned with determinations of absolute *amount of substance* (see Chapter 1 for a definition and explanation), particularly for compounds present at trace levels in complex matrices. (The only exception is the inclusion of a brief description of methods used to determine *differences* in levels of proteins in living cells or organisms subjected to different stimuli, e.g., disease state vs normal state).

The book covers analysis of 'small' (< 2000 Da) organic molecules, in environmental and biomedical matrices. The first book exclusively devoted to this subject (Millard 1977) is now rather out of date as a result of more recent spectacular advances in mass spectrometric technology. Very recently two excellent introductions to the subject have appeared (Duncan 2006; Lavagnini 2006). The present book differs from these with regard to their respective lengths; the present book is much longer, as a result of the attempt to provide a comprehensive introduction to all the many ancillary techniques and tools that must be coordinated to provide a reliable result for a trace-level quantitative analysis by mass spectrometry. Thus, many of the present chapters discuss matters that are common to any quantitative analytical method, not only to those in which mass spectrometry is the key component providing the final analytical signal

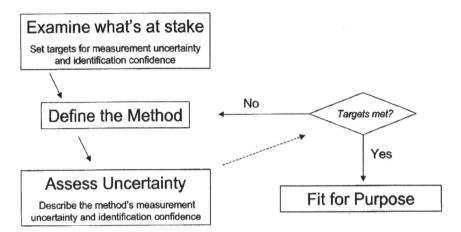

Figure P.2 Outline of a general recommended process for achieving Fitness for Purpose. Reproduced from Bethem *et al.*, *J. Amer. Soc. Mass Spectrom.* **14**, 528 (2003), with permission, copyright (2003) Elsevier.

used to estimate the concentration of the target analyte. The present book is written at a level that presupposes some basic undergraduate-level knowledge of chemistry, physics, and mathematics and statistics.

This book also treats the more recent developments of quantitative analysis of specific proteins in biological systems, even though these hardly qualify as 'small molecules'. However, it does *not* cover the important aspect of analysis of trace level metals by, e.g., ICP–MS; an excellent book covering this subject has appeared recently (Nelms 2005).

It must be emphasized that any book such as this can only be regarded as a preparation for the real learning process in this demanding practical art, namely, exposure to working on real-life problems in a real laboratory. The kinds of measurements that are addressed here really do push the various technologies involved to their current limits, and 'learning by doing' is the only truly meaningful method. This principle is well illustrated by a quotation from what might be described as 'the older literature':

Those who are good at archery learnt from the bow and not from Yi the Archer. Those who know how to manage boats learnt from boats and not from Wo (the

legendary boatman). Those who can think learned for themselves and not from the Sages. Kuan Yin Tze, 8th century.

The dangers involved in a sole reliance on 'learning by reading', in a practical discipline like analytical chemistry, are summarized in the following advice:

Il ne faut pas laisser les intellectuels jouer avec les allumettes. (Don't let the intellectuels play with the matches). Jacques Prévert, 1900–1977.

However, it is hoped that reading this book will be useful both in providing enough background information that the first exposure to the 'learning by doing' process will not seem quite so daunting, and also will provide a useful reference thereafter. The book attempts to cover a wide range of sub-disciplines, and inevitably some errors of both omission and commission will remain despite extensive checking. The authors would greatly appreciate assistance from colleagues in identifying and correcting these errors.

Robert K. Boyd, Cecilia Basic, Robert A. Bethem
October 2007

Acknowledgements

While all errors and obfuscations remain the responsibility of the authors, this book has benefited from advice and contributions generously provided by several colleagues, including Keith Gallicano, Lynn Heiman, David Heller, Bob Peterson, Eric Reiner and Vince Taguchi. We are indebted to the family of Dr A.J.P. Martin for permission to reproduce his photograph in Chapter 3 and to the Rowett Institute in Aberdeen, Scotland, for permission to reproduce the photograph of Dr R.L.M. Synge. The photographs of Dr J.J. van Deemter and Dr M.J.E. Golay were kindly provided by Drs Ted Adlard and L.S. Ettre, and the pictures of M.S. Tswett and his apparatus by Dr Klaus Beneke of the University of Kiel. The portrait of William Thomson (Lord Kelvin) was kindly provided by the Department of Physics, Strathclyde University, Scotland, and that of Joseph Black by the Department of Chemistry, University of Glasgow, Scotland. Dr Ron Majors generously sent us the original graphics from several of his articles in *LCGC* magazine.

Finally, the authors thank their families and friends for their unwavering patience, support and encouragement in the face of our often obsessive burning of midnight oil while writing this book.

1

Measurement, Dimensions and Units

Standards of Comparison

The US standard railroad gauge (distance between the rails) is 4 feet, 8.5 inches. That's a very strange number, why was it used? Because the first railroads were built in Britain, and the North American railroads were built by British immigrants.

Why did they build them like that? Because the first railways (lines and rolling stock) were built by the same companies that built the pre-railroad tramways, and they used the same old gauge. All right, why did 'they' use that gauge? Because the tramways used the same jigs and tools that had been used for building wagons, and the wagons used that wheel spacing.

Are we getting anywhere? Why did the wagons use that strange wheel spacing? Well, if they tried to use any other spacing the wagons would break down on some of the old long distance roads, because that's the spacing of the old wheel ruts.

So who built these old rutted roads? The first long distance roads in Europe were built by Imperial Rome for the purposes of the Roman Legions. These roads were still widely used in the 19th century. And the ruts? The initial ruts, which everyone else had to match in case they destroyed their wagons, were made by Roman war chariots. Since the chariots were made for Imperial Rome they were all alike, including the wheel spacing. So now we have an answer to the original question. The US standard railroad gauge of 4 feet, 8.5 inches is derived from the original specification for an Imperial Roman army war chariot.

The next time you are struggling with conversion factors between units and wonder how we ended up with all this nonsense, you may be closer to the truth than you knew. Because the Imperial Roman chariots were made to be just wide enough to accommodate the south ends of two war horses heading north.

And this is not yet the end! The US space shuttle has two big booster rockets attached to the sides of the main fuel tank. These are solid rocket boosters (SRBs) made in a factory in Utah. It has been alleged that the engineers who designed the SRBs would have preferred to make them a bit fatter, but the SRBs had to be shipped by train from the factory to the launch site. The railroad line from the factory happens to run through a tunnel in the mountains, and the SRBs had to fit through that tunnel. The tunnel is only slightly wider than the railroad track, and we now know the story behind the width of the track!

So, limitations on the size of crucial components of the space shuttle arose from the average width of the Roman horses' rear ends.

1.1 Introduction

All quantitative measurements are really comparisons between an unknown quantity (such as the height of a person) and a measuring instrument of some kind (e.g., a measuring tape). But to be able to communicate the results of our measurements among one another we have to agree on exactly what we are comparing our measurements to. If I say that I measured my height and the reading on the tape was 72, that does not tell you much. But if I say the value was 72 inches, that does provide some meaningful

Trace Quantitative Analysis by Mass Spectrometry   Robert K. Boyd, Cecilia Basic, Robert A. Bethem
© 2008 John Wiley & Sons, Ltd

information provided that you know what an inch is (tradition tells us that the inch was originally defined as the length of part of the thumb of some long-forgotten potentate but that does not help us much). But even that information is incomplete as we do not know the uncertainty in the measurement. Most people understand in a general way the concepts of accuracy (deviation of the measured value from the 'true' value) and precision (a measure of how close is the agreement among repeated measurements of the same quantity) as different aspects of total uncertainty, and such a general understanding will suffice for the first few chapters of this book. However, the result of a measurement without an accompanying estimate of its uncertainty is of little value, and a more complete discussion of experimental uncertainty is provided in Chapter 8 in preparation for the practical discussions of Chapters 9 and 10.

Actually, the only correct answer to the question 'what is an inch' is that one inch is defined as exactly 2.54 centimeters (zero uncertainty in this defined conversion factor). So now we have to ask what is a centimeter, and most of us know that a centimeter is 1/100 of a meter. So what is a meter? This is starting to sound about as arbitrary as the Roman horses' hind quarters mentioned in the text box but in this case we can give a more useful

if less entertaining answer: The meter is the length of the path traveled by light in vacuum during a time interval of 1/299 792 458 of a second. Note that the effect of this definition is to fix the speed of light in vacuum at exactly 299 792 458 meters per second, and that we still have not arrived at a final definition of the meter until we have defined the second (Table 1.1). This is the internationally accepted definition of the meter, established in 1983, and forms part of the International System of Units (Système Internationale d'Unites, known as SI for short). The SI establishes the standards of comparison used by all countries when the measured values of physical and chemical properties are reported. Such an international agreement is essential not only for science and technology, but also for trade. For example, consider the potential confusion arising from the following example:

1 US quart (dry) = 1.10122 litres
1 US quart (liquid) = 0.94635 litres
1 Imperial (UK/Canada) quart (liquid) = 1.136523 litres

(The litre is defined in the SI as 1/1000 of a cubic meter: $1L = 10^{-3} \, m^3$). Many other examples of such ambiguities can be given (see, for example, the unit conversions

Table 1.1 SI Base Quantities and Units

Quantity	Name of unit	Symbol	Definition
Length	meter	m	The meter is the length of the path travelled by light in vacuum during a time interval of 1/299 792 458 of a second.
Mass	kilogram	kg	The kilogram is the unit of mass; it is equal to the mass of the international prototype of the kilogram.
Time	second	s	The second is the duration of 9 192 631 770 periods of the radiation corresponding to the transition between the two hyperfine levels of the ground state of the caesium 133 atom.
Electric current	ampere	A	The ampere is that constant current which, if maintained in two straight parallel conductors of infinite length, of negligible circular cross-section, and placed one meter apart in vacuum, would produce between these conductors a force equal to 2×10^{-7} newton per meter of length.
Thermodynamic temperature	kelvin	K	The kelvin, unit of thermodynamic temperature, is the fraction 1/273.16 of the thermodynamic temperature of the triple point of water.
Amount of substance	mole	mol	The mole is the amount of substance of a system which contains as many elementary entities as there are atoms in 0.012 kilogram of carbon 12; its symbol is 'mol.' When the mole is used, the elementary entities must be specified and may be atoms, molecules, ions, electrons, other particles, or specified groups of such particles.
Luminous intensity	candela	cd	The candela is the luminous intensity, in a given direction, of a source that emits monochromatic radiation of frequency 540×10^{12} hertz and that has a radiant intensity in that direction of 1/683 watt per steradian.

at: http://www.megaconverter.com/Mega2/index.html). Such discrepancies may not seem to be very important when only a single quart is considered, but in international trade where literally millions of quarts of some commodity might be traded, the 19 % difference between the two definitions of the liquid quart could lead to extreme difficulties if the ambiguity were not recognized and taken into account. In a lecture on 'Money as the measure of value and medium of exchange', delivered in 1763 at the University of Glasgow, Adam Smith commented (Smith 1763, quoted in Ashworth 2004):

'Natural measures of quantity, such as fathoms, cubits, inches, taken from the proportion of the human body, were once in use with every nation. But by a little observation they found that one man's arm was longer or shorter than another's, and that one was not to be compared with the other; and therefore wise men who attended to these things would endeavour to fix upon some more accurate measure, that equal quantities might be of equal values. Their method became absolutely necessary when people came to deal in many commodities, and in great quantities of them.'

It is precisely this kind of uncertainty that the SI is designed to avoid in both science and in trade and commerce. In this regard it is unfortunate to note that even definitions of words used to denote numbers are still subject to ambiguity. For example, in most countries 'one billion' (or the equivalent word in a country's official language) is defined as 10^{12} (a million million), but in the USA (and increasingly in other English-speaking countries) a billion is used to represent 10^9 (a thousand million) and 10^{12} is referred to as a 'trillion'. In view of this ambiguity it is always preferable to use scientific numerical notation.

1.2 The International System of Units (SI)

An excellent source of information about the SI can be found at the website of the US National Institute for Standards and Technology (NIST): http://physics.nist.gov/cuu/Units/index.html

Here we shall be mainly concerned with those quantities that directly affect quantitative measurements of amounts of chemical substances by mass spectrometry. However, it is appropriate to briefly describe some general features of the SI.

Early History of the SI

There is a strong French connection with the SI, including its name and the location in Paris of the central organization that coordinates this international agreement (Bureau International des Poids et Mesures, or BIPM), and the international guiding body CIPM (Comité International de Poids et Mesures, i.e., International Committee for Weights and Measures). This connection was established at the time of the French Revolution when the revolutionary government decided that the chaotic state of weights and measures in France had to be fixed. The intellectual leader in this initiative, that resulted in the so-called Metric System, was the chemist Antoine Lavoisier, famous for his demonstration that combustion involves reaction with oxygen and that water is formed by combustion of two parts of hydrogen with one of oxygen. His efforts resulted in the creation of two artifacts

made of platinum (chosen because of its resistance to oxidation), one representing the meter as the new unit of length between two scratch marks on the platinum bar, and the other the kilogram. These artifacts were housed in the Archives de la République in Paris in 1799, and this represents the first step taken towards establishment of the modern SI.

Sadly, Lavoisier did not live to see this realization of his ideas. Despite his fame, and his services to science and his country (he was a liberal by the standards of pre-revolutionary France and played an active role in the events leading to the Revolution and, in its early years, formulated plans for many reforms), he fell into disfavour because of his history as a former farmer-general of taxes, and was guillotined in 1794. After his arrest and a trial that lasted less than a day, Lavoisier requested postponement of his execution so that he could complete some experiments, but the presiding judge infamously refused: 'L'état n'a pas besoin de savants' (the state has no need of intellectuals).

Antoine Lavoisier

Any system of measurement must decide what to do about the fact that there are literally thousands of physical properties that we measure, each of which is expressed as a measured number of some well-defined unit of measurement. It would be impossible to set up primary standards for the units of each and every one of these thousands of physical quantities, but fortunately there is no need to do so since there are many relationships connecting the measurable quantities to one another. A simple example that is of direct importance to the subject of this book is that of volume; as mentioned above, the SI unit of volume (cubic meter) is simply related to the SI unit for length via the physical relationship between the two quantities. So the first question to be settled concerns how many, and which, physical quantities should be defined as SI base quantities (sometimes referred to as dimensions), for which the defined units of measurement can be combined appropriately to give the SI units for all other measurable quantities.

At one time it was thought to be more 'elegant' to work with a minimum possible number of dimensions and their defined units of measurement, and this pseudo-esthetic criterion gave rise to the three-dimensional centimeter-gram-second (cgs) and meter-kilogram-second (MKS) systems. However, it soon became apparent that utility and convenience were more important than perceived elegance! As a simple example, consider Coulomb's Law for the electrostatic force F between two electric charges q_1 and q_2 separated by a distance r in a vacuum:

$$F = k_o.q_1.q_2/r^2$$

In the simple form of Coulomb's Law as used with the cgs system, the Coulomb's Law Constant k_o is treated as a dimensionless constant with value 1. (This is not the case

in the SI, where $k = 1/(4\pi\varepsilon_o)$ where ε_o is the permittivity of free space $= 8.854187817 \times 10^{-12}\, s^4 A^2 kg^{-1} m^{-3}$). By Newton's Second Law of Motion, force is given as (mass × acceleration), i.e., (mass × length × time^{-2}), so in the cgs system $q_1 \cdot q_2$ corresponds to (mass × length^3 × time^{-2}); thus, in such a three-dimensional measurement system, electrical charge q corresponds to (mass$^{\frac{1}{2}}$ × length$^{\frac{3}{2}}$ × time^{-1}). This very awkward (and inelegant!) result involving fractional exponents becomes even more cumbersome when magnetism is considered. Once it was accepted that usefulness was the only criterion for deciding on the base physical quantities (dimensions) and their units of measurement, it was finally agreed that the most useful number of dimensions for the SI was seven. Some of these seven are of little or no direct consequence for this book, but for the sake of completeness they are all listed in Table 1.1. Some important SI units, that are derived from the base units but have special names and symbols, are listed in Table 1.2.

The two base quantities (and their associated SI units) that are most important for quantitative chemical analysis are amount of substance (mole) and mass (kilogram), although length (meter) is also important via its derived quantity volume in view of the convenience introduced by our common use of volume concentrations for liquid solutions. (Note, however, that the latter will in principle vary with temperature as a result of expansion or contraction of the liquid).

The kilogram is unique among the SI base units for two reasons. Firstly, the unit of mass is the only one whose name contains a prefix (this is a historical accident arising from the old centimeter-gram-second system of measurement mentioned above). Names and symbols for decimal multiples and submultiples of the unit of mass are formed

Table 1.2   Some SI Derived units with special names and symbols[a]

Derived quantity	Name of unit	Symbol	Relationship to SI base units	Relationship to other SI units
Plane angle	radian	rad	$m.m^{-1}\,(=1)$	—
Solid angle	steradian	sr	$m^2.m^{-2}\,(=1)$	—
Frequency	hertz	Hz	s^{-1}	—
Force	newton	N	$m.kg.s^{-2}$	—
Pressure	pascal	Pa	$m^{-1}.kg.s^{-2}$	$N.m^{-2}$
Energy	joule	J	$m^2.kg.s^{-2}$	$N.m;\ Pa.m^3$
Power	watt	W	$m^2.kg.s^{-3}$	$J.s^{-1}$
Electric charge	coulomb	C	$A.s$	—
Electric potential difference	volt	V	$m^2 \cdot kg \cdot s^{-3} \cdot A^{-1}$	$W.A^{-1}$
Electric resistance	ohm	Ω	$m^2 \cdot kg \cdot s^{-3} \cdot A^{-2}$	$V.A^{-1}$
Magnetic flux density	tesla	B	$kg \cdot s^{-2} \cdot A^{-1}$	
Celsius temperature	degree Celsius	°C	K^b	—

a for a complete list and discussion, see Taylor (1995) and Taylor (2001).
b the *size* of the two units is the same, but Celsius temperature (°C) = thermodynamic temperature (K) − 273.15 (the ice point).

by attaching prefix names to the unit name 'gram' and prefix symbols to the unit symbol 'g', not to the 'kilogram'. (A list of SI prefixes denoting powers of 10 is given in Table 1.3). The other unique aspect of the kilogram is that it is currently (2007) the only SI base unit that is defined by a physical artifact, the so-called international prototype of the kilogram (made of a platinum–iridium alloy and maintained under carefully controlled conditions at the BIPM in Paris (Figure 1.1)). This international prototype is used to calibrate the national kilogram standards for the countries that subscribe to the SI.

Table 1.3 SI Prefixes

Factor	Name	Symbol	Factor	Name	Symbol
10^{24}	yotta	Y	10^{-1}	deci	d
10^{21}	zetta	Z	10^{-2}	centi	c
10^{18}	exa	E	10^{-3}	milli	m
10^{15}	peta	P	10^{-6}	micro	μ
10^{12}	tera	T	10^{-9}	nano	n
10^{9}	giga	G	10^{-12}	pico	p
10^{6}	mega	M	10^{-15}	femto	f
10^{3}	kilo	k	10^{-18}	atto	a
10^{2}	hecto	h	10^{-21}	zepto	z
10^{1}	deka	da	10^{-24}	yocto	y

By convention, multiple prefixes (e.g., dekakilo) are not allowed. Thus in the case of the SI unit of mass (kilogram), that for historical reasons contains a prefix in its name, the SI prefix names are used with the unit name 'gram' and the prefix symbols with the corresponding symbol 'g'.

Figure 1.1 The International prototype of the kilogram.

The addition of amount of substance as the seventh SI base unit, the 'chemical' unit, was achieved only after considerable dispute between chemists and physicists (McGlashan 1970), and was officially adopted only in 1971 about 17 years after adoption of the ampere, the kelvin and the candela (Table 1.1). Essentially the physicists felt that mass was a perfectly adequate quantity for all quantitative chemical purposes, since for all practical purposes mass is conserved in chemical reactions. Note that this can not be exactly correct since chemical reactions involve energy changes, e.g., energy loss in the case of exothermic reactions, and this energy corresponds to a change in mass via Einstein's famous relationship $E = mc^2$. However, for a typical reaction enthalpy of $10^5 \, \text{J.mol}^{-1}$, the corresponding change in mass is given as:

$$\Delta m \sim 10^5 \, \text{J}/(3 \times 10^8 \, \text{ms}^{-1})^2 \sim 10^{-12} \, \text{kg}$$

A good laboratory balance can measure mass routinely to within $10^{-7} \, \text{kg}$, and use of a microbalance with considerable precautions can lead to mass measurements to within $10^{-9} \, \text{kg}$ or so (Section 2.3). This is still three orders of magnitude larger than the mass changes equivalent to heats of reaction, so the physicists' argument is valid from this point of view. (Note that the above calculation of Δm exemplifies an important property of the SI, its coherence, by which we mean that if all quantities in a formula are expressed in SI units without prefixes the result of the calculation is also expressed in the appropriate SI unit, with no need for conversion factors).

However, the guiding principle in choice of the base quantities in any measurement system is that of usefulness and convenience, and since chemistry involves interactions among individual discrete molecules it is simply commonsense to adopt a quantity (and a corresponding unit) that reflects this reality.

The definition of the mole (Table 1.1) refers to the number of atoms in 0.012 kilogram of carbon 12 (^{12}C). This number is the Avogadro Constant $N_A = 6.0221479 \, (\pm 0.0000030) \times 10^{23} \, \text{mol}^{-1}$, formerly known as 'Avogadro's Number' but now in the SI not a dimensionless number but a quantity that must be expressed in SI units; the Avogadro Constant defines the number of molecules of a compound in 1 mol of that compound. Since different molecules interact chemically on the basis of small integral numbers of each type, it makes sense on a purely utilitarian basis to define such a base quantity and a corresponding unit, e.g., since one milligram of glucose contains 30 times as many molecules as one milligram of insulin, it makes no *physical* sense to discuss *chemical* interactions between these two compounds in terms of mass only!

The Mole and the Avogadro Constant

An interesting historical account of the origins of the concept and the name of the mole has been published by Gorin (1994).

Definition of base quantities and their respective units is a serious business, absolutely necessary for the unambiguous sharing of quantitative experimental data among scientists and engineers around the world. However, it can become a somewhat dry and even boring subject for chemists, who are never reluctant to look for ways to spice up their professional business with a little self-deprecating humour.

For example, the amazing reduction in detection limits for mass spectrometry that has been possible over the last 20 years or so has led to a proposal that a new SI prefix (see Table 1.3) will be required before long. The name of the proposed new prefix is the guaco, referring to a factor of 10^{-25}, since ambitions for guacamole sensitivity were thought to be a suitable target for instrument designers. However, this proposal was abandoned when it was realized that guacamole sensitivity was intrinsically impossible as a result of the value of the Avocado Constant.

Slightly less nonsensical is the introduction of 'Mole Day', created as a way to foster interest in chemistry. Schools throughout the United States of America and around the world celebrate Mole Day on October 23 from 6:02 a.m. to 6:02 p.m., to commemorate the Avogadro Constant (6.02×10^{23}), with various activities related to chemistry and/or moles (see www.moleday.org).

However, the physicists did have a point in their argument with respect to the importance of measurements of mass in how chemists actually set about performing quantitative analyses of amount of substance, and this will be discussed in Section 2.3.

1.3 'Mass-to-Charge Ratio' in Mass Spectrometry

How does the foregoing discussion relate to the so-called mass-to-charge ratios, universally denoted as m/z, that are used to mark the abscissa of a mass spectrum? It must be emphasized that the question of the meaning of the quantity m/z appears to be highly contentious among mass spectrometrists, and the following discussion represents only the best efforts of the present writers to devise a self-consistent interpretation that will be used where appropriate in the rest of the book.

In this book 'm/z' is best regarded as a three-character symbol, not a mathematical operation. Although no units are ever given for m/z in published spectra, within this three-character symbol 'm' does indeed denote the mass of a single atom or molecule that has been transformed into an ion so that it is amenable to analysis by the mass spectrometer. For purposes of mass spectrometry, however, it is convenient to not use the kilogram as the unit of mass, but instead to express the mass in terms of the unified atomic mass unit u (or sometimes m_u) defined as:

$$u(\text{or } m_u) = (\text{mass in kilograms of one atom of } {}^{12}C)/12$$

Since the symbol 'u' is used for other purposes in this book, 'm_u' will be used in the following discussion. The connection of this unit of mass, convenient for expressing masses of single molecules, to the definition of the mole and of the Avogadro Constant, is:

$$12 \times m_u(\text{kg}) \times N_A(\text{mol}^{-1}) = 0.012 \text{ kg mol}^{-1}$$

whence

$$m_u(\text{or } u) = 0.001/(6.02214179 \times 10^{23})\text{kg} = 1.660538782$$
$$(\pm 0.000000083) \times 10^{-27} \text{ kg}$$

Values of fundamental constants like m_u, N_A, etc., together with related information concerning their relationships, can be found at http://physics.nist.gov/cuu/Constants/index.html. Values of m for all isotopes of the elements (including radioactive nuclides) are constantly being refined; the International Union of Pure and Applied Chemistry (IUPAC) publishes frequent revised tables of isotope-averaged atomic weights (Loss 2003), and extensive updated information on individual isotopic masses can be found at http://ie.lbl.gov/toimass.html, while similarly updated information on natural isotopic abundances (Rosman 1998) is also available at www.iupac.org/reports/1998/7001rosman/iso.pdf. Such detailed high-precision information is not usually important for the kinds of measurements discussed in this book. Lists of atomic masses and isotope abundances of adequate quality for trace chemical analysis are available from several sources, e.g., http://physics.nist.gov/cgi-bin/Compositions/stand_alone.pl?ele=&ascii=html&isotype=some where the elements are listed in order of atomic number, and

Table 1.4 Relative atomic masses and relative abundances of some stable isotopes

Element (Atomic number)	Isotope	Relative atomic mass	Abundance (%)	Isotope-averaged atomic weight
H (1)	H	1.00782503	99.9885	1.00794
	^{2}H (D)	2.01410178	0.0115	
B (5)	^{10}B	10.0129370	19.9	10.81
	^{11}B	11.0093055	80.1	
C (6)	^{12}C	12.0000000 (by definition)	98.93	12.0107
	^{13}C	13.0033548	1.08	
N (7)	^{14}N	14.0030740	99.632	14.0067
	^{15}N	15.0001089	0.368	
O (8)	^{16}O	15.9949146	99.757	15.9994
	^{17}O	16.9991315	0.038	
	^{18}O	17.9991605	0.205	
F (9)	^{19}F	18.9984032	100	18.9984032
Na (11)	^{23}Na	22.9897696	100	22.9897696
Si (14)	^{28}Si	27.9769265	92.2297	28.0855
	^{29}Si	28.9764947	4.6832	
	^{30}Si	29.9737702	3.0872	
P (15)	^{31}P	30.9737615	100	30.9737615
S (16)	^{32}S	31.9720707	94.93	32.065
	^{33}S	32.9714585	0.76	
	^{34}S	33.9678668	4.29	
	^{36}S	35.9670809	0.02	
Cl (17)	^{35}Cl	34.9688527	75.78	35.453
	^{37}Cl	36.9659026	24.22	
K (19)	^{39}K	38.9637069	93.2581	39.0983
	^{40}K	39.9639987	0.0117	
	^{41}K	40.9618260	6.7302	
Br (35)	^{79}Br	78.9183376	50.69	79.904
	^{81}Br	80.9162913	49.31	
I (53)	^{127}I	126.9044684	100	126.9044684

(Note that the relative abundances are terrestrial averages; small deviations (a few parts per thousand) contain information that is valuable in several fields of science, and are measured using Isotope Ratio Mass Spectrometry.)

at www.sisweb.com/referenc/source/exactmaa.htm where the elements are listed in alphabetical order. An abbreviated list covering the elements of most interest for organic analyses in this book is given in Table 1.4.

It is important to note that m_u is the convenient unit adopted to express the mass of one unique molecule containing specified numbers of isotopes of the elements (e.g., ^{12}C^{35}Cl^{37}Cl$_2^1$H for an isotopically specified form of chloroform). The m_u unit is conventionally not used for the average mass of the molecules of the same compound, still specified as the same numbers of atoms of the elements (CCl_3H for chloroform), but now assuming the various isotopic distributions to be the average values observed on the surface of our planet Earth. (Such

quantities have been referred to in the past as the 'molecular weight' of the compound, but this usage can be misleading because the 'weight' of an object is by definition the gravitational force on that object, i.e., it depends on both the mass of the object and its position in space). Chemists and biochemists have in the past used the dalton (Da) as an atomic mass unit derived from m_u but adjusted for each element according to the average isotopic distribution of that element. This is not an official SI unit but is useful because it directly relates the mass of a macroscopic sample of a real natural compound, determined by weighing (see Section 2.3), to the molecular formula of the compound. Of course this relationship does not hold for a variant of the compound in which one or more of the atoms have been synthetically specified with a non-natural isotopic composition, for use as an internal standard in isotope dilution approaches, as discussed in Section 2.2.3). However, in actual practice the dalton is increasingly being used by mass spectrometrists as an alternative to the 'unified atomic mass unit' m_u (or u), and usually the context makes clear which interpretation of this chemical unit (i.e., the isotopic specification) is intended.

In summary, the quantity m (italicized) in m/z is related to the mass m (in kilograms) of a single isotopically-specified (ionized) molecule, and is strictly defined as:

$$m = m(kg)/m_u(kg)$$

where m is the actual mass of one of the specified molecules; m can thus be regarded as a dimensionless quantity (ratio) that requires no units, although it is intimately related to the kilogram via the Avogadro Constant and m_u.

Just as for the mass of an ion, the charge on the ion is conveniently expressed not in terms of the SI unit for electric charge, i.e., the Coulomb (C, see Table 1.2), but relative to the elementary charge (e), one of the fundamental physical constants and equal to the magnitude of the charge (i.e. without sign) on the electron:

$$e = 1.602176487 \, (\pm 0.000000040) \times 10^{-19} C$$

Then, the quantity z (italicized) in m/z is the number of elementary charges on the ion (usually quoted irrespective of sign as the context almost always makes the latter clear):

$$z = \text{magnitude of charge on the ion (C)}/e(C)$$

and is thus also, like m, a dimensionless number!

Thus, the 'mass-to-charge ratio' of mass spectrometry, conveniently denoted by the symbol m/z, is a dimensionless ratio of two dimensionless quantities that is nonetheless intimately related to both the kilogram (via m_u) and to the coulomb (via e). It is possible that the 'm' in 'm/z' could be misinterpreted as the same (but nonitalicized) symbol used in the SI to represent the base quantity 'mass' and possibly also the SI symbol for the meter (unit of length)! The letter 'm' is greatly overused in metrological notation! However, in the mass spectrometric sense, 'm' NEVER appears without '/z', so it seems that the context should never give rise to ambiguity or confusion. In fact, as mentioned above, the notation 'm/z' is best regarded as a three-character symbol for the dimensionless quantity

defined above, rather than as a mathematical operation on two different quantities. Otherwise the abscissa of a mass spectrum, invariably labeled as m/z (with no units specified!) would be labeled in units of, e.g., kilograms per coulomb (for mass:charge ratio)! In Section 6.2 we shall see how this dimensionless quantity m/z is incorporated into quantitative calculations of physical quantities that are important in describing the instruments used to separate ions according to their m/z values.

More recently, with the introduction of electrospray ionization (Section 5.3.6) to mass spectrometry, it has become much more common to observe multiply-charged ions with $z > 1$. This does not introduce any fundamental difficulty into the established measurement system for ions, but a question of convenience does arise when describing changes in, or differences between, mass spectra. For example, if it is observed that two mass spectra differ only with respect to one peak that appears at different m/z values, how does one describe the magnitude of the shift? Some authors say that the peak was shifted by 'X m/z units', which is clumsy but does transmit the desired message. It has been suggested (Cooks 1991) that the mass spectrometry community should, purely for convenience, adopt a unit for m/z defined as above, to be called the thomson (Th) in honour of J.J. Thomson; then we could speak of a peak shift by 'X Th'. This suggestion has not been approved by any international body, but has come into common use simply because it is convenient to do so under circumstances such as those mentioned above. Unfortunately a quantity (not a unit) named the Thomson cross-section (and indeed also named in honor of J.J. Thomson) already exists; this quantity describes the probability that electromagnetic radiation will be scattered by a charged particle, and its unit (symbol σ_e) is the value of this cross section for an electron ($0.665245873 \times 10^{-28} \, m^2$, see: http://physics.nist.gov/cgi-bin/cuu/Value?sigmae|search_for=atomnuc!). In the context of mass spectrometry the proposal to name a *unit* (symbol Th) for m/z in honor of Thomson (Cooks 1991) has been criticized on the basis that this would create confusion with the physical quantity (Thomson cross-section, with its own unit σ_e). This seems unlikely given the very different contexts in which the two will appear, quite apart from the fundamental difference between a physical quantity and a unit of measurement. It should also be recalled that convenience is an important criterion in deciding upon details of any measurement system, even the number of base quantities (and thus units) to be adopted! Accordingly the Thomson (Th) is used where appropriate as the unit of m/z in this book.

For convenience a list of some fundamental physical constants is provided in Table 1.5.

Table 1.5 Values of some fundamental physical constants, given to a number of significant figures sufficient for purposes of this book

Physical quantity	Symbol	Value
Elementary charge	e	1.60218×10^{-19} C
Unified atomic mass unit	u or m_u (also Da)	1.66054×10^{-27} kg
Mass of electron	m_e	9.10938×10^{-31} kg
Avogadro Constant	N_A	6.02214×10^{23} mol^{-1}
Boltzmann Constant	k_B	1.38065×10^{-23} J.K^{-1}
Gas Constant	$R(=N_A.k_B)$	8.31447J.K^{-1}.mol$^{-1} =$ 0.0820575L.atm.K^{-1}.mol^{-1}
Planck Constant	h	6.62607×10^{-34} J.s^{-1}
Permittivity of vacuum	ε_0	8.85419×10^{-12} J^{-1}.C^2.m^{-1}
Permeability of vacuum	μ_0	$4\pi \times 10^{-7}$ J.s^2.C^{-2}.m^{-1} (orT^2.J^{-1}.m^3)

(Note: T = tesla (unit of magnetic flux density, Table 1.2).)

1.4 Achievable Precision in Measurement of SI Base Quantities

The achievable precision in measuring quantities like time, length, mass and the other SI base quantities, and thus in the definitions of their units, is intimately dependent on developments in the technologies used to measure them. Once the recommendations of Adam Smith (see the introductory paragraphs of this chapter) had been adopted, and 'natural' rather than 'anthropological' units of measurement were sought, early attempts used the size of the earth (the meter was originally defined as the length of a platinum bar designed to be 10^{-7} of the length of a quadrant of the Earth), the rotation rate of the Earth to define 24 hours and thus the second, and the density of water as a link between the meter and the gram. Clearly all of these standards are subject to variation and/or uncertainty, and it became evident that standards based on atomic phenomena would be much more reproducible and constant. For example, James Clerk-Maxwell commented (Clerk-Maxwell, 1890):

'The earth has been measured as a basis for a permanent standard of length, and every property of metals has been investigated to guard against any alteration of the material standards when made. To weigh or measure anything with modern accuracy, requires a course of experiment and calculation in which almost every branch of physics and mathematics is brought

into requisition. Yet, after all, the dimensions of our earth and its time of rotation, though, relative to our present means of comparison, very permanent, are not so by any physical necessity. The earth might contract by cooling, or it might be enlarged by a layer of meteorites falling on it, or its rate of revolution might slowly slacken, and yet it would continue to be as much a planet as before. But a molecule, say of hydrogen, if either its mass or its time of vibration were to be altered in the least, would no longer be a molecule of hydrogen. If, then, we wish to obtain standards of length, time and mass which shall be absolutely permanent, we must seek them not in the dimensions, or the motion, or the mass of our planet, but in the wavelength, the period of vibration and the absolute mass of these imperishable and unalterable and perfectly similar molecules'.

An excellent review (Flowers 2004) has described the modern advances in achieving this objective. The most spectacular achievements have been in the measurement of time, for which modern cesium atom beam atomic clocks can subdivide time to better than one part in 10^{15} and thus achieve a measurement precision (and thus a definition of the second) of this order (Diddams 2004). As mentioned above, the meter is now defined as the length of the path travelled by light in vacuum during a time interval of *exactly* (1/299 792 458) of a second. The kilogram is still defined in terms of a man-made artifact (Figure 1.1), but efforts are in progress to devise a scheme of measurement and definition that will allow establishment of a unit of mass that is based on some atomic property or perhaps the mass equivalent of energy. It is interesting to note in passing that recent developments (Rainville 2004) in the measurements of cyclotron frequencies of isolated ions (either one or two ions to avoid space-charge effects, see Chapter 6) in a Penning trap have resulted in a precision of better than one in 10^{11} in measurements of ion masses. In addition to the many applications of this technology to fundamental physics (Rainville 2004), measurement precision of this order is only 1–2 orders of magnitude below that required to be able to 'weigh' chemical bond strengths to a useful degree of accuracy and precision via E = m.c^2!

These spectacular achievements in the precision of physical metrology are to be compared with the levels of precision that can be achieved in, for example, high-throughput measurements of the amounts of a target analyte present at levels of one part in $10^9–10^{12}$ in a complex matrix, a common circumstance faced in laboratories inhabited by readers of this book! In such cases, within-day and between-day precision of 10 % (one in 10^1) would be considered acceptable for organic or speciated inorganic analytes. (The precision can be improved to 1–2 % in cases where an isotope-labeled internal standard can be used with isotope dilution mass spectrometry, provided that meticulous precautions are taken as in certification of a reference material for example, see Section 2.2.2). At first sight this performance of chemical metrology would appear to be miserable compared with that of our physicist colleagues.

A first perspective on this enormous discrepancy can be found by considering the problems in measuring another physical base quantity, temperature. The definition of the kelvin, the unit of thermodynamic temperature (Table 1.1), is the fraction (1/273.16) of the thermodynamic temperature of the triple-point of water (the unique temperature at which all three phases – solid, liquid and vapor – can co-exist, see Section 4.3.2e). The implied measurement precision of $\sim$1 in 10^5 is far below that currently achievable with respect to time/frequency, for example. A major reason for this is the intrinsic complexity of chemistry! The principal barrier to improving this situation, based on the current definition, lies in uncertainties in the chemical purity and isotopic composition of the water sample used (Flowers 2004). Such complexity, that is part and parcel of chemical science, is somewhat alien to many physicists as exemplified in an extreme form by the so-called polywater scandal (see the accompanying text box).

Chemical Complexity in Physics: the Polywater Story

In the early 1960s a Soviet physicist, N.N. Fedyakin, discovered what he believed to be a new form of water with anomalous properties, e.g., higher boiling point, lower freezing point, higher viscosity, compared with 'normal' liquid water. In 1966 a senior colleague from a prestigious institute in Moscow, B.V. Derjaguin, presented a lecture on this 'anomalous water' at the Discussions of the Faraday Society in the United Kingdom.

Over the next several years, hundreds of papers from around the world, describing the properties of what soon came to be known as 'polywater', appeared in the scientific literature. Theorists developed models, supported by some experimental measurements, in which strong hydrogen bonds were causing water to polymerize. There were even some who expressed concerns that, if polywater escaped from the laboratory, it could autocatalytically polymerize all of the world's water. It is interesting that the famous American author Kurt Vonnegut published his novel *Cat's Cradle* (Vonnegut 1963) in this same time period. In this novel a substance called 'ice-nine', created by a fictitious Nobel Laureate in Physics, is an alternative structure of water that is solid at room temperature. When a crystal of 'ice-nine' is brought into contact with liquid water, it becomes a seed that indeed autocatalyzes the conversion of liquid water to the solid 'ice-nine' that unfortunately has a melting point of 45.8 °C, thus potentially ending all life on the planet. (Note that Vonnegut's fictional ice-nine is not to be confused with the real substance Ice IX, also pronounced 'ice-nine', an exotic form of solid water that exists at temperatures below 140 K and pressures between 200 and 400 MPa).

It was argued by some, early on in the story, that 'polywater' was simply impure water since boiling point elevation and freezing point depression are colligative properties characteristic of solutions. However, the flurry of papers emphasized the precautions that had been taken to avoid contamination, so for several years the existence of 'polywater' was taken for granted as a real phenomenon. However, eventually the case for 'polywater' began to crumble. Because it could only be formed in quartz capillaries of very small internal diameter, very little was available for analysis. When eventually small samples could be subjected to trace chemical analysis, 'polywater' was shown to be contaminated with a variety of substances from silica to phospholipids. Moreover, electron microscopy revealed that 'polywater' also contained finely divided colloidal particulates in suspension.

At this point the experiments that had produced polywater were repeated with extreme precautions, including rigorous cleaning of glassware. As a result the anomalous properties of the resulting water vanished, and even the scientists who had originally advanced the case for 'polywater' agreed that it did not exist. There was no question of scientific fraud. In retrospect it was simply a case where meticulous physical experiments turned out to be of no value because the subject of the experiments was not a simple pure substance, but a complex chemical mixture albeit one in which the components that caused the problem were present at ultra-trace levels.

A book describing the 'polywater' story has been published (Franks 1981).

The first step in trace chemical analyses is to determine whether or not the target analyte really is present (we might call this a 'binary quantitative analysis' with possible values of 0 or 1). An implication of this step is that we are ensuring that the signals that we eventually record using our analytical apparatus really do arise from the target analyte, and only from that compound. In order to achieve this it is necessary to selectively remove most of the other 10^9-10^{12} molecules of many kinds so that they do not interfere with our recorded signals, and do this while at the same time discarding only a small fraction (that must be measured as accurately and precisely as possible) of the analyte itself. Moreover, in real life, different methodologies must be devised for each combination of analyte and matrix, and in most real-world cases must be capable of delivering measurements that are 'fit for purpose', often on a scale of hundreds or even thousands of samples per week. An admittedly poor analogy with the approach used to achieve the spectacular successes of physical metrology would involve several generations of trace analytical chemists working on analysis of the same single analyte in the same batch of matrix (that might be a homogeneous liquid or a heterogeneous powder), in order to achieve the ultimate levels of accuracy and precision.

Naturally the likelihood of any such project being undertaken is essentially zero; in real life chemists are faced with one or several of literally hundreds of thousands of possible compounds present in a wide range of matrices; in Chapter 11 a handful of examples out of literally tens of thousands of possibilities are discussed. There is no useful purpose to a 'physics-style' effort to measure a trace level amount of substance to the utmost precision and accuracy that analytical technologies could conceivably provide. In fact, the spectacular success of chemical metrology applied to trace analysis is not the ~ 1 in 10^1 precision that is routinely achieved, but the analyte levels of one in 10^9-10^{12} for which such levels of precision (and accuracy) can be achieved on a high throughput basis for a wide range of, e.g., biomedical and environmental samples, despite their inherent complexity.

The mass spectrometer is a key component in this quest for accuracy and precision in the midst of complexity as a result of its unique combination of universality, sensitivity and selectivity for chemical analysis, but only as an important component in a series of integrated steps ranging from proper sampling of the material to be analyzed to interpretation and evaluation of the experimental data. This aspect is emphasized in Figure 1.2, and this book represents an attempt to describe the current state of the art in each step and some of the difficulties faced in

achieving a best overall compromise if optimized conditions for one particular step are not compatible with the others. But even a mass spectrometer is itself composed of several identifiable components that must be properly integrated with one another if the most reliable experimental data are to be obtained. A schematic sketch of the major components of any analytical mass spectrometer is shown in Figure 1.3, emphasizing the complexity not only the analytical samples themselves, but also of the apparatus and techniques that need to be integrated together to achieve acceptable levels of accuracy and precision in measurement of amount of substance.

1.5 Molecular Mass Limit for Trace Quantitation by Mass Spectrometry

It is important to realize that, as a consequence of the non-negligible natural abundances of higher mass isotopes of important elements (particularly carbon, Table 1.4), an upper limit of molecular mass (NOT m/z) for analytes that can be quantitated at trace levels using the techniques described here is generally around 2000 Da. One major reason for this limit arises from the natural abundance of the ^{13}C isotope (~ 0.011, Table 1.4). As an organic molecule becomes larger, the number of carbon atoms that make the largest contribution to the distribution of isotopologs (a composite word derived from "isotopic" and "analog") of a molecule (in the absence of chlorine or bromine) also becomes larger; as a result the relative importance of isotopologs containing one or more ^{13}C atoms increases (see Section 2.2.3 for a discussion of "isotopolog" vs "isotopomer"). We can get a quantitative feel for the effect by considering the probabilistic expression (binomial expansion) for the relative abundances of isotopologs $^{12}C^{13}_{(n-k)}C_k$ where n is the total number of carbon-atoms in the molecule (see isotope text box in Chapter 8 for discussion of the coefficients in the binomial expansion):

$$(a+b)^n = \Sigma_k \{n!/[k!(n-k)!]\}.a^{(n-k)}.b^k$$

where for carbon $a = 0.989$, $b = 0.011$, and n! is "factorial n", a continued product:

$$n! = n.(n-1).(n-2).—.3.2.1$$

Suppose we find the value of n for which the natural abundances of the $^{12}C_n$ and $^{12}C^{13}_{(n-1)}C_1$ isotopologs (i.e., those with k = 0 and k = 1) are equal; the corresponding molecular mass is sometimes referred to as the 'crossover' value. All we have to do is equate the first two terms

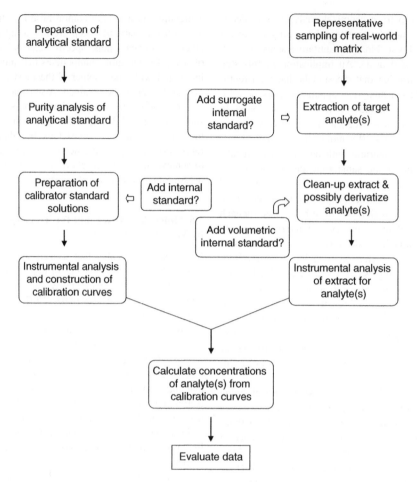

Figure 1.2 Summary of the several steps that must be integrated in an overall optimized quantitative analytical procedure. (Note that a mass spectrometer, although a key component, appears only in the 'Instrumental Analysis' steps (and to some extent in the purity analysis of the calibration standard)).

($k = 0$ and 1, corresponding to $^{12}C_n$ and $^{12}C_{(n-1)}\,^{13}C_1$) in the above expression for the binomial distribution $(a+b)^n$:

$$[n!/[0!(n-0)!].(0.989)^n.(0.011)^0$$

$$= [n!/[1!(n-1)!].(0.989)^{(n-1)}.(0.011)^1$$

Since $0! = 1 = 1!$ (by definition), and any number raised to the power zero $= 1$, cancellation of terms gives:

$$(0.989)^n = n.(0.989)^{n-1}.(0.011)$$

whence rearrangement gives $n = 0.989/0.011 \approx 90$, corresponding to a carbon contribution to this 'cross-over' molecular mass of 1080 Da. If we assume that elements other than carbon (hydrogen, nitrogen, oxygen etc.) can

contribute up to 25 % of the total molecular mass, then the 'cross-over' molecular mass is estimated as $\sim$1350 – 1400 Da.

Of course this is an approximate value since we ignored isotopic contributions from ^2H, ^{18}O, ^{15}N, etc., a reasonable first approximation even though the relative abundance of ^{15}N is a relatively high 0.368 (Table 1.4) since most organic molecules contain many fewer nitrogen and oxygen atoms than carbon. More realistic assessments of the cross-over molecular mass give a value 1700–1800 Da. The effect is illustrated in Figure 1.4 for some real examples of isotope distributions for unit-mass spectra (Section 6.2.3b) of compounds containing more than one multi-isotopic element. The restriction to unit-mass spectra implies that the small mass differences between,

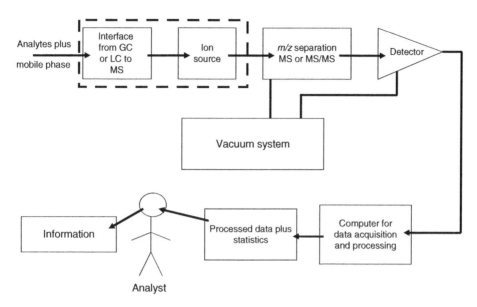

Figure 1.3 Schematic of the components integrated into a mass spectrometer used to provide selective and sensitive quantitation data. The analyte molecules entrained in the mobile phase are passed into the chromatography–MS interface and ion source where the majority of the mobile phase is selectively removed and the analyte is transformed into gaseous ions. (These two steps are sometimes incorporated into a single device, indicated by the dashed box surrounding them). The ions are then separated according to their *m/z* values (this can also include MS/MS analysis) and selected ions are introduced to the detector, usually an electron multiplier that can amplify the ion beam current by several orders of magnitude. This amplified current is transformed into a voltage signal that is digitized and fed into a computer where the raw data are logged in, stored, and processed. The processed data (usually with statistical assessments) are then outputted for interpretation and evaluation by the analyst who transforms the data into useful information.

e.g., $^{12}C_2H_5^{18}OH$ and $^{13}C_2H_5^{16}OH$ are not resolved so that a sample of ethanol will yield a mass spectrum in which the signal at *m/z* 34 contains superimposed contributions from these two species. A high resolution instrument would resolve these two signals, but for a unit-mass spectrum the combined signals would be registered and this is an underlying assumption of the calculations of isotopic distributions discussed here (Figure 1.4) and in the text box included in Chapter 8.

There are several consequences of this isotopic distribution effect for quantitative mass spectrometry. The most obvious is that sensitivity, and thus limits of detection and quantitation, will suffer as a result of the ion current arising from an analyte with naturally occurring isotopic abundances spread over several isotopologs with comparable abundances, thus lowering signal/noise (S/N) ratios (see Section 7.1) for each individual *m/z* value. This effect of isotopic distributions is in addition to that arising from the multi-charging phenomenon in electrospray ionization of larger molecules, which further dilutes the total ion current over more than one charge state (Section 5.3.6). Another obvious effect is the difficulties introduced when the higher isotopologs of the

natural analyte are isobaric with the lower isotopologs of a isotopically-labeled internal standard (see discussion of nonlinear calibrations in Section 8.5.2c).

A less obvious effect at or near the cross-over molecular mass arises for calibration of the *m/z* axis of the mass spectrometer when an automated calibration procedure is used to identify the 'molecular peak' (lowest mass isotopolog, often referred to as the 'mono-isotopic peak', i.e., ^1H, ^{12}C, ^{16}O, ^{14}N, etc. only), often specified in such automated algorithms as the most intense peak in a specified range; ambiguity in such an operational definition obviously arises near the cross-over range, potentially leading to a misidentification of the peak corresponding to the lowest-mass isotopolog (e.g., k = 0 in the $^{12}C-^{13}C$ case discussed above) and thus a *m/z* calibration that is in error (∼1 Th too high). Such uncertainties have extreme consequences in identification of target analytes. Selection of a precursor ion for MS/MS analyses (Section 6.2.9) also becomes problematic for larger molecules for which the mono-isotopic peak is no longer the most abundant natural isotopolog; selection of a higher isotopolog (containing two or more ^{13}C atoms) implies that fragment ions will also be represented by several isotopologs, further diluting

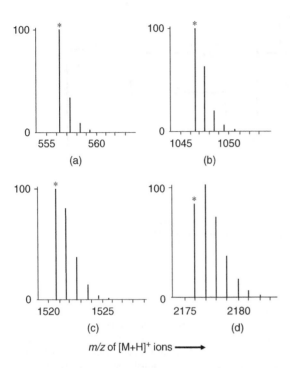

response with molecular mass observed for all ionization techniques used with current analyzers (Chapter 6) and ion detectors (Chapter 7), trace quantitative analysis is usually considered to have an upper limit of ∼2000 Da.

1.6 Summary of Key Concepts

1. The result of a measurement without an accompanying estimate of its uncertainty is of little value. The total uncertainty can be expressed in terms of **accuracy** (deviation of the measured value from the 'true' value) and **precision** (a measure of how close is the agreement among repeated measurements of the same quantity).

2. The **International System of Units** (Système Internationale d'Unites, SI) establishes the standards of comparison used by all countries when the measured values of physical and chemical properties are reported. There are **seven SI base quantities** (dimensions), for which the defined units of measurement can be combined appropriately to give the SI units for all other measurable quantities (i.e., the SI system is a **coherent system** of units).

3. The two base quantities that are most important for quantitative chemical analysis are **amount of substance** (measured in moles, mol) and **mass** (measured in kilograms, kg), although **length** (measured in meters, m) is also important via its derived quantity volume in view of the convenience introduced by our common use of volume concentrations for liquid solutions.

4. The **definition of the mole** (Table 1.1) refers to the number of atoms in 0.012 kilogram of carbon 12 (^{12}C). This number is the Avogadro Constant $N_A = 6.02214179 \, (\pm 0.00000030) \times 10^{23} \, mol^{-1}$ (formerly known as 'Avogadro's Number') that defines the number of molecules of a compound in one mol of that compound.

5. The **mass/charge ratios**, universally denoted as *m/z*, used to mark the abscissa of a mass spectrum **is best regarded as a three-character symbol, not a mathematical operation.** Within this three-character symbol, '*m*' does indeed denote the mass of a single atom or molecule that has been transformed into an ion so that it is amenable to analysis by the mass spectrometer. For purposes of mass spectrometry, it is convenient to not use the kilogram as the unit of mass, but instead to express the mass in terms of the **unified atomic mass unit** u (or sometimes m_u) defined as: u (or m_u) = [(mass in kg of one atom of ^{12}C)/12]. Thus u (m_u) is the convenient unit for the mass of one single unique

Figure 1.4 Relative abundances of isotopologs (isotopic distributions) for some peptides (containing C, H, N, S and O only), covering a range of *m/z* values for [M+H]$^+$ions including the cross-over value (∼1900Da) at which the relative abundances are equal for the lowest-mass isotopolog (the 'mono-isotopic peak' containing only ^{12}C,^{1}H,^{14}N,^{32}S, and ^{16}O, marked with an asterix in each case) and that with the next highest mass number. Relative abundances are shown normalized to 100 % for the most abundant isotopolog. (a) Leucine-enkephalin, mono-isotopic peak 556.28 Da; (b) Angiotensin II, mono-isotopic peak 1046.54 Da; (c) Fibrinopeptide A, mono-isotopic peak 1520.73 Da; (d) Fragment peptide of β-Lipotrophin, mono-isotopic peak 2175.99 Da. 'Nominal' *m/z* values, calculated using integral values for atomic masses, are (a) 556 Da, (b) 1046 Da, (c) 1520 Da, and (d) 2175 Da. 'Isotope-averaged' *m/z* values, calculated using atomic masses averaged over individual isotope masses weighted by the corresponding natural abundances within each element, are (a) 556.64 Da, (b) 1047.21 Da, (c) 1521.64 Da and (d) 2177.42 Da. Note that isotope-averaged molecular masses correspond to those used to convert weighed mass of material to moles.

the detected ion current over several *m/z* values and also introducing some ambiguity into the interpretation of the fragment ion spectrum.

For all of the reasons mentioned above (dilution of available ion current over too many *m/z* values, ambiguities in calibration and confirmation of analyte identity, choice of precursor ion in method development involving MS/MS detection), together with a significant fall-off of molecule (ion) of a specific **isotopic composition** (e.g., $^{12}C^{35}Cl^{37}Cl_2^1H$) and is not used for the average mass of the molecule (i.e., CCl_3H). Chemists and biochemists have in the past used the **dalton (Da)** as an atomic mass unit derived from u, but adjusted for each element according to the average isotopic distribution of that element; however, the Da is increasingly used by mass spectrometrists as an alternative to u.

6. Similarly, the charge on the ion is conveniently expressed *not* in terms of the SI unit for electric charge, i.e., the Coulomb (C), but relative to the **elementary charge (e)**, equal to the magnitude of the charge (i.e., without sign) on the electron; thus the quantity z (italicized) in m/z is the number of elementary charges on the ion z = magnitude of charge on the ion (C)/e (C) and, as such, is also a dimensionless number.

7. The mass spectrometer is a key component in the quest for accuracy and precision in the midst of complexity as a result of its unique combination of universality, sensitivity and selectivity for chemical analysis, but only as an important component in a series of integrated steps ranging from proper sampling of the material to be analyzed to interpretation and evaluation of the experimental data (Figures 1.2 and 1.3).

8. The upper limit on the molecular mass of analytes that are amenable to trace quantitative analysis is currently considered to be ~2000u (Da). This is the result of several factors, some of which are intrinsic to the problem, e.g., the relative abundances of higher isotopologs that increase significantly with increasing mass, and result in dilution of the total ion current derived from the analyte over too many m/z values.

2

Tools of the Trade I. The Classical Tools

2.1 Introduction

Analytical Chemists vs The Rest of the World

So Why Do You Want To Be An Analytical Chemist?

1. You wish to be held personally responsible for all environmental and ecological problems facing our planet.
2. You look forward to the opportunity to deal with irate clients demanding to know where are my results, how come you couldn't detect my compound, what do you mean my sample contained only phthalates, etc. etc.?
3. You appreciate the fact that the Psychology Department is close to the Analytical Chemistry Labs, and for a good reason.

The Wine in the Glass

Trace Quantitative Analysis by Mass Spectrometry   Robert K. Boyd, Cecilia Basic, Robert A. Bethem
© 2008 John Wiley & Sons, Ltd

(Continued)

Physicists are optimists, and say the glass is half-full.

Ecologists are pessimists, and say the glass is half-empty.

Engineers are engineers, and say the glass is the wrong size.

Analytical chemists ask 'What is that stuff in the glass made of anyway, and exactly how much of each is in there?'

In this chapter some types of apparatus are introduced that can be considered as 'classical', in contrast with the modern high technology devices like high performance liquid chromatographs (HPLC) or mass spectrometers (MS) that are the subject of later chapters. Many of these 'classical' tools are not unique to analytical strategies incorporating mass spectrometry. In particular, the general comment that all quantitative measurements are comparisons with a suitable standard (Chapter 1), includes any measurement of amount-of-substance of a target analyte (one whose identity and structure are known). Such a quantitative analysis requires comparison of an instrumentally measured signal for the unknown sample with that for an analytical standard (also called a reference or calibration standard), which is defined as a sample of the target analyte with a known degree of chemical purity (see later) and used for the preparation of stock solutions, calibration solutions and matrix-matched calibrators which will subsequently be presented to the analytical instrument as a known amount (i.e., a known volume of a solution at a known concentration). For this reason it is necessary to inquire about what exactly is meant by the concept of stock and calibration solutions, how they are made, what are the ways of using them to obtain reliable measurements on the unknown sample, and what are the particular properties of a mass spectrometer that have made it such an important component of trace analytical science.

Mass spectrometry is only the final step in an integrated procedure, involving many other kinds of apparatus, designed to provide accurate and precise quantitation of trace level analytes in complex matrices. The proper choice and use of these other apparatus, although less 'glamorous' than mass spectrometers, is equally essential for this purpose and a significant part of this chapter is devoted to this aspect.

A general scheme summarizing the steps involved in the calibration curve (or standard curve) approach, the most widely used strategy (Section 8.5), is shown in Figure 1.2. Much of the rest of this book is concerned with filling in the details and explanations that are necessary to use this scheme to obtain reliable data that are appropriate for the purpose for which the analysis was undertaken. Brief definitions of some key terms and concepts are collected together in a summary at the end of the chapter. Details of real-world practices in the selection of analytical standards, and in their use in preparing solutions for calibration of an analytical method, are presented in a later chapter (Sections 9.4, 9.5 and 9.8.1).

2.2 Analytical and Internal Standards: Reference Materials

2.2.1 Analytical Standard (Reference Standard) and its Traceability

The subject of this book is the absolute measurement of the 'amount of substance' of target analytes, i.e., of chemical substances of known molecular mass and of completely or partially known structures. To accomplish this using scientific instruments and strategies designed for relative measurements, it is essential to have available a standard sample of the analyte in question. This is known as the analytical standard (sometimes as the reference standard) and, for it to be useful, it must be of known chemical purity (as distinct from the 'isotopic purity' of isotopic variants used in isotope dilution experiments, discussed below).

The traceability of the result of a measurement is defined as the property of the result whereby it can be related to stated references, usually national (NIST, NRC, etc.) or international (CSA and EU) standards, through an unbroken chain of comparisons all with stated uncertainties, e.g., Figure 2.1. As applied to an analytical standard, distinct from a measurement of an unknown concentration, this concept implies that the purity of the standard has been determined by a traceable sequence of analytical steps, each with a known uncertainty, that permit comparison with a certified reference material (CRM) (Section 2.2.2) from a recognized supplier. This procedure results in a Certificate of Analysis, discussed in Section 2.2.2 for CRMs and also in Section 9.4.4c with respect to the analytical (reference)

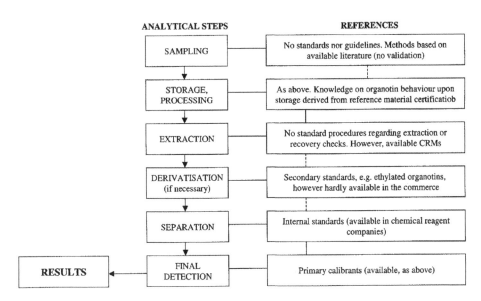

ANALYTICAL STEPS

REFERENCES

| SAMPLING | No standards nor guidelines. Methods based on available literature (no validation) |

| STORAGE, PROCESSING | As above. Knowledge on organotin behaviour upon storage derived from reference material certificatiob |

| EXTRACTION | No standard procedures regarding extraction or recovery checks. However, available CRMs |

| DERIVATISATION (if necessary) | Secondary standards, e.g. ethylated organotins, however hardly available in the commerce |

| SEPARATION | Internal standards (available in chemical reagent companies) |

| RESULTS ← FINAL DETECTION | Primary calibrants (available, as above) |

Figure 2.1 An example of a traceability chain for a measurement of an unknown concentration (organotin residues in a marine sediment in this case) in which several steps introduce appreciable uncertainty. Reproduced from Quevauviller, *Anal. Bioanal. Chem.* **382**, 1800 (2005), with kind permission from Springer Science and Business Media.

standard. The issue of traceability of a purchased analytical standard is the responsibility of the supplier and documented in the Certificate of Analysis. Nonetheless, before use the analyst is well advised to check the analytical standard for purity, as discussed below. Other precautions that should be adopted on first receipt, such as documentation, storage and stability, determination of assigned purity etc., are described in Section 9.4.4.

Measurement of chemical purity of an analytical standard is best done using a range of experimental techniques, one of which should be that used for the subsequent quantitative analysis, e.g., chromatography directly interfaced to mass spectrometry. In that example it is often considered wise to also use a different chromatography detector, e.g., a UV absorption detector, particularly a photodiode array that permits recording a wide-range UV–visible absorption spectrum. It is advisable as well to use an analytical method that operates on completely different principles from those underlying the method that will be used to analyze the real-world samples; high field nuclear magnetic resonance (NMR) would be one possibility for organic analytes.

Of course, if any of the impurities detected in the analytical standard are unknown and/or there are no suitable standards available for them, the accurate determination of their concentrations is not possible. However, if the estimates of impurity levels are clearly less than $\sim 1\%$ (estimated by assuming that, e.g., the mass spectrometric

response factors are the same for the impurities as for the analyte itself), any uncertainties thus introduced into the estimate of chemical purity will generally make an negligible contribution to the final uncertainty in the overall analytical results (Figure 1.2). If the condition of $> 99\%$ chemical purity is not met, the sample of analyte to be used as the analytical standard may have to be purified further if it is to be considered fit for purpose for the analytical task at hand. Alternatively, if the chemical purity is $< 99\%$ but is accurately known, it is sometimes acceptable practice, depending on the purpose of the analysis, to include this information in determination of the assigned purity (Section 9.4.4d) that accounts for the discrepancy between the weighed amount of analytical standard and the true quantity of analyte actually present in that weighed amount. Note that assigned purity accounts for purity aspects other than compounds that may be present as impurities in the analytical standard, e.g., corrections for the counter-ion if the standard is supplied as a salt, and (often variable) amounts of water adsorbed from the atmosphere. The practical aspects of determination of assigned purity and appropriate storage of hygroscopic materials are described in Section 9.4.4. An alternative approach, that bypasses measurements of chemical purity and conventional determinations of assigned purity, involves direct measurement of the concentration of the analyte in solution using quantitative NMR methodology in conjunction with a reference substance of some possibly unrelated compound whose purity is not subject to uncertainties arising from moisture, salts etc. This proton-counting method (Section 2.3.1) was developed (Burton 2005) for the case of an analyte that was extremely hygroscopic and difficult to separate from inorganic salt impurities, but is tedious and requires specialized equipment and operating skills and in fact the cited application involved certification of an analytical standard CRM (Section 2.2.2) where time and effort are secondary considerations.

A special kind of purity analysis for analytical standards, of growing importance, can arise for organic analytes that contain an asymmetric centre. In some cases, development of a new pharmaceutical can indicate that one of the optical isomers (enantiomers) is significantly more potent than the other, so the chiral purity of the standard must also be determined. Unfortunately, none of the chromatographic detectors with sufficient sensitivity for the kind of trace analysis of interest here (e.g., mass spectrometry and simple UV–visible absorption spectrophotometry) respond to chirality. However, chiral purity of an analytical standard can be determined by chiral chromatography (based on optically active stationary phases, Section 4.4.1d) with detection by mass spectrometry etc. Also, since by definition an analytical standard is available in macroscopic quantities (a few milligrams or more), the chiral characteristics of the standard can be investigated by the classical methods of organic chemistry (optical rotatory dispersion etc.). Racemic mixtures, e.g., samples obtained by conventional synthetic methods, can be used for quantitation since by definition they contain equal quantities of the enantiomers, but when analyzed by chiral chromatography it is essential to know which peak corresponds to which enantiomer and for this purpose a nonracemic mixture whose chiral purity is known at least approximately is essential. Practical implications of chiral purity are discussed in Section 9.4.4e.

The principles of the methods by which the analytical standard is used to obtain valid calibrations of an analytical method and then measurement of unknown concentrations are introduced in Section 2.6 and more fully described in Section 8.5.

2.2.2 Certified (Standard) Reference Materials (CRMs)

Some analytes, e.g., environmental pollutants, components or contaminants in foodstuffs, etc., are of such widespread concern that it is possible to purchase materials that are certified for the concentration of the specified analyte(s). The meaning in this context of 'certified'

requires some discussion. The following definitions are taken from the appropriate document of the International Standards Organization (ISO 2004):

Reference Material: 'A material, sufficiently homogeneous and stable with respect to one or more specified quantities, used for the calibration of a measuring system, or for the assessment of a measurement procedure, or for assigning values and measurement uncertainties to quantities of the same kind for other materials'.

Certified Reference Material (CRM): 'A reference material, accompanied by an authenticated certificate, having for each specified quantity a value, measurement uncertainty, and stated metrological traceability chain. The certificate should refer to a protocol describing the certification process'. CRMs are generally prepared in batches; for a given batch, quantity (or concentration) values and measurement uncertainties are obtained by measurements on samples representative of the batch. Note that the US National Institute for Standards and Technology (NIST) uses the terminology *Standard Reference Material (SRM)* for essentially the same thing.

Traceability: 'A property of the result of a measurement or the value of a standard whereby it can be related to stated references, usually national or international standards, through an unbroken chain of comparisons all having stated uncertainties'. (See also Sections 2.2.1 and 10.7.)

CRMs can be either analytical standards (Section 2.2.1), or calibration solutions that are solutions of an analytical standard at a certified concentration in a clean solvent, or natural material (matrix) CRMs that are prepared from material found in nature, i.e., are surrogates for the real samples to be analyzed. Matrix CRMs are invaluable as tools in quality assurance and quality control (QA/QC) in analytical laboratories, as discussed in Chapter 10. However, CRMs are generally too expensive to be used as everyday QCs and are generally used as method development tools or in occasional checks of the laboratory QCs (Section 2.5.3).

To qualify as certified, all reference materials must be prepared and characterized in accordance with a strict set of guidelines set out in ISO Guide 34-2000; a user's guide to this ISO document is available (ILAC 2000). In brief, certified calibration solutions are prepared by gravimetric methods from analytical standards whose purity (Section 2.2.1) has been assessed by at least two independent methods, and from a solvent of similar purity and known density. The concentration (expressed relative to both mass and volume of solution) of such a certified calibration solution is thus traceable to the appropriate SI quantities and units, the kilogram (and thus the mole via the known relative molecular mass) and the meter (in the case of volume concentration). For many CRMs produced by commercial companies the traceability is established, at least in part, through comparisons with CRMs produced by National Metrology Institutes (e.g., NIST, the National Research Council of Canada, the Institute for Reference Materials and Measurements of the European Union etc.). Certification of natural matrix CRMs requires even more demanding conditions, since at least two different methods of sample preparation (extraction and clean-up, Figure 1.2) and also of extract analysis, must be used and shown to agree to within stated, well-defined confidence limits (Section 8.2.5). CRMs and SRMs available for environmental analyses have been recently reviewed (Ulberth 2006; Wise 2006).

ISO Guide 31 details the requirements for preparation of the Certificate of Analysis for a CRM. These include the name and address of the certifying body and the names and signatures of the certifying officers, a general description of the CRM including its name and type with sufficient detail to distinguish it from similar materials, as well as its intended use. The code number for the material and batch number must also be included. The names and concentration(s) of the analyte(s) as well as the nature of the matrix should also be clearly listed and quantified (this might include descriptions of potential hazards). Instructions for correct use and storage of the material must be clearly specified. The listing of certified values must include their uncertainties, and where appropriate the level of homogeneity specified (the latter will determine the minimal aliquot size that must be used in order to reproduce the certified values, Section 8.6). Some analytes can be listed with uncertified values that are included for information purposes only as indications of concentration levels. The date of certification and its period of validity are required.

However, in many cases the analyst is faced with the task of quantitative analysis for an analyte for which a certified calibration solution is not available, or is too expensive and thus inappropriate for the stated purpose for which the analysis is to be conducted. In such cases it is necessary to prepare the calibration solutions from scratch from an analytical standard. Precautions in weighing the analytical standard are discussed in Section 2.3.3, and preparation of stock and spiking solutions in Section 9.5.4.

2.2.3 Surrogate Internal Standard (SIS)

Thus far discussion has been concerned with analytical standards corresponding to the target analytes of interest; such standards may be synthesized, or extracted and purified from natural sources, and are used in the preparation of solutions used as external standards for calibration of the analytical method (Figure 1.2), i.e., usually they are not added to the unknown sample (with the exception of the Method of Standard Additions, Sections 2.6.2 and 8.5.1). However, the most accurate and precise determinations also use a surrogate internal standard (SIS) (sometimes referred to as simply an 'internal standard', IS) that is added in a known amount to the unknown sample to monitor and correct for losses of the target analyte during the extraction and clean-up steps, as well as for variations in performance of the analytical instrument used. It is important to note that the SIS should be added at as early a stage of the overall procedure as possible, preferably before any extraction of target analytes (Figure 1.2), so that all uncontrolled losses of analyte are accounted for. Essentially the SIS is used in a manner such that all determinations of the target analyte (in both the actual analysis and the calibration procedure) are made relative to the (known) added quantity of the SIS (Section 8.5.2b). An obvious requirement for any SIS is that it must give rise to a signal from the analytical instrument that is unambiguously distinguished from that of the target analyte, but it must also behave as similarly as possible to the target analyte in the extraction and clean-up steps. These two requirements are clearly contradictory to some extent, and the various useful compromises between them are therefore now discussed. The ways in which an SIS is used in practice, and the care that must be taken in interpreting data thus obtained, are discussed in Section 8.5.2 and Chapter 9.

The preferred kind of SIS is a variant of the analyte in which one or more of the constituent atoms is enriched (preferably to 100 %) with a stable isotope of that atom. (Such isotopic variants of the target analyte are sometimes called 'isotopomers' of the natural analyte, but objections have been raised to this usage since the term is also used to indicate a special kind of isomer in which different 'isotopomers' are related only by exchange of different isotopes of an element among two or more positions in the molecule, e.g., $2\text{-}^{35}Cl\text{-}4\text{-}^{37}Cl\text{-phenol}$ and $2\text{-}^{37}Cl\text{-}4\text{-}^{35}Cl\text{-}$phenol are isotopomers. A suitable isotope-labeled surrogate IS for 2,4-dichlorophenol would be $3,5,6\text{-}D_3\text{-}2,$4-dichlorophenol, an example of what is sometimes called an 'isotopolog'; this nomenclature is used in this book). The most commonly used isotopes for this purpose in organic analysis are ^2H (also denoted as D, deuterium), ^{13}C and ^{15}N; these are of low natural abundance relative to

Physico-Chemical Effects of Stable Isotope Substitution: The Kinetic Isotope Effect

The observed effects of replacing a light isotope by a heavy one in a molecule are largest for H/D exchange. Some effects, e.g., diffusion rates, can be attributed directly to the resulting changes in total molecular mass. Other effects require a more subtle explanation based on changes in zero-point vibrational energy, as discussed in many texts on physical chemistry. A simple example of a C–H bond compared with a C–D bond, in the approximation of diatomic molecules, is shown in the sketch below. The potential energy curve is an electronic property determined by the nuclear charges, not their masses, and is thus the same for both; *in particular, the energy level corresponding to the dissociation limit is the same for both*. However, the quantized vibrational levels *are* a function of the masses, and the zero-point energy (ZPE) of C–X (X = H, D) is given by: $ZPE = h\nu/2 = (h/4\pi).[k(m_C + m_X)/m_C m_X)]^{0.5} \approx (h/4\pi).(k/m_X)^{0.5}$, where ν is the vibrational frequency of the oscillator (assumed to be harmonic), k is the force constant, h is Planck's Constant, and X = H or D. The further approximation that $m_C >> m_X$ gives the final result above (m_C here really denotes the mass of the rest of the ion or molecule).

For this approximate model the ZPE is lower for C–D than C–H by a factor of $\sim 2^{0.5}$, so the dissociation energies (relative to the *shared* dissociation limit) differ by the ZPE difference. This is the origin of the *primary kinetic isotope effect*, in which isotopic substitution is on an atom located on a bond that is broken at or before the transition state. The effect is by far the largest for H/D substitution in view of the large mass ratio and can lead to significant reaction rate decreases when H is replaced by D. (*Secondary kinetic isotope effects* arise when the isotopic substitution is well removed from the reaction site in the ion/molecule, and are much smaller.) *Primary kinetic isotope effects on ion fragmentations* in mass spectrometers are well studied (Cooks 1973) and are most important for ions of relatively low internal energy, not too far above that of the transition state. This *can be important for quantitation methods that use MS/MS detection* because the probability that the ionized analyte molecule will undergo a fragmentation reaction involving cleavage of a C–H bond can be significantly different from that for the corresponding deuterated internal standard! However, a full calibration method takes this into account (Chapter 8).

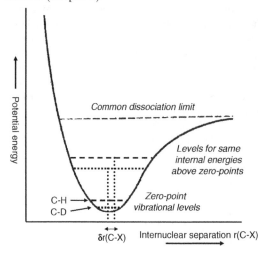

Another consequence of this ZPE difference arises for nonreacting species with the same internal energy (measured relative to the corresponding ZPEs) regardless of isotopic composition, e.g., for the thermalized molecules at the same temperature. Then the *average* internal energies *relative to the common potential energy curve* will also differ by the ZPE difference, as shown in the sketch. Because of the strongly asymmetric shape of the potential energy curve (i.e., the oscillator is *not* strictly harmonic), the D-substituted species (at lower internal energy relative to the potential curve) will have a smaller average bond length r(C–X), and thus a smaller effective molecular volume. This difference $\delta r(C–X)$ via that in molecular volume is believed to be at least partially responsible for the small but observable differences in chromatographic retention times between native and deuterium-labeled molecules (see Table 2.1).

the isotopes ^1H, ^{12}C and ^{14}N (Table 1.4) and this is the key to their use as surrogate internal standards with mass spectrometric detection, although problems can arise from the natural abundances of the heavier isotopes in the target analyte (see Sections 8.5.2c and 9.4.5b).

Similarly, for trace metal analysis the preferred stable isotope would be one that is naturally present at as low a percent abundance as possible. In general, the synthesis of an isotopic variant in which hydrogen atoms (usually at specific positions in the molecule) are replaced by deuterium atoms is the easiest to achieve, but caution is required to ensure that, under the conditions of the analytical procedure, there is no possibility of back exchange (of hydrogen for deuterium) through interactions with solvent or other constituents of the sample extract. It should also be borne in mind that, in some cases where the stable isotope substitution is deuterium for hydrogen, although the ionization efficiency (an electronic property of the molecule, essentially independent of nuclear mass) for analyte and SIS should be the same, this is not necessarily true for the fragmentation efficiencies if MS/MS (Section 6.2.2) is used for detection as a consequence of primary isotope effects (see text box on physical effects of isotopic substitution). For this reason the MS/MS responses of analyte and SIS should never be assumed to be equal; instead a full determination of the calibration curve should be undertaken, plotting response ratios for the analyte and SIS as a function of their relative concentrations (Section 8.5.2), and indeed best practice calls for this same precaution for all mass spectrometer calibrations. Moreover, the molecular volumes of analyte and deuterium-labeled SIS can be significantly different, and it is believed that this may account (at least partially) for differences in chromatographic retention times (see text box).

Preparation of an SIS labeled with one or more stable isotopes can be a demanding synthetic task in the case of organic analytes; not only is the chemical purity a concern, but also the isotopic purity, that can essentially be determined only through mass spectrometry, must be determined and preferably be as high as possible. Several methods for determining isotopic purity of a labeled compound from a mass spectrum have been published, but most of the earlier methods are subject to criticism; a critique of the earlier methods has been published (Jennings 2005), together with a viable new method that takes account of spectral noise, variable resolution etc., based on earlier work (Brauman 1966) on a related topic. This method (Jennings 2005) uses the measured mass spectral data of the unlabeled compound (i.e., natural isotopic abundances of all constituent elements) to represent any combinations of isotopologs of that compound and determines relative abundances of these isotopologs, using relative abundances obtained directly from the particular mass spectrometer for the unlabeled compound to compensate for instrumental biases as well as naturally occurring isotope biases. In the examples shown in this work (Jennings 2005), relative standard errors in the calculated relative abundances of isotopologs were < 1% if good quality mass spectral data were used. This approach not only provides a value of 'isotopic purity', but also the relative abundances of the various isotopologs

that contribute to it; it appears to offer a definitive method in the present context, though application of the method in practice does require familiarity with straightforward matrix algebra.

In the case of ^{13}C incorporation, proton NMR can be used to estimate isotopic purity by exploiting the spin–spin splittings that are zero for ^{12}C; of course this method is not applicable to carbon atoms in the molecule that are not directly bonded to at least one hydrogen atom, e.g., a carboxyl or keto group. A nondestructive method for determining isotopic purity, using the infrared thermal lens effect (Tran 1994, 1994a), has been applied to some small molecules but will require considerable development before it can be considered for more general use.

At this point it is important to emphasize that, although the chemical and isotopic purities of an isotope-labeled SIS should be known to within reasonable levels of accuracy and precision, it is not necessary to determine these purity parameters to the extent that is necessary for the analytical standard as long as the two standards are used together as described in Section 8.5.2b. In that case, as long as there are no cross-contributions between the mass spectrometric signals used to monitor the analyte and SIS (Sections 8.5.2c and 9.4.5b), and if the same amount of SIS is added to the unknown sample as to the calibration solutions, the SIS response is used simply as a normalizing factor. The response of the analyte from the sample is thus compared with those from the calibration solutions so that the SIS concentration need not be known to within the highest degree of accuracy. Of course, this is not the case if the unknown analyte concentration is estimated simply by comparison of the corresponding analytical response directly with that from an SIS solution, but this approach is seldom used in practice. (See Section 11.5.1e for an approach to analysis of endogenous compounds in which the chemical and isotopic purities of an isotope-labeled standard must be determined to high levels of accuracy and precision.)

Although an isotopolog of an organic analyte is the preferred choice as a SIS, when this is not available other less advantageous possibilities can be used, such as a simple isomer of the analyte (e.g., 3,4-dichlorophenol for 2,4-dichlorophenol), or a homolog or analog of the analyte (e.g., 3-methyl-2,4-dichlorophenol for 2,4,5-trichlorophenol). Use of chromatography directly interfaced to a mass spectrometer as the detector provides a unique advantage in an analytical method that incorporates a SIS, since it can provide an unambiguous mass spectrometric signature (for any SIS that is not a simple isomer) to complement differences in retention times in distinguishing the SIS from the analyte. This is particularly important for any SIS that mimics the analyte closely with respect to behavior in the extraction and clean-up steps (a desirable feature), since such an SIS is in principle likely to also elute with a very similar retention time to the analyte owing to the close parallels between the physical properties that control these two steps. Simultaneous elution (or nearly so) of the SIS and analyte can actually be advantageous in cases where endogenous compounds co-extracted from the sample matrix can not be removed by the clean-up procedure, and also co-elute with the analyte.

Even if such matrix interferences do not contribute signals directly to the particular m/z value(s) monitored for the analyte, they can affect the analyte signals indirectly by competing for the ionization power of the mass spectrometer ion source; this ionization suppression effect is particularly important for electrospray ionization, as discussed in Sections 5.3.6a and 9.6. Note, however, that an isotopolog SIS is doubly advantageous in this regard as it most closely resembles the analyte in the extraction and

clean-up steps (Figure 1.2) and also elutes simultaneously or very nearly so with the analyte. For a SIS that does not co-elute with the analyte, any suppression effects are potentially different at the two retention times. In such cases it is advisable, where possible, to account to some extent for these effects through use of matrix matched calibrators, i.e., extracts of aliquots of blank (control) matrix spiked with known amounts of the analytical standard that presumably contain the same endogenous interferences present in the sample to be analyzed (see Section 8.5).

The different types of SIS are compared in Table 2.1 with respect to their advantages and disadvantages in the key steps in an analytical procedure incorporating chromatography directly interfaced to the detector, especially a mass spectrometer. A brief discussion of some of the consequences of isotopic substitution that are mentioned in Table 2.1 (particularly the primary kinetic isotope effect and also the resulting changes in molecular volume) is given in the previous text box.

Table 2.1 Comparison of properties of different types of surrogate internal standard (SIS) to those of the analyte in the context of trace analysis using chromatography with mass spectrometric detection

Property compared \ Type of SIS	Isotopically Labeled	Homolog[a]	Chemical Analog[b]
Structurally identical?	Yes	No	No
Chemical properties in Derivatization	Almost identical[c]	Somewhat similar if site of homology is well removed from the reactive group	Can be somewhat similar or significantly different
Physical properties (polarity, solubility etc.)	Almost identical especially if isotope label is a heavy atom (not D/H)	Somewhat different	Can be somewhat similar or significantly different
Chromatographic and SPE properties	Almost identical but differences can be significant for D/H substitution[d,e]	Somewhat different[e]	Can be somewhat similar (for isomers) or very different[e]
Mass spectrometry properties	Identical ionization efficiency; m/z shift predictable for intact ionized molecule; m/z shifts for product ions not necessarily predictable, relative abundances can differ for D/H substitution[c]	Ionization efficiency can be similar; m/z shift predictable for intact ionized molecule; m/z shifts for product ions not necessarily predictable, can be zero or correspond to original homology	Ionization efficiency similar for isomer, can be very different for other analogs; m/z for intact ionized molecule is identical for isomer[f] and different for other analogs; m/z of product ions can be identical or different
Potential for m/z overlap between isotopic distributions of analyte and SIS[g]	Higher isotopologs of analyte can overlap lowest for SIS if degree of labelling is insufficient	Direct overlap unlikely because of m/z difference of $14n$ Da; product ion of one might overlap with that of the other	100 % overlap for isomers; for other analogs overlap can occur if mass difference is too small

| Potential for co-eluting metabolites of analyte to contribute to analytical signal for SIS[h] | Two-electron reduction can yield metabolites with molecular mass shift of +2 Da, usually too small a difference for a D-labelled SIS | Several functional groups can yield metabolites with mass shift of ±14 Da | Several functional groups can yield metabolites with mass shift of +16 Da that can possibly interfere with an analog SIS |

a Differs from analyte by n methylene groups (plus or minus).

b Chemically similar to analyte via other than homology; examples are a structural isomer (e.g. different pattern of the same substituents on a phenyl ring) or an oxidized form (hydroxyl replaces hydrogen, carbonyl replaces methylene etc.).

c Differences in chemical reactivity, whether in derivatization reactions in solution or in gas-phase ion fragmentation reactions, result from kinetic isotope effects and are significant only when a X–H/D bond is involved in the reaction.

d Differences in these properties mainly reflect differences in molecular volume (that are significant only in the case of H/D substitution) arising from the smaller average bond length of an X–D bond (X = C,N,O) that in turn reflects the lower zero-point energy (see text box).

e Differences can result in the possibility of different degrees of ionization suppression (matrix effects, Section 5.3.6a).

f If intact ionized molecules are used (SIM mode) chromatographic resolution of analyte and SIS is essential to distinguish the analyte signals of the two, but this introduces the possibility of different degrees of ionization suppression (matrix effects).

g Overlap can lead to significant cross-contributions to analytical signals for analyte and SIS and thus to a requirement for nonlinear least squares regression (Section 8.5.2c).

h See Section 9.4.7b. Such contributions to the analytical signal for the SIS apply without qualification to SIM monitoring, but only indicate potential problems depending on whether or not the m/z values of the product ions of SIS and the metabolite differ.

2.2.4 Volumetric Internal Standard (VIS)

In some cases a different kind of internal standard, that need not bear any kind of structural relationship to the analyte, is added to the extract solution to be analyzed by some form of chromatography (see Figure 1.2). Such internal standards are intended to monitor and correct for uncontrolled variations in the volume of sample extract solution actually injected into the chromatograph, and are called volumetric internal standards (VIS), (Section 8.5.2a). Obviously a SIS can monitor such variations as well as those in the efficiency of extraction and clean-up, so a VIS is not necessary in such cases. Moreover, modern loop injectors (Section 4.4.4) used with high performance liquid chromatography (HPLC) can deliver a sample solution volume with very high precision. However, this is not true of injections of liquid solutions into the hot gas flow of a gas chromatograph (GC). Accordingly, a VIS is of significant value only for analyses involving GC and for which a SIS is not available. All that is required of a VIS is that it should yield a reasonable signal in the chromatographic analysis (whether or not a mass spectrometer is used for detection) and that it also elutes well-separated from the analyte (but not too differently since the detector sensitivity can sometimes vary during the course of a chromatographic run). The signal measured for the VIS provides a normalizing (correction) factor that accounts for uncontrolled variations in the injected volume.

2.3 The Analytical Balance

Although the chemical member of the SI base quantities is amount of substance with the mole as the corresponding SI unit (Section 1.2), when analytical methods are calibrated it ultimately relies on determining the masses of aliquots of the analytical standards and internal standards and possibly of the blank matrix samples used to make up the calibrators or calibration solutions, QC samples etc. The reliability of analytical measurements depends in the first instance on the purity analyses of the analytical standard (via the potency correction as discussed above and in Section 9.4.4d), together with the accuracy and precision with which these entities can be 'weighed' (i.e., determine their masses relative to the kilogram). Indeed, it can be argued that the single most important instrumentation involved in quantitative analysis is the analytical balance. (Recall that the 'weight' of an object, unlike its mass, is a measure of the gravitational force acting on that object and thus varies with the geographical location of the object, see discussion of Equation (2.3) below; when speaking of 'weighing' an object, this means determining its mass through comparison with a standard 'weight' whose mass has been measured by a chain of experiments that are ultimately traceable to the International Prototype of the kilogram, Figure 1.1). For this reason, even though this book is devoted to a description of the use of a 'mass spectrometer' (where the 'mass' here is that of a single molecule) for determinations of amount of substance, it is essential to discuss the principles of the analytical balance and of its proper use for accurate and precise determinations of the mass of a macroscopic object (e.g., the ensemble of molecules constituting the portion of analytical standard, see Section 9.5.4a).

A brief account of the historical development of the analytical balance is included here as a text box. The following discussion deals with the practical problems in measuring mass to within the best possible accuracy and precision. Modern balances are of the single-arm type,

---

A Brief History of the Analytical Balance

(See: www.history.nih.gov/exhibits/balances)

The ability to make reliable measurements of mass has always been a fundamental aspect of experimental science. Simple balances are known to have been used as early as 5000 BC. Indeed the word 'balance' is derived from the Latin word *bilanx* which means 'two pans', as illustrated in the picture of a crude two-pan balance below. Such a balance is a comparison instrument that compares standard or known 'weights' (really standard *masses*) placed on the left-hand pan to unknown samples or objects placed on the right-hand pan; when the standard and unknown masses are equal their corresponding equal 'weights' (i.e., vertical gravitational forces) restore the indicator needle attached to the rod carrying the pans to the central position on the attached scale.

The modern analytical balance originated during the mid 18th century, when the Scottish chemist Joseph Black developed the technique of using a lightweight, rigid beam supported on a knife-edged fulcrum. (Black was a great frequenter of clubs, usually suitable for *'highly respectable literary gentlemen'*, though it was apparently not unknown for him and his companions who included Adam Smith, David Hume, Alexander Carlyle, and James Hutton, to visit *'less salubrious premises'*, see http://www.chem.gla.ac.uk/~alanc/dept/black.htm). The accuracy achieved by Black's innovation far surpassed that of any other weighing device of its time and, until the mid 20th century, further improvements were largely based on availability of better materials (strong lightweight alloys for the beam and harder materials for the fulcrum) applied to Black's original design, together with technical refinements including pan brakes, magnetic damping of beam oscillation, built-in weight sets operated by dial knobs, and microscopic or microprojection reading of the angle of beam inclination. In their most advanced form, balances of this equal-arm type could, with extreme care, achieve accuracies of less than 0.001 mg for total masses of a few grams.

However, in 1948 the design of balances moved in the direction of single pan devices. In such devices the gravitational force ('weight') of the unknown mass in the pan is balanced against an electromechanical force that is calibrated against standard masses whose own calibration can be traced back to the prototype kilogram (see Chapter 1); this design is used in analytical balances and microbalances suitable for use in analytical chemistry. In some cases requiring mass measurement of much heavier objects, the weight force is used to deform a strain gauge, thus changing its electrical resistance that is calibrated against standard masses. For an introduction to the principles involved see Berli and Bönzil (1991). A discussion of sources of uncertainty in 'weighing' unknown masses, and methods of avoiding or correcting for such uncertainties, is given in the main text.

A Crude Two-Pan Balance   Joseph Black   Modern Equal-Arm Balance

---

in which one end of the arm holds the pan carrying the unknown mass and the other is attached to a sensor of some kind that generates an electrical signal that can be readily digitized for electronic readout. Excellent introductions to the principles of electronic weighing are available (Berli, 1991; Reichmuth 2001a; Davidson 2004). One type of sensor used in balances intended for very high loads (several kilograms up to a few tons) is the strain gauge, an elastically deformable body on which are mounted strain gauges, conductive films that change their resistance on deformation by the force exerted by the weight of the unknown mass in the pan. Suitable construction ensures that this resistance change is proportional to the acting deforming force. A different approach is a modern variation of the simple 'spring balance'; in this variation a vibrating string, in which the tension is altered by the force exerted from the pan carrying the unknown mass, is used. This change in the tension applied to the string causes a change in its vibrational frequency, which is detected electrically and calibrated. Such devices are used in balances designed for a few kilograms (e.g., shop scales for butchers etc.).

However, in analytical balances of interest here, with readability in the range 10^{-4} to 10^{-6} grams (100 to 1 μg), electromagnetic compensation is used as a suitable measurement principle. The loaded weight (gravitational force) from the unknown mass in the pan at one end of the arm is compensated by an electromagnetic force generated at the other end of the arm by a compensation coil through which a permanent electric current flows, and which is inserted in a permanent magnetic field to generate a force on the same principle as that of an electric motor (or indeed of a magnetic sector mass spectrometer! See Section 6.4.4). In the unloaded condition, or in the 'tare' position in which the balance is zeroed with an empty container on the pan, regulation of the current through the coil ensures that the readout of the system is in the zero position. The coil position is controlled to an accuracy better than a thousandth of a millimeter via an opto-electronic position sensor. This information on coil position is used to generate a compensation current in the coil that returns the weighing system to zero, and is directly proportional to the weight of the load. The value of the compensation current is digitized and, through appropriate calibration using standard masses (ultimately traceable to the international prototype of the kilogram, Figure 1.1), transformed to the mass of the unknown that is shown on the electronic display.

Balances of interest in this book are often classified with respect to their resolution, i.e., the smallest mass difference that can be reliably measured. Generally, the better the resolution, the lower the maximum mass that can be reliably weighed. Thus, the term analytical balance is generally applied to a device that can measure up to a few hundred grams with an uncertainty of ±0.1 mg; a semi-microbalance can weigh up to a few tens of grams with an uncertainty of ±0.01 mg; a microbalance is limited to a few grams with an uncertainty of ±0.001 mg; and an ultramicrobalance can handle only very small loads but can provide uncertainties of as little as ±0.1 μg.

calibrated properly and whether or not adequate documentation to support the calibration can be readily retrieved upon request. In its simplest form, calibration can be defined as an operational check that generally involves the use of traceable standard materials or test instruments. Analytical balances should be calibrated and checked against an ASTM Class 1 certified set of weights, and these should be re-certified at least every year or sooner if needed by an outside contractor. All certificates should be maintained in the laboratory facility records and be readily available for routine inspections or data audits.

The balance should be checked with certified ASTM Class 1 'daily check weights' each day the balance is used for measurements. The range of weights used for the calibration check should encompass the weight of the 'substance' being measured. Typically three standard weights are used, including one below and one above the expected weight ranges. A successful calibration check of each weight to within ±1% tolerance at each weight would typically qualify as a successful daily calibration check, but any analytical laboratory should have a documented procedure that establishes calibration check acceptance for each type of balance used. The definition of the percent tolerance is:

$$\text{Percent Tolerance} = \frac{(\text{Theoretical} - \text{Measured})}{(\text{Theoretical})} \times 100$$

If the check weights do not meet the established acceptance limits, a second set of certified weights could be used. If the second weight set meets the acceptance limits, the balance may be used. Any weights from the set of check weights that are outside of tolerance after confirmation of acceptable balance performance with the new set of check weights must not be used and should be sent for re-certification. Should neither of the two weight sets meet the acceptance limits, the balance should be taken out of operation until the required maintenance and re-qualification procedures are conducted. Once again, the exact practices and allowed contingencies for daily calibration checks should be documented in the laboratory Standard Operating Procedures (SOPs) or similar document that describes the routine procedures used in the laboratory. In addition to the routine daily calibration checks described above, many laboratories will use an outside contractor to calibrate the balance at least once a year and, depending on use, more frequently than that.

2.3.1 Balance Calibration

A critical factor that must be considered before weighing an analytical standard is whether or not the balance is

2.3.2 Sources of Uncertainty in Weighing

To achieve good performance in weighing with modern balances, precautions must be taken to avoid a series of internal and external effects that can introduce error. User-friendly accounts of these effects, and of the precautions required to avoid or mitigate them, is readily available (Brunner, 2002; Davidson 2004); the present account is largely derived from these documents and from a detailed account of the underlying theory (Reichmuth 2001a, 2001b).

Some of the necessary precautions are fairly obvious. The balance should be shielded from all drafts (e.g., from windows, doors, radiators, air conditioners, fume hoods etc.) and the draft shields provided should be opened to only the minimal extent required to allow loading of the unknown object and kept closed during the actual weighing. The table or bench on which the balance is placed should transmit as few vibrations as possible (best located at the corner of the room since these are the most rigid locations in a building), should be nonmagnetic (no steel), be unsusceptible to build-up of electrostatic charge (glass or plastic can be problematic) and should be reserved for balances only. The object to be weighed should be placed centrally on the weighing pan to avoid so-called cornerload deviations arising from improper transmission of the gravitational force (weight) of the object to the balance arm.

The object to be weighed should be at the same temperature as the balance to avoid errors arising from convection currents of air. If the object is at a higher temperature the convection currents will circulate in a fashion such that they move in an upward direction near the object, thus creating a negative bias in the recorded mass, while a lower temperature for the object will have an opposite effect (Figure 2.2); this effect is sometimes known as dynamic buoyancy. The magnitude of this effect can be surprisingly large, e.g., holding the object in a hand for a minute or so can introduce a sufficiently large temperature difference to introduce a measurable negative bias in the recorded mass (despite any moisture derived from perspiration that has the opposite effect). Similarly, radiant heating from direct sunlight, or from incandescent light bulbs too close to the balance, will lead to convection air currents within the balance and give rise to false readings.

The humidity of the air in the room also can have deleterious effects on weighing accuracy. A secondary effect of a temperature difference between the weighed object and the ambient atmosphere within the balance draft screens can create a humidity effect, since water can be lost from (evaporation) or gained (condensation) by the object, depending on the direction of the temperature difference. Another humidity-related effect is concerned with electrostatic charging. Electrically insulating materials, such as glass and plastics used to fabricate weighing bottles, standard flasks and other containers, can acquire and retain

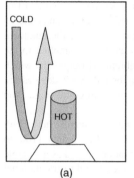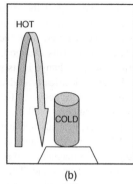

(a)   (b)

Figure 2.2 Effects of temperature differences between the object to be weighed and the balance: in (a) the object is at a higher temperature than its surroundings within the balance, and the resulting convection currents of air circulate so as to exert an upward force on the object, i.e., lead to a mass reading that is too low; in (b) the object is at a lower temperature resulting in the opposite effect.

a significant electrostatic charge primarily through friction during handling of powdered chemicals etc., if the ambient humidity is sufficiently low that the charge built up is unable to leak away quickly. A value of the ambient relative humidity of at least 40% at common room temperatures is generally required to permit such electrostatic charges to leak away to ground. (Relative humidity is the content of water vapor in the air, expressed as a percentage of the content at which liquid water starts to condense out at the specified temperature.) Modern balances are constructed so that all metallic components, particularly the weighing pan, are connected to ground via a three-pin power plug so as to minimize any Coulombic interactions between object and balance, but such effects can still occur. Use of a container made of a chemically inert metal, if available, will overcome any possibility of such problems when weighing dry powders. (While not related to humidity control, it is worth mentioning briefly here that magnetic interactions can also introduce an unpredictable bias into the recorded mass value, but again modern balances are constructed of nonmagnetic alloys in order to minimize this potential problem.)

An extreme case of problems related to humidity arises when the analytical standard to be weighed is extremely hygroscopic, to an extent that it is never possible to completely remove all the water from an otherwise chemically pure standard. A real example of this problem arose (Burton 2005) in the course of developing analytical standard solutions for saxitoxin and other shellfish toxins (Section 11.2.2). It was impossible to weigh these compounds to within any acceptable degree of either accuracy or precision as a result of their extremely hygroscopic character. The solution to the problem was to develop a method based on high field NMR. As always this quantitation procedure was based on comparison of ^{1}H NMR signals from the toxin sample with those from a standard, in this case a chemically unrelated compound (e.g., sucrose) that was chemically unrelated to the analyte but was nonhygroscopic (and thus weighable) and readily available at high purity. The method was essentially an exercise in proton counting, using well-resolved ^{1}H signals characteristic of each compound; comparison of these signal strengths, adjusted for the appropriate ratios of numbers of protons within each molecule contributing to the monitored signal to the total number of protons in that molecule, led immediately to measurement of the number of analyte molecules in the solution. To avoid contaminating the analyte it was not possible to use a mixed solution of analyte and standard in the same NMR tube, and accordingly it was unfortunately necessary to use separate tubes for analyte and standard.

Numerous factors contributing to the intensities of the NMR signals were evaluated, and it was shown that precision and accuracy of ca. 1% was obtainable for the solution concentrations using a commercial 11.7 T NMR spectrometer. Clearly, uncertainties of 1% in standard solution concentrations are larger than those that can be obtained if the analytical standard can be weighed using a modern balance. However, this approach represents a significant improvement over earlier versions of similar NMR methods that had concluded that accuracy when using external standards rarely exceeds $\pm 5\%$. An uncertainty of 1% in the standard solution concentration often makes a negligible contribution to the overall uncertainty in trace analyses of such difficult compounds in real-world samples such as shellfish tissue. This approach to quantitation has also been addressed with respect to validation (Malz 2005) and its applications (Rizzo 2005) to synthetic and combinatorial chemistry. The use of dimethylsulfone

as a potential universal reference standard for quantitative NMR has been described together with a discussion of the related uncertainty budget (Wells 2004).

A weighing perturbation that is often overlooked is that of static buoyancy (NPL 2002). The static buoyancy phenomenon is described by the famous Principle of Archimedes (see the text box dealing with this), together with the fact that the density of a chemical sample to be weighed is generally very different from that of the standard 'weights' (masses) used to calibrate an analytical balance (Section 2.3.1). Consider first an extreme example in which the calibration is done using a standard mass (e.g., 10.0000 g) on the weighing pan, with the entire balance operated in a vacuum. The parameters of the electromagnetic compensating device are then adjusted so that the balance readout is indeed 10.0000 g. Next the balance is switched off, the vacuum is broken, and the same standard mass is weighed in atmospheric air using the calibration just performed in vacuum. The indicated value will in fact be less than 10.0000 g as a result of the upward force (the buoyancy) that originally had balanced the weight of the (static) volume of air before it was displaced by the standard mass. The buoyancy can be calculated by calculating the volume of the displaced air as that of the standard mass that displaced it, for which the density of the material used is needed. Most standard masses used in such in-house calibrations are made of a high-grade nickel–chromium stainless steel, with density ~ 8 g.cm^{-3}, so the volume of the hypothetical standard mass is $(10.0000 g/8 \text{ g.cm}^{-3}) = 1.25 \text{ cm}^3$, and this is also the volume of the displaced air. An equation giving the density of air as a function of temperature and relative humidity, together with detailed tables of the parameters required, is available (Davis 1992); a typical density for air near sea level is 1.2×10^{-3} g.cm^{-3}, so the mass of the displaced air is 1.5×10^{-3} g and the weight of this mass is the buoyancy force on the standard mass; the reading on the balance will thus be 9.9985 g.

Archimedes' Principle and Chemical Analysis

Archimedes was a native of the Carthaginian (Phoenician) city of Syracuse in Sicily. He is widely regarded as one of the greatest mathematicians of all time (see: http://www-groups.dcs.st-and.ac.uk/~history/Mathematicians/Archimedes.html) However, he was also a practical inventor who devised many practical devices including *Archimedes' Screw*, a simple water pump that is still used today in some parts of the world. As a native of Syracuse he was drawn into service to devise new weapons during the Punic Wars, that ended in the final defeat of Carthage by Rome when Hannibal's famous attack via the Alps on what is now Italy ended in failure. In fact, Archimedes was killed in the fall of Syracuse to the Roman siege during the Second Punic (Phoenician) War in 212 BC.

(Continued)

Archimedes' most direct contribution to modern chemical analysis is his *Principle of Static Buoyancy* (often referred to simply as *Archimedes' Principle*), that evaluates the weight of an object when immersed in a fluid as the weight of the object measured in a vacuum *minus* the weight of the volume of fluid displaced by the object. (It is now recognized that the *weight* of an object is not its mass, but the gravitational force exerted on the object.) The reasoning is beautiful in its simplicity. Before introduction of the object, the volume of fluid that is eventually displaced by the object is in static equilibrium, so its weight must be exactly balanced by an equal and opposite upward force applied to it by the surrounding fluid that is not displaced. This balancing force will not be changed by substitution of the specified fluid volume by the object, which therefore also experiences the same upward force.

Archimedes famously discovered his Principle when he was asked to determine whether a crown had been made of pure gold or gold adulterated with silver, *probably the first documented experiment in nondestructive chemical analysis!*. The following is an account of this experiment that is directly related to the Principle and that was at least feasible using apparatus available at that time (see http://www.math.nyu.edu/~crorres/Archimedes/Crown/CrownIntro.html): *Suspend the crown from one end of a simple scale and balance it with an equal mass of gold suspended from the other end. Then immerse both the crown and gold into a container of water. If the scale remains in balance the crown and gold have equal volumes and so the crown has the same density as pure gold. However if the scale tilts in the direction of the gold, then the crown has the greater volume (experiences the greater upward force) so its density is less than gold and it must be an alloy of gold and some less dense material.* It is recognized that, in the absence of the means to create a vacuum, Archimedes compared the weights when immersed in water with those when in air, a much less dense fluid than water. (**NB** This experiment does in fact exemplify the effect of static buoyancy on the accuracy of weighing using modern analytical balances, discussed in the main text.)

Legend has it that Archimedes discovered his Principle while lying in his bath, and immediately leapt from the bath and ran naked through the building shouting '*Eureka*' ('I have found it'), a shout that echoes in our colloquial language even today. Whatever truth there may be in this story (and it certainly deserves to be true!), in fact he described the Principle in a much more extensive work titled *On Floating Bodies*, in which he also studied the stability of various floating bodies of different shapes and different densities, a subject crucial for ship design.

Of course the preceding thought-experiment, in which calibration of the balance is conducted in a vacuum, is an unrealistic example of the buoyancy effect. Consider now a realistic case where the calibration of the balance and the weighing of the unknown object are both conducted at atmospheric pressure; now the variation arises because the densities (and thus the volumes) of the standard mass and the unknown are different. The buoyancy force on the steel standard is still the same (equivalent to the gravitational force on 1.5×10^{-3} g), but in this case this buoyancy force is accounted for in the calibration procedure. Assume that the object to be weighed is a powdered or liquid chemical with density ~ 1 g.cm^{-3} and also of true mass 10.0000 g, so its volume is 10.0 cm^3; the mass of air that it displaces is thus 12×10^{-3} g. In other words, the upward (buoyancy) force on the sample is greater than that on the steel standard used to calibrate the balance by an amount equivalent to a mass of $(12 - 1.5 = 10.5) \times 10^{-3}$ g, and the indicated mass on

the balance readout will be 9.9895 g. Whether or not this is an acceptable error depends on the levels of accuracy and precision desired for the complete analytical method, i.e., on an analysis of fitness for purpose (see Preface and Section 9.2).

However, if the unknown sample has a true mass of only 1.0000 g, the buoyancy force on the steel standard mass is still equivalent to 1.5×10^{-3} g, and is accounted for in the calibration, while the corresponding value for the sample is now 1.2×10^{-3} g, i.e., smaller than that for the standard. In this case, the balance readout will be too high, but by only 0.3×10^{-3} g. (In general small discrepancies are characteristic of cases where the unknown mass is much smaller than the calibration range of the balance.) Depending on the purpose for which the weighing is being made, this kind of weighing accuracy may be acceptable, i.e., negligible compared with other sources of uncertainty in the overall analytical procedure. If not, a buoyancy correction is necessary (particularly for microbalances).

Using the same general reasoning as that used in the preceding discussion, it can be shown more generally (Brunner 2002) that:

$$m_{unknown} = R_{bal}$$
$$\times [1 - (D_{air}/D_{standard\ mass})]/[1 - (D_{air}/D_{unknown})]$$
$$[2.1]$$

where $m_{unknown}$ is the true mass of the unknown object, D represents density, and R_{bal} is the indicated reading on the balance assuming that the calibration is correct for a standard mass close to that of the unknown object. The most authoritative equation describing the density of air (Davis 1992) can be approximated as follows (Brunner 2002):

$$D_{air} = 10^{-3}$$
$$\times [0.348444P - (0.00252t - 0.020582rh)]/[273.15 + t]$$
$$[2.2]$$

where D_{air} is in $g.cm^{-3}$, P is the atmospheric pressure in mbar (hPa), t is temperature in °C, and rh is the relative humidity (%). For example, suppose an object of density $2.600\,g.cm^{-3}$ is weighed on a day when the atmospheric pressure is 1018 mbar, the relative humidity is 70 %, and the temperature 20 °C; also the calibration is assumed to be done on the same day (so D_{air} is the same for both) and that the balance readout for the unknown object is 200.0000 g. Then Equations [2.1] and [2.2] give:

$$D_{air} = 1.2029 \times 10^{-3} g.cm^{-3}, \text{ and}$$

$$m_{unknown} = 200.0625\,g,$$

representing a bias error of about 0.03 %. Again, fitness for purpose considerations will determine whether or not such corrections are necessary in any particular case. A detailed discussion (Schoonover 1981) of the calculation and application of air buoyancy corrections for single-pan direct reading analytical balances includes applications to calibration of syringes (Section 2.4.4) and to the weighing of granular or powdered materials or liquids in weighing bottles.

Another minor source of bias arises when the balance is calibrated at one location and is then used at another location of significantly different elevation, i.e., at a different distance from the centre of the Earth. Then the gravitational force on the object in the weighing pan will vary according to the inverse-square law. Suppose the calibration was done at sea level (radius of the Earth approximately 6.37×10^6 m), but is then used at a location 1000 m above sea level. Then the indicated reading R_{bal} for an object of true mass $m_{unknown}$ will be:

$$m_{unknown} = R_{bal}.(6.37 \times 10^6)^2/(6.3710 \times 10^6)^2$$
$$= R_{bal}.0.99969 \qquad [2.3]$$

This effect may not be significant for the purpose at hand, but the correction is readily made although a re-calibration at the place of use is recommended.

It is important to note that some modern analytical balances contain a microprocessor that constantly checks the calibration and corrects for some of these sources of error. Therefore, it is essential to consult the Operator's Manual for each balance to determine whether or not a given type of correction needs to be made by the operator. An interesting approach to improving the precision and accuracy that can be achieved in weighing a set of several objects, relative to the traditional method of estimating the mass of each object individually, has been described (Bzik 1998). This approach used experimental design techniques and experimental investigation to demonstrate that the precision and accuracy of such determinations can be improved by using a weighing scheme in which different sub-sets of multiple objects are on the balance simultaneously. The resulting system of linear equations is solved to yield the mass estimates for the objects, without requiring any more weighings than the number of objects provided that at least six objects are to be weighed (Bzik 1998).

This rather long discussion of the principles underlying the design and operation of a modern analytical balance is justified because all other aspects of trace quantitative analysis rely upon the accuracy and precision achieved in the weighing of standards and samples alike. A document outlining the special precautions required in weighing practices in the pharmaceutical industry (Scorer 2004) is available. In the case of reference standards, evaluation of purity (in the various senses discussed above) is equally important. The role of the mass spectrometer is now to 'simply' compare the signal from a known amount of the analytical standard with that from the analyte in the sample to be analyzed. Of course this comparison is not all that simple in practice, and the rest of this book is devoted to discussing the principles and pitfalls involved. Chapters 1–8 are devoted mainly to the more fundamental principles underlying the various methodologies that are deployed in trace quantitation using mass spectrometry, Chapters 9 and 10 are devoted to 'sharp-end' considerations of how these principles are applied in real-world situations, and Chapter 11 describes some examples selected from the literature.

2.3.3 Weighing the Analytical Standard

Since all quantitation can be traced back to a known amount (almost always weighed) of analytical standard, it is important that this weighing be as accurate and precise as possible (Section 2.3.1). Among other things this implies that some minimum amount of standard must be weighed to ensure negligible uncertainty, and in turn this will be defined by the sensitivity of the balance used. Recall that an analytical balance is a device that can measure up to a few hundred grams with an uncertainty of ± 0.1 mg, a semi-microbalance can weigh up to a few tens of grams with an uncertainty of ± 0.01 mg, and a microbalance is limited to a few grams with an uncertainty of ± 0.001 mg. Thus, for example, if quantities of the order of one milligram are to be weighed an analytical balance would be unsuitable, while careful use of a semi-microbalance would provide uncertainty of $\sim 1\%$; this may be fit for purpose in many cases of trace analysis where overall precision of 10–15% may be considered acceptable, but if not a microbalance would be used. Note that the question of how much of the analytical standard should be weighed out to make stock and spiking solutions can be determined only after the analytical strategy has been formulated (Section 9.4).

Similar comments apply to a SIS, although additional compromises can be made if extremely small amounts are weighed in order to preserve what is often a very limited and valuable commodity, as is often the case with stable isotope-labeled internal standards. The concentration of SIS generally need not be known to the highest levels of accuracy and precision since it is used (Sections 2.2.3 and 8.5.2) as a normalizing factor in the comparison of an unknown concentration with that of the analyte in the calibration solutions via the calibration curve.

2.4 Measurement and Dispensing of Volume

When an analytical standard or unknown sample is injected into a measuring instrument it is almost always in solution in some suitable solvent, so the question of measurement of liquid volume becomes an important consideration. Moreover, the analytical standard will be weighed when a series of calibration solutions is made; however, frequently in practice the solvent will not be weighed (the best practice), but will be added to make the solution volume up to some pre-determined value using a volumetric standard flask. The volume concentration of the solution is thereby established, and the volume of solution delivered by a standardized pipet, loop injector or similar device determines the amount of analyte injected.

The accuracy and precision of measurement of liquid volumes, and of their dispensing or injection into an analytical instrument, is usually appreciably worse than those obtainable by weighing. Moreover, since volume is a function of temperature via the coefficient of thermal

expansion, so is the concentration if measured relative to volume (but not if measured relative to the mass of the solution); however, this effect is usually negligible relative to other experimental uncertainties. For example, the (volume) coefficient of thermal expansion $\gamma \equiv (1/V)(\partial V/\partial T)_P$, i.e., the fractional change in volume per unit temperature change, is $2 - 3 \times 10^{-4}$ for water between $20 - 30°$ C, and 1.1×10^{-4} for ethanol at room temperature, so the fractional change in volume for a temperature change ΔT is $(\Delta V/V) \approx \gamma \cdot \Delta T$. Even for a $10°$ C change this is only 0.1–0.3% and proportionately less for a more realistic temperature variation in a laboratory. (Note that the thermal expansion coefficient for borosilicate glass is an order of magnitude lower than that of the common solvents and so makes an even more minor contribution.)

Here the properties of four important components that either measure or deliver liquid volumes and that are used in the preparation of solutions and extracts for analysis using chromatography are considered. As for the analytical balance, the proper specification and use of these volumetric devices is crucial if the ultimate quantitative analysis is to result in the desired levels of accuracy and precision. Considering the range of pipets and volumetric flasks found in a typical laboratory, there are literally thousands of combinations of glassware that may be used to dilute a stock solution in varying numbers of steps, and similarly there may be hundreds of combinations suitable for preparing a specific concentration. (See Section 9.5.4 for a discussion of practicalities involved in preparation of stock, sub-stock and spiking solutions.) A detailed paper (Lam 1980) presents tabular data on the optimum choice of volumetric glassware and the number of steps to use to minimize the final relative error (derived from propagation of error, see Section 8.2.2) in the dilution process; since there is no closed-form algebraic solution to the problem, and the number of combinations that must be considered is often very large, the results of this study (Lam 1980) are presented there as a table (too large for reproduction here) rather than as an algorithm. However, this detailed treatment accounts only for the random error contribution to the dilution procedure; the systematic errors arising from temperature variations, meniscus reading and pipet delivery errors may negate part or all of the improvements in precision and accuracy that are potentially gained by closely following this procedure (Lam 1980).

2.4.1 Standard Volumetric Flasks

An important question concerns the accuracy and precision in the total volume of a solution that can be obtained using a standard flask. The guaranteed degree of accuracy is indicated by the ' accuracy class' of the flask; the underlying definition of this concept for all measuring instruments is described in a document published by the International Organization of Legal Metrology (OIML, the acronym for the French name of the organization), and available at: http://www.oiml.org/publications/R/R034-e79.pdf

In the case of standard volumetric flasks, the accuracy class is designated by a capital (upper case) Roman letter together with the certified volume and its uncertainty at the specified temperature, all stamped on the flask; the maximum error for a volumetric flask can be given as an absolute value, so it qualifies for designation by an upper case letter. (Measuring instruments for which the maximum errors are proportional to the measured value, and are thus expressed as relative errors, are designated by numbers related to the percent maximum error.) In the case of volumetric flasks, Class A provides an accuracy ranging from 1.00 ± 0.01 mL, 2.00 ± 0.015 mL, to 10.00 ± 0.02 mL and 50.00 ± 0.05 mL. (Recall that the litre (L) is a SI-acceptable alternative name for 1 dm^3.)

The experimental precision is the more demanding parameter in the case of volumetric flasks, as a result of difficulties in filling up to the calibration mark but no further. The recommended practice is to fill until the bottom of the liquid meniscus (that is concave upwards for all solvents of interest here in glass) just touches the calibration line; there is a degree of subjectivity in this judgment, both for a given operator (e.g., on different days) and between operators. For the most demanding work, e.g., preparation of a primary stock solution (Section 9.5.4), it is thus advisable for the operator who will be using the flask to estimate both the accuracy and precision by repetitive measurements of the mass of a suitable liquid of known density in the flask that is filled to the mark according to the method used by that operator.

Once a standard solution has been made, there remains the problem of dispensing it. For example, a series of calibration solutions covering the required concentration range, or spiking solutions used to add to blank matrix in preparation of matrix matched calibrators, are often made by preparing one stock solution of known concentration and then preparing a series of appropriately diluted solutions (Section 9.5.4). The dispensing of the stock solution is often done by transferring aliquots into standard flasks using a pipet and then making up the volumes to the calibration marks.

2.4.2 Pipets

A pipet (sometimes spelled 'pipette') is a laboratory instrument used to transport a measured quantity of liquid.

It works by creating a vacuum above a liquid-holding chamber and, after the liquid-holding chamber has been brought into contact with the liquid, selectively releasing this vacuum so that the liquid is drawn up into the chamber ready to be dispensed. Medicine droppers are a type of pipet in which a rubber bulb creates the vacuum and a tapered glass tube serves as the liquid-holding chamber; classical laboratory pipets marked with a calibration line work on a similar principle (or even, in the past, via a vacuum created by sucking the liquid up by mouth!); such operation is neither accurate nor precise (nor safe!).

2.4.2a Classical Pipets

Such pipets (typically a few millilitres) are used in dilution schemes to prepare sub-stock and spiking solutions from a stock solution (Section 9.5.4), and a few words concerning their proper use are appropriate here. A TD ('to deliver') pipet is graduated in a way such that, when filled to the line (the *bottom* of the meniscus) and allowed to drain, the exact volume will be delivered while leaving a small amount of liquid in the tip of the pipet. For purposes of aspiration of liquids to be delivered by a volumetric pipet, a single-valve bulb, a three-valve bulb or an equivalent aspirating device may be used. When filling the volumetric pipet, do not allow the tip of the pipet to break the surface of the liquid while drawing in the solution; otherwise the sudden decrease in viscosity at the tip will cause a pressure surge and the inside of the rubber bulb could become contaminated.

When draining solutions into other volumetric flasks or containers, remove the volumetric pipet from the solution, transfer it to the receiving vessel and allow the solution to drain into the receiving vessel. The volumetric pipet tip should be in contact with the wall of the receiving vessel and should be angled in a way that will allow the solution to drain properly. If a single bulb pipet filler is used, draining of the solution is achieved by removing the filler from the pipet. If a three-valve bulb pipet filler is used, draining of the solution is achieved by pinching the side valve. Alternatively, the bulb may be removed from the pipet, but not until the solution has reached the bulbous portion of the pipet. When the meniscus has reached the tip of the volumetric pipet, wait approximately 10 seconds with the tip still touching the wall of the receiving vessel, then remove the volumetric pipet from the receiving vessel. The volumetric pipet should never be blown out to eject all liquid at the tip.

2.4.2b Micropipets

When small volumes (microlitres range) of liquids are to be dispensed with good accuracy and precision, the modern approach (Blues 2004) involves piston-driven pipets (Figure 2.3). In such a pipet the vacuum is generated by the vertical travel of a metal piston within an airtight sleeve, driven by the depression (either manually or by an electronically controlled electric motor) of a plunger to a first stop. In the air displacement type (see later) the air from a disposable tip attached to the bottom of the sleeve rises to fill the space left vacant in the barrel by movement of the piston, and the air is replaced by the liquid which is drawn up into the tip. The plunger is used to both draw up the liquid (as described above) and dispense it. Normal operation consists of depressing the plunger to the first stop while the pipet is held in the air. The tip is then submerged in the liquid and the plunger released in a slow and even manner; this draws the liquid up into the tip. The pipet is then moved to the receptor vessel (e.g., standard flask) and the plunger is again depressed to the first stop and then to the second

stop (dispensing) position, thus fully evacuating the tip and dispensing the liquid. The pipet tips are small disposable conical plastic tubes that attach directly to the bottom of the shaft and allow pipetting of liquids without any contact between the liquid and the pipet.

Piston driven pipets are of two main types. The first type is the air displacement pipet (Figure 2.3a) which, when filled, has a pocket of air (the 'dead air volume' representing the air originally in the pipet tip) between the head of the piston and the liquid in the cylinder. The second type (Figure 2.3b) is the positive (or direct) displacement pipet, in which the head of the piston is in direct contact with the liquid. Air displacement pipets have the advantage of minimizing the risk of contamination, since the liquid is in contact with only the disposable pipet tip, but are generally less accurate and precise than positive displacement pipets, especially for dispensing small volumes of liquid because of the compressibility of the dead air volume in the pipet. Most air displacement pipets have two stop positions for the piston travel, to allow the dead air volume to expel any liquid remaining within the tip after the main liquid volume has been dispensed. For trace analysis where cross-contamination and carryover (Sections 9.7 and 10.5.5) must be avoided, the air displacement type may be favored, although some manufacturers provide a disposable piston in addition to disposable tips so that the liquid need never be come into contact with the interior of the pipet itself. When not in use the pipets should be stored vertically in a rack.

To obtain maximum accuracy and repeatability, precautions are required when operating such a pipet (Blues 2004; Hemmings 2004). Operator consistency, in pressure applied to the plunger and operating speed, is essential particularly in view of hand fatigue resulting from prolonged operation (Martin 2002, Björksten 1994); the latter can be avoided through use of an electronically controlled motor driven pipet, sometimes operated under computer control (Dévé 2007). A dry tip should always be pre-wetted by drawing up and dispensing the liquid (either to waste or returned to the original container if the solution contains a scarce analyte) two or three times. (Use of a dry tip usually leads to delivery of too low a volume the first two or three times; pre-wetting reduces the surface tension on the inside walls of the tip and also provides the proper level of humidity inside the tip, thus reducing the possibility of evaporation of the liquid.)

When filling, the tip should be dipped well below the surface of the liquid, always at a 90° angle to the liquid surface or as close to that as possible, and removed in the same way (surface tension effects can give rise to variations in the aspirated volume if the tip is withdrawn at different angles). When dispensing into a (preferably)

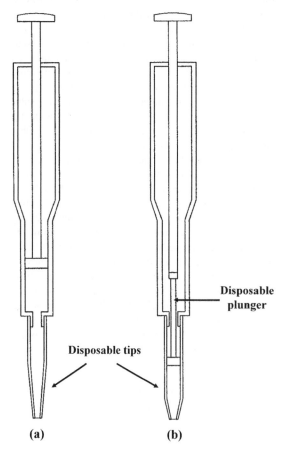

Disposable plunger

Disposable tips

(a)   **(b)**

Figure 2.3 Schematic cross-sections of (a) air displacement pipet and (b) positive displacement pipet.

glass vessel the pipet should be held at an angle between 30° and 45° and the tip placed against the side of the receiving vessel (the surface tension of the liquid on the glass provides an additional means of ensuring complete evacuation of the tip). It is important to avoid temperature variations, since the volume of a sample delivered by an air displacement pipet varies with atmospheric pressure, relative humidity and the vapor pressure of the liquid, all of which are temperature dependent; these effects can be minimized by ensuring that the liquid to be dispensed and the pipet start at the same temperature, and by avoiding excessive heating of the device by contact of the pipet body with the operator's hand.

All pipets require calibration at intervals determined by the extent of use (number of times the plunger is operated), the nature of the liquids being dispensed etc.; the international standard ISO 8655 (ISO 2002) for these calibrations recommends that all such devices should be calibrated at least once every three months, implying scheduling and record keeping (see Section 9.5.1 for further discussion). The gravimetric method is the reference method whenever possible (more difficult for micropipets, see below), and the calibration is valid only when the device is operated with the same kind of pipet tip as that used in the calibration procedure. ISO 8655 specifies the accuracy and precision limits as both absolute and relative values; the values are specified for fixed single-channel air displacement pipets, but for variable volume pipets the nominal volume is taken to be the maximum selectable volume. In such cases the required accuracy and precision apply to every selectable volume throughout the available range, e.g., for a $10-100\,\mu L$ variable volume pipet the accuracy must correspond to a maximum error of $0.8\,\mu L$ and the maximum permissible precision limit is $0.3\,\mu L$; these values are doubled for multichannel pipets.

Calibration is most often accomplished by weighing several samples of distilled water dispensed into suitable weighing bottles. This method is generally used for calibration of pipets down to $10\,\mu L$, though ultramicrobalances capable of weighing to within a few tenths of a microgram (Section 2.3) are required for the lowest volumes; such balances are not often available in busy analytical laboratories, so the pipets must be sent away for calibration (or some different method used, see below) and some less rigorous procedure used in-house as a routine check between calibrations. Water is the preferred liquid for this purpose, as its density is very well known as a function of temperature and it has a low vapor pressure at room temperature; a very complete account of this gravimetric method of pipet calibration is readily available (Rainin Instrument LLC 2005).

The accuracy and precision with which such variable volume pipets can be calibrated for a specific delivery volume varies with the volume range of that pipet. For example, a $2\,\mu L$ pipet can be calibrated for $2\,\mu L$ delivery to within 1–2 % accuracy and < 1 % precision, but it is possible to calibrate a pipet with maximum volume $20\,\mu L$ for a $2\,\mu L$ delivery to only ~8 % accuracy and 2 % precision; in contrast a $20\,\mu L$ pipet can be calibrated for $10\,\mu L$ with 1–2 % accuracy and ~0.5 % precision, but a $100\,\mu L$ pipet can be calibrated for $10\,\mu L$ to within only 3–4 % accuracy and 1 % precision (but with < 1% precision and 0.1–0.2 % precision for a $100\,\mu L$ delivery).

In the case of pipets designed to dispense very low volumes (a few microlitres), gravimetric calibration weighing with the balances commonly available in analytical laboratories can not provide accuracy and precision adequate for the most demanding purposes. A combined theoretical and experimental study (Schwartz 1989) of calibration of fixed volume pipets, that are typically calibrated gravimetrically by weighing replicate delivered aliquots, emphasized the inherent problems; when a single weighing vessel is used for this purpose and if the vessel is not tared between deliveries, the data analysis must be modified for serial correlation of errors. Appropriate statistical methods for calculating both the volume calibration and the standard deviation of the delivery volumes are illustrated (Schwartz 1989) by experimental calibration of a microlitre pipet.

If a low volume pipet is to be calibrated in-house rather than being sent away to a certified calibration service that has suitable balances available, a method based on measurements of optical density of a dye solution, before and after dispensing an aliquot of water from the pipet, is used. This approach is called the Artel PCS2 system (www.artel-usa.com); a useful description of this system is readily available (Hemmings 2004a). This system makes use of a very accurate and precise dual-wavelength UV–visible spectrophotometer that is readable to levels comparable to those of a microbalance. The method is based on principles already well established and used in microplate readers.

2.4.3 Loop Injectors for High Performance Liquid Chromatography (HPLC)

One significant advantage of high performance liquid chromatography (HPLC) over gas chromatography (GC) is derived from the reproducibility with which a volume of a solution can be injected onto the chromatographic column. In the case of GC, difficulties in reproducing the volume of a room temperature liquid injected into a gaseous environment (generally at a much higher temperature although cold on-column injection can alleviate some of the problems, see Section 4.4.4b) can lead to a requirement to include a volumetric internal standard (Section 2.2.4) to the solution to be analyzed. Here the main features are considered of modern loop injector valves designed for maximum precision (but not accuracy as will be discussed later) in the volume of a solution injected into an HPLC mobile phase stream and thus on to the column.

The most common design is based on a six-port injection valve comprising a fixed stator and a movable rotor. The stator carries six ports, i.e. connections to external components such as an injection loop, connecting tubing (mobile phase input from the HPLC pump, exits to the column and to waste) and an injection port to allow filling the loop with sample solution using a syringe. The stator is fabricated of a rugged material, usually stainless steel, while the rotor is made of a softer inert material, e.g., a fluorocarbon polymer. This facilitates formation of a liquid-tight seal between the two even while the rotor is rotated in order to change the internal connections among the ports.

The general principle on which such injectors operate is illustrated in Figure 2.4. A thorough discussion of the principles involved in achieving maximum reproducibility is provided in a Technical Note made available by the Rheodyne Company (Rheodyne LLC 2001). As shown in

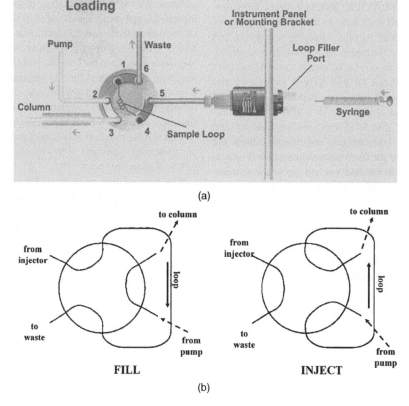

(a)

(b)

Figure 2.4   (a) Diagram of a six-port loop injector being filled by injection of the analyte solution from a syringe. In this mode the loop is bypassed and the mobile phase flows directly from pump to column through the injector valve via ports 2 and 3. In order to inject the sample in the loop onto the column the valve is rotated so that internal connections are now made between ports 1 and 2, and between ports 3 and 4. Reproduced from literature of Rheodyne Corporation, Technical Note #5 (2001), *Achieving Accuracy and Precision with Rheodyne Manual Sample Injectors*, with permission (www.rheodyne.com).

(b) Sketch of the same principles as described for (a). The sample loop is represented by the line on the extreme right, with arrows indicating the direction of liquid flow for filling the loop with sample using a syringe, and injecting the sample on the column by sweeping the sample solution out of the loop. The change in connections among the six ports (internal to the valve body) is accomplished by rotating the rotor relative to the stator (not shown in sketch).

Figure 2.4, the general idea is to configure a multi-port valve so that the flow of HPLC mobile phase from the pump can be directed to the column while the injection loop is filled from a syringe with the solution to be analyzed. Once this is done, rotation of the rotor directs the flow of mobile phase through the loop itself, thus sweeping the solution onto the column. By careful design and engineering, this can be achieved while introducing the minimum additional dead volume into the chromatography system and also the minimum carryover (memory effect from previous injections, Section 9.7). However, as always, achieving the optimum performance from such a device is somewhat more complicated. The single most important concept in this regard concerns the two methods of loading the sample solution into the sample chamber, 'complete-filling' and 'partial filling'.

To understand the differences between the relative performance of the two methods, it is necessary to discuss briefly the nature of laminar (nonturbulent) flow of a liquid in a narrow tube. Because of the frictional interactions of the liquid with the tube walls, the liquid at the centre of the tube moves faster than that near the walls, so that the flow profile is parabolic in shape. When an injector valve is in the 'load' position (Figure 2.4), the mobile phase that was in the sample chamber prior to switching is trapped and must be pushed out by the sample solution injected by syringe through the needle port. If the sample chamber (sample loop plus internal valve connections) is only partly filled with analyte solution, the loop will contain an ill-determined (approximately parabolic) profile of solution relative to the remaining mobile phase; this solution–mobile phase boundary region occupies approximately double the length within the loop compared with what would be occupied if the profile were flat instead of parabolic. As a result of this 'partial filling', the uncertainties in the repeatability (see Sections 8.1 and 10.2.5) of volume of analyte solution injected onto the column depend on the setting reproducibility of the syringe, the operator's skill etc., and is of the order of 1%. (Note that this is significantly improved if a modern autosampler is used instead of manual injection). However, the advantages of the partial filling method are that the volume injected can be varied via the volume loaded from the syringe without the need for changing the loop, and that the method is better suited to cases where the available volume of analyte solution is limited. In contrast, the 'complete filling' method requires injection of an excess of 2–5 sample chamber volumes (overfilling) to ensure that the sample chamber is filled with analyte solution that is not diluted by the mobile phase that originally filled the loop. The volume injected is then

the volume of the valve's sample chamber and the volumetric precision can be better than 0.2 % relative standard deviation (RSD, Section 8.2.1). However, the method is wasteful of analyte solution and the sample volume can be changed only by changing the sample loop. The method to be used in any particular case must be decided by consideration of fitness for purpose criteria.

The preceding discussion is a considerably abbreviated version of the much more detailed description (Rheodyne LLC 2001). In particular the advantages and limitations of the various designs of the injector valves are not discussed in detail here (but see discussion of Figure 2.5), as these vary among different manufacturers.

Thus far nothing has been said about the accuracy of volumes delivered by loop injectors. In fact, stated volumes on commercial injection loops can be in error by as much as 30 % in small loops (a few microlitres), so the actual volumes must be determined by experimental measurement. Also, the same loop on different injectors will, in general, deliver different volumes, reflecting differences in the internal injector passages that form part of the sample chamber. The actual volume injected using the 'complete filling' method could in principle be measured by some gravimetric procedure (that would, however, be difficult to design and implement), or by the following method (Rheodyne LLC 2001). The injector is operated in an HPLC system as indicated in Figure 2.5, and the observed peak heights and/or peak areas are recorded as a function of the sample volume loaded using the 'partial filling' method (these volumes must be measured using the injection syringe), covering the range from low fractional volumes to 2–5 times the nominal loop volume. A sketch of the type of curve obtained for a dual mode injector is shown in Figure 2.5a (this assumes that the HPLC detector is operating well within its linear range). The linear portion of the curve at low injected volumes corresponds to the 'partial filling' regime and the horizontal limit to the 'complete filling' regime. These two lines, when extrapolated, can be shown to intersect at the value of the loop volume actually injected onto the column. Some complications can arise for single mode injector designs, which give rise to a small offset, i.e., the linear portion does not extrapolate backwards to the origin (Figure 2.5b). This reflects a significant volume of the internal connection of the injector port to the loop itself, and this offset must be subtracted from the value obtained from the intersection of the two linear portions of the curve (Rheodyne LLC 2001).

Clearly this method of determining the absolute value of the volume dispensed to the column is unlikely to provide an accuracy of the same order as the reproducibility. Some of this uncertainty will arise from

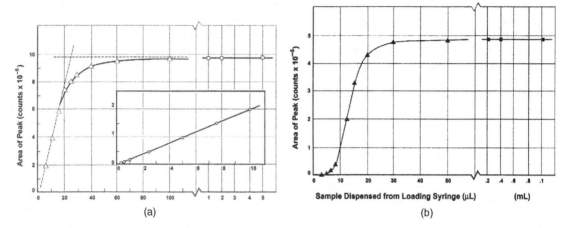

Figure 2.5   (a) Illustration of the method used to measure the actual volume delivered to the HPLC column by a sample loop of nominal volume 20 μL using a dual mode injector in which the loop is loaded from an internal passage that minimizes the amount of mobile phase that must be displaced by the analyte solution. The method uses observed chromatographic peak areas, assumed to be proportional to the actual volumes injected on-column, as a function of the volume injected into the loop by the syringe (the abscissa in μL); in this example three different syringes (a10 μL○, 100 μLΔ, and 5mL□) were used to cover the range. The linear portion corresponds to the 'partial filling' regime (see text), and the flat horizontal portion to the 'complete filling' regime. The value (μL) of the loop volume corresponds to the indicated intersection of the extrapolated linear response and the horizontal limit (note that departure from linearity starts at about 15 μL, i.e. about 60 % of total loop volume).
(b) As (a) but for a 5 μL loop using a single mode injector in which analyte solution injected by the syringe must expel mobile phase from the internal passage connecting the injection port to the loop before reaching the latter. Compared with (a) this graph is displaced to the right because of this additional volume required to displace mobile phase. Also the graph is curved at low values of volume dispensed by the syringe because these initial volumes reaching the loop are diluted because they correspond to only the leading portion of the parabolic laminar-flow profile of the analyte solution–mobile phase boundary (at the center of the tubing). For these reasons single mode injectors can achieve high precision only by using the complete filling method (see text).
Reproduced from company literature (Rheodyne 2001) with permission of Rheodyne LLC (www.rheodyne.com).

difficulties in measuring the volumes injected from the syringe; especially when operated manually, this is unlikely to provide accuracy and precision in the delivered volumes comparable with those characteristic of a modern pipet (Section 2.4.2). One approach that can alleviate this limitation is to use a gravimetric approach, e.g., weighing the filled syringe before and after each injection. If this is done, a limiting factor in these determinations of the absolute volume of the loop could be the variability in the HPLC peak heights or areas (Bakalyar 1976).

2.4.4 Syringes

Syringes of the type useful in the work discussed in this book are apparently simple devices, but appearances can be deceptive. Sketches of two types of syringe suitable for applications of the kind discussed here are shown in Figure 2.6. The volumes delivered by such syringes are certified by the manufacturer to be accurate to within ±1% of nominal volume with a precision of 1% at 80 % of the total volume. For syringes in the microlitre range

the stainless steel plunger is individually fitted to its matching borosilicate glass barrel to create a liquid-tight seal between the two. This implies that these plungers can not be interchanged or replaced if damaged, most often as a result of use with liquid samples that are heterogeneous (e.g., contain fine particulates) or homogeneous solutions that are prone to precipitation or reaction with the glass barrel, thus compromising the tight fit between plunger and barrel. The gas-tight syringes have a precision Teflon™ tip that provides a tight seal for both liquids and gases; replacement plunger assemblies are usually available for these syringes and thus they can be used with troublesome solutions. In all cases, scrupulous cleaning of the syringe after use will prolong the life of the syringe.

The best syringes have a calibrated scale containing 100 divisions in the volume markings (Figure 2.6); for example, a 10 μL syringe is readable to within 0.1 μL and the volume to be delivered can be set to within the same limit (corresponding to 1% of the total nominal volume). Checks on the accuracy and precision of volume delivery

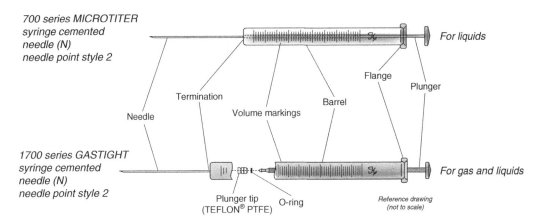

Figure 2.6 Sketches of syringe types suitable for applications discussed in this book. Reproduced from company literature (Hamilton Company 2006) with permission of the Hamilton Company.

are best done gravimetrically using distilled water at a known temperature (and thus known density). However, as discussed above for pipets, such re-calibrations of syringes of capacity below $100\,\mu L$ requires a semimicro- and/or microbalance (readable to within $10^{-5}-10^{-6}$ gram), not always available in an analytical laboratory.

A detailed discussion of the proper care and use of such syringes is available (Hamilton Company 2006). Here some of the more important points are noted. The syringe should be held via the flange (Figure 2.6) to avoid temperature variations from heating via the hand if the barrel is grasped. Air bubbles are a potential source of appreciable inaccuracies, so filling the syringe requires some care. The syringe should be successively filled and emptied several times, with the tip of the needle a few millimeters below the surface of the test solution, until no bubbles are visible in the barrel. Then the syringe should be overfilled to slightly above the desired volume marking, the needle removed from the sample solution and excess sample expelled until the desired volume is reached (hold the syringe so that it can be seen that the scale marking and the liquid meniscus are parallel). Then draw the plunger back very slightly so that the region of the needle a few millimeters from the tip contains air, and clean the external surface of the needle with a lint-free tissue; it is important here to ensure that the tissue does not come into contact with the liquid inside the syringe via the needle tip, as surface tension will draw some of the syringe contents into the tissue (a 'wick' effect). Now the sample can be injected into, e.g., an HPLC injector operated in the 'partially filled' mode or a GC injector.

As mentioned above, cleaning of these syringes is of crucial importance for good performance. Naturally, when used as described above they should also be dry before

use, so rinsing with a volatile solvent (e.g., high quality acetone) followed by drying can help ensure this.

2.5 Preparation of Solutions for Calibration

The fitness for purpose concept, introduced briefly in the Preface, will be a recurring theme throughout this book and an extensive discussion is included in Section 9.2. This concept addresses the fact of life that not all analytical problems need to be addressed using the same precautions to obtain the ultimately attainable accuracy and precision. For example, there is no point in expending large amounts of time and effort to obtain a calibration curve with $<1\%$ uncertainty if the analytical procedure yields uncertainties of 10% when applied to real-world samples.

There is no question that ultimate performance requires that all calibration solutions, calibrators and QC samples, should be made using only gravimetric procedures (Gernand 1989). This is a consequence of the foregoing discussions of accuracy and precision achievable when weighing with a modern analytical balance compared with those that are feasible using volumetric flasks and pipets. However, here trace analysis is being dealt with, implying that the microscopic amounts of analyte present in a typical analytical sample are much too small to be matched directly by weighed aliquots of the analytical standard. Accordingly it is necessary to weigh out, with suitable accuracy and precision, a macroscopic aliquot of the standard that must subsequently be sub-divided somehow to permit an instrument calibration that covers the range appropriate for the samples for analysis. The most convenient way in which such sub-division can be achieved is to dissolve the weighed standard in a suitable solvent and sub-divide the solution (see Section 9.5.4). Of course, it is in principle possible to accomplish this dilution procedure by using only weighings, but in practice some combination of weighings and volumetric dispensing is usually used as a suitable compromise between accuracy/precision and savings in time and cost. A major reason for adopting such a compromise is that the analytes of interest here are almost invariably introduced into the first stage of the analytical instrument (usually a chromatograph of some kind) in the form of a liquid solution *dispensed by volume* (e.g., via a loop injector in the case of HPLC, or directly by syringe for GC). Thus there is no point in undertaking the exacting task of making up all calibration solutions using only gravimetric methods. The only common exception is that of solutions of analytes that are certified for concentration to the highest achievable levels of accuracy and precision (e.g., CRMs).

There are two broad classes of calibration solutions used in analyses of the kind discussed here. The first class, referred to as calibration solutions, corresponds to solutions of the analyte(s) in clean solvent, possibly also containing internal standard(s); such solutions can be certified calibration solutions (Section 2.2.2) or solutions prepared in the analyst's own laboratory according to procedures determined ahead of time to be fit for the purpose for which the analysis is to be undertaken. The other class of calibration solutions, which will be referred to as matrix-matched calibrators (sometimes just "calibrators"), is prepared from aliquots of a blank (or control) matrix (identical or almost so to the matrix composing the analytical samples but devoid of the target analyte(s) to within the detection limits of the analytical method, see Section 9.4.7). When a suitable blank matrix is available, this is the preferred approach since many interferences and other effects are largely accounted for automatically.

2.5.1 Matrix-Free Calibration Solutions

The preparation of a series of solutions covering a range of concentrations of the analytical standard is almost invariably accomplished by weighing an appropriate amount (necessarily macroscopic) of the analytical standard and transferring it to a standard volumetric flask. Unless the standard flask is extremely (unrealistically) large the resulting solution is generally much too concentrated to be used directly for trace analysis. So a relatively concentrated stock solution, which can be subsequently used to prepare the calibration solutions covering the desired concentration range, is preferred (see Section 9.5.4). This range is pre-determined by consideration of factors such as regulatory limits, the range of concentrations

expected (or determined approximately by preliminary experiments) for the set of analytical samples, whether or not it is important to make measurements down to the lowest concentration possible etc. The stock solution itself can be made up by volume, i.e., adding solvent to the flask containing the weighed analytical standard up to the certified mark or, for the highest precision and accuracy, by weighing the flask before and after adding the solvent. Exactly the same comments apply to making stock solutions of internal standards.

Combining appropriate amounts of the stock solutions of analytical and internal standards, and subsequent dilution, can again be done volumetrically or gravimetrically (via a weighed syringe); the choice can only be determined by considerations of fitness for purpose, but most often careful manipulation of standard flasks and pipets (possibly re-calibrated for the purpose by weighings) is adequate for trace analysis (see Section 9.5.4 for details) as other uncertainties can considerably exceed those introduced in preparing the calibration solutions.

It is conventional practice (Sections 8.5.2b and 9.8.1) to ensure that the SIS concentration is fixed in all calibration solutions, at some value within the desired calibration range. The position of the SIS concentration within the range, e.g., near the middle or in the lower part of the range etc., has to be determined relative to the objectives of the analysis. In general, accuracy and precision are at their best when the peaks for analyte and SIS are of equal intensity, and this is related to the statistical result that accuracy and precision of a calibration curve are best at the center of the calibration curve (Section 8.4.1 and Figure 8.12).

2.5.2 Matrix Matched Calibrators

In this case (Section 9.8.1b) the solutions that are used to calibrate the analytical instrument are extracts prepared from weighed equal aliquots of blank matrix that have been spiked with the appropriate range of amounts of stock solutions of both analytical and internal standards. Here it is the amounts of the two standards added to the blank matrix that are the relevant quantities (expressed relative to the fixed mass of matrix), not the concentrations in the extracts obtained after taking the spiked blanks through the entire extraction and clean-up procedure (Figure 1.2). Again it is conventional practice to use a constant amount of SIS to spike the matrix blanks. As for calibration solutions, the amounts of analytical and internal standard stock solutions added to the matrix blanks can be determined either volumetrically (calibrated pipet) or gravimetrically (weighed syringe). It is common practice to use equal volumes of spiking solutions of different concentrations.

2.5.3 Quality Control (QC) Samples

Strictly speaking, QC samples are not used in the calibration procedure itself, but for ongoing checks on the implementation of the analytical method (including the calibration) and also in the development and validation of a new analytical method as discussed below. Routine QC samples can be CRMs, but this is rare in view of their scarcity and cost. Rather they are prepared (Section 9.8.2) from weighed aliquots of blank matrix spiked with known amounts of analytical standard, but not with the SIS since that addition is part of the overall procedure that is being checked. Note that best practice dictates that the stock solution of analytical standard used to prepare the QC samples should be prepared independently of that used for the calibrators, as a check on potential weighing errors. A more complete discussion of the preparation of QCs is given in Section 9.8.2. Typically, QCs are prepared in large batches (in bulk), aliquoted into smaller samples and then stored for subsequent use at the same conditions that are used to store the analytical samples; however, for method validation studies, QCs are sometimes prepared and 'analyzed fresh' on the same day.

2.5.3a QCs in Method Development and Validation

During method development (Chapter 9) and validation (Chapter 10), QCs are used for several purposes including checks on precision and accuracy, lower limit of quantitation (LLOQ), recovery and method robustness and ruggedness (Section 9.8.4), as well as stability studies of various kinds (Sections 10.2.7 and 10.2.8), studies of inter-day validation within a specified laboratory and cross-validations in inter-laboratory method transfer (Section 10.2.11). QC samples are also used during method development to assess the final method prior to validation; experimental runs that use QCs for this purpose are often referred to as assay pre-qualifications or pre-study assay evaluations (PSAE).

2.5.3b QCs in Sample Analysis

QCs have two main uses during sample analyses following final method validation. They are interspersed regularly among analytical samples and calibrators, blanks etc., within a sample stream to be run as a batch (e.g., overnight), in order to assess within-run acceptability. For this purpose the QCs can be either prepared fresh as a limited QC set or drawn from a bulk batch. The second main use during sample analysis is concerned with longer term tests over a complete study covering days or weeks, in which QCs are used to check that the performance of the method is within pre-specified control limits (of accuracy, precision, LLOQ etc.) and is not drifting out of control. For this second purpose the QCs must obviously be drawn from a bulk batch that has been stored under conditions that have been demonstrated to satisfy stability requirements.

The following discussion of quality control charts for lengthy analytical studies is a considerably abbreviated version of a much more complete treatment (Meier 2000). An example of a quality control plot for a method that is well under control is shown as Figure 2.7a. Two sets of

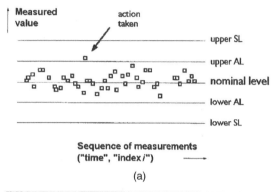

(a)

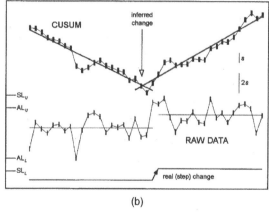

(b)

Figure 2.7 (a) Example of a QC control chart for a case in which the analytical method remains within specified limits (SL) of accuracy, precision etc. The action limits (AL) are chosen by the analyst to provide early warning that some method parameter(s) may be going out of control.

(b) As (a) but for an example in which indeed some parameter(s) did change about halfway through the time period. The point at which this change occurred is difficult to discern from the raw data but is more easily located when the data are transformed to provide the Cusum plot (see main text). The QC samples used for such long term control checks should be drawn from a single large batch stored under conditions that had been demonstrated to preserve an acceptable degree of stability. Reproduced with permission from Meier and Zünd, *Statistical Methods in Analytical Chemistry (2nd Edn)*, (2000).

limits are used, an outer set that corresponds to 'specification limits' for the permitted total variability (reflecting both accuracy and precision) determined by regulatory agencies, or by considerations of fitness for purpose when developing the analytical method (Section 9.3.3); for QC samples near the LLOQ for a bioanalytical application specification limits of $\pm 20\,\%$ might be appropriate. The upper and lower limits can sometimes be placed asymmetrically relative to the control (nominal) concentration for these QC samples. The inner limits are 'action limits' that are intended to provide early warning to the analyst that the method may be starting to drift out of control and that remedial action may be required; the action limits are determined by considering both the specification limits and the judgment of the analyst concerning the likelihood that the method will go quickly out of control once something starts to go wrong.

An example in which a significant change is observed to occur is shown in Figure 2.7b. In this example the raw data are plotted relative to the specification and action limits as in Figure 2.7a, but the point at which the change occurred is difficult to detect since in this case it is fairly small relative to the scatter of the data reflecting the analytical precision. A clever trick that can assist in such cases is to transform the raw data into a 'Cusum' chart (Doerrfel 1991), where 'Cusum' is shorthand for 'cumulative sum of deviations'. In this approach the reference concentration value C_{QC}^0 determined for the QC samples involved in the control chart is used to calculate deviations for the values $C_{QC}(i)$ determined at each time point i, and the cumulated sum of these deviations $\Sigma_i[C_{QC}(i) - C_{QC}^0]$ are calculated and plotted vs time as each new value of $C_{QC}(i)$ is measured. Since the deviations should occur symmetrically about the reference value (positive and negative deviations equally probable) the Cusum plot should be a horizontal line with smaller scatter than for the raw data (as a result of partial cancellation of positive and negative deviations). A small slope of the Cusum line suggests a degree of inaccuracy in C_{QC}^0. More important, a change in the slope of the Cusum plot will indicate the point at which the change in method performance occurred, and such discontinuities are easier to visualize than the corresponding shift in the average value of the raw data, as exemplified by Figure 2.7b. This approach is helpful when examining the method to determine which parameter has changed, since it narrows down the time points that should be examined for any abnormalities (e.g. change of solvent batch, variation in retention time etc.). If the phenomenon is one of gradual drift rather than a discontinuity of some kind in experimental parameters, this will show up in the Cusum plot as a more gradual

change of slope rather than a sharp change as exemplified in Figure 2.7b.

The use of QC control charts and subsidiary aids like the Cusum plot is an example of facing the fact that Murphy's Law predicts that something will surely go wrong if we wait long enough. These tools help in catching such problems early so as to take remedial action.

2.6 Introduction to Calibration Methods for Quantitative Analysis

This section provides a simplified introduction to the methods by which calibration solutions containing analytical (or reference) standards, with or without internal standards, are used in practice to measure amounts of analytes in unknown samples. The main deficiency of this brief account is its lack of any attempt to take into account experimental uncertainties (both random and systematic) and thus the level of confidence in the results thus obtained. While this book is directed towards analyses in which mass spectrometry is used as the chromatographic detection technique, most of the following discussion is applicable also to other detectors; the main exception concerns use of isotope-labeled surrogate internal standards, for which only mass spectrometry can provide adequate detection. A much more complete account of this material, including a discussion of the associated random and systematic errors, is given in Section 8.5.

To introduce the general concepts in this brief introduction, the only methods considered are those used to calibrate and measure the concentration of an analyte in an extract of the analytical sample (not in the original analytical sample) using calibration solutions of pure standard in clean solvent, i.e., not matrix matched calibrators. It is also assumed here that all potential sources of systematic error (e.g., analyte losses during extraction, clean-up, chromatographic analysis etc., see Section 8.5) are negligible. The notation used here is explained in Appendix 8.2. It is important to distinguish among non-primed symbols that refer to the analytical sample prior to extraction, clean-up etc.,: those marked with a single prime (′) referring to the sample extract solution (possibly spiked with internal standard), and those marked with a double prime (″) that refer to a solution of pure analytical standard used for calibration (possibly spiked with an internal standard).

2.6.1 Calibration Using an External Standard

This procedure is illustrated in Figure 2.8 and can be summarized in the following steps (recall that, to simplify this introduction to calibration and measurement only

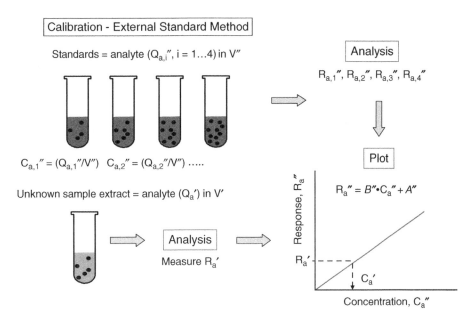

Figure 2.8   Schematic illustrating calibration of a chromatographic detector (e.g. a mass spectrometer) for quantitative analysis of a target analyte A, using external calibration with calibration solutions consisting of known amounts $Q_{a,1}''$, $Q_{a,2}''$ etc., of pure calibration standard in a fixed volume V'' of solvent i.e., calibration solutions of concentration $C_{a,1}''$, $C_{a,2}''$ etc. The detector response R_a'' is determined for each calibration solution and these data (R_a'' vs C_a'') are fitted using least-squares regression to a simple analytical equation: $R_a'' = B''.C_a'' + A''$, where B'' is the slope (sensitivity of the analytical method e.g., LC/MS) and A'' the intercept (ideally zero, that should correspond to the response of a blank solution, $C_a'' = 0$). Then the unknown solution is analyzed, and the measured response R_a'' is used to determine the unknown concentration C_a' of the extract of the analyte from the analytical sample by inversion of the experimentally determined form of the calibration equation (i.e., best-fit values of B'' and A'').

calibrations using clean standard solutions are being considered):

1. Prepare calibration solutions covering a suitable range of amounts (Q_a'', measured as mass by 'weighing' but sometimes transformed to number of moles) in volume V'', and measure the detector responses (R_a'') for these standards. Note that concentrations ($C_a'' = Q_a''/V''$) can be used instead of amounts. In principle the calibration solutions can be prepared from weighed amounts Q_a'' of pure analytical standard dissolved in clean solvent, or can be calibrators (Sections 2.5.2 and 9.8.1) prepared by extracting aliquots of a blank matrix that have been spiked with varying amounts of calibration standard; to avoid confusion in this introductory treatment only the former (simpler) case is discussed here.

Questions that often arise concern the number of different concentrations that should be used to obtain a calibration curve and the number of replicate calibration points that should be obtained at each concentration. These questions can only be answered in the spirit of fitness for purpose, but some general comments

are appropriate here. In biomedical analyses the appropriate guidelines (FDA 2001) specify that the number of standards used in constructing a calibration curve should be a function of the anticipated range of analytical values and the nature of the concentration–response relationship (e.g. linear or quadratic with respect to concentration, linear or nonlinear with respect to the fitting parameters in the calibration relationship, see Sections 8.3.6 and 8.3.8). The calibration range can be determined only after the overall analytical strategy has been determined (Section 9.4) and should include at least 6–8 different concentrations covering the expected range and including the lower limit of quantitation (LLOQ, Section 8.4). A calibration curve should also include a 'blank sample' (pure solvent in the case of calibration using clean standard solutions, or a control matrix sample processed without internal standard) and a 'zero sample' (a control matrix sample processed with internal standard).

With respect to the number of replicates at each calibration point, again this is a question to be answered in the spirit of fitness for purpose, but if three or more replicates are obtained this allows assessment of uncertainties arising from the preparation of the calibration curve independently of the repeatability of the analytical method itself. However, in turn this raises the question of what is meant by 'replicates' in this context. This could range from mere replicate injections of each calibration solution into the chromatograph on one hand, or complete replicates starting from independent weighings of the analytical standard, preparation of stock and spiking solutions, extraction and clean-up and final analysis, on the other (or possibly some intermediate case). For the most rigorously demanding tasks clearly $n \geq 3$ replicates in the latter sense are the appropriate choice (e.g. in certification of a matrix CRM), but in other less demanding applications a correspondingly less rigorous calibration procedure can be fit for purpose.

2. Plot R_a'' vs C_a'' (or Q_a'') and determine the best fit line through the points (Section 8.3). Ideally, a linear response is generated, to give a calibration equation of the form:

$$R_a'' = A + B.C_a'' \qquad [2.4]$$

where $R_a'' =$ instrument response, i.e., chromatographic peak area (counts) or peak height (counts s^{-1}) measured for the calibration solution, C_a'' (or Q_a'') $=$ analyte concentration (or amount) in the calibration solution, $B =$ instrument response factor (i.e., sensitivity, evaluated as the slope of the best-fit line, Figure 2.8), and $A =$ the intercept on the R_a'' axis at $C_a'' = 0$ (interpretable as the signal for a blank injection provided that $A \geq 0$).

3. The concentration of analyte (C_a') in the extract solution of the unknown sample is determined by measuring the response (R_a') for the extract of the unknown sample and calculating C_a' from the best-fit calibration equation (Equation [2.4]) (via 'inversion' of the calibration equation, Section 8.5):

$$C_a' = (R_a' - A)/B \qquad [2.5]$$

The amount of analyte Q_a in the original sample prior to extraction is given by ($C_a'.V'$) with a correction (to be determined experimentally) for the extraction efficiency (denoted F_a' in Section 8.5).

2.6.2 Calibration for the Method of Standard Additions

This analytical method (Section 8.5.1b) is generally used only for specialized applications as it requires large amounts of sample, analytical standard, and time. However, it does have advantages, e.g., no internal standard is required and ionization suppression effects (Sections 5.3.6a and 9.6) are automatically accounted for. The method is illustrated in Figure 2.9 and involves the following steps:

1. Varying, known amounts of analytical standard (Q_{msa}) are spiked directly into a series of aliquots of the unknown sample of fixed mass (W_S) or volume (V_S) if the sample matrix is liquid, each of which contains an unknown initial amount of the analyte (Q_a). Each spiked aliquot must be extracted and processed to give final extract solutions with a fixed final volume V'. The total amount of analyte in any given aliquot of analytical sample is then ($Q_a + Q_{msa}$).

2. Responses are measured both before (R_a') and after additions (spiking) of the analytical standard (R_{a+msa}').

3. Plot R_{a+msa}' for each spiked sample aliquot vs the amount Q_{msa} of *added* (spiked) standard in each sample and determine the best-fit line. A linear relationship between the response and the amount of analyte introduced is required for the method to be valid, i.e. the calibration equation is:

$$R_{a+msa}' = B_{msa}.(Q_{msa} + Q_a) = B_{msa}.Q_{msa} + A_{msa} \quad [2.6]$$

where the intercept on the response axis for $Q_{msa} = 0$ is $A_{msa} = (B_{msa}.Q_a)$.

4. The desired quantity Q_a is determined from the (in principle negative) intercept $Q_{msa,0}$ on the Q_{msa} axis at $R_{a+msa}' = 0$ (Figure 2.9); at this point Equation [2.6] becomes:

$$0 = B_{msa}.(Q_{msa,0} + Q_a) \text{ so that } Q_a = -Q_{msa,0} \quad [2.7]$$

Section 8.5.1b discusses the further implications of this remarkably simple result (compare Equation [8.66c]).

2.6.3 Calibration Using a Surrogate Internal Standard

This approach (Figure 1.2 and Section 8.5.2) is the most reliable analytical method used for high-throughput trace analyses, as it can account for analyte losses during extraction and processing of the sample. It also greatly alleviates the problems arising from ionization suppression (Sections 5.3.6a and 9.6) that can affect the sensitivity and response of the mass spectrometer in unpredictable

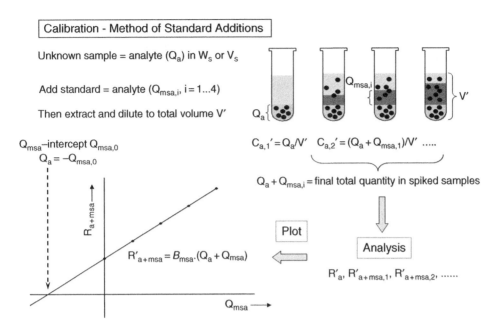

Figure 2.9 Schematic illustrating calibration and measurement using the Method of Standard Additions. A series of aliquots of the unknown sample, all the same size (e.g., a fixed mass W_S or volume V_S), are spiked with varying *known* amounts $Q_{msa,i}$ of analytical standard. The observed detector response R'_{a+msa} for each extract is then the sum of the contributions from the original unknown amount Q_a of analyte in each sample aliquot plus that from the known amount of analytical standard (Q_{msa}) spiked into that particular aliquot. These data (R'_{a+msa} vs Q_{msa}) are then fitted by least-squares regression to the simple calibration equation. The intercept on the Q_{msa} axis is necessarily a negative value $Q_{msa,0}$ corresponding to $R'_{a+msa} = 0$, so that $Q_a = -Q_{msa,0}$.

ways. The approach, illustrated in Figures 2.10 and 2.11, is summarized as follows:

1. Known *fixed* amounts of a surrogate internal standard (Q_{SIS}'') are spiked into a series of calibration solutions (varying concentrations $C_a'' = Q_a''/V''$) and also into the analytical sample that contains the unknown amount of analyte (Q_a).

2. Responses due to the analyte standard (R_a'') and those due to the internal standard (R_{SIS}'') are measured in the *same* chromatographic run for each of the spiked calibration solutions (Figure 2.10).

3. The response ratios R_a''/R_{SIS}'' are then calculated and plotted vs Q_a''/Q_{SIS}'', the ratio of amounts of analyte and internal standard in the calibration solutions (Figure 2.11). The calibration equation is:

$$R_a''/R_{SIS}'' = B''.(Q_a''/Q_{SIS}'') + A'' \qquad [2.8]$$

where, since ideally $R_a'' = 0$ if $Q_a'' = 0$ (but see Section 8.5.2b), the best-fit calibration line should include the origin (i.e., a zero intercept A''). Now the

extract solution of the analytical sample is analyzed to give the response ratio (R_a'/R_{SIS}'), and the corresponding value of Q_a'/Q_{SIS}' is calculated from the inverted calibration equation (Equation [2.8] where it is assumed here that $A'' = 0$):

$$Q_a'/Q_{SIS}' = (R_a'/R_{SIS}')/B'' \qquad [2.9]$$

The desired quantity is Q_a, not Q_a'. However, if the SIS behaves in the same fashion as the analyte in the extraction and cleanup steps (e.g., if the SIS is a heavy-atom isotopolog of the analyte, Section 2.2.3), $Q_a'/Q_{SIS}' = Q_a/Q_{SIS}$, i.e., fractional losses of analyte and SIS will be the same (see Section 8.5.2b for a discussion of this assumption), so that:

$$Q_a = Q_{SIS}.(R_a'/R_{SIS}')/B'' \qquad [2.10]$$

where Q_{SIS} is the known amount of SIS spiked into the analytical sample before extraction. This is also a very simple result, but it is valid only if some ideal conditions are met (Section 8.5.2b).

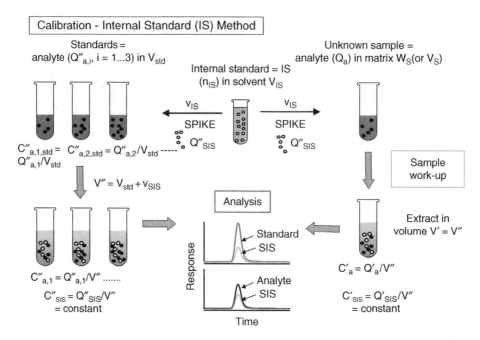

Figure 2.10   Schematic illustrating the experimental procedure used to calibrate a chromatographic detector for quantitative analysis of a target analyte using a fixed amount Q''_{SIS} of a surrogate internal standard in conjunction with calibration solutions containing known amounts $Q''_{a,i}$ of analytical standard (left). In this case the standard and the internal standard are analyzed together in the same chromatographic run (in this fictitious example the two co-elute, suggesting an isotope-labeled internal standard and thus a mass spectrometer as the detector). The same quantity Q''_{SIS} of internal standard is spiked into the unknown analytical sample, and the final extract volume of the spiked sample V' is adjusted to be the same as that of the spiked calibration solutions (V''). The extract is then analyzed as for the calibration solutions (right). The data are then processed as shown in Figure 2.11.

However the present simplified treatment does convey the general principles of the method. Compared with the method introduced in Section 2.6.1 (no SIS used) the present method has several advantages, e.g., the need for accurate and precise determination of the extraction efficiency of the analyte from the sample matrix is replaced by the considerably less challenging assumption that the analyte and SIS behave identically in that respect.

2.6.4 Curves used in Conjunction with 'Continuing Calibration Verification Standards'

The sensitivity and other parameters of a quantitative method can change with time. Such changes can be sudden (as a result of, e.g., an electrical surge or a switch to a new batch of solvent) or gradual (reflecting, e.g., a drift in room temperature or deterioration of a column or accumulation of crud in an atmospheric pressure ionization (API) interface, Section 5.3.3). A continuing calibration verification standard (CCVS) consists of either a QC sample (an aliquot of control matrix spiked with a known amount of analytical standard if the calibration

was done using matrix matched calibrators) or an analytical solution of known concentration. The purpose of the CCVS is to determine whether or not the methodology is 'in control' by verifying the linearity and slope of the calibration curve and thus assuring that the quantitative results obtained for the analytical samples reflect accurate and precise measurements. Such checks are required especially when a full curve recalibration is not performed on each analysis day. Most regulatory agencies (e.g. the US Environmental Protection Agency) stipulate upper and lower control limits for acceptance of the CCVS data relative to values predicted from the original calibration curve; these limits can vary with the nature of the analyte and matrix. (If the CCVS result lies above the upper control limit there is an implication that the method has somehow become more sensitive since the original calibration was performed.)

If the CCVS result lies outside the specified control limits, the method is said to be 'out of control'. The required remedial action will vary with the nature and purpose of the analysis, but can be as extreme as a full recalibration (the entire curve) and re-analysis

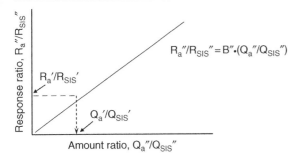

Calibration and Measurement- Internal Standard (IS) Method

$R_a''/R_{SIS}'' = B'' \cdot (Q_a''/Q_{SIS}'')$

Response ratio, R_a''/R_{SIS}''

R_a'/R_{SIS}'

Q_a'/Q_{SIS}'

Amount ratio, Q_a''/Q_{SIS}''

Figure 2.11 Schematic illustrating how data (obtained as shown in Figure 2.10) are used to obtain a calibration curve and thence determine the unknown amount Q_a. The ratios of detector responses (R_a''/R_{SIS}'') are plotted vs the ratio of known amounts (Q_a''/Q_{SIS}'') in the spiked calibration solutions. The extract of the spiked sample is the analyzed and the ratio of responses (R_a''/R_{SIS}'') is measured. The value of B'' (slope of the best least-squares fit of the calibration data to the simple calibration equation) is then used to transform the experimental value (R_a'/R_{SIS}') to (Q_a'/Q_{SIS}'). If the SIS is chosen appropriately the analyte and spiked SIS will behave identically during extraction and workup so that (Q_a'/Q_{SIS}') will be equal to (Q_a/Q_{SIS}); since Q_{SIS} is an experimentally known quantity, the value of Q_a can be calculated.

of all samples analyzed since the previous (presumably satisfactory) CCVS analysis, or as simple as re-establishing the slope of the curve by a single-point re-calibration and re-calculation of the data for the analytical samples. Even in the latter case, however, it is usually required that for all samples thus determined to be 'non-detect' relative to a minimum reporting level specified by the regulatory agency, the analyst must demonstrate that the recalibrated method is 'in control' at the method reporting level, i.e. is capable of detecting a true positive at this level. Moreover, all samples assigned as 'non-detects' using the original calibration curve must be re-analyzed using the new calibration. It is sometimes claimed that, if it is the upper control limit that is exceeded, the original non-detect results should be allowed to stand since the apparent increase in sensitivity implies that a non-detect result would not be affected by restoring the original (lower) sensitivity. However, this is generally not acceptable since the method is deemed to be 'out of control', so that the apparently higher sensitivity could reflect only a positive fluctuation rather than a negative one.

As mentioned previously, the remedial actions specified by the enforcing regulatory agencies can vary. Unless the frequency of CCVS analyses within a sample stream is specified by regulations, it becomes a matter of judgement

for the analyst to decide on the best compromise between frequent verifications (minimizes the extent of remedial action required since the previous CCVS analysis was performed) and sample throughput rate.

2.7 Summary of Key Concepts

The following list does not pretend to be exhaustive, but it does include most of the general concepts involved in **calibration and determination of unknown concentrations (amounts) of analytes**, used in this chapter and in subsequent chapters.

Absolute vs Relative Quantitation: Absolute quantitation is determination of the amount (or concentration) of an analyte; relative quantitation refers to determination of ratios of amounts (or concentrations) for two or more analytical samples, without measuring absolute values.

Analytical Sample: The material to be analyzed, comprising the target analyte(s) and the **matrix**, the mixture of often numerous and varied compounds (e.g., blood plasma, soil etc.) containing the analyte.

Analytical (or Reference) Standard: A sample of the analyte of known chemical purity, ideally the highest purity achievable and in a stable form.

Calibration (Standard) Curve : A plot of instrument response (e.g., LC/MS peak area) vs concentration or amount (mass or moles) of analyte injected. The **sensitivity** (or **response factor**) is best defined as the slope of the calibration curve (change in signal for unit change in quantity/concentration), but 'sensitivity' is often used in a colloquial sense to imply low LOD and/or LLOQ. When a SIS is used for maximum accuracy and precision, the *ratios* of instrument response (analyte:SIS) are plotted vs the corresponding concentration ratio. The standard solutions used for calibration can be clean solutions of the analytical standard (possibly plus SIS), or matrix-matched calibrators, i.e., analytical extracts of a blank matrix spiked with known amounts of analytical standard (plus possibly also SIS)

Calibration Solutions (Matrix Free): Clean solutions of the analytical standard (and SIS if appropriate) used to calibrate instrument response; use of this clean solutions can lead to failure to correct for variations in extraction efficiency, matrix effects (ionization suppression), instrument response etc. (see Chapter 8).

Calibrators (Matrix Matched Calibration Solutions): Samples of blank matrix to which known amounts of the analytical standard (and of SIS if available) have been added before extraction and clean-up; used to determine the concentration range of the complete assay and to calibrate the instrument response for analyte concentration while correcting for variations in extraction efficiency, matrix effects (ionization suppression), instrument response etc. (see Chapter 8).

Certified (Standard) Reference Materials (CRM/SRM): Real-world samples, as similar as possible to the unknown samples to be analyzed, for which the concentration(s) of analyte(s) have been certified by expert laboratories using several independent analytical methods under strict international conditions (Section 2.2.1). Such materials are rare and expensive, and if available are used only occasionally to check on the performance of the method and/or to validate **reference materials prepared in-house**.

Certified Calibration Solution: Solution of one or more analytes for which the chemical purities and concentrations are certified to within stated uncertainties; if the solution contains an isotopolog of an analyte its isotopic purity is also certified.

Dynamic Range: Range between the LLOQ and ULOQ, also known as the **calibration range** or **range of reliable response** for an analytical method. The **linear dynamic range** can refer to either the linear portion of a range of reliable response for a method, or to the linear response range for an analytical instrument obtained using clean calibration solutions of analytical standard.

Limit of Detection (LOD): The smallest signal attributable to the analyte according to specified criteria that vary from statistical definitions to a specified signal/noise ratio (Section 8.4.1) together with requirements to ensure **analyte identity**, i.e., to ensure that the measured signal can be attributed to the target analyte and only to that analyte. **Instrumental LOD** refers to detection of pure analytical standard in clean solvent, and **method LOD** to blank matrix spiked with standard analyte and subjected to the complete analytical method.

Lower and Upper Limits of Quantitation (LLOQ and ULOQ): The lowest and highest analyte amounts (or concentrations) that can be quantified to within pre-specified criteria (Section 8.4.2) for accuracy and precision; correspond to lowest and highest points on the calibration curve, the LLOQ is sometimes referred to as 'the' LOQ.

Method of Standard Additions: Aliquots (fixed mass or volume) of the untreated sample are 'spiked' with different known amounts of the analytical standard, and all aliquots (including the nonspiked raw sample) are analyzed in the same fashion. Requires neither a blank matrix nor a surrogate internal standard, but does require plentiful supplies of sample and of the analytical standard, and is time consuming.

Predicted Concentration: Estimate of concentration of analyte in the sample determined from the calibration curve; an **outlier** is an anomalously high or low value (Section 8.2.7).

Quality Control (QC) Samples: Aliquots of blank matrix spiked with known amounts of analytical standard, similar to calibrators but prepared using independent weighings. These are interspersed in a sample set to provide an independent check on the accuracy and precision of the determinations, and also in the validation of a new analytical method.

Response (Instrumental Signal): The electrical signal from the detector (e.g. mass spectrometer) used to detect the analyte; when a chromatography–MS combination is used, the response can be measured as either the chromatographic peak height or peak area, most often the latter.

Sample Work-Up: A set of procedures used to **extract** the analyte from the matrix, remove or minimize (**clean-up**) any compounds that are potential interferences in the instrumental analysis, and concentrate the analyte in the final **sample extract**, sometimes involving a solvent change for reasons of compatibility with subsequent chromatographic conditions.

Surrogate Internal Standard (SIS), often just referred to as **Internal Standard (IS):** A compound of known chemical purity that is added, in a known amount, to the sample for analysis before any extraction and work-up procedures; used to monitor and correct for analyte losses and variations in instrumental sensitivity; with mass spectrometric detection isotope-labeled versions of the analyte (**isotopologs**) are the best (this procedure is referred to as **isotope dilution mass spectrometry, IDMS**) but isomers or analogs can be used, see Table 2.1.

Target Analyte: Compound of known structure whose amount (or concentration) is required.

Validation: A test of the overall analytical method to establish that it meets pre-specified performance criteria, including analyte identity, LOD and LLOQ, dynamic range and linear dynamic range, accuracy, reproducibility (within-day precision) and repeatability (between-day precision), selectivity (freedom from interferences), sample stability under various relevant conditions etc.

Volumetric Internal Standard: A compound added, in a known amount, to the final cleaned-up sample extract just before analysis in order to monitor and correct for variability in the volume of the extract injected into a chromatograph; important for GC analyses if a surrogate internal standard is not available.

The following list summarizes other key concepts discussed in Chapter 2.

Weighing materials such as analytical standards, internal standards etc. is a crucial operation in any chemical analysis. It is therefore important to understand in a general way the principles of operation of a **modern analytical balance**, and the various problems (effects of buoyancy, elevation above sea level etc.) that can reduce the accuracy and precision of the weighed masses of these materials.

Volume is another key physical quantity since the amounts of analyte and/or internal standard that are used in trace analysis are too small to be weighed out each time. Accordingly liquid solutions of these materials at precisely and accurately known concentrations are made up and dispensed as fractional volumes of these solutions. Key pieces of apparatus in this context are the **standard volumetric flask** that is available in several grades (with respect to accuracy), and modern **pipets** used to dispense small amounts of solutions. Pipets must be carefully calibrated and operated if the best accuracy and precision are to be maintained. Similar comments apply to the **analytical syringe**. In quantitative liquid chromatography the **loop injector** is characterized by excellent precision especially in the complete-filling mode, but poor accuracy, with respect to the volume of liquid injected, and experimental methods must take this into account.

Picture of Joseph Black reproduced with permission of the University of Glasgow.

3

Tools of the Trade II. Theory of Chromatography

3.1 Introduction

Mikhail Semenovich Tswett: 'Father of Chromatography'

51. M. Tswett: Physikalisch-chemische Studien über das
Chlorophyll. Die Adsorptionen.

Eingegangen am 21. Juni 1906.

Michail Semjonowitsch Tswett

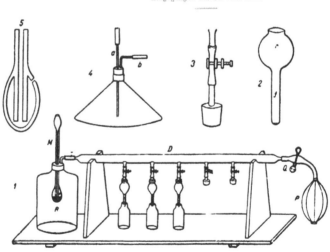

(Graphics are reproduced from Beneke (2003), with permission from Dr. Klaus Beneke, University of Kiel). M.S. Tswett was a Russian scientist who made an extensive study of plant pigments and developed the technique of chromatography to separate them. Tswett was born in Asti (Italy) in 1872 and studied at Geneva. He was appointed Professor of Botany at the University of Warsaw (then under the government of the Czar of Russia) in 1901, and during World War I organized the evacuation of the Botany Department to Moscow and Gorky.

Trace Quantitative Analysis by Mass Spectrometry   Robert K. Boyd, Cecilia Basic, Robert A. Bethem
© 2008 John Wiley & Sons, Ltd

(Continued)

In 1917 he was appointed professor of botany at Yuriev University (Estonia), but under threat of German invasion had to move once again, to Voronezh (Russia). On 21 March 1903, Tswett presented to the Biological Section of the Warsaw Society of Natural Sciences a lecture 'On a new category of adsorption phenomena and their application to biochemical analysis', now considered to have been the birth of liquid chromatography although other scientists had previously reported work on what would nowadays be called paper chromatography (Touchstone 1993). In a famous paper on 'Adsorption analysis and chromatographic method. Applications to the chemistry of chlorophyll' (*Ber. Deutsch. Bot. Ges.* **24** (1906) 384) he described the 'chromatogram', i.e., the colored zones on the 'chromatographic column' (the first time this terminology was used, albeit in German, to describe the composition of his extract from green leaves as being 'written in color'). In this work, using what now appears as a crude form of normal-phase adsorption chromatography, he detected carotenoids as yellow bands, and also invariably found two green zones for the chlorophylls (that we now call a and b). Sadly, the latter finding brought him into conflict with Richard Willstätter at München, the foremost authority on chlorophyll and a future (1915) Nobel Laureate. Willstätter, who had produced a crystalline sample of chlorophyll by classical procedures, insisted that Tswett's two zones were artifacts due to decomposition on the column and belittled the new 'chromatographic' method. Eventually Tswett was proved correct with respect to the two forms of chlorophyll, but Willstätter used his prestige to good effect with the result that chromatography was essentially ignored for the following 25 years (Engelhardt 2004). Tswett died (inflammation of the throat) in 1919 at age 47. Willstätter resigned his professorship at München in 1924 to protest the rising tide of anti-semitism. In 1938 he fled from the Gestapo with the help of a former pupil and managed to emigrate to Switzerland, losing all but a meagre part of his belongings. He died in Locarno, Switzerland, of a heart attack in 1942.

This book is about analysis of target analytes (compounds of known structure) present in complex matrices at trace levels, i.e., one molecule in 10^6–10^{12}. It is conceivable that a target molecule might possess characteristic properties (e.g., in fluorescence spectroscopy) so unique that its analysis could be successfully performed directly in its host matrix, but such examples are rare. In reality it is necessary to adopt a series of measures to selectively extract the analyte from most of the matrix components, possibly selectively concentrate the analyte in the original extract (often referred to as a 'clean-up' step), and finally to subject the sample extract to some form of high resolution separation technique before exposing the analyte to a measuring instrument such as a mass spectrometer (Figure 1.2).

By far the most common separatory technique used in this context is chromatography. This is a word that was coined from the Ancient Greek language, meaning 'writing in color' for historical reasons described in the text box that introduces this chapter; the text box also describes the contributions of M.S. Tswett, who has become known as the 'Father of Chromatography'. Interesting accounts of the history of chromatography are available (Lesney 1998, 2001; Touchstone 1993). (Note that the latter reference emphasizes that, although Tswett can be rightly regarded as the first to attempt any development of chromatography as a routinely practical

technique, other scientists had conducted experiments that are recognizably chromatographic using filter paper as stationary phase as early as 1861, and a few column separations were reported in the 1890s.) The principles underlying chromatographic separations (and also many of the clean-up procedures) are introduced below, together with a description of the most useful variants of the general principle.

There are many excellent textbooks describing the principles and practice of chromatography; an extensive listing of books and journals, together with some excellent expository material, can be found on the Internet (Kazakevich 1996). A remarkable source of information about the theory and practice of chromatography, that goes into as much depth and detail as the reader chooses, is provided by the very comprehensive and scholarly 'Chrom-Ed' books by R.P.W. Scott that are also freely available on the Internet (www.chromatography-online.org). Another invaluable source of information is available by using the search engine provided on the website of the magazine 'LC/GC' (http://www.lcgcmag.com/lcgc/).

It has been said that, in modern practice, 'the best mass spectrometrists are chromatographers'; perhaps the thought is better expressed as 'the best mass spectrometrists are *also* chromatographers'. In this spirit the present chapter on the theory of chromatography, and the following one on more practical aspects, have been included as a major portion of this book. The main content of this chapter includes a description of the *Plate Theory of chromatography*, which describes the elution times of analytes using a basic model of a chromatographic column as a series of 'theoretical plates' or compartments within each of which the analyte is fully equilibrated between the mobile and stationary phases. The Plate Theory does not account for the widths and shapes of the chromatographic peaks. These result from several kinetic processes which do not proceed at a sufficiently fast rate that the equilibrium between the phases truly is at equilibrium; this important aspect is discussed in a summary of the *Kinetic Theory of chromatography*. These two theories complement one another but, as originally formulated, they deal only with isocratic elution, i.e. elution by a mobile phase of fixed composition. This chapter also includes a summary of the extension of the Plate Theory to the case of gradient elution, i.e. elution with a mobile phase containing a mixture of solvents whose ratio is changed so that the content of the stronger solvent increases with time so as to improve separation efficiency.

For those for whom algebraic derivations of important relationships are either impenetrable or not a priority at this stage, these are presented in Appendices. The most important relationships and terms are briefly outlined in Section 3.3, and the symbols used to denote the various physico-chemical parameters involved in chromatography are listed at the end of this chapter. A comparison of parameters for isocratic and gradient elution is given in Table 3.2. A more strictly practical discussion of development of a chromatography procedure in a busy laboratory is given in Section 9.5.5. Also, unlike the other chapters, the 'Summary of Important Concepts' is included early in this Chapter rather than at the end as a guide to the rather theoretical approach taken in this Chapter.

3.2 General Principles of Chemical Separations

Almost all the separatory and clean-up methods relevant to this book are ultimately based on differences in partition coefficients of various compounds between two phases. The concept of a 'phase' in the present context (see also Section 4.3.2e) is that of a portion of matter that is separated from others by a clearly defined boundary; thus, solid, liquid and gas phases satisfy this definition, but it is possible to have two distinguishable liquid phases (e.g., water and carbon tetrachloride). A simple example of an application of the latter, that is familiar to anyone who has taken an organic chemistry laboratory course, is liquid–liquid extraction; a first step in purification of a desired synthesized compound is often to place the crude

reaction mixture in a separatory funnel together with a solvent that is not miscible with the solvent used for the synthesis. By judicious choice of solvents it is often possible to ensure that the desired compound is extracted from the reaction mixture into the second solvent much more efficiently than other components (excess reagents, side products etc.). This simple liquid–liquid procedure is still often employed with surprisingly effectiveness in modern analytical strategies, e.g., to extract pharmaceutical drugs and/or their metabolites from blood plasma. However, the same word 'phase' is also used in science to denote a difference in timing between two alternating current waveforms (see Section 6.4.5); the context makes it clear which of the two meanings is intended.

The equilibrium distribution of compound A between two phases (I and II) is usually described by the partition coefficient:

$$K^A_{I,II} = c^A_I / c^A_{II}$$

where $K^A_{I,II}$ is the partition coefficient (a special kind of equilibrium constant), and c^A_I is the (molar) concentration of compound A in phase I etc. (Strictly, the thermodynamically correct partition function is defined in terms of the *thermodynamic activities* of the compound, e.g., $a^A_I = c^A_I \cdot \gamma^A_I$, but for present purposes it is sufficient to assume that the activity coefficients γ are equal in the two phases, or are assumed to have the ideal value of unity.) Equilibrium considerations of this kind are relevant when phase partitioning is applied to batch extraction of an analyte from the matrix and in some clean-up methods; they are also the basis for the Plate Theory of chromatography, described in Section 3.4 and Appendix 3.1.

The major difference in practice between applications of phase partitioning in batch extraction (as in the simple organic chemistry experiment, i.e. not continuous or dynamic extraction) and chromatography is that one of the two phases (the mobile phase, that initially contains the liquid solution of analytes) is continuously replenished and caused to flow over and past the other (the solid stationary phase). The time dependence inevitably introduced into such a flow-past arrangement is why nonequilibrium rate effects must be incorporated into more complete theories of chromatography. As a result of differences in the thermodynamics and kinetics of partitioning between stationary and mobile phases, the various components of the analyzed solution will move at different rates when swept by the mobile phase past the stationary phase. In some methods (e.g., thin layer chromatography, TLC) the flow of mobile phase is stopped before any components of interest have been swept out of the stationary phase altogether, so the components are 'separated in space' along the flow path through the stationary phase. The separated components can then be removed (extracted) from the stationary phase and subjected to a range of analytical procedures with no inherent time constraints. More commonly the flow of mobile phase is maintained until all components of the mixture have eluted from the stationary phase; in this mode the components are 'separated in time' as they elute, so their detection and characterization (including quantitation) must by accomplished in real time 'on-the-fly'. As will be discussed later (Section 5.2.2), sometimes a chromatographic 'separation in time' is converted into a 'separation in space' by collecting discrete fractions in appropriate receptacles as they elute; this strategy combines the advantages of both approaches, i.e., the generally much superior separation efficiency that is possible with the 'separation in time' approach and the advantages arising from retaining separated fractions for multiple (or repeat) analyses, if desired.

The variety of mobile and stationary phases encountered in modern chromatography is much too extensive for a complete account in this book; indeed such a list would very quickly be out of date as a result of the intense activity in developing new and innovative phases. In gas–liquid chromatography (usually referred to as gas chromatography, GC) the mobile phase is a chemically inert gas (most often helium but sometimes hydrogen if precautions are taken). There is now a wide range of commercially available GC columns with stationary phases tailored to provide optimum separations for different classes of compounds. The most appropriate GC column currently available for a specific type of analyte is best determined by examining the literature and the information provided by column manufacturers. However, it is useful here to distinguish among some general classes of GC columns.

There are two general types of GC column: packed and capillary (also known as open tubular). Packed columns contain a finely divided, inert, solid support material (commonly based on diatomaceous earth) coated with a viscous liquid stationary phase. Most packed columns are 1.5–10 m in length and have an internal diameter of 2–4 mm. The original glass capillary columns had an internal diameter of a few tenths of a millimeter and were also one of two types, wall-coated open tubular (WCOT) and support-coated open tubular (SCOT). WCOT columns consist of a capillary tube whose walls are coated with liquid stationary phase. In SCOT columns the inner wall of the capillary is lined with a thin layer of support material such as diatomaceous earth, onto which the stationary phase has been adsorbed. Both of these types of capillary column provide greater separation efficiency than packed columns, but since 1979 they have

been largely superceded by a new type of WCOT column, the fused silica open tubular (FSOT) column. Indeed these columns have become so dominant in GC practice that they are now generally referred to simply as capillary GC columns. These fused silica columns have much thinner walls than the old original glass capillary columns, with internal diameters in the range of a few hundred micrometers, and are mechanically robust as the result of a polyimide coating on the outside walls; they are flexible and can be wound into coils. The stationary phase is chemically bonded to the inside walls (and thus provides low 'bleed' into the gas stream). The important parameters for such a column (Section 3.5.11) are its length and internal diameter, the thickness of the film of the bonded stationary phase and, of course, its chemical nature. Because GC columns are operated at high temperatures the range in which a stationary phase is thermally stable is also important.

The situation is considerably more complicated for liquid chromatography (LC), since now the composition of the mobile phase can be manipulated in a way that is not possible for GC; in later Sections of this chapter the concepts of isocratic vs gradient elution, reverse phase vs normal phase elution etc. will be described. As for GC, the internal diameters of LC columns intended for analytical work (high-performance chromatography, or HPLC) can vary from 4.6 mm to ∼100 μm, and the range of stationary phases that is available grows continuously. (Note, however, that essentially all LC columns are packed, not open tubular as in GC.) Because LC columns operate at or near room temperature, and in conjunction with a liquid mobile phase that can be tailored to suit a wide range of purposes, there is a much wider range of intermolecular interactions controlling the partitioning of analytes between mobile and stationary phases that can be exploited to provide chromatographic separations. A list of the principal types of interaction is shown in the accompanying text box; of these, essentially only the first two (adsorption and partition chromatographies) are relevant for GC. A more complete discussion of these concepts is given in Chapter 4.

Many of the 'clean-up' methods used in trace analysis exploit HPLC mobile and stationary phases in a somewhat different way. For example, a crude sample extract can be loaded onto a small cartridge packed with a suitable adsorbent under conditions such that the analyte (and its internal standard if present) are strongly retained. Washing the loaded cartridge removes many of the co-extracted compounds and the analyte plus SIS can then be eluted from the cartridge by a suitable change of solvent; in effect this is a kind of 'binary' chromatography, i.e., a compound is either retained on the stationary phase or it is not. This approach (solid phase extraction

Types of Retention Mechanism Exploited in Analytical Chromatography

Adsorption

Tswett's original experiments can probably be classified as this type. The stationary phase is a solid that can adsorb the analytes to a greater or lesser degree. The intermolecular interactions can be of many kinds (van der Waals, ion–ion, ion–dipole etc.).

Phase Partitioning

In this case the stationary phase is a substance that is covalently bonded as a thin film to a solid support. This solid support can be the internal capillary walls of fused silica capillary columns, or small silica particles that have been chemically treated to facilitate bonding of the stationary phase and are packed into the column itself. The stationary phase can be a liquid in the pure state at the operating temperature range of the column (e.g., C_8 hydrocarbon) but nonliquids are also used (e.g., C_{18}). Chromatographic separation results from the differing ratios of solubilities (expressed as their partition coefficients K^A) of the various analytes in the mobile and stationary phases.

Ion Exchange

In this case the dominant interaction is provided by coulombic forces between ionized analytes and fixed charge sites on a solid stationary phase (usually an organic resin). These fixed sites can be either negatively charged (providing cation exchange chromatography) or positively charged (anion exchange). Both types can be further sub-divided into 'strong' and 'weak'. 'Strong' exchangers retain their charge over a wide pH range ($\sim 2 - 12$); examples are $-SO_3^-$ for strong cation exchange (SCX) and $-NR_3^+$ for strong anion exchange (SAX). 'Weak' exchangers change their degree of ionization as a function of pH of the mobile phase, e.g., $-CO_2^-$ and $-NH_3^+$.

Size Exclusion (Gel Permeation)

The mobile phase passes through a porous gel that separates analytes according to their size. The pores in the gel are designed to be small relative to the larger analytes, but allow the smaller analytes to enter and thus be more retained on the column. Of course in practice there is a distribution of pore sizes. This type of chromatography is unusual in that larger analytes elute before smaller ones and is especially important for determinations of size distributions of synthetic polymers.

Biochemical Affinity

The principle underlying this type of chromatography is similar to the simplified 'lock and key' picture of enzyme–substrate interactions, i.e., strong intermolecular interactions (can be of various kinds) between the analyte and stationary phase are dependent on a close fit between molecular shapes, e.g., a concave portion of one and a matching convex one on the other. A common example would be an antibody that has been raised to some specific molecule (the analyte) and is chemically anchored to a suitable solid support. Depending on the degree of cross-reactivity of the antibody, only the target analyte (and perhaps to some degree its close analogs) will be retained.

NB: Any real column will probably to some degree exhibit characteristics of more than one kind of retention mechanism, e.g., a size exclusion column might also provide a degree of separation between molecules of the same size but of different polarities.

(SPE) Section 4.3.3) is popular because it can be fast and effective, and readily automated. Clean-up methods are crucially important especially for ultra-trace analysis; they are discussed in Section 4.3.

This book's treatment of the theory of chromatographic separations is by no means complete and rigorous, but is designed to provide some useful insight into the reasons why chromatography experiments turn out the way they do.

3.3 Summary of Important Concepts

This section contains an algebra-free summary of some of the more important concepts discussed in more detail in the following sections. Together with the list of symbols used in the algebraic development (given at the end of the chapter) this summary should hopefully provide a useful 'forest without the trees' view of chromatographic theory. A sketch of an archetypal

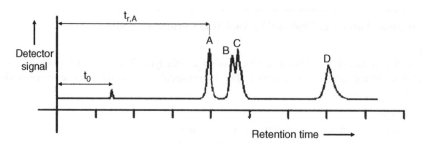

Figure 3.1 Sketch of a fictitious chromatographic separation of four components (A–D). t_0 is the time between injection of the analyte solution onto the chromatographic column and observation at the detector of any components that are not retained at all by the stationary phase, i.e. the partition coefficient K^A (Equation [3.1]) is zero. $t_{r,A}$ is the retention time of compound A on that particular column with that particular mobile phase (similarly for compounds B, C and D but these are not shown for clarity). For isocratic elution the retention time is directly proportional to the elution volume (the volume of mobile phase required to elute the compound) via the mobile phase flow rate; then $t_0 = V_m/U$, where V_m is the void volume of the column and U is the volume flow rate of mobile phase.

chromatogram that will be referred to in some of the following is shown in Figure 3.1. This summary does not include the more qualitative material (types of column, types of chromatographic retention mechanisms etc.), but concentrates on the parameters and concepts involved in the quantitative description of the chromatographic process itself. The order of discussion follows approximately that in the more detailed discussion (Sections 3.4–3.7).

K^A: *partition coefficient* for analyte A defines the ratio of equilibrium concentrations of A between stationary and mobile phases (Equation [3.1]).

Theoretical plate: a concept borrowed from distillation theory and countercurrent extraction; a chromatographic column is modeled as a series of discrete 'plates' in each of which local equilibrium of analyte partitioning between stationary and mobile phases is established. The *Plate Theory* accounts for retention of analytes, i.e., retention times, but *not* the peak shapes (widths), for isocratic elution.

Isocratic elution: elution using a mobile phase of fixed composition (compare 'gradient elution').

Laminar flow: sometimes known as 'streamline flow', this type of flow of a liquid or gas occurs when the fluid flows in parallel layers, with no macroscopic disruption between the layers (exchange at the molecular level via diffusion can still occur), in accord with the Poiseuille Equation for flow through a tube of radius r_{tube} and length l_{tube} (volume flow rate $U = \pi . r_{tube}/8.\eta . l_{tube}$), where η is the viscosity coefficient of the fluid; laminar flow is to be contrasted with turbulent flow.

Turbulent flow: in contrast with *laminar flow* at lower flow velocities, turbulent flow sets in at higher flow velocities and is characterized by eddies and chaotic motion

that do not contribute to the forward volume flow rate but instead induce mixing of the former laminar layers.

v: *volume of mobile phase* that has passed through the column at any time after sample injection, expressed not in terms of conventional volume units (mL, μL, etc., as used for V) but as the number of *plate volumes* defined as $V^A(v_m + K^A.v_s)$.

v_m and v_s: *volumes of mobile and stationary phases*, respectively, in a single theoretical plate.

V_m and V_s: *total volumes of the two phases in the column*; V_m is also called the *column void volume*, as this is the mobile phase volume that must pass through the column to elute an entirely unretained analyte (K = zero). Note that $V_m = N.v_m$ and $V_s = N.v_s$, where N is the number of theoretical plates in the column.

P: the *phase ratio* V_m/V_s for a column. Note that V_s and thus P can vary with the nature of the analyte depending on the accessibility of all binding sites on the stationary phase to analytes that can vary in molecular size etc.

Plate Theory elution equation: equation that gives the concentration of a given analyte in the last plate of the column (adjacent to the detector) as a function of the initial concentration in the first plate before elution has begun, the total mobile phase volume required for analyte elution, the number of theoretical plates N, and basic physico-chemical properties of the system (the partition coefficient); see Equation [3.11] as the Plate Theory description of a chromatographic peak like that in Figure 3.2.

V_{rA}: *retention volume* of analyte A, the volume of mobile phase that passes through the column between

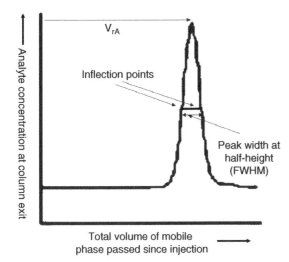

Figure 3.2 The elution curve of a single component, plotted as the analyte concentration at the column exit (proportional to the detector response R_D) as a function of V, the total volume flow of mobile phase that has passed through the column since injection of the analytical sample onto the column. (V is readily converted to time via the volume flow rate U of the mobile phase.) The objective of theories of chromatography is to predict some or all of the features of this elution curve in terms of fundamental physico-chemical properties of the analyte and of the stationary and mobile phases. Note that the Plate Theory addresses the position of the elution peak but does not attempt to account for the peak shape (width etc.). The inflection points occur at 0.6069 of the peak height, where the slope of the curve stops increasing and starts decreasing (to zero at the peak maximum) on the rising portion of the peak, and vice versa for the falling side; the distance between these points is double the Gaussian parameter σ. Modified from Scott, www.chromatography-online.org, with permission.

injection and elution (the maximum of the chromatographic peak). The theory gives $V_{rA} = V_m + K^A.V_s$ (Equation [3.12]).

t_{rA}: *retention time of analyte A* (see Figure 3.1), the time after injection at which A elutes $= V_{rA}/U$ where U is the *volume flow rate* of mobile phase..

t'_{rA}: *corrected (adjusted) retention time of A* $= (t_{rA} - t_0)$; the amount of time an eluted analyte has spent adsorbed on the stationary phase.

t_0: the *time required for an analyte that is completely unretained* $(k' = 0)$ *to elute* from the column; $t_0 = V_m/U$. Any analyte moves at the same rate as the mobile phase whenever it is in the mobile phase and is stationary when bound to the stationary phase, i.e., all analytes spend time t_0 in the mobile phase and their separation is the result of differences in the amount of time they spend bound to the stationary phase.

$\alpha_{A/B}$: *Separation ratio (selectivity factor)* for analytes A and B, a measure of chromatographic separation of two analytes using specified stationary and mobile phases that is independent of flow rate and of the volume of stationary phase (and thus of column dimensions), experimentally measured as $(t_{rA} - t_0)/(t_{rB} - t_0)$, and $\approx K^A/K^B$ (Equation [3.16]).

k'_A: *capacity factor* of analyte A is defined as $K^A.V^A_s/V_m$ (Equation [3.16]) but usually measured as $(t_{rA} - t_0)/t_0$; used to compare the behavior of an analyte on the same column when attached to two different systems so that flow rates and t_0 values may vary, but k' values should not vary. *In isocratic elution usually one aims for $3 < k' < 10$ for a good separation.* k' also indicates the number of column volumes of mobile phase used to elute an analyte, $V'_{rA} = (k' + 1).V_m$.

N: *column efficiency* defined as the *number of theoretical plates* in the column, i.e., in Plate Theory N describes the number of effective equilibrations of the analyte between mobile and stationary phases; $N = 4.(V_r/\Delta V_i)^2 = 4.(t_r/\Delta t_i)^2 = 5.545.(t_r/\Delta t_{\frac{1}{2}})^2$, where ΔV and Δt are the chromatographic peak widths in terms of elution volume and elution time, respectively, and subscripts i and $\frac{1}{2}$ refer to widths measured at the peak inflection points and at half peak height, respectively.

H: *height of the equivalent theoretical plate (HETP)* $= l/N$ (where l is the column length, strictly the length of the bed of stationary phase) $= \sigma_x^2/l$ (Equation [3.32]).

$R^{n\sigma}_{A/B}$: *chromatographic resolution*, the criterion for chromatographic separation of two components A and B by a multiple n of the Gaussian parameter σ (half the width between inflection points of the elution peak, see Figure 3.2): $R^{n\sigma}_{A/B} = (t_{rB} - t_{rA})/\Delta t_r = (V_{rB} - V_{rA})/\Delta V_r$, where the desired separation of peak maxima (Δt_r or ΔV_r) must be specified as nσ. Then $R^{n\sigma}_{A/B} = (1/N).$ $(\alpha_{A/B} - 1).N^{\frac{1}{2}}.[k'/(1 + k')]$ (Equation [3.23]) where each of N and k' is assumed to be the same for two closely-eluting analytes.

C_P: *peak capacity* of a column under given elution conditions is the number of peaks that can in principle be fitted into a chromatogram between the unretained solutes (at t_0) and the 'last peak', each peak being separated from its neighbors by twice the peak width Δt_i; it is difficult to be precise about this definition, e.g., how is 'the last peak' defined, but an ultra-simple treatment gives a lower limit for $C_P = N^{\frac{1}{2}}/4$ (Equation [3.27]).

Rate Theory of chromatography: a theory of the dispersive (nonequilibrium) processes occurring in a chromatographic column that lead to peak broadening, usually associated with the name of Van Deemter, but Giddings, Golay and Knox are also important contributors; expressed in terms of the variation of H with u.

u: *linear flow velocity* of the mobile phase, uniquely defined for a liquid mobile phase but for a gas the compressibility implies that u increases along the length of a column i.e. inversely with the pressure.

Multipath dispersion: in a packed column the molecules in the mobile phase follow meandering paths of different lengths through the gaps between the stationary phase particles, and thus elute with a range of t_r values; the corresponding contribution to H is $H_M = 2\lambda \cdot d_p$.

H_L *Longitudinal diffusion*: the effect of diffusion (random molecular movement) of a solute band in the mobile phase throughout the length of the column in accord with Fick's Law of Diffusion; the contribution to H is $H_L = (2/u) \cdot [\gamma_m D_m + \gamma_s D_s k']$

Resistance to mass transfer: a finite time is required for analyte molecules to diffuse through the mobile phase to reach and enter the stationary phase; this competes with the residence time near a given particle that is controlled by u. The contribution to H is $H_R = [f_1(k') \cdot d_p^2/D_m + f_2(k') \cdot d_f^2/D_s] \cdot u$ where the functions f_1 and f_2 are given in Equation [3.38].

Van Deemter Equation: the first theoretical expression of the dependence of H on u: $H = A + B/u + C \cdot u$, where the coefficients A, B and C were explicitly derived in terms of physico-chemical parameters. An alternative and quite commonly used formulation is that of Knox: $h = A \cdot \nu^{\frac{1}{3}} + B/\nu + C \cdot \nu$, where $h = H/d_p$ and $\nu = u \cdot (d_p/D_m)$ are the *reduced* plate height and linear velocity. The Knox equation is a convenient functional form for fitting experimental data (it accounts for a minor coupling observed between multipath dispersion and resistance to mass transfer) but does not provide expressions for the coefficients in terms of more fundamental parameters.

H_{min} (or H_{opt}): *minimum value of H*, given by the van Deemter equation as $H_{min} = A + 2 \cdot (BC)^{\frac{1}{2}} \approx 2d_p \cdot \{\lambda + [2\gamma_m \cdot f_1(k')]^{\frac{1}{2}}\}$ (Equation [3.40]); this value of H_{min} is found at the *optimum value of linear velocity* $= u_{opt} = (B/C)^{\frac{1}{2}} \approx (D_m/d_p) \cdot \{2\gamma_m/f_1(k')\}^{\frac{1}{2}}$ (Equation [3.40]).

ε: *column porosity* is the fraction of the total column volume that is not occupied by solid material; the '*pore volume*' (total volume occupied by mobile phase) includes the volume between the constituent particles plus the total volume of the 'pores' or 'cracks' within each individual particle; $\varepsilon = V_m/(V_m + V_s)$.

D Chromatographic dilution: as a result of the dispersive processes accounted for in the van Deemter equation an analyte will be progressively diluted as it proceeds along the column; *D* gives the ratio of the initial sample concentration to the concentration at the peak maximum (c_A^{pm}) as it elutes: $D \equiv c_A^0/c_A^{pm} = (\varepsilon \cdot \pi d_c^2/4V_{inj}) \cdot (1 + k_A') \cdot (2\pi l H)$ (Equation [3.50]).

Golay Equation: a modification of the van Deemter equation for a nonpacked (usually capillary) column, i.e., no multipath dispersion effect. When applied to modern capillary gas chromatography the original Golay equation should be corrected for the gas compressibility (use u_{exit} to replace u) and for the film thickness of the stationary phase (Equations [3.53–3.54]).

Gradient elution: liquid chromatography employing a mobile phase whose composition changes with time such that the solvent strength increases. A modified form of the theories developed for isocratic elution apply, leading to analogous equations for some important chromatographic parameters (Table 3.2).

Electrophoresis: motion of ions in solution under the influence of an external electric field. *Capillary electrophoresis* is electrophoresis conducted in a fused silica capillary.

Electro-osmosis: motion of a buffer solution in contact with an electrically charged surface such as fused silica at pH > 3 (deprotonated silanol groups) under the influence of an external electric field. The resulting movement of the bulk liquid is called *electro-osmotic flow (EOF)*.

Capillary electrochromatography: packed capillary column HPLC in which the mobile phase moves as the result of an external electric field (electro-osmotic flow) rather than by external pressure. Very small particle sizes can be used since there is no back pressure problem.

3.4 Plate Theory of Chromatography

There are many levels at which chromatography theory can be discussed. A very rigorous and complete discussion is freely available (Scott, www.chromatography-online.org). However, here a more intuitive and less rigorous treatment, which should be sufficient to permit informed design and use of chromatography in conjunction with mass spectrometry for trace analysis, is presented.

There are two principal approaches to chromatographic theory. Chronologically the first was the Plate Theory that models a chromatographic separation as a sequence of equilibrium partitionings of analyte between stationary and mobile phases (i.e., the theory as originally developed refers to partition chromatography, but many of the concepts derived from this theory can be transferred to other chromatographic mechanisms). The name of this theory is derived from an analogy with industrial scale distillation apparatus that sometimes does involve real 'plates' along the length of the distillation column in order to achieve chemical separations by a series of discrete liquid–vapor equilibrations. The original theory was also inspired by the popularity at the time of apparatus for countercurrent extraction that achieved separation of complex mixtures on a macroscopic scale via sequences of discrete liquid–liquid partitioning equilibrations. However, real-life chromatography involves continuous competition for an analyte between stationary and mobile phases within a limited time frame, not a series of discrete equilibrations. The inadequacies discovered in the Plate Theory could be attributed to nonequilibrium effects and led to development of the Rate Theory including the famous van Deemter Equation and its modifications (Section 3.5).

Derivation of the Plate Theory requires a fair amount of algebra that is not intrinsically difficult, but may appear confusing as a result of the terminology and many symbols used. To alleviate this situation a list of symbols is provided at the end of this chapter. Moreover, the main mathematical development is presented in Appendix 3.1, so as to not interrupt the flow of the more intuitive description in the main text.

The Plate Theory was first proposed by A.J.P. Martin and R.L.M. Synge in a paper (Martin 1941) that led to their sharing the Nobel Prize for Chemistry in 1952 'for their invention of partition chromatography'. Martin later went on to develop gas chromatography in collaboration with A.T. James (James 1952), although the idea had been suggested in the 1941 paper.

As mentioned above, the Plate Theory is based upon a concept of sequential partitioning equilibria of analyte between mobile and stationary phases, so it is convenient to define a partition coefficient for analyte A as:

$$K^A = c^A_s / c^A_m \qquad [3.1]$$

where c^A is a concentration of A and subscripts s and m denote stationary and mobile phases, respectively. K^A is assumed in what follows to be independent of total concentration, although the distribution between two such phases more closely follows the Langmuir adsorption isotherm (Langmuir 1916) that can be written as follows for our case of interest:

$$\theta = c^A_s / (c^A_s{}^*) = b.c^A_m / (1 + b.c^A_m)$$

where $(c^A_s{}^*)$ is the maximum concentration of A that can be adsorbed on the stationary phase (all adsorption sites occupied) and b is a constant. This relationship takes into account the fact that the stationary phase will have a limited capacity for the analyte and is thus subject to saturation. However, for the low analyte concentrations of interest here, far from the saturation limit ($\theta = 1$), the Langmuir isotherm reduces to a form equivalent to

Equation [3.1]. Of course, K^A does vary with temperature and the chemical compositions of the two phases. Intuitively it is obvious that if K^A increases, the analyte will take longer to elute from the column in view of the implied greater affinity for the stationary phase. This qualitative idea must now be made more rigorous. The present treatment is a simplification of a re-derivation (Said 1956) of the original theory (Martin 1941). It is emphasized that the approach assumes isocratic elution (fixed composition of mobile phase) and laminar flow (no turbulence).

It will become clear in the following that it is necessary to ensure that the sample extract to be injected on column uses a solvent that is either the HPLC mobile phase itself (the $t = 0$ composition in the case of gradient elution) or a solvent of lower elution strength (Section 4.4.2a); otherwise, the chromatographic peak is broadened by the interference with the intended separation by partitioning between mobile and stationary phases, since now K^A refers to the equilibrium distribution of A between the stationary phase and a stronger solvent than the mobile phase.

3.4.1 Elution Equation for the Plate Theory

Figure 3.2 is a diagram of the chromatographic peak for a single component, plotted as the detector response $R_D(V)$ (signal strength for the analyte, see Appendix 4.1) as a function of the total volume V of eluted mobile phase; V is readily converted to retention time via the volume flow rate of mobile phase (V). The algebra becomes simpler if V is used instead of t as the independent variable; the objective of the exercise is to derive a useful equation that describes a curve like that in Figure 3.2 in terms of fundamental parameters particularly K^A.

A schematic representation of three neighboring 'theoretical plates', all assumed to have the same size, is shown in Figure 3.3. The theory assumes that equilibrium distribution of the analyte between the two phases is achieved within each plate, thus approximately accounting for the fact that such equilibration never actually occurs at any specific point as the mobile phase flows past the stationary phase at a finite rate. The volume associated with each plate is assumed to be large enough that the residence time is long enough for equilibrium to be established. Thus, the faster the equilibration, the smaller the plate size, and the greater the number of plates in the column. The number of theoretical plates in a column (N) for a given analyte is thus directly related to the equilibration rate and thus to the column efficiency (see later).

The Birth of Modern Chromatography

A.J.P. Martin (reproduced with permission of the daughter of Dr. Martin)

R.L.M. Synge (reproduced with permission of the Rowett Institute, Scotland, UK)

Martin and Synge were awarded the Nobel Prize in Chemistry in 1952 for their invention of partition chromatography and its theoretical elaboration as the Plate Theory (Martin 1941). Martin did his graduate work at Cambridge, using countercurrent extraction to study vitamin E. (Countercurrent extraction at that time effectively chained together a series of essentially conventional separatory funnels, then repeatedly used sequential steps of shaking, settling, and separating, to increase the number of partition stages (plates) of separation.) Meantime, also at Cambridge, Synge was working on amino acids in wool, and approached Martin about how to use countercurrent extraction for this purpose. Eventually they decided to work together and moved to the Wool Industries Research Association laboratory in Leeds. After fruitless attempts to improve countercurrent extraction, Martin had his 'Eureka' moment – the idea to improve the method by moving only one phase and keeping the other stationary by coating it on silica gel particles packed in a glass tube. The mixture to be separated was injected into the mobile phase stream flowing through this column, and the various compounds were separated because of differences in partition coefficients. (Tswett's earlier work had used a solid stationary phase and relied on different degrees of adsorption of the analytes.)

Not content with inventing the practice and theory of liquid–liquid partition chromatography, Martin (in collaboration with A.T. James) went on to invent gas–liquid chromatography (James 1952), now referred to as simply gas chromatography. These remarkable accomplishments of Martin's appear to have been achieved with the aid of a dry sense of humor, as related later (James 1979):

'— I have attempted to apply the fundamental precepts I learned from A.J.P. Martin: (1) Nothing is too much trouble provided someone else does it; (2) Never answer the first letter; if it's important they'll write again; and (3) If there are twelve ways of tackling a problem, they're all wrong.'

A mathematical derivation of the Plate Theory expression for $R_D(V)$ is given in Appendix 3.1 (the equation numbers in the Appendix are in sequence with those in the main text). The final result of this derivation is:

$$c^A_{m(N)} = [c_A^0 . \upsilon^N . \exp(-\upsilon)]/N! \qquad [3.11]$$

where N is the number of theoretical plates in the column, $c^A_{m(N)}$ is the concentration of analyte A in the mobile phase in this last plate (and that is thus responsible for the observed response of the detector), c_A^0 is the concentration of A in the first plate immediately after injection of the solution of analyte solution and before any elution has

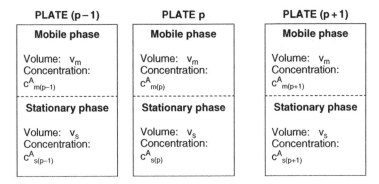

Figure 3.3 Diagrammatic representation of three consecutive theoretical plates in a column.

occurred, and υ is the so-called plate volume, a dimensionless quantity defined as $[V/(v_m + K^A.v_s)]$ where v_m and v_s are the volumes of mobile and stationary phases in one theoretical plate.

The quantity $c^A_{m(N)}$ is the concentration of analyte emerging from the last theoretical plate into the detector, so Equation [3.11] corresponds to the desired theoretical expression for $R_D(V)$ for cases in which the chromatographic detector has a concentration dependent response (Section 4.4.8 and Appendix 4.1); UV–visible absorption detectors are an important example since their response is described by the Beer–Lambert Law, but electrospray ion sources for mass spectrometers can also behave in this fashion in some circumstances (Section 5.3.6b). Electron ionization ion sources provide a response that is mass flux dependent (Section 4.4.8); however, for a fixed mobile phase flow rate U (volume per unit time), the conversion from $c^A_{m(N)}$ to the mass flow rate is trivial and this distinction is not important in the discussion of the present Section although the practical implications are discussed in Section 5.3.6b.

The right side of Equation [3.11] is a Poisson Function (see text box in Chapter 7) that for large N can be reduced to a more tractable Gaussian Function as discussed in Appendix 3.2. Note that no rate processes are explicitly considered in the Plate Theory, and the time evolution of the elution is approximated as discontinuous changes on moving from one equilibration plate to the next (Figure 3.3). However, as mentioned above, the faster the equilibration process the smaller the size of the conceptual compartments (theoretical plates) and thus the greater the number of plates (N) that are effectively acting within the column. Note also that the mathematically convenient quantity υ depends on the so-called plate volume for analyte A, $v^A_{plate} \equiv (v_m + K^A.v_s)$

and thus varies with the nature of A: it is this variation that accounts for separation of analytes in the Plate Theory. Later (Section 3.5) some of the nonequilibrium rate processes that control the peak shapes and widths are considered.

3.4.2 Retention Volume and Time

The retention volume of a solute A, V_{rA}, is that volume of mobile phase that passes through the column between injection and elution of A, the latter taken to be the maximum of the chromatographic peak (Figure 3.2); note that $V_{rA} = U.t_{rA}$ (Figure 3.1). To determine the peak maximum Equation [3.11] must be differentiated with respect to υ and the derivative set to zero. This standard mathematical procedure (not shown) proves that the peak maximum corresponds to $\upsilon_{rA} = N$, i.e., the peak maximum is reached after N plate volumes of mobile phase have passed through the column. The retention volume V_{rA}, expressed as the volume of mobile phase required to elute A, is then obtained by multiplying the number of theoretical plates (N) by the plate volume:

$$V_{rA} = N.v_m + K^A.N.v_s = V_m + K^A.V_s \qquad [3.12]$$

where V_m and V_s are the total volumes of mobile and stationary phases in the whole column; the ratio V_m/V_s is called the phase ratio of the column.

For an analyte that is not retained at all on the stationary phase $K^A = 0$ (Equation [3.1]) and $V_{r0} = V_m$, i.e., in the simplest form of the Plate Theory an unretained compound will pass through the column as if it formed part of the mobile phase itself. In practice, however, the experimental retention volume V_0 of an unretained peak (the 'void volume') will not be simply the total volume of mobile phase in the column as suggested above, but will be made up of the volume of mobile phase (V_m) plus extra-column volumes V_E that arise from sample valves, connecting tubes, unions etc. Thus:

$$V_0 = V_{r0} + V_E = V_m + V_E \qquad [3.13]$$

and an experimental retention volume can be defined as:

$$V_{r,A}{}^{exptl} = V_{rA} + V_E = V_m + K^A.V_s + V_E \qquad [3.14]$$

It is sometimes convenient to define an adjusted retention volume $V_{rA}{}'$:

$$V_{rA}{}' = V_{rA}{}^{exptl} - V_0 = V_m + K^A.V_s + V_E - (V_m + V_E)$$
$$= K^A.V_s \qquad [3.15]$$

where Equations [3.13] and [3.14] have been used.

It is useful to attribute different values of V_s to the two analytes (e.g., some analyte molecules might be too large to be able to penetrate deep within the particle pores and thus interact with all of the stationary phase that might be accessible to smaller molecules). Thus $V_{rA}{}'$ can differ from $V_{rB}{}'$ (i.e. A and B can be chromatographically resolved) not only via differences between the adsorption coefficients (K_A and K_B), but also by differences between the effective values of V_s for the two analytes. The latter corresponds to a different separation mechanism (size exclusion) from the adsorption mechanism that is the main focus of the Plate Theory.

The 'bottom line' conclusion of this Section can be stated thus: If two analytes A and B are to be chromatographically separated, their adjusted retention volumes $V_{rA}{}'$ and $V_{rB}{}'$ (Equation [3.15]), and thus the corresponding retention times t_r, must be sufficiently different. To separate two solutes, either their distribution coefficients K^A and K^B (Equation [3.1]) must be made to differ (choose appropriate phase systems) or the volumes of stationary phase with which they interact must be made to differ (choose a stationary phase with appropriate exclusion properties based on molecular size so that the effective values of V_s are different for different analytes) or a shrewd combination of both. It is appropriate to mention that if mass spectrometric detection is used, clean chromatographic separation of solutes in the analytical extract is not as crucial as for less selective detectors in view of the additional selectivity provided by the *m/z* information.

3.4.3 The Separation Ratio (Selectivity Factor) for Two Solutes

The separation ratio of two solutes A and B, denoted $\alpha_{A/B}$ and sometimes called the selectivity factor, is defined as the ratio of their corrected retention volumes:

$$\alpha_{A/B} \equiv V_{rA}{}'/V_{rB}{}' = (V_{rA}{}^{exptl} - V_0)/V_{rB}{}^{exptl} - V_0)$$
$$= (t_{rA} - t_0)/(t_{rB} - t_0) \qquad [3.16]$$
$$= K^A.V^A{}_s/K^B.V^B{}_s \approx K^A/K^B$$

where in general $t = V/U$, the second line of Equation [3.16] used Equation [3.15], and the final approximation applies when $V^A{}_s = V^B{}_s$, i.e., there are no significant differences in the physical exclusion properties of the stationary phase for the two solutes. $\alpha_{A/B}$ is independent of the mobile phase flow rate and also of the phase ratio of the column, assuming no variations in exclusionary effects for the two analytes. Thus, under this condition, the same separation ratio for two solutes would be obtained from either a packed column or a capillary column if the same phase system were used at the same temperature. In practice a separation ratio is calculated via the observed retention times as in Equation [3.16].

3.4.4 Capacity Factor (Ratio) of a Solute

The capacity factor k' is essentially a corrected retention time (Equation [3.18]) that takes into account variations in mobile phase flow rate and thus provides a more robust indicator of analyte retention for a given combination of stationary and mobile phases. As K^A is a unique property of a given solute A for a given stationary–mobile phase combination, the adjusted (corrected) retention volume $V_{rA}{}'$ (Equation [3.15] (or the corresponding retention time) can be used as a tag for analyte identification. Thus the precision and accuracy of measurement of $V_{rA}{}'$ become important and depend on those of the measurement of flow rate U since in practice retention times rather than volumes are the measured quantities. In response to this limitation of the $V_{rA}{}'$ parameter, the capacity ratio of a solute (k'_A) was defined as the ratio of its distribution (partition) coefficient to the phase ratio ($V_m/V^A{}_s$) of the column with respect to analyte A:

$$k_A{}' = K^A.V^A{}_s/V_m = V_{rA}{}'/V_m = V_{rA}{}'/(V_0 - V_E)$$
$$[3.17]$$

using Equations [3.15] and [3.13]. In practice k'_A is calculated from retention times measured on the experimental chromatogram:

$$k_A{}' = (t_{rA} - t_0)/t_0 \qquad [3.18]$$

where t_{rA} is the retention time for $A = (V_{rA}/U)$ measured from the time of injection, and t_0 is the void–volume time.

Equation [3.18] assumes that the extra-column volume V_E is negligible. However, there are two definitions of void volume, and thus also of the capacity ratio of a solute. The two void volumes are called the thermodynamic and the dynamic void volumes and they are not equal (Scott, www.chromatography-online.org); the two void volumes and capacity ratios are used for different purposes. Equations [3.16–3.18] incorporate the thermodynamic dead volume and all further discussion in this chapter assumes this definition.

Care is required when comparing k' values for the same solute measured on different columns, and for different solutes measured on the same column. Both V_m and V_s will be different for different columns and, as a result of the exclusion properties of solid stationary phases and supports, may vary between different solutes on the same column. The separation ratio (selectivity $\alpha_{A/B}$, Equation [3.16]) is a parameter that is largely independent of V_m and V_s for different solutes on the same column and is thus useful in such circumstances.

3.4.5 Column Efficiency and Height Equivalent of the Theoretical Plate

How 'efficient' is a chromatographic column operated under specified conditions? Within the Plate Theory, column efficiency is defined as the number of theoretical plates in the column (N), since that describes the number of theoretical equilibrations of the analyte between mobile and stationary phases. The measurement of N from the experimental chromatogram (Equations [3.20–3.21]) is considered here; it will become clear that N is different for different analytes and thus will vary from peak to peak in a real chromatogram, so there is no such thing as a unique value of N for a given column, rather a series of measurable quantities for specified analytes eluting from the column under specified conditions. Evaluation of N for a standard test analyte using Equation [3.21] is a common method of monitoring the performance of a column as it ages with prolonged use.

It is intuitively obvious that the greater the number of plates, i.e., the greater the number of equilibration steps, the better will be the separation, and this implies more narrow chromatographic peaks. Therefore, the strategy will be to examine the implications of the Plate Theory for peak widths. Note, however, that the Plate Theory does not pretend to explain or understand the peak widths, but only describes them. Examination of a typical peak shape (Figure 3.2) and the theoretical elution equation (Equation [3.11]) that describes its shape suggests that a

mathematically convenient approach is to define the peak width between the points of inflection since the latter correspond to the mathematical condition that the second derivative of the Poisson function in Equation [3.11] is zero. This corresponds to the intuitive notion that the inflection point represents the point at which the slope of the rising side of the peak stops increasing and starts to decrease (to zero at the peak maximum), and vice versa for the falling side of the peak (Figure 3.2). Simple differentiation and some algebraic rearrangement give:

$$d^2\{[c_A{}^0.\upsilon^N \cdot \exp(\upsilon)]/N!\}/d\upsilon^2 = (c_A{}^0/N!).\exp(-\upsilon).\upsilon^{N-2}$$
$$.[\upsilon^2 - 2N\upsilon + N(N-1)] = 0$$

so the condition giving the points of inflection is:

$$\upsilon_i{}^2 - 2N\upsilon_i + N(N-1) = 0$$

where υ_i denotes the value of υ at one of the two inflection points. This is a simple quadratic equation in υ_i with solutions given by:

$$\upsilon_i = \{2N \pm [4N^2 - 4N(N-1)]^{\frac{1}{2}}\}/2 = N \pm N^{\frac{1}{2}}$$

The difference between the two inflection points is thus $\Delta\upsilon_i = 2N^{\frac{1}{2}}$. This is the volume of mobile phase, expressed as the number of units of phase volume $(v_m + K^A.v_s)$, that elutes from the column between the two inflection points, so the actual volume (e.g., in mL, μL etc.) is:

$$\Delta V_i = \Delta\upsilon_i.(v_m + K^A.v_s) = 2N^{\frac{1}{2}}.(v_m + K^A.v_s) \quad [3.19]$$

Equation [3.19] gives the Plate Theory expression for peak width in conventional volume units. To relate this to the experimental chromatogram note that, from Equation [3.12], the retention volume V_r (maximum of the peak) is given by $V_r = N.(v_m + K^A.v_s)$, so Equation [3.19] becomes:

$$N = 4.(V_r/\Delta V_i)^2 = 4.(t_r/\Delta t_i)^2 \quad \text{(for constant}$$
$$\text{mobile phase volume flow rate U)} \quad [3.20]$$

where Δt_i is the experimental peak width measured in retention time units between the inflection points; Equation [3.20] is the relationship used to calculate N from an experimental chromatogram. Note that, as is clear from the preceding derivation, N is a function of the particular analyte via V_r and thus will vary from peak to peak in any real chromatogram, so there is no unique number of plates N for a column.

In practice it is often difficult to locate the inflection points of a real peak, particularly if appreciable noise is present, and instead the peak width measured at half the peak height (sometimes called the full width at half-maximum, FWHM) is used instead (Figure 3.2). The inflection points of any Poisson (or Gaussian) peak are located at 0.6069 of the peak height, and it follows that, if Δt_i in Equation [3.20] is replaced by Δt_{FWHM}, then:

$$N = 5.545.(t_r/\Delta t_{FWHM})^2 \qquad [3.21]$$

Equation [3.21] is the basis for routine monitoring of the performance (column efficiency) of a chromatographic column.

The column efficiency is sometimes also expressed as the height equivalent of the theoretical plate (HETP):

$$HETP = l/N$$

where l is the length of the bed of stationary phase in the column. This parameter will be important in the rate theory of chromatography, discussed in Section 3.5.

3.4.6 Chromatographic Resolution

The concept of chromatographic resolution of analytes A and B was touched on briefly in the discussion of the separation ratio $\alpha_{A/B}$, defined in Equation [3.18]. However, this did not involve any mention of criteria for successful resolution of the two peaks. The bottom line result of this section is given in Equation [3.23c], which relates the chromatographic resolution (defined as separation of peaks by two peak widths) to other chromatographic figures of merit (k', α and N). The complexity of this relationship makes it clear that arranging for an overall optimized chromatographic separation is not straightforward (Figure 3.4).

The criterion for effective separation (the resolution) is expressed as:

$$R_{A/B} = (t_{rB} - t_{rA})/\Delta t_r = (V_{rB} - V_{rA})/\Delta V_r \qquad [3.22]$$

where analyte A has arbitrarily been assigned as the first eluter and Δt_r and ΔV_r are the peak width expressed in terms of retention time and elution volume, respectively; the peak widths are assumed here to be the same for the two closely eluting peaks. The value of $R_{A/B}$, the number of peak widths (Δt_r or ΔV_r) by which the peak maxima must be separated in order that the peaks can be deemed to be resolved, must be chosen by the analyst to suit the purpose at hand. It is important to note that the present discussion refers to chromatographic resolution as detected by a nonspecific single parameter detector (e.g., a refractive index detector or a UV detector operating at

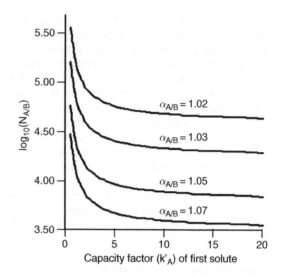

Figure 3.4 Graphs of $\log_{10}(N_{A/B})$, calculated from Equation [3.23b] for peak resolution $R^{2\sigma}_{A/B} = 2$, as a function of the capacity factor k' of the first-eluting solute A, for various values of the separation ratio $\alpha_{A/B}$. $R^{2\sigma}_{A/B}$ implies separation of peak maxima by a single peak width (measured between the inflection points), not a very stringent criterion for chromatographic separation of analytes A and B, more usually $R^{4\sigma}_{A/B}$ is used. Adapted from Scott, http://www.chromatography-online.org/, with permission.

a single wavelength at which both analytes absorb, or a flame ionization detector for GC). One great advantage of a mass spectrometer as a chromatographic detector is the detection specificity resulting from the second dimension of information (the m/z value); thus it is possible to resolve the peaks for two compounds that co-elute or nearly so if they can be characterized using different m/z values.

It is also assumed that, for two closely eluting peaks, the values of N and of $V_s (= N.v_s)$ are the same for the two analytes. In Equation [3.22], expressing V_r in units of the phase volume and using the inflection point value ΔV_i from Equation [3.19] as ΔV_r, gives:

$$V_{rB} - V_{rA} = N.[(v_m + K^B.v_s) - (v_m + K^A.v_s)]$$
$$= 2.R_{A/B}.N^{\frac{1}{2}}.(v_m + K^A.v_s)$$

as the criterion for resolution of the two peaks, where the peak width ΔV_i of A (the first-eluting peak) is measured between the inflection points (equal to 2σ where σ is the Gaussian range parameter, Equations [3.30–3.31] and Figure 3.2). Rearranging and dividing through by v_m gives:

$$N^{\frac{1}{2}} = 2R_{A/B}.[1 + (K^A v_s/v_m)]/[(K^B v_s/v_m) - (K^A v_s/v_m)]$$

However, the capacity ratio was defined in Equation [3.16] as $k' = K.v_s/v_m$, so the number of plates $N_{A/B}$ required to resolve A from B according to the criterion expressed in Equation [3.22] is:

$$N_{A/B} = 4R_{A/B}^2.(1 + k_A')^2/(k_B' - k_A')^2$$
$$= 4R_{A/B}^2.(1 + k_A')^2/[k_A'^2.(\alpha_{A/B} - 1)^2] \qquad [3.23a]$$

Recall that the values of k' can be calculated from the chromatograms using Equation [3.17], so that once the desired value of $R_{A/B}$ has been specified the necessary value of $N_{A/B}$ can be calculated. The second form of Equation [3.23a] used the definition of $\alpha_{A/B}$ in Equation [3.16] that also shows how this parameter can be evaluated from the chromatograms. Alternatively, Equation [3.23a] can be rearranged to give an expression for R:

$$R^{2\sigma}_{A/B} = 0.5.(\alpha_{A/B} - 1).N^{\frac{1}{2}}.[k'/(1 + k')] \qquad [3.23b]$$

where it is made explicit that resolution is defined here in terms of two peak maxima separated by the inflection-point peak width $\Delta V_i = 2\sigma_V$; a more commonly used definition of peak resolution requires the separation of peak maxima to be $4\sigma_V$ (two peak widths), and in such a case the factor 0.5 in Equation [3.23b] becomes 0.25 for $R^{4\sigma}_{A/B}$:

$$R^{4\sigma}_{A/B} = 0.25.(\alpha_{A/B} - 1).N^{\frac{1}{2}}.[k'/(1 + k')] \qquad [3.23c]$$

Curves relating $N_{A/B}$ and k'_A for solute pairs A and B having separation ratios ($\alpha_{A/B}$) of 1.02, 1.03, 1.05 and 1.07, calculated from Equation [3.23] using $R^{2\sigma}_{A/B}$ (not a very stringent criterion for resolution), are shown in Figure 3.4. Equation [3.23] and Figure 3.4 show that, as the separation becomes more difficult (i.e., $\alpha_{A/B}$ becomes smaller), the efficiency (value of $N_{A/B}$) required for resolution increases rapidly. In other words, as $\alpha_{A/B}$ becomes smaller the peaks become closer, so for successful resolution the peak widths must be reduced, i.e. the column must be made more efficient (larger N).

The other notable feature of Figure 3.4 is the dramatic increase in the required value of $N_{A/B}$ (note the logarithmic scale) when the capacity factor k_A' becomes small. Recall (Equations [3.17–3.18]) that a small value for k_A' implies rapid elution of A, soon after the void volume peak. The corresponding high values of $N_{A/B}$ required for successful resolution implies that short analysis times are very difficult to achieve if closely eluting compounds are to be resolved, as higher $N_{A/B}$ values imply longer

columns that in turn imply long elution times. Resolution of a solute pair with a separation ratio $\alpha_{A/B}$ of 1.02 would require $N_{A/B} = 360\,000$ for fast elution with $k_A' = 0.5$. Efficiencies of such magnitude are available with GC capillary columns, but LC columns with such efficiencies would be extremely difficult to fabricate. In LC the only practical way to resolve closely eluting compounds in a reasonable time is to choose the mobile and stationary phases so that the solutes are not eluted at low k' values. Lower $N_{A/B}$ values will be required, and thus shorter columns and consequently shorter elution times. As shown in Figure 3.4, for $k' > 10$ the required efficiency changes very little as the capacity ratio α increases. Thus, for fast analyses the phase system should provide a large separation ratio $\alpha_{A/B}$ but the first peak should elute at $k' \geq 10$.

3.4.7 Effective Plate Number

With the introduction of capillary open tube columns (Chapter 4) it became possible to obtain chromatograms corresponding to hundreds of thousands (even $>10^6$) of theoretical plates as evaluated via Equation [3.20] or [3.21]. However, as a consequence of the issues discussed in Section 3.4.6, such enormous efficiencies could be obtained only for compounds with very low k' values, i.e., those that eluted very close to the column dead volume. To provide a more realistic measure of column efficiency in such cases, the effective plate number (N_E) was defined by replacing t_r by ($t_r - t_0$) in Equations [3.20–3.21], where t_0 is the elution time after injection for unretained solutes:

$$N_E \equiv 4.[(t_r - t_0)/\Delta t_i]^2 = 5.545.[(t_r - t_0)/\Delta t_{FWHM}]^2$$
$$= N.[k'/(1 + k')]^2 \qquad [3.24]$$

The proof of the relationship between N_E and N is given elsewhere (Scott http://www.chromatography-online.org/). The correction factor (based on k') applied to N has a limiting value of 0.25 when k' is close to unity (solute is barely retained), but for large k' it approaches unity; thus N_E as defined in Equation [3.24] does account to some extent for the variation of N with k', discussed above.

3.4.8 Maximum Sample Injection Volume for a Specific Column

Details of the injection of a sample onto a column will inevitably contribute an additional source of peak broadening that will increase as the injected volume is increased. It would be useful to be able to estimate the maximum sample volume that can be injected without significantly affecting the column performance. The first question to be settled concerns the meaning of 'significantly' in this context. If it is accepted that the effect of the injected volume should be to increase the full peak width by no more than 10 %, then it is possible to show (Scott, http://www.chromatography-online.org/) that:

$$V_{inj}(max) = (1.1/N^{\frac{1}{2}}) \cdot V_r = (1.1/N^{\frac{1}{2}}) \cdot (U \cdot t_r) \quad [3.25]$$

where U is the volume flow rate of the mobile phase and N and t_r both vary with nature of the analyte. In practice we are concerned with at least one pair of analytes that elute close together, and the values of N and t_r for the first of these should be used in Equation [3.25].

It is also necessary to ensure that the sample extract to be injected on column uses a solvent that is either the LC mobile phase itself or a solvent of lower elution strength (Section 4.4.2a).

3.4.9 Peak Capacity of a Column

The peak capacity C_P of a column under stated operating conditions can been defined as the maximum number of peaks that can be fitted into a chromatogram between the unretained solutes and the 'last peak', each peak being separated from its neighbors by twice the peak width ΔV_i (or the corresponding Δt_i) given by Equation [3.19]. A lower limit estimate of C_P is given by Equation [3.27], and a less approximate approach yields the relationships summarized in Figure 3.5.

A major problem with the preceding verbal definition is the meaning of 'last peak of the chromatogram', which is vague because it varies with a number of unrelated factors such as the detector sensitivity and the column efficiency. The 'last peak' can be either theoretically (i.e. arbitrarily) defined, or somewhat less arbitrarily from properties of the chromatographic apparatus used. Limited peak capacity can be a serious problem in the analysis of multi-component mixtures if the peak capacity is insufficient to contain all the peaks as well resolved discrete entities. For example, isocratic elution (use of an LC mobile phase of fixed composition) leads to the later eluting peaks being very broad as a result of the dynamic effects discussed in Section 3.5, so that low concentration analytes which elute later can hardly be detected. If chromatographic conditions are changed so that the late peaks are eluted at lower k' values to improve peak heights and thus detection limits, the early peaks then merge together and are not resolved.

Peak capacity can be effectively improved by using temperature programming in GC (Section 4.4.3b) or

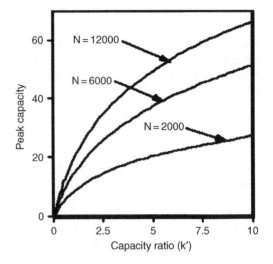

Figure 3.5 Plots of C_P (peak capacity) as a function of k' value for the last-detected peak, for several values of column efficiency N (also taken as that for the last peak), calculated from Plate Theory taking into account variation of peak width with retention time. Adapted from Scott, http://www.chromatography-online.org/, with permission.

gradient elution in LC (Section 3.6), but even this improvement is limited and may not suffice for complex mixtures. A theoretical equation describing peak capacity in terms of column and solute properties would indicate how the peak capacity might be optimized. An ultra-simple approach to calculation of peak capacity is as follows. Take the required separation of the peak for the 'last eluting' solute A from its nearest neighbours as twice the peak width ΔV_{iA} evaluated at the inflection points (Equation [3.19]):

peak separation (in terms of elution volume) $\equiv$

$$\quad [3.26]$$

$$2 \cdot \Delta V_i^A = 4 \cdot N^{\frac{1}{2}} \cdot (v_m + K^A \cdot v_s)$$

since $V_{rA} = N \cdot (v_m + K^A \cdot v_s)$ (Equation [3.12]). Adopting the approximation that, prior to the time when analyte A eluted from the column, it was possible to fit in C_P peaks with the same peak separation, so that C_P is given by:

C_P = (total time elapsed once A elutes) /

(time interval per peak)

$= t_{rA}/$(required peak separation in time units)

$= V_{rA}/$(peak separation in units of elution volume)

$= V_{rA}/2 \cdot \Delta V_{iA} = N^{\frac{1}{2}}/4$

$$\quad [3.27]$$

This estimate for C_p is a lower limit since the last detected peak will be broader than the earlier ones, so many more peaks can be fitted in than suggested by this simple expression.

It is possible to correct this ultra-simple approach with a Plate Theory model for the variation of peak width with retention time (volume), conveniently expressed via the capacity factor k'. The derivation and final result are complex (Scott, http://www.chromatography-online.org/) and are not reproduced here. Instead, Figure 3.5 shows representative plots of C_p vs k' for several values of N, calculated from this more realistic model. The values of N and k' are of course those for the last-eluting peak, but this 'last peak' will be different for different chromatographic detectors with different sensitivities. It is clear from Figure 3.5 that any chromatographic conditions that limit the k' value for the last-detected peak will thus limit the peak capacity, particularly at lower values of k'.

In a very general way, the limitations on analysis of complex mixtures imposed by peak capacity emphasize again the great advantage of a multi-parameter detector like a mass spectrometer relative to single-parameter detectors. Peaks that co-elute or nearly so can frequently be resolved if the corresponding mass spectra are sufficiently different. This is particularly important because the probability of co-elution of components of a mixture is much higher than expected from naïve considerations. The problem is analogous to a well-known trick in recreational mathematics involving calculation of the probability that two or more people (analytes) in a group (complex mixture) will have birthdays (retention times) on the same day (retention time window as assumed in calculations of peak capacity). This mathematical curiosity is described in the accompanying text box; what is remarkable is the steepness of the rate at which the probability of coincidences rises to the limiting value of unity, e.g. with a number of people (analytes) just over 10 % of the number of days (retention time windows), the probability of at least one coincidence is 90 %! The implications of the general result for proper characterization of chromatograms of complex mixtures was first pointed out (Rosenthal 1982) in the context of the requirement of mass spectrometric detection with GC in environmental analysis. Of course the birthday model described here is based on some invalid assumptions, e.g., random distribution of birthdays (analyte retention times) among the days of the year (retention time windows), and the birthday–retention time analogy itself is faulty, e.g., the size of retention time windows varies with retention time. These and other shortcomings of the model were addressed later (Davis 1983; Martin 1986). However, the general conclusion is not altered, i.e., the probability of co-elutions of

analytes in a mixture increases rapidly with the number of analytes, and a multi-parameter detector with high specificity is essential in such cases.

3.4.10 Gaussian Form of the Plate Theory Elution Equation

The Plate Theory is based on a picture of a chromatographic column as consisting of a finite number of discrete compartments, within each of which solutes establish local equilibrium between stationary and mobile phases. It is thus to be expected that a qualitative postulate of this 'discrete' kind would, when translated into mathematical language, give rise to a Poisson Function (Equation [3.11], see text box in Chapter 7) as the elution equation. It is well known that a Poisson Function becomes indistinguishable from a (more mathematically tractable) Gaussian Function when N is sufficiently large that it can be considered a continuous variable, and the purpose of the present Section is to transform the elution equation into Gaussian form, equation [3.31].

However, the elution equation in Poisson form has its zero point at the injection point ($v = 0$ in Equation [3.11]), whereas a Gaussian form as usually expressed is centred at the peak maximum. Thus conversion of Equation [3.11] to its Gaussian equivalent involves changing the origin in addition to considering the limit for large values of N. Recall that v is defined as the number of 'plate volumes' of mobile phase that have passed through the column, and in the conversion it is convenient to work with dimensionless numbers, so a new quantity ω is defined as the number of plate volumes eluted measured relative to the number required to reach the peak maximum of solute A:

$$\omega \equiv v - N \qquad [3.28]$$

More details of the conversion are given in Appendix 3.2. The final result is:

$$c^A_{m(N)} = [c_A{}^0/(2\pi N)^{\frac{1}{2}}] \cdot \exp\{-(V - V_{rA})^2/(2\sigma_V)^2\} \qquad [3.31a]$$

Equation [3.31a] is the elution equation written in Gaussian form in terms of the continuous variable V and centered at the peak maximum where $V = V_{rA}$. This is readily transformed to a form using time t as the independent variable (via U, the volume flow rate of mobile phase), replacing σ_V by $\sigma_t = (\Delta t_i)/2$:

$$c^A_{m(N)} = [c_A{}^0/(2\pi N)^{\frac{1}{2}}] \cdot \exp\{-(t - t_{rA})^2/(2\sigma_t)^2\} \qquad [3.31b]$$

The Probability of Coincident Birthdays

The proposed analogy to the possible co-elution of analytes in a complex mixture is that of coincident birthdays among a group of people. The days of the year are the analogs of the discrete retention time windows considered in the calculation of peak capacity. The problem to be considered is the calculation of the probability that two or more people (analytes) in a group (complex mixture) will have birthdays (retention times) on the same day (retention time window). It is further assumed that in both cases the distribution of analytes (people) among the windows (days of the year) is random.

At first sight this appears to be a difficult task – how to begin? However, the solution appears quite naturally when adopting a *strategy of doing the opposite of what we are asked* (a risky strategy in real life but always one worth considering if only to shed light on the problem!). Here we will calculate P_{none}, the probability that, in a group of M people, there will be NO coincident birthdays in a year of n days (we shall consider n = 365 later, but keep the discussion general for now). Then the desired probability $P = 1 - P_{none}$.

We consider *randomly* assigning the M people to the n days one-by-one. The first person is assigned to a random day, so now there are (n–1) days available to the *second* person IF there are to be no coincidences; so the probability that this will be the case is (n–1)/n. Now when the *third* person is assigned, there are (n–2) days available if no coincidences are to occur, so the probability for this is (n–2)/n. More important, the probability that both of these no-coincidence assignments will occur is the product of the separate probabilities, i.e. $(n-1)(n-2)/n^2$. It is now obvious that we can continue this process until all M people are assigned to days without coincident assignments (of course we are assuming n ≥ M), so:

$$P_{none} = (n-1).(n-2).(n-3). ----- .[n-(M-1)]/n^{(M-1)} = (n-1)!/(n-M)!.n^{(M-1)}$$

Note that if M = 1 then $P_{none} = 1$ and P = 0, as required; similarly if M = n + 1, $P_{none} = 0$ and P = 1 as required. The interesting results are those for which 1 < M < n. The first (long) form of the expression for P_{none} is convenient for exact calculation by multiplying together all the pre-calculated fractions (n–r)/n. The second (compact) form can be adapted for approximate calculation via Stirling's Theorem, e.g. $\log_e(n!) \approx n.\log_e(n) - n$.

$$\log_e P_{none} = \log_e(n!) - \log_e(M!) - (M-1).\log_e(n) \approx (n-M+1)\log_e(n) - M\log_e(M) - (n-M)$$

Evaluation of P_{none} for n = 365 as a function of M is expected to give an S-shaped curve starting at the origin and approaching the value 1.0 as M approaches 366. This is indeed what is observed, but the surprising aspect is how quickly the curve approaches the limiting value, as illustrated by these examples:

M	5	10	20	30	40	50	60
P_{none}	0.960	0.859	0.556	0.270	0.102	0.028	0.005
P	0.040	0.141	0.444	0.730	0.898	0.972	0.995

This remarkable result (e.g., for a group of only 40 people, just over 10 % of the number of days in the year, it is almost 90 % certain that there will be at least one pair of coincident birthdays) can be partly understood by considering M = 40 as an example. The continued product giving P_{none} starts at 364/365 = 0.99726, and ends at 326/365 = 0.89315, with an average value of 0.94521. Since there are 39 factors multiplied in this case we can approximate the correct answer by evaluating $(0.94521)^{39} = 0.111$. Although no one factor is small (all > 0.89), multiplying 39 of them gives the low final value ~0.1.

---

or using x (distance along the length of the column) via t = x/u(u = linear velocity of the mobile phase) and replacing σ_t by $\sigma_x = (\Delta x_i)/2$:

$$c^A_{m(N)} = [c_A{}^0/(2\pi N)^{\frac{1}{2}}].\exp\{-(x-l)^2/(2\sigma_x)^2\}$$

$$[3.31c]$$

since when a compound elutes at $t = t_r$, this corresponds to x = l.

Gaussian functions are mathematically tractable, and in fact real chromatographic peaks can be close to Gaussian. However, the Plate Theory does not address the question of the causes of peak broadening and thus limitations on

JJ van Deemter and the Kinetic Theory of Chromatography

In 1956, four years after the publication of the invention of gas chromatography (James 1952) and 15 years after the publication of the Plate Theory of chromatography (Martin 1941) by analytical chemists, our understanding of the processes that determine the widths and shapes of chromatographic peaks received an enormous boost through the efforts of three Dutch physicists and chemical engineers, reported in a paper (van Deemter, Zuiderweg and Klinkenberg (1956), *Chem. Eng. Sci.* **5**, 271) that is now a citation classic (*Citation Classics* **3**, 245: available at http://www.garfield.library.upenn.edu/classics1981/A1981KX02700001.pdf).

J.J. van Deemter (reproduced from Adlard, *LCGC North America* **24** (10), 1102 (2006), with permission from Dr. E.R. Adlard, Dr L.S. Ettre and the journal).

J.J. van Deemter, the lead author of this landmark paper that introduced the famous equation that now bears his name, worked at the time at which the work was done for the Shell Oil Company in Amsterdam, but in 1955 transferred to the branch of the same company in Houston, Texas. His expertise in transport properties in gases was applied to theories of heat and mass transfer in fluidized beds (van Deemter 1953, 1954, 1967), and also to more academic pursuits such as extension of Bernoulli's Theorem (when the speed of a fluid increases the pressure decreases) to viscous fluids (van Deemter 1952). It was this expertise in energy and mass transfer in gases in the presence of finely divided solids, originally applied to optimization of industrial processes, that was so valuable in the elucidation of the role of these fundamental phenomena in determining the shapes and widths of chromatographic peaks that were not considered in the Plate Theory (peak widths were described and correlated in that theory, not explained). According to his own account (van Deemter 1981), in the early 1950s scientists and engineers at the Shell Laboratory in Amsterdam were doing pioneering work in GC including A.I.M. Keulemans, the author of the first book on this subject (Keulemans 1959). As a physicist working in chemical engineering research, initially he had only a side interest in chromatography but realized that certain chemical engineering concepts, that had proven useful in packed column processes, could also be applied in chromatography.

The appearance of Keuleman's book made the van Deemter equation (published by a physicist in a chemical engineering journal!) much better known to chemists. As a result this paper has been cited many times (more than 370 between 1961 and 1980 alone); the author himself attributed this to the fact that the equation describes in a simple and intuitive way (that nonetheless can be expressed mathematically to permit quantitative evaluation) the role of the main design parameters of the chromatographic column. In 1978 van Deemter was one of the scientists honored by the Academy of Sciences of the USSR by presentation of a memorial medal celebrating the 75th anniversary of the discovery of chromatography by M.S. Tswett.

column efficiency, resolution etc. The reasons for such broadening and their implications must now be examined.

3.5 Nonequilibrium Effects in Chromatography: the van Deemter Equation

Separation of solutes in a chromatographic column requires that the analytes, initially injected in a common volume of solvent, will move apart from one another in the column as a result of their differences in retention characteristics, i.e., differences in the partition coefficients K (Equation [3.1]), while their dispersion (broadening) is restricted sufficiently that the analyte bands elute as resolved peaks. Until now only the retention

characteristics using the Plate Theory have been considered, but now it is necessary to consider the dispersion effects. The detailed theory of such nonequilibrium rate processes is too complex for reproduction here, so only an intuitive treatment is given. For now the discussion will be restricted (as for the Plate Theory) to isocratic elution under conditions of laminar flow, i.e., no turbulence (but see Section 3.5.9).

The rate theory of chromatography was introduced some 50 years ago by physicists and chemical engineers (van Deemter 1956). Despite all the work, both theoretical and experimental, that has been done since then on dispersion in chromatographic columns, the van

Deemter equation still provides the most commonly used description of these dispersion processes. van Deemter theory is especially useful near the 'optimum' value of the linear velocity of the mobile phase, where 'optimum' in this context implies 'maximum column efficiency'.

When discussing column efficiency in the context of the Plate Theory, the concept of the height of the equivalent theoretical plate (HETP) was introduced as $H = (l/N)$, where l is the length of the bed of stationary phase in the column and N is the number of theoretical plates (Section 3.3.5). H can conveniently be expressed as:

$$H = (l/N) = (l/4).(\Delta t_i/t_r)^2 \text{(using equation [3.20])}$$

$$= (l/4).(2\sigma_t/t_r)^2 = l.(\sigma_t/t_r)^2 \text{(using equation [3.31])}$$

$$= l.(\sigma_x/l)^2 = \sigma_x^2/l$$

$$[3.32]$$

The van Deemter approach deals with the effects of rates of nonequilibrium processes (e.g. diffusion) on the widths (σ_x) of the analyte bands as they move through the column, and thus on the effective value of H and thus of N. Obviously, the faster the mobile phase moves through the column, the greater the importance of these dispersive rate processes relative to the idealized stepwise equilibria treated by the Plate Theory, since equilibration needs time. Thus van Deemter's approach discusses variation of H with u, the linear velocity of the mobile phase (not the volume flow rate (U), although the two are simply related via the effective cross-sectional area A of the column, which in turn is not simply the value for the empty tube but must be calculated as the cross-sectional area of the empty column corrected for the fraction that is occupied by the stationary phase particles). This approach identifies the various nonequilibrium processes that contribute to the width of the peak in the Gaussian approximation and shows that these different processes make contributions to σ_x (and thus H) that are essentially independent of one another and thus can be combined via simple propagation of error (Section 8.2.2):

$$\sigma_x^2 = \Sigma_j \sigma_{x,j}^2$$

There are four independent dispersion processes operating in a packed column that contribute to the total band broadening: multipath dispersion; dispersion from longitudinal diffusion; and dispersion from resistance to mass transfer in each of the mobile and stationary phases. These are now be discussed separately in a somewhat qualitative fashion; a rigorous discussion can be found elsewhere (Scott, http://www.chromatography-online.org/). It is important to note that, in the following derivations

of Equations [3.33–3.42], it will be assumed that the compressibility of the mobile phase is negligible, i.e., as the mobile phase travels from the high pressure region at the column entrance to the low pressure at the exit, it undergoes no change in density. The implications of the breakdown of this assumption (as in GC) are discussed in Section 3.5.11a.

3.5.1 Multipath Dispersion

In a packed column the molecules in the mobile phase follow a meandering path through the gaps between the stationary phase particles; obviously some follow shorter paths than the average and some move on longer paths. Thus some molecules will elute faster than the average and some will be slower, thus leading to dispersion of an analyte band (Figure 3.6). It is intuitively obvious that particles of larger diameter d_p will cause larger deviations of molecular paths from the 'ideal' paths straight down the length of the column, and in fact it can be shown (van Deemter 1956; Scott, http://www.chromatography-online.org/),) that:

$$\sigma_{x,M}^2/l = H_M = 2\lambda.d_p \qquad [3.33]$$

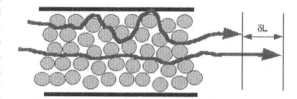

Figure 3.6 Illustration of the effect on analyte band dispersion of multiple paths through stationary phase particles, resulting in a path length difference of δL. Adapted from Scott, http://www.chromatography-online.org/, with permission.

where λ is a dimensionless parameter describing the uniformity of the particles with respect to shape and size, and how uniformly they are packed in the column. For particles of high uniformity in a well packed column (no empty gaps) $\lambda = 0.5$. The contribution H_M of the multipath effect to H can be decreased by using smaller particles (Section 3.5.7). However, this implies a need for higher pressure to force the mobile phase through the narrow interstices between the small particles and this represents a practical difficulty for such a strategy.

3.5.2 Longitudinal Diffusion

This effect is simply that of diffusion (random molecular movement) of a solute band in the mobile phase throughout the length of the column in accord with Fick's Law of Diffusion; it occurs whether or not the band is moving along the column:

$$(1/A).(\partial m^A/\partial t) = -D_m.\partial c_A/\partial x$$

where the left side gives the flux of analyte A (in mass or moles) per unit time per unit area A at a point x along the length of the column, D_m is the diffusion coefficient (diffusivity) of compound A in the mobile phase (dimensions are $length^2.time^{-1}$) and c_A is the concentration at position x. The negative sign arises because a positive flux occurs in the direction of negative concentration gradient, i.e., from high concentration to low. Conceptually a narrow band of analyte A in solution in mobile phase somewhere in the middle of the column can be imagined. If the mobile phase is not moving, diffusion of A will occur in both directions to form an ever-widening Gaussian distribution of A along the column; given enough time (and caps on both column ends to prevent escape!) there will eventually be a constant concentration of analyte A throughout the column, developing on a timescale controlled by D_m. However, in practice the mobile phase is of course flowing along the column with linear velocity u, so there is only a limited time (l/u) available for the longitudinal diffusion to act; this is illustrated in Figure 3.7. The two rate processes (diffusion controlled by the value of D_m and elution controlled by u) compete, and it can be shown (van Deemter 1956; Scott http://www.chromatography-online.org/) that the contribution of longitudinal diffusion in the mobile phase to σ_x and thus to H is given by:

$$\sigma_{x,Lm}^2/l = H_{Lm} = 2\gamma_m D_m/u \qquad [3.34]$$

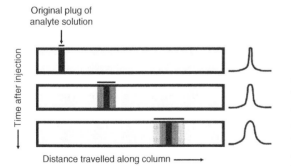

Original plug of
analyte solution

Time after injection →

Distance travelled along column ——→

Figure 3.7 Schematic illustration of peak broadening by longitudinal diffusion as the sample band is eluted along the column.

where γ_m is again a correction factor for packing inhomogeneity.

An exactly analogous effect arises from longitudinal diffusion of the analyte within the stationary phase, now controlled by a different diffusion coefficient D_s. However, now the time spent by the analyte in the stationary phase before elution is a function of the capacity ratio k' (see Equation [3.18]), and the corresponding result can be shown to be:

$$\sigma_{x,Ls}^2/l = H_{Ls} = 2\gamma_s D_s k'/u \qquad [3.35]$$

By combining Equations [3.34] and [3.35], the total contribution of longitudinal diffusion is:

$$H_L = (\sigma_{x,Lm}^2 + \sigma_{x,Ls}^2)/l = (2/u).[\gamma_m D_m + \gamma_s D_s k'] \qquad [3.36]$$

In practice, for packed columns the stationary phase is not continuous along the length of the column, as was implicitly assumed in the derivation of Equation [3.35], but is broken into discrete portions surrounding the stationary phase particles and even among different pores within a given particle. Accordingly diffusion within the stationary phase does not occur to any extent for packed columns within typical elution times, so this contribution is usually negligibly small relative to that from diffusion in the mobile phase, i.e., H_L is almost independent of k'. This is not necessarily the case, however, for capillary columns in which the stationary phase is coated as a continuous film on the walls; such columns are of course common for GC, but are almost never used for LC.

3.5.3 Resistance to Mass Transfer in the Mobile and Stationary Phases

As they pass through the column the analyte molecules are continually transferring from the mobile phase into the stationary phase and back again. In the Plate Theory this process is assumed to be equilibrated within each plate, i.e., to be effectively instantaneous. In reality a finite time is required for the molecules to diffuse through the mobile phase to reach and enter the stationary phase; molecules that are close to the stationary phase will reach it quickly, but molecules that happen to be further away will get there significantly later. However, during this diffusion time they will be swept along the column by the mobile phase and thus move ahead of those molecules that were close to the stationary phase boundary and entered it to be retained for some finite time. This leads to dispersion of the analyte band; an analogous process operates in the stationary phase.

Clearly these dispersive processes will increase in relative importance as the mobile phase flow rate increases and as the diffusion rates decrease (the opposite to the dependence for longitudinal diffusion discussed above). This is because the faster the mobile phase (and slower the inter-phase diffusion) the greater the extent to which the molecules originally far from the stationary phase interface will be swept ahead of the molecules that made it into the stationary phase and were retained there for some time. A detailed discussion (Scott http://www.chromatography-online.org/), based on that originally derived by van Deemter, of the form of this dependence of band dispersion on resistance to mass transfer, gives:

$$H_R = (\sigma_{x,Rm}{}^2 + \sigma_{x,Rs}{}^2)/l$$
$$= [f_1(k').d_p{}^2/D_m + f_2(k').d_f{}^2/D_s].u \qquad [3.37]$$

where d_p is the particle diameter of the stationary phase support, d_f is the film thickness of the stationary phase coated on the particles, and the functions of k' are:

$$f_1(k') = [1 + 6k' + 11k'^2]/24(1 + k')^2 \qquad [3.38a]$$

$$f_2(k') = k'.(8/\pi^2)/(1 + k')^2 \qquad [3.38b]$$

for resistance to mass transfer in the mobile and stationary phases, respectively.

3.5.4 Optimization to Maximize Column Efficiency

Now the physical origins of the general functional form of the van Deemter equation are understood:

$$H = \sigma_x{}^2/l$$
$$= \sigma_{x,M}{}^2/l + (\sigma_{x,Lm}{}^2 + \sigma_{x,Ls}{}^2)/l + (\sigma_{x,Rm}{}^2 + \sigma_{x,Rs}{}^2)/l$$
$$= A + B/u + C.u \qquad [3.39]$$

where the coefficients A, B and C are given by Equations [3.33–3.37] and are functions of d_p, d_f, the analyte diffusivities in both phases, of k' (and thus of the partition coefficient and the phase ratio of the column, see Equation [3.17]), and of parameters describing how well the column was packed.

To determine the 'optimum' value u_{opt}, i.e., the value of u that gives a minimum value of H and thus a maximum value of N, the hyperbolic function in Equation [3.39] must be differentiated with respect to u:

$$dH/du = -B/u^2 + C \quad \text{and} \quad d^2H/du^2 = 2B/u^3$$

Clearly $d^2H/du^2 > 0$, so the condition $dH/du = 0$ does indeed describe a *minimum* (not maximum) value for H:

$$u_{opt} = (B/C)^{\frac{1}{2}} = \{2[\gamma_m D_m + \gamma_s D_s k']/[f_1(k').d_p{}^2/D_m$$
$$+ f_2(k').d_f{}^2/D_s]\}^{\frac{1}{2}}$$
$$= (D_m/d_p).\{2[\gamma_m + \gamma_s(D_s/D_m)k']/[f_1(k')$$
$$+ f_2(k').(d_f/d_p)^2/(D_s/D_m)]\}^{\frac{1}{2}}$$
$$\approx (D_m/d_p).\{2\gamma_m/f_1(k')\}^{\frac{1}{2}}$$
$$[3.40]$$

Since generally diffusion is much faster in a mobile liquid ($D_m/D_s \ll 1$) and the film thickness is a small fraction of the particle diameter ($d_f/d_p \ll 1$), the second terms in both numerator and denominator of Equation [3.40] can be ignored relative to the corresponding first terms; thus u_{opt} is (approximately) directly proportional to the diffusion coefficient in the mobile phase (D_m) and inversely proportional to the particle diameter (d_p). Also, u_{opt} varies with the nature of the analyte via k', and to some extent on the quality of the column packing via γ_m.

Now it is possible to evaluate the minimum value for H again for $D_s/D_m \ll 1$ and $d_f/d_p \ll 1$:

$$H_{min} = A + B/u_{opt} + C.u_{opt} = A + 2.(BC)^{\frac{1}{2}}$$
$$= [2\lambda.d_p] + 2.\{2.[\gamma_m D_m + \gamma_s D_s k'].[f_1(k').d_p{}^2/D_m$$
$$[3.41]$$
$$+ f_2(k').d_f{}^2/D_s]\}^{\frac{1}{2}}$$
$$= [2\lambda.d_p] + 2.d_p.\{2.[\gamma_m + \gamma_s(D_s/D_m)k'].[f_1(k')$$
$$+ f_2(k').(d_f/d_p)^2/(D_s/D_m)]\}^{\frac{1}{2}}$$
$$\approx 2d_p.\{\lambda + [2\gamma_m.f_1(k')]^{\frac{1}{2}}\}$$

i.e. $H_{min} \approx 1.48 d_p$ \qquad [3.42]

where the very approximate expression for H_{min} in Equation [3.42] arises (Scott http://www.chromatography-online.org/) by assuming only well-retained peaks (i.e., $k' \gg 1$, elution well after the void volume) and $\lambda \approx 0.5$ and $\gamma_m \approx 0.6$ for a well-packed column. This result emphasizes the importance of particle diameter for column efficiency (Section 3.5.7) but note that use of smaller d_p values implies the need for a higher optimum linear velocity (Equation [3.40]); these criteria imply difficult demands on the pumping system.

An example of a van Deemter plot for a simple LC system is shown in Figure 3.8 (Scott, http://www.chromatography-online.org/). The major contribution to dispersion at the optimum velocity (where the value of H is a minimum) is the multipath effect. Only at much lower velocities does the longitudinal diffusion effect become significant, while the mobile phase velocity must be increased to about $0.2 \, cm.s^{-1}$ before the dispersion due to the resistance to mass transfer begins to become significant relative to that of the multipath effect. Note that the experimental minimum value of H in Figure 3.8 is $\sim 0.0015 \, cm = 15 \, \mu m$, in reasonable agreement with the value ($13.3 \, \mu m$) predicted by the very approximate Equation [3.42] for a particle size d_p of $8.9 \, \mu m$.

smaller absolute value of the slope, so that any effects on H of uncontrolled variations in u are minimized.

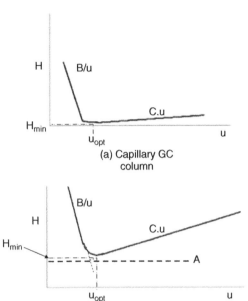

(a) Capillary GC column

(b) Packed GC column

Figure 3.9 Qualitative comparisons of van Deemter plots: $H = A + B/u + Cu$
(a) Capillary column GC. No packing so no multipath term A; low resistance to mass transfer in helium mobile phase; film thickness very low.
(b) Packed column GC (relatively large particle size d_p so multipath term A relatively large; high diffusion rates in gas).

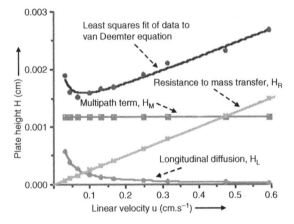

Figure 3.8 Experimental van Deemter plot of H (cm) vs u $(cm.s^{-1})$ for isocratic elution (normal phase) of hexamethylbenzene with a mobile phase of 4.8 % (w/v) ethyl acetate in *n*-decane. The column was 25 cm long, 9 mm in diameter and packed with $8.5 \, \mu m$ silica gel. The curve fitting procedure gave values for the van Deemter constants, and thus the separate contributions to the curve from the multipath dispersion, longitudinal dispersion and the resistance to mass transfer were calculated as shown. Reproduced from Scott, http://www.chromatography-online.org/, with permission.

A qualitative indication of the differences in van Deemter plots for different kinds of chromatography is shown in Figure 3.9. These differences are readily rationalized in terms of the physical meaning of the three terms in the van Deemter equation, discussed above. Note that the negative slope reflecting the B/u term is always considerably steeper than the positive slope in the region dominated by the C.u term. In practice, therefore, it is suggested that the value of u should be set just a little higher than u_{opt}, i.e., in the portion of the curve with

Recall that it was assumed in the derivations of Equations [3.33–3.42] that the compressibility of the mobile phase is negligible, i.e., as the mobile phase travels from the high pressure region at the column entrance to the low pressure at the exit, it undergoes no change in density. This is essentially true for liquid mobile phases, but not for gases. Thus in GC, as the pressure falls along the column length the pressure drops, and so the linear velocity u must increase to maintain the fixed mass flow rate of the gas. Moreover, since solute diffusivity depends on the gas pressure, this will also change along the column. The multipath effect does not depend on velocity and will be unaffected. However, the other effects are all functions of parameters that vary with gas pressure (solute diffusivity and mobile phase velocity) and therefore require modification to account for the compressibility of the gas mobile phase (Scott http://www.chromatography-online.org/). The case of gas chromatography is considered in Section 3.5.11.

3.5.5 Relationships for Estimating Optimized Conditions

One such approximate but practically useful relationship, that for the optimized theoretical plate height H_{min} in Equation [3.42], has already been seen. Note that this approximation for H_{min} clearly represents a considerable over-simplification, since it suggests that column performance for a well-packed column is independent of the nature of the analyte provided k' is sufficiently large, and is also independent of the nature of both the stationary and mobile phases. Nonetheless, this approximate relationship does emphasize the crucial importance of particle size in determining the efficiency of an LC column (see Section 3.5.7) and provides a first-order approximation to the value of H under optimum flow conditions.

Some other approximate equations are now derived; for convenience these are collected together in the accompanying text box. First the algebraic relationships are derived, then some quantitative estimates are given. It must be emphasized that the following approximate expressions are truly approximate, and that their main usefulness lies in the semi-quantitative understanding that they provide for the inter-relationships among the operating parameters. Also, the discussion thus far is restricted to isocratic elution, i.e., a mobile phase of fixed composition, and to incompressible mobile phases, i.e. not to gas chromatography.

Under the same limitations and assumptions as those that allowed evaluation of the approximation for H_{min} (Equation [3.42]), the final approximate form of Equation [3.40] for the optimum linear velocity u_{opt} gives:

$$u_{opt} \approx 1.63 D_m/d_p \qquad [3.43]$$

Now consider the maximum achievable column efficiency as measured by the number of theoretical plates N_{opt}. For this purpose help is obtained from the water engineers! An important relationship used to evaluate flow rates of water through porous media is d'Arcy's Law (see http://biosystems.okstate.edu/darcy/). The principles underlying d'Arcy's original formulation also apply to liquids other than water and the law can be expressed as follows in a form appropriate to flow of a mobile phase through a bed of stationary phase particles (Scott http://www.chromatography-online.org/):

$$u = U/A = \kappa.\Delta P.d_p^2/\eta \cdot l$$

$$\text{so that } l_{opt} = \kappa.\Delta P.d_p^2/\eta.u_{opt} \qquad [3.44]$$

where κ is the (dimensionless) d'Arcy Constant, ΔP is the pressure drop along the column and η is the viscosity of the mobile phase (dimensions = mass.length^{-1}.time^{-1}). By definition $l_{opt} = N_{opt}.H_{min}$, so that:

$$N_{opt} = (\kappa.\Delta P.d_p^2)/(\eta.u_{opt}.H_{min}) \approx (\kappa.\Delta P.d_p^2)/(2.4D_m.\eta) \qquad [3.45]$$

where the second form is obtained by using Equations [3.42–3.45] to substitute for u_{opt} and H_{min}. Equation [3.45] shows that, when operated at its optimum condition specified by u_{opt}, the column efficiency N_{opt} is proportional to the pressure drop along the column (for LC/MS using atmospheric pressure ionization sources the outlet pressure is obviously that of the atmosphere). N_{opt} is also proportional to the square of the particle diameter, i.e., larger particle diameters provide greater column efficiency N_{opt} for a column operated at a fixed pressure drop. This is apparently a counter-intuitive result in view of Equation [3.42] that indicates that decreasing d_p leads to a proportional decrease in H_{min}. The apparent discrepancy between these two measures of column performance is resolved when the column length l_{opt} is considered as the link between H_{min} and N_{opt} because an increase in d_p also increases the column permeability allowing a longer column to be used for a fixed ΔP. The permeability increases as the square of d_p but the variance per unit length increases only linearly with d_p, so that doubling d_p allows a column four times the length to be used, but the number of plates per unit length will be halved, i.e., the column efficiency N_{opt} will be increased by a factor of two. Equation [3.45] also shows that higher efficiencies are obtained using a mobile phase of low viscosity η and for analytes of low diffusivity D_m in that particular mobile phase. These two parameters η and D_m tend to be inversely proportional to one another (molecules can diffuse more easily through nonviscous liquids) and so the effect of their product in the denominator of Equation [3.45] is generally relatively small.

Within the same set of approximations as before, it is now possible to evaluate the column length that will give optimum performance when operated at the van Deemter optimum:

$$l_{opt} = N_{opt}.H_{min} \approx 0.62(\kappa.\Delta P.d_p^3)/(D_m.\eta) \qquad [3.46]$$

Another parameter of interest related to N_{opt} is the peak capacity C_p. Under the highly restrictive assumption that all peaks have the same width (taken conservatively as the width of the last-eluting peak), Plate Theory led to

---

Approximate Expressions for some Chromatographic Parameters Evaluated at the van Deemter Optimum

All of the relationships are derived from the Plate Theory with some approximations, and are evaluated for conditions at the minimum of the van Deemter Plot, usually with the following additional *approximations* appropriate for isocratic liquid chromatography:

$\lambda = 0.5$ and $\gamma_m = 0.6$; these are dimensionless parameters that have the indicated values for a well-packed column (no gaps);

k' is appreciably > 1, i.e., well-retained peaks;

Note: $f_1(k') = (1 + 6k' + 11k'^2)/[24(1 + k')^2] = 0.458$ for $k' >> 1$, and 0.400 for $k' = 10$, and 0.188 for $k' = 1$, 0.400 for $k' = 10$, 0.188 for $k' = 1$.

$D_m >> D_s$, i.e., diffusion coefficients are much larger in the mobile phase than in the stationary phase;

$d_p >> d_f$, i.e., stationary phase particle size is much greater than film thickness.

Linear velocity of mobile phase:
$u_{opt} \approx [2\gamma_m/f_1(k')]^{\frac{1}{2}}.(D_m/d_p) \approx 1.63.(D_m/d_p)$ for $k' >> 1$.

Height of theoretical plate:

$H_{min} \approx 2d_p.\{\lambda + [2\gamma_m.f_1(k')]^{\frac{1}{2}}\} \approx 1.48d_p$ for $k' >> 1$.

Column efficiency (number of plates):
$N_{opt} = (\kappa.\Delta P.d_p^2)/(\eta.u_{opt}.H_{min}) \approx (\kappa.\Delta P.d_p^2)/(2.4D_m.\eta)$ for $k' >> 1$.

Length of column:
$l_{opt} = N_{opt}.H_{min} \approx 0.62(\kappa.\Delta P.d_p^3)/(D_m.\eta)$ for $k' >> 1$.

Peak capacity:
$C_{opt} \approx N_{opt}^{\frac{1}{2}}/4 \approx 0.16d_p.(\kappa.\Delta P/D_m.\eta)^{\frac{1}{2}}$ for $k' >> 1$.

Elution time for last-eluting analyte:
$t_{opt.L} = (1 + k'_L).t_0 = (1 + k'_L).(u/l)_{opt} \approx 0.49.k'_L.[(\kappa.\Delta P.d_p^4)/(\eta.D_m^2)]$ for $k' >> 1$.

These approximations are intended to be used only for estimation of values for these parameters.

For **numerical evaluation** of these approximate expressions:

$\kappa = 35$ (d'Arcy Constant) when ΔP is in psi and all other quantities are in cgs units; for SI units $\kappa = 3.5$.
Note: $1psi = 6.895 \times 10^3 Pa = 6.895 \times 10^4$ dynes.cm^{-2}
For small molecule solutes in common solvents at room temperature, diffusion coefficients $D_m = 1 - 4 \times 10^{-5} cm^2.s^{-1} = 1 - 4 \times 10^{-9} m^2.s^{-1}$.
Values of viscosity at room temperature for common solvents are (P = poise = cgs unit): water; $\eta = 0.89 \times 10^{-2}P = 0.89 \times 10^{-3}$ (Pa.s): acetonitrile; $\eta = 0.38 \times 10^{-2}P = 0.38 \times 10^{-3}$ (Pa.s): methanol; $\eta = 0.54 \times 10^{-2}P = 0.54 \times 10^{-3}$ (Pa.s): glycerol; $\eta = 934 \times 10^{-2}P = 934 \times 10^{-3}$ (Pa.s).

Equation [3.27]; when evaluated for optimum conditions, this gives:

$$C_{P,opt} \approx N_{opt}^{\frac{1}{2}}/4 \approx 0.16d_p.(\kappa.\Delta P/D_m.\eta)^{\frac{1}{2}} \qquad [3.47]$$

Especially for high throughput applications it is extremely important to be able to estimate the time of analysis.

From the definition of the capacity factor (Equations [3.17–3.18]):

$$t_{opt,L} = (1 + k'_L).t_0 = (1 + k'_L).(u/l)_{opt}$$

where the subscript L denotes the 'last-eluting' peak and t_0 is the column dead time, i.e. the time required for nonretained solutes to exit the column along with the mobile phase. By substituting the optimum values for u and *l* from Equations [3.43–3.44], this relation becomes:

$$t_{opt,L} \approx 0.49(1+k'_L).[(\kappa.\Delta P.d_p{}^4)/(\eta.D_m{}^2)] \qquad [3.48]$$

This very strong dependence of the chromatographic run-time (when operated at u = u$_{opt}$) on d$_p$ is somewhat surprising and emphasizes the need to reduce the particle diameter as much as possible (Section 3.5.7). Of course in practical terms this means a requirement for higher ΔP as the lower particle size decreases the permeability of the stationary phase (smaller interstices between particles), but fortunately Equation [3.48] indicates that the optimum run-time varies only as the first power of ΔP.

3.5.6 Numerical Estimates for Optimized Parameters

The preceding section highlighted the role of the particle size (d$_p$) of the stationary phase as probably the most crucial single parameter in determining column performance. This treatment (Scott http://www. chromatography-online.org/) was based on an earlier seminal paper (Katz 1984) that also includes information on the d'Arcy's Law constant κ; when ΔP is given in psi (pounds per square inch, a unit of pressure still in use by engineers), the value of κ is 35 when the equations are otherwise evaluated using cgs units, and 3.5 for SI units. (The factor of 10 difference corresponds to the conversion factor between cgs and SI units of pressure: $1\,psi = 6.895 \times 10^3 Pa = 6.895 \times 10^4 dynes.cm^{-2}$; note also $1\,atmosphere = 101.325\,Pa$ and $1\,bar = 100\,Pa$.)

Since d$_p$ is so important it is interesting to evaluate, at least semi-quantitatively, some chromatographic parameters for various values of d$_p$ and for typical values of other parameters like D$_m$, η and ΔP. Until very recently, LC pumps and injector valves operated in the ΔP range of a few thousand psi; a typical working value that would ensure a long working lifetime was 3000 psi, and this value is arbitrarily chosen for these calculations. For small molecule solutes (a few hundred Da) in typical LC solvents, diffusion coefficients are in the range $1 - 4 \times 10^{-9} m^2.s^{-1}$; the viscosities η of some commonly used mobile phase solvents at room temperature are as follows (expressed in SI units, i.e., Pa.s): water, 0.89×10^{-3}; methanol, 0.54×10^{-3}; acetonitrile, 0.38×10^{-3}; glycerol 934×10^{-3}. (The value for glycerol is added for interest only.)

Representative values of D$_m$ and of $\eta(2 \times 10^{-9} m^2.s^{-1}$ and $0.4 \times 10^{-3} Pa.s$, respectively) were used in the calculations based on Equations [3.44–3.48] that led to the values shown in Table 3.1. It must be emphasized that the equations used involved several approximations, as discussed in Section 3.5.5, and so the values obtained are only approximate indications. However, the purpose of this exercise is not to be able to design LC columns, but to provide some understanding of the conflicting demands of column performance (N was chosen here as a measure of performance), physical size of the column, and the elution time, when operated at each respective van Deemter minimum and for a fixed pressure drop ΔP. Under these restrictions, use of larger particles can give very large values of column efficiency but at the expense of unmanageably long columns and unrealistically long elution times. (The latter also implies correspondingly

Table 3.1 Dependence of some performance characteristics of an LC column on the particle size of the stationary phase, for isocratic elution

Evaluated at the van Deemter optima, using the approximate expressions in Equations [3.43–3.47], for an inlet pressure of 3000 psi (20 MPa), viscosity 0.4×10^{-3} Pa.s, and diffusion coefficient of analyte in the mobile phase of $2 \times 10^{-9} m^2.s^{-1}$.

d$_p$ (microns)	N$_{opt}$	l$_{opt}$ (cm)	t$_{opt,L}$ (for k' = 10) (minutes)
1	5.5×10^3	0.8	0.6
2	2.2×10^4	6.5	9.4
5	1.4×10^5	102	368
10	5.5×10^5	814	5.9×10^3
20	2.2×10^6	6,512	9.4×10^4

NB: *The equations used for these estimates are highly approximate, but the trends are important. Note also that a different set of assumptions (e.g., a variable rather than a fixed inlet pressure) would lead to a somewhat different dependence of these performance parameters on d$_p$.*

large volumes of mobile phase, an environmental concern as well as one of cost.)

3.5.7 Ultra-Small Stationary Phase Particles

Since all of the performance characteristics are directly proportional to ΔP, if it were possible to increase ΔP by a factor of four to 12 000 psi, say, the use of two micron particles would give about 10^5 theoretical plates with a column length of 25 cm and a maximum elution time of less than 40 minutes, both much more reasonable. (Remember that for now the discussion is limited to isocratic elution.) The technical problems to be overcome in achieving such improvements include manufacture of two micron silica particles of controlled size distribution and a reliable means of packing these small particles into a column with minimal gaps (low values of λ and of γ_m, Sections 3.5.1 and 3.5.3). It is also a challenge to produce HPLC pumps that can operate reliably at up to 15 000 psi while delivering reproducible gradients (Section 3.6). Even more challenging is the development of loop injection valves that can operate at these higher pressures, and how to address the potential increased probability of clogging the column.

Quite recently all these challenges were successfully met and the resulting so-called ultra performance liquid chromatography (UPLC) system successfully interfaced to a mass spectrometer (Plumb 2004). Particles of 1.7 μm diameter with conventional C_{18} stationary phase were created, together with appropriate pumps and robust injector valves operating in the 6000–15 000 psi range. All of the advantages expected on the basis of the preceding discussion were obtained, as exemplified in Figures 3.10 and 3.11 that show some data obtained in a search for metabolites of midazolam (a drug used to produce sleepiness or drowsiness and to relieve anxiety before surgery or certain other medical procedures) in bile from dosed rats. (Note that these experiments employed gradient rather than isocratic elution, but the general trends remain unchanged.) This essentially qualitative search for drug metabolites in the complex mixture required recording full mass spectra and, because of the much more narrow chromatographic peaks obtained using UPLC, it was necessary to employ a mass spectrometer for which the spectral acquisition time was sufficiently short that several spectra could be obtained over a peak. A time of flight instrument (TOF) (Section 6.4.7) with an electrospray ionization (ESI) source (Section 5.3.6) filled this requirement, and also incidentally provided mass measurements of the ions with accuracy and precision of 10–20 ppm with resulting improvements in detection selectivity.

A further advantage of the more narrow chromatographic peaks arose from the fact that each analyte was thus presented to the mass spectrometer in a smaller volume of mobile phase, i.e., at higher concentration (see discussion of chromatographic dilution in Section 3.5.10b); this yielded a corresponding improvement in detection sensitivity since in some conditions an ESI source behaves like a concentration dependent detector over a wide range of flow rates (Section 5.3.6b). Another advantage of the increased resolution for biomedical analyses concerns the corresponding decrease in probability of co-elution with the analyte of matrix components that could lead to ionization suppression effects (Sections 5.3.6a and 9.6), and also of drug metabolites that can undergo collision-induced dissociation in an API interface (Section 5.3.3a) to create additional spurious drug ions. Moreover, as a result of the smaller particle size the minimum in the van Deemter curve is much more shallow because the C coefficient (Equation [3.37]) varies as d_p^2. This has a practical advantage of flexibility in that it is possible to increase the mobile phase flow rate (and thus decrease analysis time) without too great an increase in the theoretical plate height (i.e., decrease in N). A later publication (Yu 2006) demonstrated the advantages of the UPLC approach in quantitative analysis, exemplified by Figure 3.12. Here again, shorter elution times and more narrow peaks were obtained, together with considerable improvements in lower limits of quantitation, sensitivity (slope of calibration curve relating observed signal to analyte concentration) and some significant improvements in accuracy and precision compared with conventional reverse phase HPLC, presumably as a result of the increased concentration of analyte presented to the mass spectrometer in the more narrow chromatographic peak.

Despite these and other demonstrations of the success of very small (≤ 2 μm) particle sizes, questions remain about limitations on the improvement that can be thus obtained over the performance achievable using more conventional particle sizes. A recent investigation of this question (Butchart 2007) has emphasized that the degree to which resolution is improved by the use of particle sizes ≤ 2 μm is dependent on the quality of the column packing (i.e. accounted for by the parameters λ and γ_m incorporated in the van Deemter theory, Section 3.5). It was shown (Butchart 2007) that use of sub-2 μm particles for fast isocratic chromatography using short columns provided almost no improvement over columns packed with 3 μm particles when used with conventional HPLC pumping systems, or even with the newer systems designed to run at elevated pressures (Plumb 2004). Under isocratic conditions it was possible to achieve almost the same column efficiency as that obtained using sub-2 μm

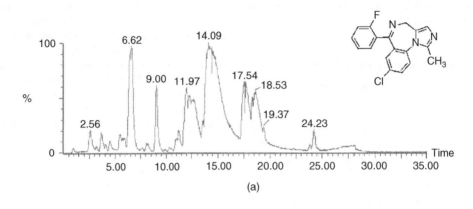

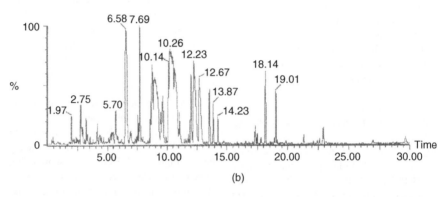

Figure 3.10 TIC chromatograms (electrospray–TOF) obtained for separations of rat bile following the administration of midazolam at 5 mgkg^{-1}: (a) 30 min separation on a conventional 2.1 × 100 mm 3.5 μmC$_{18}$ HPLC column; (b) 30 min separation on a 2.1 × 100 mm1.7μm C$_{18}$ UPLC column operated at 7000 psi. Gradient elution was used (0.1 % aqueous formic acid to 95 % acetonitrile over 30 min). The bile was diluted but not extracted or cleaned up prior to injection. The resolution is dramatically improved in (b) compared with (a), with more than double the number of identifiable discrete peaks with widths of the order of 6 s at the base, giving a peak capacity ∼300, more than double that produced by the conventional HPLC system. As a result four analyte-related peaks (metabolites) are observed in (b) compared with one (and possibly a second appearing as a shoulder at 19.37 min) in (a). Reproduced from Plumb *et al.*, *Rapid Commun. Mass Spectrom.* (2004) **18**, 2331, with permission of John Wiley & Sons, Ltd.

particle columns by using a well-packed 3 μm column but with only one-third of the required back pressure. When fast gradient conditions were used, no performance enhancement was observed with sub-2 μm particles; apparently the improvement provided by use of gradient elution (Section 3.6) swamped out any efficiency gains provided by use of sub-2 μm particles.

These disappointing observations (Butchart 2007) concerning the performance of small particle columns in fast chromatography ($\leq 3 - 4$ minute run times) were explained by considerations of the quality of column packing. The degree to which a column is well packed is affected by both the particle diameter and by the column dimensions. The smaller the particle size the greater

the difficulty in preparing a well packed column, since particle aggregation, frit blockage and particle fracture are potential problems when using the high pressures required to pack the very small particles into the column. With respect to column dimensions, shorter columns tend to have a higher proportion of poor bed quality at the bottom of the column and narrower columns are subject to a relatively greater degree of packing defects as a result of column wall effects. Another possible contribution is the effect of frictional heat that arises from the high pressures required to force the mobile phase through a packed bed of sub-2 μm particles. The thermal energy thus created can lead to radial temperature gradients and thus viscosity changes in the centre of the column more than near the

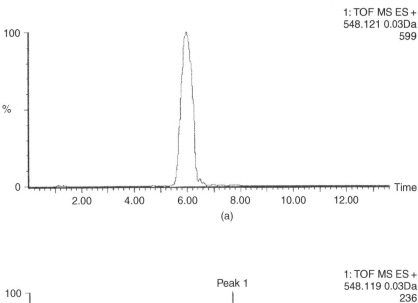

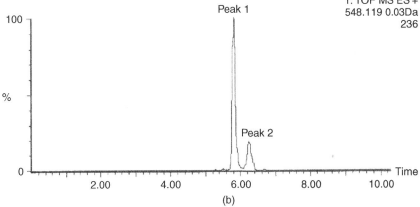

Figure 3.11 Extracted ion chromatograms for *m/z* 548.12 (glucuronides of midazolam) from the analyses shown in Figure 3.10, for (a) HPLC and (b) UPLC. Note the two metabolites, arising from glucuronidation at different hydroxylation sites in midazolam, that are well-resolved in (b) but not in (a). Reproduced from Plumb *et al. Rapid Commun. Mass Spectrom.* (2004) **18**, 2331, with permission of John Wiley & Sons, Ltd.

walls, with a resulting deterioration of column efficiency. While this thermal effect has not been investigated for the sub-2 μm particles, it has been reported for 5 μm particles in a detailed investigation (Halász 1975). In view of the much larger back pressures and resulting frictional effects that must be operative for sub-2 μm particles, it seems probable that this must contribute to the less than full realization of the theoretically expected gain in column performance for fast chromatography, particularly under conditions of gradient elution with capillary columns (Butchart 2007).

Of course these findings, which are relevant to high throughput operation, do not apply directly to other applications of sub-2 μm particle columns. For example, applications of long gradient analysis in longer column formats (≥ 100 mm), which would require the high back pressures (Plumb 2004), undoubtedly exhibit the theoretically expected advantage as exemplified by Figures 3.10 and 3.11. Moreover, if improved techniques for packing columns with the small particles could be developed, the theoretical advantage could be more nearly achieved for the fast chromatography applications also.

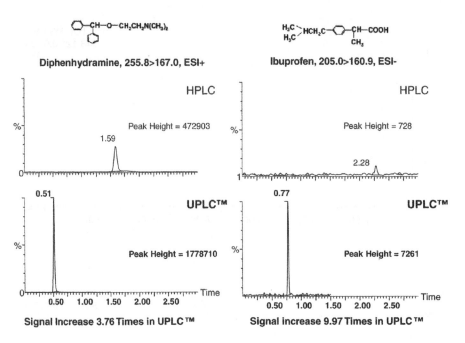

Figure 3.12 Sensitivity comparison of UPLC vs HPLC. Same sample (10 ng/mL in rat plasma) was injected twice, once for HPLC/MS/MS (3.5 μm particles) and once for UPLC/MS/MS (1.7 μm particles); both columns were 2.1 × 50 mm. Examples shown are for diphenhydramine (ESI+, left) and ibuprofen (ESI, right). Reproduced from Yu *Rapid Commun. Mass Spectrom.* (2006) **20**, 544, with permission from John Wiley & Sons, Ltd.

3.5.8 Monolithic Columns

A rather different approach to the problems associated with improving column performance, without having to greatly increase the inlet pressure or resorting to impractical column lengths or retention times, was the introduction of monolithic columns. A comprehensive short review of the development of these columns is included in one of the important papers on the subject (Cabrera 2000) and a more recent review (Ikegami 2004) describes further progress. The 'monolithic' nature of these stationary phases refers to the fact that they consist of a single extended polymer column or rod, rather than packed micron-sized particles. The fabrication of such monoliths is designed so that they contain two kinds of pores; the 'macropores' ('throughpores') are large enough (typically 2 − 6 μm) that they offer minimal resistance to hydrodynamic flow of liquid through them thus requiring low pressures, while the much smaller 'mesopores' (~120 Å) on the surface of the monolithic skeleton provide the surface area required for the chromatographic adsorption–desorption processes.

Considerable work has been devoted to development of such monoliths based on organic polymers, and their applicability to proteins and other biopolymers has been demonstrated. However, disadvantages of this approach include swelling in organic solvents, which can lead to mechanical instability, and the formation of pores of intermediate size which lead to lower column efficiency and peak symmetry (Cabrera 2000). Almost all applications of this approach to quantitative analysis of small molecules have used porous silica gel monoliths prepared using a sol–gel process based on hydrolysis and polymerization of alkoxysilanes (Nakanishi 1992). Subsequently it was demonstrated that this technology provides HPLC stationary phases with unique and useful properties (Minakuchi 1996,1997; Nakanishi 1997; Ishizuka 1998), including rather flat van Deemter curves and the need for only low back pressures that, in turn, can be exploited in two different ways; firstly, by using very high flow rates with only minor increases in plate height H for ultra-fast analyses or, secondly, by coupling several such columns in series to obtain a large number of theoretical plates N and thus greater separatory power. Quite recently, development of a monolithic column based on a capillary of only 20 μm internal diameter was reported (Luo 2005).

An example of the application of these features to separation of some closely-related marine toxins (Volmer 2002) is shown in Figures 3.13 and 3.14. Many examples of the application of monolithic columns to biomedical applications can be cited. For example, it was shown (Dear 2001) that the use of short monolithic silica columns coupled to mass spectrometry dramatically reduced analytical run times for metabolite identification from *in vitro* samples with no loss in chromatographic performance. Six hydroxylated metabolite isomers were separated in one minute, with resolution and selectivity comparable to conventional analytical chromatography resulting in reduction of analysis time per sample from 30 to 5 minutes. Similar results were reported (Wu 2001; Hsieh 2002) for rapid qualitative and quantitative analyses in drug discovery programs (i.e., not fully validated).

3.5.9 Ultra High Flow Rate Liquid Chromatography

Another strategy that appears to confound the van Deemter theory is the use of ultra high flow rate chromatography combined with relatively large particle sizes. This approach is the subject of a patent (Quinn 1997) and the patented technology is available commercially (www.cohesivetech.com). However, it is possible to exploit the principles involved using conventional HPLC pumps and injectors as the large particle sizes used (typically $50-150 \mu m$) permit use of accessible backpressures. The earliest published applications (Ayrton 1997,1998; Jemal 1998) mostly used the principle with a single column, but later work has tended to use this principle as an efficient on-line clean-up technique before passing the fraction containing the concentrated analytes to a conventional HPLC column; this aspect is further discussed in Section 4.3.3d. Here the focus is on understanding the principles, with some examples of their use in chromatographic separations.

As long ago as 1966 (Pretorius 1966) it was shown that operation of an open-tubular HPLC column under turbulent flow conditions permitted much shorter elution times than those observed using the conventional laminar (low speed) elution conditions considered by the van Deemter theory (Equation [3.39]). Extension of the turbulent flow condition to packed columns had to wait for development of HPLC pumps capable of generating sufficient pressure to drive the fast flows required. Although the approach is still referred to as 'turbulent flow' chromatography, some doubt has been expressed (Ayrton 1998) concerning

whether or not true turbulent flow conditions are actually reached under most of the flow conditions employed. Thus some of the earliest work (Ayrton 1997,1998) used a one millimeter i.d. column packed with $50 \mu m$ particles that were estimated to occupy ~40 % of the cross-sectional area (i.e., the effective cross-sectional area for flow of mobile phase $= 0.6\pi.0.05^2 = 4.7 \times 10^{-3} cm^2$), so that the volume flow rate U of $4 mL.min^{-1}$ corresponded to a linear velocity u of $[4 cm^3.min^{-1}/4.7 \times 10^{-3} cm^2] \approx 851 cm.min^{-1} \approx 14 cm.s^{-1}$. Such high flow rates could be achieved using conventional HPLC pumps only because of the large particle size ($50 \mu m$) used and the correspondingly low resistance to flow of mobile phase.

Whether or not the flow is truly turbulent, the high velocities lead to eddies around the particles that tend to equalize the flow across the column, thus creating an essentially plug flow profile rather than the parabolic profile characteristic of laminar flow. In addition, these eddies increase the rate of mass transfer between mobile and stationary phases, thus negating the increase of plate height H at higher linear velocities as expressed by the C term in the van Deemter equation. As a result it is possible to obtain much faster elution with column efficiencies that are much better than predicted for laminar flow by the van Deemter equation. An early example (Ayrton 1999) is shown in Figure 3.15. Apart from the fast elution, a notable feature is that the human plasma samples were not extracted or cleaned up in any way prior to injection. Under the ultra-fast flow conditions the large protein molecules (low values of D_m) could not access the binding sites on the stationary phase and were eluted much earlier than the small molecule analyte; by operating a divert valve prior to transfer to the mass spectrometer the protein band could be discarded to waste so as to not contaminate the mass spectrometer. It is this feature that has led to subsequent use of the ultra-fast elution strategy as an on-line clean-up method. In Figure 3.16 a similar example (Jemal 1998) is shown in which the residual effect of the macromolecules that are first eluted can be seen after the divert valve is switched over. The use of the same principle as a separate but on-line sample preparation technique is described in Section 4.3.3d.

3.5.10 Packed Microcolumns

'Standard' HPLC columns are generally of internal diameter (i.d.) 4.6 or 2.1 mm and, when packed with conventional particles (say $3 - 10 \mu m$), are operated with mobile phase flow rates of up to 1000 and $200 \mu L/min$, respectively. 'Microcolumns' are usually regarded as having

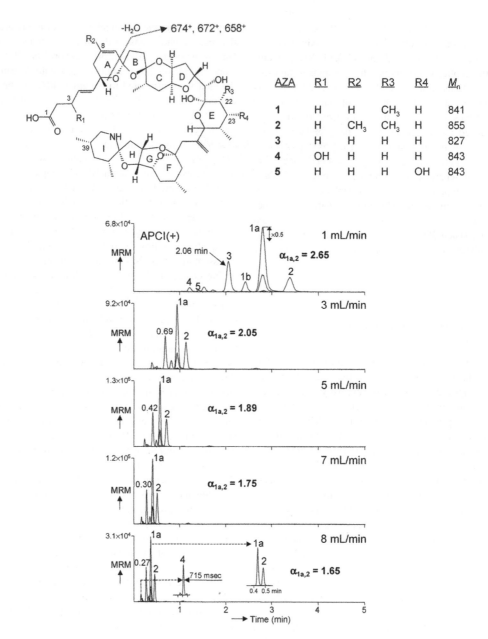

Figure 3.13 LC–MS/MS chromatograms for a mixture of azaspiracid marine toxins obtained using an isocratic mobile phase (acetonitrile/water, 66/34, pH 3.7) at different flow rates, and a 50 × 4.6 mm monolithic column (silica gel derivatized with C_{18}). APCI source on a triple-quadrupole instrument. The separation ratio α is shown for the closely-eluting pair **1a** and **2**. Reproduced from Volmer *et al.*, *Rapid Commun. Mass Spectrom.* (2002) **16**, 2298, with permission of John Wiley & Sons, Ltd.

i.d. of 0.5–1 mm (flow rates up to 50 μL/min), and when fused quartz capillaries of i.d. 0.1–0.5 mm are used they are usually referred to as capillary columns, while 10 – 100 μm i.d. columns are often referred to as nanoscale columns with flow rates in the range of a few hundred nL/min. Columns of i.d. less that one millimeter are now in wide use, but their introduction (Horváth 1967, 1969) and initial characterization was the result of pioneering work by several groups (Borra 1987; Ishii 1977, 1978a, 1978b; Kennedy 1989; Karlsson 1988;

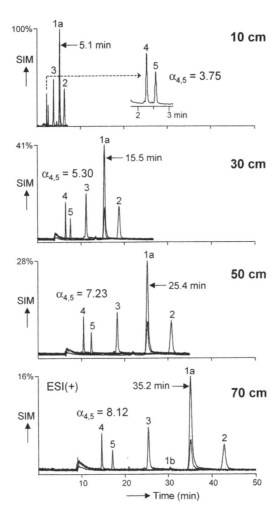

Figure 3.14 Effect of length of a 4.6 mm i.d. monolithic column (varied by connecting shorter columns together) on separation of azaspiracid marine toxins. Mobile phase as for Figure 3.13, flow rate 1 mL.min^{-1}, electrospray ionization. Peak heights are normalized to that observed for the 10 cm column as 100 %. Reproduced from Volmer *Rapid Commun. Mass Spectrom.* (2002) **16**, 2298, with permission from John Wiley & Sons, Ltd.

Kucera 1980; McGuffin 1983; Novotny 1981, 1988; Scott 1979, 1979a; Yang 1982); more recent reviews (Vissers 1997, 1999) cover this history as well as modern developments and applications. Significantly, none of the applications cited in these reviews involve quantitation, validated or otherwise.

There are advantages and disadvantages in the use of packed columns with smaller i.d. As discussed further below, the most important advantages are the small sample sizes that can be accommodated, small volume flow rates (environmental and cost considerations), reduced chromatographic dilution of the analyte

and thus enhanced detection performance, and for microcolumns and nanoscale columns an improved performance with respect to peak-broadening effects arising from dispersive mechanisms compared with similar columns of larger i.d.; these more narrow peaks can not be accounted for by the classical van Deemter theory that does not include column i.d. as a relevant parameter. However, note that parameters that are believed to largely reflect thermodynamic properties of the column, rather than the dynamic dispersive effects, and particularly the selectivity factor $\alpha_{A/B} \approx K^A/K^B$ (Equation [3.16]) and retention factor $k_A' = K^A.V^A_s/V_m$ (Equation [3.17]), are found

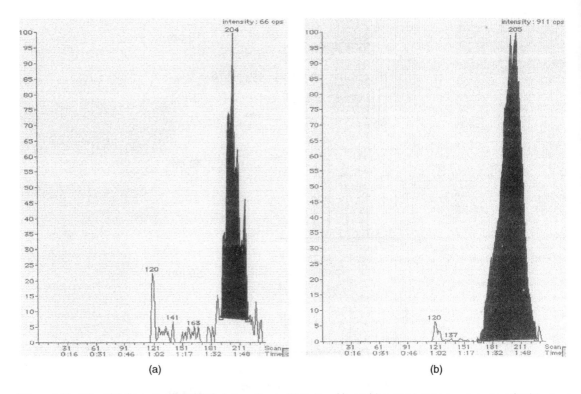

(a)   (b)

Figure 3.15 Ultra-high flow rate LC–MS/MS elution of a candidate drug (a) and (b) its D-labeled internal standard (SIS) in about 1.5 min. The drug was spiked into plasma at $0.5\,ng\,mL^{-1}$ and diluted 1:1 with aqueous internal standard solution. The capillary column was $50\,mm \times 180\,\mu m$ i.d., packed with $30\,\mu m$ particles (C_{18} silica) operated under gradient elution conditions at a flow rate of $0.13\,cm^3.min^{-1}$. Reproduced from Ayrton *Rapid Commun. Mass Spectrom.* (1999) **13**, 1657, with permission of John Wiley & Sons, Ltd.

to be only weakly affected by changes in column i.d. The main disadvantages of smaller i.d. columns arise from the low amounts of sample that can be applied before the column is overloaded (thus giving a limited dynamic range), problems with ruggedness of quantitative methods, and from difficulties in designing and fabricating HPLC components like injectors, detectors, connecting tubing etc. that do not introduce relatively large dead volumes and thus contribute significantly to peak broadening.

3.5.10a The Knox Equation

A clearcut example of the effect of column diameter on plate height (Karlsson 1988) is shown in Figure 3.17. The data were interpreted not in terms of the van Deemter equation but rather the Knox Equation (Equation [3.49]) (Kennedy 1972). This equation is very similar to the original van Deemter form, but is expressed in terms of reduced (dimensionless) variables in an attempt to provide a basis for comparison of different columns packed with

particles of different diameters. In addition, following a proposal introduced by Giddings (Giddings 1961), the A-term in the van Deemter equation was assigned a weak dependence on linear flow velocity to account for the known coupling between the A term (multipath dispersion) and the C-term (resistance to mass transfer). Accordingly, it was proposed (Kennedy 1972) that the functional form shown as Equation [3.49] would provide a more satisfactory description than the van Deemter equation (Equations [3.33–3.39]).

$$h = A.v^{\frac{1}{3}} + B/v + C.v \qquad [3.49]$$

where the reduced plate height $h \equiv H/d_p$ and the reduced linear velocity $v \equiv u.(d_p/D_m)$. A detailed description of the various rival equations to the original van Deemter form is available (Scott http://www.chromatography-online.org/); the Knox equation is widely used in the literature, but note that it was proposed only as an intelligent choice of formula for fitting experimental data, without

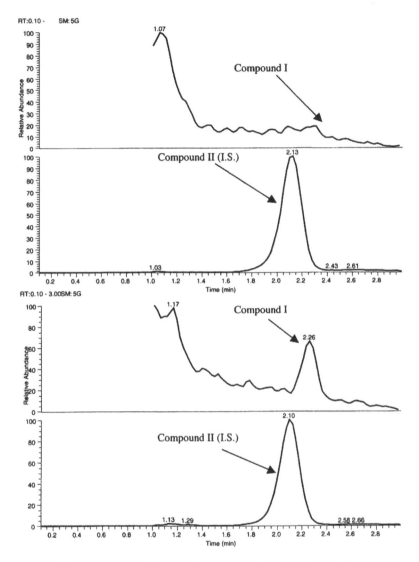

Figure 3.16 Ultra-high flow rate LC–MS/MS analysis in about 2.5 min of two candidate drug compounds in plasma. Top, compound I at zero concentration, compound II present as SIS; bottom, compound I at 1 ng mL^{-1} in plasma. The column was 50 × 1 mm packed with 30 μm particles. Gradient elution was used at a flow rate of 4 mL min^{-1}. The divert valve was switched from waste to the mass spectrometer one minute after injection. Reproduced from Jemal *Rapid Commun. Mass Spectrom.* (1998) **12**, 1389, with permission of John Wiley & Sons, Ltd.

any attempt to derive expressions for the coefficients in terms of more fundamental properties as was done for the van Deemter equation (van Deemter 1956).

Figure 3.17 clearly shows that, as the column i.d. is decreased, the reduced plate height h decreases and the optimum reduced velocity (curve minima) increases. The data shown in Figure 3.17 were fitted (Karlsson 1988) to

the Knox equation to give the following values for the coefficients:

Column i.d. 265 μm: A = 1.7, B = 1.1, C = 0.37
Column i.d. 167 μm: A = 0.71, B = 1.2, C = 0.28
Column i.d. 44 μm: A = 0.49, B = 1.4, C = 0.19

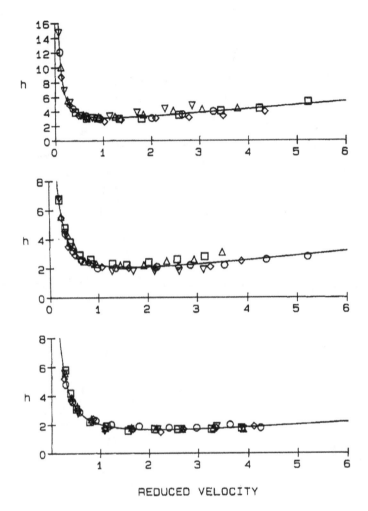

Figure 3.17 Comparison of performance for three columns of different internal diameters (from top to bottom 265, 167 and 44 μm); five independent measurements were made for each column that were all approximately one meter long. The stationary phase consisted of 5 μm C_{18} silica particles with 80 Å pore size. The mobile phase was pure acetonitrile, and the analyte was 2-methoxynaphthalene. The curves represent best least-squares fit of the data to the Knox equation (Equation [3.49]), with the diffusion coefficient of the analyte taken to be $3 \times 10^{-9} \, \text{m}^2\text{s}^{-1}$. Reprinted from Karlsson and Novotny, *Anal. Chem.* **60**, 1662. Copyright (1988), with permission of the American Chemical Society.

The most striking change is in the A-term that accounts for multipath dispersion (Equation [3.33] in the van Deemter theory) and decreases rapidly with decreasing i.d. This effect was interpreted (Karlsson 1988) as reflecting a decrease in the extent of the flow rate variation over the column cross-section as the i.d. decreases, together with shorter time for solute molecules to diffuse radially through the flow velocity profile (parabolic in the case of a nonpacked column); together these phenomena will decrease the effect of flow anisotropy that is reflected in

the multipath dispersion. The B coefficient (longitudinal diffusion, see Equation [3.34] for the van Deemter interpretation) is essentially invariant with decreasing column i.d., which makes physical sense (although there might be a slight increase as the i.d. is reduced). The C coefficient (resistance to mass transfer, Equations [3.37–3.38] for the van Deemter theoretical formulae) also decreases somewhat with decreasing i.d. but is somewhat larger than values usually observed for larger i.d. columns (0.02–0.20); this discrepancy was attributed to the unusually long columns (1 m) used in this work (Karlsson 1988). A more detailed discussion of these trends is given in a later work (Kennedy 1989) that also confirmed the general trends in the experimental data. This discussion confirms the importance of the shorter time required for solute molecules to diffuse radially between all of the possible flow paths (A-term) and retention sites on the stationary phase (C-term); in this way the effects of flow and retention inhomogeneities are averaged out by all solute molecules to a better degree than in larger columns, where the molecules have practically no chance to experience all of the flow paths and retention sites across the entire column cross-section.

3.5.10b Chromatographic Dilution

Another important phenomenon that varies strongly with column i.d. is that of chromatographic dilution. As a result of all the dispersive processes, an analyte will be progressively diluted as it proceeds along the column. The dilution factor D for analyte A can be estimated (Vissers 1997):

$$D \equiv c_A^0/c_A^{pm} = (\varepsilon.\pi d_c^2/4V_{inj}).(1 + k_A').(2\pi l/H) \qquad [3.50]$$

where c_A^0 is the initial concentration of A immediately after injection and before elution has begun (see Equation [3.7]), c_A^{pm} is the concentration of A eluting from the column at the peak maximum, l is the column length and d_c its i.d., H is the plate height and ε the column porosity. Porosity of a granular material is the fraction of the total volume that is not occupied by solid material; the 'pore volume' is the total volume between the constituent particles plus the total volume of the 'pores' or 'cracks' within the individual grains or particles. In the context of a packed chromatographic column, the porosity is the fraction of the total column volume that is available to the mobile phase, i.e. $\varepsilon = V_m/(V_m + V_s)$, see Equation [3.12].

Evaluation of Equation [3.50] shows, for example, that if all other parameters are the same, changing from a column with i.d. 2.1 mm to a 300 μm capillary will result in a 50-fold increase in the concentration of the analyte as it elutes from the column into the detector, with corresponding increases in peak height (and also peak area depending on whether the detector is concentration or mass flow dependent, see Section 4.4.8). Note, however, that such a comparison is valid only if all parameters other than i.d. are maintained the same; in particular this comparison assumes that it is possible to load the same amount of analyte in both cases, a real limitation since the sample size that can be injected without overloading a column is proportional to the amount of stationary phase

in the column. However, a major advantage of packed capillary columns is their ability to deal with very small sample sizes while still providing useful signal levels.

3.5.10c Flow Impedance Parameter and Separation Impedance

On a practical level, a parameter that is useful for evaluating columns is the flow resistance parameter ψ defined as:

$$\psi \equiv (\Delta P.d_p^2)/(u.l.\eta) \qquad [3.51]$$

This parameter relates the particle size, flow velocity, column length and mobile phase viscosity to the pressure drop required to achieve all of these characteristics (an important parameter for practical operation of capillary columns). For packed columns a useful range for ψ is 500–1000 (Knox 1980). For a series of columns that differed in d_c between 20 and 50 μm it was found (Kennedy 1989) that ψ ranged between 600–800 with a slight decrease for lower i.d. columns.

A related dimensionless parameter that is also very useful in evaluating and comparing columns is the separation impedance E (Bristow 1977):

$$E \equiv h^2.\psi = (H^2/d_p^2).(\Delta P.d_p^2)/(u.l.\eta)$$
$$= (l^2/N^2).\Delta P/(u.l.\eta)$$
$$= (\Delta P.t_0)/(N^2.\eta) \qquad [3.52]$$

since the elution time for unretained analytes $t_0 = l/u$. E is useful in assessing the practicality of a separation using a given column as it relates the required pressure drop to the column efficiency (N) and to the time required for elution of analyte A via $t_0 = t_{rA}/(1 + k_A')$, see Equation [3.18]. A useful value for E is ∼2000 for $h_{opt} = 2$ (i.e., the minimum value as a function of u) and ψ ∼500 (Knox 1980). For the same set of columns that were evaluated using ψ (Kennedy 1989) it was shown that E_{opt} for a well-retained analyte (resorcinol) decreased linearly from ∼4000 to 1500 as d_c was reduced from 50 to 20 μm. Yet again this indicates the importance of column i.d. in this range with respect to evaluating the practicality of a proposed separation.

The present treatment covers only some of the main points investigated for small i.d. columns. The major conclusions include the following. The reduction of column i.d. to one millimeter and below introduces significant effects that are not explained by van Deemter's original treatment of dispersion effects, but can still be rationalized on the basis of the same fundamental concepts.

The positive effects include a reduction in plate height (mainly via reduction in the A-term, multipath dispersion), reduction in chromatographic dilution and thus increased detection sensitivity, and reduction in volume flow rate and thus in solvent consumption. However, offsetting disadvantages include limitations on the volume of sample solution and the amount of analyte that can be injected without overloading the column, and the meticulous attention that is required to reduce the extra-column dead volume (injectors, detectors etc.) to avoid excessive band broadening. As noted above, these columns do not appear to have been used in validated quantitative methods.

3.5.11 Gas Chromatography

Gas chromatography (GC) was first proposed by A.J.P. Martin and R.L.M. Synge in their seminal paper on liquid chromatography and Plate Theory (Martin 1941). The idea was to replace the liquid mobile phase used in liquid chromatography with a suitable gas, in view of the much higher diffusivities of solutes in gases compared with liquids; they reasoned that this should ensure that the equilibration processes within a theoretical plate would be much faster, and thus the columns would be more efficient and separation times much shorter. Unfortunately, it was not until 10 years later (James 1952) that Martin and James described the first gas chromatograph. The first published GC analysis was that of some fatty acids; at the time no suitable detector was available so a microburette was used! The microburette was eventually automated and provided a remarkably effective on-line detector.

While the same general principles already discussed for liquid chromatography also apply to GC, there are some significant differences in their practical application. The most obvious difference is that, although packed GC columns are still used in some applications, trace analysis by GC (especially when interfaced to a mass spectrometer as detector) are nowadays conducted using open (unpacked) capillary columns with a stationary phase bonded in some way to the internal capillary walls; as discussed below this implies some changes to the theory of the dispersive processes and to some extent to that of the retention characteristics (Plate Theory). The other major difference arises because, unlike liquid mobile phases, gases are highly compressible; as a result, as the gas moves through the column from higher pressure to lower pressure at the exit the gas volume changes in an inverse ratio to the local pressure although the *mass* flow rate of the gas remains the same (since there is no build-up of mobile phase along the length of the column), so the linear velocity must increase correspondingly. These two effects must be accounted for if the behavior of GC separations is to be understood. A discussion of some practical implications of the following theoretical considerations is given in Section 4.4.3

3.5.11a Effect of Gas Compressibility on Elution Equation for Packed Columns

The effects of gas compressibility (change of volume as a result of pressure changes) are observed in not only the linear velocity but also the diffusion coefficient of the analyte. For a packed GC column the A-term in the van Deemter equation (multipath effect) is independent of both u and D_m, so is unaffected. However this is not the case for the B and C terms (see Equations [3.33–3.38]). In fact the van Deemter equation applies only to each point x along the length of the column if local values of the linear velocity (u_x) and diffusion coefficient ($D_{m(x)}$) are used. The ideal gas law applied to the gaseous mobile phase ($PV_{mp} = nRT$), at constant temperature and constant mass flow rate of mobile phase (dn_{mp}/dt) leads to the simple relationship for any point x along the length of the column:

$$u_x = u_{exit}.(P_{exit}/P_x)$$

Moreover, the kinetic theory of gases shows that the diffusion coefficient of a gaseous solute in a different gas (the mobile phase in our case) is inversely proportional to the pressure, i.e.:

$$D_{m(x)} = D_{m(exit)}.(P_{exit}/P_x)$$

where the subscript 'exit' refers to values at the exit of the column. A detailed treatment for a packed GC column (Scott http://www.chromatography-online.org/) then leads to the same functional form as the van Deemter equation, but with u replaced by u_{exit} and the coefficients re-defined as follows:

$$A = 2\lambda.d_p (\text{no change}) \qquad [3.53a]$$

$$B = 2\gamma_m.D_{m(exit)} \qquad [3.53b]$$

$$C = f_1(k').d_p^{\,2}/D_{m(exit)}$$
$$+ 2f_2(k').d_f^{\,2}/\{D_s.[1 + (1 + u_{exit}.\eta.l/\kappa.[P_{exit}])^{0.5}]\}$$
$$[3.53c]$$

(compare Equations [3.33–3.38]), where η is the viscosity of the mobile phase (gas) and κ is the D'Arcy Constant for the packed column (see Equation [3.44]).

This result implies that the van Deemter equation can be used to analyze the performance of a packed GC column provided that the linear velocity measured at the column exit is used. This restriction is not important for liquid chromatography since the compressibility of liquids is negligible and the linear velocity is constant along the column. For GC it is important to use u_{exit} for this purpose rather than e.g., an average linear velocity, as exemplified in Figure 3.18; the curves obtained using u_{exit} and u_{avge} are very different (Figure 3.18a), but this in itself does not tell us much. However, when the data were fitted to the van Deemter function to evaluate the coefficients (Scott http://www.chromatography-online.org/) the values obtained using u_{exit} as the fitting parameter were physically reasonable, but those obtained for the u_{avge} fit gave a negative value for the A coefficient (Figure 3.18b), which is physically impossible. Usually the GC volume flow rate is measured after the column exit anyway, perhaps at some temperature and pressure different from those that pertain to the column exit itself but the simple gas law easily allows the necessary corrections to be made (for a constant molar flow rate), and conversion from volume flow rate to linear velocity is trivial provided the column porosity ε is known or can be estimated.

3.5.11b Open Tubular Columns and the Golay Equation

Now consider the changes required if the column is not packed but is an open tubular column with the stationary phase applied to the internal walls (Figure 3.19). The equation describing the dispersive effects in such a column (Golay 1958) is still used today, but note that Golay did not take into account the mobile phase compressibility in the case of GC. Since there is now no packing material there can be no multipath effect (or its coupling to the longitudinal term as in the Knox equation), so only the B and C terms are involved. Also, the important role played by particle diameter for packed columns is now replaced by that of the column internal diameter d_c and the parameters λ and γ used to describe the quality of column packing are no longer relevant. The original Golay expression for the theoretical plate height, derived for an uncoated capillary column, is:

$$H = B/u + C.u \qquad [3.54a]$$

where:

$$B = 2D_m \qquad [3.54b]$$

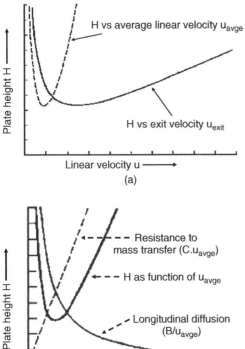

Figure 3.18 van Deemter plots of data for an analyte eluted from a GC column. (a): comparison of plots obtained using the average and exit linear velocities. (b): deconvolution of the average velocity curve into the three terms by fitting the data as a function of the average velocity (note the negative value deduced for the A-term when the average velocity is used as the fitting parameter). Reproduced from Scott, http://www.chromatography-online.org/, with permission.

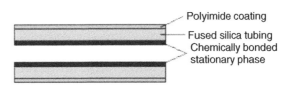

Figure 3.19 Lengthwise cross-section of a fused silica GC column; the external polyimide coating confers strength on the fragile fused silica column.

$$C = (d_c^2/24).\{(1+6k'+11k'^2)/[4D_m.(1+k')^2]$$
$$+ k'^3/[K^2.D_s.(1+k')^2]\} \qquad [3.54c]$$

As before, the partition coefficient (k) of the analyte between the two phases, as well as $k'(= K_A.V_s/V_m)$ and the two diffusion coefficients, all vary with the nature of the analyte. As for the packed column case (Equations [3.40–3.42]) the optimum (minimum) values of u and H are given by:

$$u_{opt} = (B/C)^{\frac{1}{2}} \text{ and } H_{opt} = 2(BC)^{\frac{1}{2}}$$

To apply Equation [3.54] to GC, as before u_{exit} rather than u_{avge} must be used.

Practical GC columns are coated with the stationary phase (Figure 3.19) and a wide range of such coatings is commercially available to suit different analyte types. Now an important additional parameter is the thickness of the layer of stationary phase, the so-called film thickness d_f, and the contribution to the C-coefficient from the kinetics of mass exchange from the stationary phase is a function of this parameter. The C-coefficient now becomes (Kenndler 2004):

$$C = (d_c^2/96).(1+6k'+11k'^2)/[D_m.(1+k')^2]$$
$$+ 2k'.d_f^2/[3D_s.(1+k')^2] \qquad [3.54d]$$

Thus the important properties of a real capillary GC column are the column i.d., the film thickness, and the nature of the stationary phase that determines the partition coefficients K (and thus k' values) for the range of analytes for which it is intended. So-called 'fast-GC' (Leclercq 1998) is accomplished by reducing the column internal diameter d_c and also the film thickness d_f thus reducing the C-coefficient (Equation [3.54d]) and the theoretical plate height (Equation [3.54a]); if the ratio d_f/d_c remains constant the qualitative chromatography does not change and separations in such microbore columns can be achieved much faster because of the shorter column lengths required to provide the same number of theoretical plates. The main drawback, arising from the much smaller volume of stationary phase, is a significantly lower loading capacity for such fast-GC columns.

An excellent review of modern theory and practice of GC is available on the Internet (Kenndler 2004), but unfortunately does not include a discussion of coupling to mass spectrometry.

3.5.12 Peak Asymmetry

The three dispersion mechanisms of peak broadening described by the van Deemter theory do not in themselves account for observations of asymmetric peak shapes. Sketches of so-called peak fronting (extension of a peak beyond the expected Gaussian shape on the shorter elution time side) and peak tailing (similar but on the later elution side of the peak) are shown in Figure 3.20. Peak tailing is the more common phenomenon at least for reverse phase chromatography. These peak asymmetries lead to lower peak heights (since the peak area is unaffected by the increased width) and thus possibly higher detection limits, and also poorer precision in peak areas since it is unclear where the limits of peak integration are to be located. This is one of the major reasons why it is recommended (Kipiniak 1981) that, for the most rigorous quantitative work, computer-based data analysis of chromatograms should be examined visually by the analyst to check on the reasonableness (or otherwise) of the peak-find and integration algorithms (see Section 4.4.8). This is particularly important when closely-eluting peaks are of significantly different sizes, in which case the smaller peak can appear as a badly resolved shoulder on the trailing edge of the larger.

There are two commonly used measures of asymmetry of chromatographic peaks (Dolan 2002). The peak asymmetry factor A_s is given by:

$$A_s = b/a$$

where a is the width of the front half of the peak and b is the width of the back half of the peak, measured (Figure 3.20a) at 10 % of the peak height; most analysts outside the pharmaceutical industry use A_s as the measure of peak tailing. The pharmaceutical industry generally uses the US Pharmacopeia tailing factor T_f:

$$T_f = w_{0.05}/2f_w$$

where $w_{0.05}$ is the peak width measured at 5 % of peak height and f_w is the width of the front half of the peak measured from peak apex to the (necessarily somewhat ill-defined) baseline limit of the leading edge of the peak (Figure 3.20b). For the same peak, A_s usually gives slightly larger values than T_f (Dolan 2002).

The major causes of peak tailing include: (1) injecting a sample extract in a solvent that is significantly stronger (Section 4.4.2a) than the mobile phase; (2) too large a sample injected (overload of adsorption sites on the stationary phase, this often causes sharpening of the peak front in addition to peak tailing); (3) interactions of acidic silanol groups on the silica particles with basic compounds; (4) adsorption of neutral and acidic compounds onto silica; (5) packing voids near the head of the column stationary phase; (6) degradation of silica at high pH (Section 4.4.1a) or high temperature;

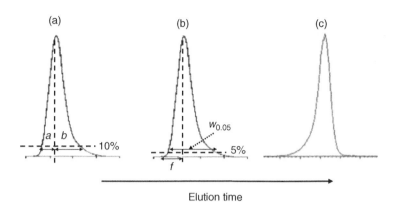

Elution time

Figure 3.20 (a) and (b): Sketch of a chromatographic peak exhibiting peak tailing (peak broadening at the later elution side only), illustrating parameters used to calculate the degree of asymmetry as (a) the peak asymmetry factor A_s and (b) as the US Pharmacopeia tailing factor T_f (see main text). (c) An illustration of peak fronting (peak broadening on the early elution side only).

and (7) unswept dead volumes in the chromatographic train. The solutions to these practical difficulties are: (1) include a solvent-switching step on the sample extract if necessary; (2) reduce the concentration and/or volume of the injected extract so that no more than $\sim 0.2 - 0.4$ mg of total adsorbable material is injected (this amount scales with the length and with the square of the column diameter); (3) use an end-capped column (Section 4.4.1b, a so-called base-deactivated column); (4) increase the total buffer concentration in the mobile phase and/or add a competing acid e.g. 1 % acetic acid or 0.1 % trifluoroacetic acid (TFA); (5) replace the column (attempts to 'fix' the column are seldom successful); (6) reduce pH or use a more completely end-capped column, reduce temperature (to $< 50\,°C$); (7) minimize number of connections and ensure that all compression fittings (including the rotor seal on the loop injector, Section 4.4.4a) are properly seated.

In addition to the foregoing strictly experimental causes of peak tailing, there are intrinsic reasons for this phenomenon that can become significant when breakdown of the simple linear adsorption equilibration relationship (Equation [3.1]) occurs due to overloading (saturation of the adsorption sites). This situation is complex and is not discussed here. An example is provided by an investigation (Fornstedt 1996) of the case of a stationary phase whose surface is covered with two different types of adsorption sites. The sites of the first type have a large specific surface area, a relatively weak adsorption energy and thus retention capability, and fast mass transfer kinetics. The sites of the second type were assumed to cover a small fraction of the total surface area, and to have a high adsorption energy and thus strong retention,

and slow mass transfer kinetics. It was shown (Fornstedt 1996) that tailing mechanisms from different origins combine when the adsorption sites with slow kinetics are subjected to a degree of overloading, i.e., under nonlinear conditions.

Peak fronting (see Figure 3.20c for an example) is less common than peak tailing. Some of the causes of peak tailing (e.g., column overloading and poor packing characteristics, see discussion of the van Deemter packing parameters λ and γ_m in Equations [3.33–3.34]) can also lead to peak fronting. In this context, poor packing characteristics correspond to less densely packed regions of the stationary phase, e.g. asymmetric voids at the head of the column, channeling within the column or a less dense bed structure along the walls off the column than in the center; if some of the analyte molecules travel through such less dense portions of the column, they will travel more quickly than the bulk of the analyte molecules that will thus be disproportionately represented in the first-eluting half of the peak, thus leading to a fronting type of distortion (as well as a slight reduction in retention time measured at the peak apex). Other causes of peak fronting include poor solubility of the analyte in the mobile phase and problems of wetting the stationary phase with the mobile phase. Solubility problems are best assessed by trying to improve the solubility of the analyte and comparing the resulting chromatogram with the original. 'Wettability' refers to the ability of the mobile phase to fully penetrate the stationary phase so that analytes interact with all of the bonded-phase (typically C_{18} for reversed phase columns). If the mobile phase contains a large percentage of water, good wetting on this highly hydrophobic material may not be achieved resulting in loss of retention and peak shape distortions including fronting. If it is not possible to increase the amount of organic solvent in the mobile phase, a different column designed for very high aqueous mobile phases should be used.

3.6 Gradient Elution

All of the theoretical considerations thus far have assumed isocratic elution, i.e., use of a mobile phase of fixed composition. This experimental approach has advantages, e.g., simplicity, highly reproducible retention times, lower susceptibility to problems, minimal need for re-equilibration of the column between injections and ease of adjustment of mobile phase composition for maximum compatibility with the ionization technique used for the mass spectrometer in LC/MS work. However, in cases where limitations of isocratic elution are disadvantageous, such as for analytes with sufficiently long retention times that the van Deemter dispersion effects result in peak widths that are too large, gradient elution is often used. Such methods simply imply that the composition of the mobile phase is progressively changed throughout the elution and appear to have been first described in 1952 (Alm 1952). This change can be done in a step-wise fashion in order to elute different fractions of the analyte mixture, either for preparative purposes or to deliver the fractions separately to a second chromatographic column. (The latter strategy is common in analysis of peptide mixtures arising from digestion of proteins by proteolytic enzymes, where the first step gradient stage could be ion exchange chromatography in which, to a first approximation, peptides are eluted according to their nett charge at a controlled pH value, and the second step is conventional reverse phase elution using a continuous linear gradient.)

Continuous gradient elution uses a mobile phase whose composition is varied (usually linearly) so as to produce a corresponding increase in the affinity of analytes for the mobile phase relative to the stationary phase. The method is most often applied in reversed phase systems, so that the ratio of organic solvent (typically acetonitrile) that is the solvent of high affinity (i.e. high elution strength, Section 4.4.2a) for reversed- phase analytes, to water, is continuously increasing. Thus, by the time the most hydrophilic analytes have eluted (or nearly so) the mobile phase composition will have changed such that somewhat less hydrophilic compounds are now more readily removed from the stationary phase and are eluted more quickly, and so on. Such an approach clearly shortens the elution times for mixtures of analytes with a wide range of hydrophilicity and thus decreases the effects of longitudinal diffusion on peak widths.

However, gradient elution also sharpens the peak for a single analyte. As a result of the various dispersive effects discussed in relation to the van Deemter equation, the (approximately Gaussian) distribution of a given analyte along the length of the column corresponds to faster-moving analyte molecules at the front of the distribution, and slower molecules bringing up the rear at the tail of the distribution. However, at any given instant during the gradient the faster molecules are in a region where the stationary phase has a greater affinity for this analyte (relative to the mobile phase) than is the case for the conditions experienced by the slower molecules, which are in a region where the mobile phase is richer in the organic solvent component. Thus the analyte molecules are always moving into a region of stronger retention by the stationary phase, so that the slower molecules can catch up as they are in a region of the column where the mobile phase has been changed so that the retention on the stationary phase is lower. Another way of expressing the same idea is that the mobile phase provides large values of k' at the beginning of the gradient and progressively smaller values as the gradient proceeds to produce a mobile phase of continuously increasing solvent strength. The practical implications of these effects include the possibility of decreasing run times while still achieving good resolution and adequate retention (k' values).

The extension of the theory of isocratic elution to the case of gradient elution is the result of work by several authors, but the most complete integration of all earlier work is contained in two landmark papers by Snyder *et al.* (Snyder 1979; Dolan 1979); other expositions and summaries are available (Snyder 1979a, 1980, 1983). Some qualitative understanding can be obtained using Figures 3.21–3.24 (Snyder 1979). Figure 3.21 is a theoretical chromatogram calculated for isocratic elution of 8 'invented' analytes, as specified in the caption. The first four or five analytes give reasonable peaks, but for the others the k' values (and thus retention times) are so large that the dispersion effects have time to make peak widths large and the peak heights correspondingly so small that the last-eluting peaks are barely detectable. Now consider gradient elution such as that described in Figure 3.22; this is an example of a linear gradient, i.e., the volume fraction ϕ of the mobile phase component of greater solvent strength (the organic component in reverse phase chromatography) increases linearly with time:

$$\phi = \phi_0 + \phi'.t \qquad [3.55]$$

Such a linear gradient, $\phi' = d\phi/dt = $ constant, with ϕ_0 the initial value of ϕ at the start of the gradient, is by far the most commonly used gradient condition as it

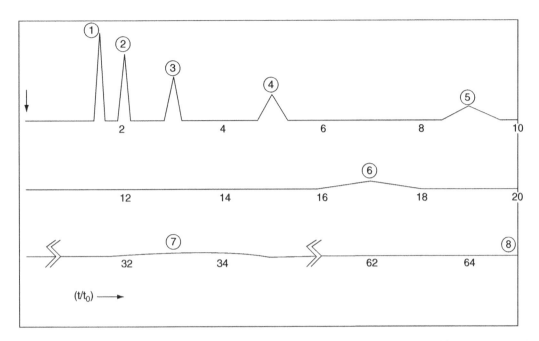

Figure 3.21 Calculated chromatogram for isocratic elution of a mixture of 8 'invented' analytes with k' values chosen to be in geometric progression (0.5, 1, 2, 4, 8 ---- 64) for peaks 1–8 and with N = 1000. The parameters are reasonable for a mobile phase of 20 % methanol in water. (t/t$_0$) values are calculated from k' (Equation [3.18]). Peaks are drawn as triangles instead of Gaussians, with the base of each peak as 4σ calculated by rearrangement of Equation [3.20] and using Equation [3.18] to give: $\Delta(t^{4\sigma}/t_0) = 2.\Delta(t_i/t_0) = (4/N^{\frac{1}{2}}).(t_r/t_0) = (4/N^{\frac{1}{2}}).(1+k')$, since Δt corresponds to 2σ. The peak heights were calculated on the assumption that the total integrated responses (peak areas) were the same for all components. Reproduced from Snyder *J. Chromatogr.* (1979) **165**, 3, Copyright (1979) Elsevier, with permission.

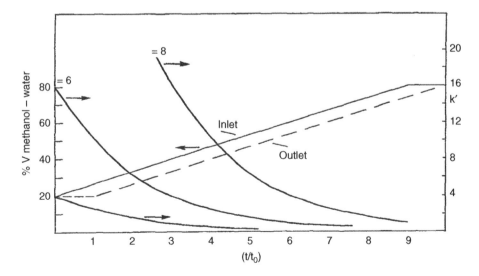

Figure 3.22 Linear solvent program for reverse phase gradient elution of the same eight components as for Figure 3.19. Note the delay in solvent composition between column inlet and outlet. The values were chosen as a reasonable representation for a program running from 20–80 % methanol in water. The k' values (right-hand abscissa) were assumed to decrease by a factor of two for each 10 % increase in methanol concentration (initial values at 20 % methanol are given in Figure 3.19); curves of k' vs (t/t$_0$) are shown for components 4, 6 and 8. Reproduced from Snyder *J. Chromatogr.* (1979) **165**, 3, Copyright (1979) Elsevier, with permission.

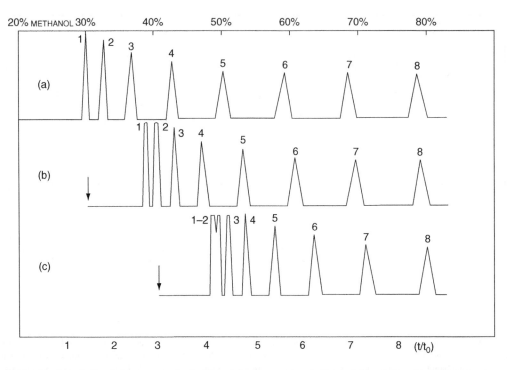

Figure 3.23 Gradient elution chromatograms for the same eight components as in Figure 3.19, calculated for linear solvent programs of methanol in water. (a) program starts at 20 % methanol (see Figure 3.20), while (b) and (c) start at 30 % and 40 %, respectively. Values of (t/t_0) and peak width calculated from gradient elution expressions (Table 3.2) with k'_{avge} taken to be the instantaneous value of k' when the analyte is halfway along the column (Figure 3.22). Reproduced from Snyder *J. Chromatogr.* (1979) **165**, 3, Copyright (1979) Elsevier, with permission.

leads to relationships among important chromatographic parameters that are remarkably similar to those that apply to isocratic elution; the most important of these comparisons are listed in Table 3.2. Note the importance in the gradient case of k'_{avge}, the average value of k' for each component during gradient elution. It turns out that a *linear* gradient yields roughly equal values of k'_{avge} for all analytes eluting at separate times, and thus roughly equal peak widths and approximately constant resolution for all analyte pairs with similar values of the separation factor α. All of these conclusions are apparent on inspection of the relationships in Table 3.2.

The full mathematical treatment of gradient elution (Snyder 1979) that led to these relationships (Table 3.2) is too complex to be reproduced here; indeed, in practice a gradient elution method is often developed by an essentially trial-and-error approach based on an isocratic method while keeping in mind the restrictions imposed by the qualitative considerations outlined above. However, it is possible to give some flavor of this theory to provide some understanding of the principles, as follows.

The theory is ultimately based on extensive experimental investigations of isocratic values of k' for a given analyte on the same column, as a function of mobile phase composition, which showed that the following relationship holds well in many cases and reasonably well in others:

$$\log(k') = \log(k'_w) - S.\phi \qquad [3.56]$$

where k'_w is the value of k' observed for isocratic elution in the pure weak solvent component (often water); this value must usually be obtained by extrapolation using Equation [3.56] rather than directly, since solvents such as water are very poor eluters for the analytes of interest. The parameter S (the slope of the function in ϕ given as Equation [3.56]) is a measure of the solvent strength of the mobile phase component of stronger elution strength; for reverse-phase chromatography, when this stronger solvent is either methanol or acetonitrile and the weaker solvent is water, the value of S is found to be about three but can

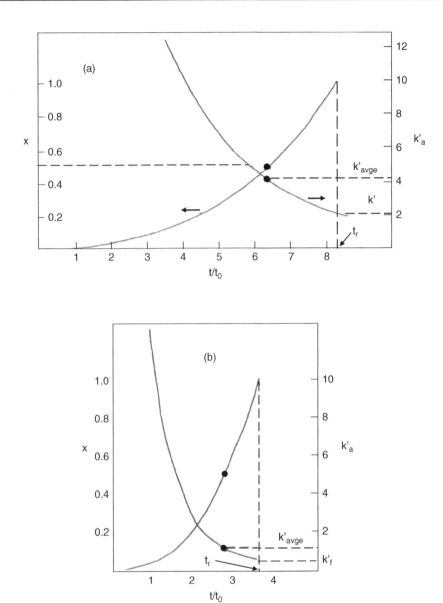

Figure 3.24 Calculated migration of compound 8 along the column under gradient elution conditions, where x is the fraction of the column length traveled and k_a is the instantaneous value of k'. k'_{avge} is taken as approximately the value of k'_a at the time when the analyte is halfway along the column (x = 0.5). (a) Gradient conditions as in Figures 3.20 and 3.21a (b = 0.2); (b) steeper gradient (b = 0.8). Reproduced from Snyder *J. Chromatogr.* (1979) **165**, 3, Copyright (1979) Elsevier, with permission.

vary from two to four, and for less polar organic solvents the value of S is larger. Combining Equations [3.55–3.56] gives:

$$\log(k_a') = [\log(k'_w) - S.\phi_0] - S.\phi'.t = \log(k'_0) - b.(t/t_0)$$
$$[3.57]$$

When this relationship is applied to gradient elution, k'_a is the 'actual' instantaneous value of k' at any point during the elution, the value that would pertain if the elution were conducted isocratically in a mobile phase of the same composition as that in which the analyte finds itself at that instant; k'_0 is the value of k' for an analyte if eluted isocratically with a mobile phase composition

Table 3.2   Basic parameters for isocratic and gradient elution (Snyder 1980, 1983)

Parameter	Isocratic	[a]Gradient
Retention time	[b]$t_r = t_0(1 + k')$	$t_{r,g} = (t_0/b).\log[2.303k'_0.b + 1] + t_0$
		$t_{r,g} = t_0.k'_{avge}.\log(2.303k_0'/k'_{avge})$
Peak width (volume)	[c]$\Delta V^{2\sigma} = V_m.(1 + k')/N^{\frac{1}{2}}$	$2\sigma_g = (V_m/2).(1 + k'_{avge})/N_g^{\frac{1}{2}}$
Resolution	[d]$R_{A/B} = (\alpha_{A/B} - 1)[k'/4(1 + k')]N^{\frac{1}{2}}$	$R_{g,A/B} = (\alpha_{A/B} - 1)[k'_{avge}/4(1 + k'_{avge})]N_g^{\frac{1}{2}}$
Capacity factor	[b]$k' = (t_r - t_0)/t_0$	$k'_{avge} = t_G/(\Delta\phi.S.t_0) = t_G.U/(\Delta\phi.S.V_m)$

(a) Approximate equations for gradient elution (Snyder 1979,1980, 1983).
(b) Equation [3.18].
(c) Equation [3.19] for $\Delta V_i = 2\sigma_i$ (i = inflection points) plus Equation [3.17] for k'.
(d) See Equation [3.23b] for R defined relative to a $\Delta t_r = 2\sigma$ definition; values given here refer to a 4σ separation.

t_G is the time from start to finish of the gradient; $\Delta\phi$ is the change in the volume fraction of the stronger solvent (organic component in reverse phase chromatography) during the gradient (e.g., $\Delta\phi = 1$ for a 0–100 % gradient); S is the slope of the curve relating $\log(k')$ to ϕ in *isocratic* elution experiments (Equation [3.56]); k'_0 is the *isocratic* value of k' for each analyte on the same column, in the gradient case with a value of ϕ used at the start of the gradient; N is the isocratic value of plate number and N_g the value for gradient elution (see Equation [3.22]); U is the volume flow rate; V_m is the volume of mobile phase in the column (thus also the *column* dead volume).

corresponding to that at the start of the gradient, and $b = (S.\phi'.t_0)$. Equation [3.57] predicts an exponential decrease with time for $k' = k_0'.10^{-bt/t_0} = k'_0.\exp(-2.303bt/t_0)$. Such exponential decreases of k' for components 4, 6 and 8 are drawn in Figure 3.22 in accord with Equation [3.57] for $b = 0.2$.

Based on the relationships in Table 3.2 (derivations are not given here), the chromatograms shown in Figure 3.23 can be calculated (Snyder 1979) for these 'invented' analytes, again assuming equal concentrations and equal total responses (peak areas) for all eight components. While it is not appropriate to reproduce the entire theoretical derivation of the equations in Table 3.2 here, it is useful to provide a sketch of the treatment sufficient for at least a qualitative understanding. The introductory description in Appendix 3.3 provides some feel for the origins of the theoretical equations in Table 3.2 that were used to calculate the chromatograms shown in Figure 3.23. The appearance of these chromatograms is very different from that of the corresponding isocratic case (Figure 3.21), particularly for the later-eluting components; their elution times are much shorter and the peak widths much narrower (so peak heights do not vary much).

The reasons for this behaviour can be understood qualitatively with reference to Figure 3.24(a). The exponentially increasing curve describes the movement of analyte 8 along the length of the column, as a function of the fractional length traveled (x) as a function of t/t_0; elution and detection occur at x = 1, i.e., $t = t_R \sim 8.4$, with a final k' value $k'_f \sim 2.1$. The exponentially decreasing curve is part of the k' curve for component 8 (it starts at the assumed value of $k' = 64$ at $t/t_0 = $ zero). At the value of (t/t_0)

for which x = 0.5, the corresponding value of k' is taken as an approximate measure of k'_{avge}. At the initial part of the gradient analyte 8 hardly moves at all along the column as a result of its very high k' value. However, as the solute strength of the mobile phase increases the k' value drops exponentially and analyte starts to move along the column with ever-increasing speed and elutes much faster than it did in the isocratic case, thus minimizing the effect of the dispersive mechanisms and accounting for the much more narrow elution peak. Very similar remarks apply to all the later-eluting bands (5–8) as the k' vs t/t_0 curves are all about the same shape as are the x vs t/t_0 curves, so these later-eluting analytes all exhibit similar values for k'_{avge} and for k'_f. This means (Snyder 1979) that all later-eluting bands exhibit similar values for resolution $R_{A/B,g}$ if the isocratic $\alpha_{A/B}$ values are similar, and the peak widths for these later-eluting bands are all rather similar. However, this is not the case for the early eluters (analytes 1–3); note that the elution times, peak widths and peak heights are all rather similar in the two cases (Figure 3.23), as a result of the fact that their k' values have not changed very much during elution from those appropriate to the initial gradient condition which, in this example, was taken to be the isocratic condition discussed for Figure 3.21. Gradient elution thus provides the possibility for elution of all components with widely varying retention characteristics in a reasonable time, with optimum resolution (narrow peaks) and increased sensitivity (again a consequence of the narrow peaks and consequent reduced chromatographic dilution, Section 3.5.10b).

The original theoretical paper (Snyder 1979) was accompanied by a paper (Dolan 1979) that discusses experimental tests of the theory and also includes advice on how best to apply this theoretical understanding to the design of practical gradient separations. It is worth repeating here that the present discussion has been restricted to *linear* gradients and has referred throughout to *reverse phase* conditions; however, these are the conditions under which gradient elution is most commonly employed.

A *ballistic gradient* is a very fast separation technique (Romanyshyn 2000; Tiller 2002a), and refers to rapid gradients using short narrow-bore columns and high flow rates leading to cycle times of two minutes or less. This approach is applicable to bioanalytical analyses in early drug discovery work, where the demands on validation are not as rigorous as in drug development (Section 11.5.1). Such rapid separations obviously increase the possibility of co-elution of matrix co-extractives that could lead to significant matrix effects (ionization suppression, Sections 5.3.6a and 9.6). A recent example of modification of a fast gradient method for pharmacokinetic studies to a ballistic method (De Nardi 2006) paid particular attention to this possibility. The fast gradient method used a $50 \times 4.6\,mm\,C_{18}(5\,\mu m)$ column at a flow rate of $1.5\,mL.min^{-1}$ and resulted in a runtime of five minutes; in contrast the ballistic method used a $30 \times 2.1\,mm$ column containing the same stationary phase at a flow rate of $1.0\,mL.min^{-1}$, and reduced the runtime to two minutes following further optimization of the gradient conditions. The effect of the ballistic gradient on chromatographic performance was assessed by comparison of peak shape, width and height for three test analytes; a decrease in peak width by up to 39 % and an increase in peak height up to 174 % were obtained (De Nardi 2006). The ballistic approach was cross-validated with the conventional method by comparison of quantitation results from several pharmacokinetic analyses. Other advantages of the ballistic approach were found to include an increased lifetime of smaller columns, their reduced cost compared to traditional $50 \times 4.6\,mm$ columns, reduction of instrument time, lower consumption of mobile phase and a less contaminated ion source as a result of the smaller injection volumes. At present it does not appear to be feasible to apply the ballistic gradient approach to the considerably more rigorous validation criteria applicable to drug development analyses.

As for isocratic elution, it is necessary to ensure that the sample extract that is injected on column uses a solvent that is either the HPLC mobile phase itself (the time-zero composition in this case) or a solvent of lower elution strength (Section 4.4.2a). It is important as well to realize that it is possible to adopt too much of a good thing

with respect to gradient elution, since if pushed too far the selectivity of separation (Section 3.4.3) can significantly deteriorate. Further, gradient elution requires a re-eqilibration time for the column to settle following completion of the gradient and return of the mobile phase composition to its initial value.

3.7 Capillary Electrophoresis and Capillary Electrochromatography

Although GC and HPLC are the high resolution techniques used most often for trace quantitative analysis, other techniques that have some potential advantages are capillary electrophoresis (CE) and capillary electrochromatography (CEC). Despite their advantages (very high separation efficiencies) these techniques also possess disadvantages that have led to their rare use for applications of interest in this book. These disadvantages include very low concentration sensitivity for CE that is only partly alleviated by sample stacking techniques (Section 4.4.6), and low reproducibility of elution times for both CE and CEC as a result of uncontrolled variations in the zeta potential (see later) that determines the electro-osmotic flow. However, this could change in future as the result of current activity in miniaturization of some or all of the steps involved in an analytical scheme and their integration on a single device. Transport and separation of analytes on such 'lab-on-a-chip' devices (Section 4.4.7) are generally achieved by exploiting the basic phenomena involved in CE and CEC, and for this reason an introduction to these general principles is given here. Excellent websites (http://www.chemsoc.org/ExemplarChem/entries/2003/leeds_chromatography/chromatography/ and http://www.chemsoc/ExemplarChem/entries/2003/leeds_chromatography/chromatography/cec.htm) give a host of references to material available on the Internet and elsewhere, and also includes a useful introduction to the subject.

Electrophoresis is defined as the migration of ions under the influence of an externally applied electric field E_{ext}. (The concept of an electric field is discussed in Section 6.3.1 and Appendix 6.1.) The resulting force F_E on an ion of overall charge $q = (\pm)z.e$, where e is the fundamental charge, z is the net number of charges (some analytes such as proteins can contain several charges, some of each polarity), and the sign of the overall charge is important in the present context and must be included in q, is given by:

$$F_E = q.E_{ext} \qquad [3.58]$$

The resulting movement of the ion under the influence of the applied field E_{ext} (usually expressed in volts.m^{-1}) is opposed by a retarding frictional force F_f that is proportional to the speed of the ion, v_{ep}, and the friction coefficient f:

$$F_f = f.v_{ep} \qquad [3.59]$$

$$f = 6\pi.\eta.r_{hd} \qquad [3.60]$$

where η is the viscosity coefficient of the medium through which the ion is moving and r_{hd} is the hydrodynamic radius of the ion that accounts for its intrinsic molecular size plus the solvation sheath around it. The ion almost instantly reaches a steady state speed velocity where the accelerating electrical force (Equation [3.58]) equals the frictional force (Equation [3.59]). Then Equations [3.58–3.60] together give:

$$v_{ep} = [q/(6\pi.\eta.r_{hd})].E_{ext} = \mu_{ep}.E_{ext} \qquad [3.61]$$

Here, μ_{ep} is the electrophoretic mobility of this ion (characterized by q and r_{hd}) in the particular medium (characterized by η); μ_{ep} is inversely proportional to r_{hd}/q, and this is the basis of the separatory mechanism in electrophoresis. Insofar as r_{hd} is related to molecular mass (both are measures of molecular size), electrophoresis is sometimes said to provide liquid phase separations in terms of mass-to-charge ratio, but this is only an approximate analogy. Nonetheless, electrophoresis has been used for many years by biochemists as the basis for separations of biopolymers (oligonucleotides, proteins) as a function of molecular mass (see Section 11.6); in that case the medium through which the ionized molecules migrate under the influence of an applied field usually involves a suitable aqueous solvent as a mobile phase and a stationary phase that can be a special paper or a gel of some kind (e.g., polyacrylamide or agarose), so the situation is more complicated than the simple picture of ions in liquid solution that has been considered thus far. In fact, traditional electrophoresis of this kind, involving migration through some porous solid medium, is more analogous to modern electrochromatography which is described briefly below.

Capillary electrophoresis (CE) is electrophoresis performed in a capillary tube, in practice almost always made of fused silica. Of the techniques available for separation of both large and small molecules in solution, CE has the highest separation efficiency. The transformation of conventional electrophoresis to modern CE was made possible by the production of inexpensive narrow-bore capillaries for gas chromatography (GC), as discussed in Section 4.4.3. The unrivalled resolution and separation efficiency of CE can be understood in terms of the van Deemter equation (Equation [3.39]). In CE, two of the three contributing band-broadening mechanisms are eliminated, namely, the multiple path (eddy diffusion) term (A term) and the mass transfer term (C term) because the separation is conducted in a single uniform liquid phase. (In contrast, for capillary GC the A term is not applicable, but a mass transfer term is operative because of the use of a stationary phase coated on to the capillary wall.) As a result, in CE the only source of band broadening under ideal conditions is longitudinal diffusion (the B term), although parameters such as initial injection volume also contribute in practice. Under typical conditions a CE separation can provide 50 000 to 500 000 theoretical plates, an order of magnitude better than competing HPLC methods. CE analyses are also usually relatively fast, use little sample and reagents, and cost much less than chromatography.

This kind of performance is possible through interaction of the differences in electrophoretic mobilities of different ions with the electro-osmotic flow in the capillary. The external electric field E_{ext} is simply applied by applying the output of a stabilized high voltage power supply along the length of the capillary, but in order for the electric field strength to be constant along the length of the capillary it is necessary that the capillary is filled with a buffer solution (the *supporting electrolyte*) of suitably high electrical conductivity. Electro-osmosis is the phenomenon whereby the buffer solution in the capillary moves under the influence of the electric field (Figure 3.25(a)). The inner surface of a fused silica capillary is covered with silanol groups (Si–OH, Figure 4.12) that are ionized to SiO$^-$ at pH $> 2 - 3$. The negatively charged surface attracts positive ions from the buffer to form a so-called electrical double layer (Figure 3.25(a)); this subject is described in any physical chemistry text book, so only a brief discussion is given here.

The negatively charged capillary wall will have a layer that is rich in positive buffer ions (called the Stern layer) immediately next to it; this layer of ions is relatively immobile as a result of the strong electrostatic attraction. The remainder of the total excess charge (equivalent to the protons removed from the silanol groups) is in the diffuse part of the double layer and extends into the bulk liquid. The concentration of ions in the solution in the double layer is relatively small compared to the total ionic concentration (as a result of the immobilized cations in the Stern layer) and falls off exponentially in a radial direction from the capillary surface; this is also the case for the corresponding electric potential (proportional to the charge density) created by this separation of charge.

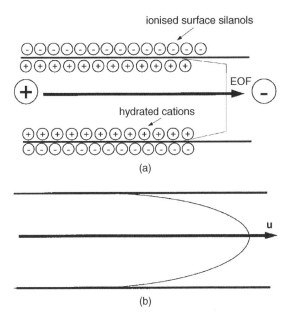

ionised surface silanols

hydrated cations

EOF

(a)

u

(b)

Figure 3.25 (a) Illustration of the principle of electro-osmotic flow. At pH > 3 the free silanol (Si–OH) groups on the wall of the fused silica capillary are ionized (deprotonated), creating a negatively charged wall that attracts a first cylindrical sheath of cations from the supporting buffer solution immediately adjacent to the wall. Under an external electric field applied along the length of the capillary, this positively charged sheath moves towards the cathode (negative electrode), creating a force that creates a flow of the bulk solution in that same direction. Because this force originates at the capillary wall there are no drag forces at the wall to slow down the motion there, so the flow profile across the capillary is essentially flat (plug flow) as illustrated, provided the capillary is sufficiently narrow. This contrasts with the flow profile (b) created in a flow driven by an external pressure difference; in that case the capillary wall provides drag forces so the flow rate is low near the walls and fastest at the centre line.

The potential at the boundary between the Stern layer and the diffuse part of the double layer is called the zeta potential (ξ) and has values ranging from 0–100 mV. Because the charge density drops off with distance from the surface, so does the zeta potential; the distance from the surface at which the potential is 0.37 times the potential at the interface between the Stern layer and the diffuse layer, is defined as the double layer thickness and is denoted δ (Figure 3.26). The equation describing δ (Knox 1987) is:

$$\delta = [(\varepsilon_r.\varepsilon_o/4\pi).RT/(2c_B.F^2)]^{\frac{1}{2}} \qquad [3.62]$$

where ε_r is the dielectric constant (relative permittivity) of the bulk solution, ε_o is the permittivity of a vacuum (Sections 1.2 and 6.3.1), ($\varepsilon_r.\varepsilon_o$) is the absolute permittivity of the solution, R is the gas constant, T is the absolute temperature (K), c_B is the molar concentration of buffer (assumed for simplicity here to be a 1:1 electrolyte) and F is the Faraday constant (the magnitude of the electrical charge equivalent to 1 mol of electrons = $N_A.e$). Evaluation (Knox 1987) of Equation [3.62] for water as solvent ($\varepsilon_r = 80$) and for a 1:1 electrolyte (e.g. sodium chloride) at a concentration c_B of 0.001M, gives $\delta = 10$ nm; for $c_B = 0.1$M, $\delta = 1$ nm. When an electric field is applied along the capillary length (parallel to the wall), positive ions in the diffuse layer that are not immobilized in the Stern layer will migrate towards the cathode and shearing will occur between these two layers (Figure 3.26). Electro-osmosis results because the core of liquid within this sheath will also be transported to the cathode along with the solvation sheaths of the cations. Since there is no charge imbalance within the sheath itself no shear takes place in this region, so the resulting flow profile is plug-like. The speed (linear velocity u_{eo}) of this electro-osmotic flow (EOF) is independent of the capillary diameter d_c provided that $d_c > 10$ δ (CE capillaries usually have $d_c > 20\delta$). If d_c has values close to δ, the double layers on diametrically opposite sides of the wall overlap and, as a result, the EOF is greatly reduced and has a parabolic flow profile like that of pressure-driven HPLC (Figure 3.25(b)). The relationship between u_{eo} and the applied electric field is given by:

$$u_{eo} = [(\varepsilon_r.\varepsilon_o/4\pi).\xi/\eta].E_{ext} = \mu_{eo}.E_{ext} \qquad [3.63]$$

where μ_{eo} is the electro-osmotic mobility of that particular buffer solution in that particular fused silica capillary. The direct dependence of u_{eo} on the zeta potential ξ (Equation [3.63]) is important. The zeta potential is dependent on the electrostatic nature of the capillary surface (density per unit area of ionized silanol groups) and to some extent on the nature of the buffer, and more particularly on the buffer concentration via the charge density; however, if the buffer concentration is too high (too high a total ionic strength), electro-osmosis can actually decrease because the double layer collapses (δ becomes too small, Equation [3.62]).

The movement of charged analytes is now considered as a consequence of the combination of their own individual electrophoretic mobilities (Equation [3.61]) and their participation in the bulk electro-osmotic flow (EOF). The net speed of motion of an analyte ion in the field direction (along the length of the capillary) is the vector sum of its electrophoretic velocity v_{ep} (Equation [3.61])

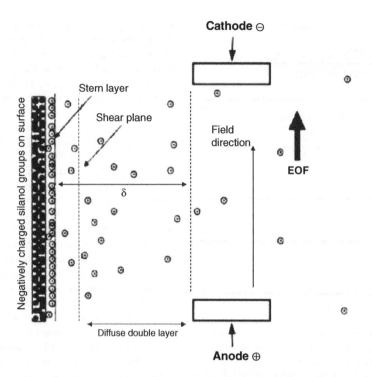

Figure 3.26 Illustrative sketch of the electrical double layer at the surface of a fused silica capillary filled with a buffer solution at pH > 3 (where the free silanol groups are deprotonated). An external electric field applied along the length of the capillary results in an electro-osmotic flow (EOF) as shown.

and the linear velocity u_{eo} of the electro-osmotic flow (Equation [3.63]):

$$u_{net} = u_{eo} + v_{ep} = (\mu_{eo} + \mu_{ep}).E_{ext} = \mu_{app}.E_{ext} \qquad [3.64]$$

where μ_{app}, defined in Equation [3.64], is the apparent mobility of that analyte under the experimental conditions. The vector nature of the addition is controlled via the sign of the electric charge q on the analyte. Neutral solutes migrate in the same direction and with the same velocity as the electro-osmotic flow and are therefore not separated from one another. Cations and anions are separated based on differences among their apparent mobilities. For cations, which under these conditions move in the same direction as the electro-osmotic flow, μ_{ep} and μ_{eo} have the same sign so $\mu_{app} > \mu_{ep}$; for anions, however, μ_{ep} and μ_{eo} have opposite signs so $\mu_{app} < \mu_{ep}$. At moderate pH values > 3, generally $u_{eo} > v_{ep}$, so anions also migrate towards the cathode, which is where the detector is typically located. Under conditions where electro-osmosis is weak, anions may never reach the detector unless the polarity is reversed so that the detector end is now the anode rather than the cathode.

The direction of the EOF can be reversed, if required, by adding a cationic surfactant such as cetyltrimethylammonium bromide $(CH_3(CH_2)_{17}N^+(CH_3)_3Br^-)$, to the separation buffer. The positively charged quaternary ammonium group at one end of this surfactant (CTAB) binds strongly via Coulombic attraction to the negatively charged silanol groups on the capillary surface while the hydrocarbon tail points away from the surface. A second layer of CTAB molecules orients itself in the opposite direction (tail-to-tail), so that the surface presented to the buffer solution effectively reverses the charge of the wall from negative to positive, resulting in reversed electro-osmotic flow for the same polarity of the applied electric field. This procedure is also known as dynamic coating. Other coating procedures have been developed but are not described here. It should be noted that such a coating in no way represents a chromatographic stationary phase analogous to those used in capillary GC; in fact the interactions of the analytes and buffer components with such a surface are dominated by electrostatic interactions similar to those with the ionized silanol groups.

There are several operational modes used in capillary electrophoresis. The main modes are capillary zone electrophoresis (CZE, the mode discussed above and frequently referred to simply as CE), micellar electrokinetic capillary chromatography, capillary isotachophoresis, capillary gel electrophoresis and capillary isoelectric focusing. From point of view of this book, CZE is the most important mode.

Again because of its applicability to lab-on-a-chip devices, capillary electrochromatography (CEC) are briefly discussed. The same websites (http://www.chemsoc.org/ExemplarChem/entries/2003/leeds_chromatography/chromatography/ and http://www.chemsoc / Exemplar Chem/ entries/ 2003/leeds _chromato graphy/chromatograpy/cec.htm) are also a useful source of information on this topic; a particularly helpful introduction is available (Smith 2001). Although CZE is a high-efficiency separation technique for charged analytes, capable of routinely generating up to ~500,000 plates.m^{-1}, it is incapable in its original form of separating neutral molecules. This problem was resolved by the introduction of micellar electrokinetic chromatography (MEKC), but this is not compatible with mass spectrometric detection. In contrast, HPLC is the most widely used separation technique and is capable of resolving a wide range of both neutral and charged analytes, offering a wide range of parameters that can be manipulated to achieve separation (Chapter 4), and can be efficiently coupled to MS. CEC is a hybrid separation method that couples the high separation efficiency of CZE with the advantages of HPLC, using an electric field rather than applied pressure to move the mobile phase through a packed bed. Since there is no back pressure in CEC it is possible to use small diameter packing materials and thereby achieve very high separation efficiencies (see Section 3.5.7). The first practical demonstration of the approach (Pretorius 1974) was followed (Jorgenson 1981) by CEC separation of 9-methylanthracene from its isomer perylene. However, the current interest in CEC is usually attributed to a later combined theoretical and practical approach (Knox 1987, 1991).

The advantages of using electro-osmosis to move the mobile phase through a packed column are the reduced plate heights as a result of the plug flow profile and the ability to use smaller particles (leading to higher peak efficiency) than is possible in pressure driven systems like HPLC. The driving force results from the electrical double layer at the liquid–solid interface between the bulk liquid and the capillary wall (in the case of CE) surface and, in CEC, also between the surface of the packing material and the mobile phase. However, because the surface areas of microparticulate silica-based stationary phases are much greater than that of the capillary walls, most of the EOF is generated by the surface silanol groups of the stationary phase (though obviously not in end capped phases (Section 4.4.1b) that have minimal levels of residual silanol groups). Thus it should be possible using an electrically driven system to use very small particle sizes and still maintain high linear velocities to yield rapid and efficient separations, provided that $d_p > 20\delta$ so that no double layer overlap occurs.

An important consideration in CEC is the relationship between the linear velocity and the concentration of electrolyte. Since u is directly proportional to the zeta potential (ξ), which itself decreases with increasing electrolyte concentration, it is an important variable. It was shown (Knox 1991) that, in a realistic test system, the reduced plate height was lowest at 10^{-3}M of the buffer (NaH_2PO_4) and that the linear velocity altered little over the range 4×10^{-5} to 2×10^{-2}M; it was thus concluded that the best overall performance (low plate height at high electro-osmotic linear velocities) would be achieved at electrolyte concentrations of ~0.002 M. Thus, although buffer concentrations used in CEC tend to be low, this might not always be the case especially if charged molecules are analyzed when it may be necessary to increase the ionic strength. Of course if CEC is to be coupled to MS, volatile buffers must be used (Section 5.3).

The heat generated per unit volume by resistive heating in an electrically driven CEC system can be considerable, emphasizing the need to use low buffer concentrations and to reduce the diameter of the capillary in order to maximize the rate of heat dissipation through the walls; capillaries with d_c up to 320 μm have been used successfully. Under conditions where self-heating is significant, serious problems can arise resulting in out-gassing within the capillary followed by a breakdown in the current. The suppression of bubble formation, when operating at high driving voltages with high concentrations of buffers and of low boiling organic solvents, requires pressurization of the mobile phase if CEC is to be performed over long periods, free from bubble formation.

Band broadening in CEC can again be described by the van Deemter equation, and in principle all three terms contribute. However, since both the A and C terms decrease when smaller particles are used (see Equations [3.33] and [3.37]), use of such particles results in smaller plate heights.

Despite all the potential advantages of both CZE and CEC, neither has been widely accepted for routine use particularly in trace quantitation; this is partly because of the limitation on the volumes of analyte solution that can be analyzed in CE without compromising separation efficiency, thus yielding poor concentration detection limits. This is to some extent a problem also for CEC although experiments using gradient elution have been evaluated (Smith 2001) and should permit larger injection volumes. Another practical limitation of CEC is the inability to analyze basic compounds (many pharmaceutical compounds contain basic nitrogenous groups) since a high pH is necessary to obtain a useful range of values of u_{eo} and thus reasonably short analysis times; in general, a CEC stationary phase should provide a good EOF across a wide pH range. Finally, both CE and CEC suffer from a lack of reproducibility of elution times relative to HPLC, a result of the strong dependence of the zeta potential on the condition of the silica surface, temperature, ionic strength etc. However, both technologies are essential for the emerging 'lab-on-a-chip' approach (Section 4.4.7) and for this reason they were discussed here.

Appendix 3.1 Derivation of the Plate Theory Equation for Chromatographic Elution

Plate Theory is based upon a approximation of discrete sequential partitioning equilibria of analyte between mobile and stationary phases (Figure 3.3), so it is convenient to define a partition coefficient for analyte A as:

$$K^A = c^A_s / c^A_m \qquad [3.1]$$

where c^A is a concentration of A and subscripts s and m denote stationary and mobile phases, respectively. The objective is to derive a theoretical expression for R_D, the response of the chromatographic detector, as a function of time following injection of the analyte onto the column (Figure 3.2); it is convenient to first consider R_D as a function of the volume V of mobile phase that has passed through the column after injection; V and t are related via U, the flow rate (volume per unit time) of mobile phase.

The differential form of Equation [3.1] is:

$$\delta c^A_s = K^A . \delta c^A_m \qquad [3.2]$$

Approximating the differentials in Equation [3.2] by discrete increments, consider a volume of mobile phase δV passing from plate (p−1) into plate (p) (Figure 3.3), at the same time displacing the same volume from plate (p) to plate (p+1). Although these volume increments must be all the same (otherwise there would be a build-up of solvent in one plate and a decrease in another, not possible for plates of equal volume), the concentrations are not (Figure 3.3), so there will be a change of mass (δm^A_p) of analyte A in plate (p), equal to the difference between the mass of A entering plate (p) from plate (p−1) and that leaving plate (p) and entering plate (p+1). Simple mass balance applied to the analyte in the mobile phase in plate (p) gives:

$$\delta m^A_p = [c^A_{m(p-1)} - c^A_{m(p)}].\delta V \qquad [3.3]$$

A basic assumption of the Plate Theory is that equilibrium is maintained within each individual plate, so that:

$$\delta m^A_p = v_m . \delta c^A_{m(p)} + v_s . \delta c^A_{s(p)}$$
$$= (v_m + K^A . v_s) . \delta c^A_{m(p)} \qquad [3.4]$$

since the theory considers that each plate has the same total volume, each containing fixed volumes v_m and v_s of the two phases. Equating Equations [3.1] and [3.2], and replacing the finite increments by differentials, gives:

$$dc^A_{m(p)} / dV = [c^A_{m(p-1)} - c^A_{m(p)}] / (v_m + K^A . v_s) \qquad [3.5]$$

It is now mathematically convenient to introduce a new dimensionless variable:

$$v \equiv V / (v_m + K^A . v_s)$$

where $(v_m + K^A . v_s)$ is called the plate volume. The physical interpretation of v is the volume of mobile phase that has passed through any point in the column, expressed not in millilitres or microlitres but in units of the plate volume. Note that both the plate volume and v vary with the nature of the solute A. Then Equation [3.5] is rewritten:

$$dc^A_{m(p)} / dv = c^A_{m(p-1)} - c^A_{m(p)} \qquad [3.6]$$

Equations [3.5] and [3.6] describe the rate of change of concentration of solute A in the mobile phase within plate p (Figure 3.2) with respect to the total volume of mobile phase that has passed through plate p. If this differential equation can be integrated it will describe the elution curve for solute A eluted from any plate p in the column; if the condition is then applied that the interest is mainly in the final plate in the column, immediately before the detector, the resulting integrated form of Equation [3.6] will indeed be the elution curve for analyte A emerging out of the column (Figure 3.2).

Firstly, it is necessary to consider the limiting case of the first plate, that for convenience will be designated plate 0 (p = 0). Assume that injection of the analyte A onto the column is restricted to this first plate to create an initial concentration in the mobile phase of $c^A_{m(0)} = c_A{}^0$ when V = 0 = v, i.e., elution has not yet started, so that $c^A_{m(p)} = 0$ for p > 0 at this initial point. Then, since there is no plate (p−1) for p = 0, for this case Equation [3.6] becomes:

$$dc^A_{m(0)} / dv = -c^A_{m(0)} \qquad [3.7]$$

Integrating this equation is straightforward:

$$\ln[c^A{}_{m(0)}] = -v + C$$

where C is the constant of integration that can be evaluated by the initial condition $c^A{}_{m(0)} = c_A{}^0$ when $v = 0$, so that:

$$\ln[c^A{}_{m(0)}/c_A{}^0] = -v, \text{ OR } c^A{}_{m(0)} = c_A{}^0.\exp(-v) \quad [3.8]$$

Now consider Equation [3.6] for the case $p = 1$ and substitute for $c^A{}_{m(0)}$ from Equation [3.8]:

$$dc^A{}_{m(1)}/dv = [c_A{}^0.\exp(-v)] - c^A{}_{m(1)}$$

i.e.:

$$[dc^A{}_{m(1)}/dv].\exp(v) + [c^A{}_{m(1)}.\exp(v)] = c_A{}^0$$

The left side of this equation is the derivative with respect to v of the product $[c^A{}_{m(1)}.\exp(v)]$:

$$d[c^A{}_{m(1)}.\exp(v)]/dv = c_A{}^0$$

that is readily integrated to give (since $c^A{}_{m(1)} = 0$ when $v = 0$):

$$c^A{}_{m(1)} = v.c_A{}^0.\exp(-v) \quad [3.9]$$

Now consider Equation [3.6] for $p = 2$:

$$dc^A{}_{m(2)}/dv = c^A{}_{m(1)} - c^A{}_{m(2)} = [v.c_A{}^0.\exp(-v)] - c^A{}_{m(2)}$$

where Equation [3.9] was used for $= c^A{}_{m(1)}$. Multiplying through by $\exp(v)$ and rearranging gives:

$$[dc^A{}_{m(2)}/dv].\exp(v) + [c^A{}_{m(2)}.\exp(v)] = v.c_A{}^0$$

Again the left side is the derivative of the product $[c^A{}_{m(2)}.\exp(v)]$, so integration of this equation gives:

$$[c^A{}_{m(2)}.\exp(v)] = v^2.c_A{}^0/2 \text{ OR}$$

$$c^A{}_{m(2)} = [c_A{}^0.v^2.\exp(-v)]/(1 \times 2) \quad [3.10]$$

Similarly:

$$c^A{}_{m(3)} = [c_A{}^0.v^3.\exp(-v)]/(1 \times 2 \times 3)$$

And in general, for $p = n$:

$$c^A{}_{m(n)} = [c_A{}^0.v^n.\exp(-v)]/n! \quad [3.11]$$

In particular, if the last plate in the column corresponds to $n = N$, it is the mobile phase that is actually delivered to the chromatographic detector that is being considered; then Equation [3.11] with $n = N$ is in fact the elution equation that describes Figure 3.2 according to the Plate Theory. If the detector in question is a concentration dependent detector (see Chapter 4) such as a photometric detector, then Equation [3.11] applies directly. However, if the detector is a mass flux dependent detector, such as a mass spectrometer operating in electron ionization or chemical ionization mode, then Equation [3.11] requires some minor modification. The right side of Equation [3.11] is a Poisson Function (see text box in Chapter 7) that for large n can be reduced to a more tractable Gaussian Function.

Appendix 3.2 Transformation of the Plate Theory Elution Equation from Poisson to Gaussian Form

Two things must be accomplished: change the origin of the Poisson distribution of analyte concentration along the column from the first theoretical plate (injection) to the last plate (detection), in particular the peak maximum; and consider the limiting case of N (number of plates) large enough that this discrete variable can be replaced by a continuous one. For mathematical convenience define ω as the number of plate volumes eluted measured relative to the number required to reach the peak maximum of solute A, reached after N plate volumes of mobile phase have eluted (see discussion leading to Equation [3.28]:

$$\omega \equiv v - N \quad [3.28]$$

Substituting for v in Equation [3.11], gives:

$$c^A{}_{m(N)} = [c_A{}^0.(\omega + N)^N.\exp - (\omega + N)]/N!$$

Substituting a version of Stirling's Approximation to N!, valid for values of N that are not very large, gives:

$$N! \approx (2\pi)^{\frac{1}{2}}.N^{(N+1/2)}.\exp(-N)$$

Then after some algebraic manipulation and approximation, this equation becomes:

$$c^A{}_{m(N)} = c_A{}^0.\exp(-\omega^2/2N)/(2\pi N)^{\frac{1}{2}} \quad [3.29]$$

This is the elution equation expressed relative to an origin located at the maximum of the elution peak, i.e., at $\omega = 0$. It has the general form of a Gaussian Function, but in order to relate it more closely to experimental observables, the elution parameter is changed from numbers of plate volumes ω to actual volumes of mobile phase (and thus making the transition from the discrete variable Poisson form to the continuous variable Gaussian form). Thus:

$$\omega = \upsilon - N = (V - V_{rA})/(v_m + K^A.v_s)$$

so Equation [3.29] can be rewritten as:

$$c^A_{m(N)} = [c_A^0/(2\pi N)^{\frac{1}{2}}].$$
$$\exp\{-(V - V_{rA})^2/[2.N^{\frac{1}{2}}.(v_m + K^A.v_s)]^2\} \quad [3.30]$$

We can recognize the quantity $[2.N^{\frac{1}{2}}.(v_m + K^A.v_s)]$ in the denominator of the exponential in Equation [3.30] as the full peak width ΔV_i measured at the inflection points (see Equation [3.19]). But this specific peak width is, for a Gaussian curve, *twice* the standard deviation parameter $\sigma_V = (\Delta V_i)/2$ (recall that V, the volume of mobile phase that has eluted, is a continuous variable). Then Equation [3.30] can be rewritten as:

$$c^A_{m(N)} = [c_A^0/(2\pi N)^{\frac{1}{2}}].\exp\{-(V - V_{rA})^2/(2\sigma_V)^2\} \quad [3.31a]$$

where V_{rA} and σ_V are defined in terms of fundamental quantities and are also measurable from the experimental chromatogram.

Appendix 3.3 A Brief Introduction to Snyder's Theory of Gradient Elution

Consider the analyte band part way along the column (fractional distance x) as it is eluted by a solvent gradient. At any such point a differential flow dV of mobile phase will cause a corresponding movement dx of the analyte band, given by (Snyder 1964):

$$dx = dV/V'_{r,a}$$

where $V'_{r,a}$ is the 'actual' (instantaneous) value of the corrected retention volume $V'_r(= K.V_s$, see Equation [3.31a] $= k'.V_m$, see Equation [3.17]). Since clearly:

$$\int_0^{elution} dx = 1$$

replacing dx by $dV/V'_{r,a}$ gives:

$$\int_0^{V_g} dV/V'_{r,a} = 1 = {_0}\!\int^{V_g} dV/(k'_a.V_m) \quad [GE.1]$$

where k'_a was defined for Equation [3.57], V_m is the volume of mobile phase in the column (and thus also the column void volume $= U.t_0$) and V_g is the retention volume of the analyte under the gradient conditions. Now, restating Equation [3.57] in terms of mobile phase volumes instead of time, i.e., $t/t_0 = V/V_m$, gives:

$$\log(k'_i) = \log(k'_0) - b(V/V_m) \quad [GE.2]$$

where k'_i at any point during a gradient is the value of k' applicable if the elution were conducted under isocratic conditions with a mobile phase composition corresponding to that just entering the column at that same instant.

Now a crucial approximation is made, namely, $k'_i = k'_a$. This implies that, at any given instant during the gradient, the composition of the mobile phase currently present within the column does not vary much as a function of x. Under this approximation we substitute k'_a in Equation [3.57] by k'_i from Equation [GE.2]:

$$(1/k'_0.V_m).{_0}\!\int^{V_g}\exp(2.303bV/V_m) \quad [GE.3]$$

where V_g is the retention volume of the solute under gradient conditions, and 2.303 is the conversion factor for conversion from 10^Y to $\exp(Y)$. Integration of Equation [GE.3] then gives:

$$V_g = (V_m/b).\log[2.303k'_0.b + 1] + V_m \quad [GE.4]$$

Equation [GE.4] gives the retention volume for the analyte under gradient conditions specified by the parameter b. It is now simple to derive a value for the corresponding elution time $t_{r,g} = V_g/U$:

$$t_{r,g} = (t_0/b).\log[2.303k'_0.b + 1] + t_0 \quad [GE.5]$$

The value of k'_f, the value of k'_a at the moment of elution from the column, is obtained by substituting V in Equation [GE.2] by $(V_g - V_m)$ from Equation [GE.4], with some simple algebraic rearrangement to give:

$$k'_f = k'_0/(2.303bk'_0 + 1) \quad [GE.6]$$

As a final brief example of the approach to gradient elution theory (Snyder 1979) it is appropriate to give some indication of the origins of the average capacity factor k'_{avge} in some of the equations in Table 3.2. Recall that in isocratic elution it is sometimes useful to define an *effective* plate number N_E given by Equation [3.24], in which N is corrected by a factor $[k'/(1+k')]^2$. It is found (Snyder 1979) that an exactly equivalent expression holds to a good approximation for the gradient elution parameter $N_{E,g}$ if k' is replaced by $k'_{avge} \approx 1/(1.13.b)$. Then use of this expression for $N_{E,g}$ instead of N_E in Equation [3.23c] is found to give an adequate description of peak resolution in gradient elution:

$$R^{4\sigma}_{A/B,g} = 0.25.(\alpha_{A/B} - 1).N_{E,g}^{\frac{1}{2}}$$
$$= 0.25.(\alpha_{A/B} - 1).N_g^{\frac{1}{2}}.[k'_{avge}/(1+k'_{avge})]$$
$$[GE.7]$$

where $N_g = G.(1+k'_f).t_0/\sigma_t^2$ and the parameter G must be obtained by numerical integration.

List of Symbols Used in Chapter 3

A	rate theory coefficient for the multipath dispersion term.
A_s	peak asymmetry factor.
b	negative of slope of plot of $\log(k')$ vs t_r/t_0 for isocratic elution.
B	rate theory coefficient for the longitudinal dispersion term.
C	rate theory coefficient for the resistance to mass transfer term.
C_P	peak capacity.
c_A^{pm}	concentration of analyte A at its peak maximum upon elution.
$c_s^A; c_m^A$	equilibrium concentrations of analyte A in stationary and mobile phases.
D	chromatographic dilution factor.
d_p	diameter of stationary phase particle.
d_f	film thickness of stationary phase.
d_c	internal diameter (i.d.) of chromatographic column.
D_m	diffusion coefficient of analyte in mobile phase.
$D_{m(exit)}$	value of D_m at column exit (important for GC).
D_s	diffusion coefficient of analyte in stationary phase.
E	separation impedance.
E_{ext}	externally applied electric field (volts.m^{-1})
f	frictional coefficient characterizing resistance to motion of an ion or molecule through a liquid.
h	reduced plate height $= H/d_p$.
H	height of the equivalent theoretical plate (HETP).
$H_{min}(H_{opt})$	optimum value of H at the van Deemter minimum.
k'_a	actual (instantaneous) value of k' at any point during gradient elution if eluted isocratically in a mobile phase of the same composition as that in which the analyte finds itself at that instant.
k'_{avge}	average value of k' over an entire gradient elution.
k_A'	capacity factor (ratio) of analyte A, isocratic elution.
k'_f	value of k' as analyte exits the column with gradient elution.
k'_i	value of k' for an analyte at any point during a gradient if eluted under isocratic conditions with a mobile phase composition corresponding to that just entering the column at that same instant.
k'_w	value of k' for isocratic elution in the pure solvent of lower solvent strength.
k'_0	value of k' for an analyte if eluted under isocratic conditions with a mobile phase composition corresponding to that used at the start of a gradient.
K^A	equilibrium partition constant expressed as the ratio of concentrations of analyte A in stationary and mobile phases.
l	length of column.
l_{opt}	value of l that gives optimum performance when operated at the van Deemter minimum.
N	number of theoretical plates for isocratic elution; column efficiency.
N_E	effective plate number for isocratic elution $= N.[k'/(1+k')]^2$

$N_{E,g}$	effective plate number for gradient elution.
P_{exit}	pressure at the exit of the column (important for GC).
P	the phase ratio V_m/V_s of a column, note this can vary with analyte.
ΔP	pressure drop across the column.
r_{hd}	hydrodynamic radius of a solvated ion in solution.
$R^{n\sigma}_{A/B}$	isocratic resolution defined relative to separation of peak maxima for analytes A and B by n times the Gaussian parameter σ, i.e., for n = 2 or 4, separation by respectively one or two peak widths measured at the inflection points.
S	slope of curve relating k' for isocratic elution vs ϕ.
I_f	U.S. Pharmacopeia tailing factor.
t_{rA}	retention time for analyte A, isocratic elution.
$t_{r,g}$	retention time for an analyte under gradient elution conditions.
$t_{opt,L}$	elution time for the last-eluting peak when operated at the optimum condition (van Deemter minimum).
t_0	'dead time' $= V_0/U$
Δt_i	elution time difference (peak width) between the inflection points of a chromatographic peak.
Δt_{FWHM}	elution time difference (peak width) between the two half-height points of a chromatographic peak.
u	linear flow velocity of mobile phase.
u_{eo}	electro-osmotic flow velocity.
u_{exit}	linear flow velocity of mobile phase as it exits the column (important for GC).
u_{opt}	optimum value of u (at the van Deemter minimum).
U	volume flow rate of mobile phase
v^A_{plate}	(plate volume for analyte A) $= (v_m + K^A.v_s)$
v_m ; v_s	volumes of stationary and mobile phases within a single theoretical plate.
v_{ep}	speed of an ion in solution under electrophoresis.
V_E	void volume contribution from components external to the column.
$V_{inj}(max)$	maximum volume of analyte solution that can be injected on a column without significantly degrading the chromatographic efficiency.
V_g	retention volume of an analyte under gradient elution conditions.
V_m ; V_s	total volumes of stationary and mobile phases within the whole column, V_m is the column dead volume.
V_{rA}	retention volume for analyte A, isocratic elution.
V_{ro}	ideal plate Theory value ($= V$ subscript m) of retention volume for an unretained analyte.
$V_r{}'$	corrected retention volume for an analyte in isocratic elution $= V_r - V_E$.
$V'_{r,a}$	actual (instantaneous) value of $V_r{}'$ at any point during gradient elution.
V_0	total void volume $= V_m + V_E$.
ΔV_i	change in total volume of mobile phase between the two inflection points of a chromatographic peak.
x	fraction of column length l that an analyte has traveled along the column.
$\alpha_{A/B}$	separation ratio for analytes A and B.
δ	thickness of an electrical double layer.
ε	porosity of column $= V_m/(V_m + V_s)$.
ε_r	relative permittivity (dielectric constant) of a bulk substance.
ε_o	permittivity of free space.
ξ	zeta potential in an electrical double layer.
η	viscosity of mobile phase.
γ_m	van Deemter parameter used to describe the effect of packing quality on the mobile phase contribution to the longitudinal diffusion term.
γ_s	van Deemter parameter used to describe the effect of packing quality on the stationary phase contribution to the longitudinal diffusion term.

κ	D'Arcy Constant.
λ	van Deemter parameter used to describe the effect of packing quality on the multipath dispersion term.
μ_{ep}	electrophoretic mobility of an ion in solution.
μ_{eo}	electroosmotic mobility of a specific buffer solution in a specific capillary.
μ_{app}	apparent mobility of an ion, the combination of its own μ_{ep} and μ_{eo} of the buffer solution.
ν	reduced linear velocity $= u.d_p/D_m$.
σ	Gaussian parameter = half the peak width measured between the peak inflection points; can be expressed in units of V, t or distance along the column.
υ	total volume of mobile phase that has passed through the column after sample injection, expressed as number of plate volumes $\upsilon^A{}_{plate}$.
ϕ	volume fraction of the mobile phase component of greater solvent strength.
ϕ_0	initial value of ϕ at the start of a gradient
ψ	flow resistance parameter.

4

Tools of the Trade III. Separation Practicalities

4.1 Introduction

Murphy's Law: The Ongoing Struggle Against the Universe

Nature always sides with the hidden flaw.

Experience varies directly with amount of equipment ruined.

Never underestimate the innate malice of inanimate objects.

Ginsberg's Theorems

1. You can't win.
2. You can't break even.
3. You can't even quit the game.

Weiler's Law

Nothing is impossible for the man who doesn't have to do it himself.

Sattinger's Law

It works better if you plug it in.

Non-Reciprocal Law of Expectations

Negative expectations yield negative results. Positive expectations yield negative results.

The Fundamental Law of Reproducibility

A man with one watch knows what time it is. A man with two watches is never sure. *Anon.* (italicized)

Nature is not human-hearted. *Tao Te Ching, Lao Tze, 604–531 BC.*

It is a capital mistake to theorize before one has data. Insensibly one begins to twist facts to suit theories, instead of theories to suit facts. *Sherlock Holmes in A Scandal in Bohemia (1891), Sir Arthur Conan Doyle.*

The machines that are first invented to perform any particular movement are always the most complex, and succeeding artists generally discover that with fewer wheels, with fewer principles of motion than had originally been employed, the same effects may be more easily produced. *Essays on the Principles which Lead and Direct Philosophical Inquiries, Adam Smith.*

(Continued)

Normal people . . . believe that if it ain't broke, don't fix it. Engineers believe that if it ain't broke, it doesn't have enough features yet. *The Dilbert Principle, Scott Adams.*

Nature has good intentions, of course, but, as Aristotle once said, she cannot carry them out. T*he Decay of Lying, Oscar Wilde.*

In the preceding Chapter theories of chromatographic separations were discussed. This Chapter is devoted to practical concerns, for both chromatography as such and also for clean-up methods; the latter can be regarded as 'binary' chromatography in the sense that in an ideal clean-up (never achieved) the target analyte(s) are either 100% eluted or are 100% retained and vice versa for the interfering endogenous compounds. Thus, in this Chapter the relatively safe world of theory is being left behind while that of wrestling with Nature and the recalcitrant instrumentation in the laboratory is being entered. As emphasized in the Preface, the only real way to understand such difficulties and acquire experience in dealing with them is to 'learn by doing'. However, hopefully the content of this chapter will help the reader avoid some of the traps into which others have fallen.

The first two sections of this Chapter deal with what is often referred to as sample preparation, which in turn can sometimes be split into extraction and clean-up. Although these two operations are often separate steps in practice, in many other cases they are considered as a whole, particularly where analytical speed and throughput are important considerations. There is a vast and sometimes overwhelming literature concerned with the practicalities of extraction, clean-up and chromatography, but an excellent

place at which to start looking for help with practical problems or for background information is the magazine *LC/GC* (http://www.lcgcmag.com/lcgc/).

Sample preparation is not as 'glamorous' a part of an integrated analytical procedure (Figure 1.2) as the actual measurement using sophisticated chromatographs and mass spectrometers, but is crucial for a reliable final result. In Figure 4.1 the findings of an intriguing analysis (Majors 1991) of the contributions to the total error in typical trace analyses for organic analytes from the four major constituent steps are summarized. While the actual percentages are, of course, somewhat variable, it is striking that well over half of the total uncertainty was found to arise from the sample preparation. Clearly this aspect must be taken very seriously.

4.2 The Analyte and the Matrix

The first things to consider, when faced with planning an analytical method for a target analyte present at trace levels in a complex matrix, are the physico-chemical characteristics of that analyte and of the matrix in which it is present at trace levels. Bear in mind that the ultimate aim of sample extraction and clean-up is to remove (extract) the analyte from the matrix, for presentation to the analytical instrument (e.g., GC or HPLC) in a 'relatively clean' solution at levels that reflect those originally present in the matrix but with any incurred losses accounted for, e.g., through use of a suitable internal standard (Sections 2.2.3, 8.5.2 and 9.4.5). The operational meaning of a 'relatively clean' final solution is one which, when analyzed by some form of high resolution chromatography (use of two orthogonal chromatographies is applied in difficult cases, Section 9.5.5c) with a suitable detector, will present the analyte to the detector free of any co-eluting matrix components that will either contribute intensity directly to the analyte signal (interference) or affect the molar response of the detector for the analyte (suppression or enhancement, Sections 5.3.6a and 9.6). Note that the latter problem, if not too serious, can be alleviated through use of an appropriate SIS and/or of so-called matrix matched calibrators (Sections 2.2, 8.5 and 9.8.1).

Some of the analyte characteristics that require attention are fairly obvious, e.g. chemical stability requires

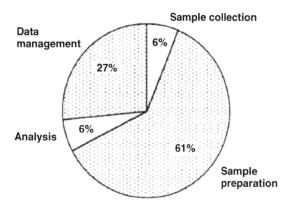

Figure 4.1 Contributions from different steps to total effort and time, and thus potential for error, in a trace organic quantitative analysis. Reproduced from Majors, *LC/GC North America* **9** (1991), 16, with permission from Dr. Ron Majors and the journal.

simple precautions and checks within the validation procedure (Sections 10.2.8 and 10.4.1). Highly polar analytes will have low volatility and thermal stability and so will require either suitable derivatization (Section 5.2.1b) to be compatible with GC, or separation by HPLC with electrospray ionization (ESI) for compounds with high polarity, or atmospheric pressure chemical ionization (APCI) or photoioinization (APPI) for those that are moderately polar (see Sections 5.3.3–5.3.6). The problem of chemical artifacts refers to analytes containing highly reactive or labile functional groups that may, if precautions are not taken, react or decompose in an uncontrolled fashion during the extraction and clean-up steps. Another parameter for which the implications are reasonably obvious is the amount of available sample; this involves two parameters, the total amount of analytical sample (matrix containing the analyte) and the concentration of analyte in the matrix, that taken together determine the number of moles of analyte available. Together with the number of replicate analyses required (i.e., the degree of analytical precision required), the amount and concentration of the analytical sample are obviously important in determining the analytical method to be adopted. Limitations imposed by molecular mass are not quite as apparent and have been discussed in Section 1.5.

There are also issues with the matrix that can mostly be related to the physical state (gas, liquid or solid) and thus the degree of homogeneity of the sample. Thus gases are intrinsically thoroughly mixed and aliquots can be removed and analyzed directly, possibly involving an enrichment procedure of some kind before the final GC/MS analysis. Liquids are also essentially homogeneous and can generally be extracted directly and the extracts cleaned-up to prepare a solution for injection into the chromatograph; the only concerns here are those involved in accounting for losses of analyte during these steps and ensuring removal of matrix components that could potentially interfere with the subsequent analysis, e.g., via ionization suppression. Cleaner extracts always lead to more rugged analytical methods! However, analyte losses can be addressed through addition of an SIS before extraction, since the extraction efficiencies of a good SIS (Table 2.1) and of the native analyte from a homogeneous liquid matrix (e.g., urine, lake water) will be the same. The suppression phenomenon is discussed later (Section 5.3.6a).

However, solid samples are often highly inhomogeneous (e.g., contaminated soils), leading to a sampling problem (Section 8.5) before any kind of analytical procedure is even started. This problem refers to selection of aliquots of matrix plus analyte (e.g. soil samples plus contaminant) that are of suitable size for the analytical procedure and that are truly representative of the entire analytical sample (e.g. the contaminated field or garden). As an additional problem, frequently addition of a SIS before extraction does not correctly account for analyte extraction efficiencies since part of the analyte can be occluded in small pores within the solid and is inaccessible to the extraction medium. An example where both of these problems arose involved analysis for polycyclic aromatic hydrocarbons (PAHs) of a contaminated harbor sediment. In this case the principal source of inhomogeneity arose from differences among the sediment particles with respect to size, shape and PAH concentration; this source of uncertainty could be reduced to manageable size by specifying a minimum mass of sediment (i.e., number of sediment particles) that should be analyzed. The problems arising from occlusion of analyte, and thus unequal extraction efficiencies for analyte and surrogate internal standard that was spiked into the raw sample (Section 8.5.2b), were much more difficult and it was never really certain that all of the analyte had been extracted (Boyd 1996). The best that could be done was to subject the residue remaining after the first extraction (solvent extraction in a Soxhlet apparatus, Section 4.3.2a) to additional extraction methods such as use of a different Soxhlet solvent and/or supercritical fluid extraction (Section 4.3.2e), and combining the resulting extracts. Less extreme cases of this general phenomenon are also found, e.g., in analysis of blood or plasma for a small molecule that is strongly bound to a protein (Ke 2000). It is important to document all details, no matter how apparently unimportant, of the extraction procedure.

4.3 Extraction and Clean-Up: Sample Preparation Methods

Although extraction and clean-up are often conducted as separate steps (Figure 1.2), in many cases they are considered together as 'sample preparation', particularly where analytical speed and throughput are important considerations. Here, the methods in current use for both are discussed together for convenience as most of them overlap. This stage in planning an analytical method is where the 'fitness for purpose' concept (Preface and Section 9.2) is particularly important. The extent and rigor of the combined extraction–clean-up procedures must be considered in the light of required performance criteria such as LOD and LLOQ (Section 8.4), dynamic range, specificity, speed and/or throughput etc., as well as considerations of the available analytical equipment (chromatography, mass spectrometry etc.) Some of these criteria can be mutually incompatible if ultimate performance is required in all of them, so a suitable compromise must be reached that will best fit the intended purpose of the analyses. As a general point, however, the lower the extent of clean-up, the more susceptible will be the final mass spectrometric step to interferences and ionization suppression (Section 5.3.6a), and also the greater the degree of fouling of the instrument by large quantities of involatile matrix components that were not removed. For the ultimate achievable accuracy, precision and selectivity, the extraction–clean-up procedures should be as rigorous as possible, but this choice may not be the most appropriate one given the actual purpose of the analysis. For example, if it is believed that local groundwater may have become contaminated as the result of an environmental accident, speed to a preliminary estimate of the extent of pollution becomes a dominant criterion in determining the analytical strategy. Another example concerns analysis of blood plasma for a candidate new drug and its metabolites. Depending on the stage of the drug discovery–development process (Section 11.5) at which the analyses are being done, a simple clean-up step (e.g., precipitation of plasma proteins by addition of acetonitrile, Section 4.3.1f) may suffice for the purpose at hand, or a much more rigorous clean-up procedure may be adopted based on one or more of the other approaches described in this chapter. In all methods for quantitative trace analysis, it is crucial to monitor and control contamination and carryover in general; this practical concern is discussed in detail in Section 9.7.

Before starting the descriptions of these approaches, it seems worthwhile to try to put their relative importance in current analytical practice into some perspective. An interesting article (Majors 2002) has described the results of a survey of analytical chemists working in a wide range of application areas, concerning the sample treatments used in their laboratories. The results are summarized in Figure 4.2. (These methods are not mutually exclusive, i.e., more than one of the methods listed might be used in any given analytical method.) Note the popularity of column chromatography (presumably including flash chromatography, Section 4.3.3a), sonication assisted liquid extraction (Section 4.3.2c), liquid–liquid extraction (LLE, Section 4.3.1) and solid phase extraction (SPE, Section 4.3.3c), the moderate usage of Soxhlet extraction (Section 4.3.2a) and purge-and-trap methods (Section 4.3.3b), and the very low use of supercritical fluid extraction (SFE, Section 4.3.2e), microwave assisted extraction (MAE, Section 4.3.2d) and pressurized solvent extraction (PSE, Section 4.3.2b) although it appears that PSE was gaining in popularity at the time of this survey. Of course, as in all surveys, the results are a function of the interests of the analysts who responded and may not correspond to an entirely representative

summary; also, even the least popular methods are useful for small groups of analysts undertaking specialized work. However, the overall picture (Majors 2002) is unlikely to be misleading. With regard to SPE, more than 50 % of those respondents to the survey who work in the pharmaceutical sector favored C_{18} reverse phase materials, but this number dropped to about 25 % in a more general survey (but was still the most popular choice, the next most popular being C_8 phases).

The following discussion of methods used for sample extraction and clean-up attempts to cover the more commonly used techniques as well as those that appear to offer some potential for the future but are not yet routine. A review of microextraction procedures used in analytical toxicology (Flanagan 2006) is mainly concerned with extraction of analytes from plasma with particular emphasis on LLE and protein precipitation; however, miniaturized LLE and SPE are also described as promising ancillary methods.

4.3.1 Liquid–Liquid Extraction (LLE)

Most scientists with even minimal exposure to practical chemistry will be familiar with liquid–liquid extraction (LLE) in a separatory funnel. The general principle is simple. A target compound (traditionally a desired product of a chemical synthesis in the raw reaction mixture, or a target analyte in a liquid matrix) is mixed with a second liquid that is essentially immiscible with the liquid in which the target compound is originally dissolved. The conditions are manipulated so that the target compound is transferred from its original solution into the second liquid, with as high a degree of selectivity as possible. A common circumstance would involve an organic molecule containing an acidic (or basic) functional group, present in an original matrix that is highly aqueous (e.g., urine, blood plasma); by adjusting the pH of the matrix to a low (or high) value, the target compound can be rendered electrically neutral and thus more likely to have a high solubility in the second liquid. The latter will (if immiscible with water) be a nonpolar solvent (e.g. dichloromethane). Sometimes two LLE steps are performed, for example back extraction into a clean aqueous phase. It is good practice, where feasible, to profile different extraction solvents with different strengths (Section 4.4.2a) to determine the best choice to minimize co-extractives.

This simple principle is still applied in trace analysis, often in miniaturized form, as either an extraction procedure *per se* or as a clean-up method (or both). Classical LLE uses large amounts of organic solvent (an environmental concern) and is slow, while the various miniaturized methods use much smaller amounts of solvent, are much faster and more amenable to automation, and

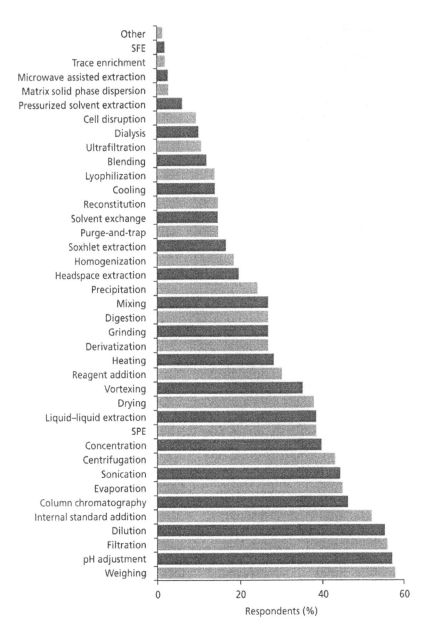

Figure 4.2 Summary of sample preparation methods used (alone or in combination with others) as reported in a survey of analysts covering a range of application areas. Reproduced from Majors, *LC/GC* **20** (2002) 1098, with permission from Dr. Ron Majors and the journal.

can also result in better sample enrichment (higher selectivity). The following discussion is based on a review of miniaturized LLE (Majors 2006).

However, it is necessary to first consider a concern for the design of integrated extraction–clean-up–reversed phase HPLC analytical methods that arises from

the elution strength (Section 4.4.2a) of the solvents commonly used in LLE from aqueous matrices. The elution strength of the LLE solvents is usually appreciably higher than that of the HPLC mobile phase, so that direct injection of such LLE extracts can significantly distort the HPLC peak shape (Section 3.5.12). This degrades the chromatographic resolution and limits the precision of measurement of peak areas. Frequently evaporation of the LLE extract to dryness (e.g., under a stream of dry nitrogen) and re-dissolution in a solvent that is compatible with reversed-phase HPLC can solve this problem, but at the expense of an additional step and thus lower throughput. However, use of 96-well plates (Venn 2005) and evaporators can minimize this effect for low volume bioanalytical assays, and efficient modern rotary evaporators can do the same for larger volume extracts from e.g. environmental, food and similar samples. More recently the introduction of hydrophilic interaction chromatography (HILIC, Section 4.4.2c) has circumvented this problem (Weng 2004) for the highly polar compounds amenable to this technique, since the mobile phases used in this chromatographic method have high elution strength.

The remainder of this section on LLE describes some variations that are either seldom used or are in the experimental stages of development.

4.3.1a Solid-Supported Liquid–Liquid Extraction (SLE)

This technique is a modification of traditional LLE that lends itself to miniaturization and automation; it involves immobilization of an aqueous sample matrix (e.g., urine, diluted plasma) on a chemically inert solid material with high surface area and strong affinity for polar molecules (diatomaceous earth is the most common material used). The immiscible organic solvent is then passed over the thin film of immobilized aqueous phase and, thanks to the high surface area, can efficiently extract the analyte. This method can be implemented on a millilitre scale using SLE cartridges available commercially, and all the subsidiary approaches (e.g. pH control) that have been used in traditional LLE can also be employed here without risks associated with emulsion formation on shaking, and with shorter overall times. With regard to miniaturization and high-throughput applications, 96-well plates with each well packed with a suitable solid support phase are available for extraction of $100-200\,\mu L$ of aqueous sample. A brief report of the advantages of SLE over LLE in a biomedical analysis, particularly with respect to speed and precision, has been published (Shapiro 2005).

4.3.1b Single Drop Microextraction (SDME)

The ultimate in miniaturization of LLE uses a single drop of immiscible liquid. The seminal work on this approach to LLE (Jeannot 1996, 1997) was based on earlier work (Liu 1995) on partitioning of gas molecules into liquid

droplets; the latter work has developed into an approach to headspace analysis (Wood 2004). In SDME small droplets $(1-2\,\mu L)$ of an organic solvent are suspended from the tip of a GC syringe and exposed to the aqueous sample matrix. After exposure the entire droplet is withdrawn back into the syringe and then injected directly into the GC. It was found (Jeannot 1997) that equilibration of the liquid–liquid partitioning can not be established in the time available before significant loss of the organic solvent occurs as a result of the small but nonzero aqueous solubility of the solvent. For this reason the procedure involves kinetics not equilibration, so good precision requires precise timing of the exposure but standard deviations of 1.5 % were reported (Jeannot 1997) even when the amount of analyte extracted was only 38 % of that expected if equilibrium had been established. Combination of SDME with HPLC has been reported (Gioti 2005), but in that case the original droplet (an octanol–chloroform mixture) was transferred to a microvial and diluted with the water-soluble solvent methanol before injection into the loop injector of a reverse phase HPLC; this paper provides a detailed description of the precautions required to obtain high quality data by SDLE–HPLC. However, in its present state of development SDME is not widely used in trace analysis as a result of the difficulties in its implementation, especially with respect to automation.

An intriguing variation of single droplet extraction uses droplets $(0.1-2\,\mu L)$ levitated in the nodal points of a standing ultrasonic wave. The idea is to levitate an aqueous droplet containing the target analyte, then add the organic solvent to the levitated droplet using a micropipette and allow the LLE to proceed. Removal of the organic phase for subsequent analysis is again achieved by micropipette. The motivation for developing this approach (Welter 1997) was to avoid the effects of glass or silica surfaces on very small liquid volumes and the subsequent uncontrolled losses of analyte by adsorption. This technique is very much in its infancy and is subject to difficulties such as solvent evaporation. However, some impressive results have been reported by specialists (Santesson 2004, 2004a).

While not thus far exploited for LLE, it is interesting to note experiments (Bogan 2002, 2004) in which charged droplets ($\sim0.27\,\mu L$) are produced from a piezo-electric atomizer and are levitated in the electric field of a modified quadrupole ion trap (Section 6.4.5) operated at atmospheric pressure. Thus far the main analytical application of this technique (Bogan 2004) has involved wall-free preparation of micrometer-sized sample spots for fmol detection limits of proteins by MALDI–MS, but extension to LLE in such levitated droplets is a possibility.

4.3.1c Dispersive Liquid–Liquid Microextraction (DLLE)

A new approach to miniaturized LLE (Ahmadi 2006, Rezaee 2006) from aqueous solutions avoids some problems of SDME, including droplet break up during stirring and the possibility of air bubble formation. This new approach is based on the dispersion of tiny droplets of the extraction liquid within the aqueous liquid, rather than a single droplet. A ternary solvent system is used; a typical example would use ~5 mL of aqueous solution in a vial, rapidly mixed with ~1 mL of acetone (disperser solvent) and ~8 μL of tetrachloroethylene (TCE) as the extraction solvent. The apparatus is very simple, incorporating a syringe for rapid injection of the disperser and extraction solvents into the aqueous solvent. With gentle shaking a cloudy suspension of minute droplets of TCE in the aqueous sample is formed, and equilibration of the analyte(s) between aqueous and organic phases is effectively instantaneous (a few seconds) as a result of the extremely large interface area between the two phases. A brief centrifugation (1–2 minutes) forces the droplets to coalesce at the bottom of the vial; this organic phase can then be removed for analysis. This extraction procedure was used in proof-of-principle determinations of polycyclic aromatic hydrocarbons in surface, river and well water. The principal advantages of this innovative method include the low cost of common solvents and the very small amounts used, the use of simple equipment, high recoveries and enrichment factors, and very short extraction times. It remains to be seen whether or not this approach (Ahmadi 2006, Rezaee 2006) can be developed into a robust analytical technique.

4.3.1d Flow Injection Liquid–Liquid Extraction

A continuous flow LLE technique (Karlberg 1978) involves flow injection of sample solution and organic solvent alternately into a flowing liquid stream. The plugs of the two immiscible liquids are mixed by inducing turbulence in a specially designed mixing chamber, and the two phases are subsequently separated again and the organic phase removed for subsequent analysis. The original application of this approach was to the analysis of aspirin tablets for caffeine, but since then the main use of the method appears to have been on-line enrichment of trace level metal ions for analysis by atomic absorption spectrophotometry or inductively-coupled plasma mass spectrometry (ICP–MS). This approach is designed for high throughput screening but seems unlikely to be adaptable to validated quantitative methods.

4.3.1e Membrane Extraction

As emphasized previously (Majors 2006), it is surprising that LLE methods using a microporous membrane have not been exploited more often. Such methods, pioneered in the 1980s (Audunsson 1986), potentially provide good clean-up efficiency, high enrichment factors (i.e. ratio of concentrations in the two phases) and ease of automation and of on-line connection to HPLC (Jönsson 2001, 2003; van de Merbel 1999). Most membrane-assisted LLE methods use flowing systems, the simplest involving aqueous and organic phases separated by a suitable microporous hydrophobic membrane; this approach has been called microporous membrane liquid–liquid extraction (MMLLE). Such a membrane, e.g., a polypropylene membrane with pore size of 0.2 μm and a thickness of 25 μm as used in analysis of pesticides in wine (Hyötyläinen 2002), has essentially no affinity for water so the organic solvent (*cyclo*hexane in this example) fills the membrane pores and thus allows direct contact of the two phases. A diagram of the conceptually simple extractor unit that permits continuous extraction of the nonpolar analytes from the flowing aqueous matrix into the organic solvent (toluene) is shown in Figure 4.3. The strengths and limitations of the MMLLE approach as presently developed are well exemplified in this work (Hyötyläinen 2002); the extracts were very clean thus permitting unambiguous analyses by GC with a single-parameter detector (flame ionization detector), but extraction required about 40 minutes. More recently (Hyötyläinen 2004) this application was extended by devising a means of using MMLLE on-line with GC.

A different approach that has turned out to be more versatile is supported liquid membrane extraction (SLME), a three-phase method in which analytes in an aqueous matrix are extracted into a clean aqueous solvent through an organic phase held by capillary forces in the pores of a hydrophobic membrane. This is analogous to the back-extraction approach in classical LLE, and is superior as it involves two independent partitioning equilibria that can be made complementary by appropriate choices of the three phases. Detailed theory for SLME, incorporating several physico-chemical parameters as well as the sample flow rate, has been worked out (Jönsson 1993). An interesting application of both SLME and MMLLE, on-line with HPLC (Sandahl 2000), permitted analysis of a wide range of compound types (permanently charged, ionizable and nonpolar) arising from metabolism and breakdown of thiophanate-methyl, a pesticide widely used for fungal control.

These membrane-based approaches do not appear to have been commercialized as yet and undoubtedly this is a factor in the current rarity of their use. However, another

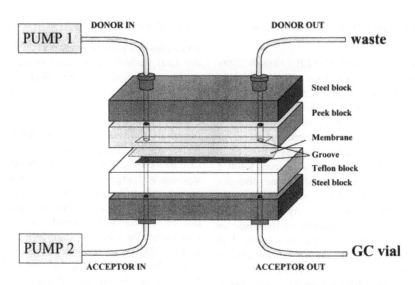

Figure 4.3 Diagram of an extractor unit used in microporous membrane liquid–liquid extraction (MMLE) analysis of pesticides in wine. Reproduced from Hyötyläinen *et al., Anal. Bioanal. Chem.* **372** (2002), 732, copyright 2002, with kind permission from Springer Science and Business Media.

disadvantage appears to be susceptibility to memory effects particularly in on-line versions. An off-line variant that avoids memory effects (Pedersen-Bjergaard, 1999) has been termed liquid phase microextraction (LPME). This approach uses a membrane in the form of a disposable hollow fibre impregnated with a solvent that is immersed in the liquid matrix to be sampled, in a fashion similar to that used in solid phase microextraction (SPME, Section 4.3.3f). An extensive review of developments of this technique, together with a comparison of its advantages and disadvantages relative to other microextraction methods, has been published (Psillakis 2003).

A more recent variation of LPME has been named solvent bar microextration (SBME), in which a 2 cm length of a polypropylene hollow fibre membrane (inner diameter 600 μm, wall thickness 200 μm, wall pore size 0.2 μm) was sealed at both ends, filled with suitable organic liquid (~3 μL) as the extraction phase using a microsyringe, and placed in the aqueous sample to be extracted (Jiang 2004). The aqueous sample was then agitated with a conventional magnetic stirrer and the small solvent-filled membrane device tumbled through the sample thus affording faster extraction compared with the static LPME method. (Note that this approach should not be confused with 'stir-bar microextraction, SBME', discussed in Section 4.3.3g as an example of extraction from a liquid into a solid phase.) After extraction the liquid inside the membrane bar is extracted with a microsyringe and injected into a GC. The original work (Jiang 2004) used pentachlorobenzene and

hexachlorobenzene as model compounds to investigate the extraction performance, yielding high enrichment (~100-fold) in 10 minutes and good reproducibility (RSD < 4%, $n = 6$). Since the hollow fiber membrane was sealed it could be used for extraction from 'dirty' samples such as soil slurries (Jiang 2004). A later application (Chia 2006) investigated organochlorine pesticides in wine, and demonstrated very high enrichments (1900–7100-fold) at an aqueous sample/organic solvent volume ratio of 20 mL/2 μL with good repeatabilities (RSD < 12.6%). A more recent modification of SBME (Melwanki 2005) is analogous to SLME, described above, and involves forward- and back-extractions across an organic film immobilized in the pores of the porous polypropylene hollow fiber. Four chlorophenoxyacetic acid herbicides were chosen as model compounds. They were extracted from the stirred acidified aqueous sample (donor phase) through a thin film of organic solvent in the pores of the hollow fibre and then finally extracted into another aqueous phase containing sodium hydroxide (acceptor phase); the acceptor phase was analyzed by HPLC. Good enrichment (438–553-fold) and repeatability (RSD 4–6%) were obtained in 40 minutes. The analytical potential of the method was demonstrated by applying the method to a spiked river water sample (Melwanki 2005), and comparisons with SPE, SPME and in-tube SPME (Section 4.3.3f) were described. Like all of these innovative membrane-based extraction methods, the range of applicability and general robustness of this approach remain to be established.

One last variant of membrane-based LLE will be mentioned briefly. Microdialysis corresponds to a miniaturized version of dialysis techniques routinely used to remove small ions and molecules from biological extracts. At this time it is clear that there are many obstacles to be overcome before routine use of this approach in trace quantitative analysis can be established (Torto 2001). An example of successful application of the method to a specialized biological problem is the determination of basal acetylcholine in microdialysate from the brain striatum of awake and freely moving rats, in which the miocrodialysis probe was interfaced to HPLC with tandem mass spectrometric detection (Zhu 2000).

It is emphasized again that the LLE variants described in Sections 4.3.1a–e are either early-stage developments or approaches that have not found wide use in validated analytical methods.

4.3.1f Protein Precipitation from Biological Fluids

A method of sample preparation that is not strictly liquid–liquid extraction, and indeed is not really an extraction method at all, concerns the addition of an organic solvent (typically acetonitrile) to analytical samples that are biological fluids (plasma, urine). The purpose of this clean-up step is to simply precipitate the endogenous proteins that would otherwise interfere with subsequent LC/MS analyses. A typical manual procedure for protein precipitation might involve spiking $50\,\mu L$ of plasma with $10-50\,\mu L$ of a solution of surrogate internal standard in a solvent containing $<10\,\%$ organic if possible (to avoid precipitating the proteins at this stage), adding $200\,\mu L$ of acidified (sometimes neutral) acetonitrile, mixing in a vortex mixer, and final centrifugation in order to make a pellet out of the precipitated proteins. An aliquot of the supernatant is then removed for further analysis; in screening procedures used in drug discovery (Section 11.5) direct LC–MS analysis of such supernatants is frequently fit for purpose. However, manual implementation of protein precipitation is much too slow to provide the desired high throughput screening, and automation based on 96-well plates (Venn 2005) and laboratory robots (Watt 2000) is nowadays an industry standard. However, note that, as for LLE in general, the solvent of the supernatant extract is generally of significantly higher elution strength (Section 4.4.2a) than that of reverse-phase HPLC mobile phase, and direct injection of such solutions can lead (Section 3.5.12) to significant deterioration of chromatographic peak shape, resolution etc. However, this problem can sometimes be addressed by simple dilution, since modern LC/MS instruments now afford enough sensitivity to allow for dilution of the supernatant in an aqueous solvent, resulting in a solution compatible with reversed phase LC/MS with ample sensitivity to provide a lower limit of quantitation (LLOQ) of $1\,ng.mL^{-1}$. Otherwise a solvent-switching step is introduced or (where appropriate) HILIC separations (Section 4.4.2c) are used. Protein precipitation is also used for assays for multiple analytes that are not readily extracted using LLE or SPE (Section 4.4.3), e.g. ester and acid metabolites of drug candidates.

4.3.2 Liquid Extraction of Analytes from Solid Matrices

As for LLE, extraction of analytes by a liquid solvent from solid (or semi-solid) matrices is a long established approach. A well known example is the extraction of lipids from biological tissues by well designed solvent mixtures (Folch 1957; Bligh 1959).

4.3.2a Soxhlet Extraction

A classic approach uses freshly distilled solvent to continuously extract analyte from the matrix in a continuous cyclic procedure (Soxhlet 1879). This simple piece of apparatus, originally designed as a tool for study of lipids in milk (see accompanying text box), is still a standard item in many laboratories today. Modern versions of the Soxhlet apparatus, offered by several suppliers, consist of variations including micro or macroscale extractors in automated or semi-automated format. The common theme is the continuous use of freshly distilled solvent to extract analytes from the solid (or semi-solid) matrix in a closed cycle operation, with the potential to recover and repurify most of the solvent after use. Disadvantages of the Soxhlet method include the need for sizable quantities of sample, rather long extraction times (hours to days) and thus extended exposure of the analyte to boiling solvent, and use of large amounts of solvent. However the method is still used in applications where these disadvantages are not severe, e.g. analyses of environmental samples for thermally stable pollutants that require a minimum sample size to avoid statistical sampling errors (Section 8.6).

4.3.2b Pressurized Solvent Extraction

It is a common experience in even introductory practical chemistry that, in most cases, the solubility of compounds in a given solvent increases with temperature. For example, the organic chemists' use of recrystallization to purify compounds relies on selective dissolution

The Soxhlet Extraction Apparatus

1. Boiling flask.
2. Extraction chamber in which the solid sample is contained in a thimble made of a material porous to the solvent (e.g. cellulose). The tube on the right side of 2 is a siphon that returns the contents of 2 to the boiler 1.
3. Solvent recovery vessel (note stopcock between items 3 and 2 open during extraction, closed for solvent recovery).
4. Condenser.

Franz von Soxhlet was born in 1848, the son of a Belgian immigrant, in Brünn (Brno) in what was part of Bohemia at that time, and died in 1926 in Munich (Germany). The following account of his life's work, together with the portrait and sketch of his famous apparatus, are taken from: http://www.cyberlipid.org/extract/soxhlet.htm

After obtaining his Ph.D. he was appointed assistant at the agricultural research station in Vienna in 1873, and in 1879 became Professor of Animal Physiology and Dairy at the Agricultural College in Munich; he also earned a medical degree in 1894 from the University of Halle. His life's work was devoted to improvement of the health of infants via studies of the chemistry and biochemistry of milk; invention of his extraction apparatus (Soxhlet 1879) was motivated by a need to efficiently extract lipids from milk and dairy products. Other contributions to milk chemistry included analysis of sugars (he appears to have been the first to characterize lactose as the sugar present in milk) and acidity. His work on the sterilization of milk for infants led him to describe a simple household device to sterilize (pasteurize) feeding bottles. He also described the chemical differences between human and cow milk. Another first was his fractionation of milk proteins into casein, albumin, globulin and lactoprotein. He investigated the relationship between the calcium content in milk and the prevalence of rachitis (rickets). One of his last papers (in 1912) investigated the connection between the iron content of human and cow milk and infant anemia.

Although Soxhlet is nowadays chiefly remembered for his invention of the apparatus that bears his name, his life and work remain an exemplary case of the application of creative chemistry to the improvement of human health.

of the desired compound from a mixture using a heated solvent, followed by removal of residual solid and slow cooling of the solution to allow crystallization of the compound in a manner that hopefully excludes most other dissolved compounds. This phenomenon is the basis for pressurized solvent extraction; extraction at temperatures well above the solvent's normal boiling point (i.e., that at atmospheric pressure) is achieved by raising the pressure in the extraction vessel. The risk of decomposition of analytes at these higher temperatures is partly alleviated by the fact that the rate of extraction from a solid matrix also substantially increases with temperature; this is at least partly the result of decreases in surface tension and viscosity that allow the solvent to more readily wet the solid and also to penetrate small pores where some of the analyte is occluded. Nonetheless, this approach is generally restricted to analytes with appreciable thermal stability, e.g. environmental pollutants that are amenable to GC analysis; this aspect was recently reviewed (Schantz 2006).

The most common form of pressurized solvent extraction at present is the so-called accelerated solvent extraction (ASE®) approach, for which an instructive summary is available (Dionex Corporation 1998). For all techniques involving solvent extraction of solids the ideal sample is a dry, finely divided solid, and this is certainly true of ASE. Thus, initial grinding to particle sizes < 0.5 mm, addition of an inert dispersing agent (e.g. sand) to prevent re-aggregation of sample particles, and addition of a drying agent such as anhydrous sodium sulfate in ratios up to 1:2 by weight, are all advisable or essential for ASE. Solvent type and temperature are the most important experimental parameters, while pressure is important only to the extent that it must be sufficient to maintain the solvent in the liquid state at the temperature used. ASE is conducted in static mode as it is difficult to arrange a dynamic (continuous flow) method while maintaining the pressure at a suitably high value, but it is possible to arrange for several cycles of static extraction for the same sample. The chief advantages of ASE as an off-line extraction method are its speed and its need for only modest amounts of organic solvent.

4.3.2c Sonication Assisted Liquid Extraction (SAE)

SAE exploits the irradiation of a liquid with ultrasound to extract an analyte from a matrix. There is a growing body of literature on sonochemistry, the application of sonication to chemistry (Suslick 1994; Maynard 2000); the following account draws heavily from these references. Ultrasonic irradiation at moderate intensities can produce bubbles of vaporized liquid that oscillate in size as they experience the expansion and compression phases of the sound wave. However, in addition to this oscillation the bubbles gradually increase in size because they grow a little more during the expansion phase than they shrink during the compression phase. Under appropriate conditions the bubbles can grow to such an extent (150–200 μm) that they undergo a violent and very fast (< 1 ns) collapse, thus generating very high pressures and temperatures from the rapid (adiabatic) compression of the vapor; this process is called cavitation. Since the surrounding liquid is cold it quickly quenches the heated cavity. However, the short-lived (< 1 μs) localized hot spot can reach a temperature of ~5000 °C and a pressure of about 1000 atmospheres (~14 000 psi); for a rough comparison (Suslick 1994) these are, respectively, the temperature of the surface of the sun and the pressure at the bottom of the ocean, lasting for the lifetime of a lightning strike.

This concentration of the diffuse energy of sound (the energy density in an ultrasonic field that produces cavitation is about 10^9 times less than that in the collapsed cavitation bubble) into a chemically useful form is the basis of sonochemistry, including SAE. In the latter case, however, there is an added complication arising from the presence of the solid matrix to be extracted. In liquids the cavity remains spherical during collapse but, if close to a solid surface, the cavity collapse can be very asymmetric and create a new range of phenomena involving generation of high-speed liquid jets that hit the surface with tremendous force. This process can cause severe damage at the point of impact and can produce newly exposed highly reactive surfaces. It is believed (Suslick 1994) that this phenomenon is important for understanding the corrosion and erosion of metals observed in propellers, turbines and pumps where cavitation is a continual technological problem; cavitation is also responsible for the noise generated in fast heating of water in a (tea) kettle, for example. It is conceivable that the erosion phenomenon could contribute to the efficiency of SAE from solids, but these asymmetric distortions of bubble collapse occur only near surfaces several times larger than the size of the bubble (Suslick 1994), i.e., the presence of fine powders (few hundred micrometers) does not induce jet formation, though coarser solid particles might do so.

Commercially available high-intensity ultrasonic probes (10 to 500 W cm^{-2}) are the most effective sources for laboratory scale sonochemistry. A typical system operates at 24 kHz with an adjustable total power output of up to 200 W and also adjustable irradiation times per pulse of a few tenths of a second. Lower intensities can often be used in liquid–solid heterogeneous systems of interest here because of the reduced liquid tensile strength at the liquid–solid interface, and a common ultrasonic cleaning bath (about 1 W cm^{-2}) can often be adequate for SAE.

In summary, SAE is a rather special case of off-line liquid extraction at higher temperatures, possibly assisted in some cases by the erosion effect on larger solid particles. Increased extraction speed and efficiency of conventional solvent extraction procedures are the main advantages. However, chemical decomposition or other reactions of the analytes is always a possibility to be borne in mind. The chemical consequences of ultrasonic irradiation of organic liquids have been studied much less that those for water, but it has been established that when cavitation occurs almost all organic liquids will generate free radicals.

4.3.2d Microwave Assisted Extraction (MAE)

This method bears some superficial resemblance to SAE, but the mechanism whereby energy is transferred into the solvent is quite different. As discussed above, SAE produces highly localized 'hot spots' within which highly efficient extraction from a solid matrix can occur. For MAE the process is exactly the same as that of cooking food in a domestic microwave oven. Microwaves cover the electromagnetic frequency range from about 300 to 300 000 MHz and are used for radar and telecommunications; to avoid interference with these operations, domestic and industrial microwave devices are required to operate at either 900 or 2450 MHz. Devices of interest here use the larger frequency (2450 MHz) corresponding to a photon energy of 0.9 J mol^{-1}, far too small to lead to any chemical reaction; indeed, the only mechanisms responsible for coupling the electromagnetic energy to the thermal energy of the solvent are rotation of molecular permanent dipoles and oscillatory translational motion of ions, following the oscillating electric field.

MAE is thus another special case of liquid extraction accelerated by increasing the temperature of the solvent. It differs from traditional solvent extraction (including the Soxhlet method) by the fact that the heat is generated internally within the solvent–solid matrix system wherever the microwaves can interact with solvent molecules, including within any pores in the solid. MAE differs from SAE by the fact that no spectacular localized 'hot spots' are involved. In addition, because of the mechanism involved, different solvents have different properties with respect to microwave radiation and this provides a degree of selectivity for MAE by adjusting the nature of the solvent. The principal molecular properties involved in response to a microwave field are the dipole moment of the solvent and the extent of nonbonding interactions

of a solvent molecule with its neighbors, which determines how free it is to rotate and how efficiently the rotation energy is transferred to random (thermal) energy of the solvent as a whole. On the macroscopic level these concepts appear as the components of the dielectric permittivity ε of the solvent:

$$\varepsilon = \varepsilon' - i.\varepsilon''$$

where $i = (-1)^{\frac{1}{2}}$, ε' is the dielectric constant that describes how well the solvent absorbs the microwave radiation (analogous to an extinction coefficient in UV–Visible spectroscopy) and ε'' is the dielectric loss factor that expresses the efficiency of conversion of the microwave energy into thermal energy of the entire solvent and thus the extent of cooling of localized zones where appreciable energy absorption has occurred. The dissipation factor, $\delta = \varepsilon''/\varepsilon'$, is a figure-of-merit that roughly corresponds to the fraction of the locally absorbed microwave energy that can be dissipated to the surrounding medium. Values of these related constants for several commonly used solvents are listed in Table 4.1. Note the large difference between the values for water and ice, reflecting the much more restrictive movement of water molecules in the solid state; the increased dissipative factor when ions are dissolved in the water (Table 4.1) reflects the additional mechanism associated with oscillation of the ions which is readily converted to random thermal energy, and the very low values for the nonpolar solvents as expected. Water, as usual, is somewhat unique with a very high ε' value but only a moderate ε'' value compared with those for other polar solvents.

In practice, different configurations of extraction vessel and microwave oven are employed. In one approach the vessel is open to the atmosphere and the microwave energy is delivered in a focused manner via a waveguide

Table 4.1 Dielectric constants, loss factors and dissipation factors for some common solvents[a] (3 GHz microwave radiation, 25° C)

Solvent	Dielectric Constant, ε'(Fm^{-1})[b]	Loss Factor, ε''(Fm^{-1})[b]	Dissipation Factor, $\delta (\times 10^4)$
Water	77	12	1570
Ice	3.2	0.029	9
0.1 M NaCl in water	75.5	18	2400
Acetone	20.7	11.5	5555
Methanol	23.9	15.2	6400
Ethyl Acetate	6	3.2	5315
CCl$_4$	2.2	0.00088	4
Hexane	1.9	0.00019	1

a data from Zlotorzynski 2005.
b F = farad (unit of electrical capacitance) = m^{-2}.kg^{-1}.s^4.A^2 = C.V^{-1}

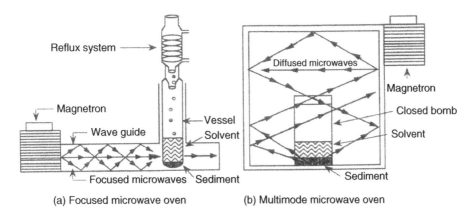

Figure 4.4 Schematic diagrams of different MAE configurations used to extract environmental pollutants from a sediment. (a) Open extraction vessel heated by focused microwaves. (b) Closed extraction vessel heated by microwaves that establish several modes within the reflective walls of the microwave oven. Reproduced from Letellier, *Analusis* **27**, 259 (1999), with permission.

(Figure 4.4(a)); the temperature that can be reached in this way should be the normal boiling point of the solvent but a degree of superheating (a few degrees) is observed with some solvents. Higher temperatures can be obtained by pressurizing in a closed system so as to reach temperatures that can be $50-100\,°C$ higher than the normal boiling point, thus effecting more efficient extraction but with the risks associated with heating liquids in closed vessels; a multimode microwave technique (as used in domestic microwave ovens) is used for closed MAE systems.

An excellent review (Letellier 1999) describes many of the principles of MAE and its applications. The nature of the heating mechanism ensures a high degree of temperature uniformity within the solvent–sample system as long as no shielding of solvent molecules from the microwaves (e.g., by the sample matrix) occurs. This advantage over external heating methods (e.g. Soxhlet) is shared by SAE, as is the appreciably smaller volume of solvent required (although it is possible to recover solvent in modern Soxhlet apparatus, see text box). The main disadvantage of MAE is its restriction to fairly polar solvents that can absorb microwave radiation. Nonpolar solvents like hydrocarbons can not be used without adding some polar solvent to act as absorber, but this can result in co-extraction of undesired compounds in addition to the target analyte(s) and thus require extensive clean-up before the final analysis. Use of a secondary absorber consisting of a solid liner made of a chemically inert fluoropolymer has been described (Shah 2002). This approach certainly overcomes the need for a polar co-solvent but at the price of localizing the heating, although this is inside the extraction vessel in immediate contact with the solvent. Good performance of this approach for extraction

of polycyclic aromatic hydrocarbons (PAHs) in polluted soil is reported (Shah 2002) and the use of MAE in environmental analysis has been recently reviewed (Bélanger 2006).

Of course, thermal stability of analytes becomes a concern for all extraction methods (Sections 4.3.2a–d) that involve extended exposure to boiling solvents. A somewhat different approach is discussed below.

4.3.2e Supercritical Fluid Extraction (SFE)

This is a book about analytical chemistry but, as can be seen in other Chapters, most of the fundamental concepts underpinning the techniques and methods that are used are derived from classical (i.e., non-quantum) physical chemistry, possibly the least favorite topic of science students who require some chemistry courses in their degree program.

'Phys. Chem. is TWO four-letter words!' (Bumper sticker presented to one of the authors.)

To understand the nature of a supercritical fluid it is necessary to briefly discuss the famous Gibb's Phase Rule (see accompanying text box):

$$F = C - P + 2$$

where C is the number of independent chemical components (i.e., those that are not in chemical equilibrium with one another), P is the number of phases (distinguishable states of matter such as gas and immiscible liquids and solids with well defined physical boundaries between them), and F is the number of degrees of freedom (the

Josiah Willard Gibbs and his Phase Rule

Gibbs was a distinguished mathematician and physicist (1839–1903) who was born in New Haven, Connecticut, and spent his entire career at Yale University where he had been a student and won prizes for excellence in Latin and Mathematics. He later obtained his doctorate at Yale, the first doctorate of engineering to be conferred in the USA. After some years in Europe he was appointed as Professor of Mathematical Physics at Yale in 1871. In addition to his seminal work on the theory of physico-chemical equilibria, he made many contributions in physics and pure mathematics including the development of modern vector calculus.

Gibbs' work on the theory of chemical equilibria was based on his concepts of Gibbs Free Energy and its derivative, the *chemical potential*, of a chemical compound. His derivation of his famous Phase Rule was based on the concept of chemical potential, but the brief derivation sketched below avoids details at the expense of rigor.

A *phase* is a part of the system that is immiscible with the other parts (e.g. solid, liquid, or gas); P is the number of phases present. Each phase may contain several *constituents*, simply the distinct chemical compounds in the system; C is the number of these. (If some constituents remain in equilibrium with each other at all times they should be counted as a single constituent.) An *intensive variable* is a property of the system whose value is independent of the size (e.g., temperature, pressure, and chemical composition i.e. *relative* quantities of the constituents). The number of *degrees of freedom* (F) is the number of intensive variables that can be varied without restriction without changing the number of phases present, and thus whose values need to be specified to fully describe the system.

If a multiphase multicomponent system is to be at equilibrium (no change with time of the intensive variables) obviously temperature and pressure must be the same for all phases and also the chemical compositions (mole fractions of each constituent). In any given phase there are $(C-1)$ independent mole fractions (their sum is unity by definition), so there are $P.(C-1)$ composition variables involved and thus $[P.(C-1)+2]$ intensive variables in total. But if *chemical* equilibrium in all phases simultaneously is to hold, the *chemical potential* of each constituent (a function of the composition) must be the same in each phase; thus there are $C.(P-1)$ independent constraints on the composition variables arising from the equilibrium condition (the chemical potential in one of the phases is used as the reference standard for the other phases). Thus: $F = [P.(C-1)+2]-[C.(P-1)] = C-P+2$. This is the famous Gibbs' Phase Rule.

number of independent variables like temperature, pressure, chemical composition, that are intensive, i.e., whose values do not depend on the size of the system). For present purposes the system considered is composed of a single pure chemical compound, i.e., C = 1, so F = 3−P (Figure 4.5). Note in passing that, for three phases (i.e. P = 3) to co-exist at equilibrium, F = 0, i.e. this condition can be found at only one unique combination of temperature and pressure, the so-called triple point that was mentioned in Chapter 1 in the context of the definition of the SI unit of temperature in terms of the triple point of water. Figure 4.5 is a general phase diagram for any

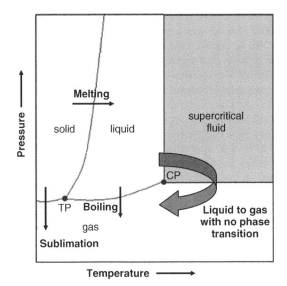

Figure 4.5 Phase diagram for a single pure compound, i.e., C = 1 in the Gibbs' Phase Rule so F = (3–P). The curved lines indicate the temperatures and pressures at which two phases (P = 2) can co-exist in equilibrium, i.e., F = 1, so once either temperature or pressure is specified for such a system the other is fixed at an equilibrium value. The areas marked 'gas', 'liquid' and 'solid' are single phase regions (P = 1) so F = 2, i.e., both temperature and pressure can be varied within each of these regions without changing the number of phases present. If three phases are to co-exist at equilibrium (P = 3) then F = zero, i.e., there is only one unique combination of temperature and pressure at which this is possible; this is the triple point (TP) for this compound. The point CP is the critical point; for temperatures and pressures that are both greater than the critical values there is no longer any distinction between gas and liquid phases.

pure substance, but of course the values of temperature and pressure involved vary from substance to substance.

The important aspect for present purposes concerns the critical point, a combination of temperature and pressure above which there is no distinction between gas and liquid (P = 1 so F = 2) and thus no phenomena like evaporation and condensation. A compound in this condition is said to be a supercritical fluid (SCF) and possesses physical properties intermediate between those of a gas and a liquid (Table 4.2). A SCF retains solvent power properties approximately the same as those of the corresponding liquid, but simultaneously has the transport properties (viscosity, diffusivity) of the gaseous form. This combination makes an SCF ideal for extracting compounds that are soluble in the fluid in its liquid form, from pores and other occlusion sites within a solid matrix that are relatively inaccessible to the liquid.

Table 4.2 Comparison of typical values for some physical properties for a substance in gaseous, liquid and supercritical fluid (SCF) states

	Density ($kg.m^{-3}$)	Viscosity $(cP)^a$	Diffusivity $(m^2.s^{-1} \times 10^6)$
Gas	1	0.01	1–10
SCF	100–800	0.05–0.1	0.01–0.1
Liquid	1000	0.5–1.0	0.001

a $1cP = 1$ centipoise $= 10^{-2}$ poise $= 10^{-3}$ Pa.s $= 10^{-3} kg.m^{-1}.s^{-1}$

In practice, carbon dioxide is the most commonly used SCF because of its moderate critical temperature and pressure (Table 4.3), its nontoxicity, low cost and nonpolluting properties (if we ignore global climate change!). Unfortunately carbon dioxide is highly nonpolar (zero dipole moment as a result of its molecular symmetry), so is limited in the range of compounds that it can extract. This drawback is overcome by judicious use of low levels of additives, typically methanol, to the carbon dioxide; this permits the solvent power to be increased in a somewhat selective manner. On an industrial scale supercritical carbon dioxide with appropriate additives is used for a wide range of applications, e.g., decaffeination of coffee (the so-called 'naturally decaffeinated' coffee). The other increasingly common SCF, particularly for industrial rather than laboratory scale extractions, is water. Its relatively high critical point parameters (Table 4.3) present a considerable barrier for routine laboratory use, but supercritical water has the unexpected and unique property that it is an excellent solvent for organic compounds and a rather poor one for inorganic salts, i.e., the exact converse of the situation for liquid water.

The advantages of SFE over the various liquid extraction methods include the facile 'tuning' of its solute

Table 4.3 Critical point values for some commonly used supercritical fluids

Fluid	Critical Temperature (K)	Critical Pressure (bar)^a
Carbon dioxide	304.1	73.8
Ethylene	282.4	50.4
Ammonia	405.5	113.5
Water	647.3	221.2
Pentane	469.7	33.7
Toluene	591.8	41.0

a 1 bar $= 10^5$ Pa $= 0.98692$ atmospheres $= 14.5038$ psi.

power and selectivity by varying temperature and/or pressure and/or additives, the ready removal of the SCF following extraction by changing conditions so that it enters the gaseous state, absence of residues from toxic solvents in the extract, the possibility of extracting thermally labile components at low or moderate temperatures, and the ability to access occluded sites (pores etc.) as a result of its gas-like transport properties. Disadvantages include the need to maintain high pressures in a controlled manner (requiring complex apparatus) that mitigates against its use in high throughput and/or on-line modes, as well as a limitation on the range of compounds that can be efficiently extracted, particularly for carbon dioxide.

Supercritical fluids are also sometimes used as a chromatographic mobile phase; an example is discussed in Section 11.5.2a.

4.3.3 Solid Phase Extraction from Liquids and Gases

There is a wide range of methods falling under the general heading of solid phase extraction (SPE). Such methods are those most closely related to chromatography in the sense that many of the solids used for selective extraction and concentration of analytes are also used as chromatographic stationary phases. Such phases and associated analytical methods are in constant development, so the present discussion will focus on more general principles; the text box on retention mechanisms in Chapter 3, and later discussion in this Chapter, cover the main classification of retention mechanisms used in SPE and chromatography.

4.3.3a Flash Chromatography

This is a method for preparative column chromatography that is well suited to high throughput clean-up, though it was originally introduced (Still 1978) as a method in synthetic organic chemistry. Flash chromatography differs from conventional gravity-driven column chromatography in two related ways. Slightly smaller particle sizes (250–400 mesh) are used (mesh size refers to the sieve used to separate the silica particles into size ranges, specifically the number of holes per unit area of the mesh, i.e., the larger the mesh size the smaller the particles that pass through). As a result of the lower porosity resulting from the more closely packed small particles, a low pressure drop (10–15 psi), easily applied via laboratory compressed air or water-based vacuum, is required to drive the solvent through the column. The net result is rapid ('over in a flash') and moderate resolution chromatography. An

excellent detailed description of procedures used (Rubin 2000) also includes a reprint of the original paper (Still 1978). In practice most applications of flash chromatography have used normal phase conditions (Section 4.4) with silica or alumina stationary phase (there is a close relationship of normal phase flash chromatography to thin layer chromatography (TLC) separations), but reverse phase packings are increasingly used. Flash chromatography is an exception to the binary 'all-or-nothing' view of SPE clean-up methods as it is a true chromatography requiring fraction collection. However, it is seldom if ever used in trace quantitative analysis methods, fully validated or not, presumably because it is less versatile and efficient than solid phase extraction (Section 4.3.3c).

4.3.3b Purge-and-Trap Analysis for Volatile Organic Compounds

Volatile organic compounds (VOCs) are organic compounds with sufficiently high vapor pressures under ambient conditions to enter the atmosphere in significant amounts. The term (and its abbreviation VOCs) is also sometimes used in a more narrow legal or regulatory sense to include only those compounds that have been shown to interact with nitrogen oxides in sunlight to produce ozone that can cause serious respiratory problems. To confuse the terminology even further, the acronym 'VOC' is also used as an abbreviation for 'volatile organic carbon', i.e., the total amount of carbon present in a volatile form, especially in some biological and ecological contexts. Volatile organic compounds in the more general sense are emitted from a wide variety of products, many of which are widely used household items; indoor concentrations of many VOCs are consistently higher (up to ten times) than outdoors.

Purge-and-trap techniques are designed to evaluate the levels of VOCs present in contaminated water, soil etc. and involve forcing a gas (usually helium) through a sample of water or soil slurry to entrain ('purge') the VOCs. For the best quantitative analyses the water or slurry samples are first spiked with known amounts of surrogate internal standards. The flow of gas containing the entrained volatiles can be fed directly into an analytical instrument (almost always a GC), or can be used with a 'trap' containing a suitable solid sorbent to concentrate samples that are later desorbed thermally into the GC. Commonly used sorbents for this purpose are Tenax™ (a polymer resin based on 2,6-diphenylene oxide), silica gel, and charcoal preparations; a good example of a purge-and-trap method is the United States Environmental Protection Agency (EPA) method for VOCs in aqueous samples (EPA 2003). Similar sorbent tubes are used to concentrate

VOCs from ambient air (or other gaseous environments) by pumping large volumes of gas through the sorbent for specified times. Many rugged validated methods using the purge-and-trap approach have been developed and approved by the EPA.

One of the major challenges in VOC analysis using purge-and-trap methods together with GC/MS involves is the connection of the concentrator to the GC inlet, a result of the vastly different flow rate requirements of the purge and trap apparatus, the capillary GC column and the mass spectrometer (Boswell 1999). Some laboratories choose wide bore (0.53 mm) capillary GC columns with a jet separator (Section 5.2.1) to enrich the GC effluent with respect to the analytes and thus decrease the total carrier gas flow to levels that can be handled by the mass spectrometer vacuum system. This approach is readily adaptable to the high flow rates required to efficiently desorb the trap but has some significant problems, including susceptibility to column contamination by high concentration samples, poor chromatographic behavior of early eluting and closing eluting compounds, long analysis times (run times approaching 40 minutes) and frailty of the jet separator (Boswell 1999). Narrow bore (0.25 and 0.32 mm) GC capillaries, which potentially provide better separations, cannot easily handle the relatively high flow rates from the purge-and-trap concentrator. One way to handle this problem is to insert a cryogenic trap between the concentrator and GC, so as to condense the thermally desorbed analytes from the concentrator before subsequently desorbing them into a much lower flow of GC carrier gas. This cryfocusing approach for narrow bore columns works well, but represents an additional capital expense and adds to the total analysis time. More modern purge-and-trap concentrator designs exploit a conventional split/splitless GC injector (Section 4.4.4b) that is usually installed as a standard item on commercial GC/MS systems and can be connected in series with the purge-and-trap concentrator. The excess flow coming from the concentrator is thus vented at the column inlet, resulting in a reduction in carrier gas flow rate to a value consistent with high resolution (narrow bore) gas chromatography. This approach has been shown (Boswell 1999) to be simple to implement in practice and, combined with optimization of the concentrator desorb time and the GC oven temperature program, it can produce significant improvements in overall performance, reduce susceptibility to column contamination by high level samples, improve chromatographic behavior of early eluting and closely eluting compounds, and permit shorter analysis times. A more recent review (George 2001) describes further developments in this technique.

4.3.3c Solid Phase Extraction (SPE)

In its most widely used sense, SPE refers to one of two complementary extraction–clean-up strategies: (i) selective extraction and concentration of target analytes from a liquid matrix on to a suitable solid phase that does not bind matrix interferences followed by elution of the analytes using a stronger solvent; (ii) selective elution of the analytes through the solid phase with strong retention of impurities on the solid phase. The first strategy is usually chosen when analytes of interest are present at low levels or when the analytes have widely differing polarities, and provides an efficient means of enrichment of analytes present at very low concentrations. The second strategy is normally used when the desired analyte is present at relatively high concentrations. When the sample is particularly complex, a combination of the two strategies is sometimes used. Note also that solutions subjected to SPE can either be raw liquid extracts of, e.g., a solid or liquid matrix, or sometimes the untreated liquid matrix itself.

As mentioned previously this process is essentially 'binary' liquid chromatography, a slightly different exploitation of the same competition for solutes between mobile and solid phases; the ideal 'either 100 % retained or eluted' SPE method corresponds to perhaps $\sim$50 theoretical plates compared with $\sim$10^4 for an HPLC column. Other differences include larger particle size ($\sim$5–10 μm for HPLC, 40–80 μm for SPE), column length (usually 5–30 cm for HPLC although some applications have used shorter columns, and $\sim$1cm for SPE) and degree of concern (essentially zero for SPE) with quality of the packing of the stationary phase (compare the λ and γ parameters in the nonequilibrium theory of HPLC, Section 3.5), extra-column dead volumes etc. The large particles and short lengths imply high porosity and low resistance to liquid flow, so large pressure drops are not required for elution. On the other hand, the low column efficiency means that SPE requires very high (≥ 4) values of selectivity ($\alpha_{A/B} = K^A.V_s^A/K^B.V_s^B$, equation [3.16]) for effective separation of solutes A and B, so that impurities that are somewhat similar to the analyte will not be depleted or removed by SPE. In terms of capacity factors k' (equation [3.17]), selective capture of the target analyte while eluting the matrix interferences implies that k' should be large for the analyte and $\sim$0 for the interferences, and vice versa when impurities are strongly retained while the analyte is eluted.

Of course such ideal conditions are seldom if ever fully satisfied and SPE can give incomplete recovery of analytes (monitored by use of internal standards) and incomplete removal of potential interferences. These problems can be alleviated by adjusting the conditions

such as solvent composition and/or pH, or using a second
SPE phase with different retention properties, e.g., ion
exchange (Section 4.4.1c) followed by a reversed-phase
packing (Section 4.4.1b). So-called mixed mode SPE
combines more than one retention mechanism within
the same stationary phase. For example, a mixed mode
stationary phase based on silica derivatized to provide
both C_8 (reversed phase) and $-NR_3^+$ (strong anion
exchange) functionalities provides selective retention of
analytes containing both nonpolar (hydrophobic) and
acidic (negatively charged) functional groups; an analogous stationary phase with C_8 and $-SO_3^-$ (strong cation
exchange) functional groups is selective for analytes with
both hydrophobic and basic functional groups. An excellent introduction to SPE is available on the internet (Levin
2002) and a recent review (Poole 2003) covers advances
in all aspects together with theoretical approaches
to designing new sorbents. Also, some suppliers of
chromatographic and SPE materials make available
some useful background and specific information (e.g.,
http://www.sigmaaldrich.com/Graphics/Supelco/objects/
4600/4538.pdf).

Ready-made SPE devices are intended for single use
and can be supplied in different formats. (It is important to
be aware, however, that lot-to-lot variation of SPE packings and devices can occur and can lead to problems with
reproducing sample preparation procedures.) The most
conventional format is the SPE tube (Figure 4.6) designed
for manual or partially automated operation. SPE disks
have the sorbent phase embedded in an inert matrix such
as glass fibres, and are recommended for large volume
samples or when high elution flow rates are required.
However, these devices are subject to clogging, a potential
problem for all SPE methods that can lead to channeling
of the stationary phase and loss of separation efficiency.
It is also possible to have short cartridges that are essentially short LC columns, intended to be used on-line with
HPLC columns together with multiport valves that can
derict different solvents through the cartridge and thence
either to waste (for the discarded matrix components) or
on to the analytical HPLC column.

More sophisticated fully automated multichannel
versions of this on-line approach are also available from
several manufacturers. The current ultimate in automated
multiplexed SPE for high throughput operation uses 96-
well plates (or even 384-well plates) with robotic addition of extracts and addition/removal of solvents; a recent
article (Majors 2004) describes developments in this
technology. However, in most instances of rigorously
validated methods, all critical pipetting steps (sample
aliquoting and addition of standards) is done manually

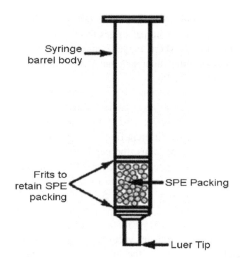

Figure 4.6 Sketch of a typical SPE tube. Various sizes are
available; for extract volumes $\sim$1 mL, 100 mg of sorbent in a
1 mL tube is appropriate (typical values for trace analysis) but
larger sizes are available (e.g., 12–20 mL for fast extraction of
volumes up to 250 mL). The Luer tip is for easy attachment to a
vacuum manifold.

followed by liquid transfer with 96-well devices, otherwise it is necessary to demonstrate adequate precision and
accuracy for each of the individual 96 channels. Also,
well-to-well contamination can be a major problem and
must be assessed and monitored (Section 9.7.1). As for
any technology, the 96-well plate approach must be carefully assessed to determine fitness for purpose.

The practical aspect of a SPE clean-up that should
not be overlooked is the multistep nature of any such
procedure designed to simultaneously achieve concentration of the target analytes, selective removal of interfering matrix contaminants, and preparation of a final
solution that is more concentrated than the pre-SPE solution and suitable for injection into an analytical instrument (chromatography–MS in the context of this book). These
steps are outlined in a generalized way in the accompanying text box. A typical cleaned-up extract would
provide concentration of $\sim$10-fold for the analyte(s) in
addition to the clean-up (selective complete or partial
removal of matrix components). Note that, in the case
of analytes to be analyzed by reverse phase HPLC, the
SPE extract is generally in a solvent with higher elution
strength (Section 4.4.2a) than that of the HPLC mobile
phase; to maintain optimum chromatographic performance, the solvent in the injected extract should have
elution strength no higher than that of the mobile phase,
and this requires that the elution solvent used in SPE must

Generalized Procedure for Solid Phase Extraction (SPE)

Regardless of whether SPE is conducted in high throughput mode using 96-well plates, or using standard cartridges with either manual manipulations or some semi-automated version, the general approach requires several distinct steps to be optimized and coordinated to achieve the overall objective. This involves selectively depleting matrix components that potentially could interfere with the final analysis while increasing the analyte concentration compared with that in the raw (pre-SPE) extract (or sample), in a solvent that is compatible with subsequent chromatography.

Pre-conditioning. Wash the SPE phase (cartridge, disk, plate wells) with several 'hold-up volumes' of the solvent to be used to elute the analyte, followed by the same volume of the noneluting solvent. (Hold-up volume is the volume of noneluting solvent required to flush a nonretained compound from the SPE phase). If the procedure is intended to first retain and then elute the analyte, rather than vice versa, the first wash would use a solvent of high elution strength (e.g., methanol or acetonitrile for reverse phase systems) and the second wash would use the original extract solvent (usually aqueous buffer with low organic content).

Loading. An 'appropriate' volume of raw extract (or analytical sample), preferably spiked with internal standard at an earlier stage (Figure 1.2), is loaded on the SPE phase. The major question concerns what volume is appropriate in any given case, as it is important to not overload the SPE phase with strongly retained solutes (not necessarily restricted to the target analyte(s)). A commercial SPE device will usually specify the recommended maximum loading, but if a bulk phase is used to construct in-house devices the supplier should specify the total surface area of the phase (e.g. $3 \times 10^5 \, \text{m}^2.\text{kg}^{-1}$) and the coverage of active partitioning agent (e.g. tethered C_{18} content, $3 \times 10^{-6} \, \text{mol.m}^{-2}$); the product of these two parameters gives the total concentration of sites ($0.9 \, \text{mol.kg}^{-1}$), so a 100 mg cartridge would contain 9×10^{-5} mol of partitioning sites and this represents the theoretical maximum loading of strongly retained compounds. It is a good rule of thumb to load no more than $\sim 5\,\%$ of this theoretical maximum to ensure that all of the target analyte molecules are retained. Most ion exchange SPE phases contain $\sim 0.2 \, \text{mol.kg}^{-1}$ of active sites (for singly-charged ions, this value must be divided by the number of charges per ion).

Cleaning Wash. To selectively remove the nonretained compounds, wash the SPE phase with the same volume of noneluting solvent (e.g. aqueous buffer for reverse phase systems) as that used in the pre-conditioning step. Some analyte might be washed out also in this step so this eluant can be retained for secondary SPE treatment if desired; however, a suitable SIS (Table 2.1) will correct for any such losses so this eluant is often discarded.

Elution. To recover the analyte (inevitably combined with any other strongly retained compounds) wash the SPE phase with the same volume of the strong elution strength solvent as was used in the pre-conditioning step, and retain all of this eluant.

Preparation of Final Cleaned-Up Extract. The objective of this final step is to prepare the resulting solution of target analyte (plus any other components that were strongly retained on the SPE phase) in as concentrated a solution as possible (using gentle solvent evaporation, e.g., under a stream of nitrogen) without risking any solubility issues, and in a solvent compatible with the subsequent chromatography–MS analysis. For GC/MS this solvent need only be volatile and be essentially non-retained on the GC column. For HPLC–MS this final solvent should have an elution strength no higher than that of the HPLC mobile phase in order to maintain the separation efficiency, and for SPE (also for LLE) this usually implies that the elution solvent used for SPE must be completely removed and replaced by, e.g., the HPLC mobile phase itself.

Naturally details will vary with the nature of both analyte and matrix, and with the overall purpose of the analysis, but the principles underlying these five steps are important in all cases.

---

be completely evaporated and the dry residue re-dissolved in, e.g., the HPLC mobile phase itself. (However, for highly polar analytes for which hydrophilic interaction chromatography (HILIC, Section 4.4.2c) is applicable, direct injection of the primary SPE extracts is not a problem (Weng 2002).) The solvent-switching step using gentle evaporation under a nitrogen stream is similarly required for LLE (Section 4.3.1).

SPE procedures are generally time consuming compared with, e.g., simple liquid–liquid extraction, but this disadvantage is partly compensated by the fact that SPE is easier to automate for simultaneous batch processing of multiple samples. However, the multiple steps involved in SPE (see text box) do increase the chances for errors and, as a result, the training of operators for reliable and reproducible use of SPE methods is more demanding than for LLE. SPE avoids the potential emulsion problems that can arise with LLE and also requires use of much lower quantities of organic solvent; the latter is an advantage with respect to safety and environment, but not necessarily with respect to cost of consumables. Because of the large variety of stationary phases available (see later) SPE offers greater flexibility than LLE in designing clean-up strategies via greater selectivity.

Miniaturization in analytical chemistry has prompted development of novel formats for sample preparation, including SPE. The micropipet tip format permits handling of sub-μL volumes of samples such as biological fluids (Majors 2005). Various SPE phases are available packed, embedded or coated on the walls of a conventional pipet tip (see Chapter 2), although at present the C_{18} tip is probably the most common; however, these devices are not generally used in validated quantitative analyses but can be useful for exploratory work. Liquid samples can be moved and transferred without plugging or requiring a significant pressure drop. In addition to SPE, micropipet tips have also been used to miniaturize dialysis and enzyme digestion. One of the main advantages of using micropipet tips is that they can be used with single or multichannel manual micropipettors or in liquid-handling robots, and this has resulted in routine use of micropipet tip-based SPE in bioanalytical laboratories for high throughput applications that do not demand rigorous validation (Majors 2005).

4.3.3d Turbulent Flow Chromatography

A new on-line approach to SPE is the use of so-called turbulent flow chromatography which combines rapid mobile phase linear velocities with larger particle sizes; this approach was discussed in Section 3.5.9 in the context of breakdown of the conventional (van Deemter) rate theory. Whether or not the flow can truly be described as turbulent (Ayrton 1998), there is no question that eddies are formed that enhance the interactions of smaller molecules with the stationary phase. In contrast the large proteins and other biopolymers have rates of mass transfer from the liquid to stationary phase (C term in the van Deemter equation) that are too

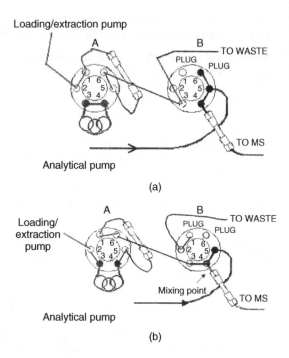

Figure 4.7 Configuration of a typical dual column, two valve, on-line extraction LC/MS system: (a) loading samples and (b) eluting, mixing and analyzing extracted compound(s). Reproduced from Zhou *et al., Rapid Commun. Mass Spectrom.* **19** (2005) 2144, with permission of John Wiley & Sons, Ltd.

low to compete with the high linear velocity are therefore selectively swept out of the column. This serves as a useful clean-up procedure for small molecule pharmaceuticals and their metabolites in plasma, urine etc. Figure 4.7 shows a typical arrangement of multiport valves and columns that allows selective removal of large proteins on-line with conventional reverse phase HPLC analysis of the small molecule analytes (Zhou 2005). Figure 4.8 illustrates the quality of the performance obtained (Zhou 2005) for analysis of plasma for three related compounds in which the 'turbulent flow' on-line extraction was combined with a monolithic reverse-phase analytical column, resulting in extremely low background and very short analysis times. An example of an entirely different application is illustrated in Figure 4.9, where successful analysis of pesticides at $ng.L^{-1}$ levels in river water was achieved using a similar approach (Asperger 2002). A list of publications describing a wide range of applications can be found at the website of a company that supplies equipment and columns for this approach   (http://www.cohesivetech.com/technologies/turboflow/index.asp).

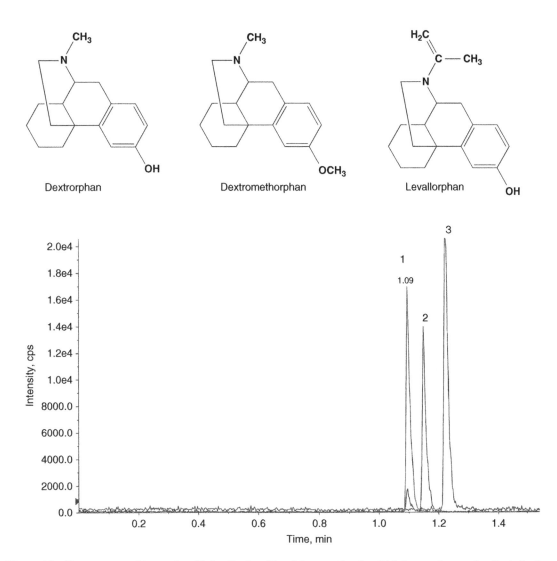

Figure 4.8 Chromatogram of dextrorphan (1), levallorphan (2) and dextromethorphan (3) in human plasma using the dual column on-line extraction LC/MS/MS arrangement shown in Figure 4.7. The extraction column was a large-particle C_{18} column, and the analytical column a monolithic column. During the analysis step, the extraction loop on valve A (Figure 4.7) was refilled with 0.1 % formic acid in methanol in readiness for the next sample. Reproduced from Zhou *et al.*, *Rapid Commun. Mass Spectrom.* **19** (2005) 2144, with permission of John Wiley & Sons, Ltd.

There are two different modes in which on-line extraction/clean-up by turbulent flow chromatography is performed (Zhou 2005). In the simpler mode, single column extraction, the analytical sample or primary extract is injected directly onto the extraction column and extracted directly onto the analytical column by the turbulent flow mobile phase that has high eluent strength (Section 4.4.2a), i.e., high organic content in the case of analytes to be analyzed by reverse phase HPLC. This approach permits very high throughput and could possibly reduce carryover (Section 9.13), but the use of strong solvents in both extraction and HPLC steps implies that the chromatographic resolution is low. For applications in which this is potentially a problem, e.g., if analytes must be chromatographically separated from other constituents that can cause ionization suppression (Section 5.3.6a), or from metabolites whose ions can be converted to the analyte ions by in-source CID (Section 11.5), a dual

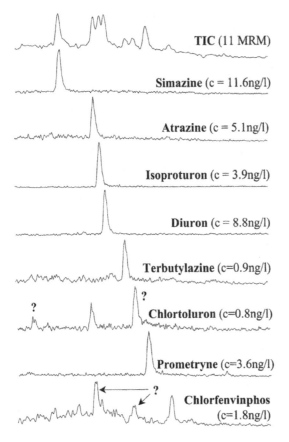

TIC (11 MRM)

Simazine (c = 11.6ng/l)

Atrazine (c = 5.1ng/l)

Isoproturon (c = 3.9ng/l)

Diuron (c = 8.8ng/l)

Terbutylazine (c=0.9ng/l)

?

?   Chlortoluron (c=0.8ng/l)

Prometryne (c=3.6ng/l)

?   Chlorfenvinphos (c=1.8ng/l)

Figure 4.9 Total ion current chromatogram and extracted MRM chromatograms for eight pesticides in 10 mL of river water (River Parthe, February 10, 2001, Leipzig, Germany), analyzed by on-line 'turbulent flow' extraction and analysis using a monolithic column. Reproduced from Asperger *et al., J. Chromatogr. A* (2002), **960**, 109, copyright (2002), with permission from Elsevier

column mode must be used (Zhou 2005). In this mode the analytes are extracted with solvent stored in an extraction loop, and are then transferred to an analytical column containing a weak mobile phase flowing at a relatively high rate in order to dilute the strong extraction solvent and focus the analytes onto the analytical column. The analytical separation can then proceed with minimal distortion from the strong extraction solvent, but at the expense of significantly longer analysis times.

4.3.3e Molecularly Imprinted Polymers (MIPs)

A type of SPE stationary phase, still in its experimental development phase, that is not often used in chromatography is the molecularly imprinted polymer. This approach is described here rather than in the general discussion of stationary phases. Useful reviews of the subject (Ensing 1999, 2001; Stevenson 1999; Owens 1999; Ramström 1997; Spégel 2002) cover developments of this technology as they apply to SPE and to other applications such as biosensors, highly selective chemical catalysis (analogous to enzymatic catalysis) etc.

The basis of the concept is derived from what has come to be known as the Pauling theory of formation of an antibody to a specific antigen. (Although Pauling's theory is now known to be incorrect, and has been supplanted by the clonal-selective theory, it still provided the inspiration for the MIP concept.) According to Pauling's theory (Figure 4.10), those protein chains that are the nascent antibodies encounter an antigen and, guided by their characteristic structure or partial structure (the epitope), fold in such a manner as to closely fit their own three-dimensional shapes to that of the epitope thus allowing strong

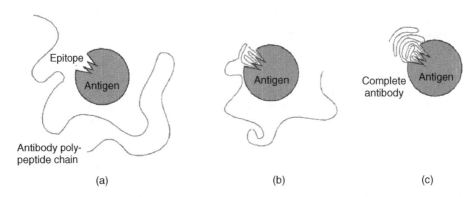

Figure 4.10 Pauling theory of formation of an antibody to an antigen (now known to be incorrect, replaced by the 'clonal selective' theory). (a) The polypeptide chain that will eventually form the antibody encounters the antigen. (b) The peptide chain starts to fold in a manner guided by the three-dimensional characteristics of the antigen. (c) The fully folded antibody is completed. Reproduced with permission from Dr. Olof Ramström (1997), The Royal Institute of Technology, Stockholm.

binding of the antigen. (Antigens are usually proteins or polysaccharides but can be small molecules (haptens) coupled to a carrier protein.) Regardless of the mechanism of formation of antibodies, the strong binding of antigen to antibody is the principle underlying immunoaffinity columns for SPE and chromatography (Section 4.4.1e).

The MIP approach uses the same 'lock-and-key' concept, but is based on polymerization chemistry conducted so as to be adapted to the three-dimensional shape and chemical functionalities of the analyte of interest. MIPs can be produced using a covalent (Wulff 1972) or a noncovalent (Arshady 1981) strategy (Figure 4.11). The great advantage of MIPs in SPE is the very high selectivity. However, they do have disadvantages. Large quantities of analytical standard are required for the MIP synthesis, and the high selectivity means that for multiple analytes a MIP has to be produced for

each. Another problem (Owens 1999) arises from template (analyte) species remaining in the MIP after polymerization, even with careful washing (Figure 4.11), which reduces column capacity (not a major problem for trace analysis) and leakage of remaining template species to contaminate the sample. The latter problem can be alleviated by employing a chromatographically distinguishable structural analogue of the analyte as the template molecule, sacrificing some selectivity for analytical accuracy and precision. For small molecule analytes, SPE using a MIP is about as effective as that obtained using immunoaffinity, but suitable antibodies are often difficult to produce and are considerably less robust than MIPs (Ye 2000). The same problems apply to use of MIPs as chromatographic stationary phases, but they do appear to hold some promise for enantioselective separations of racemic mixtures of optically active compounds (Section 4.4.1d).

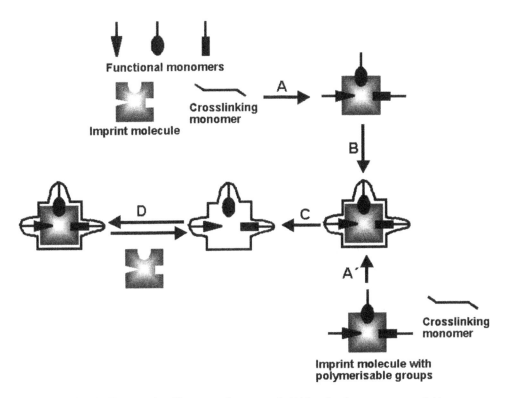

Figure 4.11 Schematics of MIP preparation. *The noncovalent approach:* (A) functional monomers, cross-linking monomers, radical initiator and imprint molecule are mixed in a suitable solvent to allow noncovalent complexes to form between the functional monomers and the imprint molecule; (B) the functional monomers are locked in position by the polymerization reaction; (C) after having extracted the imprint molecule (D) the MIP is able to again recognise the imprint molecule. *The covalent approach:* (A') imprint molecules with covalent substitutions of polymerizable groups are mixed with cross-linking monomers and radical initiator in an appropriate solvent and polymerization is initiated; (C) after polymerization is complete the covalent bonds between the imprint molecule and the MIP are broken; (D) the MIP is able to re-bind the imprint molecule using covalent bonds. Reproduced from Spégel *et al., Anal. Bioanal. Chem.* **372** (2002), 37, copyright (2002), with kind permission from Springer Science and Business Media.

Their use in validated quantitative methods will have to wait for these problems to be addressed.

4.3.3f Solid Phase Microextraction (SPME)

SPME is a very simple and efficient sample preparation method, requiring no solvent (at least for GC applications). Following its introduction (Belardi 1989; Arthur 1990), SPME has been widely used in different fields of analytical chemistry since its first applications to environmental and food analysis. A recent extensive review (Vas 2004) covers the principles and applications of the technique. Other reviews (Queiroz 2004; Hinshaw 2003; Wang 2004; O'Reilly 2005; Ouyang 2006) describe special aspects. All conventional steps of extraction, concentration, possibly derivatization, and transfer to a GC are integrated into one step and one device, considerably simplifying the sample preparation procedure. (If HPLC is to be used, SPME more closely resembles conventional SPE in a special miniaturized format.)

SPME uses a fused silica fiber that is coated on the outside (occasionally internally if a capillary is used) with an appropriate stationary phase, typically an immobilized polymer, a solid adsorbent, or a combination of the two. A wide range of coatings is available, but probably the most widely used is polydimethylsiloxane (PDMS), well known as a GC stationary phase that is thermally stable (can be used in a temperature range $20-320\,^\circ$C). The analytes in the sample (liquid or gas) are directly extracted to the fibre coating, and are then thermally desorbed directly into the injection port of a GC or are extracted from the fiber using a suitable solvent for HPLC analysis; this process provides high sensitivity because the complete extract can be analyzed.

It is important to note that SPME, like conventional SPE, is based upon distribution of analytes between the analytical sample (gas or liquid) and the fiber coating and that the distribution process proceeds at a finite rate. Thus the extraction efficiency has a theoretical maximum determined by the distribution coefficient K (Chapter 3) and, if extraction time must be minimized for reasons of desired throughput so that equilibrium is not reached, precise timing of the extraction process is crucial particularly if chosen to correspond to the steep initial rise before approaching the equilibrium plateau. The extraction kinetics are strongly influenced by sample size and fiber parameters such as diameter and length as well as coating thickness; the extraction time for liquid samples can be decreased by stirring, ultrasonic assistance etc. Probably the most important feature determining the analytical performance of SPME is the type and thickness of the coating material; a thicker coating requires a longer extraction time but the recoveries are generally higher. The time of extraction for a specified fractional recovery is independent of the absolute concentration of analyte in the sample, and the relative number of molecules extracted at a given time is also independent of the concentration of analyte (Vas 2004). Originally SPME was mainly used for qualitative or semiquantitative work, but better quantitation is possible with use of appropriate internal standards. Most sample preparation methods attempt to achieve 100 % extraction efficiency, but SPME is an exception and improved reproducibility and precision require careful control of extraction time and temperature. For this and other reasons it will be difficult to incorporate SPME into rigorously validated quantitative methods.

In-tube SPE is a variant, developed for coupling with HPLC, that uses an open tubular capillary column instead of a fiber; it can continuously perform extraction, desorption and injection using a standard autosampler, i.e., is suitable for automation, and commercial apparatus for this purpose is available from several suppliers. Organic compounds in aqueous samples are directly extracted into the internal coating of the open capillary column and then desorbed by introducing a stream of mobile phase (or a static desorption solvent for analytes that are more strongly adsorbed to the coating). With the in-tube SPME method it is necessary to remove particulates from the sample before extraction to prevent plugging, in contrast with fiber SPME where particles can be removed after extraction by washing the fiber with water before insertion into the SPME/HPLC interface. However, a disadvantage of SPME fibers is their fragility, as they easily break and/or the fiber coating can be damaged; a recent method for overcoming this limitation (Azenha 2006) uses sol-gel deposition on titanium wires. Furthermore, high molecular mass compounds from biological matrices can be adsorbed irreversibly on the fibers, thus changing their properties. An advantage of in-tube SPME over fiber SPME for use in combination with HPLC is the possible decoupling of desorption and injection. In the fiber SPME method analytes are desorbed during injection as the mobile phase passes over the fiber directly into the HPLC column; in contrast, in the in-tube SPME method analytes are first desorbed from the SPME device by mobile phase or other desorption solvent, and then transferred to the HPLC column by mobile phase flow. As a result, peak broadening is low relative to that obtained with a fiber because the analytes are completely desorbed before injection; as an added bonus the carryover (Section 9.7.2) for this 'flushing' approach to injection can be 0.1 % or less (Vas 2004).

Another variant of SPME is designed to overcome one of its fundamental limitations particularly in its fiber implementation, namely, the actual amount of sorption medium (e.g., PDMS) available for extracting the analyte. For a typical $100\,\mu$m fiber the volume of extraction phase in the coating is approximately $0.5\,\mu$L. As emphasized in a recent review (David 2003), this means that if a large volume of aqueous sample is to be extracted and/or if the analyte has a reasonably high solubility in the aqueous phase, or has a high affinity for the glass sample container or other components in contact with the sample, the fiber will not compete well for the analyte as a result of the limited amount or sorption phase. This can be demonstrated by the following simple considerations of the appropriate distribution equilibrium between the aqueous and adsorption phases; note that this simple approach is applicable (with appropriate modifications) to all SPE procedures but is particularly relevant for SPME:

$$K^A_{p/w} = c^A_p/c^A_w = (m^A_p/V_p)/(m^A_w/V_w)$$

so that:

$$m^A_p/m^A_w = K^A_{p/w}.(V_p/V_w)$$

where $K^A_{p/w}$ is the distribution coefficient for analyte A between the aqueous and SPME adsorptive phases, c^A_p and c^A_w are the equilibrium concentrations of A in the adsorptive and aqueous phases, respectively, V_p and V_w are the volumes of the two phases, and m^A_p and m^A_w are the masses of A found in the two phases at equilibrium. (Strictly, the thermodynamic activity coefficients required for an accurate description of the equilibrium are being ignored, but this is not important for the present purpose of drawing generalized qualitative conclusions.) The extraction efficiency $(E_{p/w})$ can be defined as the fraction of the total amount of A that is found in the SPME sorptive phase:

$$E_{p/w} = m^A_p/(m^A_w + m^A_p)$$
$$= (m^A_p/m^A_w)/[1 + (m^A_p/m^A_w)]$$
$$= K^A_{p/w}.(V_p/V_w)/[1 + K^A_{p/w}.(V_p/V_w)]$$

This expression for efficiency is the theoretical maximum, assuming that the extraction procedure is allowed to proceed to equilibrium (see above). If it is sought to increase $E_{p/w}$ by increasing V_p, in general the equilibration rate will be slower.

4.3.3g Stir-Bar Sorptive Extraction (SBSE)

To address the problems intrinsic to SPME, i.e., to increase V_p and thus $E_{p/w}$ while avoiding long extraction

times, SBSE was developed (Baltussen 1999). The sorptive phase is coated on the glass envelope of a magnetic stir-bar with volumes V_p that are 50–250 times greater than those on a typical SPME fiber, giving correspondingly higher sensitivities. The extraction is done by spinning the bar within the aqueous sample using a conventional laboratory magnetic stirrer; the stirring speed has a strong effect on the extraction time. A comprehensive review (David 2003) provides further details and a review of applications of the technique to environmental, bioanalytical and food analysis problems. Even nonvalidated quantitation again relies on use of suitable internal standards. A major disadvantage of SBSE is the requirement for a relatively complex apparatus for thermal desorption of the analyte from the stir-bar into a GC or for extraction of analyte into a solvent suitable for HPLC analysis, and consequent difficulties in arranging for automation.

4.4 Chromatographic Practicalities

Chapter 3 was devoted to acquiring some understanding of the physico-chemical theories of chromatography, although some practical aspects were discussed for cases with particular relevance to limitations of the theory (e.g. 'turbulent flow' chromatography). Here, some real-world issues are described. The following account is clearly not exhaustive, but hopefully it will provide some help and guidance to readers who are about to start, or have already started, the process of learning by doing, as already mentioned several times in this book. One of the freely available internet resources already mentioned (Kazakevich 1996) is particularly helpful in these practical aspects, though it is now a little out of date in some cases.

4.4.1 Stationary Phases for SPE and Liquid Chromatography

The present section adds more detail to the classification of retention mechanisms in HPLC (Chapter 3). Recall that an important distinction in HPLC is that between normal phase chromatography (in which the mobile phase is less polar than the stationary phase) and reverse(d) phase (both usages are found in the literature) chromatography (mobile phase is the more polar phase); the same general discussion applies also to SPE, although larger particles and particle pore sizes are used in that case. This distinction does not arise in GC, for which the mobile phase is almost always helium (occasionally hydrogen or other gases). The concern here is mainly with HPLC column packings that are suitable for retention of types of analytes commonly subjected to trace quantitative analysis (molecular mass generally $< 1000\,$Da). Currently the main analyte type that is larger than this limit is proteolytic peptides that have been determined to be surrogates for their parent proteins (Section 11.6). However, in that case the chosen surrogate peptides are always sufficiently polar that they are separated using appropriate reverse phase techniques, and they are generally ionized as the doubly charged forms that mostly fall into the optimum m/z range of the common mass spectrometers. Larger biopolymers can not be quantitated directly and are analyzed qualitatively using gel permeation (GP) and size exclusion chromatography (SEC), which separates molecules based on their sizes (more accurately, their hydrodynamic volumes). SEC is widely used in purification (and to some extent analysis) of biopolymers using a gel stationary phase, usually polyacrylamide, dextran or agarose, under low pressure conditions (gravity-fed flow). GP chromatography is a term usually used to describe separation and analysis of synthetic polymers dissolved in an organic solvent, typically using a silica or crosslinked polystyrene stationary phase and somewhat higher pressures.

A more recent development that has yet to be widely accepted in the trace quantitation community is that of HPLC analysis at higher temperatures using stationary phases based on metal oxides (mainly zirconia and some titania). The many experimental adjustments that must be made to achieve the potential advantages (faster separations, partial or complete elimination of the need for organic solvents in the mobile phase etc.) are described in a recent review (McNeff 2007). However, no account of interfacing this novel HPLC approach to mass spectrometry appears to have been published as yet.

4.4.1a Alumina and Silica Particles

Normal phase (sometimes called 'adsorption') chromatography for nonpolar solutes is conducted on stationary phases of specially prepared silica (either unmodified or derivatized silica gel), alumina, magnesium or aluminum silicate, or hydrophobic cross-linked polymers (e.g. copolymers of styrene with divinylbenzene to provide the cross-linking). Alumina has the advantage that it is stable over a wide pH range, whereas silica becomes increasingly soluble above about pH 6 with a dramatically steep rise in both solubility and dissolution rate above pH 7. However, it is difficult to chemically derivatize alumina, a result of only a few free hydroxyl groups on the solid surface, thus restricting its range of applications. Alumina is also sensitive to the amount of water bound to it; the higher its water content, the fewer polar sites, and thus the less retention for organic compounds; the activity range is designated I, II or III, with I being the most

active. Alumina is often purchased with activity I and subjected to controlled deactivation with water before use according to specific procedures. Alumina also comes in three chemical forms, acidic, neutral and basic. Of course the (nonpolar) mobile phase solvents used with alumina must be carefully dried in order to obtain reproducible behavior.

There are two main types of silica, both formed as silica gels of general formula $SiO_2.(H_2O)_x$, i.e. water is bound in a nonstoichiometric ratio. The so-called sol-gel type is made from a silica sol, a colloidal suspension of particles of size 1–100 nm formed by polycondensation of solutions of soluble silicates brought to pH ~8; these colloidal particles are nonporous and amorphous (i.e. noncrystalline). Manipulation of the sol then leads to aggregation of the colloidal particles into a silica hydrogel, a three-dimensional structure of sol particles linked together and containing considerable amounts of water. This hydrogel is washed to remove metal ions etc. and is then heated and treated further to drive off the water and form a xerogel, a rigid form of silica of high porosity as a result of its formation from small nonporous sol particles linked together. Formation of silica particles suitable for stationary phases is done by milling the xerogel with subsequent sieving to obtain various particle sizes. This process leads to irregularly shaped particles of moderate particle porosity that, when used as a stationary phase, are found to provide lower chromatographic efficiency than comparable spherical particles. The latter are usually more expensive and are sometimes confusingly referred to as the sil-gel type; they are manufactured by emulsifying a silica sol by rapid mixing in an immiscible organic liquid and conversion of the droplets thus formed into beads of gelatinous hydrogel. The size of the droplets is controlled via the viscosity of the liquid, and the spherical hydrogel beads are formed into solid particles of high porosity by appropriate heat treatment. Another difference between the two types concerns the nature of the free hydroxyl (silanol) groups on the surface (Figure 4.12), mostly isolated (nonbonded) in the sil-gel type and associated (bonded) in the sol-gel type.

The importance of particle size in packed column chromatography was discussed in Chapter 3. Only the surface of the stationary phase participates in the retention process, so chromatography (and SPE) are dynamic interface (adsorption) phenomena. As a result, other important parameters are the pore size and pore size distribution, since these determine the adsorbent surface area available to those analyte molecules that can penetrate inside the particle. This is important because the ratio of the outer particle surface to its inner one can be about 1:1000, i.e., the surface molecular interaction with analytes mainly

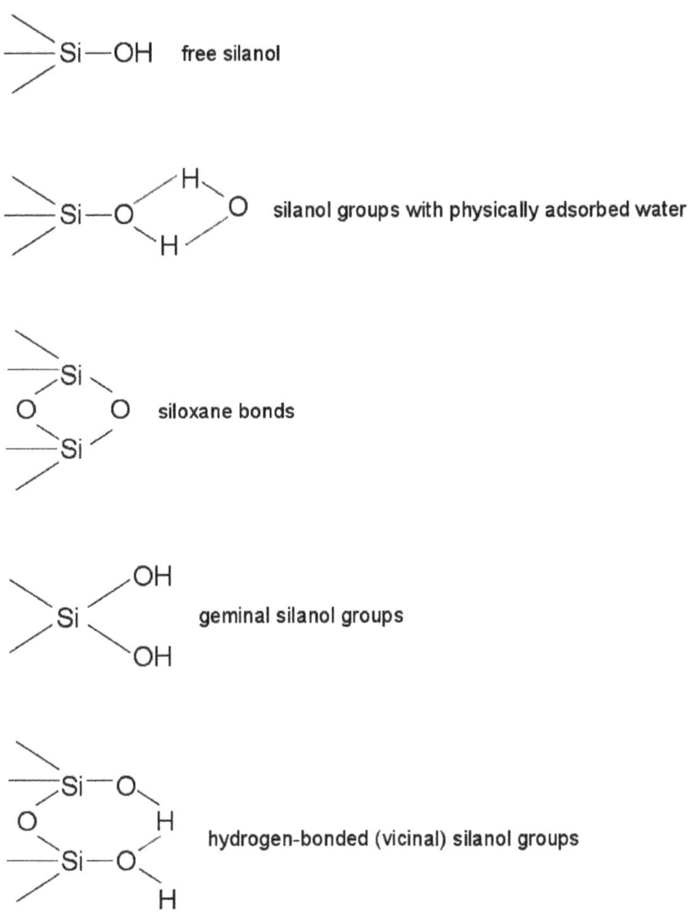

Figure 4.12 Structures of various silanol groups on the surface of silica gel particles.

occurs on the inner particle surfaces. Pore size distribution is a secondary parameter; for HPLC the most important aspect is the absence of micro and mesopores, arbitrarily defined as pores with diameters less than 10 Å and 50 Å, respectively. Micropores are generally too small to be accessible to analyte molecules while mesopores are partially accessible, but this access is sterically hindered and as a result the rate of mass transfer (accounted for by the C term in the van Deemter and Knox equations, see Chapter 3) is significantly slowed, and column efficiency is thus decreased. Other relevant parameters are the specific pore volume V_{pore} (the sum of volumes of all pores in one gram of adsorbent, i.e. only the internal volume inside the adsorbent particles), the specific surface area S, and mean pore diameter D_{pore}. These parameters are correlated to one another in a way dependent on the pore shape. Typical values are

$V_{pore} \approx 0.5–1.0\,\mathrm{mL.g}^{-1}$, $D_{pore} \approx 100–300\,\text{Å}$, $S \approx 100–300\,\mathrm{m}^2.\mathrm{g}^{-1}$.

These unmodified silica particles are used as stationary phases in normal phase chromatography and SPE but, as for alumina, the role of water adsorbed on the strongest sites must be kept in mind. Most of the normal phase chromatographic properties of the silica surface are related to the interactions with silanol groups, although siloxane and hydrogen bonded silanols (Figure 4.12) may also contribute. In keeping with the highly polar nature of this stationary phase compared with typical normal phase solvents (see discussion of eluotropic strength, Section 4.4.2a), retention mechanisms are based on dipole/induced-dipole forces and dispersion forces (i.e. fluctuating dipoles arising from molecular polarizability). As a result a typical elution order in normal phase chromatography is found to be: saturated hydrocarbons < olefins < aromatic hydrocarbons ≈ organic halides < sulfides < ethers < nitro compounds < esters, aldehydes and ketones < alcohols and amines < amides < carboxylic acids.

4.4.1b Derivatization of Silica for Normal and Reverse Phase Chromatography

An important property of silica particles for chromatographic stationary phases is the reactivity of the hydroxyl groups on the surface, which permits derivatization to provide a wide range of properties. The silanol groups are weakly acidic, with a concentration on the surface of porous silica $\approx 8\,\mu mol.m^{-2}$, i.e. $\approx 5\,Si-OH$ groups per $100\,\mathring{A}^2$, corresponding to an average inter-silanol separation $\approx 4-5\,\mathring{A}$. Derivatization is almost always achieved by reaction of the hydroxyl groups with organosilanes:

$$-Si-OH + X_n SiR(CH_3)_{3-n}$$

$$\rightarrow \quad -Si-O-SiX_{n-1}R(CH_3)_{3-n}(+HX)$$

where X is a reactive group (usually chlorine) and R is the functional group to be attached to the silica surface. The reaction indicated above is the only one possible when $n = 1$, e.g., a monochlorosilane, leading to just one alkyl-silane ligand per hydroxyl group (monomeric bonding). So-called polymeric phases can be produced in the presence of trace amounts of water when $n = 2$ or 3; such phases correspond to formation of tree-like structures on

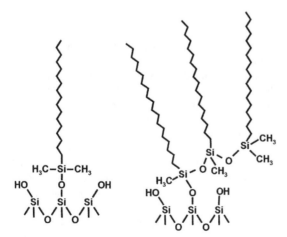

Figure 4.13 Examples of monomeric and polymeric phases formed by derivatizing silanol groups on a silica surface with organochlorosilanes.

a single silanol group (Figure 4.13). Such bulky organosilane groups can not react with all the silanols on the surface as a result of steric effects ($\approx 4-5\,\mathring{A}$ inter-silanol separation).

Carbon loading in a sample of silica-based stationary phase is the percentage by mass of carbon atoms in the derivatized sample (usually measured by combustion elemental analysis); typical values for $R = octyl(C_8)$ are 6–9 %, and for $R = octadecyl(C_{18})$ 8–12 % with some phases as high as 18 %. Some of the remaining under-ivatized silanols are accessible to analytes and these highly polar sites can seriously affect the chromatographic performance in reverse-phase mode. Accordingly such phases are subjected to end capping whereby trimethylchlorosilane, a molecule small enough to minimize steric hindrance effects, is used to react with those of the previously underivatized silanols that this reagent can access. Although end capping still leaves many free silanol groups, these are usually not accessible to analyte molecules that are generally at least as large as trimethylchlorosilane. As a result the reverse phase chromatographic properties are significantly improved, with less peak tailing (Section 3.5.12) and more predictable and reproducible retention of polar analytes.

A great deal of creativity is still devoted to devising new HPLC/SPE stationary phases with unique properties, and here only those that are currently most commonly used are referred to. For polar functional groups R in the organosilane, the resulting derivatized silicas are suitable for normal phase separations. The most common such phases are 'cyano' (nitrile) with $R = (CH_2)_3CN$, 'amino' with $R = (CH_2)_3NH_2$, and 'diol' with $R = (CH_2)_2OCH_2CH(OH)CH_2OH$. These phases provide normal phase separations somewhat similar to those of unmodified silica though usually permitting shorter column equilibration times in gradient elution. Diol columns are slightly polar and exhibit less retention for polar compounds than silica itself, and are thus useful for complex mixtures of analytes with a range of polarities. Amino columns are useful for normal phase separations and also have weak anion exchange properties. Cyano phases are slightly polar and are typically used for fast separations of mixtures of analytes with very different retention times on most other stationary phases; they are also sometimes used for reverse phase separations.

For nonpolar functional groups R, the most commonly used choices for reverse-phase separations are alkyl-bonded phases with $R = C_8H_{17}$ and $C_{18}H_{37}$, though $R = CH_3$ and C_4H_9 have applications for protein separations. (Note that the $-Si(CH_3)_2C_{18}H_{37}$ functional group is frequently referred to as octadecylsilyl and abbreviated ODS). So-called phenyl columns with $R = (CH_2)_3$–phenyl show lower reverse phase retention than C_8 or C_{18}, but also exhibit weak dipole-induced dipole interactions with polar analytes and can thus be used in normal phase conditions also.

Reverse phase HPLC is by far the most widely used HPLC method, a result of its ability to separate a wide variety of organic compounds, often including compounds of closely similar structures. For this reason it is worthwhile to add a few comments concerning some of the stationary phase parameters that are important in this context (Kazakevich 1996). Pore size (D_{pore}) is important because, if it is too small, larger analyte molecules can not access the in-pore surface (by far the greatest contributor to the total surface area); in particular derivatization with bulky C_{18} groups decreases the effective pore size considerably, so for larger analytes it is advisable to choose a C_{18} column with larger pore size. The pore volume (V_{pore}) is important insofar as it is closely related to the surface area (S); the higher the values of V_{pore} and S, the more mechanically fragile the particles but the higher the retention and the loading capacity of the column. The carbon loading by itself is not very informative, but together with S and the molecular formula of R it determines the bonding phase density, i.e., the number of R groups per unit area of adsorbent surface. Molecular models indicate that an ODS group is about 22 Å long and has a cross-sectional area of about $42 Å^2$. Thus under optimum conditions (ODS groups essentially vertical to the silica surface) a maximum of about 2.5 ODS groups can be accommodated in $100 Å^2$, leaving the same number of unreacted silanol groups some of which can be end-capped. This corresponds to about $4 \times 10^{-6} \, mol.m^{-2}$ as the maximum bonding density. Compare this theoretical limit with a typical practical C_{18} material with a carbon loading of 15% and a specific surface area of $300 m^2.g^{-1}$. For a C_{18} substituent the mass of carbon per mole is 216 g, so a carbon loading of 15% corresponds to 0.15 g of $C.g^{-1}/216 g.mol^{-1} = 7 \times 10^{-4} \, mol.g^{-1}$; then the bonding density is $7 \times 10^{-4} \, mol.g^{-1}/300 \, m^2.g^{-1} = 2.3 \times 10^{-6} \, mol.m^{-2}$, well below the theoretical maximum. Thus in such an adsorbent the silica surface as 'seen' by the analytes is a mixture of hydrophobic (C_{18}) zones and polar zones (unmodified silica); as a result, specific interactions and thus increased column selectivity are possible, whereas an adsorbent carrying the (sterically determined) maximum bonding density will present a purely hydrophobic surface to the analytes so only dispersion and dipole-induced dipole interactions are possible.

4.4.1c Ion Exchange Media

In ion exchange chromatography the dominant interaction is provided by coulombic forces between ionized analytes and fixed charge sites on the stationary phase (usually an organic resin, e.g., suitably derivatized styrene-divinylbenzene cross-linked copolymers). These fixed sites can be either negatively charged (providing cation exchange chromatography) or positively charged (anion exchange). Both types can be further sub-divided into 'strong' and 'weak'. 'Strong' exchangers retain their charge over a wide pH range ($\sim$2–12); examples are $-SO_3^-$ for strong cation exchange (SCX) and $-NR_3^+$ for strong anion exchange (SAX). 'Weak' exchangers change their degree of ionization as a function of pH of the mobile phase, e.g., $-CO_2^-$ and $-NH_3^+$ for WCX and WAX, respectively; by controlling pH it is then possible to obtain a wide range of conditions for selective adsorption of ionized or ionizable compounds in both SPE and HPLC. Obviously, for coulombic interactions the pH must be such that both the analyte of interest and the active site on the stationary phase are ionized. Separation of analytes is the result of their differences in the competition for the charge sites with the original counter-ions supplied with the adsorbent (usually Cl^- for SAX and Na^+ for SCX), and is influenced by the pH and ionic strength of the mobile phase. Alternatively, a solution that has a high ionic strength, or that contains an ionic species that is highly competitive with the analyte for the stationary phase, can be used to elute the analyte in ion exchange (either SPE or HPLC). Ion exchange columns are not much used in organic trace analysis, although SPE clean-up to remove ionic contaminants can be useful. In some strategies employed in proteomics investigations an ion exchange column is used as the first in a two-column separation together with a conventional reverse phase C_{18} column as the second stage. This strategy can achieve good separation of proteolytic peptides derived from a complex protein mixture; usually a stepwise variation of ionic strength is used to deliver peptide fractions to the second column. A recent review of this technology (Pohl 2004) also shows an example of fast separation of small organic carboxylic acids.

4.4.1d Chiral Separations

It appears (Barron 2007) that Kelvin was the first to coin the word 'chirality' to represent the concept of right- or left-handedness in science: 'I call any geometrical figure or group of points chiral, and say that it has chirality if its image in a plane mirror, ideally realized, cannot be brought into coincidence with itself ' (Kelvin 1904). This definition is essentially that still used in modern stereochemistry. The original discovery in the early 19th century, that some natural product compounds could cause the rotation of the plane of polarization of polarized light, reached its climax in the remarkable discovery by Pasteur. At the age of 26, in a paper presented to the Paris Academy of Sciences, he showed that mirror-image enantiomers generate equal and opposite optical-rotation angles. In this famous work he investigated properties of tartaric acid $HOOC-C^*H(OH)-C^*H(OH)-COOH$, an acid formed in grape fermentation, where the carbon atoms marked with an asterisk are what we now recognize as asymmetric chiral centers (all four bonds are attached to different groups). A solution of this compound, when derived from a natural source (wine byproducts), rotated the plane of polarization of light passing through it. However, if prepared by chemical synthesis using no natural products, the solution exhibited no such effect although the chemical reactivity and elemental composition were identical for the two forms. When the synthetic sample was crystallized from solution, Pasteur noticed that the macroscopic crystals came in two asymmetric forms that were mirror images of one another. He then tediously sorted the crystals by hand into the two types; when dissolved separately, one form rotated polarized light clockwise while the other form rotated light counterclockwise, but an equal mix of the two had no polarizing effect on light. Pasteur correctly deduced that the tartaric acid molecule is intrinsically asymmetric and could exist in two different forms (enantiomers) that are related to one another in the same way as left- and right-hand gloves, and that the naturally occurring form of the compound consists of only one type. This seminal discovery led to modern stereochemistry, although the 'chiral' terminology was not introduced till later (Kelvin 1904).

Chiral analysis, requiring the distinction between enantiomers of compounds containing at least one asymmetric carbon atom, represents an increasingly important class of analytical problems. Developments in this area have been largely driven by the pharmaceutical industry since many pharmaceutical drugs contain one or more asymmetric carbon centers but are synthesized by achiral methods as racemic mixtures. Often one of the enantiomers is much more physiologically active than the other, and in some cases one of the enantiomers can have detrimental side effects or other undesirable characteristics. Therefore, it is essential to investigate the characteristics of each of the enantiomeric forms and their absorption, distribution, metabolism and excretion (ADME) properties for successful drug development. Accordingly, the

policy statement of the United States Food and Drug Administration (FDA) policy statement for development of new stereoisomeric drugs (FDA 2005) requires pharmaceutical manufacturers to develop quantitative assays for the individual enantiomers in *in vivo* samples early in the drug development process (an example is discussed in Section 11.5). This subject appears likely to become increasingly important in trace quantitative analysis. At present the state of the art is such that chiral columns can be easily fouled if precautions are not taken (e.g., if extracts to be analyzed are too 'dirty', i.e. contain too many co-extractives), leading to a problem with method robustness (Section 9.8.4). Also, run times required for full separation of the enantiomers are generally very long.

One approach to chiral separations of racemic mixtures is to devise a chemical reaction with an optically active reagent to produce diastereomeric derivatives that have different physical properties depending on the original enantiomers; these can then be separated by conventional (achiral) chromatographic or SPE approaches, such as a recent demonstration (Xia 2006) of the utility of porous graphitic carbon for this purpose. However, this approach is not always possible and, even when it is, complex derivatization procedures are not desirable in trace analytical procedures and are avoided if possible.

Development of chiral stationary phases for HPLC separations of enantiomeric mixtures is a highly active field. A useful review (Däppen 1986) covers earlier work, and more recent overviews (Beesley 2004; Bojarski 2005) of such developments are supplemented by a review of chiral LC/MS quantitation in the pharmaceutical industry (Chen 2005). Chiral stationary phases have specific requirements concerning mobile phases and this raises nontrivial questions concerning the ionization technique (Chapter 5) that is best suited to interfacing with a mass spectrometer in any particular case.

Chiral stationary phases (CSPs) developed on the basis of the strategy devised by Pirkle *et al.* (Pirkle 1980, 1984, 1988, 1992; Wolf 2002) depend on an explicit recognition of the three-dimensional fit between the CSP molecule and the enantiomers of the analyte. This strategy is based on Pirkle's *Principle of Reciprocity* (Figure 4.14), namely, that a single enantiomer of a racemate which separates well on one CSP will, when used to produce a second CSP, usually afford separation of the enantiomers of analytes that are structurally similar to the chiral selector of the first CSP. Figure 4.14 also shows two chiral selectors that do exhibit reciprocity, as well as a representation of how the specific interaction between the (S)-form of the first CSP can select for one enantiomer of the second. This is a typical example of the Pirkle-type chiral selectivity,

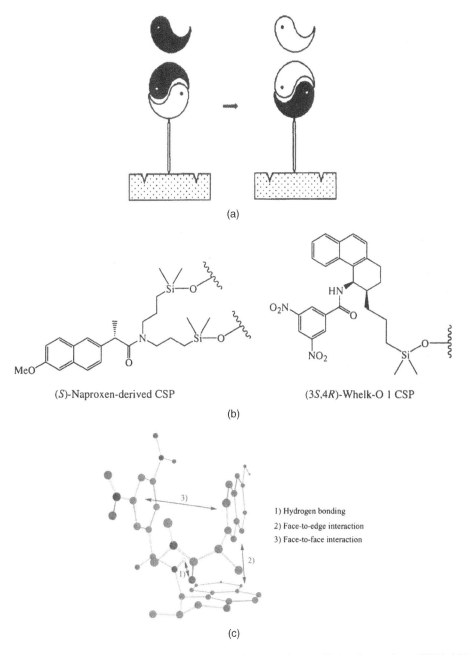

(a)

(S)-Naproxen-derived CSP

(3S,4R)-Whelk-O 1 CSP

(b)

1) Hydrogen bonding

2) Face-to-edge interaction

3) Face-to-face interaction

(c)

Figure 4.14 Schematic illustration of the principles underlying design of Pirkle-type chiral stationary phases (CSPs). (a) Illustration of the concept of reciprocity: a single enantiomer of a racemate which separates well on the CSP shown on the left, when used to produce a second CSP shown at the right, will usually afford separation of the enantiomers of analytes that are structurally similar to the chiral selector of the first CSP. Reproduced from Pirkle *et al.*, *J. Org. Chem.* **57** (1992), 3854, Copyright (1992), with permission of the American Chemical Society. (b) Two CSPs that exhibit reciprocal behavior, and (c) enantiomeric recognition model for the more stable diastereomeric complex between (S)-naproxen dimethylamide and the Whelk-O-1 (3R,4R) analog. Note that hydrogen atoms bonded to carbons are omitted for clarity. Reproduced from Wolf and Pirkle (2002), *Tetrahedron* **58**, 3597, copyright (2002), with permission from Elsevier.

depending on a combination of three different intermolecular interactions (in this case $\pi-\pi$, charge transfer and hydrogen bonding) that must be active for binding (i.e. retention) to occur. The number of CSPs that have been developed using the Pirkle strategy is quite large. Enantiomer discrimination mechanisms in Pirkle-type CSPs are thus well characterized and their chromatographic behaviors are reasonably predicable. Pirkle-type columns also appear to be more robust than polysaccharide CSPs (see below). Since $\pi-\pi$ and charge transfer interactions as well as hydrogen bonding are stronger under normal phase conditions, the majority of chiral separations on

Pirkle-type columns are performed in normal phase mode, and this in turn determines the type of LC/MS interface that is most appropriate (Chen 2005).

Another well established type of CSP is that based on the properties of cyclodextrins, a family of cyclic oligosaccharides containing five or more α-D-glucopyranoside (glucose) units linked 1→4. Very large macrocycles of this type are known, but the most commonly used are those containing six, seven or eight glucose monomers, known as α-, β- and γ-cyclodextrins, respectively. The structure of β-cyclodextrin is shown in Figure 4.15(a), and a sketch of the three-dimensional

(a)

(b)

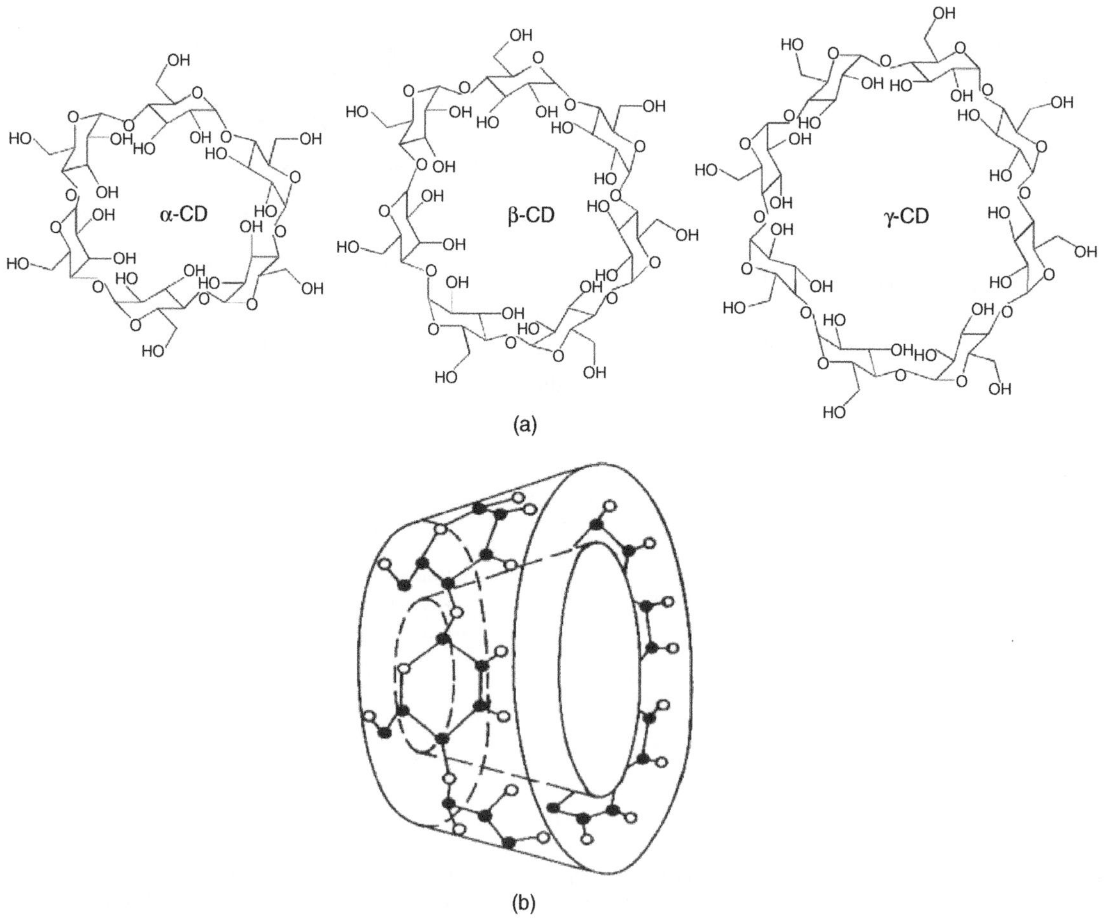

Figure 4.15   (a) Chemical structures of α-, β- and γ-cyclodextrins (derivatization of hydroxyl groups can create a range of retention mechanisms), consisting of 6, 7 and 8 α-1,4-linked glucopyranose units, respectively. (b) Sketch of the three-dimensional shape of β-cyclodextrin, a hollow cone with a hydrophilic external surface (secondary hydroxyl groups around the larger rim, primary hydroxyls, i.e., $-CH_2OH$ groups, are round the smaller) and a lipophilic cavity in the centre. The height of the cone is 7.2 Å, and the internal diameters are 4.7–5.2, 6.0–6.5 and 7.5–8.5 Å for α-, β- and γ-cyclodextrin, respectively (Szejtli 1988).

hollow-cone structure, with its hydrophilic external surface and somewhat hydrophobic internal cavity, is shown in Figure 4.15(b). Their use as HPLC or SPE stationary phases was first introduced in the form of the cyclodextrin ring bound to a silica support via a 6–10 carbon spacer (Armstrong 1984, 1985, 1986); the selectivity arises from the requirement that a molecule must satisfy requirements of shape and size to fit inside the cavity, the goodness-of-fit determining the association constant. The cyclodextrins are also chiral molecules (e.g., β-cyclodextrin contains 35 chiral centres), so that enantiomeric analytes will satisfy the shape requirements for best-fit in the internal cavity to different degrees. Cyclodextrin-based CSPs have been used in both normal and reverse phase modes.

CSPs based on larger derivatized noncyclic polysaccharides, e.g., cellulose tribenzoate and cellulose trisphenylcarbamates, have also been widely studied and used extensively in the pharmaceutical industry (Yashima 2001). A wide range of derivatized polysaccharide CSPs is available and is normally used in normal phase mode although some have been designed to operate in reverse phase mode (Tachibana 2001; Franco 2001). The popularity of these polysaccharide-based CSPs is based on their versatility, ruggedness when used under appropriate conditions, and loading capacity, although their enantiomeric discrimination mechanisms are not well understood and consequently their chromatographic behavior is not predictable.

Another commonly used type of CSP is based on macrocyclic antibiotics (Ward 2001). The macrocyclic antibiotics used for this purpose include the ansamycins, the glycopeptides and the polypeptide antibiotic thiostrepton; some representative structures are shown in Figure 4.16. More chiral analytes have been resolved by HPLC using the glycopeptides than with all the other macrocyclic antibiotics combined. The macrocyclic antibiotics have been used as CSPs when covalently bound to silica gel via linkage chains employing a variety of chemistries, as first demonstrated using vancomycin, rifamycin B and thiostrepton (Armstrong 1994). Several CSPs based on macrocyclic antibiotics and chemically derivatized forms have been commercialized (Ward 2001); they behave somewhat similarly to those based on proteins (see later), but exhibit higher loading capacities and are more stable. In particular they can be used with normal phase solvents without denaturation and loss of functionality and enantioselectivity. Many variations of these basic types have been devised (Ward 2001).

CSPs based on proteins are not as widely used as the polysaccharide types although they can provide enantiomeric separations of a wide range of analytes as a result of their multiple binding interactions and/or multiple binding sites. The proteins used include albumins such as bovine serum albumin and human serum albumin, glycoproteins such as α_1-acid glycoprotein, ovomucoid, ovoglycoprotein, avidin and riboflavin 1 binding protein, enzymes such as trypsin, α-chymotrypsin, cellobiohydrolase I, lysozyme, pepsin and amyloglucosidase, and other proteins such as ovotransferrin and β-lactoglobulin. A review (Haginaka 2001) has described the properties and binding mechanisms of protein-based CSPs, that can generally be used only with reverse phase solvents because of denaturation in normal phase conditions.

Synthetic optically active polymers used as CSPs have been classified into three major categories, i.e. addition polymers, condensation polymers and cross-linked gels (molecularly-imprinted polymers, Section 4.3.3e). Thus far most emphasis has been devoted to polymethacrylate addition polymers with helical conformation, synthesized using asymmetric anionic or radical polymerization techniques (Nakano 2001); such polymers exhibit enantiomeric separations for a wide range of racemic analytes. While considerable progress has been made, future progress in preparation of intrinsically asymmetric polymer chains will depend (Nagano 2001) on invention of a new stereo-regulation method for polymerization reactions because the degree of stereo-regularity of a polymer chain often significantly affects its physicochemical properties in the solid state. Many of the addition polymers prepared to date have been synthesized by free radical polymerization, which is versatile and inexpensive but generally provides rather poor stereocontrol. Desirable new methods would alter both the main chain stereochemistry and also the higher order structure of the polymer, thus improving the chiral recognition capability of the polymer. A different related approach is to use acrylate monomers with pendant chiral groups attached.

The other approach to preparation of CSPs based on synthetic polymers is that based on molecular imprinting (MIP), discussed in Section 4.3.3e in the context of achiral separations but here involving chiral templates using the same general approaches (Nakano 2001; Sellergren 2001; Turiel 2004). This approach clearly has considerable potential but is currently faced with several problems (Sellergren 2001; Turiel 2004; Kandimalla 2004). The template molecule must be available in appreciable amounts and also be stable under the polymerization conditions, so a close analog of the target analyte is sometimes used. The cross-linked gel matrix must also possess several properties (preparation as small spherical particles with sufficient porosity that the analyte can access the imprinted vacancies with sufficiently fast mass transfer) if high performance HPLC separations are to be achieved.

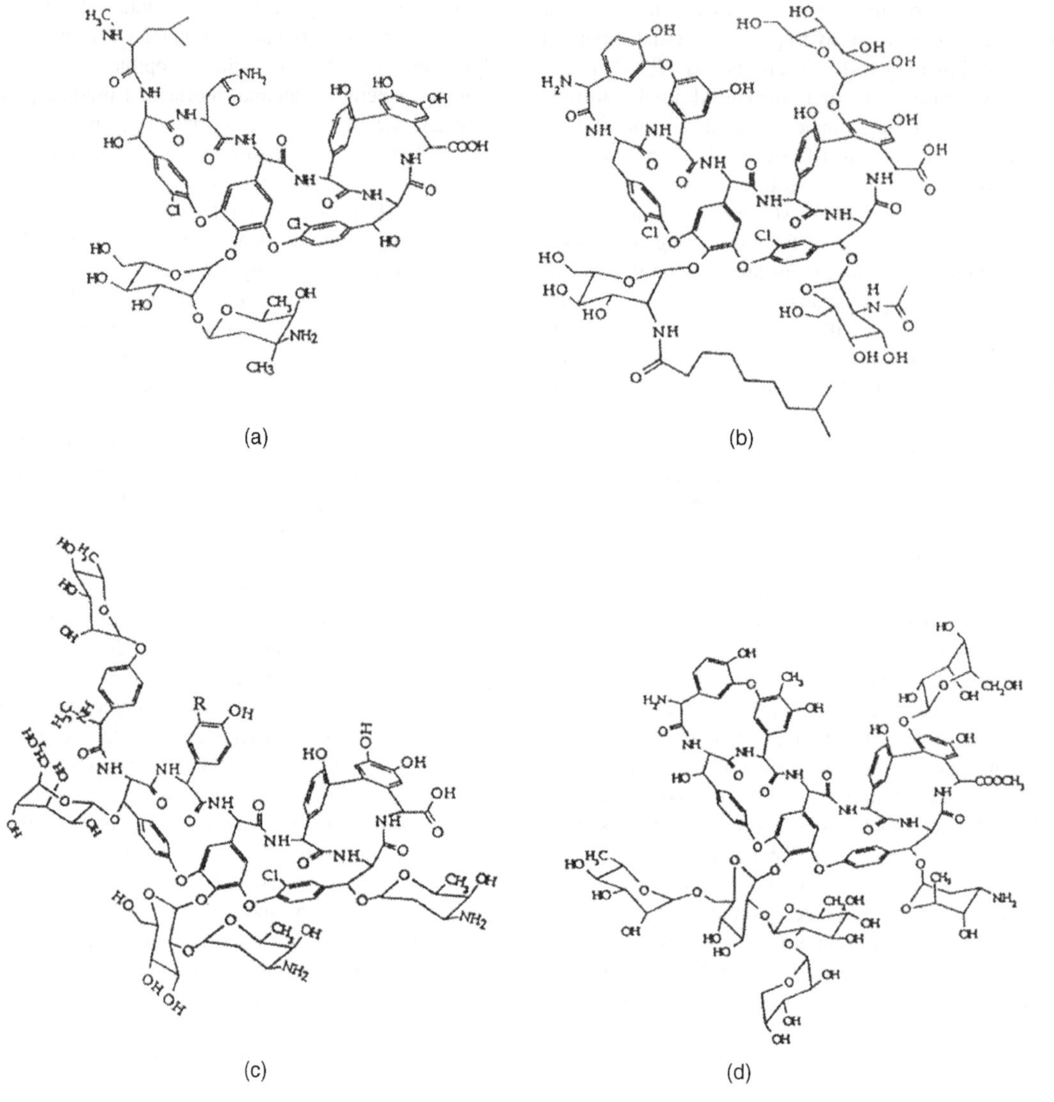

Figure 4.16 Chemical structures of the macrocyclic glycopeptide antibiotics: (a) vancomycin, (b) teicoplanin, (c) avoparcin, (d) ristocetin A, that have been used as chiral selectors in CSPs for HPLC. Reproduced from Ward and Farris, *J. Chromatogr. A* **906** (2001), copyright (2001), with permission from Elsevier.

At present the approach tends to give broad asymmetric peak shapes and limited loading capacities as a result of heterogeneous distributions of binding sites and slow mass transfer (C term in the van Deemter equation). If these problems can be overcome the MIP approach could become a method of choice in the future. In fact, MIP stationary phases can be considered as a rather special case of restricted access media (Cassiano 2006); this term is somewhat different as it is intended to describe

a group of phases that limit the accessibility of macro-molecular compounds to the adsorption sites of porous supports as a result of their molecular size. Analytes are then separated through a combination of size exclusion and conventional hydrophobic and/or ion exchange inter-actions that result in elution of macromolecules within or close to the void volume of the column but can achieve separation of small molecule analytes. This approach can lead to direct and repetitive injection of untreated biological samples (e.g., blood plasma) into an otherwise conventional reversed phase HPLC system without a need for a separate clean-up step. However, at present the most demanding analyses of this kind still require a rigorous clean-up as described (Section 3.4).

4.4.1e Affinity Media

Affinity chromatography is widely used in biochemistry for purifications, and sometimes in pharmaceutical and biomedical trace analysis though not often where high throughput is an important criterion. Of the two approaches discussed here, immobilized metal affinity chromatography (IMAC) is mainly used for purification of proteins and selective extraction of phosphopeptides and phosphoproteins from complex protein mixtures. The underlying chemical principle is the selective chelation by transition metal ions of molecules containing a phosphate group and of certain amino acid residues, particularly histidine (His) but also cysteine, lysine and tryptophan. Efficient chelator molecules are immobilized on Sepharose beads and doped with metal ions such as copper (Cu^{2+}), nickel (Ni^{2+}), zinc (Ni^{2+}) and cobalt (Co^{2+}), to give metal chelates which have one or more 'free' coordination sites (occupied by solvent molecules) that are available to retain the target amino acid residues. In the case of selection for His residues the best selectivity is provided by immobilized chelated metals with just one free site, since otherwise the specificity is reduced by other amino residues attaching to the last site on the central metal ion (see www.affiland.com/imac.htm). This is particularly important when a target protein has been His-tagged (genomically coded to have an additional His_6 moiety) to enable selective extraction.

In the present context, immunoaffinity chromatography and application of the principle to SPE are more important than IMAC. A comprehensive review (Delaunay-Bertoncini 2004) describes the principles of the immunoaffinity SPE approach for pharmaceutical and biomedical analysis, including its coupling with HPLC and LC/MS as well as examples of other applications, e.g., to environmental analysis. An immunoaffinity sorbent contains antibodies that are specific to the target analytes and are immobilized on a solid support. An excellent free-access source for detailed information on all aspects of modern biology (Kimball 2006) includes informative descriptions of antibodies, particularly monoclonal antibodies (see below).

An antibody is a protein (one of the immunoglobulins) used by the immune system of a mammal to identify and neutralize antigens (foreign objects like bacteria, viruses and others). Each antibody 'recognizes' a specific antigen via its 'epitope', a portion of the molecular structure of the antigen that has a unique shape and other molecular characteristics (e.g., distribution of hydrogen bonding sites). This results in a 'lock-and-key' match of antibody and antigen (see the discussion of Figure 4.10 that emphasizes the modern view of this interaction and is more complex than the preceding simple picture). Thus, the first step in making an immunosorbent is to produce antibodies that recognize either one specific analyte or a group of analytes. This book is concerned with relatively small analytes (up to 2000 Da or so) and such molecules are unable to evoke an immune response in a mammal (rabbits are frequently used); accordingly they must be converted into a hapten, a covalent complex of the analyte with a carrier protein. Of course this means that at least some of the antibodies raised by the test animal against this hapten are specific for epitopes of the carrier protein rather than the analyte, and this must be accounted for in the subsequent purification. The biological response of the mammal after injection of the hapten involves the β-lymphocyte cells (an important class of white blood cells responsible for immune response) to produce immunoglobulin (IgG) molecules that 'fit' the various epitopes on the hapten. Even after purification from those antibodies that respond to the carrier protein, the resulting fraction is generally a mixture of antibodies raised to different epitopes of the analyte by different cell lines of β-lymphocytes.

For present purposes such polyclonal antibodies might be desirable if the ultimate intention is to selectively retain a class of analytes, e.g. an antibody raised against a drug that would also interact with its metabolites, or the class of coplanar polychlorobiphenyls that mimic the chlorinated dibenzodioxins. On the other hand, if maximum specificity is required one must resort to a monoclonal antibody, so-called because they were produced by just one type of immune cell as clones of a single 'parent' cell. The procedure to form monoclonal antibodies is even more complicated, involving fusion with myeloma tumor cells that are effectively 'immortal', i.e. can multiply indefinitely in culture (myeloma is a B-cell cancer). This is the trick used to enable just a single member of the polyclonal cell population, one that has been shown to respond to just one epitope of the analyte, to be amplified sufficiently to yield a useful quantity of the desired monoclonal antibody. The detailed procedure is much more complex (Kimball 2006) but this brief summary suffices for present purposes. Although monoclonal production is considerably more lengthy (and expensive), the production of the 'immortal' fusion cells guarantees long term availability without the need for immunizing additional animals.

Once the antibodies are available they are immobilized on a solid support that must be biologically inactive (no specific interactions with the analyte), chemically inert (but readily activated to allow covalent attachment of the antibody), and hydrophobic in order to limit nonspecific interactions in the aqueous environment within which the antibodies operate. Modified agarose gel or silica particles are the common choices. Such immunoaffinity columns are stored in a suitable solution, often a phosphate buffer saline (PBS) solution containing sodium phosphate and sodium chloride at pH 7.4. Before use such a column must be conditioned by replacing the storage solution with one that favors the specific interactions, generally aqueous solutions at an appropriate pH possibly containing a small percentage of an organic solvent (but not so much as to denature the antibody). The result is a SPE clean-up similar to those already discussed but many times more specific than those achievable using conventional SPE sorbents. This is nicely illustrated by the example shown in Figure 4.17. More details of the use of such immunoaffinity SPE columns and their interfacing to HPLC and LC/MS are given in the cited review (Delaunay-Bertoncini 2004).

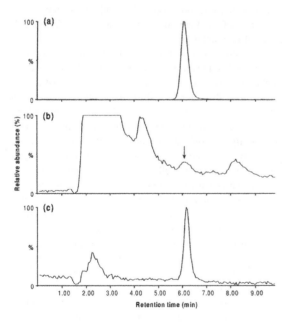

Figure 4.17 Comparison of HPLC–MS analyses of (a) melatonin standard, (b) a human serum sample processed by conventional SPE (C_{18} sorbent, the arrow indicates the melatonin peak), and (c) by an anti-melatonin immunosorbent. Reproduced from Rolčík *et al.*, *J. Chromatogr.* B **775**, 9 (2002), copyright (2002), with permission from Elsevier.

4.4.2 Mobile Phases Used in SPE and Liquid Chromatography

In HPLC (and SPE) the type and composition of the mobile phase is one of the important variables influencing the separation. Several criteria are important for all mobile phases. Purity must be sufficiently high that impurity molecules do not interact so strongly with the active sites on the stationary phase that the analyte molecules can not compete; also impurities must not interfere significantly with the HPLC detector, e.g., strongly absorbing chromophores for UV–visible detectors and highly surface-active impurities for electrospray mass spectrometry (see Section 5.3.6). The solubility of the analytes in the mobile phase must obviously be sufficiently high that the analyte remains in solution right up until detection, no matter how low the chromatographic dilution factor (D) (equation [3.50], Section 3.5.10b). The mobile phase must also be chemically inert under the conditions of the separation. Viscosity is an important parameter, since it partly determines the back-pressure that must be maintained by the pumps to sustain the desired flow rate. Polarity of the mobile phase is extremely important for reversed phase operation. For LC/MS it is important that volatile buffers (usually based on ammonia plus acetic or formic acid, rather than the chromatographers' traditional phosphate buffers) should be used to control pH, ionic strength etc., to avoid fouling of the mass spectrometer. A useful rule-of-thumb is that, to obtain good separations in a reasonable time, the mobile phase composition should be adjusted so that k' values for the analyte(s) should be between 1 and 10, and preferably > 3. And finally, the cost of purchasing and disposal of solvent must always be considered.

Dissolved gas is a special case of purity considerations for HPLC mobile phases. If the mobile phase contains excess dissolved gas which remains in solution at the high pressures produced by the pump within the column, once the liquid is exposed to room pressure at the column exit, or even within the detector, the gas may come out of solution resulting in sharp spikes in the chromatogram. These spikes are created by microscopic bubbles that make the flowing stream heterogeneous (the same effect causes 'the bends' in deep sea divers who come to the surface too quickly). The main problem appears to be oxygen from the air dissolving in polar solvents, particularly water. Solvent degassing can be an important precaution and may be accomplished by subjecting the liquid to a vacuum, or boiling the liquid or subjecting it to sonication, or bubbling helium through the liquid, or a combination of these. For high pressure mixing for gradient elution (Section 4.4.2a) an on-line degasser is usually sufficient to eliminate this problem.

Another problem with even chemically pure solvents is that of particulates, which can clog up the column inlet or other components. Sometimes solvents are filtered before use but most commercial HPLC systems have a line filter installed between the pump and the sample injector. These filters are generally of porous stainless steel, with pore size of $\sim 2\,\mu m$ for columns packed with larger particles, but $\sim 0.5\,\mu m$ pore-size filters are recommended for use with columns packed with particles of much less than $10\,\mu m$.

4.4.2a Solvent Polarity and Elution Strength

There are several measures of solvent polarity, e.g., the dipole moment of the solvent molecule, and the dielectric constant and Snyder polarity index of the liquid. The following abbreviated account draws from a detailed discussion of solvent properties in the context of their use as HPLC mobile phases (Snyder 1974). A selection of data for a few commonly used mobile phase components is included here as Table 4.4. The elution strength parameter (sometimes called eluotropic strength) is a measure of the ability of the solvent to remove an analyte from the active sites of the stationary phase; these values are based on the adsorption energy of the solvent molecules on the active sites (the so-called Hildebrand parameter), so the higher this value the greater the affinity of the solvent for the active sites and the greater the ability of the solvent molecules to displace an analyte molecule and thus contribute to its elution. There are a few peculiarities in the data in Table 4.4; thus, although acetonitrile has a higher dipole moment (3.87 debye) than water (1.84 debye), liquid water has a higher dielectric constant

$(80\,F.m^{-1})$ than acetonitrile $(37\,F.m^{-1})$ and also higher effective polarity (Snyder Polarity Index P'); this is a result of the highly ordered structuring of water molecules in the liquid state compared with the single molecules in the dilute gas state (where dipole moments are measured). Extensive lists of values of elution strength are available for solvents on the highly polar active sites of under-ivatized alumina and silica, which are useful for normal phase separations (values on alumina are generally some 20 % higher than on silica). However, only a few values have been estimated for revered phase separations on C_{18} stationary phases; in that case elution strength is a measure of the affinity of the solvent for the hydrophobic C_{18} chains, so organic solvents have a higher elution strength than water, the opposite trend to that for normal phase operation.

For gradient elution, solvent miscibility over the entire range used is an important consideration. As a rough guide, solvents are completely miscible if they are on the same half of the elution strength scale, i.e. solvents on the upper half of the scale are completely miscible, or all on the bottom half are miscible with each other, and all solvents from the center half ($\frac{1}{4}$ to $\frac{3}{4}$) are mutually miscible. There are several 'universal' solvents, such as tetrahydrofuran (THF) and acetonitrile (ACN), which are miscible with almost all others except hexane and pentane. The most useful 'cleanout' solvent (for removing strongly retained compounds from the column) is iso-propanol, which is miscible with all others at all concentration levels. It is also important to be aware of the potential for precipitation of analytes out of solution if the organic content gets too high.

Table 4.4 Values of physico-chemical properties of some liquids commonly used as mobile phase components in HPLC and SPE.

Solvent	$\varepsilon^*_{alumina}$	ε^*_{silica}	ε^*_{C18}	$(P')^{**}$	η^{***}	Boiling point (°C)
Hexane	< 0.01	< 0.01	n/a	0	0.31	69
CCl₄	0.17	0.11	n/a	1.7	0.97	76
Toluene	0.2–0.3	0.22	n/a	2.3	0.59	101
CH₂Cl₂	0.4	0.32	n/a	3.4	0.44	40
THF	0.45	0.35	3.7	4.2	0.55	66
Acetonitrile	0.6	0.5	3.1	6.2	0.37	82
Methanol	0.95	0.73	1.00	6.6	0.6	65
Water	Very large	Large	Small	9	0.9	100

* Solvent (elution) strength parameters ε are based on adsorption energy on active sites of the indicated stationary phase; the higher this energy, the greater the affinity of the solvent for the active sites and the greater its tendency to elute the analytes off these sites. Values for alumina and silica are defined relative to a value of 0.00 for pentane; values for C_{18} are relative to that for methanol defined as 1.00.
** Polarity Index P' (Snyder 1974). Rule of thumb: a 2 unit change in P' gives a $\sim$10 unit change in k' (larger with increase in P' for reverse phase, smaller with increase in P' for normal phase chromatography).
*** Viscosity at $20\,°$C, in centipoise (cP): $1 cP = 10^{-3} N.s.m^{-2} (kg.m^{-1}s^{-1})$.

A practical question in gradient elution concerns the way in which the two (or more) solvents are to be mixed, before or after the high pressure pumping stage. In high pressure mixing each solvent is delivered by its own high pressure pump (Section 4.4.5) to a mixing chamber located between the pumps and the column. Among the advantages of high pressure mixing is the ability to achieve precise control and repeatability of mobile phase mixtures to ~0.1% levels within the medium composition range (much worse at either end of the composition range), and reasonably rapid response to changes in composition. The disadvantages include difficulties with the mixing efficiency, significant changes in total volume on mixing, i.e., the volume of the mixture is not simply the sum of the volumes of the two individual solvents, although this potential problem can be avoided through use of a good low void volume mixing tee. In low pressure mixing systems the overall flow rate is controlled by a single high pressure pump and mixing is accomplished prior to the pump at its low pressure side. This has been a popular approach to gradient formation, although the precision of delivery of eluent composition for the medium concentration range is generally not as good as for high pressure mixing and there is a greater potential for cavitation than for high pressure mixing.

4.4.2b Reverse Phase Chromatography

In practice, reverse phase separations are conducted far more frequently than normal phase for quantitative analysis of small analytes. The solvents most frequently employed are water, acetonitrile and methanol, especially for LC/MS applications. It is often important to add buffers to these solvents to control the pH so that the analytes remain in the desired form (electrically neutral for stronger interactions with the hydrophobic C_{18} etc.). As mentioned above, it is essential for interfacing to a mass spectrometer that the buffer components be volatile so as not to foul the instrument with precipitated salts. This precludes the use of alkali phosphates, which are the traditional buffers used by chromatographers with other detectors, but volatile buffers formed from ammonia and formic and acetic acids (and their ammonium salts) are generally found to work equally well. Sometimes ion-pairing reagents are used to help ensure that the analyte is in a neutral form when interacting with the hydrophobic sites. However, some of these, such as the perfluoroalkanoic acids (and particularly trifluoroacetic acid (TFA) commonly used to improve chromatographic performance), can have a strong deleterious effect on the sensitivity of electrospray mass spectrometry (Section 5.3.6). TFA is also notorious for being extremely difficult to completely flush out of a LC/MS system; this

special case of a carryover effect (Section 9.7.2) is particularly important if the system is to be used for negative ion operation since the TFA residues can dominate the negative ionization at the expense of the analytes.

4.4.2c Hydrophilic Interaction Chromatography (HILIC)

Use of a high organic:low water mobile phase (~80 : 20), with conventional normal phase adsorbents (usually silica, either underivatized or derivatized to give amino or diol phases though some others have been reported), provides a new form of normal phase chromatography suitable for polar compounds (Alpert 1990). A comprehensive review (Grumbach 2004) provides full details. The HILIC approach to retention of more polar compounds uses normal phase adsorbents with eluants typical of reverse phase separations but with unusual composition, e.g. 80:20 acetonitrile:water. Ion exchange, and reverse phase chromatography with ion pairing or mobile phase pH manipulation are techniques traditionally used for retention of polar analytes. However, each of these techniques has drawbacks since they are applicable only if the analytes are ionizable. Moreover ion exchange and ion pairing reverse phase chromatographies are unsuitable for detection by mass spectrometry because ion exchange involves high concentrations of salts that can contaminate the spectrometer, while ion pairing reagents cause signal suppression in electrospray ionization (Section 5.3.6a). Several retention mechanisms have been shown to be operating in HILIC on silica columns; a combination of hydrophilic interaction, ion exchange, and reverse phase retention result in a unique selectivity that allows for retention of more highly polar analytes. Retention is approximately proportional to the polarity of the solute and inversely proportional to that of the mobile phase (Alpert 1990).

In addition to improved retention for more polar analytes, HILIC also provides improved response in electrospray mass spectrometry. This is a consequence of the high volatility and low surface tension of the organic component relative to water, leading to electrosprayed droplets that are smaller and that evaporate faster (Section 5.3.6). HILIC is thus well suited to LC/MS analysis of polar pharmaceuticals and their even more polar metabolites. Another advantage arises from the relatively low back pressure generated in HILIC, a consequence of the low viscosity of the high organic mobile phase plus the absence of any bonded phase (e.g. C_{18}) on the silica particles forming the stationary phase; this feature permits faster chromatographic separations with no loss of chromatographic resolution (Shou 2002). An excellent example of application of HILIC to high throughput analysis of drugs in urine (Weng 2002) also exploited the high organic eluant composition in HILIC to eliminate the usual tedious steps of evaporation and reconstitution of C_{18} SPE extracts in a suitable solvent with elution strength weaker than that of the reverse phase solvent (Section 4.4.2a); use of a SPE extraction solvent compatible with HILIC enabled a considerable time saving in fully validated SPE–LC/MS procedures. Another example of the exploitation of the unique properties of HILIC is described in Section 11.2.2.

The wide variety of modes of use in HPLC, including the choice of packed microcolumns (Section 3.5.10), is one of its principal advantages over, e.g., GC, leading to its applicability to a much wider range of analytes, but the wide range of choices can also be confusing. Practical experience ('learning by doing') is the best guide to designing a good separation by combining appropriate stationary and mobile phases with suitable column dimensions and flow rates. A useful starting point is to search the literature for separations (whether HPLC or SPE) of analytes that are structurally similar to those of interest, and use the conditions described there as a starting point for modification via the very general principles outlined here.

4.4.3 Mobile and Stationary Phases for Gas Chromatography

A useful compendium of terms and techniques used in GC is freely available on the internet (Hinshaw 2002a). The range of choices for mobile and stationary phases for GC is much less extensive than for HPLC, quite apart from the requirement that analytes be volatile and thermally stable.

4.4.3a GC Mobile Phase

The GC mobile phase (carrier gas) is almost always helium, nitrogen or hydrogen, with helium by far the most commonly used in applications of interest here. However, despite this relative lack of complications there are a few aspects that require attention. Helpful details of the proper use of carrier gases are available on the internet (www.chem.agilent.com/cag/cabu/carriergas.htm). The most obvious requirement is that the gases be pure and chemically inert, and to this end various traps for water, oxygen and other impurities are available to reduce total impurity levels to < 1 in 10^5 (10 ppm) and levels of each of water and oxygen to < 2 ppm. Other precautions include the use of connecting tubing, valves (including pressure reducing valves on gas cylinders) and other

in-line components that contain no organic polymers, as these contain volatile components and do not have sufficient mechanical strength to withstand the high pressures involved; gas regulators with stainless steel diaphragms should be used and connecting tubing should be stainless steel or copper.

Most GC analyses, and almost all that are concerned with trace quantitative analysis, are nowadays conducted using unpacked (open tube) capillary columns; the following discussion will be mainly concerned with these. In this regard, a great deal is owed to Golay (see the accompanying textbox), who was responsible for both the underlying theory and the first practical demonstrations.

Flow rate is an important parameter in GC, as for HPLC, but in this case the linear flow rate (u) varies with pressure drop along the column as a consequence of the compressibility of gases (see discussion of equation [3.53] and Figure 3.18). While the use of u_{exit} allows use of a modified Golay equation that accounts well for this effect (Scott, http://www.chromatography-online.org/), in practice the average value u_{avge} provides a useful approximate (Figure 3.18) indication of relative properties, and is more readily estimated directly via:

$$u_{avge} = l/t_0$$

where, as before, l is the column length and t_0 the elution time of an unretained peak. Flow rate is one of the parameters that determine the choice of carrier gas via the van Deemter plots exemplified in a qualitative way in Figure 4.18. As shown in Section 3.5.11, the minima in

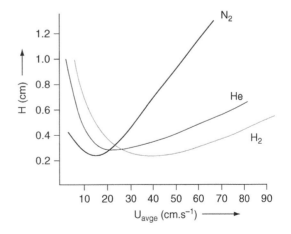

Figure 4.18 Sketch of van Deemter curves for a capillary GC column as a function of average linear flow rate u_{avge}, showing a qualitative comparison between nitrogen, helium and hydrogen. Modified from Scott, http://www.chromatography-online.org/, with permission.

Marcel Golay: Electrical Engineer, Mathematician and Inventor of the Open Tubular Capillary Column

Marcel J.E. Golay was educated in his native Switzerland as a mathematician and electrical engineer, and later received his Ph.D. in nuclear physics from the University of Chicago. He spent 25 years at the Engineering Laboratories of the US. Army Signal Corps, where he made distinguished contributions to information theory as applied to transmission of information along a telegraph line. In 1955 he joined the Perkin–Elmer instrument company where he made several important contributions to analytical chemistry, including the famous Savitzky–Golay smoothing algorithm, introduced in a paper (Savitzky 1964) that is one of the most widely accessed papers of all time

(Photograph reproduced from Adlard, *LCGC North America* **24** (10), 1102 (2006), with permission from Dr E.R. Adlard, Dr L.S. Ettre and the journal).

This algorithm represents a signal peak (e.g., from chromatography) as a series of equally spaced points and obtains a local smoothed value for each point in a fashion that preserves features such as relative maxima and minima as well as width, that are usually 'flattened' by other averaging techniques.

On joining Perkin–Elmer, Golay became interested in the new technique of gas chromatography (James 1952); the history of this work that led to development of the wall-coated capillary column for gas chromatography has been well described (Ettre 1987, 2001). Interestingly his first attempt at a theory of gas chromatography (Golay 1957) was based on his experience with signal transmission via the so-called 'Telegrapher's Equation', in which he viewed the chromatographic process as an analogy to the system of capacitors and resistors used as the idealized model for a telegraph line. The analogy was somewhat forced but even so predicted several results such as the existence of an optimum gas flow velocity. His next attempt viewed a GC column as a bundle of capillary tubes each representing a passage through the particles in the packed column. The progression to his realization of the advantages of using real (as opposed to imagined) capillaries was remarkably fast and led to his famous paper (Golay 1958) in which he described his complete theory (essentially identical to that independently proposed by van Deemter at about the same time, but without accounting for multiple path dispersion that applies only to packed columns). To demonstrate the validity of his new approach he had to devise a GC detector of sufficiently small volume that it did not introduce excessive extra-column band broadening. Subsequent developments were concerned with new materials more suitable for fabrication of the columns themselves, and with new stationary phases and how to stabilize them on the capillary walls (Ettre 2001).

these plots, defined as the optimum values of u giving a minimum theoretical plate height, are given by:

$$u_{opt} = (B/C)^{\frac{1}{2}} \text{ and } H_{opt} = 2(BC)^{\frac{1}{2}}$$

where the B and C coefficients are given by equation [3.54]. For well-retained peaks ($k' \gg 1$), and for the usual condition that the column diameter is much larger than the stationary phase film thickness ($d_c \gg d_f$), the optimum parameters for an open tubular GC column with wall-coated stationary phase, become:

$$H_{opt} \approx d_c/3.5 \text{ and } u_{opt} \approx 14D_m/d_c$$

Under this approximation the optimum plate height is independent of the nature of stationary and mobile phases but in practice this is not strictly correct (Figure 4.18). Nitrogen provides slightly better efficiencies (lower plate heights) than the other gases for a given column. However, the van Deemter minimum occurs at an appreciably lower value of u_{opt}, a consequence of lower values of diffusion coefficients of analyte molecules in an atmosphere of nitrogen molecules, since the latter are heavier and also 'stickier' (larger intermolecular forces dependent on polarizability) than for helium or hydrogen. This implies significantly longer elution times (about double those for helium) for little gain in column efficiency. In addition, the dominance of the term involving d_c^2 in the C-term (equation [3.54]) means that the positive slope section of the van Deemter plot (dominated by the C-term, see Figures 3.8–3.9) is considerably larger for nitrogen than for helium or hydrogen as a result of the smaller value of D_m in the denominator. Thus small variations in u can lead to large variations in retention time (Hinshaw 2004), much larger than the case for the much flatter curves for helium and hydrogen (Figure 4.18).

If the GC is equipped with some type of electronic pressure or flow control system, the flow rate variations can be reduced to negligible values. In such devices the column length and diameter, type of carrier gas and desired average linear velocity are inputted into the GC software. The GC automatically determines the current column temperature during temperature programming (see below) and adjusts the column head pressure so that the desired u_{avge} is obtained. A more rigorous approach to measuring volume flow rate and linear velocity (Hinshaw 2002) explicitly considers the effects of gas compressibility.

Thus flow rate considerations provide an argument against the use of nitrogen, but there are other considerations, e.g., cost (nitrogen is the cheapest). A less obvious criterion is the compatibility of the carrier gas with the GC detector. For interfacing to mass spectrometry there are at least two considerations. The first is the pumping speed of the mass spectrometer vacuum system that is strongly dependent on the molecular mass of the gas being pumped (Section 6.6). Basically this is because the action of the pump in moving the gas molecules in the desired direction (out of the mass spectrometer) has to compete with the random thermal motion of the molecules, and the latter varies inversely with molecular mass. Because of the more limited pumping speed of a typical vacuum system for helium or hydrogen, the volume flow rate must be limited to $\sim 1 \ cm^3.min^{-1}$ for direct GC/MS coupling in order to not allow the vacuum to deteriorate to an extent that the performance of the instrument is jeopardized. Fortunately for a typical capillary GC column this is entirely compatible with an optimum value of u_{avge} for helium of about $0.25 m.s^{-1}$. Then the corresponding volume flow rate is given by:

$$U = \pi.(d_c/2)^2.l/t_0 = \pi.(d_c/2)^2.u_{avge}$$

where the numerator is the total volume of the column ignoring the small contribution of the film of stationary phase. For a typical value for d_c of $0.3 \ \mu m$ and $u_{avge} = u_{opt} = 0.25 m.s^{-1}$, this gives $U = 1.77 \times 10^{-8} m^3.s^{-1} = 1 cm^3.min^{-1}$.

The other limitation placed on the nature of the carrier gas by use of a mass spectrometer as detector is the extent to which the carrier gas might overwhelm the ionization capacity of the ion source. For trace quantitative analysis by GC/MS, an electron ionization (EI) source (Section 5.2.1) is the most common and the 70 eV electrons normally used are quite capable of ionizing the carrier gas molecules that are present in great excess over the analyte even though the probability of ionization is less for helium (ionization energy 24.5 eV) than for nitrogen and hydrogen (14.5 and 13.6 eV, respectively). This is not usually a limitation on the limits of detection and quantitation, but can become so for truly ultra-trace analytes; in that case (e.g., for dibenzochlorodioxins, Section 11.2) the electron energy is sometimes lowered to decrease the cross-section for ionization of helium much more than that of organic analytes (ionization energy usually $\sim$10eV), and this can sometimes yield improved analytical performance. This approach does not work for nitrogen and hydrogen as their ionization energies are not much larger than those of typical organic analytes.

Hydrogen is sometimes regarded as too dangerous for use as a GC carrier gas but in fact it can be used safely if minimal precautions are taken. The diffusion coefficient of hydrogen in air is very high (low molecular mass), so if any does escape from the venting provided it does not have much chance to build up to levels within the explosion limits (that actually cover a surprisingly narrow concentration range). Also, modern GCs have safety features incorporated to minimize any such risks of creating an explosion. Hydrogen provides the highest value of u_{opt} of the three common carrier gases, resulting in the shortest analysis times; also, its very flat van Deemter curve (Figure 4.18) provides a wide range over which high efficiency is obtained, so that hydrogen is the best carrier gas for samples containing compounds that elute over a wide temperature range using temperature programming (see below).

4.4.3b Temperature Programming

There is no analogy to HPLC gradient elution in GC; varying the composition of a carrier gas is not a realistic option and in any case the differences in elution power between e.g., nitrogen and helium, are unlikely to yield much useful effect. The only analogous variable available in GC is column temperature and indeed temperature programming is widely used since k' decreases strongly with increasing temperature. In principle it would be a good idea to arrange a temperature gradient along the GC column so that analytes always are moving into a region of greater retention, by analogy with gradient elution in HPLC. In fact attempts to do this have been made (Rubey 1991, 1992), but have failed to be widely adopted because of difficulties in finding a cladding material with a thermal conductivity sufficiently low that the length-wise gradient can be maintained, but that is at the same time sufficiently high that heat transfer to the fused silica capillary is effectively instantaneous. Accordingly, temperature programming of the entire column in a well stirred oven is employed to permit analytes with a wide range of volatilities to be analyzed in one chromatographic run, all with k' values sufficiently low that diffusion broadening of the peaks is not too serious. In fact, the vapor pressures of the analytes turn out to be one of the key parameters in GC since there are no specific interactions between mobile phase and analyte. Thus, as before the distribution coefficient between the two phases is defined as (see equation [3.1]):

$$K^A = c^A_s/c^A_m$$

where now the mobile phase is the carrier gas and $c^A_m = p^A/RT$, where p^A is the partial pressure of analyte A present at equilibrium above the stationary phase containing a concentration c^A_s of analyte. But according to Raoult's Law of Partial Pressure:

$$p^A = \gamma^A_s.X^A_s.p^A_0$$

where X^A_s is the mole fraction of A in the stationary phase, p^A_0 is the equilibrium vapor pressure of pure A at temperature T, and γ^A_s is the thermodynamic activity coefficient of A in the stationary phase accounting for nonideal behavior of the solution (unequal intermolecular forces between A–A, A–S and S–S). Taking X^A_s as an appropriate measure of c^A_s gives:

$$K^A = X^A_s/c^A_m = (p^A/\gamma^A_s.p^A_0)/(p^A/RT)$$
$$= RT/(\gamma^A_s.p^A_0)$$

At first sight this result suggests that K^A should increase with temperature, but the linear increase indicated in the numerator is swamped by the exponential increase of p^A_0 (the vapor pressure of pure A described by the Clausius–Clapeyron equation), so that in fact K^A decreases strongly with increasing temperature and therefore so does k_A'

(see equation [3.17]). Then, for example, the selectivity coefficient for analytes A and B (equation [3.16]) is given for GC by:

$$\alpha_{A/B} = K^A/K^B = k'_A/k'_B = (p^B_0/p^A_0).(\gamma^B_s/\gamma^A_s)$$

Thus selectivity in GC is controlled by the ratios of vapor pressures of the pure analytes at the working temperature (instantaneous temperature if a temperature program is used) and by the corresponding ratio of the activity coefficients in the stationary phase; it is the latter that account for differences in selectivity on different stationary phases.

The strong dependence of k' on temperature (in fact log(k') ~1/T) means that temperature settings must be precisely and reproducibly controlled if retention times are to be reproducible (Hinshaw 2004, 2005). Thus the oven is a crucial component of any GC system (Hinshaw 2000). Recently the use of resistive heating, rather than a heated oven with rapid mixing of the air around the column, has been used to obtain heating rates of $20\,°C.s^{-1}$ with resulting very short elution times (Bicchi 2004).

4.4.3c GC Stationary Phases

Unlike the mobile phases there is a wide selection of GC stationary phases available, sufficiently so that the inexperienced analyst may feel overwhelmed. A helpful guide is available on the internet at www.chem.agilent.com/cag/cabu/gccolchoose.htm. The first decision concerns whether a packed or open tube capillary column is to be used, but in modern practice the only common application of packed columns is for analytes that are gases at room temperature. The absence of column packing means that the multiple-path term (A-term) in the van Deemter equation is not operative, and this accounts in large part for the improved separation efficiency of capillary GC (Ettre 2001). The introduction of fused silica capillary columns (Dandeneau 1979), strengthened mechanically by an external coating of polyimide polymer (Dandeneau 1990) and with internal diameters in the range 100 – 530 μm, affords superior column efficiency and (for 530 μm columns) an upper limit to sample capacity that is adequate for most purposes. (Imides are compounds containing a –CO–NR–CO– functional group). Usually 15 m long columns are used for fast screening and 30 m columns for most analyses, with 50–100 m columns available for extremely complex samples. However, column length does not affect column performance very much and does so at the expense of much longer analysis times. It is usually preferable to deal with complex samples by other means, e.g., use of a thinner film of stationary phase, optimization of flow rate and temperature program etc. The choice of internal column diameter (d_c, Section 3.5.11b) is a compromise between achievable resolution (better on narrow columns) and sample capacity (larger on wider columns because of increased surface area and thus more stationary phase, but note that this leads also to greater column bleed so upper limits to operating temperature must be lowered). Additionally the volume flow rate of carrier gas increases as the square of the diameter and this may cause problems for the vacuum system of a directly coupled mass spectrometer.

Another general parameter for which a decision must be made is the film thickness (d_f, Section 3.5.11b). As a general rule of thumb, thinner films elute analytes faster, with better resolution, and at lower temperatures, than thicker films. For this reason thin films are well suited to analysis of analytes that are high-boiling and/or thermally sensitive, and for closely-eluting analytes (similar values of p^A_0). The standard range of film thickness is $0.25 - 0.5\,\mu m$, and this suffices for most samples that elute at up to ~300 °C. Thin films (0.1 μm) are available for analytes that elute at > 300 °C. Conversely, to obtain sufficient retention for low boiling analytes (i.e., with appreciable values of p^A_0 at lower temperatures) that elute in the range 100–200 °C, thicker films (1.0–1.5 μm) are available. Wide-bore columns ($d_c = 530$ μm) are usually available only with thicker films as this is necessary to maintain resolution and reasonable retention times.

GC capillary columns are mostly the WCOT (wall-coated open tubular) type with a viscous liquid film coated (sometimes chemically bonded) to the deactivated silica wall. So-called PLOT (porous layer open tubular) columns have a solid substance (molecular sieves, alumina, some polymers) coated on the wall, but are used for only a few special purposes (analytes that are gases at room temperature, or resolution of small aliphatic hydrocarbons up to ~C_{10}). Most WCOT stationary phases are based on thermally stable polysiloxanes with different substituents on the alternating silicon atoms (Figure 4.19) and are usually cross-linked and sometimes also bonded to the column walls for extra stability. The polysiloxane phases possess different polarities depending on the nature of the R_1 and R_2 groups (Figure 4.19), and this property determines the types of compounds that are best retained since nonpolar phases tend to retain nonpolar analytes, and similarly for the polar case. (Note that compounds of higher polarity tend to have lower vapor pressures, see the preceding discussion of the relationship between analyte vapor pressure and k').

Polydimethylsiloxane (all substituents CH_3) is highly nonpolar, is usable up to ~350 °C and is widely used as a general purpose phase for e.g., aliphatic and aromatic hydrocarbons, PCBs, steroids and less polar drugs. Incorporation of increasing percentages of phenyl groups replacing methyl increases the polarity and polarizability of the phase and allows separation of correspondingly polar analytes; however, a higher percentage of phenyl groups tends to lower the maximum usable temperature to ~250 °C for 50 % phenyl. Introduction of short chain cyano groups (propyl or allyl) confers still higher polarity on the polysiloxane, but again lowers the maximum temperature to ~250 °C. Use of polyethylene glycol phases (the so-called carbowaxes) allows interactions via hydrogen bonds so that analytes like free fatty acids, glycols etc. can be separated at temperatures up to ~250 °C. For operation at temperatures up to ~400 °C, polydimethylsiloxanes modified with phenyl groups incorporated directly into the siloxane chain (Figure 4.19) provide phases with somewhat variable polarity depending on the percentage of phenyl groups.

(a)

(b)

(c)

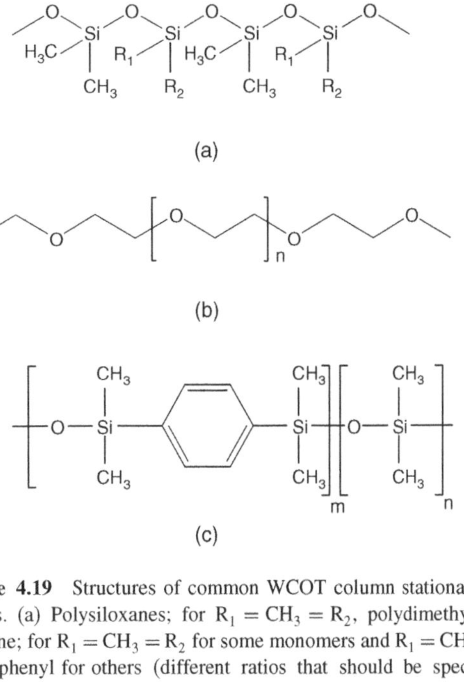

Figure 4.19 Structures of common WCOT column stationary phases. (a) Polysiloxanes; for $R_1 = CH_3 = R_2$, polydimethylsiloxane; for $R_1 = CH_3 = R_2$ for some monomers and $R_1 = CH_3$, R_2 = phenyl for others (different ratios that should be specified), methylphenylpolysiloxanes; for $R_1 = CH_3 = R_2$ for some monomers and $R_1 = CH_3$, R_2 = phenyl for some others and $R_1 = CH_3$, $R_2 = C_3H_5CN$ or C_3H_7CN for yet others (different ratios that should be specified), methylphenylcyanopolysiloxanes. (b) Polyethyleneglycol. (c) Silylarene polymer.

As mentioned previously for HPLC packings, development of creative new stationary phases is an extremely active and competitive area, and websites and catalogs

of column suppliers contain up-to-date information. For
example, chiral phases have been developed for enantiomeric separations by GC (Schurig 2001), and fluoro-
alkyl stationary phases can provide increased selectivity
compared with their nonfluorinated analogs although the
interaction mechanisms are unclear (Przybyciel 2004).
However, the foregoing summary covers the main types
that are used in current practice.

4.4.4 Sample Injection Devices for Chromatography

4.4.4a Automated Loop Injectors for HPLC

Loop injectors for HPLC based on six-port valves were
discussed in Section 2.4.3. The main point that bears
repetition is that in usual practice the loop volume is
seldom known to within any useful degree of accuracy, but the reproducibility is excellent ($\sim$0.2 % for the
'complete filling' mode of operation). Use of a surrogate
internal standard provides the accuracy required in quantitative methods. For high throughput operation, loop
injectors are combined with an autosampler in which
a robot is used to continuously complete the cycle of
filling the syringe, injecting into the loop and flushing
the syringe. An excellent description of the design principles used to construct autosamplers (Dolan 2001a)
is accompanied by a discussion of common problems
encountered with these devices (Dolan 2001b), particularly carryover (Dolan 2001, see also Section 9.7.2).
Sample carryover exists when a peak for a previously
injected sample appears in the chromatogram for a blank
(clean solvent) injection, and is particularly important
for LC/MS methods that are usually validated over a
dynamic range of 500–1000. The most common sources
of carryover are sample residue left in the autosampler
(needles and valves), and on frits at the column inlet.
Adjustment of the material used to fabricate tubing and
fittings, selection of a better wash solvent, addition of
additional wash cycles etc. can solve most carryover
problems (Dolan 2001). The same general comments
with appropriate changes apply equally to autoinjectors
for GC (these do not employ a six-port valve).

The optimum volume of sample extract injected varies
with the HPLC column dimensions and the mobile phase
flow rate. Essentially the initial 'plug' of sample extract
swept onto the column from the loop should not exceed
a few per cent of the column void volume, since the
injection volume then becomes a significant fraction of
the elution volume resulting in broad chromatographic
peaks (Section 3.4.8). It was shown in the case of
column chromatography (not HPLC in that case) that
peak width was directly proportional to injection volume

for isocratic elution (Huber 1972). Similarly the injection
volume also affects the height of the chromatographic
peak (Karger 1974) and thus the detection limit. A typical
injection volume for an HPLC column of 1 mm i.d. and
150 mm length, with a flow rate of $50 \,\mu L.min^{-1}$, would
be $2 \,\mu L$, while for a 2×50–100 mm column operated at
$300 \,\mu L.min^{-1}$ an injection volume of $20 \,\mu L$ would be
appropriate. In addition, overloading the column (injection of too much analyte to be accommodated on the
adsorption sites of the stationary phase) can lead to peak
asymmetry (Section 3.5.12). There are additional complications arising from the choice between overfilling and
partial filling the HPLC injection loop (Section 2.4.3).
Partial filling has the disadvantage that the repeatability of
the injection volume is degraded, but is often used when
the total extract volume is limited; however it is then
important to optimize the loop volume since too large a
value can result in a significant contribution to the dead
volume of the HPLC system. For example a $25 \,\mu L$ loop
would be a good choice for a $10 \,\mu L$ injection volume
using partial filling. Of course, as in all such decisions,
less than ideally optimal choices might still be fit for
purpose in any particular application.

4.4.4b GC Injectors

Because of the difficulties inherent in the injection of
liquid samples into a stream of hot carrier gas, injectors
for GC are not capable of providing good accuracy or
precision of injected volumes. For this reason, if a surrogate internal standard is not used, a volumetric internal
standard (Sections 2.2.4 and 8.5.2a) is added to the sample
extracts and calibration solutions just before injection to
correct for variations in the effective volume injected.
The analyte solution, of the order of one microlitre, is
injected from a syringe (Section 2.4.4) and completely
evaporated at a high temperature that is usually well above
that of the column. This process results in a volume of
vaporized solvent plus analyte ($\sim$10$^3 \,\mu L$) that is up to
10^3 times greater than that of the liquid, depending on
the temperature and pressure; this should be compared
with the internal volume of a capillary column 30 m long
and $300 \,\mu m$ internal diameter, that is only about twice
as large. Clearly if such a volume of vapor (including
analyte vapor) were injected directly into such a column
it would fill about half the column before any retention
could occur, resulting in zero separation; thus something
must be done to reduce the total volume reaching the
column. This problem is not as severe for packed columns
with significantly larger internal volumes, but even there it
is usually necessary to preferentially discard a large fraction of solvent vapor while retaining most of the analyte

by using a jet separator (Section 5.2.1a); however, packed column GC is not very relevant for most trace analyses. The following description of the solutions used for capillary columns is a modified version of a more detailed discussion (Kenndler 2004).

Two approaches are used to overcome the problem of large vapor volumes for capillary columns. With the split–splitless injector (Figure 4.20), the sample is introduced via a syringe into the heated injector and evaporated. Then either only part of the evaporated sample is allowed to enter the capillary column (the so-called split mode), or alternatively most of the solvent is separated in the injector from the analytes (splitless mode). In the latter mode the analytes (usually considerably less volatile than the solvent) are re-condensed at the entrance of the column and the more volatile solvent is swept away; the analytes are then flash heated to start the elution. The alternative to the split–spitless injector does not use an injector that is already heated, but rather one at around ambient temperature which, following sample injection, is heated in a controlled fashion so as to preferentially evaporate the solvent. An advantage of the heated injector is that it can be used in either mode to best suit the sample under analysis at the time. A more complete comparison of the two approaches is given below.

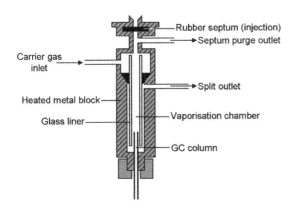

Figure 4.20 Sketch of a split–splitless injector for GC. The fractions of the inlet gas flow that exit through the septum purge outlet and the split outlet are controlled by needle valves.

In the split–splitless injector (Figure 4.20) the carrier gas flow is divided initially to permit a small fraction to continually flush the injection septum (maintained at a cooler temperature) to minimize carryover effects. The sample extract is rapidly injected from a syringe and evaporated in the liner of the heated injector; the resulting gas mixture (carrier gas plus analytes plus solvent) can either be completely directed into the column (i.e. split-less mode), or else the majority of the gas mixture is directed to waste and only a small fraction is allowed to flow into the separation capillary (i.e. split mode). A needle valve at the split exit (Figure 4.20) can be either closed completely (splitless) or opened partially to control the split ratio; another needle valve is sometimes used to control the fraction of carrier gas flow used to flush the septum.

The split mode has the advantage that the injected zone is narrow and the small sample aliquot entering the capillary avoids overloading the column. Although very flexible in practice, this mode has a number of disadvantages. In many cases mass discrimination among analytes is observed, especially when their volatilities are very different, and this can lead to systematic errors in quantitative analysis. Another disadvantage, especially in trace analysis, arises from the (sometimes small) fraction of the analytes that is transferred into the capillary i.e. the majority exits via the split outlet.

Splitless mode does not significantly suffer from these problems; in this case the injector is run in the split mode before injection, i.e., only part of the carrier gas flow is directed through the capillary column. Directly prior to sample injection the needle valve controlling the split ratio is closed completely (zero split). The sample is then slowly injected into the heated injector and analytes plus solvent are evaporated at a relatively low temperature, lower than the boiling point of the solvent so that the solvent vapor condenses near the entrance of the column and forms a kind of stationary phase there. Volatile sample components, also evaporated in the injector, are re-dissolved in the re-condensed solvent and thus become re-focused ('solvent trapping'). Less volatile sample components, which were also eventually evaporated in the hot injector, are re-condensed and re-focused on the colder part of column ('cold trapping'). After these two processes have finished (usually 30–90 s) the split is opened and the rest of the solvent is flushed via the split exit. The solvent that initially formed a liquid sheath at the entrance of the column evaporates gradually, with progressive evaporation from the side of the hot injector to the detector side, thus amplifying the refocusing of the analytes on the column. Finally the analytes are evaporated by application of a temperature program (obligatory for this injection technique).

The entire (apparently complicated) procedure avoids any large solvent peak tail and allows transfer of the majority of the analytes into the column, and thus to the detector. Splitless mode is, therefore, the better technique for minor components in trace analysis. However, the re-condensation of the solvent at the first portion of the capillary can damage the stationary phase, so only columns with chemically-bonded phases should be used in this mode. Also, the stationary phase must be wettable by the condensed solvent, otherwise droplets are formed rather than a liquid layer. Both of these problems can be overcome by use of a short length of empty capillary (usually widebore, ~500 μm i.d. and 20–200 cm long, without stationary phase), between the injector and the GC column, in which both re-focusing processes (solvent trapping and cold trapping) occur. Such 'retention gaps' are used in splitless mode and more commonly with the on-column injection technique described below, and have been discussed in more detail elsewhere (Hinshaw 2004a).

A variation of the split–splitless injector, the programmed temperature vaporizer (PTV) injector, alleviates the problem of analyte discrimination via volatility when injected into the hot injector; this device can be used in both split and the splitless modes. Injection is made into the cooled injector, initially at a temperature slightly above the boiling point of the solvent but below those of the most volatile sample components; the majority of the solvent is removed via the open split, and after thus flushing the solvent the PTV injector is heated rapidly so that the sample is transferred to the top of the column as a reasonably narrow band. This approach permits injection of much larger sample volumes (up to 50 μL), thus lowering detection limits for ultra-trace analytes; for this reason such devices are also referred to as large volume injectors. The approach also allows use of solvents like dichloromethane that can seriously damage a stationary phase if large amounts are injected.

An excellent account of practicalities in optimizing split–splitless operation (Reedy 2006) includes a description of how pulsing the pressure of the carrier gas supply can improve the overall performance. The relative advantages and disadvantages of split vs splitless operation can be summarized as follows (Kenndler 2004). The split mode yields a narrow sample zone on the column before elution starts, thus optimizing separation efficiency; it also allows optimization of the split ratio to avoid overloading the column. However, analyte discrimination via differing volatilities can lead to systematic errors in quantitation, and the levels of the minor components that reach the column can be reduced to below their detection limits. Splitless mode provides transfer of the majority of the analytes onto the column (and is thus best for low-level components) and also minimizes the long tail of the solvent peak. Disadvantages of splitless mode include possible damage to the stationary phase by the re-condensed solvent and the requirement for wettability,

although both of these problems are alleviated through use of a solvent gap at the expense of increased operational complexity.

The on-column injection technique avoids problems arising from the large volumes of solvent vapor. In this approach the sample solution is directly inserted into the column with the aid of a special syringe with a long, narrow needle, while maintaining all components at low temperature. The syringes used must be such that the outer diameter of the needle is less than the inner diameter of the capillary column and the needles thus have limited mechanical strength; for this reason a conventional GC septum cannot be used and is replaced by a special design. Usually a retention gap is interposed between injector and column and within the column oven, whose temperature must therefore be adjusted to the boiling point of the solvent. If the temperature is below the boiling point, solvent trapping (see above) takes place, and if it is selected slightly above the boiling point, cold trapping of the sample components occurs. In both cases the elution of the analytes requires an appropriate temperature program, an essential step when using this injector type. Its advantages include the complete transfer of analytes onto the column with no volatility discrimination effects, and minimal thermal stress on thermally labile analytes. However, the fragile nature of the syringe needle implies that development of an autoinjector is difficult.

4.4.5 Pumps for HPLC

The driving force producing mobile phase flow in GC is simply the cylinder of compressed gas delivered via a pressure regulator, although modern electronic flow controls introduce a useful level of sophistication (Hinshaw 2002b). However, HPLC pumps must provide a continuous constant flow of the eluent through the injector, column and detector, and are a crucial component of any HPLC system (Kazakevich 1996).

The most common requirements of an HPLC pumping system are delivery of a constant flow of mobile phase in both isocratic and gradient modes in the range from a few mL.min^{-1} to ~10 μL.min^{-1}, with an inlet pressure generally up to 5000 psi (~35 MPa) though considerably higher values (up to ~15 000 psi) are required for columns packed with ultra-small particles (Section 3.5.7); moreover, pressure pulses from piston-driven pumps must be no larger than 1% of the total flow rate for normal and reverse phase separations.

Constant-flow pumping systems are of two basic types, reciprocating piston and positive displacement (syringe) pumps. Both can provide reproducible elution times and peak areas regardless of viscosity changes or column blockage, up to the pressure limit of the pump. Piston pumps can maintain a liquid flow for indefinitely long times, while a syringe pump has to be refilled after it displaces the syringe volume. However, a syringe pump does not create any flow and pressure pulses while the piston pumps do, and syringe pumps are often used for microHPLC applications that require a constant flow in the $\mu L.min^{-1}$ range.

The basic principle of operation of reciprocating piston pumps is shown in Figure 4.21 for a single-piston pump. The pump rate is adjusted by controlling the distance the piston retracts, thus limiting the amount of liquid pushed out by each stroke, and/or by variation of the rotation speed of the eccentric cam. Such a pump can deliver a constant flow rate with a high back pressure and with effectively no limitations on the total amount of mobile phase that can be pumped without interruption since the reservoir is external to the pump itself (unlike a syringe pump). The main disadvantage of this simple type of piston pump is the sinusoidal pressure pulses and corresponding sinusoidal variations in flow rate. Better performance is obtained when two single-piston pumps like that illustrated in Figure 4.21 are arranged on opposite sides of the eccentric cam, so that when one pump is emptying the other is filling. This simple arrangement doubles the

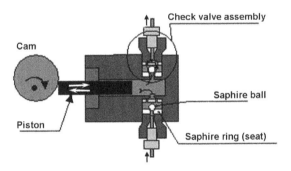

Figure 4.21 Diagram illustrating the operation of a single piston HPLC pump. A circular cam with an eccentric shaft is held tight against the piston. As the cam rotates it pushes the piston inwards as the larger side of the cam moves around to contact the end of the piston, and then withdraws the piston as the smaller dimension moves around. When the piston is moved inwards it creates a positive pressure in the mobile phase liquid in the body of the pump, and this opens the spring-loaded sapphire ball in the exit check valve and simultaneously forces closed that in the entrance check valve. In the next half cycle when the piston moves backwards, a negative pressure is created and the exit valve is forced to close while the entrance valve is opened allowing additional mobile phase into the body of the pump.

frequency of the pulses and decreases their amplitude for a given flow rate since now each pump has to deliver only half of the total demand; actually in practice the result is even better than this since the two out-of-phase pulses of pressure and flow rate tend to cancel one another out. The main operational disadvantage of these dual-head piston pumps arises from the role of the check valves as the weakest link since they can be contaminated or clogged, although use of volatile buffers (see Section 4.4.2) greatly alleviates this problem; reduction of the number of check valves while still maintaining the benefits of the dual-head pump is achieved by some ingenious designs (Kazakevich 1996).

Nonetheless, for the most demanding work, with minimal pulses of pressure and thus flow rate, a pressure pulse damper is used to 'smooth out' the pulses. The simplest such device is simply a coil of narrow bore tubing inserted between the pump and the injector. As the pump strokes the coil flexes, absorbing the energy of the pressure pulses. This type of pulse damper holds a large amount of liquid that must be purged during solvent changes and is not well suited to gradient elution. The most usual damper type is a low volume chamber (less than 0.5 mL), separated into two by a membrane. One half forms part of the flow path of the mobile phase between pump and injector while the other half is filled with an inert liquid (heptane) whose compressibility is large enough to absorb the energy associated with the positive pressure pulses sensed through the membrane. Such a damper is very effective in compensating for the pulses created by a dual-head pump with each piston volume around $100\,\mu L$.

Syringe pumps are generally larger scale versions of manual syringes (Section 2.4.4) and consist of a cylinder containing the mobile phase that is expelled by a piston driven by a motor connected through worm gears to provide smooth pulse-free flow. They have a number of other advantages; the mechanical aspects are simple leading to low maintenance requirements, and they are capable of generating quite high pressures (tens of thousands of psi). The disadvantages are the limited reservoir capacity and difficulties in programming two such pumps to provide reproducible gradients, especially fast gradients. They are still sometimes used in their original implementation for microcolumn chromatography at flowrates of $1-100\,\mu L.min^{-1}$.

Recently, a different approach to HPLC pumping was developed (Jensen 2004). Pressure in the system is generated by connecting laboratory air or nitrogen to a pneumatic amplifier that produces an amplification factor for pressure values up to 36; for example, a nitrogen supply at 100 psi can be amplified to deliver pressures in the range 0–3600 psi. These systems also incorporate microfluidic flow control (MFC) to generate precise LC gradients at both nanoscale and conventional capillary flow rates. This combination can provide highly precise gradients at flow rates in the nL.min⁻¹ range (important for proteomics research) without the post-pump flow splitting (see later) that is generally required

with conventional HPLC pumping systems; it can also respond extremely rapidly to set-point changes. Flow meters in the line following the pneumatic pumping stage for each mobile phase continuously monitor flow rate and feed a proportional signal back to a microprocessor (Figure 4.22), which in turn sends out a voltage signal to the controller at the pressure source for each mobile

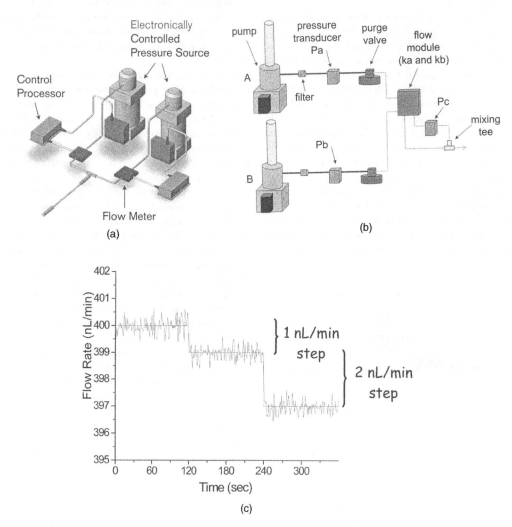

Figure 4.22 Schematic of a pumping system based on a pneumatic gas pressure amplifier with microfluidic flow control via feedback from a sensitive flowmeter. In this way the flow rate is maintained regardless of changes in system back pressure or mobile phase viscosity, and changes in flow rates can be established rapidly and accurately. (a) A gradient system in which the mobile phase composition is controlled via flow rates of both mobile phase solvents. (b) A gradient system in which both back pressures and flow rates are monitored; volume flow rates $U_A = k_A.(P_C - P_A)$ and $U_B = k_B.(P_C - P_B)$ where k_A and k_B are calibration constants. (c) A demonstration of the precision and accuracy with which controlled flow rates can be changed rapidly at total flow rates in the nL.min⁻¹ range, suitable for packed capillary HPLC. Reproduced from company literature (Eksigent 2005, 2006) with permission from Eksigent LLC.

phase. This signal is proportional to the pressure required in each mobile phase to achieve the desired flow rate or gradient. The MFC controller regulates this pressure to generate the required flow rate (Jensen 2004). For high pressure mixing for gradient operation, a microscale mixer with a volume of 300 nL was developed, resulting in a mixing delay volume of only two seconds at $10\,\mu L.min^{-1}$. This rapid mixing virtually eliminates the mixing delays required in conventional pumping systems at the start of the method and during column re-equilibration. The rapid mixing plus the feedback control lead to extremely precise gradients and thus reproducible retention times, rapid system response and excellent gradient linearity for low flow rates compatible with those in packed capillary columns. Such performance is not possible with conventional piston pump systems, which in many cases must create the gradient at a considerably higher flow rate that is then split before entering the capillary column, leading to wastage of large amounts of solvent.

Flow splitters are available commercially for pre-column splitting in order to match a higher flow rate pump to the flow appropriate for a small diameter column, and also for post-column splitting to match the flow rate of an HPLC eluent to the value that optimizes the efficiency of e.g., an ESI source. For example a typical eluent flow rate from a 1 mm i.d. column is $50\,\mu L.min^{-1}$, and this matches quite well with the optimum flow for many conventional ESI sources. However, if a 2 mm i.d. column is used at a flow rate of $0.2-1\,mL.min^{-1}$, a post-column split of ~4–20 is required to optimize the ESI source. Under the conditions where an ESI source provides a response that is a function of concentration (Section 5.3.6b) rather than of mass flux of analyte (and this is the case for many practical experimental conditions), such diversion of up to 95 % of the eluent away from the mass spectrometer makes little difference to the mass spectrometric response. The diverted 95 % can be directed to a fraction collector or to a UV or other detector. Flow splitter designs can be fairly simple, e.g., essentially a zero dead volume T-piece with the perpendicular arm holding a length of capillary whose length and internal diameter are adjusted to provide the appropriate split via the flow resistance of the capillary relative to that of the tubing connecting the T-piece to the ion source. More complex designs incorporate a valve whose flow resistance can be continuously adjusted to provide a correspondingly adjustable flow. In either case, the arrangement is analogous to an electrical circuit incorporating two resistors connected in parallel, one of which is adjustable to control the current that flows through the other fixed resistor (that corresponds to the flow path from HPLC to the ion source).

4.4.6 Capillary Electrophoresis and Electrochromatography

As mentioned in Section 3.7, CE and CEC are not used on a routine basis when ultimate performance (accuracy and precision etc.) is required for true trace level quantitative analysis of small molecules using mass spectrometry. Direct coupling of bench scale CEC (and CE to some extent) to MS is not straightforward. However, both CE and CEC are key technologies for development of integrated 'lab-on-a-chip' devices (Section 4.4.7) and this is the reason for the following discussion.

Applications of CE with UV detection in the pharmaceutical industry (Altria 1998) are mainly concerned with analysis of pharmaceutical preparations for the active ingredient(s) and impurities, not for the truly challenging analyses of drugs and metabolites in plasma, urine etc. However, a more recent review (Ohnesorge 2005) of quantitation using CE with mass spectrometric detection has indicated that many of the intrinsic problems are being alleviated significantly. Helpful hints on troubleshooting a CE system (though not CE/MS) are available from Biorad Inc. (Darling 2005), and at the website http://www.chem.agilent.com/scripts/literaturepdf.asp?iWHID=15854. An excellent introduction to chiral analysis by CE is available (Fanali 1995) and a recent review (Van Eeckhaut 2006) covers both theory and applications.

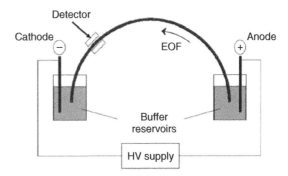

Figure 4.23 Diagram of a simple system for capillary zone electrophoresis (CZE, the most commonly used version of CE) using an integrated spectroscopic detector (UV–visible absorption or fluorescence). The example illustrated is set up for analysis of compounds that are cations under the buffer conditions used.

A simple diagram illustrating a typical CE apparatus is shown in Figure 4.23. The intrinsic simplicity is one of the main advantages of CE over HPLC. The fused silica capillary itself is one of the most important components;

internal diameters d_c of 20–100 μm are normally used, with lengths that are limited by the available output from the high voltage power supply (typically up to ~30 kV) and the desired electric field (a few hundred V.cm^{-1}). The power supply is operated in constant voltage mode and must be capable of supplying an output current consistent with the applied voltage and the electrical conductivity of the buffer solution in the capillary. An uncoated fused silica capillary is usually prepared for its first use in CE by rinsing it with 10–15 column volumes of 0.1 M sodium hydroxide followed by 10–15 column volumes of water and finally 5–10 column volumes of the separation buffer. For a coated capillary (Section 3.7) the preparation procedure is similar except that the 0.1 M sodium hydroxide is replaced by methanol.

Sample injection involves extremely small volumes (picolitre to nanolitre range) so as to not occupy more than a few percent of the total capillary volume (a 1 m capillary with d_c of 50 μm contains a volume of 2.5 μL). The two commonly used injection methods for CE are hydrodynamic (pressure) and electrokinetic. Each involves temporarily removing the buffer reservoir from the inlet end (the cathode in the case illustrated in Figure 4.23) and replacing it by a vial containing the solution to be analyzed. Hydrodynamic injection involves application of a pressure difference between the two ends of the capillary. The volume amount of sample solution injected can be calculated using the Poiseuille equation:

$$V_{inj} = (\Delta P/l).(d_c^4.t_{inj})/(128\eta)$$

where V_{inj} is the calculated injection volume, ΔP is the pressure difference between the ends of the capillary of length l, d_c is the inner diameter of the capillary, t_{inj} is the injection time and η is the sample viscosity. Electrokinetic injection is performed by simply turning on the voltage for a period. The moles of each analyte injected, $Q(i)$, are determined by the net electrophoretic velocity, u_{net}, of each analyte (equation [3.71]), the injection time, t_{inj}, and the ratio of electrical conductivities of the separation buffer and the sample solution, k_b/k_s:

$$Q(i) = u_{net}(i).C(i).(k_b/k_s).t_{inj}.(\pi d_c^2/4)$$

where C(i) is the molar concentration of the *i*th analyte. Because each analyte has a different mobility and thus $u_{net}(i)$, electrokinetic injection provides a biased injection and the composition of the injected sample is different from that of the original sample; this problem can be overcome in quantitative analysis only through use of carefully chosen surrogate internal standards. It is here that mass spectrometric detection becomes essential since an SIS

that behaves the same as the analyte in the electrokinetic injection will also by definition behave very similarly in the electrophoresis itself. Despite this drawback, electrokinetic injection is useful for capillary gel electrophoresis in which the polymeric gel inside the capillary is too viscous for hydrodynamic injection.

The major limitation of CE, poor concentration sensitivity, is a consequence of being able to load a sample volume no more than a few percent of the total capillary volume without seriously degrading the separation efficiency. This limitation can be overcome in principle by field amplified sample stacking (Figure 4.24); this technique uses sharp gradients in electrolyte conductivity to subject sample ions to nonuniform electric fields. Typically, sample ions are dissolved in an electrolyte buffer of relatively low concentration and thus low electrical conductivity, i.e., high electrical resistance. After injection this sample region is surrounded by zones containing buffer of more usual (higher) conductivities, a step that requires careful preparation of the capillary and subsequent analysis before longitudinal diffusion reduces the gradient to unusably low values. When the high voltage is applied, since the resulting electrical current must be the same at all points along the capillary, there is a high potential drop across the high-resistance sample region and thus a higher effective value of E_{ext} than exists in the low resistance regions (Figure 4.24). In turn, this leads to a higher local electrophoretic velocity (equations [3.66–3.68]) than in the neighboring regions with high buffer concentrations. Sample ions stack up as they move from the high field, high velocity region to the low field, low velocity region. In this way the analyte can be concentrated by large factors in some cases. However, the sample stacking process is tedious and time consuming, and frequently the sample solution is instead pre-enriched by, e.g., SPE. More recently (Ptolemy 2006), strategies have been described that permit on-line sample pre-concentration together with chemical derivatization in a CE capillary for high throughput metabolite screening, though at present the methodology is not directed at quantitative analyses.

Capillary electrochromatography (CEC) usually employs capillaries packed with 3–5 μm particles with C_{18} or C_8 stationary phases held in place with two retaining frits. As for CE the optical detection window is made by removing a small region of the polyimide external coating, thus allowing on-column detection with little if any extra-column band broadening; in CEC this window is best located immediately after the second frit. Good capillaries packed with 3 μm particles should generate at least 150 000 plates per metre (Smith 2001). The driving force in CEC is the electro-osmotic flow (EOF) that is highly

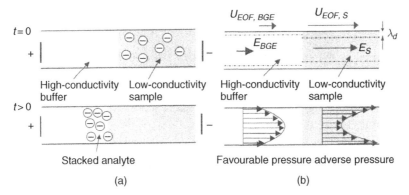

Figure 4.24 (a) Schematic showing field-amplified sample stacking (FASS) of *anionic* species *in the absence of electro-osmotic flow (EOF)*. A gradient in the background electrolyte ion concentration (and thus electrical conductivity) is established and the sample is placed in the low conductivity zone at t = 0. Upon application of an electric field in the direction shown (t > 0), this axial gradient in conductivity results in an electric field gradient. Since area-averaged current density is uniform along the axis of the channel (no build-up of overall charge anywhere), the low conductivity (high resistance) section (containing the analyte) is a region of high electric field and thus of high electrophoretic velocity, and the region of high conductivity (low resistance) is a region of relatively low electric field. As analyte anions exit the high field/high electrophoretic velocity region and enter the low velocity region, they accumulate locally near the region boundary and thus increase in concentration. Cations can be similarly stacked by reversing the field direction. (b) Stacking in the presence of an electro-osmotic flow (EOF). Gradients in conductivity generate axial variations in both electric field and electro-osmotic mobility, thus generating internal pressure gradients that tend to disperse the sample. Reproduced from Bharadwaj and Santiago, *J. Fluid Mech.* **543**, 57 (2005), with permission of Cambridge University Press.

dependent on pH, the buffer concentration, the organic modifier and the type of stationary phase. The concentration of deprotonated silanol groups present under the operating conditions largely determines the EOF, that drops off almost linearly between pH 10 and pH 2 by as much as a factor of three; for this reason most CEC is performed at pH > 8. The most common stationary phase used in CEC is $3\,\mu C_{18}$; depending on the manufacturer the silica can be very acidic and thus promote EOF even at low pH. Some stationary phases developed specifically for CEC contain both hydrocarbon substituents and sulfonic acid groups on each particle, resulting in a sizable EOF even at pH 2 because the sulfonic acid groups are very acidic and provide a negative charge on the particles at these low pH values.

CEC is essentially reverse phase HPLC modified to promote EOF rather than external pressure as the driving force for the mobile phase. The nature of the organic component of the mobile phase has a significant effect on the separation in CEC. With acetonitrile as organic component the EOF increases almost linearly with increasing acetonitrile concentration, but passes through a minimum with methanol (Smith 2001). The effect of electrolyte concentration on both EOF and separation efficiency was investigated (Knox 1987,1991) and shown to have little effect on EOF over the range 4×10^{-5} to

2×10^{-2}M; however, efficiency increased significantly with electrolyte concentration over this range.

Bubble formation is one of the common problems arising in practical implementation of CEC and was originally believed to be mainly caused by the heat generated by passage of the electro-osmotic current. However, in CEC the overall buffer concentrations used are low relative to those in CE, resulting in operating currents much lower ($\leq 10\,\mu$A) than those in CE, and it is now believed (Smith 2001) that bubbles are mainly formed at the exit frit as a result of the change in electro-osmotic flow as the solvent passes from the packed bed into the unpacked region of the capillary through this frit. Bubble formation can cause breakdown of the current and thus of the EOF, and it is recommended (Smith 2001) that the capillary should be pressurized at both ends during each run to suppress bubble formation (this is difficult to achieve when coupling CEC to mass spectrometry). Frit blockage by solid microparticles can be alleviated by pre-filtering solvents and sample solutions (as in HPLC the latter should be made using mobile phase or a solvent of lower elution strength).

There is considerable ongoing research and development in CEC; currently its main interest from the viewpoint of this book is its importance for on-chip devices (Section 4.4.7). Instrumentation for conventional

CEC, including pressure assisted CEC, has been recently reviewed (Steiner 2000). The question of gradient elution in CEC is a complex one (Rimmer 2000) and approaches include use of pH gradients as initially developed for CE. Step gradients are the most easily implemented in practice but provide less flexibility in separation than continuous gradients. Pressure assisted CEC is most easily adapted for gradient elution, but does not offer the very high column efficiency provided by totally electro-osmotically driven mobile phases. The development of flow-injection interfaces allows a true solvent gradient to be generated by micro LC pumps, with the mobile phase drawn into the CEC separation capillary by pure electro-osmotic flow. Finally voltage gradients provide a mobile phase velocity gradient (Rimmer 2000).

Although most stationary phases used in CEC are still the traditional reverse phase silica particles (mostly C_{18}), other novel phases are being developed (Pursch 2000), including ion exchange phases, sol–gel approaches to monolithic columns, organic polymer continuous beds etc. A very recent review of monolithic stationary phases for CEC (Kłodzińska 2006) emphasizes their applicability to microfabricated devices, especially as such phases do not require retaining frits. However, currently the reproducibility of the fabrication process is still problematic and an alternative approach of entrapping stationary phase particles in an organic polymer has been investigated (Chirica 2000, Xie 2005), a combination of the packed and monolithic methods. It was shown (Xie 2005) that beds of commercially available $3 \mu C_{18}$ silica beads could be trapped in a capillary column by photo-initiated polymersization of a suitable monomer, such that the entrapping polymer 'glued' the beads together at specific bead–bead and bead–capillary contact points permitting a no-frit fabrication technology that could withstand applied pressures > 4400psi. In addition to the applicability of this fabrication protocol to SPE, nanoHPLC and CEC columns, it is particularly promising for 'lab-on-a-chip' systems since a bed may be precisely positioned through photo-patterning.

Fundamental studies of CEC include an investigation of the electrochemical parameters in CEC columns (Henry 2005). Most CEC columns consist of a packed segment and an open (but buffer filled) segment; the two segments differ in two respects, their electrical resistivity and their zeta potentials at a multitude of solid–liquid interfaces. Measurements of resistivity and zeta potentials of an entirely open unpacked column can be used together with those of the CEC column itself to determine the electrochemical properties of both segments, and the properties of each segment separately can be combined to give the same properties of the CEC column as a whole (Henry 2005).

4.4.7 Micro Total Analysis Systems (Lab-on-a-Chip)

The lab-on-a-chip concept is undergoing a rapid development phase at present, in contrast with the maturity of many of the other technologies that are discussed in this book; indeed, the Royal Society of Chemistry (UK) has started publishing a new journal with this very title (http://www.rsc.org/Publishing/Journals/lc/index.asp). Biennial reviews of this field (Reyes 2002; Auroux 2002; Vilkner 2004; Dittrich 2006) provide ample evidence that any attempt here to describe the current state-of-the-art would be rapidly out of date, so the main thrust will be focused on more fundamental concepts; additional problems associated with interfacing such devices to mass spectrometry are discussed in Chapter 5. It should be mentioned that the current flurry of activity in this field has been driven mainly by applications to proteomics and other areas of modern molecular biology, rather than to trace quantitative analysis of small molecules; some examples of the latter applications, of direct interest here, are also given in Section 5.2.3.

The need for rapid, on-line quantitative measurements of target analytes at low concentrations in complex mixtures requires a high level of discrimination of the analyte from potential interferences. A chemical sensor that responds to only the target analyte, with zero response from any of the uncountable numbers of other chemical compounds that could be present, would be the ideal solution if such a thing could be found. The more realistic approach, that is the underlying concept of this book, is to incorporate sample preparation steps before the actual measurement, thus decreasing the selectivity demands on the detector and reducing limits of detection and quantitation as a result of reduced background. The concept of a total analysis system (TAS) involves a series of steps in which sample extraction and concentration, analyte separation and detection, are conducted in an automated fashion, originally using conventional laboratory equipment like that discussed in this book, to transform chemical information into electronic information (Kopp 1997). Modification of the TAS approach, by miniaturization and integration of the multiple steps into a single device, can yield a system that can be viewed as a complete chemical sensor with fast response time, low sample consumption and high stability which is readily transportable for on-site work.

The basic principles of scaling down bench scale apparatus to a size that can be accommodated on a chip, using technology developed for microfabrication in the electronics industry, were described in remarkable work by Manz and his collaborators (Manz 1990). The concept of a micro total analysis system (μ-TAS) is far from trivial and requires a fundamental understanding of how to scale physical dimensions (downward in this case) while maintaining the inter-relationships that describe the operating principles of the various steps that are to be integrated on the device. The original paper (Manz 1990) provides a very general approach, using reduction of these physico-chemical relationships to a form involving only dimensionless variables by multiplying each physical variable by an appropriate constant and by an appropriate power of d, an appropriate length variable (e.g., the diameter of a tube or a particle diameter etc.). Instead of reproducing these very elegant general arguments here, the flavor of the approach can be appreciated from the ad hoc treatment of a special case (the analysis time and separation efficiency for CE, i.e. the number of theoretical plates) originally given in an excellent exposition of μ-TAS (Jakeway 2000).

The time t_A for analyte A to emerge from a CE column is simply:

$$t_A = l/u_{net,A} = l/(\mu_{app,A}.E_{ext}) = l^2/(\mu_{app,A}.V_{ext})$$

where l is the length of the capillary between injection and detection, $u_{net,A}$ and $\mu_{app,A}$ are the net speed of A and its apparent mobility under CE conditions (equation [3.71]) and $E_{ext} = V_{ext}/l$ is the external electric field obtained by applying a potential drop, V_{ext}, across the length of the capillary. Assuming that the same external field can be applied, the analysis time should therefore decrease proportionally with the capillary length, and if the same potential drop can be sustained t_A decreases as l^2.

This is good news for miniaturization, but does not address the important question of separation efficiency N_A (do not confuse this with the Avagadro Constant) for analyte A. It is a very general result (equation [3.20]) that:

$$N_A = 4.(t_A/\sigma_{t,A})^2 = 4.[(t_A.u_{net,A})/(\sigma_{t,A}.u_{net,A})]^2$$
$$= 4.(l/\sigma_{x,A})^2$$

where $\sigma_{x,A}$ is the half-width of the peak (assumed to be Gaussian) measured at the inflection points along the main axis (x) of the capillary as analyte A reaches the detector. In the case of CE the only cause of peak dispersion, in addition to extra-column contributions from injector and detector, is longitudinal diffusion (no contributions from multiple path or mass transfer effects). The extra-column contributions can be arranged experimentally to be much

less than that from longitudinal diffusion, which is given (equation [3.34]) as:

$$\sigma_{x,A}^2 = 2D_m^A.l/u_{net,A}$$

where D_m^A is the diffusion coefficient of A in the mobile phase solvent. Then by combining the last two equations, N_A can be expressed as:

$$N_A = (2.l.u_{net,A})/D_m^A = (2.l.\mu_{app,A}.E_{ext})/D_m^A$$
$$= (2.\mu_{app,A}.V_{ext})/D_m^A$$

This last result is important because it tells us that, provided that the device can withstand the same potential drop, V_{ext} as the length, l, of the CE device is decreased, the separation efficiency is not affected by miniaturization so theoretically it is possible to obtain a significant decrease in analysis time (see above) without a loss in separation efficiency.

However, the optimistic assumption concerning V_{ext} must be examined further. Apart from the possibility of electrical breakdown (arcing) if too large a V_{ext} is applied over too small a length, the heat generated by the electrical current passing through the CE electrolyte must be considered since this can lead to temperature gradients that increase $\sigma_{x,A}$. Then H, the amount of heat generated per second (in watts, $W = J.s^{-1}$) generated by an electrical current I (in amps, A), flowing through a conductor of resistance R (in ohms, Ω) as a result of a potential drop V_{ext}, is given by:

$$H = V_{ext}.I = V_{ext}^2/R \text{ (by Ohm's Law)}$$

In CE the conductor is the supporting electrolyte solution, characterized by its molar conductivity, k_{molar}, defined as the electrical conductance (Ω^{-1}, reciprocal of resistance) of a solution of the electrolyte in a specified solvent that contains one mole of electrolyte and is of unit length (m) and unit cross-sectional area (m^2) relative to the direction along which V_{ext} is applied. The value of R is directly proportional to the length, l, of the electrolyte solution and inversely proportional to its cross-sectional area, A, and to its molar concentration, C ($mol.m^{-3}$), with proportionality constant k_{molar}:

$$R = l/(k_{molar}.A.C)$$

where k_{molar} has SI units ($m^2.\Omega^{-1}.mol^{-1}$). Note that k_{molar} itself is a function of C as a result of inter-ion electrostatic forces (see any textbook of physical chemistry) but it is not necessary to go into this level of detail for present purposes. Then for the supporting electrolyte solution in CE:

$$H = V_{ext}^2 . A . k_{molar} . C/l$$

Since this rate of heat production occurs uniformly throughout the electrolyte, and since cooling through the walls is the only means of dissipating this heat, its deleterious effects on CE performance can be limited by using a large surface:volume ratio, i.e. the capillary in conventional CE and a microchannel for an on-chip device. This wall-cooling effect gives rise to a radial temperature gradient perpendicular to the walls which in turn gives rise to local convection effects that add to the effect of longitudinal diffusion to increase $\sigma_{x,A}^2$ and thus decrease N_A. Thus, since H varies as l^{-1}, miniaturization (decrease of l) increases the contribution of this band broadening effect. However, in turn this disappointing result can be alleviated in the case of a microchannel by manipulation of the cross-sectional area, A. Such microchannels are usually fabricated with a rectangular cross-section so $A = h_c^2 . a_c$, where h_c is the height (depth, measured in the vertical direction) of the channel and a_c is its aspect ratio (its width divided by its height, e.g., the aspect ratio of a traditional television screen is 4:3, and that of a high definition television screen is 16:9). Since thermal convection is a gravitational effect, i.e., it occurs in the vertical dimension, the channel height is a much more important parameter than the width; indeed, further analysis (Jakeway 2000) has shown that in such channels the effective value of N_A is essentially independent of channel width while the contribution of thermal convection to band broadening varies as h_c^2, so use of microchannels with small h_c and large aspect ratio can alleviate the deleterious effect of miniaturization on N_A via the reduction of l. (Of course such considerations do not apply to conventional capillaries which have circular cross-sections as a result of the way in which they are fabricated.) In summary, CE analysis times can be reduced by decreasing channel length and/or increasing the applied potential drop, but separation efficiency (plate number) is to a first approximation independent of CE channel length and can be increased by an increase in potential drop; degradation of separation efficiency by heat generation can be alleviated by reducing channel heights.

The purpose of the foregoing analysis of a simple example is merely to emphasize that the process of scaling down laboratory scale apparatus to on-chip dimensions is far from trivial but is amenable to theoretical analysis. Thanks to the enunciation of the general principles (Manz 1990, Janasek 2006) it is now possible to approach this task in a rational manner. Nonetheless, the

first microfabricated device for chemical analysis predated publication of the theoretical principles by over 10 years. A gas chromatograph fabricated on a planar silicon wafer (Terry 1979) was fabricated using standard photolithographic and wet-etching techniques; it incorporated a sample injection system, a 1.5 m column and a thermal conductivity detector (Figure 4.25). An interesting discussion of this and subsequent work on chip-based chrom-

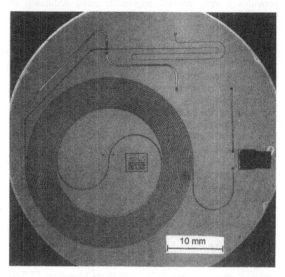

Figure 4.25 Photograph of the first GC on a chip. Reproduced from de Mello, *Lab Chip* **2**, 48N, (2002), with permission from the Royal Society of Chemistry.

atography systems (de Mello 2002) has described this first remarkable achievement that was essentially ignored for several years possibly because the GC separation, although fast, did not provide separation efficiencies equivalent to those achievable using bench scale apparatus. A very recent (Agah 2005) development has described a microfabricated GC with integrated heaters and temperature sensors for temperature programming and integrated pressure sensors for flow control. In 1990, the first on-chip HPLC (Manz 1990a) was described (see Figure 4.26), followed two years later by the first on-chip CE (Manz 1992).

As mentioned above, design, microfabrication, testing and application of devices of this kind is undergoing an explosive growth phase (Reyes 2002; Auroux 2002; Vilkner 2004; Dittrich 2006; Whitesides 2006). Some authors prefer to distinguish between μ-TAS and lab-on-a-chip systems. The former correspond to chemical sensors requiring no external ancillary devices other than

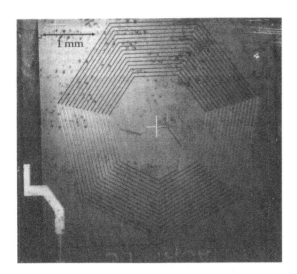

Figure 4.26 Photograph of the first HPLC chip (Manz 1990a). Reproduced from de Mello, *Lab Chip* **2**, 48N, (2002), with permission of the Royal Society of Chemistry.

sample introduction and electronic read-out, and the latter to systems incorporating at least two of the analytical steps in Figure 1.2, thus excluding multi-well sample plates, for example; however, this distinction is not universally accepted. Another excellent published overview of μ-TAS systems (Lee 2004) is complemented by engineers' views of this subject (Gale 2001) and of integrated microfluidics systems (Erickson 2004); in addition the complete Proceedings of the 7th International Conference on micro-Total Analysis Systems is available on-line from the University of Alberta (http://www.chem.ualberta.ca/~microtas/Start.pdf). An extensive review of sample pre-treatment on microfabricated devices (Lichtenberg 2002) addresses one of the more difficult problems still facing true μ-TAS and lab-on-a-chip systems; a more recent contribution (Gong 2006) has addressed the theory and practice of on-chip field-amplified stacking. Another continuing concern is the cost of these devices and their useful lifetime (number of analyses). Fabrication using plastic substrates rather than glass or silica is a favorite approach at present (Kameoka 2001). One promising approach (Koerner 2005; Brown 2006) involves transformation of a (positive or negative) image of the proposed microstructures micromachined in glass (the master) into a (negative or positive) image made of epoxy resin that can be used as a robust stamping tool for hot embossing the microstructures in a cheap polymer like poly(methylmethacrylate).

A recent review (Lazar 2006) has described the state-of-the-art in microfabricated devices with special emphasis on their use in sample introduction for mass spectrometric applications to proteomics. The potential importance of such devices for reducing carryover (Section 9.7.2) for bioanalytical work on small molecules has been described (Dethy 2003). The coupling of these devices to mass spectrometry is discussed in Section 5.2.3.

4.4.8 General Comments on Detectors for Chromatography

This book is concerned with the unique benefits of mass spectrometry for trace level quantitative analysis, so no detailed discussion of other chromatographic detectors is included except for occasional comparisons. However, some general aspects applicable to all detectors are discussed here, with special emphasis on how these generalizations apply to mass spectrometry (see also Section 6.2.3).

The ideal chromatographic detector for trace analysis should satisfy all of the following requirements (of course this is impossible): low intrinsic detector noise, low background signal and low long term drift; low detection limits; high sensitivity (slope of response curve); low limits of quantitation and detection; large linear dynamic range; low dead volume with negligible remixing of separated bands; high selectivity and universality; not susceptible to minor variations in temperature, flow rate, solvent composition and other operating parameters; operationally simple and reliable; nondestructive. Not all of these 'figures of merit' are completely independent of the others but it is convenient to consider them separately; it may be helpful in the following discussion to refer to Figure 4.27.

Noise is the erratic variations in detector output that occur rapidly relative to the timescale of a chromatographic peak width, regardless of whether or not there is any analyte present. The *intrinsic noise* level of a detector arises from random electrical fluctuations within the detector and its associated circuitry. Intrinsic noise should be distinguished from the so-called *chemical noise*, better described as 'chemical background', an additional contribution observed in a mass spectrometer that arises from small and rapid variations in the chemical composition of the mixture flowing into the spectrometer, and/or from ubiquitous contaminants from laboratory air or from plasticizers in plastic components etc. that can give rise to the 'peak at every m/z value' which is observed at high magnification of many mass spectra (Aebi 1996). In most practical applications to trace analysis, chemical background is the more important contribution and as a result can be alleviated through use of detection methods with increased selectivity, e.g. MS^n or high mass resolution, although the

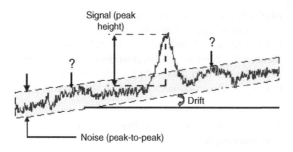

Figure 4.27 Sketch of portion of a chromatogram illustrating the relationship among short term, high frequency noise and background, long term drift, a peak that is above the limit of detection (LOD) based on a signal:noise criterion of ≥ 2, and two possible peaks (marked with ?) that are nonetheless below the LOD and are assigned as chemical background. Note that in this example the total 'noise' level (random noise plus possible chemical interferences) has been defined as the peak-to-peak envelope rather than the root mean square (RMS) average or some multiple of standard deviation of the noise, and that the drift has been accepted as a nonhorizontal baseline rather than as a contribution to the total 'noise'. Modified from Scott, www.chromatography-online.org, with permission.

intrinsic noise defines the 'statistical limit' beyond which these strategies break down (see Figure 6.1). Measurement of the level of 'noise' plus 'background' is done in several different ways and should be specified when a value of S/B and/or S/N is quoted. The peak-to-peak envelope of all contributions to background on either side of a signal peak is sometimes used (Figure 4.27) but other definitions include the square root of the average of the squares of deviations from the average baseline (so-called root mean square or RMS) and some multiple of the standard deviation of the background measured relative to the average baseline. Such calculations are available as data processing options in most mass spectrometer data systems but it is the responsibility of the user to determine which definition of 'noise' is used in these algorithms since the values can differ by 50–100 %. These concepts are discussed in more depth in Sections 7.1 and 9.5.6c.

Drift is a longer term variation corresponding to a baseline signal (no analyte present) with a nonzero slope (Figure 4.27); drift is usually associated with the warmup process when the instrument is first switched on after a significant shut-down period, or it could arise in a chromatogram obtained using gradient elution such that the progressively stronger mobile phase dissolves more and more material from the stationary phase and/or plastic components.

The *detection limit* (LOD) of a detector for a specified analyte is the smallest amount or concentration of that analyte that produces a signal that is clearly

distinguishable from the average noise (background) level (Section 8.4.1). A common working definition of LOD uses a S/B value background B ratio (S/B) of 2–5 (three is the most commonly quoted value) as the specification of 'clearly distinguishable'; the signal S is the signal produced by the specified amount or concentration of analyte (peak height in the case of chromatographic detection), and the background B is usually the peak-to-peak amplitude of the total background B observed in the absence of analyte (close to the analyte chromatographic peak, see Figure 4.27) though sometimes the root mean square average of the background B is used. The main disadvantage of such a definition of LOD is that it does not readily permit an evaluation of the statistical probability that the assignment ('peak' or 'no peak') is not a false positive or false negative (see Section 8.2.4). Section 8.4.1 discusses statistically defensible definition and measurement of LOD values.

The *sensitivity* of the detector for a specified analyte is strictly defined as the slope of the plot of detector signal as a function of the amount or concentration of that analyte (the calibration curve). For a given noise (background) level, a higher sensitivity does permit a lower detection limit, since the higher slope of the signal–concentration curve implies that the signal level necessary to satisfy the S/B criterion will be observed at a lower concentration. However, these two parameters are not the same, since the LOD varies with sensitivity but also with the background level (see Section 8.4.1). Nonetheless, sometimes the word 'sensitivity' is used in a more colloquial sense than includes the general concept of detection limit. It is important to distinguish the *instrumental* 'sensitivity' and 'detection limit' measured using purified analyte, from the 'method detection limit' (Section 8.4.1), which refers to analysis of the specified analyte following a particular *integrated analytical method* (Figure 1.2) for the analyte in a complex matrix.

The *lower limit of quantitation* (LLOQ, Section 8.4.2) is strictly a figure of merit for an integrated analytical method. It refers to the lowest concentration of analyte in a particular complex matrix that can be analyzed quantitatively by the analytical method with specified levels of accuracy and precision. Sometimes, if not critical for the end purpose of the analysis, the LLOQ can be estimated as the lowest concentration that provides a S/B ratio of 10:1 (Section 8.4.2).

The *linear dynamic range* (sometimes referred to as the *range of reliable response*, or simply the *calibration range*) is the range of concentration or amount of analyte between the LLOQ and the higher point at which the calibration curve used to determine sensitivity (see above) deviates from linearity. This quantity can refer to either the instrument (pure analyte) or to a specified analytical procedure and is important because a linear calibration simplifies the quantitation process (Section 8.4).

The *dead volume* refers to any void volume in the detector (or anywhere else in the chromatographic system) in which mixing can occur, thus causing deterioration of chromatographic resolution and selectivity (Section 3.5.12).

Detector selectivity measures the degree to which the detector can yield a signal arising from an analyte (or group of analytes) and only from that/these analyte(s) even when present in a mixture of compounds (e.g., co-eluting compounds in chromatography). In other words, selectivity indicates the extent to which the detector (or overall analytical method) can yield final signal levels that characterize the concentrations of the target analytes *without direct interferences* from other compounds. (This does not include indirect interferences such as ionization suppression, Sections 5.3.6a and 9.6, but may include metabolites that co-elute with the analyte and are subject to dissociation in the API interface (Section 5.3.3a) to produce ions indistinguishable from those used to quantitate the analyte, see Sections 9.3.4 and 11.5. Refractive index detectors are highly nonselective, UV–visible detectors are somewhat selective if the entire absorption spectrum is recorded (diode array detectors) and fluorescence detectors are more so (two wavelength parameters are selected and monitored). Mass spectrometric detection can have varying (but comparatively high) degrees of selectivity depending on the operating mode (tandem mass spectrometry is extremely selective, Section 6.2.2).

The term '*specificity*' is frequently used as an synonym for 'selectivity' in analytical chemistry but this usage can lead to ambiguity if applied in any but the most general sense (Vessman 2001). If a detector (or analytical method) is to be designated as 'specific' for a target analyte, the implication is that *only* this analyte can give rise to a signal with no possibility of interferences from other compounds; in other words, a detector (or method) is either specific or it is not (i.e., there is a binary decision to be made). Of course, chemical complexity being what it is, no detector or method is specific in this sense, so that the strictly defined concept of specificity is meaningless in chemistry (Vessman 2001). In contrast, 'selectivity' is not a binary property but could assume a range of values (though difficult to quantify, see Section 9.3.3b). Nonetheless, the ambiguity in the use of 'selectivity' vs 'specificity' does appear elsewhere in other disciplines in which 'specificity' is defined as a statistical measure of how well a binary classification test (i.e., 'yes' or 'no') correctly identifies the negative cases that do not meet the criterion of the test. An example might be a medical

test designed to determine whether or not a patient has a certain disease; in this context the 'specificity' of the diagnostic test to this particular disease can be quantified as the probability that the test indicates 'negative' if the patient truly does not have the disease, i.e., the 'specificity' of the test in this context is the fraction of true negatives in all of the 'negative' decisions indicated by the test results for a significantly large number of cases. This definition of 'specificity' appears to be more consistent with our foregoing discussion of 'selectivity', since within the same strict definition of 'specificity' even a medical diagnostic test is either truly 'specific' for the disease or it is not!

Universality of a detector, i.e., the ability to respond to all compounds to some degree, appears at first sight to be inconsistent with selectivity but in fact a very high degree of selectivity is perfectly compatible with a high level of universality, as exemplified by tandem mass spectrometry. Perhaps it should be mentioned that, to partially offset its otherwise high scores on these desirable properties, mass spectrometry is a 100 % destructive technique, which can be a problem for scarce samples if the analysis does not work well the first time.

A distinction between chromatographic detectors that should be kept in mind when planning an analytical method is that between those detectors that *respond to analyte concentration* independently of flow rate of amount of analyte, and those which respond only to the *flow rate of amount of analyte* independent of analyte concentration. Some appropriate units for the sensitivity (strictly defined as the slope of the calibration curve, see above) of a *concentration dependent detector* using peak height as the measure of signal strength would be $V.mol^{-1}.m^3$, $mV.g^{-1}.L$, $mV.\mu g^{-1}.\mu L$ etc. If peak height is replaced by peak area (signal integrated over the chromatographic peak) these units would be, e.g., $(Vs).mol^{-1}.m^3$, and so on. Examples of sensitivity units for a *mass flow dependent detector* using peak height would be $V.mol^{-1}.s$, $mV.mg^{-1}.min$, or $mV.\mu g^{-1}.min^{-1}$, and for peak area $(Vs).mol^{-1}.s = V.mol^{-1}.s^2$, and so on. In general this distinction is important because it affects the choice of peak height or peak area as most appropriate for calibration and quantitation. The importance of the distinction appears to have been first realized for GC detectors (Halász 1964). In that case the example used for a concentration dependent detector was the conductivity detector, in which the resistance of an electrically heated metal wire is monitored to provide a measure of its temperature; in turn the wire temperature varies as the result of differences in thermal conductivity of the flowing GC eluant as the concentration of analyte vapor varies with time. The original example of a mass flow dependent detector (Halász 1964) was the flame ionization detector, in which the electrical conductivity of a hydrogen–oxygen flame varies when an organic analyte enters it and burns to produce amounts of gaseous ions (e.g., CHO^+ and e^-). Here the UV–visible absorption detector is chosen as the archetype for concentration dependent detectors and the electron ionization (EI) mass spectrometer as that for mass flow dependent detectors.

Theoretical expressions for both peak height and peak area, using simplified models for a UV–visible absorption detector and for an electron ionization (EI) mass spectrometer, are derived in Appendix 4.1 together with an attempt to clarify the fundamental reasons underlying the difference. This treatment considers highly idealized and approximate models for the detectors themselves and for the nature of the time profile of the analyte as it is presented to the detector, for reasons of algebraic simplicity only; these approximations do not invalidate the following qualitative conclusions that are generally valid for the limiting cases of detectors that are 'truly' concentration dependent or mass flow dependent:

> The chromatographic peak height for a concentration dependent detector varies only with analyte concentration and is independent of flow rate; peak height for a mass flow dependent detector is not simply dependant on analyte concentration in the mobile phase as such, but is directly proportional to the mass (molar) flow rate into the detector. (Note that mass (molar) flow rate can be varied by changing one or both of analyte concentration and volume flow rate of the mobile phase.)
>
> Chromatographic peak area for a concentration dependent detector is inversely proportional to volume flow rate and that for a mass flow dependent detector is independent of flow rate.

Yet again, the intrinsic complexity of chemistry guarantees that most chromatographic detectors do not exactly fit either of these idealized models and electrospray ionization for mass spectrometry (Section 5.3.6a) is a prime example of this 'grey area'. For very low flow rates (approaching zero), at a fixed analyte concentration, the signal level (peak height) for a 'true' concentration dependent detector (a UV detector approximates this very closely) remains unchanged and the peak area increases because the signal is integrated over a correspondingly longer time. In comparison, as flow rates approach zero with a mass flow dependent detector the peak height decreases towards zero and the peak area remains constant since the decrease in signal is balanced by the longer integration time; however, in reality the peak becomes indistinguishable from noise at low mass

flow rates and so the peak area is effectively zero also. If the mobile phase flow rate is subject to uncontrolled variations during a run, the foregoing conclusions suggest that better precision in quantitation would be obtained using peak height for a concentration dependent detector and peak area for a mass flow detector. Indeed an extensive cooperative study (McCoy 1984) of quantitation by LC–UV (the ideal case of concentration dependent detection) did find that use of peak height provided higher precision.

The practical problems involved in actual measurements of peak heights and areas for poorly resolved peaks (e.g., smaller peaks observed as shoulders on larger ones, or badly defined due to significant noise and/or drift) have been considered (Kipiniak 1981) with special emphasis on computer-based methods. Uncertainties exist when peaks are not well resolved or are asymmetric (Section 3.5.12), or where background noise and drift are significant, e.g., is the area of a small peak observed as a shoulder on a larger peak defined by 'tangent skimming' or by 'dropping perpendiculars'. The closing comments of this paper reflect the state-of-the-art in 1981 but are still a relevant warning to those who use computer-based algorithms to analyze chromatograms and evaluate peak heights and areas in an automated high-throughput mode: 'In situations as complex as these, there is no single right answer. A computer program, no matter how powerful, sophisticated and well designed, cannot take the place of the watchful eye of the chemist. The computer can carry out analysis and do so by several procedures within its repertoire, but it has to show the user, preferably graphically, what it did, how it did, and what are the consequences of its assumptions. Be it in QC or research environment, the final responsibility for the accuracy and reliability of analysis rests with the human. Many of those who have intimate experience with computers think that perhaps that is just as well.' (Kipiniak 1981). An additional discussion relating to some of the practical aspects around this topic as they relate to data review and analytical run acceptance for routine quantitative analyis can be found in Section 10.5.2.

4.5 Summary of Key Concepts

To analyze for a specific target analyte by chromatography, even using high specificity mass spectrometric detection, it is necessary to **extract** the analyte from the matrix and prepare it (usually) in a liquid solution. This **sample extract** should contain the analyte at a concentration compatible with the instrument; it should also contain a sufficiently low concentration of co-extracted compounds that they do not interfere with the final analysis. It is sometimes necessary to conduct a separate **clean-up** step as part of this critical **sample preparation**.

The complexity of sample preparation depends on the **nature of both analyte and matrix**. The nature of the **solvent** used for the final extract is also important if LC–MS is used insofar as it must have a **solvent strength** that is no higher than that of the mobile phase into which the extract solution is injected in order to maintain chromatographic efficiency. Sample preparation methods can be categorized based on **analyte stability** and on the **nature of the matrix**, particularly its **physical phase** (liquid, solid or gas). In the majority of cases the final prepared sample is presented to the chromatograph as a liquid.

For liquid matrices, a common method of extraction or overall sample preparation is **liquid–liquid extraction (LLE)**. The solvents used in LLE usually have an appreciably higher solvent strength than the mobile phases used in conventional reversed phase HPLC, so that a solvent switching step is often required. **Protein precipitation** from biological fluids through addition of an organic solvent (typically acetonitrile) is widely used in bioanalytical applications. Direct injection of the protein-depleted fluid (even plasma) into an HPLC is often demonstrated to be fit for purpose in bioanalytical applications, but here again it is necessary to be careful about incompatibility of solvent strengths.

Methods for extraction of analytes contained in **solid matrices** into a liquid solvent almost all involve subjecting matrix and analyte to more or less robust conditions, so that care must be taken to ensure that the analyte is not subjected to extremes of temperature etc. The classic **Soxhlet extraction** method involves boiling the extraction solvent with the solid matrix dispersed in it, with an arrangement to condense the solvent vapor and continuously return it to the boiling flask. While **pressurized solvent extraction (accelerated solvent extraction)** can decrease extraction times, restrictions to thermally stable analytes are even more severe for such methods. **Sonication assisted liquid extraction (SAE)**, in which irradiation with ultrasound creates microscale bubbles of vapor within the solvent that eventually collapse, creates highly localized hot spots. **Microwave assisted extraction (MAE)** uses microwave irradiation as an external energy source and produces more uniform heating but is restricted to polar solvents that can absorb microwaves. Neither SAE nor MAE has been commonly adopted by analytical chemists. **Supercritical fluid extraction (SFE)** has many advantages in cases in which analyte molecules are occluded in small pores and crevices of a solid matrix, but is again not widely used by analysts. **Selective extraction by adsorption** on to a solid surface – of an analyte present in solution in a liquid matrix or as a minor component in a gaseous mixture – has been developed into several widely used sample extraction methods. Such **solid phase extraction (SPE)** can be regarded as **'binary chromatography'**, since the objective is to either adsorb (or not) a component of the solution. The solid phases used in SPE are generally closely related to those used as chromatographic stationary phases. SPE is most often used in **extraction of analytes from liquid matrices** as either a **primary extraction** step or as a **clean-up technique** for crude extract solutions. A variant of SPE, **solid phase microextraction (SPME)**, differs from conventional SPE mainly with respect to the amount of solid adsorbent phase used resulting in greater speed and convenience. **Turbulent flow chromatography** is a special form of SPE that can provide an online clean-up procedure (selective removal of large biomolecules) for HPLC analysis of small-molecule pharmaceuticals and their metabolites in plasma and urine. An exception to the 'binary chromatography' generalization is **flash chromatography**, which is essentially fast preparative chromatography requiring fraction collection. **Purge-and-trap** techniques are designed for analysis of volatile analytes usually with GC/MS, but difficulties arise in connection of the trap to the GC inlet, especially if a capillary column is used.

The modern practice of analytical chromatography for nonvolatile analytes often involves the use of **high performance liquid chromatography (HPLC)**. **Particles used for HPLC packings** are almost invariably made of **silica or alumina**. The **pore size** and **pore size distribution** of these particles are important since interaction with analytes mainly occurs on the inner particle surfaces. **Derivatization of silica particles** to replace the surface silanol (Si–OH) groups with (usually) $Si - O - Si(CH_3)_2 - R$ groups allows a wide range of stationary phases with tailored retention properties to be created for both normal and reversed phase chromatography. In **normal phase chromatography** the mobile phase is *less* polar than the stationary phase, and in **reversed phase chromatography** these relative polarities are reversed! For polar functional groups R the derivatized silicas are suitable for normal phase separations; phases with $R = (CH_2)_3CN$ ('cyano') and $R = (CH_2)_3NH_2$ ('amino') are common examples. For reversed phase chromatography the most common phases have $R = C_8H_{17}$ or $C_{18}H_{37}$, which can be either **monomeric or polymeric**. **End capped phases** have any free surface silanol groups that did not react with the desired organosilane converted to $Si - O - Si(CH_3)_3$. Figures of merit frequently quoted for derivatized silica phases include the **carbon loading**, the **surface area** and the **bonding phase density**.

HPLC methods allow **isocratic elution** where a single mobile phase composition is used, or **gradient elution** where variation of the composition of the mobile phase during elution is used to control the retention properties. **Hydrophilic interaction chromatography (HILIC)** is a good example of exploitation of the versatility of HPLC in that it uses a high organic:low water mobile phase ($\sim$80:20), together with stationary phases typically used for normal phase separations, to provide efficient normal phase chromatography suitable for highly polar compounds.

Stationary phases used for **ion exchange chromatography or SPE** are organic resins, e.g., suitably derivatized styrene-divinylbenzene cross-linked copolymers containing fixed charge sites on the stationary phase. 'Strong' exchangers retain their charge over a wide pH range ($\sim$2 – 12), e.g. $-SO_3^-$ for **strong cation exchange (SCX)** and $-NR_3^+$ for **strong anion exchange (SAX)**. **'Weak' exchangers** change their degree of ionization as a function of pH of the mobile phase, e.g., $-CO_2^-$ and $-NH_3^+$ for **WCX and WAX**, respectively, so by controlling pH it is possible to obtain a wide range of conditions for selective separation of ionized or ionizable analytes.

Chiral separations depend on significant differences in the K^A values for the enantiomers of the analyte on an optically active stationary phase. Many of the stationary phases have restrictive conditions on the nature of the mobile phase and this in turn can lead to problems with compatibility of the mobile phase with the ionization source of the mass spectrometer. **Immunoaffinity chromatography** is based on a stationary phase that contains **antibodies** that are highly specific for the target analyte(s) and are immobilized on a solid support. The antibodies are difficult and costly to prepare and are susceptible to denaturation. Those using **polyclonal antibodies** are not usually 100 % specific for the target analyte and are useful if selectivity for a class of compounds is desired. **Monoclonal antibodies** are more difficult to make but have much greater selectivity for the target analyte.

In gas chromatography (**GC**) the **mobile phase** is almost always helium, although nitrogen and hydrogen can be used. Gas flow rate is a crucial parameter for optimum performance and use of an **electronic pressure or flow control system** can reduce flow rate variations to negligible values. **Temperature programming** is also used to improve chromatographic performance. There is a wide selection of GC stationary phases available even for the **wall-coated open tube (WCOT) capillary columns** that nowadays dominate the technique. The **internal column diameter (d_c)** and **film thickness (d_f)** are crucial parameters in determining the column performance. Most **WCOT stationary phases** are based on **thermally stable**

polysiloxanes with different substituents on the alternating silicon atoms and are usually cross-linked and sometimes also bonded to the column walls for extra stability.

Other components of a chromatographic train, such as **sample injectors**, are also important for best performance. The **HPLC loop injector** is capable of delivering volumes of analyte solution into the mobile phase flow with very high repeatability (precision) but significantly poorer accuracy; highest precision is obtained in the overfill mode and with automated injection (the **autosampler**). **GC injectors** are not capable of providing good accuracy or precision of injected volumes and, as such, use of a **volumetric internal standard** is required. In addition, vaporization of one microlitre of analyte solution in the hot GC carrier gas creates a volume of vaporized solvent plus analyte that is up to 10^3 times larger, and that can be as much as half of the total internal volume of the capillary column. The **split–splitless injector** evaporates the analyte solution and is then arranged to transmit either only part of the evaporated solution to the capillary column (**split mode**), or to separate most of the solvent from the less volatile analytes within the injector (**splitless mode**). The **programmed temperature vaporizer (PTV) injector** is a variant of the split–splitless injector that allows much larger volumes of analyte solution (up to $\sim$50 μL) to be injected. The **on-column injection technique** avoids problems arising from the large volumes of solvent vapor by injecting the solution directly onto a cool portion of the column and applying temperature programming to selectively remove the more volatile solvent well before the analytes.

The pressure of the GC mobile phase (helium) can be readily controlled, preferably electronically. But for HPLC the back pressure driving the mobile phase must be created by **liquid pumps**. The most common type are **reciprocating piston displacement pumps** whose main disadvantage arises from the sinusoidal pressure pulses and corresponding sinusoidal variations in flow rate that can be minimized by using a **pressure pulse damper**. **Syringe pumps** are larger scale versions of manual syringes, driven by either a mechanical device or more recently by **amplified pneumatic pressure with electronic control**, which can deliver highly reproducible mobile phase gradients at flow rates compatible with packed microcolumns (i.d. $\sim$300μm). **Flow splitters** are devices used either for pre-column splitting to match a higher flow rate pump to the flow appropriate for a small diameter column, or for post-column splitting to match the HPLC eluant flow rate to a value that optimizes the efficiency of, e.g., an ESI source. A question for gradient elution concerns whether to use **high pressure mixing** (each solvent is delivered by its own pump to a mixing chamber located between the pumps and the column) or **low pressure mixing**, in which mixing is accomplished prior to a single pump at its low pressure side. Both have their advantages and disadvantages but low pressure mixing is probably the more common.

An entirely different concept in analytical separations is provided by **capillary electrophoresis (CE)** in which the flow of liquid is generated by electro-osmotic flow (EOF) driven by an external electric field. The major advantage of this approach is the essentially flat plug flow profile that leads to intrinsically more narrow elution peaks than the parabolic flow profiles characteristic of pressure-driven viscous flows. In **capillary zone electrophoresis (CZE)** separation is achieved by superimposing the different electrophoretic mobilities of the solutes on to the EOF. In **electrochromatography** the separation is achieved as in packed column HPLC but using an EOF to generate flow of the mobile phase past the stationary phase particles. The importance of these EOF-based techniques is their application to miniaturized devices, '**lab-on-a-chip**' or '**micro total analysis**' systems. Such devices that can be directly interfaced to a mass spectrometer via an ESI source are currently under intense development.

Multidimensional separations use more than one separation mechanism applied to a sample, with each considered as an independent separation dimension. Multidimensional chromatography includes both **heart-cutting methods**, in which fractions collected from the first separation are collected, stored and re-analyzed by the second method, and also the more ambitious **comprehensive multidimensional chromatography**, which attempts to subject each and every component in the sample extract to the full combined separatory power of different chromatographic dimensions. Comprehensive strategies are denoted using an 'x' notation, e.g., LCxLC, GCxGC, LCxGC. Equipment required for multidimensional chromatography is specialized and complex, with the interface (the 'modulator') between the two dimensions as a key component.

Development of **fast digital electronics and computers** has made it possible to perform chemical analyses that would have been unthinkable previously. However, the algorithms used to acquire the raw data and process it into a form convenient for the analyst necessarily contain assumptions and approximations, in addition to errors that inevitably arise when, e.g., a chromatogram contains features that are not accounted for in the algorithms used to find peaks and determine their areas. Visual inspection of processed data by a trained analyst is highly recommended to minimize such errors.

Appendix 4.1 Responses of Chromatographic Detectors: Concentration vs Mass–Flux Dependence

The electrical signal generated by a concentration dependant UV-visible detector is given by:

$$S_{uv} = I_{tr}.G_\lambda = I_0.G_\lambda.\exp[-c^A_{uv}.\varepsilon_{A\lambda}.L_{uv}]$$
$$\approx I_0.G_\lambda.[1 - (c_A^{uv}.\varepsilon_{A\lambda}.L_{uv})] \qquad [4.1]$$

where I_{tr} is the light intensity transmitted through the detector containing a molar concentration c^A_{uv} of analyte A to the detector, I_0 is the value of I_{tr} when there are no absorbing compounds in the cell, G_λ is the combined amplification gain of the photomultiplier plus its electronic amplifier for light of wavelength λ, $\varepsilon_{A\lambda}$ is the molar extinction coefficient of A at wavelength λ, and L_{uv} is the path length of light in the cell. The second form of equation [4.1] introduced the Beer–Lambert Law, and the third (approximate) form is valid when $c^A_{uv}.\varepsilon_{A\lambda}.L_{uv} << 1$, so that higher order terms in the power series for $\exp[-(c^A_{uv}.\varepsilon_{A\lambda}.L_{uv})]$ can be ignored. Note that this approximation is not essential for derivation of the desired results but it is made here to simplify the treatment. Now, in a well designed HPLC system in which the size of the UV cell is properly matched to the internal diameter of the column, connecting tubing etc., $c^A_{uv} = c^A_m$, the concentration of A in the mobile phase entering the UV cell (see Chapter 3). So finally:

$$S_{uv} = I_0.G_\lambda.[1 - (c_A^m.\varepsilon_{A\lambda}.L_{uv})] \qquad [4.2]$$

Equation [4.2] indicates that the chromatographic peak height for a UV detector is independent of flow rate (all the way down to zero!).

The corresponding treatment for the (mass flux dependant) EI mass spectrometer is similar but contains an additional complication. Now, instead of a beam of light, the analytical probe is a beam of energetic electrons that interact with the analyte molecules in a variety of ways, one of which is to create ions of a particular m/z value chosen as characteristic of the analyte and selectively transmitted to the detector by the mass selection portion of the instrument. Then the observed signal is given by:

$$S_{EI} = i_{m/z}.G_{m/z} \qquad [4.3]$$

where $i_{m/z}$ is the electrical current corresponding to the beam of m/z ions reaching the electron multiplier detector of the mass spectrometer, and $G_{m/z}$ is analogous to G_λ in equation [4.1]. To relate $i_{m/z}$ to concentration of A, the equation analogous to the Beer–Lambert Law can be written:

$$i_{el} = i^0_{el}.\exp[-c^A_{EI}.\sigma_A.L_{EI}] \approx i^0_{el}.[1 - c^A_{EI}.\sigma_A.L_{EI}]$$
$$[4.4]$$

where i_{el} is the electron beam current that survives the passage through the gas in the EI source, i^0_{el} is the value of i_{el} with no gaseous A in the cell, c^A_{EI} is the concentration of A in the EI source, σ_A is the total cross section for removal of electrons from the beam via interactions with molecules of A, and L_{EI} is the length of the path of the electron beam within the source volume that contains A. The second (approximate) form of equation [4.4] is valid when $(c^A_{EI}.\sigma_A.L_{EI}) << 1$, which is generally true for normal operation of an EI source. Now:

$$i_{m/z} = (i^0_{el} - i_{el}).F_{m/z}.T_{m/z} \quad [4.5]$$

where $F_{m/z}$ is the fraction of all the electron-A interactions that lead to formation of m/z ions, and $T_{m/z}$ is the transmission efficiency for these ions between their formation within the EI source, through the mass-selector portion of the instrument, and finally to the detector. Combining equations [4.3–4.5] then gives:

$$S_{EI} = i^0_{el}.G_{m/z}.F_{m/z}.T_{m/z}.c^A_{EI}.\sigma_A.L_{EI} \quad [4.6]$$

Thus far the only difference between the UV and EI detectors is that the observed signal for the former is a measure of transmitted intensity I_{tr}, while that for the latter is a measure of the absorbed intensity $(i^0_{el} - i_{el})$, but it turns out that is of little consequence for the present purpose. Thus, for example, a fluorescence detector records a signal that is a measure of the absorbed intensity of the exciting radiation like the EI case, but it is also a concentration dependent detector like the simple UV absorption detector. The difference between the two types arises rather in the relationship between the analyte concentration delivered by the mobile phase (c_A^m) and that within the absorption cell or EI source; in the former case $c_A^{uv} = c_A^m$ (see equation [4.2]), but the situation is very different in the EI case where the mass spectrometer vacuum pumps continuously remove the analyte from the EI source. In fact c^A_{EI} represents an 'instantaneous' steady state value, a compromise between the flow rate of A into the source and the pumping rate out of the source; here 'instantaneous' means simply that the establishment of the steady state value c^A_{EI} occurs on a timescale appreciably shorter than that of the chromatographic peak. Then at this steady state:

$$dc^A_{EI}/dt = (Q_A^{in} - Q_A^{out})/(N_A.V^{EI}) = 0 = [(U.c_A^m)/V^{EI}]$$
$$- [C_{gasA}^{EI}.(P_A^{EI} - P_{vac})]/(V^{EI}.RT) \quad [4.7]$$

where Q_A represents the flow rate of A expressed as (molecules.time^{-1}) as a special case for analyte A of Q_{gas}

(see equations [6.46–6.50]), U is the mobile phase volume flow rate (Chapter 3), V^{EI} is the volume of the EI source, C_{gasA}^{EI} is the gas flow conductance (Section 6.6.1) of the dilute gas A out of the ion source (proportional to the total area of apertures in the ion source and inversely proportional to molar mass $M_A^{\frac{1}{2}}$ via the average molecular speed), P_A^{EI} is the steady state pressure of A in the EI source $(= c^A_{EI}.RT)$, and P_{vac} is the pressure maintained in the surroundings of the EI source by the vacuum pumps of the mass spectrometer. In any EI mass spectrometer $P_{vac} << P^{EI}_A$. Equation [4.7] can be solved for $c^A_{EI} = P_{EI}^A/RT$:

$$c^A_{EI} = (U.c_A^m)/C_{gasA}^{EI} = (Q_A^{in}/N_{AV})/C_{gasA}^{EI} \quad [4.8]$$

since the molar flow rate of A into the EI source is simply the product of the mobile phase volume flow rate and the concentration of A in the mobile phase. The symbol N_{AV} is used here for the Avagadro constant to avoid confusion with N_A, the column efficiency for analyte A. Then combining equations [4.6–4.8] gives:

$$S_{EI} = Q_A^{in}.[(i^0_{el}.G_{m/z}.F_{m/z}.T_{m/z}.\sigma_A.L_{EI})/(N_{AV}.C_{gasA}^{EI})] \quad [4.9]$$

This looks extremely complicated but the expression within the square brackets contains only the Avogadro constant N_{AV} and purely instrumental parameters, apart from σ_A and C_{gasA}^{EI} which are functions of the chemical nature of A, independent of concentration and flow rate of A. Thus the fact that the ion source of a mass spectrometer pumps away the remaining analyte that is not ionized (i.e., is a destructive detector) transforms the signal's original concentration dependence (equation [4.6]) into a dependence on the molar flow rate of analyte (equation [4.9]). Of course the molar flow rate (Q_A^{in}/N_{AV}) is readily transformed into a mass flow rate by multiplying by the molar mass M_A, so equation [4.9] indicates that the EI signal strength (i.e., chromatographic peak height) is directly proportional to flow rate.

To determine the corresponding relationships for peak area (PA), it is necessary to integrate the signal over the time-width of the peak, i.e., it is necessary to know how the signal varies with time across the peak. For the UV-viscase:

$$PA_{uv} = \int(S_{uv})dt$$

(where the integration is over the entire peak)

and similarly for the EI case. It is known (see discussion of equation [3.31] in Chapter 3) that chromatographic peak shapes can approximate well to a Gaussian form. However, to keep the algebra simple here an ultra-simple case is considered, that of a rectangular plug flow, i.e., the analyte is contained within a fixed length plug as the mobile phase carries it along. This is an unrealistic model even for flow injection analysis in which the analyte solution is injected directly into the mobile phase without eluting through a chromatographic column since longitudinal diffusion will smear out the boundaries of the plug, but it allows a simple integration that will lead quickly to the desired final qualitative result. In algebraic terms, this simplifying assumption is:

$$c_A^m = \text{constant for } t_1 < t < t_2 \quad \text{and}$$

$$= \text{zero everywhere else}$$

where $t_1 = (t_{plug} - \Delta t/2)$ and $t_2 = (t_{plug} + \Delta t/2)$, $t_{plug} =$ the time at which the centre of the plug enters the detector, and Δt is the time-width of the plug passing through the detector. Also, the simplifying assumption that the length of the plug L_{plug} is constant (independent of flow rate) means that:

$$\Delta t = (t_2 - t_1) = L_{plug}/U = L_{plug}.\pi r^2_{tube}/u = V_{plug}/U$$

where r_{tube} is the internal radius of the tubing leading into the detector, u and U are the linear velocity and volume flow rate of the mobile phase, respectively, and V_{plug} has been assumed to be constant. Then, from equation [4.2]:

$$PA_{uv} = \smallint(S_{uv})dt = \smallint_{t1}^{t2}\{I_0.G_\lambda.[1 - (c_A^m.\varepsilon_{A\lambda}.L_{uv})]\}dt$$

$$= I_0.G_\lambda.[1 - (c_A^m.\varepsilon_{A\lambda}.L_{uv})].(t_2 - t_1)$$

$$= \{V_{plug}.I_0.G_\lambda.[1 - (c_A^m.\varepsilon_{A\lambda}.L_{uv})]\}/U \quad [4.10]$$

Equation [4.10] indicates that the chromatographic peak area for a UV detector is inversely proportional to the volume flow rate.

A similar integration of equation [4.9] over the time-width of the chromatographic peak, can be performed to give:

$$PA_{EI} = Q_A^{in}.[(i^0_{el}.G_{m/z}.F_{m/z}.T_{m/z}.\sigma_A.L_{EI})/(N_A.C_{gasA}^{EI})]$$

$$\times (V_{plug}/U)$$

$$= (U.c_A^m).[(i^0_{el}.G_{m/z}.F_{m/z}.T_{m/z}.\sigma_A.L_{EI})/(C_{gasA}^{EI})]$$

$$\times (V_{plug}/U)$$

$$= [c_A^m/(C_{gasA}^{EI}.V_{plug})].[(i^0_{el}.G_{m/z}.F_{m/z}.T_{m/z}$$

$$\times \sigma_A.L_{EI})$$

$$[4.11]$$

where $Q_A^{in} = (U.c_A^m/N_A)$ is used so that U cancels, i.e., for an EI–MS detector (or any similar destructive detector) the chromatographic peak area is independent of flow rate.

Of course, the entire treatment of this problem has been highly idealized and approximate models for the detectors themselves and for the nature of the time profile of the analyte as it is presented to the detector (the latter was for reasons of algebraic simplicity only and does not invalidate the qualitative conclusions). These theoretical conclusions are considered further in Sections 4.4.8 and 5.3.6a.

5

Tools of the Trade IV. Interfaces and Ion Sources for Chromatography–Mass Spectrometry

---

Joseph John Thomson: Discoverer of the Electron and Inventor of Mass Spectrometry

Thomson was born in Manchester, England, of Scottish parentage on his father's side. His father wanted him to be an engineer but a waiting list for apprenticeships at the locomotive works led to his attendance at a local college where his intellectual brilliance became apparent. He was able to continue his studies after his father's early death by winning scholarships, culminating in one that enabled him to study mathematics and physics at Cambridge. A detailed account of his life and career was published (Griffiths 1997) to celebrate the centenary of his classic experiment that measured the ratio of electrical charge to mass of the electron and marked the birth of mass spectrometry. The apparatus he used is shown in the accompanying sketch. Electrode C was the cathode and electrode A the anode, maintained at a high positive potential relative to C so that a discharge of 'cathode rays' (now known to be electrons) moved from A to C. A small hole in C passed a narrow beam of rays that was further collimated by electrode B to a well defined beam that fell on the fluorescent screen S.

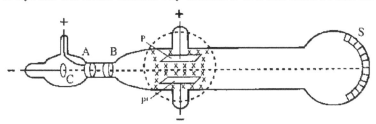

A deflecting electric field was applied between parallel plates P and P1 in the vertical direction, and a magnetic field in the horizontal direction (into the plane of the page), indicated by the small crosses.

His extension of the same general approach to positive ions used a somewhat different apparatus, reproduced below from his famous book *Positive Rays and their Application to Chemical Analysis* (Longmans, London, 1st Ed. 1913, 2nd Ed. 1921). Here a different combination of electric and magnetic fields (parallel in this case) was used to produce different parabolas on a photographic

---

(Continued)

different mass:charge ratios. Part of this success is attributed to his glassblower (E. Everett) who joined him in 1896 and whose skills greatly extended Thomson's experimental range.

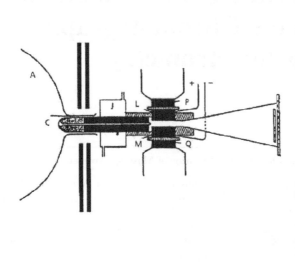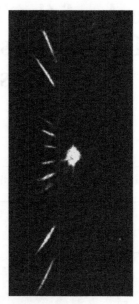

It is interesting that Thomson's son, G.P. Thomson, won a Nobel Prize for demonstrating that electrons behave not only as charged particles as his father had shown, but also as waves that undergo diffraction.

5.1 Introduction

The material covered in this book could not have developed without the invention and development of mass spectrometry. There have been suggestions that mass spectrometry may have been known over four centuries ago, based on passages in the 'old literature', e.g.:

> *'Witness this army of such mass and charge'*
> Shakespeare (1601).

However, it is generally recognized that the discipline originated in 1897, when J.J. Thomson measured the ratio of the electrical charge to the mass of the electron (see the accompanying text box), which resulted in his Nobel Prize for Physics in 1906. Thomson later extended his work to positive ions and several of his students, including F.W. Aston (also a Nobelist), laid the foundations for mass spectrometry as we now know it, an essential analytical technique in all areas of chemistry and biochemistry.

This chapter and the following one will discuss modern mass spectrometry only in the context of quantitative analysis of organic compounds present at trace levels in complex mixtures. A diagram of the components of

the components of a modern mass spectrometer was shown as Figure 1.3; for convenience these will be considered separately in Chapters 5–7, though proper integration of all components is essential for an efficient system. Instruments and techniques that are not best suited for purposes of trace analysis will be mentioned only briefly. Clearly, a major part of this focused approach must concentrate on methods for interfacing mass spectrometry on-line with chromatography. The major question concerns how to selectively discard the mobile phase while preserving as much as possible of the analyte; this has been much more difficult for HPLC than for GC. An indication of this intrinsic difficulty is the fact (van der Greef 1992) that, of the > 25 methods that have been developed for coupling HPLC online to MS, only about five have been used to any significant extent for difficult analyses.

The problem is two-fold; firstly, the total molecular flow rate into the mass spectrometer must be sufficiently low that economically feasible vacuum systems (Section 6.6) can maintain a sufficiently good vacuum that the ion trajectories through the m/z selection stage are not significantly perturbed via ion–molecule collisions (this problem applies also to operation of the ion detectors used in mass spectrometry, Chapter 7). Different *m/z* analyzers have different vacuum requirements, see Chapter 6. Since they are more susceptible to perturbations from ion collisions with background gas molecules if they are to achieve their design performance with respect to high resolution, analyzers such as double focusing magnet systems, reflectron time of flight, and Fourier transform ion cyclotron resonance (FTICR) instruments require a higher vacuum (10^{-6}–10^{-8} torr, even lower for FTICR, note 1 torr = 133.3 Pa and 1 atm = 760 torr) than those that provide lower resolution. Linear quadrupoles and Paul ion traps operate at 10^{-5}–10^{-6} Torr, except inside a trap that requires ~10^{-3} torr of helium to damp the motion of the ions for efficient trapping. The second aspect that demands removal of most of the mobile phase is that the ionization capacity of the ionization technique must not be saturated by a large excess of mobile phase molecules, thus suppressing ionization of the analytes. The ionization method to be used is also intimately connected with the selective removal of mobile phase to manageable levels and, indeed, the two aspects are conveniently considered together since in some important cases they are in practice integrated into a single device. A recent review (Abian 1999), subsequently updated (Gelpi 2002), has described the history of the development of chromatography–MS coupling, but here the focus is mainly on those devices that are currently in common use; these include atmospheric pressure ionization sources (API), which have also been extensively reviewed (Niessen 1998).

In the great majority of cases, detection of ions following *m/z* separation by the mass analyzer (Figure 1.3) uses some form of electron multiplier (requiring a good vacuum for efficient operation) together with some form of current-to-voltage amplifier (Chapter 7). The principal exception in current practice is the use of image current detection with FTICR, but since such instruments are seldom used for absolute quantitation they will be discussed only briefly here. The volume of raw data generated by today's instruments is sufficiently large that it can only be stored for later processing by a computer, so the raw signal from the detector must be digitized in real time (i.e., on-the-fly during acquisition). The computer also is programmed to process the raw data by a variety of algorithms, to generate processed data usually with appropriate statistical analysis. The final and most important component of the chromatography–mass spectrometry system is the human analyst (Figure 1.3), who is responsible not only for the design of the entire analytical protocol but also for applying his or her expertise and judgment to interpret the processed data and transform it into usable information.

It is important to appreciate the strengths and weaknesses of mass spectrometry (MS) as a chromatographic detector with respect to important criteria (see Section 4.4.8). These include degree of universality (applicability to all analyte types), the sensitivity and the intrinsic noise levels (that together determine the 'instrumental detection limit' (LOD) for pure analyte as distinct from the 'method detection limit' for the entire integrated analytical method applied to the analyte in its complex matrix, see Section 8.4), linear dynamic range, specificity, response speed (must provide sufficient data points across a chromatographic peak that the peak can be properly characterized and its area determined), duty cycle of the *m/z* analyzer (fraction of the total ion beam current generated specifically from the analyte that is actually transmitted to the detector) etc. Clearly many of these parameters are inter-related in some way, and they also vary somewhat from case to case, but in general MS is regarded as providing by far the most useful combination of properties. Its strengths greatly outweigh its shortcomings, which include a limited degree of specificity, e.g., the amount of chemical information in a mass spectrum compared with an NMR spectrum. However, use of NMR as an on-line detector on the chromatography timescale is not compatible with LODs that are anywhere near adequate for trace quantitation, and in any case operation of an NMR instrument to provide quantitation data of acceptable accuracy and precision is far from straightforward, even for isolated fractions where acquisition time is not an issue (Burton 2005). Hopefully, Chapters 5–7 will provide an understanding of the origins and limitations of all the strengths and weaknesses of mass spectrometry in this context.

5.1.1 Matrix Effects

Some important issues peculiar to the use of mass spectrometry as a chromatographic detector in quantitative analyses have come to be referred to in general as matrix effects. These effects are all related to the presence of compounds other than the target analyte(s) in the extract solution that is injected into the chromatograph, and that co-elute with one or more of the analytes and interfere with the mass spectrometric analysis. Such interfering compounds can be endogenous (originally present in the sample matrix and co-extracted with the analytes) or exogenous (introduced to the solution during the sample preparation stage, e.g., compounds introduced from plastic tubing etc.).

The simplest interferences to recognize are those in which a co-eluting compound yields a mass spectrometric signal that coincides or overlaps with that used to monitor the concentration of an analyte; such effects are sometimes referred to as direct interferences and the spurious signals may be observed even when no analyte is present. In contrast, indirect interferences are more subtle and do not involve overlap of mass spectrometric signals. Rather, the presence of the interfering compound together with an analyte in the ion source either suppresses or enhances the ionization efficiency of the latter in an unpredictable fashion (suppression is the more common). These effects present a problem if the instrument calibration is acquired using clean solutions (no potential interfering compounds) of the analytical standard and is then applied to quantitation of the analyte in extracts of the matrix, since the ionization efficiency of the analyte might be quite different in the two circumstances. Use of a suitable co-eluting internal standard (SIS), preferably a stable-isotope-labeled version of the analyte, can greatly alleviate the situation (Section 8.5.2b) but it is always preferable to be aware of any potential problems of this kind and take steps to reduce or eliminate them.

Suppression of ionization efficiency is important when the total ionizing capability of the ionization technique is limited, so that there is a competition for ionization among compounds that are present in the ion source simultaneously. In principle such a saturation effect must be operative for all ionization techniques, but in practice it is most important for electrospray ionization (Section 5.3.6), slightly less important for atmospheric pressure chemical ionization (Section 5.3.4), atmospheric pressure photoionization (Section 5.3.5) and matrix assisted laser desorption ionization (Section 5.2.2); it does not appear to be problematic under commonly used conditions for electron ionization and chemical ionization (Section 5.2.1) or thermospray (Section 5.3.2). Enhancement of ionization efficiency for an analyte by a co-eluting compound is less commonly observed and is, in general, not well understood.

The present chapter covers the more fundamental aspects of these ionization techniques, since some appreciation of these principles is essential for an understanding of their respective strengths and limitations, particularly with respect to matrix effects. The most complete discussion of the fundamental aspects is that given in Section 5.3.6a, where the main focus is on electrospray, and their practical implications are described in Sections 9.6 and 10.4.1d.

To simplify discussion, it is convenient to categorize the eight ionization techniques of interest here according to whether or not the chromatographic eluant can be introduced directly into the ion source. If not, a discrete device of some sort, distinct from either the chromatograph or the ion source, must be interposed between the

two. Such a device can be an on-line enrichment apparatus (e.g. a jet separator for packed column GC with an electron ionization source), or involve off-line collection of eluant fractions that are later introduced separately into the ion source. As with all attempts at categorization there are examples that do not fit comfortably, e.g., fast atom bombardment (Section 5.3.1) was originally an off-line method but was later adapted for online use. Similarly, a simple flow splitter is sometimes used between HPLC and one of the atmospheric pressure ionization techniques.

5.2 Ion Sources that can Require a Discrete Interface Between Chromatograph and Source

5.2.1 Electron Ionization and Chemical Ionization

Until the 1980s the only ionization techniques used for trace analysis on-line with chromatography were electron ionization (EI, formerly known as electron impact ionization) and chemical ionization (CI). These closely related ion sources are discussed below but for now it is sufficient to emphasize that both sources require introduction of analytes in the gas phase, and that the sources are located within the high vacuum chamber of the mass spectrometer; thus it is essential to severely limit the amount of mobile phase that can enter the mass spectrometer in order to maintain the necessary vacuum conditions. As a result, for quantitative analyses it is necessary to either restrict the flow rate of the mobile phase or to develop external on-line devices that remove the bulk of the mobile phase prior to introduction into an EI or CI source. This is fairly easy to achieve for GC, but remains difficult for HPLC (see below). The restriction to vapor phase analytes implies that EI and CI are applicable to only thermally stable and volatile compounds, and sometimes chemical derivatization is necessary to achieve this (Section 5.2.1b).

Electron ionization is the classical ionization technique in mass spectrometry and yields essentially only positive ions; the original design (Dempster 1921, 1921a) was subsequently improved by several workers, but the design (Nier 1940, 1947) that forms the basis of modern EI sources has not changed much since its first development. Inside the ion source (typically $\sim 10^{-4}$ torr, 200–250 °C), the gaseous sample is bombarded with fast electrons (the standard value corresponds to acceleration of the electrons through 70 V, i.e., electron energy 70 eV, see Section 6.3.1) usually generated from an electrically heated tungsten or rhenium filament (Figure 5.1). Because the pressure is kept low, ion–molecule reactions do not occur to any significant extent and $[A + H]^+$ ions characteristic of chemical ionization (see later) are not observed unless the sample is

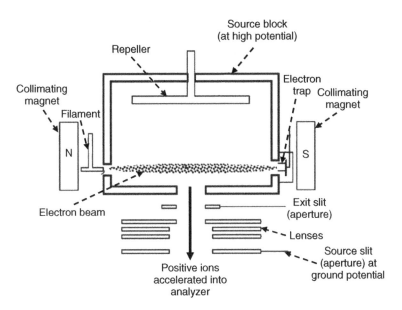

Figure 5.1 Sketch of a cross-section through an electron ionization source. The gaseous sample enters the source in a direction perpendicular to the plane of the figure. The electrons are emitted from the electrically heated filament and accelerated through a potential difference between filament and trap (standard value is 70 V). The electron trap current is monitored and used in a feedback control circuit to control the filament heating current to provide a constant trap current (typically 100 μA). The magnetic field collimates the electron beam and simultaneously causes the electrons to follow a spiral path through the source, to increase the total path length and thus the probability of an ionizing collision with an analyte molecule. The potential on the repeller is adjusted relative to that of the source block so that the electric field in the source maximizes the efficiency of positive ion extraction through the exit slit; counter-intuitively, this 'tuned' potential difference can be negative for positive ions since the electric field lines of force (see text box in Chapter 6) then focus the ions at the source slit. The entire assembly up to and including the ion exit slit is maintained at an electrical potential higher than that of the source slit and analyzer (kV range for magnetic sector and time of flight, a few volts for quadrupoles and ion traps) to accelerate the ions to an appropriate translational energy to match the requirements of the analyzer. The source is heated (usually to 200–250 °C) and is located inside the high vacuum chamber of the mass spectrometer to ensure efficient pumping of the source.

introduced in excess (so-called 'self CI'). The application of EI is restricted to thermally stable analytes with relatively low molecular masses ($< \sim 1000$ Da) with sufficient volatility that they are completely vaporized at the source temperature. The EI process for an analyte molecule A can be described as in Equation [5.1]:

$$A(g) + e^-(g, \text{fast}) \rightarrow (A^{+\bullet})^* + 2e^-$$

$$\downarrow$$

$$(F_1{}^+ + N_1^\bullet) + (F_2{}^{+\bullet} + N_2) + \dots$$

$$[5.1]$$

The superscript $^+$ denotes a net positive charge, $^\bullet$ denotes an unpaired electron (such ions are radical species referred to as odd-electron ions, as opposed to stable even-electron species in which electrons are paired up with opposite spins to form chemical bonds or electron lone pairs); the $A^{+\bullet}$ ion is conventionally referred to as the molecular

ion of analyte A, though for complete lack of ambiguity it should be called the molecular radical cation since many other ionized forms of intact molecules are observed in mass spectrometry and are sometimes referred to as 'molecular ions'. In Equation [5.1] the notation * denotes a species with a large amount of internal energy such that it fragments within its residence time in the source (~ 1 μs) to form fragment ions F^+ and complementary neutral fragments N (net charge and electron count must each balance in Equation [5.1] like the total chemical composition). The array of ions thus formed, including their relative abundances, constitutes the EI mass spectrum of compound A. By controlling the source temperature and the energy of the ionizing electrons, it is found that such mass spectra are remarkably reproducible and comparable among different m/z analyzers, sufficiently so that libraries of such 'fingerprint' spectra are maintained and used for compound identification.

The most reliable library of such spectra is maintained by the US National Institute for Standards and Technology (Stein 2004); the NIST/EPA/NIH Mass Spectral Library is the product of a multiyear, comprehensive evaluation, and currently (version NIST05) includes ~160 000 compounds with evaluated spectra, retention times etc. The library plus associated search engine is included in virtually every standard software package for GC/MS instruments and can also be purchased separately from several licensed vendors listed on the NIST website (http://www.nist.gov/srd/mslist.htm). The 2007 version of the Wiley Mass Spectral database (http://www3.interscience.wiley.com/cgi-bin/mrw home/114177614) contains over 399 000 spectra and over 182 000 chemical structures, covering compounds monitored by the Organisation for the Prevention of Chemical Weapons (OPCW) and by governmental bodies such as the US EPA, FDA etc., often also including their metabolites and precursors. There are several specialized libraries available; these have been conveniently summarized (Halket 2005). For unknown compounds for which no good match can be found in the library, *a priori* interpretation of the spectrum in terms of chemical structure partakes of the nature of both a science and an art (McLafferty 1993). For 'soft' ionization methods that yield few if any fragment ions directly, such structural information can be obtained using collision induced dissociation (CID) with tandem mass spectrometry (MS/MS), as discussed in Section 6.2.2.

Qualitative identification is important in quantitative analysis of target analytes because there is no point in assigning a concentration or amount to a component in a complex mixture if the analyst has no confidence that the chromatographic peak in question does indeed contain the target analyte; moreover assurance is needed that the signal used for quantitation does not contain contributions from co-eluting compounds (so-called peak purity). This interplay between qualitative and quantitative analysis is discussed further in Section 9.3.3. Here some aspects of EI that pertain directly to the quantitation aspect are examined.

As discussed in Section 4.4.8 and Appendix 4.1, the EI source is an archetypal example of a mass flow dependent detector for chromatography, as a result of the high level of pumping applied to the source that removes the analyte almost as soon as it enters. Further, the characteristics that led to the high level of spectral reproducibility, mentioned above, also lead to excellent quantitative precision with a wide linear dynamic range and low detection and quantitation limits. The main drawback of the EI source is its limitation to thermally stable and volatile analytes; also, in a significant number of cases it does not provide reliable information on molecular mass (abundance of the $A^{+\bullet}$ ion can be zero or very low), a disadvantage with respect to analyte verification. Thus EI is typical of a 'hard' ionization technique, i.e., the initial ionization event deposits a large quantity of internal energy into the $A^{+\bullet}$ ion (Equation [5.1]) leading to formation of an array of fragment ions (the mass spectrum) that in some cases depletes the $A^{+\bullet}$ ion abundance to zero. The extent of fragmentation in EI can be reduced by using a much lower electron energy ($\sim 20\,eV$) than the standard 70 eV, but this is achieved at the expense of a significant decrease in the overall ionization efficiency to the detriment of performance in trace analysis. An alternative approach (Fialkov 2006) interposes a supersonic nozzle device between GC and EI source so that the analyte molecules entering the EI source are vibrationally 'cold', i.e., possess only low internal energies prior to ionization. In this way the $A^{+\bullet}$ ions initially formed have less internal energy than those formed in a conventional GC–EIMS arrangement, where the pre-ionization internal energies can correspond to temperatures of 500 K or more. This lowers the average fragmentation rate constants and correspondingly increases the $A^{+\bullet}$ relative abundance.

Note that negative ions are never observed in any significant abundance using EI, unless the pressure inside the ion source chamber is allowed to build up to values $\gg 10^{-4}$ mbar. At such pressures, in addition to self CI processes, some of the ionizing electrons are cooled by collision to energies that are sufficiently low that they can be captured by analyte molecules to form $A^{-\bullet}$ radical anions that can either be detected as such or react further. When such processes are allowed to proceed in a controlled manner, as in negative ion chemical ionization (see below), extremely high sensitivities for electronegative analytes can be obtained.

The search for a 'soft' ionization technique that would ionize a molecule without inducing extensive fragmentation was historically first satisfied with the invention of field ionization (Beckey 1977, Prokai 1990). However, this ionization method is not suitable for trace quantitation in the present context, although its extremely soft nature still finds application in analysis of complex mixtures such as fossil fuels. The first technique of interest in the present book was chemical ionization, developed into a useful analytical technique in the 1960s (Munson 1966) based on earlier physical chemistry studies of ion–molecule reactions (Tal'roze 1952). Significant development of this approach followed and has been admirably summarized (Harrison 1992).

As its name suggests, chemical ionization (CI) involves ionization of an analyte molecule A by chemical reaction with a 'reagent ion' formed by EI of a reagent gas added to the source in large (10^3–10^4) excess. The occurrence of bimolecular ion–molecule reactions during the residence time ($\sim$1 μs) of the reactants in the source requires a much higher pressure within the source than is used with EI, to permit sufficient collisions for reaction to be efficient. This is achieved by adding a suitable reagent gas to an EI source that has been engineered to have a much lower total area of apertures to reduce the pumping speed (Section 6.6.1) out of the source, so that when reagent gas is bled into the source the total internal pressure is $\sim$10^{-1} torr. (The pressure indicated on an ion gauge located near the ion source inside the vacuum chamber will be orders of magnitude less, depending on the gas conductance of the ion source and on the pumping speed in the source region, see Section 6.6.1). The large excess of reagent gas effectively ensures that the vast majority of ionizing collisions with the electrons do not involve the analyte molecules, so that there is a negligible contribution from EI in the mass spectrum of the analyte. This relatively high pressure within a CI source means that few if any of the electrons ever reach the electron trap (Figure 5.1), so the trap current can not be used in a feedback loop to control the electron emission. Instead the current of electrons scattered to the ion source walls is measured and used in this way. The potential difference between the source chamber and the filament is usually raised above the standard EI value to 100 V or more. It should be mentioned here that CI shares the same restrictions as EI with respect to volatility and thermal stability of analytes, and also the requirement to severely limit the flow rate of mobile phase into the source.

Because CI involves chemical reactions it is intrinsically more complicated than the relatively simple electron–molecule collisions involved in EI. For this reason it is more difficult to control and replicate experimental conditions, and as a result there are no extensive libraries of CI mass spectra and quantitation with CI usually demands use of a co-eluting internal standard. To illustrate the complexity, consider the list of ion–molecule reactions occurring in a CI source when methane is used as the reagent gas:

$$CH_4 + e^- \rightarrow (CH_4^{+\bullet})^* + 2e^-$$

$$\downarrow$$

$$CH_3^+, CH_2^{+\bullet}, CH^+, C^{+\bullet}, H_2^{+\bullet}, H^+$$

$$CH_4^{+\bullet} + CH_4 \rightarrow CH_5^+ + CH_3^\bullet$$

$$CH_3^+ + CH_4 \rightarrow C_2H_7^+ \rightarrow C_2H_5^+ + H_2$$

$$CH_2^{+\bullet} + CH_4 \rightarrow C_2H_4^{+\bullet} + H_2$$

$$CH_2^{+\bullet} + CH_4 \rightarrow C_2H_3^+ + H_2 + H^\bullet$$

$$C_2H_3^+ + CH_4 \rightarrow C_3H_5^+ + H_2$$

$$C_2H_5^+ + CH_4 \rightarrow C_3H_7^+ + H_2$$

and the reactions that lead to ionization of analyte molecules A are:

$$CH_5^+ + A \rightarrow CH_4 + [A+H]^+ \text{ (protonation)}$$

$$C_2H_5^+/C_3H_5^+ + A \rightarrow (A+C_2H_5^+/C_3H_5^+)$$
$$\text{(electrophilic addition, adduct formation)}$$

$$C_2H_5^+/C_3H_5^+ + A \rightarrow (A-H)^+ + C_2H_6/C_3H_6$$
$$\text{(hydride abstraction)}$$

Clearly, in view of the complexity of this ion–molecule chemistry in the CI plasma ('plasma' is a gas in which the total concentration of ions is high), the conditions inside the CI source must be carefully controlled if any level of reproducibility is to be achieved. Fortunately, at the methane pressures in the source, $>$0.1 torr, the abundance ratios of the methane-derived ions are essentially independent of pressure but do vary with temperature. At $\sim$200 °C typical relative abundances might be: m/z 17 (CH_5^+), 100; m/z 29 ($C_2H_5^+$), 85; m/z 41 ($C_3H_5^+$), 15 %; however, these abundance ratios do vary with temperature and with levels of impurities (oxygen, water, etc.) such that, at even quite low water concentrations, H_3O^+ becomes the dominant positive ion in a methane CI plasma. The relative efficiencies of the three analyte ionization processes depend strongly on the chemical nature of A but in most cases protonation is the major contributor.

The apparently pentavalent carbon atom in CH_5^+ has drawn much attention and in terms of simple intuitive chemical structures this species is best thought of as a CH_3^+ ion solvated by a hydrogen (H_2) molecule (Heck 1991). The tendency of a reagent ion BH^+ like CH_5^+ to donate a proton to a gas molecule is conveniently expressed in terms of the proton affinity (PA) of a species B:

$$B(g) + H^+(g) \rightarrow BH^+(g); PA(B) = -\Delta H° \qquad [5.2]$$

PA is intrinsically a positive quantity for such proton capture reactions that are exothermic ($\Delta H°$ negative); the more exothermic (i.e., energetically favorable) is this PA-defining reaction, the higher is PA(B), and thus the lower is the tendency of BH^+ to give up its proton, i.e., BH^+ is a weaker gas phase acid. As might be expected, methane has a very low PA (552 kJ.mol^{-1}) and thus CH_5^+ is a strong gas phase acid that readily protonates most analyte molecules A. The overall energy balance for the gas phase protonation of A by a BH^+ ion (i.e., proton transfer reaction) is:

$$\Delta H^\circ = PA(B) - PA(A) = 552 - PA(A)\,kJ.mol^{-1}$$
$$\text{for } B = CH_4 \qquad [5.3]$$

Methane has one of the lowest PA values known, only hydrogen has a lower value ($424\,kJ.mol^{-1}$) among common molecules; the larger alkanes have higher values, e.g., ethane $(C_2H_6)\,596\,kJ.mol^{-1}$, *iso*butane $678\,kJ.mol^{-1}$), while values for molecules containing a heteroatom are much larger (e.g., ethanol $(C_2H_5OH)\,776\,kJ.mol^{-1}$, dimethyl ether $(CH_3\text{-}O\text{-}CH_3)$, $792\,kJ.mol^{-1}$. A very complete database of carefully evaluated PA values and ionization energies, together with a great deal of other thermodynamic data, is freely available from the NIST Chemistry WebBook: http://webbook.nist.gov/chemistry/.

Then, for example, gas phase protonation of ethanol by CH_5^+ will proceed with ΔH° given by:

$$\Delta H^\circ = PA(CH_4) - PA(C_2H_5OH) = 552 - 776$$
$$= -224\,kJ.mol^{-1}.$$

This large exothermicity will reside as internal energy of (mostly) the reaction product (protonated ethanol) rather than methane, so these highly energized ions will have a high probability of dissociating in the ion source to produce a range of fragment ions (in this case dominated by water loss to give $C_2H_5^+$). In contrast, if *iso*butane were used as CI reagent gas instead of methane:

$$\Delta H^\circ = PA(iC_4H_{10}) - PA(C_2H_5OH) = 552 - 678$$
$$= -126\,kJ.mol^{-1}.$$

Thus CI with *iso*butane deposits much less internal energy in the reaction products so that much less fragmentation will be observed in the mass spectrum, i.e., *iso*butane CI is much 'softer' than with methane. This is just one simple example of how the energetics of CI reactions can be tailored to fit the purpose at hand by judicious choice of reagent gas (Harrison 1992). For purposes of quantitation, it is advantageous to arrange that all the ion current generated from an analyte is concentrated in just one or a few m/z values in order to keep LOD and LLOQ values low.

In negative ion mode a range of ion–molecule reactions can be used (Harrison 1992) to provide 'true' CI (involving chemical reactions). However, for trace quantitative analysis, electron capture ionization is particularly important in the case of highly electronegative compounds, including those containing halogens (pesticides etc.) and other electronegative groups (e.g., nitro). When electron energies in a CI source operated in negative mode are reduced to 1–2 eV by collisions with a buffer gas that does not capture electrons (methane is frequently used), the following processes can be observed:

$$A + e^-(<2\,eV) \rightarrow A^{-\bullet}\,(\text{resonance electron capture})$$

$$A + e^-(\sim2\text{-}10\,eV) \rightarrow (A^{-\bullet})^* \rightarrow \text{fragment ions}$$
$$(\text{dissociative electron capture})$$

$$A + e^-(>10\,eV) \rightarrow F_1^+ + F_2^-\,(\text{ion pair formation,}$$
$$\text{one of } F_1^+ \text{ and } F_2^- \text{ is odd-electron})$$

The primary electron capture event, forming the molecular radical anion $A^{-\bullet}$, is almost always exothermic. The energy quantity for negative ions that is analogous to proton affinity for positive ions is the electron affinity (EA) of the corresponding neutral molecule, defined as $EA = (-\Delta H^\circ)$ for the resonance electron capture process, i.e., EA is a positive quantity since this process is exothermic. Large values of EA tend to favor dissociative electron capture if the molecule can not accommodate large amounts of internal energy and/or if the ion source temperature is too high. But for suitable (highly electronegative) analytes, resonance electron capture can proceed with extremely high efficiency to provide the highest sensitivity and lowest LOD values achievable for any ionization technique. In cases where the analyte is not very electronegative, a variety of derivatization reagents (Section 5.2.1b) which introduce suitable groups such as pentafluorophenyl that can greatly improve limits of detection and quantitation, is available, but at the expense of an additional step in the sample preparation scheme that can introduce additional losses of analyte and reduce throughput. It must be mentioned that negative ion CI (often abbreviated NICI) is particularly irreproducible (Stöckl 1982), sometimes introducing ambiguity into analyte identification, e.g. some halogenated analytes readily undergo dissociative electron attachment to give halide anions X^- that can yield $(M + X)^-$ ions in variable yields; moreover precise and accurate quantitation using NICI definitely demands an internal standard.

5.2.1a Discrete Chromatograph-Ion Source Interfaces

Packed column GC involves carrier gas flow rates of 10–20 $mL.min^{-1}$, far too much for any affordable vacuum pump system to handle while providing a sufficiently good vacuum for EI–MS. Several devices have been used to selectively discard the bulk of the helium (Abian 1999) but by far the most commonly used has been the jet separator (Becker 1957; Ryhage 1964). This device (Figure 5.2) exploits the large difference in atomic/molecular mass between a helium atom ($M_{He} = 4\,Da$) and a typical GC analyte A (M_A up to $\sim$400 Da). The pressure of the GC effluent behind orifice 1 (Figure 5.2) is sufficiently high that it is forced through this orifice into the evacuated body of the separator as a supersonic jet. The atoms and molecules in this jet move with velocities of the order of their average (root mean square) thermal values c_{avge}, given by the kinetic theory of gases as $(3RT/M)^{\frac{1}{2}}$.

The important aspect here is the inverse relationship between the average velocity and the square root of the molar mass M (R is the gas constant and T the absolute temperature). This relationship holds for both the axial velocity component (that carries the atoms and molecules between the two orifices) and the radial component (that tends to cause them to diverge sufficiently that they miss

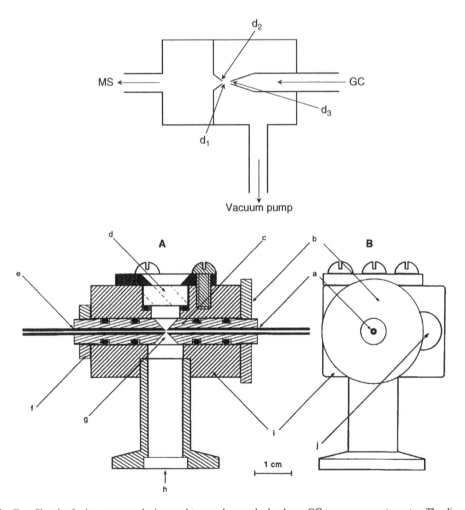

Figure 5.2 *Top*: Sketch of a jet separator device used to couple a packed column GC to a mass spectrometer. The dimensions are typically $d_1 = 100\,\mu m$ and d_2 and d_3 both 250–300 μm. The two tubes leading to the MS and from the GC, and drawn down to small apertures, must be accurately aligned. Such devices are normally fabricated of an inert material such as borosilicate glass. *Bottom*: Schematic diagram of an experimental jet separator designed with an adjustable inter-jet gap, in (A) cross-section and (B) axial view. (a): delivery capillary connected to transfer line from GC; (b) gap adjustment threaded disk; (c) nozzles; (d) window; (e) receiving capillary connected to the ion source; (f) gap zero-setting threaded disk; (g) expansion chamber; (h) vacuum port; (i) brass body of device; (j) cartridge heater (not visible in cross-section). Reproduced from Pongpun, *J. Mass Spectrom.* **35**, 1105 (2000), with permission of John Wiley & Sons, Ltd.

orifice 2 and are pumped away). Since M_{He} can be up to 100 times less than M_A its velocity components can be 10 times greater. This difference is of no consequence for the axial velocity components, since these carry the atoms and molecules between the two orifices and into the MS anyway. However, the difference in the radial velocity components results in a strong discrimination against helium, since they travel further in a radial direction than the analyte molecules during their passage through the distance d_3 (Figure 5.2(a)). This simple device is effective and introduces only minimal peak broadening but does suffer from discrimination among analytes on the basis of molecular mass as a result of its principle of operation. Packed GC columns are very seldom, if ever, used nowadays for true trace analysis although investigations of their performance for fast GC have been reported (von Lieshaut 1999).

Wall-coated capillary GC columns (Section 4.4.3) operate with helium flow rates in the range 1–2 mL.min^{-1} (measured at room temperature and pressure) and can thus be introduced directly into a mass spectrometer without overloading the vacuum pumps. With appropriate heating of the (short) capillary that emerges from the GC oven and enters the EI or CI ion source, such an arrangement is extremely efficient with minimal extra-column broadening and discrimination effects. This approach is extremely important for trace analysis in several fields of application, particularly environmental analysis; in some cases extremely fast separations (a few seconds up to $\sim$2 minutes) are possible but such 'fast GC' (Section 3.5.11) places extreme demands on the response time characteristics of the mass spectrometer (Leclercq 1998) with modern TOF analyzers the likely best choice (Section 6.4.7). However, when multicapillary GC systems are employed to increase the permissible sample loading on-column (Baumbach 1997; von Lieshaut 1999; Pongpun 2000) the helium flow rate must be reduced before introduction into an EI source, and for this purpose a sophisticated design of a jet separator with an adjustable inter-jet gap has been described (Pongpun 2000, Figure 5.2b). A recent book (Niessen 2001) contains specialist chapters on most aspects of GC/MS analysis.

It is much more difficult to couple HPLC on-line with EI and CI sources (Ardrey 2003) since the volume of vaporized mobile phase is up to 10^3 times larger than that of the liquid, far beyond the pumping capabilities of any realistic vacuum system unless the flow rate is restricted by a post-column split to a few μL.min^{-1}, as in the so-called direct liquid introduction interfaces (Abian 1999) for CI and (less commonly) EI; these interfaces

are never used in modern practice for quantitative analysis. For more realistic flow rates the first practical interfaces were transport systems that carry the analytes via a series of heaters and pumps that remove the mobile phase before flash vaporization of the analytes into the EI or CI source. The moving wire (Scott 1974) and moving belt (McFadden 1976) interfaces are examples of such devices, but these are never used nowadays since a combination of GC/MS with LC/MS using atmospheric pressure ionization can analyze all analytes that could be analyzed using these transport interfaces that were infamously liable to breakdown.

Another type of discrete LC/MS interface that now appears to be almost obsolete is the particle beam interface (Takeuchi 1978; Willoughby 1984; Winkler 1988) based on the same general principle as the jet separator. In this interface the effluent emerging from the LC is first nebulized into a fine aerosol spray, which facilitates evaporation of solvent from the small aerosol droplets, to leave microparticles that consist mostly of relatively involatile analyte. These high mass microparticles can be efficiently separated from solvent molecules in a jet separator, sufficiently so that ionization by either EI or CI is possible following flash evaporation of the microparticles in the ion source. Again, thermal stability and volatility are obviously prerequisites for analysis of compounds using this interface (Abian 1999). The advantage of this approach over the transport devices was the lack of a complex mechanical system incorporating vacuum locks that permitted the moving belt or wire to enter the mass spectrometer vacuum. However, even for highly stable and reasonably volatile analytes like polycyclic aromatic hydrocarbons (PAHs), it was found that use of an atmospheric pressure chemical ionization (APCI, Section 5.3.3) source with no pre-ionization removal of mobile phase was much preferable to both the moving belt and particle beam interfaces (Anacleto 1995).

5.2.1b Chemical Derivatization for EI and CI

The limitation on applicability of both EI and CI to analytes that are thermally stable and volatile can be alleviated by chemical derivatization to protect any highly polar functional groups within the analyte molecule (Knapp 1979; Drozd 1981). A more recent book describing this approach (Blau 1993) has been updated and extended in a series of publications (Halket 2003, 2004, 2004a, 2005, 2006; Zaikin 2003, 2004, 2005). The structures of some common derivatization reagents, suitable for transforming involatile compounds into derivatives amenable to analysis by GC and EI or CI, are shown in Figure 5.3. A paper (Little 1999) describing artifacts that can arise in derivatization by silylation

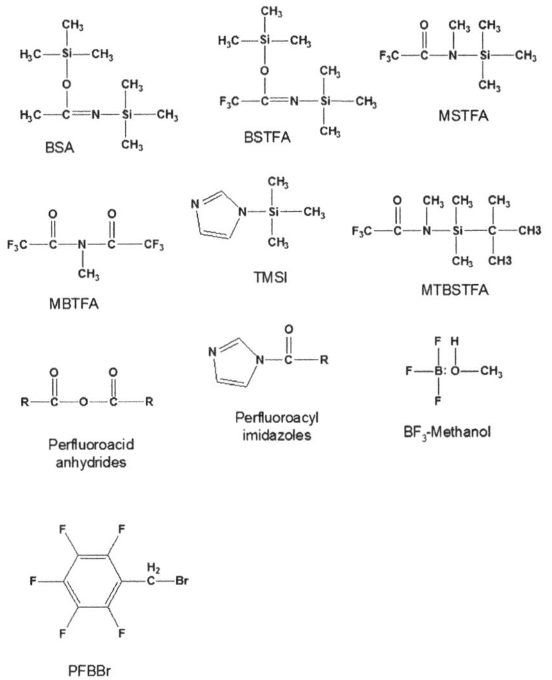

BSA

BSTFA

MSTFA

MBTFA

TMSI

MTBSTFA

Perfluoroacid anhydrides

Perfluoroacyl imidazoles

BF₃-Methanol

PFBBr

Figure 5.3 Structures of some popular derivatization reagents for GC. Butylsilylacetamide (BSA) is a silylation reagent that reacts quantitatively under relatively mild conditions with a wide variety of compounds to form volatile, stable trimethylsilyl (TMS) derivatives. *N, O*-bis(trimethylsilyl)trifluoroacetamide (BSTFA) is a powerful TMS donor with highly volatile byproduct more easily separated than that from BSA. *N*-methyl-*N*-trimethylsilyltrifluoroacetamide (MSTFA) is also a powerful TMS donor with a byproduct (*N*-methyl-*N*-trifluoroactamide) that is so volatile that it often coelutes with the solvent. *N*-methyl-*bis*(trifluoroacetamide) (MBTFA) trifluoroacylates primary and secondary amines, and hydroxyl and thiol groups, under mild non-acidic conditions. *N*-trimethylsilylimidazole (TMSI) is the strongest silylation reagent for hydroxyl groups, reacts quickly and smoothly with hydroxyl and carboxyl groups, but does not react with amines or amides making it possible to do selective multi-derivatization of compounds containing both hydroxyl and amine groups. Dimethyl-*tert*butylsilyltrifluoroacetamide (MTBSTFA) yields stable TBDMS (*tert*-butyldimethylsilyl) derivatives of hydroxyl, carboxyl, thiol and primary and secondary amine groups that are much more stable towards hydrolysis than TMS ethers. Perfluoroacid anhydrides (R = CF_3, C_2F_5, C_3F_7 are common alternatives) react readily with alcohols, phenols and amines, producing stable, volatile, highly electronegative acyl derivatives suitable for sensitive detection by electron capture CI. Perfluoroacylimidazoles (typically R = CF_3 or $C_4H_2F_7$) also provide effective acylation of hydroxyl groups and primary and secondary amines, but have the advantage over the anhydride reagents that produce no acidic byproducts that must be removed before GC injection (the principal byproduct is the relatively inert imidazole). BF_3-methanol is a convenient methanol-catalyst system that quickly and quantitatively converts carboxylic acids to methyl esters (note that BF_3 is a strong Lewis acid so must be used with caution for compounds that can undergo reactions or rearrangements under acidic conditions). Pentafluorobenzylbromide (PFBBr) converts carboxylic acids, phenols, and mercaptans to electronegative fluorinated derivatives via loss of hydrogen bromide that are highly suitable for analysis by electron capture CI more commonly known as negative ion chemical ionization (NICI).

(one of the most commonly used derivatization reactions) is updated by the author on a website (http://users.chartertn.net/slittle/) that contains other helpful information of interest to readers of this book. A great deal of useful information is available on the websites of companies that supply derivatization reagents. For example, a convenient brief guide to selection of derivatization reagents is available at http://www.piercenet.com/files/TR0037-GC-reagent-guide.pdf and a

brief introduction to derivatization is freely available at http://www.sigmaaldrich.com/img/assets/4242/fl_analytix 3_2002_new_.pdf.

Silylation is the most widely used derivatization technique for GC. Some functional groups are problematic for GC as a result of acidic or basic properties that promote intermolecular hydrogen bond formation and thus involatility; thus hydroxyl, carboxylic acid, amine, thiol, even phosphate ester groups, can be derivatized by converting the acidic hydrogen into an alkylsilyl group (e.g., $-SiMe_3$, where $Me = CH_3$). Such derivatives are more volatile and more thermally stable than the original molecules and also usually have more favorable characteristics for EI–MS via more diagnostic ion fragmentation patterns of use in structure investigations and/or characteristic ions ideal for trace quantitative analyses. *Acylation*, an alternative to silylation, converts compounds with active hydrogens (OH, SH and NH groups) into esters, thioesters and amides, respectively; as for silylation, acylation also provides characteristic fragmentation patterns and intense specific fragment ions with EI-MS. A useful example involves insertion of perfluoracyl groups to enable negative ion CI via electron capture (Section 5.2.1). *Alkylation* is the replacement of an active hydrogen in R–COOH, R–OH, R–SH, and $R-NH_2$ with an alkyl (or sometimes aryl) group. *Esterification*, the condensation of the carboxyl group of an acid and the hydroxyl group of an alcohol with the elimination of water, is the first choice for derivatization of acids and is also used for other acidic functional groups.

An extensive report (Birkemeyer 2003) has described a painstaking evaluation of 17 different derivatization reactions covering trifluoroacetylation, pentafluorobenzylation, methylation and trimethylsilylation. It was concluded that, for the authors' analytes of interest (phytohormones), the N-methyl-N-(*tert*-butyldimethylsilyl) trifluoroacetamide (MTBSTFA) reagent (Figure 5.3) provided the best derivatization for GC/MS analysis with EI. This paper contains much useful information on derivatization reagents, reaction conditions etc.

Derivatization can also be an enabling technology that makes it possible to conduct an analysis with improved sensitivity and selectivity. Recently (Gao 2005) the use of derivatization to improve sensitivity and selectivity in LC–MS using atmospheric pressure chemical ionization (APCI, Section 5.3.4) and electrospray ionization (ESI, Section 5.3.6) has been emphasized. For example, introduction of permanently charged moieties (e.g., quaternary ammonium) or readily ionized species can sometimes dramatically improve the ionization efficiency for ESI, while introduction of moieties with high proton affinity or electron affinity in positive or negative ion modes, respectively, can improve sensitivity for ESI and APCI. However, it is worth repeating that introduction of such an additional step into the overall analytical method (Figure 1.2) can introduce analyte losses and will also reduce throughput in cases where the latter is an important criterion. As always, it is necessary to adopt the compromise that best fits the purpose at hand.

In the context of trace quantitative analysis, GC coupled with EI or CI and with many types of mass analyzer (Chapter 6), is a highly mature technique that has been, and continues to be, applied to wide range of analytical problems involving volatile and thermally stable analytes. Matrix effects are usually not a serious problem in practice. For these reasons, the foregoing discussion is significantly shorter than the following treatments of the more recent ionization techniques used for quantitative analyses.

5.2.2 Matrix Assisted Laser Desorption/Ionization (MALDI)

MALDI is one of the newer ionization methods (other than ESI) that has propelled mass spectrometry into a leading role as an essential technique in modern molecular biology and led to a Nobel Prize for Koichi Tanaka in 2002. The technique bears some similarity to the earlier fast atom bombardment method (Section 5.3.1), in that the sample is dissolved or embedded in a suitable matrix, introduced into a (usually) evacuated ion source and activated by an external high energy agent (keV atoms or ions for FAB, laser photons for MALDI). Until relatively recently, MALDI was considered to be unsuitable for trace quantitative analysis of 'small' molecules in combination with chromatography as it is an off-line technique. However, the intrinsically high ionization efficiency of MALDI, with its potential for improved limits of detection and quantitation as well as high throughput, has led to considerable activity to the point where the technique is now a potentially important contender in this area. For this reason a more complete discussion is provided compared with that of EI/CI in Section 5.2.1

The history of the development of MALDI has been admirably described (Beavis 1992) and this account will be briefly summarized here. The Tanaka approach, first described in 1987 in the proceedings of a conference that were not widely available to the international scientific community, became well known in a later publication (Tanaka 1988). This approach, referred to by the authors as the 'ultra fine metal plus liquid matrix' method, involved suspending a fine cobalt powder (particle size about 30 nm) in liquid glycerol to which a biopolymer or synthetic polymer sample was added in nmol quantities. Irradiation of this mixture (held on a sample holder inserted into the vacuum chamber of the mass spectrometer) by a pulsed laser (nitrogen laser emitting at 337 nm, pulse width 15 ns, pulse energy up to 4 mJ, no information given on laser spot size at the sample) then yielded mass spectra of polymers with molecular masses in the tens of kDa (Figure 5.4) using a time of flight mass spectrometer. (Such analyzers are an ideal fit for pulsed ion sources like MALDI, since the laser pulse can be used as the start signal for timing the ion's flight time,

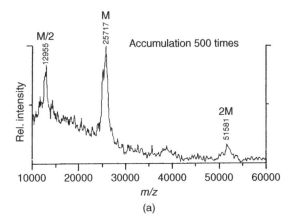

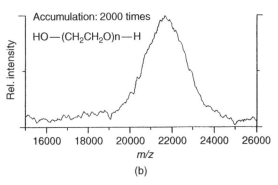

Figure 5.4 Mass spectra obtained using the 'ultra fine metal plus liquid matrix' method and a time of flight mass spectrometer. (A): chymotrypsinogen (protein of calculated molecular mass 25 717 Da) showing both singly- and doubly-protonated molecules plus a weak peak corresponding to a proton-bound dimer. (B): sample of polyethylene glycol (PEG20K) with nominal average molecular mass 20 kDa. Reproduced from Tanaka, *Rapid Commun. Mass Spectrom.* **2**, 151 (1988), with permission of John Wiley & Sons, Ltd.

Section 6.4.7.) No attempt was made to explore the mechanism(s) responsible for these observations, although the role of the glycerol matrix (that is essentially completely transparent at 337 nm) was assumed to be that of replenishment of analyte to the region of the surface irradiated and ablated by the laser. The fine metal particles were essential for the success of the method, ascribed to the high photo-absorption, low heat capacity and extremely large ratio of surface area to volume of these particles, which were assumed to couple the laser photons to the analyte via the 'rapid heating effect' (see the text box in this Chapter).

Despite its initial success, this approach to laser mass spectrometry has not been pursued. Instead, the approach developed by Hillenkamp and Karas (Karas 1987, 1988) has come to dominate the technique now universally referred to as MALDI. This approach uses a matrix that is almost always an organic solid that absorbs the laser radiation; early examples of protein mass spectra obtained in this way are shown in Figure 5.5. Although the m/z resolution evident in all these early laser mass spectra (Figures 5.4 and 5.5) is very poor (~20–40), the MALDI method required sample loadings of only ~1 pmol, about 1000 times less than the 'ultra fine metal plus liquid matrix' method. This, together with the experimental inconvenience associated with using a liquid matrix in an evacuated ion source, has led to intense development of all aspects of the MALDI method and almost complete abandonment of the rival approach.

The mechanism(s) underlying MALDI are still a matter of considerable investigation and debate, as exemplified by a special issue of *Chemical Reviews* (2003, **103**, Issue 2); two of the contributions (Karas 2003; Knochenmuss 2003) are particularly relevant to concerns of direct interest here, although it must be added that in both cases the main thrust is towards qualitative MALDI–MS analysis of macromolecules, the area where the technique has enjoyed its major success.

A major problem in understanding MALDI concerns how the technique produces any ionization at all, in view of the fact that the most commonly used laser wavelengths (337 nm, 355 nm) correspond to photon energies of only 3.5–3.7 eV, far below the ionization energy of any organic molecule (all the commonly used MALDI matrices have ionization energies >8 eV). Indeed, infrared lasers (e.g., carbon dioxide lasers emitting at 10.6 μm) can yield MALDI mass spectra with photon energies of only 0.12 eV. The primary ionization event clearly can not correspond to direct single photon photoionization of isolated matrix (or analyte) molecules, and the lasers commonly used in MALDI experiments do not deliver sufficient power densities (equivalent to number of photons per unit volume per unit time) for multi photon photoionization. The primary contender for the primary ionization mechanism at present (Karas 2003) involves formation of gas phase clusters as a result of explosive ablation of matrix (plus entrained analyte) by the laser pulse; this picture can account for the laser power threshold observed (see later discussion of Figure 5.8) for the production of ions by MALDI, interpreted as the point at which the accumulated excitational energy density at the irradiated area of the matrix solid reaches a critical value that can induce explosive cluster formation.

Typical sample preparation protocols for MALDI (see below) involve additives such as acids for analytes with basic functionalities like proteins and peptides, or metal

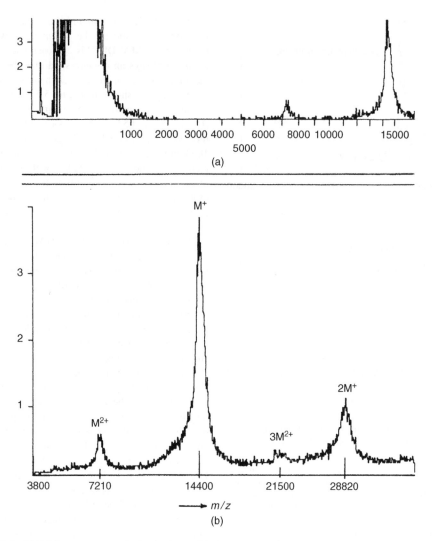

Figure 5.5 Early MALDI mass spectra obtained using solid nicotinic acid (3-carboxyl pyridine) as matrix and a Q-switched frequency-quadrupled Nd YAG laser (266 nm, 10 ns). Mass spectrum of porcine trypsin (protein of molecular mass 23 463 Da); (a) full range spectrum; (b) expanded region showing protein peaks. Reproduced from Karas and Hillenkamp, *Anal. Chem.* **60**, 2299 (1988), copyright (1988), with permission of the American Chemical Society.

ions (often alkali metal ions) for synthetic polymers with no acid–base properties, so there are many pre-formed ions in the solid matrix-plus-analyte. As a result of statistical formation of such clusters within the initial few nanoseconds following arrival of the laser pulse, some clusters will carry an excess positive charge and others a net negative charge. Considerable experimental evidence exists (Karas 2003) for formation of such clusters, and indeed the ubiquitous 'chemical noise' ('peak at every m/z value') observed in MALDI mass spectra has been shown to correspond to clusters of matrix plus other molecules

such as water or other solvent used in the sample preparation. The clusters then proceed to give single molecule ions by processes somewhat related to those invoked in mechanisms of ESI (Section 5.3.6).

It must be emphasized that most MALDI experiments are conducted with an evacuated ion source, so that cluster (droplet) evaporation might not seem to be possible in the same way as in ESI, but there are many more ($\times 10^4$–10^7) neutral species than ions in the dense plume ejected from the solid surface so the local pressure is high. Although the cluster ionization mechanism is consistent with a great deal of experimental evidence (Karas 2003), it can not account for some clearcut observations such as the observation of abundant radical ions like [Ma–2H]$^{-\bullet}$ (where Ma = matrix) in negative mode. One possible additional mechanism could involve energy pooling of several photo-excited matrix molecules within a cluster to an extent that results in ionization of one matrix molecule (a very special mechanism of multi photon photoionization). Despite some high level physical chemistry and theoretical investigations, there is still considerable uncertainty concerning all the mechanisms responsible for the primary ionization in MALDI; clearly infrared excitation need not follow all the same pathways as UV MALDI.

Regardless, there is extensive evidence (Knochenmuss 2003) that secondary reactions within the desorbed plume, occurring on a timescale of tens or even hundreds of nanoseconds, are the main determinant of the final detected MALDI mass spectrum. Again this hypothesis is supported by a wide range of observations. As one example, it was shown (Papantonakis 2002) that a wide variety of excitation methods, with UV wavelengths ranging from 266 to 400 nm, pulse durations from 0.12 to 2ns, and energies per pulse 0.25 to 5 mJ, yielded qualitatively identical mass spectra (including both matrix and analyte peaks) for three test analytes in a matrix of 2,5-dihydroxybenzoic acid. Even more striking, this spectral similarity was maintained when the more usual UV laser was replaced by a free-electron infrared laser, continuously tunable over the range 2–10 μm and with energy per pulse of ~2 μJ in micropulse mode (one picosecond pulse duration) and 0.6 mJ in macropulse mode (100 ns pulse duration). This evidence (Papantonakis 2002) is consistent with the view (Knochenmuss 2003) that secondary reactions in MALDI, in principle predictable from thermochemistry, are the dominant factor in determining the appearance of the final spectrum so that the primary ionization events are simply not reflected in the final ion distribution.

From the point of view of the present book, some understanding of the mechanism(s) underlying MALDI is important because of effects of ionization suppression (and possibly enhancement, though the latter has not been widely observed in MALDI) that must be taken into account if MALDI is to be used for quantitation; an extended discussion of suppression and enhancement effects is given in Section 5.3.6a for the case of electrospray ionization. MALDI suppression effects have been extensively investigated (Knochenmuss 1996, 1998, 2000, 2003) and correlated with the detailed theory of in-plume reactions controlled largely by thermochemical considerations. Examples of the suppression of matrix ions by relatively large amounts of analyte are shown in

Figure 5.6. In general it was found that the matrices tested showed suppression in only one polarity, or not at all. In those cases where suppression was observed, the matrix-to-analyte signal ratios follow a typical curve with increasing analyte concentration (Figure 5.6c); the intermediate plateau region (between about 200–600 mole ratio in Figure 5.6c) is not well understood. The evidence suggests (Knochenmuss 1998) that, at high matrix:analyte ratios, the gas phase reactions responsible for the final observed mass spectrum continue in the desorption plume; at low matrix:analyte ratios the primary ions are depleted before significant plume expansion occurs. This can be understood as a situation in which one reactant is limiting (Knochenmuss 2003). When secondary reactions of primary matrix ions with analyte are thermochemically favorable, they can proceed to equilibrium; when analyte is in sufficiently low concentration it is limiting and some matrix ions survive to appear in the spectrum, but when analyte is abundant the matrix ions are limiting and analyte ions are the only observed products. Therefore, above a certain concentration, analyte signal cannot be increased by adding more analyte to the MALDI sample, with obvious implications for dynamic range.

A notable and perhaps unexpected observation (Figure 5.6) is that, when matrix suppression occurs, all matrix ions are suppressed, e.g., an analyte that appears in the spectrum as the protonated molecule can suppress matrix alkali ion adducts [Ma+Na]$^+$ as well as protonated matrix [Ma+H]$^+$. This observation has been interpreted (Knochenmuss 2000) in terms of quantitative considerations of secondary plume reactions among matrix ion species that can be interconverted by reactions with neutral matrix. These reactions are sufficiently close to isoenergetic that they should proceed rapidly under plume conditions, so that when a highly efficient matrix–analyte reaction depletes one matrix species (e.g. [Ma+H]$^+$), all others (e.g., [Ma+Na]$^+$) will also be efficiently depleted via this interconversion reaction channel.

Even more important for present purposes than the effect of ionization suppression of the MALDI matrix is the suppression of one analyte by another; examples are shown in Figure 5.7. This effect is intrinsically more complex than ionization suppression of the matrix; it may be the result of competition among analytes for the same or different primary matrix ions, or via direct reactions among analytes, so the concentration dependence can become highly complicated with serious implications for quantitation experiments. Inter-analyte suppression has been observed (Knochenmuss 2000) for both similar (e.g., both protonated) and dissimilar (e.g., protonated vs sodiated) analytes (Figure 5.7); again, the dissimilar case must presumably involve efficient interconversion reactions among the ionized forms of the suppressed analyte in the MALDI plume.

The foregoing discussion of the present understanding of MALDI mechanisms is a much abbreviated and simplified version of the detailed considerations (Karas 2003; Knochenmuss 2003) of extensive experimental evidence. However, it should provide sufficient understanding for an appreciation of the difficulties faced in using MALDI–MS in trace quantitative analysis. Before moving on to discuss this in detail, some practicalities of MALDI in general will be addressed.

The properties of the matrix material are critical, but only a few compounds are useful and very little is known about what makes a material a 'good' matrix for any given analyte type; some common qualities of 'good' matrices are the solubility of the matrix in the appropriate solvents, the absorption spectrum of the matrix, and its reactivity. Matrix solubility is found to be necessary so that the analyte and matrix material can be dissolved in the same solution; in practice this condition means that the matrix must dissolve sufficiently in a suitable solvent to make a ∼5 mM solution. Exceptions to this general rule do exist; for analyte/matrix combinations that

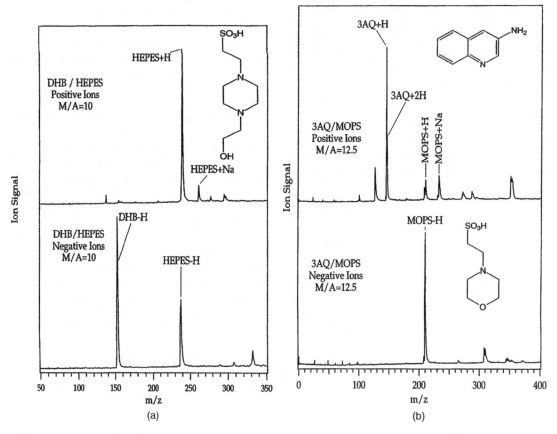

Figure 5.6 (a) and (b): Examples of suppression of matrix ionization by analytes at low matrix:analyte ratios. DHB = 2,5-dihydroxybenzoic acid; HEPES = 4-(2-hydroxyethyl)-piperazine-1-ethansulfonic acid; 3AQ = 3-aminoquinoline; MOPS = (3-morpholino-propanesulfonic acid. Note that both analytes are sulfonic acids expected to give strong MALDI signals in negative mode, but HEPES suppresses the acidic matrix DHB only in positive mode while MOPS suppresses the basic matrix 3AQ in negative mode. The polarity in which matrix suppression occurs is thus determined by both the matrix and the analyte. (c): Matrix to analyte ion intensity ratios (negative ions) vs matrix:analyte mole ratio (M/A). The prominent drop as M/A is decreased to a plateau at M/A ∼ 1000 is followed at M/A < 50 by a final drop to full matrix suppression. The inset shows the low M/A region with an expanded scale. Each point represents a single measurement. Reproduced from Knochenmuss, *Rapid Commun. Mass Spectrom.* **12**, 529 (1998), with permission of John Wiley & Sons, Ltd.

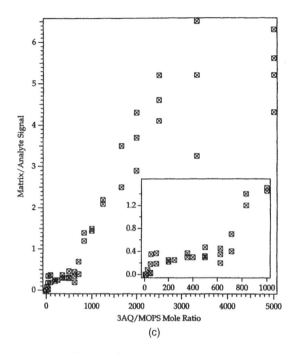

Figure 5.6 (Continued)

can not be co-dissolved to suitable concentrations, it was shown (Skelton 2000) that finely ground materials can be mixed as solids then pressed together in a fashion similar to preparation of an alkali halide disk for infrared spectroscopy to give useful MALDI spectra.

The condition of efficient absorption of laser radiation by the matrix restricts the initial energy to the matrix, rather than the analyte that might be photochemically altered. The range of values of extinction coefficients for 'good' matrices at the commonly used UV laser wavelengths is 3000–16 000 L mol^{-1} cm^{-1} (Beavis 1992); in practice this condition is quite restrictive and in effect limits the choice of matrix to aromatic compounds (Table 5.1). The condition of low reactivity ensures that the matrix does not chemically modify the analyte. Even when a candidate matrix material satisfies these three requirements, it is still necessary to test the material in MALDI practice, and indeed the vast majority of materials that meet the criteria do not turn out to be 'good' matrices. Even closely related compounds can behave very differently as MALDI matrices, e.g. 4-hydroxycinnamic acid is a reasonably good matrix for proteins, but its close isomer 3-hydroxycinnamic acid is a poor matrix despite having very similar solubility properties and absorption spectra.

Preparation of the analyte/matrix combination for MALDI–MS is also crucial and also something of an art

rather than a science. A summary of popular methods is freely available at: http://www.sigmaaldrich.com/Brands/Fluka____Riedel__Home/Literature/AnalytiX/AnalytiX_2001.html and a more detailed version at: http://www.chemistry.wustl.edu/~msf/damon/samp_prep_dried_droplet.html. Note that most descriptions of MALDI preparation methods are strongly slanted towards analysis of pre-purified samples of macromolecules (particularly biopolymers), rather than of small molecules eluting as fractions from an HPLC column. Nonetheless, the experience accumulated is useful for present purposes and an abbreviated discussion is provided as Appendix 5.1.

Another experimental factor governing the efficiency of MALDI production of ions is the nature of the laser irradiation; as mentioned above in the discussion of MALDI mechanisms, the wavelength of the laser radiation has little or no influence on the MALDI process (Papantonakis 2002). The important parameters are the total amount of energy (J) deposited per pulse by the laser, the fluence (energy per pulse per unit irradiated area, often expressed in J.cm^{-2}), the laser power (energy per pulse divided by the pulse time-width, in J.s^{-1}, i.e., in W), and the irradiance (energy per pulse per unit irradiated area divided by the pulse time-width, J.cm^{-2}.s^{-1}). The dependence of the analyte ion yield on the pulse irradiance is highly nonlinear (Figure 5.8); below a threshold value (that varies considerably with experimental conditions) no analyte ions are produced and above this threshold the ion yield increases rapidly with increasing irradiance to a plateau. The irradiance is the parameter intuitively expected to be most important for MALDI efficiency but in order to measure irradiance it is necessary to know the total pulse energy (relatively straightforward) and also the temporal and spatial profiles of the beam at the sample (more difficult), and to account for the variations over these profiles. Usually only an approximate value of the irradiance, averaged over the irradiated area and pulse time-width, is estimated. However, for a given MALDI–MS apparatus operating under conditions fixed except for the pulse energy, the latter is proportional to the irradiance. Thus it is straightforward to determine the ionization threshold for a given analyte–matrix combination in a given MALDI–MS instrument but inter-instrument comparisons of thresholds etc. are more difficult. For example, although spatial profiles of laser beam intensities perpendicular to the beam direction are generally Gaussian, it was recently shown (Holle 2006) that by altering the spatial profile of a Nd:YAG laser the MALDI efficiency could be significantly improved.

From the point of view of interfacing mass spectrometry to high resolution separation methods, MALDI is

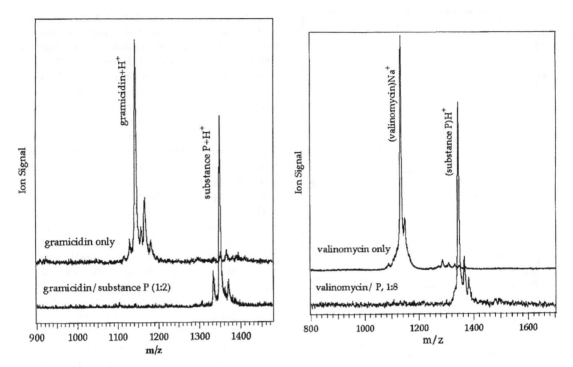

Figure 5.7 Examples of MALDI suppression of one analyte by another. *Left*: MALDI mass spectra of gramicidin S and of a mixture of gramicidin S and substance P in DHB matrix (mole ratio 1 : 2 : 2000), illustrating the similar analyte suppression effect (both analytes appear as protonated molecules). The gramicidin S concentration was the same in both spectra, and the spectra were recorded under identical conditions. When sufficient substance P is present, the gramicidin S signal disappears. *Right*: MALDI mass spectra of valinomycin and of a mixture of valinomycin and substance P in DHB matrix (mole ratio 1 : 2.5 : 1000), illustrating the dissimilar analyte suppression effect (one analyte appears as the Na^+ adduct, the other as the protonated molecule). The valinomycin concentration was the same in both spectra, and the spectra were taken under identical conditions. When sufficient substance P is present forming abundant protonated molecules, the valinomycin Na^+ adduct signal almost completely disappears. Reproduced from Knochenmuss, *J. Mass Spectrom.* **35**, 1237 (2000), with permission of John Wiley & Sons, Ltd.

Table 5.1   Common matrices used for UV-MALDI of macromolecules

Name(S)	Structure	Formula and Monoisotopic Molecular Mass (Da)	Most common applications
3-Hydroxypicolinic acid (3-hydroxy-2-pyridinecarboxylic acid		$C_6H_5NO_3$ 139.035	Nucleic acids
Ferulic acid (4-hydroxy-3-methoxycinnamic acid)		$C_{10}H_{10}O_4$ 194.066	Nucleic acids

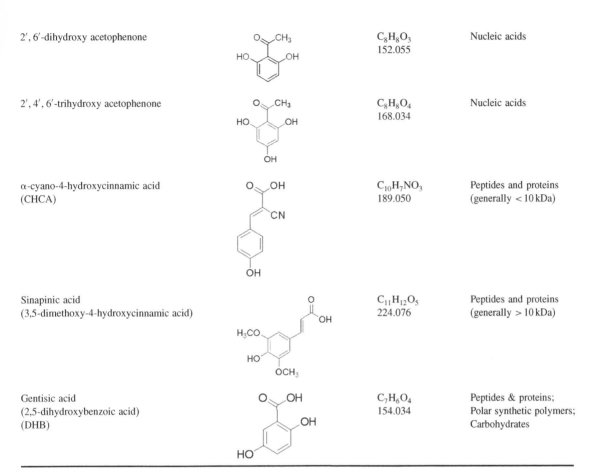

2′, 6′-dihydroxy acetophenone		$C_8H_8O_3$ 152.055	Nucleic acids
2′, 4′, 6′-trihydroxy acetophenone		$C_8H_8O_4$ 168.034	Nucleic acids
α-cyano-4-hydroxycinnamic acid (CHCA)		$C_{10}H_7NO_3$ 189.050	Peptides and proteins (generally < 10 kDa)
Sinapinic acid (3,5-dimethoxy-4-hydroxycinnamic acid)		$C_{11}H_{12}O_5$ 224.076	Peptides and proteins (generally > 10 kDa)
Gentisic acid (2,5-dihydroxybenzoic acid) (DHB)		$C_7H_6O_4$ 154.034	Peptides & proteins; Polar synthetic polymers; Carbohydrates

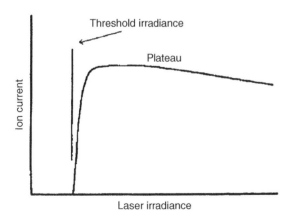

Figure 5.8 A sketch of a typical dependence of MALDI ion current as a function of laser irradiance (total laser energy per unit area divided by pulse time width), showing the characteristic threshold for ion formation. Reproduced from Beavis, *J. Mass Spectrom.* **27**, 653 (1992), with permission of John Wiley & Sons, Ltd.

typically an off-line technique. Attempts at on-line interfacing have been reviewed (Murray 1997; Gelpi 2002) and these are not described in detail here; the basic incompatibility between the two technologies has thus far frustrated all attempts at on-line coupling in a manner suitable for trace analysis. For example, a continuous vacuum deposition interface for continuous introduction of liquid samples (from a CE in this case) into a MALDI–TOFMS instrument (Preisler 1998, 2000) is really a modern version of the moving belt interface developed for LC–EI–MS (McFadden 1976). In this device the CE eluant was premixed with a suitable matrix and deposited on a rotating quartz wheel inside the vacuum region of a TOF instrument. Rapid solvent evaporation produced narrow sample trace segments on the rotating wheel that transported them to the ion source for irradiation by the laser beam. Good reproducibility and mass sensitivity (down to 2 amol for the peptide angiotensin I) were obtained, with general results similar to those obtained using an off-line dried droplet approach. The later version

(Preisler 2000) used a disposable moving tape instead of the rotating wheel to avoid the need for regular cleaning of the transport medium. However, no attempts were made at quantitation of small molecule analytes at trace levels in complex matrices, and limitations of the approach with respect to flow rate, throughput and general robustness have not encouraged further development.

Off-line coupling of LC or CE to MALDI–MS may seem like an approach that is unlikely to provide performance competitive with that of on-line LC–API methods (Sections 5.3.3–5.3.6), but in fact this approach is now undergoing intensive development. By effectively decoupling the liquid phase separation from the MALDI–MS analysis, this approach has potential for very high throughput analysis, and also in principle provides a semi-permanent 'chromatogram' of the deposited eluant that can be revisited for further characterization. A review of small molecule analysis by MALDI–MS (Cohen 2002) summarizes earlier work on this approach, but these early developments focused on qualitative characterization of analytes, not on trace quantitation of analytes extracted from a complex matrix.

A major barrier to development of MALDI as an off-line interface between HPLC and MS for trace quantitation has been the 'chemical noise' (peak-at-every-m/z background, particularly at low m/z < 500) where matrix-derived peaks can dominate the spectrum. However, the overall problem can be resolved into constituent parts, namely, development of efficient deposition of eluant to provide adequate representations of the chromatographic separation, and efficient MALDI–MS analysis of the deposited analytes on a timescale that should certainly be no longer than the chromatographic time, and also methods to provide acceptable accuracy, precision, repeatability, LOD, LLOQ and dynamic range of the quantitation.

At the time of writing this chapter, such developments are at an early stage and incorporate several complementary innovations. The development (Krutchinsky 1998) of so-called orthogonal MALDI (o-MALDI) separates operation of the mass spectrometer analyzer from the laser pulse timing, by providing a high pressure quadupole cell between the MALDI source and analyzer for efficient ion transmission that also transforms the MALDI source into what is effectively a continuous ion source (Section 6.4.2a). This conversion to a quasi-continuous ion source is assisted by the use (McLean 2003; Hatsis 2003; Gobey 2005) of high-frequency lasers (~1 kHz) rather than the nitrogen lasers (5–10 Hz) used in conventional MALDI–MS instruments. These high frequency lasers allow signal averaging of a large number of measurements (laser pulses) over a short amount of time

and thus have the potential for acquisition of quantitative data at useful high throughput rates. In addition, use of these lasers can lead to essentially complete ablation of a representative portion of a matrix–analyte spot (Sleno 2005), thus alleviating any sampling errors arising from inhomogeneities in the matrix–analyte solid mixture. Ionization suppression effects in the desorption step of MALDI imply that use of appropriate internal standards is essential (Gobey 2005; Sleno 2006) to account for competition in the plume reactions (see above). An early investigation of MALDI–TOF–MS in quantitative analysis (Bucknall 2002) used the Method of Standard Additions (Section 8.5.1b), but this is unsuitable for high throughput analysis and requires larger amounts of sample and of the analytical standard.

Most conventional MALDI mass spectrometers use a time of flight (TOF) m/z analyzer (Section 6.4.7), originally as a result of the natural fit between the pulsed nature of the two devices (now superceded by the introduction of o-MALDI), but partly because of the very high m/z range that can be covered in macromolecular analyses and the moderately high resolution and accuracy/precision of mass measurement that TOF–MS can provide. However, TOF analyzers are not ideal with respect to trace quantitation of small molecules despite their potential for fast acquisition of complete mass spectra, as a result of their limited dynamic range (~2 orders of magnitude) and low (~10–20 %) duty cycle (the fraction of all ions created in the source that are recorded by the detector, Section 6.2.3e). Tandem mass spectrometry using a quadrupole-TOF combination (Section 6.4.7) can increase the specificity to alleviate the chemical noise problem but at a cost of lower ion transmission efficiency. Ion trap instruments (Section 6.4.5) also provide complete MS/MS spectra of m/z selected precursor ions at unit mass resolution but the time per spectrum is relatively long (~100 ms) as a result of poor duty cycle (Section 6.2.3e). Moreover, all ion trap devices are subject to severe space–charge effects (deterioration of performance by Coulombic repulsive effects among the trapped ions) if the total ion population becomes too high, although the newer linear traps (Section 6.4.5a) are much less susceptible to this problem.

Triple quadrupole instruments (Section 6.4.3) have not previously been considered suitable for use with MALDI sources in view of their nature as scanning analyzers (serial recording of a mass spectrum resulting in very low duty cycle in full spectral acquisitions) with a limited m/z range. However, for the present purpose of small molecule quantitation by MALDI, neither of these limitations applies. In particular, if the so-called multiple reaction monitoring (MRM) mode of tandem mass spectrometry is used (Section 6.2.2), the duty cycle can approach 100 % depending on how many precursor–fragment ion pairs are monitored in the experiment in order to confer detection selectivity together with adequate confirmation of analyte identity. In addition, the ion transmission efficiency from ion source to detector is very high in a triple quadrupole instrument, with the potential to provide very low limits of detection and quantitation.

Recently (Corr 2006) a detailed account of design considerations for, and initial performance characteristics of, a MALDI–MS/MS instrument for high speed quantitative analysis of small molecules has described incorporation of most of the considerations of the preceding paragraph. The use of a triple quadrupole analyzer in MRM mode with a high repetition rate laser avoids the disadvantages associated with use of an o-MALDI–TOF

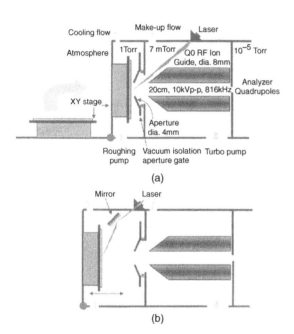

Figure 5.9 Sketches of experimental designs of MALDI ion source intended for use with a triple-quadrupole analyzer for high-throughput trace quantitation. (**a**) Initial source design showing MALDI source at ∼1 torr pressure to cool internal energies of ions, make-up gas flow to improve ion transfer into wide (4 mm diameter) sampling aperture, and a differentially pumped RF-only quadrupole ion guide to transfer ions into the mass spectrometer. (**b**) Modification to (a) to permit experimental flexibility in the laser angle of incidence via a mirror and spacing of the MALDI target from the aperture. Reprinted by permission of Elsevier from "Design consideration for . . . ", by Corr *et al.*, Journal for the American Society for Mass Spectrometry, 17, 1129 Fig.1, p.1130. Copyright 2006 by the American Society for Mass Spectrometry.

instrument. Figure 5.9 shows sketches of experimental MALDI sources designed (Corr 2006) for use with triple quadrupole analyzers, and Figure 5.10 shows images of a MALDI sample spot after a high repetition rate laser had been rastered across it (Sleno 2005), illustrating the virtually 100 % ablation of material.

As mentioned previously, this technology is still at the development stage and it is unclear what its niche will be once it has matured. It does seem to be promising for higher bioanalytical throughput in early discovery ADME (absorption–distribution metabolism–excretion) studies (Section 11.5.1) for pharmaceutical companies (Gobey 2005). Figure 5.11 shows a comparison of the overall efficiencies of MALDI–MS/MS analysis of a cleaned-up (SPE) serum extract and LC–ESI–MS/MS analysis of the same extract, using the same triple quadrupole instrument; the MALDI analysis delivered about 10 times fewer ion counts than LC–ESI–MS/MS, but with the consumption of 1000 times less sample, i.e., the combined efficiency of ionization plus ion transport through the analyzer to the detector was about 100 times greater for MALDI in these experiments. A similar investigation of natural toxin levels in marine phytoplankton (Sleno 2005a) is illustrated in Figure 5.12, demonstrating that the two techniques gave comparable quantitative results. It should be noted that neither of these examples (Gobey 2005; Sleno 2005a) used LC with MALDI–MS/MS but rather only extraction plus some form of sample clean-up.

The advantage of the MALDI approach lies in its potential for much higher throughput than LC–API–MS. It was shown (Corr 2006) that, when MALDI–MS/MS is operated so as to optimize the throughput advantage, the LOD is about 1–15 times worse (higher) than for LC–ESI-MS basically as a result of lower sample utilization efficiency (that could be improved by spending more time on laser ablation of the MALDI sample spot, i.e., with lower throughput). If the MALDI sampling efficiency could be improved by ensuring that the sample plus analyte are deposited in a much smaller spot size, the same number of laser shots should provide a correspondingly greater sample utilization efficiency. This was achieved (Corr 2006) by using HPLC to deposit the analyte in a fashion that ensured that spot spreading before solvent evaporation was minimized. Figure 5.12b compares results obtained using the two techniques, with essentially the same S/B values. Of course in this case the throughput was limited by the chromatography (same for both), but the deposited MALDI trace could be 'read' by the mass spectrometer in only a fraction of a second so that the same mass spectrometer could 'read' as many as 20 chromatograms in the same time as that required for a single LC–ESI–MS/MS experiment. Since

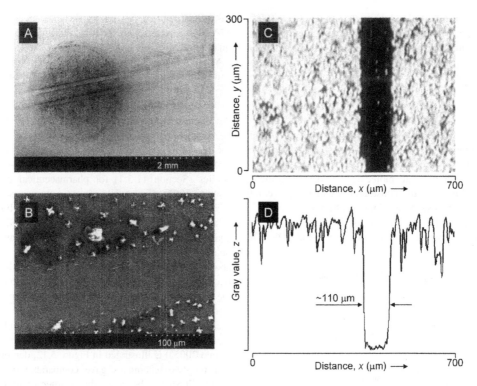

Figure 5.10 Images of a MALDI sample spot on an uncoated stainless steel target following ablation due to a high repetition rate laser tracked across the spot. (A) SEM image after MALDI analyses. (B) Higher magnification SEM image showing that the laser ablated virtually the entire sample material in its path. (C) Light microscope image of a section of the spot. (D) Cross-section of the same area as in (C) obtained from image analysis (ImageJ software, NIH, Bethesda, MD, USA). Note that in (C) and (D) only the x,y-plane is calibrated; the z-axis is expressed in contrast (gray) values from the digital image, but does not have specific physical meaning describing the actual dimension in the z-direction. The actual depth of the channel is much smaller than implied by the gray values. Reproduced from Sleno and Volmer, *Rapid Commun. Mass Spectrom.* **19**, 1928 (2005), with permission of John Wiley & Sons, Ltd.

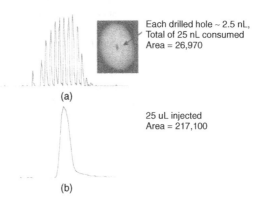

Figure 5.11 Comparison of combined efficiency of ionization-ion transmission for the same serum extract of a drug candidate, using (a) MALDI-MS/MS without LC, showing variation of MRM peak height for different spots as the laser is rastered across the deposited matrix-analyte sample, and (b) an LC–ESI–MS/MS peak obtained using the same triple-quadrupole mass spectrometer. Note the difference between the volumes of analyte solution consumed ($\sim$25 nL vs 25 μL, i.e., factor of 1000) for peak areas that differ by a factor of only eight. Reproduced from Gobey, *Anal. Chem.* **77**, 5643 (2005), copyright (2005), with permission from the American Chemical Society.

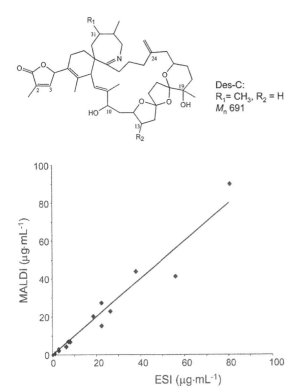

Des-C:
$R_1 = CH_3$, $R_2 = H$
M_n 691

Figure 5.12 Comparison of quantitative data for 13-desmethyl spirolide C in an extract of marine phytoplankton obtained using conventional LC–ESI–MS/MS vs MALDI–MS/MS. Reproduced from Sleno and Volmer, *Anal. Chem.* **77**, 1509 (2005a), copyright (2005), with permission from the American Chemical Society.

HPLC equipment is much less expensive than mass spectrometers, chromatographic systems could be parallelized for preparation of MALDI targets that could also possibly be archived for future re-examination.

This ongoing development of instrumentation for trace quantitation using MALDI illustrates well an intrinsic shortcoming of any book such as this; the pace of development is such that the book is liable to become outdated in many respects and offers only a temporary snapshot of the situation. For example, a recent paper (Kovarik 2007) describes a rigorous evaluation of the LC-MALDI-MS/MS approach to quantitation of pharmaceutical compounds in plasma, with generally encouraging results particularly with respect to throughput. For this reason the present book seeks to emphasize the more fundamental underpinnings of the various technologies involved, at least to a degree sufficient that readers can understand the basics of how their instrumentation works

(or does not!), and can then progress to more complete discussions of these principles if interested.

Although the majority of MALDI–MS experiments have followed the originators (Tanaka 1988; Karas 1987, 1988) by inserting the matrix–analyte mixture into the high vacuum chamber of the mass spectrometer before irradiation, development of MALDI sources that operate at atmospheric pressure has been described (Laiko 2000; Moyer 2002). Such methods obviously require efficient interfaces between the atmospheric pressure ion source and the mass spectrometer vacuum as for the more usual API sources (Section 5.3.3). Currently this approach does not appear to have been exploited for trace quantitative analyses of small molecules.

5.2.3 'Lab-on-a-Chip'

A special case of discrete interfaces between a separation system and mass spectrometry is that of a 'lab-on-a-chip' (Section 4.4.7). The mass spectrometer and the microchip are well matched, despite the gross mismatch in physical dimensions, as a result of the similarity in flow rates generated by CE or CEC (or even HPLC) on a microchip with those best suited to electrospray ionization (Section 5.3.6) and its variants. Note that miniaturization of mass spectrometers to on-chip dimensions while retaining performance characteristics does not appear to be possible at present, despite some remarkable achievements (Taylor 2001a; Geear 2005) in this area. Early approaches to achieving the nontrivial coupling of microchip devices with mass spectrometers have been reviewed (Oleschuk 2000), and recently (Yin 2005) a commercially available chip device was developed; this device, fabricated from polyimide, incorporates enrichment (SPE) and separation columns packed with conventional reverse phase materials, and a nanospray tip (Section 5.3.6) on a single chip, eliminating the need for conventional LC connections. The device is pressure driven rather than electrically driven (i.e. not CEC, Section 4.4.6) because in the latter case the stationary phase must generate the electro-osmotic flow by having a high surface charge, yet at the same time must be capable of providing good chromatographic separation. It is difficult to find an optimum surface to achieve both, especially for reversed phase separations where changing solvent composition causes a change in the flow rate. This device was designed primarily for proteomics applications and, indeed, remarkable performance has been reported in this context (Fortier 2005). Thus far little has been reported on its performance in applications to trace quantitative analysis of small molecules, although one brief report (Gauthier 2006) describes chip-assisted gradient

LC–MS/MS determination of sulfadimethoxine in human serum with on-chip sample enrichment and retention time < 7 minutes. The only other details provided are an enhanced sensitivity that enabled quantitation down to 10 fg with a linear dynamic range of 10^3 and precision in the range 2.6–10 %.

Most earlier reports of true 'lab-on-a-chip' devices (integrating more than one function on-chip) have been directed at proteomics applications (e.g. Wang 2000). Most applications to small molecule analysis have been the work of Henion and his collaborators (Deng 2001, 2001a; Kameoka 2001; Tan 2003; Benetton 2003) and have mostly involved on-chip CE or SPE integrated with an interface to an on-chip electrospray device. Two aspects of this work are worth special mention; one is the introduction of a special plastic (Zeonor™) that can be used for chip fabrication by hot-embossing and that can also support electro-osmotic flow for on-chip CE as a result of its chemical properties (Kameoka 2001), and the other is on-chip metabolism by P450 enzymes for rapid metabolite screening (Benetton 2003). Some other publications (Dethy 2003; Yang 2004) describe the use of an on-chip electrospray emitter for direct analysis of small molecules in biomedical extracts but with no other on-chip functionalities, and thus do not qualify as true 'lab-on-a-chip' devices; nonetheless, the potential for such devices to alleviate or remove carryover effects (Section 9.7.2) is important in the context of the book. These contributions, and those from other groups that were primarily directed at developing effective on-chip ESI emitters, will be described later in the section on electrospray ionization (Section 5.3.6). Similarly, on-chip devices for atmospheric pressure photoionization (APPI), (Kauppila 2004) and atmospheric pressure chemical ionization (APCI), (Östman 2006) have been described.

Although use of lab-on-a-chip devices for trace quantitative analysis has not yet been widely accepted, it seems likely that this will occur at some time in the future.

5.3 Ion Sources not Requiring a Discrete Interface

5.3.1 *Flow Fast Atom Bombardment (Flow-FAB)*

The fast atom bombardment (FAB) ionization technique is of great historical importance in mass spectrometry, as it provided the first facile approach to applications to fragile biomolecules. When first introduced (Barber 1981), FAB created a sensation within the mass spectrometry community but was supplanted after a few years by the newer techniques of electrospray ionization (ESI) and matrix assisted laser desorption ionization (MALDI). FAB is still used for some specialized applications. It was first developed as a means of overcoming an intrinsic limitation of

conventional secondary ion mass spectrometry (SIMS), used to characterize molecules on solid surfaces by sputtering with an energetic (typically 10–30 eV) primary beam of ions such as $Xe^{+\bullet}$ (although a variety of primary ions has been investigated). The problem with SIMS is that it destroys the surface under investigation and the yield of secondary ions sputtered from the surface thus falls off rapidly. This is avoided in FAB by dissolving the analyte in a suitable liquid matrix so that the surface to be sputtered can be continuously replenished. At first a fast beam of neutral atoms (typically Ar° or Xe°) was used as the primary beam, but later the fast ion sources used in SIMS were applied to FAB; this variant is sometimes called liquid assisted secondary ion mass spectrometry (LSIMS).

Selection of the FAB matrix is not an exact science and is most often done by trial and error as for the solid matrices used in MALDI. However, there are some requirements for a suitable FAB matrix, including solubility of the analyte(s) without matrix–analyte reactions, low volatility in the vacuum conditions of the source, high viscosity (to prevent dripping of the liquid matrix from the probe tip!) and matrix ions that do not overlap with those of the analyte(s) in the resulting mass spectrum. Glycerol was the first matrix used but several others were also developed. The general concept is illustrated in Figure 5.13

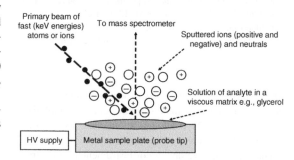

Figure 5.13 Sketch of a fast atom bombardment (FAB) ion source. The potential difference between the probe tip and extraction grid (not shown) accelerates the ions (from both the liquid matrix and analyte) into the mass spectrometer. Since a large quantity of neutral matrix is also sputtered by the fast primary beam, the source region must be provided with adequate pumping and the m/z analyzer region differentially pumped. The fast atom beam can be replaced by a beam of fast primary ions (often Cs^+) and the technique is then sometimes referred to as liquid assisted secondary ion mass spectrometry (LSIMS).

From the viewpoint of trace analysis by mass spectrometry interfaced online with HPLC, flow-FAB interfaces were used for a short time before introduction of the atmospheric pressure ionization methods ESI and APCI. In a flow-FAB interface the column effluent is mixed post-column with a FAB matrix (generally aqueous glycerol) and is continuously deposited on the FAB probe tip situated inside the evacuated ion source; the liquid mixture is transported to the tip through a fused silica capillary. Two such interfaces were developed, the frit-FAB interface (Ito 1985) and continuous flow FAB (Caprioli 1986). The subject has been extensively reviewed (Caprioli 1990). These LC/MS interfaces were difficult to operate for a long period, and could handle only very low flow rates $(1–10\,\mu L.min^{-1})$ due to limitations of vacuum pumping capacities, thus requiring flow spitting or use of capillary chromatography; flow-FAB has been almost entirely supplanted by ESI and APCI.

An investigation of matrix effects in FAB (ionization suppression, again enhancement has not been reported) caused by ionic surface active compounds has rationalized the unexpected finding that, while cationic surfactants do suppress ionization of glycerol matrix in positive mode this is not true in negative mode, and vice versa (Kosevich 2007). This paper is an extension of previous work (Kosevich 2003) that rationalized many otherwise puzzling features of FAB ionization through the realization that a FAB matrix in the ion source vacuum is a superheated liquid.

5.3.2 Thermospray Ionization

The thermospray ionization interface was the first device to provide a viable means of coupling HPLC to MS for quantitative analysis, particularly for relatively involatile analytes. In its first form (Blakley 1978), HPLC eluant

was sprayed as a liquid jet from a steel capillary into the path of an infrared laser beam sufficiently powerful to completely vaporize the eluant at flow rates up to $1\,mL.min^{-1}$. The resulting beam of vaporised eluant then intersected an electron beam for ionization (via CI since large amounts of solvent vapor were present). Later it was found that, if the laser was used to heat the sprayer tip instead of the liquid jet, the same results were obtained and the system was therefore simplified by replacing the laser with an oxygen–hydrogen flame (Blakley 1980). Somewhat later it was accidentally discovered that, by not applying the electron beam, ions were still observed, thus demonstrating that another ionization mechanism method was operating (Blakley 1980a).

This extended series of developments culminated in the thermospray design (Blakley 1983) that was subsequently commercialized as an LC/MS interface. A solution of the analyte (e.g., HPLC eluant) and a volatile buffer (typically 0.1M ammonium acetate added post-column) was evaporated from a heated capillary at a flow rate of up to 1.5 ml/min into a heated chamber (whence the name 'thermospray'), forming a mist of droplets containing relatively involatile analytes and solvent vapor; as the solvent evaporated the analyte formed adducts with ions from the added salt. It is believed that the formation of free gaseous ions from the microdroplets then proceeds in a manner similar to that discussed for ESI in Section 5.3.6. Most of the neutrals are removed by a vacuum pump and the ions are extracted orthogonally by some electrostatic lenses and a repeller through a pinhole (restricted to $\sim\!25\,\mu m$ to protect the vacuum in the m/z analyzer, typically a quadrupole because of its better tolerance to poor vacuum). Such an arrangement is found to be efficient

---

Improved Vaporization of Involatile Analytes By Rapid Heating of Microparticles

The advantage of rapid heating for volatilization of thermally labile compounds (Beuhler 1974, Daves 1979) can be understood by assuming that both proceed with unimolecular rate constants k_r. At lower temperatures (high

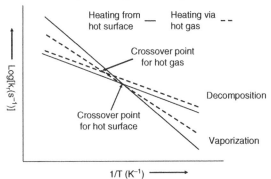

1/T), k_r for decomposition is larger than for vaporization and, on heating, mainly decomposition products are formed in the vapor.

However, if the activation energy for vaporization (breaking relatively strong intermolecular bonds, including those with a strongly adsorbing surface of a sample container) is larger than that for decomposition (breaking relatively weak intramolecular bonds of a labile molecule, partly compensated by new bond formation), the two Arrhenius plots must cross as indicated. If the compound can be heated quickly to above this crossing point, k_r for vaporization will now be

(Continued)

the larger and the intact molecule will appear in the vapor. Further, in cases where the analyte is heated via the walls of its container, e.g., a quartz sample cup used in an insertion probe, the adsorption energy can be high. By judicious choice of material for the sample container such strong adsorption energies can be avoided and the overall level of k_r for vaporization will increase as will the activation energy (dashed lines).

The same result can be obtained by nebulization of an analyte solution into small droplets that are then flash heated by exposure to hot gas, as in thermospray and APCI sources. Thus, it is common to find that a compound that will not yield a meaningful mass spectrum using conventional CI (solid sample heated in an insertion probe) will do so when subjected to flash heating in APCI. Another physical phenomenon that contributes to this effect is the increase of vapor pressure P_A of an analyte as the sample size becomes very small, as in a nebulized solution. This increase is described by the Kelvin equation (Thomson 1871, Moore 1972):

$$P_A/P_A^o = \exp[(4.M_A.\xi_A)/(RT.\rho_A.d_p)]$$

where P_A is the vapor pressure for a bulk sample of A (i.e., particle diameter d_p effectively infinite), M_A is the molar mass of A, ρ_A its density, and ξ_A its surface energy (a measure of intermolecular forces at the surface, the energy required to increase the surface by unit area, e.g. $J.m^{-2}$; the equivalent for liquids is the surface tension). Thus P_A increases exponentially as d_p decreases, and the magnitude of this increase is greater for particles with large values of ξ_A, exactly the case of interest here (solid analytes with low values of P_A^o resulting from strong intermolecular forces).

only when using highly polar solvents (not a problem for reverse phase HPLC) in the presence of a buffer salt. This limitation was overcome in some versions of the interface by addition of two modifications, one using an electrical discharge in the vapor phase (similar to APCI, Section 5.3.3) and the other an electron-emitting filament (as used for EI and CI). Under these conditions thermospray closely resembles APCI, although the pressure within the thermospray source is considerably lower (~8 torr) than for APCI but still much higher than for conventional CI.

The success of the thermospray interface can be attributed to a combination of physico-chemical phenomena. Apart from the direct production of gaseous ions directly from charged microdroplets as in ESI, it is likely that increased volatilization of analytes is produced by a combination of the rapid heating effect on the small droplets and the inverse dependence of vapor pressure on the radius of a particle (see the accompanying text box for a discussion of these phenomena). The underlying mechanisms were discussed in a remarkable early effort (Vestal 1983) that sought to harmonize theoretical understanding of ion emission from liquids in a wide range of different experimental configurations, including thermospray, FAB etc.; this work was later extended to include effects of superheating of liquids in a vacuum (Kosevich 2003). Detailed reviews of the theory and practice of thermospray (Yergey 1990) are complemented

by extensive reviews of practical applications (Arpino 1990, 1992).

For several years in the 1980s thermospray dominated LC/MS analyses. However, problems of clogging the spray tip, reflecting the high buffer concentration and rapid heating, are a major disadvantage relative to ESI and APCI. Also, sensitivity and response were found to vary widely and in many cases unpredictably, and the technique has been largely abandoned although it still appears occasionally in the literature.

5.3.3 Atmospheric Pressure Ionization (API)

Ionization at atmospheric pressure was for many years considered to be an experimental curiosity despite excellent early work (Horning 1974, 1974a; Carroll 1975). However, API techniques nowadays dominate on-line LC–MS work. There are currently three kinds of API sources in common use for trace quantitative analyses, namely, atmospheric pressure chemical ionization (APCI), atmospheric pressure photoionization (APPI) and electrospray ionization (ESI). In addition, atmospheric pressure MALDI has been investigated but not as an on-line approach to quantitative LC–MS. Since the highly practical discussions of Chapters 9 and 10 will be concerned mainly with LC–MS analyses (since GC–MS is now almost routine), an extended discussion of API methods is appropriate here.

The major problem with all API techniques for mass spectrometry concerns the transfer of the ions from the atmospheric pressure ion source into the vacuum required for operation of the m/z analyzer itself, a pressure drop by a factor $\sim 10^9$! Such a transfer involves a sudden expansion of the gas at some stage and this tends to enhance the condensation of solvent molecules (particularly water) on the ions to produce clusters of various sizes that redistribute the total ion current among several species thus complicating the spectra and reducing S/B values. An interface between any atmospheric pressure ionization (API) source and a mass spectrometer must be able to deal with the pressure ratio (pumping speed is a crucial factor here, Section 6.6.1) and the de-clustering of the analyte ions before m/z analysis and detection.

Despite these difficulties (discussed in Section 5.3.3a) API sources are currently used widely in combination with a wide range of liquid chromatography methods, i.e. normal and reverse phase, isocratic and gradient, normal bore (4.6 mm i.d.) and capillary columns, as well as ESI with capillary electrophoresis and APCI with GC. In the context of trace level quantitation, API techniques are most often used in combination with reverse phase HPLC, and it is this combination that will be the main focus of discussions of matrix effects (Sections 5.1.1 and 5.3.6a) and of the more practical aspects in Sections 9.6 and 10.10.4.1d. In this regard it is worth noting here that in reverse phase chromatography using e.g., a C_{18}-derivatized silica

phase, the more polar (hydrophilic) components elute earlier than the more hydrophobic compounds, so that the elution time of the analyte relative to those of matrix components that are potential causes of matrix effects becomes an important issue.

A phenomenon common to all API techniques is the production of ionized intact analyte molecules in a variety of forms of the type $[A + X]^\pm$, where X can be one of several ionic species; the most common is $X = H^+$, but others are commonly observed (see Table 5.2 for those observed in APCI and ESI). It is possible to use any of the adduct ions listed in Table 5.2 as the species monitored in quantitative LC–MS, but it is important to ensure that only one form (one adducting ion X) contributes significantly to the ion current derived from the analyte (to maximize sensitivity) and also that the concentration of X is constant under the conditions used (to maintain a constant sensitivity for $[A + X]^\pm$).

5.3.3a Coupling of API Sources to Mass Spectrometers

Coupling an API source (APCI, ESI or APPI) to a mass spectrometer, so as to maximize the analyte signal while disposing of the large amount of vaporized LC mobile phase, is an essential and important component of any API–MS system. The following discussion is based upon two excellent early reviews (Bruins 1991; Niessen 1995)

Table 5.2 Common adduct ions observed in LC–MS with ESI or APCI[a]

Chemical species forming adduct	Formula of adduct ion	Cause of adduct formation	POS and/or NEG	m/z shift of adduct ion relative to $[A+H]^+$ (POS) or $[A-H]^-$ (NEG)
Alkali metal ions Me^+ (Li, Na, K)	$[A + Me]^+$	Alkali metal salts	POS	+6 (Li); +22 (Na); +38 (K)
Ammonia	$[A + NH_4]^+$	Ammonium salts and/or ammonia	POS	+17
Chloride	$[A + Cl]^-$	Chlorinated solvents; chloride salts	NEG	+34 (^{35}Cl); +36 (^{37}Cl)
Water	$[A \pm H + H_2O]^\pm$	Aqueous solvent or water as impurity	POS/NEG	+18
Methanol	$[A \pm H + CH_3OH]^\pm$	Methanol in solvent	POS/NEG	+32
Acetonitrile	$[A \pm H + CH_3CN]^\pm$	Acetonitrile in solvent	POS/NEG	+41
Formate	$[A + HCOO]^-$	Formic acid and/or formates	NEG	+46
Acetate	$[A + CH_3COO]^-$	Acetic acid and/or acetates	NEG	+60
Trifluoroacetate	$[A + CF_3COO]^-$	Trifluoroacetic acid (TFA)	NEG	+114

a Note that the m/z shifts are calculated relative to the 'default' ions expected, i.e. $[A+H]^+$ and $[A-H]^-$, and *not* relative to the molecular mass of the neutral analyte. Ammonia, formate and acetate are often present as buffer components in the LC mobile phase, and TFA as an ion-pairing reagent. (NB It is *not* recommended to use TFA if the instrument is to be used in negative ion mode any time soon afterwards, as it is difficult to flush completely from the system and trifluoroacetate at m/z 113 can swamp out other negative ion signals). The adduct ions with ammonia, water, methanol and acetonitrile are relatively easily converted to $[A \pm H]^\pm$ in the API interface by raising the 'cone voltage'. The adduct ions with the organic anions can also be dissociated in this way to yield $[A-H]^-$.

and will also build upon the concepts of pumping speed, gas flow and conductance introduced in Section 6.6.1.

A quadrupole m/z analyzer (Section 6.4.5) and its electron multiplier detector (Chapter 7) are generally operated at a pressure P_{MS} of 10^{-6}–10^{-5} torr ($\sim 10^{-4}$ Pa). A good turbomolecular or oil diffusion pumping system backed by an adequate rotary pump (Section 6.6.2) can provide an effective pumping speed S_{pump} of $\sim 2\,m^3.s^{-1}$, so the maximum gas leak Q_{gas} (including that from an API source) that the system can handle while maintaining adequate vacuum is $P_{MS}.S_{pump}$, i.e. $2 \times 10^{-4}\,Pa.m^3.s^{-1}$. If this gas load is derived directly (no differential pumping) from a gas supply at atmospheric pressure (1.01×10^5 Pa), the gas conductance C_{or} of the orifice and associated components connecting the API source to the MS vacuum chamber is given by Equation [6.48] as:

$$Q_{gas}/(P_{source}-P_{MS}) \approx Q_{gas}/P_{source}$$
$$= (2 \times 10^{-4}\,Pa.m^3.s^{-1})/(1.01 \times 10^5\,Pa)$$
$$= 2 \times 10^{-9}\,m^3.s^{-1}$$

The volume leak rate of gas at atmospheric pressure that is compatible with maintaining the MS vacuum ($\sim 2 \times 10^{-9}\,m^3.s^{-1}$) corresponds to $2\,\mu L.s^{-1}$ or $120\,\mu L.min^{-1}$. This should be compared with the volume of eluent from e.g., a 1 mm i.d. HPLC column, typically $100\,\mu L.min^{-1}$ of liquid that gives approximately $10^5\,\mu L.min^{-1}$ of vapor, about 1000 times greater than the permissible leak rate into the MS vacuum system. While this calculation should be regarded as only an order of magnitude indication of the extent of the problem (the details will vary with the available pumping speed, provision of additional stages of pumping, i.e., differential pumping of intermediate regions, tolerance of the MS to somewhat poorer vacuum etc.), clearly only a very small fraction of the HPLC effluent can be accepted from the API source into the MS, so that some means of discriminating strongly in favor of gaseous ions and against nonionized mobile phase must be provided. For these reasons early designs (Horning 1974, 1974a; Carroll 1975) were restricted to orifices of $25\,\mu m$ diameter, but improvements in vacuum systems since then have permitted use of orifices up to 10 times larger with corresponding increases in sensitivity that in principle should increase as the square of the orifice diameter.

Expansion of a higher pressure gas into a vacuum via a small orifice produces a molecular beam. This is a well developed field with an extensive literature in many fields of science and engineering e.g., Campargue (1984), but the present simplified account (Bruins 2001) suffices for present purposes; a simple scheme of expansion through a nozzle is shown in Figure 5.14(a). Immediately behind

(downstream from) the nozzle (background pressure $\sim 10^{-3}$ torr), the gas molecules originating from the atmospheric pressure gas acquire a high velocity exceeding the local speed of sound, essentially by transferring internal (vibrational–rotational) energy into translational energy; it is this internal cooling that leads to formation of ion–solvent clusters that must be broken up if meaningful mass spectra are to be observed (see later). This rapid

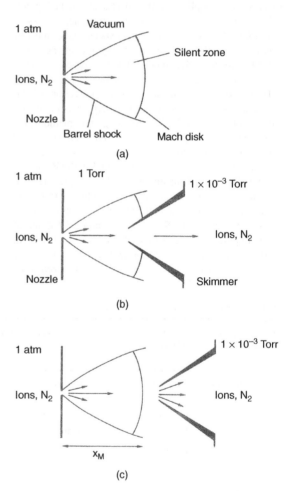

Figure 5.14 Expansion of gas and ions from atmospheric pressure into a vacuum. (a) simple case showing shockwaves (barrel shock and Mach disk) and zone of silence. (b) generation of a beam of gas and ions by sampling from the silent zone via a skimmer penetrating through the Mach disk. (c) sampling of gas and ions with a skimmer located downstream from the Mach disk, leading to the beam of gas and ions being scattered by passage through the Mach disk. Reproduced from Bruins (1991), *Mass Spectrom. Revs.* **10**, 53, with permission of John Wiley & Sons, Ltd.

expansion also creates a conical shock wave (the so-called barrel shock, Figure 5.14). A shock wave is not a sound wave, but rather involves a very sharp change in the gas properties over a distance corresponding to a few mean free paths. To produce a shock wave some initial disturbance (the sudden expansion of the gas in this case) has to be travelling faster than the local speed of sound; the gas molecules near the disturbance cannot 'get out of the way' before the disturbance arrives. A useful picture of a shock wave is as the furthest point downstream of a supersonic disturbance that 'knows' about the approach of the disturbance (via local increases in collision frequencies and energies). On this view the shock wave position is the boundary between the zone that has no 'information' or 'forewarning' about the disturbance that is moving faster than the the local speed of sound (i.e. the speed of information transfer about its approach), and the zone that is 'aware' of the approach of the disturbance. The energy of a shock wave dissipates relatively quickly with distance and terminates at a so-called Mach disk as the initially supersonic gas slows down and the molecules change from predominantly directed to random motion.

The region inside the barrel shock wave (Figure 5.14) is known as the 'zone of silence' or 'silent zone', where ions and gas move at equal speeds in the same direction and undergo strong and rapid internal cooling (it is here that solvated clusters can be formed). The distance x_M from the orifice of the sampling nozzle to the Mach disk is important for the design of an API interface and is given by:

$$x_M = 0.67.d_{or}.(P_{source}/P_1)^{\frac{1}{2}} \qquad [5.4]$$

where d_{or} is the diameter of the sampling orifice, P_{source} is the pressure on the high pressure side of the sampling nozzle (for an API source $P_{source} = 1$ atm) and P_1 is the pressure in the region immediately downstream of the nozzle. For $d_{or} = 100\,\mu m$ and a well pumped expansion region ($P_1 \sim 10^{-5}$ torr), as found in some commercial instruments, evaluation of Equation [5.4] predicts that the Mach disk is located at $x_M = 670$ mm, i.e., beyond the dimensions of the system. At the other extreme, in an expansion into a region pumped by only a rotary pump, with $d_{or} = 300\,\mu m$ and $P_1 \approx 1$ torr, as used in some experimental systems (Bruins 1991), the Mach disk is located at $x_M = 5$ mm. In general, for API sources x_M is largely determined by the pumping speed (Section 6.6.1) available in the region near the nozzle orifice.

Often the gas flow from the atmospheric pressure side through the nozzle is too large for the vacuum pumps of the mass analyzer. In such cases differential pumping (Section 6.6.2d) can be applied, such that the central portion of the molecular beam is sampled into an intermediate vacuum stage by means of a skimmer cone (Figures 5.14(b) and 5.14(c)). In the arrangement shown in Figure 5.14(b) the core of the beam is sampled from the silent zone and the transmitted ions and neutral molecules continue their movement in straight lines. The advantage of this arrangement is that collection of ions from the gas flow through the skimmer should be efficient because of the directional effect of the molecular beam. However, since the gas in the molecular beam has strongly cooled upon expansion (other than the motion in the beam direction of course), there will be a strong tendency for ions and mobile phase molecules to form clusters in addition to those already formed within the ion source itself (e.g. Equation [5.5] below). In the alternative arrangement (Figure 5.14(c)) a sample of the ions plus gas is collected by the skimmer cone from behind (downstream from) the Mach disk, where ions and molecules have undergone extensive scattering; as a result, extraction and focusing of ions are more difficult, but in compensation the ions and molecules have warmed up as a result of collisions in the Mach disk so that cluster formation is much less of a problem. Most commercially available instruments sample the molecular beam from the silent zone and deal with cluster formation later.

The choice made between these two possibilities (Figures 5.14(b) and (c)) dictates other requirements for the API–MS interface with respect to the arrangements for dealing with clustering of solvent molecules on the ions. All practical designs of API instruments are directed towards prevention of clustering and/or 'fixing' the problem by breaking up the clusters. In early work (Siegel 1976) a combination of higher temperature and careful pre-drying of the gas in an APCI source was used to reduce clustering. Collision induced dissociation (CID, see Section 6.5) was used (Kambara 1976, 1976a) to strip water molecules from various molecular ions by applying accelerating fields of about 10 V.cm^{-1} in a differentially pumped region (background pressures $\sim$1 torr) between the ion source and the mass spectrometer.

All of these strategies are used in modern API interfaces. Some designs use a heated capillary to connect the atmospheric pressure source to the mass spectrometry vacuum, thus restricting the gas flow and at the same time de-clustering solvent molecules from the ions, together with facilities for CID induced by applying a user-controlled potential to the final skimmer cone, the so-called 'cone potential' or 'skimmer potential'. (The cone potential can also be increased well above the values needed to break the noncovalent bonds responsible for cluster formation, and instead create fragment ions by breaking covalent bonds within the ionized molecule to provide additional chemical and diagnostic information; this technique is known as in-source CID, see Section 6.5.)

Other early work (Buckley 1974, 1974a; French 1977) developed a 'gas curtain' or 'gas membrane' between the ion source and vacuum expansion, consisting of a flow of carefully dried nitrogen; most of this flow is directed outwards (Figure 5.15), thus maintaining the cleanliness of the vacuum system since neutral molecules are thereby impeded from entering the orifice connecting the source to the vacuum while the ions are directed through the orifice against the bulk flow by appropriate electrical fields. By aerodynamic design of the gas flows within the 'curtain', all ions and molecules that survive the passage through the initial part of the curtain (that sweeps back toward the

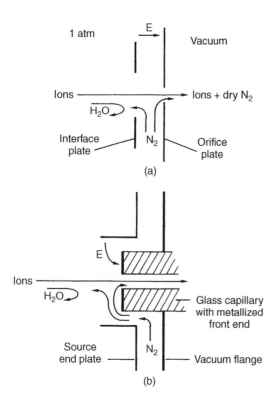

(a)

(b)

Figure 5.15   (a) Sketch of a gas curtain interface for API–MS coupling (Buckley 1974, 1974a; French 1977). The ultra-dry gas (N_2) curtain separates the ionization chamber (atmospheric pressure) from the orifice leading to the skimmer cone and thence to the mass spectrometer vacuum. (b) Sketch of an API–MS interface based on a heated glass capillary that connects the atmospheric pressure source to the low vacuum region preceding the sampling cone (Figure 5.17). In both cases an electric field **E** helps direct the ions into the sampling orifice. Reproduced from Bruins, *Mass Spectrom. Revs.* **10**, 53 (1991), with permission of John Wiley & Sons, Ltd.

atmospheric pressure region) are swept forward toward the differentially pumped region leading to the mass spectrometer. The use of ultra-dry nitrogen in the gas curtain strongly shifts the cluster equilibria towards species with lower numbers of solvent molecules and also eliminates (via nonavailability of water molecules) the possibility of additional clustering during the rapid cooling during expansion and formation of the molecular beam (Figure 5.14). This system also uses CID to complete the de-clustering and (if required) to provide facilities for producing fragment ions in-source. These considerations for API–MS coupling apply equally to APCI, atmospheric pressure photoionization and electrospray (see later).

The ions emerging from the skimmer cone must be refocused in order to efficiently transmit them into the high vacuum chamber containing the *m/z* analyzer. A highly efficient method of accomplishing this is by exploiting the phenomenon of collisional focusing in quadrupoles (Douglas 1992). This subject is discussed in Section 6.4.2a in the context of quadrupole *m/z* analyzers and collision cells for CID in tandem mass spectrometers. Here it is sufficient to mention only that a short RF-only linear quadrupole device, arranged to have an internal pressure of several mtorr, acts to constrain ion trajectories through the device to close to the central axis, thus reducing ion losses via radial scattering. The only drawback arises from the simultaneous damping of the motion in the axial direction, thus increasing the ions' transit times to what can be unacceptably large values if appropriate precautions are not taken (e.g. provision of a linear acceleration field). In the context of efficient transmission of ions from the skimmer cone into the *m/z* analyzer, several devices involving variations on the collisional focusing principle have been described and are incorporated into commercially available instruments.

The intrinsic difficulty in transferring gaseous ions from atmospheric pressure to the mass spectrometer vacuum leads to a disappointingly low overall transfer efficiency (<1 %), expressed as the ratio of number of ions detected to the number of analyte molecules consumed. In addition to methods used to improve ion transmission on the low vacuum side, discussed in the preceding paragraph, there have been many investigations of methods applied on the atmospheric pressure side to improve the transfer efficiency. These methods have included manipulation of nebulizing gas (Covey 1995) that yielded disappointing results. However, more recently a Venturi device based on a commercially available 'air amplifier', which generates a concentric high velocity converging gas flow around the electrospray needle to reduce spreading of the electrospray plume and thus improve conduction of ions to the sampling orifice, was shown (Zhou 2003;

Hawkridge 2004) to provide order-of-magnitude increases in ion transmission especially when used in conjunction with a focusing potential applied to the device. However, this approach does not appear to have been incorporated into analytical instrumentation as yet, for reasons that are not clear. Other approaches applied on the atmospheric pressure side have involved electrostatic focusing (Shaffer 1997; Him 2000; Schneider 2002), but none of these approaches have turned out to be suitably versatile and robust to deliver a reliable advantage in other than specialized instruments.

The discussion in this section applies equally to all API–MS techniques, so that the components of the interface (nebulization, heating, drying, molecular beam skimmers, etc.) are also common to all. This commonality results in ease of swapping one kind of API source for another on a given analyzer and also in development of combined sources (e.g., electrospray–APCI) in which only minor adjustments are required in order to switch between the two ionization techniques. Of course different manufacturers do not all use exactly the same combinations of these elements for their API–MS interfaces and such differences can account for sometimes subtle differences in the characteristics of such devices from different suppliers for the same target analyte. Even differences in susceptibility to matrix effects can be attributed to such variations in interface design; for example, this could arise if the degree of evaporation of the electrosprayed microdroplets (resulting from combined nebulization and heating) is an important factor in determining the competition between analyte and the interfering compound for the available sites on the droplet surface, and thus the magnitude of the matrix effect.

5.3.4 Atmospheric Pressure Chemical Ionization (APCI)

The first API technique to be successfully commercialized as an LC/MS interface was APCI particularly when combined with a deceptively simple device, the 'heated pneumatic nebulizer' (HPN). The increase in ease of solute vaporization due to pre-nebulization of the LC eluent combined with rapid heating, discussed in Section 5.3.2 with respect to thermospray, probably contributes also to the success of the HPN interface when combined with APCI. To appreciate the strengths and limitations of the HPN–APCI approach to LC/MS coupling, the two aspects will be discussed separately.

Conventional chemical ionization (CI, Section 5.2.1) is initiated by a beam of fast electrons in a modified EI source operated at internal pressures ($\sim$0.1 torr) that are much higher than those in an EI source ($\sim 10^{-3}$ torr) or

than the analyzer vacuum (10^{-5}–10^{-8} torr but $\sim 10^{-3}$ torr in Paul ion traps, Section 6.4.5). CI can be extremely efficient at creating analyte ions because the ion–molecule reactions responsible are generally (for exothermic reactions) simply the molecular collision rates between analyte molecules and reagent ions. One of the motives for development of APCI ion sources was the realization that, at even higher pressures and thus collision rates, the ionization efficiency could be much greater.

APCI can be regarded as CI using ambient air as the reagent gas, initiated not with a beam of fast electrons since these would not penetrate far into atmospheric pressure air, but by using a high potential (several kV) applied to a sharp needle in the atmospheric pressure gas (a so-called corona discharge to distinguish it from a glow discharge that operates at much lower pressures). Thus the ionization reactions again proceed in the gas phase, hence the importance of the efficiency of vaporization of analytes. The components of air that are important here are water vapor, nitrogen, oxygen and nitric oxide. The cascade of processes (positive ions only) occurring when air is subjected to an ionizing agent of some kind is illustrated in Figure 5.16 (see below); the primary ions $N_2^{+\bullet}$, $O_2^{+\bullet}$, $H_2O^{+\bullet}$ and NO^+ are initially formed, and these react extremely rapidly ($\sim 10^{-6}$ s) at atmospheric pressure to form the terminal equilibrium set of proton hydrates (Figure 5.16) that dominate the positive ion population and are thus the most likely to react with trace level analytes within the source residence time of 10^{-3}–10^{-4} s; these conclusions are the result of studies by many groups that have been evaluated and summarized (Huertas 1975). In negative ion mode, the high energy free electrons initially formed are rapidly thermalized by collisions with the high pressure gas, thus acquiring an energy distribution characterised by the ambient temperature of the gas, and are captured by oxygen as the most electronegative component to form the superoxide anion $O_2^{-\bullet}$, its hydrates $[O_2^{-\bullet}(H_2O)_n]$, and clusters $[O_2^{-\bullet}(O_2)_n]$, that are the major negative ions in the APCI plasma.

The APCI reactant ions can analyze trace analytes A by two general mechanisms, R = reactant neutral or ion:

$$A + RH^+ \rightarrow AH^+ + R; AH + R^- \rightarrow A^- + RH$$
$$\text{(proton transfer)}$$

$$A + R^+ \rightarrow A^+ + R; A + R^- \rightarrow A^- + R$$
$$\text{(charge transfer, really electron transfer)}$$

The energetics of these processes are related to the values of one or more of proton affinity (PA), electron affinity (EA) and ionization energy (IE). In general it is possible to predict whether or not a proposed reaction

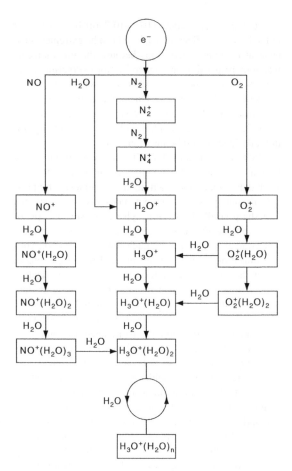

Figure 5.16 Positive ion processes occurring within an APCI source using atmospheric air. Reproduced from MDS–Sciex literature (*The API Book*), with permission.

will proceed by evaluating $\Delta H°$ for the reaction (negative values promote reaction) via tabulated values of these quantities (http://webbook.nist.gov/chemistry/). Also the degree of 'hardness' or 'softness' (Section 5.2.1) of an APCI reaction can be qualitatively predicted based on the absolute magnitude of $\Delta H°$; this aspect is fully discussed elsewhere (Harrison 1992). An example important for APCI is the protonation reaction of an analyte by proton hydrates:

$$A + [(H_3O)^+(H_2O)_n] \rightarrow [(AH^+)(H_2O)_m]$$
$$+ (n+1-m)H_2O \qquad [5.5]$$

Reaction [5.5] will generally proceed only if the PA of A is greater than that of water ($691 \, kJ.mol^{-1}$), a condition that is met for most organic compounds containing oxygen or nitrogen. For analytes for which this condition

is not satisfied, charge exchange by $O_2^{+\bullet}$ and its hydrates is the alternative ionization mechanism in positive mode APCI. However, as indicated in Figure 5.16, these species are very quickly converted to proton hydrates under APCI conditions and as a result the sensitivity of APCI for positive ions in charge exchange mode is far less (by a factor of up to 10^3) than for proton transfer. It is possible to improve this situation by doping the APCI plasma with a compound like benzene that can form a stable radical molecular cation (Lane 1980; Ketkar 1991; Anacleto 1992) that can ionize compounds with lower ionization energies (9.2 eV for benzene). However, this approach introduces additional variability and is seldom used in quantitative trace analysis, although its application in analysis of polycyclic aromatic hydrocarbons (PAHs) on-line with HPLC and supercritical fluid chromatography has been demonstrated (Anacleto 1991, 1995). In negative APCI mode the main reagent ion $O_2^{-\bullet}$ has a fairly low EA value and thus readily transfers an electron to a molecule of higher EA, particularly one containing electronegative functional groups. Also $O_2^{-\bullet}$ has a fairly high PA and can accept a proton from analytes A that contain acidic protons, such as carboxylic acids and amides, phenols etc., to form $(A–H)^-$ ions.

The instrumental practicalities of APCI, and its exploitation as an LC/MS interface, are discussed next. Most reverse phase HPLC mobile phases contain mixtures of water with acetonitrile (PA $779 \, kJ. \, mol^{-1}$) or methanol (PA $754 \, kJ.mol^{-1}$). These PA values are considerably higher than that for water, so that the corresponding RH^+ ions will protonate a correspondingly more narrow range of analytes; as a result the proton hydrates must be responsible for ionizing analytes with lower PA values. The volatile buffers (generally formic or acetic acid and their ammonium salts) used for LC/MS also must be taken into account for APCI; if there is a high ammonia content, only analytes with PA values higher than that of ammonia ($854 \, kJ.mol^{-1}$) would be ionized.

A word of caution is in order with regard to all such predictions, since in fact the reactions in an APCI source almost all involve solvated (usually hydrated) species, exemplified by the characteristic APCI protonation reaction described in Equation [5.5]. For example, benzene has a higher PA ($750 \, kJ.mol^{-1}$) than water ($691 \, kJ.mol^{-1}$), but is not protonated in positive ion APCI conditions because neither the neutral nor ionized form likes to be solvated (after all benzene is essentially immiscible with water); thus benzene is ionized preferentially by charge exchange under positive APCI conditions.

Modern APCI–MS interfaces incorporate a heated pneumatic nebulizer (HPN, Figure 5.17) that pre-nebulizes the solution of analyte and subjects the droplets

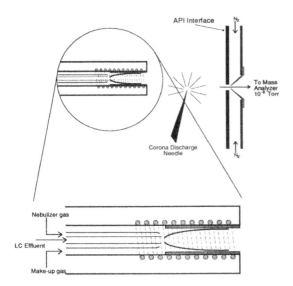

Figure 5.17 Sketch of a heated pneumatic nebulizer interface for APCI–MS. The LC effluent is nebulized by a fast gas stream and subjected to rapid heating via a heated make-up gas (typically N_2 at 120–200 °C, although the heater temperature can be much higher). The vaporized effluent is then introduced into the APCI plasma activated by the corona discharge, all at atmospheric pressure. The gaseous solvated ions are then introduced into the mass spectrometer vacuum via an API interface that must discriminate in favor of ions vs neutrals and also de-solvate the ions (see text). Reproduced from MDS-Sciex literature (*The API Book*) with permission.

to rapid heating to exploit the advantages arising from flash heating and the increase of vapor pressure for small particles described above for thermospray (see text box). Moderate heating (120 °C) at atmospheric pressure facilitates evaporation of solvent from the nebulized droplets; note that temperatures recorded at the heater itself can be several hundred degrees centigrade but the nebulized droplets encounter much cooler gas. The HPN approach to coupling LC with APCI–MS was introduced in the early 1980s (Henion 1982; Thomson 1983) building on earlier work (Horning 1974) that did not, however, incorporate pre-nebulization of the LC eluate. The modern HPN interface (Covey 1986; Huang 1990) has proved to be a remarkably useful workhorse in high throughput quantitative analyses of analytes of moderate molecular masses and low to moderate polarities. In early work the APCI process was activated by a ^{63}Ni β-particle emitter, but is now almost always enabled using a corona discharge; this avoids the necessity for hot filaments as a source of electrons as are used in conventional EI and CI sources, with their requirement for thermal stability of analytes.

In a recent innovation (Östman 2006) the nebulization and evaporation steps, but not the corona discharge and APCI itself, were accomplished on a microchip to provide an efficient low volume interface between GC and APCI–MS.

As emphasized in Section 5.1.1, the possibility of ionization suppression or enhancement of the analyte(s) by co-eluting compounds arising from the sample matrix or elsewhere must be considered for APCI. At one time it was thought that APCI was not seriously affected by matrix effects, but it is now clear that the extent of the problem is compound dependent. In a thorough investigation of the effect of ionization suppression on quantitative bioanalytical methods based on LC–MS/MS (Matuszewski 2003) it was found that, for the particular test compound used in that study, there was significant suppression in electrospray but none in APCI (enhancement was not observed in that case). However, other examples in which APCI suppression was clearly observed (van Hout 2003; Shou 2003; Sangster 2004) show that, although the problem is much less serious and prevalent in APCI than in electrospray, the possibility should not be ignored. Mutual suppression of ionization efficiencies of an analyte and its stable-isotope-labeled internal standard (Liang 2003) was demonstrated in electrospray (Figure 5.18(a)), but in most cases the same analyte–internal standard pairs exhibited enhancement (Figure 5.18(b)) of APCI ionization efficiencies. It was confirmed (Mei 2003) that suppression in APCI is compound dependent but that its extent also varied significantly with the design of the APCI source (extent and method of heating etc.); for one commercial API source, ionization suppression was found to be more serious with APCI than with electrospray! Moreover, the same study showed that suppression can arise not only from endogenous components co-extracted from the sample extract and co-eluting with the analyte but also from exogenous compounds such as plasticizers used in plastic components used in the analytical procedure and from additives commonly used to stabilize samples during storage before analysis, e.g., Li-heparin commonly used as an anti-coagulant for blood samples.

The mechanism of suppression/enhancement of ionization in APCI has received much less attention than that of the much more serious effect in electrospray ionization, and is currently not well understood. It is likely that a major source of these effects is the complex ion–molecule chemistry established in the APCI plasma (Figure 5.16). The example shown in Figure 5.19 is consistent with this idea, particularly the ionization enhancement observed on HPLC injection of a pure water sample into the mobile phase that was 80 % acetonitrile.

5.3.5 *Atmospheric Pressure Photoionization (APPI)*

Photoionization (PI) has been used for many years in physical chemistry studies with mass spectrometry, e.g., in determination of ionization energies, but until very recently the technique was not suitable for trace analysis with MS. For many years PI was used as the basis for a GC detector in which the electrical current generated by ionization of components with ionization energy (IE) lower than the photon energy is detected and recorded. Since most organic molecules have IE energies in the range 7–10 eV while the common GC carrier gases have much higher IE values (24.6 and 15.4 eV for helium and hydrogen, respectively), low background currents are obtained for this single parameter GC detector.

Most early reports of PI detection for LC were simply adaptations of the PI GC detector, i.e., simple detection only, with no complementary MS information. The only exception (Revel'skii 1991) involved some experiments that did not explore direct LC–MS coupling, but rather its feasibility in which vaporized samples were transported to an atmospheric pressure photoionization (APPI) source interfaced to a quadrupole MS. Encouraging results were obtained with the exception that, when abundant solvent vapor (water or methanol) was added, a significant decrease in the APPI response for the model analytes was observed. This phenomenon is still important for modern LC–APPI, as discussed below.

The recent activity in development of practical LC–APPI–MS can be attributed to a detailed study (Robb

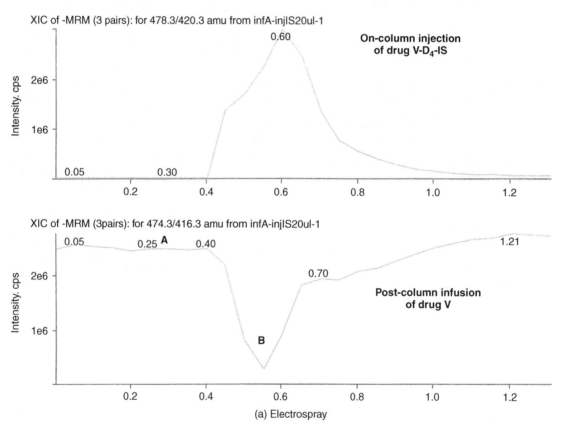

Figure 5.18 (a): Top trace shows the LC–ESI–MS peak for the D_4 analog of a proprietary drug V, observed using an arrangement in which a post-column T-junction allowed simultaneous infusion at a fixed rate of drug V itself so that the LC eluant was mixed with the solution of drug V before entering the ion source. The lower trace in (a) shows the time response for this constant infusion of drug V, indicating $\sim$80 % suppression by its D_4-internal standard during the elution window (top trace) of the latter. Concentrations of drug V and its D_4 analog were $10\,\mu g.mL^{-1}$. (b): same as (a) except that the APCI response is enhanced $\sim$7 times by the D_4 analog. Extent of suppression or enhancement (%) was calculated as $100.\{(A-B)/A\}$ or $100.\{(B-A)/A\}$. Reproduced from Liang (2003), *Rapid Commun. Mass Spectrom.* **17**, 2815, with permission of John Wiley & Sons, Ltd.

XIC of -MRM (2 pairs): for 474.3/416.3 amu from APCI- infA-injIS-20ul

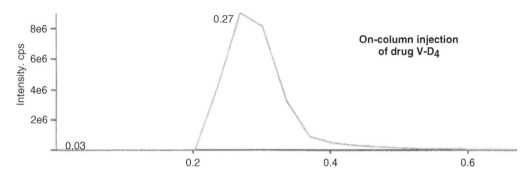

XIC of -MRM (2 pairs): for 474.3/416.3 amu from APCI- infA-injIS-20ul

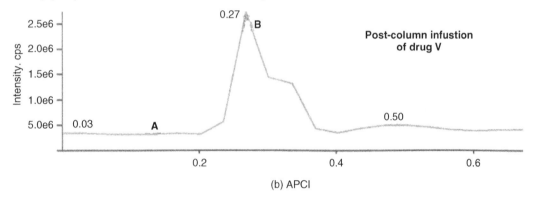

(b) APCI

Figure 5.18 (Continued)

2000) of the performance of a commercial APCI source modified by replacing the corona discharge needle by a photoionization lamp originally designed for PI detectors for GC; this lamp was filled with krypton that emits sharp lines in the vacuum UV at 10.0 and 10.6 eV, below the PI values of the solvents commonly used in reverse phase HPLC (12.6, 10.8 and 12.2 eV for water, methanol and acetonitrile, respectively). Thus, direct PI of organic molecules with IE < 10 eV was expected to be favored, but the very low concentrations of analytes in realistic LC–MS quantitation procedures meant that the total ionization was very low. This situation was remedied (Robb 2000) by adding a dopant that could be photoionized by 10 eV photons; in these initial APPI experiments toluene and acetone (IE values 8.8 and 9.7 eV, respectively) were tested as dopants. A sketch of this APPI interface (Robb 2000) is shown in Figure 5.20(a).

This work has initiated a large amount of activity in APPI development and applications in a relatively short time, but fortunately two excellent reviews have been published (Raffaelli 2003; Bos 2006). Here the focus is on some more fundamental aspects and on illustrative applications that are pertinent to the subject of this book. On the instrumentation side a somewhat different approach to APPI source design (Syage 2001; Hanold 2004) uses an orthogonal spray geometry with illumination from the PI light source directly in the spray region (Figure 5.20(b)). The coupling of APPI to flow rates of analyte solutions in the range 50 nL–5 µL.min^{-1}, typical of microfluidic separation systems or micro and nano HPLC, has been reported (Kaupilla 2004); this approach employed essentially the same microfabricated heated nebulizer chip as that used for coupling GC to APCI–MS (Östman 2006). More recently (Constapel 2005; Droste 2005) the use of a laser light source for APPI (leading to the acronym APLI) has been described. The KrF* excimer laser used in this work emits photons ($h\nu$) at 248 nm (∼5 eV), much too low for direct single photon PI of any organic molecule, so the laser irradiance, i.e., energy (or number of photons) per unit area per unit time, must be sufficiently high for

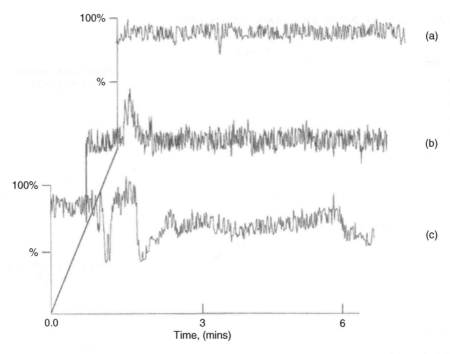

Figure 5.19 Effect of HPLC injections on the APCI signal from the analyte infused post-column at a constant rate; the isocratic mobile phase was 80 % acetonitrile and 20 % aqueous ammonium acetate (1 % w/v), and the analyte solution that was infused post-column was 1 μg.mL^{-1} in the mobile phase. (a) Infusion of parent compound with no injection on the HPLC (mobile phase only); (b) enhancement effect of injection of water on the signal from infused analyte; (c) suppression effect of injection of extract of a blank plasma sample on the signal from the infused analyte (see Section 5.3.5a). The diagonal line is drawn to connect the injection times in the different chromatograms. Reproduced from Sangster (2004), *Rapid Commun. Mass Spectrom.* **18**, 1361, with permission of John Wiley & Sons, Ltd.

multiphoton absorption to occur with sufficiently high probability that useful ion currents are produced from the analyte. With modern lasers this is not difficult to achieve and there is a considerable potential advantage to be gained with respect to ionization selectivity since dopants are not a requirement, especially if the mechanism is resonance enhanced multiphoton ionization (REMPI):

$$A + mh\nu \rightarrow A^*$$

$$A^* + nh\nu \rightarrow A^{+\bullet} + e^-$$

where h is Planck's Constant. These processes describe a simple (m+n) single color REMPI scheme that relies on the existence of a sufficiently long lived electronically excited state A^* of the neutral analyte to boost the multiphoton ionization efficiency. As emphasized in the original APLI paper (Constapel 2005), for applications of REMPI to trace analysis overall-resonant $(1 + 1)$ REMPI approaches are likely to be the most useful. In these initial experiments single color REMPI (fixed wavelength photons from a single laser) were used, and this essentially limited the application to polyaromatic compounds.

However, use of multicolor REMPI (more than one laser emitting different wavelengths) should permit extension of the approach to analytes with more complex UV absorption behavior. Coupling of both conventional and capillary HPLC and also CEC to MS via the APLI source was reported (Constapel 2005; Droste 2005), with detection limits for polycyclic aromatic compounds in the low fmol range. This approach is still in its early development stage but appears to have considerable potential.

At present the great majority of APPI work employs rare gas discharge lamps as light source, together with judicious choice of dopants. It turns out that the mechanism of action of dopants in APPI is not as simple as first believed, i.e., is more complicated than the simple charge exchange mechanism. A series of papers (Marotta 2003; Tubaro 2003; Basso 2003) has described this complexity. An investigation (Marotta 2003) of APPI of some furocoumarins, dissolved in dry acetonitrile and infused into the source, showed surprisingly that the majority of the ionized analyte was formed as $[A+H]^+$ rather than the $A^{+\bullet}$ expected to be formed by direct photoionization (the ionization energies of

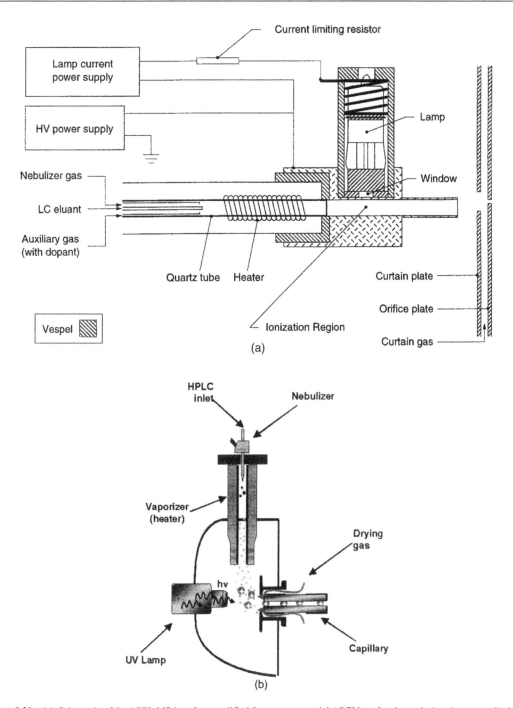

Figure 5.20 (a). Schematic of the APPI–MS interface modified from a commercial APCI interface by replacing the corona discharge needle by a PI lamp, showing the heated pneumatic nebulizer and introduction of the dopant in the auxiliary gas line. Reproduced from Robb (2000), *Anal. Chem.* **72**, 3653, copyright (2000), with permission of the American Chemical Society. (b). Sketch of the APPI–MS interface based on an orthogonal spray geometry and with a capillary-based connection between the API source and the MS vacuum. Reproduced from Hanold (2004), *Anal. Chem.* **76**, 2842, copyright (2004), with permission of the American Chemical Society.

these analytes should be well below 10 eV). Also, formation of protonated acetonitrile molecules $(C_2H_4N^+)$ and of their hydrate $(C_2H_6NO^+)$ was observed (the latter presumably arising from traces of water).

The origins of these ions were unclear as the ionization energy of acetonitrile (12.2 eV) is well above the photon energy, and the same is true of water (12.6 eV). To clarify the nature of these species, ^{13}C and ^2H labeling experiments were performed. The data suggested that either neutral acetonitrile together with some other species (probably water) formed a complex with ionization energy < 10 eV, or that photon irradiation led first to isomerization of acetonitrile molecules to give species with IE < 10 eV. Once the acetonitrile radical molecular cations were formed, typical APCI reactions would lead naturally to the $C_2H_4N^+$ species; the $C_2H_6NO^+$ species had already been detected in the stratosphere and in drift tube experiments. Both of these species can be considered as protonating agents, thus accounting for the preponderance of $[A+H]^+$ ions in the APPI spectra.

In the case of benzene and toluene dopants (Tubaro 2003), the formation of $A^{+\bullet}$ analyte ions from the simple charge exchange mechanism was indeed observed to predominate when the analyte was dissolved in pure dopant, but protonation reactions were also observed. It was shown by deuterium labeling and MS^n experiments (Section 6.2.2) that, when benzene is irradiated in the APPI source, the odd-electron molecular ions of phenol, diphenylether and phenoxyphenol are produced in high yield, together with some protonated diphenylether and protonated phenoxyphenol. A possible mechanism based on radical attack by benzene molecular ions on oxygen molecules present (although at low concentration) in the APPI source was proposed. Similar results were obtained with toluene, demonstrating that APPI is able to activate a series of ion–molecule reactions within the APPI plasma leading to highly reactive species that are effective in promoting protonation of those molecules that have ionization energies higher than the photon energy. A further experiment, in which 10 % of phenol was added to the benzene solution of analyte that was infused into the APPI source, led to a dramatic increase in the yield of $[A+H]^+$ ions, suggesting the use of phenol as an efficient dopant.

Anisole (methylphenylether) can be regarded as methylated phenol and has been proposed (Kauppila 2004a) as a suitable dopant for APPI of analytes with low proton affinity (PA) values in reverse phase HPLC solvents in which benzene and toluene are unsuitable. Previously only acetone was used in these conditions but its fairly high PA ($812 \, kJ.mol^{-1}$) meant that it was suitable for only analytes with even higher PA values. It was shown

that a high abundance of the $A^{+\bullet}$ ion of anisole can be produced upon APPI in reverse phase solvents, and that as a result samples with low IE and PA values can be ionized efficiently by charge exchange. Formation of negative ions from electronegative analytes has been shown (Basso 2003) to arise from electron attachment following photoemission of electrons from metal surfaces followed by thermalization of the photoelectrons in the API conditions.

The take-home message from all of this work is the unsurprising one that chemistry, including that in an APCI plasma however activated (corona discharge or photoionization), is inherently complex and needs to be borne in mind when exploiting these techniques for trace quantitative analysis. For example it has been shown (Kauppila 2002) that buffers commonly used in LC/MS (e.g., ammonium acetate) can severely suppress ionization by APPI. Nonetheless, the technique does appear to have considerable potential as a rival and/or complementary ionization method for LC/MS applied in trace quantitation. For example, an investigation (Hsieh 2003) of APPI in LC–MS/MS analysis of small molecules in rat plasma demonstrated good sensitivity and fast turnaround, with data quality equivalent to that obtained by this group's standard HPLC–APCI–MS/MS approach. Several major experimental parameters, including the delivery rate of dopant solution, the composition and flow rate of the LC mobile phase, and nebulizer temperature, significantly affected the ionization efficiency. Also, APPI was shown in this study (Hsieh 2003) to suffer from only minimal matrix suppression or enhancement of ionization for the two test compounds employed.

A more recent study by the same group (Wang 2005) compared the performance of APCI, APPI and ESI (electrospray) for quantitation of two drug molecules (Cyclosporin A and Lornafarib) in drug plasma. All three interfaces were compared using reversed phase HPLC (using toluene as dopant for APPI) and also with normal phase APPI (iso-octane/ethanol as isocratic mobile phase) both with and without dopant. All combinations provided good quality data for concentrations above $5 \, ng.mL^{-1}$ in rat plasma, although ESI gave detection limits four times lower than APCI and 10 times lower than APPI. No statistically significant differences were observed among the concentrations determined by the various methods in the range $5–1000 \, ng.mL^{-1}$. A considerable effort was expended (Wang 2005) in investigating ionization mechanisms in APPI with normal phase HPLC, where it was surprising to observe that the dopant (toluene) made no difference to the ionization efficiency. It was shown using D-labeling experiments that the proton transfer process was predominantly from proton bound dimers of ethanol $[C_2H_5OH–H^+–HOC_2H_5]$, with iso-octane (IE 9.89 eV) acting as a 'self dopant' for the normal phase mixed solvent. A somewhat different comparison (Cai 2005) investigated the relative performance of APPI, APCI and ESI as 'universal' ionization methods in support of drug discovery. In all, 106 standard compounds and 241 proprietary drug candidates, representing a wide range of chemical types, molecular masses, polarities etc., were tested for detectability at concentrations in methanol in the range $0.01–10 \mu g.mL^{-1}$ (depending on the compound). APPI was superior to APCI and ESI in detection of less polar compounds and was about the same with respect to polar compounds; when positive and negative detection modes were combined, APPI, APCI and ESI detected 98 %, 91 % and 91 % of the compounds, respectively.

While these results do suggest that APPI–MS more closely approaches a 'universal' detector, it should be pointed out that in this test APPI used a dopant (toluene) that was not used in the corresponding APCI tests so that this particular comparison is not straightforward. Also, simple detection is a much easier criterion to satisfy than acceptable accuracy and precision in quantitation. At present the characteristics of APPI with respect to important figures of merit such as repeatability, precision, susceptibility to matrix effects etc. remain unclear.

5.3.6 Electrospray Ionization (ESI)

Electrospray ionization (ESI) is one of the two newer ionization techniques (the other being MALDI) that has revolutionized the application of mass spectrometry to biochemistry and molecular biology, and may fairly be said to have revolutionized these disciplines also. This is the result of the ability of ESI–MS to provide molecular mass information of hitherto unthinkable accuracy and precision for fragile biopolymers, particularly proteins, and to provide amino acid sequence information for specific peptides present at trace levels in complex mixtures characteristic of biological extracts. The discipline of proteomics would not exist without the development of ESI and MALDI. However, this book is concerned with quantitation of small molecules present at trace levels in complex matrices, rather than the essentially qualitative data typically acquired in proteomics experiments, although application of the approach to quantitation of proteins and peptides is discussed in Section 11.6. Nonetheless, development of methods to produce and characterize gaseous ions from macromolecules was very important in the development of ESI–MS, and this history will be briefly described here. Excellent reviews of this history as it pertains to macromolecule characterization (Fenn 1990; Smith 1991, 1992; Fernandez de la Mora 1992) are available, while

an interesting account of this history from the viewpoint of a physicist (Hamdan 1991) adds a different perspective.

At its most basic level, ESI involves production of gaseous ions at atmospheric pressure from analytes in liquid solution by nebulizing the solution in such a way as to produce small droplets that carry a net electrical charge. It is convenient to consider the overall process as two steps, the first leading to production of the charged droplets and the second to the formation of gaseous ions from these droplets.

There is a long history of investigations of production of electrically charged droplets by spraying or shattering liquids in impact with surfaces, including early work on electrically charged spray from waterfalls. This work culminated in publication of a comprehensive book on the subject (Loeb 1958) and some of it is summarized in one of the recent reviews of ESI (Hamdan 1991).

The phenomenon has many practical implications, including build up of charge on aircraft in flight (with no means to allow the charge to escape to ground) as a result of exposure to heavy rain. However, for present purposes a good starting point, which led directly into one of the series of experiments resulting in modern ESI–MS, is provided by an investigation conducted at the Ford Motor Company (Hines 1966) into the physics of spray painting using electrostatic atomization. A practical system of this kind consists of a knife edge held at a potential of about 100 kV and held about 30 cm away from the target workpiece that is electrically grounded. Paint is fed to the knife edge and becomes electrically charged (see below), so that the strong electric field around the knife edge pulls the charged paint away in the form of numerous jets; a photograph of two such jets produced in a model system designed for fundamental studies of the process is shown in Figure 5.21 (Hines 1966). The excess electrical charge tends to break up such a jet into droplets via Coulombic repulsion that must, however, be sufficiently strong to overcome the surface tension of the liquid that tends to hold the jet together; the size of the resulting droplets is a result of the balance between these two opposing forces. Because the droplets are charged the electric field created by the high potential on the knife edge ensures that virtually all the droplets land on the grounded workpiece, a major advantage of this painting method. As a result of these detailed experimental and theoretical studies (Hines 1966) it was shown that the experimental values of the charge transported by the jet and of the drop size agreed with theoretical values calculated from the magnitude of the electric field and the physical properties (electrical conductivity, viscosity, density, surface tension and dielectric constant) of the sprayed liquid.

This excellent work (Hines 1966), in which fundamental principles were applied to improve the understanding and performance of a strictly practical device,

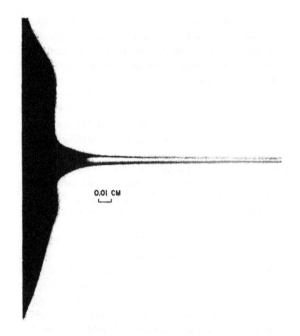

Figure 5.21 Photograph of jets formed from an oil–ethanol mixture in a laboratory scale model of an electrostatic paint sprayer. Reproduced with permission from Hines, *J. App. Phys.* 37, 2730 (1966). Copyright (1966) American Institute of Physics.

caught the attention of a materials science research group interested in producing intact single molecules of synthetic polymers in the gas phase, and resulted in the first experiments (Dole 1968) using a device that is recognizable as the precursor to electrospray ion sources in use today with mass spectrometers (Figure 5.22). This

work involved Hines as a collaborator, and again created droplets from the bulk liquid via electrification as in the spray painting experiments. The results obtained by this first attempt at creation of gaseous single ions were unfortunately ambiguous since characterization of the ions produced was possible only by a very crude method employing application of retarding electric fields to distinguish ions produced from the solvent from the desired polymer ions. However, this work is important partly for its historical significance and mostly for its introduction of the atmospheric pressure bath gas–supersonic nozzle combination (Section 5.3.3a) still used today to connect an atmospheric pressure ion source to a detection system that operates in a vacuum.

This work (Dole 1968) also introduced the 'charge residue mechanism' proposed to account for the formation of gaseous ions from the charged droplets (this mechanism is discussed below). Further work on both synthetic polymers and biopolymers (Mack 1970; Clegg 1971) produced similar results, but in later work (Gieniec 1984) the electrosprayed ions were characterized via their drift times through an atmosphere of nitrogen under the influence of an applied electric field, what would nowadays be called their ion mobility spectra. The data thus obtained for a solution of lysozyme (a protein of molecular mass 14,307 Da) are shown in Figure 5.23; peaks 4 and 5 are clearly due to ions produced from the solvent, and peaks 1, 2 and 3 were assigned as singly, doubly and triply charged lysozyme, respectively, appreciably lower than charge values established later.

It is interesting that this paper (Gieniec 1984) mentions that attempts to characterize these ions using a time

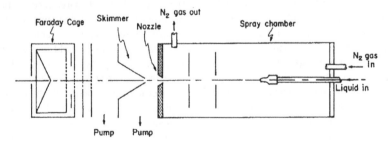

Figure 5.22 Sketch of an electrospray ionization apparatus constructed to investigate the possibility of creation of single molecules of polystyrene (molecular mass ~50 000 Da) in the gas phase. The polymer solution (0.01 wt % of polymer) was infused at a rate of $300\,\mu L.min^{-1}$ and the needle voltage was generally $-10\,kV$ with $-3\,kV$ on the first collimating plate (preceding the nozzle) and $-1.4\,kV$ on the second. The ion current transmitted by the supersonic nozzle-skimmer system was detected by a simple Farady cage that collected all the ions are fed to resulting current to a sensitive electrometer. To distinguish the polymer ions from ions produced from solvent, it was assumed that the nozzle-skimmer behaved perfectly to produce a beam in which all molecules and ions traveled with exactly the same velocity so that their kinetic energies were proportional to their masses. By applying retarding potentials to the grids placed in front of the Faraday cage it was possible to screen out the low mass ions produced from the solvent. Reproduced from Dole, *J. Chem. Phys.*, **49**, 2240 (1968a). Copyright (1968), with permission of the American Institute of Physics.

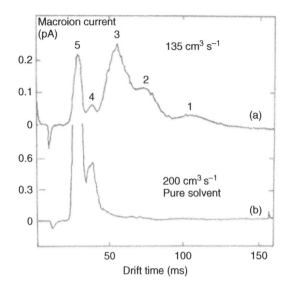

Figure 5.23 Current–drift time curves (ion mobility spectra) recorded using a drift voltage of 4.0 kV, for positive ions formed from a 0.0053 wt% solution of lysozyme in 90 % alcohol, and for the pure solvent, at carrier gas flow rates of 135 and 200 mL.s^{-1}, respectively. Peaks 4 and 5 clearly arise from the solvent, but peaks 1, 2 and 3 were assigned as different charge states of lysozyme. Reproduced from Gieniec, *Biomed. Mass Spectrom.* **11**, 259 (1984), with permission of John Wiley & Sons, Ltd.

of flight mass spectrometer were unsuccessful. Practically simultaneously with this last publication of Dole on the subject (Gieniec 1984), Fenn and his collaborators (Yamashita 1984) achieved the successful marriage of the same electrospray–atmospheric pressure bath gas–supersonic nozzle combination pioneered by Dole with mass spectrometry. The ions thus detected and measured for *m/z* were mainly clusters of solvent molecules on protons and alkali metal ions, although a few small organic molecules were also successfully ionized; this limitation was imposed by the *m/z* working range of the quadrupole analyzer used. Subsequently (Whitehouse 1985) a quadrupole with a range up to *m/z* 1500 was used and permitted acquisition of mass spectra of biomolecules with molecular masses up to almost 1400 Da. Even more significant was the finding that, in the case of each of the peptide analytes tested, the *m/z* value of the dominant peak in the spectrum corresponded to the doubly-protonated form [A + 2H]$^{2+}$; this demonstration of the multi-charging phenomenon led to the applications to protein analysis (Fenn 1989) that resulted in a Nobel Prize for Fenn in 2002.

In view of the importance of the multi-charging phenomenon for biological applications, it is worth noting that the work of Dole on synthetic polymers culminated (Wong 1988) in observation by ESI–MS of ions bearing up to 23 charges (due to associated sodium ions) from polyethylene glycol preparations with average molecular masses up to 17 500 Da. Even more significantly, in the same year Fenn *et al.* reported ESI mass spectra of multi-protonated intact proteins with molecular masses up to 40 000 Da, bearing up to 45 charges (Meng 1988). This report sparked immediate follow-up by other groups (Loo 1988; Covey 1988); it is a measure of the importance of this discovery that such spectra are now almost common-place.

However, from the viewpoint of this book it is actually more important to note that, under the experimental conditions used (Whitehouse 1985), the signal strength was independent of the flow rate of the same solution through the electrospray needle in the range 6.5–14.5 µL.min^{-1} (Figure 5.24). This finding is of considerable significance for the use of ESI–MS as a detector for quantitative HPLC analysis, as it suggests that the response of ESI–MS is concentration dependent rather than mass flow dependent like EI–MS (Section 4.4.8 and Appendix 4.1); this is discussed further in Section 5.3.6b.

The foregoing greatly abbreviated account of the progression from an investigation of a paint sprayer to a device used to reveal fine details of the properties of proteins and other biopolymers is an excellent example of the purely artificial distinction between 'pure' and 'applied' science discussed in the Preface. The same is true, though perhaps to a lesser extent, of another series of investigations that owes its roots to the very early work on static electrification (Loeb 1958) but progressed from studies in cloud physics to a device designed to link HPLC on-line with MS with particular emphasis on trace quantitation. For present purposes this trail can be said to have begun with a series of papers (Thomson 1975, 1979; Iribarne 1976) designed to investigate charged species obtained by spraying liquids via their ion mobility spectra, and in particular to investigate whether the high mobility species thus observed corresponded to singly charged small ions or multiply charged microdroplets. At this stage of this work the charged droplets were formed without use of an applied electric field and the work was thus a direct extension of the classical work on static electrification (Loeb 1958).

For present purposes this particular series of papers is important because the authors later used a mass spectrometer for the first time (Thomson 1979) to characterize the charged species produced by their spray technique (Figure 5.25). Equally important, the authors introduced the 'ion evaporation theory' for the mechanism whereby single molecular ions are formed from charged droplets (see later). It was soon realized that this

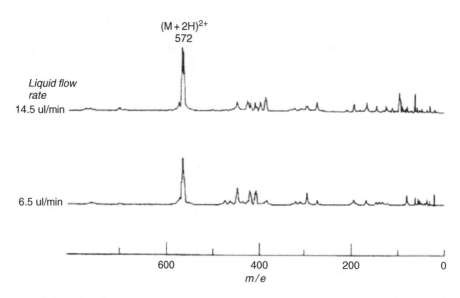

Figure 5.24 Demonstration of the independence of ESI–MS signal intensity on solution flow rate (see Section 5.3.6b). The analyte was the cyclic peptide gramicidin S dissolved in methanol:water (1:1); this solution was infused into the ESI source using a syringe pump. Reproduced from Whitehouse, *Anal. Chem.* **57**, 675 (1985), copyright (1985), with permission of the American Chemical Society.

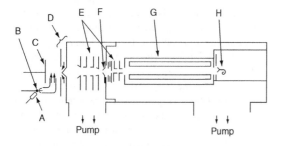

Figure 5.25 Sketch of the ion evaporation–MS apparatus of Thomson and Iribarne. For convenience the ion source section (items A–D) is shown at 90° (horizontal cross section) to the MS vacuum section (items F–H). A, high voltage electrode (±3.5 kV); B, sprayer nozzle; C, deflecting electrode to steer ions towards the orifice; D, gas curtain plus plate containing 25 μm orifice leading to skimmer cone; E, ion lenses; F, plate separating low-vacuum chamber (10^{-4} torr) from high-vacuum chamber (10^{-6} torr) containing the quadrupole mass filter (G) and continuous dynode electron multiplier (H). The sprayer was a pneumatic type, and the spray was directed through a 90° bend to eliminate the larger droplets. Reproduced from Thomson, J. Chem. Phys., **71**, 4451 (1979). Copyright (1979), with permission of the American Institute of Physics.

combination was ideal for the analysis of polar involatile compounds and also that the extremely soft nature of the ionization process would limit the chemical information to molecular mass, so the single quadrupole MS was

replaced (Thomson 1982) by a triple-quadrupole instrument; Figure 5.26 shows the first tandem mass spectra of ions produced by an ionization technique that is a recognizable version of ESI. Note that the electric field strength produced at the spray tip by the external electrode (A, Figure 5.25) is much less than that created by a high potential applied directly to the spray needle, as introduced by Dole (Figure 5.22), so the extent of multiple charging is much lower than was observed later using the latter method.

Later work on this 'ion evaporation' source (Iribarne 1983) surveyed its applicability to a wide range of compounds, and also for the first time coupled it on-line to an HPLC operating at a flow rate of 1 mL.min^{-1}. However, this 'ion evaporation' source did not directly result in practical LC–MS technology but was supplanted (Bruins 1987) by an electrospray-type source (labeled 'ionspray' that was later used as a trademark of the MDS-Sciex company) in which the best features of the Thomson–Iribarne approach (e.g., pneumatically assisted nebulization) were combined with those of the Dole–Fenn method (direct application of the high potential to the sprayer itself). This arrangement and some of the data obtained are shown in Figure 5.27. The advantage of pneumatically assisted nebulization is that it decouples the nebulization from the electrification of the spray, thus permitting operation at higher flow rates characteristic of HPLC, and also is much more suitable for handling

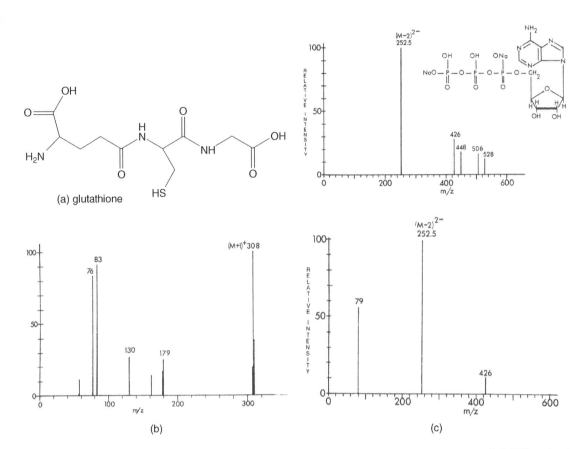

Figure 5.26 (a) Structure of glutathione. (b) Fragment ion (MS/MS spectrum of singly protonated glutathione (*m/z* 308) produced by ion evaporation; the labeled fragment ions correspond to simple backbone cleavages. (c). Ion evaporation mass spectrum of adenosine triphosphate di-sodium salt (top) and MS/MS fragment ion spectrum (bottom) of the di-anion at *m/z* 252.5; the fragment ions correspond to PO_3^- and to the singly-charged anion formed through loss of PO_3^- from the selected precursor ion. Reproduced from Thomson, *Anal. Chem.* **54**, 2219 (1982), copyright (1982), with permission of the American Chemical Society.

gradient elution where the physical properties (vapor pressure, surface tension, viscosity etc.) of the mobile phase can vary significantly through the gradient and thus affect the overall efficiency of the ESI process itself.

However, unknown to these two groups working in North America, an entirely independent development of LC/MS using ESI was underway in what was then the Soviet Union. In 1984 Aleksandrov and his colleagues demonstrated the use of ESI as an LC–MS interface (Aleksandrov 1984). This was a doubly heroic achievement since the mass spectrometer used was a magnetic sector instrument (Section 6.4.4) for which the ion source must be at several kV relative to the analyzer that is usually held at ground, and of course the ESI needle must be another several kV relative to the ion source. Even today, operation of an ESI source with a magnetic sector analyzer is difficult and indeed is seldom attempted. Figure 5.28 shows a sketch of this apparatus and a mass spectrum of erythromycin (Aleksandrov 1984) obtained

using their method, which they named EDIAP (extraction of dissolved ions at atmospheric pressure). Subsequently, this group applied ESI–MS to oligosaccharides, peptides with molecular mass up to 1500 Da including observation of dominant doubly charged ions (Aleksandrov 1984a, 1985), and later (Aleksandrov 1988) reported the ESI mass spectrum of a small protein containing intense peaks corresponding to doubly and triply charged ions. These remarkable achievements went largely unnoticed outside the Soviet Union (even though an English translation of the first paper was available in 1985) until publication of an excellent early review (Smith 1991). However, from the point of view of the present book, it was the use of ESI as an on-line LC–MS interface (Aleksandrov 1984) that is of major interest; it was essentially contemporary with the demonstration (Iribarne 1983) of LC–MS coupling using the ion-evaporation interface.

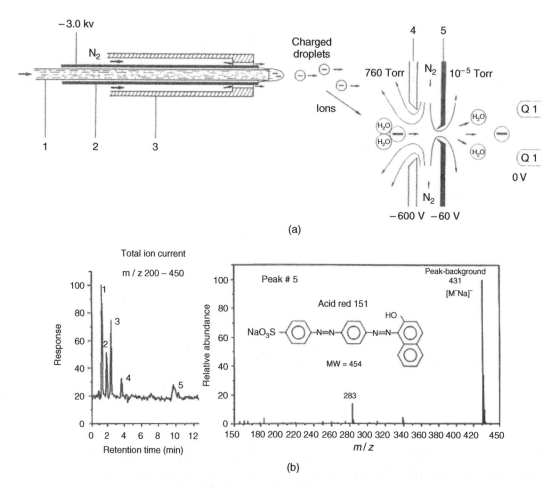

Figure 5.27 (a) Pneumatically assisted electrospray source and API interface designed for LC/MS applications. (b) Total ion current (*m/z* 200–450) chromatogram observed in LC/MS analysis of a mixture of five monosulfonated azo dyes ($\sim$20 ng of each injected on-column), and ESI mass spectrum of the last-eluting component. Separation was achieved using a 1 × 100 mm C$_{18}$ column, operated in reverse phase isocratic mode with 40 μL.min^{-1} of a 30:70 mixture of acetonitrile and aqueous ammonium acetate (10^{-3} mol.L^{-1}). Reproduced from Bruins, *Anal. Chem.* **59**, 2642 (1987), copyright (1987), with permission of the American Chemical Society.

It is now clear that all three approaches exploited the same physico-chemical phenomena and that differences among them reflect differences in the electric field strength applied to the flowing solution of analyte (the ion evaporation approach used a relatively low value) and the decoupling of nebulization from droplet electrification in the ion spray source. All three can be considered together as examples of electrospray ionization (ESI). The physico-chemical mechanisms involved have been the subject of some high level investigations using classical (i.e., nonquantum) physical chemistry. A recent summary of this work is contained in a series of articles published back-to-back in 2000 (Cole 2000; Van Berkel 2000; Amad 2000; Gamero-Castaño 2000; Kebarle 2000), and in other

key papers (Van Berkel 2000a; King 2000; Cech 2001); references to earlier work on ESI mechanisms are given in these papers. This book is not the place to attempt a full summary of this work; instead the following discussion concentrates on those aspects that are most relevant to applications of ESI to quantitative analysis of small molecules, i.e., ionization suppression or enhancement (Section 5.3.6a), and the variation of ionization efficiency with flow rate of the analyte solution (Section 5.3.6b).

The general overall ESI process is illustrated in Figure 5.29. Here it is assumed that the electric field alone is responsible for the nebulization of the flowing solution of analyte (no pneumatic assistance). The field will be strongest near the edges of the tip of the ESI

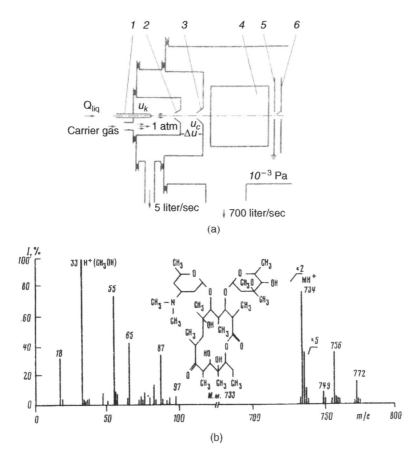

Figure 5.28 (a) Sketch of the electrospray ion source and MS interface of Aleksandrov *et al.* 1: metal capillary held at 2–4 kV relative to the first of the two differentially pumped diaphragms (2 and 3); the potential difference between 2 and 3 was varied between 0 and 900 V to study solvent declustering and in-source collision induced dissociation. 4: ion lenses. 5: entrance slit of magnetic sector mass spectrometer. 6: total ion current monitor plate. (b). Mass spectrum of erythromycin obtained using the apparatus shown in (a). Reproduced from Aleksandrov *Doklady Phys. Chem.* **277**, 572 (1985), copyright (1985), with permission from Springer Science and Business Media.

capillary (needle) and will polarize the emerging solution by Coulombic forces in such a way that the liquid just emerging from the tip will contain an excess of ions of the same polarity (positive in the example shown in Figure 5.29) as the applied potential at the needle (Coulombic repulsion). Note that the solution must contain a significant concentration of ionic species in the first place in order to permit charge separation at the tip. These ionic species may be analyte, added volatile electrolytes such as acetic or formic acid or their ammonium salts, and/or charged products of the electrochemical reaction at the needle (see below); if insufficient ions are provided it is impossible to maintain a stable electrospray.

It was shown some time ago (Taylor 1964) that such an emerging liquid from a narrow tip at a high potential adopts the shape of a cone as a result of a balance between Coulombic repulsion at the surface and the surface tension of the liquid; this is known as a 'Taylor cone'. At the point where Coulombic repulsive potential energy starts to exceed the surface tension (energy), a condition known as the Rayleigh Limit (Rayleigh 1882), the cone breaks up into droplets each of which carries an excess charge (positive in the example shown in Figure 5.29). This flow of net positive charge away from the ESI needle in the gas phase must be compensated by an equal flow of (negative) electrons away from the needle through the external circuit, and this can only come about if oxidation processes occur at the needle, such as:

$$2H_2O \rightarrow O_2 + 4H^+ + 4e^-$$

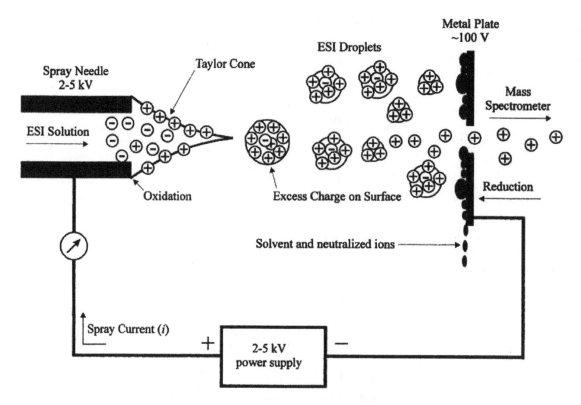

Figure 5.29 Schematic of the electrospray ionization process. The analyte solution is pumped through a needle to which a high electrical potential (positive in this example) is applied. A Taylor cone with an excess of positive charge on its surface forms as a result of the electric field gradient between the ESI needle and the counter-electrode (i.e., the metal plate that contains the orifice leading into the mass spectrometer). Charged droplets are formed from the tip of the Taylor cone (see text); these droplets evaporate and, if the Rayleigh limit is reached (see text), split into smaller droplets as they move towards the mass spectrometer orifice. Gas phase charged analyte molecules are formed from the final charged microdroplets. The transport of charge (positive in this example) in the gas phase must be balanced by a flow of electric current in the external circuit that maintains the potential difference between needle and counter-electrode; this spray current is formally a flow of positive charge in the external circuit from counter-electrode to needle as shown, but is actually a flow of electrons in the opposite direction (that can be measured using the meter in the diagram). This implies that electrochemical oxidation (loss of electrons) occurs at the needle, and reduction at the counter-electrode. Reproduced from Cech and Enke, *Mass Spectrom. Revs.* **20**, 362 (2001), with permission of John Wiley & Sons, Ltd

The view of an ESI source as an electrochemical cell is important and this aspect has been discussed in some detail (Van Berkel 2000, 2000a). Note that for analytes A with oxidation potentials <2 eV or so, and depending on the nature of the solvent, electrochemical oxidation at the ESI needle can produce molecular radical cations $A^{+\bullet}$ instead of $[A+H]^+$.

As these primary charged droplets drift through the atmospheric pressure gas under the influence of the applied field, the solvent evaporates, the droplet size decreases and the concentration of excess charge at the droplet surface increases until the Rayleigh limit is again reached, whereupon the droplet spits out 'offspring'

droplets that on average contain ∼2 % of the original mass and ∼15 % of the original charge. This process has been captured on film in some remarkable photographs (Gomez 1994). The process then starts over again, with droplet evaporation alternating with Coulombic fission at the Rayleigh Limit (Figure 5.30). The general features briefly outlined thus far appear to be well established both experimentally and theoretically, as summarized recently (Cole 2000; Kebarle 2000). The only additional minor comments concern the early ion evaporation experiments (Thomson 1975, 1979, 1982; Iribarne 1976) where the electric field experienced by the liquid emerging from the needle tip was considerably lower than that operating

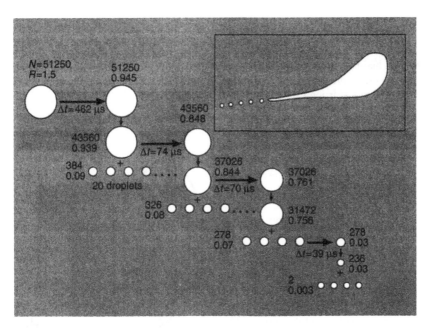

Figure 5.30 Sketch based on experimental observations (Gomez 1994) describing the evolution of charged droplets of heptane, moving through atmospheric pressure gas under an applied electric field, due to alternating steps of solvent evaporation at constant charge and of Coulomb fission at the Rayleigh Limit. The first droplet shown is at the Rayleigh Limit and produces 20 offspring droplets that carry off ~2% of the mass and ~15% of the charge. N = number of elementary charges, R = radius in μm, and Δt is the time (μs) required to reach the next fission from the evaporating droplet. Note that these experiments (Gomez 1994) used heptane and thus are only approximately applicable to liquids with a high aqueous content. The insert at the top right is a sketch of a photograph (Gomez 1994) of the fission of a droplet at the Rayleigh Limit. Reproduced from Kebarle and Tang, *Anal. Chem.* **64**, 972A (1993), copyright (1993), with permission from the American Chemical Society.

when the high potential is applied directly to the needle, and the improved efficiency of droplet formation (but not charging) made possible by pneumatic assistance (Bruins 1987) for solutions of higher surface tension or for higher HPLC flow rates.

The final step to be accounted for is the production of single analyte ions from the final sub-μm charged droplets (Figure 5.30). Two theoretical approaches have been proposed. The 'charge residue model' (Dole 1968) proposes that the gas phase ions will be produced when the cascade of droplet fissions (Figure 5.30) continues to the point where no further evaporation of the solvent can occur; what is left behind constitutes a charged residue containing analyte molecule(s) present as a cluster with residual solvent molecules and possibly other constituents of the original solution, e.g., buffer components. The clustered molecules of solvent etc. are removed by collisions in the API–MS interface mediated by the 'cone potential' (Section 5.3.3a) and the bare analyte ions are detected by the mass spectrometer. This concept is consistent with the original motivation of Dole *et al.* who wished

to produce single polymer molecules in the gas phase for fundamental studies, as discussed above; it is now generally accepted (Kebarle 2000) that this mechanism is the major contributor to successful ESI production of gaseous ions of macromolecules, particularly the globular proteins. However, the ion evaporation mechanism, originally proposed (Thomson 1975, 1979; Iribarne 1976) in the context of the physics of electrified clouds, appears to be responsible for ESI of small molecules as recently demonstrated experimentally (Gamero-Castaño 2000) and summarized (Kebarle 2000).

It is necessary for the purposes of this book to describe the general features of the ion evaporation theory without going into too much detail. The following discussion is a much simplified abbreviated version of a recent expert summary (Kebarle 2000). When solvent evaporation and Coulomb droplet fissions have reduced the size of the charged droplets to a critical radius r_{drop} (~10–20 nm, Figure 5.30), direct emission of ions from the droplet to the gas phase begins. According to the quantitative evaluation of the theory by its originators (Iribarne

1976; Thomson 1979), the charge on droplets of this size that is required for this ion evaporation is lower than that required for Coulomb fission, i.e., ion evaporation replaces Coulomb fission as a means of relaxing the Coulombic repulsion at the droplet surface. The equation predicting the rate constant k_I for ion evaporation from the droplets was derived on the basis of the transition state theory (see any text on physical chemistry):

$$k_I = (k_B T/h).\exp(-\Delta G^{\#}/RT) \qquad [5.6]$$

where k_B is the Boltzmann Constant ($= R/N_A$), h is the Planck Constant, R the universal gas constant, and $\Delta G^{\#}$ is the difference in (molar) Gibbs Free Energy between the transition state and the starting point (the ion at the droplet surface). The assumed transition state (i.e., the maximum in the curve relating the potential energy to the distance of the evaporated ion from the droplet surface) is a 'late' transition state, i.e. the transition state is located after the ion has escaped from the droplet surface, at a distance ~ 0.5 nm between the ion and the surface. The maximum of the potential energy barrier, located at this point, is due to a balance between the strong attractive force between the ion and the polarizable droplet; this attractive force decreases rapidly ($\sim r^{-6}$) with distance r from ion to droplet centre, and acts to oppose the longer range repulsive Coulombic forces ($\sim r^{-1}$) between the ion and the remainder of the charged droplet. In this theoretical model $\Delta G^{\#}$ contains contributions from both this electrostatic term $\Delta G^{\#}(E)$ (where E is the electric field perpendicular to the droplet surface, readily evaluated as a result of the spherical symmetry) and a solvation free energy $-\Delta G^{\circ}_{solv}$ corresponding to the free energy required to transfer the ion from a neutral droplet of the same size to the gas phase

$$\Delta G^{\#} = \Delta G^{\#}(E) - \Delta G^{\circ}_{solv} \qquad [5.7]$$

Individual properties of specific ions and solvents are included in k_I mainly via the ΔG°_{solv} term. The preceding account does not do justice to the sophistication of the ion evaporation theory or its recent experimental tests (Gamero-Castaño 2000), but will suffice for present purposes.

In addition to the summary and assessment of theories of all the various aspects of ESI (Kebarle 2000), an account of their practical implications (Cech 2001) is available; the following discussion draws in part on this expert review and emphasizes aspects of importance for trace quantitative analysis rather than those that control ESI–MS of proteins and other biopolymers. A sketch illustrating some of the processes occurring within a

charged microdroplet that compete with analyte ion evaporation, and lead to several limitations of ESI, is shown in Figure 5.31. One of these limitations concerns the linear dynamic range, as illustrated by the log–log calibration curve shown in Figure 5.32.

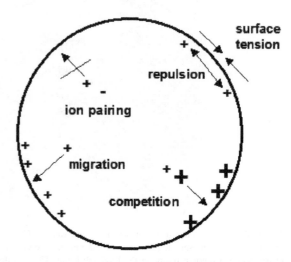

Figure 5.31 Sketch illustrating processes within a positively charged microdroplet sufficiently small that ion evaporation is possible. The small crosses represent analyte cations, and the large crosses are other cations that compete with the analyte for positions on the droplet surface.

At the low end of the concentration range, positive deviations from the expected linear relationship with slope of unity is the result of spurious additional signal intensity arising from chemical background. Some of this background occurs at well defined m/z values and can be assigned to cluster ions formed from solvent, mobile phase buffer components etc. This type of interference can be minimized by heating the ESI source or by appropriate adjustment of the cone potential to break down these clusters to very small ions. However, close examination of any ESI mass spectrum reveals some random background ('chemical noise') at virtually every m/z value. The sources of chemical noise in ESI–MS have been a matter of debate (Cech 2001), although recent work (Guo 2006) has identified the main signals as corresponding to either ions of contaminants or their degradation fragments, or cluster-related ions. Significant contributions were demonstrated to arise from contamination, either airborne or from tubing and/or solvents, from plasticizer additives (phthalates, phenyl phosphates, sebacates and adipates etc.) and silicones. These contaminant ions also

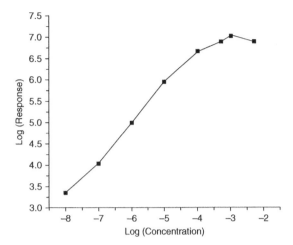

Figure 5.32 Typical ESI calibration curve generated for a concentration range covering six orders of magnitude. The analyte was the surface active trimethyldecylammonium cation, dissolved in 50:50 water:methanol containing 0.5 % acetic acid. If response (observed signal intensity) is directly proportional to concentration, such a log–log plot should have a slope of 1.0, and this is indeed observed over the concentration range $10^{-7} - 10^{-5}$ mol.L^{-1}. Positive deviations from linearity at the low end are attributed to chemical background interference, and negative deviations at the high end to saturation in ESI response. Reproduced from Cech and Enke, *Mass Spectrom. Revs.* **20**, 362 (2001), with permission of John Wiley & Sons, Ltd.

serve as nuclei for clustering of HPLC solvent or additive molecules, e.g., water and acetic acid. This study (Guo 2006) accounts for the persistence of at least some chemical background, in both ESI and APCI spectra, even under fairly strong declustering conditions.

From the viewpoint of this book, this phenomenon results in a limitation on the limit of detection (LOD) and lower limit of quantitation (LLOQ) (see Section 8.4 and Figure 8.10). The current best way to overcome this limitation is to use tandem mass spectrometry (MS/MS) for more selective detection (Section 6.2.2), since the ion(s) contributing to the chemical noise at the m/z value of the analyte ion are unlikely to give rise to fragment ions at the same m/z as those derived from the analyte. Thus, although the absolute signal intensity always is less for MS/MS detection, the chemical noise generally decreases by a much larger factor (see Figure (6.1), leading to an overall increase in S/B values and thus to improved (lower) LOD values and to extension of the linear dynamic range to lower values.

The other approach to minimizing the effect of chemical background is to increase the sensitivity, i.e., the slope of the (nonlogarithmic) plot of the response vs

concentration calibration curve. A significant increase in the slope would raise the analyte signal well above the chemical background at correspondingly lower concentrations (Figure 8.10), but this approach is easier said than done even for pure analytes in clean solutions. The problem of ionization suppression and enhancement in real-life extracts of complex matrices is discussed in Section 5.3.6a.

Extension at the high end of the linear dynamic range (Figure 5.32) is highly desirable for quantitative analysis with ESI–MS, but the reason(s) for a limitation in the ESI response at high concentrations is(are) still not fully understood (Cech 2001). This limitation is most likely due to a combination of instrumental factors and some fundamental limitation in the ability to produce a charged analyte in the ESI process. One such fundamental limitation could be that applicable in principle to all ionization techniques, a limited capability to provide a charge to all the molecules presented to the ion source (charge saturation). In the case of ESI this amounts to recognition of the maximum achievable excess charge available on a microdroplet at the ion evaporation limit. However, experimental evidence suggests (Cech 2001) that limited charge considerations can not account quantitatively for the ESI saturation effect observed at higher concentrations (Figure 5.32). Another fundamental parameter that could account, at least in part, for the saturation effect is the available area of droplet surface from which ion evaporation can occur.

Whatever the reasons might be (note that this discussion involves the case of pure analytes in clean solvents), no experimental method of extending the linear dynamic range for ESI to concentrations $> 10^{-5}$ M or so has been discovered. It is worth noting that APCI typically provides a linear dynamic range extending to appreciably higher concentrations than ESI. In a properly operated APCI source the analyte and co-eluting volatile components are completely vaporized prior to ionization (so no problems arising from limited surface area of droplets arise), and the quantities of APCI reagent ions are generally in large excess relative to the quantities of volatile sample derived components (Bruins 1999) so that the fundamental limitation of limited ionizing power is not reached in practice.

5.3.6a Ionization Suppression/Enhancement: Matrix Effects

For real-world extracts of complex matrices, the situation is even more complicated and the phenomenon of ionization suppression or enhancement (i.e. matrix effects, Sections 5.1.1 and 9.6) becomes extremely important for ESI. Detailed accounts of this topic have been published

(Mei 2004; Jessome 2006). Ion suppression (or occasionally enhancement) is a form of matrix effect that negatively affects LC–MS, regardless of the type of the mass spectrometer analyzer (Section 6.4), with respect to LOD, LLOQ, precision, accuracy etc. The importance of this phenomenon in the validation of analytical methods is emphasized by the extensive discussion in the *Guidance for Industry on Bioanalytical Method Validation* (FDA 2001). Figure 5.31 illustrates some of the possible causes of ionization suppression; these include competition between analyte and other matrix components for either the total available charge or the available surface area of the droplet, as discussed above with respect to the dynamic range.

Another possible mechanism involves increases in the surface tension of the droplets as a result of high concentrations of co-eluting compounds (Mallet 2004). This effect would reduce the rate of solvent evaporation, and thus also the probability that the droplet will reach a sufficiently small size that ion evaporation can occur, and also increase the difficulty of ion evaporation itself via the ΔG°_{solv} term in the rate constant, see Equations [5.6–5.7]. Another effect that seems to be particularly important for biological sample extracts arises from the presence of involatile solutes, believed (King 2000) to cause ionization suppression not only via a strong effect on surface tension but also via co-precipitation of analyte as the droplets shrink in size.

Ion pairing of analyte ions with mobile phase additives (Figure 5.31) is a well established cause of ionization suppression; this effect arises most often in LC–ESI–MS experiments when a strong ion pairing agent like trifluoroacetic acid (TFA) is used to improve chromatographic efficiency. If the additive anion (or cation in the case of negative ion operation) forms ion pairs with the analyte cations (or anions), the analyte ions become effectively neutralized and are insensitive to the electrical field at the droplet surface that leads to ion evaporation. Buffers composed of weak acids and bases (acetic or formic acid and/or their ammonium salts), that are volatile and therefore can be pumped away from the API source, are commonly used at fairly low concentrations for LC–MS. Strong acids like TFA (pK$_a$ 0.23) are notoriously effective in suppressing ESI of positively charged analytes, an effect attributed to their ion pairing properties; this hypothesis is supported by the fact that weak acids like acetic acid (pK$_a$ 4.72) are much less effective in ionization suppression, and by the finding (Kuhlmann 1995) that strongly basic analytes that were most readily protonated in positive mode ESI, and were thus expected to pair most strongly with TFA anions, were strongly suppressed while weakly basic analytes were hardly

affected. Finally, analyte ions that have successfully evaporated from the droplets can be neutralized (via deprotonation in the case of cations) by gas phase reactions with highly basic compounds (Amad 2000) and are thus not detected by the mass spectrometer. For any given matrix–analyte combination it is likely that some or all of these effects operate to suppress the ionization efficiency of the analyte.

However, methods have been developed to test for the presence of ionization suppression or enhancement of an analyte extracted from a complex matrix, to take its effects into account and thus hopefully alleviate the problem or even eliminate it altogether. An early systematic approach (Buhrman 1996), directed at detection and quantification of ionization suppression and other parameters, was later modified (Matuszewski 2003) to take account of the possibility of ionization enhancement, and this modification is adopted here. The approach compares the results obtained for the mass spectrometric responses (peak areas) in three different LC/MS experiments for a target analyte, using the same known amount of analytical standard at the same final concentration in each case:

(a) pure analyte dissolved in clean solvent, peak area *A*;
(b) pure analyte dissolved in extract of blank matrix after extraction and clean-up, peak area *B*;
(c) pure analyte added to blank matrix before extraction and clean-up, peak area *C*.

Then the following figures of merit are defined as:

Matrix Effect: $ME\,(\%) = (B/A) \times 100$
Extraction Efficiency: $RE\,(\%) = (C/B) \times 100$
Process Efficiency: $PE\,(\%) = (C/A) \times 100$
$$= (ME \times RE)/100$$

The generalized matrix effect can take account of ionization suppression ($ME < 100$) or enhancement ($ME > 100$). If the extraction and clean-up procedures were 100 % efficient then *C* and *B* would be equal regardless of any matrix effect since both are susceptible to exactly the same effects; clearly if *RE* is significantly > 100, something is wrong with the experiment. The process efficiency *PE* measures the net effect of extraction efficiency plus matrix effect on the final response of analyte extracted from the matrix.

An important distinction was drawn (Matuszewski 2003) between an absolute matrix effect, defined as *ME* above, and a relative matrix effect that refers to the comparison of *ME* values determined using different sources (batches) of blank matrix (e.g., biofluids such as plasma or urine for bioanalysis, or environmental matrices such as soil or water). Experience has shown that the matrix derived co-eluting interferences that are responsible for *ME* values significantly different from 100 % can vary from batch to batch of blank matrix. Thus, if calibration of the analytical method uses matrix-matched calibrators obtained from a single batch of blank matrix, the *ME* value for the batch used for calibration may differ significantly from those relevant to the actual unknown samples to be analyzed, i.e., a significant relative matrix effect may be operating. More recently (Matuszewski 2006) it was shown that the precision of the slopes of calibration curves, expressed as the coefficient of variation (CV %, Equation [8.4]) determined using LC–ESI–MS and different lots of a blank matrix (biofluid in this example), can serve as a good indicator of a relative matrix effect. For cases in which an isotope-labeled internal standard was not used, it was suggested (Matuszewski 2006) that this precision of slopes should be $\leq$ 3–4 % for the method to be considered free from the relative matrix effect. However, when an isotope-labeled SIS was used it was found that the CV of calibration curve slopes in five different lots of a biofluid was $\leq$ 2.4 % for both ESI and APCI, and this was taken as an indication that (at least for the examples studied) the use of such an SIS reduced the relative matrix effect to negligible values.

It was originally recommended (Matuszewski 2003) that five different batches of blank matrix (biological fluid in their case) should be tested for *ME* and *RE* values in order to assure reliable validation of a bioanalytical method. In the later work (Matuszewski 2006) it was further proposed that assay precision and accuracy should be determined in five different lots of a blank matrix (e.g., a biofluid) instead of repeat ($n = 5$) analyses using a single lot of blank matrix. Then if an isotope-labeled SIS is not available, determination of the slopes of the five calibration curves and their precision, together with use of <3–4 % (CV %) as a criterion, provides a useful guideline for assessment of the presence or otherwise of a significant relative matrix effect. Of course such a procedure is not always possible, especially for environmental samples; a recent example of this problem (Stüber 2004) involved environmental samples (river water).

In general for the most reliable quantitative methods it is necessary to eliminate matrix effects as far as possible; the practicalities are addressed in Sections 9.6 and 10.4.1d. The most obvious method is to improve the extraction and clean-up procedure to the point where the interfering compounds are reduced to insignificant levels. This is never a bad idea, but the associated increase in analysis time and related decrease in sample throughput can be a serious disadvantage depending on circumstances, so alternative and complementary approaches have been developed that can be fit for purpose.

One approach is to change the chromatographic conditions so that the matrix components that are the source of the suppression/enhancement no longer co-elute with the analyte. To achieve this, the most widely used approach (Bonfiglio 1999) involves continuous monitoring of the MS signal specific for the analyte that is continuously infused via a T-connection inserted between the HPLC and MS (Figure 5.33). Extracts of a blank matrix sample (plasma in the original experiments) are injected on-column, so that deviations of the analyte-specific signal (MS/MS in these experiments) from its pre-injection level are attributed to matrix effects, i.e., either ionization suppression or enhancement by extracted components. Figure 5.33 shows examples for three different analytes for injections of the same blank plasma that had been treated only by protein precipitation by acetonitrile addition (Section 4.3.1f); each of the infusion chromatograms thus obtained is compared with that obtained by on-column injection of pure mobile phase. In each case a strong ionization suppression effect is observed.

This original work (Bonfiglio 1999) was undertaken to determine which extraction/clean-up procedure yielded the smallest matrix effect; the results obtained using six different methods for phenacitin as test analyte are shown in Figure 5.34. In this case simple liquid–liquid extraction gave good results (panel (c), Figure 5.34), so there was no need to change the chromatographic conditions (these experiments actually used isocratic elution) in order to chromatographically separate the analyte from the interfering compounds. If gradient elution is used, the two major 'hot spots' for interferences are generally the solvent front where unretained compounds elute and the end of the gradient where strongly retained compounds are by design washed from the column, so it is a good rule-of-thumb that the gradient should ensure that the analyte elutes in the middle of the gradient. A change of column is a fairly drastic step but switching MS polarity (e.g., from positive ions to negative) can be successful if the analyte gives good response in both modes, since many fewer compounds give a strong response in negative mode compared with positive.

A modification of the post-column infusion method was used (Mallet 2004) to evaluate the ion suppression/enhancement effects on 16 different pharmaceutical compounds covering a range of acidity vs basicity. A range of different SPE extracts of a blank plasma sample was compared with those prepared by simple protein precipitation by acetonitrile; in addition, the effects of a range of mobile phase additives were also evaluated. The modification consisted of doing away with HPLC separation and using the HPLC pumping system to deliver a constant flow of a solution containing all of

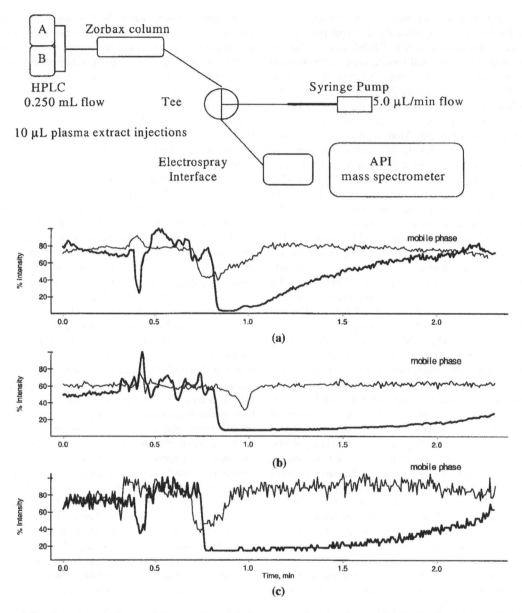

Figure 5.33 *Top*: schematic diagram of the post-column infusion apparatus introduced to establish the time dependence of ionization suppression and other matrix effects in LC/MS analyses. Extracts of blank matrix or of pure mobile phase are injected into the HPLC and a standard solution of pure analyte is infused post-column while the mass spectrometer is set to continuously monitor the signal specific for the analyte.

Bottom: Effect of on-column injections of blank plasma treated by protein precipitation only, with post-column infusion of analytes. The three panels are for (a) a proprietary Merck compound, (b) phenacetin, (c) caffeine. In each case the upper chromatogram is for an injection of clean mobile phase and the lower chromatogram is for an injection of blank plasma after protein precipitation. Reproduced from Bonfiglio, *Rapid Commun. Mass Spectrom.* **13**, 1175 (1999), with permission of John Wiley & Sons, Ltd.

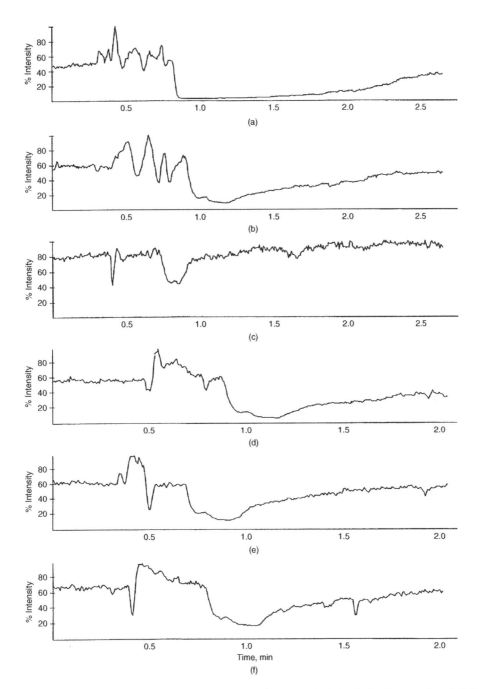

Figure 5.34 Infusion chromatograms covering the LC/MS assay time (2.5 minutes), obtained using the post-column infusion method shown in Figure 5.33, comparing the ability of different sample preparation methods to remove endogenous sample matrix components that interfere with the ionization of phenacetin. Panels (a)–(f) show the variation with time of the MS signal specific for the infused standard (phenacitin) following on-column injection of 10 μL of a blank plasma sample prepared by one of the tested sample preparation methods. (a) Protein precipitation. (b) Oasis SPE. (c) Methyl-*tert*butyl ether (MTBE) liquid-liquid extraction. (d) Empore C2 disk SPE. (e) Empore C8 disk SPE. (f) Empore C18 disk SPE. Reproduced from Bonfiglio, *Rapid Commun. Mass Spectrom.* **13**, 1175 (1999), with permission of John Wiley & Sons, Ltd.

either the acidic or basic test analytes; the post-column infusion pump was used to introduce the blank plasma extract or solution of mobile phase additive at a constant rate. The mass spectrometer was operated so as to record full mass spectra covering the range required to detect all the ionized test compounds, and absolute responses were recorded. This arrangement permitted determination of ion suppression or enhancement arising from a wide variety of SPE plasma extracts and HPLC additives, for a useful range of compound types within the pharmaceutical area of interest, and provides a useful guide and source of information for method design.

A review of the nature of matrix effects in LC–MS analyses of pesticide residues in various matrices, and of methods used to alleviate or remove them altogether (Niessen 2006), also discusses these problems in the context of bioanalytical work. More recently, the concept of 'ruggedness' for API–LC/MS methods was introduced (Heller 2007); this concept may be defined as the absence of significant variation in analytical results as a function of the amount of co-injected matrix. Thus a nonrugged API–LC/MS method may give consistent results only if a fixed amount of a fixed matrix is co-injected on a specific instrument. While such a nonrugged method may be adequate for a limited series of analyses, it can not be extended to more general cases without further testing and possible modification. Extensive experiments on LC–ESI–MS analysis of a new candidate veterinary drug in animal organ tissues (Heller 2007), using a surrogate internal standard that differed from the analyte by one methyl group and eluted very close to the analyte, indicated not only a marked effect of matrix:analyte ratio but also notable differences between mass spectrometer designs with respect to their susceptibility to suppression or enhancement effects. Graphical presentations of the experimental data, effectively 'matrix effect maps', help in visualization of the impact of various experimental parameters on these matrix effects. This approach differs significantly from those proposed previously (e.g., Bonfiglio 1999; Mallet 2004), which evaluated the behavior of a given method with a fixed amount of co-injected matrix, and will be important in development of rugged methods for trace analyses conducted for regulatory purposes such as those that motivated this work (Heller 2007).

Even if the extent of ionization suppression/enhancement is appreciable but considered fit for purpose, e.g., in order to satisfy a requirement for high throughput, it is possible to obtain robust and reliable data if care is taken to employ an appropriate calibration procedure. The most reliable method is the Method of Standard Additions (Section 8.5.1b) in which different known amounts of pure (unlabeled) analytical standard are added to fixed-mass aliquots of the sample to be analyzed before extraction, clean-up and LC/MS analysis; a simple extrapolation procedure then yields the original analyte concentration. However, this method requires ample supplies of all of sample, analytical standard and time, and this is seldom the case. Use of isotope-labeled surrogate internal standards that co-elute with the analyte (^{13}C and ^{15}N labels are preferable to ^{2}H as the latter can lead to significant variations in retention times and thus also in degrees of matrix effects, see Section 2.2.3) can usually provide reliable data provided that appropriate precautions are taken. The mutual interference of analyte and co-eluting SIS was already mentioned with respect to Figure 5.20 and a more complete discussion of this problem (Liang 2003; Sojo 2003) is necessary at this point.

The nature of the problem is well exemplified by the data (Sojo 2003) shown in Figure 5.35. The signal from the fixed quantity of SIS (D_6-labeled analyte) is clearly suppressed as the concentration of analyte increases and subsequently recovers as the latter decreases back to zero; moreover, this effect is much less evident when the flow rate was reduced from 600 to 60 μL.min^{-1}, an example of a more general effect discussed in Section 5.3.6b. Although the converse effect (suppression of analyte signal by the internal standard) is not known from these experiments as the actual analyte concentrations are not known *a priori*, presumably this also occurred. This conjecture is supported by the data shown in Figure 5.36 for a different analyte; the raw data acquired for calibration purposes again clearly show suppression of the signal for the fixed quantity of the D_4-labeled internal standard, while the presence of the converse effect is suggested by the highly nonlinear data for the analyte itself. However, when the calibration curve is plotted in the usual way as peak area ratio vs concentration ratio, a linear calibration plot is obtained (Figure 5.36). This suggests that in this case the mutual suppression effects of analyte and its D_4-SIS are equal and that the phenomenon does not affect the validity of the quantitative analysis, although it may adversely affect limits of detection and quantification as a result of the depressed signal levels. It is instructive to examine this phenomenological conclusion in terms of a simple theory (Tang 1993; Kebarle 2000) of competition for available excess charge at the droplet surface.

The basic assumption of this approach is that ion evaporation occurs from the surface layer of the droplet and that the concentrations of the various species in the surface later are related to those in the bulk of the droplet by a simple equilibrium distribution coefficient:

$$[A^+] = K_{si}^{\;A}.C_i^{\;A} \qquad\qquad [5.8]$$

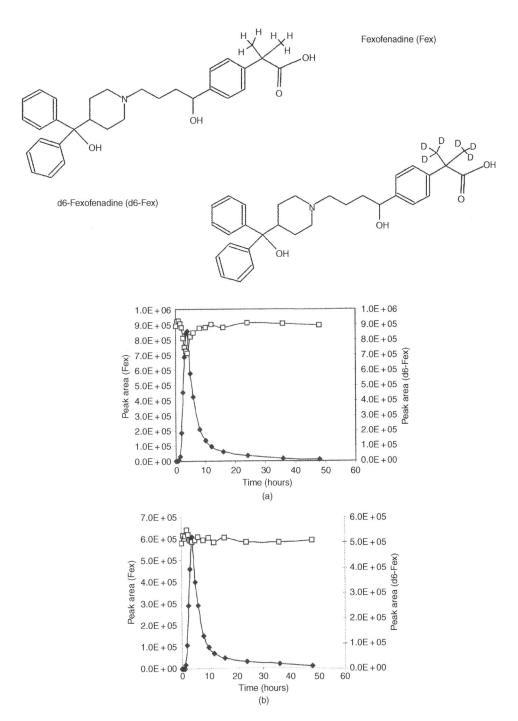

Figure 5.35 Pharmacokinetic curves (LC–MS peak areas) for fexofenadine in human plasma analysed by LC–ESI–MS at (a) $0.6 \, \mathrm{mL.min^{-1}}$ and (b) $60 \, \mathrm{\mu L.min^{-1}}$. Internal standard ($D_6$-analogue) response (for constant concentration) shown as open symbols, and analyte as closed symbols. Note that the suppression of the signal for the internal standard at the maximum concentration of analyte is much reduced at the low LC flow rate. Reproduced from Sojo, *Analyst* **128**, 51 (2003), by permission of the Royal Society for Chemistry.

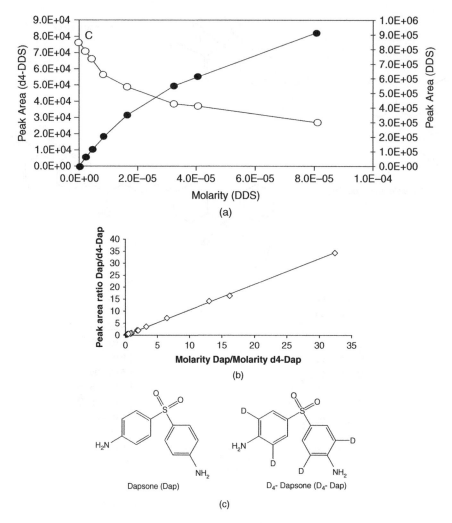

Figure 5.36 (a): LC–ESI–MS response curves for dapsone (closed symbols) as a function of concentration, in the presence of dapsone-D_4 (internal standard, open symbols) present at fixed concentration. (b): Calibration curve for dapsone plotted as peak area ratio relative to that of the D_4-analog (present at fixed concentration) vs the concentration ratio. Note that the slope is very close to 1.0. (c) Structures of dapsone and of D_4-dapsone. Reproduced from Sojo, *Analyst* **128**, 51 (2003), by permission of the Royal Society for Chemistry.

where $[A^+]$ is the surface concentration of ionized analyte A (assumed to be positively charged for convenience) and C_i^A is the concentration of A^+ in the interior of the droplet. Similar equations can be written for a co-eluting internal standard B, and for a composite species E representing mobile phase additives, co-eluting matrix components etc. If I is the total electrospray current leaving the electrospray needle (that is readily measured, see Figure 5.29), f is the fraction of the total charge on the droplets that is converted to gaseous ions of all kinds, and p is the sampling efficiency of the mass spectrometer for the gaseous ions

formed in the API source, then the ion current I_{A+} carried by the analyte ions into the mass spectrometer is given by the appropriate fraction of $(I.p.f)$:

$$I_{A+} = (I.p.f).\{k_I^A.[A^+]\}/\{k_I^A.[A^+]+k_I^B.[B^+]$$
$$+k_I^E.[E^+]\} \qquad [5.9]$$

where the k_I factors are the ion evaporation rate constants (Equations [5.6 and 5.7]). It is now possible to use Equation [5.8] and its analogous versions for B and E to replace the surface layer concentrations with those in the droplet interior:

$$I_{A+} = (I.p.f).\{k_I^A.K_{si}^A.C_i^A\}/\{k_I^A.K_{si}^A.C_i^A + k_I^B.K_{si}^B.C_i^B$$
$$+ k_I^E.K_{si}^E.C_i^E\}$$
$$= (I.p.f).\{k_I^A.K_{si}^A.C^A\}/\{k_I^A.K_{si}^A.C^A + k_I^B.K_{si}^B.C^B$$
$$+ k_I^E.K_{si}^E.C^E\} \qquad [5.10]$$

where C^A etc. (without subscripts) represent the bulk solution (pre-spraying) concentrations and the second form of Equation [5.10] is obtained from the first since $C_i^A = C^A.K_{evap}$, where K_{evap} is the ratio of the volume of the newly formed electrosprayed droplet to that of the same droplet after evaporation and fission to the final droplet at the ion evaporation limit (Figure 5.30); of course this ratio is the same for all solutes and so cancels in Equation [5.10].

An exactly analogous equation can be derived for I_{B+}, so the ratio of the detected ion currents (i.e., signal strengths) for analyte and internal standard is:

$$I_{A+}/I_{B+} = (k_I^A/k_I^B).(K_{si}^A/K_{si}^B).(C^A/C^B) \qquad [5.11]$$

This is the desired relationship between the ratio of MS responses and the bulk solution concentration ratio. Its derivation did involve several assumptions, e.g., that the factor p is the same for A and B, and more fundamentally that the competition between solutes is mainly for the available excess charge and not for space at the droplet surface. In the special case that B is sufficiently similar to A that their k_I values are equal, and the same is true of the K_{si} values, Equation [5.11] predicts that such a calibration curve should have a slope of 1.0, and indeed this is very nearly so in the example of mutual suppression of an analyte and its deuterated internal standard shown in Figure 5.36. The internal standard B that would most closely resemble the analyte A in this sense, and still be distinguishable by MS, is a heavy atom (^{13}C, ^{15}N, ^{18}O) labeled isotopolog, as such internal standards are much less susceptible to kinetic and/or equilibrium isotope effects than are D-labeled versions (see the text box on this subject in Chapter 2). However, when the fraction of H/D labeling is not large, as in the example shown in Figure 5.36, the effects are not large if the H/D atom is not directly involved in a chemical reaction and also the molecular volume effect on retention times is not significant. Other internal standards B that are not isotopologs of the analyte, even if they could be arranged to co-elute with the analyte (so that the nature and concentration of the E components and their k_I and K_{si} values are the same for analyte and internal standard and thus cancel when

the ratio of the two forms of Equation [5.10] is taken to give Equation [5.11]), are likely to have values of k_I that differ from that for A (via ΔG_{solv}°, Equation [5.7]). Similarly the K_{si} values (Equation [5.8]) are likely to differ significantly and indeed it is well known that more surface active analytes (e.g. detergent molecules with a charged head group and a hydrophobic tail that prefers to stick out from the surface of an aqueous droplet) give much higher ESI responses than most other analytes.

To summarize the situation with regard to matrix effects (suppression or enhancement of ionization efficiency by co-eluting components), it is always advisable to be aware of their importance in any proposed analytical method and to minimize them as far as possible. Use of an appropriate isotope-labeled standard in conjunction with matrix matched calibrators should give reliable results if all other appropriate precautions are taken (see Section 8.5.2), but it is important to investigate the possibility of relative matrix effects (Matuszewski 2003) if it is suspected that the matrix in the analytical sample might differ appreciably from that used to make the matrix matched calibrators.

An excellent example of this problem concerns an SPE–LC–ESI–MS/MS method for determination of mevalonic acid ($CH_2OH–CH_2–C(OH)(CH_3)–CH_2–COOH$, an intermediate in the biosynthesis of cholesterol) in human plasma and urine (Jemal 2003); a deuterated version of the analyte was used as the SIS. Extracts of some batches of urine showed severe ionization suppression (matrix effect) when a larger sample volume was extracted for analysis. Under some conditions, the analyte:SIS response ratio changed from the expected value, indicating that the responses of the analyte and the SIS were not affected to the same extent by the matrix effect in this case. This is an example where use of a stable isotope labeled SIS did not turn out to be a panacea for the problems with accuracy and precision resulting from matrix effects. It was recommended (Jemal 2003) that this unexpected kind of relative matrix effect should be checked for during method development, again by using several batches of the biological matrix (Matuszewski 2003).

Of course, if a suitable isotope-labeled internal standard is not available so that other approaches must be used, it is best to reduce any matrix effects as much as possible at the probable expense of a more complicated and lengthy clean-up procedure. A structural homolog, or close structural analog of the analyte, can be arranged to elute closely and, provided that the chemical differences are not too large, the k_I and K_{si} values may not be very different from those of the analyte and the slope of the calibration plot will be close to 1.0. An interesting alternative approach when an isotope-labeled internal standard is not available was described (Alder 2004) as the 'ECHO' technique. In this method two injections are made separated by about 50 seconds; the first is a solution of the (unlabeled) calibration standard in mobile phase and this is closely followed by the second – the extract of the unknown sample. In this way the unknown analyte peak elutes very close to, but chromatographically slightly resolved from, the 'echo peak' corresponding to the pure standard; an example is shown in Figure 5.37. The time delay of 50 seconds was chosen as a compromise between the need to resolve the two peaks and the desire to expose both injections to matrix components that are as similar as possible, so that they are affected somewhat equally by ionization suppression (or enhancement). The expected compensation for matrix effects on the analyte was generally observed. However, when both injections were pure standard in solvent, i.e., no co-eluting matrix components, the expected peak area ratio differed significantly from the expected value (1.0), at least in part because of imperfect resolution so that the first (echo) peak contributed signal to the second when the integration algorithm was applied. This is an intrinsic problem with this otherwise clever approach, since steps taken to improve the resolution will inevitably lead to differences between the two peaks with respect to the matrix effects to which they are subjected.

As a final comment in this section, it is worthwhile mentioning that ionization suppression effects are generally known to become progressively less important as the solution flow rate is decreased (e.g. Figure 5.35).

Discussion of this phenomenon is more conveniently postponed until the next section, but this observation provides another weapon in the analyst's war against matrix effects (ionization suppression or enhancement).

5.3.6b ESI–MS: Concentration or Mass Flow Dependent?

A frequent comment in the LC–MS literature, to the effect that ESI–MS is a concentration dependent detector, clearly can not be true in the same sense as for a UV detector. As discussed in Section 4.4.8 and Appendix 4.1, an essential feature of a concentration dependent detector is that the signal strength (corresponding to peak *height* in LC–MS experiments) for an analyte of fixed concentration is independent of flow rate of analyte solution and, in particular, remains constant when the flow rate is reduced to zero (see below for the corresponding predictions for peak areas). These criteria are fulfilled by a UV absorption detector, but if the flow rate is reduced to zero in an LC–ESI–MS experiment the signal is also zero, since an ESI source is, like an EI source, a destructive detector in which analyte is continuously removed simultaneously with its introduction from the HPLC (Appendix 4.1). Nonetheless, the common belief that LC–ESI–MS response exhibits concentration dependence must obviously have some validity and indeed an early thorough investigation of the question (Hopfgartner 1993) established the point. For example, it was demonstrated that post-column splitting of the HPLC effluent to the ESI–MS did not affect the peak heights, demonstrating that, in the flow rate regime and other conditions used, the ESI–MS peak height response was independent of the

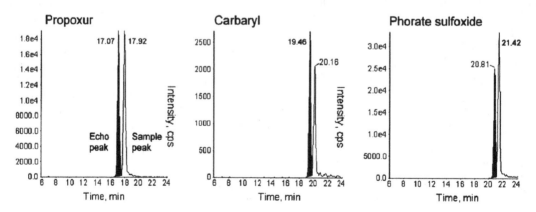

Figure 5.37 Examples of LC–MS/MS multiresidue analysis for pesticides spiked into lemon at 0.1 mg.kg^{-1}, using the ECHO technique. The first (black) peaks are the 'echo' peaks corresponding to injection of calibration standards in solvent at 0.1 µg.mL^{-1}, and the second peaks correspond to the injection of the extract of spiked lemon. Reproduced from Alder, *J. Chromatogr. A* **1058**, 67 (2004), copyright (2004), with permission from Elsevier.

mass flow rate for a fixed analyte concentration. Actually this was established very early (Whitehouse 1985) in the development of ESI (see Figure 5.24).

However, things are not always so clearcut, as exemplified by Figure 5.38 for experiments in which peak *areas* were measured using flow injection (no HPLC column) of fixed volumes of a solution of diltiazem. For a true concentration dependent detector the peak areas should be inversely proportional to the flow rate (Equation [4.10]) and this is approximately true over a limited low flow rate range (Figure 5.38a), but at higher flow rates the peak areas fall increasingly below the linear extrapolation. This apparent paradox is examined in the following.

It will be convenient to first consider the ESI–MS response as a function of flow rate for constant infusion of an analyte solution, i.e., with no time dependence of the rate of analyte introduction within any given experiment such as that arising as a result of the chromatographic peak shape. Under such circumstances the concept of peak area does not arise and only signal intensity (analogous to peak height) is relevant. There has been considerable interest in the development of low flow rate ESI–MS since the demonstration (Wilm 1994, 1996) that decreasing flow rates could provide significantly higher signal strengths (equivalent to peak heights in LC–MS experiments); ESI at these low flow rates ($nL.min^{-1}$ instead of $\mu L.min^{-1}$) was achieved by using an ESI needle with a tip diameter of the order of $1\,\mu m$, with no external forced pumping. This approach (nano-electrospray, sometimes simply nanospray) results in much smaller initial charged droplets, estimated (Wilm 1996) as 180 nm compared with the micrometer range in conventional ESI. It has been emphasized (Juraschek 1999; Schmidt 2003) that, in view of some unique features, nanospray should be regarded as more than simply low flow ESI; in addition to the generally higher ionization efficiencies compared with forced flow ESI, nanospray significantly reduces interference effects from salts and other species present in the analyte.

These attributes of nanospray can be understood (Juraschek 1999; Schmidt 2003) in terms of the order of magnitude smaller charged droplets originally formed; as a consequence, the asymmetric droplet fission process (see preceding discussion of Figure 5.30) for nanospray starts off one generation later than for conventional ESI. Furthermore, the smaller initial size also implies a higher initial charge density on the droplet surface, leading to droplet fissions without the need for extensive solvent evaporation that leads to increased concentrations of analyte and of salts and other components in the sprayed solution. It was shown (Juraschek 1999) that this viewpoint is consistent with observations that

the qualitative nature of mass spectra obtained using conventional ESI (and ionspray) closely resembles that of nanospray spectra from analyte solutions that are similar to those used for conventional ESI, except that they contain salt concentrations about an order of magnitude greater, thus illustrating the relative insensitivity of nanospray to ionization suppression by other components in solution (Section 5.3.6a). Subsequently (Schmidt 2003), these experiments were extended by studying variations of ionization efficiency and other parameters with flow rate in the $nL.min^{-1}$ range. It was found that ionization suppression effects essentially vanished at very low flow rates ($< 20\,nL.min^{-1}$), even for compounds in the sprayed solution that were of strongly different physicochemical properties, but were observed to occur at flow rates as low as $50\,nL.min^{-1}$ with quite an abrupt onset when the flow rate was gradually increased (Schmidt 2003).

These findings implied that nanospray–MS responses might be effectively equal for all analytes at flow rates $< 20\,nL.min^{-1}$. At first sight it seems unlikely that these ultra-low flow rate phenomena could be of any direct practical consequence for real-life analytical methods of interest in this book, but an example where the equimolar ionization efficiencies for different compounds was exploited to assist with a problem in a practical analytical method is described later (Section 11.5.1e). It is also of interest that a specially designed device for post-column splitting of HPLC effluent ($200\,\mu L.min^{-1}$) which allowed the introduction of flow rates down to $100\,nL.min^{-1}$ to the ESI source, importantly with no degradation in HPLC performance, was shown to significantly reduce ionization suppression (Gangl 2001) as well as improve absolute response.

Other developments of the first nano-ESI experiments (Wilm 1994, 1996) have improved on the overall absolute ionization efficiency ε:

$$\varepsilon = (\text{number of ions detected per second})/$$

$$(\text{number of sample molecules consumed per second}) \quad [5.12]$$

Thus ε measures the product of the ionization efficiency (fraction of analyte molecules delivered as ions to the entrance from the API source into the MS) with the transmission efficiency of ions from the MS entrance aperture to the detector. The first experiments, conducted at flow rates of tens of $nL.min^{-1}$, gave ε values of about $1/1000$ (Wilm 1994), later improved to $1/390$ (Wilm 1996) using a quadrupole mass spectrometer operated with very high peak widths to improve transmission efficiency; the sampling efficiency

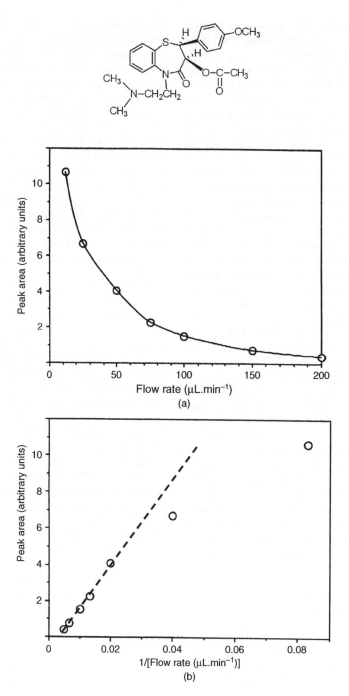

Figure 5.38 Electrospray ionization peak areas measured as a function of flow rate for a solution of diltiazem injected (no HPLC column) into a flow of 50:50 acetonitrile water with 1 % acetic acid. A true mass flow dependent detector is predicted (Equation [4.5], Appendix 4.1) to yield peak areas independent of flow rate but this is clearly not observed in (a). A concentration dependent detector should give peak areas proportional to flow rate^{-1}, and this is observed in (b) at lower flow rates but the peak area response falls below the extrapolated prediction as flow rate increases. Thus in this flow rate range the ESI mass spectrometer did not behave in accord with either of these idealized models.

was approximately 0.5–1 %. A very thorough investigation (Geromanos 2000) using much lower flow rates $(1–2 \, nL.min^{-1})$ achieved ε values of 1/20, corresponding to sampling efficiencies of 10–15 % with transmission efficiencies of $\sim 50 \%$. More recently (El-Faramawy 2005) very high sampling efficiencies (up to 33 %) were obtained for a quaternary ammonium compound (well known to have very high efficiencies in ESI as a result of the permanently charged nitrogen and the high surface activity conferred by the long hydrophobic tail). Another notable feature of this work was the observation of wide variations in the performance of different nano-ESI needles, all nominally identical from the same supplier.

Most recently (Schneider 2006) an experimental source was used to conduct studies under conditions of 'total solvent consumption', with pneumatically assisted nebulization to stabilize the ESI process, a heated laminar flow chamber to enhance desolvation and ion production, and various atmosphere-to-vacuum aperture diameters to maximize ion transfer. The motivation for these experiments was to investigate the proposal that the reason for the much lower ionization sampling efficiencies at higher flow rates $(\mu L.min^{-1}$ and above) is that the electrosprayed droplets are much larger in view of the much larger ESI needle tip diameters required to maintain flow rates in this regime, and thus are much less efficiently evaporated down to the Rayleigh and/or ion evaporation limits than the droplets formed from the $\sim 1 \, \mu m$ diameter tips used in nano-ESI (Juraschek 1999; Schmidt 2003).

However, it was also suspected that an increased degree of spray divergence with larger flow rates could also contribute to the lower efficiencies. As the flow rate is increased, the sprayer plume diverges into a wider region of the API source and an increased fraction of the spray escapes the viscous drag forces into the vacuum system. For this reason, a heated laminar flow chamber was used to transport the spray from the needle to the entrance orifice of the mass spectrometer (Figure 5.39) to improve both droplet evaporation and spray transport efficiency. Careful measurements of ε (Equation [5.12]) and of transmission losses through the MS/MS system to the detector allowed reliable values of the ionization sampling efficiency to be obtained under conditions where all droplets were evaporated (i.e., 'total solvent consumption') and where spray transport efficiency was very high.

Under these ideal conditions, favorable compounds (e.g., a long chain quaternary ammonium compound) showed ionization sampling efficiencies of 70–85 % at flow rates in the range $50–500 \, nL.min^{-1}$, although compounds with less favorable characteristics or solutions with high aqueous content gave lower efficiency values. However, all sampling efficiencies were far superior to

those obtained with conventional ESI sources operating at higher flow rates. Most significantly in the present context, in all cases for which measurements were made in the ideal 'total solvent consumption' flow regime in the range $50–500 \, nL.min^{-1}$, the ion count rate was always directly proportional to the absolute quantity of analyte molecules entering the source. That is to say, this nano-ESI source behaved strictly as a mass flow dependent detector! This conclusion related to absolute responses for a given analyte is related to, but not quite the same thing as, the conclusion (Schmidt 2003) that at very low flow rates $(< 20 \, nL.min^{-1})$ the response becomes essentially independent of the physico-chemical nature of different analytes with little or no mutual suppression of ionization efficiency. However, the two conclusions presumably both reflect the same underlying phenomena relevant to total droplet consumption; an example of practical application of this concept is discussed later (Section 11.5).

Thus, it seems probable (Schneider 2006) that losses arising from incomplete droplet evaporation and from incomplete spray transport efficiency can account for observations of concentration dependent rather than mass flow dependent response of conventional electrospray at higher flow rates; bear in mind that the claimed concentration dependence was based only on the observed dependence of ESI response on flow rate over a limited range. As the flow and thus mass flux into the source are increased at a fixed analyte concentration, the ionization sampling efficiency will drop as a result of these two loss mechanisms; if the two effects (increased mass flux and increased losses) approximately balance one another there will be no net signal increase. Thus, this observed concentration sensitive response of ESI–MS is probably an artifact of what is truly a mass flow sensitive system but with ionization sampling efficiencies that vary significantly under different flow conditions (Schmidt 2003; Schneider 2006). By capturing and desolvating the entire spray, the expected mass flow sensitive response that is also significantly higher than that obtained without these precautions should be expected. For such high efficiency operation, the nanosprayer must be located within the vacuum drag region to ensure capture of the entire spray and thus close to the heated interface (required for desolvation); unfortunately this could lead to boiling within the nanosprayer with its low flow rate, so a cooling nebulizer gas is required.

Thus far the discussion has involved mainly constant infusion experiments rather than LC/MS. In the latter case, flow rate is generally varied via the diameter of the HPLC column or by post-column splitting of the effluent before transport to the ESI sprayer. In the case of conventional ESI sources and mobile phase flow rates, the apparent concentration dependent response will apply

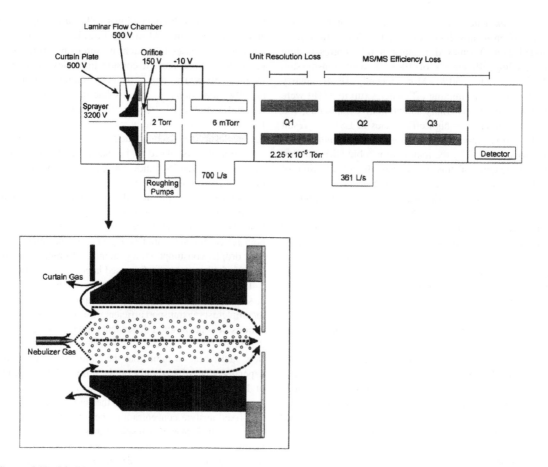

Figure 5.39 Modified nanoelectrospray source used to study ionization sampling efficiencies under conditions of 'total solvent consumption'. The inset shows the modified curtain gas and heated laminar flow particle discriminator interface with the nebulizer assisted nano-ESI arrangement. Reproduced from Schneider, *Rapid Commun. Mass Spectrom.* **20**, 1538 (2006), with permission of John Wiley & Sons, Ltd.

(Hopfgartner 1993). However, the situation in comparisons of different diameter columns is more complicated because the intrinsic effects of flow rate on ESI efficiency, discussed above, are convoluted with the increased analyte concentration by smaller diameter columns (a consequence of more narrow peak widths and thus decreased chromatographic dilution, see Equation [3.50] and associated discussion in Section 3.5.10b). However, the benefits thus obtained must be balanced against limitations on column loading (amount of analyte that can be injected on-column without overloading). In LC–ESI–MS experiments the best choice of HPLC column size (and thus flow rate, with or without post-column splitting) depends on the circumstances, including the design features of the ESI source itself.

Despite all the theoretical advantages of low flow rates, including large reductions in matrix effects, there are well known practical problems with nanosprayers. One problem was already mentioned, the wide variations in performance of sprayers that are nominally identical. Another common problem, which is probably related to the first, is their susceptibility to blockage of the micron-size tips. A novel approach to this problem (Lee 2005) involves photo-initiated fabrication of a small region containing a porous polymer monolith (PPM) at the end of a fused silica capillary (Figure 5.40). Such devices facilitate a stable electrospray over a wide range of flow rates ($10–1200\,\mathrm{nL.min^{-1}}$) with only a modest increase in back pressure caused by the PPM plug. The PPM assisted electrospray emitter compared well with a commercial conventional nanosprayer, in terms of sensitivity, stability and robustness. A PPM filled electrospray tip produced a day-to-day signal variation of 23 % (RSD) over a three-day period when spraying a $1.0\,\mathrm{\mu M}$ peptide

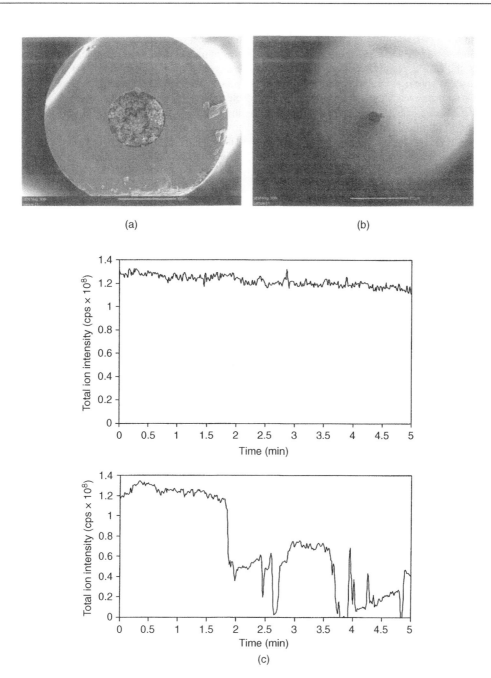

Figure 5.40 Scanning electron micrographs of (a) a PPM filled capillary (100 μm i.d., 360 μm o.d.) and (b) a commercial 360 mm o.d. nanospray capillary with a 30 μm tip opening. A comparison of the robustness of the two nanospray devices after four hours of use is provided by the total ion current traces recorded over a 5 minutes period, (c) for the PPM device and (d) for the conventional nanospray tip. Reproduced from Lee, *Rapid Commun. Mass Spectrom.* **19**, 2671 (2005), with permission of John Wiley & Sons, Ltd.

solution, and three different PPM capillaries produced a signal variation of 17 % RSD, indicating good fabrication reproducibility. The multiple flow paths through the pores of the PPM appear to provide multiply-redundant nanosprayers so that, even after accumulation of debris after prolonged use, a stable spray could still be generated with the PPM filled capillary while the commercial nanosprayer ceased to function properly. The PPM assisted electrospray showed superior ESI response at infusion flow rates between 100 and 1000 nL.min^{-1}, but the conventional nanospray tips performed slightly better at flow rates below 100 nL.min^{-1}. Moreover, the PPM also provided a means of sample purification by using the PPM as a stationary phase to desalt and pre-concentrate samples prior to mass spectrometric detection. The same approach has also been applied to fabrication of robust electrospray emitters for both glass (Koerner 2005a) and polymeric ((Bedair 2006) microfluidic chips.

Finally in this section, it is appropriate to summarize the practical implications of the foregoing discussion of flow rate dependence of ionization efficiency in ESI. With respect to use of peak height vs peak area in LC–MS analyses, in practice there is no theoretical reason to prefer one over the other as long as the same flow conditions are used for the analyses as for the calibration. The only meaningful criterion concerns which of the two measures provides the better accuracy and precision under the conditions used. However, it is necessary to be aware of the implications of the distinction between mass flow vs concentration dependence during method development, when flow rates are varied in order to search for optimum conditions. The marked improvement in ESI ionization efficiency at lower flow rates is another aspect that should be kept in mind but must be balanced against chromatographic run time (and thus throughput), permissible sample loading etc. The same physical phenomena that result in higher ESI efficiencies at low flow rates also account for the marked reduction in the importance of matrix effects. Nonetheless, it is unlikely that nanospray methods will be adopted for routine LC–MS analyses in the near future, although their use in obtaining a calibration when pure analytical standards are not available (Section 8.5.1e), e.g., drug metabolites, could prove to be useful.

5.3.7 Atmospheric Pressure Desorption Methods

As mentioned in Section 5.2.2, MALDI operated at atmospheric pressure (and thus requiring an API–MS interface) has received some attention (Laiko 2000; Moyer 2002) although no applications to trace quantitative analysis appear to have been published. An entirely different approach to analyte desorption/ionization from surfaces was initiated (Takats 2004) by the introduction of desorption electrospray ionization (DESI). In this approach an aqueous solvent is subjected to pneumatically assisted electrospray (essentially IonSpray™, Section 5.3.6) and the resulting ions and charged droplets are directed towards a solid surface containing the target analyte(s). These charged particles entrained in the stream of nebulizing gas act as primary projectiles and desorb ions from the surface as a result of electrostatic and pneumatic forces (Takats 2004). The gas phase ions desorbed from the surface are transferred to a mass spectrometer via an API–MS interface (Section 5.3.3a). This technique has proved to be capable of rapid (no sample preparation required) qualitative analysis of analytes on surface areas $\leq 2\,mm^2$, and a wide range of applications has appeared in the literature since its introduction. With respect to quantitative analysis, the original publication (Takats 2004) reported attempts to obtain quantitative results by using appropriate internal standards in experiments in which the analyte was deposited in known amounts on a target surface. However, when the internal standards were added to the surface by a spraying method, the results obtained by DESI were described (Takats 2004) as semi-quantitative (e.g., relative standard deviation ∼30 % for spiked plant tissue samples, no information provided on accuracy). It was noted (Takats 2004) that quantitation by any method is intrinsically difficult in the analysis of natural surfaces, and indeed it is difficult to find any mention of quantitation in the many publications describing applications of DESI.

The APCI version of DESI, labeled desorption atmospheric pressure chemical ionization (DAPCI), was described shortly afterwards (Takats 2005). In this approach analytes are desorbed from the surface to be analyzed by exposure to a heated stream of air and thus transported into the corona discharge region of a conventional APCI source. The stream of air can be doped with a suitable APCI reagent gas if desired. This approach has received much less attention than DESI, possibly because of concerns about thermal degradation of analytes. Note that the physical effects minimizing such degradation in the heated pneumatic nebulizer (Section 5.3.4) are not exploited in the DAPCI approach. No mention of quantitation was made (Takats 2005). An essentially identical apparatus, labeled an atmospheric pressure solids analysis probe (ASAP), was published practically simultaneously (McEwen 2005), but again no attempts at quantitation were reported.

A related but somewhat different method (Cody 2005), labeled direct analysis in real time (DART), uses a flow of chemically inert gas (helium, sometimes nitrogen) that is subjected to a high voltage discharge that creates electronically excited species together with ions and electrons. (In the case of nitrogen, vibronic excitation is probably involved.) It is believed that the electronic and/or vibronic excited state species (metastable helium atoms or nitrogen molecules) are the active ionizing agents, although protonated water clusters derived from trace moisture appear to be responsible for analyte protonation. DART was shown (Cody 2005) to be successfully applied to analysis of gases, liquids and solids but its main attraction is its ability to direct detection of chemicals on surfaces without requiring any sample preparation, such as solvent extraction or wiping on a swab. Again the main emphasis has been on qualitative identification of analytes. The only reference to quantitative analyses by DART (Cody 2005) concerned observation of relative concentrations of capsaicin (a compound producing the 'burning' sensation on eating a chili pepper) in different parts of a hot pepper.

A comparison of the relative abilities of DESI, DAPCI/ASAP and DART in qualitative analysis of a range of analytes on the surfaces of a range of matrices (Williams 2006) also investigated the advantages and disadvantages of adding solvent to the surface before exposure to the ionizing agent. Very recently (Venter 2007) DESI was made easier to implement, more robust, and the exposure of the analyst to the sample was reduced, by a modification in which a small pressure-tight enclosure with fixed spatial relationships between the sprayer, surface and sampling orifice (a heated capillary in this case) was used. A different recent variant of DESI (Cotte-Rodríguez 2007) describes the use of an ion transport tube to transport analyte ions from ambient surfaces located up to three meters from the mass spectrometer. Yet another approach (Ratcliffe 2007) uses plasma assisted desorption/ionization (PADI) that involves generating a nonthermal radio frequency driven atmospheric pressure plasma and directing it onto the surface of the analyte without charged particle extraction. However, none of these more recent innovations describe quantitative applications.

5.4 Source–Analyzer Interfaces Based on Ion Mobility

Both APCI and ESI produce low intensity background signals of the 'peak-at-every-mass' type (Guo 2006) that can limit the S/B value and thus ultimately limits of detection and quantitation when MS^1 monitoring is used (full scan or selected ion recording, SIR). One way to improve this situation is to use MS^2 monitoring (MRM, Section 6.2.2), since it is unlikely that the background

ions in the MS^1 spectrum will yield an abundant fragment ion at the m/z value(s) characteristic of the analyte specific precursor ion selected from the MS^1 spectrum. This works reasonably well for known target analytes since the precursor–fragment ion m/z relationships characteristic of the analyte can be pre-programmed into the mass spectrometer acquisition file. However, in environmental or toxicological analyses, for example, where the analytes of interest are not necessarily known ahead of time, initial identification (as opposed to quantitation) by automated MS/MS requires the LC–MS system to isolate the appropriate precursor ions and trigger the MS/MS acquisition. The effectiveness of this approach is highly dependent on the chemical noise (background), appearing as a 'peak at every mass' in ESI/APCI. Indeed the background signals can sometimes be larger than those of analytes of interest so the precursor selection algorithms that rely on relative intensities of signals can end up recording MS/MS of background cluster ions.

An alternative approach is to use post-acquisition data processing to filter out background interferences; various software packages have been described for this purpose (Muddiman 1995; Windig 1996; Andreev 2003). On-line software approaches (Ramsey 1993; Fleming 1999; LeBlanc (2004); Kohli 2005), which essentially detect and remove periodic background noise during acquisition, have also been described and are effective to some degree. However, none of these approaches is completely satisfactory and it would be useful if some way could be found to selectively remove the background ions before the mass spectrometric analysis. A somewhat similar problem is encountered in inductively coupled plasma (ICP) mass spectrometric analysis of elemental ions at low mass resolution, in which polyatomic isobaric interferences limit the performance for some elements; these interferences are removed through chemical reaction with gases such as oxygen, ammonia etc. (Eiden 1997; Tanner 1999) before introduction into the mass spectrometer. Development of an analogous chemical reaction interface for organic analytes, to selectively remove the various background ions produced by API sources (Guo 2006), would be highly advantageous.

Another recent approach has been to insert an ion mobility device between the ion source and mass spectrometer. The additional separation power added by the differences in ion transit times through a buffer gas can often distinguish between analyte ions and the background ions that originate in the ion source (and thus can not be chromatographically separated). This multidimensional approach is feasible because of the differences in timescales among the HPLC peaks (seconds), ion mobility (tens of milliseconds) and SIR or MRM dwell times (a few milliseconds per channel). Several devices based on this concept have been described (Hoaglund 1998; Henderson 1999; Merenbloom 2006; Pringle (2007); Loboda 2006) and are conveniently categorized by the gas pressures at which they operate; generally high pressure (∼1 atm) cells provide better drift time resolution, but low pressure cells (∼1 torr) are more conveniently incorporated into mass spectrometers operating at high vacuum. Recently (Loboda 2006) a novel low pressure mobility cell was described that achieved higher resolution by providing a counterflow of gas to counteract the force exerted by the axial electric field, so that ions become trapped at the front portion of the mobility cell and can only pass to the cell exit when the electric field force exceeds the drag force from the gas flow; this idea is somewhat analogous to the situation in CZE (Section 3.6) when the electro-osmotic flow opposes the ion motion induced by the electric field applied along the capillary. However, most of these ion mobility interfaces appear to have been designed specifically for applications limited thus far to peptides and proteins.

A rather different way of exploiting differences in ion mobilities through a buffer gas is embodied in so-called High Field Asymmetric Waveform Ion Mobility Spectrometry (FAIMS), in which the physical property that is exploited to separate ions is the variation of ion mobility with the strength of the applied field. The principles of the method (Buryakov 1993) were subsequently developed (Purves 1998; Guevremont 1999, 2004) into a device suited to reduction of chemical background in ESI mass spectra. The website www.faims.com is a source of information, references etc. In essence, a periodic electrical waveform (Figure 5.41(a)) is applied to conductive surfaces about 2 mm apart; this waveform must be asymmetric in the sense that there is a significant difference between the positive and negative peak voltages (each in the kV range, either may be the higher). Because the high field mobility is larger than the low field value (Figure 5.41(b)), an ion injected axially into such a device will be progressively moved to one of the plates and eventually be lost. By applying a DC 'compensation voltage' the ion trajectory through the device can be maintained near the central axis so that it is transmitted. Because different ions have different mobility-field curves (Figure 5.41(b)), the values of the compensation voltage required for transmission will differ and the device can be used as a filter for ions that have different discrepancies between high- and low field mobilities (Figure 5.41(c)). FAIMS devices have been successfully applied to discrimination among protein and peptide ions but also have been used to assist small molecule analyses. An example (Kapron 2006) is illustrated in Figure 5.42; it exemplifies

the dramatic reduction of chemical background and consequent loss of ambiguity in defining the LC peak for integration that can sometimes be achieved. Other examples of application of FAIMS–MS to trace analysis can be cited (Barnett 1999; Handy 2000; Ells 2000; McCooeye 2001, 2002, 2003; Kapron 2005, 2006a).

5.5 Summary of Key Concepts

Qualitative identification is important in quantitative analysis of target analytes because there is no point in assigning a concentration or amount to a component in a complex mixture if there is no confidence that the chromatographic peak in question does indeed contain the target analyte. Moreover, assurance is needed that the signal used for quantitation does not contain contributions from co-eluting compounds (so-called peak purity).

The variety of mass spectrometer ion sources that can readily be coupled to GC and LC allows **retention times to be correlated with the mass spectra** (MS and MS/MS), providing an unprecedented approach to requisite compound identification for quantitation.

The major question concerning methods for interfacing mass spectrometry on-line with chromatography is **how to selectively discard the mobile phase** while preserving as much as possible of the analyte, which is much more difficult for HPLC than for GC.

The design of these interfaces is dictated, in large part, by the **method used to ionize the molecules**, which in turn is dictated by the nature of the molecules themselves, i.e. volatile, nonvolatile, polar, nonpolar, low mass, high mass etc., as well as the **need to couple chromatographic techniques that operate at or near atmospheric pressure to the vacuum requirements of the MS.**

The successful **on-line interfacing** of several ion sources has made them dominant players in quantitative analyses using mass spectrometry. These include: electron ionization (EI) and chemical ionization (CI) both coupled to GC, and the atmospheric pressure ionization (API) methods of atmospheric pressure chemical ionization (APCI) atmospheric pressure photoionization (APPI), and electrospray ionization (ESI) coupled to LC. In addition, matrix assisted laser desorption ionization (MALDI) is seeing increased application in **off-line** LC/MS applications.

Electron ionization (EI) is the classical ionization technique performed by bombarding a volatilized sample with a fast moving (typically 70 eV) electron beam. The application of EI is restricted to thermally stable analytes with relatively low molecular masses (<∼1000 Da) with

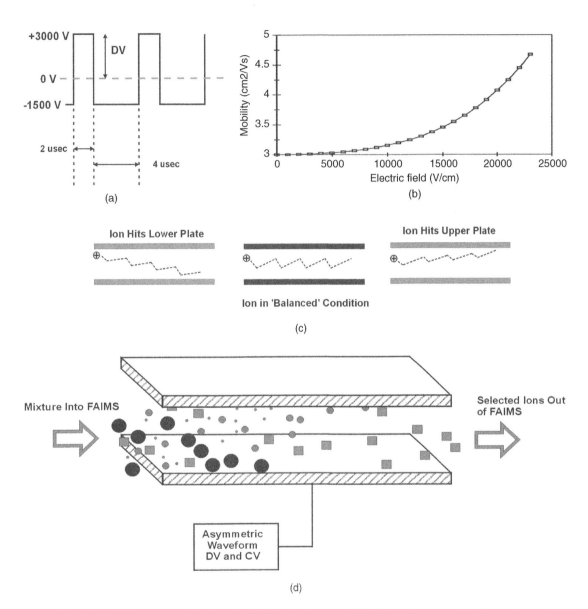

Figure 5.41 High Field Asymmetric Waveform Ion Mobility Spectrometry (FAIMS), (a) The asymmetric RF waveform such that the product of peak voltage (DV = dispersion voltage) and half-cycle time is the same (apart from sign) for positive and negative half-cycles. (b) Variation of sodium ion mobility in a buffer gas with applied field strength. (c) Illustration of interplay between the effects of the asymmetric RF field and the applied DC field ('compensation voltage', CV) on ion path stability. In the left-hand diagram the compensation voltage is too low and the asymmetric RF field drives the ion to the lower plate. In the right-hand diagram the compensating DC field is too large; in the centre diagram the DC field has been tuned so as to just compensate for the effect of the asymmetric RF field and the ion is transmitted through the device. Because different ions have different mobility field curves (as in (b)) the values of the compensation voltage required for transmission will differ and the device can be used as a filter for ions of varying differences between high and low field mobilities, as illustrated in (d). These figures are reproduced from the website www.faims.com

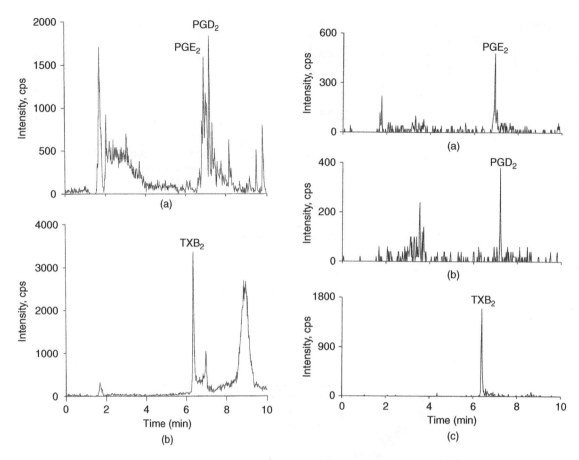

Figure 5.42   Left panel: Representative LC–MRM chromatograms obtained for target prostanoids in a guinea pig spinal cord homogenate sample. (a) PGE2 and PGD2 (m/z 351 → 315) showing peaks for the analytes at retention times 6.9 and 7.2 min, respectively, plus considerable interference signals; and (B) TXB2 (m/z 369 → 169) showing the mixture of anomeric species between 6.4 and 7.0 minutes together with a broad interference between 8 and 10 minutes. Right panel: Representative LC–FAIMS–MRM chromatograms obtained for the same guinea pig spinal cord extract. (a) MRM transition and FAIMS compensating voltage (CV) set for PGE2 (the predominant peak at 6.9 minutes) showing selective removal of most of the interferences and of PGD2 (different CV value). (b) MRM transition and CV set for PGD2 (peak at 7.2 minutes), with most of the interferences and PGE2 removed by the selectivity of FAIMS. (c) MRM transition and CV set for TXB2, resulting in only the predominant TXB2 species emerging from the FAIMS device and elimination of the interference between 8 and 10 minutes. Reproduced from Kapron, *Rapid Commun. Mass Spectrom.* **20**, 1504, (2006) with permission of John Wiley & Sons, Ltd.

sufficient volatility that they are completely vaporized at the source temperature. In the positive ion mode, EI results in the formation of odd-electron, molecular radical cations $A^{+\bullet}$ whose m/z value is essentially equal to the molecular mass of the neutral molecule.

EI mass spectra are remarkably reproducible and comparable among different m/z analyzers, leading to excellent quantitative precision with a wide linear dynamic range and low detection and quantitation limits. However, the 'hardness' of this ionization technique can result in extensive fragmentation thus depleting the $A^{+\bullet}$

ion abundance to zero. In such cases a possibly less characteristic fragment ion must be selected for quantitation, preferably with m/z as close as possible to that of $A^{+\bullet}$.

Chemical ionization (CI) is a 'soft' ionization technique that can ionize a molecule without inducing extensive fragmentation via a chemical reaction with a 'reagent ion' formed by EI of a reagent gas added to the source in large excess. CI shares the same restrictions as EI with respect to volatility and thermal stability of analytes, and also the requirement to severely limit the flow rate of mobile phase into the source. A wide variety of CI reagent gas systems can be used resulting in formation of either positive ($[A+H]^+$) or negative ($[A–H]^-$ or $A^{-\bullet}$) ions, though occasionally adduct ions are also observed (e.g. $[A+NH_4]^+$ if ammonia is used as reagent gas). In addition, the availability of different reagent gases allow selective tuning of the exothermicity of the CI reactions, effectively controlling the degree of molecular ion fragmentation in the source.

CI is intrinsically more complicated than the relatively simple electron–molecule collisions involved in EI. For this reason it is more difficult to control and replicate experimental conditions, and quantitation with CI usually demands use of a co-eluting internal standard. Negative ion CI (NICI) is particularly irreproducible and quantitation definitely *requires* an internal standard.

The limitation on applicability of both EI and CI to analytes that are thermally stable and volatile can be alleviated by **chemical derivatization** to protect any highly polar functional groups within the analyte molecule. However, introduction of such an additional step into the overall quantitative method can introduce analyte losses and will also reduce throughput in cases where the latter is an important criterion.

Classical EI and CI sources can be readily coupled to GC and much less readily to liquid flows using particle beam (PB) interfaces. Packed column GCs require the used of a jet separator for successful coupling, while wall-coated capillary GC columns operating with helium flow rates in the range 1–2 mL.min^{-1} can be introduced directly into a mass spectrometer without overloading the vacuum pumps.

Matrix assisted laser desorption ionization (MALDI) is performed by mixing the analyte with a suitable (usually solid) matrix, spotting the sample onto a sample holder (a multi-well plate in the case of high throughput analyses) that is then inserted into the vacuum of the MS. Selected regions of the spots are then bombarded with a laser beam resulting in desorption of intact ionized molecules (typically $[A+H]^+$ or $[A–H]^-$ ions) from the 'plume' formed above the droplet surface. Analytes with molecular masses greater than 20 000 Da can be ionized with MALDI making it a leader in the qualitative and quantitative analysis of large molecules. A number of matrices and spotting techniques are available, most developed for large molecule analyses although techniques for high throughput quantitation of small molecules are currently under development.

The mechanism of ion formation in MALDI is not completely understood; however, it is known that **ionization suppression effects** must be taken into account if MALDI is to be used for quantitation. Also, above a certain concentration analyte signal cannot be increased by adding more analyte to the sample, with obvious implications for dynamic range. These effects indicate the essential use of appropriate internal standards to account for competition.

MALDI is typically an off-line technique, with its basic incompatibility with chromatography technologies thus far frustrating all attempts at on-line coupling in a manner suitable for trace analysis. However, off-line coupling of LC or CE to MALDI–MS effectively decouples the liquid phase separation from the MALDI–MS analysis and has potential for very high throughput analysis (relative to LC–MS). In addition, the intrinsically high ionization efficiency of MALDI, with its potential for improved limits of detection and quantitation, has led to considerable activity in its application to small molecule quantitative analyses.

Fast atom bombardment (FAB) (also known as liquid assisted secondary ion mass spectrometry, LSIMS) also makes use of a spotted sample present in a matrix to generate gas phase ions from fragile biomolecules. Flow-FAB interfaces capable of on-line interfacing to LC have been developed for trace quantitative analyses. The **thermospray ionization interface (TS),** performed by directly flowing sample through a heated region to effect ionization, was the first device to provide a viable means of coupling LC to MS for quantitative analysis, particularly for relatively nonvolatile analytes. For several years in the 1980s TS dominated LC/MS analyses; however, as with flow-FAB, TS has been largely abandoned in favour of the much more facile and reliable atmospheric pressure ionization (API) methods.

There are **three types of API sources** that are readily coupled on-line to LC: atmospheric pressure chemical ionization (APCI), atmospheric pressure photoionization (APPI) and electrospray ionization (ESI). While the three techniques differ in the physico-chemical processes that lead to ion formation, they all readily accommodate liquid flow by making use of heating and/or nebulization aids, often involving the flow of inert gases to aid desolvation. As such the **three sources have similar design features** that also include optics for transferring ions from the atmospheric pressure ion source into the vacuum required for operation of the *m/z* analyzer, a pressure drop by a factor $\sim10^9$.

Atmospheric pressure chemical ionization (APCI) can be regarded as CI using ambient air as the reagent gas, initiated not with a beam of fast electrons, since these would not penetrate far into atmospheric pressure air, but by using a high potential (several kV) applied to a sharp needle in the atmospheric pressure gas (a so-called corona discharge). The APCI method is usually used in combination with a heated pneumatic nebulizer

(HPN), which achieves nebulization of the liquid sample plus rapid vaporization of the droplets in a heated region prior to the corona discharge. APCI primarily results in the formation of $[A+H]^+$ or $[A-H]^-$ ions as a result of the complex ion–molecule chemistry that occurs in the plasma formed following the discharge. Because of the requisite heating, the technique is best suited to analytes that possess some thermal stability (not as demanding as for EI and CI) and are of moderate polarity (and thus volatility).

Atmospheric pressure photoionization (APPI) is performed in a manner similar to APCI, by replacing the corona discharge needle with a photoionization lamp. Unlike APCI, however, APPI requires the addition of a dopant to aid in the ionization process. The mechanism of action of dopants in APPI is not as simple as first believed, i.e., it is more complicated than a simple charge exchange mechanism as evidenced by the fact that the majority of the ionized analyte is formed as $[A+H]^+$ rather than the $A^{+\bullet}$ expected to be formed by direct photoionization. Developments in APPI for LC/MS are ongoing; however, APCI and ESI currently dominate trace quantitative applications requiring LC and complement one another.

Electrospray ionization (ESI) at its most basic level involves production of gaseous ions at atmospheric pressure from analytes in liquid solution by nebulizing the solution in such a way as to produce small droplets that carry a net electrical charge. This is achieved by flowing the sample through a narrow bore capillary that is held at a high voltage (volts to several kV) relative to the ion sampling orifice of the mass spectrometer. While ESI primarily forms $[A+H]^+$ and $[A-H]^-$ for small molecules, the multiple charging phenomenon that readily produces $[A+nH]^{n+}$ and $[A-nH]^{n-}$ ions for large molecules is what revolutionized the application of mass spectrometry to biochemistry and molecular biology.

ESI can be performed over a wider range of flow rates than APCI, i.e. from nl/min (using so-called nanospray) to ml/min, thus allowing facile coupling to normal bore, narrow bore and capillary LC systems. It should be noted that studies have shown that at **higher flow rates, ESI behaves as a concentration dependent detector**, while at **lower flow rates (nl/min) it behaves as a mass flow dependent detector**. This change from a true mass flow dependent detector to a concentration dependent detector at higher flow rates is attributed to greater analyte losses arising from incomplete droplet evaporation and from incomplete spray transport efficiency at higher low rates.

API techniques can be readily coupled to a wide range of LC systems, across a range of flow rates, with reverse phase LC, with or without gradient elution, and are widely employed in trace quantitative analyses. In addition to the formation of $[A+H]^+$ and $[A-H]^-$ ions, all three of these API techniques can also lead to the formation of **adduct ions**, e.g. $A+NH_4^+$. The nature of the specific adduct ions observed is a function of the analyte structure and of the composition of the mobile phase as well as the presence of trace impurities, usually salts. These adduct ions can be used in quantitative method development.

All ionization techniques used for quantitative analysis present the possibility of ionization suppression (and sometimes enhancement) of the analyte(s) by co-eluting compounds arising from the sample matrix or elsewhere; this phenomenon is a form of **matrix effect**. This problem is much more serious for API techniques than the others presented in this chapter, with ESI presenting the highest occurrence of matrix effects. Matrix effects are also of concern in MALDI analyses, especially in the quantitation of small molecules. **It is always advisable to be aware of the importance of matrix effects** in any proposed analytical method **and to minimize them as far as possible**. Use of an appropriate isotope-labeled standard in conjunction with matrix matched calibrant standards should give reliable results if all other appropriate precautions are taken, but it is important to investigate the possibility of **relative matrix effects** if one suspects that the matrix in the analytical sample might differ appreciably from that used to make the matrix matched calibrants.

5.1 Appendix 5.1: Methods of Sample Preparation for Analysis by MALDI

Dried droplet method: The dried droplet method is the oldest (Karas 1988) and is probably still the most commonly used. A drop of matrix solution is mixed with analyte solution and allowed to dry (usually in ambient air), resulting in a solid deposit of analyte-doped matrix crystals that is introduced into the mass spectrometer for analysis. The analyte/matrix crystals may be washed to remove the nonvolatile components of the original solution. This method tolerates salts and buffers relatively well, but this tolerance has its limits. Washing the sample can help but if signal suppression is suspected a different approach should be tried (e.g., crushed crystal, see below). The method is usually a good choice for samples containing more than one analyte; thorough mixing of the matrix and analyte prior to crystallization usually provides the best possible reproducibility for mixtures. A common problem in the dried droplet method is the aggregation of higher amounts of analyte/matrix crystals in a ring around the edge of the drop. Normally these crystals are inhomogeneous and irregularly distributed, which is the reason for searching for 'sweet spots' on the sample surface. Another problem that is often observed during crystallization is segregation, i.e., as the solvent evaporates and the matrix crystallizes, the salts and some of the analyte are excluded from matrix crystals, resulting in an inhomogeneous mixture of analyte throughout the sample and thus highly variable analyte ion production as the laser is moved across the sample surface. This is a major problem that must be addressed in quantitative analysis.

Vacuum drying method: this method is a variation of the dried droplet method in which the final analyte/matrix drop applied to the sample stage is rapidly dried in a vacuum chamber. When it works, vacuum drying provides uniform crystalline deposits with small crystals and increases crystal homogeneity, thus reducing the segregation effect; by thus greatly improving spot-to-spot reproducibility it minimizes the need to search for 'sweet spots'. The main disadvantages of vacuum drying are that it is not guaranteed to work better than dried droplet in all cases, and it requires accessory vacuum hardware.

Crushed crystal method: the crushed crystal method was developed to allow for the growth of analyte-doped matrix crystals in the presence of high concentrations of involatile solvents and additives commonly encountered in protein preparations (e.g., glycerol, urea, dimethyl-sulfoxide) without pre-purification. A dried droplet of matrix-only solution is first applied to the MALDI sample stage and a clean glass slide (or the flat end of a glass rod) is placed on the deposit and pressed down on to the surface with an elastic rod such as a pencil eraser to smear the deposit into the surface. The crushed matrix is then brushed with a tissue to remove any excess particles, and a one microlitre drop of the analyte/matrix solution is then applied to the smeared matrix material. An opaque film forms quickly over the crushed matrix. After about one minute, the sample is immersed in room temperature water to remove involatile contaminants, the film is blotted with a tissue to remove excess water and allowed to dry before loading into the mass spectrometer. Results obtained using this method can be very different from those from the dried droplet method, particularly in the presence of involatile solvents and contaminants. In this method rapid crystallization directly on the metal surface is seeded by the nucleation sites provided by the smeared matrix bed crushed on the sample holder plate; crystal nucleation shifts from the air/liquid interface to the surface of the substrate, so that microcrystals form inside the solution where the concentrations change more slowly. The films produced are more uniform than dried droplet deposits with respect to spot-to-spot reproducibility. The disadvantages of the crushed crystal method are the increase in sample preparation time and a strict requirement for

control and removal of particulates in solution preparation to eliminate undissolved matrix crystals that can shift the nucleation from the plate surface to the bulk of the droplet.

Fast evaporation: this simple sample preparation method, in which matrix and sample handling are completely decoupled, was introduced with the main goal of improving the resolution and mass accuracy in of MALDI analysis of macromolecules. A matrix-only solution is prepared in acetone containing 1–2 % pure water or water containing 0.1 % acid such as trifluoroacetic acid; the matrix concentration can range between saturation and 30 % of that value. A $0.5\,\mu L$ droplet of the matrix-only solution is applied to the sample stage, whereupon the liquid spreads quickly and the solvent evaporates almost instantly (1–2 seconds). A drop ($1\,\mu L$) of sample solution (0.1–$10\,\mu M$) is applied on top of the matrix bed and allow to dry either by itself or in a flow of nitrogen; the analyte can be dissolved in any solvent provided that this does not completely re-dissolve the matrix crystals. The sample is allowed to dry in air and is then introduced into the mass spectrometer. This method provides polycrystalline surfaces with roughness 10–100 times smaller than that of comparable dried droplet deposits, and with more uniform distribution of analyte across an entire sample deposition area. There are some disadvantages associated with this method. It does not provide reproducible sample-to-sample data for analyte mixtures. Also, because the layer of analyte-doped matrix on each crystal is usually very thin, ions are produced for only a few laser shots from a fixed irradiated spot. The use of a highly volatile solvent like acetone makes it difficult to make reproducible sample spots, as some varying amount of solvent is always lost to evaporation before the matrix-only droplet is delivered to the sample plate. Moreover, acetone has low surface tension and so spreads uncontrollably along the sample plate surface.

Overlayer (also Two Layer and Seed Layer) Method: this method (several variants have been developed by different research groups) is a combination of the crushed crystal and fast evaporation methods. Fast solvent evaporation is used to form the first layer of small matrix-only crystals, followed by deposition of a solution of matrix plus analyte on top of the crystal layer. The difference between the fast evaporation and the overlayer methods is in the second layer solution, where the addition of matrix to the latter is believed to provide improved results particularly for mixtures. The method has several convenient features that make it a popular approach. It inherits all the advantages of the fast evaporation method and avoids some of its limitations, while providing enhanced sensitivity and better spot-to-spot reproducibility.

Sandwich method: this method is derived from the fast evaporation and overlayer methods. The sample analyte is not premixed with matrix; instead a sample droplet is applied on top of a fast evaporated matrix-only bed as in the fast evaporation method, followed by deposition of a second layer of matrix in a nonvolatile solvent so that the sample is sandwiched between the two matrix layers. This method is not widely used.

Electrospray deposition: sample deposition by electrospray for MALDI–MS involves spraying a small amount of matrix–analyte mixture from a stainless steel capillary (3–5 kV) onto a grounded metal sample plate mounted 0.5–3 cm away from the tip of the capillary. It creates a homogenous layer of equally sized microcrystals and the guest molecules are evenly distributed in the sample, achieving fast evaporation and effectively minimizing sample segregation effects. Electrospray deposited samples have been shown to give several advantages over traditional droplet methods. In particular, the reproducibility from spot-to-spot within one sample desposit, and from sample-to-sample for multiple depositions, is much improved (typical sample-to-sample variations are 10–20 %); the correlation between analyte concentration and signal is also much improved, an important consideration for quantitative analysis. Obviously the characteristics of this method offer a possible path for interfacing MALDI sample preparation to liquid chromatography. The method does suffer from some disadvantages. It is slow, requiring 1–5 minutes to build up a useful deposit; pneumatic nebulization has been suggested as an alternative and faster spraying method, with potential to combine pneumatic nebulization with an electric field (as in ionspray ionization, see above) to create charged droplets that can be efficiently guided to the grounded sample plate.

Quick and Dirty: this aptly named sample preparation technique is widely used to obtain fast results without consideration of signal optimization. It is simple, fast and broadly applicable, and completely decouples matrix handling from sample handling with little or no sample purification prior to analysis. Typically, a one microlitre drop of analyte solution $(0.1–10 \mu M)$ is deposited on the sample plate and a one microlitre drop of matrix solution $(1–10 \mu M)$ is immediately added on top. The two solutions are mixed thoroughly with the pipet tip before the mixture is dried in a stream of air or nitrogen and introduced into the mass spectrometer. When signal supression is suspected, or if no signal is observed, the dried sample spot can be rinsed once or twice with 2–5 μL of 4°C water. There are several advantages of this method: it is fast, it decouples the sample and matrix preparation steps, and it is very easy to add a calibration standard to the sample before the matrix is added. The main disadvantage of the method is that it provides the least controlled conditions of all sample preparation procedures, yielding results that are considerably less reproducible and consistent than those obtained using the somewhat similar dried droplet method. This method is clearly not of interest for quantitative analyses.

Matrix precoated targets: the use of matrix precoated targets, particularly for MALDI analysis of peptides and proteins, has been investigated by several research groups. The advantages of a sample preparation method reduced to straightforward deposition of a single drop of sample solution to a precoated target spot would likely include increased speed and sensitivity relative to the methods described above, but would also include a means to directly interface MALDI sample preparation to the output of LC and CE columns.

Most early efforts focused on the development of thin layer matrix precoated membranes, including those made of nylon and other synthetic polymers, nitrocellulose, anion- and cation-modified cellulose and regenerated cellulose. Several manufacturers currently offer 96-well plates with proprietary coatings for LC–MALDI–MS applications. This approach was described in more detail above. As mentioned previously, all of these sample preparation methods were originally developed for prepurified samples or fractions, not for analytes eluting from an HPLC or capillary electrophoresis column.

6

Tools of the Trade V. Mass Analyzers for Quantitation: Separation of Ions by m/z Values

6.1 Introduction

In Chapter 5 the methods of transforming analyte molecules in liquid solution into gaseous ions in a vacuum enclosure were discussed. This chapter is concerned with devices used to separate a mixture of ions according to their m/z values by subjecting them to electric and/or magnetic fields under conditions in which ion collisions with background gas molecules are either negligible or are carefully controlled. Such devices are commonly referred to as 'mass analyzers', although they truly separate ions according to their m/z values; this usage is consistent with that of 'mass spectrum' rather than 'm/z spectrum'. To understand the basic principles of how this is achieved, it is necessary to discuss some concepts of electromagnetic fields (Section 6.3.1 and Appendix 6.1) and of vacuum systems (Section 6.6 and Appendix 6.2). As an aid, a list of the symbols used in this chapter is given in Appendix 6.3.

Historically many different mass analyzers have been invented (Beynon 1960; Farmer 1963), each exploiting its own blend of electric and/or magnetic fields, but today only a few of these are in common use particularly for quantitative analyses. A useful categorization is that of ion beam vs ion trap analyzers. Ion beams are collections of ions all moving in well defined directions and were so named because of their analogy to beams of light. Indeed, the various devices used to steer and control ion beams are referred to as 'ion optics' that include focusing elements called 'ion lenses' (see Section 6.3.1). Historically the first ion beam instrument was the magnetic sector analyzer (Section 6.4.4); the double focusing version of this analyzer is still extensively used today for quantitative

analyses in which a combination of high resolution and high sensitivity are required (Section 11.4). However, in the context of trace level quantitative analysis of small (a few hundred Da) organic molecules, the linear quadrupole mass filter (Section 6.4.2) is by far the most commonly used. Another ion beam instrument that has recently been developed from its earliest forms into a version that has shown some promise for quantitative analysis is the time of flight analyzer (Section 6.4.7). In contrast, ion trap analyzers manipulate groups of ions within a limited volume, where they undergo oscillatory motions under the influence of time dependent electric fields, and are subsequently either extracted from the trapping volume in a m/z dependent manner for detection or are directly detected by exploiting the m/z dependent nature of their oscillatory motions.

Some additional general discussion of these types of mass analyzer is given later but it is worthwhile noting that, for quantitative analyses that demand the most rigorous validation requirements (Chapter 10), ion beam analyzers (especially linear quadrupole and magnetic sector instruments) are predominantly used in current practice. While there are many reasons for this distinction, space charge effects and their implications are probably the most important. This phenomenon refers simply to the importance of electrostatic (Coulombic) interactions (Section 6.3.1) among ions that are confined to a limited space. The ions are thus sufficiently close together on average that the forces arising from these interactions are sufficiently large to yield significant perturbations on the controlling forces applied via the time dependent electric fields that trap the ions in the first place. The same number of ions in an ion beam instrument is strung

Trace Quantitative Analysis by Mass Spectrometry   Robert K. Boyd, Cecilia Basic, Robert A. Bethem
© 2008 John Wiley & Sons, Ltd

out along the length of the beam so that the average inter-ion distance is much larger and the inter-ion electrostatic interactions are correspondingly small. Ion trap instruments can be used for quantitative analyses that are performed for purposes that do not require the most rigorous levels of validation. This fundamental limitation of traps relative to beam instruments is related to the ease with which the latter can switch quickly between transmission of ions of different *m/z* (e.g. ions characteristic of an analyte and of an internal standard) while excluding all other *m/z* values. However in a stand-alone trap ions of all *m/z* values are allowed to enter the trap at the same time during the 'ion fill' time, thus requiring some (admittedly ingenious) resonance ejection techniques (described later) to eject unwanted ions (including matrix ions), thus mitigating the space charge effects but with penalties elsewhere (e.g. reduced duty cycle and variable scan cycle times, see Section 6.4.5).

To judge whether or not any particular mass analyzer is fit for purpose, some key figures of merit must be considered relative to the needs of the specific analytical problem. These figures of merit are discussed in more detail later (Section 6.2.3) but are listed here for convenience in the introductory sections: accessible *m/z* range, resolving power, accuracy and precision of mass measurement, transmission efficiency (for beam analyzers), duty cycle, sensitivity, data acquisition rate, dynamic range (particularly linear dynamic range), versatility for tandem mass spectrometry (MS/MS), capital and maintenance costs, and ease of use. It should be noted that many of these parameters are inter-related, e.g., transmission efficiency (ratio of detected signal to the number of analyte ions supplied to the analyzer from the ion source) and resolving power are generally inversely related for a given instrument type, while resolving power and cost are usually positively correlated among different instrument types. As a matter of nomenclature, 'resolving power' is strictly a (generally adjustable) ability of an analyzer to separate ions with closely similar *m/z* values, while 'resolution' refers to the mass peak widths actually observed in the mass spectrum, but this distinction is not always followed. It is clear from this long list of figures of merit that choosing an ideal instrument that is fit for purpose could be a complicated task but in fact, as discussed below, these decisions have already been worked out through the experience of analytical colleagues over several years. It is usually more problematic to determine whether or not an existing instrument is fit for purpose with respect to some new quantitative analytical problem. In this regard, instruments capable of tandem mass spectrometry are almost always superior in terms of the selectivity–sensitivity compromise

(Section 6.2.1) Attempts to compare different analyzer types with respect to some key figures of merit are made in Table 6.1, but it should be borne in mind that such attempts at a succinct comparison are both somewhat subjective and also are likely to become outdated in view of the rapid pace of instrumental development.

A stand-alone magnetic sector analyzer of the type universally used today (2007) provides separation of ions with moderate mass resolution (a measure of the ability to separate ions with closely similar *m/z* values). To provide higher mass resolution, a magnetic sector is combined with an electric sector to form a combined analyzer that simultaneously provides high resolution (to provide a high degree of detection selectivity, i.e., the analyzer's ability to differentiate the analyte in the presence of other components of closely similar *m/z* in the sample) and high ion transmission efficiency (through the concept of double focusing Section 6.4.4). As a convenient shorthand, a magnetic sector is often denoted as 'B' (from the SI notation for magnetic flux density, Table 1.2) and an electric sector as 'E'; a double-focusing combination can be either EB or BE. In the context of trace quantitative analyses, these analyzers are generally used only with EI sources (Section 5.2.1) together with GC separation (Section 4.4.3), i.e., for volatile and thermally stable analytes (most notably polychlorodibenzodioxins and related compounds, Section 11.4.1). It is difficult to use the modern API sources with sector analyzers (Sections 5.3.3–5.3.6), so unfortunately rigorous quantitation by LC/MS is not feasible.

The linear quadrupole mass filter, frequently referred to as a 'linear quadrupole' or even simply a 'quadrupole', is compatible with all ion sources in current use for quantitative analysis (EI, CI, MALDI and the API methods, Chapter 5) and is frequently fit for purpose for even rigorously validated analyses (usually EI or API sources). However, a linear quadrupole has limited resolving power so that increased detection selectivity, when required, must be achieved by means of tandem mass spectrometry (MS/MS) (Section 6.2). For this purpose three quadrupoles are linked together in series to form a triple quadrupole instrument usually designated QqQ (Section 6.4.3), where lower-case 'q' (*not italicized*) denotes a quadrupole collision cell rather than a mass analyzer. Such instruments are currently the gold standard for bioanalytical and indeed many other quantitative applications due to the relative ease with which they can be interfaced to chromatography systems in general, ease of use, selectivity, and their ability to enable validation of rugged and robust methods with the sensitivity, precision and accuracy required by the end user.

Table 6.1 Comparison of some figures of merit for tandem mass spectrometers used in quantitative analysis. Double focusing magnetic instruments are not included here as their range of applicability is somewhat limited and they are never used in tandem mode for quantitation. The stand-alone linear quadrupole is not included in this Table (no MS/MS capabilities) although it is widely used in quantitative applications by GC/MS. The stand-alone TOF analyzer also has potential in this regard.

Feature	QqQ	3D Trap	2D (linear) trap	QqTOF	$q_0Q_1q_{2/T}Q_{3/T}$ ("QTRAP")
highest m/z	2000 to 4000	2000 to 4000		20 000 (full scan) 4000 (MS/MS)	2800
resolving power	$\Delta m = 1$ (normal) $\Delta m = < 0.01$ (enhanced-res)*	$\Delta m = 1$ (normal) $\Delta m = 0.1$ (high resolution)	$\Delta m = 1$ (normal) $\Delta m = 0.05$ (high resolution)	5000 to 10 000 FWHM	> 3000 (FWHM) at m/z 609 at 250 Th.s^{-1}
accurate mass	no (normal) limited (enhanced-res)	no	no	yes	±0.03 Th in some scan modes
instrumental sensitivity	5 to 10 times greater with QqQ vs 3D trap in MRM	5 to 10 times greater	5 times greater than 3D trap	very sensitive	Provides QqQ sensitivity in MRM and 2D trap in full scan
MS/MS?	yes tandem-in-space	yes tandem-in-time	yes tandem-in-time	yes tandem-in-space	yes (tandem-in-time and/or in-space)
MS/MS scan modes control over CID conditions**	all full	product ion, MRM limited	production, MRM limited	production full	all full
MS^n? (other than using in-source CID)	no	yes MS^{10} (MS^3 on-the-fly)	yes MS^{10} (MS^5 on-the-fly)	in theory MS/MS/MS (using PSD in reflectron-TOF)	yes
MS/MS excitation	nonresonant, i.e. both precursor and product ions undergo CID	resonant, i.e. precursor only undergoes CID	resonant, i.e. precursor only undergoes CID	nonresonant, i.e. both precursor and product ions undergo CID	can be resonant and/or nonresonant
interfacing to LC and GC	facile	facile	facile	facile	facile
preferred MS/MS operating mode for quantitation	MRM	limited scan	limited scan	full scan	MRM
duty cycle in preferred MS/MS mode for quantitation	high	low	medium	low	high
size and cost***	bench-top and floor-standing (medium cost range)	bench-top (low cost range)	bench-top (low–medium cost range)	bench-top and floor-standing (high cost range)	floor-standing (medium–high cost range)

* This mode implies limited mass range.

** In trap instruments the conditions for optimal trapping efficiency limit the possibilities to optimize collisional activation.

*** This category is particularly subject to rapid changes as instrumentation is developed further.

The time of flight (TOF) analyzer is most easily interfaced to EI, CI and MALDI ion sources, though it can also be used with the API sources through use of an intervening specialized (RF-only) quadrupole (not a quadrupole mass filter). The TOF analyzer is unique among mass analyzers in that its principle of *m/z* separation does not involve externally applied electric or magnetic fields, but rather the flight of ions through a field-free region. However, most modern TOF analyzers (Section 6.4.7) do incorporate an additional 'reflectron' device that imposes electric fields in order to improve the resolving power. Although the literature contains descriptions of examples in which TOF analyzers were fit for purpose in some quantitation applications covering both small and large molecule analytes, at the present state of development they do not appear to be suitable for the most rigorously validated analytical methods (Chapter 10).

Three-dimensional quadrupole ion traps (Section 6.4.5, often referred to as Paul traps in honor of their inventor, or as 3D traps) have been shown to be compatible with all of the important ionization techniques including MALDI. Although these analyzers have been fit for purpose for a wide range of quantitative methods, they appear to be intrinsically incapable of meeting the conditions for the more rigorous validation requirements (Chapter 10). As mentioned above, their most important limitations in this regard are consequences of space charge effects. These effects are considerably mitigated in the more recent linear ion traps (2D traps) (Section 6.4.5) that are essentially modified linear quadrupole analyzers operated in ion trap modes. As a result of their much larger internal volume, which permits larger average ion–ion distances, space charge effects are much less serious than in the 3D traps.

Two other types of ion trap currently used in chemical analysis are the Fourier Transform Ion Cyclotron Resonance (FTICR) and the Orbitrap analyzers. These analyzers share the distinctions of providing the possibility of very high resolving power and of using a unique method of ion detection (image current detection). These analyzers have seldom been used for trace level quantitation of small molecule analytes, and certainly not for fully validated methods. Accordingly, they will be described only briefly in this book (Section 6.4.8) although they are increasingly used for quantitative measurements of larger molecules.

The preceding brief introduction to the mass analyzers commonly used for demanding quantitative analyses is intended to provide context for the more detailed discussions in subsequent sections. Note that these comments have referred largely to the various designs as standalone analyzers although a brief reference was made to the importance of triple quadrupole instruments and of

tandem mass spectrometers in general (Table 6.1). Other combinations that are increasingly being used as fit for purpose in some quantitative analyses include an ion trap (usually 3D but increasingly also linear) followed by a TOF analyzer, the QqTOF design, especially those in which some means of ion trapping is included before final TOF analysis, and a triple quadrupole design in which the final quadrupole can also be operated as a linear trap. Several other such 'hybrid' combinations comprising two different analyzer types that can facilitate tandem mass spectrometry have been described, but the QqQ and QqTOF are presently the most important in the context of this book.

This chapter and Chapter 7 are concerned with the mass spectrometer hardware, i.e. components fabricated from stainless steel and other metals, plastics, ceramics etc. The associated electronics and computer-based algorithms (software) that control the instrument and process the data are not discussed in depth in this book (but see Section 7.4.3). The 'firmware' is a set of computer programs that are resident on dedicated microprocessors. Like software, these are computer programs executed by a computing device of some kind, but are also an essential and integral part of a piece of hardware and are generally of no use beyond their purpose for that particular hardware, e.g. the instructions on a microprocessor for communication between a GC or HPLC with a mass spectrometer that synchronizes initiation of data acquisition that is triggered by a new sample injection and start of a chromatographic elution.

It should be emphasized again that the current pace of instrumentation development implies that at least some of the statements made above will become outdated sooner rather than later. This is one of the reasons why this book has included an introduction to the more fundamental aspects of mass spectrometer design and of the physical and chemical processes that occur within them. The other important reason is that, without at least some understanding of these principles, the analyst must treat his/her instruments as 'black boxes' with no appreciation of the intrinsic strengths and limitations that control the validity of the data generated through their use.

6.2 Mass Analyzer Operation Modes and Tandem Mass Spectrometry

This is an important section, as it introduces important concepts that will be referred to in subsequent more detailed sections. Thus some nomenclature will be established and some general comments on operational modes of mass analyzers, including tandem mass spectrometry, are intended to provide a context for the more detailed discussions in later sections of this chapter. The figures of merit relevant to comparisons of different types of mass analyzer are then discussed in more detail than in the preceding brief introduction. Firstly, however, a major reason why mass spectrometry is *the* enabling technology in trace level quantitative analyses is discussed, namely, the unique combination of selectivity (via combinations of m/z values) and sensitivity (in the generalized sense of the term) that is unmatched by any other chromatographic detector (compare Section 4.4.8).

6.2.1 The Selectivity–Sensitivity Compromise

It is appropriate to start by revisiting what is meant here by 'selectivity' and 'sensitivity', see Chapters 2, 5 and 8. The definition of selectivity in the FDA Guidance for Industry on Bioanalytical Method validation (FDA 2001), is ' — the ability of an analytical method to differentiate and quantify the analyte in the presence of other components in the sample'. Other components include metabolites, impurities, degradation products or matrix components. While this definition was written so as to apply to a complete integrated analytical method (Figure 1.2), it can readily be rewritten to apply to the chromatographic detector (mass spectrometer in this case). In particular, the ideal mass spectrometric detection technique would be sufficiently selective that interfering components that co-elute with a target analyte do not in any way affect the integrity of the analytical signal chosen to monitor the target analyte. Such interferences can be direct (contribution to the intensity of the analytical signal) or indirect ('matrix effects', Sections 5.3.6a and 9.6). The term 'specificity' is sometimes used as a synonym for selectivity and appears to be used twice in this way in the cited regulatory document (FDA 2001), but specificity is probably best regarded as the property of an analytical method for which only the target analyte gives a response, i.e., as the extreme case of selectivity. (It seems unlikely that a truly specific method exists for any analyte, and in any event it is not a useful concept since it is not feasible to test the signal response for every conceivable potential interfering compound.)

The term sensitivity is widely used in two rather different ways. Firstly, it is widely used colloquially in a somewhat loose undefined manner to refer to (but not evaluate) the magnitude of the analytical signal produced by introduction of some (usually unspecified) amount of analyte into the mass spectrometer. Alternatively, this colloquial usage sometimes implies a reference to the smallest quantity of analyte that can be detected or quantitated (i.e., the LOD and LLOQ, Section 8.4). However, in a more rigorous usage, the sensitivity of either an

integrated analytical method or of a mass spectrometric detector is best defined as the slope (S) of the calibration curve (i.e., the change in analytical signal for unit change in quantity/concentration of analyte supplied to the detector, Sections 2.6, 8.3 and 9.8.1). As illustrated in Figure 8.10, for a given noise (or background) level a higher value of sensitivity S will lead to a lower uncertainty in the amount of analyte deduced from an observed analytical signal via the calibration curve, thus allowing smaller changes in analyte amount to be detectable and, by extension, lead to a lower LOD. The various usages of 'sensitivity' are obviously closely related but the context in which the term is used usually makes it clear which interpretation is appropriate.

In this present section the more colloquial sense is intended and applied to the sensitivity of a mass spectrometer rather than to that of an integrated analytical method. The sensitivity of a mass spectrometer refers to its ability to create ions from analyte molecules supplied to it (e.g., from a chromatograph), transmit these ions to the analyzer where they are separated according to m/z value, and thence to the ion detector where the current of ions is transformed into electrical signals that can be manipulated by detection electronics and computers. Each of these steps is associated with its own efficiency rating, i.e., ionization efficiency (ensure that the most appropriate ionization method is used for the target analyte, see Chapter 5), the ion transmission efficiency from ion source to analyzer to detector (that can be optimized by 'instrument tuning', see Section 6.3.2), the CID efficiency (Section 6.5) if tandem mass spectrometry MS/MS is used (see Section 6.2.2), and ion detector efficiency (that can vary with age of the detector and also with the electrical potentials applied to it, an aspect of instrument tuning). Note, however, that instrument tuning (Section 6.3.2) mainly affects the magnitude of the signal, whereas it is S/B value that ultimately determines important performance criteria like detection limits. Since increased selectivity can reduce 'chemical noise' (Aebi 1996), i.e. low level chemical background rather than random noise that can be electrical in nature, it is important to remember that detection and quantitation limits are a function of both sensitivity and selectivity.

It is now possible to move on to discussion of general parameters of a mass spectrometer that are relevant to the selectivity–sensitivity compromise that is ubiquitous in analytical chemistry. 'Full scan' acquisition mode refers to operation of a mass analyzer so as to acquire the entire spectrum of ions (or a major portion thereof) introduced into the analyzer. In the case of a pure compound, full scan mode delivers the maximum chemical information that the mass spectrometric experiment (MS or MS/MS) can provide. However, this is not necessarily true of chromatographic peaks observed in analysis of a trace level analyte in a complex matrix, where it is never possible to be certain which of the ions observed in the full scan spectrum were derived from the analyte of interest and which from co-eluting matrix components. This is an example of the fundamental difference between 'data' (always maximized in full scan mode) and 'information' (can be compromised by too much confusing data), see the final step in the overall process summarized in Figure 1.3.

For analysis of complex mixtures for a target analyte selectivity can in fact be increased by monitoring only those m/z values that have been determined to be characteristic of that analyte while ignoring all the rest. This strategy also has the advantage of simultaneously increasing sensitivity by devoting most of the available instrument time over the chromatographic peak to recording signal(s) characteristic of the analyte. (This latter advantage mainly applies to beam instruments, as will become clear later in the discussion of selected ion monitoring (SIM) and multiple reaction monitoring (MRM) modes.) For the important ion beam analyzers (the magnetic sector and linear quadrupole) in which the m/z spectrum is obtained sequentially by scanning the electric and/or magnetic fields, full scan mode is characteristically of low sensitivity since all ions other than those of the m/z value being transmitted and detected at any given point in the sequential detection are lost and wasted. The TOF analyzer acquires a full mass spectrum practically instantaneously (but not by 'scanning' since no time variation of electric or magnetic fields is used to separate ions). The TOF and ion trap analyzers (both 3D and linear) are more sensitive than the magnetic sector and linear quadrupole in full scan mode.

In contrast, various techniques of selected ion recording (SIM) are used to increase the detection sensitivity (and possibly selectivity, see preceding paragraph) of ion beam instruments. The most straightforward of such methods does not involve tandem mass spectrometry (MS/MS, Section 6.2.2), but instead involves operation of a mass analyzer so as to record signals corresponding to only selected m/z values (or small m/z ranges), rather than scan continuously over the entire relevant m/z range. The specified m/z values are chosen to be characteristic of (i.e. selective for) the analyte(s) of interest, such that the ions that are discarded and not recorded do not carry the majority of the potential chemical information. (Note that a stand-alone TOF analyzer is unique in this regard as it intrinsically must record the full mass spectrum.) This approach, in the absence of MS/MS, is usually referred to by the acronyms 'SIM' or 'SIR' that are interpreted in the literature as either 'single' or 'selected'

ion monitoring/recording, depending on whether only one or several m/z values (or small ranges) are used. Usually the context of the published method resolves the single/selected ambiguity. In this book SIM will be used in preference to SIR for reasons of consistency with the corresponding acronyms used in MS/MS detection (see below), but SIR is found in many publications.

To increase detection selectivity in SIM mode, e.g., to screen out contributions to the observed signal from matrix components that co-elute with the target analyte, the mass resolution can be increased so as to enable discrimination among ions with smaller and smaller differences in their m/z values. A trivial example that illustrates the principle is that of molecular nitrogen (N_2), carbon monoxide (CO) and ethylene (C_2H_4), all with nominal molecular masses of 28 Da if expressed to the nearest integral value only. However, the m/z values of the monoisotopic molecular ions (Section 1.5) can be calculated from the atomic masses of ^{12}C, ^{16}O and ^{1}H (Table 1.4) as 28.0061, 27.9949 and 28.0313 Da, respectively, so that an analyzer capable of cleanly separating these three species could be set up in SIM mode to selectively detect any one of them. Of course, this principle becomes increasingly difficult to apply as the m/z value increases since there are many more possible isotopic combinations to be distinguished. Magnetic sector instruments are the preferred choice for this strategy for trace level quantitation with chromatographic pre-separation, although modern TOF analyzers can provide resolving power that is fit for purpose in some cases (Table 6.1).

However, it is very difficult to reliably operate a magnetic sector analyzer together with an API source and this is one reason why linear quadrupoles have come to dominate trace quantitative analyses. Unfortunately, linear quadrupoles have a limited upper limit to resolving power while still providing useful ion transmission efficiency (and thus sensitivity), so selectivity increases must be obtained by some other means. It is here that tandem mass spectrometry (MS/MS) takes over. The principles of this approach to increasing detection selectivity are discussed in Section 6.2.2 but for now it can be mentioned that it is possible to obtain full scan MS/MS spectra as well as the MS/MS counterparts of SIM. The latter are usually referred to by the acronyms 'SRM' and 'MRM' denoting single and multiple reaction monitoring, respectively, and we shall use 'MRM' to avoid any possible confusion with 'SIM'.

Finally in this section, it can be mentioned that there is a tendency in the literature to use the terminology 'full scan mode' to refer only to acquisition of a complete mass spectrum, i.e., ignoring the fact that it is also perfectly feasible to obtain a complete spectrum in MS/MS mode (e.g. a product ion spectrum). Ion traps and TOF analyzers are the most effective in any 'full scan' mode, but note that TOF analyzers do not 'scan' a spectrum in the usual sense.

6.2.2 Tandem Mass Spectrometry (MS/MS)

This section deals with some general principles of tandem mass spectrometry and its applicability to quantitative analysis. The MS/MS acronym is used in this book as a general term for all tandem mass spectrometry techniques. More detailed descriptions of how the principles are exploited in practice for the various instrumental types are given in later sections of this chapter. The general concept of tandem mass spectrometry in qualitative (structural) analysis is that additional chemical information, over and above that contained in a conventional one-dimensional mass spectrum, can be obtained by examining the connectivity relationships among some or all of the ions in that mass spectrum. The connectivities arise as a result of the dissociation reactions that lead to the fragment ions in a mass spectrum, e.g.:

$$[A^+] \rightarrow F_1^+ \ (+N^o_1)$$
$$\downarrow$$
$$F_2^+ \ (+N^o_2)$$
$$\swarrow \quad \searrow$$
$$F_3^+ \ (+N^o_3) \quad F_4^+ \ (+N^o_4)$$
$$\downarrow$$
$$F_5^+ \ (+N^o_5)$$
$$\downarrow$$
$$F_6^+ \ (+N^o_6)$$

where F and N denote ionic and neutral fragments, respectively; for convenience, only singly charged positive ions are considered in this mythical example in which $[A^+]$ represents the ionized analyte molecule, e.g., $A^{+\bullet}$, $(A + H)^+$, $(A + Na)^+$ etc. A conventional 'full scan' mass spectrum contains all the ions, with no information about their hierarchy. The most common type of MS/MS experiment used in quantitative analysis is that in which a precursor ion (most often the ionized intact analyte $[A^+]$ or $[A^-]$) is mass selected (strictly via *m/z*), caused to fragment and one or more of its product ions detected. Note that the F^+ ions in a conventional mass spectrum are conventionally called 'fragment ions' to distinguish them from 'molecular ions', but when these same ions are observed in an MS/MS experiment in which a mass selected precursor

ion is isolated and caused to dissociate, they are called 'product ions'. Indeed an MS/MS experiment designed to reveal all product ions of a selected precursor ion is called a 'product ion scan' and the corresponding spectrum a 'product ion spectrum' (or sometimes a 'full scan MS/MS spectrum'). Ion traps have a unique advantage in this respect since they can readily provide spectra that follow the fate of the product ions through several steps, e.g., $[A^+] \rightarrow F_1^+ \rightarrow F_2^+ \rightarrow F_4^+ \rightarrow F_5^+$. This operating mode is referred to as MS^n, where superscript n denotes the number of the 'generation' of the product ion, e.g. in this example F_1^+ would appear in an MS^2 scan, F_2^+ in an MS^3 scan, and F_3^+ and F_4^+ would be observed in an MS^4 scan, and so on. (This nomenclature is related to that used in some older literature in which precursor ions were referred to as 'parent ions' and product ions as 'daughter ions'; this latter nomenclature is no longer used.) These MS^n experiments with n > 2 are difficult or impossible with ion beam (scanning or time of flight) analyzers unless these are combined in multi-analyzer combinations. The simple product ion scan, discussed above, corresponds to an MS^2 experiment in this notation, and the conventional mass spectrum (or SIM experiment) is MS^1.

In the context of quantitative analysis, the usefulness of MS/MS modes relative to SIM is related to the consequent reduction of background to a greater extent than of the signal, in cases where the background consists of chemical background. Such chemical background is not really 'noise' in the strict sense, since it is not random but well defined and unidirectional so that it can not be minimized by averaging replicate experiments that can lead to partial cancellation of truly random fluctuations (see Figure 4.27). This general concept of minimization of chemical background 'noise' is exemplified by a recent proposal (de Silva 2006) for identification of individual members of a large population of related molecules (e.g. prepared by some combinatorial synthesis approach) using a series of single input 'logic gates'. This example (de Silva 2006) involved fluorescent dyes that respond differently to chemical structural differences under various solution conditions (pH etc.) and whose combined response is unique to a particular molecule within the population, but the general concepts can clearly be applied to MS^n analysis.

In a somewhat different theoretical approach based on information theory, and using an example directly pertinent to the concerns of this book (Fetterolf 1984), the 'logic gates' concept is replaced by that of 'informing elements' such as clean-up procedure, chromatographic retention time, ionization efficiency with a particular technique, selection of *m/z* values of a precursor ion and of one or more product ions (possibly via more than one stage of

MS^n with n ≥ 2). Such 'informing elements' are not open-or-shut ('Yes' or 'No') logic gates as in the first example (de Silva 2006); however, the two approaches share the same essential idea of greatly increasing the confidence in identification of compounds via the combined effect of a series of experimental 'filters' (Figure 6.1). Note also that the informing power varies quite widely among the 'informing elements' (Fetterolf 1984), e.g., survival of an analyte in a particular clean-up method is much less informative than specification of *m/z* values of precursor and product ions. Modern developments (Section 9.4.3b)

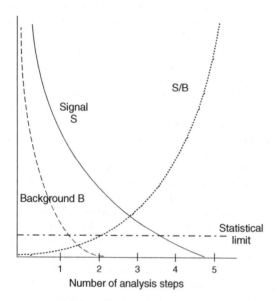

Figure 6.1 Illustration of the effect of increasing the number of analysis steps or 'informing elements' (e.g., value of n in MS^n analyses) on absolute signal level S for the target analyte and on the level of 'chemical background' (interference on the analyte signal channel from co-eluting components). It is the signal/background (S/B) ratio that determines limits of detection and of quantitation provided that ion statistics are acceptable. (Note that it is common practice to refer to this ratio as 'signal/noise' although this terminology is misleading in the present context, see Section 7.1.) In this example the chemical background level decreases more rapidly than the analyte signal as the number of analysis steps is increased, so S/B increases. However, this strategy can not be extended indefinitely since eventually the background will be reduced to insignificant levels and the analyte signal S will approach the level of intrinsic random (nonchemical) noise, the statistical limit at which the absolute number of ions detected is too small for a reliable quantitative measurement of adequate precision (see the discussion of 'shot noise' in Section 7.1.1a). In the hypothetical case illustrated here the signal level S reaches the statistical limit once the number of analysis steps exceeds three.

have attempted to place some approximate general rankings of informing power on these experimental filters. In theory the 'informing power' can be increased indefinitely by adding yet more 'informing elements' (e.g., MS^n with n >> 2), but in practice a limit is reached because of inevitable losses of target analyte or of ions derived from it; such losses accumulate until insufficient ions are actually detected for a reliable quantitative measurement to be possible (the so-called 'statistical limit', Figure 6.1, also see Section 7.1.1). Also, it should be added that higher MS^n techniques using ion traps are currently seldom if ever used in trace quantitation, despite their higher informing power, since the disadvantages of ion traps in this context (e.g. low duty cycle, various consequences of having to deal with the effects of space charge) are even more severe with n > 2.

Mass analyzers capable of MS/MS experiments have been classified (Johnson 1990) as either 'tandem-in-space' or 'tandem-in-time'. Tandem-in-space experiments are performed using ion beam instruments (typically a triple quadrupole or a QqTOF) in which the ions are transported from one analyzer or other device to another, each of which is responsible for one of the precursor selection, ion fragmentation and product ion analysis functions. In contrast, ion traps conduct precursor ion selection plus fragmentation and product ion analysis by a tandem-in-time sequence of events all within one spatial trapping region.

The product ion scan and MRM operating modes are important because of increased selectivity compared with SIM, since matrix compounds that co-elute with the target analyte and that happen to yield ions with the same *m/z* as the target-derived ion selected for SIM are unlikely to yield product ions with the same *m/z* as the analyte. Even though the absolute signals are lower in MS/MS spectra than in the conventional mass spectrum, the S/B values (see Section 4.4.8) are increased since the increased selectivity ensures that the background decreases even more and thus yields lower detection limits at the end of the day (Figure 6.1). One reason why the QqQ instrument is so predominant in trace quantitation work is that it is capable of monitoring several precursor–product ion 'reaction channels' in MRM mode (e.g. for additional product ions monitored to confirm analyte identity and/or for multiple analytes). However, it is important to obtain a sufficient number (~10) of data points across chromatographic peaks in order to define the peak sufficiently well for integration to give a peak area with precision and accuracy that are fit for purpose (see Section 6.2.3f and Figure 6.7). This aspect is a reflection of the excellent duty cycle (Section 6.2.3e) of the

QqQ in MRM mode and enables use of shorter chromatography columns and shorter run times (and thus increased throughput), while the additional selectivity provided by MRM detection can permit simplification of sample preparation procedures (Chapter 3). In practice, usually for each target analyte just one reaction channel (sometimes referred to as an MRM transition) is used to provide the quantitative data while one or two others are sometimes monitored simultaneously in order to provide confirmation of analyte identity via the relative responses (essentially a check on the selectivity of the analytical method, Section 9.4.3b). In contrast with the QqQ, the 3D ion trap has a poor duty cycle in MRM mode and the full scan product ion scan method is the method of choice if this analyzer is used for quantitation, since it is often possible to acquire 10 such scans across a chromatographic peak with adequate S/B values. Additional post-acquisition data processing is required to obtain quantitation data from such full scan MS/MS experiments,

e.g., decide which ions in the product ion spectrum will be included in a summed intensity to give the final analytical signal.

Other MS/MS operating modes can be useful in screening experiments. The most important of these are the precursor ion scan (searching for precursor ions for a mass selected product ion) and the constant neutral loss scan (searching for precursor ions that fragment to yield a neutral fragment of user selected molecular mass). The former can help identify compounds in a mixture that contain a common partial structure contained in the mass selected product ion, while the latter mode can provide information on components that contain a particular functional group that is characterized by the molecular mass of the chosen neutral fragment; extensive lists of correlations (McLafferty 1993) can provide invaluable help in interpreting these screening MS/MS scans. The relationships among the various MS/MS scan modes are illustrated in Figure 6.2(a) (Schwartz 1990).

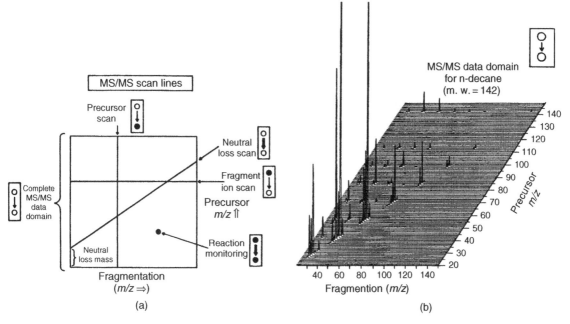

Figure 6.2 (a) Diagram illustrating the inter-relationships of the one-dimensional spectra that can be extracted from the complete MS/MS domain. Full circles denote ions whose m/z values are fixed by the conditions of the experiment, and open circles denote ions whose m/z values are to be obtained from the appropriate one-dimensional scan; this convention is also readily extended to MS^n experiments with n > 2. A conventional (MS^1) mass spectrum corresponds to a scan along the 'Precursor m/z' axis. (b) An example of an experimental MS/MS domain obtained for EI of decane using a triple quadrupole mass spectrometer; the three-dimensional plot (signal intensity is plotted vertically) is shown in perspective view. Since all ions in this spectrum were singly charged, all fragment ions have m/z values less than those of their respective precursors. (If doubly charged ions were present some fragmentation reactions could result, instead of one fragment ion plus a neutral fragment, in two singly charged fragments one of which could have a m/z value (but *not* mass!) greater than that of its precursor). Adapted from Schwartz, *Anal. Chem.* **62**, 1809 (1990), copyright (1990), with permission from the American Chemical Society.

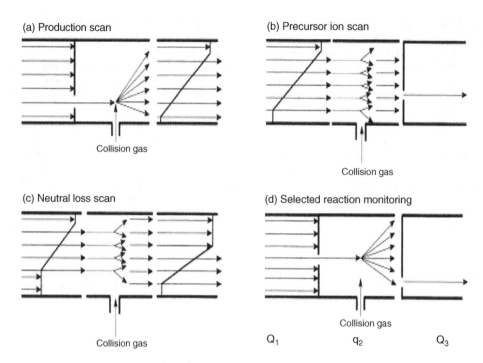

Figure 6.3 Sketches illustrating different MS/MS experiments implemented using a triple quadrupole mass spectrometer $Q_1q_2Q_3$ (Section 6.4.3). Such instruments are particularly suitable for this purpose since the linear quadrupoles respond directly to m/z values of ions (i.e., are m/z filters) rather than to, e.g., ratio of ion momentum to charge. In these sketches a diagonal line indicates a scan of m/z, while a small aperture indicates that the m/z filter is set to transmit only ions with specified m/z values. Note that in all cases the species detected after the final quadrupole Q_3 are *fragment ions* formed by collision induced dissociation (CID) in the central quadrupole q_2 (no mass filtering in this component). Selected reaction monitoring (d) is a special case of the fragment ion scan in which Q_3 is set to transmit only ions of a selected m/z rather than scanning the entire fragment ion spectrum; this operating mode is extremely important for quantitative analyses. The precursor ion scan (b) is simply the converse of the fragment ion scan (a). The neutral loss scan determines the (user-selected) mass of the neutral fragment formed in the CID only indirectly by scanning Q_1 (determines m/z of precursor ions) and Q_3 (determines m/z of fragment ion) in a linked fashion such that the transmitted m/z values are separated by a user-selected difference.

Once again the triple quadrupole instruments shine in this respect, in that they can be readily operated in all of these screening scan modes in addition to MRM (Figure 6.3). Ion traps and QqTOF instruments can not be operated to provide precursor or constant neutral loss scans, although the corresponding information can be obtained by extensive post-acquisition processing of product ion scan spectra obtained for all precursor ions in the MS[1] spectrum that leads to a three-dimensional map like the example shown in Figure 6.2(b) (Schwartz 1990).

Double focusing magnetic sector analyzers were very important in the initial development of MS/MS experiments. Historically, the first information of the MS/MS type was provided by the so-called 'metastable peaks' (Hipple 1945), observed as broad peaks at nonintegral values of m/z in magnetic sector MS[1] spectra. These broad signals correspond to fragment ions formed from

ions extracted from the source but that dissociate spontaneously (and are thus 'metastable') while traversing the field-free region immediately upstream of the magnetic sector; 'metastable ions' dissociate as a consequence of excess internal energy acquired during the ionization event itself. Subsequently techniques of so-called ion kinetic energy spectrometry were developed (Cooks 1973), of which the best known was mass analyzed ion kinetic energy spectrometry (MIKES) that was later shown (Boyd 1994) to be a special case of the family of 'linked scans' that can provide all three of the important MS/MS scan modes (and, with some additional difficulty, MRM). However, such operating modes of double focusing analyzers are never used for serious quantitative analyses and very seldom any more for screening experiments (precursor and constant neutral loss scans). Similarly, while it is possible to operate a stand-alone

TOF analyzer equipped with a reflectron (Section 6.4.7) in an equivalent 'metastable ion' mode (so-called post-source decay, PSD), this operating mode is too tedious and insensitive to be useful for trace quantitation.

In addition to the potential of MS^n to increase selectivity in quantitative analyses (Figure 6.1), as noted above this technique is also an aid to deduction of chemical structures of unknown compounds from mass spectrometric data using first principles, as much an art as a science (McLafferty 1993). In addition, a mass spectrum of all the ions observed in any mass spectrometric experiment can also be used as an uninterpreted fingerprint for the compound that was ionized, as in library matching methods. For electron ionization (EI) mass spectra (MS^1 spectra), the reproducibility achievable using different instruments under well specified conditions (Section 5.2.1) is sufficient that universal (instrument independent) spectral libraries are feasible (Stein 2004, Halket 2005). For soft ionization techniques such as the API methods, which produce only ionized intact molecules in the MS^1 spectra, MS^n spectra are required to provide sufficient diagnostic product ions for usable chemical information. For many years it was believed that lack of reproducibility in MS^2 spectra precluded construction and use of libraries, but recently this view has been challenged and interesting results obtained (Baumann 2000, Bristow 2004, Halket 2005, Milman 2005, Mueller 2005). However, spectral library matching is not considered to provide a definitive proof of structural identity in validated quantitative analyses, since the scores describing the similarity of library and unknown spectra can only be interpreted on the basis of comparisons with the compounds that happen to be represented in the library used (see Section 9.3.3b for further discussion).

Thus far very little has been said about how intact ionized molecules, e.g. those formed by 'soft' ionization techniques like the API methods (Sections 5.3.3–5.3.6), can be induced to dissociate for subsequent MS/MS analysis. (In fact any ion produced in an ion source can be m/z selected and subjected to MS/MS analysis). Soft ionization does not produce metastable ions (see above) in any abundance if at all. Historically the most common method of ion activation has been collisional activation (CA), whereby ions are accelerated through a defined potential drop to transform their electrical potential energy into kinetic (translational) energy, and are then caused to collide with gas molecules that are deliberately introduced into the ion trajectory; the history of this approach has been described in an excellent overview (Cooks 1995). This involves conversion of part of the ions' kinetic energy into internal energy that in turn leads to fragmentation. It is still by far the most commonly used method.

The overall process of CA plus dissociation is known as collision induced dissociation (CID), first introduced in the very earliest days of mass spectrometry, but occasionally CAD is used to denote the same overall process. More detailed discussion of CA and CID is given in Section 6.5.

6.2.3 Figures of Merit for Mass Analyzers

The concept of figures of merit for mass analyzers was briefly introduced in Section 6.1. Further, sensitivity and selectivity were introduced early in Section 6.2 since they were necessary for the ensuing discussion of the various operating modes (including MS/MS) of mass analyzers. (Note that these two figures of merit can be applied to an overall analytical method as summarized in Figure 1.2, as well as to mass analyzers alone.) Here other properties of mass analyzers that must be considered in determinations of fitness for purpose for any specific analytical problem are outlined. Although it is convenient to discuss these figures of merit separately, many of them overlap or interact with one another, e.g., absolute response (ratio of detected signal to the number of analyte ions supplied from the ion source) and resolving power are generally inversely related for a given instrument type, as are absolute response and m/z range, while resolving power and cost are usually positively correlated among different instrument types.

6.2.3a Accessible m/z Range

The 'mass range' parameter quoted by a manufacturer for a particular instrument generally refers to the highest mass of singly charged ions for which the (mass) peak height is significantly greater than the background instrumental noise for some (generally unstated) conditions. As a general rule, and to varying extents and for different reasons, the signal level recorded by the ion detector for a given number of ions extracted from the ion source decreases as the mass and/or m/z of the ions increases. This fall-off can be the result of transmission losses through the m/z analyzer and the characteristics of the detector (Chapter 7). However, in the context of trace level quantitative analysis there is an effective upper limit of ~2000 Da (Section 1.5), so the mass range is not usually very important except when the target analytes are larger peptides (Section 11.6). In any case, the upper limit of the accessible mass range is not as important *per se* as the resolution–transmission curve for the analyzer at each m/z value of interest (see Section 6.2.3d).

There is usually less concern with the low end of the accessible m/z range since ions in this range are generally of low diagnostic value (do not provide sufficiently high selectivity), as they can be formed from a wide variety of compounds including solvents and other components of mobile phases. However, ion traps do have significant limitations with respect to the lowest m/z value that can be trapped and stored, particularly in MS/MS mode (a consequence of the q value, see Section 6.4.5).

6.2.3b Resolving Power

Resolving power (RP) is a measure of the ability of a mass analyzer to separate ions of different m/z values. In this book 'resolving power' is taken to refer to a property of an instrument, while 'resolution' refers to the separation between similar m/z values actually achieved in a real mass spectrum. The resolving power of any given analyzer can be set by the user as a tuning parameter, up to the practical limit that is determined by the analyzer type and specific design. In practical terms, setting the RP value is a very important aspect in development of a quantitative analytical method, since it is inevitably involved (Chapter 9) in finding a best compromise between sensitivity (inversely related to RP) and selectivity (increases with RP).

Unfortunately, several different definitions of resolving power are used in mass spectrometry. (Recall that 'resolving power' is taken to refer to a property of an instrument, while 'resolution' refers to the separation between similar m/z values actually achieved in a real mass spectrum.) 'Unit resolution' (sometimes 'unit mass resolution' or 'Δmass = 1') is a somewhat vague term but generally means that a peak at any chosen integer value m/z is 'clearly' separated from peaks at adjacent integral m/z values, e.g., m/z 50 can be unambiguously distinguished from m/z 49 and 51, and m/z 1000 from 999 and 1001, and so on. This somewhat loose definition is most often used with respect to linear quadrupole mass filters and ion trap mass spectrometers that can provide approximately constant peak widths across an appreciable m/z range. For this reason, for quadrupoles and ion traps a working definition of unit mass resolution is that the full peak width at half-maximum *(FWHM)* should be 0.7 Th (Figure 6.4).

The RP for other analyzer types that do not provide approximately constant peak widths across the m/z range is conventionally defined by Equation [6.1]:

$$RP = (m/z)/\Delta(m/z) = m/\Delta m \qquad [6.1]$$

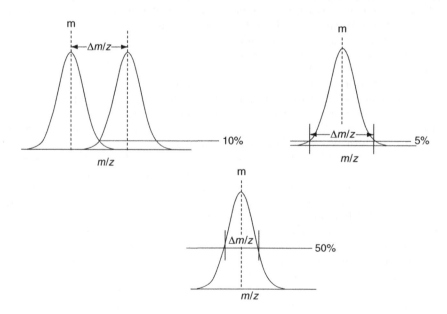

Figure 6.4 Illustration of the different definitions of resolving power for mass spectrometers; $RP = (m/z)/\Delta(m/z)$. In the 10 % valley definition $\Delta(m/z)$ refers to the separation of two peaks that is required in order that the height of the valley between them is only 10 % of the average height of the two peaks. The peak maxima are denoted as 'm'. In the other two definitions $\Delta(m/z)$ is measured on a single peak as the full peak width measured at a specified % of the peak height; the 5 % peak height definition is approximately equivalent to the 10 % valley definition, and the 50 % peak height case is frequently referred to as RP based on FWHM (full-width-half-maximum) of a specified peak.

where $\Delta(m/z)$ is the m/z difference between two neighbouring peaks that can be distinguished; usually these two peaks correspond to ions with the same charge (value of z) so the second form of Equation [6.1] is commonly used. This definition of RP was introduced for magnetic sector analyzers (that dominated mass spectrometry until the 1980s) since they can provide approximately constant RP over a significant m/z range. The other important point about Equation [6.1] concerns the definition and measurement of Δm (Figure 6.4). For most analyzers (e.g. FTICR and time of flight) the FWHM definition of Δm is used. However, for magnetic sector mass spectrometers adjacent peaks are said to be resolved if there is a 10 % valley between them, i.e., if Δm is such that the height of the valley is 1/10 of that of the higher of the two peaks (Figure 6.4). (Such a value of Δm can be measured approximately from a single peak as the peak width at 5 % of peak height). Simple trigonometry applied to an (assumed) triangular peak shows that the FWHM definition of RP gives a value approximately double that for the 10 % valley definition for the same peak. Magnetic sector analyzers provide RP values (Equation [6.1]) that are approximately independent of m/z (i.e. Δm is proportional to m), but this is not always true for other mass analyzers, so in such cases it is important to know the value of m/z at which a reported RP value was obtained.

Conventionally, analyzers are often classified with respect RP as low (RP~100–1000), medium (RP~2000 – 10 000) and high (RP > 10 000). Currently quadrupole mass filters and ion traps are generally classified as low, time of flight analyzers as medium, double focusing instruments straddle the medium-to-high RP categories, while FTICR and Orbitraps have high RP, although RP can vary strongly with the m/z value of the ions.

An obvious question concerns the reason why an observed peak in a mass spectrum, corresponding to detection of ions that all have identical m/z values, should have a finite width on the m/z axis at all. One of the main reasons in the case of ion beam instruments concerns the less-than-ideal characteristics of the ion beams extracted from the ion source and presented to the analyzer. Ideally the ion beam would be extremely narrow, containing ions all moving in the same direction and with the same kinetic energy (i.e., the same velocity for ions of a given m/z value). In reality the ion trajectories cover an appreciable width and diverge from one another; also their kinetic energies are distributed over a range of values for various reasons (see in particular the discussion of time of flight analyzers in Section 6.4.7). These distributions of the ions' properties in space and velocity affect different ion beam analyzer types to different degrees, but always contribute to broadening of the mass spectral

peak recorded by the ion detector following the analyzer to an approximate Gaussian shape (Figure 6.5). The real-life spectral peaks can be recorded in continuum mode (full peak shape is preserved) or in centroided mode (the acquisition computer processes the originally acquired continuum data to calculate and record only the centroid of the peak, i.e., a special average of the m/z values across the peak such that if the peak shape were printed on paper and cut out, the peak would balance on a knife-edge situated along the centroid line, see Equation [7.4] in Section 7.4). Peak broadening in ion trap instruments is the result of other physical processes that can still be traced to distributions of ions of the same m/z with respect to both spatial position and velocity.

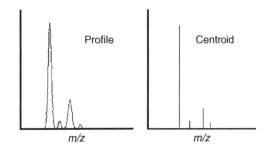

Figure 6.5 Ions extracted from an ion source and passed to a mass analyzer have a range of speeds and directions that lead to m/z peaks of finite width. When the full peak shapes are recorded the resulting spectrum is said to be in 'profile (sometimes continuum) mode'. Such spectra require much more disk storage than those that are converted to peak centroids (see main text) that can be done 'on-the-fly' i.e., during data acquisition, or post-acquisition.

In the present context of quantitative analysis of target analytes present at trace levels in complex matrices, the importance of high resolving power lies in the additional degree of detection selectivity that it provides. A particularly important example where this is essential is that of PCDDs ('dioxins') discussed later in Section 11.4.1. It is also true that higher RP implies that the positions of the peaks along the m/z axis are more precisely defined and, in turn, this can facilitate more precise measurements of the m/z value; this in turn can provide additional confirmation of analyte identity in cases where this is an important issue (Sections 6.2.3c and 9.4.3b).

6.2.3c Accuracy and Precision of Mass Measurement

In qualitative (structural) analysis, so-called 'accurate' (sometimes 'exact', although that can be a misleading term) mass measurements are important contributions of mass spectrometry. The 'accurate mass' value and its associated uncertainty are used to calculate all possible elemental compositions that have calculated masses falling within the specified uncertainty of the measurement, an important contribution to qualitative (structural) analysis by mass spectrometry (Beynon 1960; Gross 1994).

To be useful these mass measurements (strictly of m/z) should have uncertainties (combined accuracy and precision) in the parts per million (ppm) range (or sometimes with uncertainties <10 mDa), or even better, and are normally made using a high resolving power instrument. This ensures both that the mass peak of interest does not correspond to a multiplet of unresolved ions and also that the position of the peak on the m/z axis is more precisely defined; instrument stability is a crucial concern in this regard. The term 'high resolution' is often used interchangeably with 'accurate mass' in the literature but the terms really refer to two different things, since mass measurements of a useful degree of accuracy and precision can, in some cases, be obtained using even unit mass resolving power (Tyler 1996; Kostianen 1997; Paul 2003; Grange 2005; Gu 2006). However, as the sample becomes more complex (e.g. increased risk of unresolved multiplets, and/or higher mass compounds that require smaller uncertainties in measured mass in order to pin down a unique molecular formula), very high resolving power (FTICR or Orbitrap) becomes essential if useful results are to be obtained (Sleno 2005b). This figure of merit is important for quantitative analysis via its close relationship to resolving power (Section 6.2.3b) and also if it is important to confirm analyte identity (Section 9.4.3b).

6.2.3d Transmission Efficiency

This parameter is important as a result of its direct influence on the sensitivity of an integrated analytical method, and thus on the limits of detection and quantitation (Section 8.4). Usually the concept is applicable to ion beam analyzers and generally refers to the fraction of ions injected from the ion source that are successfully transmitted through the analyzer to an ion detector (Figure 1.3). More generally, transmission efficiency describes the fraction of the ions in a beam, injected into one component of a mass spectrometer, that are successfully passed on (transmitted) to the next component; for example, designers of QqQ instruments are concerned with the transmission efficiency from one quadrupole component to the next. Modern analyzers of all types can now be designed and constructed so that transmission efficiencies are very high. For example, it was shown (El-Faramawy

2005) that the SIM transmission efficiency for a range of ions of molecular mass from a few hundred Da to ~16.9 kDa (but all with $m/z < 2000$), through a linear quadrupole mass filter operated at strictly controlled unit-mass resolving power, was ~33 % based on a comparison with the same quadrupole operated in ion funnel (RF-only) mode (see Section 6.4.5). In all beam-type analyzers transmission efficiency is inversely related to resolving power (not usually linearly) and the only truly meaningful representation of this figure of merit is the graph relating these two quantities for a specified m/z value; Figure 6.6 shows an example, in this case for a single linear quadrupole analyzer. Transmission efficiency is one of the instrument performance parameters that are optimized during tuning (Sections 6.3.2 and 9.5.2).

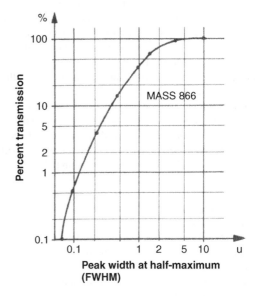

Figure 6.6 Example of the dependence of transmission efficiency on resolving power (measured here on the FWHM definition). This example is for a linear quadrupole mass filter. Reproduced from Feser and Kögler, *J. Chromatogr. Sci.* **17**, 57 (1979), copyright (1979), with permission from Elsevier.

The transmission efficiency of the analyzer is just one component of the overall sampling efficiency of the instrument, defined as the fraction of analyte molecules delivered to the ion source that are ionized and delivered to the detector; obviously the ionization efficiency of the ion source must be considered in addition to the transmission efficiency of the analyzer. Even this does not fully describe the overall efficiency of production of analyte derived signal, since detector response (a function of m/z, see Chapter 7) and duty cycle of the instrument analyzer (Section 6.2.3e) are also important.

6.2.3e Duty Cycle

This figure of merit is closely related to others, e.g. it contributes to detection sensitivity, and is related to the time required to acquire mass spectral data with adequate ion statistics to account for shot noise (Section 7.1.1). Ideally all ions injected into the analyzer from the ion source should reach the detector and thus contribute to the observed signal but in reality this is not the case due, in part, to transmission efficiencies <100%. However, the duty cycle of the analyzer can be the major contributor to wastage of ions. An important example arises when a complete mass spectrum of ions from a continuous ionization source (e.g., APCI, or ESI and its variants) is recorded by serial detection, i.e., by scanning the m/z value transmitted to the ion detector, as is done with, e.g., magnetic sector and linear quadrupole analyzers. In contrast, parallel detection arranges to simultaneously record signals for all ions, regardless of m/z value; this is a property of TOF, FTICR and Orbitrap spectrometers, and of magnetic sector instruments fitted with focal plane detector arrays or even photographic plates. For serial spectral scanning, during the time that the detector is recording the signal intensity at one particular m/z value all other ions are lost to the walls or other components of the analyzer. The fraction of the total scan time during which ions of a particular m/z value, extracted from the ion source, actually reach the detector is called the duty cycle. For such instruments, when operated so as to record a full mass spectrum, the duty cycle can be only a few per cent depending on the range to be scanned. Ion trap instruments operated in full-spectrum mode partake of the characteristics of both serial and parallel detection, since the ions to be analyzed are first efficiently trapped and the trapped ions are then scanned out to the detector. The time required to scan out the ions (and thus the time during which ions are not trapped and are thus wasted) can be appreciably less than the time devoted to ion accumulation in the trap, but nonetheless a low duty cycle is one of the disadvantages of ion trap analyzers (see discussion in Section 6.4.5). For time of flight analyzers (Section 6.4.7) operating with continuous ion sources, although these are not 'scanning' instruments in the usual sense, the duty cycle is limited to perhaps 10–20% (depending on the m/z range) as a result of the wastage of ions arriving from the ion source during the flight time of the previous 'packet' of ions that was accelerated into the flight tube. For time of flight instruments operated with pulsed ion sources like MALDI, such that the ionization pulse also

provides the 'start' signal for the recording of flight time, the duty cycle is close to 100%. However, more recent developments (Brenton 2007) suggest that this limitation can be overcome in future if the m/z range of interest can be specified in advance (see Section 6.4.7).

Note that the preceding comments apply to analyzers operating in full spectrum acquisition ('full scan') mode. However, in the context of quantitative analyses of target analytes, scanning instruments like linear quadrupoles and magnetic sectors have a significant advantage in that they can be operated in SIM or MRM mode (Section 6.2.1) to monitor only one or a few m/z values characteristic of the analyte. If the number of m/z values monitored in this way is n, then the duty cycle can be close to $(100/n)\%$, ignoring the (generally short) switching times among the m/z values.

6.2.3f Data Acquisition Rate

The time-widths of some chromatographic peaks can be as low as 1–2 seconds, so characterization of such peaks requires a commensurately shorter (perhaps 10 times) data acquisition time. Generally a minimum of 10 data acquisitions is required to define a chromatographic peak sufficiently well that precise and accurate measurements of chromatographic peak area can be made by integrating the peak area, or to measure the peak height for purposes of quantitation (Figure 6.7). Here emphasis is on quantitation of target analytes using ion beam instruments operated in SIM or MRM mode, and for this purpose a minimum of 8–10 of the SIM or MRM cycles, each cycle monitoring n different m/z channels, can provide a representation of the chromatographic peak that is adequate for peak area integration together with some degree of analyte identification via relative intensities at the n m/z values. Here again, the linear quadrupole and the triple quadrupole are instruments of choice as only electric fields need to be switched in order to monitor discrete m/z channels, so that interchannel switching times are negligible and the duty cycle (%) approaches $100/n$. Magnetic sector instruments can also be operated in SIM mode using switching of only electric field fields (potentials applied to the ion source and electric sector, see Section 6.4.4) but transmission efficiency falls off rapidly if too large an m/z range is covered. In such circumstances the magnetic field must also be switched (See e.g. Section 11.3.1a) and magnet hysteresis (Section 6.4.4) limits the speed and increases the difficulty of such SIM switching.

Operating parameters that affect the rate of acquisition of data points in SIM and MRM modes are the switching time (an intrinsic property of the analyzer) and the dwell time per SIM or MRM channel (specified m/z value for

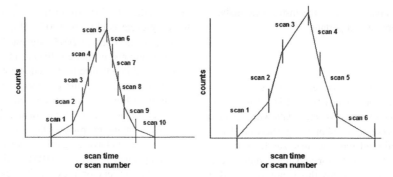

Figure 6.7   Sketch illustrating the importance of the number of mass spectrometric data points (either full spectra or SIM or MRM) acquired across a chromatographic peak, for adequate representation of the shape of the peak in order to calculate accurate peak areas and/or heights. Usually 10 scans (or SIM or MRM acquisition windows) are considered necessary to provide an adequate representation of the peak centroid (and thus retention time) and peak area.

SIM and precursor product ion m/z values for MRM). The dwell time is the time spent accumulating ion signals in a particular SIM or MRM channel before switching the instrument to the next, and can be selected within limits by the analyst. The importance of dwell time selection arises from the need to compromise between signal:noise ratio and the number of data points recorded across a chromatographic peak. In this context the 'noise' component that it is wished to minimize is mainly that arising from ion statistics (see discussion of Figure 6.1 and Section 7.1.1) in truly trace level analyses, where individual ions are detected as independent events using ion counting detection techniques. Such 'shot noise' can be minimized by detecting and recording a sufficiently large number of ions that the time variation of the rate of ion arrival at the ion detector is averaged out; this can be achieved in practice, for a given number of ions generated by the ion source, only by increasing the dwell time. (However, in the vast majority of cases of interest in this book chemical background is by far more important than random noise, see Section 7.1.) On the other hand, the dwell time must be sufficiently short that the data acquisition rate is sufficient that enough data points are collected across a chromatographic peak (Figure 6.7). The overall sensitivity of the ion source–analyzer–detector combination (Figure 1.3) is also important here, since a higher sensitivity implies a larger rate of ion counting and thus a less difficult compromise between signal:noise and dwell time. The situation is further complicated by the time required for the raw data to be processed by the detection electronics and written to the disk of the computer.

In the final reckoning it is the 'total data acquisition time' that must be considered when ensuring at least 10 data points are acquired across a chromatographic peak when using ion beam instruments in SIM or MRM mode:

total data acquisition time per SIM/MRM cycle = [(dwell time per SIM/MRM channel) + (total switching time) + (data transfer time)] × [number of channels monitored per cycle]

A need for more than one SIM or MRM channel ($n > 1$) could arise if there is a requirement to analyze for more than one analyte and/or to monitor more than one channel (m/z value) per analyte in order to provide some degree of confirmation of analyte identity via relative intensities of the signals recorded in the different channels. If the dwell time per channel must be increased for reasons of signal:noise (assuming the limiting noise is ion statistics and not chemical background, see Figure 6.1 and Section 7.1.1), then to maintain a data acquisition time per cycle that is compatible with the chromatographic peak width the only practical recourse is to reduce n, the number of channels monitored per cycle. If some of these channels were intended to monitor analytes that are chromatographically well separated, time-scheduling of different sets of channels (via the mass spectrometer data system) can often be used to accommodate acquisition of all the required data.

While the concept of duty cycle also applies to ion traps, if an attempt is made to produce a similar simple relationship the situation becomes immediately more complicated as a result of the need for such instruments to deal with the implications of space charge effects. For example, only a limited total number of ions (analyte plus matrix) can be accumulated in a trap at any one time, so the number of microscans (fill-scanout cycles, Section 6.4.5), which must be accumulated and averaged to provide adequate ion statistics for the analyte ions, is highly variable depending on the analyte:matrix ratio. If selective ejection techniques (Section 6.4.5) are used to eliminate matrix ions this also increases the time per microscan in a variable fashion.

6.2.3g Dynamic Range (Range of Reliable Response)

This parameter is strictly a property of the complete ion source-analyzer-detector combination, rather than of the analyzer alone, and indeed the concept can be applied to an entire integrated analytical method (Figure 1.2); however, the instrument only version is conveniently described here. The linear dynamic range of an instrument refers to the range over which the observed signal is directly proportional to the amount (or concentration) of analyte supplied to the ion source. In most cases, the lower limit of this range is defined by the instrumental lower limit of quantitation, the smallest amount (concentration) of pure analyte introduced into the ion source that provides a signal that can be measured with sufficient precision that it can meaningfully be said to follow the same linear signal concentration relationship as the data obtained for higher values (to within experimental uncertainty limits that must be specified in advance). The upper limit of the instrumental linear dynamic range is defined by the amount (concentration) where significant deviations from the linear relationship start to appear, sometimes as a result of saturation of the ion detector (Chapter 7). It is possible to conduct quantitative measurements when this relationship is nonlinear, although less easily (Section 8.3.6), and the more general term 'dynamic range' is used in such cases.

The same terminology is also applied to a complete integrated analytical method for a target analyte in a particular type of matrix. If the distinction between instrumental and method dynamic ranges is not made explicit, the context usually makes clear which of the two is intended.

6.2.3h Versatility for Tandem Mass Spectrometry

This figure of merit was discussed in Section 6.2.2, where it was emphasized that the triple quadrupole (QqQ) instrument is by far the most versatile in this respect (Figure 6.3). The scans for precursor ions of selected product ions or neutral fragments are never used directly in quantitative trace analysis, since they have intrinsically very low duty cycles, but can be useful in detection of analytes with chemical structures related to that of the target analyte.

6.2.3i Ease of Use

This criterion refers not only to the ease (or otherwise) of operating an analyzer so as to meet its performance specifications, but also to the longer term difficulties of maintaining the instrument in a condition that is fit for purpose (Section 9.5.1). As a broad generalization, experience indicates that ease of use is inversely related to the resolving power that can be obtained at a specified transmission efficiency (Section 6.2.3d). For example, double focusing magnetic instruments do not score high on ease of use since they require maintenance of a high or very high vacuum to achieve their designed resolving power-transmission characteristic curve. Also, they are particularly sensitive to effects of contamination of the internal metal components by organic (nonconducting) layers that can charge up electrostatically in an uncontrolled fashion. At the high end of the ease of use scale, linear quadrupoles and 3D and 2D ion traps can tolerate relatively poor vacuum ($\sim 10^{-5}$ torr) and are significantly less susceptible to the effects of such contamination.

6.2.3j Capital and Maintenance Costs

As a broad generalization, of the analyzer types currently used in quantitative analysis, double focusing magnetic instruments are the most expensive in terms of both capital and maintenance costs (the latter are related to the comments under ease of use, Section 6.2.3i). Linear quadrupole analyzers and ion traps are usually regarded as being at the low end of this scale, with time of flight analyzers (particularly those fitted with reflectrons) in the middle. The foregoing comments refer to stand-alone single analyzers. Instruments for MS/MS operation are more difficult to fit into a generalized cost ranking, e.g., stand-alone ion traps provide MS^n facilities (Section 6.2.2) at no extra cost, while multi-analyzer ion beam instruments such as the QqQ and QqTOF are necessarily much more expensive than single analyzer instruments (Table 6.1).

Clearly, these figures of merit (others could be included) are sufficiently numerous and their inter-relationships sufficiently complex that the proposed purchase of a new instrument requires careful consideration of the purposes for which it is required. The same complex web of inter-relationships also applies to the converse case, in which one is considering whether an existing instrument can provide the data required for a new application. In later sections of this chapter the more important analyzers currently in use for quantitative analyses are discussed with these criteria in mind.

6.3 Motion of Ions in Electric and Magnetic Fields

All *m/z* analyzers rely on application of electric and/or magnetic fields to separate ions according to their *m/z*

Michael Faraday and the Concept of a Field of Force

Michael Faraday (1791–1867) was a British chemist and physicist (he considered himself a natural philosopher) who made enormous contributions to electromagnetism and electrochemistry. He is widely regarded as the greatest experimentalist in the history of science. It was largely due to his efforts that electricity became viable for use in technology. The SI unit of capacitance (the farad) is named after him, as is the Faraday Constant (the charge on a mole of electrons, about 96 485 coulombs). He made many discoveries in chemistry, including benzene, and invented a system of oxidation numbers of the elements.

Faraday was born in south London to a poor family; his father was a Yorkshire blacksmith who suffered ill-health throughout his life. In the rigidly class-conscious England of that day, a poor lad like Faraday had no chance of much of a formal education and indeed in his early years he suffered considerably from intolerance of this kind, particularly from the wife of the scientist Humphrey Davy who employed Faraday as laboratory assistant. He appears to have borne no rancor as a result; he was a devout member and elder of the small Sandemanian denomination, an offshoot of the Church of Scotland. During his lifetime, Faraday rejected a knighthood and twice refused to become President of the Royal Society. There is a plaque in his memory in Westminster Abbey near Newton's tomb, but he refused to be buried there and is interred in the Sandemanian plot in Highgate Cemetery in London.

Fortunately for science, Faraday caught the eye of John 'Mad Jack' Fuller, who became his sponsor and mentor and set him on the road to his brilliant career. Of his many contributions to science, his demonstration that a changing magnetic field produces an electric field became known as Faraday's Law of Magnetic Induction, which was later incorporated in much more sophisticated mathematical form as one of the four fundamental equations of electromagnetic theory by James Clerk Maxwell. These equations in turn evolved into the generalization we know today as *field theory*. Faraday used the principle to construct the first electric 'dynamo', the ancestor of modern power generators. He proposed that electromagnetic forces extended into the empty space around the conductor; this concept of lines of force emerging from charged bodies and magnets provided a way to visualize electric and magnetic 'fields of force', a model that was crucial in the development of electromechanical devices like generators and electric motors. In this context a beam of ions is an electric current and has forces exerted on it by magnetic and electric fields in exactly the same way as on the wires carrying an electric current in Faraday's experiments. (Note, however, the historical accident that electric current in a wire was originally thought to be a flow of positive charge, but we now know it is a flow of negative electrons in the opposite direction, with implications for the *direction* of the forces.)

---

values and the concept of a 'field' (has been used several times in preceding chapters with little or no explanation. In the 18th and 19th centuries physicists were concerned with the phenomenon of 'action at a distance'. When a hammer strikes a nail, it is understood intuitively that the motion of the nail was a result of the force exerted by the impact of the hammer directly on the head of the nail. But how can we understand the phenomenon whereby the ground on which we are standing attracts the hammer if we happen to drop it? A great advance in

the scientific treatment of such 'action at a distance' was the invention of the concept of a 'field of force', or more briefly a 'field'. This concept allowed the quantitative mathematical treatment of 'action at a distance', not only by Newton for gravitational fields acting on hammers, planets etc., but also (originally by Michael Faraday, see the accompanying text box) for electrical and magnetic fields acting on objects carrying electrical charges.

In very general terms, a field (e.g. gravitational, electric or magnetic) at a point in space describes the force (a vector with both magnitude and direction) on a particle possessing unit magnitude of the appropriate property (mass or electric charge or electric current) that is situated at that particular point. Such electrical and magnetic interactions through space are actually different manifestations of a more generally inclusive electromagnetic field, as first described for classical (nonquantum) and nonrelativistic phenomena by Clerk Maxwell in the 19th century, but it is convenient here to consider them separately (see accompanying text box on the concept of physical fields). These fields are the physical basis of machines that exploit electric currents (flows of electrically charged particles) within a magnetic field to create mechanical forces that are used to do useful work (i.e., electrical motors and magnetic sector analyzers), and also the 'converse' machines that use mechanical forces from other sources (falling water, steam turbines, etc.) to produce the electric currents that are used at work and home (i.e., generators). In the present context, these same fundamental effects underlying the operation of electrical motors are responsible for the motions of ion beams (i.e., electric currents of a special kind) in mass spectrometers that are exploited for separation of the ions according to their *m/z* values.

Various types of electric field are used in different mass analyzers. From this perspective the simplest example is provided by the constant field (DC) strengths used to accelerate ions, e.g. out of an ion source into an analyzer, or from an intermediate component into the flight tube of a time of flight analyzer etc. More elaborate fields whose strength is varied in time at frequencies in the range of a few MHz (within the radiofrequency (RF) range) are used in linear quadrupoles and ion traps. However, the same principles of physics are exploited in all cases, and the following section is an introduction to these principles.

6.3.1 Introduction to Interactions of Electric and Magnetic Fields with Ions

The interaction of electrical and magnetic fields with charged particles (ions and electrons in the present context) is fundamental to the understanding of how mass spectrometers separate ions according to their ratios of mass to charge. A brief introduction to this subject is given in Appendix 6.1. For convenience the equation numbers in the Appendix continue those in the main text.

The relationships discussed in Appendix 6.1 that are most important for the following discussion concern the relationships among electric field strength E_X along a spatial coordinate X (see accompanying text box), electrical potential Φ and potential energy ($q_{ion} \cdot \Phi$) of an ion with charge q_{ion} located at a position X:

$$E_X = d\Phi/dX \qquad [6.9]$$

The SI unit for electrical potential is the volt (V); unfortunately the same symbol (V) is frequently also used instead of Φ to denote the physical quantity (electrical potential) for which the volt is the SI unit, but in practice the context does not usually lead to confusion and this book will follow the literature by using both symbols occasionally. The SI unit for the magnitude of an electric field (the electric field strength) E is thus ($V.m^{-1}$).

A consequence of these relationships together with the principle of conservation of energy will play an important role in this book. In the absence of any additional energy loss/gain arising from collisions with background gas molecules, when an ion moves from a point with electrical potential Φ_1 to another with Φ_2, i.e., changes its potential energy from $q_{ion} \cdot \Phi_1$ to $q_{ion} \cdot \Phi_2$, the translational (kinetic) energy T of the ion must also change so that the total energy is unchanged:

$$\text{change in } T = (m_{ion}/2) \cdot [v_{ion,2}{}^2 - v_{ion,1}{}^2] =$$
$$- (\text{change in potential energy})$$
$$= -q_{ion}(\Phi_2 - \Phi_1) = q_{ion}(\Phi_1 - \Phi_2) \quad [6.10]$$

Note that q_{ion} represents both the magnitude and the sign of the charge. Then, for example, a positive ion ($q_{ion} > 0$) moves so as to minimize its potential energy between two positions (1 → 2) such that $\Phi_2 < \Phi_1$, thus acquiring additional translational energy $q(\Phi_2 - \Phi_1)$. For a negative ion ($q_{ion} < 0$) this will happen if $\Phi_2 > \Phi_1$. This convention about the signs of electrical charges results ultimately from the historical accident that it was originally believed that electrical current through a wire corresponds to a flow of positive charge, whereas we now know (thanks to J.J. Thomson, see text box in Chapter 5) that in fact the flow is of negatively charged electrons in the direction opposite to that of the 'conventional' (positive) electric current. A gravitational analogy to Equation [6.10] might involve a hammer slipping from a hand at a meter or so above the corresponding feet; the hammer will then spontaneously seek out a position of lower gravitational potential (shorter distance from the centre of the planet) while transforming its gravitational potential energy to kinetic energy (that in turn is transformed into heat energy when it is brought to rest by the foot!).

Equation [6.10] and its variants account for many processes occurring in a mass spectrometer; these include the transport of ions from ion source to analyzer to ion

Non-Relativistic Classical Fields Theory

Analytical mass spectrometers operate on the principles of classical (nonquantum) physics without relativistic effects. A physical force field can be thought of as a 'map' in which, at each point of space and time, the value of the force of interest on a specified 'test object' is assigned a value (both magnitude and direction). A simple example is a maritime weather map showing the wind velocity at various locations in a region of ocean (with implications for the resulting forces on a 'test sailing boat'). Historically, the first classical field theories were those describing electric and magnetic fields, first introduced in an intuitive way by Faraday; the 'test object' for an electric field is a body of negligible dimensions carrying a unit positive charge (one coulomb in SI), so *the magnitude of the electric field strength at a point is the electrical (Coulombic) force per unit charge*. A similar definition applies in a less intuitively obvious way to magnetic fields. Faraday and others showed experimentally that these two fields are related, and are in fact two aspects of the same field, the electromagnetic field. The interaction of charged matter (e.g., ions) with this field in the classical non-relativistic limit was fully described in the 19th century in Maxwell's electromagnetic theory, that uses vector fields to describe the electric and magnetic fields and their interactions. This classical theory is still widely used today within its sphere of validity.

electric field lines of force

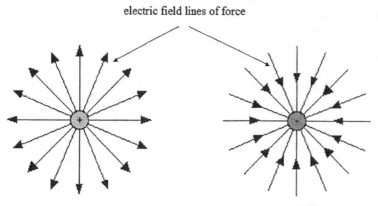

electric field arising from an
isolated positive charge

electric field arising from an
isolated negative charge

An electric field can be visualized by drawing 'lines of force' or 'field lines, that start on positive charges and end on negative charges and indicate both the strength and direction of the field. The 2D fields surrounding *isolated* charges are illustrated here; by convention the field is highest where the field lines are closest. When there is more than one charge present the electric field lines will not be straight lines but will curve in response to the different charges. The *magnitude* of the forces is ultimately derived from Coulomb's Law for the force operating between two charges of magnitude q_1 and q_2 and separated by a distance d_X in a vacuum: $F = q_1.q_2/(4\pi\varepsilon_0.d_X^2)$, where ε_0 is a constant called the *permittivity of free space* ($= 8.85 \times 10^{-12}$ $C^2.N^{-1}m^{-2}$ in SI). Then as a simple example, the vector field $\mathbf{E}$ at a specific point X a distance d_X from a single charge Q is: $\mathbf{E_X} = \mathbf{i_{QX}}.Q/(4\pi\varepsilon_0.d_X^2)$, where $\mathbf{i_{QX}}$ is a dimensionless vector of unit magnitude linking Q to X, and the corresponding *force* experienced by an object at X and carrying charge q is: $\mathbf{F_X} = q.\mathbf{E_X}$, where the *sense* of the vector $\mathbf{F}$ (i.e., Q→X vs X→Q, i.e., repulsion vs attraction) is determined by whether Q and q have the same or opposite signs. Vector signs are shown in bold font.

Magnetic fields are generally similar, but the magnetic forces of interest in mass spectrometry are exerted on charged objects (ions) that are *in motion* (i.e., on electric currents), so the intuitive understanding of such fields is more difficult to explain in simple terms.

detector (via tuning of the ion optics, Section 6.3.2), control of the collision energy of precursor ions with collision gas molecules to promote collision induced dissociation (CID, Section 6.5, note that ions of the same mass m_{ion} but different charge $q_{ion} = \pm z.e$ acquire different kinetic energies when accelerated through the

same change $(\Phi_2 - \Phi_1)$ in electrical potential) and the principle of operation of magnetic sector fields as *m/z* analyzers (Section 6.4.4).

Note also that a commonly used unit of energy, the electron volt (symbol eV), is defined as the difference in potential energy of a unit elementary charge *e* when moved between two points whose electrical potentials differ by one volt. Thus:

$$1\,eV = e.1V = [1.60217653(\pm 0.00000014) \times 10^{-19}\,C] \times$$
$$(1V) = 1.6022 \times 10^{-19}\,J$$

This calculation exemplifies the coherence of the SI system of units (Section 1.2), i.e. if the quantities appearing in an expression for another quantity are expressed in their base SI units, the result will automatically be in expressed in its SI base unit (joules in this case). The eV is a unit of energy per ion or molecule; if it is wished to convert to energies per mole, it is necessary to multiply by the Avogadro Constant $= 6.02205 \times 10^{23}$ molecules.mol^{-1}, i.e., $1\,eV = 9.6485 \times 10^4$ J.mol^{-1}.

In a typical mass spectrometer the range of potentials relative to 'ground potential' (the electrical potential of the surface of our planet, including ourselves!) ranges from a few volts to a few tens of thousands of volts. In this regard, it is conducive to the pursuit of a long career to memorize the First Law of Mass Spectrometry: 'A metal object at 10 000 volts looks exactly the same as one at ground potential'.

6.3.2 Ion Optics and Lenses: Instrument Tuning

The remainder of this chapter is mainly devoted to a description of some mass analyzers that are important in quantitative trace analysis. Other important components in mass spectrometers are the devices that maximize the efficiency of transfer of ions, e.g., from ion source to analyzer. Such devices exploit properties of electric (and occasionally magnetic) fields as expressed via Equation [6.10] and are referred to as 'ion lenses', since their properties resemble those of thick lenses in light optics. The arrangements of ion lenses used for this purpose are referred to as 'ion optics'. These are devices whose electrical potentials are changed during 'tuning' of a mass spectrometer; note that tuning for optimum combinations of sensitivity and resolving power can be extremely important for success of trace quantitative analysis. It can also be mentioned that if a mass spectrometer is to be used for negative ions, in accord with Equation [6.10] all tuning potentials should be reversed and possibly re-adjusted.

Details of instrument tuning vary widely among instrument types and recommended tuning procedures are highly manufacturer dependent. Most modern mass spectrometers provide the facility to store values of potentials applied to the various ion optical elements in a 'Tune File' in the computer, which can be called up for an analytical method. Note, however, that the optimum tuning potentials (especially for optical elements close to the ion source) can vary somewhat with time as a result of accumulation of the mass spectrometrist's worst enemy, variously referred to as 'dirt', 'crud', 'ion burn' etc. It is also important to realize that some tuning parameters are m/z dependent and others are not, and it is important to know which are which. In some instruments some tuning parameters (mostly potentials applied to the various lenses) can be optimized via an automated procedure, while others can be adjusted for optimum performance following such an 'auto-tune' procedure. The latter usually correspond to lenses closest to the ion source and are sometimes referred to as 'compound dependent lenses'; these secondary adjustments are generally more common for LC/MS than GC/MS instruments, since the effluents from the latter are usually 'cleaner' resulting in less build-up of the above mentioned dirt.

Although they are extremely important for the overall efficiency of a real mass spectrometer, detailed discussion of ion lenses and optics is outside the scope of this book. However, as an example Figure 6.8 shows a sketch of one of the most commonly used ion lenses, the Einzel Lens, an arrangement of three closely spaced metal tubes or pairs of plates, designed to focus a diverging ion beam. The two outermost elements of the lens are at ground potential (or alternatively at the same potential as that from which the ion beam originated), while the central element is held at a high potential (adjustable as a tuning parameter). This arrangement produces electric field lines that result in ion trajectories within the lens such as those indicated in Figure 6.8. (The polarity of the potential applied to the central element is the same as that of the charge on the ions but, as shown in Figure 6.8, the lens action is not a straightforward matter of repulsive forces exerted on the ions by the central element.) Because the arrangement of the lens elements is symmetrical, the ions regain their initial kinetic and potential energies on exiting the lens; this is important property for instrument design.

It should also be noted that the RF-only quadrupoles (as well as the related RF-only hexapoles and octapoles) can be operated as ion lenses, often referred to as RF-only ion guides or simply ion guides. RF-only quadrupoles (denoted as lower case 'q', nonitalicized, to distinguish them from the closely related quadrupole mass filters Q) also act as efficient collision cells for collision induced dissociation in triple quadrupole analyzers (QqQ, Section 6.4.3) and hybrid tandem quadrupole–time of flight instruments (QqTOF, Section 6.4.7). Discussion of these devices must be postponed until the principles of all linear quadrupoles have been considered (Section 6.4.2).

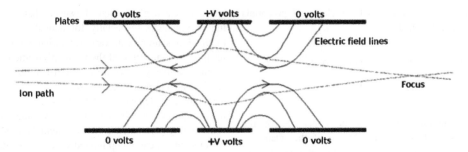

Figure 6.8   Sketch of the mode of action of an Einzel lens. This is a simple example in which the three lens elements are flat plates providing focusing in only one of the directions perpendicular to the main ion beam axis. The ions are assumed to be positively charged, and thus experience forces in the directions indicated on the field lines, i.e., contrary to naïve expectations, these positive ions are first deflected *towards* the central element held at a positive potential, but are subsequently re-directed to the central axis. It is the shapes of the field lines that determine the ion optical properties.

6.4 Mass Analyzers

In this section the mass analyzers that are important in current practice of trace level quantitation are described in some detail. Before starting the descriptions of specific analyzers, a topic of general applicability (calibration of the m/z axis) is discussed. In addition, it should be mentioned here that an extremely important component of any mass spectrometer is the vacuum system; this is discussed separately (Section 6.6). Another topic that is important for tandem mass spectrometry is that of collisional activation of ions, and discussion is also deferred till later (Section 6.5).

6.4.1 Calibration of the m/z Axis ('Mass Calibration')

As for any physical quantity (Section 1.1), measurement of m/z values of ions is ultimately a comparison of the unknown with a standard of some kind. In the case of a mass spectrometer, the raw data transferred to the computer do not involve m/z values at all but are specifications of the times at which electrical signals (arising from arrival of ions at the detector) are transferred to the computer during a scan of the electric and/or magnetic fields that the particular analyzer employs to separate the ions. (In the case of time of flight analyzers the relevant quantity is the time elapsed since the ions were pulsed into the flight tube, see Section 6.4.7.) To transform these times of detection into m/z values, it is necessary to obtain a calibration of time vs m/z obtained using mass calibration standards that yield ions of known m/z calculated from their molecular formulae and values of atomic masses (Table 1.4). It can be mentioned that

calibration of a mass analyzer can drift with time as a result of temperature changes, contamination build-up on lenses etc. and should therefore be done only after tuning.

Proper mass calibration of the analyzer, and regular checks on its validity, are important aspects of method development and validation (Section 9.5.1b). In most modern computer-controlled instruments, mass calibration is automated like instrument tuning (Section 6.3.2) and both 'Tune' and 'Calibration' files can be stored and called up from the computer for specific applications. For standard calibrant ions for which the analyzer can cleanly resolve the isotopologs, the monoisotopic ion (the lowest mass isotopolog containing only the lowest mass isotopes of the constituent atoms, i.e., ^{12}C, ^{1}H, ^{16}O, ^{14}N, ^{35}Cl etc.) is used, but note that at higher m/z values this is not necessarily the isotopolog of highest abundance and this can sometimes create confusion in automated calibration procedures and erroneous calibrations (see the discussion of the cross-over value in Section 1.5). When the analyzer can not resolve the isotopologs of a calibrant ion (usually at high m/z), the isotope averaged m/z value of the calibrant must be used. For analyzers that can cover a wide m/z range, e.g., TOF analyzers, often two mass calibrations, covering low and high m/z ranges, must be used.

Common volatile mass calibration compounds for EI and positive and negative ion CI include perfluorokerosene ('PFK') for the m/z range 30–1000, perfluorotributylamine (commercially available as 'PTA' or 'FC-43') for m/z 30–650, perfluorotripentylamine ('FC-70') for m/z 30–780 and perfluoroamines. For ESI and APCI a mixture of fluorinated phospazenes ('Ultramark 1621') is suitable for positive and negative ion calibration in the range m/z 900–2200. Cesium iodide cluster ions (both positive and negative) provide good calibrations for ESI for m/z 500–2100 (this is also a good approach for MALDI) and mixtures of quaternary ammonium salts do the same in positive ion mode only for m/z 70–500. Polypropyleneglycol (PG) mixtures with different ranges and distributions of molecular mass are popular choices for both ESI and APCI. Many other mass calibration standards are used, e.g. protonated water clusters $[(H_2O)_nH^+]$ for APCI. Manufacturers of mass spectrometry systems provide details on recommended calibrants that are incorporated into automated calibration routines.

Finally in this section, it is worthwhile to emphasize again that careful mass calibration is an essential feature of validation of any analytical method. Automated calibration routines are convenient and easy to use, but should be checked before use.

6.4.2 Quadrupole Mass Filters

Radiofrequency (RF) electric quadrupole mass (really m/z) filters represent a considerable majority of analyzers in current use, particularly in trace quantitative analysis; for this reason the operating principles of these devices will be discussed in some detail to emphasize their advantages and limitations. (The RF range corresponds to a few MHz.) These analyzers are used as stand-alone (non-tandem) MS detectors, as the components of the workhorse QqQ tandem instruments and as the first analyzer in the QqTOF hybrid tandem instrument. Quadrupole mass filters (Q) are essentially the same device as the RF-only collision cells and ion guides (q) discussed later in this section and are intimately related to the RF ion traps described in Section 6.4.5. In this regard, it can be mentioned that Q and q devices are not called 'quadrupoles' because they are constructed of four electrodes (rods), but because a quadrupolar electric field (see Equation [6.11]) is formed in the space between the rods; indeed the three-dimensional (Paul) trap (Section 6.4.5) creates a quadrupolar field using just three electrodes!

Before starting the discussion of the physical principles underlying the Q and q devices, it is convenient here to summarize the figures of merit for the linear quadrupole m/z filter. As discussed later, it is possible to trade-off the m/z range against RP, and indeed instruments (so-called 'enhanced resolution' quadrupoles) have been described that sacrifice performance in m/z range for significantly improved RP. Most commercially available instruments provide unit mass RP over an m/z range that covers the range of 'small molecule' analytes of interest here, i.e., molecular masses up to ∼2000 Da. Ion transmission efficiency is another figure of merit that is closely related

to RP (Figure 6.6); modern quadrupoles operating at unit mass RP have transmission efficiencies estimated experimentally to be ∼30 %. It is possible to obtain m/z measurements of useful degrees of accuracy and precision using either unit mass RP or enhanced RP instruments provided that the mass peaks in question are known to not contain unresolved mass multiplets. In both cases the accuracies in molecular masses thus determined are in the 1–2 mDa range, with similar values for precision for ions in the range m/z 300–500.

As for all scanning instruments the spectral acquisition rate depends on the m/z range required and on other factors, but a good quadrupole can typically scan and record a spectrum covering a m/z range of several thousand Th with unit mass RP in ≤ 1s, certainly adequate for full spectral characterization of conventional chromatographic peaks. As for all scanning instruments the duty cycle is very poor in full spectral acquisition mode, but a quadrupole can be operated in selected ion monitoring (SIM) mode; if only one m/z value is monitored the duty cycle is 100 %, but this value is decreased proportionately with the number of m/z values that are successively monitored in a SIM cycle. Switching between different SIM channels is fast for a quadrupole, so the switching time has only a small effect on the duty cycle.

When operated in single channel SIM mode the dynamic range of a quadrupole analyzer can be as high as 10^5, but this is decreased (at the lower concentration end) by a factor corresponding to the duty cycle if more than one channel is used or for full scan recording. A single linear quadrupole provides no MS/MS capabilities other than those provided by exploiting in-source CID (Section 6.5), but a triple quadrupole instrument is the most versatile of all tandem instruments. Other major advantages of quadrupole instruments are their ease-of-use (high) and cost (low). Most of these figures of merit are discussed in some detail later in this section.

Historically, the quadrupole mass (m/z) filter actually arose out of an apparently unrelated field of study, that of focusing charged particles in high energy accelerators or storage rings (see the accompanying text box). The 'strong focusing alternating gradient' principle employed a sequence of fixed magnetic fields, oriented in alternating perpendicular directions, applied to the beams of very fast charged particles during their trajectories within the apparatus. In contrast, in the linear RF quadrupole device the ions move relatively slowly along the main axis; the accelerating electric fields that impart the kinetic energy that cause the ions to move through the device are

---

The Strong Focusing Principle: from Giant Mile-Sized Particle Accelerators on the GeV Scale to Mass Spectrometers on the Scale of a Few cm and A Few eV. *(Also a story of the very best of the true scientific spirit)*

In the early 1950s high energy physicists were considering problems in building particle accelerators much more powerful than the then-best of a few GeV to probe even deeper into the structure of the atom. These machines were based on the cyclotron principle by which charged particles, injected into a magnetic field perpendicular to their direction of motion, would describe a circular path (exactly the same principle as that exploited in FTICR mass spectrometers). Calculations showed that, using existing technology, building a proton accelerator ten times more powerful than the 3.3 GeV machine would require 100 times as much steel to construct the more powerful magnets required and would weigh 200 000 tons! A group of four physicists at the US Brookhaven National Laboratory in Long Island, NY, overcame this barrier by co-inventing the *alternating gradient* or *strong focusing principle* (Courant 1952; Blewett 1952); an oversimplified description of this discovery is that, by alternating the directions of the focusing magnetic fields in the two directions perpendicular to the particle trajectory, a 'strong focusing' of the proton beam could be achieved using much smaller magnets. Since its discovery, strong focusing has been one of the guiding principles behind the design of every new high energy particle accelerator in the world.

BUT, were the Brookhaven physicists the true discoverers? Nicholas C. Christofilos was born in Boston in 1916 but was taken by his parents to Greece at an early age. He received electrical and mechanical engineering degrees from the National Technical University in Athens, was employed by a Greek elevator firm, and later branched out on his own. At this time he developed the study of physics as his hobby, with a particular interest in particle accelerators. In 1950 he invented the strong focusing principle and that same year filed for a patent (US Patent No. 2,736,799) that was eventually issued in 1956. This achievement was all the more remarkable in that he conceived and worked out these ideas while in almost complete isolation from any modern active scientific community (Moir 1989).

Meanwhile, the Brookhaven physicists eventually learned of the independent work of Christophilos. In the very best of the true scientific tradition, they published a letter stating that '– *it is obvious that his proposal pre-dates ours by over two years. We are therefore happy to acknowledge his priority*'. It seems fitting to acknowledge in turn the exemplary behavior of these four distinguished gentlemen by spelling out the reference in full: Ernst D. Courant, M. Stanley Livingston, Hartland S. Snyder and John P. Blewett (1953), *Phys. Rev.* **91**, 202. As a result, Christophoulos was employed as a physicist at Brookhaven in 1953 and subsequently had a highly productive research career (Moir 1989).

The strong focusing principle is also applied, with two major differences, in modern quadrupole mass spectrometers. Firstly, the focusing magnetic fields are replaced by electric fields and, secondly, instead of transmitting an ion beam very fast through static fields arranged along the ions' trajectory, the ions move rather slowly and the fields themselves are made to alter their direction *with time* at a rate (MHz range) that is fast relative to the ions' transit time (see the main text).

---

of the order of 10 V, corresponding to linear velocities v_{ion} of an ion given by Equation [6.10] as:

$$v_{ion}^2 = 2.q_{ion}.(\Phi_1 - \Phi_2)/m_{ion} = 2.ze.(\Phi_1 - \Phi_2)/(m.m_u)$$

$$= (2e/m_u).(\Phi_1 - \Phi_2)/(m/z)$$

Evaluation of this rearranged form of Equation [6.10] with $(\Phi_1 - \Phi_2) = 10$ V and $m/z = 500$ gives $v_{ion} \approx 2000$ m.s^{-1}, so for quadrupole rods that are 10 cm long such an ion with traverse the device in about 5×10^{-5} seconds. If a perpendicular alternating electric field of frequency 1 MHz is applied to such an ion beam, each ion will experience 50 cycles of this RF field and will thus experience the same strong focusing effect (although using electric rather than magnetic fields in this case). Note that this calculation is intended only to provide a 'feel' for the magnitudes of the parameters, not as a description of a real instrument.

In the linear quadrupole the focusing field is created by applying an RF potential to an assembly of four metal rods arranged to lie parallel to one another, symmetrically around a central axis (Figure 6.9(a)). The general features of the end result are the same as far as the original strong ion focusing is concerned, i.e., the restoring force on an ion increases with its distance from the central axis

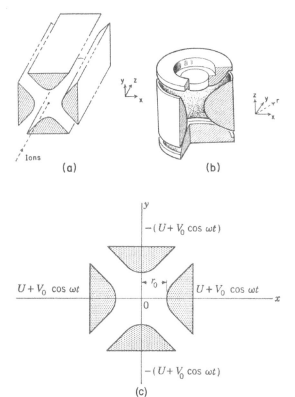

$$U + V_0 \cos \omega t$$

$$-(U + V_0 \cos \omega t)$$

$$U + V_0 \cos \omega t$$

$$-(U + V_0 \cos \omega t)$$

(c)

Figure 6.9 Sketches of the two commonly used types of RF quadrupole devices: (a) the linear *m/z* filter (two-dimensional quadrupole field), and (b) the Paul ion trap (a three-dimensional quadrupole field). Reproduced from Dawson, *Mass Spectrom. Revs.* **5**, 1 (1986), with permission of John Wiley & Sons, Ltd. (c) Diagram of the cross-section of a linear quadrupole *m/z* analyzer at z = 0, showing the potentials applied to the two pairs of electrically connected rods; the rod spacing is 2r₀ at the closest approach, as shown. Reproduced from Farmer, in '*Mass Spectrometry*' (CA McDowell, Ed), McGraw-Hill (1963), with permission of John Wiley & Sons, Ltd.

(see Equation [6.12] below). Also the time dependent field alternately 'squeezes' and 'relaxes' the ion beam in each of the field directions to create a net focusing effect (see discussion of Figure 6.17). The action of such devices as a selector (filter) for *m/z*, rather than as a focusing lens, depends on the combination of the RF field plus a superimposed DC field, as described below. However, RF-only quadrupoles (q) are also used extensively in mass spectrometers as lenses and as gas cells for CID for tandem mass spectrometry.

The realization that the strong focusing effect could be transferred from high energy physics to other areas, including chemical applications of mass spectrometry, was taken up very quickly at the University of California

Radiation Laboratory in Berkeley (Shewchuck 1953). Two physicists (R. Post and L. Henrich) developed the basic theory for a linear *m/z* filter based on the principles of strong focusing, but investigated what is now referred to as the second stability region whereas modern instruments exploit the first stability region (see later). A prototype instrument with hyperbolic rods was in fact constructed and found to behave as predicted (Shewchuck 1953) but no further work on this approach appears to have been reported. The first published reports of the full theory and its practical realization were the result of the work of Paul and his collaborators (Paul 1953, 1953a, 1955, 1958; Von Busch 1961). A popular account of this early work (Dawson 1968) was followed by a classic text (Dawson 1976) that was re-issued in 1995 by the American Institute of Physics. The following account is based largely on a later review (Dawson 1986).

The 'ideal' quadrupole *m/z* filter consists of four rods of hyperbolic cross section, arranged symmetrically around a central axis with spacing defined by r₀, the distance from the central axis to the apex of each hyperbola (Figure 6.9(c)). A hyperbola is a type of curve in a plane, described algebraically by an equation of the form:

$$Ax^2 + Bxy + Cy^2 + Dx + Ey + F = 0$$

such that $B^2 > 4AC$, and where all of the coefficients are real-valued. In the case of interest here $F = r_0^2$, $D = 0 = E$, and either $(A = -1, B = +1, C = +1)$ or $(A = +1, B = -1, C = -1)$, corresponding to the two pairs of opposite rods (Figure 6.9(a)); note that indeed $4AC = -4 < B^2$. The specification of hyperbolic rods is 'ideal' in that the mathematical description of ions moving through such a device is straightforward, and is also 'ideal' in that a hyperbola by definition stretches to infinity. A real device is thus based on rods with cross-sections that are truncated hyperbolae, although in practice round rods are most commonly used. Figure 6.9 shows a sketch of such a device and the relationship of its fields and coordinate systems to those of the three-dimensional ion trap (see Section 6.4.5). In the linear quadrupole the two pairs of opposite rods (maintained strictly parallel to the central z-axis and centred on the x- and y-axes, Figure 6.9(a)) are connected electrically. A potential difference is applied between the two pairs of rods; this potential difference is the sum of a time independent (DC) potential U (measured relative to the potential at the z-axis) and a dynamic RF component $V_0.\cos(\omega t)$, where ω is the angular RF frequency (ω = 2πf, where f is the frequency in Hz) and V_0 is the zero-to-peak amplitude (in volts). A sketch of the arrangement is shown as Figure 6.9(c). The detailed mathematical derivation of the equations of motion of an ion moving in the z-direction through such a field have been fully described (Dawson 1976; Campana 1980, 1980a; March 2005), and here only enough detail is discussed to permit an understanding of the operation of such devices and of the terminology used to describe them.

To write down the equations of motion of ions within such a device, it is necessary to know the magnitude and direction of the forces acting on them. For this purpose the electrical potential at any point (x,y) within a two-dimensional field (i.e., independent of z) created by a symmetrical arrangement of four rods of hyperbolic cross-section must be known. This expression is not derived here, but is given (Dawson 1976, March 2005) by:

$$V_{xy} = [U + V_0.\cos(\omega t)].(x^2 - y^2)/r_0^2 \qquad [6.11]$$

where $2r_0$ is the electrode spacing (Figure 6.9(c)). Now it is known (Equation [6.9], see Appendix 6.1) that the potential V_{xy} leads to the corresponding field $\mathbf{E}_{xy}$ and thus to the electrical force on an ion of mass m_{ion} and charge $q_{ion} = \pm z.e$ (the sign of the charge is not included in z but is essential in q_{ion}), i.e. $\mathbf{F}_{xy} = q_{ion}.\mathbf{E}_{xy}$. The components of this force vector (F_x etc.) can then be incorporated into the corresponding Newtonian equations of motion (Equation [6.2], Appendix 6.1):

$$F_x = m_{ion}.a_x = m_{ion}.(\partial^2 x/\partial t^2)$$
$$= -q_{ion}.[U + V_0.\cos(\omega t)].(x/r_0^2) \qquad [6.12a]$$
$$F_y = m_{ion}.a_y = m_{ion}.(\partial^2 y/\partial t^2)$$
$$= +q_{ion}.[U + V_0.\cos(\omega t)].(y/r_0^2) \qquad [6.12b]$$
$$F_z = m_{ion}.a_z = m_{ion}.(\partial^2 z/\partial t^2) = 0 \qquad [6.12c]$$

Note that in Equations [6.12] m_{ion} and q_{ion} denote the actual mass (in kilograms) and charge (in C and including the sign) of the ion, respectively, and the 'a' parameters are accelerations (not the italicized 'a' parameters used to characterize the ion motion in such an RF field, see Equation [6.14]). Conversion of these equations of motion to include the mass spectrometrists' familiar *m/z* is described below. Since there is no electric field acting in the z-direction, the motion of the ion in this direction is not affected and is controlled only by the fields and forces that injected it into the device in the first place or by any additional fields that can be imposed (see later). On the central z-axis $x = 0 = y$ and, similarly, $F_x = 0 = F_y$, i.e., Equations [6.12a–c] indicate that the magnitude of the field strength in any direction is zero for ions on the z-axis and increases with increasing x and y. It is also important to note that, in this idealized mathematical model, the motions of the ions in the x- and

y-directions are completely independent of one another and are identical in form. So, as an example, consider the motion in the x-direction only, and for convenience rewrite Equation [6.12a] as:

$$(\partial^2 x/\partial \tau^2) = -[a_x + 2q_x.\cos(2\tau)].x \qquad [6.13]$$

where $\tau = (\omega t/2)$ (i.e. time measured in units related to the RF frequency), and:

$$a_x = (4/\omega^2.r_0^2).(q_{ion}/m_{ion}).U$$
$$= (4e/\omega^2.r_0^2.m_u).[U/(m/z)] \qquad [6.14]$$
$$q_x = (2/\omega^2.r_0^2).(q_{ion}/m_{ion}).V_0$$
$$= (2e/\omega^2.r_0^2.m_u).[V_0/(m/z)] \qquad [6.15]$$

To relate the quantities m_{ion} and q_{ion} (non-italicized) to the dimensionless *m/z* used to describe mass spectra (see Section 1.3), recall that $m \equiv m_{ion}/m_u$ where m_u is the unified atomic mass unit = [(mass in kg of one atom of ^{12}C) / 12] (see Table 1.5), and z is the (integral) number of elementary charges e, so that $q_{ion} = \pm z.e$ where the polarity (+ or −) of the ion is not included in z but must be in q_{ion}.

It is important to distinguish the following somewhat similar symbols: a non-italicized lower-case 'q' denotes a linear quadrupole device operated using only RF fields, an upper-case Q denotes a similar device operated with both RF and DC fields, an italicised *q* denotes the trajectory stability parameter described above, and q_{ion} represents the charge on an ion (in coulombs and containing the sign indicating the polarity).

An analogous transformation of Equation [6.12b] for the y-dimension gives a result similar to Equation [6.13] except that now a_x and q_x are replaced by $a_y = -a_x$ and $q_y = -q_x$. This relationship between the x- and y-directions, in which only the sign changes, corresponds in a real device to the fact that while the rods on the x-axis are at potentials corresponding to the positive half-cycle of the RF potential $V_0.\cos(\omega t)$, simultaneously the y-axis rods are at the negative half-cycle (see Figure 6.9c), and vice versa.

Equations of the form [6.13] had been examined in 1868 by the French mathematician Émile Mathieu in connection with vibrations of an elliptically shaped membrane (actually a drumhead!). Solutions to such a differential equation (i.e. expressions describing x or y for an ion as functions of time t) were thus known. These mathematical solutions (Dawson 1976; March 2005) are not derived here. An important feature of the solutions is that they can be divided into two classes, 'stable' and 'unstable'. Stable solutions for x (or y) as a function of time are those in which the value of x (or y) does oscillate but within well defined limits, thus predicting that ions will survive under such conditions. In contrast, unstable solutions are those in which the coordinates increase without limit, thus predicting that ions will be lost by hitting one of the rods. The mathematically stable solutions are conventionally plotted on a stability diagram (Figure 6.10(a)) that relates the values of the *a* and *q*

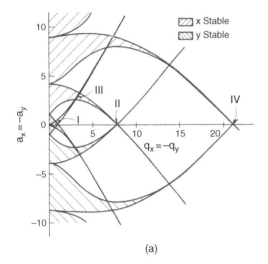

(a)

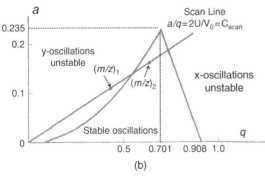

(b)

Figure 6.10 (a) The stability diagram for a two-dimensional quadrupole field in which the x- and y-directions differ by a factor of −1. Reproduced from Dawson, *Mass Spectrom. Revs.* **5**, 1 (1986), with permission. (b) Expanded view of the first stability region I (positive half only) in which both x- and y-motions are stable. The slope of the scan line $a/q = C_{scan}$ is determined by adjusting the ratio $2U/V_0$ (Equation [6.16a]). In this diagram $(m/z)_1 < (m/z)_2$. The larger intercept on the *q*-axis is 0.908, and this corresponds to the *minimum* value of *m/z* that undergoes stable oscillations, i.e., is transmitted by the quadrupole. Modified from Farmer, in *Mass Spectrometry* (C.A. McDowell, Ed), McGraw-Hill (1963), with permission of John Wiley & Sons, Ltd.

parameters that yield stable solutions. Of special interest here is the so-called first stability region at low values of *a* and *q* (Figure 6.10(b)), within which ions have stable trajectories in both the x- and y-directions. The precise descriptions of the curves encompassing this region are defined by the mathematical solutions (Dawson 1976; March 2005) of the Mathieu-type Equation [6.13]. (It is convenient from here on to omit the distinction between a_x and a_y, q_x and q_y in the discussion.) It is important to note that these stability diagrams are completely general, limited only by the extent to which the idealized mathematical model provides an adequate description of a real device. In particular, because the convenient dimensionless quantities *a* and *q* (Equations [6.14–6.15]) involve only combinations of the experimental variables ω, r_0, U and V_0, it is possible to devise particular combinations that will optimize the performance of a device for a specified purpose. For example, if the wish is to fabricate a miniature quadrupole (i.e. significantly reduce r_0), appropriate adjustment of the RF frequency ω and/or the voltages U and V_0, so as to provide values of *a* and *q* that still fall into the stability region, will be necessary.

Thus far only the conditions under which ions possess stable trajectories have been discussed, but not the means by which a real device can be made to work as a mass spectrometer. To achieve this some additional constraint must be placed on operation of the device and the choice that is both experimentally feasible and achieves the desired result is to operate the device by varying U and V_0 such that:

$$a/q = 2U/V_0 = C_{scan} \qquad [6.16a]$$

where C_{scan} is a constant, as illustrated in Figure 6.10(b). That is to say, the device is operated as a mass spectrometer by scanning U and V_0 together, in a linked fashion, in accord with Equation [6.16a]. This strategy, of scanning the voltages up from low to high values, results in production of a mass spectrum as a result of successively passing increasing *m/z* values from the y-unstable region, through the stability region (so that these ions successfully pass through the device to the detector) and finally into the x-instability region (Figure 6.10(b)). The only alternative strategy would require scanning the frequency ω but this is not technically feasible. To understand the general principle a little better, recall that the scanning strategy constrains all ions emerging from the ion source and entering the quadrupole to lie on the scan line (Figure 6.10(b)) whose slope is determined by the operator adjusted constant C_{scan} (Equation [6.16a]). Because *a* and *q* are both inversely proportional to *m/z*, the ions are lined up in order of decreasing *m/z* with increasing *q*. As

U and V_0 are scanned upwards, so are the values of m/z for which the corresponding values of a and q fall within the range of the scan line that lies within the stability region and are therefore transmitted to the detector; all ions with temporary m/z values that correspond to (a,q) points lying outside this limited stability region at the instantaneous values of U and V_0 during the scan have unstable trajectories in either the x- or y-direction, and are thus lost on the rods (Figure 6.11). In this way the mass spectrum is recorded.

The value of q corresponding to the maximum of the stability region in Figure 6.10(b) turns out to be $q_{max} = 0.701$ and the corresponding value of a is 0.2353; this means that the slope $C_{scan,\infty}$ of the 'theoretical infinite resolution' scan line that passes exactly through this point

is 0.33568 and the corresponding value of U/V_0 (Equation [6.16a]) is 0.16784. The value of q at which the boundary line, separating the stability and x-instability regions intersects the q-axis (Figure 6.10(b)), is 0.908; this determines (via Equation [6.15]) the minimum value of m/z that can be transmitted through a quadrupole of fixed r_0 and operating with a fixed value of ω. Note that this minimum value is achieved only when $a = 0 = U$, i.e., the device is operated in RF-only mode, as in the central collision cell q in a triple-quadrupole tandem instrument. (This parameter that determines the 'low mass cutoff' for a linear RF-only quadrupole is also important for the three-dimensional quadrupole trap, discussed in Section 6.4.5).

The width of the stability region intersected by the scan line determines the resolving power via the corresponding

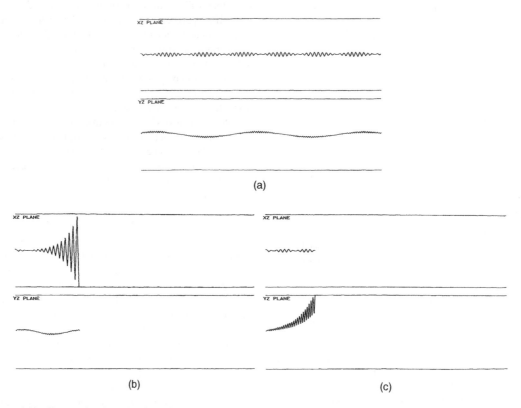

Figure 6.11 Computed trajectories through a linear quadrupole mass filter for ions of m/z (a) 100, (b) 99 and (c) 101. In all cases the top trajectory is for the xz plane and the other for the yz plane. All trajectories were calculated for the same initial conditions (ion energy and point and angle of entry) under (a, q) conditions chosen such that ions of m/z 100 are close to the apex value (0.701,0.2353) within the stability range (Figure 6.15(b)) and are thus transmitted as shown in (a). However, the scan-line slope $C_{scan} = 2U/V_0$ was set so that the corresponding (a, q) values for m/z 99 and 101 lie outside this intersected stability region at the instantaneous values of U and V_0 that led to stable trajectories for the m/z 100 ions, and are thus not transmitted but are lost on the quadrupole rod as shown in (b) and (c). (Recall that the z-direction is the main axis of the quadrupole, i.e., the horizontal direction in this sketch). Reproduced from Campana, *Int. J. Mass Spectrom. Ion Phys.* (1980) 33, 101, copyright (1980), with permission of Elsevier.

m/z stability range that is transmitted (Figure 6.10(b)), and in turn this width is determined by the slope C_{scan}, twice the ratio of U to V_0 (Equation [6.16a]). In this mode of operation the resolving power is essentially independent of *m/z* and thus minimizes *m/z* discrimination (variable transmission efficiency). However, most applications of such devices, of interest in the present context, require that RP should increase with *m/z* to provide a constant peak width $\delta(m/z)$ corresponding to unit mass resolving power. This is achieved (Dawson 1986) using a scan law relationship such as:

$$U = 0.16784.V_0 - U' \qquad [6.16b]$$

where 0.16784 is half the value of $C_{scan,\infty}$ discussed above and U' is a fixed voltage that determines the value of $\delta(m/z)$. As in most ion beam instruments there is a resulting complicated trade-off between transmission efficiency and RP, but clearly the use of the constant $\delta(m/z)$ mode results in progressively lower transmission as *m/z* increases. Some instruments arrange for the U/V_0 ratio to be changed throughout a scan in order to provide some flexibility in the trade-off between RP and transmission efficiency.

In actual practice, in order to distinguish an ion with an unstable trajectory from one with a stable trajectory and thus permit the *m/z* filter action, it is necessary to arrange for the ions to experience a minimum number n_{min} of cycles of the RF field so that ions with unstable trajectories are in fact lost on the rods. Theoretically (Dawson 1986) it can be shown that:

$$(n_{min})^2 > h . RP \qquad [6.17]$$

where the required value of *h* depends on the required value of RP and also on other instrumental conditions; a value of $h = 10 - 20$ is typical for RP determined based on width at half-height (FWHM, Figure 6.4). In turn, the value of *n* is simply determined by the transit time of the ion t_{ion} through the length 1_{quad} of the quadrupole device:

$$t_{ion} = 1_{quad}/v_{ion} = 1_{quad}/[2.e.V_{acc}/(m/z).m_u]^{\frac{1}{2}}$$

so that

$$n = t_{ion}.(\omega/2\pi)$$
$$= (\omega/2\pi).1_{quad}/[2.e.V_{acc}/(m/z).m_u]^{\frac{1}{2}} \qquad [6.18]$$

where V_{acc} is the potential drop responsible for injecting ions into the quadrupole (compare Equation [6.10]). These relationships show that there is a trade-off between scan

speed (and thus spectral acquisition rate) and RP since these parameters are limited in opposite directions by t_{ion}. Modern quadrupoles can maintain unit mass resolving power while scanning a *m/z* range of several thousand Th in < 1second. Practical limitations on 1_{quad} arise not only from scan time considerations but also from difficulties in maintaining sufficiently accurate alignment of the rods; this limitation affects other figures of merit also.

For example, from knowledge of the stability diagram and the limitations on *n* imposed by the finite length 1_{quad}, the maximum value of *m/z* that can be transmitted by a realistic practical device can be deduced (Dawson 1986):

$$(m/z)_{max} \propto V_0/(f.r_0)^2 \qquad [6.19]$$

where $f = \omega/2\pi$. Similarly, the minimum achievable peak width is given by:

$$[\delta(m/z)]_{min} \propto h.V_{acc}/(f.1_{quad})^2 \qquad [6.20]$$

Note that increasing the upper limit to the *m/z* range can be achieved by lowering the frequency (Equation [6.19]), but this change simultaneously increases the peak width; the converse is also true of course, and quadrupoles designed to provide lower peak widths must accept a correspondingly lower *m/z* range (Schoen 2001; Yang 2002). The practical limit to RP is given by combining Equations [6.19–6.20]:

$$RP_{max} = (m/z)_{max}/\delta(m/z)_{min}$$
$$\propto (V_0/V_{acc}).(1_{quad}/r_0)^2 \qquad [6.21]$$

Reductions in r_0 in order to increase RP will result in decreased the ability to accept wide ion beams, thus requiring additional demands on the design of the ion source and of the ion optics that transfer the ion beam to the analyzer. Reductions in r_0 will also increase susceptibility to contamination of the rods and at some point will introduce increased fabrication deviations from the required tolerances that will eventually limit performance. Similarly, increases in 1_{quad} are limited by the required tolerances in maintaining constant r_0 along the length of the rods. In modern instruments with good performance, stability in each of U, V_0 and f of the order of one in 10^5 is required and the mechanical tolerance for parallelism of the rods (i.e. variations in r_0) is ~1 in 10^4 (implying a tolerance of ~0.5 μm for $r_0 = 5$ mm). Typical values of key parameters would be $r_0 = 5 - 10$ mm, $1_{quad} = 200$ cm (longer rods require more RF power), $f = 2$ MHz, $V_{acc} = 5$V, V_0 up to 4 kV.

Although some commercially available quadrupole rods are fabricated with cross-sections that conform closely to the 'ideal' hyperbola (see discussion of Equation [6.11]), most such instruments are for practical reasons constructed using rods of circular cross-section with radius r_{rod}. Theory and experiment agree that best performance is obtained if $r_{rod} \approx 1.13 - 1.14.r_0$. There has been much debate over the relative advantages of rods with truncated hyperbolic cross-sections compared with round rods. Theoretical investigations (Dawson 1986) suggest that limitations of round rods will become apparent only at RP values appreciably greater than unit mass RP, but that if round rods are incorrectly aligned the deterioration of performance is likely to be appreciably more serious than that for a similar problem with truncated hyperbolic rods. In practical applications of the kind discussed in this book, it appears that other factors are more important than this distinction.

It is necessary to discuss briefly the arrangement of the electron multiplier detector (Chapter 7) for a quadrupole analyzer. To maintain a low background in the mass spectrum, it is necessary to position the detector well off-axis (i.e. removed from the linear line-of-sight from the ion source), with appropriate deflector electrodes to guide the ions onto the detector. Some sources involving high energy ionizing agents (EI at 70 eV, FAB, MALDI etc.) produce photons that are transmitted through the quadrupole without hindrance and would be registered by a line-of-sight detector. Also, almost all ion sources can produce fast neutral species that would yield similar erratic and spurious signals; these fast neutrals are produced from ions I^+ that have been accelerated out of the source and that, while in flight, undergo charge transfer collisions with background gas molecules G, e.g.:

$$I^+(\text{fast}) + G(\text{thermal}) \rightarrow I(\text{fast}) + G^+(\text{thermal})$$

Finally, the effects of fringe fields at the entrance and exit of the quadrupole must be considered. It was tacitly assumed in the theoretical treatment summarized above that the ions enter the quadrupole field through a sharp boundary in which there are no fields in the x- or y-directions. The on-axis accelerating field in the z-direction affects only t_{ion} (Equation [6.18]). However, in reality the RF and DC fields within the quadrupole 'spill out' into the fringe regions at either end. These fringe fields produce lines of force that are curved in all three dimensions rather than strictly perpendicular to the z-axis, and as a result the x-, y- and z-motions are coupled. Various approaches to alleviating the deleterious effects of this coupling have been proposed (Dawson 1986). The effects are particularly important for operation of triple quadrupole instruments in which the ion beam is transmitted through three quadrupoles (Section 6.4.3); transmission losses can be large unless

appropriate precautions are taken, usually involving close coupling of the RF fields within the three quadrupoles (Dawson 1986). On the other hand, these fringe field effects can be exploited in operation of a linear ion trap (Section 6.4.5).

The figures of merit for the quadrupole m/z filter that were summarized in the second paragraph of this section are discussed in more detail here. As discussed above (Equations [6.19–6.20]), it is possible to trade off the m/z range against RP, and indeed instruments have been described that sacrifice performance in one of these parameters to improve the other. Most commercially available instruments provide unit mass RP (FWHM = 0.7 Th, so that adjacent peaks are almost resolved at the baseline) over an m/z range that covers the range of 'small molecule' analytes of interest here, i.e. molecular masses up to 2000 Da. Ion transmission efficiency is also closely related to RP (Figure 6.6); modern quadrupoles operating at unit mass RP have transmission efficiencies estimated experimentally to be $\sim$30 % (El-Faramawy 2005).

It is possible to obtain m/z measurements of useful degrees of accuracy and precision using either unit mass RP (Tyler 1996, Kostiaian 1997) or so-called enhanced RP instruments (Paul 2003, Grange 2005) in which m/z peak widths (FWHM) as low as 0.1 Th were used. (This enhanced resolution version of the quadrupole mass filter was actually incorporated as the first analyzer in a triple quadrupole instrument but this is irrelevant to its operation to achieve more precise mass measurements.) In both cases the accuracies in molecular masses thus determined were in the 1–2 mDa range, with similar values for precision (standard deviations of replicate measurements), for ions in the range m/z 300–500. It is somewhat surprising that the enhanced resolution quadrupole did not significantly out-perform the unit mass instrument in this respect and presumably other factors, perhaps instrumental linearity of scan function ($2U/V_0$) with respect to m/z, were limiting the performance. Of course, a precondition for a successful mass measurement using a low resolution mass spectrometer is that the analyte ion must be completely separated from background ions, since even very low levels of unresolved ions can interfere significantly with the accuracy even though the precision of the measurement can remain excellent! From the point of view of trace quantitative analysis, however, such mass measurements are of value only to confirm the identity of the target analyte.

However, an enhanced resolution quadrupole (Schoen 2001; Yang 2002) can in principle provide increased selectivity in quantitative analyses using SIM or MRM. An example in which this advantage is clearly evident is shown in Figure 6.12, where the SRM chromatogram obtained using a RP yielding a peak width (FWHM) of

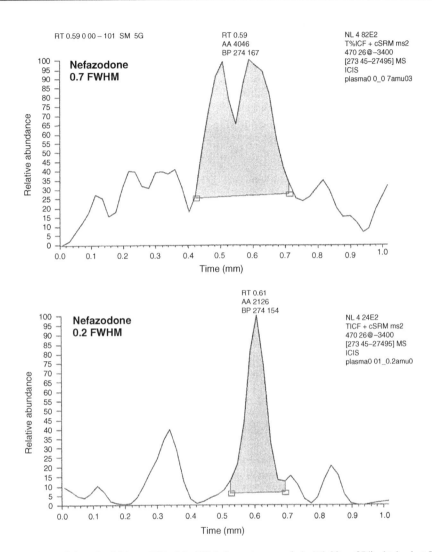

Figure 6.12 Comparison of the selectivities exhibited in SRM chromatograms (*m/z* 470.22 → 274) obtained at FWHM settings of 0.70 (top) and 0.20 (bottom) Th for Q_1 (precursor ion selection) in LC–MS/MS analyses using the same enhanced-resolution triple quadrupole instrument. In each case the sample was human urine spiked at 30 pg.mL^{-1} of nefazodone, with 60 fg injected on column. Reproduced from Jemal, *Rapid Commun. Mass Spectrom.* **17**, 24 (2003), with permission of John Wiley & Sons, Ltd.

0.7 Th for the precursor ion indicates interferences sufficiently serious that it is not clear where the analyte peak boundaries should be set in order to measure peak area; in contrast, use of a peak width of 0.2 Th gave a much cleaner, well defined peak (Figure 6.12). This example is taken from an extensive examination (Jemal 2003a) of the performance of this enhanced resolution quadrupole system in the context of high throughput bioanalytical assays. While the increased selectivity was clearly established, it was also found that the enhanced resolution mode of operation requires considerably more attention

to detail than conventional 'unit mass' RP mode using the same instrument. Most of the variability appeared to arise from drift of the *m/z* axis with small temperature variations, which implied a need for required periodic checks and adjustment of the SRM *m/z* settings (Jemal 2003a). This creates difficulties in validation (Chapter 10) of large sample LC–MS runs (e.g. multiple samples loaded in an autosampler to be analyzed overnight).

As for all scanning instruments, the spectral acquisition rate depends on the *m/z* range required and on other factors, but a good quadrupole can typically scan and record a spectrum covering a *m/z* range of several thousand Th with unit mass RP in ≤ 1 second. This is certainly adequate for full spectral characterization of conventional chromatographic peaks but for fast chromatography giving peaks only a few seconds wide this performance becomes marginal. However, very recently a fast-scanning quadrupole (Perkin-Elmer Clarus 600 GC/MS instrument) was introduced that can provide good-quality mass spectra covering a *m/z* range of 300 at an acquisition rate of 23 Hz, and in selected ion monitoring mode (actually narrow scan range ~1 Th) an acquisition rate of 90 Hz can be achieved when monitoring a single *m/z* value. An account of the performance of this instrument in GCxGC–MS analyses (Korytár 2005) emphasizes the operational compromises that must be made among spectral acquisition rate, sensitivity etc.

This increased acquisition rate does not help a fundamental limitation of all scanning instruments, including quadrupoles, namely, that the *duty cycle* is very poor in full spectral acquisition mode since all ions of *m/z* values other than that instantaneously being transmitted at any given time during the scan are wasted. However, a quadrupole can be operated in SIM mode in which only one or a few *m/z* values, characteristic of a target analyte, are monitored. To operate a quadrupole in SIM mode, U and V_0 are set so that the *m/z* of interest yields the *a* and *q* values at the upper apex of the stability diagram (0.701 and 0.253, respectively, see above), but possibly using an offset U' to ensure constant peak widths across the *m/z* range (Equation [6.16b]). U and V_0 are then 'switched' together so that ions of different selected *m/z* values are successively arranged to satisfy the apex (*a*, *q*) values.

If only one *m/z* value is monitored the duty cycle is 100 %, but this value is decreased proportionately with the number of *m/z* values that are successively monitored in a SIM cycle. Switching between different SIM channels is fast for a quadrupole, so the switching time has only a small effect on the duty cycle. The choice of number of SIM channels represents a balance between sensitivity, limit of detection etc., on the one hand, and confidence in the identity of the compound being monitored and/or number of analytes monitored, on the other. Similar comments apply to a triple quadrupole tandem instrument, where a number of fragment ions arising from a selected precursor ion can be chosen in multiple reaction monitoring (MRM) mode (Section 6.2.1). When operated in single channel SIM/MRM mode the dynamic range of a linear quadrupole can be as high as 10^5, but this is limited at the lower concentration end by a factor corresponding to the duty cycle if more than one channel is used as a consequence of inadequate ion statistics (shot noise, Section 7.1.1).

A single linear quadrupole provides no MS/MS capabilities but a triple quadrupole instrument is the most versatile of all tandem instruments (Section 6.2.2). Other major advantages of quadrupole instruments are their ease-of-use (high) and cost (low), as well as their small size. Indeed, a remarkable miniaturized quadrupole etched on a silicon chip (Geear 2005), with rods of diameter 0.5 mm and length 30 mm, yielded unit mass RP up to *m/z* 400.

6.4.2a RF-Only Quadrupoles

Linear quadrupole devices are also used in RF-only mode (U = 0), i.e. only the *q*-axis in Figure 6.10(b) is considered so that a range of *m/z* values will correspond to stable trajectories. Such devices (denoted as nonitalicized q) are used as ion lenses (guides) to transport ions from one part of an apparatus to another (e.g., from ion source to analyzer) and as collision cells to effect collision induced dissociation; an extensive review (Douglas 2005) covers the literature on all these topics up to 2003. The emphasis is placed on linear quadrupole devices, essentially the same as those discussed above in the context of quadrupole mass filters.

Arrays with six or even eight rods have been used less often for several reasons. In these higher multipoles, unlike in quadrupole devices, the ion motions in the x- and y-directions are strongly coupled and the frequencies of ion oscillation and the stability of the trajectories depend strongly on the initial conditions. As a result no general stability diagram can be given for motion in higher multipoles, although a stability diagram can be calculated for a specified set of initial conditions. However, even in this case the boundaries between stable and unstable regions are not well defined, so that there is no sharp cutoff in transmission vs *m/z* for ions close to such a stability boundary. This restricts the usefulness of higher multipole linear traps as mass analyzers or as ancillary lenses and traps for mass spectrometers.

When operated as ion guides, RF-only quadrupoles (q-devices) benefit from collisional cooling to achieve collisional focusing and thus high transmission efficiencies. This counter-intuitive method arose from studies of ion transmission through a RF-only quadrupole as a function of collision gas pressure ($5 \times 10^{-4} - 1 \times 10^{-2}$ torr), originally undertaken (Douglas 1992) to study the extent of ion losses as a result of collisional scattering! It was shown that, provided the initial injection energy of ions into the q device was low (1–30 eV), the ion transmission observed through a small aperture at the exit actually first increased as the gas pressure increases, reaching a maximum at ~8×10^{-3} torr, before decreasing at higher pressures. This observation was in direct contrast to the expectations of classical scattering theory and experiment, and appears to be analogous to the effect of helium damping gas in Paul traps (Section 6.4.5) although damping gases such as nitrogen or air can be used rather than low mass gases (helium) required in ion traps. The extent of the observed collisional focusing increased with the mass of the ion (not *m/z* value) up to at least 16 950 Da and, significantly, was accompanied by significant losses of axial kinetic energy. In fact, all ions (both surviving precursors as well as the full range of product ions) were found to have axial energies close to zero at the exit of the q-device, thus greatly facilitating control of ion transmission between such an ion guide and the following ion optical element in the instrument.

When the same q-devices are employed such that the axial injection energy of the ions is sufficiently high to produce collision induced dissociation (Section 6.5), the device is referred to as a collision cell. The most common example of this application is that of triple quadrupole mass spectrometers (Section 6.4.3), and also in QqTOF analyzers (Section 6.4.7). Other aspects of this application are discussed in more detail in Section 6.4.3. (It is interesting that the focusing properties of RF-only collision cells allow the latter to be either straight as implied thus far, or slightly bent as an aid to preventing fast neutrals and/or photons from the ion source from reaching the detector, or strongly curved to also reduce the overall space requirement (footprint) of the instrument.)

Finally in this section, it is noted that single quadrupole mass filters (as opposed to triple quadrupole instruments) are generally used in trace quantitative analyses only for GC–EI(CI)–MS methods for thermally stable analytes (e.g. pesticides in various matrices) in SIM mode. (It is interesting that GC/MS instruments still appear to account for a large fraction of mass spectrometer units sold worldwide.) The same linear quadrupole devices can also be operated as 2D ion trap mass spectrometers, discussed in Sections 6.4.5 and 6.4.6.

6.4.3 Triple Quadrupole Instruments

The triple quadrupole (QqQ, sometimes denoted for convenience as $Q_1q_2Q_3$) is the workhorse of trace level quantitative analysis, especially for analytes that must be separated by HPLC. This design was first described as an analytical instrument by Yost and Enke (Yost 1978, 1979), although versions designed for fundamental studies of ion dynamics had been described previously (Vestal 1974; McGilvery 1978). Its advantages for quantitative analysis by MS/MS with all the selectivity advantages of the latter (Section 6.2.1) arise from its conceptual simplicity based on the function of Q_1 and Q_3 as *m/z* filters, the high CID efficiency obtainable through low

energy collisional activation in the RF-only quadrupole q_2, the high transmission efficiency through all three quadrupoles and the very high duty cycle when operated in selected ion monitoring (SIM) or multiple reaction monitoring (MRM) modes (although the duty cycle is extremely low in full scan mode). As a tandem-in-space beam instrument it is effectively immune from limitations arising from space charge effects when dealing with trace level analytes. Since its first commercialization in the early 1980s, for many years only technological improvements were made to the basic design, as the principles underlying efficient coupling of the RF fields in the three quadrupoles in order to minimize the effects of fringe fields (Section 6.4.2) were well understood (Dawson 1982). Some manufacturers use stacks of Einzel lenses (Figure 6.8, Section 6.3.2) to achieve this coupling, while others use very short RF-only quadrupole lenses ('Brubaker rods') to achieve the same purpose.

However, a problem associated with efficient analysis of product ions emerging from the collision cell q_2 was solved only relatively recently. The problem involves arranging for all product ions entering Q_3 from q_2 to have an axial velocity that is low enough to allow these ions to experience a sufficient number of RF cycles that they are properly resolved (see discussion of Equation [6.18], Section 6.4.2), but not so low that they have insufficient kinetic energy to successfully penetrate the fringe fields at the entrance and exit of Q_3. The problem arises (Boyd 1994) because of the uncertainty in the axial kinetic energies (and thus velocities) of fragment ions formed under the multiple collision conditions normally used in triple quadrupoles (pressure of collision gas within $q_2 \sim 10^{-3}$ torr). If there were no collisions the velocities of the initially formed product ions would all be identical to those of the precursors (see text box) and thus the kinetic energies $(m.v^2/2)$ would be proportional to the masses of the ions. Then arranging for suitable velocities of entry into Q_3 would be a relatively simple matter of providing appropriate potential offsets $(\Phi_1 - \Phi_2)$ and $(\Phi_2 - \Phi_3)$ between $Q_1 - q_2$ and $q_2 - Q_3$. However, because of the uncontrolled losses of kinetic energy of both precursor and fragment ions in 'conventional' multiple collision conditions (pressure not so high as to create excessive ion scattering and thus sensitivity loss), the product ion velocities entering Q_3 are essentially unpredictable and also exhibit wide distributions. Figure 6.13 illustrates this difficulty for a simple example (Shushan 1983) in which optimum fragment ion signal intensities and resolution were obtained only by actually measuring the kinetic energies of the product ions.

Relationship Between Velocities of Precursor and Product Ions in Tandem Mass Spectrometry

The archetypal ion fragmentation reaction is: $P^{p+} \rightarrow F^{f+} + G^{g+}$, where conservation of charge requires $p = f + g$ and conservation of mass requires $m_P = m_F + m_G$ (where the m's denote actual masses (kg) of the various species, *not m/z* values). If this reaction occurs in field free space, so that no external forces are exerted during the reaction, the linear momentum and kinetic energy are also conserved in each of the three mutually orthogonal directions. It is convenient to label the original direction of the precursor P^{p+} as the Z-axis. If velocity components measured relative to the laboratory are denoted by v, the momentum conservation conditions are:

$$m_P.v_{PZ} = m_F.v_{FZ} + m_G.v_{GZ}; \quad 0 = m_F.v_{FX} + m_G.v_{GX}; \quad 0 = m_F.v_{FY} + m_G.v_{GY}$$

since axes were chosen so that $v_{PX} = 0 = v_{PY}$. Similarly, for kinetic energy conservation:

$$m_P.v_{PZ}^2 + 2.\Gamma_Z = m_F.v_{FZ}^2 + m_G.v_{Gz}^2; \quad 2.\Gamma_Y = m_F.v_{FX}^2 + m_G.v_{GX}^2; \quad 2.\Gamma_Y = m_F.v_{FY}^2 + m_G.v_{GY}^2$$

where the Γ quantities (so-called 'kinetic energy release') represent the amount of internal energy in P^{p+} that is transformed into kinetic energy components of the two products F^{f+} and G^{g+}. For present purposes it is sufficient to consider the case where the Γ values are negligible relative to $m_P.v_{PZ}^2$ etc. (see also below). If now all three conservation equations are considered as simultaneous equations in three 'unknowns' m_P, m_F and m_G, there is a restriction on all the various v_Z values because *all of these mass values must be real and positive*. Similar conclusions hold for the various values of v_X and v_Y. To discover the nature of this restriction, it is *convenient* to consider the conservation equations in a reference frame fixed on the original direction of P^{p+} before the fragmentation event instead of fixed on the laboratory; if the velocities measured relative to this new set of axes are denoted u, then: $u_{IJ} = v_{IJ} - v_{PJ}$, where I = P, G or F and J = X, Y or Z (for all combinations of I and J). Then, substitution of the u_{IJ} for the corresponding v_{IJ} quantities in the kinetic energy conservation equations (with $\Gamma_J = 0$) gives:

$$0 = m_F.u_{FJ}^2 + m_G.u_{GJ}^2$$

for each of J = X, Y and Z.

Now each of the u values can be positive or negative, but they must be real-valued so their squares are necessarily positive; but so are each of m_F and m_G. So the energy conservation equation can be satisfied only if *either* $m_F = 0 = m_G$ (a trivial case of no fragmentation) *or* if $u_{FJ} = 0 = u_{GJ}$ for each of J = X, Y, Z. Converting this nontrivial condition back to the coordinate system fixed in the laboratory gives the final condition imposed by the conservation equations:

$$v_{FJ} = v_{PJ} = v_{GJ}, \text{ for each of } J = X, Y, Z$$

That is to say, for the special case where the Γ values are negligible, the two products (F^{f+} and G^{g+}) formed in the reaction continue with the same velocity components as those of the precursor ion P^{p+}. It can be shown (e.g., Boyd 1994) that when the Γ values are significant, the same result holds to an excellent approximation if now the v_{IJ} values are taken to be the median values of the resulting velocity distributions. Note also that if P^{p+} is induced to dissociate by collision with a gas molecule, the relevant value of v_{PZ} is the value *after* the collision but *before* the fragmentation event. Further, the restriction to zero external fields strictly applies only to velocity *components* in directions in which no fields are applied, e.g., the axial Z-direction in an RF-only quadrupole collision cell, but in that case obviously *not* to the radial components (X and Y, see Equations [6.12]).

Then an appropriate linked-scan law relating Φ_3 to *m/z* of the product ion was applied (the latter instrument control information was obtained from the instantaneous U/V_o ratio of Q_3, Equation [6.16(a) and (b)]).

Such an ad hoc procedure is obviously impractical for, e.g., LC–MS/MS analyses, and manufacturers of commercial instruments have devised useful compromises to permit acceptable performance. For example, the

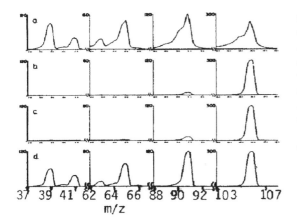

Figure 6.13 Portions of product ion spectra obtained by CID of the molecular ion $C_8H_{10}^{+\bullet}$ of p-xylene in a $Q_1q_2Q_3$ instrument. In all cases the collision energy of the precursor ions entering q_2 was $e.(\Phi_{source} - \Phi_2) = 66\,eV$, the collision gas was argon maintained at a fixed pressure $\sim 10^{-3}$ torr, Φ_{source} was +66V, Φ_1 was 55V, and Φ_2 was zero (ground potential). (a) $\Phi_3 = $ zero $(= \Phi_2)$; (b) $\Phi_3 = 55V\ (= \Phi_1)$; (c) $\Phi_3 = [(m/z)_{frag}/(m/z)_{prec}.\ (\Phi_{source} - \Phi_2)]$, i.e., the 'ideal no-collision' limit; (d) $\Phi_3 = [(m/z)_{frag}/(m/z)_{prec}]^2.$ $(\Phi_{source} - \Phi_2)$, the linked scan law experimentally determined for this particular example by measuring kinetic energies of the product ions. Reproduced from Shushan, *Int. J. Mass Spectrom. Ion Phys.* **46**, 71 (1983), copyright (1983), with permission from Elsevier.

following semi-empirical linked-scan relationship has been used to relate the offset potential Φ_3 of Q_3 to the value of $(m/z)_{prod}$ determined by Q_3 itself:

$$\Phi_3 = [(m/z)_{prod}/(m/z)_{prec}].(\Phi_s - \Phi_2)$$
$$+ \Phi_2 - [\Phi_{min} + k.(m/z)_{prod}]$$

where k and Φ_{min} are tuning constants, Φ_2 is the offset potential of the collision cell q_2, and Φ_s is the potential of the source of ions injected into Q_1. This point of origin of precursor ions is the ion source itself for e.g. EI or CI, but in cases of ions extracted from an ESI or APCI source the point of origin would be the ion guide (often a short RF-only quadrupole denoted q_0) connecting the API source to Q_1 and which is maintained at sufficiently low pressure that the mean free path is long enough that the ions can be accelerated by the applied field with minimal perturbation by collisions.

Some modern instruments avoid the problem altogether by operating at much higher pressures of collision gas within q_2, taking advantage of the collisional focusing effect (Douglas 1992) discussed in Section 6.4.2a. In this way all ions (both surviving precursors as well as the full

range of product ions) were found to have axial energies close to zero at the q_2 exit. As a result, all uncertainties about axial velocities of ions were removed and it became easy to operate Q_3 so as to simultaneously obtain optimum transmission and resolution. Some early results obtained (Morris 1994) using this collisional focusing mode in q_2 are compared in Figure 6.14 with those obtained using the same apparatus but at lower pressures in q_2 corresponding to 'conventional' multiple collision conditions. The improvement in performance is striking.

However, this new mode of operating q_2 had a serious problem in all scan modes in which Q_1 was scanned as in precursor or neutral loss scans (or switched between m/z settings as in MRM). Because of the reduction of ion axial velocity to almost zero the transit time of the ions through q_2 was greatly increased. As a result, peak shapes were significantly distorted in scans for precursors of selected ionic or neutral fragments, and in MRM mode product ions arising from the preceding precursor ion were detected during that part of the MRM cycle that was arranged to detect the products from the current precursor (determined by the Q_1 setting). The resulting cross-talk among MRM channels could be reduced to negligible values by using sufficiently long dwell times per MRM channel, but it turned out that the dwell times required were too long relative to the chromatographic timescale (Mansoori 1998). This problem was solved by applying a small axial DC field to impart sufficient velocity to the ions at the q_2 exit that the transit time was sufficiently short that MRM operation was compatible with fast chromatography, and could still benefit from all the advantages of the collisional focusing mode (Mansoori 1998). A detailed experimental and simulation study of one such q_2 design (Lock 1999, 1999a) clarified the interplay among the experimental parameters.

The advantages of RF-only quadrupoles as collision cells thus include high efficiency in collision induced dissociation (CID, Section 6.5), efficient transmission of surviving precursor ions plus product ions to the following m/z analyzer (a quadrupole Q in QqQ instruments, a time of flight analyzer for QqTOF instruments, Section 6.4.7), flexibility in choice of collision gas (and thus the centre of mass collision energy, Section 6.5) and its pressure (and thus the average number of collisions experienced by ions during their transit through the device), and flexibility in selecting the laboratory collision energy of precursors as they enter the cell (via adjusting the potential offset ($\Phi_s - \Phi_2$), see above). The only significant limitation of these devices from an analytical point of view is an upper limit of $\sim 100V$ for ($\Phi_s - \Phi_2$) imposed by the need to subject the ions to a minimal number of RF cycles (discussion

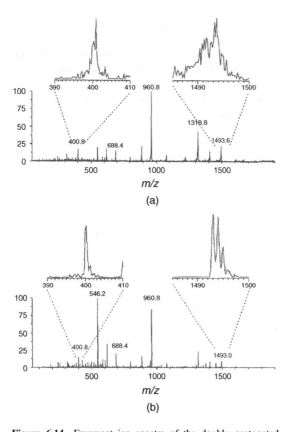

m/z

(a)

m/z

(b)

Figure 6.14 Fragment ion spectra of the doubly protonated peptide AGEGLSSPFWSLAAPQRF (monoisotopic molecular mass of peptide 1920.0 Da) obtained by flow injection of $2\,\mu l$ of a $1\,mg.mL^{-1}$ solution (50:50 aqueous acetonitrile with 15 % acetic acid) into a $10\,\mu L\ min^{-1}$ solvent stream of 50:50 aqueous acetonitrile that contained 0.1 % trifluoroacetic acid. The precursor ion analyzer Q_1 was operated with a 4 Th window to transmit the most intense isotope peaks corresponding to the $[A+2H]^{2+}$ species, and the fragment ion analyzer Q_3 was operated to give peaks approximately 1.5 Th wide near the base. Q_3 was scanned in steps of 0.1 Th, with a dwell time of 1 ms per step. Argon was used as collision gas. (a) Conventional collision cell, 140 eV collision energy, and collision gas thickness 38×10^{13} atoms cm^{-2}. (b) High pressure collision cell, 60 eV collision energy, and collision gas thickness 350×10^{13} atoms cm^{-2}, equivalent in this q_2 cell to $\sim 8 \times 10^{-3}$ torr. (Collision gas thickness $= (N_{gas}/V).1_2 = (P_2.N_A/RT).1_2$, a measure of the total number of collision gas atoms with which an ion can collide in its passage through q_2 of length 1_2, see Equation [6.43]). Reproduced from Morris, *J. Amer. Soc. Mass Spectrom.* **5**, 1042 (1994), copyright (1994), with permission of Elsevier.

of Equations [6.17–6.18]) and a corresponding limit of $ze(\Phi_s - \Phi_2)\,eV$ for the laboratory-frame collision energy (thus placing these devices in the 'low collision energy' category, compared with ultra-low energies in ion traps

and high energy collisions (keV range) in magnetic sector analyzers and TOF-TOF instruments).

A different and relatively minor limitation of triple quadrupole instruments, that of limitations on the linear dynamic range at the upper end, was addressed recently (Fries 2007) in order to quantify, within a single analysis, the concentrations of a potential pharmaceutical drug in a multiple dose (Section 11.5.1) rodent study that previously had required a dilution scheme in order to keep the observed concentrations within the linear range. The alternative strategy involved simultaneously acquiring data at varying sensitivities achieved via collision energy adjustment, for the same precursor-to-product ion transition within a single MRM method. In the application described (Fries 2007) the linear dynamic range was extended in this way from $2.0 - 10\,000\,ng.mL^{-1}$ to $0.50 - 100\,000\,ng.mL^{-1}$, while still satisfying fitness for purpose criteria within a drug discovery context.

In summary, triple quadrupole analyzers are the gold standard workhorse of quantitative trace analyses requiring MS/MS detection as a consequence of their favourable combination of figures of merit. They share all the advantages and disadvantages of stand-alone quadrupole mass filters in this regard, despite the additional complexities of their three-component construction, and extend these with the additional selectivity acquired through use of MS/MS techniques. They can be operated as single quadrupole analyzers by using either one of Q_1 or Q_3 in RF-only mode so that only one of the two acts as a m/z filter. In this latter regard they are the only instrument capable of providing all four MS/MS operating modes (Figure 6.3) in a facile fashion. For quantitation, however, the SIM and MRM modes are the key to their dominance in trace quantitation and for this reason the practical aspects of quantitative analysis discussed in Chapter 9 (e.g. dwell times for SIM or MRM, number of MRM channels monitored etc.) will concentrate on triple quadrupole instruments.

6.4.4 Magnetic Sector Analyzers

Magnetic sector instruments are still used in trace analysis of analytes (generally of environmental concern) that are sufficiently stable that they are amenable to GC separation followed by ionization by EI or CI. Double focusing instruments provide an unmatched combination of high transmission efficiency at high resolving power, up to 10 % transmission at a resolving power of 10 000 (on the 10 % valley criterion, equivalent to $\sim 20\,000$ on the FWHM definition of RP, Section 6.2.3b). Coupled with excellent duty cycle in SIM mode, this has led to their continued use (Section 11.4) in difficult analyses, e.g., of trace level chlorinated dibenzodioxins ('dioxins') in the presence of large quantities of polychlorobiphenyls ('PCBs').

In fact, chemical applications of mass spectrometry were dominated by magnetic sector instruments well into the 1980s, although linear quadrupole mass filters had become important for GC/MS applications. All of the basic principles of qualitative structural analysis, and many of those of quantitative analysis, were established during this period. However, the introduction of new atmospheric pressure ionization techniques such as APCI and ESI, which facilitated direct coupling of HPLC to MS, was difficult for magnetic sector instruments because of their intrinsic requirement to inject the ions at high energies (keV range) from source to analyzer in a high vacuum. Although ingenious attempts have been made to provide robust interfaces between API sources and magnetic sector instruments, such combinations have proved generally unsuitable for high throughput trace quantitative analysis.

Other disadvantages of these instruments include the fact that they are intrinsically scanning instruments with a very low duty cycle except in selected ion monitoring (SIM) mode, and are generally costly to install and to maintain. Moreover, the electromagnets used in such instruments are constructed with a ferromagnetic core that is subject to hysteresis, a property of systems that do not instantly follow the forces applied to them (in this case the magnetic lines of force created by the current flowing in the coils surrounding the iron core), but respond slowly, or do not return completely to their original state. Hysteresis is the property of systems whose states do not depend only on the instantaneous values of the parameters that define them (e.g., electric current in the electromagnet coils) but also on on their immediate past history. As a simple example, if a piece of putty is pushed it will assume a new shape but when the pressure is removed it will not return to its original shape, or at least not immediately and not entirely. In the case of an electromagnet, when the external field is applied by the current flowing through the coils the iron core absorbs some of the external field so that, even when the external field is removed, the magnet will retain some field, i.e., it has become internally magnetized. This hysteresis property implies that such a magnetic field can not be switched sufficiently rapidly to be compatible with fast chromatographic timescales and with a repeatability sufficient to provide high resolution monitoring without some additional aid (usually an internal lock-mass standard, see below). The hysteresis problem can be greatly alleviated by use of an electromagnet with a laminated iron core

(Bill 1983), which permits rapid switching between SIM channels with minimal magnet settling time.

A good historical perspective on their development (Beynon 1960) can be complemented by a detailed treatment of the underlying physics (Kerwin 1963; Duckworth 1963). These instruments exploit properties of so-called sector fields, both electric and magnetic, that are analogous to those of achromatic lenses in light optics. Such a lens does indeed spatially refocus a divergent beam of light to an image of the source of the light but simultaneously disperses the light according to wavelength, thus forming an array of images of the light source that are separated in space according to their wavelengths (colours). Similarly, a magnetic sector field refocuses a spatially divergent beam of ions of the same *m/z* to form an image and simultaneously disperses (in space) ions of differing momentum/charge ratios (including ions of the same *m/z* but different axial velocities).

These ideas are better explained by referring to Figure 6.15; consider here only the dispersion properties of these devices (that lead to their use as mass spectrometers) since the ion optical theory of the direction focusing action is complex (Kerwin 1963). A magnetic sector is an electromagnet shaped like a slice of cake with the narrow point cut off that is then sliced horizontally to allow the ion flight tube to pass through the gap between the two halves (the magnetic poles). Figure 6.15(a) is a sketch of such a device with the magnet pole faces above and below the plane of the page, so that the magnetic field lines of force are perpendicular to the page. If now a beam of ions (effectively an electric current) is passed through this perpendicular magnetic field, an electromagnetic force is created as described by the Lorentz Force Law (Equation [6.3], Appendix 6.1), which causes motion exactly as in a DC electric motor. In this case it is the beam of ions that is moved by this force that is perpendicular to both the original direction of the ions and to the magnetic field lines, i.e., in the plane of the page in Figure 6.15(a) but always perpendicular to the direction of the ion trajectory. As a result the ions follow a circular path with a radius r_{ion} such that the magnitude of the magnetic force F_{mag} (Equation [6.3], Appendix 6.1) just balances that of the centrifugal force F_{cen} associated with the circular motion:

$$F_{mag} = q_{ion}.v.B = F_{cen} = m_{ion}.v^2/r_{ion} \qquad [6.22]$$

where q_{ion} and m_{ion} here denote the actual charge (in coulombs) and mass (in kilograms) of the ion, r_{ion} is in meters, and the magnetic flux density B in tesla

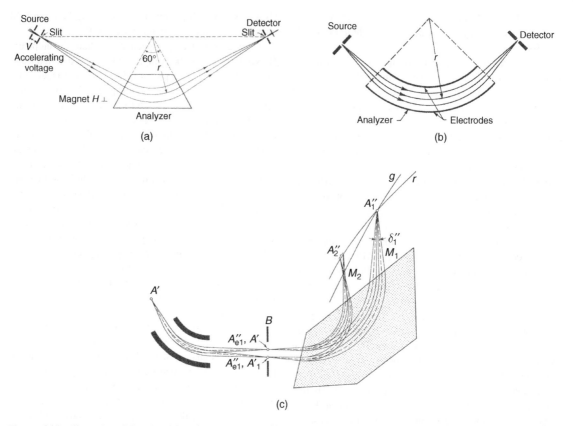

Figure 6.15 Illustration of directional focusing properties of (a) a magnetic sector field and (b) an electric sector field. Reproduced from Farmer in *Mass Spectrometry*, C.A. McDowell (Ed.), McGraw-Hill, 1963, with permission. (c) Schematic diagram of a double focusing instrument. Each of the electric and magnetic sectors acts as a lens to produce an image of the source of the ions, usually an optical-quality slit. The image formed by the electric sector is at the intermediate slit B and this acts as the 'ion source' for the magnetic sector that forms its own image at a point on the direction-focusing curve r. However, each sector also separately disperses ions of a given *m/z* according to their axial velocities. Velocity focusing is achieved, for ions of the same *m/z* but with some variation in velocity, by arranging that the velocity dispersion caused by the electric sector at the intermediate directional focal point B is exactly cancelled by an equal but opposite velocity dispersion produced by the magnetic sector. The resulting velocity focusing along the curve g is thus a consequence of appropriate arrangement of the properties of both sector fields. The double focusing point A_1'' is where the direction and velocity focusing curves intersect, so that ions of the same *m/z* but with moderate variations in direction and velocity all arrive at this same point, where a slit accepts these ions and rejects neighbouring ions. Different *m/z* values are detected at this point by varying the magnetic field strength. Reproduced from Duckworth, in *Mass Spectrometry*, C.A. McDowell (Ed.), McGraw-Hill, 1963, with permission of the McGraw-Hill Companies.

($T = kg.s^{-2}.A^{-1}$, see Table 1.2). Rearrangement of Equation [6.22] gives the fundamental equation of magnetic sector mass spectrometry:

$$B.r_{ion} = m_{ion}.v/q_{ion} \qquad [6.23]$$

The physical meaning of Equation [6.23] is that an ion moving in a sector magnetic field will follow a circular trajectory with a radius r_{ion} determined by the magnetic flux density and by its own ratio of linear momentum ($m_{ion}.v$) to charge (q_{ion}).

Operation of such a device as a 'mass spectrometer' (i.e., as a *m/z* analyzer) requires two further restrictions. A magnetic sector is constructed with its own circular radius r_M, and selection and detection of ions of a specified momentum/charge ratio is achieved by placing an appropriate aperture (or 'image slit') at the point of direction focusing lying on the main axis defined by r_M (Figure 6.15(a)). Then the ions actually transmitted and detected are those defined by Equations [6.22–6.23] with $r_{ion} = r_M$. The second condition concerns the value of the velocity v of the ions. In practice the ion beam is created from an ion source at a high potential $\Phi_1 = V_{acc}$ that accelerates the ions towards the magnetic sector through a source slit held at ground potential (like the rest of the analyzer, i.e. $\Phi_2 = 0$). Then, as discussed in Equation [6.10], the kinetic energy acquired by the ion is equal to the drop in potential energy:

$$T = m_{ion}.v^2/2 = q_{ion}.(V_{acc} - 0)$$

so that:

$$v = (2.q_{ion}.V_{acc}/m_{ion})^{\frac{1}{2}}$$
$$= [(2e.V_{acc}/m_u)/(m/z)]^{\frac{1}{2}} \qquad [6.24]$$

since $q_{ion} = ze$ and $m_{ion} = m.m_u$. Substitution of Equation [6.24] into Equation [6.23] with $r_{ion} = r_M$, gives:

$$m_{ion}/q_{ion} = (r_M^2/2.V_{acc}).B^2$$
i.e., $m/z = [(e/m_u).(r_M^2/2)].(B^2/V_{acc})$ $\qquad [6.25]$

Equation [6.25] defines which ions (m/z values) will be transmitted through a magnetic sector with slits placed as indicated on the main axis, when accelerated into the sector through a potential drop V_{acc}, for a given setting of the magnetic flux density. For values of B that can be achieved with an electromagnet constructed with separation between its poles sufficiently large to accommodate an evacuated flight tube for the ions, values of V_{acc} in the kV range are required. The device can be scanned to give a mass spectrum by scanning B or V_{acc}, usually the former since, if V_{acc} is changed too much from the optimized tuned conditions for the EI ion source, a significant loss of sensitivity is observed (so-called 'source detuning'). For SIM mode, usually V_{acc} is switched at fixed B since an electric field is not subject to hysteresis effects and can therefore be switched fast and reproducibly in a manner that can provide SIM at high resolving power. If the m/z values chosen for the SIM mode are too different, to avoid excessive source detuning the value of B must also be switched despite hysteresis problems. Usually it is arranged for an ion of known m/z (the 'lock-mass' ion) to be present in the ion source at all times and, when the value of B is changed, an automated routine is invoked whereby V_{acc} is scanned over a small range until the lock-mass ion is located and the system 'locks-on' to the tuning condition thus re-established.

The foregoing describes the basic features of the dispersive properties of a magnetic sector; as mentioned above,

the directional focusing properties are taken as given. The main limitation on resolving power of such a stand-alone magnetic sector arises from the fact that the ion velocity given by Equation [6.24] is actually the median value of a range of velocities, since not all ions experience exactly the same accelerating potential drop V_{acc} due to the range of starting (pre-acceleration) positions of the ions within the EI source. Thus ions of the same m/z value will require slightly different values of B to allow them to follow circular trajectories of radius $r_{ion} = r_M$, and thus be transmitted through the image slit and detected. This corresponds to a broadening of the observed peak for these ions, i.e., a decrease of resolving power. This limitation is addressed by the application of velocity focusing, achieved by incorporating an electric sector.

Figure 6.15(b) shows a sketch of an electric sector, usually comprising parallel part-cylindrical plates (electrodes) arranged symmetrically along the central axis that is in the form of a circle with radius r_E and plate separation d_E. A DC electrical potential difference V_E between the two plates creates a radial electric field V_E/d_E between the two plates, thus exerting a radial force $q_{ion}.(V_E/d_E)$ on an ion beam that is again balanced by the centrifugal force corresponding to the induced circular motion, as for the magnetic sector:

$$q_{ion}.(V_E/d_E) = m_{ion}.v^2/r_{ion},$$
i.e. $r_{ion} = 2.(d_E/V_E).(m_{ion}.v^2/q_{ion})$

Thus, in contrast with the magnetic sector, the electric sector disperses ions according to their ratio of kinetic energy $(m_{ion}.v^2/2)$ to charge. Again, appropriate slits on the main axis ensure that the only ions that are transmitted are those following a circular trajectory of radius r_E, and the velocity v is still given by Equation [6.24], so substitution for v gives:

$$r_{ion} = r_E = 2.(d_E/V_E).V_{acc} \qquad [6.26]$$

Thus, under the condition of all ions having kinetic energy/charge defined by V_{acc}, an electric sector alone can not separate ions according to their m/z values since r_{ion} is independent of m/z. However, an electric sector does disperse ions according to their velocity via variations in their kinetic energy/charge ratio arising in turn from variations in the effective value of V_{acc}, as discussed above for a magnetic sector.

The concept of velocity focusing in an instrument that combines magnetic and electric sectors in series (Figure 6.15(c)) involves choosing the dimensions (r_M and r_E) and sector angles, as well as distances between ion source and the first sector and that between sectors, in such a way that the velocity dispersion created by the electric sector is equal and opposite to that created by the magnet. This is the role of the electric sector, i.e., not as a kinetic energy filter as is sometimes stated (that would result in rejection of a large fraction of the ions) but as a compensator for variations in kinetic energy.

The term double focusing refers to such an arrangement, in which both directional and velocity focusing are achieved simultaneously (Figure 6.15(c)), thus leading to very narrow peaks, i.e., high resolving power mass spectrometry (HRMS). Traditionally, magnetic sector instruments specify resolving power using the 10 % valley definition of Δm (Figure 6.4). For demanding trace analyses such as 'dioxin' analyses (Section 11.4.1), the benchmark data are provided using such instruments operated with RP $\geq 10\,000$ (equivalent to $\sim 20\,000$ on a FWHM definition of Δm) with still excellent transmission efficiency (up to 10 % of that when operated at RP $\sim 10^3$), and with high duty cycle in SIM mode (usually achieved by switching V_{acc} and V_E together so that Equation [6.26] is satisfied). Such data obtained at RP values of this order are generally free from interferences from co-eluting compounds that yield ions with the same nominal m/z value (nearest integer or one additional significant figure) and thus require high RP for unambiguous analyses.

For many years double focusing sector instruments were the instruments of choice for highly accurate and precise (within $\sim 1\,ppm$) measurements of molecular mass, used to determine molecular compositions of unknown compounds or at least limit the possibilities to a small number. This role is now usually filled by upper-end TOF analyzers (less expensive, superior ease of use, but lower accuracy and precision) and by FTICR and Orbitrap instruments (more expensive but higher RP possible). Magnetic sector instruments are still used by the petroleum industry in type analysis (see Preface), and also for GC/HRMS quantitation as mentioned above, since TOF, FTICR and Orbitrap instruments do not posses the necessary combination of figures of merit for such trace quantitative analyses.

6.4.5 Quadrupole Ion Traps

Quadrupole ion traps are now available in two geometries, the recently introduced 2D linear ion trap and the (now classical) 3D quadrupole ion trap (Paul trap). The operating principles of 3D ion traps are closely related to those of the quadrupole mass filter but are considerably more complex and thus demand more space for explanation. However, they are discussed here in some detail both because of their popularity and because the basic principles underlying their operation are shared by the more recently introduced linear (two-dimensional) ion traps. Another reason for the relatively lengthy treatment of these devices is that stand-alone versions can be operated both as conventional mass spectrometers (MS^1) and in MS^n mode (see Section 6.2.2), whereas the quadrupole mass filter (Section 6.4.2) must be used in tandem-in-space combinations (Section 6.4.3) to provide MS/MS capability. The more recently developed linear (two-dimensional) ion traps are much less susceptible to space charge effects and have faster spectral acquisition rates, and may in future be shown to be key players in trace quantitative analyses. For convenience the two types are discussed separately.

3D ion traps have probably been used in as many quantitative application areas as triple quadrupoles. Much like the single quadrupole mass filter, 3D traps are popular because of their small size (and cost), their ease of use and the facility with which they can be coupled to both GC and LC. However, because 3D traps store ions for subsequent m/z analyses, this leads to some inherent limitations that have prevented them from achieving the high levels of precision and accuracy required in the most demanding high throughput trace quantitative environments, such as bioanalytical applications in the drug development phase (Section 11.5.2). The need to first fill the trap with ions before m/z analysis, and at the same time limit the number of ions present in the trap at any one time to avoid space charge effects (Section 6.1), leads to most of the important limitations of ion traps in trace quantitation, e.g. without ion number control, space charge would lead to strong limitations on dynamic range at the high end. Further, this fill/analysis cycle, coupled with the need to control ion numbers, results in total scan times that vary in length across the chromatographic peak as the rate of analyte derived ions from the ion source rises and falls thus leading to poorer integration precision (see later discussion of Figure 6.24). Also, in SIM and MRM modes data acquisition times are longer relative to those for mass filters and triple quadrupoles, which can switch rapidly from one ion (or ion pair) to another. Strangely, full scan product ion acquisition is faster for ion traps than the rather complex series of operations required to operate these devices in SIM or MRM mode, so the selectivity increase afforded by MS/MS usually exploits the former rather than the latter. Another disadvantage of ion traps in MS/MS mode is the low mass cutoff (analogous to that for quadrupole mass filters, see discussion of Figure 6.10b in Section 6.4.2) that can sometimes prevent characteristic fragment ions from being stored for analysis and subsequently observed.

A more subtle disadvantage of 3D ion traps in particular, arising from the need to control space charge effects

(2D traps are considerably less susceptible), concerns the conventional use of a co-eluting isotopically-labeled SIS at a fixed concentration. In MRM mode using a triple quadrupole, during one dwell time the RF/DC voltages are set to transmit the precursor–product ion pair for the analyte, and in the next dwell time the SIS ion pair, and this cycle is repeated, i.e. during any one dwell time only ions from one or other of the analyte or SIS are transmitted and counted. This reduces the duty cycle from its theoretical best value of 100 % but the independence of the monitoring for analyte and SIS is an offsetting advantage. In contrast, in ion trapping devices the trap must first be filled before it can scan or go through the ion isolation procedures required for SIM or MRM (see later). When the trap is filled all ions are initially trapped, i.e. matrix and SIS ions as well as analyte ions. At the low end of the calibration curve (this also applies to matrix free solutions) because there are more SIS ions relative to analyte ions the trap is predominantly filled with SIS ions, i.e. the ion fill time (determined by automated controls on total ion numbers) is shorter than what it would be if only the (lower concentration) analyte ions were present. This can lead to intrinsic limitations on the LOD and LLOQ relative to what would be possible using a triple quadrupole. Of course this particular limitation does not apply when a (non-coeluting) analog SIS (Table 2.1) is used.

The preceding comments were made with no detailed justifications. These can only be made via a reasonably detailed consideration of the fundamental principles underlying operation of 3D quadrupole ion traps. The limitations outlined above were in fact the motivation behind the recent development of the newer 2D ion traps. The remainder of this section attempts to provide a description of these fundamental principles at a level that at least permits an understanding of the strengths and limitations of ion traps in chemical analysis, particularly when compared with triple quadrupoles in the context of trace level quantitation.

6.4.5a Three-Dimensional (Paul) Traps

In 1989 the Nobel Prize in physics was shared by Wolfgang Paul (for development of the three-dimensional quadrupole ion trap as an extension of the linear quadrupole mass filter) and Hans Dehmelt (for spectroscopic studies of ions suspended in ion traps of various kinds, including the Paul trap; the Nobel award lectures (Paul 1990; Dehmelt 1990) incidentally also provide accounts of their work that are interesting historically and also lucid and accessible to nonexperts. Other early work on development of the same general principles for ion trapping (Good 1953; Wuerker 1959) should also be

mentioned. The first published account of the use of such a device as a mass spectrometer (Fischer 1959) led ultimately to the development of the sophisticated devices in common use today for trace analytical chemistry.

An early account (Dawson 1976) of the bridge between the subsequent developments of the device by physicists (who refer to the device as the 'Paul Trap') and chemists (who have used several names, most often simply 'ion trap' but sometimes 'quadrupole ion storage trap, QUISTOR', '3D trap' and, in one commercially available form, 'ion trap mass spectrometer'), was followed by extensive reviews written by and for chemists (e.g., Todd 1991; March 1992) and a three-volume set (March 1995) covering theory, practicalities and applications. More recently an excellent first introduction for chemists (March 1997) was updated (March 1998) and followed by a comprehensive treatise on the subject (March 2005). An interesting personal perspective by one of the leading contributors to the field (Stafford 2002) describes the additional problems faced in producing a commercial instrument.

A sketch of the device is shown in Figure 6.9(b) as a comparison with the quadrupole mass filter but a better representation is shown in Figure 6.16. The quadrupole ion trap consists essentially of three electrodes that are arranged to create a quadrupolar electric field; two of the three (called end cap electrodes) are virtually identical and resemble small inverted saucers that are hyperboloids of revolution, each conceptually created by rotating one half of a two-dimensional hyperbola around the z-axis (Figure 6.16(b)). Each end cap has a central small aperture, one for admitting ions into the trap and the other to permit ions to be transmitted to a detector. In ion traps intended for use in GC/MS analyses with EI, one of the end caps has two apertures, one to admit the GC eluant and the other to admit a 70 eV electron beam used to create ions within the trap itself. The third electrode is also of hyperboloidal geometry, but in this case can be thought of as resulting from rotation around the z-axis of both halves of a hyperbola; this 'ring electrode' resembles a serviette holder (March 1997) and is positioned symmetrically between the two end caps. When an appropriate RF potential (usually ~1 MHz) is applied to the ring electrode with both end caps held at ground potential, any cross-section containing the z-axis (Figure 6.16(b)) shows a field of force that is identical (apart from the overall physical dimensions) with that produced in a quadrupole mass filter (compare Figures 6.9(c) and 6.16(b), note that in both cases the electrodes are truly truncated hyperbolae for practical reasons). The distance from the central axis of cylindrical symmetry (z-axis) to the ring electrode,

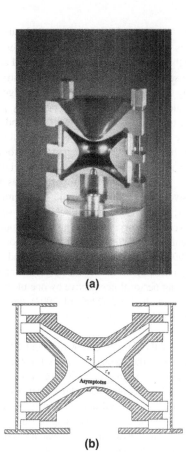

(a)

(b)

Figure 6.16   Representations of a Paul ion trap (quadrupole ion trap). (a) Photograph of a trap cut in half through a plane containing the axis of cylindrical symmetry (z-axis). The physical size of the device is generally of the order of 1 cm within the trap cavity. (b) Schematic diagram of the cross-section. Reproduced from March, *J. Mass Spectrom.* **32**, 351 (1997), with permission of John Wiley & Sons, Ltd.

measured in the central plane of symmetry that is perpendicular to the z-axis (Figure 6.16(b)), is r_0; the separation of the two end caps measured along the z-axis is $2.z_0$. Originally the following relationship was thought to be necessary in order to provide an ideal field within the trap:

$$(r_0/z_0)^2 = 2 \qquad [6.27]$$

but, as discussed below, this is not an absolute requirement and ratios other than 2 are now widely used. However, whatever the value of this ratio, once the value of r_0 has been chosen (most traps use $r_0 \leq 1$ cm), essentially all other dimensions of the trap are defined. The small size is dictated by the requirement for sufficiently

strong RF fields ($V.cm^{-1}$) within the device; by keeping the dimensions small the RF amplitude V_0 required can be kept down to a few kV.

The theory of operation of a Paul trap is very similar to that for the linear quadrupole mass filter, discussed above, and the following treatment will build on that as well as on a more detailed introduction (March 1997). The mechanism of the device as an ion trap (as distinct from a mass spectrometer) can be understood in qualitative terms through a mechanical analogy (Paul 1990) illustrated in Figure 6.17. (With appropriate modifications this analogy also applies to ion motion in the

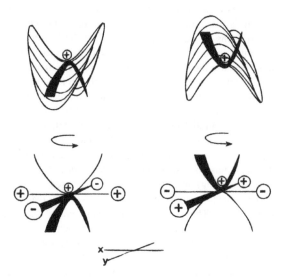

Figure 6.17   Top: Schematic of a ball-on-a-saddle model (operating in a gravitational field) of ion trapping in a quadrupole electric field. Top left: the ball will tend to roll down in the y-direction, but if the saddle is rotated quickly enough by 90° (top right) before the ball moves too far, the ball will now roll back downhill to the centre; this will continue indefinitely if the frequency of rotation is correctly matched to the rolling speed of the ball (a real physical model of this gravitational analog of an ion trap was constructed and is described by Paul (1990), *Rev. Modern Phys.* **62**, 531). Bottom: Similarly, a positive ion submitted to a quadrupolar electric field will move towards a negative electrode (vice versa for a negative ion), but if the field polarity changes sufficiently quickly relative to speed of motion of the ion, the latter will be brought back to the centre of the structure by the field that is now applied in the opposite direction. This trapping action will continue indefinitely if the frequency of the RF field is correctly matched to the speed of motion of the ion under the influence of the field. This is the principle underlying ion trapping in the quadrupole ion (Paul) trap and of establishment of stable trajectories through a linear quadrupole. Reproduced from de Hoffmann, *J. Mass Spectrom.* **31**, 129 (1996), with permission.

xy-plane of a linear quadrupole operated in the RF-only mode, i.e., with $U = 0$.

It can be shown (March 2005) that the electrical potential Φ at any point (x,y,z), within a hyperboloid trap satisfying condition [6.27], is given by:

$$\Phi = 0.5(U + V_0.\cos\omega t).[(x^2 + y^2 - z^2)/r_0^2] \quad [6.28]$$

The ion trajectory within such a potential is described by the Newton equations of motion (Equation [6.2], Appendix 6.1), one for motion in the z-direction and one for each of the x- and y-directions. The component of force exerted on an ion with charge magnitude ze in the x-direction, for example, is $ze.E_x$ where E_x is the electric field component (Equation [6.9]):

$$E_x = \partial\Phi/\partial x = (x/r_0^2).(U + V_0.\cos\omega t)$$

At the exact geometrical centre of the trap where $x = y = z = 0$, so also $E_x = E_y = E_z = 0$, i.e., there is zero electric field at the centre of the trap and the field strength increases linearly with distance from the centre. As for the linear quadrupole, the equations of motion have the algebraic form of Mathieu equations. As a result of the cylindrical symmetry of the device (Figure 6.16) the x- and y-directions are entirely equivalent (unlike the case for the linear quadrupole where the two are related via a simple sign change, see discussion of Equations [6.14–6.15]) and only one equation of motion in the generalized r-direction is required. The solutions to the Mathieu equations for a trap are expressed in terms of a and q parameters analogous to those defined in Equations [6.14–6.15]:

$$a_r = 8/[\omega^2.(r_0^2 + 2z_0^2)].(q_{ion}/m_{ion}).U$$
$$= 8e/[\omega^2.(r_0^2 + 2z_0^2).m_u].[U/(m/z)] \quad [6.29]$$
$$q_r = -4/[\omega^2.(r_0^2 + 2z_0^2)].(q_{ion}/m_{ion}).V_0$$
$$= -4e/[\omega^2.(r_0^2 + 2z_0^2).m_u].[V_0/(m/z)] \quad [6.30]$$
$$a_z = -16/[\omega^2.(r_0^2 + 2z_0^2)].(q_{ion}/m_{ion}).U$$
$$= -16e/[\omega^2.(r_0^2 + 2z_0^2).m_u].[U/(m/z)] \quad [6.31]$$
$$q_z = 8/[\omega^2.(r_0^2 + 2z_0^2)].(q_{ion}/m_{ion}).V_0$$
$$= 8e/[\omega^2.(r_0^2 + 2z_0^2).m_u].[V_0/(m/z)] \quad [6.32]$$

Clearly, $a_z = -2\,a_r$ and $q_z = -2\,q_r$, i.e., the stability parameters for the z- and r-directions differ by a factor of -2. This differs from the case of the linear quadrupole for which the stability parameters for the x- and y-directions differ by a factor of -1 (see discussion of Equations [6.13–6.15]).

The original relationship between r_0 and z_0 (Equation [6.27]) is dramatically changed in most modern traps in order to alleviate the effects of the truncation of the hyperbolic electrodes on the field inside the trap as well as field perturbations arising from the holes in the end caps. The most troublesome of these effects were unpredictable chemical mass shifts in the observed m/z values (Traldi 1992, 1993; Bortolini 1994; Syka 1995); errors in m/z as large as 0.7 were observed, even when the measured m/z was in good agreement with the true value for other ions in the same spectrum. Even more puzzling, such shifts could be observed for one positional isomer (typically a substituted aromatic compound) but not for another isomer investigated under identical conditions. These compound-specific chemical mass shifts were greatly alleviated, or even eliminated, by adjusting the electric field in the trap by 'stretching' the trap. In such 'stretched' ion traps the value of z_0/r_0 is increased from the original specification (Equation [6.27]) of $(\frac{1}{2})^{\frac{1}{2}}$ by $\sim 10\%$ up to as much as 57% (March 1997). An investigation of the physical-organic chemical aspects of the 'chemical mass shift' phenomenon (Peng 2002) contains a good review of earlier work and confirms the conclusion (Londry 1995; Yost 2000) that such shifts are caused by dissociation of fragile ions during ejection in a mass selective instability scan (see below). In an ideal Paul trap the electric field varies linearly from the center of the trap to the end cap electrode (see above) where it abruptly drops to zero, but the end cap holes lead to nonlinearities in the electric field along the z-direction, which in turn cause increased levels of collisional activation (Section 6.5) and thus the uncontrolled fragmentation that leads to the chemical mass shifts. All ion trap mass spectrometers now use some form of geometric modification to adjust the field to correct for the perturbations on the ideal trap potential distribution by end cap apertures, and the chemical mass shift problem is now not a significant problem in operation of a Paul trap as a low resolving power mass spectrometer.

Exactly as for the linear quadrupole, the Mathieu equations for a Paul trap provide solutions to the ion's equations of motion that are of the 'stable' and 'unstable' types. The details of derivation of these solutions (March 2005) are omitted here. The 'stability' and 'instability' solutions are again conveniently represented on a stability diagram in which the stability regions of interest are those in which the motions in both the z- and r-directions are stable. The double stability region that is exploited in almost all traps is shown in Figure 6.18, plotted in (a_z, q_z) space for convenience (the radial diagram is inverted and compressed by a factor of two since the corresponding stability factors differ by a factor of -2, see above).

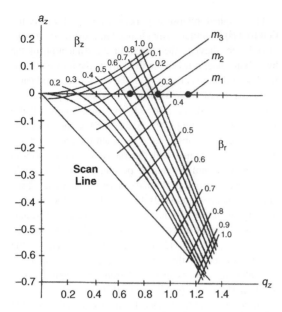

Figure 6.18 Stability diagram in (a_z, q_z) space for the region of simultaneous stability in both the r- and z-directions near the origin for a Paul ion trap. (Recall that $a_z = -2 a_r$ and $q_z = -2 q_r$, so the corresponding diagram in (a_r, q_r) space is inverted and compressed but otherwise identical.) The β parameters are functions of a and q, and describe the secular frequencies of the ion motion in the trap. The 'scan line' represents the locus of all possible m/z values for a fixed ratio U/V_0 and mapped on to the stability diagram as a single straight line through the origin with a slope $= -2U/V_0$; this line refers to the *mass selective storage* mode of operation. In the *mass selective instability (ejection)* mode with $U = 0 = a_z$, the points marked m_1, m_2, and m_3 represent the (a_z, q_z) coordinates of three singly-charged ions of the indicated masses ($m_1 < m_2 < m_3$), such that m_1 has already been ejected by scanning V_0 upwards (and thus detected), m_2 has $q_z \sim 0.908$ and is about to be ejected, and the species with mass m_3 is still trapped. Reproduced from Todd, *Mass Spectrom. Revs.* **10**, 3 (1991), with permission of John Wiley & Sons, Ltd.

The parameter β, that is included in Figure 6.18 as lines of either fixed B_z or B_r, are complex functions of a and q that describe the natural resonance frequencies of the oscillatory motion of the ion; these natural frequencies are analogous to those of a stretched string in a musical instrument and are called the secular frequencies of the ion's motion. Just as for a musical instrument, the secular frequencies include the fundamental frequency plus so-called overtone frequencies. In the case of a musical instrument it is the particular mix of these overtone frequencies with the fundamental frequency that confer the distinctive tone of a violin, say, compared with that of a guitar or a piano playing the same note (fundamental frequency). In the

case of an ion oscillating in the RF field of a Paul trap, the complete mathematical theory gives the spectrum of secular frequencies as:

$$\omega_{u,n} = (n + 0.5\beta_u).\omega \quad \text{for } 0 \leq n;$$

$$\omega_{u,n} = -(n + 0.5\beta_u).\omega \quad \text{for } n < 0 \quad [6.33]$$

where $\omega_{u,n}$ is the n'th overtone (angular) frequency for both $u = r$ and z. The fundamental frequency is for $n = 0$, i.e., $\omega_{u,0} = \beta_u.\omega/2$. Although β_u is strictly defined by a complicated function (March 1977), the so-called Dehmelt Approximation is useful for sufficiently small values of q:

$$\beta_u^2 \approx a_u + 0.5.q_u^2 , \quad \text{for } q_r < 0.2 \text{ and } q_z < 0.4 \quad [6.34]$$

Note that the boundaries of the doubly stable region (Figure 6.18) are defined by the curves for $\beta_u = 0$ or 1, for $u = r$ and z. The $\beta_z = 1$ curve intersects the q_z axis at $q_z = 0.908$; since in practice ion traps are usually operated without a DC potential difference U, i.e., with $a_r = a_z = 0$, the point with $q_z = 0.908$ represents the ion of lowest m/z value that can be stored in the trap (the so-called low mass cutoff).

Before proceeding to a discussion of how a Paul trap can be operated as a mass spectrometer, some further discussion of the trapping action is necessary. Figure 6.19(a) is a famous photograph of a small ($\sim$20 μm) particle of aluminum that had been electrically charged and injected into an ion trap (Wuerker 1959). This remarkable early paper described many of the principles used in operating such traps even today, including the use of higher pressures in the trap to 'cool' the ion motions (see below). In the case of the metal particles that were sufficiently large to scatter light, their motion could be observed directly and photographed to show the low amplitude direct response of the driving field ($\omega/2\pi = 200$ Hz in this case), superimposed on the secular motion with frequency determined experimentally to be $\omega_{z,0}/2\pi = 163$ Hz. Since the trap was operated with $U = 0 = a_z$, Equation [6.34] gives $\beta_z = 2\omega_{z,0}/\omega = 0.163$, and Equation [6.34] then gives $q_z \approx \beta_z.2^{\frac{1}{2}} = 0.231$. Of course no such experimental visualization is possible for organic ions but Figure 6.19(b) shows computed three-dimensional trajectories for an ion with m/z 105 and $q_z = 0.3$ (Nappi 1997); the general features of the ion trajectory are similar to those of the metal particle (shown as a two-dimensional projection on the r-z plane in Figure 6.19(a)). A detailed understanding of such ion motions in 3D (and 2D) traps, and their associated frequencies, leads to techniques for manipulation of ions for MS and MS/MS analyses.

(a)

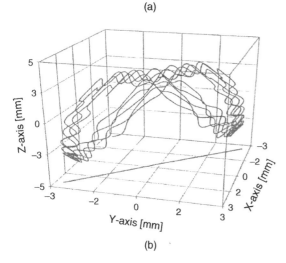

(b)

Figure 6.19 (a) Photomicrograph of the motion of a single electrically charged aluminum particle ($\sim$20 μm) viewed in an r-z plane of a Paul trap, obtained using an exposure time of 0.1–0.2 s. (The motion is three-dimensional; the trajectory from top right to bottom left lies *behind* the trajectory from top left to bottom right.) The large scale motion corresponds to the secular motion of the particle and the small superimposed oscillations to the direct response of the particle to the RF field. The observations led to an estimate of the charge/mass ratio of the particle of 189 C.kg^{-1}. Reproduced from R.F. Wuerker, *Journal of Applied Physics*, 30, 342 (1959). Copyright (1959), with permission from the American Institute of Physics.
(b) Computed trajectory in a Paul trap of an ion with m/z 105, $q_z = 0.3$, and zero initial velocity. The trajectory develops a shape that resembles a flattened boomerang, and again shows the low amplitude oscillations at the RF frequency ω superimposed on the secular frequency of the motion. Reproduced from March, *J. Mass Spectrom.* **32**, 351 Copyright (1997), with permissions from Elsevier

Space charge effects can present problems for all mass spectrometers but are particularly important for traps because of their small dimensions. All the preceding discussion has implicitly assumed that only a single ion (or charged particle) is present in the trap but of course in the context of chemical analysis this is an absurd restriction. If a collection of ions is present in sufficient numbers in a restricted space, such that the average inter-ion distance is sufficiently small that ion–ion electrical forces (Equation [6.4], Appendix 6.1) are significant relative to the forces exerted by the external RF field, the ion trajectories will deviate from those predicted and the performance of the trap will deteriorate. In general, the effects of significant space charge effects in a Paul trap include a broadening of the distribution of secular frequencies of ions of a given m/z value and a lowering of the mean values of these distributions relative to the single ion values (Equation [6.33]). There is also a decrease in ion detection efficiency because the ions are now more widely distributed spatially and thus no longer line up as well with the holes in the exit end cap during the mass selective instability scan mode (see later).

The intrinsically small dimensions of a Paul trap (imposed by the practical requirement to apply RF field strengths of the order of 10^3 V.cm^{-1}) implies that there is an upper limit to the number of ions that can be accommodated before space charge effects become important; depending on circumstances and the purpose of the particular experiment, this number can be as low as 10^5, or even lower. Space charge effects in a Paul trap have been extensively investigated (Todd 1980). Note also that it is the total number of ions present that controls space charge effects, not just the number of ions produced from the target analyte, so that all solvent derived ions from ESI or APCI or matrix ions from MALDI, potentially many times more abundant than analyte ions, can contribute. The methods that have been devised to minimize this problem are extremely important for quantitative applications and will be described later together with the methods used to generate a mass spectrum with a Paul trap.

Collisional focusing of ions is now an extremely important phenomenon in the operation of modern Paul traps as analytical mass spectrometers. A very early observation (Wuerker 1959) described the beneficial effects of a background pressure of a few mtorr on the trapping characteristics. However, realization of the importance of the effect for mass spectrometry can be attributed to investigations (Bonner 1976) of some apparently anomalous results of studies of charge exchange of argon ions in an excess of argon atoms. The experimental observations were interpreted with the aid of theoretical modeling of the motion of the ions using Monte-Carlo methods and suggested that, as a result of ion–atom collisions, the ions migrated towards the center of the trap device with a corresponding increase in extraction efficiency and a decrease in the ions' kinetic energy. The full potential of the phenomenon was not realized until several years later (Stafford 1984) in a landmark paper that also introduced the mass selective instability scan mode.

It was shown that the presence of a gas of low molecular mass (in practice helium), at a relatively high pressure ($\sim 10^{-3}$ torr) compared with that usually required in a mass analyser ($\leq 10^{-5}$ torr), significantly enhanced the resolution, sensitivity and detection limit. (Note here a significant difference between this collisional focusing in a 3D trap with that in an RF-only linear quadrupole, where gases of much higher molecular mass can be used and are indeed preferred.) The most likely causes of the observed behavior were discussed briefly (Stafford 1984). A collision between a trapped ion and a low mass atom or molecule results in only a small change in the momentum (speed and direction) of the heavier ion (compare a collision between a bowling ball and a ping-pong ball), and thus generally does not produce significant scattering of the ion away from its former trajectory so that the ion motion is mostly under the expected control of the external field. The net result of a large number of such collisions will thus tend to maintain the directional property of the original ion trajectory (while decreasing the ion velocity), since all changes in direction will be equally probable as the initial parameters describing the collision are randomly sampled so that directional changes will tend to average out to zero. In this way the trapping efficiency is not significantly reduced by ion scattering. However, if the neutral gas molecules have molecular masses similar to or greater than that of the ion, the collision can result in significant scattering of the ion away from the predicted trajectory, sufficiently large that the deviation can not be corrected for by the applied field or cancelled by scattering in the opposite direction by the next collision, so that such ions are lost to the trap walls. Collisions with very low mass atoms like helium, however, also result in transfer of kinetic energy from the ion to the neutral gas atom/molecule. Since kinetic energy depends on the square of the velocity, i.e., all directional effects are irrelevant, no similar cancellation of the effects of multiple collisions on kinetic energy can occur (no 'positive' and 'negative' effects that can mutually cancel). Therefore, the kinetic energy is progressively reduced to values ~ 0.1 eV. (Compare with the case in RF-only quadrupole cells in which ion kinetic energies of tens of eV can be accommodated. It is this large difference in collision energies, together with the difference in molecular mass of the collision gas molecules, that leads to

inherently different CID conditions (Section 6.5) and thus appearance of MS/MS spectra for traps compared with RF-only linear quadrupole q_2 cells.) This 'cooling' of the ion motion in the trap is analogous to viscous damping and causes the orbits of the trapped ions to collapse close to the center of the trap (radii of trajectories of the order of one millimeter) where field imperfections, caused by fabrication errors in the hyperbolic electrode structure (or by truncation of the hyperbolae or by holes drilled in the end-caps), are minimized. It is well known (Dawson 1976) that several mechanisms exist for loss of trapped ions as a result of field distortions, so that such ion losses will be minimized at the center of the trap. The collisional damping improves not only the trapping properties but also the mass selective instability scan mode (Stafford 1984) that is described below.

Resonant excitation is an important technique in modern ion trap mass spectrometry; it is central to using a trap for MS/MS and for improving detection efficiency by axial modulation (see below). This technique exploits one or both of the two secular frequencies, axial and radial (Equation [6.33]) to excite ion motion by irradiation at either or both of these frequencies. Axial irradiation is widely used, via application of a small supplementary oscillating potential of a few hundred mV amplitude between the end cap electrodes, i.e. along the z-axis (so-called dipolar mode excitation). Recall that the secular frequencies $\omega_{z,n}$ are functions of m/z via the dependence (Equation [6.34]) of β_z on q_z. Thus it is possible to apply user-selected waveforms composed of specified frequencies or frequency ranges as the basis for techniques that discriminate among m/z values. Prior to dipolar resonant excitation of trajectories of ions of selected m/z values, the ions are allowed to collisionally cool and focus near the center of the trap. Then resonant excitation of the cooled ions, at the axial secular frequencies for those ions whose m/z values match those within the excitation waveform, causes these ions to move away from the ion trap center where they experience a larger trapping field (recall that the field, i.e., the potential gradient, is zero at the exact centre of the trap, see discussion of Equation [6.28]). This method of ion excitation is often referred to as 'tickling'. The ions that have been excited away from the center of the trap are further excited by the trapping field and can thus achieve kinetic energies of tens of eV.

Resonant excitation is used to remove ions from the trap in order to isolate a range of m/z values by increasing the kinetic energy of the unwanted ions so that they escape from the trapping potential and are ejected. This mode can be used to eject unwanted ions and isolate in the trap only ions with a selected m/z value (e.g., for subsequent MS/MS experiments, see later) or a range of m/z values (e.g., exclusion of matrix or mobile phase ions), or to eject ions selectively and successively by *m/z* by scanning the applied frequency. Resonant excitation is also used (see later) in a less extreme mode to increase the collision energy of the ions with the helium bath gas atoms, thus increasing the internal energy of the ions and create collision induced dissociation (CID). The same dipolar resonant excitation technique is also used in the so-called axial modulation method of operating a Paul trap as a mass spectrometer (see below).

The effective well depth in a Paul trap is another concept that is frequently met in the literature. The hyperboloid shapes of the electrodes were chosen so as to produce an ideal quadrupole electric field when an RF potential is applied to the ring electrode and the two end cap electrodes are grounded. Such a field in turn produces a (truncated) parabolic potential well ($V \propto r^2$ or z^2) for the confinement of ions (recall that the field strength $E_r = dV/dr$ and similarly for z); thus hyperboloid shapes were chosen because the motions of particles in parabolic potential wells is a very well understood problem in mechanics. The wells are truncated at the physical boundaries of the trap to define well depths for the electrical potential given by:

$$D_z \approx q_z.V_0/8 = 2.D_r \qquad [6.35]$$

The practical importance of the potential well depth concept arises because it defines the maximum amount of kinetic energy that an ion can acquire through resonant excitation before it overcomes the potential energy well and is lost to the walls or ejected from the trap. However, note that the concept was developed for collision free conditions for one or a few trapped ions (negligible space charge) and thus gives only a rough indication of the maximum achievable ion kinetic energy in an ion trap mass spectrometer.

Before proceeding to describe how a Paul trap is operated as a mass spectrometer, it will be useful (March 1997) to indicate some typical magnitudes of the various parameters discussed thus far. Consider a typical stretched trap with $r_0 = 1\,cm = 10^{-2}\,m$, $z_0 = 0.783\,cm = 0.783 \times 10^{-2}\,m$, with RF amplitude $V_0 = 757V$ (zero-to-peak) at $f = 1.05\,MHz$ so that $\omega = 2\pi f = 6.60 \times 10^6\,radians.s^{-1}$. From Equation [6.32], by substituting the proposed values for r_0, z_0, V_0 and ω, and $e = 1.602 \times 10^{-19}$, $m_u = 1.660 \times 10^{-27}\,kg$:

$$q_z = 8e/[\omega^2.(r_0{}^2 + 2z_0{}^2).m_u].[V_0/(m/z)]$$
$$= 0.0796 \times [V_0/(m/z)] = 60.27/(m/z) \quad [6.36]$$

Then for $m/z = 100$, $q_z = 0.603$, and for $m/z = 1000$, $q_z = 0.060$ for $V_0 = 757\,V$.

An important quantity in all electric quadrupole devices (linear and three-dimensional) is the low mass cutoff, the minimum value of *m/z* that has stable trajectories in the device under the stated operating conditions, i.e., the ion that has q_z just less than 0.908 (see discussion of Figures 6.10 and 6.18). Then for this trap operated at $V_0 = 757\,V$, $(m/z)_{min} = 60.27/0.908 = 66.4$, i.e., in practice the low mass cutoff for trapping of ions would be *m/z* 67. The converse question, (i.e. what change would have to be made to the operating conditions to permit a specified low mass cutoff) amounts to determination of the required value of V_0 since the RF frequency is considerably more difficult to adjust once fixed. Then the required value of V_0 (in volts) for this particular hypothetical trap (i.e., r_0, z_0 and ω all fixed) is given by Equation [6.36] as $[(0.908/0.0796).(m/z)_{min}] = 11.41.(m/z)_{min}$. Calculations of this latter kind are particularly important for MS/MS experiments in which a precursor ion is stored at a particular q_z value and then subjected to internal excitation (e.g., by resonance excitation), since only fragment ions with $m/z > (m/z)_{min}$ will be stored and thus detected successfully.

This is the basis of the so-called 'one-third rule for ion traps'; this rule-of-thumb states that if a precursor ion is isolated and trapped at a q_z value of about 0.3 (a typical value although there is some variability in this value), since the ion of $(m/z)_{min}$ will correspond to the q_z of 0.908, the $(m/z)_{min}$ for product ions will be about one-third of the *m/z* of the selected precursor ion, because q_z is inversely proportional to *m/z* if all else is held constant (Equation [6.36]) and $0.3/0.908 = \frac{1}{3}$. This is a significant difference between an ion trap and a triple quadrupole for MS/MS experiments; the latter instrument is not subject to this limitation because precursor ion selection, fragmentation and analysis of fragment ions are all conducted in different quadrupole devices. (Some newer generation 3D traps have improved on this one-third rule somewhat, effectively by lowering the q_z value used to initially isolate the precursor ions.) This difference can provide a serious obstacle to transference of an analytical method originally developed using a triple quadrupole to an apparatus using a 3D ion trap, since the product ion used for the QqQ method might not be observable with a trap. (Moreover, even if the product ion does in fact have *m/z* greater than the low mass cutoff, it may well be present at very different relative abundance in view of the very different CID conditions mentioned above.)

Similarly, Equation [6.36] can be used to calculate the upper limit of the *m/z* range that can be ejected at $q_z = 0.908$ from this trap (and thus recorded in a mass spectrum) when operated in the scan mode whereby V_0 is scanned upwards to the maximum value $V_{0,max}$ permitted by the circuitry (see below); then $(m/z)_{max} = (0.0796/0.908).V_{0,max} = 0.0877 \times V_{0,max} = 650$ if the trap has $V_{0,max} = 7.4\,kV$ (zero-to-peak). The ingenious way in which this upper limit can be greatly extended without resorting to impractically high values of V_0 is described below.

First, however, consider the evaluation of the parameters β_z, $\omega_{z,0}$ and D_z. From the Dehmelt Approximation (Equation [6.34]) $\beta_z \approx q_z/2^{\frac{1}{2}}$ for $a_z = 0 = U$; hence if the trap is operated so that an ion of interest has $q_z = 0.908/2 = 0.45$, then $\beta_z \approx 0.318$ (actually this approximate value is ~5 % high in this case as the upper limit for the approximation is $q_z = 0.4$). The corresponding value of the fundamental secular frequency $\omega_{z,0}$ is given by Equation [6.33] as $\beta_z\omega/2 = 0.318 \times 1.05 \times 10^6 \times 2\pi/2$ radians.s^{-1} or $f_{z,0} = 2\pi.\omega_{z,0} = 167\,kHz$ (again ~5 % high). The reason for this choice of q_z (and thus of β_z) as an example will be explained below. Equation [6.35] gives $D_z \approx q_z.V_0/8$, so for an ion operating at our chosen q_z of 0.45 at $V_0 = 1\,kV$, $D_z \approx 56\,V$ (incidentally Equation [6.36] tells us that this ion must be m/z 177). Under exactly the same operating conditions, an ion of m/z 354 (double the previous case) will have $q_z = 0.225$ (Equation [6.35]) and $D_z = 28\,V$. This exemplifies a general point, i.e., at any specified operating point of the trap (i.e. value of V_0) ions of low m/z are in the deepest potential energy wells.

Operation of a Paul trap as a mass spectrometer has undergone several fundamental changes (Todd 1991). The preceding discussion focused on the trapping properties of the device such that, for successful trapping, the working (a_u, q_u) coordinates (Equations [6.29–6.32] must lie within the boundaries of the stability diagram (Figure 6.18). In addition, the trajectories consist of a superimposition of the secular components (low frequency and high amplitude) plus the components that are the direct responses (low amplitude) to the driving RF frequency. Operation of the ion trap as a mass spectrometer exploits one or other of these properties of the trajectories of the trapped ions.

The earlier techniques are now of mainly historical interest and therefore are described only briefly here; they are described in detail elsewhere (Todd 1991). The original method for performing m/z analysis was the resonance detection technique that was described in the original patent (Paul 1960) and realized experimentally soon after the patent submission (Fischer 1959). The mass selective storage technique for operation of a Paul trap as a mass spectrometer (Dawson 1969, 1970) involves trapping only one m/z value at a time and then ejecting these

ions from the trap by applying a voltage pulse between the end caps, thus forcing the trapped ions through holes in one of the end caps on to an electron multiplier. In practice the DC and RF amplitudes were scanned slowly relative to the time required for the sequence of creating, storing, ejecting and detecting the ions.

Both of these early techniques were too slow and inefficient for use of a Paul trap as an on-line detector for GC or HPLC, and indeed this restricted their use in chemical applications to more fundamental studies of ion–molecule reactions. This situation changed dramatically with the key invention of the mass selective instability scan mode (Stafford 1984). The theoretical principles underlying this technique are clearly explained in the original paper and the following description follows the original closely. The trap is operated with $U = 0$ (i.e. $a_z = 0$) and ω fixed (~1 MHz), i.e. only the q_z axis is considered in the stability diagram (Figure 6.18), and V_0 is the only parameter that is scanned. Initially V_0 is set at a value required to establish the desired low mass cutoff $(m/z)_{min}$, using Equation [6.36] with $q_z = 0.908$ as described above; then all ions with $m/z > (m/z)_{min}$ are initially trapped (in the original experiments ions were formed by EI inside the trap itself).

The sequence of manipulations is illustrated in Figure 6.20; ions are created in (or nowadays often introduced into) the trap. After this relatively short period,

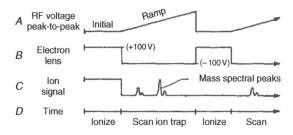

Figure 6.20 The original timing sequence employed for operation of a Paul ion trap in the mass selective instability mode. In this case the ions were created by electron ionization of a gaseous sample directly inside the trap using a pulse of electrons controlled by switching the potential on a lens, but could have been created in an external ion source and pulsed into the trap using a similar lens. Some of the key operating parameters used were: $\omega = 1.0\,MHz$; $r_0 = 0.9252\,cm$; ionization time $= 1.0\,ms$; V_0 (maximum) $= 6\,kV$ (zero-to-peak); m/z scan range ~60–600 at a scan speed of $6000\,s^{-1}$, i.e., the scan time in this example was ~100 ms; helium pressure $= 10^{-3}$ torr; pressure of sample (perfluorotributylamine, i.e. FC143 mass calibration standard) $= 5 \times 10^{-6}$ torr. The figure shows one complete microscan plus part of the next. Reproduced from Stafford, *Int. J. Mass Spectrom. Ion Proc.* **60**, 85 (1984), copyright (1984), with permission from Elsevier.

V_0 is ramped up so that trapped ions of successively larger values of m/z become unstable in turn since Equation [6.36] shows that $(m/z)_{min} \propto V_0$. As the field conditions reach the point where ions of a particular m/z value become unstable, all such ions develop trajectories that exceed the boundaries of the trapping field (only in the axial z-direction in this case). These ions pass out of the trapping field through a perforation in one of the end caps and are detected using an electron multiplier. The ion current signal intensity, detected and recorded as a function of V_0 (i.e. scan time) corresponds to the mass spectrum of half of the ions that were initially trapped (the other half are excited in the opposite z-direction and are thus lost on the other end cap). The V_0 scan rate is chosen so that ions of consecutive values of m/z are not made unstable at a rate faster than the rate at which unstable ions depart the trapping field region. A single such scan is usually completed in a few tens of milliseconds depending on the m/z range and is referred to as a *microscan*. To record a mass spectrum of acceptable signal:noise ratio (ion statistics noise) the results of several ionization/trapping–microscan cycles are accumulated and recorded, requiring a suitable compromise between S/N and the timescale imposed by the time-widths of the chromatographic peaks that are to be analyzed.

Like most successful clever ideas, the mass selective instability scan mode seems rather simple once someone had conceived it. However, it turned out (Stafford 1984) that full experimental realization of the potential of the method required another innovation, the exploitation of the collisional cooling/focusing effect of a low pressure of helium, discussed above. It was found (Stafford 1984) that the presence of helium at a relatively high pressure of approximately 10^{-3} torr significantly enhanced the mass resolution, sensitivity and detection limit of a Paul ion trap operated as a mass spectrometer using the mass selective instability scan to obtain the mass spectrum.

The foregoing discussion of the stability diagram characterizes the conditions under which ions within a real electrode structure can have stable trajectories but said nothing about the amplitudes of these stable trajectories. These amplitudes are a function of the initial position and velocity (i.e. magnitude and direction) of the charged particle, together with the phase of the applied RF field at the instant of the introduction of the ion into the trap field. Ions created within (or introduced into) a trap, even with (a_z, q_z) values lying within the stability region, may or may not be trapped successfully; such ions will remain trapped within a real electrode structure only if they possess certain preferred initial conditions. For example, suppose a positive ion is created with (a_z, q_z) values well inside the stability region but spatially very close to the ring electrode and with a velocity that would cause it to collide with the ring if no RF were applied. If the phase of the RF waveform at this instant is such that the ring electrode is in its positive half cycle, the ion will be repelled by the electrode and still has a chance of being trapped. But if the RF phase is in its negative half cycle, the positive ion will be attracted to the electrode and this attractive force will assist the initial velocity in causing the ion to strike the ring electrode and be lost. Thus, possession of (a_z, q_z) values corresponding to the stability region is a necessary, but not sufficient, condition for an ion to be successfully trapped. (However, all ions with unstable trajectories will necessarily lead to collisions with the trapping electrodes.)

One result of the collisional focusing caused by the low pressure of helium is to initially restrict ions to orbits in the trap that have maximum displacements from the centre of about $0.1 \, r_0$ (~ 1 mm) or less (i.e. well away from the electrodes); this effect directly reduces ion losses arising from initial positions and velocities of the ions, and thus also the detection limit. A less direct but still significant effect of the collisional focusing arises because, during the mass selective instability scan, ions of a given m/z spend most of their lifetime in the center of the trap where the theoretical field is zero (see discussion of Equation [6.28]) so that field imperfections, caused by mechanical errors in the fabrication of the electrode structure, are at a minimum.

Further, if collisional damping is operating, displacements of the trapped ions in both the axial and radial directions will remain small throughout the V_0 scan, up until ions of a particular m/z become theoretically unstable (q_z increases to 0.908). At that point the z-displacements of such an ion increase rapidly (in a few tens of RF cycles) to an extent that they reach the end caps and can be transmitted through apertures to a detector. In contrast, without a background damping gas the maximum displacements of trapped ions grow relatively gradually (since $E_z \propto z$) as the V_0 scan brings the ions close to, and eventually into, instability. This results in a distribution of z-amplitudes of the excited ions, so some of them will have sufficiently large amplitudes that they are transmitted through the end caps before the time at which V_0 has reached the value at which the q_z value of these ions reaches the theoretical instability value (0.908). This results in m/z peaks that are broad and have poor peak shapes, resulting in low resolving power; Figure 6.21 illustrates the dramatic improvements made possible through use of helium as damping gas. Of course further increase of the helium pressure will eventually cause significant scattering of the ions, and indeed the peak widths were observed (Stafford 1984) to reach a minimum (i.e. resolving power at a maximum) at $\sim 10^{-3}$ torr.

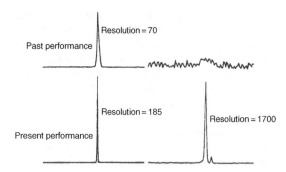

Figure 6.21 Partial spectra illustrating the dramatic effect of use of helium as a damping gas in a Paul trap on its performance as a mass spectrometer using the mass selective instability scan mode. Peaks on the left are for m/z 69, those on the right for m/z 502. The top spectra were obtained without damping gas, and the bottom spectra by using 10^{-3} torr of helium. Reproduced from Stafford, *Int. J. Mass Spectrom. Ion Proc.* **60**, 85 (1984), copyright (1984), with permission from Elsevier.

The mass selective instability scan represented an enormous advance in the exploitation of the Paul trap as a mass spectrometer but was limited with respect to the available $(m/z)_{max} \approx 650$ (see above) for a trap of reasonable physical dimensions and maximum RF amplitude $V_0 \sim 7.4\,kV$. This value of $(m/z)_{max}$ is the value at which $q_z = 0.908$ at $V_0 = 7.4\,kV$ (see above discussion of Equation [6.36]) and is adequate for a trap to be used as a GC detector since most analytes amenable to GC analysis have molecular masses less than 650 Da. However, the analytical applications of interest in this book can require m/z values up to ~ 2000. To address this problem of limited m/z range, an ingenious application of the resonance excitation technique (see above) was developed (Tucker 1988; Weber-Grabau 1988). A simple example can be used to explain this approach (March 1997). At the point when V_0 has been scanned to its maximum value and ions with m/z 650 are ejected, ions with m/z 1300 ($= 2 \times 650$) will still be trapped as they will have a q_z value of ~ 0.45 ($= 0.908/2$), corresponding to $\beta_z \approx 0.318$ and secular frequency $f_{z,0} = 2\pi.\omega_{z,0} = 167\,kHz$ (see above discussion of the implications of Equation [6.36]). Such ions can be ejected from the trap if resonantly excited by applying an AC waveform between the end caps at this frequency (an amplitude of $\sim 6\,V$ is found to be sufficient) with V_0 scanned to its maximum value. In practice, in order to double $(m/z)_{max}$ to 1300, following introduction of the ion pulse the mass selective instability scan would be initiated as originally developed (Figure 6.22), but simultaneously the resonant excitation waveform is applied; this scanning method is referred to as axial modulation. As a result ions are now ejected when V_0 has been

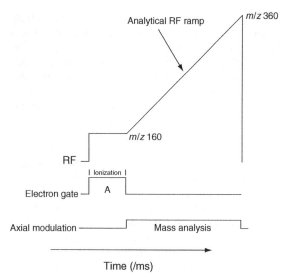

Figure 6.22 Scan function used to obtain an EI mass spectrum (one microscan) using a Paul trap with mass selective instability scanning with concurrent axial modulation. The scan function shows the ionization period A (this could be replaced by a period of ion injection from an external source), followed immediately by activation of the axial modulation waveform and the analytical ramp of V_0. Note that the pre-scan for the automatic gain control function is not shown. Reproduced from March, *J. Mass Spectrom.* **32**, 351 (1997), with permission of John Wiley & Sons, Ltd.

scanned up such that the q_z value reaches ~ 0.45 instead of ~ 0.90, as a result of the choice of resonant excitation frequency of 167 kHz (that was itself determined as a consequence of the choice of q_z value, see above). Similarly, a different choice of the working q_z value as 0.045 implies $\beta_z = 0.032$ and $f_{z,0} = 1.67\,kHz$ (Equations [6.33–6.34]), and a theoretical $(m/z)_{max} = 13\,000$. While truly high mass operation is not of direct importance for this book, it is worth noting that this general principle has been applied with great effect (Kaiser 1989, 1991) such that ions of m/z 70 000 were experimentally recorded.

In addition, use of axial modulation scanning was also found to increase resolving power; this effect is believed to arise as follows (March 1998). Immediately prior to ion ejection, as V_0 is scanned upwards the axial secular motions of ions of a particular m/z value become resonant with the supplementary potential applied between the end caps, so that the axial excursions of the resonantly excited ions (and only these ions) increase in magnitude. As a result these ions escape from the space charge effects of the cloud of ions of higher m/z while the latter are still collisionally cooled at the center of the trap. In this manner the ions that are resonantly excited become tightly bunched as they exit the ion trap, i.e., the range of times over which the ions of any given *m/z* value, when in resonance, move from the central cloud through the exit aperture to the detector is much shorter than in the case of the simple V_0 scan, so the observed peaks (detected ion counts as a function of V_0 scan time) are more narrow. Indeed, when combined with very slow scanning speed (so that ions experience many more RF cycles resulting in greater discrimination among closely-spaced *m/z* values) a resolving power $> 10^6$ at *m/z* 3510 (FWHM peak width 3.5×10^{-3}) was achieved (Kaiser 1991) using an axial modulation scan. However, this remarkable achievement is of little consequence for trace quantitative analysis.

Some ion trap mass spectrometers use variants of the original axial modulation principle. For example, some instruments use an auxiliary AC potential angular frequency $\omega/3$, applied to the end cap nearer to the detector. In this case, when V_0 is ramped, ion ejection occurs at a fixed working point well removed from the $\beta_z = 1$ boundary of the stability diagram.

As mentioned above, the upper limit to the number of ions that can be stored in a Paul trap without significant deterioration of performance (as a result of space charge effects) must be addressed in order to permit a useful dynamic range. For a stand-alone trap the solution that is adopted is to insert a pre-scan function ahead of the main analytical scan, with the objective of determining only the total number of ions present (i.e. ultimate trap performance is not required for the pre-scan). The instrument software determines the ratio of the acceptable ion count to this experimental ion count and reduces the time of ion creation or introduction by this ratio, in a manner exemplified by Figure 6.23.

This approach is referred to by several names, e.g. ion population control, automated gain control, ion control time, trap fill time etc., and can be combined with selective ejection by resonance excitation of abundant unwanted ions (see above). The quantitative ion count value thus obtained must be multiplied by the ion fill time ratio to give the true value. In this way the dynamic range can be increased upwards by several orders of magnitude, with levels of accuracy, precision and reproducibility that are adequate for many purposes but are not for the most demanding applications. Limitations at the lower end of

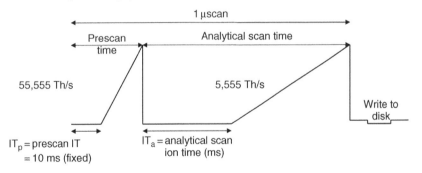

Determining ion time (IT):

let P_c = number of counts determined during prescan = 10^6
and T = target value set in software = 10^7

then for P_c to equal T, must multiply P_c by a factor $f = T/P_c = 10^7/10^6 = 10$

The ion fill time IT_a required for the analytical scan to be conducted on the target number of ions T is therefore also given by the ratio $f = T/P_c$

i.e. $IT_a = f \times IT_p = 10 \times 10\,ms = \underline{100\,ms}$

Figure 6.23 Illustration of an example of determination of appropriate 'ion time (IT)', i.e., trap fill time to minimize space charge effects in a MS[1] full scan acquisition. The fast pre-scan is used to first estimate ion counts and this is compared with T, the target value for number of trapped ions suitable for the analytical scan. The ratio thus determined also fixes the ratio of ion fill times required to ensure that T ions are trapped for the analytical scan. Of course this assumes that the *rate* of ion introduction does not change between the prescan and the analytical scan fill time.

the range are not an issue when running a pure analytical standard, and indeed a 3D trap can analyze fmol (some claim amol) amounts of material. However, as discussed in Section 6.4.5, when a co-eluting isotope-labeled SIS is used there are critical limitations at the low end of the range because now the trap will preferentially fill with the more abundant SIS ions that are now the limiting factor in determining the total number of ions, thus effectively decreasing the number of analyte ions stored compared with what would be the case if only analyte were present.

Note also that an ion trap can be operated in a mode analogous to the selected ion monitoring (SIM) mode for linear quadrupoles (see above), by simply filling the trap and then ejecting all the unwanted ions, i.e. mass selecting the ion of interest before scanning it out thus alleviating the space charge problem. There is an appreciable time penalty for this approach relative to a quadrupole mass filter, and it can become impractically long if more than one m/z value is to be monitored since the trap–isolation–scan-out cycle must be repeated for each m/z value. Note that ion traps can scan at around 5000 Th.s^{-1}, so a scan from m/z 160 to 360 (Figure 6.22) would require about 40 ms. This scan-out time would be long relative to the ionization (trapping) time (as suggested in 'A' of Figure 6.22) if the latter were the lowest it could be, perhaps ~0.1 ms. However, the ionization time 'A' could be up to 1000 ms based on the rate at which ions enter the trap, i.e. the time 'A' in Figure 6.22 is variable. This Figure does not include any contribution to data acquisition time arising from ion number control; in older traps this requires about 10 ms although newer traps (both 3D and 2D) can achieve this much faster. (This is a good example of how a book can become obsolete with respect to the performance characteristics of instrumentation almost as soon as it is written when the pace of development is as rapid as is currently the case for mass spectrometry; and this also emphasizes the benefits of understanding the fundamental principles.) Note that if full scan mass spectra (or MS/MS spectra) are recorded rather than SIM or MRM data, post-acquisition data processing allows construction of so-called extracted ion chromatograms in which intensities of signals for only characteristic m/z values are summed and used as the basis for quantitation. This approach takes advantage of the fast full scan acquisition properties of an ion trap (see above).

Another reason for limited performance of 3D traps in quantitation can also be traced to this same need to control space charge. The same number of ions is always allowed into the trap (with differences in the ion trap fill time corrected for after the fact), and this leads to longer scan times at the beginning and the end of chromatographic peaks where the numbers of ions are lower,

with shorter data acquisition times near the maximum of the peak (Figure 6.24). This ultimately results in poorer precision and accuracy provided by ion traps in quantitative analyses because the 'start' and 'end' limits of the chromatographic peaks (used to define integration limits in peak area determination) are poorly defined because of the required longer intervals between data acquisitions. In contrast, a linear quadrupole has minimal or zero problems arising from space charge effects and can use fixed data acquisition intervals (Figure 6.24) that define the chromatographic peak just as well at its extremities as near the peak maximum. In this way the use of ion number

single scan time

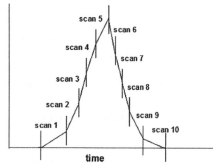

differing scan time with AGC:

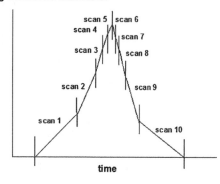

Figure 6.24 Schematic illustration of the difference with respect to accurate definition of a chromatographic peak between fixed acquisition times ('scan times') as used by a linear quadrupole, and variable acquisition times dictated by automated gain control (AGC) applied in order to control space charge effects in a Paul ion trap. Since the rate of ion arrival is lower at the extremities of the peak, the acquisition times for the trap are longer than near the peak maximum, so that the 'start' and 'end' points for peak area integration are less well defined that in the case of fixed acquisition intervals (can be scan times or SIM/MRM channel dwell times).

control on a trap leads to greater variability (lower precision) in the ability of the peak integration algorithm to define the integration limits of the peak.

The preceding discussion has emphasized the various reasons why ion traps (particularly 3D traps) perform less well than quadrupole mass filters and triple quadrupoles in trace level quantitation. Nonetheless, their performance in this regard can often be fit for purpose. One major reason for this relatively poor performance is the time penalty required in achieving SIM or MRM using a trap, which generally means that traps have problems with providing at least 10 data points across a chromatographic peak, but more modern traps (3D and 2D) appear to be doing better in this regard. However, it is difficult to see how the problems arising from space charge effects can be mitigated any better than at present. The data acquisition time problem with a stand-alone trap can be overcome by combination with a linear quadrupole operated in RF-only mode (Cha 2000), which allows for storage of ions from a continuous ion source (e.g. ESI) and also m/z-selective ejection of unwanted ions (e.g. matrix or solvent ions) in the linear quadrupole, while each microscan function is in progress in the 3D trap. This 'proof of principle' approach greatly improves the performance and alleviates the space charge problem at the expense of a significant increase in the complexity and size of the instrument. The principles of operation of a linear quadrupole as a linear trap are discussed in Section 6.4.5b.

Most of the early advances in use of a Paul trap as a mass spectrometer used electron ionization of gaseous analytes inside the trap volume. The need to use the trap as detector for ionization sources like ESI, APCI and even MALDI, required development of interfaces that permit efficient injection and trapping of ions from an external source. Externally generated ions are pulsed into a trap through apertures in an end cap, or in the ring electrode, or through the gap between the truncated ring electrode and truncated end cap (diagonal lines in Figure 6.16(b)). However, this choice turns out to be less important than the method used to slow down and trap the incoming ions. During ion injection the amplitude V_0 of the RF drive voltage may either be held constant or changed. If it is held constant at a value that generates an active trapping field, the kinetic energy of the injected ions must be dissipated in collisions with buffer gas atoms (McLuckey 1994). Unfortunately, when applied with the normal operating background pressure of 10^{-3} torr of helium, this method exhibits a trapping efficiency that is both low and mass dependent. The most efficient ion injection methods exploit a time of flight effect whereby a pulse of ions extracted from the external ion source, or from an intermediate RF-only quadrupole (Cha 2000), is strung out in

time as a function of m/z; then the trapping RF amplitude V_0 can be synchronized with ion arrival time and matched to ion energy and m/z value (Doroshenko 1993; Eiden 1993,1994; Garret 1994; Qin 1996). This ramped V_0 trapping method provides high trapping efficiency with essentially no mass dependence and has been extensively investigated using both experimental and simulation studies (Doroshenko 1994, 1996, 1996a, 1997).

A different disadvantage of a Paul trap operated as a mass spectrometer appears to be the limited accuracy and precision in measuring m/z values. This limitation arises from a combination of factors including the stability of RF amplitude, buffer gas pressure, space charge effects and the phenomenon of chemical mass shifts (although all modern 'stretched' traps do not suffer from this last phenomenon). Accuracy of m/z measurement in commercial traps is currently sufficient to allow m/z assignment to the nearest integral value up to m/z ~5000 (Stafford 2002). However, careful operation of a modified trap (not commercially available) was found to permit m/z measurements with ~10 ppm accuracy and ~4 ppm precision over an m/z range ~540–1350 (Stafford 2002); the modifications that led to this improved performance included reducing the internal operating pressure to 4.4×10^{-6} torr and a somewhat reduced stretch of the end cap separation (i.e. a smaller increase in the z_0/r_0 ratio) from that employed in the commercial trap investigated.

A Paul trap has unique properties with respect to tandem mass spectrometry in the context of product ion spectra, particularly MS^n, although it can not perform precursor scans for a fixed product ion m/z or neutral mass loss (Figure 6.3) and is subject to a significant limitation with respect to the low mass cutoff for product ions (see discussion of the 'one-third rule' following Equation [6.36]). Efficient isolation of precursor ions of a user-selected m/z value is clearly an essential step that is essentially the same as SIM (although the latter is seldom used on a trap in view of the combination of the limited selectivity and the time penalty involved in ion isolation). The MRM mode requires two ion isolation steps, improving selectivity while significantly increasing the data acquisition time. As a result the operating mode of choice for MS/MS quantitation using a trap is full scan product ion spectrum acquisition (MS^2) that requires just one ion isolation step plus a fast scan step.

Several methods for ion isolation exploit relatively straightforward combinations of DC voltages, RF ramps and tickle frequencies to eject unwanted ions. An example is the so-called 'DC isolation method', which is based upon an earlier method used for selective ion storage (Bonner 1977; Fulford 1978). This isolation method uses a DC component to bring the working point near the upper apex of the stability diagram (Figure 6.18); this can be done by separating the DC and RF components by applying the former to the end caps and the latter as usual to the ring electrode, or by applying both to the ring. After creation or injection of ions with a range of m/z values into the trap, the DC and RF components are adjusted so that only the user-selected m/z value is stable in a manner analogous to the scan function for a linear quadrupole. Exploitation of this method for isolation of precursor ions in MS/MS experiments (Strife 1988) has been critically examined (Louris 1990). This and similar methods have limitations in cases where the target precursor ions are present at low abundance in the presence of high abundances of unwanted ions such as matrix or solvent cluster ions. Although effective in removing unwanted ions, the resulting efficiency of isolation of the low-abundance target precursor ions may be too low to provide adequate ion statistics (S/N) in the subsequent MS^n experiments. Even more important from the point of view of trace level quantitation, this ion isolation method is far too slow to be compatible with the chromatography timescale.

A clever alternative technique (Buttrill 1992) was designed to overcome such problems in isolating ions present in trace quantities in the presence of abundant unwanted ions with m/z values close to that of the desired precursor. This technique exploits resonance excitation (rather than moving the selected ions to the boundaries of the stability diagram) using a complex broadband waveform to eject unwanted ions during ionization/injection. This waveform contains many frequencies whose amplitudes can be varied to optimize ejection of unwanted ions but also contains one or more frequency windows of zero amplitude, corresponding to resonances of the m/z-selected precursor ions. Thus the unwanted ions are continuously ejected so that their abundances do not have a chance to build up to levels at which space charge is important, but the abundance of the desired precursor ions can build up.

A related type of precursor isolation technique uses multiple frequency resonant ejection with a 'noise field' to excite and eject ions. The filtered noise field (FNF) approach (Kelley 1992; Goeringer 1994) uses an excitation waveform containing all frequencies (the 'noise field') that will excite all of the ions in the trap; selected secular frequency components, corresponding to a single m/z value or multiple values (or even m/z bands) are removed by digital filtering ('notches') before the waveform is converted to an analog signal that drives the circuit that generates the noise field. In this way the FNF waveform removes all of the unwanted ions from the trap. An alternative procedure, originally developed

for Fourier Transform ion cyclotron resonance mass spectrometry, uses a stored waveform inverse Fourier Transform (SWIFT) that can be tailored (Marshall 1985; Chen 1987; Guan 1993) to eject selected bands of m/z values; early applications of the SWIFT approach to a Paul trap (Julian 1993; Mordehai 1993) achieved selective elimination of matrix ions. Details of the practicalities involved in these and related approaches to isolation of precursor ions (March 1995) are not given here and tend to be specific to the manufacturer of the trap.

Collision induced dissociation (CID, Section 6.5) is essential for any MS/MS detection method. Ion dissociation in a Paul trap is induced via collisions with the helium bath gas. The following account of the methods used to do this is an abbreviated version of an excellent review (March 1998). As a result of application to the end cap electrodes of a supplementary auxiliary RF potential (sometimes called a 'tickle voltage'), with amplitude only a few tenths of a volt, the collisionally cooled ion cloud at the center of the trap can be expanded both radially and axially, with m/z dependence via the frequency of the applied voltage, i.e. this is an example of resonant excitation. Ions are thus moved away from the center of the trap and are accelerated by the increasingly larger RF field (recall that the field strength, the gradient of the electric potential, is zero at the trap center, see discussion of Equation [6.28]). Such resonant excitation, if conducted in the presence of helium buffer gas atoms so that a large number of ion–atom collisions occur, results in translational energy being transformed incrementally into vibrational excitation. As a result of the discrepancy in mass (not m/z) between helium (4 Da) and precursor ion (can be larger by a factor ≥ 100), the resulting conversion of kinetic into internal energy in a collision between the two is limited to low values (see discussion of center-of-mass collision energy in Section 6.5). As a result, collisional activation in an ion trap is an example of a 'ladder climbing' ('slow heating') activation process (McLuckey 1997) that tends to access the dissociation channels with the lowest critical energies. The time allowed for this process controls the average number of collisions experienced by the precursor ion and thus the average degree of internal excitation, and the amplitude of the resonant excitation waveform can increase the average collision energy. The ion cloud can then be refocused to the center of the trap, so that the ion translational energies are again reduced to ca. 0.1 eV; however, the internal (vibrational) energy distribution remains essentially unchanged since loss of the newly gained vibrational energy requires many more collisions than the translational cooling. It is possible to further excite these ions upon an additional cycle of resonant excitation, and several such cycles can result in the excited ions accessing dissociation channels requiring relatively high ($\sim$5–6 eV) critical energies; however, this would increase yet further the data acquisition time for on-the-fly chromatographic applications.

This relatively straightforward resonant excitation process is complicated somewhat in a 'stretched' ion trap (see above). Since ion axial secular frequencies increase as an ion approaches an end cap electrode, it is necessary to adopt one of several strategies to maintain resonance and thus ion excitation. There are three such methods of resonant excitation of ions of a selected m/z value isolated within an ion trap, namely, single frequency irradiation (SFI), secular frequency modulation (SFM) and multi-frequency irradiation (MFI).

SFI is simple in principle, requiring application of the resonant excitation voltage at the axial secular frequency for the isolated ion of selected m/z value. While this frequency may be readily calculated theoretically, the actual secular frequency of ions focused near the center of the trap in a relatively dense ion cloud is subject to frequency shifts arising from space charge effects (that vary with the concentration of analyte in the ion source and thus the number of ions in the trap), nonideal trap geometry, and the inadequacies in automated frequency calibration. Thus, in practice, SFI involves a laborious procedure in which an applied frequency waveform is matched empirically to the actual secular frequency of the isolated ions.

SFM involves modest variations in the amplitude V_0 of the RF drive potential applied to the ring electrode, such that ions move into and out of resonance with the applied single frequency tickle waveform; such slight modulation of V_0 changes q_z and β_z so as to produce a sweep of the axial secular frequency $\omega_{z,0}$ over a narrow range of some 1–2 kHz. Thus the resonance condition for ion activation is guaranteed to be met for some fraction of the time over which the resonant excitation voltage is applied.

In contrast, MFI involves the application of a resonant excitation waveform consisting of several frequency components while the value of q_z is held constant. The frequency components of the waveform bracket the entire anticipated range of the secular frequency so as to compensate for frequency shifts. Other variants of multi-frequency irradiation (March 1998) include random noise, swept frequency and broadband excitation. A rather different ion activation method (Paradisi 1992, 1992a; Curcuruto 1992) is accessible with ion trap instruments equipped with a DC power supply so that non-zero values of U and thus a_z are available. The method involves moving the working point of a given ion species to either the β_r or the β_z boundary of the stability diagram

(Figure 6.18). Ions are efficiently excited resonantly at such working points; moreover, the resulting product ions remain stored in the ion trap, available for subsequent further examination.

Although incapable of scans that reveal the precursor ions of user-selected product ions or neutral fragments, traps can follow fragmentations of m/z selected precursor ions through several steps. Such MS^n experiments are discussed in Section 6.2.2. A typical scan cycle for a MS^2 microscan is shown in Figure 6.25 for application to analysis of tetrachlorodibenzodioxin (TCDD), though this approach provides only a convenient screening technique and can not satisfy the requirements of fully validated analyses (Section 11.4.1).

All of the methods of collisional activation of m/z selected precursor ions isolated in a trap, discussed above, are effective for MS^n experiments but carry a time penalty (resulting from multiple ion isolation steps) that may be incompatible with fast chromatography or electrophoresis separations. Figure 6.26 illustrates qualitatively the relative scan cycle time differences between a full scan MS^1 acquisition, a full scan MS^2 experiment and an SRM (MRM) acquisition in which both precursor and fragment ions are isolated using a selective ejection technique. Note that, as discussed above, ejection and detection of the fragment ion still requires a V_0 scan, so the duty cycle is actually decreased in MRM mode compared with full scan MS^2 for a Paul trap! This is in marked contrast with the large increases in duty cycle that result from use of MRM mode with a triple quadrupole instrument. Moreover, for MS^n with n > 2 the price that must be paid for increased selectivity in terms of penalties in data acquisition time (via longer microscan cycle times) and sensitivity is generally too large for trace analysis. This time penalty arises because many microscans may be necessary at each stage (each value of n) in order to accumulate sufficient ions for the subsequent stage in the sequence. In addition, the manipulations required to isolate each generation of product ions that are themselves to be fragmented subsequently require time. However, MS^3 experiments are often desirable in cases where the 'slow heating' nature of the collisional activation in a MS^2 experiment with a Paul trap (see above) accesses only fragmentation reactions that are not sufficiently specific, usually losses of small molecules like water, carbon dioxide etc. An interesting approach to this problem of specificity vs analysis cycle time (Volmer 2000) exploits FNF as an excitation (rather than isolation) method by deleting all frequencies in the noise field except for a window centered at the excitation frequency of the selected precursor ion (a 'reverse notch'). Sequential MS^n experiments (n > 2) can be achieved, in a single excitation period, by simultaneously applying

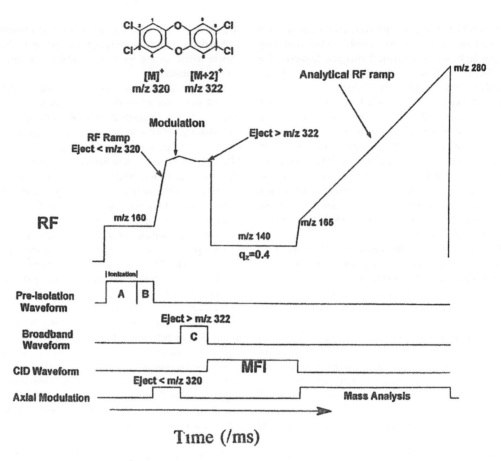

Figure 6.25 Scan function for MS^2 spectrum of the molecular ions of tetrachlorodibenzodioxin (TCDD, or 'dioxin'); *m/z* 320 and 320 correspond to the $^{35}Cl_4$ and $^{35}Cl_3$$^{37}Cl_1$ isotopologs, respectively. MFI refers to resonance activation using multi-frequency irradiation. The pre-scan for AGC is not shown. Reproduced from March, *J. Mass Spectrom.* **32**, 351 (1997), with permission of John Wiley & Sons, Ltd.

multiple reverse notch bands corresponding to the tickle frequencies for the *m/z* selected precursor ion and each of the intermediate anticipated product ions that are themselves to be fragmented. This approach is a somewhat more selective version of CID in a triple quadrupole instrument where all first-generation fragment ions might be subject to subsequent dissociation in an uncontrolled fashion.

The practicalities of Paul traps operated in MS^n modes (n ≥ 1) are quite complex, exemplified by the length of this section (much longer than that for e.g., time of flight analyzers), which was necessary to present even this abbreviated version of the underlying principles and their experimental realization. Many of the scan functions are fairly complex and the ease-of-use of Paul traps for such experiments relies heavily on computer control with

a user-friendly software interface. Ion traps are available from several manufacturers and their software interfaces do indeed greatly simplify the operations. However, such heavily automated assistance comes with a danger that an operator can use the instrument without understanding the limitations of the various modes made available and thus draw unjustified conclusions from the experimental results; hopefully the present discussion will help to fill this gap. Paul traps are generally of moderate cost and these attributes together with the MS^n capability has led to their popularity for structural analyses, although accuracy and precision of mass measurements is limited. They are also widely used for many quantitative analyses, although problems of limited duty cycle, very low energy collisions in MS/MS operation, low dynamic range unless automated gain control methods are applied, limited precision

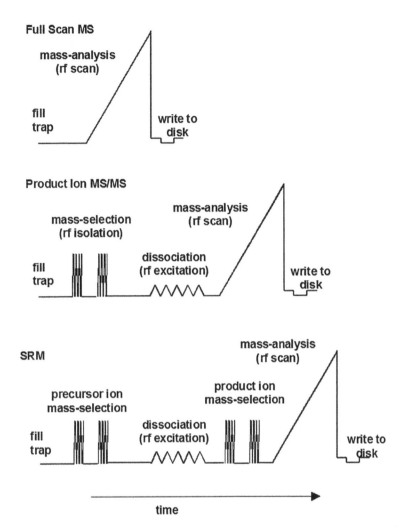

Figure 6.26 Comparison of scan cycles (and cycle times) used in operating a Paul ion trap in full scan MS[1] mode, in full scan MS[2] mode, and in MRM mode in which both precursor and fragment ions are isolated using selective injection techniques (but observation and detection of the fragment still requires a scan function).

in replicate measurements etc., result in triple quadrupole instruments (in MRM mode) being the instruments of choice for ultimate performance in trace analysis.

6.4.5b Two-Dimensional (Linear) Traps

Linear (2D) ion traps have more recently been introduced for use as ion storage devices, as stand-alone mass spectrometers and as components of hybrid tandem instruments. Although no example of their use in rigorously validated analytical methods has been published to the best of the knowledge of the present authors, there are reasons to believe that this situation might change. For

example, as a result of their larger physical dimensions, linear traps do not exhibit operating problems arising from space charge effects until much larger ion populations are reached compared with 3D traps, for which space charge is a serious problem. Linear quadrupole devices are also used as ion guides to transport ions from one part of an apparatus to another (e.g., from ion source to analyzer) and as collision cells to effect collision induced dissociation; an extensive review (Douglas 2005) covers the literature on all these topics up to 2003 and is the basis for the following discussion. The emphasis is placed on linear quadrupole devices, essentially the same as those discussed above in the context of linear quadrupole mass filters. As discussed in Section 6.4.2a, arrays with six or eight rods have been used less often.

A linear quadrupole is readily converted into a linear ion trap by (usually) operating in RF-only mode while applying DC stopping potentials to electrodes at the entrance and exit. Ions are trapped axially (z-direction, Figure 6.9(a)) by these stopping potentials, in contrast with the 3D Paul trap in which ions are confined axially by RF potentials. Of course in a linear trap the ions are still confined radially (x- and y-directions) by the RF potential, as in a 3D trap. Linear traps are thus often referred to as two-dimensional (axial and radial) traps, since the equations of motion in the two radial directions (x and y) are identical apart from a sign change (see discussion of Equations [6.13–6.15], Section 6.4.2). When operated in this way a linear quadrupole rod assembly controls the radial ion motions in exactly the same way as that described in Section 6.4.2 for the quadrupole mass filter, except that now (usually) $U = 0 = a$ so that the RF-only device generally operates along the q-axis only (Figure 6.10).

When operated as ion guides, RF-only linear quadrupoles benefit from collisional cooling (Douglas 1992) to achieve high transmission efficiencies, but do not require very low molecular mass buffer gases like helium as in a 3D trap (air or nitrogen are often used). When the same devices are employed such that the axial injection energy of the ions is sufficiently high to produce collision induced dissociation, the device is referred to as a collision cell q (Section 6.4.2a).

It is appropriate to repeat here the following clarification regarding notations used that appear to be similar (see Section 6.4.2): a nonitalicized lower-case 'q' denotes a linear quadrupole device operated using only RF fields; an upper-case Q denotes a similar device operated with both RF and DC fields; an *italicised q* denotes the trajectory stability parameter described above; and q_{ion} represents the magnitude of the charge on an ion (in coulombs and containing the sign indicating the polarity).

When used as a linear ion trap (2D trap) many of the same techniques used with Paul traps can be used with only minor modifications. Trapping of ions with selected m/z values can be achieved, as in 3D traps, by resonant excitation methods or by methods that exploit the boundaries of the stability diagram. Selective ion trapping is necessary if the trap is to be used also to create fragment ions by in-trap collision induced dissociation for subsequent analysis by a Paul trap or by FTICR or time of flight analyzers. Also, particularly in the case of downstream trapping analyzers, excess unwanted ions that cause space charge problems can be selectively ejected. Resonant excitation at the ions' secular frequencies (Equation [6.33])

can be achieved using either dipolar excitation (an auxiliary potential at the secular frequency is applied between one pair of opposite rods) or quadrupole excitation (the auxiliary potential is superimposed on the main RF waveform applied between pairs of opposite rods). Either the excitation frequency or the RF amplitude can be manipulated to eject (radially) all unwanted ions with specified m/z values. Alternatively, application of a broadband waveform, with a notch (zero amplitude region) in the frequency distribution at the secular frequency of the desired m/z value that is to be selectively trapped, can be used. This approach was used quite early for linear traps (Langmuir 1967) but all such approaches developed initially for Paul traps (Section 6.4.2) can be adapted to linear quadrupole traps. The same general principles of resonant excitation developed for Paul traps can also be applied to effect collision induced dissociation in linear traps.

Again, as for 3D traps, boundary ejection methods for ion isolation exploit unstable ion motion at the boundaries of the stability diagram of a linear quadrupole (Figure 6.10). In this case both RF (V_0) and nonzero DC (U) potentials can be manipulated so that ions with the m/z value to be selectively trapped have (a,q) values near a stability boundary such that ions of neighboring m/z values fall into instability regions and are ejected. Specific methods that have been shown to achieve this have been summarized (Douglas 2005).

As mentioned above, the 2D trap has an advantage over the Paul trap with respect to the relative limits on trapping capacities imposed by space charge effects. It has been shown (Douglas 2005) that, in agreement with naïve intuition, these limiting ion densities (as opposed to absolute ion numbers) are essentially the same for linear traps as for Paul traps. In fact the trapping capacity (maximum number of stored ions) of any trap is approximately proportional to the internal volume of the trap, about one order of magnitude larger for a linear trap than for a Paul trap. Thus an estimate of the ratio of trapping capacities of a linear trap to that of a 3D trap has been given (Campbell 1998) as $(r_0^2 . l_{quad}/z_0^3)$, where all quantities were defined previously; for example, typical values of $r_0 = 0.4$ cm, $l_{quad} = 20$ cm and $z_0 = 0.707$ cm, gives a trapping capacity ratio of $\sim$9:1. The higher ion capacity increases the concentration linear dynamic range of the linear trap relative to the 3D trap. This situation is more complicated if buffer gas is present to create collisional cooling thus shrinking the sizes of the respective ion clouds, particularly since it was shown (Collings 2003) that a linear trap can provide ion fragmentation at pressures $< 5 \times 10^{-5}$ torr (and thus without collisional focusing) whereas a Paul trap requires collisional cooling for effective operation (Section 6.4.5a), but the general conclusion is still valid.

However, this is too simplistic a view if the purpose is not simply to store the maximum number of ions, but rather to be able to manipulate the ions by appropriate application of external fields in order to achieve, e.g., selective isolation and/or excitation of m/z-selected ions. In such cases the deleterious effects of space charge become evident at ion densities that are orders of magnitude lower than those estimated for simple maximum trapping capacity, e.g. compared with an estimated maximum trapping capacity of ~8×10^8 ions.cm^{-3}, significant effects of space charge were observed (Hager 2002) for trapping and excitation at much lower ion densities ~5×10^5 ions.cm^{-3}. The 2D trap advantage of one order of magnitude increase in the effective ion capacity is thus considerably accentuated at such low absolute numbers of ions.

Linear ion traps are often used as interfaces between continuous ion sources (e.g., APCI or electrospray and its variants) and m/z analyzers with limited intrinsic duty cycle and/or susceptibility to space charge effects, e.g. Paul traps, time of flight analyzers and FTICR. Duty cycle limitations with all of these latter analyzers with a continuous ion source arise because, while the analyzer is processing the preceding ion injection, the ions produced during that processing period are wasted. An intermediate linear trap can store these otherwise wasted ions while the processing is in progress, thus (theoretically) increasing the duty cycle to 100 %. A simple example of a 2D trap used between a continuous source and a 3D (Paul) trap (Douglas 1993) can illustrate the principle, but applies more generally to other analyzers. For a 3D trap operated alone as a mass spectrometer the duty cycle is given by Equation [6.37a], where $t_{fill,3D}$ and $t_{A,3D}$ are the times required to fill the trap with ions from the continuous source and the time required for m/z analysis. For a hypothetical case (Douglas 1993, 2005) where a continuous ESI source supplies 6.2×10^8 ions.s^{-1} and the trap can accommodate 1×10^6 ions without space charge effects, $t_{fill,3D} \approx 1.6$ ms. The value of $t_{A,3D}$ can vary quite widely but if it is 100 ms the duty cycle of the stand-alone trap is only ~1.6 %, i.e., more than 98 % of the ions extracted from the ESI source are wasted.

$$(\text{Duty Cycle})_{3D} = t_{fill,3D}/(t_{fill,3D} + t_{A,3D}) \qquad [6.37a]$$

$$(\text{Duty Cycle})_{L-3D} = t_{fill,L}/(t_{fill,L} + t_{A,L} + t_{tr}$$
$$+ t_{A,3D}) \qquad [6.37b]$$

$$(\text{Duty Cycle})_{L-3D,opt} = (t_{fill,L} + t_{A,3D})/(t_{fill,L}$$
$$+ t_{A,L} + t_{tr} + t_{A,3D}) \qquad [6.37c]$$

With a linear trap used as an intermediate stage between source and 3D trap the duty cycle is given by Equation [6.37b], where $t_{fill,L}$ is the fill time used for the linear trap in which the ions are processed (e.g., high abundance matrix ions are ejected) for a time $t_{A,L}$, and t_{tr} is the time required to transfer the ions from the linear trap to the 3D trap. If it is assumed that the unwanted matrix ions represent 90 % of the total and are ejected during $t_{fill,L}$ by resonant ejection, in order to reach the same limiting number of ions admitted to the 3D trap $t_{fill,L} \sim 10.t_{fill,3D}$, and this alone represents a 10-fold increase in the duty cycle. The optimum situation requires that ions can continue to be accumulated in the linear trap while the 3D trap is performing the m/z analysis, and in that case the duty cycle is given by Equation [6.37c] and in favorable cases can approach 100 %. During the storage period the linear trap can also be operated so as to eject unwanted ions (see above), thus alleviating the space charge problem though at some cost to the duty cycle, and can also induce ion fragmentation of isolated precursor ions.

Of course this discussion of a 3D trap using pre-processing by a 2D trap is not intended to promote a particular linear trap–analyzer combination, and indeed the linear trap–3D trap serves only to illustrate the general principle that applies to any other analyzer with duty cycle problems. Examples in which this principle has been applied to time of flight and FTICR instruments, as well as Paul traps, have been reviewed in detail (Douglas 2005). An example in which a linear trap is used together with a time of flight analyzer (Campbell 1998) is illustrated in Figure 6.27. In this case the linear trap could be operated so as to store and process the ions from the ESI source (i.e. isolate and fragment the selected precursor ions), thus providing MS/MS capability. The duty cycle for MS/MS operation in this instrument was ~25 % (see ratio of injection time to total MS/MS time, Figure 6.27). In a later improved version of this apparatus (Collings 2001), it was possible to store ions also in the small quadrupole ion guide q$_0$ connecting the ion source to the 2D trap (Figure 6.27) for periods during which the user could program the main 2D trap to implement ion manipulations (including two possible stages of ion fragmentation that permitted MS^3 operation), thus maintaining an improved duty cycle. It was also found that a reduction in the cooling gas pressure from 7.0 to 1.8 mtorr resulted in an increase of the effective m/z resolving power in the excitation step from 75 to 240, a trend that was pushed even further (Collings 2003) when it was discovered that, contrary to initial expectations, ions can be fragmented in a linear trap with high efficiency and high resolving power at relatively low pressures (< 0.05 mtorr). This pressure range is compatible with operation

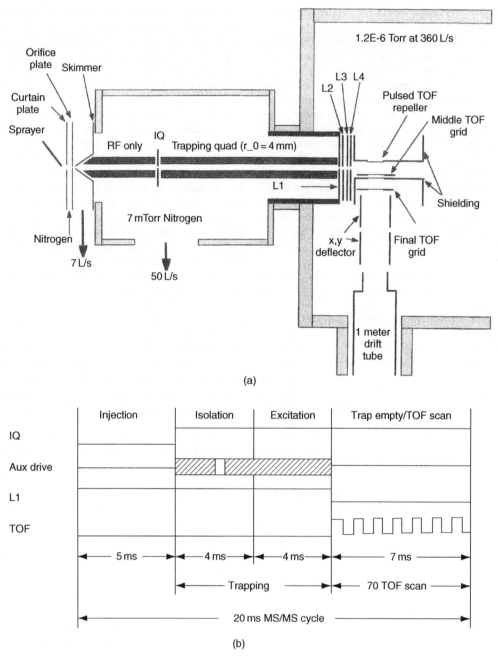

(a)

(b)

Figure 6.27 (a) Schematic diagram of a linear trap used as an intermediate between an electrospray ion source and an orthogonal-acceleration time of flight instrument. The short RF-only quadrupole assembly (q_0) located before the trapping quadrupole is used as an ion guide. IQ is the inter-quadrupole aperture and serves also as one of the axial trapping plates when an appropriate DC potential is applied to it. The ion lens L1 serves also as the trapping plate at the exit of the trapping quadrupole. (b) The timing parameters for a typical 20 ms MS$_2$ experiment in which the precursor ion is selected, isolated and subjected to collision induced dissociation in the storage quadrupole; all the resulting ions are then transferred to the time of flight analyzer. The horizontal lines represent levels of the potentials (not shown to scale) applied to the corresponding ion optical elements in the different time periods; the auxiliary drive is the resonance excitation waveform (note the 'notch' in the isolation period). Reproduced from Campbell *et al.*, *Rapid Commun. Mass Spectrom.* **12**, 1463 (1998), with permission of John Wiley & Sons, Ltd.

of a linear quadrupole as a m/z analyzer (mass spectrometer) and this has implications for the new tandem instrument ($q_0 Q_1 q_{2,trap} Q_{3,trap}$) described in Section 6.4.6. However, relatively long excitation times ($\sim$100 ms) were required (Collings 2003) and this places strong demands on the cleanliness of the instrument to avoid formation of adducts of the trapped ions with background impurities like phthalates (Collings 2001). In addition, such long excitation times could make the total data acquisition time too long for compatibility with on-the-fly chromatographic detection.

Similarly to Paul traps, 2D traps are generally operated using no DC component (U = 0), i.e., RF-only plus auxiliary AC potentials for purposes of resonance excitation. It is convenient here to briefly mention operation of a linear quadrupole device in RF-only mode in a fashion that still provides m/z analysis. The first description of this mode of operation (Brinkmann 1972), and its subsequent development by several authors, is described in a more recent investigation (Hager 1999). While this device is not widely used in practice, and never in trace quantitative analysis, the general principles of operation are important for one of the methods of using a linear quadrupole trap as a m/z analyzer. The following discussion is a considerably abbreviated version of a much more detailed treatment (Hager 1999) and is directed principally towards the later description (Section 6.4.6) of a 2D trap mass spectrometer based on the same general principles.

The equations of motion used to describe the trajectory of an ion in a linear quadrupole (Equation [6.12], Section 6.4.2) are strictly valid only well inside the rod assembly, well removed from the entrance and exit. At each of these ends the ideal quadrupole field (Equation [6.11]) terminates abruptly, but in any real device is affected not only by the RF and DC potentials applied to the rods but also by the potentials applied to nearby ion optical elements (lenses etc.). Moreover, the field lines created by the potentials applied to the rods 'spill out' for some distance outside the theoretical boundaries. These curved fringe fields (Section 6.4.2a) distort the ideal quadrupole field such that the ion motions in the x- and y-directions that are independent of one another in the main quadrupole field (Equation [6.11]) become coupled as a consequence of mixing radial and axial potentials, i.e. the electrical force exerted on an ion in the z-direction can be a function of the time dependent potentials applied in the x- and y-directions (but now curved in three-dimensions), and vice versa. These effects of fringe fields are important in the following discussion.

In RF-only mode (U = 0) the scan line corresponds to the q-axis (i.e., $a = 0$), so that the quadrupole is operated as a broadband high-pass filter that transmits only ions with $q < 0.908$ (see discussion of low mass cutoff following Equation [6.36], Section 6.4.2). As the RF amplitude V_0 is scanned up, ions of a particular m/z are transmitted with almost 100 % efficiency until they approach this limiting q value when the ions gain significant radial amplitude until they hit the rods. However, this simple picture applies only to ions within the interior of the quadrupole, well removed from the entrance and exit. Within the fringing field region at the exit, some of the increased radial energy associated with the ions' transfer into an instability region is converted into axial energy as a result of the coupling of radial and axial fields, mentioned above. Importantly, ions with large radial displacements within the exit fringing field (i.e. those with q values close to 0.908) receive a proportionately greater axial kinetic energy boost from the radial–axial coupling than ions with small radial displacements (those with m/z values corresponding to q values < 0.908). Thus, ions on the verge of instability (high m/z) can be distinguished from those with stable trajectories (lower m/z) by virtue of their excess axial kinetic energy. In practice, exploitation of this effect to produce a mass spectrum requires use of simple energy filtering techniques, e.g. application of variable retarding potentials to metal grids located between the quadrupole exit and the detector. The results of applying different retarding potentials to ions injected into an RF-only quadrupole are exemplified (Figure 6.28) by some of these early results (Brinkmann 1972).

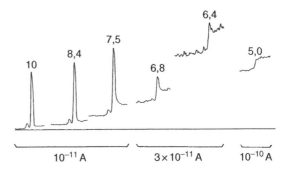

Figure 6.28 Appearance of peaks at m/z 414 (larger peak) and 415 (from EI of perfluorobutylamine) obtained using an RF-only linear quadrupole with different retarding potentials applied between the quadrupole exit and the detector (numerical values ranging from 5–10 V are annotated in each case). The original axial energy of the injected ions was 6 eV. The relative intensities of the peaks are indicated by the full scale settings of the ion current amplifier. Reproduced from Brinkmann, *Int. J. Mass Spectrom. Ion Phys.* **9**, 161 (1972), copyright (1972), with permission from Elsevier.

One problem of RF-only linear quadrupoles operated in this way as mass spectrometers is their susceptibility to background ion signals originating from ions with low q-values but higher radial energies that can therefore successfully pass through the applied retarding potential. This effect is evident in Figure 6.28 and results in a continuum background signal and thus degraded signal-to-noise. This problem can be particularly severe with ESI sources since fast ions can be formed from multiply-charged ions that subsequently undergo charge reduction processes in post-acceleration collisions with background gas. Indeed, one of the major features described in a paper on this topic (Hager 1999) was the development of techniques to reduce this background to negligible levels. This advance was essential for the operation of a linear quadrupole trap as a m/z analyzer with axial ion ejection, discussed below, and this is the main reason for the foregoing discussion of the RF-only quadrupole operated as a mass spectrometer in transmission mode (i.e. not in trap mode), although such instruments do have potential as small, cheap mass spectrometers (Hager 1999). Indeed, the poor resolving power evident in Figure 6.28 was improved by unbalancing the RF (unequal RF amplitudes applied to the x- and y-rods) and by applying a small resolving DC potential between the pairs of rods. Also the sensitivity and resolving power were further improved (Hager 1999) by applying an auxiliary waveform at the frequency of the quadrupole RF to the exit aperture.

Two somewhat different approaches to exploiting a linear quadrupole trap as an m/z analyzer have been described; one uses m/z selective radial ejection of ions (Schwartz 2002) and the other uses axial ejection (Hager 2002; Londry 2003). The motivation for these developments was, of course, to alleviate the limitations on dynamic range and duty cycle (with respect to ions generated externally by a continuous ion source) imposed by space charge effects in the small dimensions of a Paul trap. In addition, because ions from an external source can be injected along the z-axis of a linear quadrupole with a consequent long path to facilitate cooling, the initial injection energy, capture and trapping of such ions is much less problematic and more efficient than in the small volume of a Paul trap, for which a complicated trapping method, e.g. ramping V_0 synchronized with ion arrival times, is necessary (see Section 6.4.5).

Figure 6.29 shows the linear quadrupole mass spectrometer designed (Schwartz 2002) to exploit radial ejection by appropriate resonance activation techniques, which are entirely analogous to those employed in Paul traps. A conventional linear quadrupole rod set ($r_0 = 4$ mm) was cut into three sections 12, 37 and 12 mm long. The central section has a 30×0.25 mm ejection slot cut in one of the x-rods and this section is shielded from fringe field distortions of the trapping and excitation fields by the two shorter sections on either end. Ions are confined radially in this central section by the trapping RF with V_0 up to 5 kV zero-to-peak at 1 MHz, corresponding to $(m/z)_{max} \sim 2000$ for resonant ejection at $q = 0.88$; axial trapping is effected by DC potentials applied to the two 12 mm sections. To compensate for the field distortions arising from the slot in the x-rod, both x-rods were moved out an extra 0.75 mm from the center (z-axis). The ejected ions are detected by an ions-to-electrons conversion dynode (Section 7.3) held at -15 kV for positive ions, and an electron multiplier. Dipole excitation with auxiliary AC voltages on the x-rods is used to isolate, excite and eject ions. For tandem mass spectrometry, precursor ions are isolated by a broadband waveform with a notch at the frequency of the precursor ions, typically at $q = 0.83$. Collisional activation of ions by resonant excitation is done at $q = 0.25 - 0.35$ as a compromise between fragmentation efficiency and trapping efficiency. The storage capacity (the maximum number of ions stored, limited by space charge) for this device was estimated (Schwartz 2002a) to be 7×10^6, but the spectral space charge ion limit, i.e., the maximum number of stored ions that can provide a mass spectrum with specified resolving power and accuracy and precision of m/z measurement, was $\sim 2 \times 10^4$, more than ten times higher than for a typical Paul trap.

Trapping of ions in a linear trap requires that the injection energy must be sufficiently large to overcome the trapping barrier at the trap entrance and also that the ions must subsequently lose excess injection energy during a single pass through the trap. The dominant energy loss mechanism is collisional, and indeed the trapping efficiency in the radial ejection linear trap was found (Senko 2002) to be dependent on both the pressure of helium in the trap and on the mass of the ion, as expected for this mechanism, varying in the range 55–70 % at a helium pressure of 1 mtorr. However, for all ions tested the trapping efficiency was found to approach 100 % at helium pressures of 4 mtorr (Senko 2002). A slightly different version of this radial ejection linear trap, with 3 mtorr of helium (Schwartz 2002), gave trapping efficiencies ~ 29 %, still much better than typical values (< 5 %) for 3D Paul traps. Note that the trapping efficiency must be established as a compromise with achievable resolving power, since increasing the pressure above ~ 1 mtorr was found to increase the former but degrade the latter.

Overall detection efficiencies were determined (Senko 2002) by comparison of a direct measurement of the ion current delivered to the trap (using a conventional detector positioned axially at the trap exit) with an estimate of the number of ions observed during a m/z scan

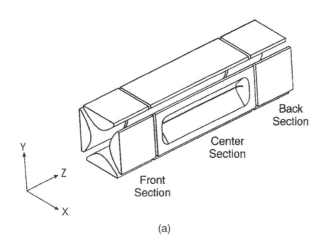

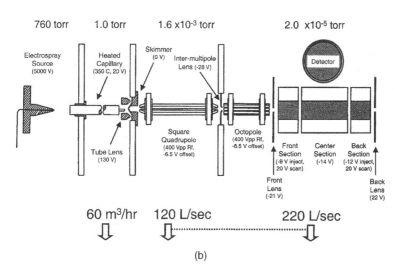

Figure 6.29 A linear quadrupole ion trap for use as a mass spectrometer using radial ion ejection. Top: schematic of the trap showing the ejection slot along the length of one of the x-rods. Bottom: an overall view of the complete instrument showing typical potentials and pressures. The first (square) quadrupole is an ion guide to transport ions from the ESI source into the higher vacuum region and the function of the small octapole is similar but to facilitate ion transfer into the trap. Reprinted by permission of Elsevier from "A Two – Dimensional Quodrupole Ion Trap . . . ", by Schwartz, *et al.*, Journal of the American Society for Mass Spectrometry, **13**, p. 659–669, 2002, Fig 1, p. 660; Fig 5, p. 662, by the American Society for Mass Spectrometry.

to the radial detector. Above 1.5 mtorr the detected ion current on the radial detector did not continue to increase as expected from the axial detector data, indicating some loss in radial ejection efficiency at higher pressures particularly for low mass ions, most likely an indication of scattering. At 1 mtorr of helium the radial ion ejection efficiency was estimated to be 50–100 % (Senko 2002) and 44 % (Schwartz 2002). By combining the estimates of trapping and ejection efficiencies the overall detection efficiency was estimated as ~13 % (Schwartz 2002), but

if it is assumed that half of the ions are neutralized on the x-rod opposite the ejection slot, use of a second slot on the opposite rod in combination with a second detector will in principle double this value. The increased ion capacity and the higher ion injection efficiency, compared to those possible for a Paul trap, significantly improved detection limits. For example, a 500 fg sample of alprazolam, analyzed by LC/MS/MS (m/z309 → 274), gave a S/N ~20 with a detection limit five times lower than that obtained using a 3D Paul trap (Schwartz 2002). Unit mass resolving power was obtained up to m/z 2000 at scan speeds > 5000 Th.s^{-1}, and MS^4 acquisition was demonstrated (Schwartz 2002).

Axial detection of ions from a linear quadrupole trap is a less obvious strategy from the point of view of extension of techniques developed for Paul traps, but follows naturally from work on operation of an RF-only linear quadrupole as an m/z analyzer by exploiting the fringe fields at the quadrupole exit (Brinkmann 1972; Hager 1999), summarized above. These same principles were used (Hager 2002) to design a linear trap mass spectrometer using axial ejection; a sketch of the experimental apparatus (a modified triple quadrupole mass spectrometer that can be described as a $q_0Q_1q_{2/T}Q_{3/T}$ combination) is shown in Figure 6.30. The various scan modes accessible using this new instrument are described in Section 6.6 and the present discussion is restricted to the more fundamental aspects of operation. Ions could be trapped in either the collision cell $q_{2/T}$ (at a pressure of 1×10^{-4} torr) or $Q_{3/T}$ (3×10^{-5} torr) by applying stopping potentials to aperture plates at the ends of the quadrupoles. Ions were excited at their resonant frequencies in the region near the trap exit by applying an AC potential to the exit aperture either as dipole excitation between a pair of rods or as quadrupole excitation. A mass spectrum can be obtained either by scanning the excitation frequency at fixed amplitude of the drive RF, or vice versa, to bring ions of different m/z values into resonance; scanning the RF amplitude was found to work better.

When in resonance with this auxiliary waveform applied in the xy-plane, ions of the appropriate m/z accumulate kinetic energy as usual but now the fringing fields mix the lines of force, and thus the ion motions, in the x-, y- and z-directions. As a result these resonant ions gain sufficient axial kinetic energy to overcome the stopping potential at the trap exit. The first experiments (Hager 2002) were performed with quadrupole excitation of ions trapped in $q_{2/T}$; unit mass resolution was obtained at m/z 609. Trapping efficiencies varied with pressure and with the q value of the trapped ions; $q = 0.4$ gave the highest trapping efficiencies (58–95 %). The extraction efficiency also varied with pressure, ranging from 2–18 % for ions of m/z 609. Note that in this case of axial ejection by the fringing fields, only ions in the last $\sim 5r_0$ of the length of the quadrupole are ejected since this is the limit of penetration of the fringe fields into the quadrupole array; the remaining ions are lost to the rods.

For MS/MS the linear trap $q_{2/T}$ could, in principle, be operated as in the linear trap–TOF system (Campbell 1998; Collings 2000) mentioned above (see discussion of Figure 6.27) using ion injection, ion isolation with tailored waveforms, excitation to induce dissociation and axial ejection of fragment ions to produce a mass spectrum. However, because the instrument is based on a triple quadrupole system, new scans and methods of producing MS/MS spectra are possible. Precursor ions can be selected by Q_1 and injected into $q_{2/T}$ with sufficient energy to cause fragmentation; surviving precursors and

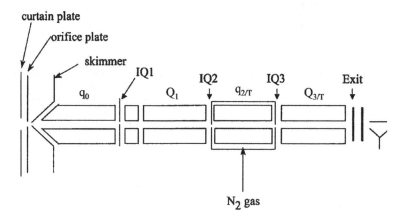

Figure 6.30 Sketch of a triple quadrupole instrument adapted to demonstrate use of a linear quadrupole as a 2D trap mass spectrometer with axial ejection. Q_1 was operated as a conventional linear quadrupole m/z filter. Either the pressurized (≥ 0.1 mtorr) RF-only quadrupole $q_{2/T}$ or the evacuated ($\sim 10^{-5}$ torr) $Q_{3/T}$ could be used in trapping mode. The RF-only quadrupole $q_{2/T}$ could be operated as a conventional collision cell, and $Q_{3/T}$ could also be used as a conventional RF-DC quadrupole m/z analyzer. The ion guide quadrupole q_0 is 12 cm long and is maintained at ~ 7 mtorr. All of Q_1, $q_{2/T}$ and $Q_{3/T}$ are 127 mm long with $r_0 = 4.17$ mm, and the RF drive frequency was 1 MHz. IQ1 – IQ3 are differential pumping apertures that can also act as ion lenses. Adapted from Hager, *Rapid Commun. Mass Spectrom.* **16**, 512 (2002), with permission of John Wiley & Sons, Ltd.

fragment ions can then be ejected axially (with Q_1 operated in RF-only mode as an ion funnel) to produce a MS^2 spectrum. The advantages of such a Q_1–$q_{2/T}$ arrangement include an increased duty cycle, MS/MS spectra similar to those provided by a triple quadrupole in that higher energy fragmentation reactions can be accessed that are not possible with a 3D Paul trap, and significant reduction of space charge problems because only precursor ions selected in Q_1 are injected into the trap so that no ion isolation procedure is necessary within the trap itself (Hager 2002). The sensitivity of the Q1–$q_{2/T}$ system for product ions of reserpine $[M+H]^+$ ions (m/z 609) was ~12 times greater than for conventional triple quadrupole MS/MS operation (i.e. with $q_{2/T}$ operated as a conventional collision cell), reflecting the improved duty cycle. The resolving power for the Q_1–$q_{2/T}$ system with axial ejection was typically 1000 (based on FWHM) measured at m/z 609 with a scan rate of 1000 Th.s^{-1}.

Some surprising results were obtained (Hager 2002) using $Q_{3/T}$ as the ion trap, i.e., with $q_{2/T}$ operated as a conventional collision cell at a pressure ~5 mtorr. A pulse (variable from 5–1000 ms) of product ions plus surviving precursor ions (pre-selected for m/z in Q_1) was then trapped in $Q_{3/T}$ with surprising efficiency considering the appreciably lower pressure (~3×10^{-5} torr). With a $q_{2/T} \rightarrow Q_{3/T}$ potential drop (for positive ions) fixed at 8 V, the trapping efficiency at m/z 609 varied from ~45 % for $Q_1 \rightarrow q_{2/T}$ energies of 5–15 eV to ~30 % at 45 eV, surprisingly insensitive to this collision energy. It was proposed (Hager 2002) that, because the pressure within $q_{2/T}$ was well within the collisional focusing regime for this quadrupole, the ions exiting $q_{2/T}$ were already collisionally cooled with little or no 'memory' of the original axial energies of the ions injected from Q_1 into $q_{2/T}$. Furthermore, there will be a pressure gradient along $Q_{3/T}$ resulting from collision gas escaping from the aperture connecting $q_{2/T}$ to $Q_{3/T}$ (the quoted pressure of 3×10^{-5} torr represents an estimated average value), and this will tend to promote collisional trapping at the front end of $Q_{3/T}$ while minimizing collisional effects at the exit, as required for optimum coupling of the motions in the fringe fields at the exit. Other observations (Hager 2002) included considerably improved resolving power vs scan speed characteristics compared with the $Q_1q_{2/T}$ mode of operation, not unexpected as a result of the lower pressure in $Q_{3/T}$.

However, a truly surprising result was a remarkable insensitivity to space charge for mass spectra obtained by axial injection from $Q_{3/T}$ in 2D-trap mass spectrometer mode. For example, a spectrum of the $[M+H]^+$ ions of reserpine (m/z 609) obtained using $q_{2/T}$ as a collision cell and $Q_{3/T}$ in 2D-trap MS mode at an ion

count rate of 2.27×10^6 counts.s^{-1}, was indistinguishable (m/z assignments changed by only 0.02 Th) from that obtained under identical conditions but with a count rate of 1.93×10^2 counts.s^{-1}. (This instrument used ion counting detection, see Section 7.3.) This insensitivity to anticipated space charge effects is dramatically different from that observed using the same instrument but operated using $q_{2/T}$ operated in 2D-trap MS mode with a pressure ~1×10^{-4} torr, for which the spectral space charge limit was observed, as predicted, at an ion density equivalent to that typical of 3D Paul traps. While not properly understood, preliminary model calculations (Hager 2002) suggested that the observed effect (relative insensitivity to space charge effects in $Q_{3/T}$ when operated in 2D-trap MS mode) was related to the fact that the axial ejection mechanism is most efficient for ions with large radial displacements, which will arise in the exit region of $Q_{3/T}$ as a result of the lower pressure and thus lower collisional cooling effect. These conditions also correspond to those where the ion density, and thus space charge effects, are minimized. Whatever the correct explanation may be, this represents an unexpected advantage of the $Q_1q_{2/T}Q_{3/T}$ instrument when operated with $Q_{3/T}$ in 2D-trap MS mode.

In summary, it appears that linear quadrupole trap mass spectrometers can be fabricated and operated with all the expected advantages over 3D Paul traps. These advantages mainly arise from the larger ion volumes that alleviate space charge problems but also include more efficient (factors of 10–20) trapping of ions injected from external sources, faster scan speeds (~3 times faster) and much improved duty cycle. It is too soon to tell whether or not linear traps will dominate the trap market in future, since Paul traps do have an advantage with respect to size and cost; if linear traps can be shown to be capable of better performance in trace level quantitation, this would be a decisive advantage. With respect to comparisons between radial and axial ejection in linear traps, it is difficult to make any judgments at present since the radial ejection trap has been evaluated as a stand-alone device or as a component of hybrid instruments incorporating either an FTICR or Orbitrap analyzer (Section 6.4.8), while the axial ejection version was developed as a component of a modified triple quadrupole configuration (Section 6.4.6).

6.4.6 The QqQ_{trap} Analyzer

The experimental apparatus incorporating an axial ejection linear trap (Hager 2002), introduced in Section 6.4.5b, has appeared on the market (Hager 2003) as a fully engineered version. This new QqQ_{trap} instrument (Figure 6.31) is more fully described as $q_0Q_1q_{2/T}Q_{3/T}$, where the subscript 'n/T' with n = 0 – 3 indicates that the

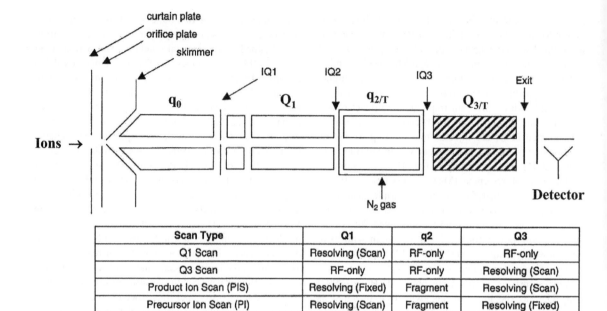

Scan Type	Q1	q2	Q3
Q1 Scan	Resolving (Scan)	RF-only	RF-only
Q3 Scan	RF-only	RF-only	Resolving (Scan)
Product Ion Scan (PIS)	Resolving (Fixed)	Fragment	Resolving (Scan)
Precursor Ion Scan (PI)	Resolving (Scan)	Fragment	Resolving (Fixed)
Neutral Loss Scan (NL)	Resolving (Scan)	Fragment	Resolving (Scan Offset)
Selected Reaction Monitoring mode (SRM)	Resolving (Fixed)	Fragment	Resolving (Fixed)

	Q1	q2	Q3
Enhanced Product Ion (EPI)	Resolving (Fixed)	Fragment	Trap/scan
MS3	Resolving (Fixed)	Fragment	Isolation/frag trap/scan
Time delayed frag capture Product Ion (TDF)	Resolving (Fixed)	Trap/No frag	Frag/trap/scan
Enhanced Q3 Single MS (EMS)	RF-only	No frag	Trap/scan
Enhanced Resolution Q3 Single MS (ERMS)	RF-only	No frag	Trap/scan
Enhanced Multiply Charge (EMC)	RF-only	No frag	Trap/empty/scan

Figure 6.31   Top: diagram of a $q_0Q_1q_{2/T}Q_{3/T}$ hybrid instrument (triple quadrupole/linear trap, similar to Figure 6.29); IQ1, IQ2 and IQ3 denote inter-quadrupole ion optical lenses used to guide and/or trap ions. Q_1 and $Q_{3/T}$ are operated at pressures $\sim 10^5$ torr, q_0 and $q_{2/T}$ at $\sim 8 \times 10^{-3}$ torr. Bottom: summary of operational modes in conventional triple quadrupole configuration (upper table) and using $Q_{3/T}$ as a linear trap (lower table). Reproduced from Hopfgartner, *J. Mass Spectrom.* **38**, 138 (2003), with permission of John Wiley & Sons, Ltd.

device can be operated either in the mode typical of a traditional triple quadrupole or in linear trap mode. It provides not only the usual high sensitivity–high duty cycle MRM operation of a triple quadrupole for quantitative analyses, but also a range of enhanced scan modes using the linear trap features. These new scan modes allow characterization of other compounds (e.g. metabolites) related to the original target analyte, with subsequent MRM quantitation using so-called 'information dependent acquisition', all within the chromatographic timescale (Figure 6.31).

'Enhanced Product Ion' (EPI) spectra are produced when a selected precursor resolved by Q_1 is subjected to CID in q_2 in the usual way, but the fragment ions and

surviving precursors are then transferred to $Q_{3/T}$ where they are trapped by the ion optical lens between $q_{2/T}$–$Q_{3/T}$ and the exit lens following $Q_{3/T}$; the $Q_{3/T}$ quadrupole is first operated in linear trap (RF-only) mode, but after for a user-specified time the ions are scanned out axially as described previously (Hager 1999, 2002). The resulting spectra are qualitatively similar to conventional triple quadrupole fragment spectra but the sensitivity is higher and the scan speed is increased (up to $4000\,\mathrm{Th.s^{-1}}$), though the resolution is best at lower scan speeds (e.g. $250\,\mathrm{Th.s^{-1}}$). This mode permits quantitative analysis via full scan MS^2 spectra at collision energies characteristic of triple quadrupoles but actually acquired as in 3D (Paul) ion traps without the usual low *m/z* cutoff and without the need for selection of precursor ions via tailored isolation waveforms with narrow *m/z* isolation widths. The latter point is an advantage since it has been shown (McClellan 2002) that use of narrow (and thus specific) isolation widths can result in CID of 'fragile' ions (e.g. glucuronide conjugates of analytes or their metabolites), leading to reduced intensity or complete elimination of the precursor ion.

The 'Enhanced MS' mode uses Q_1 operated in RF-only mode so that the final $Q_{3/T}$ linear trap spectrum contains all ions produced in the source (down to the low *m/z* cutoff of Q_1); this mode resembles a MS^1 scan of $Q_{3/T}$ operated in conventional RF/DC resolving mode except that the scan cycle time is much faster. To generate MS^3 spectra, first generation fragment ions are produced in $q_{2/T}$ from precursor ions selected by Q_1 and are transmitted and accumulated in $Q_{3/T}$ operated in linear trap mode; isolation and CID of *m/z*-selected first generation fragment ions are then achieved as described above for linear traps (Hager 2003).

Several examples of the use of this instrument in identification and quantitation of analytes of interest have been published (Hager 2003a; Hopfgartner 2003, 2004; King 2003; Xia 2003). In some cases, conventional triple quadrupole MRM mode and Enhanced Product Ion full scan spectra gave comparable results for target analyte quantitation, with similar values of sensitivity, reproducibility and linear dynamic range; use of the Enhanced Product Ion scan mode for quantitation was shown to be useful (Xia 2003) in selecting MRM transitions during the method development stage. The several other scan modes that are possible using this new instrument are summarized in Figure 6.31. It remains to be determined whether the new 'enhanced' scan modes are fit for purpose in rigorously validated quantitative methods.

6.4.7 *Time of Flight and QqTOF Analyzers*

Time of flight analyzers (TOFs) are important in two main areas, as high *m/z* analyzers for ions produced by MALDI and as analyzers of medium resolving power that can provide mass measurements for low mass ions to within a few ppm. They are available either in stand-alone single TOF arrangements, or hybrid tandem QqTOF or TOF–TOF arrangements. Use of any of these instruments for quantitation is really their secondary application, with the exception of GC/TOF instruments that were specifically designed for fast capillary GC.

The TOF analyzer was first described by Cameron and Eggers (Cameron 1948), but useful performance characteristics were not obtained till the work of Wiley and McLaren some years later (Wiley 1955). In the context of trace level quantitation it is handicapped by poor duty cycle and low dynamic range, but has advantages arising from its ability to provide appreciable selectivity through a FWHM resolving power $\sim 10^4$ and also full mass spectral detection at an acquisition rate compatible with even the fastest chromatography. However, SIM mode is not possible for TOF analyzers. A stand-alone TOF fitted with a reflectron can provide MS/MS data with some difficulty (the so-called post-source decay (PSD) method) and this limitation has led to development of the hybrid QqTOF combination described later.

Two detailed accounts of the basic principles involved in modern TOF instruments (Guilhaus 1995; Weickhardt 1996) are the basis for the following simplified version. A sketch of an ultra-simple system that will serve to introduce the concept is shown in Figure 6.32(a). This system consists of an evacuated flight tube with a detector at the end, and a much smaller 'acceleration' region from which ions can be accelerated into the flight tube under the influence of an electric field. These ions can be formed within this acceleration region (as is the case with conventional MALDI–TOF instruments, for example) or can be injected from an external ion source (as is the case with electrospray–TOF, or QqTOF instruments, or orthogonal MALDI–TOF discussed in Section 5.2.2). In any event, it is important to keep in mind that these ions will be accelerated into the TOF analyzer in a nonideal fashion, with distributions of both direction and kinetic energy leading to spectral peaks of finite width.

As a first simple example, consider a group of analyte molecules all at rest relative to the instrument and located over a very narrow range of distances along the main axis X (the flight path for the ions). An experimental realization of this situation is provided by a unimolecular film of analyte on a metal surface such as those studied by so-called 'static secondary ion mass spectrometry' (Adriaens 1999, Van Vaeck 1999). Further in this idealized situation, the surface carrying these analyte molecules is held at a high potential (typically several kV) relative to the rest of the apparatus, which is at ground potential (other than the front end of the detector, see Chapter 7), and are 'instantaneously' ionized by some means. In the case of static SIMS this is provided by a short burst of high energy primary ions from an ion gun, with a time-width that must be as short as possible relative to the subsequent flight time of the ions. The ions (assumed here to be positively charged, for convenience in discussion) are thus accelerated from the sample holder (at potential $\Phi_1 = +V_{acc}$) towards an extraction grid (a fine metal mesh) held at ground potential ($\Phi_2 = 0$); for now consider the distance of this grid from the sample surface to be negligible relative to the distance X_d from ion source

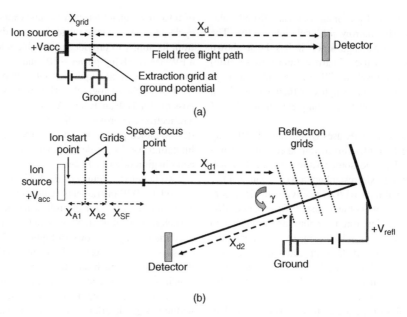

Figure 6.32 Sketches of (a) an ultra-simple linear time of flight analyzer and (b) a modern design with two extraction grids (permitting the space focus point to be moved to a reasonably large distance X_{SF}) and a reflectron (the grids are at increasing electrical potentials from the first entrance grid that is at ground), see main text. Not drawn to scale (in particular the angle γ should be small $\sim$1–2°).

to detector. The assumption that all analyte molecules are at a fixed distance from the accelerating grid means that they are all accelerated through the same potential drop $(V_{acc} - 0) = V_{acc}$. Then, within all these restrictions, ideally the ions acquire translational energy in the X-direction given by Equation [6.10] as:

$$T = ze.V_{acc} = m_{ion}.v_X^2/2$$

where m_{ion} is the actual mass (in kg) of the ion, e is the magnitude (without sign) of the electron charge and z is the number of charges on the ion ($= q_{ion}/e$), so that:

$$v_X = [2.e.V_{acc}/m_{ion}.z)]^{\frac{1}{2}}$$
$$= [(2.e.V_{acc}/m_u)/(m/z)]^{\frac{1}{2}} \qquad [6.38]$$

where v_X is the (mass-dependent) speed of the ions in the X-direction and $m/z = (m_{ion}/m_u)/z$. Then the time t_f required for an ion to travel the distance X_d to the detector is:

$$t_f = X_d/v_X = (m/z)^{\frac{1}{2}}.X_d/[2.e.V_{acc}/m_u]^{\frac{1}{2}} \qquad [6.39]$$

Thus if the flight time t_f, can be measured, a simple transformation (or more usually in practice, a calibration) will

provide the value of m/z. The uncertainty δt_f in measuring t_f (see below) will control the resolving power of this analyzer, i.e. the corresponding uncertainty in m/z is given by simple differential calculus applied to Equation [6.39]:

$$\delta t_f = X_d/[2.e.V_{acc}/m_u]^{\frac{1}{2}}.[\delta(m/z)/2.(m/z)^{\frac{1}{2}}]$$

so that the resolving power is given by:

$$RP = [(m/z)]/[\delta(m/z)] = t_f/(2.\delta t_f) \qquad [6.40]$$

i.e. the precision of the flight time measurements (δt_f) determines the resolving power. In this highly idealized model of a linear TOF analyzer, the value of δt_f is determined by the detector response time and by the time-width of the pulsed ionization event in cases where the voltage is applied continuously, or of the accelerating potential pulse applied to the pre-formed ions from an external source. These parameters are fixed for a given instrument, so that the RP must be improved by increasing t_f via either using a low value of V_{acc} or a long flight path X_d.

However, when applied to chemical analysis conditions other than those approximating static SIMS, such a primitive instrument gives very poor resolving power in practice as a result of a distribution of values of v_X. In most cases the neutral analyte molecules and/or ions occupy a range of positions within the ionization–acceleration region prior to acceleration, and are thus subjected to different effective accelerating potentials V_{eff} depending on the pre-ionization distance X_{mol} of the molecule from the point at which V_{acc} is applied relative to that (X_{grid}) to the grounded extraction grid:

$$V_{eff} = V_{acc}.[1 - (X_{mol}/X_{grid})]$$

Thus ions that are accelerated from initial points with small values of X_{mol} have further to travel to the detector than those that start out closer to the acceleration grid but are accelerated through larger values of V_{eff} and thus have higher values of v_X. Analysis of these competing effects (Guilhaus 1995; Weickhardt 1996) shows that there is a so-called 'space focus' at $X = X_{SF}$, where all the ions of fixed m/z but originating from different values of X_{mol} arrive at the same time (the faster ions from smaller values of X_{mol} have caught up with the slower ions that started off closer to the grid). X_{SF} is the same for all values of m/z but the time of arrival at the space focus varies with m/z. Placing the detector at the space focus would thus improve the resolving power since the contribution to δt_f from the spatial dispersion within the acceleration region is thereby reduced to zero.

However, for a simple single stage acceleration field, such as that considered so far, it can be shown (Guilhaus 1995; Weickhardt 1996) that $X_{SF} = 2.X_{grid}$ and this length of flight path is much too short to be useful. (The value of X_{grid} is kept small otherwise the corrections necessary to the simple time-to-mass conversion (Equation [6.39]), to account for the time spent during acceleration, become so large as to be unmanageable.) The situation is saved by use of a two stage acceleration region (Wiley 1955) using two grids instead of one; the first grid is usually called the extraction grid and the second the acceleration grid (Figure 6.32(b)). By appropriate choice of the positions of the grids and the potentials applied to them (the potential drop to the extraction grid is usually ~10% of the total drop to the acceleration grid), the value of X_{SF} can be increased to much larger values resulting in correspondingly larger values of t_f and thus improved RP (Equation [6.40]). This famous Wiley–McLaren condition is now used in all TOF instruments. Later it was shown (Weinkauf 1989) that the Wiley–McLaren condition could be optimized even further to provide so-called second order energy focusing (Weickhardt 1996). Linear TOF instruments, even with all these optimizations, have limited RP (up to a few thousand in the best cases but much lower in common practice) but also better ion transmission than the high resolving power designs described below.

The Reflectron Principle: An Ultra-Simple Example

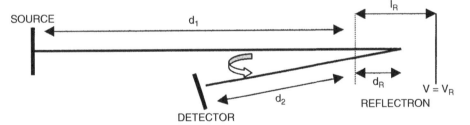

The following is intended only to illustrate the principles of velocity focusing of ions in a time of flight analyzer with a single stage reflectron, using an unrealistically simple example that avoids all the complications of real working systems. It requires the condition that the total flight time is independent of the initial speed of the ion along the X-axis, and prefers that this condition should apply to ions with all values of m/z. If v_X represents the 'ideal' speed set by the ion source, the corresponding kinetic energy T_X is simply $(m_{ion}/2)v_X^2$, where $m_{ion} = m.m_u$ is the actual mass (kg) of the ion. Also, it is assumed that $T_X = ze.V_{acc}$, ignoring all effects of acceleration time etc. The reflectron is modeled as a simple linear electric field with potential $V = 0$ at the entrance ($X = d_1$) and V_R at the far end ($X = d_1 + l_R$); thus at any point a distance d_R from the reflectron entrance the potential is $V_R.(d_R/l_R)$. An ion entering the reflectron with kinetic energy T_X will reach a depth d_R where the speed is reduced to zero, i.e., where the potential energy $zeV_R = T_X$. Substitution and simple algebraic rearrangement gives:

$$d_R = (m_u/2e)(m/z)(l_R/V_R)v_X^2.$$

(Continued)

Now make another simplifying assumption, i.e., that the average speed of the ion during its journey into the reflectron is the simple average of its initial (v_X) and final (0) values, i.e., $v_{X,avge} = v_X/2$. Then the flight time *into* the reflectron is:

$$t_{R,in} = d_R/v_{X,avge} = (m_u/e)(1_R/V_R)(m/z).v_X = t_{R,out}$$

where the equality $t_{R,in} = t_{R,out}$ is exact if the reflection angle γ (curved arrow) = 0 since the reflected ion will have a speed equal but opposite to v_X, but for $\gamma \sim 2°$ the error is small. The flight times through d_1 and d_2 at speed v_X are (d/v), so the total flight time is:

$$t_{total} = t_1 + t_{R,in} + t_{R,out} + t_2 = [(d_1 + d_2)/v_X] + (2m_u/e)(1_R/V_R)(m/z).v_X$$

Now consider an ion of the same m/z but with a different speed $v_X' = v_X(1 + a + b.v_X + c.v_X^2 + -)$; in this simple example only the first term is considered, i.e., $v_X' \approx v_X (1 + a)$. Such an ion penetrates the reflectron to a different depth and of course the times required to travel the various distances are also changed, but the derivation is exactly analogous and gives:

$$t'_{total} = (d_1 + d_2)/[v_X(1 + a)] + (2m_u/e)(1_R/V_R)(m/z).v_X.(1 + a)$$

We are seeking the condition for $t_{total} = t'_{total}$, i.e., for $t_{total} - t'_{total} = 0$:

$$(d_1 + d_2)/\{(1/v_X) - 1/[v_X(1 + a)]\} + (2m_u/e)(1_R/V_R)(m/z).v_X.[1 - (1 + a)] = 0$$

Simple algebraic rearrangement and cancellation gives the velocity focusing condition:

$$(d_1 + d_2)/(1 + a) - (2m_u/e)(1_R/V_R)(m/z).v_X^2 = 0$$

For small spreads in speed $a << 1$. Also in our highly ideal model, $T_X = zeV_{acc}$ so $(m_u/e)(m/z).v_X^2 = 2V_{acc}$, and the final highly over-simplified condition for velocity focusing for all m/z values is: $(d_1 + d_2) = 41_R(V_{acc}/V_R)$. Design of a velocity focusing instrument based on this ultra-simple model must satisfy this condition.

An important contribution to development of modern TOF analyzers was the invention (Alikanov 1957; Mamyrin 1973) of the reflectron, a device that compensates for the flight time differences of ions of the same m/z but different translational energies in the flight path (X) direction (Figure 6.32(b)). (The energy spread of the ions can arise from various sources, including the effect of the spatial position of the ions at the initiating pulse in the acceleration region, and also angular divergence of the accelerated beam since only the velocity component along the X-axis determines the flight time.) This device introduces an electric field designed to slow down and stop the ions and then reverse their flight direction (and is thus sometimes referred to as an 'ion mirror'). Ions with higher translational energies penetrate deeper into the retarding field in the reflectron than the slower ones and thus have a longer total path to reach the detector. The

theory (Guilhaus 1995; Weickhardt 1996) allows prediction of the appropriate combination of reflectron field and the distances from source to reflectron and from the latter to the detector, to permit ions of the same m/z but different translational energies to arrive at the detector simultaneously. In a fashion somewhat analogous to that described above for the acceleration region, a two-stage reflectron (Mamyrin 1973) produces a smaller time-width δt_f of the energy focused ions at the detector, and can correct for larger energy widths, than the original single-stage version (Alikanov 1957).

Naturally the reflectron must be tilted at a small angle (γ) in order that the reflected beam does not simply return to the acceleration region (Figure 6.32(b)), but the effect on δt_f of a small value of $\gamma(\sim 1 - 2°)$ is negligible. A highly simplified discussion of the reflectron principle is given in the accompanying text box. A very complete treatment of velocity focusing in time of flight analyzers (Doroshenko 1999) takes into account many variables totally ignored in this ultra-simple model, including ion source acceleration regions, field free regions, reflectron (not necessarily single stage), energy discrimination filters, post-analyzer acceleration before ion detection etc.

Originally the reflectron fields were created by grids of very fine metal wires to which the appropriate potentials were applied, but these wires also act as electrostatic deflectors that perturb the ions from their ideal trajectories. Grid-free reflectrons use instead a series of metal discs with appropriately sized apertures, for each of which the potential can be controlled independently. The problem with this concept is that the electric fields thus produced are no longer simple functions of position and cannot be described by analytic equations, so that numerical simulations must be used to calculate the fields and the resulting ion trajectories to find the conditions for optimum energy compensation. However, there is an advantage to this approach in addition to the freedom from uncontrolled perturbations of the trajectories by the potentials on the grid wires, in that the electric fields also have a focusing effect perpendicular to the flight path (analogous to that of an Einzel lens, Figure 6.8) which increases the transmission of the reflectron. Various 'tailored' fields within the reflectron have been described.

There are other effects contributing to δt_f that can not be compensated for by a reflectron. One of these is the 'turnaround time', the result of the distribution of initial ion velocities prior to acceleration. As a simple example, consider an ion that is moving in the negative X-direction at the moment when the acceleration field pulse is applied. It will require a certain 'turnaround time', depending on the magnitude of this pre-acceleration velocity, before starting its flight to the detector. An ion at roughly the same initial position, but that happened to be already moving in the positive X-direction, will have a correspondingly shorter flight time than an ion that was at rest pre-acceleration. This effect is an intrinsic time spread that can not be corrected by a reflectron, but a number of techniques involving time dependent acceleration fields have been developed to alleviate the effect. The first of these techniques was so-called 'time-lag focusing' (Wiley 1955), introduced for a linear TOF instrument in which the ions were created in the acceleration region by pulsed EI of gaseous samples. A delay was interposed between ionization and acceleration of the ions so that the spread in initial velocities was converted into a spatial spread that could be corrected for as described above. Unfortunately the optimum delay time is dependent on *m/z* and thus the correction can only be applied

for a small part of a spectrum in any one pulse acquisition. More recently the principle of time-lag focusing (now termed 'delayed extraction') has been applied to MALDI–TOF instruments, for which the turnaround time problem reflects a large momentum and energy spread of the nascent ions in the plume produced by the laser pulse, resulting in disappointingly poor RP in MALDI–TOF even with a reflectron; appropriate modification of the time-lag focusing principle led to greatly improved performance (Vestal 1995).

Another phenomenon, whose effect on δt_f (and thus RP) is somewhat analogous to that of turnaround time, is that of space charge, i.e. the mutual electrostatic repulsions among the ions if they are bunched together so closely that the resulting Coulombic forces (Equation [6.4], Appendix 6.1) are significant relative to those arising from the externally applied fields. There is no real solution to this problem other than ensuring that the ion beam current is kept sufficiently low that the effect of the ion–ion repulsion forces is negligible during the ion flight time.

Another effective approach to minimizing the effect of velocity distributions to is to minimize pre-acceleration velocities in the X-direction. The most common method is the use of orthogonal acceleration of collimated ion beams extracted from a continuous ionization source and traveling perpendicular to the TOF flight path (Figure 6.33). This approach (Dawson 1989; Dodonov 1991) is the subject of an extensive review (Guilhaus 2000); it is often referred to as oa–TOFMS and minimizes the pre-acceleration velocity spread in the X-direction of the TOF by cleanly separating the ionization event from the acceleration, and can provide RP values up to $\sim 20\,000$ (FWHM definition) when combined with electrospray ionization, appropriate time-lag focusing and a reflectron (Dodonov 2000). In an ingenious application of the oa–TOF principle (Krutchinsky 1998) a conventional MALDI source was converted into an essentially continuous source by collisional cooling in a RF-quadrupole ion guide. In addition to the resulting improvement in RP, an important advantage relative to conventional MALDI–TOFMS resulted from the nearly complete decoupling of the ion production from the TOFMS analysis since the latter was independent of source conditions. This allows much greater flexibility to optimize the matrix, substrate (including even insulating substrates) and the wavelength, fluence and pulse widths and frequencies of the laser.

For an oa–TOF, in the simplest approximation the instrumental sampling duty cycle for ions with a particular *m/z* is approximately given by the ratio l_p/l_b (Figure 6.33), where l_p is the length of the ion packet

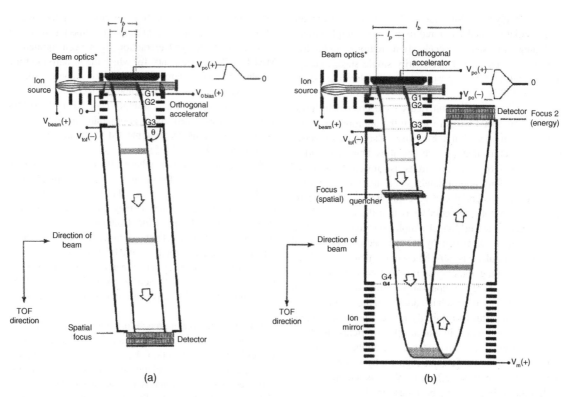

Figure 6.33 Typical configurations of oa–TOF systems with (a) a linear TOF and (b) a single stage reflectron–TOF. The ion beam enters from a continuous source at the left in each case. For positive ions, the beam is formed literally or as though it originated from a source held at a positive potential of V_{beam} and accelerated to enter the orthogonal accelerator (oa) held at 0 V. The beam optics make the beam more parallel before it enters the oa, thus minimizing pre-acceleration velocities in the acceleration direction. The beam is allowed to fill the oa at 0V until a monopolar pushout pulse (a) or bipolar pushout pulse pair (b) is applied. A packet of ions of length l_p is thus sampled and accelerated through grids to enter the drift region at a potential of V_{tof} (equivalent to V_{acc} in the previous discussion); note that the velocity in the original beam direction is maintained so the post-acceleration ions move at an angle as shown. Conventional linear (a) or reflecting (b) TOF optics are used to bring the ions to a space or space–time focus on the detector, respectively. During the time that the ions are in the TOF drift region (and ion mirror in (b)) the oa is refilled with new ions. Within the TOF drift time for an ion of a particular *m/z* value, the same ions in the from the source beam travel a distance l_b determined by the beam velocity (varies with *m/z* for a fixed extraction potential from the ion source). The distances l_p and l_b determine the sampling efficiency of the oa-TOF and thus the analyzer duty cycle. Reproduced from Guilhaus, *Mass Spectrom. Revs.* **19**, 65 (2000), with permission of John Wiley & Sons, Ltd.

that is accelerated into the TOF drift region, and l_b is the distance traveled in the original beam direction by the new ions of the same *m/z* during the time t_{max}. The time t_{max} is the reciprocal of the repetition frequency of the push-out acceleration pulse and is set to be sufficiently long that ions at the top of the *m/z* range reach the detector before the next acceleration pulse. This instrumental duty cycle (ion sampling efficiency) is appreciably higher for an oa–TOF with continuous ion sources than for conventional in-line TOF analyzers with the same sources. Bear in mind that the overall analyzer efficiency is given by multiplying the duty cycle by the ion transmission efficiency and by the detector efficiency; for the overall instrumental efficiency (relative to the amount of sample introduced to the ion source), the ionization efficiency must also be considered.

Nonetheless, the introduction of the oa–TOF principle has meant that use of a TOF analyzer on-line with chromatography for trace analysis is now a viable option. The advantages of such an arrangement include the medium-high resolution (10 000–15 000) and reasonably good accuracy/precision of *m/z* measurement (Sleno 2005b) that can be obtained, thus providing increased detection selectivity compared with unit mass RP analyzers as well as some confirmation of analyte identity. (A more complete discussion of limits on precision of *m/z* measurement is given later with respect to QqTOF instruments.) Furthermore, a TOF intrinsically records full mass spectra at a relatively fast spectrum acquisition rate, so that fast chromatography (narrow peaks) can be accommodated (Zhang 2000).

On the other hand, the linear dynamic range for quantitation that can be obtained with all TOF analyzers is appreciably less than that possible for quadrupole or triple quadrupole analyzers in SIM or SRM modes, and the precision is also significantly worse (Zhang 2001). At the low end of the range it is often found that the lower limit of quantitation (and the detection limit) are worse by an order of magnitude for a TOF analyzer (reflecting a low overall duty cycle with continuous ion sources), while the upper limit of the linear range is determined by the inherent characteristics of the TOF detector. These detectors are ion counting devices in which arrays of continuous channel multipliers (see Section 7.3.2b) convert an ion that strikes one of the channels into an avalanche of electrons; this pulse of electron current is converted to a voltage pulse across a resistor attached to the output of the multiplier channel and, provided that the amplitude of this pulse is greater than the user-set background level, the associated electronics assigns such a pulse as arrival of one ion within the appropriate time window following the pulsed acceleration of the ions into the TOF (that also starts the timer of the electronics). The width of this time window is determined mainly by the flight time of the electrons through the multiplier channel and the resistor, and considerable development effort has been devoted to reducing this time mainly through reduction in the diameter and spacing of the channels (Section 7.3). This time window is the detector 'dead time'; if a second ion strikes the detector during the dead time it can not be recorded. It has been well documented that these dead time effects influence the *m/z* assignment via a distorted peak shape, and obviously also record too low an ion count thus limiting the upper end of the dynamic range.

While the limitations of TOF analyzers (particularly of those employing orthogonal acceleration and for low *m/z* values for which the flight times are longest) with respect to duty cycle have been emphasized in all preceding discussions, very recently (Brenton 2007) a method has been described whereby duty cycle and *m/z* range can be traded off between each other. This method relies on one or more ion gates that can switch from the open state to the closed state in tens of nanoseconds, placed in the beam path to transmit or stop the beam and thus select

portions of the *m/z* range that are passed to the CEMA detector. (All other ions are deflected from the beam and are lost, and thus do not impose a wait-time penalty on the next pulse.) Because it is possible in this way to select a limited number of *m/z* values (or regions) for detection, an oa–TOF analyzer can now be operated in what is effectively SIM mode (or MRM if it is a component of a QqTOF instrument, see below) with a duty cycle that can in principle approach 100 %. This technique is very new (Brenton 2007) and has not yet been tested for trace quantitation, but has the potential to greatly expand the usefulness of QqTOF instruments in this area by essentially mimicking a triple quadrupole in MRM mode but with appreciably higher resolving power for fragment ions. Also, in full scan acquisition mode (MS^1, or MS^2 with a QqTOF) this approach should at least rival trap instruments in data acquisition time.

A linear TOF analyzer can not provide any MS/MS information but a reflectron instrument can provide fragment ion spectra of metastable ions that fragment (either spontaneously or via laser photodissociation) while in flight between the final acceleration grid and the reflectron entrance (Weickhardt 1996). However, this 'post-source decay' technique has never been used for quantitative trace analysis.

These disadvantages have limited the use of stand-alone TOF and oa–TOF analyzers in quantitative trace analyses requiring the best possible performance with respect to accuracy and precision, dynamic range etc. However, clearly for some purposes the advantages of TOF technology will outweigh the disadvantages, particularly when used as the second analyzer in a tandem mass spectrometer. The ease-of-use of TOF analyzers is nowadays quite favorable and the cost of a stand-alone instrument can vary considerably depending on the type of application for which it is designed (e.g. a detector for fast chromatography, or one for MALDI analyses etc.) but generally falls in the medium range for single analyzer instruments.

To add readily usable MS/MS capability to the TOF analyzer, the QqTOF hybrid design was introduced; a diagram of one such instrument is shown in Figure 6.34. Although originally introduced (Morris 1996; Shevchenko 1997) as tools for sequencing peptides and proteins, they have attracted some attention (Clauwaert 1999; Jeanville 2000; Stolker 2004, 2004a) with regard to applications in trace analysis.

The QqTOF concept is applicable to ion sources that supply continuous ion beams (e.g. ESI, APCI); special arrangements must be made for pulsed ion sources (e.g., MALDI) to convert them into quasi-continuous sources (Krutchinsky 1998; Loboda 2000). A QqTOF instrument can be regarded either as addition of a mass filter Q and

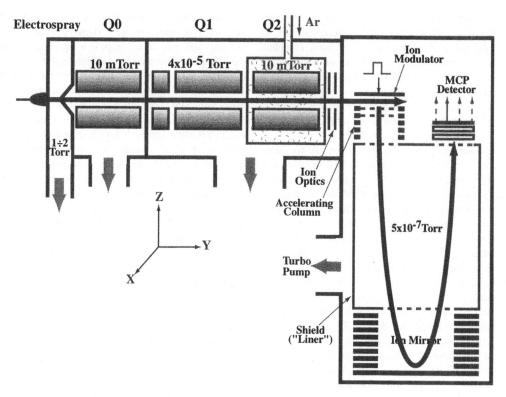

Figure 6.34 Schematic diagram of a QqTOF hybrid mass spectrometer. Reproduced from Chernushevich, *J. Mass Spectrom.* **36**, 849 (2001), with permission of John Wiley & Sons, Ltd.

collision cell q to a TOF analyzer, or as replacement of the third quadrupole (Q_3) in a triple quadrupole by a TOF. The advantages (that come with a significantly higher capital cost) include high sensitivity in full spectral acquisition and moderately high resolution as well as accuracy and precision in *m/z* measurement (Sleno 2005a); these advantages apply to both MS[1] spectra (Q_1 in RF-only mode) and MS[2] fragment ion spectra acquired using the TOF of precursor ions selected with unit-mass resolution (Q in RF/DC mode). The advantage of full scan sensitivity over a wide *m/z* range is a result of the quasi-parallel detection property of a TOF analyzer. However, note that a QqTOF can not readily provide an operating mode analogous to the precursor ion scan for a *m/z*-selected fragment ion, although an ingenious method has been described (Chernushevich 2000) that involves trapping ions in q and then ejecting them in short bursts synchronized with the TOF acceleration pulse. More importantly from the present viewpoint, this q trapping mode increases the (normally low) duty cycle of a TOF analyzer to higher values (see below) and in principle offers an acquisition

mode truly analogous to MRM. (The term 'scan' cannot strictly be applied to a TOF instrument, but this terminology is in common use).

The general principles underlying the coupling of the constituent QqTOF components have been reviewed (Chernushevich 2001) and the following account is a much-abbreviated version of that review with particular emphasis on the topics of interest in this book. Use of sufficiently high pressures in q to effect collisional focusing is particularly important for efficient operation of the TOF component of a QqTOF because an orthogonal–TOF places much more demanding restrictions on the properties of the input ion beam than a Q_3 mass filter. Ions are accelerated out of the high pressure q cell through several tens of volts and are focused by ion lenses into a parallel beam (thus minimizing initial velocity spread in the TOF acceleration direction) that enters the 'ion accelerator' (Figure 6.33). For most of the time the accelerator region is field free so that ions move in their original direction. However, a short electric field pulse (several kV at a frequency of several kHz) is applied across the accelerator in a perpendicular direction into the field free drift space where m/z separation is achieved via TOF. The ratio of the (m/z dependent) velocity in the original direction within the field free accelerator to the post-acceleration velocity in the perpendicular (TOF) direction must be selected so that the width of the ion beam matches that of the detector. A single stage reflectron is sufficient to compensate for the initial energy and spatial spreads of the ions, provided that the collimation (degree of parallelism) of the beam within the accelerator (determined by appropriate operation of q in the collisional focusing regime) is sufficient; thus ions originating from different depths in the accelerator are focused on to a plane at the detector entrance. The detector usually consists of two microchannel plates in a chevron configuration (see Section 7.3.2b). An important difference from stand-alone TOF analyzers results from the need to maintain the Qq section at low (near ground) potentials. Thus e.g., positive ions are accelerated from ground to high negative potentials in the TOF section of a QqTOF instrument, rather than from high positive potentials to ground as in e.g., MALDI-TOF mass spectrometers. This requires an extra shield (liner) within the TOF chamber (Figure 6.33) to create a field free drift region floated at high potential, but this shield must be designed carefully since even a small penetration of an external field (e.g. from the ground potential vacuum housing) can seriously damage the TOF performance.

For reliable measurement of the low ion currents reaching the detector in such an instrument, an ion counting technique using a time-to-digital (TDC) detector (Section 7.3) is essential. Such detectors place intrinsic limits on the upper end of the dynamic range as a result of the detector 'dead time' following detection of an ion, together with the probability of one or more additional ions arriving within that dead time (and that are thus not counted). The low end of the dynamic range is mainly limited by the duty cycle, as discussed above with respect to stand-alone orthogonal–TOF analyzers. In most orthogonal–TOF instruments the duty cycle is between 5 % and 30 %, depending on m/z range (and of course on the instrumental parameters).

As mentioned above with respect to obtaining precursor ion spectra using a QqTOF instrument (Chernushevich 2000), trapping ions in q while the previous pulse of ions is traversing the TOF analyzer, and then gating them into the modulator synchronized with the next TOF acceleration pulse, significantly increases the duty cycle to values dependent on the m/z range to be recorded. If combined with appropriate stepping of Q to transmit only precursor ions characteristic of the target analyte(s), the duty cycle for this mode could in principle come close to that for the MRM mode for a triple quadrupole. However, no such

experiments appear to have been reported in the literature and all applications of QqTOF instruments in trace analysis of small molecules reported thus far have used conventional full spectral recording. The advantages of a QqTOF over a triple quadrupole in this regard include increased confidence in analyte identification and in chromatographic peak purity, arising from full product ion spectra obtained with m/z uncertainties of a few ppm.

A thorough investigation of accuracy and precision of m/z measurement using a QqTOF instrument (Wolff 2003) confirmed and expanded earlier work (Blom 2001) that considered the limitations on the measurement arising from limited numbers of ions detected for any mass peak in a spectrum, which in turn limits the definition of the peak shape and position on the m/z axis. This ion statistics aspect is considered in Chapter 7. Similarly, the limitation on dynamic range arising from detector dead time can be improved by several measures, including post-acquisition processing that exploits the Poisson distribution of ion arrival times to determine an appropriate correction for the dead time effect (see Chapter 7).

The published reports of trace quantitative analysis using QqTOF instruments generally confirm the generalizations discussed in the preceding paragraph. The small number of such reports presumably reflects the generally excellent performance of the considerably less costly and more user-friendly triple quadrupole instruments. Thus QqTOF instruments are unlikely to be used for small molecule quantitation except in unusual circumstances requiring additional specificity (provided by the better resolution and m/z measurement). A growing exception to this conclusion involves absolute quantitation of proteins in cell extracts via quantitative analysis of characteristic small proteolytic peptides, discussed in Section 11.6, where the greater mass range of the TOF provides an advantage over a quadrupole Q as product ion analyzer. The same conclusion is likely valid for TOF–TOF tandem instruments (Cornish 1993; Medzihradszky 2000) and magnetic sector–TOF hybrids (Bateman 1995) that exploit high energy (keV range) collisional activation. Early TOF–TOF instruments exhibited only poor resolution for precursor ions but more recent improvements have increased this parameter to ~400 (based on peak widths measured as FWHM). However, in proteomics applications the QqTOF instruments must compete with FTICR analyzers provided with linear traps to increase the duty cycle (Schrader 2004), and similarly with the new Orbitrap instruments (Hu 2005).

A summary of figures of merit for some tandem mass spectrometers currently used for quantitative analysis is shown in Table 6.1. Of course such comparisons are highly time dependent in view of the pace of mass spectrometry development and in any case no such summary table can possibly provide a complete comparison that would require multi-parameter considerations of fitness for purpose, but Table 6.1 provides a starting point.

6.4.8 FTICR and Orbitrap Analyzers

The double focusing magnetic sector mass spectrometer was for many years considered to provide the ultimate in resolving power obtainable using reasonable acquisition times. However, the application (Comisarow 1974, 1974a, 1974b) of Fourier transform pulse techniques, which had already been successfully introduced to nuclear magnetic resonance spectroscopy, transformed the old mass spectrometric technique of ion cyclotron resonance into what is now an important analytical tool for proteomics (e.g., Bogdanov 2005). This is a consequence of the unparalleled resolving power and accuracy of mass measurement (Sleno 2005b), combined with high sensitivity, of the technique. Excellent tutorials and reviews of the subsequent development of the FTICR technique for mass spectrometry (Marshall 1992, 1998; Amster 1996; Heeren 2004), however, do not mention trace quantitative analysis at all. Indeed, an attempt to search the literature for any publications describing such applications turned up only one short paper (Choi 2001) that compared quadrupole, time of flight and FTICR analyzers for LC/MS applications, and even this paper did not comment on the use of FTICR for quantitation. A review of methods and applications of HPLC–FTICR (Schrader 2004) does mention that applications of this combination to quantitative analysis are 'extremely rare'. There are several reasons for this, including cost as well as a low duty cycle arising from the time penalty imposed by the unique detection technique, which involves recording of image currents induced in the FTICR cell walls by the cyclotron motions of the trapped ions.

The same lack of application to trace quantitative analysis appears to be true of a more recent innovation, the Orbitrap analyzer (Makarov 2000); indeed a recent extensive review of this device (Hu 2005) also does not mention quantitation at all. Like the FTICR analyzers, the Orbitrap operates under very high vacuum ($\sim10^{-10}$ torr) to achieve its ultimate performance; it also uses image current detection. At this time neither of these analyzers appears to be suitable for trace level quantitation experiments of the kind discussed in this book.

6.5 Activation and Dissociation of Ions

Thus far very little has been said about how precursor ions, including the intact ionized molecules formed by 'soft'

ionization techniques such as ESI or APCI (Chapter 5), can be induced to dissociate for subsequent MS/MS analysis. The ideal activation method for trace quantitation of a target analyte using MS/MS is different from that for structural elucidation of an unknown compound. In the latter case it is desirable to observe a large number of fragment ions that convey structural information, with relative and absolute abundances that are not of primary importance as long as acceptable mass measurements are possible. In contrast, an ideal situation for a quantitative analysis involves production of a relatively small number (two or three) of product ions so that the ion current is concentrated at these m/z values, thus permitting lower limits of detection and quantitation. (Of course, neither of these ideal situations is ever attained in practice.) Moreover, since the structure of a target analyte is known (by definition), structural information is not directly of primary importance; rather, uniqueness of the mass of the corresponding neutral fragment, and thus the detection selectivity, is the more important criterion. For example, a neutral mass loss of 18 Da (certainly water) tells us that the molecule contains an oxygen atom most likely present as a hydroxyl group (and is thus structurally informative); however, 18 Da is such a commonly observed neutral mass loss that its use to provide an MS/MS signal for quantitation is likely to have low selectivity and thus to provide only marginal improvement in S/B value ratio (Figure 6.1) in many cases.

Historically, from the earliest days of mass spectrometry the most common method of ion activation has been collisional activation (CA), whereby ions are accelerated through a defined potential drop and caused to collide with gas molecules that are deliberately introduced into the ion trajectory; the history of this approach has been described in an excellent overview (Hu 1995). This involves conversion of part of the kinetic energy of the ion into internal energy (that in turn leads to fragmentation), and is still by far the most commonly used method.

In general the timescale for the activation itself is appreciably shorter than that for the subsequent dissociation, and this two-step model appreciably simplifies theoretical descriptions of the phenomenon. This can be understood to some degree for the case of collisional activation by estimating the time during which the fast ion interacts with its neutral collision partner. The velocity of an ion that has been accelerated through a potential difference from V_{acc} to zero was given previously (Section 6.4.4) as:

$$v = (2.q_{ion}.V_{acc}/m_{ion})^{\frac{1}{2}}$$
$$= \{2.e.V_{acc}/[m_u.(m/z)]\}^{\frac{1}{2}} \qquad [6.24]$$

Using SI units, this can be evaluated for an ion with $m/z = 400$ as $(6.944 \times 10^2 . V_{acc}{}^{\frac{1}{2}})$ m.s^{-1} (with V_{acc} in volts); for $V_{acc} = 100V$, $v = 6.944 \times 10^3$ m.s^{-1}. Now if the effective distance over which the ion and collision gas molecule interact with one another for collisional activation to occur is taken as ∼5 Å (0.5 nm), the corresponding interaction time is $(5 \times 10^{-10}/6.994 \times 10^3) \approx 7 \times 10^{-14}$ seconds. Of course this estimate varies somewhat with m/z, V_{acc} and the interaction length, but the important point is that the collision interaction time is shorter than that required for a single vibration of all but the fastest molecular vibrations(∼10^{-13} seconds), and shorter by several orders of magnitude than the average time required for a collisionally activated molecule to undergo the complex combination of coordinated atomic rearrangements involved in a dissociation reaction of even minimal complexity. Thus the timescale for collisional activation is much shorter than that for the subsequent dissociation; the latter occurs after the two collisional partners have well separated from one another and thus dissociation following activation is a truly unimolecular event. No chemical change of the collision gas molecule is involved in 'collision induced dissociation (CID)', a term that was introduced very early in the development of the approach (Hu 1995) to describe the overall process of activation plus dissociation. Below, this widely used method of ion activation is discussed in the context of other activation methods that are not often used at present in trace quantitative analysis.

This book is not concerned with detailed theories of dissociation of activated ions but a limited discussion is appropriate in order to understand some experimental trends that can be important in designing an analytical strategy. The first important aspect is that mass spectral fragmentations represent a problem in chemical kinetics, so that appropriate theories must attempt to predict the first-order rate constants for unimolecular dissociation of activated but isolated molecules (ions in this case). Most theories of ion fragmentation are so-called 'statistical theories', in which the total internal energy ε_{int} of the internally activated ion (or molecule) is assumed to be rapidly funneled into vibrational energy of the ground electronic state, and further to be rapidly distributed in an entirely random (statistical) fashion among the vibrational modes. Both of these physical steps are assumed to be 'rapid' relative to the subsequent chemical processes (i.e., breaking and/or rearrangements of chemical bonds) but slower than the collisional activation. The random distribution assumption has led to adoption of the unfortunate term 'quasi-equilibrium theory, QET' for such theories applied to ion fragmentations (Rosenstock 1952; Vestal 1962). It is further assumed that dissociation occurs

when a sufficient fraction of ε_{int} is randomly accumulated in the crucial vibrational mode(s), localized in specific bonds, that are involved in a given fragmentation reaction and together describe the 'reaction coordinate'. The reaction occurs when ε_{int} exceeds the minimum energy $\varepsilon_{int}{}^*$ required for that reaction; the simplest example is a single bond cleavage in which case the relevant vibrational mode (reaction coordinate) is the bond stretching vibration and $\varepsilon_{int}{}^*$ is the bond dissociation energy. In the language of the transition state concept, $\varepsilon_{int}{}^*$ corresponds to the ground state energy of the transition state for that particular reaction. (Note, however, that the transition state theory of chemical reaction rates assumes that the reactant is at thermal equilibrium with a high pressure of molecules, thus providing sufficient collision rates that a unique temperature can be established; this is not the case for an isolated ion about to dissociate in a mass spectrometer vacuum, which is why the terminology 'quasi-equilibrium' theory can be misleading.)

On the basis of these fundamental assumptions, statistical rate theories calculate the probability that this critical amount of energy is in fact accumulated in the vibrational modes involved in the 'reaction coordinate'. Obviously this probability will increase as the excess internal energy ($\varepsilon_{int} - \varepsilon_{int}{}^*$) increases. This energy probability requirement is a necessary but not sufficient condition for reaction to occur. In addition, the motions (vibrations) of the atoms involved in this particular reaction coordinate must be synchronized in such a way that, when sufficiently energized, the resulting extreme vibrations do actually lead to dissociation. This requirement for appropriate 'coordination' of the atomic vibrations is loosely related to the 'entropy' of activation in a transition state theory formulation (but this analogy should not be pursued too closely since entropy is a concept restricted to systems of molecules in thermal equilibrium). This requirement for a specified synchronization of the relative vibrations of the atoms that must move appropriately for this reaction to occur is trivial for a simple cleavage of one bond, since only one vibrational mode is directly involved (i.e., other than to compete for the available pool of internal energy ε_{int}). However, the synchronization requirement can be demanding for a complex reaction that involves rearrangement of the original chemical structure in some way; these are sometimes referred to as 'steric requirements'.

Both of these factors are contained in the following ultra-simple expression for the unimolecular dissociation rate constant k_{uni} for isolated ions (molecules) (i.e. no thermal equilibrium among large numbers of such species):

$$k_{uni}(\varepsilon_{int}) = v^* . [(\varepsilon_{int} - \varepsilon_{int}{}^*)/\varepsilon_{int}]^{s-1} \qquad [6.41]$$

This expression arises as a result of a full treatment of the statistical theory plus several highly approximate additional simplifications. Here ν^* has the dimensions of frequency (units s^{-1}) and is often referred to as the 'frequency factor'; it incorporates the steric (synchronization) requirements for reaction to proceed. The quantity s roughly represents the number of vibrational modes that are sufficiently closely coupled that they can efficiently share internal energy with those modes that are directly involved in the reaction process, within the experimental timescale defined by the nature of the apparatus. For example, consider the molecular ion $M^{+\bullet}$ formed by EI of a phenol that also has a long saturated aliphatic substituent on the phenyl ring, i.e., $[C_nH_{2n+1}-C_6H_4-OH]^{+\bullet}$; the characteristic fragmentations of such ions are expulsion of a molecule of either water (H_2O) or carbon monoxide (CO), and it is chemically reasonable to suppose that the vibrational modes involving the carbon and hydrogen atoms at the tail of the aliphatic chain do not readily participate in energy exchange with those of the aromatic ring that are intimately involved in these two reactions.

However, the parameters ν^* and s in Equation [6.41] are not readily predictable or calculable from the properties of the reactant ion (molecule) and the ultra-simplified Equation [6.41] is useful only for understanding some broad trends in a qualitative fashion. A sketch of the dependence of k_{uni} on ε_{int} for two types of fragmentation reaction, requiring different degrees of structural rearrangement for fragmentation to occur, is shown in Figure 6.35. Simple bond cleavages have high values of ε_{int}^* (the bond dissociation energy) but higher values of ν^* (corresponding to the vibration of that one bond with no requirements for synchronization with other modes). In contrast, rearrangement reactions also involve bond breaking but at least some of the energy required is compensated for by simultaneous formation of new bonds in the rearrangement, a process with high synchronization requirements and low ν^* values. At values of $\varepsilon_{int} >> \varepsilon_{int}^*$, Equation [6.41] gives $k_{uni} \approx \nu^*$, so the simpler reaction (typified by a direct cleavage of just one bond) has the higher value of k_{uni} and the corresponding product ion will dominate the product ion spectrum. However, at values of ε_{int} below the curve-crossing (Figure 6.35), and thus at lower values of k_{uni}, i.e., for slower reactions and longer reaction times, the rearrangement product ions will dominate.

In practice the situation is more complex. It is necessary to take into account the average value of ε_{int} reached by the particular ion activation method used, and also the timescale available for reaction to occur (in general longer for a tandem-in-time (trapping) instrument than for a tandem-in-space (beam) instrument), which in turn determines the range of values of k_{uni} that can produce

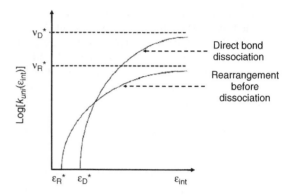

Figure 6.35 Qualitative depiction of the dependence of the unimolecular rate constant k_{uni} on internal energy ε_{int}, for a simple 'direct' bond cleavage (or more generally a fragmentation reaction requiring minimal structural rearrangement), compared with a fragmentation that requires prior extensive rearrangement. The critical energy for the latter (ε_R^*) is typically the lower as a result of new bonds being formed simultaneously with those being broken to produce the fragments, while no such energy compensation is possible for simple cleavage of one bond. However, the steric (synchronization) requirements for the fragmentation reaction that requires extensive rearrangement are more severe so $\nu_R^* < \nu_D^*$. When ε_{int} is very large, Equation [6.41] predicts that k_{uni} approaches ν^*, as shown; as a result the two curves must cross, i.e., rearrangements tend to dominate at low internal energy and direct bond dissociations at higher ε_{int} values.

observable quantities of product ions. Nonetheless, the essentially qualitative picture provided by Equation [6.41] and Figure 6.35 does permit some understanding of broad trends observed in differences in fragment ion spectra obtained using different instrumental methods. This can be important when product ion spectra obtained using very mild collisional activation (i.e., low values of ε_{int}) as in a 3D Paul trap, are compared with those obtained using activation methods (e.g., higher collision energies) that yield higher average ε_{int} values. Frequently, product ion spectra obtained using a Paul trap can be qualitatively different from those obtained using a triple quadrupole, for example! (It is this basic difference that partly accounts for difficulties in transferring an analytical method developed on one of these instrument types to the other.)

The efficiency with which the translational (kinetic) energy of the ion/molecule collision is transformed into internal energy in the collisional activation step is a highly complex problem but it is possible to understand some practically important aspects in a relatively simple fashion. The objective is to increase ε_{int} to a value sufficiently greater than ε_{int}^* that fragmentation proceeds at a rate sufficient to provide a useful yield of the product

Center of Mass Collision Energy: The Maximum Fraction of the Relative Kinetic Energy Between Two Objects that can be Transformed into Internal Energy

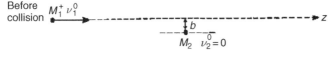

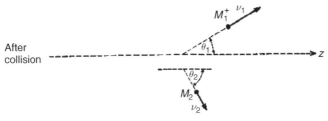

The above scheme illustrates a collision between a fast ion M_1^+ (mass m_1, initial velocity v_1°) and a thermal gas molecule M_2 (mass m_2, effectively stationary, $v_2^\circ \approx 0$). The 'impact parameter' b is the miss-distance if there were no M_1^+–M_2 intermolecular forces. If there are no *external* forces acting perpendicular to the page, all velocity vectors 'before' and 'after' the collision lie in the same plane. 'Before collision' quantities are written with superscript $\circ$. Also for convenience we can write:

'after collision' momentum of M_1^+ in z-direction $= m_1.v_1.cos\theta_1 = (2.m_1.T_1)^{\frac{1}{2}}.cos\theta_1$ since $T_1 = m_1.v_1^2/2$;

'after collision' momentum of M_1^+ in perpendicular direction $= (2.m_1.T_1)^{\frac{1}{2}}.sin\theta_1$

'after collision' momentum of M_2 in z-direction $= (2.m_2.T_2)^{\frac{1}{2}}.cos\theta_2$

'after collision' momentum of M_2 in perpendicular direction $= -(2.m_2.T_2)^{\frac{1}{2}}.sin\theta_2$

Then linear momentum is conserved in the two independent directions (z and the in-plane perpendicular direction), and the total energy is also conserved:

$$T_1^\circ = T_1 + T_2 + \varepsilon_{int}$$

$$(2.m_1.T_1^\circ)^{\frac{1}{2}} + 0 = (2.m_1.T_1)^{\frac{1}{2}}.cos\theta_1 + (2.m_2.T_2)^{\frac{1}{2}}.cos\theta_2$$

$$0 = (2.m_1.T_1)^{\frac{1}{2}}.sin\theta_1 - (2.m_2.T_2)^{\frac{1}{2}}.sin\theta_2$$

where ε_{int} is the amount of translational energy converted into internal energy as a result of the collision. Since m_1, m_2 and T_1° are known experimental parameters, we essentially have three equations relating the five unknowns T_1, T_2, θ_1, θ_2 and ε_{int}. (The two missing equations are the equations of motion in the two in-plane directions for which we need to know the intermolecular force fields, see Equation [6.8].) However, we can eliminate two of the unknowns (choose T_2 and θ_2) from these three equations, to give:

$$(1 + m_2/m_1)(T_1/T_1^\circ)^{\frac{1}{2}} = cos\theta_1 \pm [cos^2\theta_1 - (1 - m_2^2/m_1^2) - (m_2/m_1).(1 + m_2/m_1).(\varepsilon_{int}/T_1^\circ)]^{\frac{1}{2}}$$

Recall that we seek the *maximum* possible value of $\varepsilon_{int}/T_1^\circ$. The expression inside the square brackets appears as the square root and so can not be negative (must be ≥ 0); moreover the maximum value for $cos\theta_1$ is 1.0 (corresponding to $\theta_1 = 0$, i.e., head-on collision with $b = 0$). Algebra leads to the desired result: $(\varepsilon_{int}/T_1^\circ)_{max} = m_2/(m_1 + m_2)$ The physical meaning is that the molecule M_2 has to balance the momentum changes in M_1^+ by acquiring a velocity v_2 that has to be larger as θ_1 increases well above zero; however this condition also increases T_2 even faster ($\sim v_2^2$) and the value of ε_{int} must eventually be limited by the need to also satisfy the energy conservation condition.

ions on the instrumental timescale. It is intuitively obvious that, with a larger initial kinetic energy T_1° of the ion (see the accompanying text box), the average amount of kinetic energy converted to internal energy (ε_{int}) should be increased and indeed this is observed experimentally. (Of course there is always an instrumental upper limit to the value of T_1°, e.g. a few tens of eV in q collision cells and perhaps a few eV in traps.) Also, by using noble gases (most often argon or helium) as the collision partners, all of ε_{int} is essentially guaranteed to reside in the ion since the first excited states of these inert gas atoms lie at several eV above their ground states and are thus not readily accessible in most analytical mass spectrometers. A result that is not immediately obvious is that there is an upper limit to the value of ε_{int} for a given projectile ion M_1^+ and collision gas molecule M_2:

$$\varepsilon_{int,max} = [m_2/(m_1 + m_2)].T_1^\circ \qquad [6.42]$$

This upper limit arises quite generally from the need to conserve both energy and linear momentum (see the accompanying text box); $\varepsilon_{int,max}$ is often referred to as the center of mass collision energy, since it arises quite naturally if the problem is expressed in terms of a coordinate system that moves with the centre of mass of the M_1^+–M_2 system, rather than relative to a coordinate system fixed relative to the laboratory. However, it is not necessary for present purposes to further investigate this aspect (Boyd 1984), or any other of the consequences of the problem summarized in the accompanying text box. However, it is important to note that the simple treatment leading to Equation [6.42] contains zero information about the probability of observing a collision in which $\varepsilon_{int} = \varepsilon_{int,max}$ or indeed any other value. To calculate the relative probabilities for different values of ε_{int} would require a full treatment of the motion based on the laws of motion and a known or assumed potential energy function for the interaction of the M_1^+–M_2 collision partners. Nonetheless, the dependence of the upper limit for ε_{int} (Equation [6.42]) on the mass of the collision gas molecule M_2 is reflected in practical implementations of CA. It is well known that CID in ion traps, based on collisions with the helium damping gas ($m_2 = 4\,\mathrm{Da}$), can access only the fragmentation reactions with the lowest values of ε_{int}^*, frequently the less specific reactions such as loss of' a neutral molecule of water. In contrast, CID in linear quadrupole collision cells using argon ($m_2 = 40\,\mathrm{Da}$) can access many more reactions to give a richer fragment ion spectrum, consistent with a $\varepsilon_{int,max}$ value that is almost ten times higher. (Note that strictly the quantity ε_{int} that appears in Equation [6.42] for k_{uni} is the sum of the value

acquired in collisional activation plus ε_{int}°, the initial (pre-collision) value arising from the ionization event itself.) The mandatory use of a low mass collision gas such as helium as the damping gas in a Paul trap is a consequence of the low scattering angles (θ_1 in the accompanying text box) for projectile ions M_1^+ that are generally much heavier than helium atoms. This choice of collision gas permits the ions' trajectories to be largely controlled by the applied electric fields with only minor perturbations from collisional scattering (see Section 6.4.5).

In fundamental studies of the CA phenomenon it is important to distinguish between 'single collision' and 'multiple collision' conditions, since in the latter case any conclusions drawn about the CA mechanism will be blurred by the effect of superposition of several different collisions on the observed outcome. However, this aspect is of little consequence for present purposes, far less important than optimization of both the yield and specificity of the observed fragmentations. Of course CA in a Paul trap is inherently a multiple collision phenomenon that involves climbing the internal energy ladder (McLuckey 1997). As discussed in Section 6.4.2a, use of relatively high pressures of heavier collision gas in a linear RF-only quadrupole collision cell can yield unexpected improvements in overall performance.

The most important distinction within collisional activation methods is between so-called 'low energy CA' (up to $\sim$100 eV collision energy measured relative to the laboratory) and 'high energy CA' (laboratory collision energies in the keV range). The former applies to trap instruments and to linear quadrupole devices, and the latter to magnetic sector and TOF analyzers (including TOF/TOF tandem instruments). Low energy CA in linear quadrupole cells involves laboratory-frame collision energies of a few tens of eV with argon, nitrogen etc. as collision gas. The collision energy regime in a Paul trap is often referred to as very low energy CA, a consequence of the low laboratory-frame ion energies (a few eV) and even lower centre of mass collision energies with helium (see text box). As a result of the combination of appreciably higher collision energies (and thus higher $\varepsilon_{int,max}$ values) and shorter times available for fragmentation reactions to occur, the high energy CID spectra tend to contain more product ions of the simpler bond cleavage type relative to those requiring rearrangement of the ion structure, and vice versa for the high energy CID instruments. This trend (not a black-and-white distinction) is consistent with the preceding discussion of Figure 6.35.

All CA methods share the disadvantage that they lead to purely statistical dissociation, i.e., they can not be 'tuned' to yield only a few specific fragment ions that meet the ideal criteria for use in quantitation by MS/MS. The same is true of infrared multi-photon dissociation (IRMPD) conducted in trap instruments with low power continuous wave infrared lasers, another example (McLuckey 1997) of activation by 'slow heating' (usually for irradiation times 10–100 ms) that also accesses the fragmentation reactions with lowest $\varepsilon_{int}{}^*$ values. A significant advantage of IRMPD over CID in a Paul trap is that no resonant excitation (see Section 6.4.5a) of precursor ions is required, i.e., the activation procedure is much simpler and subject to fewer complications. In fact, CID in a Paul trap usually requires values of q_z for the precursor ion that have negative consequences for the low mass cutoff for storage and detection of the fragment ions. In contrast, photodissociation can proceed efficiently at low q_z values, thus allowing storage of a wide *m/z* range with less discrimination of the fragment ions. However, the cost of IRMPD is a significant factor and it does not appear to have been exploited in trace quantitative analysis as its advantages over CID appear to be greatest for Paul traps that have other disadvantages in the context of quantitative analysis.

Another ion activation method that has received considerable attention is surface induced dissociation (SID), in which a solid surface is used as the collision partner M_2 instead of a gas molecule. A principal motivation for development of SID (Cooks 1990; Grill 2001) was the effectively infinite value of m_2 in Equation [6.42], so that the theoretical value of $\varepsilon_{int,max}$ is now the kinetic energy of the ion relative to the laboratory ($T_1{}^\circ$). It turns out that at least some of the ion collisions involve mainly adsorbed molecules on the surface, i.e., although the effective value of m_2 is larger than that of the commonly used gas target molecules, it is not always as large as originally hoped. However, it is also found that the range of values of ε_{int} resulting from ion–surface collisions is appreciably less than that resulting from gas molecule CA, thus increasing somewhat the selectivity of the activation process although the subsequent dissociation again appears to be entirely statistical (i.e., no possibility of 'tuning' the overall SID process to produce just a few highly specific fragment ions). No tandem instrument incorporating SID has been commercialized thus far, a consequence of several factors including the all-too-ready contamination of the surface that changes the balance of the competition among SID, neutralization, chemical sputtering from the surface, and ion–surface complex formation. The SID approach, together with other ion activation methods, has been extensively reviewed (Sleno 2004).

The one activation method that appears to provide some practical degree of non-statistical dissociation is so-called electron capture dissociation (ECD) (Zubarev 1998). As developed and investigated (Zubarev 2002), and reviewed (Zubarev 2003), the method involves capture

of low energy electrons in an ion trap by multiply-protonated proteins or peptides. This converts the even-electron precursor ions into odd-electron ions in which the newly formed localized radical site is highly activated by some 5–7 eV (the energy required to remove the negative electron from the multiply-protonated protein), and appears to initiate simple cleavage dissociation reactions at that site that are faster than the internal energy redistribution that would otherwise result in purely statistical dissociation as for CID, SID etc. ECD is performed most often using an FTICR instrument since the RF fields in a Paul trap tend to heat up the electrons to a point where they are not captured efficiently. The ECD method is a preferred technique for sequencing peptides and proteins, but its current restriction to FTICR instruments accounts in part for its absence from the literature on trace quantitative analysis of small molecules. In the case of singly-protonated proteins and peptides (such as those produced by MALDI), to which the ECD technique is not applicable, activation by UV lasers emitting in the 266–157 nm range also appears to yield non-statistical (bond specific) fragmentation (Cui 2005; Devakumar 2005; Kim 2005; Moon 2005; Oh 2004, 2005). However, again the current cost and difficulty of use of such apparatus precludes any consideration of its use for trace quantitative analyses.

In summary, although collisional activation using gas targets is far from the ideal solution to ion activation for MS/MS analyses, it is the method that currently offers the best compromise in terms of cost, ease of use and effectiveness for trace level quantitation of small molecules. For this reason it is the method that is almost universally used for this purpose. It is important to note that, even when using a specified instrument, the optimum conditions for CID can vary widely among different precursor ions and should be optimized for each case to whatever extent is possible given the limitations of the instrument used. Some manufactures of triple quadrupoles have provided the option of a rapid collision energy ramp over the course of a MS/MS scan so that the optimum collision energy for a particular precursor → fragment transition is always sampled for at least part of the time.

In practical terms in the context of quantitative analyses, three parameters affect the nature and efficiency of the CID process within a specified apparatus, namely: target gas type (usually one of helium, nitrogen, argon); target gas pressure (and thus average number of collisions); and initial ion kinetic energy. With respect to 3D Paul traps and the current version of 2D traps employing radial ejection, helium is the only realistic choice in order to achieve the necessary collisional cooling of the ions prior to mass selective instability scanning of the spectrum (see discussion of Figures 6.20–6.21). In tandem-in-space instruments employing a quadrupole collision cell, including the $q_0Q_1q_{2/T}Q_{3/T}$ instrument (Section 6.4.6), nitrogen or argon are best. With regard to collision gas pressure, the helium pressure in a 3D Paul trap can be varied over only a relatively small range ($\sim 10^{-3}$ torr) in order that the essential collisional cooling effect can be maintained without too great a loss of ions as a result of scattering at higher pressures. In contrast the gas pressure in a quadrupole collision cell can be varied over a wide range to provide the optimum conditions for the particular analysis at hand. Some instrument manufacturers specify 'collision gas thickness (CGT)' for a q_2 collision cell, rather than simple pressure of collision gas, in an attempt to take into account the fact that the number of collisions experienced by an ion during its flight through the cell is a function of both the pressure P_2 and of the length of the cell (l_2):

$$CGT \ (molecules.cm^{-2}) = (N_{gas}/V_g).l_2$$
$$= (P_2.N_A/RT).l_2$$
$$= 3.2 \times 10^{16}.P_2(torr).l_2(cm)$$

$$[6.43]$$

where (N_{gas}/V_g) is the number density (*ND*, see Equation [6.46]) of collision gas molecules so that] $(N_{gas}/V_g).l_2]$ is the number of gas molecules per unit area 'seen' by the ion as it 'looks' through the cell at the start of its journey; the number of collisions is then the product of the CGT and the collision cross-section (also an area) for that ion with the collision gas molecules. With regard to the ion kinetic energies used to effect collisional activation, a Paul trap can access a limited range of low values via the tickle voltage amplitude, compared with those accessible (up to ~ 100 eV) in a quadrupole collision cell by adjusting the potential offset between Q_1 and q_2. This difference is further accentuated when the center-of-mass collision energies, and thus values of $\varepsilon_{int,max}$, are considered in view of the order of magnitude difference between the molecular masses of helium and argon (see text box).

So-called 'in-source CID' occurs in the vicinity of the skimmer cone in API interfaces (e.g. Figure 6.40(b)) where the background pressure is comparable to that in a q collision cell (mtorr range). A potential is normally applied to the skimmer cone to permit gentle collisional activation that leads to stripping of clustered solvent molecules from the analyte ions (see Section 5.3.3a), but by increasing this potential (referred to as the 'cone potential' or 'skimmer potential') it is possible to increase the

collision energy sufficiently that covalent bonds in the analyte ion are broken, i.e. create CID. This in-source CID is difficult to control and reproduce but it does permit acquisition of MS^2 using a stand-alone quadrupole mass filter and of so-called quasi-MS^3 information using a tandem-in-space instrument, e.g. a QqQ or QqTOF design. Because of its somewhat irreproducible nature, this phenomenon is very seldom exploited in quantitative methods.

6.6 Vacuum Systems

The vacuum system is another component of an analytical apparatus that is less 'glamorous' than some others but is essential for mass analyzers and detectors to function. To separate a mixture of ions according to their *m/z* values they are subjected to electric fields of one kind or another, sometimes in combination with a magnetic field. The ions follow trajectories under the influence of these fields in a fashion arranged such that they are presented to the MS detector at different times that are a function of the *m/z* values. If time dependent applied fields are to be able to separate the ions by *m/z* to the best possible resolution that the design of the device can permit, each ion must be able to travel the total distance of its sometimes meandering trajectory (i.e., not necessarily simply the distance 'as the crow flies' between the point at which the ion enters the analyzing field and the detector) without interference from collisions with background gas molecules.

Suppose this total trajectory path length is L. Moreover, suppose that an ion is deflected from its desired trajectory by interaction with a molecule of background gas only if their centers approach one another to within a distance $\leq r_{coll}$; the quantity πr_{coll}^2 is called the collision cross-section for interaction between this particular kind of ion and background gas molecule. During its trajectory the ion will sweep out a volume $\pi r_{coll}^2.L$ that should remain free of gas molecules if the trajectories are to remain undeflected. The ideal gas laws give the number of moles of background gas in a volume V_g at pressure P as $n = PV_g/RT$, so the number N_{coll} of interfering molecules with which the ion will collide in its trajectory is:

$$N_{coll} = (N_A.P.\pi r_{coll}^2.L/RT) \qquad [6.44]$$

where N_A is the Avogadro Constant (Chapter 1). Strictly, the fact that the background gas molecules are moving around in constant thermal motion should be taken into account, but in all mass spectrometers the ions are moving so much faster than the average thermal speed that the molecules can be regarded as effectively motionless for the purposes of this approximate calculation. Evaluation of Equation [6.44] using SI units, with T = 298 K, r_{coll} in nm, L in m, and P in Pa, gives:

$$N_{coll} = 763.7r_{coll}^2.L.P = 10^5r_{coll}^2.L.P(torr)$$

since 1 torr = 133.3 Pa. Since r_{coll} = 1 nm and L = 1 m are reasonable orders of magnitude for these quantities (that of course vary with the nature of the ion and background gas molecules, and with the instrument design, respectively), a pressure of 10^{-6} torr maintained within the *m/z* analyzer would reduce the number of perturbing collisions to less than one (physically this means that such events become highly improbable). Such a vacuum is nowadays routine in a mass spectrometer and in what follows the technologies that permit maintenance and measurement of vacuum of this quality are discussed briefly.

6.6.1 Pumping Speed, Conductance and Gas Flow

Because analytes plus residual mobile phase (and often also collision gas in the case of tandem mass spectrometry) are being injected continuously into a mass spectrometer, the vacuum pumps must be capable of continuously evacuating significant input gas flows in order to maintain a sufficiently low pressure within the analyzer and detector. Before describing the types of pump commonly employed, it is necessary to discuss the concepts of gas flow and its measurement. In this regard two helpful websites are recommended for both introductory and more detailed discussions: www.newequation.net and www.vacuumlab.com.

In the context of gas flow there are three different flow regimes to consider, namely, viscous, molecular and intermediate. The transitions among these states depend primarily on the pressure, which in turn determines λ, the mean free path of the gas molecules, i.e., the average distance traveled by a molecule before it collides with another. The kinetic theory of gases (see any textbook of physical chemistry) gives:

$$\lambda = RT/[2^{\frac{1}{2}}.\pi.r_{coll}^2.N_A.P] \qquad [6.45]$$

where all symbols were defined for Equation [6.44], so that:

$$\lambda(m) = 9.26 \times 10^{-4}/[r_{coll}^2.P]$$
$$= 7.0 \times 10^{-6}/[r_{coll}^2.P(torr)]$$

Thus at a pressure of 10^{-6} torr and for r_{coll} = 1 nm, the mean free path λ ≈ 7 m, considerably larger than any reasonable dimension within an analytical mass spectrometer. Of course λ decreases in inverse proportion to P.

The value of λ is very small at atmospheric pressure ($\sim9 \times 10^{-9}$ m = 9 nm) so that the flow of the gas is limited by the 'sticky' interactions between the molecules, i.e., by its viscosity. At the low pressures characterized as 'high vacuum', i.e., $\sim10^{-6}$ torr, it was shown above that λ ≈ 7 m, and the flow of the gas is essentially that of each molecule moving independently and is thus called molecular flow. At the intermediate values that the pressure must go through when the system is being pumped down from atmospheric pressure to high vacuum, the gas flow is governed by a combination of viscosity and molecular phenomena, and is termed intermediate flow. In the viscous flow range the flow can be laminar (the flowing gas layers are parallel, their velocity increasing from ~zero at the walls to a maximum at the central axis of the flow) or turbulent (at high linear gas velocities, the layers are not parallel). As an approximate guide for air at room temperature, the condition for viscous flow is (D.P) > 5×10^{-1} cm.torr, and that for molecular flow is (D.P) < 5×10^{-3} cm.torr, where D is the diameter of the pipe or orifice (cm) and P the average pressure (torr).

These distinctions are important when designing a vacuum system to best match the expected gas flows. It is not appropriate to go into this in detail here, but it is appropriate to present some rudimentary definitions and relationships that should help with understanding how and why a mass spectrometer vacuum system behaves as it does. The first quantity to be considered is pumping speed, defined as the volume of gas per unit time (dV_g/dt) removed from an orifice (e.g., high pressure entrance to a pump, or an orifice in an enclosure such as an EI ion source or collision cell that is at a higher internal pressure than its surroundings). In this definition the volume is measured at the high pressure side (e.g., at the pump inlet or at the interior of the ion source). In molecular terms pumping speed can be related to Q_N, the number of molecules N passing per unit time through the cross-section of a pipe, orifice in an enclosure etc. Consider two orifices 1 and 2 in series (or entrance and exit ends of a pipe) of cross-section areas A_1 and A_2, and such that all the gas molecules leaving 1 also enter 2; the numbers of molecules N crossing these orifices per unit time are given by:

$$Q_{N1} = ND_1.(u_1.A_1) = ND_1.S_1 \text{ and}$$
$$Q_{N2} = ND_2.(u_2.A_2) = ND_2.S_2 \qquad [6.46]$$

where *ND* = number density = number of molecules per unit volume, Q_N is the number of molecules per unit time and S is the pumping speed (volume per unit time);

u denotes the average speed of a molecule given as $\sim(RT/M_g)^{\frac{1}{2}}$ for an ideal gas, so that $(u.A)$ is the volume of gas that is transported through the orifice with cross-section area A per unit time, i.e., the pumping speed S. Then, by combining the two forms of Equation [6.46], assuming $u_1 = u_2$ (constant T) and as a result of the restriction that all molecules leaving 1 will enter 2, i.e., $Q_{N1} = Q_{N2} = Q_N$ (no leaks in the system), simple algebraic rearrangement gives:

$$Q_N = (ND_1 - ND_2)/(S_1^{-1} - S_2^{-1}) \qquad [6.47]$$

In contrast with these molecular considerations, from the purely phenomenological point of view the gas flow from orifice 1 to orifice 2 occurs because the pressure $P_1 > P_2$, so it is convenient to define the conductance C_{gas} of the pipe (www.newequation.net) as:

$$Q_{gas} = C_{gas}.(P_1 - P_2) \qquad [6.48]$$

where the gas throughput Q_{gas} is measured in units of (pressure.volume.time^{-1}) so that C_{gas} has units (volume.time^{-1}).

The molecular and phenomenological descriptions must be equivalent. Q_N (number of molecules.time^{-1}) and Q_{gas} (pressure.volume.time^{-1}) are easily related via the ideal gas law in the form:

$$P.V_g = N.(RT/N_A)$$

where R is the universal gas constant and N_A is the Avogadro Constant, so that:

$$Q_{gas} = Q_N.(RT/N_A) \qquad [6.49]$$

Similarly $P_1 = (N_1/V_g).(RT/N_A) = ND_1.(RT/N_A)$, and similarly for P_2. Then, comparison of Equations [6.7] and [6.49] provides the relationship:

$$C_{gas} = (S_1^{-1} - S_2^{-1})^{-1} \text{ or } C_{gas}^{-1} = (S_1^{-1} - S_2^{-1}) \quad [6.50]$$

To summarize, consider the gas flow Q_{gas} (sometimes referred to as the mass flow or throughput) that enters a pipe or vacuum enclosure of conductance C_{gas} (volume.time^{-1}) at pressure P_1. If no additional gas leaks into or out of the pipe, this same quantity of gas Q_{gas} exits the pipe at pressure P_2; thus, provided the system is isothermal, Q_{gas} is the same all through the system. The pumping speed at any point of such a vacuum system is $S = Q_{gas}/P$ where S and P are defined at the point in question. The pumping speed at any point in the system can be obtained using Equation [6.50] from the known pumping speed at some other point and the total conductance of the components of the system (pipes, orifices, valves etc.) between the two points.

As a meaningful example, consider a vacuum pump with a pumping speed S_{pump} measured at its inlet (Figure 6.36). Then the pumping speed S_{MS}, that can be obtained at a mass spectrometer vacuum housing that is

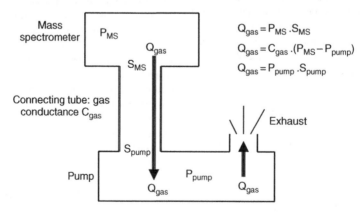

Figure 6.36 Illustration of the effect of gas conductance (C_{gas}) of a connecting tube between pump and mass spectrometer on the pumping speed available at the mass spectrometer. If there are no leaks in the system the mass flow of gas (Q_{gas}) is the same throughout. By algebraic manipulation of the three expressions for Q_{gas} the following relationship is obtained: $S_{MS}/S_{pump} = (C_{gas}/S_{pump})/[1 + (C_{gas}/S_{pump})]$, i.e., the pumping speed delivered to the mass spectrometer is only a fraction of that provided by the pump if C_{gas} is too small (see main text).

connected to the pump via a pipe plus valves with total conductance C_{gas}, is given by rearranging Equation [6.50]:

$$1/S_{MS} = 1/S_{pump} + 1/C_{gas}, \text{ i.e.,}$$

$$S_{MS}/S_{pump} = (C_{gas}/S_{pump})/[1 + (C_{gas}/S_{pump})]$$

The fractional decrease of pumping speed S_{MS}/S_{pump} is a function of the ratio C_{gas}/S_{pump}. When the value of the conductance is equal to that of the pumping speed of the pump, only 50 % of the pumping speed is available at the vacuum chamber; in order to use 80 % of the pumping speed to evacuate the mass spectrometer the ratio C_{gas}/S_{pump} must be 4. Clearly there is no point in increasing the speed of the pump (size and cost) if the conductance of the connecting pipe and valves is inappropriate. In the design of any vacuum system for a mass spectrometer a balance is struck between the sizes of the conductance elements (largely dictated by the size of the mass spectrometer analyzer components which differ in size for different analyzers) and the speed of the pumps; this balance aims to minimize costs and also the footprint of the instrument.

6.6.2 *Vacuum Pumps*

A large number of vacuum pumps is available, each one meeting some specific requirement. They can be categorized into three main types. Positive displacement pumps operate in a cyclic fashion to expand a cavity, allow gas to flow in from the vacuum chamber to be evacuated, seal off the cavity, compress the trapped gas and push it out to the atmosphere, then repeat the cycle; the example considered here is the commonly used rotary vane pump. Such pumps are effective for low vacuum (down to $\sim 10^{-3}$ torr) but are subject to high backstream flows through mechanical seals and this limits their usefulness in producing a high vacuum. In contrast, momentum transfer pumps arrange for the gas molecules that are to be evacuated to receive large momentum transfers in an appropriate direction that will cause them to move to the high pressure exhaust side of the pump. The momentum transfers are derived from high speed jets of vapor of large (heavy) molecules as in a diffusion pump, or from high speed rotating blades as in a turbomolecular pump. The final category can be labeled entrapment pumps since, instead of expelling gases from the vacuum chamber to the ambient atmosphere, they immobilize the gases on a surface within the chamber; examples are cryopumps, which freeze down the gases as solids, and getter pumps, which provide a highly reactive surface to trap the gas molecules as nonvolatile chemical compounds.

Here the three pump designs that currently dominate applications to mass spectrometry are considered; such pumping systems must handle significant gas loads while producing pressures as low as 10^{-7}–10^{-8} torr (1 torr = 1.333 mbar = 133.3 Pa).

6.6.2a *Rotary Vane Pumps*

These pumps are ubiquitous in mass spectrometry vacuum systems; the mode of operation is illustrated in Figure 6.37. It comprises a cylindrical cavity with entrance and exit ports used to draw in the gas to be

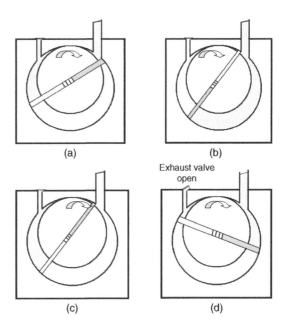

Figure 6.37 Diagram indicating the operation of a rotary vane pump. The electrically-driven rotor (rotates clockwise in this diagram) is mounted eccentrically within the main pump cylindrical cavity (shown here in cross-section) and carries two spring-loaded vanes (blades, white and darker grey) that are forced to fit tightly against the inner cavity wall by the action of the spring. In (a) the pump cycle has just begun and the grey vane starts to increase the volume into which gas from the vacuum chamber can expand. In (b) the rotor has moved by almost 180° and the gas from the chamber is about to be trapped between the two vanes. In (c) the rotor has moved enough that the white vane can now compress the trapped gas. In (d) this process is almost complete, the white vane has compressed the gas sufficiently that the pressure causes the exhaust valve to open and the grey vane has started to pull in the next volume of gas from the vacuum chamber. The pump oil that both seals and lubricates the pump is not shown for clarity, but must be filled to a level to cover the exhaust valve.

pumped and subsequently exhaust it to atmosphere. An off-center rotor is driven electrically inside the cylinder (Figure 6.37). One end of each spring-loaded vane (blade) is fitted into a slot on the central rotor and the other end slides along the cylinder wall, thus dividing the inner space into two sickle-shaped working chambers. A special lubricating oil (with low vapor pressure!) is used to seal the operating part of the pump from the atmosphere and to minimize frictional wear between the vanes and the cylinder wall. Regular attention to changing this pump oil is one of the low grade but essential tasks involved in maintaining an effective vacuum system for a mass spectrometer.

During one full rotation of the rotor each chamber volume increases from zero (Figure 6.37(a)) to the maximum volume (Figure 6.37(b)) and then decreases until it reaches the minimum value (Figures 6.37(c) and (d)). The compression of the gas in this latter half of the cycle is sufficient to open the exhaust valve so that the gas can be expelled to atmosphere. Older models drive the rotor via a drive shaft between the electric motor and the rotor itself, and this shaft must be introduced into the pump via a rotor seal. Modern designs eliminate this potentially troublesome feature by using magnetic coupling between motor and rotor through the external wall of the pump that can thus be hermetically sealed. A high vacuum safety valve is usually inserted between the pump and the vacuum chamber in case of uncontrolled shut-down (power failure etc.). The safety valve prevents the rise of the pump oil into the vacuum system (mass spectrometer) under the action of atmospheric pressure that can access the pump when the rotor stops.

A rotary vane pump can produce a vacuum of about 10^{-2}-10^{-3} torr at the pump input, with pumping speeds that are readily increased by increasing the size of the pump cavity. Apart from the limited ultimate vacuum, their main drawback involves the trapping in the pump oil of corrosive compounds that have been pumped from the vacuum chamber; particularly in the case of LC/MS the residual mobile phase reaching the vacuum pumps can be a major problem in this regard requiring regular changes of pump oil. The problem can be alleviated by a combination of special pump oils that can withstand operation at higher temperatures, thus increasing the vapor pressure of the contaminants making it easier to expel them, plus use of a gas ballast. The latter is simply a small valve that admits a small quantity of ambient air to the pump cavity in the compression phase (e.g., between vane B and the pump outlet in step (c) in Figure 6.37); this has a 'carrier' effect and improves the probability that the corrosive vapors will not condense but be expelled instead. This positive effect comes at the expense of a slight reduction in the ultimate vacuum and also with a risk of producing a mist of fine oil droplets (that can be caught in an oil trap). If the pump oil does become contaminated it is common practice to close off the rotary pump inlet from the vacuum chamber and open the gas ballast valve wide to allow the pump to flush out the contaminants; of course it is wise from a safety point of view to vent the exhaust gases safely, e.g., into a fume hood or suitable trap.

6.6.2b Diffusion Pumps

To achieve the desirable vacuum $\leq 10^{-6}$ torr in the vacuum chamber the rotary vane pump must be used in conjunction with a different type; the oldest type of high vacuum pump is the so-called diffusion pump (Figure 6.38(a)). Such a pump starts pumping only at pressures around 10^{-3} torr, which is fortunately the pressure range that can be maintained by a rotary vane pump used as forepump. Diffusion pumps are so-called because part of their mode of action involves diffusion of some of the air or other gas from the vacuum chamber into the jets of oil vapor. Since these pumps involve no moving parts they are highly reliable and can deliver a vacuum down to 10^{-8} torr. Their disadvantages include a significant warm-up time on starting from cold and susceptibility to potentially disastrous backflow of air through the pump from the low vacuum side, e.g., after failure of the backing (rotary vane) pump or from a power failure; such 'back-streaming' can lead to a massive oil contamination of the high vacuum side of the equipment (the mass spectrometer in our case). A passive oil-trap buffer or shield is often included on the high vacuum side to help reduce the effects of such accidents, but at the cost of reduced pumping speed. Such events can also lead to significant degradation of the hot oil, although modern pump oils are more resistant to oxidation.

6.6.2c Turbomolecular Pumps

A more modern alternative to the diffusion pump is provided by the turbomolecular pump (Figure 6.38(b)). This pump was patented by the Pfeiffer company in 1957; a full account of the principles and use of these pumps is freely available at: http://www.pfeiffer-vacuum.com/ent/en/652. The general principle on which such pumps are based is that of vectorial addition of velocities. When a molecule strikes a metal surface at rest it will bounce off at an angle to the surface that is (ideally) equal and opposite to its angle of incidence. However, if the surface is moving fast the angle at which the molecule bounces off will change as a result of vector addition of

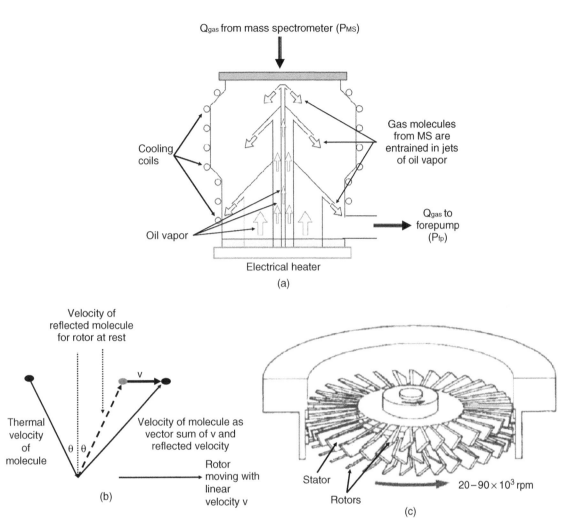

Figure 6.38 (a) Schematic diagram showing operation of a diffusion pump. A heater evaporates the special low vapor pressure diffusion pump oil in the boiler. The oil vapor rises through each of the concentric vertical tubes and then shoots at high speed (up to > 700 miles per hour!) out of a system of concentric ring-shaped nozzles (one such system is connected to each of the vertical tubes carrying the oil vapor) in the shape of a cone. The oil then condenses against the water-cooled wall of the pump and then returns into the boiler as a thin film along the wall. The air or other gas from the vacuum chamber is entrained and moves along with the high speed oil vapor in the three stages of cone-shaped jets, exits the pump on the low vacuum side ($\sim 10^{-3}$ torr), and is pumped away by the forepump (rotary vane pump). The cooled baffle condenses any oil that escaped from the body of the pump. Modified from literature of Varian Vacuum Inc., with permission. (b): principle of operation of a turbomolecular pump; a molecule approaches a fast rotor and instead of reflecting (bouncing) from the surface at the same angle θ to the vertical as the incident trajectory, its final velocity is the vector sum of that reflected velocity and the rotor velocity, so the molecule is directed towards an orifice in the stator and enters the next stage (this process is continued through the series of rotor–stator stages until the molecules are delivered to the forepump and removed). (c): sketch of a single rotor–stator stage for a turbomolecular pump.

the velocities of the original molecule and the surface, and the final magnitude of the molecular velocity (its speed) will be correspondingly increased. If now a second fast moving metal surface is arranged relative to the first

such that it accepts only molecules moving in a specified direction or small range of directions, the random thermal motion of the molecule before it struck the first surface is converted into a directional component (with increased magnitude, i.e., speed), so that a pumping process has been achieved. By arranging a cascade of such stages a highly effective high vacuum pump can be constructed. However, this effect can only operate if there are no collisions of the 'pumped' molecules with other molecules during their passage between the stages, so such a pump only works at pressures low enough that the gas flow is in the molecular region (see above), i.e., the mean free path is larger than the separation between the pumping stages (high speed rotors). Therefore, as for diffusion pumps, turbomolecular pumps require a forepump to provide a backing pressure of no more than 10^{-2}–10^{-3} torr, conveniently provided by a rotary vane pump.

In practice a turbomolecular pump consists of a vacuum-tight casing containing a series of high speed rotors (20 000–90 000 rpm) each moving relative to its own stator disk. As gas molecules enter from the vacuum chamber through the inlet the angled blades on the rotor impact on the molecules and their newly acquired momentum allows the molecules to enter the gas transfer holes in the stator and proceed to the next stage. This process is continued through the series of rotor–stator stages until they are delivered to the forepump and removed. To collect as much as possible of the highly dilute gas at the pump entrance (the vacuum chamber itself) the first rotors have a larger radius, and consequently must withstand a higher centifugal force requiring fabrication from stronger materials. Turbomolecular pumps are very versatile and can be configured to generate degrees of vacuum from intermediate ($\sim 10^{-4}$ torr) to high vacuum ($\sim 10^{-6} - 10^{-7}$) up to ultrahigh vacuum ($\sim 10^{-10}$ torr). Unfortunately, the compression ratio (ratio of outlet to inlet pressures) is found to vary exponentially with the square root of the molecular mass of the gas, so heavy molecules are pumped much more efficiently than light ones, and it is difficult to pump hydrogen and helium efficiently. Qualitatively these differences in performance can be understood by considering the average gas speeds of gas molecules given by the kinetic theory of gases as:

$$u = (3RT/M)^{\frac{1}{2}} \qquad [6.51]$$

where R is the gas constant and M the molar mass of the gas. The values of u at 25 °C for hydrogen (1700 m.s^{-1}) and nitrogen (400 m.s^{-1}) should be compared with the tangential velocity of a typical rotor, e.g., for a 15 cm diameter rotor rotating at 36 000 rpm the outer tangential velocity is $\sim$280 m.s^{-1}. Thus the fraction of the lighter gas that will 'backstream' through the rotors (without blades striking it) into the vacuum chamber will be higher because of the faster speed. By using a smaller turbo pump

backed by a rotary pump as the forepump combination for the primary turbomolecular pump, good overall compression ratios for hydrogen and helium can be obtained (at higher cost!).

Another drawback of turbo pumps is the requirement for very high-grade bearings for the high speed rotors, which increases the cost. Lightweight ceramic ball bearings with lubrication by low vapor pressure grease is common. More recent introduction of magnetically levitated bearings, either permanent magnets for smaller pumps or a combination of permanent and dynamic magnetic fields, permits supporting the rotor shaft without contact, resulting in a turbomolecular pump that has truly zero oil vapor backstreaming with bearings that in principle never wear down. Although modern designs are now quite robust, when they do fail (breakage of bearings or rotors) they can do so spectacularly and replacement can be costly. However, the fact that they do not require pump oil of any kind is a great advantage with respect to risk of contaminating the vacuum chamber and its contents. In addition, turbomolecular pumps reach full operating speed within a few minutes, much faster than warm-up for diffusion pumps, and are nowadays more commonly used in mass spectrometer vacuum systems than diffusion pumps. The pressure ranges at which some common pumps are effective are shown in Figure 6.39.

6.6.2d Differential Pumping

An important concept for vacuum systems in mass spectrometers is that of differential pumping (Figure 6.40).

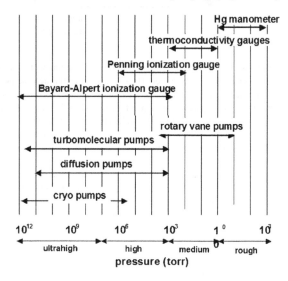

Figure 6.39 Operating pressure ranges for some commonly used vacuum pumps and gauges.

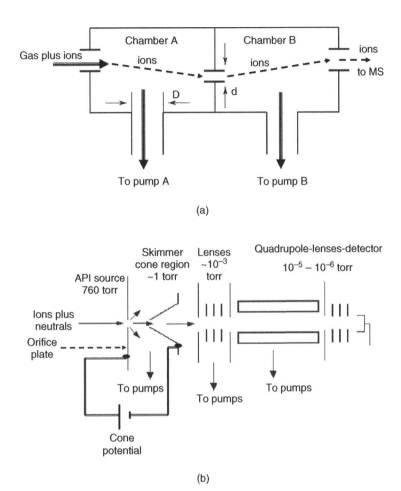

(a)

(b)

Figure 6.40 (a) Schematic illustration of the principle of differential pumping. The pressure at the inlet is typically too high for vacuum pump B to operate efficiently to maintain a high vacuum in chamber B, so an intermediate pump A is used to decrease the pressure part-way. The orifice connecting chambers A and B is much smaller than those connecting the chambers to their respective pumps, and indeed need only be large enough to permit efficient transmission of ions moving under the influence of electric fields. (b) Sketch of an API ion source coupled to a quadrupole mass spectrometer operating at $\sim 10^{-6}$ torr (a pressure ratio of 10^8–10^9) via three stages of differential pumping. Note that neutral molecules and ions are both subjected to the pumping action, but the ions are controlled by the much stronger forces from the applied electric fields and continue their trajectories through the instrument while the neutrals are pumped away. The skimmer cone selectively transmits ions and the cone potential (potential difference between orifice plate and skimmer cone) can be adjusted to strip solvent molecules from ions or to larger values to effect in-source CID.

This term refers simply to situations in which different regions within the vacuum chamber are to be operated at different pressures while still permitting free passage of ions from one region to the other. The most common examples involve ion sources that operate at appreciably higher pressures (up to atmospheric) that must be coupled with analyzers operating at high vacuum ($\leq 10^{-5}$ torr).

This coupling is achieved via a series of differentially pumped stages (the number will depend on circumstances) in which the pressure is progressively lower. Differential pumping is achieved by a combination of restricting the apertures connecting the different regions to sizes that efficiently trasmit the ions (controlled by electric fields) but minimize the gas conductance C_{gas}, with provision of effective pumping speeds on either side of the connecting aperture that can maintain the desired pressures in the two regions.

6.6.3 Vacuum Gauges

Measurements of pressure within the vacuum systems of mass spectrometers used in organic trace analysis are not usually required to be of high accuracy. Rather they are used to monitor the pressure in the various regions, particularly the high vacuum region containing the analyzer and detector, to ensure that the mean free path is sufficiently large. In addition, it is important to be able to monitor the pressure at the inlet of the forepump to ensure that this is sufficiently low to allow the high vacuum pump (diffusion or turbomolecular) to operate. Sometimes it is also desirable to monitor pressure in intermediate differentially pumped regions. Therefore, the requirement for vacuum gauges in a mass spectrometer system is to provide reasonably accurate indications of pressure in a cost-effective manner for a range of pressures covering the range from atmospheric ($760\,\mathrm{torr} = 1.013 \times 10^5\,\mathrm{Pa} = 1.013\,\mathrm{bar}$) to $< 10^{-6}$ torr. No single type of gauge can cover this range (see Figure 6.39) and in the following the types of gauge most commonly used in mass spectrometers are briefly described.

6.6.3a Capacitance Manometer

None of the vacuum gauges commonly used in mass spectrometers measure pressure directly, but rather they measure some physical property that varies with pressure and, unfortunately, also with the chemical nature of the gas. For this reason gauges are usually delivered with a calibration for air, but if some other gas is present the indicated pressure values will be systematically incorrect. On the few occasions where direct pressure measurements are required with good accuracy, a capacitance manometer is used. This device operates by measuring the change in electrical capacitance that results from the movement of a sensing diaphragm relative to fixed capacitance electrodes that are located in a separate chamber maintained at some reference pressure (often a high vacuum). The lower the unknown pressure, the farther it will pull the diaphragm away from the fixed capacitance plates. In some versions the diaphragm is allowed to move and the resulting change in capacitance is recorded, and in others a measured voltage is applied to keep the capacitance constant. Accuracy is typically 0.25–0.5 %; thin diaphragms can measure down to 10^{-5} torr, while thicker diaphragms can measure in the low vacuum to atmospheric pressure range. They

can be constructed from corrosion-resistant materials and are widely used in circumstances where this is important, e.g., the semiconductor industry. However, for purposes of semi-quantitative monitoring required in mass spectrometers, less accurate but cheaper types of gauge are used.

6.6.3b Pirani Gauge

For the pressure range appropriate to monitoring the forepump vacuum (atmospheric to $\sim 10^{-3}$ torr during pumpdown), the most common type is the Pirani gauge. This is an example in which the pressure of a gas is measured indirectly via its thermal conductivity. In the pressure range in which the mean free path is larger than the dimensions of the container, the thermal conductivity of a gas varies with its pressure. If an electrically conducting element (e.g., a metal wire) is placed in a gas, the surface temperature of the element will represent a compromise between the electrical heating and the cooling effect of the gas transporting heat to the container walls (higher pressures imply more molecules to conduct the heat). Hot wire gauges of suitable physical dimensions can be used down to $\sim 10^{-3}$ torr. In Pirani gauges the pressure of the gas is determined by measuring the current needed to maintain the wire at a constant temperature (sensed via the electrical resistance of the wire). A simple Pirani gauge can not detect pressure changes above about one torr because above these pressures the thermal conductivity of a gas is not determined by molecular properties since the mean free path is too small, and bulk properties of the gas dominate the cooling effect (by convection). Simple Pirani gauges are linear in the range $10^{-2} - 10^{-4}$ torr and are inexpensive, convenient and sufficiently accurate for many purposes. Their main limitation is their lack of sensitivity to pressure changes much above 10^{-2} torr (response is roughly logarithmic).

The convection enhanced Pirani gauge extends the usable range of the same principle by exploiting the convection phenomenon that limits the simple Pirani design. At pressures $< \sim 10^{-2}$ torr the convection enhanced design responds to pressure exactly like the original version, but at pressures greater than one torr and up to atmospheric pressure the response depends on convective cooling by the bulk gas. The convection enhancement is achieved by using a larger internal volume for the gauge; within the pressure range where the convective heat loss mechanism is significant, such gauges must be oriented according to the manufacturer's instructions (horizontal or vertical) since convection is controlled by gravitational forces acting on heated gas.

6.6.3c Thermocouple Gauge

The thermocouple gauge is another common type in which the pressure of a gas is measured indirectly via its thermal conductivity, but in this case the temperature of the wire is measured directly by a thermocouple spot-welded onto the wire rather than via the electrical resistance. The thermocouple generates a DC voltage signal of up to $\sim$20 mV DC for its usable pressure range of $\sim 10^{-3} - 1$ torr. This range can be increased by use of an electronic controller with a digital/analog converter and digital processing that can calibrate the highly nonlinear response at higher pressures, thus extending the range up to atmospheric pressure. This design competes with the convection enhanced Pirani gauge but is cheaper; however, the response time of the thermocouple gauge ($\sim$5 seconds) is appreciably longer that that of the Pirani gauge ($\sim$50–800 ms).

6.6.3d Ionization Gauge

Pressure measurement in the high vacuum regions of a mass spectrometer most often uses an ionization gauge that measures pressure via the current carried by ions formed in the gas by the impact of electrons. Two types are available; hot cathode ionization gauges produce the ionizing electrons by thermal emission from a heated filament, while in the cold cathode type the electrons are extracted from a metal electrode by a high electrical potential. In both cases the signal generated is a function of not only the pressure of the gas but also of its chemical nature via the ionization efficiency (roughly inversely related to the ionization energy, see Table 6.2). In the hot cathode ionization gauge, also known as a Bayer-Alpert gauge, the electrons emerging from the hot filament (cathode) are accelerated towards a grid held at $\sim$150 V positive

Table 6.2 Approximate ion gauge (Bayer–Alpert gauge) sensitivities (S_{IG}) relative to nitrogen, and ionization energies (IE), for some representative gases

Compound	IE (eV)	S_{IG}
Acetone	9.7	3.6
Air	–	1.0
Argon	15.8	1.2
Carbon dioxide	13.8	1.4
Helium	24.6	0.14
Hydrogen	15.4	0.44
Methanol	10.8	1.8
Nitrogen	15.6	1.0
Oxygen	12.1	1.0
SF_6	15.3	2.3
Water	12.6	1.0

with respect to the cathode; in their flight towards the grid they collide with some of the gas molecules to form positive ions (cations) that are accelerated away from the grid towards the collector (anode) held at a negative relative potential ~ -30 V. The electrons can not escape from the positive potential of the grid and do not reach the anode, so the anode current measures the number of positive ions produced from the gas. Most hot cathode ion gauges are usable in the range $\sim 10^{-2} - 10^{-9}$ torr and are calibrated for air. Some designs have the electrode assembly mounted within its own envelope made of glass or metal, with a flange that allows connection to a mating flange on the mass spectrometer vacuum chamber. The so-called 'nude' designs do not have an independent envelope; instead the electrode assembly is mounted directly on its flange and is inserted directly into the chamber vacuum.

A common feature of all types of cold cathode ionization gauges is that they contain just two unheated electrodes (cathode and anode) between which the discharge of electrons is induced by a DC voltage of about 2 kV and maintained so that the discharge continues at very low pressures. This is accomplished by use of a magnetic field to make the paths of the electrons so long that their collision probability with gas molecules is sufficiently large to maintain the discharge, i.e., create the required number of charge carriers. (A common design of gauges of this kind is the Penning gauge.) The magnetic field is arranged so that the magnetic field lines of force cross the electric field lines, thus confining the electrons to spiral paths superimposed on oscillations between the two cathode plates (top and bottom) held at a negative potential (to repel the electrons). The positive ions are collected by the cylindrical anode and form the pressure dependent discharge current indicated on the meter as a measure of the pressure of the gas (and also of its chemical nature). The upper pressure limit of $\sim 10^{-2}$ torr is due to the change of the cold cathode discharge to a glow discharge with intense light output in which the current depends to only a small extent on the pressure.

A disadvantage of the cold cathode type is a consequence of sputtering of material from the cathode surface by the ions produced in the discharge that were accelerated by the high potential drop within the device. The sputtered cathode material condenses on the tube walls, forming a highly reactive film that can react with components of the gas whose pressure it is desired to measure, thus potentially leading to highly inaccurate results (up to $\pm$ 50%). However, this is not necessarily unacceptable for a mass spectrometer where the purpose is monitoring to ensure that the mean free path is sufficiently long.

The cold cathode ionization gauge has offsetting advantages relative to the hot cathode type. It is insensitive to mechanical vibrations, is not susceptible to an inrush of air in the event of a vacuum accident (that can destroy the healed filament in the hot cathode type) and is less susceptible to harm from contaminants.

No book that claims to provide practical guidance to anyone using a mass spectrometer can avoid the necessity of dealing with the question of leaks in the vacuum system. Some information on leak detection is given in Appendix 6.2

6.7 Summary of Key Concepts

All mass spectrometers can perform quantitative analyses that are fit for purpose in some applications. However, of the many types of mass spectrometers available, distinguished in large part by the type of **mass analyzer(s)** used to separate ions of different m/z, several have risen to the forefront in quantitative applications.

Key among these are **quadrupole based instruments**, single quadrupoles and triple quadrupoles as well as the closely related quadrupole ion trap instruments (3D and 2D ion traps). Their widespread use in trace quantitative analyses has been facilitated by the fact that these instruments can also be readily coupled to GC and/or LC because of inherent features of the mass analyzer fields, i.e. they operate at or near ground potential.

Of the quadrupole-based instruments available, including single quadrupole (Q), triple quadrupole (QqQ, sometimes $Q_1q_2Q_3$) and hybrid quadrupole time of flight (QqTOF), the QqQs are today's gold standard for rigorous, robust, high throughput quantitative analyses.

QqQs have attained their 'workhorse' status in quantitative analyses due to their unique combination of ruggedness and reliability, **data acquisition rate**, **duty cycle** (fraction of total operating time spent measuring the ions of interest in selected ion modes) and **tandem MS capabilities**, notably the ability to perform **multiple reaction monitoring (MRM)**, also called **selected reaction monitoring (SRM)**.

In SRM a structurally significant ion of the target analyte (**precursor ion**, usually one of the possible ionized forms of the intact molecule loosely referred to as 'molecular ions') are **mass (really m/z) selected** (first mass analysis step, MS^1, performed by Q_1 on a QqQ), then induced to dissociate (fragment) via **collision induced dissociation** (CID, also called collisionally activated dissociation, CAD) in a collision region of the mass spectrometer (q_2 in the case of a QqQ), and then a specific, structurally distinct fragment ion (also called a **product ion**) is mass selected in a second mass analysis

step (MS^2, performed by Q_3 on a QqQ) and detected (see Chapter 7).

Compared to non-tandem mass spectrometers (e.g. a single Q where only a single MS^1 mass analysis step is possible) the **MRM mode provides enhanced selectivity in quantitative analysis**, leading to increased confidence that the analyte of interest is being monitored. With a MS^1 scan there is an increased possibility that **isobaric interferences** (non-analyte related ions of the same m/z) will be detected in addition to (or instead of!) the target analyte ions, especially in assays with complex matrices. Because **MRM mode** involves two distinct mass analysis steps, this leads to a selective decrease (filtering) of the chemical background, leading to **improved S/B** ratios compared to those possible with single Q instruments.

In addition, as a tandem mass spectrometer containing two coupled analyzers of the same type (both Qs), facile linked changing of applied analyzer voltages is possible providing **rapid switching between the precursor/product ion pairs** monitored at any one time in SRM mode. This performance feature is critical to quantitative analyses using **stable isotope labeled surrogate internal standards that co-elute** with the target analyte, i.e. the QqQ is able to rapidly switch between SRM pairs for analyte and SIS and still provide the requisite 10 data points across a chromatographic peak for both (i.e. 20 data points total for a single analyte and its SIS). It is also possible to apply MRM for multi-analyte assays, although the number of analytes that can be monitored while still providing 10 data points across the chromatographic peak is a function of the width of the latter. If this becomes problematic it is possible to program the QqQ to monitor different sets of MRM ion pairs as a function of retention time, so that the question of 'how many analytes can be monitored' involves not only dwell times but also chromatographic peak widths and retention times.

As with all Q-based instruments, it is also possible to increase the time that each of Q_1 and Q_3 is set to transmit the precursor/product ion pairs (so-called **dwell times**) while still maintaining data acquisition rates that can provide the requisite number of data points across the peak. Longer dwell times result in detection of larger numbers of ions (at trace analyte levels the detected ion beams are of very low intensity so the arrival of discrete ions at the detector is statistically distributed in time, see discussion of ion beam shot noise in Chapter 7), and thus enhanced sensitivity. In addition, once dwell times are set, **SRM dwell times for a QqQ are constant across the time frame of a chromatographic peak.** These constant dwell times are a distinguishing feature of QqQs (and Qs) relative to quadrupole ion traps (see below).

Single quadrupoles (Qs) are non-tandem mass spectrometers that provide a single stage of mass (m/z) analysis for quantitation, most often using **selected ion monitoring mode SIM** (sometimes called selected ion recording SIR) where only ions of a selected few m/z values are transmitted through the analyzer to the detector in a cyclic fashion as in MRM. Typically these ions are chosen to be structurally characteristic of the target analyte and internal standard, most often the 'molecular' ions. SIM can also be performed using a QqQ by operating one of the Qs in RF-only (ion funnel) mode. While not providing the same degree of selectivity as MRM since there is no filtering of chemical background, **SIM increases the duty cycle of a single Q relative to full scan mode** thus detecting many more of the ions of interest and increasing sensitivity. As for the QqQs, single Q dwell times can be increased to enhance sensitivity while still maintaining the requisite total number of data points across a chromatographic peak even while monitoring three ions per analyte when using relative abundance ratios for confirmation purposes (the 'three-ion rule'). This attribute, coupled with their **facile coupling to GC** and **their small footprint**, combine to make single quadrupoles the largest selling quantitative GC/MS instrument, with extensive applications in environmental analysis. In the latter context, single quadrupoles operated in full scan mode and with library search capabilities essentially provide the selectivity component of a US EPA environmental assay.

Double focusing magnetic sector mass spectrometers make use of electric and magnetic fields that allow the separation of ions with **resolving power (RP) of up to 100 000**, though transmission efficiency falls off rapidly for RP > 10 000. This can be compared to quadrupoles (Qs and QqQs) which operate at unit mass resolution, i.e. ions with m/z values that differ by no less than 1 Th can be resolved. Once dominant in quantitative MS analyses, sectors now play a niche, albeit important, role in analyses that require high RP in order to provide needed selectivity when low RP MS/MS can not provide enough reduction in chemical background; GC/MS analyses of 'dioxins' and coplanar PCBs are the main examples. The waning popularity of sectors is due not only to their large footprint and higher maintenance needs, but also to the **inherent difficulty of coupling atmospheric pressure ionization (API) sources** used for LC to high voltage (kV) ion sources required for sector fields.

3D quadrupole ion traps and 2D linear quadrupole ion traps are closely related in operating principles to quadrupoles and as such share many of the same performance features, such as unit mass resolution and near-ground potentials that allow facile coupling to both GC

and LC. However, unlike their cousin the quadrupole, **3D and 2D traps are intrinsically tandem mass spectrometers**, offering **MS^n analyses** (where n = 1–10) in a single analyzer device. This high qualitative informing power distinguishes ion traps from single quadrupoles and QqQs, which are capable only of MS/MS (MS^n with n = 2. Note, however, that QqQs can also perform precursor and neutral loss scans used in the rapid screening of complex samples for compounds of related structures; ion traps can not perform these qualitative scans.

As **trapping mass analysers**, ions are either formed within the volume of the trap (as in GC/MS) or injected into the trap (as in LC/MS) where they are **trapped and stored** for m/z analysis. The various stages of the MS^n analyses are performed by applying appropriate voltages to the three trap 'electrodes' as a function of time. Thus, ion traps are often called **tandem-in-time instruments**, vs QqQs which are called **tandem-in-space instruments**, i.e. the first mass analyses, dissociation and second mass analysis steps are performed in different spatial regions of the instrument.

Unlike QqQs, **quantitative analyses in ion traps (usually) make use of the product ion scan (also called full scan MS/MS)** mode of detection because of the shorter times required to perform full scan MS/MS relative to SRM. Full scan MS/MS on a trap requires that the trap first be filled with ions, followed by isolation (m/z selection) of the target analyte ions, followed by dissociation of these precursor ions (usually via resonant CID with the **helium buffer gas** in the trap), and finally scanning the product ions out of the trap to the detector. If MRM were to be performed, considerable additional time would be required to subsequently m/z select the selected product ion prior to final detection.

While the inherent trapping capability provides enhanced sensitivity, and the tandem-in-time mode of operation provides enhanced ion structure informing power (MS^n) relative to linear Q-based instruments, these same features also lead to **decreased quantitative performance of 3D traps compared to QqQs in MRM mode**. This is largely due to the fact that the **total scan times associated with full scan MS/MS on a trap are inherently longer than those required for MRM on a QqQ**. In addition, unlike a QqQ that provides constant scan times over the elution of the chromatographic peak, **scan times on an ion trap vary** due to the need to control the population of ions stored in the trap to avoid the decreased performance (resolution and sensitivity) that comes with space charge effects. These variations in scan times leads to poorer precision of chromatographic peak areas across replicate injections. Further, should **a stable isotope labeled SIS be used** in the analysis, the 3D ion trap will preferentially 'fill' with either SIS ions (at the low end of the calibration curve) or analyte ions (at the high end of the curve) leading to an **inaccurate sampling of the ion ratios,** which in turn leads to flattening of the curve at the low and high ends, resulting in a narrower linear dynamic range and higher values for detection limits (LODs) and lower limits of quantitation (LLOQs) relative to those achievable using a QqQ in MRM mode.

In recent years **newer generation 2D linear ion traps** have been introduced. As with a single quadrupole mass analyzer, 2D ion traps consist of four rods (of circular or truncated-hyperbolic cross-sections) with trapping potentials applied to either end. The resulting **linear trap has a higher internal volume,** and thus is able to trap and store more ions than a 3D trap before falling victim to space charge effects. In addition, the **2D traps have faster spectral acquisition rates than 3D traps,** allowing acquisition of a greater number of scans across the peak, albeit still of varying lengths. Detailed comparisons of the quantitative abilities of 2D traps relative to QqQs are only now underway, but these preliminary studies point to a potentially important niche for the new generation 2D traps in quantitative analyses.

Another relatively new instrument seeks to leverage the storing and structural informing power of the 2D ion traps with the performance features offered by the QqQ geometry. These instruments can be represented as $Q_1q_{2/T}Q_{3/T}$ and can be used with q_2 and/or Q_3 operating either each in its own conventional mode or in ion trap mode. These capabilities are enabled by the similar fundamental operating principles of Q and 2D ion trap analyzers. As with the stand-alone 2D ion traps, studies evaluating the quantitative abilities of these instruments with Q_3 operating in trap mode are just appearing and the quantitative role of these enhanced performance instruments remains to be evaluated (note that **such instruments can always be operated in conventional QqQ mode**).

Another design that seeks to leverage the performance features of two types of mass analyzer coupled together (often called **hybrid mass spectrometers**) is the **quadrupole time of flight** or **QqTOF instrument. Time of flight (TOF) mass analyzers** have been used for many years and, more recently, have seen widespread application in large molecule analyses, notably coupled to matrix assisted laser desorption ionization (MALDI). Single TOF analyzers are now available coupled to GC or LC, where the reflectron type is **primarily used as accurate mass analyzers,** where their **resolving power and extremely rapid spectral acquisition rates** (thousand of full scan

spectra per second) can provide m/z values with uncertainties as low as 10 ppm.

While this level of mass measurement accuracy is rarely necessary in quantitative analyses, the fast spectral acquisition rates of TOFs have been used to advantage in methods employing extremely fast GC separations. Stand-alone TOF analyzers cannot readily mass select ions, thus precluding the use of SIM mode. As such, **quantitative assays performed on TOFs must currently make use of full scan MS mode.** Moreover, TOFs have a low duty cycle when operated with continuous ionization sources since the slowest (highest m/z) ions in one ion packet accelerated into the flight tube must be allowed to reach the detector before the next packet is selected. At the other extreme, TOF analyzers have an essentially 100 % duty cycle when directly interfaced to pulsed ionization sources like MALDI if the ionization pulse is synchronized with the TOF acceleration.

QqTOF tandem-in-space mass spectrometers were developed largely for qualitative applications requiring the selectivity and ion structural capabilities of full scan MS/MS, as well as more accurate m/z measurements for the product ions than those provided by a unit mass resolution quadrupole. Coupling of these two different mass analyzers proved most successful with respect to achievable RP when the flight path of the **TOF analyzer is placed orthogonal to the main axis of the Qq combination, the so-called oa–QqTOF** (although the duty cycle problems remain). The main advantage of the QqTOF in quantitative assays arises from the increased selectivity of the high resolution detection of product ions. In addition to the low duty cycle the main disadvantage is an intrinsic limit on the upper limit of quantitation (ULOQ) imposed by the detector dead time that can lead to nondetection of those ions that reach the detector during the dead time caused by an earlier ion arrival.

The current pace of instrument development is such that these summarized comments will probably have to be rewritten sooner rather than later. The potential role of the 2D traps in quantitative analyses was already mentioned and very recently a modification to a QqTOF instrument that permits SIM/MRM operation with high duty cycle has been described; the higher product ion resolution could be useful in bioanalytical and similar assays where isobaric interferences from unknown metabolites could be a factor. However, it seems likely that, for the applications of interest in this book, the triple quadrupole instrument will remain the workhorse instrument of choice for the foreseeable future. The recent introduction of a triple quadrupole design in which Q_1 is capable of enhanced RP (peak widths as low as 0.1–0.2 Th) permits increased selectivity via precursor ion selection.

Appendix 6.1 Interaction of Electric and Magnetic Fields with Charged Particles

Note that equations in this Appendix are numbered as if they formed part of the main text starting from where the Appendix is first mentioned.

Two fundamental equations of classical physics provide convenient mathematical summaries of these experimentally observed effects, namely, Newton's Second Law of Motion and the Lorentz Force Law, written here as they apply to an ion of mass m_{ion} (kg) and charge q_{ion} (coulombs C):

$$\mathbf{F} = m_{ion}.\mathbf{a} \text{ (Newton's Second Law)} \qquad [6.2]$$

$$\mathbf{F} = q_{ion}.(\mathbf{E} + \mathbf{v} \times \mathbf{B}) \text{ (Lorentz Force Law)} \qquad [6.3]$$

where quantities in bold font are vectors (possess magnitude and direction), $\mathbf{F}$ is the force applied to the ion, m_{ion} is the mass of the ion (in kilograms units in SI units) and q_{ion} its charge (including the sign, coulomb units in SI), $\mathbf{a} = d\mathbf{v}/dt$ is the acceleration produced by the applied force, $\mathbf{E}$ is the electric field, $\mathbf{v}$ is the ion velocity vector and $\mathbf{B}$ is the magnetic flux density (magnetic field strength); $\mathbf{v} \times \mathbf{B}$ is the *vector* cross product of the ion velocity and the applied magnetic field, i.e., another vector quantity whose magnitude and direction are determined by combining those of its two constituents). Note that the Lorentz Law (Equation [6.3]) tells us that there is no effect of a magnetic field on an ion unless the latter is in motion relative to the magnetic field lines, i.e., a magnetic field interacts only with an ion beam, not with an ion at rest. This law also makes it clear that the magnitude of the electromagnetic force applied to an ion is proportional to the *charge* on the ion, while Newton's Law tells us that the result (the acceleration) of applying this force to an ion is dependent on the ion *mass*. This is the origin of the fact that so-called 'mass spectrometers' do not actually measure the mass of an ion, but rather the ratio of its mass to its charge (see the discussion of the symbol *m/z* in Section 1.3).

Magnetic fields are of prime importance in the history of mass spectrometry, since essentially all such instruments used until the second half of the 20th century relied on the q.$\mathbf{v} \times \mathbf{B}$ interaction (Equation [6.3]) to achieve *m/z* spectra (Beynon 1960). However, for the purposes of quantitative measurements of trace level components eluting from an HPLC column, magnetic field instruments have been largely superseded by spectrometers that rely on only electric fields of various kinds. The main exception to this generalization is the use of double focusing magnetic sector instruments in analysis of chlorinated pollutants (e.g. dibenzodioxins) eluting from a capillary GC column, where this analyzer's ability to provide high transmission at high resolving power makes it the gold standard for this application. Fourier Transform Ion Cyclotron Resonance (FTICR) instruments also incorporate a magnetic field but are seldom used in quantitation and are discussed only briefly in this book.

It is convenient to briefly define the meaning of *electrical potential*, a concept that was used (e.g. in Section 4.4.6) without proper discussion. In the present context 'potential' is closely related to 'potential energy', i.e., the energy possessed by an object by virtue of its *position* only; of course 'position' is intrinsically a relative term, it is necessary to specify 'position relative to what', and usually in definitions of potential energy the 'what' is conventionally defined as a position where the forces acting on the object are zero. The relationship between electric field strength and electrical potential is derived for a simple special case although this relationship is generally valid.

Consider the electric field arising from a generalized charged particle bearing charge Q' (the symbol includes both the magnitude and sign of the charge). Coulomb's Law (an experimental result!) gives the *magnitude* of the force exerted by this field on another particle (an ion in this case) with test-charge q_{ion} located in a vacuum a $d_X = \infty$ from Q' (here *X denotes any general spatial coordinate*):

$$F_X = q_{ion}.Q'/(4\pi\varepsilon_0).(1/d_X{}^2) \qquad [6.4]$$

where F_X is the force acting in the X-direction connecting Q' and the test-charge q_{ion}. In SI, ε_0 is a constant called the *permittivity of free space* with value 8.8542×10^{-12} $C^2.N^{-1}.m^{-2}$ ($= A^2.kg^{-1}.s^4m^{-3}$ in SI base units, see Chapter 1); if q_{ion} and Q' are located in a medium other than a vacuum ε_0 must be corrected by a factor characteristic of that medium (the 'dielectric constant'), but ions in a mass spectrometer are separated by vacuum. Then Equation [6.4] indicates that the state of zero force on q_{ion} due to Q' is at $d_X = \infty$, so the potential energy of q_{ion} is evaluated relative to this condition.

The incremental work energy ∂W required to move the particle with charge q_{ion} from a distance d_X from Q' to an incrementally larger distance $[d_X + \partial(d_X)]$ is defined as the product of the force acting along coordinate X and the distance moved:

$$\partial W = F_X.\partial(d_X)$$

so the total work energy required to move q_{ion} from ($d_X = \infty$) to distance d_X is:

$$W_X = \int F_X.\partial(d_X) = q_{ion}.Q'/(4\pi\varepsilon_0). \int [1/d_X{}^2]\partial(d_X)$$

where the integration is over the range d_X to ∞. Trivial integration then gives:

$$W_X = q_{ion}.Q'/(4\pi\varepsilon_0).[-1/d_X]_X^\infty$$
$$= q_{ion}.Q'/(4\pi\varepsilon_0).[-1/d_X] \qquad [6.5]$$

since $1/\infty = $ zero. Now the quantity W_X in Equation [6.5] is simply, by definition, the potential energy of q_{ion} at distance d_X from Q', and is to be compared with the expression for the *magnitude* E_X of the corresponding electric field along the coordinate X (see the accompanying text box):

$$E_X = Q'/(4\pi\varepsilon_0.d_X{}^2) \qquad [6.6]$$

Comparison of Equations [6.5 and 6.6] gives:

Force (on test-charge q_{ion} at distance d_X from Q')

$$= q_{ion}.E_X = -d(W_X)/d(d_X) \qquad [6.7]$$

obtained by straightforward differentiation of Equation [6.5]. Equation [6.7] is a simple special case of a general result, i.e., *the force is the derivative of the field with respect to distance*.

Rather than using W_X explicitly, it is convenient to use a quantity Φ_X, the *electrical potential at point X*, defined as the electrical potential energy *per unit charge* at X:

$$\Phi_X = W_X/q_{ion} \qquad [6.8]$$

so that:

$$E_X = -d\Phi_X/dX \qquad [6.9]$$

where *for convenience d_X has been replaced with the coordinate itself (X)*. The field strength along any coordinate X at a specified point in space is thus the derivative of the electric potential with respect to X. The SI unit for electrical potential is the volt (V); *unfortunately the same symbol (V) is frequently also used instead of Φ to denote the physical quantity (electrical potential) for which the volt is the SI unit*, but in practice the context does not usually lead to confusion and this book will follow the literature by using both symbols occasionally. The SI unit for the magnitude of an electric field (the *electric field strength*) E is thus ($V.m^{-1}$).

In the discussion of the principles of operation of *m/z* analyzers Equations [6.2–6.9] will be referred to with respect to the motion of ions subjected to electromagnetic fields. In addition, a consequence of these relationships together with the principle of conservation of energy will

play an important role; in the absence of any additional energy loss/gain arising from collisions with background gas molecules, when an ion moves from a point with electrical potential Φ_1 to another with Φ_2, i.e., changes its potential energy from $q_{ion}.\Phi_1$ to $q_{ion}.\Phi_2$, the translational (kinetic) energy T of the ion must also change so that the total energy is unchanged:

change in $T = -$(change in potential energy)

$$= -q_{ion}(\Phi_2 - \Phi_1) = q_{ion}(\Phi_1 - \Phi_2) \qquad [6.10]$$

Note that q_{ion} represents both the magnitude and the sign of the charge.

Appendix 6.2 Leak Detection

Leak detection can be a very frustrating task, requiring patience and perseverance. In particular, except as an extreme last resort, do not panic and tear the system apart, replacing every gasket and seal; such an approach often makes matters worse. A comprehensive guide to systematic leak detection, together with a great deal of other valuable information about vacuum systems and their components, is freely available on the website of the Kurt J. Lesker Company: http://www.lesker.com/newweb/menu_techinfo.cfm. In addition, useful hints provided by experienced operators can be found at: http://www.sisweb.com/index/referenc/tip19.htm. What follows represents an abbreviated combination of these two helpful contributions.

In general it is much easier to avoid leaks than to find them. There are several relatively easy and inexpensive things that can be done to minimize the risk of leaks. If the system contains flanges requiring metal gaskets (usually only on instruments that require prolonged heating or 'baking' of the vacuum chamber to remove molecules adsorbed on the walls and on other metal components), always replace the metal gaskets when such a flange is disassembled and reassembled; copper gaskets are cheap, and in the case of gold gaskets it is sometimes possible to obtain a leak-free seal by putting two used gold gaskets together in the flange. For flanges that will not be heated and that use synthetic polymer O-rings, it is usually helpful to apply a very light film of low vapor pressure lubricant (e.g., diffusion pump oil) before reassembly. Make sure that mating flange surfaces are kept scrupulously clean and free of scratches.

If a leak is suspected, it should first be determined whether a leak really exists or if the vacuum gauges have become dirty or are malfunctioning in some way and are giving false readings. If the vacuum measurements are not at fault, determine whether the leak is real (i.e., atmosphere is being sucked into the mass spectrometer) or virtual (e.g., too much solvent and/or sample has been introduced into the instrument and is now outgassing, or the rotary vane pump oil is contaminated, or the high vacuum pump has failed etc.). No method of leak testing will find virtual leaks and if the high vacuum pump is confirmed as operable the pump oil should be changed; if that fails to solve the problem (remember new pump oil will need some time to degas itself), the instrument must be cleaned (or baked if appropriate). If the mass spectrometer can still be operated despite the leak, and if an electron ionization source (see below) is available the intensity ratio for m/z 28 (nitrogen) to m/z 32 (oxygen) will be 4:1 if the leak is indeed real. If it finally seems clear that there is a real leak, it usually occurs at the last point(s) where the vacuum system was disassembled and reassembled, and the most likely culprits here are metal gaskets and O-ring seals on flanges. Other likely possibilities are electrical feedthroughs, welds and shut-off valves (valves that can isolate one part of the vacuum system from the rest) that have developed a through-valve leak.

The general principles of vacuum leak detection (as distinct from the practice!) are simple. The atmosphere side of a suspected leak location is sprayed with a gas or volatile liquid that gives a response on the detection device that is different from that of air. The gas or vapor diffuses through the leak, displacing the air, and the output of the 'detection device' changes. The 'detection device' can be the human ear if the forepump is running and the leak is large enough to create a whistling sound as the air rushes through. If the mass spectrometer has an electron ionization (EI) source and the vacuum is just good enough that it can be operated (but beware risk of damage to the electron multiplier detector) the mass spectrometer can be used as the detection device; in such a case acetone is a commonly used probe vapor as it gives a characteristic EI mass spectrum, m/z 43 (100), 15 (70), and 58, 28, 27 and 14 (all ~15), where the numbers in parentheses are the relative abundances of the ions. (Note, however, that acetone is recommended only for use on flanges sealed by metal gaskets rather than polymer O-rings, as acetone can permeate the polymer and slowly outgas into the vacuum system over a significant time.)

Alternatively, the vacuum gauges can be used for leak detection. If the vacuum is good enough that the ion gauge will come on (and stay on), helium is a good choice as a probe gas since its ion gauge response is so much smaller than that of air (Table 6.2). However, things are not quite as simple as this might suggest, since helium diffuses much faster than air (a consequence of its much smaller mass and thus much higher u value) so more helium is

likely to diffuse through the leak than air and this will tend to *increase* the ion gauge reading; the net effect on the ion gauge reading will thus be a case-by-case compromise between the two effects, but the direction of the change on the ion gauge is less important than the fact that it will change if helium is sprayed over a leak. If the achievable vacuum is not good enough to allow the ion gauge to switch on, the forepump gauge (Pirani or thermocouple) can be used if the vacuum is in the $10^{-2} - 10^{-3}$ torr range. In this case if helium displaces air through the leak the faster diffusion will increase the pressure and thus the thermal conductivity of the gas in the gauge so the gauge reading will increase.

If none of these approaches succeeds in locating the leak, the next step is to use a *helium leak detector*, a self-contained EI mass spectrometer tuned to respond only to helium (m/z 4) and complete with its own vacuum system with sufficient pumping speed that it can handle inlet pressures (i.e., from the system with the leak) of up to ~10 torr. These items are fairly expensive and are certainly not available in every laboratory, so it may be necessary to borrow or rent one. Another possibility is to use a *residual gas analyzer*, a small EI mass spectrometer mounted on a vacuum flange that can be inserted into the vacuum chamber, e.g., to temporarily replace the ion gauge or other component. These devices record complete EI mass spectra typically in the range m/z 1–~200, but usually are not supplied with their own independent vacuum system and in fact require a vacuum of better than 10^{-4} torr for operation (and at this pressure can operate only with a Faraday cup detector instead of an electron multiplier, see Chapter 7).

An alternative to acetone that is sometimes used is sulfur hexafluoride (SF_6), a gas that also has a ion gauge response (Table 6.2) that is significantly higher than that of air (but of course diffuses through a leak much more slowly). Finally, it is sometimes advantageous to use plastic sheets to 'bag' small portions of a system where the leak is suspected to localize the availability of the helium or other test gas.

Appendix 6.3 List of Symbols Used in Chapter 6

a_x (italicized): a mathematically convenient dimensionless quantity, defined as $(4/\omega^2.r_0^2).(q_{ion}/m_{ion}).U = (4e/\omega^2.r_0^2.m_u).[U/(m/z)]$, used to describe the nature of the motion in the x-direction of an ion in a linear quadrupole electric field. An analogous expression holds for a_y that characterizes motion in the y-directiom. Sometimes the subscripts x and y are omitted since $a_x = -a_y$ as a result of the symmetry of the device and the different half-cycles of the RF waveform applied to the two pairs of rods.

a_r (italicized): a mathematically convenient dimensionless quantity, defined as $8/[\omega^2.(r_0^2 + 2z_0^2)].(q_{ion}/m_{ion}).U = 8e/[\omega^2.(r_0^2 + 2z_0^2).m_u].[U/(m/z)]$, used to describe the nature of the motion in the radial direction of an ion in a 3D (Paul) ion trap. An analogous expression holds for a_z that characterizes motion in the axial (z-) direction. Note that $a_z = -2a_r$ as a result of the symmetry of the device and the different half-cycles of the RF waveform applied to the endcap and ring electrodes.

B (bold font): a vector quantity, the magnetic flux density (magnetic field strength); B_X represents the magnitude of the component of **B** in the X-direction.

C_{gas}: the conductance of a pipe or orifice with respect to flow of gas through it as a result of a pressure difference, in units (volume.time^{-1}).

e: the elementary unit of charge, corresponding to the magnitude of the charge carried by an electron (negative) or proton (positive), 1.60217653 ($\pm 0.00000014) \times 10^{-19}$ C.

E (bold font): a vector quantity, the electric field strength; E_X represents the magnitude of the component of **E** in the X-direction (units of V.m^{-1}).

F (bold font): a vector quantity representing the force experienced by a particle; F_X represents the magnitude of **F** in the X-direction.

k_{uni}: a unimolecular rate constant for an isolated ion with internal energy ε_{int}.

l   : the axial length of any quadrupole device.

m: this symbol is used as a symbol for several different quantities depending on the context, i.e., the SI unit for length (meter) or the SI base quantity 'mass'; when used to indicate the mass of a particular object it is best to denote the latter with a subscript, e.g., m_{ion}, m_{molec} etc.

m_u: a unit of mass convenient for describing individual molecules etc., defined as 1/12 of the mass (in kilogram) of one ^{12}C atom; in the convention used by physicists this quantity is denoted 'u'. The value of m_u is 1.66053886 ($\pm 0.00000028) \times 10^{-27}$ kg.

m/z (italicized): in this book this is regarded as a three-character symbol denoting a characteristic property of an individual ion, defined as the ratio of $m = [m_{ion}(kg)/m_u(kg)]$ to $z = [|q_{ion}|(C)/(e(C)]$ where only the *magnitudes* of the charges are considered (i.e., no signs).

M: molar mass of a compound ($= N_A.m$); sometimes also used as a unit of concentration in solution ('molarity') = n/V$_{solution}$.

n: number of moles (sometimes number of replicate experiments, depending on context).

N: number of molecules $= n.N_A$.

N_A: Avogadro Constant $= 6.022 \times 10^{13}$ molecules.mol^{-1}.

ND: number density of gas molecules $= N/V_g$.

q_{ion}: the electric charge (including the sign) carried on a physical object; not to be confused with the operating parameter q (italicized) or the notation q conventionally used to denote an RF-only quadrupole used as a collision cell.

q_0: notation used to denote a linear quadrupole device used as an ion guide only.

q_2: notation conventionally used to denote an RF-only quadrupole used as a collision cell as in triple quadrupole ($Q_1q_2Q_3$) or QToF (Q_1q_2TOF) instruments.

$q_{2/T}$: notation conventionally used to denote an RF-only quadrupole that can be operated either as a collision cell or as a 2D trap.

q_x (italicized): a mathematically convenient dimensionless quantity, defined as $(2/\omega^2.r_0^2).(q_{ion}/m_{ion}).V_0 = (2e/\omega^2.r_0^2.m_u).[V_0/(m/z)]$, used to describe the nature of the motion in either the x- or y-direction of an ion in a quadrupolar electric field. Sometimes the subscripts x and y are omitted, since $q_x = -q_y$ as a result of the symmetry of the device and the different half-cycles of the RF waveform applied to the two pairs of rods. Not to be confused with the symbol q (nonitalicized) used to denote the charge (including the sign) on a particle or the notation q_2 conventionally used to denote an RF-only quadrupole used as a collision cell.

q_r (italicized): a mathematically convenient dimensionless quantity, defined as $-4/[\omega^2.(r_0^2 + 2z_0^2)].(q_{ion}/m_{ion}).V_0 = -4e/[\omega^2.(r_0^2 + 2z_0^2).m_u].[V_0/(m/z)]$, used to describe the nature of the motion in the radial direction of an ion in a 3D (Paul) ion trap. An analogous expression holds for a_z that characterizes motion in the axial (z-) direction. Note that $q_z = -2q_r$ as a result of the symmetry of the device and the different half-cycles of the RF waveform applied to the endcap and ring electrodes.

Q, Q_n: notation conventionally used to denote a linear quadrupole m/z filter; subscripts (usually n = 1 or 3) are sometimes used to distinguish between two such devices, e.g., as incorporated into a triple quadrupole instrument.

$Q_{n/T}$: notation conventionally used to denote a linear quadrupole device that can be used either as a conventional m/z filter or as a 2D trap mass spectrometer.

Q_N: the number of molecules of a gas pumped through a specified location in a vacuum system per unit time.

Q_{gas}: the so-called 'mass flow' of a gas through a pipe or orifice in units of (pressure.volume.time^{-1}) = $Q_N.(RT/N_A) = C_{gas}.$(pressure drop across pipe or orifice).

r_0: the characteristic dimension in a linear quadrupole (radial distance between the main axis and each rod) or in a 3D (Paul) ion trap (radial distance between the geometrical centre of the trak and the ring electrode).

R: universal gas constant, as in the ideal gas equation $P.V_g = nRT$.

s (italicized): represents the number of vibrational modes in an activated ion that are sufficiently closely coupled that they can efficiently share internal energy with those modes that are directly involved in the reaction process. Not to be confused with s (nonitalicized), the SI symbol for the second.

S: pumping speed for a gas at a specified location in a vacuum system; volume per unit time, e.g., L.s^{-1}, measured on the high pressure side of the specified location.

T: translational (kinetic) energy = $m_{ion}.v_{ion}^2/2$ for an ion.

u: average speed of a gas molecule = $(3RT/M)^{\frac{1}{2}}$ for an ideal gas; also used by physicists to denote m_u (the 'unified' atomic mass unit, see above).

U: DC potential applied between the two pairs of opposing rods in a linear quadrupole mass filter.

v: when used in bold font this represents the vector quantity *velocity* (magnitude and direction), but in normal font is often used to represent the *scalar* magnitude of the velocity (the *speed*).

V: symbol for the volt, the SI *unit* for electrical potential; often used (strictly incorrectly) as a replacement

symbol for the *physical quantity* (Φ, electrical potential) itself.

V_0: The amplitude (zero-to-peak, in volts) of the RF waveform applied between the two pairs of opposing rods in a linear quadrupole mass filter.

V_g: volume of a gas at a specified pressure P and absolute temperature T.

z_0: distance along the central z-axis of a Paul trap from the centre to one of the end cap electrodes.

z (italicized): the number of elementary charges e on an electrically charged object, regardless of the sign of the charge; for an ion $q_{ion} = \pm z.e$ where the sign corresponds to the polarity of the charge on the ion.

ε_{int}: the internal energy of an ion, e.g. that leads to unimolecular dissociation if sufficiently large, i.e., $\geq \varepsilon_{int}^*$, the critical energy for reaction.

ε_0 : the *permittivity of free space*, a constant = 8.8542×10^{-12} A^2.kg^{-1}.s^{4}m^{-3}

Φ_X: the electrical potential at point X (in volts V), defined as the electrical potential energy per unit charge at X and frequently replaced by the symbol V (the SI unit for the physical quantity Φ); related to electric field strength by $E_X = d\Phi_X/dX$, and to the potential energy W_X of a particle carrying a charge q (including the sign) by $W = q. \Phi_X$.

τ : ($\omega t/2$), i.e. a mathematically convenient measure of time in units related to the RF frequency.

v^* : an effective molecular frequency that describes the rate at which the constituent atoms of an activated ion can reach the critical configuration of both positions and velocities for unimolecular reaction to occur.

ω : the angular frequency (radians.s^{-1}) of an AC or RF waveform, related to the frequency f (in Hz) by $\omega = 2\pi.f$.

Q, the number of molecules of a gas pumped through a specified location in a vacuum system per unit time

Q, the so-called mass flow of a gas through a pipe, which is in units of pressure-volume/time.

Q_s (PTA_s) = Q_{in} (pressure drop across type of conditions)

Tools of the Trade VI. Ion Detection and Data Processing

7.1 Introduction

One of the important advantages of mass spectrometry over other analytical methods is its combination of applicability, sensitivity and selectivity. This combination results from the integrated mass spectrometry system (Figure 1.3) but the excellent sensitivity of mass spectrometry is largely the result of extremely efficient ion detection systems located immediately after the mass analyzer (Figure 7.1). These detection systems are particularly important since the ratio of the number of ions delivered to the mass spectrometer to the number of analyte molecules consumed in the ion source is always disappointingly low, in the range $10^{-3} - 10^{-5}$. The same conclusion does not hold for the ratio of the number of electrons created by the ion detector (and delivered to the detection electronics that create the observed electrical signal) to the number of analyte ions arriving from the analyzer. This is thanks to modern developments in the design of m/z analyzers and, particularly, to the remarkable properties of electron multiplier detectors. A recent review (Koppenaal 2005) has emphasized the importance of this aspect.

Here, only the detectors currently used in instruments designed for trace analysis will be described in any detail. With the exception of TOF analyzers, which are a special case in this respect, virtually any of the electron multipier detectors described in this Chapter are compatible with any of the analyzers discussed in Chapter 6. (Note that a mass spectrometer is often referred to as a 'detector' in the context of chromatography, but that a mass spectrometer itself includes an ion detector, often referred to only as 'the detector' or 'the multiplier'; the context should always make clear in which sense the term 'detector' is used.) These ion detector systems all give rise to a digitized time series of measures of ion beam intensity, such

as 'counts per second', that correspond to the ordinate (vertical axis) in the chromatograms that are integrated to give quantitative data.

An ideal ion detector for mass spectrometry would possess at least the following attributes (Koppenaal 2005): 100 % detection of all ions reaching the detector, low or no intrinsic noise, high stability (both short and long term), parallel (simultaneous) detection of ions over a wide m/z range, response (signal intensity) that is independent of m/z, a wide linear dynamic range, fast response time (i.e., short time delay between ion arrival and recording of corresponding signal), short time required for recovery (i.e., short 'dead time' following arrival of the previous ion) and high saturation level (i.e., a large value for the upper end of the linear dynamic range). In addition to these performance criteria, the ideal detector would have a long operational life, require low maintenance and be easy and affordable to replace. Of course no real-world detector can satisfy all these criteria and the compromise choices adopted are those that are most fit for the purpose at hand. It should also be mentioned that the following discussion is not directed towards image current detection (Marshall 2002) used in FTICR and Orbitrap instruments (Section 6.4.8) since the latter are not used in trace quantitation in the sense of this book.

The ion detectors used in practice are discussed in more detail later in this chapter but are introduced in general terms here. All analyzers based on quadrupolar electric fields (including 3D and 2D traps) use electron multiplier detectors, usually channel electron multipliers (CEMs) but in some cases an external photomultiplier connected optically to a phosphor target. Magnetic sector instruments also use electron multipliers, including CEMs, photomultipliers and the discrete dynode type; some

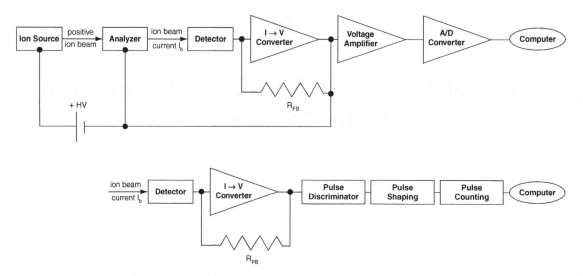

Figure 7.1 Diagrams of a generalized mass spectrometer system emphasizing the detector chain for a beam current I_b of (assumed) positive ions. The top diagram shows the features of a system in which the output signal is recorded as a continuous voltage (possibly digitized for computer storage and data processing) and the bottom diagram is for an ion counting detection system. Recall that an electrical current is conventionally (by historical accident) regarded as a flow of positive electricity but is in fact a flow of negatively charged electrons in the opposite direction. The electrical current delivered by the detector is $G_D.I_b$, where G_D is the (current) gain of the detector and the output voltage from the I → V converter is $G_D.I_b.R_{FB}$, where R_{FB} is the value of the feedback resistor. The (voltage) signal supplied to the digitizer (if used) is then $G_D.I_b.R_{FB}.G_V$, where G_V is the (voltage) gain of the amplifier. In the case of the ion counting system, pulse discrimination rejects voltage pulses below a user-set level, while pulse shaping conditions the pulses to a constant height and shape to permit optimized operation of the pulse counter.

specialized sector instruments designed for accurate measurements of isotope relative abundance ratios use Faraday cups that efficiently collect ions from an ion beam but do not 'multiply' them in any way. Time of flight (TOF) analyzers are somewhat unique in that, unlike quadrupoles and sector instruments that first sort out ions according to their m/z values and then pass them on to the detector, in a TOF instrument the detector (a channeltron multiplier array, or CEMA) is also a crucial component of the m/z analysis (see Section 7.3.2b).

All electron multipliers degrade with use and must be replaced periodically; this is described in more detail later, but in general quantitative applications place more demands on them than qualitative analyses. Remember that mass spectrometry is a destructive technique and the detector is the final destruction point, i.e., analyte ions are ultimately converted into neutral molecules embedded in the multiplier surface thus creating a flow of electrons emerging from the multiplier that is amplified and processed and recorded (Figure 7.1).

This chapter focuses on CEMs (including CEMA plates), as these detectors currently provide the best combination of ruggedness, gain, dynamic range, response time and possibility for optimized design for use in either analog detection (for high level signals) or pulse counting mode (for low level signals such as those encountered in true trace level quantitative analyses). For circumstances in which detector time response is a critical parameter, as in TOF analyzers, CEMA plates (usually in a chevron configuration) are required. In future the CEM detectors may be supplanted by new concepts (Koppenaal 2005) but at present they represent the best available performance except for a few specialized applications.

Diagrams of a generalized mass spectrometer with particular emphasis on the detection chain are shown in Figure 7.1. Two cases are illustrated; the first for detection of higher ion currents using a current-to-voltage converter, a voltage amplifier and possibly an analog-digital (A/D) converter for ease in electronic recording of the data. For very low ion currents, such as those that are likely to occur in trace analysis, some form of direct ion counting (pulse counting) is used in place of current-to-voltage conversion; a crucial parameter for ion counting is the detector dead time, i.e., the time delay between detection (and counting) of one ion and that of the next, imposed by the finite time required for the detector to amplify and transmit this signal to the associated electronics. In the case of secondary electron multiplier detectors (Section 7.3), an important contributor to dead time is the transit time of the electron avalanche through the detector itself. As discussed later (see, e.g., Section 7.4.2), this phenomenon imposes an intrinsic upper limit to the dynamic range of any such detector.

In Section 7.1.1 the general problems of signal:noise (S/N) and signal:background (S/B) ratios, and their improvement for data sets consisting of discrete data points (as obtained in chromatography with mass spectrometric detection) by smoothing techniques are discussed. Before proceeding, however, it is appropriate to make a few comments about the chromatograms that are integrated in quantitative analyses. In the case of SIM or MRM chromatograms (obtained using stand-alone or triple quadrupole instruments and also magnetic sectors) the data are acquired as discrete data points of different kinds (e.g. one or more m/z values or MRM channels for each of an analyte and SIS), separated in time by an amount determined by the dwell times and the number of m/z values monitored in the experiment. It is this discrete quality of the data that leads to the need for smoothing to obtain adequate precision, as discussed later. However, if the raw data are acquired as full scan spectra (generally either MS^1 or MS^2 obtained using TOF analyzers or ion traps), one way to simplify the resulting data set is to decide on one or a few m/z values from all those observed in the spectra that are selective for the analyte of interest, and instruct the data system computer (post-acquisition) to construct chromatograms for only these m/z values or combinations thereof; such chromatograms are referred to as extracted ion chromatograms and are often denoted by the acronyms EIC or XIC.

A somewhat different distinction is that between the more selective SIM/MRM/EIC chromatograms and the total ion current (TIC) chromatogram in which the total signal at all detected m/z values (whether full scan or SIM/MRM) is summed and plotted as a function of time. TIC chromatograms can provide some idea of the complexity of the sample especially if the raw data are of the full scan MS^1 type but are often sufficiently complex to be of little use for quantitation because of inadequate peak resolution, which can introduce appreciable uncertainty into peak integration. In Figure 7.2(a) a simple example of an analyte and its SIS that almost co-elute, so that the TIC chromatogram can not be used to provide peak areas for them separately (this mimics what would be observed with a single parameter detector such as a fixed wavelength UV detector), is shown. However, thanks to the extra dimension of information provided by the m/z axis of a mass spectrometer, it is easy to separate the two

peaks into the two components so that integration can proceed with minimal uncertainty arising from overlapping peaks. Figure 7.2(a) is a little misleading in that, for purposes of illustration, the chromatographic peaks are represented as continuous smooth curves. As mentioned above, the peaks are actually generated as discrete data points with blank spaces between them, illustrated in a rather extreme fashion in Figure 7.2(b) for the case of a single analyte and its SIS.

7.1.1 Signal:Noise vs Signal:Background

In Section 4.4.8 an important distinction was drawn between 'noise' and 'background', and the corresponding ratios to signal. 'Signal-to-noise ratio, S/N' is a concept introduced in electrical engineering and explicitly refers to the degradation of the signal (that carries the desired information) by superposition of statistical (random) fluctuations. In the engineering context these random fluctuations are mostly derived from the random thermal motion of electrons in components of an electrical circuit, so-called 'Johnson noise'; this effect can in principle be decreased by reducing the temperature (e.g. of a large resistor used as the feedback resistor of an electron multiplier, see Figure 7.1, to convert the output current into a voltage that can then be processed by well designed electronic circuits). More important in the present context is the random nature of such noise. This means that such fluctuations can be averaged out by spending a longer time on the measurement, either in a single observation interval or by averaging several independent measurements, since positive and negative fluctuations are equally likely. However the signal is unidirectional, so the effects of replicate measurements on signal strength are additive, not canceling. Another example of this phenomenon is ion beam shot noise, discussed in Section 7.1.1a.

In contrast, in applications of mass spectrometry to analysis of analytes present at trace levels in complex matrices, so-called 'chemical noise' (better described as 'chemical background') is by far the most important in determining the limit at which a chromatographic peak signal can be meaningfully claimed to be observable (the LOD, Section 8.4.1). Under these conditions it is better to use the term 'signal-to-background ratio (S/B)'. However, there is a limit to the extent to which additional 'informing elements' (Section 6.2.2) can improve signal:background ratios; this is referred to as the 'statistical limit' at which the absolute number of ions detected is too small for a reliable quantitative measurement of adequate precision (see Figure 6.1). At this stage (e.g. using MS^n with n >> 2 to reduce the background to negligible levels) the essentially random arrival of ions at the ion detector (ion beam shot noise) can then become the limiting factor in determining the LOD. This statement of fact actually carries a hidden message concerning the almost zero contribution to noise from modern electron multiplier detectors and their associated electronics (Sections 7.3 and 7.4). Even when chemical background is the more important, as is almost always the case for experiments of interest in this

book, shot noise in the ion beam is still a concern in that it determines the SIM/MRM dwell time required to obtain a statistically relevant sampling of the ions arriving at the detector. The challenge arises when this push for longer dwell times conflicts with the need to shorten them sufficiently that at least 10 data points are acquired for each SIM/MRM channel monitored over the chromatographic peak. The key difference is that when chemical background dominates this statistically meaningful collection

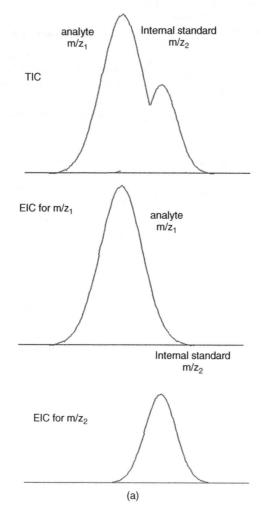

(a)

Figure 7.2 (a) Idealized illustration of how poorly resolved peaks in a total ion current (TIC) chromatogram can be resolved using the additional information provided by a mass spectrometric detector via construction of extracted ion chromatograms (EICs). The illustration is idealized in the sense that the chromatograms are drawn as continuous peaks. For mass spectrometric detection completely continuous curves would be observed only if just one single SIM or MRM channel were monitored, in which case the TIC would in effect be a single EIC with no additional information acquired.

(b) A more realistic illustration of the same principle that takes into account that mass spectrometric data are generally obtained as discrete separated segments. Segments are illustrated here as narrow MS^2 scans but could also represent SIM/MRM dwell times. The segmented nature of the data for each channel reflects the need to switch the analyzer between the channels for the analyte and SIS.

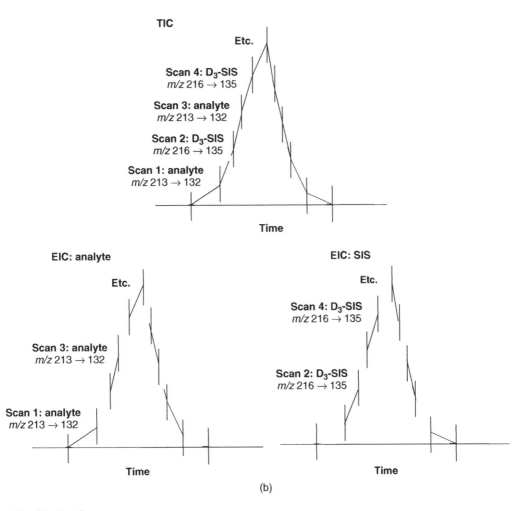

Figure 7.2 (Continued)

of ions includes both analyte-derived signal and the background (both unidirectional), but when background is reduced to negligible proportions the signal is still unidirectional but the shot noise provides positive and negative fluctuations with equal probability.

Another way of looking at this distinction is to compare the situation for a single SIM or MRM dwell time with the chromatogram that corresponds to stitching all of these dwell times together. Within a single dwell time window (t_D) the background signal provides a bias error to the signal that is attributed to the analyte only within that single dwell time window, and both are subject to ion beam shot noise fluctuations; the effect of the latter can be minimized by ensuring that the dwell time is sufficiently long to average out the bi-directional fluctuations. However,

from the viewpoint of the complete chromatogram, when it is the objective to identify the peak arising from the analyte, define its integration limits and actually conduct the integration, it is the time variation of these shot-noise-averaged chemical background signals among the various dwell times that lead to uncertainties and lack of precision (illustrated in Figure 4.27). If the background bias errors were constant within all dwell time windows they could easily be accounted for as a raised baseline in peak integration. Similarly, if they varied in a monotonic continuous manner the result would be a sloping baseline (compare the 'drift' in Figure 4.27) that could again be accounted for in the integration procedure. Although the background signal is recorded as a constant shot-noise-averaged value within each individual dwell time, the rise and fall of these background signals with time gives the appearance of 'noise' on the chromatogram.

The fact that so-called 'chemical noise' (better described as 'chemical background') is by far the most important in determining limits of detection is the reason why so much attention is paid to increasing the detection selectivity using MS^n (see discussion of Figure 6.1 in Section 6.2.2). The 'peak at every m/z value' phenomenon that characterizes mass spectra obtained at high magnification (Aebi 1996) is unidirectional like the signal and so can not be mitigated by averaging of replicate measurements or by increasing the time over which a signal (and its superimposed background) are observed. However, the analytical literature is full of misleading references to 'signal:noise ratio' where the limitation is truly the chemical background; this is not merely a matter of pedantic terminology since the experimental strategies available to mitigate the two phenomena are very different. Unlike random noise, chemical background can only be filtered out by increasing the number of 'informing elements' (see the discussion of Figure 6.1 in Section 6.2.2).

7.1.1a Shot Noise in the Ion Beam

As noted above, in practical applications 'shot noise' ('statistical noise') in the ion beam arriving at the ion detector is important mainly in the context of determining minimum dwell times for SIM or MRM; analogous statistical noise arising from the detector and its associated electronics is relatively unimportant. This reflects the fact that the arrival of ions at a detector is a random process, i.e. the numbers of ions arriving within a series of sampling time intervals (dwell times) t_D are not constant but show statistical variations within each time window; the 'ion current' is simply the rate of arrival of ions. A detailed discussion of in measurement of S/N values for low ion currents, and the individual contributions to the final overall noise from the various components of detection chains (Harris 1984), is the basis for the present abbreviated discussion.

The shot noise can be considered first for an ion counting detection system, universally used for trace analyses by mass spectrometry; suppose the average number of ions arriving at the detector within a sampling window t_D is N_D. Simple statistical theory (not included here, but see the accompanying text box discussing Poisson statistics) gives the result that the standard deviation (σ_N) of the individual observations of N_D would be $N_D^{\frac{1}{2}}$, and since 95 % of these observations would fall within the range $N_D \pm 2\sigma_N$ (see Section 8.2) it is reasonable to define the S/N for ion beam shot noise as $N_D/2\sigma_N = N_D/2\ N_D^{\frac{1}{2}} = N_D^{\frac{1}{2}}/2$. To transform this expression for S/N in the number of ions arriving within a specified time

---

The Poisson Distribution

The statistical distribution of rare events, such as the probability that an ion in a low intensity ion beam will strike a detector within a short sampling time interval, follows a distribution law first derived by the famous French mathematician Siméon-Denis Poisson (1781–1840). Despite his many official duties, he found time to publish more than 300 works, several of them extensive treatises most of which were intended to form part of a great work on mathematical physics, which sadly he did not live to complete.

Poisson distributions model discrete random variables, i.e., variables that can assume only non-negative integral values with no correlation among the observations of these values. A typical Poisson random variable is a count of the number of events that occur in a specified time interval, e.g., the number of coins I will find on my walk home on any given day, or the number of cars passing a fixed point in a five-minute interval. The Poisson distribution applies under the conditions that the length of the observation period is fixed, the events occur at a constant average rate and the numbers of events occurring in different intervals are statistically independent (no correlations among these values). The Poisson distribution also appeared in a different context in the Plate Theory of chromatography (Chapter 3), but the following short discussion is more directly related to detection of particles (e.g., ions). It describes the behaviour of a large number of independent experiments of which only a very small fraction is expected to result in the events of interest. The example of interest is the number of ions N_D in a low intensity beam that strikes a mass spectrometer detector in a (fixed) short time interval. The probability for k such events to occur is given by the Poisson probability that is stated here without proof:

$$P(k) = N_D{}^k . \exp(-N_D)/k!, \text{ for } k = 0, 1, 2, \dots \text{etc.}$$

where N_D is the average number of ions detected for a large number of sampling intervals (and is thus not necessarily integral). Unlike the Gaussian distribution the Poisson is not symmetrical, as shown in this example for $N_D = 3$:

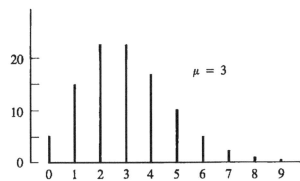

$\mu = 3$

An important property of the Poisson distribution is σ_k, the standard deviation of k:

$$\sigma_k{}^2 \equiv \; <(k - N_D)^2> \; = \Sigma[(k - N_D)^2 . P(k)] = N_D$$

where the notation $<>$ denotes an average and the summation is over all values of k. The final result $\sigma_k = N_D{}^{\frac{1}{2}}$ follows from the above definition by extensive but simple algebra.

'Johnson noise', arising from random thermal motion of electrons in a component of an electrical circuit, also follows Poisson statistics.

window (e.g. an SIM or MRM dwell time) into S/N for the corresponding ion beam current I_b, note that I_b is just the average number of ions per second arriving at the detector multiplied by the charge per ion, i.e., $I_b = N_D . e / t_D$. Then the standard deviation in I_b measured by an ion counting method and resulting from only shot noise in the beam itself, is given by $\sigma_I = \sigma_N . e / t_D = N_D{}^{\frac{1}{2}} . e / t_D = (I_b . e / t_D)^{\frac{1}{2}}$, and the corresponding S/N is given by:

$$(S/N)_I = I_b/2\sigma_I = (I_b . t_D/2e)^{\frac{1}{2}} \qquad [7.1a]$$

For a measurement of continuous beam current (i.e. for rates of ion arrival that are too large for ion counting detection as a result of detector dead time, see below), it is appropriate to replace the ion counting time window t_D with a more suitable parameter; the appropriate parameter is the time constant (τ_D) of the detection chain (usually dominated by the RC time constant of the current-to-voltage converter, see Section 7.4). If the I_b value is 'instantaneously' increased to a new value, the time

required for the output (voltage) signal to reach 95 % of the final value is $\sim 3\tau_D$; this provides an appropriate analogy for t_D so that for continuous current measurement $\sigma_I = (I_b . e / 3\tau_D)^{\frac{1}{2}}$ and the S/N is:

$$(S/N)_I = I_b/2\sigma_I = (I_b . 3\tau_D/2e)^{\frac{1}{2}} \qquad [7.1b]$$

These expressions for the S/N, corresponding to only the shot noise in the ion beam itself (i.e. ignoring contributions from the components of the detection chain, Figure 7.1), reflect the intuitive appreciation that these statistical variations should become relatively less important as the total number of ions actually recorded increases, i.e. as either or both of the average ion current and time window of detection are increased. Of course the value of I_b is primarily determined by the size of the original analytical sample and the concentration of analyte within it, and is thus not within the control of the analyst other than via efficient extraction and clean-up procedures (Chapter 4). Increasing $(S/N)_I$ by increasing

t_D (or τ_D) has an upper limit imposed by requirements that the sampling (or response) time should be sufficiently short relative to the chromatographic timescale that an accurate representation of the chromatogram is obtained. Moreover, in the case of mass spectral recording for a TOF analyzer using ion counting detection, increasing t_D limits the m/z resolution since now the t_D windows correspond to flight time windows (see Equation [6.19]).

In the remainder of this chapter it is mostly S/B values that are referred to, with S/N used only on the few occasions when required. Unfortunately, the analytical literature does not uniformly make this distinction and 'noise' is frequently used when 'background' is appropriate.

7.1.1b Data Smoothing Before Integration

When raw data are acquired as discrete digitized data points, as in SIM or MRM data obtained across a chromatographic peak using ion counting detection, some degree of uncertainty is inevitably introduced into integration of the peak area since the peak shape between adjacent data points must be 'guessed' in some fashion. (This is generally true of any digitized data.) This is the reason for the frequent references throughout this book to a need to acquire at least 10 data points across a chromatographic peak. Even when this is done to minimize the degree of 'guesswork' involved, however, some level of uncertainty remains. For example, should the data points be connected by straight lines before integration, or should some form of curvature be introduced to mimic the 'real' peak shape that might be revealed by some continuous detection method, e.g. a fixed wavelength UV detector with analog electronics (fictitious in the sense that the sensitivity would be far too low), or a single fixed SIM/MRM channel.

Guessing a digitized peak shape in the 'blank' regions by introducing some notion of curvature is known as 'smoothing'. Even in the fictitious case of a continuous detector with analog electronics, the true signal would be 'smoothed' by the time response characteristics (the effective RC time constant, see Section 7.4) of the analog circuitry that would discriminate against high frequency components of the response. This is simply 'smoothing' the signal in the analog world. For discrete or digitized signals, smoothing can be accomplished by using post-acquisition algorithms that connect the discrete data points in a well defined transparent manner. Note that this type of smoothing is a necessary prerequisite for integration of discrete data, since some kind of decision must be made as to how to account for the blank

segments (Figure 7.2(b)). In contrast, smoothing is sometimes applied even to continuous data in order to 'make the data look better', i.e. for purely cosmetic reasons.

Smoothing algorithms are provided in any mass spectrometer software suite; these are often proprietary, each manufacturer creating a partially unique algorithm. However, all are designed using the same basic principles to accomplish the same objectives, i.e. estimate the true signal shape (chromatographic peak shape in our case) in a more realistic manner than simply joining up data points with straight lines, and apply some degree of averaging to the fluctuations superimposed on the signal regardless of their fundamental nature (true noise or background signals). Smoothing is an example of data enhancement methods that in general intend to improve the quality of data without using any knowledge or presumed knowledge about the nature of the superimposed effects (noise, background) that are degrading the data. (This is in contrast with restoration or deconvolution methods that use real or presumed knowledge about the degrading phenomena to correct for their effect.) Since in chromatography–MS data the degradation of signal is overwhelmingly chemical background rather than Johnson-type random noise, no detailed understanding of the phenomenon is available that would permit its description in mathematical terms to develop an algorithm that could be used to restore the 'true' signal. Accordingly it is necessary to resort to data enhancement in the form of digital filtering (i.e. smoothing) techniques.

A wide variety of digital filters and their combinations can be used to smooth data and are too complex to discuss in detail here. However, the following discussion provides a brief insight into how three of the most commonly used methods work, namely, box-car averaging, moving average and polynomial smoothing with Savitzky–Golay filtering. Note that a smoothed signal can also be smoothed, i.e. the smoothing process can be repeated.

Box-car averaging (also called rectangular or unweighted sliding average smoothing) is conceptually the simplest approach. All values falling within a given range of the independent variable x (time or MS scan number in the case of a chromatogram) are averaged and this average 'bin' value is then assigned to the center point of this range of x. The range of discrete x values chosen for averaging is called the smooth width, m, where m is an odd positive integer. (For all smoothing methods the larger the smooth width m, the greater the noise reduction, but the greater the possibility of peak distortion, see below.) The box-car algorithm continually performs this form of bin averaging as it steps through the sets of m discrete data points across the chromatographic peak.

Moving average is performed similarly to the box-car averaging except that, instead of providing a single averaged signal value for each smooth width (bin) and then stepping to the next bin, the algorithm moves forward by only one data point thereby capturing all but one of those used to determine the previous average plus one new one. This recalculated average then is assigned to the new center point of the new smooth width (bin), and so on. This approach provides a larger number of smoothed points, but for signals (chromatographic peaks) defined by only a few data points there is a trade-off between increase of the S/N or S/B and distortion of the peak that can lead to artifactual shifting of the position of the peak on the x-axis (time or scan number).

Polynomial smoothing with Savitzky–Golay (SG) filtering operates by the same basic mechanism as moving average smoothing but with the addition of a polynomial weighting function applied to the data points within each bin (smooth width) that discriminates in favor of those data points near the center of the bin at the expense of those at either end. For example, if x_0 represents the central data point in a bin, the m values of (signal + interference) before averaging would be weighted by a n'th order polynomial, i.e. by factors $[1/(x - x_0)^n]$. This provides the desired result while minimizing the accompanying disadvantage of peak shape distortion. By adjusting values of m and n, the procedure for the set of data at hand can be optimized. As with any form of signal averaging, the S/N (or S/B) improves with the square root of m, e.g. a 25-point smoothing gives about a five-fold improvement.

In view of the limitations imposed by the need to acquire a minimum number of mass spectrometric data points across a chromatographic peak (in turn determined by a combination of factors, e.g. in SIM or MRM by the need to use dwell times sufficiently long that ion beam shot noise is effectively averaged out), analysts almost always smooth the chromatograms before integration in order to achieve the required precision. It is a good rule to compare chromatographic peaks before and after smoothing to ensure that the process has not introduced too much distortion.

7.1.1c Integration and Experimental Determination of Signal:Background

The preceding discussion has attempted to summarize some general fundamental concepts underlying the practical goal of integrating appropriate chromatographic peak areas (e.g. EIC vs TIC). Modern mass spectrometer data systems include integration routines and these are briefly discussed in Section 7.4.3 but practical issues concerning peak integration are discussed in more detail in Section 10.5.3a. Similarly, practical measurement of S/B values are important in method development since it is good practice to ensure a S/B of at least five at the LLOQ; this practical aspect is discussed in Section 10.2.2.

7.2 Faraday Cup Detectors

The first component in the detection chain is the detector itself; this is always a converter of ion beam current to a conventional (electron) current that can be manipulated by subsequent components (Figure 7.1). The gold standard for accuracy in quantitative ion detection is the Faraday Cup detector (another of the contributions of Michael Faraday, see the text box in Chapter 6) although practical concerns with sensitivity and speed of response result in its restriction to specialized applications in highly precise determinations of isotope abundance ratios. In its simplest form it consists of a conducting metallic chamber or cup that intercepts the charged particle beam; a suitable bias potential is applied either to the cup itself or to a metal grid in front of the cup, to prevent secondary emission (mostly electrons) from escaping and thus distorting the measurement (the preferred interior lining material is carbon because it produces few secondary ions). A high ratio of cup depth to width also helps to efficiently capture ions while minimizing scattering losses. When the cup intersects a beam of ions it acquires the charge of the ions while the latter are neutralized by acquiring electrons from the cup. The Faraday Cup can be regarded as part of an electrical circuit in which the electrical current is carried from the ion source through the mass spectrometer vacuum by the ions, and the detector provides the interface to the solid state external circuit in which electrons act as the charge carriers (Figure 7.1).

Because of the simple correspondence between the numbers of ions and of electrons in the detector circuit, a direct absolute measurement of the ion current reaching the detector is possible with Faraday Cups by measuring the current in the external circuit. However, they have severe limitations with respect to both detection limits and response time, and as a result are not used in trace quantitation procedures in which the mass spectrometer is interfaced to a high resolution separation technique such as GC or HPLC. Ion beams of interest in this context carry currents in the range of nA to < fA (i.e., $10^{-9} - < 10^{-12}$ A, i.e. < 6000 ions s^{-1}). For a current of 10^{-12} A to provide a voltage signal of one volt for signal handling by typical electronic circuits, amplification with a high input impedance and large feedback resistance ($R_{FB} \sim 10^{12}$ ohms, Figure 7.1) is required (Koppenaal

2005). This represents the practical limit for such detection schemes, so Faraday Cup detectors are relatively insensitive compared with electron multipliers (see below) especially when used in pulse counting mode.

In addition, the Johnson noise (arising from random thermal motions of charge carriers, mainly electrons,in the large feedback resistor R_{FB},) significantly restricts the achievable S/N and detection limits. The stray capacitance (C_{stray}) in the circuit can be reduced to the pF range but, with values of $R_{FB} \sim 10^{12}$ ohms, the time constant $\tau_D \sim R_{FB} \cdot C_{stray}$ (Equation [7.1]) becomes too long for detector response to be much faster than the chromatographic timescale, and much too long for a Faraday Cup to be used in pulse counting mode. Accordingly, Faraday Cup detectors are generally used nowadays for isotope ratio mass spectrometry with high resolution magnetic sector instruments, for which high signal stability is important for precise ratio measurements that can use long observation times and appropriate signal integration techniques, i.e. the response time constraint is not relevant. Note that, since electron multipliers deliver a signal current greater than the ion beam current by factors of 10^4–10^6, the values of R_{FB} required to convert the current signals to voltages are correspondingly smaller, as are the intrinsic $R_{FB} \cdot C_{stray}$ time constants, so that pulse counting detection is limited by other factors (see Section 7.3).

Recently (Knight 2002; Barnes 2002, 2004) an array detector based on microFaraday Cups was developed for use with magnetic sector mass spectrometers that are designed to focus the mass spectrum in a plane (so-called Mattauch–Herzog design (Beynon 1960; Farmer 1963)) and was used as a charge-integrating detector in trace elemental analyses using laser ablation of samples. Conventional Faraday Cups are also sometimes used with other instrument types to measure an ion beam current directly so as to provide the reference value required to determine the (current) gain of an electron multiplier, although other indirect methods are available to do this (Fies 1988).

7.3 Electron Multipliers

The secondary electron multiplier (SEM) detector is the key to the role of mass spectrometry as an extremely sensitive analytical technique with wide dynamic range and compatibility with on-line coupling to fast chromatographic separations. The SEM was a natural development from the invention of the photomultiplier (Zworkin 1936, 1939), in which photoelectrons produced by photons falling on a conversion dynode with a photo-sensitive surface are amplified in an 'avalanche' fashion by accelerating the original (first strike) photoelectrons on to a

second dynode constructed of a material that efficiently yields several secondary electrons for each initial electron that strikes it. In turn, these secondary electrons are passed on to a second dynode, and so on through as many as 20 stages, till the final electron avalanche reaches the anode that is connected to the appropriate circuitry. A schematic of such a device containing only three dynodes is shown in Figure 7.3. The current gain of such a device depends on the photo-emission efficiency of the initial photo-sensitive surface of the conversion dynode, on the secondary electron emission efficiency of the material used for the second and later dynodes (most often a beryllium–copper alloy that has been activated by an oxidation procedure), on the number of dynodes and on the total potential applied across the device via the resistor chain (Figure 7.3). Gains higher than 10^6 can be obtained for a new high grade photomultiplier, with a linear range covering several orders of magnitude and response times of the order of nanoseconds or less. As discussed in Section 7.3.1, photomultipliers are sometimes used as electron multiplier detectors for mass spectrometry through use of a phosphor screen as the initial ion detector inside the mass spectrometer vacuum chamber, optically coupled to the external photomultiplier through a quartz window.

7.3.1 Discrete Dynode Secondary Electron Multipliers

The mass spectrometric detector that corresponds most closely to the photomultiplier is the discrete dynode SEM, in which the photo-sensitive conversion dynode is replaced by a conversion dynode of a different kind (see later) that converts the ions from the mass analyzer into electrons that are subsequently amplified by the same avalanche effect. Although discrete dynode SEMs are not much used in modern analytical mass spectrometers because of their susceptibility to degradation by chemical contamination and response time limitations in ion counting (see below), they are of historical importance and provide a useful starting point for discussion of SEMs in general. Initially the case of positive ions striking the SEM will be discussed; additional problems arise for negative ion detection, since the initial negatively charged particles (the negative ions) carry a charge of the same polarity as the secondary electrons, so that any potential difference designed to accelerate the primary particles onto a dynode will automatically retard the secondary electrons. These problems are discussed later.

The (current) amplification gain (G_{SEM}) of a discrete dynode SEM, for ions of a specified *m/z* value and chemical type accelerated on to the first (conversion) dynode

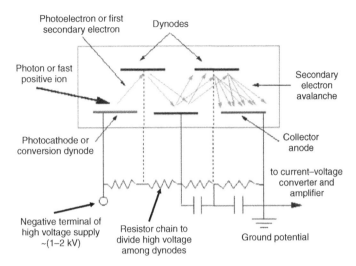

Figure 7.3   Schematic representation of a simple electron multiplier with three dynodes, applicable to either a photomultiplier or a detector for positive ions. The photocathode/conversion dynode is maintained at a high negative potential so that the positive ions are accelerated into it and the photoelectrons/secondary electrons are accelerated away from it towards the first dynode, which is held at a somewhat less negative potential (of the order of 100 V less negative than the photocathode). The secondary electrons thus produced from the first dynode are in turn accelerated towards the second dynode that is maintained at ~100 V less negative than the first, and so on, via the voltage divider chain (each resistor ~10^5–10^6 ohms). In this sketch the amplification factor per dynode (ratio of number of secondary electrons to the number of electrons striking the dynode) is two, so that the total current gain of this fictitious three-dynode device is $2^3 = 8$. For negative ions a more complex arrangement is required (Figure 7.5) since the negative potential on the conversion dynode as shown here repels negative ions.

through a fixed potential drop (V_{acc}), is defined as the ratio of the final output current of secondary electrons to the primary ion beam current (from the mass spectrometer) striking the first dynode. It can be expressed as:

$$G_{SEM} = A.\gamma_m.V_{SEM}{}^{\chi(n-1)}$$

where V_{SEM} (Figure 7.2) is the total potential difference between the first (conversion) dynode to the collector anode that controls the kinetic energy to which secondary electrons are accelerated through the device (Equation [6.16]), γ_m is the ion-to-electron conversion ratio for the first (conversion) dynode (see below), χ is a constant related to the efficiency of secondary electron emission for the material used for the other dynodes and also their design structure, n is the number of dynodes and A is a constant that varies from device to device and (sadly) with time-of-use and operating conditions for a given detector. The multiplier gain G_{SEM} is adjusted by the user via V_{SEM} and can thus be considered a tuning parameter of a special kind. A typical discrete dynode SEM with n = 18 will have a gain of 10^3 at $V_{SEM} = 1200$ V, increasing linearly in a log–log plot to ~10^7 at $V_{SEM} = 2000$ V. Such performance is typical of a new SEM, but G_{SEM} at a fixed V_{SEM} decreases with time as a result of contamination

(from the incident ions) and radiation damage in the first (conversion) dynode, and general chemical contamination on all dynode surfaces that leads to deterioration of the secondary emission characteristics. (It is true that mass spectrometry is a destructive analytical technique, in that analyte is consumed and lost, but the procedure is also somewhat destructive to the instrument itself!)

The dependence of G_{SEM} on the chemical nature of the incident ions can be illustrated by some early experiments (Stanton 1956) in which G_{SEM} was compared for hafnium ions ($Hf^{+\bullet}$) and molecular ions of anthracene ($C_{14}H_{10}{}^{+\bullet}$), both with m/z 178; the value for the organic ion was 2.2 times greater than for the monatomic ion. This and similar observations are usually interpreted as indicating that the atoms constituting a polyatomic ion can, to some extent, be considered to act independently when they strike the surface of the conversion dynode. Nonetheless, for ions accelerated on to a conversion dynode through a given potential drop V_{acc} (see discussion of Equation [6.10] in Section 6.3.1 and Appendix 6.1), the value of G_{SEM} falls off dramatically with the m/z of the incident ions. (Note that part of this acceleration can arise from extraction from the ion source and part from additional acceleration within the detector assembly, see below.) This effect reflects the *m/z* dependence of the ion–electron conversion efficiency at the conversion dynode. In an extensive investigation (Beuhler 1977) of this phenomenon in the V_{acc} range 21–84 kV, the average secondary electron yields from the conversion dynode per ion impact, for atomic and polyatomic ions (γ_a and γ_m, respectively), were found to be functions of ion velocity with an apparent threshold value. Furthermore, γ_m was found to be an additive property; for an ion accelerated on to the conversion dynode through a total potential drop V_{acc} (not to be confused with V_{SEM}) the experimental values of γ_m for small organic ions (largest were $C_{10}F_{17}^+$ and $C_7H_{16}^{+\bullet}$) were found (Beuhler 1977) to be well described by the relation:

$$\gamma_m = \Sigma_{atoms}\{(\gamma_{a,s}/26.8).[(V_{acc}/m/z)^{\frac{1}{2}} - 3.2]\} \quad [7.2]$$

where the summation is over all atoms in the ion, and $\gamma_{a,s}$ is the value of γ_a for the atom under consideration determined at a standard reference velocity chosen to be 4.17×10^5 m.s^{-1}. Representative values of $\gamma_{a,s}$ at this standard velocity for a copper conversion dynode were reported as hydrogen = 0.45, carbon = 2.15, nitrogen = 2.37, oxygen = 2.37, argon = 4.1 and xenon = 7.8. Equation [7.2] can not be expected to hold exactly for all experimental devices and all ions but it does emphasize the dependence of γ_m on V_{acc} and on the *m/z* value and chemical composition of the incident ion. Note that Equation [7.2] predicts a threshold at which $\gamma_m = 0$, given by $[(V_{acc}/m/z)^{\frac{1}{2}} - 3.2] = 0$; this can be transformed to the threshold velocity for secondary emission from the ion–electron conversion dynode since (see Equation [6.24]):

$$m_{ion}.v_{ion}^2/2 = z.e.V_{acc}, \text{ i.e., } v_{ion} = (2.z.e.V_{acc}/m_{ion})^{\frac{1}{2}}$$
$$= (2e/m_u)^{\frac{1}{2}}.(V_{acc}/m/z)^{\frac{1}{2}}$$

where m_{ion} is the mass of the ion in kg and $e = 1.609 \times 10^{-19}$ C, $m_u = 1.661 \times 10^{-27}$ kg. The threshold value of $v_{ion} = v_{ion}^0$ at which $\gamma_m = 0$ is given by Equation [7.2] as that defined by $(V_{acc}/m/z)^{\frac{1}{2}} = 3.2$, so that $v_{ion}^0 = 4.5 \times 10^4$ m.s^{-1}. This threshold prediction represents a considerable extrapolation of the experimental data on small organic ions that led to formulation of Equation [7.2] (Beuhler 1977) and thus can not be expected to provide an accurate description of the threshold region for all ions, but it does provide a semi-quantitative appreciation of the nature of the significant drop in G_{SEM} as the size (*m/z*) of the incident ions increases. The use of post-acceleration detector designs to circumvent this problem is discussed below.

Another important characteristic of any SEM is the time response. The definitions of the main time response parameters are illustrated in Figure 7.4. For the discrete dynode case, depending on the design of the dynodes, the rise time will vary over a range of a few nanoseconds down to a few femtoseconds; the fall time is usually 2–3 times longer, and this can place a limit on the dead time of such a detector in pulse counting applications. As a result the pulse width (FWHM) of the SEM output can vary from one to as high as 20 ns. The transit time of electrons through the device (Figure 7.3) can vary up to ~50 ns and this is clearly a major consideration with respect to detector dead time in pulse counting; SEM response times can be decreased by increasing V_{SEM} (and thus G_{SEM}) since the average electron speed will vary with $(V_{SEM})^{\frac{1}{2}}$ (recall that the kinetic energy (m.v^2/2) varies with V_{SEM}, Equation [6.16]). Time response in the nanosecond range is adequate for many applications of interest in this book, but for ion counting detectors and for time of flight mass spectrometers a faster response is required (see the discussion of channel electron multiplier arrays in Section 7.3.2).

Delta function light

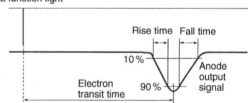

Figure 7.4 Definition of time response characteristics of an electron multiplier detector. The initiating event is here assumed to be an effectively instantaneous ('delta function') flash of light, but could equally well be a single charged particle. Reproduced from *Photomultiplier Tubes: Basics and Applications (3rd Edn)*, Hamamatsu Corporation, with permission.

The so-called 'dark noise' introduced by a discrete dynode SEM when operated under appropriate conditions is largely Johnson noise, the result of random thermionic emission of electrons from the dynode surfaces, and is minimized by appropriate choice of materials. When SEMs are used appropriately the dark noise is generally low ($< 10^{-12}$A, negligible in applications of interest here) even when operated to provide $G_{SEM} = 10^6$. (The maximum output current from such devices can be as high as 10^{-5}A before saturation occurs.) In addition, some dark current can arise from leakage through the ceramic supports for the dynodes, particularly when they become chemically contaminated thus decreasing their electrical resistance. A very important requirement in this regard is that of background pressure; if the pressure within the SEM is too high ($> 10^{-3}$ Pa, i.e. $\sim 10^{-5}$ torr), the fast electrons moving through the device can collide with background gas molecules resulting in sporadic scattering and even ionization of these molecules. Usually the background pressure for a discrete dynode SEM is maintained at $< 10^{-6}$ torr, not only to minimize noise arising from this effect but also to avoid contamination of the device.

Linear dynamic range is another important consideration for any detector and in the case of a discrete dynode SEM this amounts to the range of input beam current over which G_{SEM} is constant. (Note that the linear range discussed here refers to the output:input ratio of the detector only and not to that of an overall quantitation method, although it can contribute to the latter as a result of saturation of the SEM, see below.) The lower end of the linear range ($\sim 10^{-12}$A) is determined by the dark current and the shot noise in the incident ion beam (Section 7.1.1a). The upper limit is determined by the (reverse) current flowing through the voltage-divider resistors that control the potential differences between dynodes (Figure 7.2) under dark current conditions within the SEM itself. This current corresponds to a flow of electrons through the resistor chain from the first (conversion) dynode to the anode (recall that electric current through a conductor is conventionally assigned as a flow of positive charge as the result of a historical accident, whereas we now know that it really comprises a flow of electrons in the opposite direction). Since these resistors are generally ~ 1 MΩ each, for a 20-stage SEM the total resistance is 20 MΩ and this background current for $V_{SEM} = 2000$ V will be 10^{-4} A. This background current through the resistor chain is in the direction opposite to the signal current arising from the electron avalanche through the SEM itself; the latter electron flow is from the first dynode to the anode within the SEM vacuum, i.e., the current through the return half of the complete circuit (the resistor chain, Figure 7.3) is from the anode to first dynode. Thus the signal and background currents oppose one another and become equal when the signal current $\sim 10^{-4}$ A so that, as this limit is approached, the output electron current is no longer directly proportional to the input ion beam current (i.e., G_{SEM} is no longer constant) and this saturation phenomenon places an upper limit (on the output current) of $\leq 10^{-5}$ A to the linear range of such a detector. Note that in pulse counting mode the linear range is also controlled by the time response of the detector (dead time).

7.3.1a Off-Axis Conversion Dynodes

For a variety of reasons virtually all SEM detectors are now used in an off-axis mode (Beuhler 1977; Stafford 1980; Rinn 1982), in which both the SEM itself and a first conversion dynode are located on an axis perpendicular to the ion beam direction (Figure 7.5). The off-axis conversion dynode can be maintained at a high potential. One reason for this strategy is that SEM detectors located in a 'line-of-sight' from the ion source can be struck by UV photons from some types of ion source, and also by fast neutrals formed in electron transfer collisions of fast ions with background gas molecules; such rogue particles give rise to spurious signals (amounting to nonchemical background) from the SEM. Another important advantage of the off-axis location for the conversion dynode is that it facilitates the efficient detection of very large ions by maintaining this dynode at much higher potentials (up to 10–20 kV) than can be applied to the SEM itself, increasing V_{acc} and thus γ_m to useful values (Equation [7.2]). A third advantage is that negative ions can be detected much more efficiently by holding the off-axis conversion dynode, rather than the SEM itself, at a high positive potential to accelerate the negative ions into the conversion dynode to produce positively charged

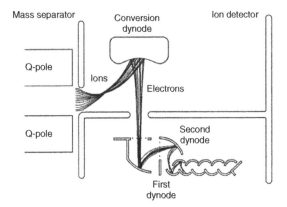

Figure 7.5 Schematic diagram of an off-axis detector for a mass spectrometer. The conversion dynode is maintained at a high potential (up to 10–20 kV) thus accelerating the ions to high velocities to improve the secondary emission efficiency (the diagram is drawn to illustrate the arrangement for positive ions). The secondary electrons are then accelerated to the first dynode (maintained at ~ 2 kV) and the SEM then amplifies the secondary electron current as usual. For negative ions the conversion electrode is maintained at a high positive potential and secondary positive ions are accelerated on to the first dynode. This example portrays a quadrupole analyzer with a discrete dynode SEM. Reproduced from *Photomultiplier Tubes: Basics and Applications (3rd Edn)*, Hamamatsu Corporation, with permission.

secondary partices (see next paragraph); this allows the detector itself to be operated in exactly the same way as for positive primary ions, so that the final dynode (anode) of the SEM can be held at ground potential with consequent ease of connection to the signal processing electronics (also at ground), rather than having to deal with complications involved in coupling an SEM at several kilovolts from ground to the electronics.

An obvious question concerns the nature of the secondary particles produced from the conversion electrode and accelerated to the first dynode. This question was investigated by some elegant experiments (Wang 1986) that demonstrated that, when positive sample ions are detected, the secondary particles from the conversion dynode consist of electrons as well as up to 50 % of negative ions; these negative ions may be the original (charge switched) sample ions or their fragments, or may originate from the conversion electrode surface, either from the electrode material itself or surface contamination. When negative sample ions strike the conversion electrode surface, positive ions are ejected and are accelerated to the SEM; again, these positive ions may be derived from the original negative sample ions or from the electrode material or surface contamination. Some X-ray photons are also probably produced in both positive and negative cases but they appear (Wang 1986) to be of only minor importance in generating the SEM output.

This work was conducted using primary ions of moderate size produced by caesium ion (Cs^+) bombardment of caesium iodide (CsI) (to produce positive ions) and Ultramark 1621 (a mixture of fluorinated alkylphospazines, commercially available from PRC Inc., Gainesville, Florida). Later it was shown (Spengler 1990) that, for large singly charged protein ions (up to > 10^5Da,) produced by MALDI, with an off-axis conversion electrode maintained at up to 20 kV, secondary electron emission was a less important contributor with increasing molecular mass of the ions. Instead, small secondary ions appeared to be the dominant secondary charge carriers for both positive and negative primary ions. Regardless of the composition of the primary ions (and to some extent of their polarity), the mass spectra of secondary ions of a given polarity were similar; therefore sputtering of surface adsorbates was assumed (Spengler 1990) to be the major mechanism of secondary ion formation, rather than surface induced dissociation of the primary ions.

A different kind of discrete dynode electron multiplier detector using an off-axis conversion dynode is exemplified by the Daly detector (Daly 1960), developed in response to reduced lifetimes of conventional discrete dynode SEMs resulting from chemical contamination and/or from venting an instrument to atmospheric pressure to effect repairs then pumping down to vacuum again etc. This device (Figure 7.6) also introduced the concept of using a photomultiplier as the electron multiplier via optical linking to a phosphor plate located inside the mass spectrometer vacuum chamber. The Daly detector consists of a metal 'doorknob' off-axis conversion electrode maintained at a high potential, a scintillator (phosphorescent screen) and a photomultiplier located outside the vacuum chamber. Ions that strike the 'doorknob' release secondary electrons that are accelerated to the scintillator (maintained at ground potential) where they are converted to photons. These photons are detected by the photomultiplier through the quartz window in the vacuum chamber. The main advantages of the Daly detector are those common to all off-axis conversion dynodes, including increased detection efficiency for high m/z ions (although the original design was not applicable to negative ions), together with avoidance of contamination of the electron multiplier and easy replacement of the latter when required. The reason for the peculiar 'doorknob' shape of the conversion electrode was the focusing of the secondary electrons on the phosphor slab by the shape of the electrostatic lines of force created by this shape. Currently some manufacturers of mass spectrometers use modifications of the Daly concept (use of a phosphor plate and an external photomultiplier) with a design of the conversion dynode and electron optics different from the 'doorknob' design. More recently (Pramann 2001) the Daly concept was modified for negative ions by accelerating the negative ions to an on-axis metal grid with 80 % transmittance and held at +12kV. The secondary electrons generated from this grid then pass to an (on-axis) aluminum-coated scintillation crystal held at a potential between +110 and 114 kV, where each electron induces a light flash. The scintillation crystal is mounted on a cylindrical light guide that seals the vacuum chamber and that directs the light flashes to a fast (2.6 fs rise time) photomultiplier.

Discrete dynode SEMs have excellent dynamic range and gain (up to 10^7, see above) but have some significant disadvantages. The emissive properties of the oxidized beryllium–copper (Be–Cu) dynodes are highly susceptible to chemical contamination, they are rather expensive to fabricate and have limited lifetime even when maintained clean, and their time response characteristics (rise times and peak widths of several nanoseconds, electron transit times of tens of nanoseconds, also inconveniently large pulse height distributions) are inadequate for modern ion counting applications particularly in recording time of flight spectra.

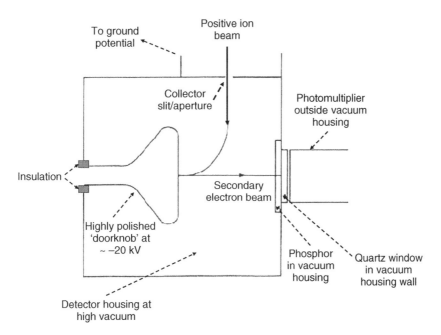

To ground potential

Positive ion beam

Collector slit/aperture

Photomultiplier outside vacuum housing

Insulation

Secondary electron beam

Highly polished 'doorknob' at ~ −20 kV

Phosphor in vacuum housing

Quartz window in vacuum housing wall

Detector housing at high vacuum

Figure 7.6 Schematic of a Daly detector for positive ions.

7.3.2 Channel Electron Multipliers

Most of the disadvantages of discrete dynode SEMs are overcome by the development of channel electron multipliers (CEMs) (sometimes also referred to as continuous dynode electron multipliers, CDEMs) that are by far the most widely used ion detectors in analytical mass spectrometers.

7.3.2a Single Channel Electron Multipliers

These devices use a continuous tube fabricated from lead glass (and more recently of ceramic) made conductive by alkali-doping but with controllable resistance (usually 10^8–10^9 ohms) to suit the purpose at hand (Kurz 1979). (Such devices are often referred to as 'Channeltrons' but this is actually the trademarked name for the version marketed by BURLE ElectroOptics Inc.) These devices maintain their secondary emission characteristics even when cycled to atmospheric pressure and back to vacuum. In addition they exhibit dark counts $< 0.1s^{-1}$, low power demands on the high voltage supply and can be engineered (see later) to provide much faster response characteristics and more narrow distributions of pulse heights than discrete dynode SEMs. While it is always best to operate any SEM at as low a pressure as possible to avoid contamination of the emissive surface, CEMs can be successfully operated with good lifetimes at pressures

upto discrete dynode $\sim 10^{-5}$ torr without the arcing that can destroy multipliers. For these reasons CEMs currently dominate all other types of SEM as detectors for analytical mass spectrometers. In future, CEM detectors may be supplanted by new concepts (Koppenaal 2005) but at present they represent the best available performance in terms of compromises among gain, dynamic range, useful lifetime and response time.

Schematics illustrating the principles underlying CEMs are shown in Figure 7.7. Unlike the discrete dynode case, the potential gradient that accelerates the electrons along the device is not created in discrete steps using a resistor chain (Figure 7.3) but rather by adjusting the intrinsic resistance of the interior wall of the device such that a continuous potential gradient is provided; this also results in a low background current that replenishes the surface charge lost to the secondary electron avalanche.

Straight tube CEMs can not be operated at gains $> 10^4$ as a result of instability arising from ionization of residual background molecules by fast electrons within the tube and acceleration of the positive ions towards the input end. Some of these ions strike the wall, creating secondary electrons that are multiplied as usual, thus causing spurious output pulses not directly related to the ion current input (i.e. sharp spikes in the detector output). The probability of such events obviously increases with both a rise in background gas pressure and with the

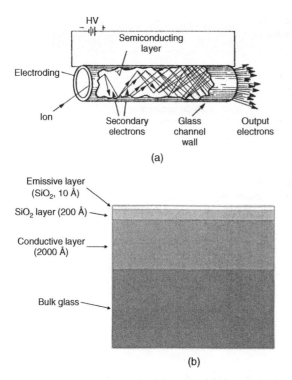

(a)

(b)

Figure 7.7 Schematic diagrams of a channel electron multiplier. (a): principle of operation of a CEM for detection of positive ions; the 'electroding' at each end consists of a thin band of metal deposited to provide electrical contact, and the potential gradient along the length of the device is developed from the external high voltage supply applied along the intrinsic resistance of the doped lead glass. Reproduced from Wiza, *Nucl. Instrum. Methods* **162**, 587 (1979), copyright (1979), with permission from Elsevier. (b): A cross-section of the surface structure of a CEM. Reproduced from literature provided by Burle ElectroOptics Inc, with permission.

number of secondary electrons (i.e. the gain G_{SEM}). To minimize this effect the CEM tube is curved, thus limiting the distance that a positive ion can travel towards the input end. Since the highest probability of such ion creation is at the output end (highest electron current), a high degree of curvature near this end severely limits the distance over which such ions can be accelerated and thus the energy with which they strike the surface. In this way CEMs can be fabricated with gains up to 10^7 when operated in analog mode and 10^8 in pulse counting mode (more on this distinction later).

It is important to note that the gain of a CEM is a function of both channel length and diameter, but not independently. Rather the gain is found to be a function of the aspect ratio, defined in this case as the ratio of the length

to the diameter. It is this fact that allows miniaturization of CEMs without loss of gain and thus fabrication of CEM arrays with many applications, including sensitive detection with fast response time characteristics (that do scale with CEM length) suitable for TOFMS, for example (Section 7.3.2b).

Some CEMs are operated using an external bias resistor in series with the power supply (Figure 7.8(a)) to ensure that the collector electrode for the secondary electrons is maintained at the most positive potential in the detector circuit. Some more modern CEMs provide an integrated bias resistor by also making a portion of the outside wall resistive together with appropriate electroding of the exterior wall, thus simplifying the circuitry. The optimum value of the bias resistor depends on the internal CEM resistance that can vary fairly widely in the manufacturing process. Also, as the CEM ages a higher voltage is must be applied along the CEM to obtain the desired gain, and as a result a higher potential drop is also applied across the gap between the exit of the CEM itself and the electrically-isolated collector (Figure 7.8(a)). This CEM collector potential difference can have an effect on gain and other operational parameters and it is preferable to maintain it at a constant value, so a Zener diode is sometimes used in place of the bias resistor to 'clamp' the bias voltage (Figure 7.8(b)). Operation as a detector for negative ions is optimized, as for discrete dynode SEMs, by combining with an off-axis conversion dynode (Figure 7.5).

The collection efficiency of any SEM measures the ability to detect an input event (e.g. an ion from the mass spectrometer) and convert it into an output signal. CEMs are constructed with collector 'cones' or 'horns', with a larger cross section than that of the CEM channel, to permit efficient collection of beams of primary positive ions (or of secondary ions or electrons from an external conversion dynode). All secondary electrons arising from the first impact of these input ions/electrons should travel down the channel and create an electron avalanche, but some of the 'first strike' electrons can escape from the cone. A grid carrying a moderate negative potential can be placed in front of the cone to act as a shield from any outside potentials that may attract these electrons, so forcing them to travel down the channel. Another option is to add focusing plates to direct the input ion beam so that the first strike event will occur at the apex of the cone; the secondary electrons will then be created deep enough in the voltage gradient that they will preferentially travel down the channel.

For CEMs the linear dynamic range is limited at the upper end by the maximum count rate capability for versions designed for use in pulse counting mode, or by maximum linear output current for those designed for

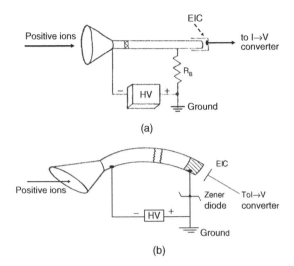

Figure 7.8 (a): Positive bias provided for a CEM with an electrically isolated collector using an external bias resistor. It is also possible to provide the electrical potential gradient along the length of the CEM by incorporating a conductive coating of variable resistance on the external wall as well as the interior wall. The output from the EIC is passed to a current–voltage converter plus an analog amplifier (analog mode) or pulse counting device (see Figure 7.1). (b): Provision of positive biasing of a CEM using a Zener diode. The 'horn'-shaped inlet end permits collection of an ion beam with large cross-section. Modified from literature provided by Burle ElectroOptics Inc., with permission.

analog mode. (This can contribute to limitations in ULOQ for an integrated analytical method.) Both of these limiting parameters for CEMs are dependent upon the bias current at the operating voltage (the current flowing through the CEM's internal resistive layer when the high voltage is applied, see the corresponding discussion for a discrete dynode SEM). CEMs designed for continuous current operation will produce a linear output up to a value equal to 10–20 % of the bias current; this limit arises because the bias current supplies the electrons necessary to replenish the surface charge and thus sustain the secondary emission process, as noted above. Typical bias currents for CEMs designed to operate in analog mode are up to $30-40\,\mu A$, so the maximum linear output is $\sim5-10\,\mu A$. Since typical dark currents in a analog mode CEM are $< 1pA$, the usable dynamic range of such a detector is thus $\sim10^6$.

For a CEM designed for pulse counting detection the maximum count rate capability depends not only on bias current, but also on channel capacitance (and thus the response time) and the condition of the channel surface. As the count rate is increased beyond a limiting point the pulse amplitudes begin to decrease, and eventually

fall to a level below the threshold at which genuine ion detection pulses can be successfully distinguished from spurious noise spikes (via a pulse height discriminator, see Section 7.4). The upper limit on count rate is ultimately determined by the output pulse width, about 10–20 ns for standard single channel CEMs, as this defines the detector dead time during which no additional ions can be counted; the maximum count rates in the linear range can be as high as 10^7-10^8 counts s^{-1} for modern CEMs designed for this purpose. The lower limit on the dynamic range for a pulse counting CEM is determined by the dark count rate, which can be as low as 0.05 counts s^{-1}, generally far too low to have any significant effect on the LLOQ of a complete analytical method where chemical background is far more important.

The primary difference between CEMs designed specifically for pulse counting and analog modes is that CEMs designed for pulse counting operate at the condition where space charge effects within the electron avalanche cloud cause saturation (plateau) in the gain vs V_{CEM} curve. The important consequence is that, at this condition, pulse counting CEMs produce output pulses with a characteristic narrow amplitude distribution whereas analog CEMs (operating in the region where gain increases with V_{CEM}) have a very wide distribution of output pulse amplitudes and thus detector dead times. The high (saturated) gain of pulse counting CEMs allows setting a discriminator level to reject low level intrinsic noise (not chemical background) while passing essentially all the fixed amplitude signal pulses, resulting in very good S/N performance (Section 7.4). Note that further increase of CEM voltage above the space charge plateau results in significant undesirable contributions from positive ions created from background gas molecules, discussed above. This phenomenon implies an operational disadvantage of using very high V_{SEM} values to maximize CEM gain in order to meet sensitivity targets. To produce the high gain optimized for pulse counting operation the channel length-to-diameter ratio must be $> 50:1$, whereas for CEMs intended for analog operation this ratio needs to be only $\sim30-35:1$. A longer channel results in a higher channel resistance and thus lower bias current (and thus lower upper limit to the linear range if operated in analog mode, see above) for a given V_{SEM} value. Mass spectrometers designed for true trace analytical applications will have the CEM detector selected for pulse counting mode.

As mentioned above, the ruggedness of a CEM represents an advantage over the older discrete dynode SEMs. However, as for the latter, the lifetime of a CEM can be significantly affected by the environment in which it is operated; chemical contamination together with intense electron bombardment can result in a change in surface composition that in turn results in decreased gain. In cases where an insulating organic layer has built up on the emissive surface, washing with 1M sodium hydroxide solution has been reported (Gibson 1984) to restore the original performance but, in general, successful cleaning of a contaminated CEM is difficult to achieve. For a newly installed CEM, proper pre-conditioning is critical to remove any loosely adsorbed gases on the surface prior to full operation; this is generally done by operating the CEM at some fixed voltage with an input signal for a short period under a good vacuum to degas the device. Most CEMs used in mass spectrometry applications are nowadays pre-conditioned at the factory and are shipped in hermetically sealed containers; storage for more than six months is generally recommended to be done in dry nitrogen or in a vacuum.

The two areas of a CEM that are most susceptible to degradation are the output and input areas of the inner channel wall. For the output end the high density cloud of fast electrons results in physical and chemical changes to the surface that reduce the secondary emission characteristics, thus reducing the gain. Under very clean vacuum conditions where chemical contamination is not a problem (i.e., not in an analytical mass spectrometer!), the normal life of the output end surface can be measured in terms of total output charge (in coulombs) and is ~30C. CEM useful lifetime varies with the specific application and environment but is typically in the order of about 2000 hours operation. In mass spectrometry applications the input region of a CEM is important in determining useful lifetime. If used without an off-axis conversion dynode (Figure 7.4) the impact of high molecular mass ions at high energies can result in chemical changes in the glass surface and even physical sputtering. When an external conversion dynode is added to perform ion–electron conversion, the primary ions from the mass spectrometer do not strike the CEM directly. The CEM emissive surface is much less affected by electrons than by much more massive and chemically active ions, and positive secondary ions from the conversion dynode are less likely to damage the CEM surface since they are usually small ions rather than heavy organic ions.

One method to improve lifetime of the CEM output end involves increasing the output surface area. This can be accomplished by using a microchannel plate with thousands of channels (Section 7.3.2b). A less drastic approach using the same principle is to use a CEM with more than one channel (four or six channel versions are commercially available) that are twisted in a barber's pole fashion around a solid central support, resulting in effectively 4–6 times the output surface area and about double the usable lifetime. An additional advantage of this approach results from the tight spiral construction that severely limits the distance that secondary positive ions produced near the output end can travel back up the tube before striking the surface, so that their kinetic energy and thus ability to create spurious secondary electrons is minimized. As a result it is possible to operate such multichannel CEMs at appreciably higher pressures (down to $\sim 5 \times 10^{-3}$ torr), an advantage for design of portable field-deployable mass spectrometers since much less demanding vacuum pumps are required.

Conventional CEMs, either single or multi channel, are perfectly adequate for ion counting detection for weak mass spectrometry signals of interest in this book (unless TOF analyzers are used). The time characteristics, typically pulse risetimes of 3–5 ns and pulse widths (FWHM) of ~20ns, permit detection of single ions every 50 ns or so, i.e., 2×10^7 ions s^{-1} corresponding to an ion beam current $\sim 3 \times 10^{-12}$ A (assuming singly charged ions each carrying a charge 1.602×10^{-19}C). Such a low beam current is clearly susceptible to significant shot noise (Section 7.1.1a), leading to a theoretical best S/N value of ~ 0.7 (given by Equation [7.1a]). The ion sampling time (dwell time) t_D (Equation [7.1a]) can be lengthened (or alternatively the recorded count numbers over several sampling times can be averaged) to improve this figure of merit. (Note, this is not the signal:background problem (Section 7.1) which can not be addressed in this fashion). As noted previously, the experimental challenge is to reach a compromise between increasing the ion sampling time t_D (experimentally corresponds to SIM/MRM dwell time) while still providing sufficient data points for an adequate record of the rise and fall of beam current as a chromatographic peak passes through the ion source.

7.3.2b Channel Electron Multiplier Arrays

For modern time of flight (TOF) analyzers that can provide resolving power (FWHM definition) $\geq 10^4$ via incorporation of the ion detector as an intrinsic component of the m/z analysis, such CEM response time characteristics are inadequate. This is easily illustrated by an ultra-simplified calculation for a typical case in which the spatial and time focusing properties of the TOF analyzer (Section 6.4.7) are sufficient that the resolving power (RP) is limited by the time response of the detector. The time of flight of an ion is given by Equation [6.39]:

$$t_f = (m/z)^{\frac{1}{2}} . X_d / [2.e.V_{acc}/m_u]^{\frac{1}{2}} \qquad [6.39]$$

Consider an ion with m/z 1000 accelerated through $V_{acc} = 20$ kV in an instrument with a flight path length

$X_d = 2$ m. Then evaluation of Equation [6.39] gives $t_f = 32\,\mu$s. For this example with fixed X_d and V_{acc}, differentiation of Equation [6.39] and simple algebraic rearrangement gives:

$$\delta t_f = (t_f/2)/[m/z/\delta(m/z)] = (t_f/2)/[RP]$$
$$= (32 \times 10^{-6}/2)/[10^4] = 1.6 \times 10^{-9}\,\text{s} \qquad [7.3]$$

This represents the difference in flight times between ions of m/z 1000 and m/z 1000.1, i.e., the difference between the peak maxima in the raw flight time spectrum, for an instrument with $RP = 10^4$ at m/z 1000. However, for these two peaks to be clearly defined and resolved from one another, it is possible to estimate that the flight time spectrum in this inter-peak region should be represented by 5–10 time sampling 'slices', so the time resolution should be $\sim$5–10 times smaller than δt_f, i.e., of the order of 0.2 ns (200 ps). Although this is only an order of magnitude estimate for purposes of illustration, it is far outside the capability of even the best modern single channel CEM.

As discussed above, the time resolution of a CEM based ion detector is ultimately limited by the electron transit time (Figure 7.4) through the channel and the anode impedance. Obviously the former can be decreased by shortening the channel length, but to maintain the channel gain (controlled by the ratio of the length to the diameter, i.e., the aspect ratio) the channel diameter must be reduced proportionally. This requirement places severe demands upon fabrication technology. These were successfully met in the channel electron multiplier array (CEMA, Figure 7.9, also known as channeltron plate detectors) originally developed (Wiza 1979) as an amplification element for image intensification devices ('night vision' devices in which the primary particles are photons rather than ions or electrons). For this original purpose it was necessary to provide acceptable spatial resolution and this in turn required that the channels should have small diameters (and thus lengths to maintain the high gain) with similarly small center-to-center inter-channel spacings (the 'pitch' of the CEMA plate).

The first CEMAs exploited as detectors for TOFMS had pore diameters of 25 μm and time resolution of two nanoseconds, appreciably better than that of single channel CEMs but still a factor of 10 worse than the target value estimated above. Modern CEMAs designed for TOFMS detection have channel diameters $\sim$2 μm (Figure 7.9) with proportionally short channel lengths (aspect ratios typically 70:1) in order to provide current gain (see above) of about 10^4 at $\sim$2 kV across the plate. This is not sufficient for the most demanding applications, so a 'chevron' composed of two CEMA plates in series (Figure 7.10) is used to increase the gain

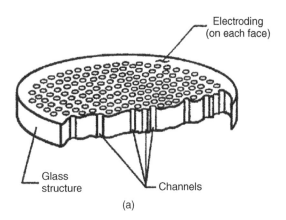

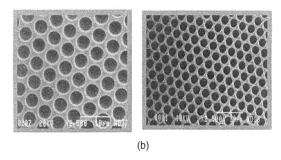

(b)

Figure 7.9 (a): Sketch of a cutaway view of a channel electron multiplier array plate. Nickel–chromium electroding is applied to both surfaces of a microchannel plate to provide electrical contact and also penetrates into the channel. The penetration depth is minimized on the input face (0.3–0.7 of the channel diameter) to maximize the first strike conversion efficiency of incoming ions/elecrons channel. Reproduced from Wiza, *Nucl. Instrum. Methods* **162**, 587 (1979), copyright (1979), with permission from Elsevier.

(b): electron micrograph images of portions of the surface of CEMA plates with 5 μm and 2 μm pore diameters, taken at the same magnification. Reproduced from literature provided by Burle ElectroOptics Inc, with permission.

to $> 10^6$. Of course this coupling of two plates as a chevron increases the electron transit time and thus the response time, but even so impressive performance can be achieved with rise times $\sim$250 ps and peak widths $\sim$350 ps (Figure 7.10). Reduction of the channel diameter also permits an increase of the percentage of the entire area of the CEMA plate that consists of open pores (the 'open area ratio'), and thus increases the number of channels covering a given area of a CEMA plate. In turn, this reduces the detector dead time that is partly determined by the number of missed events (incident ions or electrons), which arise because of a second event entering

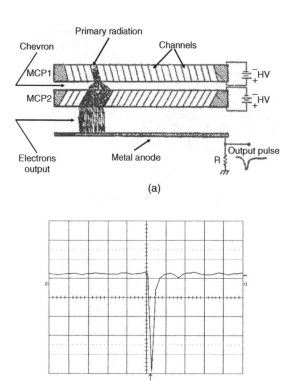

(a)

(b)

Figure 7.10 (a): Sketch of a double-plate chevron detector showing the angled channels designed to ensure that incident particles strike the interior walls at an optimum angle. Reproduced with permission from Wiza, *Nucl. Instrum. Methods* **162**, 587 (1979), copyright (1979) Elsevier.
(b): Typical pulse width measurements (1 ns/division) for a chevron comprised of two CEMA plates with $2\,\mu m$ diameter channels; risetime $\sim$250 ps, peak width (FWHM) $\sim$350 ps. Reproduced from literature provided by Burle ElectroOptics Inc, with permission.

a specific channel within the dead time of that particular channel. For example, if a larger diameter channel (e.g., $25\,\mu m$) is replaced by a number of smaller pores (e.g., 2–$5\,\mu m$), only the individual 2–$5\,\mu m$ channel struck by the first ion would be dead for its characteristic dead time period, and a second ion or electron entering another 2–$5\,\mu m$ channel, displaced from the first by as little as 2–$5\,\mu m$, would still be detected and amplified. This effect can significantly improve the dynamic range of a pulse counting CEMA chevron at the top end and greatly alleviate the limitations on ULOQ values for TOF analyzers that are often imposed by the dead time characteristics of the ion detector.

A wire mesh is usually placed in front of the entrance of a CEMA plate, typically 20×20 or 30×30 wires

per centimeter, providing ion or electron transmission through the grid of approximately 90 %. Such grids are used to create a uniform field across the detector entrance and/or to apply accelerating or retarding voltages. The 'bias angle' of a CEMA plate is the angle between the channels and the perpendicular to the CEMA surface, and is typically 5–12° (Figure 7.9). This optimizes the performance since it ensures that incident primary charged particles strike the internal emissive surface at an appropriate channel depth (close to the channel entrance). It is found that reversing the direction of the bias angle in a two-plate chevron (Figure 7.10) optimizes the combined performance.

The dynamic range of a CEMA chevron at the lower end (detection of extremely low signal levels) is limited by the dark current (dark count) and/or the background noise inherent in the multiplier. The background count rate of an ion detector can be ultimately limited by the cosmic ray background found here on planet Earth, but in practice the limit for a CEMA is usually controlled by the amount of radioactive decay of trace impurities in the glass used to fabricate the device; specialized low noise glass formulations are now used to minimize this problem. The upper limit of dynamic range is ultimately limited by charge saturation effects within the multiplier itself (see above for a discussion of this effect in single channel CEMs, recall that any CEM in pulse counting mode is operated at the space charge limit). As mentioned above, increasing the channel redundancy per unit area (i.e., reducing channel diameter) increases the upper limit of the count rate within the linear dynamic range.

The useful lifetime of any CEM detector can be expressed in terms of the total accumulated electric charge associated with the electron avalanche emerging from the output end. For a CEMA chevron, up to $\sim$40 coulombs per cm^2 can be extracted in a clean working environment without significant gain degradation (typical CEMA active areas are circular with 1.8–2.5 cm diameters, i.e., $\sim$3 cm^2). Considerable care must be taken to avoid contact with corrosive gases. CEMA plates readily absorb moisture from the environment as they are highly hygroscopic (even storage with a typical silica gel dessicant is dangerous because the CEMA will actually pull moisture out of the dessicant!) but the hydration process is reversible.

In summary, single channel CEMs (or variants with four or six channels 'twisted' in a barber's pole configuration) are currently used as mass spectrometric detectors and are designed for either analog (current measurement) mode for higher level signals or for pulse counting mode for low level signals. For circumstances in which detector time response is a critical parameter, as in TOF analyzers,

CEMA plates (usually in a chevron configuration) are required and such an application inherently demands some kind of pulse-counting strategy.

7.4 Post-Detector Electronics

Post-detection electronics are now briefly considered (Figure 7.1). Most mass spectrometrists do not concern themselves with details of the various circuits used for this purpose, as these represent the specialized expertise of electrical engineers. However, these circuits are important components of the overall measurement process (Figure 7.1), so it is helpful for users to have at least a general understanding of their principles of operation.

7.4.1 Analog Signal Processing

In the case of input ion beam currents too high for ion counting techniques (as a consequence of detector dead times), the CEM used is selected to be optimized for current mode (i.e. in the operating range where output current is linearly related to the input, well below the space charge limited condition for CEMs, Section 7.3.2a). This is the approach used in traditional mass spectrometry (Harris 1984) and is only briefly discussed here. The first step in such cases is to use a current-to-voltage converter to transform the detector output current to a proportional voltage across a resistor R_{FB} (Figure 7.1). The main sources of noise with a current-to-voltage amplifier are associated with the fluctuating thermal current ('Johnson noise') in the feedback resistor, which gives rise to corresponding noise on the output voltage signal of the converter; this signal is proportional to R_{FB} since the noise current is a function of temperature only, while the noise varies as $R_{FB}^{\frac{1}{2}}$, so the S/N value (not signal:background) increases with $R^{\frac{1}{2}}$. However, the value of R_{FB} also determines the minimum time response constant of the system, since there are inevitably stray capacitances (of the order $C = 1pF$) associated with any resistor and the leads connected to it, thus giving an RC time constant of $10^{-12}R_{FB}$ s. This introduces the need for a compromise between arranging for a reasonably high output signal from the current–voltage converter via high R_{FB} values, and a time response that is fast enough for faithful recording (without too much 'smoothing' by this analog circuitry, Section 7.1.1b) of rapidly varying signals. In addition, the value of R_{FB} must be at least 10^3 times lower than the leakage resistance to ground via the material used to construct the structures used to physically support the CEM, R_{FB} etc.; 10^{14} ohms is about the upper limit for ground leakage resistance, so an absolute upper limit to R_{FB} is 10^{11} Ω, implying a time response constant of 0.1 seconds that is clearly far too long for applications as a chromatography detector. Most current–voltage converters are provided with the facility to switch to lower values of R_{FB} to provide a suitable time response, but at the cost of lowering the voltage signal delivered as input to the variable gain voltage amplifier (Figure 7.1).

Amplification of low level DC voltage signals (DC signals are here defined as signals whose time variation is much slower than the time response characteristics of the processing electronics) is difficult because DC amplifiers are notoriously susceptible to drift and other instabilities. A common method used to overcome this problem is to use some form of chopper amplifier, in which the direct current input is converted into a square-wave alternating current signal by either one or two choppers that alternate between the signal and ground; this square-wave is then amplified by a high gain AC amplifier, then converted back to a DC signal (by the integrated 'demodulation' circuit).

In modern practice final recording of even analog mode mass spectrometer signals uses a computer based system rather than the chart recorders that were formerly used, so that the data always end up as a time series of digital measures of ion beam intensity. The final analog output voltage signal from the chopper amplifier must therefore be converted to digital form. Analog-to-digital converters (ADCs) are now ubiquitous in laboratories as well as in everyday life and only a few comments are offered here. The signal resolution of an ADC simply indicates the number of discrete values it can produce over the maximum range of input signal voltage and is usually expressed in bits, e.g. a converter that can transform a input analog signal into one of 256 discrete values (0 to 255 inclusive) has an ideal resolution of eight bits since $256 = 2^8$, while 12-bit and 16-bit converters assign digital numbers (0 to 4095) and (0 to 65 535), respectively. (These are values commonly used for mass spectrometer systems.) The corresponding voltage resolution is simply given by the maximum permissible input signal voltage (from the detector) divided by the number of discrete values, e.g., for an input signal voltage range of 0 to 10V (a common choice), a 12-bit converter will provide voltage resolution of $10V/4096 = 2.4\,mV$.

The so-called 'quantization error', the difference between the analog input voltage and its digital representation, reflects this intrinsic finite resolution of an ADC and is measured in a unit called the LSB (least significant bit); for a 12-bit converter an error of one LSB is 1/4096 of the full signal range. If the input signal is much larger than one LSB, the quantization errors are independent of the actual magnitude of the signal and are found to have a distribution (equivalent to noise) over many sampling periods with a standard deviation given by $LSB/(12^{\frac{1}{2}}) = 0.289 \times LSB$. For a 12-bit converter this corresponds to a quantization error of 0.007 % of the full signal range, entirely insignificant compared with other sources of error and noise in the overall system.

However, at lower signal levels (the regime of interest here) the quantization error becomes a function of the input signal and can become significant; in such cases the effective resolution of the converter is limited by the S/N value of the input signal and it is impossible to accurately resolve beyond a certain number of bits, the so-called 'effective number of bits'. In such cases the lower bits are simply measuring electrical noise. It is possible in principle to increase the signal resolution of an ADC by increasing the number of bits but this can only be done at the cost of a significant decrease in the time resolution.

7.4.2 Digital Electronics for Ion Counting Detectors

The most accurate and precise quantitative measurements do not employ the current measurement methods described thus far, but instead use ion counting techniques. The use of ion counting techniques in mass spectrometric analyses of very small samples was first intoduced (Barton 1960; Ihle 1971) for determination of relative isotope abundances in radioactive metals, but the first published report of application to trace level quantitation of organic analytes by GC/MS appears to have been somewhat later (Picart 1980); modern practice uses the same fundamental principles but with greatly improved detector and electronics technology.

Unfortunately, as mentioned previously, such techniques can be applied only to low level ion beam currents, such that the probablity of a second ion striking the detector during the dead time is sufficiently low, thus limiting the upper limit of the linear dynamic range of the detector system. A problem common to all analog voltage digitizers at low ion counting rates is that the digital output is not simply the probability of a primary particle (ion, electron etc.) striking the detector but only some function of that desired information distorted by the rise and fall times of the electronics, electronic noise spikes etc.; somehow there has to be some means of discriminating against the artifacts and recording only the part of the signal that can be attributed to the primary particles. Thus it makes sense to eliminate such electronic artifacts by recording only digital 'yes or no' responses as to whether or not a primary particle has arrived.

However, there are practical difficulties in the application of this pulse counting principle for mass spectrometric detection. One of these problems arises from the fact that the number of electrons at the multiplier output, arising from the arrival of individual ions, is not the same for each ion, as a result of the distribution of secondary electrons in each step of the multiplication avalanche (this is equivalent to saying that G_{SEM} for individual ions is not a constant but exhibits a distribution of values). Although CEMs give an appreciably more narrow distribution than discrete dynode SEMs, this variation in the amplitudes of the single ion voltage pulses to be counted creates problems for the pulse counter (Figure 7.1). Moreover, as discussed above, spurious pulses can arise from sputtering of ions from the CEM surface that are accelerated back towards to the inlet end, and also from radioactive impurities in the glass and from cosmic radiation. However, particularly for CEMs the majority of these spurious pulses are significantly smaller than those produced by the arrival of a fast heavy ion at the inlet of the detector, and these sources of error (noise) can be reduced to small values by introducing a pulse discriminator between the multiplier and the pulse counter (Figure 7.1) that rejects pulses below a user-set amplitude level. (These discriminator levels are often referred to as 'ion threshold levels', or some similar phrase, in mass spectrometer control software.) In practice only a small fraction of the true signal pulses are rejected at a discriminator level that cuts out most of the undesired background. Such pulse height discriminators usually also adjust the peak heights and shapes to constant values to optimize the performance of the pulse counter.

Practical ion counting detector systems for mass spectrometers, requiring a precise record of counting rates that vary with time, are often based on the multichannel scaler (MCS) concept. Even when recording full scan spectra an MCS behaves like a series of (typically) 65 536 counters. When the measurement is started by an appropriate start signal, the first counter counts the pulses (corresponding to ions striking the detector) for a counting dwell time t_D (see Equation [7.1]). When the first counter stops counting after the user-set dwell time, the second counter counts the pulses for the second time interval t_D, and so on until the last counter has completed its counting interval. At the end of the 'scan' the contents of the 65 536 counters show the variation of count rate with time over a total time period of $65\,536 \times t_D$, and these data are transferred to a computer as a series of ion counts for the 65 536 time 'bins'. Although this same purpose could be achieved by computer-based automation of a single counter, the dead time contributed by computer read-out of the counter leaves significant gaps in the data when the counting intervals are short. An MCS is an efficient solution for recording a spectrum of count rate vs the time after the start signal pulse. For example, in scanning instruments a mass calibration file of m/z vs time (a 'look-up table') is acquired using a standard sample yielding a range of ions of known m/z values (Section 6.4.1), and this calibration is used by the computer to assign m/z values to each of the counters in the MCS. In the case of TOF instruments, the relationship between m/z and the time at which each counter starts counting (i.e., the flight time) is particularly direct. It is possible, by carefully synchronizing the start of the instrument scan with that of the MCS, to repeat the scan several times and accumulate the results in a single 'spectrum', in order to reduce the statistical scatter and improve the S/N value by a factor $\sim n_S^{\frac{1}{2}}$, where n_S is the number of scans thus accumulated.

Some form of time-to-digital converter (TDC) is generally used for TOF instruments in view of the common use of time measurement. In a TDC a counter is set up to count at a constant clock rate and the time difference is measured as the number of clock counts (not ion counts) between a start signal (provided by the laser pulse in a conventional MALDI–TOF instrument, or by the orthogonal acceleration pulse in the modulator of an oa–TOF apparatus, see Section 6.4.7) and the stop signal provided by a single ion pulse from the CEMA detector. This time difference, corresponding to time of detection of a single pulse following the start pulse that sends the ion on its flight through the TOF analyzer, is sent to the data processing unit as the measurement of the flight time t_f (Section 6.4.7). In an alternative method, the counter is normally reset to zero after sending the previous data, then starts counting at the time when the start signal arrives at the TDC and stops counting when the stop signal arrives, and the value of the counter is recorded as before. Since the clock frequency of the TDC is known, the time at which the pulse arrived is easily derived and can be related to the m/z value of the ion via the measured flight time. In such a 'single stop' TDC only the time difference between the start signal and the first stop signal is measured, so that only the pulse that first arrived at the ion detector can be measured in any one measurement. In modern practice a 'multi stop' TDC, which can output a plurality of counter values in response to a corresponding plurality of stop pulses (and thus to corresponding detection times and thus to m/z values), is used.

CEMA dead times, typically ~ 1ns (i.e. of the same order as the width of a TOFMS mass peak but usually less than the spacing between mass peaks 1 Th apart), provide a limit to the upper end of the dynamic range for any ion counting strategy. However, since ion arrival times are described by Poisson statistics (see the text box in this chapter), saturation effects on beam current measurements can be corrected to some extent by applying a correction factor derived from the known mathematical characteristics of the Poisson distribution to each recorded count rate for each bin (Ihle 1971). The precise way in which such a correction calculates the fraction of hits that were not detected during a dead time period (and applies a corresponding correction) varies with the particular application but generally it can provide extension of the linear dynamic range by a factor of up to 10. Algorithms that apply such correction factors are generally included in the software package supplied with an instrument that uses ion counting detection.

Somewhat different specifications apply to different applications. Thus, for accurate and precise m/z measurements using a TOF analyzer operating with mass resolution $\sim 10^4$, if the uncertainty required is 5 ppm the corresponding maximum uncertainty in the measured flight time is half of this value (see Equation [7.3]), i.e., 2.5 ppm. For example, for ions with flight times of 20 μs, the uncertainty in flight time measurement is required to be < 50 ps. Such performance can be achieved with modern CEMAs with multi stop TDCs. On the other hand, for scanning mass spectrometers the time accuracy/precision requirements are generally less demanding; an MCS capable of higher count rates (up to $\sim 10^8$ s^{-1} is possible) with excellent time resolution (~ 15 ppm) is usually fit for purpose.

Thus far the discussion of ion counting detection has emphasized application to full spectrum recording. For TOF and ion trap analyzers this is currently the only realistic option anyway (but see the recent innovation for TOF analysers (Brenton 2007) described in Section 6.4.7), but for quadrupole mass filters and magnetic sector instruments the SIM and/or MRM modes are appropriate for trace quantitative analyses. In this case, an MCS can be used in a slightly different mode in which the dwell time per bin is increased to permit statistical averaging of the signals; of course there is an upper limit on the MCS dwell time imposed by the need to provide enough data points across a chromatographic peak that it can be sufficiently well defined for integration.

7.4.3 Computer-Based Data Systems

The final step in the detection and measurement scheme (Figure 7.1) is to transfer the data to a computer for storage and also processing. The many algorithms used for data processing in commercial instruments are generally opaque to users, so it is appropriate to repeat the cautionary words that were quoted already in Chapter 4:

'In situations as complex as these, there is no single right answer. A computer program, no matter how powerful, sophisticated and well designed, cannot take the place of the watchful eye of the chemist. The computer can carry out analysis and do so by several procedures within its repertoire, but it has to show the user, preferably graphically, what it did, how it did, and what are the consequences of its assumptions. Be it in QC or research environment, the final responsibility for the accuracy and reliability of analysis rests with the human. Many of those who have intimate experience with computers think that perhaps that is just as well.' (Kipiniak 1981).

This quotation emphasizes the importance of the judicious conversion of data (however processed by a computer) into information, the final step in any mass spectrometric analysis (Figure 1.3), or indeed in any scientific experiment. In general, the calibration algorithms for converting detection time to m/z value perform extremely well. The only cautionary words here concern the associated 'm/z-peak-find' algorithm that identifies features in the mass spectrum as peaks and sets rules to define the beginning and end of the m/z peak (see discussion of peak shapes and widths with reference to Figure 6.5 in Section 6.2.3b); if the associated parameters are properly set relative to the noise level, the subsequent assignment of an m/z value by calculating the time centroid of the identified peak and transforming to m/z via the calibration file is straightforward. The time centroid C_{peak} of a peak recorded as a function $I(t)$ of time t, where I denotes the intensity, is defined as:

$$C_{peak} = \int I(t).t.dt / \int I(t)dt \qquad [7.4]$$

which amounts to determination of the t = constant line (parallel to the I-axis) along which the peak would 'balance' if held horizontal. The problem arises in determining the limits of the m/z peak over which the (numerical) integration is to be performed; many algorithms ignore the lower intensity edges of the peak. Modern computer-based data systems can calculate centroids on-the-fly, so that the only data actually stored are the time centroid and its corresponding m/z value, the peak area $\int I(t)dt$ and possibly the peak maximum I_{max} and its corresponding time t_{max}. This represents a great saving in the required storage capacity on-disk, although for some purposes the complete time series of digital values $I(t)$ from the ADC or ion counter are stored (often referred to as 'continuum acquisition'). Usually the user does not have to adjust the parameters used in this on-the-fly m/z peak integration and determination of centroids.

The same general problems of peak identification and integration (determination of peak area) are apparent in processing of chromatograms (see Figure 4.27). It is here that the warning that the computer should 'show the user, preferably graphically, what it did, how it did it and what are the consequences of its assumptions' (Kipiniak 1981) most directly affects trace level quantitation. Most data systems do provide a facility whereby the user can examine the chromatographic peak, together with the assumed baseline and limits of integration (limits with respect to both t and I), and even permit replacement of the automatically-determined limits by user-determined values based on human judgement. This facility can be particularly important for peaks with low S/Nvalues (see e.g. Section 10.5.3a). A generalized logic sequence such as that used for automated determinations of chromatographic peak positions and their areas is shown in Figure 7.11. These algorithms identify the 'start' and the 'end' of a peak, based on criteria built into the algorithm, to determine when to start and stop counting ions. As a result of this feature, together with the fact that 3D ion traps operate with longer and less well defined reproducible scan times at the beginning and the end of the peaks (because of the decreased ion populations and thus longer ion fill times, Section 6.4.5), these integration algorithms lead to poorer precision in quantitation for traps than for constant dwell time instruments such as quadrupoles.

Note that the logical test as to whether or not the S/B value of the putative chromatographic peak is above a specified level (Figure 7.11) is analogous to the use of a pulse height discriminator to determine whether or not a pulse height from a CEM detector in ion counting mode arose from a genuine ion detection event. However, Figure 7.11 captures only the general principles of a peak identification algorithm. There are additional features in such algorithms that can handle peak splitting, fronting, tailing etc.(Section 3.5.12) with various degrees of success. When additional features such as these must be employed, it is especially important to visually inspect the integrated chromatograms to check on the reasonableness (or otherwise) of what the automated algorithms have decided to do!

7.5 Summary of Key Concepts

With the exception of FTICR and Orbitrap instruments, all mass spectrometers in current use for quantitative trace analysis detect and 'count' ions using **ion detectors based on the electron multiplier** concept. The first such devices (not for mass spectrometric use) were **photomultipliers** designed to detect and measure very low intensity

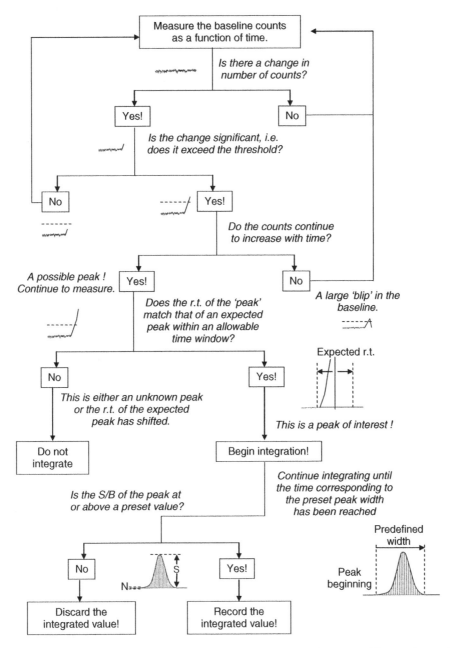

Figure 7.11   Illustration of a general logic sequence for automated determination of the presence/absence of a genuine chromatographic peak (clearly evident within the background, see Figure 4.31) and the limits over which a genuine peak should be integrated for purposes of quantitation. Note that the step that queries whether the S/B is above a pre-set value corresponds in a general way to use of a pulse height discriminator in testing a pulse height from a CEM detector in ion counting mode to determine whether or not the pulse arose from a genuine ion detection event.

UV-visible radiation. These devices create an **electron avalanche** by accelerating **secondary electrons** from one metallic dynode to the next using externally applied potential differences between adjacent dynodes. The same **discrete dynode design** was adapted (with appropriate changes in the materials used in their construction) to ion detectors operated within the mass spectrometer vacuum system. While these discrete dynode ion detectors can provide gains in electric current (ratio of output electron current to the incoming ion beam current) of up to 10^6 with remarkably little contribution to noise, they are disappointingly susceptible to chemical contamination, particularly on the first dynode under direct ion beam bombardment, leading to degradation of gain and other characteristics. Moreover, their response time is too slow for modern ion counting applications.

These problems can be solved by directing the ion beam onto a **phosphor block** that is **optically coupled** via a quartz window in the vacuum enclosure wall **to a modern fast response photomultiplier**. Such ion detectors are in use in some modern mass spectrometers, but the most common detectors are some form of **channel electron multiplier (CEM)**, in which discrete metallic dynodes are replaced by a **continuous tube fabricated from lead glass or special ceramic made conductive by alkali doping** with controllable resistance; in this way a large potential difference can be maintained along the channel length (to accelerate the electrons) without drawing too large a current. CEMs are **remarkably resistant to chemical contamination** from within the mass spectrometer vacuum enclosure and **can be operated at pressures as high as 10^{-5} torr** (compared with 10^{-6} torr for discrete dynode detectors).

CEMs can be engineered either for **analog operation (continuous measurement of ion beam current)** or to provide very short response times to single ion events, which is important for **ion counting detection strategies** that are **relevant to trace level analyses** that are the subject of this book. CEMs designed for pulse counting operate at the plateau in the curve of multiplier gain vs applied accelerating potential to provide output pulses with a narrow distribution of pulse amplitudes, a desirable feature for pulse counting electronics. The CEM response time (really the '**dead time**' between detection of one ion and recovery of the device such that it is ready to detect the next ion) can be decreased by controlling the physical dimensions of the channels. **Channel electron multiplier arrays (CEMAs)** can be manufactured with channel diameters down to a few micrometers, with response times sufficiently short to meet the needs of time of flight analyzers, at the expense of somewhat lower gain than that of larger CEMs.

The **upper limit for ion beam currents that can be measured using ion counting detection** (for trace quantitative analysis) is **determined by the dead time of the ion detector**, since ions arriving during a detector dead time do not give rise to an output pulse and are thus not counted. The **time distribution of individual ions arriving at the detector is essentially random** but is well **described by Poisson statistics**; this can be exploited in application of a correction factor applied to ion counts to extend the linear range by up to a factor of 10 above the limit imposed by the detector dead time.

This random (irregular) rate of ion detection also corresponds to so-called '**ion beam shot noise**' that could, in principle, limit the signal:noise ratio (S/N) in measurements of the kind discussed in this book. However, time widths of chromatographic peaks are almost always sufficiently long that the electronics can be set for **sampling (ion counting) periods that are long enough to average out the irregular ion arrival rate**. In this context, a 'sampling period' is most easily visualized as the dwell time for a specific MRM channel. It is important to arrange for a sufficient number (typically 9–12) of ion counting sampling periods across a chromatographic peak so that the peak area can be determined with appropriate accuracy and precision. Under such conditions the **S/N value of the chromatographic peak area is limited not by ion beam shot noise but by the chromatographic background** ('chemical noise') that is, in turn, determined by the selectivity of the overall analytical method. If the acquisition involves full scan spectra, the scan speed must be fast enough to provide 9–12 spectra across the chromatographic peak but sufficiently slow that ion beam shot noise is not a limiting factor.

With the notable exception of CEMA arrays in time of flight instruments, ion detectors are now used in an **off-axis configuration**, usually in conjunction with an additional **conversion dynode that can be held at a high potential** independent of that applied to the detector itself. Such an arrangement has three **practical advantages**: it avoids detection of **spurious signals such as UV photons or fast neutral species** arising from the ion source; the high potential applied to the conversion dynode can **significantly increase the overall detector gain;** and the polarity of the potential applied to the conversion dynode can be changed to permit **detection of either positive or negative ions**.

The electronics associated with electron multiplier detectors always include a **high stability power supply** that provides the accelerating potential to create the electronavalanche. Processing the output signal from the electron multiplier is accomplished in **analog mode for high**

ion beam currents by special amplifiers such as a chopper amplifier, but this is not particularly relevant for instruments designed for trace analysis. **Digital electronics are required in ion counting mode**. In either case a digitized signal is supplied to a **computer-based system designed to record and process the raw data**, which consist of values of ion beam current reaching the detector from the mass analyzer as a function of (chromatographic) time. Modern data systems do an excellent job of recording the raw data but it is advisable to **treat with caution the decisions made by automated** **routines** such as those used to detect a chromatographic peak, determine the start and stop points, and evaluate the corresponding peak area. Complications arising from poor chromatographic resolution, peak tailing/fronting, chemical background etc. are not always properly dealt with by computer-based algorithms. Inspection of individual chromatograms by an experienced analyst is essential in such cases. Ideally, peaks should be integrated consistently throughout the run, prior to and independent from evaluation of the calibration curve and of concentration values for QCs or samples.

8

Tools of the Trade VII: Statistics of Calibration, Measurement and Sampling

Some Thoughts on Errors, Statistics and Real Life

This famous photograph of the tragic result (several lives were lost) of a human error at a railway station of the 'Chemin de Fer de L'Ouest' (Western Railway) in 1895 reminds us that everything we do, including within the laboratory, is prone to error and uncertainty. It is unlikely that statistical methods would have been of any practical assistance in this example, but this is certainly not the case for our analytical measurements, despite

this equally well known quotation usually attributed to Benjamin Disraeli, a British Prime Minister of the 19th century: 'There are three kinds of lies: lies, damned lies and statistics.' One might question whether this distinguished politician knew as much about statistics as the other two categories he mentioned?

A more candid view of how politicians view the science of statistics is attributed to Winston Churchill, another British Prime Minister :

> "I gather, young man, that you wish to be a Member of Parliament. The first lesson that you must learn is, when I call for statistics about the rate of infant mortality, what I want is proof that fewer babies died when I was Prime Minister than when anyone else was Prime Minister. That is a political statistic."

Unfortunately this kind of comment has led to a degree of cynicism about statistical methods in the general public. For example:

> "It is now proved beyond doubt that smoking is one of the leading causes of statistics".

> "97.2516 % of the people who use statistics in arguments make them up".

Nonetheless, in view of the uncertainties inherent in any quantitative scientific measurement, statistical treatment of the raw data is essential. However, the necessity of application of mathematics and statistics to chemical measurements in particular has not always been enthusiastically endorsed:

Trace Quantitative Analysis by Mass Spectrometry   Robert K. Boyd, Cecilia Basic, Robert A. Bethem
© 2008 John Wiley & Sons, Ltd

(Continued)

"Every attempt to employ mathematical methods in the study of chemical questions must be considered profoundly irrational and contrary to the spirit of chemistry . . . . if mathematical analysis should ever hold a prominent place in chemistry – an aberration which is happily almost impossible – it would occasion a rapid and widespread degeneration of that science." Auguste Comte, *Cours de philosophie positive*, 1830.

Comments of this kind tend to confirm the Murphy Law of Statistical Distribution:

'Whatever it is that hits the fan will not be evenly distributed'.

8.1 Introduction

All measurements made by human beings are subject to error and uncertainty. This is neither a 'good' nor a 'bad' thing but a fact of life that calls for an intelligent and balanced approach to interpretation of the results of the measurements. Statistical methods provide an approach that yields quantitative estimates of the uncertainties in the raw measurements themselves and also in the conclusions drawn from them. The theory and practice of statistics are themselves demanding scientific disciplines with a considerable literature, despite many jokes about statistics and statisticians (see text box). Many excellent texts and study courses on general statistics are available but only a few take into account the complexity (and messiness) of practical analytical chemistry, particularly at the trace level. For example, the real-world analytical chemist usually does not have available a large population of analytical samples that can be analyzed quickly (and at negligible cost) to provide a large number of measurements that can be interpreted in terms of some idealized statistical theory; this also applies to proposals to conduct large numbers of replicate measurements on a few samples. Practical considerations, such as imperfect sampling of heterogeneous matrices like soils or solid biological tissues, qualitative confirmation of analyte identity, chemical interferences of various kinds, plus nonstatistical considerations (cost, time pressures etc.) that can affect decisions on experimental strategy, are usually not considered at all in general introductions to statistical methods. Further, analytical chemists are regularly faced with a wide range of circumstances that call for different strategies. For example, a very large effort is currently devoted to development, validation and use of analytical methods designed for routine repetitive use in all stages of discovery, development and manufacture of new pharmaceutical drugs; a very different example might be the need to urgently develop a new analytical method to address the extent of an unexpected contaminant in a food product that has already been released to the public. The criteria to be applied in determining an optimum analytical method

that is fit for purpose are obviously different in these two cases, and so are the statistical approaches to evaluation of the measurements.

As mentioned above, statistical methods yield quantitative estimates of the uncertainties in the raw measurements and also in the conclusions drawn from them. Any statistical evaluation of quantitative data involves a mathematical model for the theoretical distribution of the experimental measurements (reflecting the 'precision' of the measurements, Section 8.1.1), a set of hypotheses, and a user-selected criterion for making a decision concerning the validity of any specific hypothesis (e.g., the 'confidence level', see Appendix 8.1). The outcome of a statistical test is never a 'hard fact' but rather a statement about the probability of the validity of a formulated hypothesis concerning the experimental measurements (e.g., is the 'best' experimental value for a pollutant concentration, however determined, greater than a specified regulatory limit). Because of this, statistical considerations do not, in principle, give a clear-cut answer to the apparently simple question as to whether or not a particular experimental result is 'acceptable'. Rather, it will be seen that the analyst must ultimately determine whether an experimental result is valid, or not, and that there are quantitative statistical tests to help guide these decisions.

The most widely used theoretical model for the distribution of experimental measurements remains the Gaussian distribution, also known as the 'normal' distribution, an unfortunate label in view of the common colloquial meaning of 'normal'. Three major assumptions are made when the Gaussian model is applied to quantitative analyses: i) that the number of replicate measurements of x conducted is representative of a 'normal' distribution, (ii) that the experimental mean value $\bar{x}$ provides a valid estimate of the 'true' value, together with the assumption that the distribution is centered on the experimental mean value $\bar{x}$ (this amounts to assuming that systematic errors are negligible), and (iii) that the experimental standard deviation s_x [Equation 8.2c] is a good estimate for the theoretical (Gaussian) counterpart σ_x (Section 8.2.3). In real life, however, the number of replicate measurements of some trace level target analyte will generally be in the range 3–6, dictated by considerations of time, cost, availability of the analytical sample etc.; as a result, estimates of confidence levels obtained on the basis of the three assumptions can only be regarded as approximate indications. Clearly, a better mathematical model for the statistics of small numbers of replicate measurements is required and it will be seen how the 'normal' (Gaussian) model has been modified for application to small data sets (particularly the Student t-distribution, Section 8.2.5).

Several excellent texts describing general statistical methodology are available (e.g., Taylor 1982) and are certainly valuable, but the present account relies heavily on texts that are directed specifically to, and written by, analytical chemists (Meier 2000; Brereton 2003), as well as on some admirably concise but complete articles (Miller 1988, 1991; Lavagnini 2007). The first part of the chapter (Sections 8.1–8.3) should be regarded as a brief introduction to the general concepts and language of statistics, with an emphasis on their application in trace quantitative analysis; a glossary of some terminology used by statisticians is included as Appendix 8.1. Section 8.4 deals with statistically defensible definitions and evaluation of the limit of detection (LOD) and of the lower limit of quantitation (LLOQ, often denoted simply as LOQ). Section 8.5 discusses various real-life circumstances that dictate the theoretical models (equations) that describe expectations of how calibration should behave, i.e., the models that form the reference point for statistical testing. For example, what is the appropriate calibration equation if no internal standard or blank matrix are available, and what are its intrinsic limitations. Section 8.6 introduces the difficult problem of statistically valid sampling of heterogeneous matrices (particularly solids and particulate solids such as soils) although in some cases, e.g. when the analytical sample is a well mixed homogeneous liquid, choosing a representative aliquot for analysis is essentially trivial. This last aspect (the 'sampling problem') might

logically be expected to be considered first, since selection of a representative aliquot of a heterogeneous sample to be taken through the analysis is necessarily always the first step (Figure 1.2). However, since some fundamental statistical concepts must be established first, discussion of the sampling problem is deferred till Section 8.6.

In general throughout the chapter, although some of the algebraic notation may appear unfriendly, in fact the algebra itself is rudimentary.

8.1.1 Systematic and Random Errors: Accuracy and Precision

It is convenient to classify experimental errors into two categories which can be illustrated by the commonly used analogy with hits on a target or dartboard (Figure 8.1), such that the center of the target represents the 'true' or 'reference' value of the measured quantity. Illustrations of various combinations of high/low accuracy with high/low precision are readily accommodated within this simple model.

'Systematic error' describes the deviation between the (usually unknown) 'true' value (often referred to more realistically as the 'reference' value, see later for experimental estimates of this value) and the value determined by a specified analytical method. Systematic errors are conveniently further sub-divided into 'bias' errors (constant offset of the experimental value from the 'true' value) and 'proportional' errors (systematic differences between experimental and 'true' values with a magnitude that is proportional to that of the 'true' value). The effect of a systematic error is to lead to measured values that are either all larger or all smaller than the 'true' value or its best experimental estimate (although simultaneous random errors, if sufficiently large, might possibly cancel out this effect of the systematic error and result in a few measured values deviating in the opposite direction). The operational meaning of 'true value' varies with

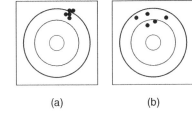 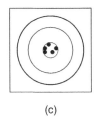 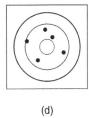

(a)   (b)   (c)   (d)

Figure 8.1 Simple visualization of meanings of accuracy and precision in experimental measurements, using darts thrown at a dartboard as an analogy in which the center of the dartboard represents the 'true' value. (a) Good precision, low accuracy. (b) Poor precision, poor accuracy. (c) Good accuracy and precision. (d) Poor precision, and therefore accuracy is indeterminate (quantitative statistical evaluation is essential).

circumstances; in some cases the analytical method can be checked for systematic errors by applying it to a suitable certified reference material (CRM, Section 2.2.1), but in other circumstances the CRM is replaced by a blank matrix spiked with known amounts of analytical standard (Section 8.5). 'Accuracy' is the term used to describe the magnitude of systematic errors in application of a specified analytical method to a particular type of sample (high accuracy implies low systematic errors). An example of bias might be a constant contribution to the instrumental signal (that was chosen to monitor the analyte) from some unsuspected contaminant arising from the solvent used in the analytical extract and that is thus present at constant amount for a fixed injection volume.

An example of a proportional systematic error would arise if the unsuspected interfering contaminant were present in the analytical standard used to create the calibration curve used to quantify the analyte; thus the amount of contaminant contributing to the calibration curve would be proportional to the amount of standard analyzed. Another example of a proportional error would arise if the extraction efficiency of analyte from the sample matrix was determined by spiking a blank matrix sample with pure standard, and then assuming that the extraction efficiencies are the same for both the added standard and the analyte originally present. This condition cannot be assumed to be valid for many environmental samples, in which occlusion of the analyte in the sample matrix can cause a constant fraction of the native analyte to be unavailable to the chosen extraction method, a situation that is difficult to detect (Boyd 1996, Ke 2000). Methods that permit separation of systematic error into its bias and proportional components (Cardone 1983, 1986, 1990; Ferrus 1987) are available and are discussed later in the context of the Method of Standard Additions (Section 8.5.1b). For now, it is important to note that systematic errors can not be detected nor determined by the extensive statistical methods that have been developed to deal with random error, and which are a major topic of this chapter. In general, when systematic errors are detected or suspected it is best, whenever possible, to investigate the origins of these errors and eliminate or minimize them. Unlike random errors, systematic errors can not be reduced by increasing the number of measurements performed, but if identified and quantified they can in principle be used to correct the experimental values.

Random errors, unlike systematic errors, can lead to measured values that are either larger or smaller than either the 'true' value or the best experimental estimate of the true value. Such errors can arise from various sources, some of which are unavoidable in principle (e.g.,

fluctuations that are the result of shot noise (Section 7.1.1) in a low intensity ion beam), while some correspond to experimental parameters that could in principle be better controlled (e.g., temperature of an extraction solvent or of a chromatography column). 'Precision' is the term used to describe the magnitude of random errors (high precision implies low random errors); as explained later, the precision can be improved by increasing the number of replicate measurements, but estimates of the random error (again unlike systematic error) can not be used to correct the experimental values, even in principle.

In analytical practice, precision is often sub-divided into repeatability and reproducibility. Repeatability refers to the spread in values obtained on multiple repeats of the same analysis of the same sample, performed consecutively by the same operator using the same apparatus; repeatability thus measures the degree of short term control of the analytical method. In contrast, the reproducibility of the analytical procedure refers to comparison of experimental estimates of repeatability obtained some time later by the same analyst using the same method and equipment, or perhaps in a different laboratory and/or by a different analyst using similar equipment on another aliquot of the same analytical sample or a similar sample; thus defined, the experimental reproducibility clearly involves both random errors and possibly uncontrolled systematic errors (e.g., arising from temperature variations, differences in clean-up efficiency etc.).

Some conventions (Ellison 2000) restrict use of the term 'error' to describe systematic error, and 'uncertainty' to refer to random fluctuations in the experimental values. Thus, the best result of an analysis after appropriate correction may by chance be very close to the 'true' value, i.e., have a negligible 'error', but the 'uncertainty' may still be large, reflecting the effects of uncontrolled random variations in experimental parameters reflected in the spread of the measured values. However, note that the usefulness of the concepts of accuracy and precision can break down in some cases (Figure 8.1(d)) in which the 'best' value derived from the replicate measurements could conceivably be assigned as close to the 'true' value but with a large uncertainty. In other words, a claim of accuracy is not sustainable if the precision is very low, since the apparent accuracy could have arisen purely by chance; it is here that the statistical concepts of confidence limits and confidence levels are crucial. (A short glossary of statistical terminology is included as Appendix 8.1.) Figure 8.2 shows a different illustration of some of the same concepts, for a case in which the number of replicate measurements is sufficiently large that the distribution of measured values can usefully be described by a continuous curve.

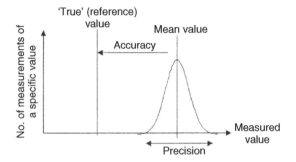

Figure 8.2 Graphical illustration of results of a sufficiently large number of replicate measurements of the same quantity that the distribution of measured values can be represented by a continuous curve, here assumed to be a 'normal' or 'Gaussian' distribution (see text). The accuracy (measured as the systematic error) is intrinsically a uni-directional (vector) parameter, i.e., must be assigned a sign (+ or –), while the precision is a spread (a scalar parameter, represented here by the total range of the measured values). The 'reference value' can be interpreted as the generally unknown 'true' value or a best estimate of the latter, e.g., the certified value documented for a certified reference material (CRM). The case corresponding to Figure 8.1(d) would be represented here by a much more broad distribution that encompasses the reference value.

8.2 Univariate Data: Tools and Tests for Determining Accuracy and Precision

Although not stated as such, the discussion thus far has implicitly concerned univariate data, i.e., replicate measurements of a single parameter under closely controlled conditions. A simple example might be a series of weighings to determine the mass of an object. Of course, the fact that a spread of experimental values is always obtained indicates that some of the experimental conditions are not completely under control. However, this class of measurements is usefully contrasted with bivariate and multivariate data (we shall be mainly concerned with the bivariate case, Section 8.3). Experimental measurements become two-dimensional under various sets of circumstances (Meier 2000). The case of main interest in this book corresponds to cases in which measured values (e.g., mass spectrometric signal intensities) are considered as functions of an experimental parameter (e.g., concentration or amount of a specified analyte injected into the instrument), as in acquisition of a calibration; determination of the functional relationship between the two parameters is called regression. A related but somewhat different case concerns correlation analysis between two experimentally observable quantities (e.g., signals from a mass spectrometer and from a UV absorbance detector). The correlation behaviour is tested

to discover whether one of the parameters (e.g., UV absorbance) can be used as a surrogate for the other (e.g., the more specific but also more expensive mass spectrometer). A third example of a bivariate data set arises from a 'round robin' (laboratory inter-comparison) exercise in which two samples are tested for the same parameter (e.g., concentration of a target analyte) by each of several laboratories, using a fixed experimental method. A variant of this example arises in tests of the uniformity of many nominally similar samples (e.g., pharmaceutical pills) by a single laboratory that measures two different parameters (e.g. amount of the active ingredient and the total mass of each pill). Such data sets are normally interpreted using some version of *ANOVA* (*AN*alysis *Of VAri*-*ance*) that tests the significance of deviations among mean values obtained for two or more groups of measurements (Section 8.2.6). However, a logical progression in this introduction to statistical methods for quantitative analysis requires that concepts applicable to univariate data should be considered first.

8.2.1 Mean, Median, Standard Deviation, Standard Error

When summarizing the results of a set of n replicate measurements x_i ($i = 1, 2, \ldots, n$) of a single experimental measured quantity x, it is necessary to deduce both some typical value representing all the measured values and also an indication of the spread of values obtained (i.e., how reliable is the quoted typical value). The most commonly used quantity describing the typical value is the mean (arithmetical average) $\bar{x}$ defined as:

$$\bar{x} = (1/n).\Sigma_i(x_i) \qquad [8.1]$$

The experimental mean $\bar{x}$ is taken as the best available approximation to the 'true' mean, usually denoted μ. When only a small number n (< 10 or so) of values is available, the median x_{med} is sometimes a more appropriate measure, particularly if there appear to be observed values that are possible outliers (values that are far removed from all the others, probably as the result of some uncontrollable factor such as a power surge, see Section 8.2.7 for methods to determine whether or not it is justifiable to discard an experimental value as an outlier). The median is defined as the value that divides all the experimental values into equal halves of 'high' and 'low' values. If n is odd, there are $(n-1)/2$ values of x_i smaller than the median so the next higher value, x_j with $j = (n+1)/2$, is taken as x_{med}; if n is even, the average of the middle two observations ($i = n/2$ and $i = (n+1)/2$) is taken as x_{med}.

For example, for the set of $n = 7$ values [$x_i = 2, 3, 5, 5, 6, 6, 7, 15$], $x_{med} = 5.50$ and $\bar{x} = 6.125$. However, if the possible outlier ($x_7 = 15$) is omitted, the median is barely changed to 5.00 while the mean $\bar{x}$ is shifted to 4.857 (Figure 8.3). Typically the median value is a more robust indicator of a typical value of a set of univariate data than is the mean, e.g., in comparisons of family incomes in two different countries if a small fraction of families can have very high incomes well removed from the vast majority this can lead to misleading conclusions based on the mean values. In analytical chemistry, the main value of comparisons exemplified by the fictional data in Figure 8.3 lies in their ability to highlight 'suspicious' values x_i that should be examined as possible outliers whose exclusion can be justified by appropriate statistical tests (Section 8.2.7).

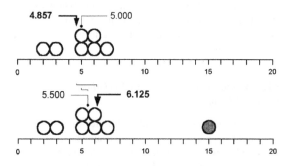

Figure 8.3   Graphical representations of a set of $n = 8$ experimental values [$x_i = 2, 3, 5, 5, 6, 6, 7, 15$]. Top: data omitting the possible outlier $x_8 = 15$, represented by $\bar{x} = 4.857$ and $x_{med} = 5.0$. Bottom: all data included, represented by $\underline{x} = 6.125$ and $x_{med} = 5.5$. Note the relative robustness of the median value towards inclusion or not of the possible outlier. Reproduced from Meier and Zünd, *Statistical Methods in Analytical Chemistry*, 2nd *Edition* (2000), with permission of John Wiley & Sons Inc.

The other property of a set of univariate data that must be specified to give an adequate summary description is the dispersion, the extent of the spread (scatter) of the n values around the mean reflecting the extent of random error, i.e., the precision. The simplest parameter describing the dispersion is the range, the difference ($x_{max} - x_{min}$) between maximum and minimum values. Obviously this parameter is extremely susceptible to distortion from extreme values (possible outliers). The most commonly used measure of dispersion is the standard deviation of the data set s_x defined as follows:

$$SS_x = \Sigma_i(x_i - \bar{x})^2 = \Sigma_i(x_i^2) - (\Sigma_i x_i)^2/n \quad [8.2a]$$

$$V_x = SS_x/(n-1) \quad [8.2b]$$

$$s_x = (V_x)^{\frac{1}{2}} \quad [8.2c]$$

where the summation giving SS_x is over $i = 1$ to n. SS_x is the sum of the squares of the deviations of the x_i from $\bar{x}$ (the second form of Equation [8.2a] is derived by simple algebra and is more convenient for calculation). Note that the sum of the deviations of the x_i values from the mean would be zero by definition and is thus not a very useful parameter for describing the dispersion; the squares of the deviations are all positive and so do not cancel like the actual values do. V_x is the variance of the data set and is a measure of the average squared deviation using $(n - 1)$ instead of n as the divisor since, based on the definition of SS_x, there are only $(n - 1)$ independent values of x_i, i.e., once $\bar{x}$ and $(n - 1)$ values of x_i are defined the final (n^{th}) value is fixed. Thus $(n - 1)$ is the number of degrees of freedom (*dof*) in this simple example. In principle, the standard deviation can be properly interpreted and used in statistical tests only if the data set meets the criteria for a 'normal' (or Gaussian) distribution (Section 8.2.3), but it is widely used in less stringent cases, usually without serious consequences for the ultimate conclusions drawn from its use.

It is helpful to distinguish between the experimentally derived parameter s_x (Equation [8.2]) and the theoretical quantity σ_x that appears in mathematical models of experimental distributions, e.g., the 'Gaussian' or 'normal' distribution. This mathematical model for the distribution function for measured values of a quantity subject to random error (systematic errors are not accounted for) is discussed more fully in Section 8.2.3, but it is convenient to specify its form here:

$$G_{\mu,\sigma}(x) = [\sigma_x.(2\pi)^{\frac{1}{2}}]^{-1}.\exp[-(x-\mu)^2/2\sigma_x^2] \quad [8.3]$$

where $G_{\mu,\sigma}(x)$ gives the probability of obtaining a particular measured value x, μ is the theoretical 'true' value (estimated experimentally as $\bar{x}$), and σ_x is the theoretical counterpart for s_x; the function varies with the values of μ and σ so these parameters are appended as subscripts as in Equation [8.3]. This equation is the normalized Gaussian ('normal') function. It is unfortunate that the same terminology is used for two different purposes, as exemplified by the preceding sentence. The 'normal' distribution function $G_{\mu,\sigma}(x)$ is so-called because it describes the situation for a large number of real-world circumstances, but this same 'normal' function can be expressed in either 'normalized' or 'non-normalized' form; in general mathematical terminology, a function or quantity is said to be 'normalized' if multiplied or divided by some constant factor in order to make the resulting value(s) comparable with some other data or satisfy some theoretical requirement. In the present case, inclusion of the factor

$[\sigma_x.(2\pi)^{\frac{1}{2}}]^{-1}$ in Equation [8.3] guarantees that the integral of G(x) over all values of x is unity, reflecting the obvious fact that any measured value must lie between $-\infty$ and $+\infty$. The 'normal' distribution, less ambiguously termed the 'Gaussian function', corresponds to the 'bell shaped curve' all too familiar to some students; Figure 8.4 shows some examples of Gaussian distribution curves for different combinations of μ and σ.

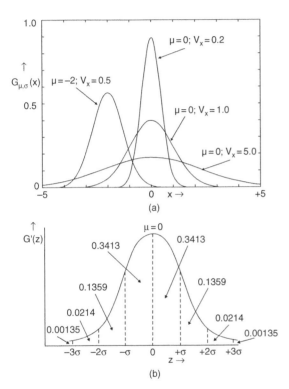

(a)

(b)

Figure 8.4 Normal (Gaussian) distribution curves, normalized so that total area under each curve is unity. (a) Three normalized curves for $G_{\mu,\sigma}$ (x) vs x (Equation [8.3]), with $\mu = 0$ and $V_x = \sigma_x^2 = 0.2$, 1.0 and 5.0 (top to bottom), plus a normalized curve with $\mu = -2$ and $V_x = \sigma_x^2 = 0.5$. (b) Normalized curve for $G'(z)$ vs z, where $z \equiv (x - \mu)/\sigma$ (Equation [8.14]). The numerical values are the contributions to the total area under the curve (normalized to unity) of the indicated areas.

The relative standard deviation (RSD), also known as the coefficient of variation (COV or CV) and usually expressed as a percentage, is given by:

$$RSD(\%) = 100.(s_x/\bar{x}) \qquad [8.4]$$

The RSD is useful in comparing reproducibilities. The standard deviation s_x should not be confused with the

standard deviation of the mean (sometimes called the standard error of the mean, or simply standard error), denoted as $s_{x,mean}$ or SE_x, and defined by:

$$s_{x,mean}(= SE_x) = V_x^{\frac{1}{2}}/n^{\frac{1}{2}} = (V_x/n)^{\frac{1}{2}} = s_x/n^{\frac{1}{2}} \qquad [8.5]$$

Essentially s_x provides a measure of the average uncertainty of each of the n individual values x_i, while $\bar{x}$ is the best experimental estimate of the (generally unknown) true value μ and $s_{x,mean}$ indicates the uncertainty in $\bar{x}$. This makes sense because, particularly when n is increased to large values, it is clear that the uncertainty in $\bar{x}$ is appreciably less than that in any one of the n individual values x_i. Indeed, for a distribution of x_i values that corresponds to a normal (Gaussian) distribution, the standard error $s_{x,mean}$ (SE_x) can be shown theoretically to be the appropriate measure of uncertainty in $\bar{x}$. Note that, in view of its definition (Equation [8.5]), SE_x decreases for increasing values of n even if the width of the distribution (measured as s_x) does not change appreciably; this is consistent with our intuitive understanding of how the reliability of $\bar{x}$ should vary with increasing numbers of measurements. Note, however, that this reliability indicated by the value of SE_x refers only to the random error; increasing the number of measurements does not in any way reduce the uncertainty associated with any systematic errors that might be present.

As a simple example, consider the set of fictitious measurements [2,3,5,5,6,6,7] considered above (see Figure 8.3 without the potential outlier). Then, the appropriate quantities are readily calculated as: $\Sigma_i x_i = 34$; $(\Sigma_i x_i)^2 = 1156$; $\bar{x} = \Sigma_i x_i/n = 4.857$; $\Sigma_i(x_i^2) = 184$; $SS_x = \Sigma_i(x_i^2) - (\Sigma_i x_i)^2/n = 184 - (1156/7) = 18.857$; $V_x = SS_x/(n-1) = 18.857/6 = 3.143$; $s_x = V_x^{\frac{1}{2}} = 1.77$; $SE_x = s_x/n^{\frac{1}{2}} = 1.77/2.646 = 0.67$. Thus, the most appropriate summary of the set of seven measurements (no outliers) is $\bar{x} \pm SE$, i.e., 4.86 ± 0.67; in view of recommendations regarding the number of significant figures (Section 8.2.2) this should strictly be expressed as 4.9 ± 0.7. A more complete description would make explicit what are the confidence limits (confidence interval) and confidence level for this result (see Section 8.2.4).

8.2.2 Significant Figures and Propagation of Error

Note that the number of figures quoted in the preceding simple example has been truncated from that displayed on an electronic calculator, in accord with the rules for number of significant figures. Since experimental measurements with precision expressed as, e.g., s_x or SE_x, are being dealt with here, it is obvious that numerical values of such parameters should not be quoted with too high a level of precision (as reflected in the number of figures quoted). It follows that this must also apply to $\bar{x}$, the best experimental estimate of the 'true' value, given by $34/7 = 4.857142857$ on the particular calculator used, clearly a ridiculous expression of a value for $\bar{x}$. The working rule for the number of significant figures (digits) that should be used when expressing the result of an experimental measurement, can be stated thus:

> The last significant digit in any quoted measured value should usually be in the same decimal position as the uncertainty.

As with all rules, this one is subject to interpretation in unusual circumstances and in such cases it is prudent to quote one additional digit in addition to those prescribed by the rule. Thus, returning to the above simple example, it could be argued that the best expression of the result is 4.9 ± 0.7. However, inclusion of one additional significant digit could be useful, e.g., if this experimental value of x were to be combined with measured values of other parameters y, z etc. to calculate a new quantity W via a mathematical function $W = f(x,y,z, \ldots)$; the resulting uncertainty in W must be calculated from those in x,y,z etc. to account for propagation of error, and it is best to include additional significant figures in such calculations and apply the significant figures rule to the end result.

The calculation of s_W from s_x, s_y, s_z etc. is an example of evaluation of the propagation of error. The accepted rules for this procedure are derived from the theory of the normal distribution (Equation [8.3]) and are given below without detailed proof. However, some insight into the origins of these rules can be obtained by considering an ultra-simple example, in which $W = f(x,y) = (x+y)$, where x and y are parameters that are measured separately and independently of one another. The 'best' experimental estimates for x and y are $\bar{x} \pm SE_x$ and $\bar{y} \pm SE_y$, i.e., it is believed that there is a high probability (see Section 8.2.4 for a discussion of confidence limits on such probabilities) that the true value of x lies between $\bar{x} + SE_x$ and $\bar{x} - SE_x$, and similarly for y. Then it might be considered reasonable to suppose that the true value of W probably lies between the corresponding highest and lowest limits for probable values of (x+y), i.e.:

highest probable value of $W = \bar{x} + \bar{y} + SE_x + SE_y$
lowest probable value for $W = \bar{x} + \bar{y} - SE_x - SE_y$

On the basis of this simple analysis, it might be expected that the best estimate of the true value of W can be expressed as:

$$\bar{W} = (\bar{x} + \bar{y}) \pm (SE_x + SE_y), \text{ i.e., } SE_W = (SE_x + SE_y)$$

That is to say, it is assumed that the true value of W lies between these two limits to within the same level of confidence (thus far not quantified) as that applied to x and y. Incidentally, the corresponding result for $W = (x-y)$ is also $\bar{W} = (\bar{x} - \bar{y}) \pm (SE_x + SE_y)$, since a worst-case scenario in which uncertainties are considered to be additive is being assumed.

However, this approach is too simple, for the following reason. If the experimental quantities x and y are measured independently, in the sense that the corresponding uncertainties do not depend on one another and are not correlated with one another in any sense, then the proposed upper limit for the probable range for $\bar{W}$, i.e., $(\bar{x} + \bar{y}) + (SE_x + SE_y)$, assumes that simultaneous observation of the highest probable values for x and y will occur with a probability equal to that of cases of partial cancellation of positive deviations from the mean value in one and negative deviations in the other (similarly for the lower limit). This assumption is simply not valid and, when the possibility of mutual cancellation of errors is taken into account within the theory of the normal distribution (Section 8.2.3), the appropriate formula for combining uncertainties in simple sums and/or differences of several independently variable measured quantities, can be shown to be:

$$\delta W_{\pm} = [\delta x^2 + \delta y^2 + \delta z^2 + \ldots]^{\frac{1}{2}} \qquad [8.6]$$

where δx etc. denote any appropriate measure of uncertainty arising from random error, usually s or SE, so δx^2 etc. represent the respective variances (Equation [8.2]). Equation [8.6] was derived (proof not given here) by taking into account the relative probabilities of combinations of random errors of opposite sign from the separate quantities x, y, z, … etc.

Similarly, it can be shown that the appropriate measure of uncertainty for a quantity $W_{x \div}$ calculated as a simple series of multiplications and divisions:

$$W_{x \div} = (x.y \ldots .z)/(u.v \ldots .w)$$

is given by:

$$(\delta W_{x \div}/W_{x \div}) = [(\delta x/\bar{x})^2 + \ldots + (\delta z/\bar{z})^2 + (\delta u/\bar{u})^2$$
$$+ \ldots + (\delta w/\bar{w})^2]^{\frac{1}{2}} \qquad [8.7]$$

i.e., take the square root of the sum of squares of the relative uncertainties of the independent experimentally measured quantities.

Equation [8.7] cannot be applied to the special case where $x = y = z = \ldots$, i.e., x^m (a power of x), since the m variables (all x) are obviously not independent of one another, a necessary condition for validity of Equations [8.6–8.7]. Instead it is necessary to revert to the general case in which W is a general algebraic function, and it is convenient to consider first a function of a single experimental quantity x, i.e., $W = f(x)$. Then simple differentiation gives:

$$\delta W = (dW/dx).\delta x \qquad [8.8]$$

Simple examples of application of Equation [8.8] are the following:

$$\text{For } W_m = f(x) = x^m, \ \delta W_m = m.x^{(m-1)}.\delta x, \text{ or}$$

$$\delta W_m/W_m = \delta W_m/x^m = |m|.(\delta x/x) \qquad [8.9]$$

There are no restrictions on the value of m (it can be positive or negative, integral or not) and this accounts for the

---

A Simple Example of Propagation of Error: Determination of the Dissociation Constant of a Weak Acid

To measure the acidic dissociation constant K_a of a weak acid AH, available as a pure solid, the plan is to make up an aqueous solution of known concentration and measure its pH.
 The following series of operations would be performed:

(1) Weigh a clean dry weighing bottle: mass of empty bottle $m_1 = 9.8916$ g, with an estimated uncertainty $\delta m_1 = \pm 0.0005$ g (an analytical balance, Section 2.3).
(2) Weigh the bottle plus a small amount of the acid, mass $m_2 = 10.0827$ g, with $\delta m_2 = \pm 0.0005$ g.
(3) Then mass of acid $m_{AH} = (m_2 - m_1) = 0.1911$ g, with $\delta m_{AH} = [(\delta m_1)^2 + \delta m_2)^2]^{\frac{1}{2}}$ (using Equation [8.6]) $= [2 \times (5 \times 10^{-4})^2]^{\frac{1}{2}} = \pm 7.1 \times 10^{-4}$ g, since m_1 and m_2 were measured in independent experiments with no correlation of any kind between them.
(4) Suppose the relative molecular mass of AH (i.e., relative to m_u) is known from its molecular formula to be 202.0956, i.e., the molar mass is 202.0956 g.mol^{-1}, with essentially zero uncertainty. Then the weighed sample of AH contains a number of moles $n_{AH} = 0.1911$ g/202.0956 g.mol$^{-1} = 9.4559 \times 10^{-4}$ mol, with $\delta n_{AH} = \delta m_{AH}/202.0956 = \pm 3.51 \times 10^{-6}$ mol.
(5) This weighed sample of AH is transferred into a 10.00 mL Class A standard flask and the volume made up to the mark with distilled water, i.e., final volume of solution $V = 10.00$ mL$(= 1.000 \times 10^{-2}$ L) with $\delta V = \pm 0.02$ mL (see Section 2.4.1).
(6) Then the total concentration of AH in this solution is $C_{AH}° = n_{AH}/V = 9.4559 \times 10^{-2}$ mol.L^{-1}, with (from Equation [8.7]) $[\delta(C_{AH}°)/C_{AH}°] = [(\delta n_{AH}/n_{AH})^2 + (\delta V/V)^2]^{\frac{1}{2}} = [1.378 \times 10^{-5} + 4.0 \times 10^{-6}]^{\frac{1}{2}} = \pm 4.216 \times 10^{-3}$; thus $\delta(C_{AH}°) = 4.216 \times 10^{-3} \times 9.4559 \times 10^{-2} = \pm 3.99 \times 10^{-4}$ mol.L^{-1}.
(7) The desired quantity is $K_a = [H^+].[A^-]/[AH]$ at equilibrium, where $([AH] + [A^-]) = C_{AH}°$. The strictly correct treatment takes into account the amount of $[H^+]$ contributed by the autoprotolysis of water, but here we assume that $[A^-] = [H^+] << C_{AH}°$ (in fact this introduces negligible systematic error in this case), so that $K_a \approx [H^+]^2/C_{AH}°$.
(8) The value of $[H^+]$ was measured using a pH meter graduated in units of 0.05, i.e. could be read to $\delta(pH) \sim \pm 0.025$ pH units. The measured value was pH $= 2.95$, so $[H^+] = 10^{-2.95} = 1.1220 \times 10^{-3}$ mol.L^{-1} and $(\delta[H^+])/[H^+]$ is given by Equation [8.11] as $2.30258 \times \delta(pH) = \pm 5.76 \times 10^{-2}$ (i.e. $\sim 6\%$ uncertainty). Thus $\delta[H^+] = 5.76 \times 10^{-2} \times 1.1220 \times 10^{-3} = \pm 6.463 \times 10^{-5}$ mol.L^{-1}.
(9) To calculate K_a we need $[H^+]^2 = 1.2589 \times 10^{-6}$ mol^2.L^{-2} and, using Equation [8.9], $\delta([H^+]^2) = 2.\{\delta[H^+]/[H^+]\}.[H^+]^2 = 2 \times 5.76 \times 10^{-2} \times 1.2589 \times 10^{-6} = \pm 1.4502 \times 10^{-7}$ mol^2.L^{-2}.
(10) Finally, we can now calculate $K_a = [H^+]^2/C_{AH}° = 1.2589 \times 10^{-6}/9.4559 \times 10^{-2} = 1.331 \times 10^{-5}$ mol.L^{-1}. Using Equation [8.7], since $[H^+]$ and $C_{AH}°$ were measured independently, we calculate $(\delta K_a)/K_a = \{(\delta[H^+])/[H^+])^2 + (\delta C_{AH}°/C_{AH}°)^2\}^{\frac{1}{2}} = \{1.3271 \times 10^{-2} + 1.78 \times 10^{-5}\}^{\frac{1}{2}} = \pm 0.1153$ (i.e., $\sim 11\%$). Thus $\delta K_a = 0.1153 \times 1.331 \times 10^{-5} = \pm 0.153 \times 10^{-5}$ mol.L^{-1}, so finally our 'best' result for K_a is $1.331(\pm 0.153) \times 10^{-5}$ mol.L^{-1}, i.e., $1.33(\pm 0.15) \times 10^{-5}$ mol.L^{-1}.

Note that we worried about significant digits only at the very end of this somewhat long (but straightforward) calculation and expressed our final result retaining two digits after the decimal, although some might argue that only one was justified.

use of |m|, the absolute value of m, in Equation [8.9]. As another simple example, consider the exponential function $W_{exp} = \exp(x)$:

$$\delta W_{exp} = \exp(x).\delta x, \text{ i.e. } \delta W_{exp}/W_{exp} = \delta x \qquad [8.10]$$

since $(d/dx)[\exp(x)] = \exp(x)$. Similarly if $W_{10} = 10^x = \exp(2.30258x)$ where $10 = \exp(2.30258)$:

$$\delta W_{10} = 2.30258.10^x.\delta x, \text{ i.e. } \delta W_{10}/W_{10} = 2.30258.\delta x \qquad [8.11]$$

A simple (fictitious) chemical example can be used to illustrate these principles (see accompanying text box). Suppose a pure sample of an organic acid AH has been isolated, and indications of a surprisingly high acidity led to a plan to measure the acidic dissociation constant (K_a). The apparently complex (but essentially simple) series of steps involved in making the necessary measurements, estimating the uncertainty in each, and calculating both the value of K_a and its uncertainty, is described in the text box. Note that in this simple example the final uncertainty was almost entirely due to that in reading the pH meter, and it would have been possible to use a much less precise balance and lower grade standard flask without any noticeable negative effect on the final result; such a choice would have provided a simple example of the 'fitness for purpose' principle in practice (see Chapter 9). Finally, what exactly is meant by the 'best' result has still not been explained, and, in particular, what is really implied by the value quoted for the uncertainty in K_a (δK_a), and now attention must be turned to a quantitative interpretation of indicators of precision such as s_x and SE_x.

8.2.3 Normal (Gaussian) Distribution

The kind of questions that have not been addressed thus far can be summarized by asking whether the experimental values for $\bar{x}$ and s_x (or SE_x) correspond to expectations, i.e., what are the confidence limits (Section 8.2.4) within which the best measured value $\bar{x}$ can be regarded as being an 'acceptable' estimate of the 'true' value μ (note that only random error is considered here). To introduce some important concepts, particularly what we mean by 'acceptable', we consider first the question of whether or not a 'normal' (Gaussian) distribution (Equation [8.3]) provides an adequate model for a set of replicate measurements; as we shall see (Section 8.2.5), modifications to the Gaussian theory are required to account for real-world complications, especially the limited number of replicate measurements that can realistically be made by the analytical chemist.

The discussion in the remainder of the present section is devoted to cases in which the number of replicate measurements of a single parameter (e.g. concentration of an analyte in a given matrix) is sufficiently large that the unmodified Gaussian theory is applicable. While this condition is only infrequently met in practice in analytical chemistry, a brief discussion is presented here as an introduction to the more useful statistical models (Section 8.2.5).

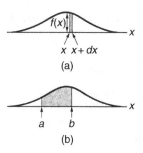

Figure 8.5 Graphical illustration of an arbitrary distribution function f(x), i.e., the fraction of all measurements yielding a particular value for x, for a very large number of replicate measurements of x (large enough that f(x) can be considered without error to be a continuous function). (a) The area of the strip between x and x+dx becomes f(x).dx for infinitesimally small values of dx(such that f(x) does not change significantly over this range of x). (b) The fraction of all measured values falling between x=a and x=b is represented by the shaded area shown; this area is given by the sum of all appropriate micro-areas f(x).dx, i.e. by the integral I_a^b f(x).dx. Reproduced from Taylor, *An Introduction to Error Analysis*, University Science Books (1982), with permission.

Consider any normalized (see Section 8.2.1) distribution function f(x) that describes the fraction of all measured values that fall between x and x+dx as f(x).dx, where dx is a very small increment in x; graphically this can be represented as the shaded 'incremental' area under the curve illustrated in Figure 8.5(a). By extension, the larger area under the curve between x=a and x=b (Figure 8.5(b)) gives the fraction of all measurements obtained between these two values of x. Another way of saying the same thing is that the probability $P_{a,b}$ of observing a measured value of x between these two values is:

$$P_{a,b} = \int_a^b f(x).dx \qquad [8.12]$$

In our case we are assuming for now that f(x) is given by the Gaussian function $G_{\mu,\sigma}(x)$, normalized (Equation [8.3]) so that the integral for $a = -\infty$ and $b = +\infty$ is unity; the Poisson function considered in Sections 3.4.1

and 7.1 is another example of f(x). Note that this assumption includes, among other things, the specific supposition that there are negligible systematic errors so that our experimental mean value x̄ will provide a valid estimate of the parameter μ of the theoretical distribution function. Purely for algebraic convenience, we define a new variable z that describes the function in terms of deviations of individual values of x from μ, expressed in units of the dispersion parameter σ:

$$z \equiv (x - \mu)/\sigma, \text{ so } dx = \sigma.dz \text{ for given values of } \mu \text{ and } \sigma \quad [8.13]$$

Then the distribution function $G_{\mu,\sigma}(x)$ becomes:

$$G'(z) = (2\pi)^{-\frac{1}{2}}.\exp(-z^2/2) \quad [8.14]$$

This function (Equation [8.14]) is independent of particular values of μ and σ and is thus convenient for constructing tables of values of $G'(z)$ that are universally applicable. Actually the values of $G'(z)$ as a function of z are generally of less practical use than the evaluated integrals (Equation [8.12]) of this function. As an example, one important class of such integrals corresponds to integration limits of z that are symmetrical about $x = \mu$, corresponding (Equation [8.13]) to values of $|x - \mu|$ that are the same multiple m of σ (where the notation $|x - \mu|$ denotes the absolute value of $x - \mu$). Such integrals P_m (area under the distribution curve) are given by:

$$P_m = \int_{-m}^{+m} G'(z).dz = \int G_{\mu,\sigma}(x).dx \quad [8.15]$$

where the integration limits for $G_{\mu,\sigma}(x)$ are $x = \mu - m\sigma$ to $\mu + m\sigma$.

Extensive tables of numerical values of the integral P_m in Equation [8.15] are given in any statistics textbook; it is important to understand that P_m represents the probability that, for a very large number of measurements that conform to the mathematical model described by the 'normal' distribution $G'(z)$, a particular value for z (and thence for x = μ + zσ) will fall within a range ±m on either side of z = 0 (i.e., of x = μ). Figure 8.6 shows values of P_m calculated for small integral values of m and also one example of the converse case, i.e., a value of m shown to correspond to a user-selected value of P (= 0.5 in this case, i.e., to the case where 50 % of all measurements will fall within the calculated range of z and thus x).

The value of m that we choose when expressing a summary of our experimental results (as a mean value plus an indication of the width of the distribution) will depend on the use to which the width is to be put. Thus, the common practice of expressing the results of a number of replicate measurements of x as x̄ ±s_x (as

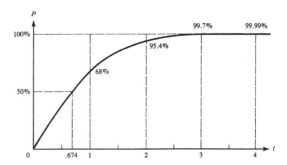

Figure 8.6 Graph of the Gaussian probability P_m vs m (Equation [8.15]) that a measured value x will fall in a range centered by the mean x̄ (best experimental estimate of the 'true' value μ, i.e. systematic errors are assumed to be negligible) and with a width given by a multiple m of the standard deviation s_x (best experimental estimate of σ), i.e. in a range (x̄ ± m.s_x). It is assumed that a very large number of replicate measurements results in a normal distribution $G_{\mu,\sigma}(x)$. Note that a range of 1σ gives a probability of only 68 % that one specific measurement will fall in that range but this probability increases rapidly with the value of m. For example 95 % of measured values that obey a Gaussian distribution fall within a range x ± 2σ. As an example of the converse calculation, if a 50 % probability is desired it turns out that the required value of m is 0.674. Reproduced from Taylor, *An Introduction to Error Analysis*, University Science Books (1982), with permission.

our best estimate for μ ± σ) corresponds to $P_m = 0.68$ (Figure 8.6), i.e., only 68 % of all measurements are expected to fall within one standard deviation of the mean. (Note that though this is indeed common practice, it is not necessarily the best practice; SE_x is a better estimate of the uncertainty of the mean value, Equation [8.5].) This choice may be fit for some purposes, but note (Figure 8.6) that a choice of m = 2 in Equation [8.15] means that >95 % of measured values should fall within two standard deviations on either side of the mean. Of course, the probability that a measured value will fall outside the stated range is simply $P'_m = (1 - P_m)$. Recall (Equation [8.5]) that in general s_x gives the best experimental estimate of the uncertainty in each individual measurement of x, but SE_x gives the corresponding estimate of the uncertainty in the mean value x̄.

As an example of how the foregoing discussion might be used in practice, consider a situation in which a certified reference material (CRM) is being analyzed in order to test the ability of a laboratory to reproduce the certified value x_{cert} for the concentration of a target analyte in the CRM matrix; assume that a sufficiently large number of replicate analyses was performed to justify use of the unmodified Gaussian model. This is a case where the value of μ (the 'true' value of x) can reasonably be claimed to be known (= x_{cert}) and the test amounts to deciding whether or not the laboratory's best experimental value $\bar{x}$ is in 'acceptable' agreement with this 'true' value, i.e. are there any systematic errors, where the quantitative implications of 'acceptable' have yet to be specified. Start by assuming:

(i) that the replicate measurements of x are representative of a normal distribution, i.e., can be considered to be a (large) set of values randomly selected from this theoretical distribution;

(ii) that this theoretical distribution underlying the set of replicate measurements is centred on the certified value (this amounts to assuming that systematic errors are negligible); and

(iii) that the experimental standard deviation s_x (Equation 8.2c] is a good estimate for the theoretical (Gaussian) counterpart σ_x.

We shall comment on these three assumptions later. However, we first calculate the absolute value of the experimental discrepancy $|\bar{x} - x_{cert}|$ and compare this with the value of σ_x (taken to be s_x); the experimental ratio $\bar{z} = (|\bar{x} - x_{cert}|/s_x)$ (Equation [8.13]) can then be interpreted in terms of the probability P_m that the value $\bar{x}$ obtained by a laboratory would be found to lie within $(m = \bar{z})$ standard deviations of the 'true' value, for the (assumed) normal distribution defined by $(\mu = x_{cert}; \sigma = s_x)$. For this purpose the tabulated values of P_m are essential, a few are shown in Figure 8.6. P_m is an example of a confidence level, here referring to situations in which the data set includes a sufficiently large number of data points that a use of a Gaussian distribution is justifiable; other definitions applicable to small data sets are discussed in Section 8.2.5.

Usually the interpretation of the experimental data uses $P'_m = (1 - P_m)$. For example, suppose that the experimental value of $(\bar{z} = m)$ was 0.54, i.e., the absolute value of $(\bar{x} - x_{cert}) = 0.54.s_x$; the table (not shown) of the numerically calculated integrals defined by Equation [8.15] then gives $P_{0.54} = 0.4108$, or $P'_{0.54} = 0.5882$. That is to say, 59% of experimental determinations of x would be theoretically expected (on the basis of our assumed normal distribution for random errors as defined in points (i)–(iii)) to fall outside the range $x_{cert} \pm m.s_x$; therefore, the fact that this laboratory obtained a best value $\bar{x}$ that falls inside the range implies that their performance would rank better than significantly more than half of such attempts. If the value of $\bar{x}$ was much closer to x_{cert}, e.g., $m = \bar{z} = 0.15$, then $P_{0.15} = 0.1192$ and $P'_{0.15} = 0.8808$, i.e., 88% of such measurements would be expected on the basis of statistical probability to fall outside the range defined by the experimental results obtained by our particular laboratory, which therefore performed extremely well (within the top

12%). At the other extreme, if the measured values of $\bar{x}$ and s_x gave m $(= \bar{z}) = 2.87$, then the table gives $P_{2.87} = 0.9959$, indicating a probability $P'_{2.87}$ of only 0.4% that a measured value would not fall inside the range $x_{cert} \pm m.s_x$, and therefore that one or more of the assumptions (i)–(iii) are (almost) certainly invalid. In such a case the laboratory would be well advised to examine its procedures carefully, particularly assumption (ii), since some undetected systematic error would be a common cause of such a large discrepancy even though the experimental standard deviation s_x seemed reasonable; such a case, in which systematic errors are possible even if the precision is good, is an example of the situation depicted graphically in Figure 8.1(a).

This is as far as our statistical method can take us in this test using a CRM and a large set of replicates, but it does provide a quantitative tool on the basis of which we can decide whether or not the laboratory's performance was 'acceptable' under any selected meaning of 'acceptability', i.e., value of P'_m. This quantitative prescription of 'acceptability' is usually expressed as the selected value of P_m given as a percentage. For example if we choose $P'_m = 0.05$, this corresponds to $P_m = 0.95$, often given as a percentage (95% in this case). The corresponding range of values between $\pm m.s_x$ are described as the confidence limits. This concept is discussed more fully in Section 8.2.4.

The selection of the appropriate value of P'_m depends on the details of the purpose for conducting the test in the first place; the choice of $P_m = 95\%$ is used quite commonly, since then an experimental value corresponding to m = 2, i.e., a difference from the 'true' value $= 2.s_x$ is only just unacceptable (since $P_2 = 0.954$, see Figure 8.6). In fact any experimental value corresponding to m $(= \bar{z}) \geq 1.95$, corresponding to $P_m \approx 5\%$, is unacceptable on this criterion of acceptability. The higher the specified value of P_m' (i.e., the lower the value of P_m) the more stringent is our criterion of acceptability. For example, if our acceptability criterion is $P_m' \leq 0.02$ (i.e. $P_m = 98\%$) any experimental discrepancy m ≥ 2.32 is unacceptable, while if we decide that an appropriate measure of acceptability is $P_m' \leq 0.10$ (i.e. $P_m = 90\%$), then any experimental m ≥ 1.65 must be considered unacceptable.

This is a simple example of the general rule that statistical considerations do not, in principle, give a clearcut answer to the apparently simple question as to whether or not a particular experimental result is 'acceptable'. In quantitative science we can not allow ourselves to get away with purely qualitative concepts like 'acceptability' that we use in everyday life to evaluate, e.g., some aspect of human behaviour. It is the essentially probabilistic nature of quantitative statistical tests that give rise to the frustrations of people unfamiliar with the concepts when faced with a proper presentation of results of an opinion poll, for example, or the risk/benefit analysis for a particular pharmaceutical etc.

The foregoing simple example serves to introduce the main concepts of statistical analysis of experimental data but is inadequate in dealing with real-world data from chemical analyses. Assumptions (i) and (iii) together imply that the separate experimental values ($\bar{x}$ and x_{cert}) obtained for x are examples of values randomly drawn from an assumed normal distribution with standard deviation σ_x well estimated by the experimental parameter s_x. This combined assumption is reasonable (but even then not necessarily valid) if the number of replicate measurements is sufficiently large, as exemplified by the (fictitious) data used to obtain the histograms shown in Figure 8.7; even for 1000 data points the distribution is not well defined and certainly the width parameter s_x will itself be subject to considerable uncertainty. In real life

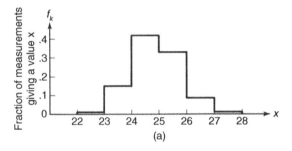

(a)

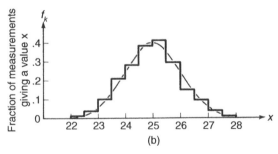

(b)

Figure 8.7 Fictitious results of replicate measurements of a quantity x, generated using an appropriate random number generator. The histograms show the fractions of measured values of x that fall within each bin. The bin width is chosen in each case so that several of the central bins contain *some* measured values. (a): results of 100 simulated experiments. (b): results of 1 000 simulated experiments (the dashed curve represents the theoretical normal distribution $G_{\mu,\sigma}(x)$ that was used to generate the plotted data by using it as a weighting function for a random number generator). Reproduced from Taylor, *An Introduction to Error Analysis*, University Science Books (1982), with permission.

the number of replicate measurements of some trace level target analyte will generally be in the range 3–6, dictated by considerations of time, cost, availability of the analytical sample etc.; as a result, estimates of confidence levels obtained on the basis of assumptions (i)–(iii) can only be regarded as approximate indications. Clearly a better mathematical model for the statistics of small numbers of replicate measurements is required (Section 8.2.5). First, however, some other concepts and definitions must be established.

8.2.4 Hypothesis Tests: False Positives and False Negatives

A statistical test provides a mechanism for making quantitative decisions about a set of data. Any statistical test for the evaluation of quantitative data involves a mathematical model for the theoretical distribution of the experimental measurements (reflecting the precision of the measurements), a pair of competing hypotheses and a user-selected criterion (e.g., the confidence level) for making a decision concerning the validity of any specific hypothesis. All hypothesis tests address uncertainty in the results by attempting to refute a specific claim made about the results based on the data set itself.

As mentioned above, in order to conduct meaningful statistical tests on the data sets typically generated by analytical chemists, the 'normal' (Gaussian) model requires some modification if it is to be applied to such small data sets. We shall see that we commonly use the Student's-*t* distribution (Section 8.2.5), as it is universally accepted as the best available model for assessing the variability of mean values in the experimental data generated by analytical chemists. (Note, however, that this model assumes an underlying normal distribution for the individual measurements themselves, that would have been observed for a very large number of measurements.) We also make use of the Fisher *F*-distribution (Section 8.2.5) as well as the chi-squared (χ^2) distribution (Section 8.2.6) in our hypothesis tests. The need to specify a confidence level that is appropriate for the purposes of the user of the data with all of these tests is a reminder that statistics provides a means of organizing and assessing raw experimental data to provide information in a user-defined, unambiguous and transparent manner.

A crucial first step in any statistical test is the formulation of a hypothesis, i.e., an appropriate wording of the question we wish to ask about the data. The outcome of a statistical test is never a 'hard fact' but rather a statement about the probability of the validity of a formulated hypothesis. A null hypothesis (H_o) is a hypothesis that corresponds to a presumed 'default' state, e.g. that an average value obtained from replicate analyses of CRM is not significantly different from the certified value, or that two analytical results from different laboratories agree, or that a measured pollutant level does not exceed a regulatory limit, or that the concentration of a disinfectant used in a drinking water plant is above a required minimal level etc. Null hypotheses are formulated in such a way as to be capable of being refuted (i.e. 'nullified') in order to support the alternative hypothesis (H_1) that states the opposite, i.e., the experimental mean value differs from the certified value by an amount determined to be statistically significant at the user-selected confidence level (the accuracy is not acceptable), or the two analytical results are significantly different, or the pollutant level does exceed the regulatory limit, or that insufficient disinfectant is present to make the water supply safe. The objective of the statistical test is to determine whether or not H_0 can justifiably be discarded in favor of H_1 on the basis of available experimental data, and is conducted to test the explicit presumption that the null hypothesis H_0 (the 'default state') is 'probably' true, where we are careful to specify quantitatively what level of probability we are implying (see below).

Clearly, the formulation of the null hypothesis is a crucial step and must reflect the objectives of the entire exercise. For the example of comparison of an experimental mean value $\bar{x}_{expt}$ for a CRM with the certified value x_{cert}, in order to determine whether or not the difference between them is statistically significant, H_0 (the 'default state') would presume that the accuracy of the experimental method was adequate, i.e., that $\bar{x}_{expt}$ and x_{cert} represent mean values of two sets of data points belonging to the same underlying ('normal') distribution with a fixed mean value $\mu = x_{cert}$ and are thus are statistically indistinguishable within user-selected statistical criteria for such decisions (see the discussion of significance level, p-value and confidence level, in the following paragraph). The alternative hypothesis H_1 would then claim that $\bar{x}_{expt}$ and x_{cert} belong to populations with different mean values μ_1 ($= x_{cert}$ in this example) and μ_2. In this case H_0 would be tested using a two tailed (or two sided) significance test (see below), since here we are interested in differences that can be either positive or negative. Very similar comments apply to testing the mean values of two sets of replicate measurements of a target pollutant in a water sample, reported by two different laboratories, in order to determine whether or not the difference between them is statistically significant; H_0 (the 'default state') would presume that the two results $\bar{x}_1$ and $\bar{x}_2$ represent mean values of data points belonging to the same underlying ('normal') distribution with a fixed (though usually unknown in this example) mean value μ, and are thus are

statistically indistinguishable within user-selected statistical criteria. As for the accuracy test using a CRM, the alternative hypothesis H_1 would then claim that $\bar{x}_1$ and $\bar{x}_2$ belong to populations with different mean values μ_1 and μ_2 and a two tailed significance test is again appropriate. If the objective was to decide whether or not the pollutant concentration fell below a specified regulatory limit x_{reg} (i.e., we are interested in a uni-directional difference), a single tailed significance test of the null hypothesis [H_0: $\bar{x} \le x_{reg}$] would be appropriate. In contrast, for the example in which the concern was about whether the disinfectant concentration met or exceeded the minimum specified concentration x_{min}, the null hypothesis [H_0: $\bar{x} \ge x_{min}$] would again require a single tailed significance test. As we shall see, because the Student's-t distribution is symmetrical the same table of single tailed critical test values of t applies to both 'upper' and 'lower' significance tests.

Note that, in general, if a statistical test of the experimental data results in H_0 being rejected, this means either that H_0 is false or that the measured value represents a case of observation of experimental data with low (user-selected) probability. But this does not imply that H_1 is necessarily true. The intrinsically probabilistic nature of statistical testing (no cut-and-dried 'yes or no' answers) is emphasized by considerations of false positive and false negative conclusions (sometimes referred to as type I and type II errors, or α and β errors). A false positive is the error of rejecting H_0 when subsequent (usually more extensive) testing indicates that it is in fact 'true', i.e., H_0 was wrongly rejected relative to the user-selected statistical criteria (e.g., p-value, see below) on the basis of the first (limited) set of experimental data. If this initial finding resulted in acceptance of the validity of H_1, rather than a conclusion that the initial finding is a low probability result consistent with the 'default' hypothesis H_0, we say that a type I error (a false positive conclusion with respect to H_1) has been made. In contrast, a type II error (false negative, or β error) arises when H_0 is accepted when further evidence shows that the alternative hypothesis H_1 is 'true' within the statistical probabilities; again, a type II error is a false negative conclusion with respect to H_1.

This is an appropriate point at which to describe the relationship between the various parameters used in practice to specify the statistical criteria used in decisions on hypothesis tests on data sets that are too small to justify using the unmodified Gaussian model. (The parameters P_m and P'_m that were introduced in Section 8.2.3 can be regarded as special cases of the parameters discussed in the following, but will not be referred to again in this book.) The significance level (α) of a statistical hypothesis test and the corresponding probability value (*p*-value) and confidence level $100(1-p)$ are all widely used in the present context (Appendix 8.1). (Note that the *p*-value is not related to P_m, Section 8.2.3; P_m is an integrated probability while the *p*-value is a critical value selected as the cutoff for 'acceptability', see below). The significance level α is a user-selected probability (area under the appropriate distribution curve, an example is shown in Figure 8.8(a)) of wrongly rejecting the null hypothesis H_0 and accepting H_1 i.e. the probability of a false positive, and is selected by the investigator in the context of the consequences of such an error. Thus, if the purpose of the investigation demands it, the value of α is chosen to be as small as possible in order to protect the null hypothesis and to minimize the chance that the investigator will inadvertently make a false positive conclusion (about H_1); a common choice is $\alpha = 0.05$. The probability of not reaching a false positive conclusion $(1-\alpha)$ is called the confidence level and is often expressed as a percentage. Thus if we present a quantitative result X (e.g., the result of a poll of a representative sample of the population, or of analysis of a pollutant in a soil sample) as $X \pm \delta X$ at a 95 % confidence level (sometimes stated as '19 times out of 20' in poll results), where δX is a measure of the random uncertainty of the measurement, we mean that we are 95 % confident that the stated interval $X \pm \delta X$ contains the 'true' value. Another way of looking at such a statement is that 95 % (the confidence level) of all such 'confidence intervals' ($\pm \delta X$), determined from different samples of the population or of the polluted soil, will include the 'true' value. In some cases we are interested in determining whether or not a measured value X is either above or below some critical value X_{crit}, i.e., our statistical test needs to be uni-directional. In such cases we must consider the probability of either a positive deviation ($+\delta X$) alone or a negative deviation ($-\delta X$) and then we distinguish between the two integrated 'tails' of the *t* distribution, illustrated in Figure 8.8. In practice this distinction between uni-directional and absolute deviations is expressed in single tailed vs two tailed *t*-tests (see below), and also has implications for the concept of false positive vs false negative conclusions. An example of how a hypothesis test might be applied in practice is illustrated in Figure 8.8(b).

The probability value (*p*-value) of a statistical hypothesis test is closely related to the significance level α; the *p*-value is the critical value of the statistical parameter (z or m in the Gaussian case in Section 8.2.3, or the Student-*t* introduced in Section 8.2.5) for which H_0 would be only just rejected for a given value of α. That is, the *p*-value is the threshold value for false positives (type I

errors) whereas α is the total probability of making type I errors, i.e., the integrated probability of observing an experimental value of the test statistic as extreme as, or more extreme than, the value predicted by chance alone if the null hypothesis H_0 were true. If the investigator had decided that H_0 should be rejected at $\alpha = 0.05$, this would be reported as '$p < 0.05$' (note the crucial difference between '=' and '>'). For calculation of a confidence level, $(1-p)$ and $(1-\alpha)$ are equivalent despite this distinction between their meanings.

Note that selection of small *p*-values tends to favor conclusions that H_0 is unlikely to be true; the smaller the *p*-value, the more convincing is the rejection of H_0. Thus this approach indicates the strength of evidence for e.g., rejecting H_0, in contrast with a simple binary choice between 'reject H_0' and 'do not reject H_0'.

For any given set of data, the probabilities of false positives and false negatives (type I and type II errors) are inversely related, i.e., the smaller the risk of one, the higher the risk of the other, and these two risks must be carefully balanced against one another when selecting the statistical criteria to be used. A false positive (with respect to H_1) is in many circumstances considered to be more serious and, therefore, more important to avoid than a false negative; in such cases the hypothesis test procedure is therefore adjusted so that there is a guaranteed 'low' probability of wrongly rejecting the null hypothesis (of course this probability is never exactly zero); this is done by judicious choice of the significance level α. The probability of a false negative (with respect to H_1) is generally not known to within the same degree of confidence, a result of the asymmetric way in which statistical hypotheses are generally formulated.

8.2.5 Student's t-Test and Fisher F-Test for Comparison of Means and Variances: Applications to Two Data Sets

The universally accepted approach to statistical evaluation of small data sets is that originated by William Gosset (Gosset 1908) and developed and publicized by Richard Fisher (Fisher 1925, 1990). The so-called Student's *t*-distribution (see the accompanying text box) addresses the problem of estimating the uncertainty in the mean of an assumed normal distribution based on a small number of data points assumed to be random samples from that distribution. It forms the basis of *t*-tests commonly used for the statistical significance of the difference between the means of two small data sets believed to be drawn from the same normal distribution, and for determination of the confidence interval for the difference between two such data set means.

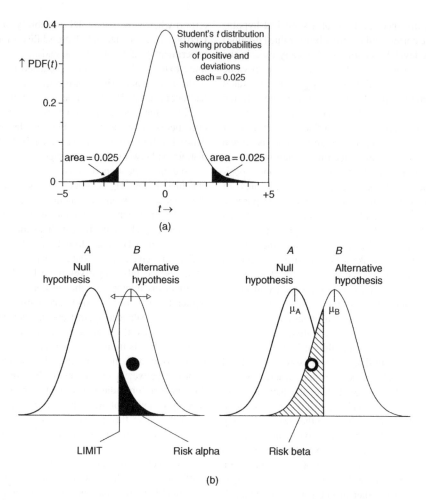

(a)

(b)

Figure 8.8   (a) The probability density function (PDF) for the Student's-t test statistic, plotted as a function of $t \equiv (\bar{x}_n - \mu)/(s_x/n^{\frac{1}{2}})$ (Equation [8.16]), i.e., the difference between the experimental mean value for n replicate measurements and the reference value μ, normalized relative to the standard error (standard deviation of the mean) $SE = (s_x/n^{\frac{1}{2}})$. The shaded areas each correspond to values of t for which the significance levels $\alpha = 0.025$ accounting for both positive and negative deviations, respectively, i.e., are relevant for the two kinds of single tailed t-tests (e.g., is a measured value above or below a specified critical value, [H_o: $\bar{x} \geq x_{crit}$] or [H_o: $\bar{x} \leq x_{crit}$]). Usually for a test of a null hypothesis in which the sign of a deviation is irrelevant, e.g. [H_o: $\bar{x} = \mu$], a two-tailed test with $\alpha = 0.05$ is used. The vertical boundaries of these shaded regions correspond to the critical value t_c for a 95 % confidence level, i.e., for $p = 0.05$ for a two tailed test. (b) An example of application to hypothesis testing. It is assumed that two data sets A and B are to be compared, such that μ_A (taken to be the experimental value $\bar{x}_A$) is a reference value (e.g., a certified value for a CRM or a regulatory limit) and the data obtained as data set B are to be compared with those for A to determine whether or not $\bar{x}_B > \bar{x}_A$. The curves show the distributions for the two data sets. Then we set up [H_o: $\bar{x}_B = \bar{x}_A$] and [H_1: $\bar{x}_B > \bar{x}_A$] and choose a critical value (shown as the 'LIMIT' in the Figure) of the experimental measurement x that determines the value of α, the error probability associated with H_o (this implies that a single tailed test is the appropriate choice in this case). Then the probability of a false positive conclusion (applied to H_1) is given by the dark shaded area (fractional value of the total area under the H_o curve is α) in the left panel, exemplified by the hypothetical result shown as the dark circle that has been assigned to the population characteristic of H_1 but turns out on further experimentation to really belong to that for H_o. The right-hand panel illustrates a case where the result represented by the open circle is assigned to the H_o distribution but is later shown to actually belong to H_1, thus representing a false negative conclusion with probability β. The ability of the test to discriminate between the two hypotheses obviously increases as ($\bar{x}_B - \bar{x}_A$) increases (in this particular example $\bar{x}_A$ is fixed) and/or by narrowing the distribution for dataset B by increasing the number of replicate determinations. Reproduced with permission from Meier and Zünd, *Statistical Methods in Analytical Chemistry*, 2nd *Edition* (2000), John Wiley & Sons Inc.

Twentieth Century Statisticians: Gosset and Fisher

William Sealy Gosset

William Sealy Gosset (1876–1937) was an English chemist and statistician. After graduating from Oxford University in 1899 he joined the Dublin brewery of Arthur Guinness & Son, which was even then a progressive agro-chemical business. He applied his statistical knowledge (acquired from time spent with the famous statistician Karl Pearson at King's College London) in the brewery and also on the farm to the selection of the best varieties of barley, a basic ingredient in the best known Guinness product!

Because of the nature of his work Gossett was forced to develop statistical theory for small data sets, since then-current theories dealt only with very large sets. Unfortunately another researcher at Guinness had published a paper containing Guinness trade secrets so the company prohibited its employees from publishing *any* papers and Gosset was unable to publish under his own name. He therefore used the pseudonym *Student* for his publications to avoid detection and his most famous achievement is thus referred to as the *Student-t distribution*. However, the *t*-statistic was actually introduced by RA Fisher; Gosset's original statistic *z* was slightly modified by Fisher to the *t*-form in accord with his theory of degrees of freedom. Gosset was said to be a modest man who disarmed praise with the comment that 'Fisher would have discovered it all anyway.'

Sir Ronald Aylmer Fisher (1890–1962) was an English statistician and geneticist. Because of poor eyesight he was taught mathematics without pen and paper, which developed his ability to visualize problems in geometrical as opposed to algebraic terms. He was famous for producing mathematical results without writing down the intermediate steps. Among his many achievements he introduced the concept of *analysis of variance*, the foundation ot modern ANOVA that still employs Fisher's F-statistic. He also introduced the related concept of *experimental design* permitting e.g., quantitative testing of hypotheses by statistical comparison of results for test and control groups. As a first illustration of the concept, Fisher described how to test quantitatively the hypothesis that a certain lady could distinguish by taste alone whether the milk or the tea was first placed in the cup. While this appears to be a piece of English whimsy, in fact it represented a major breakthrough in the general scientific method that is nowadays taken for granted.

Sir Ronald Aylmer Fisher

The basis of the approach rests on Gosset's proof that, for a theoretical normal distribution of a quantity x with standard deviation σ, the means of random samples of n independent measurements of x are also normally distributed but with standard deviation $\sigma/n^{\frac{1}{2}}$, i.e., the standard deviation of the mean, or standard error SE (Equation [8.5]). Moreover, if it turns out that the true distribution is not truly Gaussian, i.e., is not well described by Equation [8.3], the distribution of the mean values of several small sub-sets does tend to be well approximated by its own normal distribution. The mathematical theory underlying the Student's-*t* approach is well outside the purposes of the present book and we shall only discuss briefly the aspects that are of direct importance for its judicious application.

The Student's-*t* parameter (sometimes referred to simply as *t*) is formally defined by:

$$t \equiv (\bar{x}_n - \mu)/(s_x/n^{\frac{1}{2}}) = (\bar{x}_n - \mu)/SE_x \qquad [8.16]$$

where μ is again the mean of the theoretical normal distribution from which the *n* experimental values are assumed to have been drawn (sometimes replaced by, e.g., an independent experimental estimate $\bar{x}$, see later), $(s_x/n^{\frac{1}{2}})$ is the standard error SE_n (Equation [8.5]) for this small data set of *n* values for x, and we also have emphasized by using the symbol $\bar{x}_n$ that this is the mean of the small subset of *n* values of x. (Thus *t* is somewhat related to the parameter z that was introduced for convenience in Equation [8.13], but note the important differences; z was used as an algebraic convenience defined relative to the entirely theoretical Gaussian function (Equation [8.3]), while t incorporates the experimental quantities $\bar{x}_n$ and s_x.) The mathematical theory and the explicit form of the distribution function $T_{dof}(t)$ for t are not described here; *dof* is the number of degrees of freedom and is given by $(n-1)$ for a data set with a known mean value. However, it is important to note that $T_{dof}(t)$ is independent of μ and σ (it is this property that has led to its universal applicability), but varies with the value of *dof* although for *dof* larger than ~ 100 the distribution becomes essentially a normal (Gaussian) distribution independent of *dof*. The graph of $T_{dof}(t)$ vs t (Figure 8.8) is generally lower and wider than its 'parent' normal distribution of $G_{\mu,\sigma}(x)$ vs x.

In exactly the same way as for the normal (Gaussian) distribution function (see Equation [8.15]), it is actually the corresponding probability distribution function for t, calculated as integrals of $T_{dof}(t)$, that is more useful than $T_{dof}(t)$ itself (Figure 8.8). These integrals are usually listed (Table 8.1) as the analogues of P'_m, discussed above for the case of a normal distribution (Section 8.2.3); recall that P'_m gives the probability that a measured mean value will fall **outside** the prescribed limits and the same is true of the Student's-t analogy.

The Student's-t test provides a statistical test for significant differences (where 'significant' must be explicitly defined, see below) between the means of two data sets that ostensibly represent the same quantity x. All statisticians agree that t-tests on two mean values $\bar{x}_1$ and $\bar{x}_2$ are valid if the variances $V_{x,1}$ and $V_{x,2}$ (Equation [8.2b]) of the two data sets to be compared are statistically indistinguishable; when this is not the case, the $t-$ test result can be questioned in borderline cases.

Thus, in order to apply the t-test, we must first test for differences between the two variances; Fisher's F-test is used for this purpose. Again we shall omit all reference to the derivation of this test statistic and the rationale for statistical tests based on it. Instead we focus here on its application to testing two data sets to determine the confidence level at which we can assert that the variances $V_{x,1}$ and $V_{x,2}$ (Equation [8.2b]) of the two are indistinguishable (the null hypothesis H_0 for the test). The value of the experimental test statistic F is calculated simply as:

$$F = (V_{x,1}/V_{x,2}) = (s_{x,1}/s_{x,2})^2 \geq 1.00 \qquad [8.17]$$

where the ratio is always taken such that $F \geq 1.00$. The critical value F_c is taken from a table for the chosen p-value (usually 0.05, Table 8.2) and for the two values of the degrees of freedom dof_1 and dof_2 where $dof = (n-1)$. (Note that dof_1 and dof_2 must be assigned consistent with the designation of data sets as 1 and 2 such that $F > 1$.) Then if $F \leq F_c$, the null hypothesis H_0 is retained (i.e., $V_{x,1}$ and $V_{x,2}$ are indistinguishable at this confidence level), but if $F > F_c$ then H_0 is rejected, i.e., the variances (and thus standard deviations, Equation [8.2c]) of the two data sets must be regarded as statistically different at the chosen confidence level.

As a simple example, suppose that two analytical methods are compared by replicate ($n_1 = 6 = n_2$) analyses of a homogeneous sample, giving $s_{x,1} = 0.50$ and $s_{x,2} = 0.35$. Then from Equation [8.17] $F = (0.50/0.35)^2 = 2.041$. For $dof_1 = 5 = dof_2$ and $p = 0.05$, Table 8.2 gives $F_c = 5.050 > F$, so H_0 is retained, i.e., the difference between these standard deviations is 95 % likely to be consistent with the variation expected for such small data sets selected randomly from the (assumed normal) distribution of all such measurements; then we can proceed with confidence to apply a t-test to the means of the two data sets in this case. However, suppose that $s_{x,1} = 0.60$ and $s_{x,2} = 0.26$, so that $F = 5.325$; now since $F > F_c$ we must reject H_0 and the s_x values for the two data sets can not be considered indistinguishable at this confidence level; as mentioned above, application of a t-test to two mean values for data sets that failed the F-test can give questionable conclusions about the equivalence of the two mean values in borderline cases (see below).

Once the variances of the two data sets to be compared have been shown to be indistinguishable to within the limits implied by the selected value of F_c, the Student's-t

Table 8.1 Values of critical Student's-t factors t_C for three commonly used values of the error probability p as a function of the number of degrees of freedom (*dof*) for both one and two sided cases. Note that for a given value of *dof*, the value of t_C for the two sided case at a given p-value (e.g., $p = 0.1$) is equal to the corresponding t_C-value for the one sided case but at half the p-value (e.g., $p = 0.05$).

(a) Two sided				(b) One sided		
$p = \rightarrow$	0.1	0.05	0.01	0.1	0.05	0.01
$dof = \downarrow$						
1	6.314	12.706	63.66	3.078	6.314	31.821
2	2.920	4.303	9.925	1.886	2.920	6.965
3	2.353	3.182	5.841	1.638	2.353	4.541
4	2.132	2.776	4.604	1.533	2.132	3.747
5	2.015	2.571	4.032	1.476	2.015	3.365
6	1.943	2.447	3.707	1.440	1.943	3.143
7	1.895	2.365	3.499	1.415	1.895	2.998
8	1.860	2.306	3.355	1.397	1.860	2.896
9	1.833	2.262	3.250	1.383	1.833	2.821
10	1.812	2.228	3.169	1.372	1.812	2.764
12	1.782	2.179	3.055	1.356	1.782	2.681
20	1.725	2.086	2.845	1.325	1.725	2.528
∞	1.645	1.960	2.576	1.282	1.645	2.326

Table 8.2 Critical F-values for $p = 0.05$ and for several values of the degrees of freedom f for the data sets 1 and 2 to be compared. Note that by definition $F = V_2/V_1$ is > 1.0, so data sets 1 and 2 are chosen such that this condition is satisfied; in turn this specifies which of the degrees of freedom is dof_1 and which is dof_2 for purposes of the F-test using this table

dof_2/dof_1	1	2	3	4	5	6	7	8	9	10	20	30	Inf
1	161.4476	199.5000	215.7073	224.5832	230.1619	233.9860	236.7684	238.8827	240.5433	241.8817	248.0131	250.0951	254.3144
2	18.5128	19.0000	19.1643	19.2468	19.2964	19.3295	19.3532	19.3710	19.3848	19.3959	19.4458	19.4624	19.4957
3	10.1280	9.5521	9.2766	9.1172	9.0135	8.9406	8.8867	8.8452	8.8123	8.7855	8.6602	8.6166	8.5264
4	7.7086	6.9443	6.5914	6.3882	6.2561	6.1631	6.0942	6.0410	5.9988	5.9644	5.8025	5.7459	5.6281
5	6.6079	5.7861	5.4095	5.1922	5.0503	4.9503	4.8759	4.8183	4.7725	4.7351	4.5581	4.4957	4.3650
6	5.9874	5.1433	4.7571	4.5337	4.3874	4.2839	4.2067	4.1468	4.0990	4.0600	3.8742	3.8082	3.6689
7	5.5914	4.7374	4.3468	4.1203	3.9715	3.8660	3.7870	3.7257	3.6767	3.6365	3.4445	3.3758	3.2298
8	5.3177	4.4590	4.0662	3.8379	3.6875	3.5806	3.5005	3.4381	3.3881	3.3472	3.1503	3.0794	2.9276
9	5.1174	4.2565	3.8625	3.6331	3.4817	3.3738	3.2927	3.2296	3.1789	3.1373	2.9365	2.8637	2.7067
10	4.9646	4.1028	3.7083	3.4780	3.3258	3.2172	3.1355	3.0717	3.0204	2.9782	2.7740	2.6996	2.5379
12	4.7472	3.8853	3.4903	3.2592	3.1059	2.9961	2.9134	2.8486	2.7964	2.7534	2.5436	2.4663	2.2962
14	4.6001	3.7389	3.3439	3.1122	2.9582	2.8477	2.7642	2.6987	2.6458	2.6022	2.3879	2.3082	2.1307
16	4.4940	3.6337	3.2389	3.0069	2.8524	2.7413	2.6572	2.5911	2.5377	2.4935	2.2756	2.1938	2.0096
18	4.4139	3.5546	3.1599	2.9277	2.7729	2.6613	2.5767	2.5102	2.4563	2.4117	2.1906	2.1071	1.9168
20	4.3512	3.4928	3.0984	2.8661	2.7109	2.5990	2.5140	2.4471	2.3928	2.3479	2.1242	2.0391	1.8432
30	4.1709	3.3158	2.9223	2.6896	2.5336	2.4205	2.3343	2.2662	2.2107	2.1646	1.9317	1.8409	1.6223
inf	3.8415	2.9957	2.6049	2.3719	2.2141	2.0986	2.0096	1.9384	1.8799	1.8307	1.5705	1.4591	1.0000

test statistic can be used for statistical comparisons of the two mean values (see Section 8.2.6 for cases where more than two are involved). A variety of circumstances that must be carefully distinguished from one another can arise; the following discussion is an abbreviated version of one given previously (Meier 2000), where further details are provided. It is convenient to consider separately three categories of Student's *t*-tests:

(A) One of the values to be compared is an experimental value known to high precision (e.g., the certified value for a CRM), or is a definite prescribed value such as a specification or regulatory limit, denoted μ.

(B) The two data sets are comparable experimental results that yield indistinguishable values of s_x (as determined by the *F*-test at a specified *p*-value); this category includes several sub-cases, that differ, e.g., as to whether the two data sets are wholly independent or involve comparisons of data sets that are correlated in some manner.

(C) Same as (B) but the two experimental values of s_x are shown (*F*-test) to be different.

In categories (B) and (C) the difference of the mean values $d_{1,2} = (\bar{x}_1 - \bar{x}_2)$ and the appropriate number of degrees of freedom *dof* must be calculated (see later). We now discuss these different cases in turn.

(A) In such cases we are comparing the experimental mean $\bar{x}_2$ (with $dof_2 = (n_2 - 1)$ and $SE_2 = s_{x,2}/n_2^{\frac{1}{2}}$, Equation [8.5]) with the fixed comparison value μ (defined with zero or negligible uncertainty, e.g., x_{cert} for a CRM), using the test statistic *t* (Equation [8.16]) calculated here as:

$$t = (\bar{x}_2 - \mu)/SE_2$$

where we retain the sign of *t* for purposes described below. Now the absolute value of this experimental value of *t* is compared with the critical value t_c taken from a table of *t*-values (Table 8.1) for the particular value of dof_2 and the confidence level CL $= (1 - p).100$, selected by the investigator. This procedure is best illustrated using a simple example (Meier 2000). (Note that we are not stating the units of the measured quantities in all the following examples, since the focus is on the numerical values.)

Suppose $\bar{x}_2 = 12.79$, $n_2 = 7$, $s_{x,2} = 1.67$, $\mu = 14.00$, and that the chosen *p*-value is 0.05 (95 % confidence level); then $SE_2 = 1.67/\sqrt{7} = 0.631$, and $dof_2 = 6$, and the experimental value of $t = (12.79 - 14.00)/0.631 = -1.918$. At this point it is essential to explicitly state the hypotheses being tested. If the purpose of the test is to determine whether or not the null hypothesis [H_o: $\bar{x}_2 = \mu$] is valid (i.e., the alternative hypothesis is the simple

inequality [H_1: $\bar{x}_2 \neq \mu$], regardless of the sign of the difference $(\bar{x}_2 - \mu)$), then a two sided t-test (Figure 8.8) is applied since we are here making no distinction between $\bar{x}_2 > \mu$ and $\bar{x}_2 < \mu$; for the same reason we use only the absolute value of $t = 1.918$. From Table 8.1 we find $t_c = 2.447$ for ($p = 0.05$, $dof = 6$, two sided test). Since here $t < t_c$, we do not reject H_o, i.e., in this case H_1 can not be substantiated within the stated confidence level, and within this restriction $\bar{x}_2$ is statistically indistinguishable from the fixed reference value μ.

However, if μ is a specified regulatory limit, e.g., a maximum allowed concentration of a compound in a foodstuff, we probably wish to test the null hypothesis [H_o: $\bar{x}_2 \leq \mu$] against the alternative hypothesis [H_1: $\bar{x}_2 > \mu$], i.e., H_1 proposes that the regulatory limit has been exceeded. Here, since we are interested only in a one-way deviation, we use a one sided t-test for which the appropriate $t_c = 1.943$ (Table 8.1), which is very close to the experimental *t*-value of 1.918. Nonetheless, by applying the appropriate one sided *t*-test for this experimental *t*-value to our chosen null hypothesis [H_o : $\bar{x}_2 \leq \mu$], since the condition for rejection of H_o is $t \geq t_c$ which is not true in this case, we can not reject H_o, i.e. the regulatory limit was not breached (though it was close!).

(B) This category of *t*-tests examines the statistical significance of the difference $(\bar{x}_1 - \bar{x}_2)$ between the means of two experimental data sets for which the *F*-test has shown that the two standard deviations are indistinguishable. However, application of the *t*-test again varies somewhat with the circumstances, and three sub-cases must be considered (Meier 2000).

(B1) Each of the two data sets whose means are to be compared contains the same number of replicate measurements, i.e., $n_1 = n_2 = n$, but without pairing of the measurements (see (B2) below); examples include a single analytical method applied *n* times to each of two different analytical samples (e.g., calibrating a laboratory QC standard against a CRM), or two different analytical methods each applied *n* times to the same analytical sample (e.g., comparison of a new LC–UV analytical method against a fully validated LC–MS method). In this case, if the *F*-test found the two variances to be indistinguishable within a user-chosen confidence level, the standard deviation (or standard error) of the difference $d_{1,2}$ is given by Equation [8.6] (propagation of error) as e.g., $SE_d = [SE_{x,1} + SE_{x,2}]^{\frac{1}{2}}$, where $SE_{x,1}$ is given by Equation [8.5] as $s_x/n_1^{\frac{1}{2}}$ with $n_1 = n$, so the experimental $t = d_{1,2}/SE_d$; also $dof_d = (2n - 2)$. The two-sided Student's-*t* test is then applied since we are interested in any deviation (positive or negative) of $\bar{x}_2$ from $\bar{x}_1$.

An example should help to clarify the procedure. Suppose $\bar{x}_1 = 17.4$, $s_{x,1} = 1.30$, $\bar{x}_2 = 19.5$, $s_{x,2} = 1.15$,

$n_1 = n_2 = n = 6$, and we choose to test the null hypothesis [$H_o : d_{1,2} = 0$]. We apply the F-test for $F = (s_{x,1}/s_{x,2})^2 = 1.278$; the degrees of freedom within each data set $dof_1 = dof_2 = (n-1) = 5$, and thus Table 8.2 gives $F_c = 5.050$ for $p = 0.05$, so $F < F_c$ and $s_{x,1}$ and $s_{x,2}$ are indistinguishable, so we can proceed to the t-test. Then $SE_d = [(s_{x,1}^2 + s_{x,2}^2)/n]^{\frac{1}{2}} = 0.709$, and our experimental t-value $= d_{1,2}/SE_d = 2.1/0.709 = 2.964$. To find the critical value t_c we need $dof_d = (2n-2) = 10$; testing our chosen H_o (a simple claim that the two means are indistinguishable, regardless of the sign of the difference) requires a two sided test for $p = 0.05$, and Table 8.1 then gives $t_c = 2.228$. Since $t > t_c$, our two mean values are distinguishable at the 95 % confidence level, i.e., the experimental value of $d_{1,2}$ is indeed significantly different from zero at this confidence level.

(B2) This case of pairwise testing differs from (B1) in a fashion that appears to be subtle but in fact is important. In (B1), the comparison of two analytical methods was done by applying method 1 to just one set of n replicate samples, and independently method 2 to a different set of n replicates, giving $2n$ measurements on $2n$ samples. However, in (B2) we consider a case in which the $2n$ measurements are performed on just n replicate samples, each replicate sample being analyzed by both methods in parallel. An example might be a comparison of the LC–UV and LC–MS methods but done by subjecting each and every one of the n samples to both methods. The two data sets are thus different (distinguished by the analytical method) but they are not completely independent of one another since any variations arising from extraction, clean-up etc. are now the same for both methods. The difference is well illustrated by an instructive example suggested previously (Meier 2000). Suppose that the following paired raw data were obtained for five replicate samples using methods 1 and 2:

$$x_1(1) = 1.73, x_2(1) = 1.61; x_1(2) = 1.70, x_2(2) = 1.58;$$

$$x_1(3) = 1.53, x_2(3) = 1.41; x_1(4) = 1.78, x_2(4) = 1.64;$$

$$x_1(5) = 1.71, x_2(5) = 1.58$$

where subscripts 1 and 2 denote the experimental method and numbers in parentheses the identification number of the replicate sample. It is easy to calculate $\bar{x}_1 = 1.690$, $s_{x,1} = 0.0946$, $\bar{x}_2 = 1.564$, $s_{x,2} = 0.0896$, and to show by the F-test that $s_{x,1}$ and $s_{x,2}$ are indistinguishable at the 95 % confidence level ($p = 0.05$), so we can proceed to the appropriate t-test. If we ignore the fact that the two sets of values were experimentally paired, and proceed as described for case (B1), we have $d_{1,2} = 0.126$, $n_1 = n_2 = n = 5$, and $SE_d = 0.0583$. Then the experimental t_d-

value $= d_{1,2}/SE_d = 2.162$. Under this assumption we have $dof_d = (2n-2) = 8$, and Table 8.1 then gives $t_c = 2.306$ for a two sided test at $p = 0.05$. Since $t_d < t_c$, on this basis we would conclude that the two analytical methods were equivalent since the difference $d_{1,2}$ is statistically indistinguishable from zero at the 95 % confidence level.

However, this (B1) approach ignores the pairwise nature of the raw experimental data. To take account of this additional information, we must first calculate the absolute values of the individual differences $d_{1,2}(j)$ for $j = 1-5$:

$$d_{1,2}(1) = 0.12; d_{1,2}(2) = 0.12; d_{1,2}(3) = 0.12;$$

$$d_{1,2}(4) = 0.14; d_{1,2}(5) = 0.13$$

We now calculate the mean and standard deviation for these differences using Equation [8.2], to give $\bar{d}_{1,2} = 0.126$, $SE_d = 0.00894$. This gives an experimental t_d-value for the five individual paired differences of $\bar{d}_{1,2}/s_d = 14.1$. Now we test the null hypothesis [H_o: $\bar{d}_{1,2} = $ zero] with $dof = (n-1) = 4$ for $p = 0.05$, for which $t_c = 2.776$ (Table 8.1). Clearly $t_d > t_c$, and the two analytical methods are definitely not statistically indistinguishable at the 95 % confidence level, unlike the conclusion drawn from the (B1) evaluation of the same data. This is an excellent example of the need to fit the specifics of a t-test to include all of the available information about the raw experimental data.

(B3) This case is a slight variation of (B1), i.e., not pairwise testing, but the case in which the numbers of replicate analyses are different in the two data sets, i.e., $n_1 \neq n_2$; however, we still restrict the discussion to two data sets with indistinguishable standard deviations (via the F-test). The degrees of freedom $dof = (n_1 + n_2 - 2)$, an obvious extension of (B1). Now, however, calculation of s_d, the standard deviation of the difference of the two means, is somewhat more complicated, and indeed there is no universally agreed formula among statisticians (Meier 2000). The method described here appears to be the most robust:

$$V_d = [(n_1-1).V_{x,1} + (n_2-1).V_{x,2}].[(1/n_1) + (1/n_2)]/dof$$

We proceed as in (B1) to calculate the t_d value for the difference between the two means calculated as $t_d = d_{1,2}/\sqrt{V_d}$, and compare with the critical value t_c from Table 8.1 for the null hypothesis [H_o: $d_{1,2} = $ zero] at the 95 % confidence level ($p = 0.05$). The following illustrative example is discussed in detail elsewhere (Meier 2000). Suppose $\bar{x}_1 = 109.73$, $s_{x,1} = 4.674$, $n_1 = 8$, $\bar{x}_2 = 101.26$, $s_{x,2} = 7.328$, $n_2 = 6$. It is easy to calculate $F = 2.46$, and the value of F_c for $dof_1 = 7$, $dof_2 = 5$, is 3.972 (Table 8.2) so the standard deviations are indistinguishable at $p = 0.05$. The two variances are $V_{x,1} = 21.846$ and $V_{x,2} = 53.700$, with $d_{1,2} = 8.47$ and $dof = 12$. Substitution of these values in the expression for V_d in this case gives the experimental $t_d = d_{1,2}/\sqrt{V_d} = 2.65$. From Table 8.1, for $dof = 12$ and $p = 0.05$, the two sided $t_c = 2.179$. Here, $t_d > t_c$ so the null hypothesis H_o is rejected at the 95 % confidence level ($p = 0.05$) and [H_1: $d_{1,2} \neq$ zero] is accepted, i.e., the two mean values are indeed distinguishable, with the proviso typical of all statistical tests that there is a small chance (determined by the chosen p-value) that our data set corresponds to a false positive (with respect to H_1).

(C) There appears to be no robust mathematical theory applicable to the case where $s_{x,1} \neq S_{x,2}$ (Meier 2000), i.e., when the F-test is failed. However, the following alternative sets of working relationships can be used with caution. In both cases V_d, the variance in the difference between the means $d_{1,2}$ is approximated by the simple expression $V_d = [(V_{x,1}/n_1) + (V_{x,2}/n_2)]$, and $t_d = d_{1,2}/\sqrt{V_d}$, as before. In contrast, two alternative complicated expressions for dof have been proposed:

$$dof = \{(Q_1 + Q_2)^2/[Q_1^2/(n_1 - 1) + Q_2^2/(n_2 - 1)]\}$$
$$- 2, \text{with}$$

$$Q_j = V_{x,j}/n_j, \text{ for } j = 1, 2$$
$$dof = 1/[k^2/(n_1 - 1) + (1 - k)^2/(n_2 - 1)],$$

with $k = (n_2 \cdot V_{x,1})/(n_2 \cdot V_{x,1} + n_1 \cdot V_{x,2})$

These alternative expressions can give significantly different values for dof and thus for t_c. We use the same illustrative example as for case (B3) above, except that now $s_{x,1} = 3.674$ and $s_{x,2} = 8.328$, and it is easy to show that the F-test is now failed, as might have been guessed from the wide divergence between these two values of s_x. We evaluate V_d as $[(s_{x,1}^2/n_1) + (s_{x,2}^2/n_2)] = 13.247$, and thus t_d as $d_{1,2}/\sqrt{V_d} = 8.47/3.640 = 2.33$. To estimate t_c we need to evaluate the two rival expressions for dof as 4.4 and 5.0, respectively, which we round off to the nearest integers as $f = 4$ or 5. We are testing the null hypothesis [H_o: $d_{1,2} =$ zero] so the corresponding values of t_c are for a two sided test at the 95 % confidence level, i.e., 2.776 and 2.571 (Table 8.1); in both cases $t_d < t_c$ so we can not reject H_o at the 95 % confidence level. However, since the difference between t_d and t_c is not that large, in view of the uncertainty (Meier 2000) concerning the range of validity for such a test with $s_{x,1} \neq s_{x,2}$, if this were an important decision it would be advisable to acquire more data to settle the question.

The analyst frequently is faced with assessing more than two data sets with respect to equality (or not) of

the mean values, e.g., concentration of the active ingredient in several batches of a pharmaceutical formulation, or of a pollutant in soil samples taken from several locations in a polluted site, or of analytical data reported by several laboratories for the same analytical sample in a round robin exercise. In general, such a situation involves m_d independent data sets, each containing n_j replicate measurements where the data set sizes $n_j (j = 1, 2, ..., m_d)$ are not necessarily all equal. To determine whether or not the resulting mean values $\bar{x}_j$ are all equivalent at a chosen confidence level, it would be possible to conduct appropriate F-tests and t-tests, as described above, for all possible pairings of the m_d data sets. This is highly inefficient for all but small values of m_d. Moreover, it does not answer the question as to whether or not all m_d data sets can be regarded as belonging to the same overall normal distribution; multiple t-tests can not achieve this because, as the number of data sets increases, the number of required pairwise comparisons grows quickly. For m_d data sets there are $m_d(m_d - 1)/2$ pairs, e.g., if $m_d = 8$ there are 28 pairs, and if we test all these pairs at the 95 % confidence level we should not be surprised to observe things such as nonequality of means and/or variances that happen in 5 % of all cases. Thus, if the significance level α is set at 0.05 for each t-test, for 28 such tests the new effective p-value is simply the product of the p-values for the 28 pairwise comparisons, i.e. $28 \times 0.05 = 1.4$ (an example of multiplying individual probabilities for a series of events to obtain the the overall probability for all events combined); thus within the total 28 pairwise comparisons we have a much higher probability of making a false positive decision (type I error, i.e., [H_o: all eight means are equal to within $p = 0.05$] is wrongly rejected) than that applicable to any individual pairwise comparison. In other words, within all these 28 tests there is a non-negligible probability that we would find some significant differences between data sets in any event, though this probability would decrease if the value of p was increased. An ANOVA test (Section 8.2.6) controls the overall error by testing all eight means against one another simultaneously, so the effective p-value remains at 0.05, and uses just one test statistic (the F-value for the set of means) to test the null hypothesis [H_o: all means are equal].

8.2.6 Statistical Testing of More Than Two Data Sets: Bartlett Test and ANOVA

Comparisons of multiple data sets ($m_d > 2$) require examination of three questions, two of which are obvious extensions of those pertinent to the simple t-test applied to $m_d = 2$ data sets.

Question 1. Do all m_d data sets have the same standard deviation (or variance), indistinguishable within a stated confidence level? For the $m_d = 2$ case (Section 8.2.5) this question was answered by the F-test, but for $m_d > 2$ the Bartlett test is used (see below). If at least one data set variance is thus found to be significantly different from the rest, it is advisable to investigate the reasons for the discrepancy before continuing the statistical testing and/or to omit that particular dataset. On the other hand, if the variances are found to be indistinguishable they can be pooled as described below.

Question 2. Do all m_d data sets have the same mean within the stated confidence level? The appropriate test here is a simple analysis of variance (ANOVA) test that detects variability among the data set means in excess of that expected on the basis of chance alone. Note that if the m_d data sets are confirmed to have indistinguishable variances (Question 1), the mean values $\bar{x}_j$ are predicted by theory (not given here) to conform to a normal (Gaussian) distribution with a standard deviation equal to the average of the within-group standard deviations $s_{x,j}$ divided by the square root of the average group size, i.e.:

$$\bar{s}_{md} = (\Sigma_j s_{x,j}/m_d)/(\Sigma_j n_j)/m_d)^{\frac{1}{2}}$$

where the summations are over j (the label for the data sets) from $j = 1$ to m_d. If the ANOVA test detects a variance of the means greater than this prediction, one or more of the data set means differs significantly (as defined by the selected confidence level) from the rest, and this has to be investigated (see Question 3). On the other hand, if no excess between data set variance of the means is detected by ANOVA (following demonstration in Question 1 that the standard deviations are indistinguishable), the m_d means can be pooled because, at the specified confidence level, they can all be considered to belong to the same overall normal distribution and no further testing is required.

Question 3. If the ANOVA test in Question 2 indicates that not all of the m_d groups have means that are consistent with the rest at the stated confidence level, it is important to identify these particular data sets. Two approaches are described in Section 8.2.7, the Multiple-Range Test that effectively combines several t-tests among pairs of data sets into one simultaneous test, and a test for 'outliers' (see e.g. Figure 8.4).

In the following discussion of this three-step process, numerical examples are generally not provided as they occupy too much space; indeed these tests are usually conducted using a computer algorithm (Meier 2000). A more complete discussion, together with an example of a fully worked-out procedure, are presented elsewhere (Meier 2000, Section 4.4). Here we sketch out the basic procedures without referring at all to the mathematical theory that justifies the various

tests, and provide some relevant tables of critical test statistics.

To answer Question 1, the Bartlett test (Bartlett 1937) is applied to the null hypothesis $[H_o : s_{x,1} = s_{x,2} = \ldots = s_{x,md}]$, with $[H_1:$ same as H_o BUT at least one of the equalities is replaced by an inequality]. As before, within each data set j, $dof_j = (n_j - 1)$, and summations over j are over the range $j = 1$ to m_d. The following computational steps are performed:

$$A_B = \Sigma_j(dof_j.V_j); B_B = \Sigma_j(dof_j.\ln V_j); C_B = \Sigma_j(1/dof_j);$$
$$D_B = \Sigma_j(dof_j).$$

Then

$$E_B = D_B.\ln(A_B/D_B) - B_B; G_B = [C_B - (1/D_B)]/$$
$$(3m_d - 3) + 1; H_B = E_B/G_B$$

The quantity H_B is the Bartlett test statistic, and is to be compared with tabulated values of the χ^2 ('Chi-squared', pronounced 'ky-squared') parameter (Table 8.3); if the

Table 8.3 Critical values of Chi-Squared χ^2 for several values of p and *dof*, for a two sided test

Degrees of Freedom (*dof*)	Probability *p*				
	0.99	0.95	0.05	0.01	0.001
1	0.000	0.004	3.84	6.64	10.83
2	0.020	0.103	5.99	9.21	13.82
3	0.115	0.352	7.82	11.35	16.27
4	0.297	0.711	9.49	13.28	18.47
5	0.554	1.145	11.07	15.09	20.52
6	0.872	1.635	12.59	16.81	22.46
7	1.239	2.167	14.07	18.48	24.32
8	1.646	2.733	15.51	20.09	26.13
9	2.088	3.325	16.92	21.67	27.88
10	2.558	3.940	18.31	23.21	29.59
11	3.05	4.58	19.68	24.73	31.26
12	3.57	5.23	21.03	26.22	32.91
13	4.11	5.89	22.36	27.69	34.53
14	4.66	6.57	23.69	29.14	36.12
15	5.23	7.26	25.00	30.58	37.70
16	5.81	7.96	26.30	32.00	39.25
17	6.41	8.67	27.59	33.41	40.79
18	7.02	9.39	28.87	34.81	42.31
19	7.63	10.12	30.14	36.19	43.82
20	8.26	10.85	31.41	37.57	45.32
22	9.54	12.34	33.92	40.29	48.27
24	10.86	13.85	36.42	42.98	51.18
26	12.20	15.38	38.89	45.64	54.05
28	13.57	16.93	41.34	48.28	56.89
30	14.95	18.49	43.77	50.89	59.70

value of $H_B > \chi^2_c$ for $dof_{md} = (m_d - 1)$ and the selected value of p, then H_o must be rejected in favour of H_1, i.e., there is at least one data set with a standard deviation that is significantly outside the range of the others at the stated confidence level $100(1 - p)$. As a simple example, consider the following dataset (these values of y_i, $V_{y,i}$, and f_i actually correspond to the data set means $\bar{x}_j$, variances V_j and degrees of freedom dof_j used to illustrate the one way ANOVA procedure in the ANOVA text box):

$$y_i = 6.08, 7.01, 5.91, 7.09, 5.51; V_i = 1.103, 1.588, 2.161,$$
$$1.210, 0.314; dof_i = 7, 4, 5, 7, 7.$$

Straightforward but tedious hand-calculation then gives:

$$A_B = 35.546; B_B = -0.3852; C_B = 0.8786; D_B = 30;$$
$$E_B = 5.4741; G_B = 1.0704; H_B = 5.114$$

We now use Table 8.3 to find the critical value $\chi_c^2 = 9.49$ for $p = 0.05$ and $dof = 4$ (one less than the number of tested V_i values). Since $H_B < \chi_c^2$ we can not reject [H_o: all V_i values are equal] and thus can proceed on the assumption that the variances (and thus standard deviations) all belong to the same normal distribution at the 95% confidence level.

To answer Question 2 once Question 1 has been decided, an analysis of variance (ANOVA) is used. ANOVA consists of a collection of theoretical statistical models and and their associated procedures that compare means of several data sets. This approach was pioneered by R.A. Fisher (see text box); indeed ANOVA is sometimes known as 'Fisher's analysis of variance' and uses Fisher's F-test (Section 8.2.5) as the test of statistical significance. In this book we will consider only the so-called 'simple ANOVA' or 'one-way ANOVA', that tests for the existence of differences among a number of mean values of a single dependent variable (e.g., concentration of one target analyte). These mean values represent several data sets that are distinguished from one another by only one independent variable, e.g., a comparison among analytical results obtained by different laboratories for the same analytical sample and all using the same analytical method. If, in addition, some laboratories used analytical methods different from the others, i.e., if there are now two independent variables, a so-called 'factorial ANOVA' would be applied. If the experiment involves more than one dependent variable (e.g., concentrations of more than one target analyte), a 'multivariate analysis of variance' (MANOVA) is applied. Neither of these two latter approaches is described here.

A simple (one way) ANOVA is straightforward in principle but can be cumbersome to calculate by hand if m_d is too large and accordingly such statistical tests are usually conducted using a computer algorithm. Here we describe only the general principles and parameters used, illustrated with a simple example (Meier 2000) without attempting to discuss any of the underlying theory. A basic assumption of the model is that the total variance of the combined data of all data sets (i.e., $\Sigma_j n_j$ individual data) can be split into a simple sum of the variance within a single data set plus the variance between data sets; this will be clearer later. Then, to evaluate and test the between data set variance (the objective of the exercise), it is necessary to obtain a characteristic measure of within data set variance. If the null hypothesis (H_o: all $\bar{x}_j$ values are statistically indistinguishable) is to be valid within the selected confidence level, e.g. p-value of 0.05, then within data set and between data set variances should be similar and all the data can be pooled because they have been shown to belong to the same underlying normal distribution from which all the m_d small data sets are assumed to have been randomly drawn. On the other hand, if the between data set variance is significantly larger than its within data set counterpart, we must accept [H_1: at least one of the $\bar{x}_j$ values is significantly different from those of the group as a whole]. Note that, although this is a powerful statistical test, the result does not tell us which of the data set means is (are) significantly different.

The parameters that must be evaluated for a one way ANOVA, and their interpretations, are as follows:

n_T: $\Sigma_j n_j$ for $j = 1$ to m_d; total number of measurements in the m_d data sets;

x_{ij}: i'th measurement in the j'th data set;

$u_j = \Sigma_i(x_{ij})$; sum of all measured values in the j'th data set;

$u_{nT} = \Sigma_j u_j$: sum of all n_T measured values;

$\bar{x}_j = u_j/n_j$: mean of the n_j measured values within data set j;

$\bar{x}_{GM} = u_{nT}/n_T$: 'grand mean' of all n_T measured values;

$dof_W = (n_T - m_d)$: degrees of freedom *within* data sets;

$dof_B = (m_d - 1)$: degrees of freedom *between* data sets;

$dof_T = (n_T - 1)$: total degrees of freedom $= (dof_W + dof_B)$;

$r_{ij} = (x_{ij} - \bar{x}_j)$: residual of i'th value in data set j relative to mean of that data set;

$S_W = \Sigma_j[\Sigma_i(r_{ij})^2]$ for $j = 1$ to m_d, $i = 1$ to n_j: sum of squares of all residuals in all m_d data sets;

$S_B = \Sigma_j[n_j.(\bar{x}_j - \bar{x}_{GM})^2]$: sum of squares of residuals of the data set means relative to $\bar{x}_{GM}$, weighted according to the number of measurements in each data set (n_j);

$S_T = S_W + S_B$: total sum of squares of residuals;

$V_W = S_W/dof_W$: variance within data sets;

Example of a One Way ANOVA Calculation (Meier 2000)

Data index i	Data set x_j				
	1	2	3	4	5
1	7.87	6.35	4.65	8.74	5.29
2	6.36	7.84	5.06	6.02	6.03
3	5.73	5.31	6.52	6.69	6.06
4	4.92	6.99	6.51	7.38	5.64
5	4.60	8.54	8.28	5.82	4.33
6	6.19	-	4.45	6.88	5.39
7	5.98	-	-	8.65	5.85
8	6.95	-	-	6.55	5.51

This table presents five data sets ($m_d = 5$) with $n_j = 8, 5, 6, 8$, and eight determinations, respectively, of the same target analyte using the same method, i.e., only one controlling variable so a one way ANOVA is appropriate. The number in the i'th row and the j'th column is the value of x_{ij} (see main text). These simulated data (without units!) are used to illustrate the computations in a simple ANOVA test, after demonstrating that the variances all belonged to the same normal distribution (see Bartlett test sample calculation in the main text).

The definitions of all of the following parameters are given in the main text.

n_T = total number of measurements = 35;
$u_j = \Sigma_j(x_{ij})$ are the sums of all entries in the j'th column = 48.60, 35.03, 35.47, 56.73, 44.1;
u_n = sum of all measured n_T values = 219.93;
$\bar{x}_j$ = 6.08, 7.01, 5.91, 7.09, 5.51, for j = 1 to 5;
V_{xj} = 1.103, 1.588, 2.161, 1.210, 0.314 (calculated using Equation [8.2c]);
dof_j = 7, 4, 5, 7, 7;
$\bar{x}_{GM}$ = grand mean = 6.284;
dof_W = degrees of freedom *within* data sets = 30;
dof_B = degrees of freedom *between* data sets = 4;
dof_T = total degrees of freedom = 34;
S_W = 35.52; S_B = 13.76; S_T = 49.28 (intermediate calculations not shown)
From which the variance within (V_W) and between (V_B) data sets can be determined, as well as the total variance (V_T): V_W = 1.184; V_B = 3.441; V_T = 1.450

The experimental value of the *F* test statistic is thus $V_B/V_W = 2.91$. From Table 8.2, the critical value F_c for [$p = 0.05$ and $dof_1 \equiv dof_B = 4$, $dof_2 \equiv dof_W = 30$,] is 2.69.

Thus, $F > F_c$, so the null hypothesis [H_o: all $\bar{x}_j$ values are equal] must be rejected at the 95 % confidence level, although the discrepancy between *F* and F_c is not that large.

This is important information but in itself the ANOVA does not identify in any way how many of the data set means $\bar{x}_j$ differ significantly from the grand mean $\bar{x}_{GM}$, and certainly provides no information on which of the means is/are divergent, i.e. which data set(s) is/are possible outliers. The statistical tools used to clarify these questions are the Multiple-Range Test and/or a test for outliers (see main text).

$V_B = S_B/dof_B$: variance between data sets;
$V_T = S_T/dof_T$: total variance.

V_W and V_B are then subjected to an *F*-test with $F = V_B/V_W$ so that, if $F > F_c$ (from Table 8.2 for the selected *p*-value, usually 0.05), H_o must be rejected. Note that, as a result of their definitions, V_B must be greater than V_W (proof not given here); as a result of computational artifacts it can sometimes happen that $V_B \leq V_W$, in which case H_1 can be rejected without the *F*-test.

The meaning of an acceptance of H_o is that each and every data set mean $\bar{x}_j$ is judged to be statistically indistinguishable from the grand mean $\bar{x}_{GM}$; then the data are pooled, with V_T as the variance of $\bar{x}_{GM}$ with $dof_T = (n_T - 1)$ degrees of freedom. In the worked example shown in the accompanying text box, it turned out that the experimental $F > F_c$, so H_o had to be rejected (i.e., there is at least one 'outlier' among the means) although the difference $(F - F_c)$ was not very large. (Another indication of the marginality of the decision that at least one outlier must exist is described in Section 8.2.7.) In any case this ANOVA conclusion provided no information as to which of the $\bar{x}_j$ values were divergent 'outliers', i.e., differed significantly $(p = 0.05)$ from $\bar{x}_{GM}$.

8.2.7 Multiple-Range Test and Huber–Davies Test for Outliers

The problem of identifying 'outliers', i.e., experimental results among a set of replicate data that appear to be sufficiently different from the majority that some unusual event can be assumed to have interfered with the measurement, can be addressed in several ways (Miller 1993, Meier 2000), of which two will be discussed here. The first of these is the Multiple-Range Test (Duncan 1955) that effectively combines several t-tests into one combined test. The procedure is summarized and applied to our simple example in the accompanying text box, and indicates that $\bar{x}_5 = 5.51$ is the experimental mean that caused the ANOVA test to suggest that not all of these experimental means of small data sets could be assigned to the same normal population at the 95 % confidence level. It has been pointed out that the Multiple-Range Test is especially protective against false negatives for H_1 at the expense of a greater risk of false positives (Type I errors, i.e. rejecting H_o when it is in fact true). In our case the Multiple-Range Test suggested (see text box) that H_o is not true ($\bar{x}_5$ inconsistent with $\bar{x}_2$ and $\bar{x}_4$). However, this conclusion is marginal since $\bar{x}_5$ is in fact consistent with each of $\bar{x}_1$ and $\bar{x}_3$ both of which are consistent with $\bar{x}_2$ and $\bar{x}_4$, and thus this conclusion might be an example of a false positive (for H_1). This ambiguous indication is at least consistent with the somewhat marginal rejection of H_o in the ANOVA test (see above).

An alternative approach to identifying the (somewhat ambivalent) rejection of H_o in the ANOVA test is to use a different test for outliers. Several such tests are available but one of the most robust appears (Meier 2000) to be Huber's elimination rule that was not originally formulated as an outlier test (Davies 1988). Moreover, this Huber–Davies test is extremely easy to apply. The first step is to rearrange the test measurement values in increasing order (as for the Multiple-Range Test), thus $[\bar{x}_5 = 5.51; \bar{x}_3 = 5.91; \bar{x}_1 = 6.08; \bar{x}_2 = 7.01; \bar{x}_4 = 7.09]$. This allows selection of the median value $\bar{x}_{med} (= \bar{x}_1 = 6.08$ in our case). The absolute values of the differences $\Delta_{i,med} = |\bar{x}_i - \bar{x}_{med}|$ are then calculated and arranged also in

increasing order, thus $[\Delta_{1,med} = 0; \Delta_{3,med} = 0.17; \Delta_{5,med} = 0.57; \Delta_{2,med} = 0.93; \Delta_{4,med} = 1.01]$, allowing selection of the median value of $\Delta_{i,med} = \Delta_{5,med} = 0.57$ (this median value of $\Delta_{i,med}$ is known as the 'median absolute deviation, or 'MAD'!). The test statistics k_i are then calculated as $k_i = \Delta_{i,med}/MAD$, thus $[k_1 = 0; k_2 = 1.63; k_3 = 0.30; k_4 = 1.77; k_5 = 1.0]$. The critical value k_c against which the k_i values are compared is usually chosen to be in the range 3–5, as this has been found to provide the most robust test. In our example none of the k_i values is greater than k_c, indicating that none of the $\bar{x}_i$ values should be considered to be an outlier in this case. This result supports the suggestion that the tentative rejection of the null hypothesis [H_o: all means are equal] in the Multiple-Range Test could have been a Type I error (false positive, see above). It would probably be advisable in such cases to try to obtain more experimental data to clear up the matter.

In our simple example, all of these calculations could be done by hand in a reasonably short time, but for cases involving many more data sets ($m_d > 5$) and total numbers of measurements ($n_T > 35$), this would be impractical. There are many statistics software packages available that will perform these calculations. The purpose of working through our simple example explicitly is to emphasize what goes into these sometimes opaque algorithms (although we did not attempt to discuss the mathematical theory underlying any of these tests), and that the output from these calculations often requires some interpretation.

8.3 Bivariate Data: Tools and Tests for Regression and Correlation

Although the simple (one way) ANOVA test was described as a topic in univariate statistics, in fact it could also be regarded as a case of bivariate statistics since the single experimentally determined parameter (e.g., concentration of a target analyte in a given sample) was determined under more than one set of circumstances (e.g., different laboratories using the same analytical method). The two cases of bivariate statistics that are most relevant to this book are correlation and regression, that are related but different; correlation quantifies how closely two variables are connected with one another, while regression finds the algebraic relationship that best predicts the value of an experimental observable from values of an experimentally controlled parameter. An example of correlation analysis, discussed in Section 8.3.1, might involve an experimental test of a new instrumental method (e.g., LC with UV detection) as a cheaper alternative to a proven method (e.g., LC/MS). Regression analysis is vital for

Example of a Multiple-Range Test Calculation

u (N_u) → v (N_v) ↓	1 (8)	2 (6)	3 (8)	4 (5)	5 (8)
1 (8)	–	0.40 **0.963** *2.888*	0.57 **1.482** *3.035*	1.50 **3.420** *3.131*	1.58 **4.107** *3.199*
2(6)		–	0.17 **0.409** *2.888*	1.10 **2.361** *3.035*	1.18 **2.840** *3.131*
3(8)			–	0.93 **2.120** *2.888*	1.01 **2.625** *3.035*
4(5)				– **0.182** *2.888*	0.08
5(8)					–

In each box in this Table the first entry (normal font) is $(X_u - X_v)$, the second (bold font) is the corresponding value of $q'_{u.v}$ and the third entry (italics) is the critical value $q'_c(u, v)$ taken from Table 8.4 for $dof_{MRT} = 30$. Diagonal elements (u = v) are obviously zero and the empty boxes correspond to negative values of $(X_u - X_v)$ with magnitude equal to that of the diagonally opposite box.

The procedure is as follows:

(1) Arrange experimental values from small → large, i.e., $X_1 = \bar{x}_5 = 5.51$; $X_2 = \bar{x}_3 = 5.91$; $X_3 = \bar{x}_1 = 6.08$; $X_4 = \bar{x}_2 = 7.01$; $X_5 = \bar{x}_4 = 7.09$, with corresponding values of n_u and n_v (given in brackets in the Table).

(2) Then calculate values of $(X_u - X_v)$ (positive values only, converting subscript i/j to u/v for clarity), and tabulate. Convert each $(X_u - X_v)$ value to the corresponding value of the test statistic:

$$q' = (X_u - X_v) \cdot [n_u \cdot n_v / (n_u + n_v)]^{\frac{1}{2}} \cdot [2/V_W]^{\frac{1}{2}}$$

where V_W is the same 'variance within data sets' as in the corresponding ANOVA calculation (in this case $V_W = 1.184$).

(3) Then compare with critical values $q'_c(u, v)$ from Table 8.4 for the particular value of $dof_{MRT} = \Sigma_u(n_u) - m_d = 35 - 5 = 30$, as in the ANOVA calculation; note that the appropriate entry in the $q'_c(u, v)$ Table is in the column such that $\delta u = [|u - v| + 1]$, e.g., for (u = 2, v = 1) the appropriate entry in Table 8.4 is for $dof_{MRT} = 30$ and $\delta u = 2$, and for (u = 5, v = 3) the appropriate entry is for $dof_{MRT} = 30$ and $\delta u = 3$.

The only cases for which $q'_{u.v} > q'_c(u, v)$, thus indicating that the corresponding (X_u, X_v) values do not belong to the same population, are for (u = 4, v = 1) and (u = 5, v = 1); all other pairwise comparisons indicate consistency. This result indicates that $X_1 = \bar{x}_5$ is the problematic value that led to the ANOVA test to indicate (though somewhat marginally, see main text) that not all $\bar{x}_i$ values could be assigned to the same underlying normal distribution at the 95 % confidence level. This multiple-range test result is also marginal since $X_1 = \bar{x}_5$ is indicated to be inconsistent with both $X_4 = \bar{x}_2$ and $X_5 = \bar{x}_4$, BUT is also consistent with $X_2 = \bar{x}_3$ and $X_3 = \bar{x}_1$, each of which is consistent with both $X_4 = \bar{x}_2$ and $X_5 = \bar{x}_4$. This is a typical example of a statistical result requiring interpretation (see Meier 2000, Section 4.4 for a more complex fully worked example).

Table 8.4 Critical q' values ($p = 0.05$) for two means with index number differences $\delta u = [|u - nv| + 1]$; Multiple-Range Test

$\delta u \rightarrow$ $dof \downarrow$	2	3	4	5	6	7	8	9	10	11	12	14	16	18	20
1	17.97	17.97	17.97	17.97	17.97	17.97	17.97	17.97	17.97	17.97	17.97	17.97	17.97	17.97	17.97
2	6.085	6.085	6.085	6.085	6.085	6.085	6.085	6.085	6.085	6.085	6.085	6.085	6.085	6.085	6.085
3	4.501	4.516	4.516	4.516	4.516	4.516	4.516	4.516	4.516	4.516	4.516	4.516	4.516	4.516	4.516
4	3.926	4.013	4.033	4.033	4.033	4.033	4.033	4.033	4.033	4.033	4.033	4.033	4.033	4.033	4.033
5	3.635	3.749	3.796	3.814	3.814	3.814	3.814	3.814	3.814	3.814	3.814	3.814	3.814	3.814	3.814
6	3.460	3.586	3.649	3.680	3.694	3.697	3.697	3.697	3.697	3.697	3.697	3.697	3.697	3.697	3.697
7	3.344	3.477	3.548	3.588	3.611	3.622	3.625	3.625	3.625	3.625	3.625	3.625	3.625	3.625	3.625
8	3.261	3.398	3.475	3.521	3.549	3.566	3.575	3.579	3.579	3.579	3.579	3.579	3.579	3.579	3.579
9	3.199	3.339	3.420	3.470	3.502	3.523	3.536	3.544	3.547	3.547	3.547	3.547	3.547	3.547	3.547
10	3.151	3.293	3.376	3.430	3.465	3.489	3.505	3.516	3.522	3.525	3.525	3.525	3.525	3.525	3.525
11	3.113	3.256	3.341	3.397	3.435	3.462	3.480	3.493	3.501	3.506	3.509	3.510	3.510	3.510	3.510
12	3.081	3.225	3.312	3.370	3.410	3.439	3.459	3.474	3.484	3.491	3.495	3.498	3.498	3.498	3.498
13	3.055	3.200	3.288	3.348	3.389	3.419	3.441	3.458	3.470	3.478	3.484	3.488	3.490	3.490	3.490
14	3.033	3.178	3.268	3.328	3.371	3.403	3.426	3.444	3.457	3.467	3.474	3.482	3.484	3.484	3.484
15	3.014	3.160	3.250	3.312	3.356	3.389	3.413	3.432	3.446	3.457	3.465	3.476	3.480	3.480	3.480
16	2.998	3.144	3.235	3.297	3.343	3.376	3.402	3.422	3.437	3.449	3.458	3.470	3.476	3.477	3.477
17	2.984	3.130	3.222	3.285	3.331	3.365	3.392	3.412	3.429	3.441	3.451	3.465	3.472	3.475	3.475
18	2.971	3.117	3.210	3.274	3.320	3.356	3.383	3.404	3.421	3.435	3.445	3.460	3.469	3.473	3.474
19	2.960	3.106	3.199	3.264	3.311	3.347	3.375	3.397	3.415	3.429	3.440	3.456	3.466	3.472	3.474
20	2.950	3.097	3.190	3.255	3.303	3.339	3.368	3.390	3.409	3.423	3.435	3.452	3.463	3.470	3.473
21	2.941	3.088	3.181	3.247	3.295	3.332	3.361	3.385	3.403	3.418	3.431	3.449	3.461	3.469	3.473
22	2.933	3.080	3.173	3.239	3.288	3.326	3.355	3.379	3.398	3.414	3.427	3.446	3.459	3.467	3.472
23	2.926	3.072	3.166	3.233	3.282	3.320	3.350	3.374	3.394	3.410	3.423	3.443	3.457	3.466	3.472
24	2.919	3.066	3.160	3.226	3.276	3.315	3.345	3.370	3.390	3.406	3.420	3.441	3.455	3.465	3.472
25	2.913	3.059	3.154	3.221	3.271	3.310	3.341	3.366	3.386	3.403	3.417	3.439	3.454	3.464	3.471
26	2.907	3.054	3.149	3.216	3.266	3.305	3.336	3.362	3.382	3.400	3.414	3.436	3.452	3.463	3.471
27	2.902	3.049	3.144	3.211	3.262	3.301	3.332	3.358	3.379	3.397	3.412	3.434	3.451	3.463	3.471
28	2.897	3.044	3.139	3.206	3.257	3.297	3.329	3.355	3.376	3.394	3.409	3.433	3.450	3.462	3.470
29	2.892	3.039	3.135	3.202	3.253	3.293	3.326	3.352	3.373	3.392	3.407	3.431	3.448	3.461	3.470
30	2.888	3.035	3.131	3.199	3.250	3.290	3.322	3.349	3.371	3.389	3.405	3.429	3.447	3.460	3.470
32	2.881	3.028	3.123	3.192	3.243	3.284	3.317	3.344	3.366	3.385	3.401	3.426	3.445	3.459	3.470
34	2.874	3.021	3.117	3.185	3.238	3.279	3.312	3.339	3.362	3.381	3.398	3.424	3.443	3.458	3.469
36	2.868	3.015	3.111	3.180	3.232	3.274	3.307	3.335	3.358	3.378	3.395	3.421	3.442	3.457	3.469
38	2.863	3.010	3.106	3.175	3.228	3.270	3.303	3.331	3.355	3.375	3.392	3.419	3.440	3.456	3.469
40	2.858	3.005	3.102	3.171	3.224	3.266	3.300	3.328	3.352	3.372	3.389	3.418	3.439	3.456	3.469
60	2.829	2.976	3.073	3.143	3.198	3.241	3.277	3.307	3.333	3.355	3.374	3.406	3.431	3.451	3.468
120	2.800	2.947	3.045	3.116	3.172	3.217	3.254	3.286	3.313	3.337	3.358	3.394	3.423	3.446	3.466
240	2.786	2.933	3.031	3.103	3.159	3.205	3.243	3.276	3.304	3.329	3.350	3.388	3.418	3.444	3.466
Inf	2.772	2.918	3.017	3.089	3.146	3.193	3.232	3.265	3.294	3.320	3.343	3.382	3.414	3.442	3.466

quantitative chemical analysis as it permits statistically defensible calibration of detector signal as a function of concentration (or amount) of analyte, and use of this calibration to determine unknown concentrations.

For an experiment intended to test for correlation it makes no difference which variable is designated as the controlled parameter and which as the experimental observable, but this is definitely not the case for an experiment designed to determine a regression relationship; in the latter case it is important to be explicit about which is the independent (controlled) parameter (usually designated x, e.g. analyte concentration in a calibration solution) and which is the dependent variable (usually designated Y, e.g., LC–MS peak area). Linear regression calculations (Section 8.3.2) are not symmetrical with respect to x and Y, i.e., if we exchange the designations (dependent vs independent) for the two variables, a different regression equation will be obtained; in contrast correlation calculations are symmetrical with respect to x and Y, i.e., if we exchange the labels we will still obtain the same correlation coefficient (Section 8.3.1).

The main example of bivariate statistics in this book involves simple linear regression and its application to calibration of the response of an analytical instrument with respect to concentration (or amount) of a target analyte; values of the dependent variable (e.g., instrument response) are recorded as a function of an experimental (independent) variable (e.g., amount of analyte injected) under fixed instrumental conditions; the most useful circumstance involves a linear relationship between the two, and the relevant statistical approach is linear regression analysis (Sections 8.3.2 and 8.3.5). Another example would involve a series of repeated analyses of the same sample (typically a laboratory QC sample) as a function of the time (date) of analysis, as a monitor of stability and reproducibility of the instrument or complete analytical method.

8.3.1 Correlation Analysis

An example in which correlation analysis could be useful in an analytical laboratory involves an experimental test of a different instrumental method (e.g., LC with UV detection) as a cheaper alternative to a proven method (e.g., LC/MS). The objective is to provide a measure of the 'goodness of fit' (Section 8.3.3) of the relationship between the two data sets. However, in correlation analysis the nature of the relationship (e.g. linear, quadratic etc.) is not determined; in the simple example one would measure the responses of the two instruments for a series of test solutions covering the desired concentration range and compare the two to determine whether there is a strong correlation. The statistical measures most often employed for such purposes are the correlation coefficient R and the coefficient of determination R^2, defined by:

$$R = S_{xy}/(S_{xx}.S_{yy})^{\frac{1}{2}} = \Sigma_i[(x_i - \bar{x}).(y_i - \bar{y})]/[(\Sigma_i(x_i - \bar{x})^2.$$
$$(\Sigma_i(y_i - \bar{y})^2]^{\frac{1}{2}} \qquad [8.18]$$

where (x_i, y_i) are the paired responses of the two analytical methods for the same test solution i, and $\bar{x}$ and $\bar{y}$ are the mean values of the two sets of experimental responses. The range for R is $-1 \le R \le +1$, where positive/negative signs for R indicate positive/negative linear correlations, respectively. An R value of ± 1 indicates a perfect fit, i.e., all experimental points would lie exactly on a (undetermined) curve. The value of R^2 (range 0 to $+1$) gives the proportion of the variance (fluctuation) of one variable that is predictable from the other variable, thus denoting the strength of the correlation between x and y. For example, if $R = 0.922$, then $R^2 = 0.850$, indicating that 85 % of the total variation in y can be explained by the relationship between x and y (as described by the regression equation, see later); the other 15 % of the total variation in y remains unexplained (random 'noise' and/or uncontrolled parameters such as temperature etc.).

Although high values for R or R^2 do indeed indicate a high degree of correlation note that, with modern analytical instrumentation, coefficients of determination > 0.999 are common, leaving only the 4th digit after the decimal as the only meaningful test statistic to determine the degree of correlation. As an alternative the residual standard deviation s_{res} (Equation [8.25]) is recommended (Meier 2000) as a better measure of goodness of fit in correlation analysis, because it is a quantity that has the same physical dimensions and units as the measured quantity Y, thus providing a better 'feel' to the analyst than the dimensionless quantities R and R^2. Note however that we are dealing here with essentially the same situation as that discussed in case B2, Section 8.2.5, where statistical measures of degree of agreement between the two paired data sets, based on appropriate *t*-tests with well defined confidence levels, were described; the latter approach is thus much superior to considerations of R or R^2 in this case.

8.3.2 Simple Linear Regression for Homoscedastic Data

In statistical theory regression analysis refers to procedures used to model relationships between variables and determine the magnitude of those relationships so that the quantified models can be used to make predictions. Regression analysis models the relationship between one or more dependent variables (sometimes referred to as response variables) and one or more independent variables (sometimes called control variables). Simple linear regression and multiple linear regression are related statistical methods for modeling the relationship between two or more random variables using a model equation that is a linear function of the fitting parameters; it is important to understand that this definition of 'linear' regression does not necessarily imply linearity with respect to the independent variables.

Simple linear regression refers to a regression on only two variables (Y and x denote the dependent and independent variables), while multiple regression refers to a regression on more than two variables. Equations [8.19a–8.19b] are thus both examples of simple linear regression models, although Equation [8.19b] is quadratic with respect to x (and will be met again later in the context of calibration curves that are nonlinear with respect to (x_i, Y_i), Section 8.3.6). Equation [8.19c] is an example of a multiple linear regression model and Equation [8.19d] exemplifies a non-linear regression model (non-linear with respect to at least one of the fitting parameters):

$$Y = A + B.x \qquad [8.19a]$$

$$Y = A + B.x + C.x^2 \qquad [8.19b]$$

$$Y = A + B.x_1 + C.x_1^2 + B'.x_2 + C'.x_2^2 \qquad [8.19c]$$

$$Y = A + B.\exp(C.x) \qquad [8.19d]$$

where the coefficients A, B, C are the fitting parameters that must be determined from experimental data in order that the regression equation can be used predictively. Note that Equation [8.19d] is non-linear in C (and thus requires non-linear regression); it is incidentally also non-linear in x, but that is not the distinguishing criterion here.

In analytical chemistry we always try to arrange that Equation [8.19a] provides an adequate model for the relationship between the instrumental response (Y) and the concentration (or amount) of analyte (x) injected directly into the instrument (instrumental calibration) or used to spike a blank matrix (method calibration, see Section 8.5). When analytical chemists speak of a linear calibration equation they refer to Equation [8.19a], a simple linear regression model that is linear in both the fitting parameters and also in the independent variable; Equation [8.19b] might be referred to as a non-linear calibration equation by a chemist, although to a statistician it is an example of a simple linear regression model, i.e., it is linear in all of the fitting parameters.

Equation [8.19a] is desirable for several reasons: only two fitting parameters need to be calculated, it is straightforward to invert the equation so as to calculate an unknown value of x (e.g., analyte concentration) from a measured value of Y (e.g. mass spectrometric signal intensity) as $x = (Y - A)/B$, and relatively few experimental measurements (x_i, Y_i) are required to establish reliable values of A and B. In principle it would be possible to calculate values for A and B from just two such experimental measurements, requiring algebraic solution of just two simultaneous equations. However, this procedure takes no account of experimental uncertainties in the measurements, and instead we prefer to over-determine the two simultaneous equations by incorporating many more than two (x_i, Y_i) points on the curve $(n \gg 2)$. The least squares model, developed independently by two famous mathematicians Gauss and Legendre in the 19th century (see the text box), applies the constraint given by Equation [8.20] to the model equation $Y = f(x)$ (Equation [8.19a] in our preferred case):

$$\Sigma_i[r_i^2] = \Sigma_i[(Y_i - Y_{i,pred})]^2 = \Sigma_i[Y_i - f(x_i)]^2$$
$$= \Sigma_i[Y_i - (A + B.x_i)]^2 = \text{minimum} \qquad [8.20]$$

where $Y_{i,pred}$ is the value of Y_i predicted for a particular value of x_i from Equation [8.19a] plus the values of the fitting parameters A and B, and r_i is the residual associated with the i'th measurement, always defined as (observed – predicted) values of Y_i, and the summation over i is from 1 to n (number of (x_i, Y_i) data points) This constraint implies that we are seeking values of A and B in Equation [8.19a] that minimize differences r_i regardless of their sign, achieved by considering the squares of these deviations (that are necessarily positive). (Use of $\Sigma_i[r_i]$ instead of $\Sigma_i[r_i^2]$ would lead to mutual cancellation of positive and negative deviations and the procedure would not yield useful results.) This definition is appropriate when the experimental uncertainty in x is much smaller than that in Y, a condition that is almost always fulfilled when x is the independent (experimentally controlled) variable and Y is the observed response, although it has been suggested that this is not necessarily always the case when x corresponds to concentration of a standard solution and Y to its mass spectrometric response. Alternative procedures that use functions other than $\Sigma_i[r_i^2]$ as the quantity to be minimized have been described (York 1966), but we shall not discuss these here.

One other restriction applies to the following expressions for A and B and related parameters, namely, the data are assumed to be homoscedastic, i.e., the reproducibilities of Y (given by standard deviations $s(Y_i)$, Equation [8.2c])

Legendre and Gauss: the Birth of Least-Squares Regression

Adrien-Marie Legendre (1752–1833) was a distinguished French mathematician whose many contributions are still in active use by mathematicians and theoreticians today. He was born into a wealthy family and was given a top quality education in mathematics and physics in Paris. He decided to enter for the 1782 prize on projectiles offered by the Berlin Academy. The actual task was stated as follows: *Determine the curve described by cannonballs and bombs, taking into consideration the resistance of the air; give rules for obtaining the ranges corresponding to different initial velocities and to different angles of projection.* Legendre won the prize and this marked the start of his research career.

As a result of the French Revolution in 1793, Legendre had difficulties as he lost the inherited capital that had provided him with a comfortable income. In 1794 Legendre published *Eléments de géométrie*, the leading elementary text on the topic for ~100 years. In 1806 he published a book on determining the orbits of comets. His method involved three observations taken at equal intervals; he assumed that the comet followed a parabolic path so he ended up with more equations than unknowns and, in an appendix, he described the *least-squares method* of fitting a curve to the data. However, Gauss published his version of the least-squares method in 1809 and, while acknowledging it had appeared in Legendre's book, claimed priority for himself (see below and www-history.mcs.st-andrews.ac.uk/Biographies/)

Johann Carl Friedrich Gauss (1777–1855) was a child prodigy who continued his remarkable creativity all his life. He started elementary school at age seven in Brunswick (now in Germany), and his teacher was amazed when he instantly mentally summed the integers from 1 to 100 by realising that the sum was 50 pairs of numbers with each pair summing to 101. In January 1801 an astronomer published the orbital positions of Ceres, a new 'small planet' (comet), but unfortunately only nine degrees of its orbit were observed before it disappeared behind the Sun. Several predictions of its position were published, and that by Gauss differed greatly from the others. When Ceres re-emerged on 7 December 1801 it was almost exactly where Gauss had predicted. Although he did not disclose his methods at the time, Gauss had used his least-squares approximation method. In retrospect it is thus a matter of opinion who had priority.

are assumed to be essentially constant over the experimental range of values for x. We shall consider the case where this condition is not fulfilled later (Section 8.2.5).

The strategy to apply the constraint (Equation [8.20]) to determination of the fitting parameters involves taking partial derivatives of $\Sigma_i[r_i^2]$ with respect to each of the N parameters in f(x) (N = 2, i.e., A and B in the case of Equation [8.19a]), setting each partial derivative to zero (the condition for a minimum in $\Sigma_i[r_i^2]$), and solving the resulting system of N equations for the N unknowns.

This part of the derivation is straightforward and can be sketched in as follows. Firstly, fully multiply out the expression in Equation [8.20]:

$$\Sigma_i[r_i^2] = \Sigma_i[Y_i^2 - 2Ay_i - 2Bx_iY_i + A^2 + 2ABx_i + B^2x_i^2]$$

Now treat x_i and Y_i as given values, and differentiate with respect to A and B regarded as variables that are to be adjusted so as to minimize $\Sigma_i[r_i^2]$, a condition given by zero values for each of the partial derivatives:

$\partial(\Sigma_i[r_i^2])/\partial B = \Sigma_i[-2x_iY_i + 2Ax_i + 2Bx_i^2] = 0$

i.e., $B.\Sigma_i(x_i^2) = \Sigma_i x_i Y_i - A.\Sigma_i x_i$ [8.21a]

$\partial(\Sigma_i[r_i^2])/\partial A = \Sigma_i[-2Y_i + 2A + 2Bx_i] = 0$

i.e., $A = (\Sigma_i Y_i - B\Sigma_i x_i)/n$ [8.21b]

since the summation is over i = 1 to n so $\Sigma_i(A) = n.A$. Now subsitute A from Equation [8.21b] into Equation [8.21a]:

$B.\Sigma_i(x_i^2) = \Sigma_i x_i Y_i - \Sigma_i x_i(\Sigma_i Y_i - B\Sigma_i x_i)/n = \Sigma_i x_i Y_i$

$- (\Sigma_i x_i)(\Sigma_i Y_i)/n + B(\Sigma_i x_i)^2/n$

Solving for B gives:

$B = [\Sigma_i x_i Y_i - (\Sigma_i x_i)(\Sigma_i Y_i)/n]/[\Sigma_i(x_i^2) - (\Sigma_i x_i)^2/n]$ [8.22]

It is convenient to rewrite Equations [8.21b and 8.22] as:

$B = S_{xy}/S_{xx}$ and $A = \bar{Y} - B.\bar{x}$ [8.23]

where the average values $\bar{Y} = (\Sigma_i Y_i)/n$ and $\bar{x} = (\Sigma_i x_i)/n$, and:

$S_{xx} = \Sigma_i(x_i - \bar{x})^2 = \Sigma_i(x_i^2) - (\Sigma_i x_i)^2/n$ [8.24a]

$S_{YY} = \Sigma_i(Y_i - \bar{Y})^2 = \Sigma_i(Y_i^2) - (\Sigma_i Y_i)^2/n$ [8.24b]

$S_{xY} = \Sigma_i(x_i - \bar{x}).(Y_i - \bar{Y})$

$= \Sigma_i(x_i.Y_i) - (\Sigma_i x_i).(\Sigma_i Y_i)/n$ [8.24c]

In each of Equations [8.24a–8.24c] the two expressions given are algebraically equivalent, e.g.:

$S_{xx} = \Sigma_i(x_i^2 - 2x_i.\bar{x} + \bar{x}^2)$

$= \Sigma_i(x_i^2) - 2(\Sigma_i x_i).(\Sigma_i x_i)/n + \Sigma_i[(\Sigma_i x_i)/n]^2$

$= \Sigma_i(x_i^2) - (\Sigma_i x_i)^2/n$

where the summations are over i = 1 to n. In both cases (Equations [8.24b] and [8.24c]) the second expression is more convenient for calculation than the first, and these expressions are widely available as pre-programmed parameters in scientific calculators (but take care that the calculator retains a sufficient number of significant digits throughout all calculations!). Note that Equation [8.23] implies that the average point ($\bar{x}$, $\bar{Y}$) of the experimental data always satisfies the least-squares regression equation and thus forms the 'center of gravity' of the corresponding

least-squares regression line. This is a consequence of the choice of $\Sigma_i[r_i^2]$ (Equation [8.20]) as the quantity to be minimized in order to provide the 'best' possible summary of the data by regression analysis using Equation [8.19a] as the fitting equation. Some alternatives to $\Sigma_i[r_i^2]$ as the quantity to be minimized (York 1966) do not carry this implication.

The preceding part of the derivation is easy, but the calculation of the corresponding uncertainties is considerably more complicated and is not given here. The important definitions and relationships are:

$V_{res} = s_{res}^2 = \Sigma_i(Y_i - Y_{i,pred})^2/dof = (S_{YY} - B.S_{xY})/(n-2)$ [8.25]

$V_B = s_B^2 = V_{res}/S_{xx}$ and $V_A = s_A^2 = V_{res}.[(1/n) + (\bar{x}^2/S_{xx})]$ [8.26]

$V_{Y,i,pred} = (s_{Y,i,pred})^2 = V_{res}.[(1/n) + (x_i - \bar{x})^2/S_{xx}]$ [8.27]

$s_{Y,i,k} = V_{res}^{\frac{1}{2}}.[(1/k) + (1/n) + (x_i - \bar{x})^2/S_{xx}]^{\frac{1}{2}}$ [8.28]

$CL(Y_{i,pred}) = (A + B.x_i) \pm [t(dof, p).(V_{Y,i,pred})^{\frac{1}{2}}]$, with $dof = (n-2)$ [8.29a]

$CL(A) = A \pm [t(dof, p).(V_A)^{\frac{1}{2}}]$, with $dof = (n-2)$ [8.29b]

$CL(B) = B \pm [t(dof, p).(V_B)^{\frac{1}{2}}]$, with $dof = (n-2)$ [8.29c]

$\bar{Y}_i^* = \Sigma_j(Y_{i,j}^*)/k^* = $ mean of k^*

replicates of Y_i^* (j = 1 to k^*)

measured for an unknown concentration x_i^* [8.30]

$V^*_{x,i} = (s^{*2}_{x,i}) = (V_{res}/B^2).[(1/n) + (1/k^*) + (\bar{Y}_i^* - \bar{Y})^2/(B^2.S_{xx})]$ [8.31]

$CL(x_i^*) = [(Y_i^* - A)/B] \pm [t(dof, p).(V^*_{x,i})^{\frac{1}{2}}]$ [8.32]

Equations [8.25–8.29] are the statistical quantities relevant to a simple linear regression (Equation [8.19a]), e.g., as in a calibration experiment for an analytical method that determines instrument response (Y) as a function of analyte concentration or amount (x); note again that these equations are appropriate for homoscedastic data for which the variances in Y_i ($V_{Y,i}$) are constant and independent of x_i, i.e., all Y_i values are of equal statistical weight. V_{res} is the variance of the residuals, i.e. the portion of the variance of the complete dataset that can not be accounted for by the predicted (from Equation [8.19a]) variation of Y resulting from variations in x; V_{res} includes possible contributions from unrecognised deviations of the (x_i, Y_i) calibration data from the simple linear relationship (Equation [8.19a]) as well as from breakdown of the homoscedacity assumption (see Section 8.3.3 for discussion of this point and Equation [8.33] for an alternative formulation of Equation [8.25] in the case where several replicate determinations of Y_i are made for each x_i value). Note also that s_{res} (Equation [8.25]), the standard deviation of the residuals, is the quantity recommended (Section 8.3.1) as an alternative to R or R^2 to assess the degree of correlation of two data sets. V_B and V_A are the variances in the least-squares estimates of B and A (s_B and s_A are the corresponding standard deviations).

$s_{Y,i,k}$ (Equation [8.26]) is the standard deviation of k values of Y_i predicted by Equation [8.19a] for a known x_i and $V_{Y,i,pred}$ is the variance of the normal (Gaussian) distribution of replicate determinations assumed to underlie the small data sets usually obtained in analytical practice, as predicted from Equation [8.19a] for a chosen value of x_i. Clearly $(s_{Y,i,k})^2 = V_{Y,i} = (s_{Y,i})^2$ in the limit k $\to \infty$, i.e. $V_{Y,i,pred}$ is the variance of the normal distribution assumed to describe the determinations of Y_i. Later the quantity $s(Y_i)$ is used to denote a simple experimental determination (not prediction as for $s_{Y,i,k}$) of the standard deviation of a set of replicate determinations of Y_i for a fixed x_i in the calibration experiments (Equation [8.2] with Y replacing x).

$CL(Y_{i,pred})$ gives the corresponding confidence limits for $Y_{i,pred}$ using a two sided Student's-t value corresponding to $dof = (n-2)$ and the user-selected value of p and thus of t (Table 8.1). Confidence limits for the intercept and slope, $CL(A)$ and $CL(B)$, are calculated similarly. Of course these definitions refer to small data sets for which the Student-t is the appropriate statistical test parameter.

In contrast, Equations [8.30–8.32] are the quantities used to assess the statistical reliability of the inversion of the simple linear regression equation, e.g., determining the concentration (or amount) x_i^* of an anayte from a measured average response $\bar{Y}_i^*$ by inverting the calibration equation. The superscript * in Equations [8.30–8.32] is used to denote quantities relevant to experiments in which an unknown value of x is obtained

from a measured value of Y via inversion of the least-squares regression calibration equation that was determined separately. Thus $V^*_{x,i}$ is the variance of a value x_i^* predicted by inverting the calibration equation (Equation [8.19a]) and inserting an experimental value of the analytical signal $Y = Y_i^*$. Note that $\bar{Y}$ in Equation [8.31] is still the mean value of Y_i in the calibration experiments, not their inversion. Then $CL(x_i^*)$ (Equation [8.32] represents the corresponding confidence limits for x_i^* that vary with the details of both the determination and calibration experiments, as they should. Recall that the experimental standard deviation of the k^* measurements of Y_i^* corresponds to the value of $s(Y_i)$, which was assumed to be independent of x_i in our assumption of homoscedacity (see text box on heteroscedacity), and is not the same thing as $s_{Y,i,k}$ in Equation [8.28], which refers to the standard deviation of values of Y_i predicted by Equation [8.19a] for a known value of x_i.

A simple example of a regression calculation, performed by hand to make clear what is involved, is shown in the accompanying text box. In practice, such calculations are usually performed using a computer algorithm or a scientifc calculator; in the latter case it can be important to retain many more digits in the intermediate steps of the calculation than are required for the final result, especially when small differences of two large numbers are involved (see e.g., the calculation of V_{res} in the text box).

A question that arises frequently, and has been the subject of considerable discussion (Strong 1979, 1980; Ellerton 1980; Schwartz 1986), concerns the advisability or otherwise of constraining a calibration curve to pass through the origin (x = 0 = Y), i.e. setting $A = 0$, if it is expected for nonstatistical theoretical reasons that this should be the case. It was shown (Schwartz 1986) that, even if the experimental data can be shown to be consistent with such a constraint by, e.g., an appropriate Student's t-test at a chosen confidence level (i.e. does $CL(A)$ in Equation [8.29] encompass the origin?), the imposition of such a constraint has a marked effect on the uncertainty estimate of the analysis by inversion of the calibration equation; that this uncertainty is lower than for the unconstrained model ($A \neq 0$) in the vicinity of the constraint (the origin in this case). However, imposition of this constraint implies that the point (x = 0 = Y) is both accurate and has zero variance, an assumption that ignores the grim reality of trace level analysis (lack of selectivity, noise and drift, irreproducibilities in extraction and clean-up procedures etc. that are generally more critical for near-zero values of x). It

Example of a Simple Linear Least-Squares Regression

x_i	Y_i	$Y_{i,pred}$	r_i	$(x_i - \bar{x})^2$	$V_{Y,i}$	$t.(V_{Y,i})^{\frac{1}{2}}$	$CL(Y_{i+})$; $CL(Y_{i-})$
1.0	5.3	5.54	−0.24	6.2500	0.0265646	0.4525	5.99; 5.09
2.0	10.9	10.78	+0.12	1.3225	0.0122849	0.3077	11.09; 10.47
3.0	16.2	16.03	+0.17	0.2500	0.0091769	0.2659	16.30; 15.76
4.0	21.3	21.27	+0.03	0.2500	0.0091769	0.2659	21.54; 21.00
5.0	26.7	26.51	+0.19	1.3225	0.0122849	0.3077	26.82; 26.20
6.0	31.5	31.76	−0.26	6.250	0.0265646	0.4525	32.21; 31.31

The first two columns represent a fictitious calibration experiment; the concentrations x_i (no units!) were chosen to be integers to emphasize the constraint that assumes negligible uncertainty in the independent variable.

Some initial calculation steps are: $\Sigma_i(x_i) = 21.00$; $\Sigma_i(x_i^2) = 91.00$; $\Sigma_i(Y_i) = 111.90$; $\Sigma_i(Y_i^2) = 2568.17$; $\Sigma_i(x_i.Y_i) = 483.40$; $n = 6$; $\bar{x} = 3.500$; $\bar{Y} = 18.650$

Some additional calculation steps required for the regression are:

$S_{xx} = 17.500$ (Equation [8.24a]); $S_{YY} = 481.23500$ (Equation [8.24b]); $S_{xY} = 91.7500$ (Equation [8.24c])

from which:

$B = +5.242857143$ (Equation [8.23]); $A = +0.3000$ (Equation [8.23]); $V_{res} = 0.050714$ (Equation [8.25]);
$V_B = 0.0028980$ Equation [8.26]; $V_A = 0.0439524$ Equation [8.26]

(*This is an example of the importance of carrying more digits in intermediate steps than are significant in the final result*);

$V_{Y,i}$ values from Equation [8.27] and the $t(V_{Y,i})^{\frac{1}{2}}$ values give the confidence limits around the value $Y_{i,pred}$ predicted from the best-fit line for a given x_i, i.e. $CL(Y_{i,pred}) = (A + B.x_i) \pm [t(dof, p).(V_{Y,i})^{\frac{1}{2}}]$, with $dof = (n - 2) = 4$ (Equation [8.29a]). For a chosen value of $p = 0.05$ and $dof = 4$, $t_C = 2.776$ (Table 8.1) for a two sided test since we are interested in both upper and lower 95 % confidence limits.

Now consider the application of our calibration ($A = +0.300$, $B = 5.24286$) to determine unknown concentrations x_j^* of the target analyte from values of $Y_{mean,j}^*$ representing mean values of $k^* = 5$ replicates measured for x^* at each j:

$$x_j^* = (Y_{mean,j}^* - A)/B;$$

where the uncertainties in x_j are given as $V_{x,j}^*$ (Equation [8.31]) determined using the values $\bar{Y}$, n, V_{res}, S_{xx} and $t(0.05)$ from the *calibration* experiments.

Again we use a fictitious set of data summarized in the following Table. Note that the uncertainties $\pm(t.V_{x}j^*)^{\frac{1}{2}}$ estimated for the values calculated for x_j^* are largely determined by the *calibration* experiments, together with the number of replicates k^* of Y_j in the replicate measurements of the unknown sample, in accord with intuition.

$Y_{mean,j}^*$	x_j^*	$(Y_{mean,j}^* - \bar{Y})^2$	$V_{x,j}^*$	$t.(V_{x,j}^*)^{\frac{1}{2}}$
6.42	1.1673	149.573	0.0012502	0.098
12.76	2.3765	34.692	0.0008096	0.079
19.18	3.6011	0.2809	0.0006776	0.072
26.04	4.9095	54.612	0.0008860	0.083

These uncertainties in x_j^* (corresponding to the chosen p value used to determine t) are smaller ($\sim \pm 0.07$) near the 'centre of gravity' ($\bar{x}, \bar{Y}$) of the calibration experiments, and increase to $\sim \pm 0.1$ at the extremities of the calibration range (compare Figure 8.12B).

is appropriate to investigate whether or not the value obtained for A is zero to within some selected confidence level but it is recommended (Schwartz 1986) to use the unconstrained model (Equation [8.19a]) for the most reliable inversion of the calibration to obtain estimates for x.

8.3.3 Test for Goodness of Fit: Irreproducibility, Heteroscedacity and Nonlinearity

The example calculation of a simple linear regression, shown in the text box, assumed that the six calibration experiments involved a single determination of Y_i at each of the different concentrations x_i of the calibration standard. Of course this is seldom the case in standard practice; fortunately, the procedure described applies equally well if the n_C total calibration experiments involved several replicates at each concentration, i.e. using a fixed number k_C of replicate analyses at each of m_C different concentrations in the calibration experiment, so that $n_C = m_C.k_C$. This approach facilitates tests of the calibration line (Equation [8.19a]) for goodness of fit that are more meaningful than the correlation coefficient R and the coefficient of determination R^2 (Section 8.3.1), and for the reasons in cases of failure of such tests (Analytical Methods Committee 1994). Thus we use Y_{ij} to denote the j'th determination of Y at the i'th concentration x_i in these calibration experiments, where $i = 1$ to m_C and $j = 1$ to k_C. Now the calculation of the coefficients A and B proceeds exactly as before (Equations [8.22–8.24]), treating all n_C data points on an equal basis under the same set of assumptions (homoscedastic data, the least-squares line includes ($\bar{x}, \bar{Y}$)). Where the analysis differs is in the analysis of the residuals. Now Equation [8.25] still applies but can be re-written as:

$$V_{res} \equiv MS_{res} = SS_{res}/dof_{res} = \Sigma_i[\Sigma_j(Y_{i,j} - Y_{i,pred})^2]/dof_{res};$$
$$dof_{res} = k_C.m_C - 2 \qquad [8.33]$$

where dof_{res} is the number of degrees of freedom associated with V_{res} (the total variance of all n_C values of $Y_{i,j}$ relative to those predicted by Equation [8.19a]), and SS_{res} and MS_{res} are the 'sum of squares' and the 'mean square' parameters used in the quoted reference from the UK Royal Society for Chemistry (Analytical Methods Committee 1994). The objective now is to sub-divide SS_{res} into a contribution SS_{repr} from intrinsic irreproducibility in the measurements themselves (obtained from consideration of the k_C replicate measurements at each different concentration), and a contribution SS_{lof} from lack-of-fit of

the raw data to the model fitting function (Equation [8.19a] in this case); the latter can correspond to nonlinearity of the experimental data and from other miscellaneous reasons such as significant deviations from homoscedasticity:

$$SS_{res} = \Sigma_i[\Sigma_j(Y_{i,j} - Y_{i,pred})^2] = SS_{repr} + SS_{lof} \qquad [8.34]$$
$$SS_{repr} = \Sigma_i[\Sigma_j(Y_{ij} - Y_{i,avg})^2]; \ dof_{repr} = k_C(m_C - 1) \qquad [8.35]$$
$$SS_{lof} = SS_{res} - SS_{repr}; \ dof_{lof} = k_C - 2 \qquad [8.36]$$

where $Y_{i,avg}$ is the average instrumental response over the k_C measurements for the i'th concentration, so that SS_{repr} is an experimental indication of the intrinsic reproducibility of the measurements. Note that $(dof_{repr} + dof_{lof}) = dof_{res}(= k_C m_C - 2$, as required).

We now wish to test a null hypothesis (H_o: the chosen model (Equation [8.19a]) provides an adequate description of the experimental data), by applying an F-test (Section 8.2.5). In this case we must convert SS_{repr} and SS_{lof} into corresponding values of variance (i.e. MS values in this case, e.g. Equation [8.33]) to take account of the different degrees of freedom in the two cases:

$$V_{repr} \equiv MS_{repr} = SS_{repr}/dof_{repr} = SS_{repr}/[k_C(m_C - 1)] \qquad [8.37a]$$
$$V_{lof} \equiv MS_{lof} = SS_{lof}/dof_{lof} = SS_{lof}/(k_C - 2) \qquad [8.37b]$$

We now construct our test statistic:

$$F = MS_{lof}/MS_{repr} \qquad [8.38]$$

bearing in mind that F is defined so as to always be ≥ 0 (Section 8.2.5) and that this in turn defines which parameter is assigned to dof_1 and which to dof_2 when reading the critical value F_c from Table 8.2. If $F < F_c$, H_o is accepted for this set of data, i.e. the simple linear regression is adequate, but if $F \geq F_c$, H_o fails at the level of cofidence used to determine F_c now the reason for the poor goodness of fit must be sought. Again a numerical example should help to clarify the procedure from the user's point of view and this is provided in the accompanying text box. (Note that once again no attempt has been made to summarize the underlying theory justifying this procedure.)

For the example worked through in the text box it turned out that indeed H_o failed and the next step must be to investigate the reason. In general this is the result of either heteroscedastic data or intrinsic nonlinearity of the functional relationship between x and Y (i.e., inadequacy of Equation [8.19a] as a model fitting function),

Example of an F-Test for Goodness of Fit for a Simple Linear Least-Squares Regression for Replicate Determinations

x_i	$Y_{i,1}$	$Y_{i,2}$	$Y_{i,avg}$	$|Y_{i,1} - Y_{i,2}|$	$Y_{i,pred}$	$r_{i,1} =$ $(Y_{i,1} - Y_{i,pred})$	$r_{i,2} =$ $(Y_{i,2} - Y_{i,pred})$
3.2	0.276	0.284	0.280	0.008	0.271	+0.005	+0.013
6.4	0.560	0.564	0.562	0.004	0.563	−0.003	+0.001
9.6	0.846	0.846	0.846	0.000	0.854	−0.008	−0.008
12.8	1.130	1.142	1.136	0.012	1.145	−0.015	−0.003
16.0	1.440	1.436	1.438	0.004	1.437	+0.003	−0.001
19.2	1.740	1.732	1.736	0.008	1.728	+0.012	+0.004

The first three columns represent fictitious experimental data for duplicate measurements ($j = 1 \ldots m = 2$) of instrument response Y for each of six concentrations x_i ($i = 1 \ldots k = 6$).

We first must establish that the data are homoscedastic in order to apply the simple (unweighted) linear regression formulae. In this case simple examination of the data shows that there is no detectable trend of the absolute values $|Y_{i,1} - Y_{i,2}|$ with increasing x_i, as we would expect since the range of values of Y_i would increase with x_i if heteroscedasticity were present, compare Figure 8.9. (This preliminary conclusion of homoscedacity is confirmed after the event by examining the residuals calculated on the basis of this assumption (columns 7 and 8) that demonstrate no trend of $|r_{i,j}|$ with x_i; the implications of these residuals for the goodness of fit are discussed later.)

Now we calculate the linear least-squares parameters A and B as before, treating all $n = k.m = 12$ data points individually. This case is slightly more complicated than the simple linear regression case considered in an earlier text box since now, although there are only six different values of x_i, each value appears twice so we have to be sure to take this into account.

We have:

$$\bar{x} = \Sigma_i x_i / k = 11.2000 \text{ and } \bar{Y} = \Sigma_i \Sigma_j (Y_{i,j})/n = 0.99967$$

where we are carrying more significant figures than are strictly justifiable in order to avoid rounding-off errors; $S_{xx} = \Sigma_l (x_l{}^2) - (\Sigma_l x)^2/n$ (Equation [8.24a]) where l enumerates *all* x-values, i.e., although there are only six *different* x-values each one appears twice in the data set ($m = 2$) so the summation over l is from 1 to $n = m.k(= 2.6 = 12)$; we replace summation over l by double summation over i and j then:

$$S_{xx} = \Sigma_j \Sigma_i (x_{i,j}^2) - (\Sigma_j \Sigma_i x_{i,j})^2/n$$

where the summation over j is 1 to m ($= 2$ in this case) and that over i is 1 to k ($= 6$ in this case); i.e.

$$S_{xx} = m.\Sigma_i(x_i^2) - (m.\Sigma_i x_i)^2/(k.m) = m.[\Sigma_i(x_i^2) - (\Sigma_i x_i)^2]/k$$

where the summation over i is 1 to k. Then $S_{xx} = 2x[931.840 - (67.20)^2/6] = 358.400$.

From Equation [8.24b]: $S_{YY} = [\Sigma_i \Sigma_j (Y_{i,j}^2) - (\Sigma_i \Sigma_j Y_{i,j})^2/n] = \Sigma_l(Y_l^2) - (\Sigma_l Y_l)^2/n]$ (where the summation over l is 1 to $n = 12$) $= 14.964898 - (11.996^2)/12 = 2.97290$

From Equation [8.24c]:

$$S_{xY} = \Sigma_l(x_l.Y_l) - (\Sigma_l x_l).(\Sigma_l Y_l)/n \text{ where } l = 1 \text{ to } n$$

$$= \Sigma_i[\Sigma_j(x_i.Y_{i,j})] - [(\Sigma_i \Sigma_j x_i).\Sigma_i \Sigma_j(Y_{i,j})]/n, \text{ where } i = 1 \text{ to } k(= 6) \text{ and } j = 1 \text{ to } m(= 2)$$

$$= 32.63360.$$

Then $B = S_{xY}/S_{xx} = 0.09105357$, and $A = \bar{Y} - B.\underline{x} = -0.02013334$, where we are again temporarily carrying too many digits. These values for A and B lead to the $Y_{i,j,pred}$ values (same for all j) in the Table, and thus to the residuals in columns 7 and 8.

To test the goodness of fit (see Equations [8.33–8.36]), the sum of the squares of the residuals is: $SS_{res} = \Sigma_i\Sigma_j(r_{ij}^2) = 7.35522 \times 10^{-4}$ with $dof_{res} = m(k-1) = 2.(6-1) = 10$ degrees of freedom. This sum of squares is split into two parts, one of which gives an estimate of the intrinsic reproducibility of the raw measurements:

$$SS_{repr} = \Sigma_i[\Sigma_j(Y_{i,j} - Y_{i,avg})^2] = 1.5200 \times 10^{-4} \text{ with } dof_{repr} = k(m-1) = 6$$

Then the contribution to the total SS_{res} from lack of fit to the proposed simple linear model function (Equation [8.19a]) is:

$$SS_{lof} = SS_{res} - SS_{repr} = 5.8452 \times 10^{-4} \text{ with } dof_{lof} = k - 2 = 4.$$

Now we calculate the corresponding 'mean squares' MS parameters (equivalent to V values):
$MS_{res} \equiv V_{res} = SS_{res}/dof_{res} = 7.35522 \times 10^{-5}$,
$MS_{repr} = SS_{repr}/dof_{repr} = 0.25333 \times 10^{-4}$, and
$MS_{lof} = SS_{lof}/dof_{lof} = 1.46130 \times 10^{-4}$.

Then, finally, our experimental test statistic $F = MS_{lof}/MS_{repr} = 5.768$, where we chose MS_{lof} as the numerator (corresponding to $dof_1 = dof_{lof}$ in the table of F_c values) since by definition $F \geq 1$.

Then from Table 8.2, $F_c = 4.5337$ for ($dof_1 = 4$, $dof_2 = 6$, $p = 0.05$). Since our experimental $F > F_c$ the null hypothesis [H_0: the chosen model (Equation [8.19a]) provides an adequate description of the experimental data] **fails!**

This statistical test tells us whether or not our fitting equation (Equation [8.19a] in this case) provides an adequate description of the experimental data, but if not the test itself gives no indication of the reason. We can eliminate heteroscedasticity in this case (see above) and thus suspect a degree of nonlinearity in the raw data, confirmed by plotting the residuals r_{ij} vs x_i (compare the accompanying plot of these residuals with Figure 8.9). A slightly better fit would be obtained by using a quadratic fitting equation (Equation [8.19b], see Section 8.3.6).

Note that this nonlinearity would *not* have been detected by examination of the *correlation coefficient* $R = S_{xY}/(S_{xx}.S_{YY})^{\frac{1}{2}}$ (Equation [8.18]) $= 0.999876$, which tells us only that Y is well correlated with x but says very little (actually nothing) about the adequacy of the fitting function.

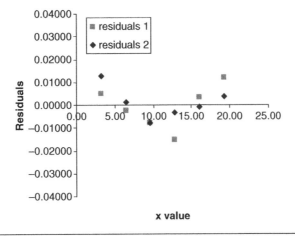

Residuals

| Reproducibility | Nonlinearity | Heteroscedacity |

Figure 8.9 Plots of values (including the sign) of residuals $(Y - Y_{pred})$ vs x, where Y_{pred} is the value of Y predicted from the simple linear model $(Y = A + B.x)$ with A and B determined by simple linear regression, for three important causes of significant values of the residual variance V_{res}. Note that in the case of nonlinearity *of the data* (i.e., inapplicability of Equation [8.19a] to this data set) the plot could just as well be concave upwards. A sufficiently high degree of irreproducibility can mask the other two if present. It is possible to observe nonlinearity of heteroscedastic *data* if the latter effect is not too extreme. Reproduced from Meier and Zünd, *Statistical Methods in Analytical Chemistry*, 2nd *Edition* (2000), with permission of John Wiley & Sons Inc.

or possibly both. A simple plot of the residuals $(Y_{i,j} - Y_{i,pred})$, including the sign, vs x_i, can usually indicate the reason(s) for such a failure of H_o, as shown in a general way in Figure 8.9. For the example in the text box, the failure was clearly due, at least in part, to nonlinearity of the data, i.e., breakdown of Equation [8.19a] for this data set (see plot of the residuals in the text box). In this case the next step would likely be to perform linear regression on the quadratic fitting function (Equation [8.19b]), see Section 8.3.6. If instead the residuals plot indicated heteroscedasticity, a weighted regression would be required (Section 8.3.5).

8.3.4 Inversion of the Calibration Equation

We discuss here the use to which calibrations obtained by least-squares regression are put, i.e., their use in determining an unknown value (x^*) of concentration x from a measured instrumental response Y^*; moreover, importantly we can also obtain a value for the corresponding uncertainty (variance) in x^*. Such determination of x^* involves inversion of the calibration fitting function, Equation [8.19a] in our case; inversion of this equation gives:

$$x^* = f^*(Y^*) = (Y^* - A)/B$$

Thus for the simple fitting function Equation [8.19a] inversion is easy, and this simple evaluation of x^* from Y^* is one of its major advantages.

However, it is equally important to also obtain confidence limits for values of x^* thus calculated from measurements of Y^*, and indeed these were already stated (again without proof) in Equations [8.30–8.32] (Meier 2000); some example calculations for $k^* = 5$ are shown in the

linear regression text box, but here we will analyze these a little more closely in order to establish the main sources of uncertainty in x^* derived from Y^*. The relevant information from the text box, concerning the calibration data, includes the following:

$$n_C = 6; S_{xx} = 17.500; V_{res} = 0.050714; \bar{Y} = 18.650;$$

$$B = 5.24286; dof = (n_C - 2) = 4$$

with $t = 2.776$ for $p = 0.05$ (95 % confidence limits) and $dof = 4$ (Table 8.1).

Now we evaluate Equations [8.31–8.32] for two values of $Y_{mean,j}{}^*$, one far from the mean value $\bar{Y}$ of the calibration experiments and one close to $\bar{Y}$; we also consider two values of k^*, the number of replicate measurements that were averaged to give $Y_{mean,j}{}^*$.

By inserting the values of the quantities from the calibration experiments into Equation [8.31], we obtain for this example:

$$V^*_{x,j} = 0.001845[(1/6) + (1/k^*) + (Y_{mean,j}{}^* - 18.650)^2$$

$$/481.0327] \qquad [8.39]$$

For $Y_{mean,j}{}^* = 6.42$ and $k^* = 5$, Equation [8.39] gives:

$$V^*_{x,i} = 0.001845[0.16667 + 0.2000 + 0.31094]$$

$$= 1.2502 \times 10^{-3}$$

one of the examples already given in the linear regression text box. Now, however, we can see the origins of the range of the confidence limits $CL(x^*_j) = \pm t.(V^*_{x,j})^{\frac{1}{2}}$ (Equation [8.32]). Increasing both n_C (total number of independent data points obtained for the calibration) and k* to very large values would reduce the value of the expression in the square brackets in Equation [8.32] by a factor of about 2, and thus $(V^*_{x,j})^{\frac{1}{2}}$ by a factor of about 1.4, but at the expense of a greatly increased amount of time and effort; a more fruitful approach in this case would involve decreasing V_{res} (and thus the pre-multiplier of the expression in square brackets) by increasing n_C and/or improving the analytical method so that more precise data are obtained thus minimizing the $(Y_i - Y_{i,pred})^2$ terms (Equation [8.25]). On the other hand, the value of a reasonable number of replicates k* becomes apparent if we evaluate Equation [8.39] for the same $Y_{mean,j}^*$ and for k* = 1 (the absolute minimum value):

$$V^*_{x,j} = 0.001845[0.16667 + 1.0000 + 0.31094]$$

$$= 2.726 \times 10^{-3}.$$

Thus use of a moderate number (k* = 5) of replicates provides in this example an improvement in $V^*_{x,j}$ by a factor of $(2.726/1.250) = 2.2$, and in $(V^*_{x,j})^{\frac{1}{2}}$ by a factor of 1.5, relative to a single determination. The appropriate compromise between more narrow confidence limits and amount of experimental time and effort (and cost) required, depends on the purpose for which the analyses are to be conducted. Even an approximate evaluation such as that given above can be valuable in deciding the fit for purpose compromise; however, as always, it is strongly recommended to fully document the details of all such evaluations and decisions before proceeding to perform the analyses themselves.

The preceding example used a value of $Y_{mean,j}^*$ far from $\bar{Y}$; it is of interest to compare the same evaluation for a case in which $|Y_{mean,j}^* - \bar{Y}|$ is small and for this purpose we choose another example summarized in the linear regression text box, $Y_{mean,j}^* = 19.18$, i.e., $|Y_{mean,j}^* - \bar{Y}| = 0.530$. For this example, Equation [8.31] becomes:

$$V^*_{x,j} = 0.001845[(1/6) + (1/k^*) + 0.000584]$$

Here it is clear that the 3rd term is negligible relative to the others and that large increases in n_C and k* would indeed reduce $V^*_{x,j}$ to very low values in a narrow range of $Y_{mean,j}^*$ around $\bar{Y}$. Again, the narrow range of x* values over which the range of CL(x*) is also very narrow must be determined on a case by case basis, e.g., by preliminary rough experiments if sufficient unknown sample is available. Note in passing that the reason for this effect can be traced ultimately in the supporting theory to the constraint placed on the least-squares linear regression that the regression line always includes the average point ($\bar{x}$, $\bar{Y}$) of the calibration data (Section 8.3.2).

8.3.5 Weighted Linear Regression

We must now address the additional complications arising when the raw calibration data are clearly heteroscedastic, i.e., the standard deviations $s(Y_i)$ are clearly a function of x_i. The best diagnostic test for heteroscedacity is to first fit the raw data by a simple linear least-squares regression, then plot the residuals (including the sign) as a function of x (Figure 8.9). There are two general approaches to dealing with this situation, namely, transformation of the raw data in such a way that the transformed data are homoscedastic, or use of statistical weights in the procedure to evaluate the least-squares estimates of the parameters in the model function $Y = f(x)$ to account for the unequal standard deviations. Both approaches can work well in different situations and indeed can often give very similar values for the parameters (A and B in our simple linear case); however, the confidence limits can behave in an undesirable manner if the data are transformed. For example, if a logarithmic transformation is used the confidence limits become unsymmetrical, e.g., $Y_i = 18.4 \pm 1.1$ (experimental confidence limits are symmetrical) becomes $\ln(Y_i) = (2.9124 + 0.056)$ to $(2.9124 - 0.0617)$. If the transformation approach is to be used, one must find a transformation that equalizes the variances in Y' (the transformed Y values) across the experimental range, then also (if necessary) find a transformation for the values of $x \rightarrow x'$ that provides a simple functional form (preferably Equation [8.19a]) relating Y' to x', then fit this new fitting function to the transformed (hopefully now homoscedastic) variables, check that the residuals of the Y'_i do indeed indicate no heteroscedastic trend, then transform the predicted values $Y'_{i,pred}$ back into $Y_{i,pred}$ using the inverse of the transformation applied for $Y_i \rightarrow Y'_i$. This is very complicated and subject to error; data transformations should be used with care, if at all.

In analytical chemistry, heteroscedastic data are usually handled using weighted regression so that data points measured with greater precision (smaller variance) influence the values assigned to the parameters in the model fitting function more than do data measured with lower precision (Schwartz 1979; Garden 1980). For this purpose, the quantity that is minimized in order to provide least-squares estimates of the parameters in the fitting function is no longer that given in Equation [8.20], but is instead:

$$\Sigma_i[w_i.(r_i^2)] = \Sigma_i[w_i.(Y_i - Y_{i,pred})^2]$$

$$= \Sigma_i\{w_i.[Y_i - (A_W + B_W.x_i)]^2\} = minimum$$
$$[8.40]$$

where the w_i are the statistical weights. An important result (not proved here) is that the optimal results, i.e., those that minimize the uncertainties in the estimates thus obtained for the parameters (A_W and B_W in the simple linear case), are obtained when the w_i are inversely proportional to the variances at each experimental value of the independent variable x_i:

$$w_i = K/[s(Y_i)]^2; K = \text{normalizing constant often}$$

$$\text{chosen so that } \Sigma_i[w_i] = n \qquad [8.41]$$

where n is again the total number of data points to be fitted. The $s(Y_i)$ values (do not confuse these with the $s_{Y,i}$ parameter defined in Equation [8.27]) can be obtained experimentally by performing large numbers of replicate measurements of Y_i for each x_i, and this approach is sometimes used (e.g., for strict validation experiments). A less stringent approach that is often fit for purpose is to determine estimates of $s(Y_i)$ from relatively small numbers of determinations of Y_i for a few selected values of x_i covering the desired range, then fitting these estimates of $s(Y_i)$ to a simple function of x_i often chosen to be a version of Equation [8.19a]:

$$s(Y_i) = a_s + b_s.x_i \qquad [8.42]$$

using simple (unweighted!) linear regression (Equations [8.23–8.24]). Such a procedure effectively averages out uncertainties in the individual $s(Y_i)$ values resulting from the small numbers of replicates used to determine each one. It is always advisable to make sure that the model function used to describe the $s(Y_i)$ values (Equation [8.42]) appears to fit the observed data reasonably well; however, to be effective the goodness of fit of this model function (Equation [8.42]) often does not have to meet the usual standards.

By setting $\partial[\Sigma_i(w_i.r_i^2)]/\partial B_w = 0 = \partial[\Sigma_i(w_i.r_i^2)]/\partial A_w$ and solving for B_w and A_w as before, the following relationships are derived to define A_w and B_w in the weighted version of Equation [8.19a] :

$$\bar{x}_w = \Sigma_i(w_i.x_i)/n; \quad \bar{Y}_w = \Sigma_i(w_i.Y_i)/n \qquad [8.43]$$

$$B_w = S_{xY,w}/S_{xx,w} \text{ and } A_w = \bar{Y}_w - B_w.\bar{x}_w \qquad [8.44]$$

$$S_{xx,w} = \Sigma_i[w_i.(x_i - \bar{x}_w)^2] \qquad [8.45a]$$

$$S_{YY,w} = \Sigma_i[w_i.(Y_i - \bar{Y}_w)^2] \qquad [8.45b]$$

$$S_{xY,w} = \Sigma_i[w_i.(x_i - \bar{x}_w).(Y_i - \bar{Y}_w)] \qquad [8.45c]$$

Equations [8.44–8.45] are analogous to Equations [8.23–8.24] in the unweighted case. Equations [8.25–8.32] remain unchanged, with $S_{xx,w}$ replacing S_{xx} etc. Once the raw

data (x_i,Y_i) have been fitted to the model function using weighted least-squares regression, the validation of the model includes examination of the weighted residuals as a function of x_i:

$$r_{i,w} = w_i.(Y_i - Y_{i,pred}) \qquad [8.46]$$

to check that the heteroscedacity has been adequately accounted for and also to examine whether any nonlinearity can be detected (Figure 8.9).

A numerical example of weighted linear regression is worked out in the accompanying text box and compared with unweighted regression of the same heteroscedastic data. This example is typical in that only small differences result with respect to the slope and intercept, but the weighting results in an appreciable decrease in V_{res} that is reflected in corresponding decreases in $s_{Y,i,k}$ and $CL(Y_{i,pred})$ (Equations [8.28 and 8.29]), and thus also in $V^*_{x,i}$ and $CL(x^*_i)$ (Equations [8.31 and 8.32]). Furthermore, the appreciable movement of the average data point ($\bar{x}_w, \bar{Y}_w$) away from ($\bar{x}, \bar{Y}$) in cases where $s(Y_i)$ increases with x_i (the usual case) leads to further reductions in $V^*_{x,i}$ and $CL(x^*_i)$ for small values of Y_i^* via the term $(Y_i^* - \bar{Y})^2$ in Equation [8.31]. This represents a considerable advantage with respect to justification of lower limits of detection and quantitation (see later).

In the spirit of so-called 'robust regression' (Phillips 1983; Thompson 1989), weighted least-squares regression would employ an iterative procedure in which the weighting coefficients w_i are re-evaluated after each cycle and the least-squares regression then repeated using the revised weights until no significant change is observed. This approach is similar in intent to the rejection of outliers (Section 8.2.7) but proceeds by progressive decreases in the weights assigned to the 'dubious' data rather than their outright rejection. Such procedures have been shown to be relatively insensitive to large deviations in a small number of data points that are difficult to justify as representatives of an underlying normal distribution (a necessary assumption for the validity of the conventional least-squares approach such as Equations [8.23–8.32]). The complexity of such iterative calculations, however, means that computer-based algorithms are the only realistic approach.

At the other end of the scale of degree of rigor vs expenditure of time and effort, many analysts assume that their data are heteroscedastic with w_i values that vary with $1/x_i$ or $1/x_i^2$; these assumptions amount to assuming (Equation [8.41]) that $s(Y_i)$ varies as x_i or as $x_i^{\frac{1}{2}}$. It is difficult to justify such arbitrary assumptions nowadays when

Example of Linear Regression on Heteroscedastic Data

x_i	$Y_{i,1}$	$Y_{i,2}$	$Y_{i,3}$	$s(Y_i)$	$s_{fit}(Y_i)$	w_i
1	30.7	32.1	31.4	0.700	0.7240	2.4174
2	65.5	63.0	64.1	1.253	1.2067	0.8702
3	96.1	94.0	97.3	1.670	1.6894	0.4439
4	126.9	128.6	131.2	2.170	2.1721	0.2685

This set of fictitious data is intended to illustrate the case where the data are clearly heteroscedastic, as demonstrated here by the values of $s(Y_i)$ (the standard deviation of the three replicate values determined for Y_i from Equation [8.2c]) that steadily increase with x_i; do not confuse this purely experimental quantity with $s_{Y,i}$ defined in Equation [8.27], the standard deviation of the value of Y_i *predicted* from Equation [8.19a] for a chosen value of x_i, as the mean value of the Gaussian distribution assumed to underly the observed values.

For purposes of comparison we first subject the $n = 12$ data points to simple linear regression as if they were in fact homoscedastic, i.e., we assign equal statistical weights to all data.

Thus we obtain $\bar{x} = 2.500$ and $\bar{Y} = 80.0750$. Then, from Equations [8.24a to 8.24c]:

$S_{xx} = 15.00$,
$S_{YY} = 15,776.36$ and
$S_{xY} = 486.15$.

Equation [8.23] then gives $B = 32.410$ and $A = -0.950$, while Equation [8.25] gives $V_{res} = 2.0241$.

A plot of the 12 residuals $r_{i,j} = (Y_{i,j} - Y_{i,pred})$ vs x_i confirms that the data are indeed heteroscedastic (compare Figure 8.9).

To deal with the heteroscadacity we must assign statistical weights w_i to the data; it would be possible to use values of w_i given by Equation [8.41], where the $s(Y_i)$ values are those calculated directly from the raw data (see Table). However, in this case only three values of Y_i were determined for each of the four values of x_i, so the $s(Y_i)$ values are themselves subject to appreciable statistical uncertainty; in such cases the $s(Y_i)$ values are in turn fitted by a least-squares method to a suitable fitting function.

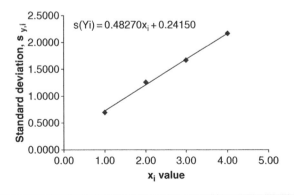

$s(Yi) = 0.48270x_i + 0.24150$

(Continued)

Simple linear regression was used here, giving $s_{fit}(Y_i) = 0.24125 + 0.48270x_i$, where the notation $s_{fit}(Y_i)$ denotes a value of $s(Y_i)$ calculated from the linear regression equation for one of the four specific values of x_i (see Table for these values); $s_{fit}(Y_i)$ is an experimental counterpart of $s_{Y,i,k}$ (Equation [8.28]) that refers to the standard deviation of values of Y_i *predicted* by Equation [8.19a] for a given x_i for a small number (k) of experimental determinations.

Then the four w_i values in the Table were calculated from Equation [8.41] from the corresponding values of $s_{fit}(Y_i)$. The question now arises of how to normalize these w_i values. Since we intend to treat each of our $n = 12$ data points (three Y_i values for each x_i) individually in our final weighted linear regression, the sum of all 12 statistical weights should be 12 $(= n)$, so the four independent w_i values in the Table should be weighted such that their sum $= 4$, as shown. The weighted linear regression then proceeds using Equations [8.43–8.45], giving the following results where the subscript $_w$ is used to distinguish these parameters from those obtained above using unweighted least squares:

$\bar{x}_w = 1.6409$ (remember there are three data points for each of the four x_i values so $n = 12$);

$\bar{Y}_w = 52.2779$;

$S_{xx,w} = 10.2591$;

$S_{YY,w} = 10,794.612$;

$S_{x,Y,w} = 332.6134$;

$A_w = -0.9722$;

$B_w = 32.4213$;

$V_{res,w} = 1.0854$ (from Equation 8.25).

This example is typical of a comparison of weighted linear regression with a corresponding simple linear regression for heteroscedastic data in that only small differences result with respect to the slope (32.421 vs 32.410) and intercept (–0.972 vs –0.950), but are accompanied by an appreciable decrease in V_{res} (a factor of ~2 in this case).

Such decreases in V_{res} are reflected in corresponding decreases in $s_{Y,i,k}$ and $CL(Y_{i,pred})$ in Equations [8.28 and 8.29], and also in $V_{x,i}{}^*$ and $CL(x_i{}^*)$ in Equations [8.31 and 8.32].

Furthermore, the appreciable movement of the average data point $(\bar{x}_w, \bar{Y}_w) = (1.641, 52.278)$ from $(\bar{x}, \bar{Y}) = (2.500, 80.075)$ leads to further reductions in $V_{x,i}{}^*$ and $CL(x_i{}^*)$ for small values of $Y_i{}^*$ via the term $(Y_i{}^* - \bar{Y})^2$ in Equation [8.31], a considerable advantage when reductions are required in the limits of detection and quantitation.

computer algorithms (and even the more expensive scientific hand calculators) permit rapid evaluation of experimental values of $s(Y_i)$ and their variation with x_i, using the data that must be acquired for the calibration in any case. Unfortunately, these arbitrary weighting options are provided in many computer-based data processing algorithms and it is easy to select one or other of these options in attempt to make the data appear 'better' without any experimental justification or examination of the implications for confidence limits etc.

The procedure for testing 'goodness of fit' for heteroscedastic data also requires modification of that used in the homoscedastic case (Equations [8.33–8.38]). For heteroscedastic data the following equations apply (Analytical Methods Committee 1994):

$$V_{res,w} \equiv MS_{res,w} = SS_{res,w}/dof_{res}$$
$$= \Sigma_i\{w_i[\Sigma_j(Y_{i,j} - Y_{i,pred,w})^2]\}/dof_{res} \qquad [8.47]$$

$$SS_{res,w} = SS_{repr,w} + SS_{lof,w} \qquad [8.48]$$

$$SS_{repr,w} = \Sigma_i\{w_i[\Sigma_j(Y_{ij} - Y_{i,avg})^2]\}; \; dof_{repr} = k(m-1) \qquad [8.49]$$

$$SS_{lof,w} = SS_{res,w} - SS_{repr,w}; \; dof_{lof} = k-2 \qquad [8.50]$$

with $dof_{res} = dof_{repr} + dof_{lof} = (k.m - 2)$ as before. Note that $Y_{i,avg}$ is the average value of the j determinations of Y_i. The statistical *F*-test of the null hypothesis [H_o: the chosen model (Equation [8.19a] with weighted linear regression) provides an adequate description of the experimental data] now proceeds as before:

$$V_{repr,w} \equiv MS_{repr,w} = SS_{repr,w}/dof_{repr} = SS_{repr,w}/[k(m-1)] \qquad [8.51a]$$

$$V_{lof,w} \equiv MS_{lof,w} = SS_{lof,w}/dof_{lof} = SS_{lof,w}/(k-2) \qquad [8.51b]$$

$$F = MS_{lof,w}/MS_{repr,w} \qquad [8.52]$$

If $F < F_c$ then H_o is accepted for this set of data, i.e. the weighted linear regression is adequate; however if $F \geq F_c$ H_o fails, and the reason for the poor goodness of fit must be sought. A worked numerical example (Analytical Methods Committee 1994) is available. Frequently such lack-of-fit arises because of an intrinsic departure from linearity in the raw data and this is the next complication that we must face.

8.3.6 Regression for Nonlinear Data: the Quadratic Fitting Function

A statistically significant lack-of-fit is often associated with significant nonlinearity in the raw data (x_i, Y_i), though it may still be true that some fitting function other than Equation [8.19a] that is still linear in the fitting parameters can provide an adequate model and is thus amenable to least-squares regression. Such nonlinearity of the data can be indicated by a plot of residuals (weighted or unweighted) vs x (Figure 8.9). If this is found to be the case, it is important to acquire many data points evenly distributed across the concentration range of interest in order to properly describe the $Y = f(x)$ function that is to be fitted by least-squares regression. One approach to subsequent regression analysis of the data is to transform the Y data (and possibly also the x data) in such a way that the transformed data $(x_{tr,i}, Y_{tr,i})$ display a linear dependence. A common transformation is the logarithmic function, often used to transform both x and Y in cases where the experimental ranges in each of x and Y cover several orders of magnitude. While such a transformation can be useful in exhibiting the full range of data on a single graph (although the same end result can be accomplished using two plots of nontransformed data with an expanded range for data near the origin), linear regression of the transformed data $(x_{tr,i}, Y_{tr,i})$ followed by back-transformation to the original variables (x_i, Y_i) can lead to undesirable consequences (Meier 2000). For example, if both x_i and Y_i are transformed logarithmically and fitted by least-squares regression to a simple linear model $(Y_{tr} = P + Q.x_{tr})$, then back-transformation to the raw experimental variables leads to $(Y = 10^P.x^Q)$, so that the intercept A for the original data is always predicted to be zero (that may or may not agree with experimental observation), and any uncertainty in the intercept P for the log-transformed data translates into a more significant uncertainty in the slope $(B = 10^P)$ for the raw data. Moreover, inversion of $(Y = 10^P.x^Q)$ to give $(x^* = [Y^*/10^P]^{1/Q})$ is not straightforward and the confidence levels $CL(x_i^*)$ are difficult to

determine. Of course, depending on the particular application, such a procedure could be determined to be fit for purpose, but this would have to be considered carefully (and the reasons documented ahead of time).

Fortunately, in most cases of interest in this book the deviations of the raw data from linearity are not very large even with extensive ranges for x_i and Y_i and can usually be handled by using a simple quadratic relationship instead of Equation [8.19a]:

$$Y = A' + B'.x + C'.x^2 \equiv \bar{Y}_w + B'(x - \bar{x}_w) + C'(x^2 - \bar{x}_w^2) \qquad [8.53]$$

which is still linear in the coefficients A', B' and C'. The first form of this equation is equivalent to Equation [8.19b] and the second form is deduced from the first by eliminating A from Equation [8.53] and the special version of that equation applicable to $(\bar{x}_w, \bar{Y}_w)$, the weighted average point (Equation [8.54]). The same quantity as in Equations [8.20] or [8.40], depending on whether the data have been shown to be homoscedastic or heteroscedastic, is now minimized with respect to variations in the coefficients by setting each of the (now three) partial derivatives equal to zero and solving the resulting three simultaneous linear equations for the coefficients that minimize the sum of squares of residuals. This problem has been addressed (Schwartz 1977) and leads to the following solution, given here for the general case where the data are weighted by a factor $W_i = k_i.w_i$, with k_i the number of replicate measurements at $x = x_i$ (i = 1 to m_C = number of concentration levels investigated) and w_i as before the statistical weight assigned to the measurements of Y_i (Equation [8.41]). For homoscedastic data all $w_i = 1$ and it is common in analytical chemistry to use k_i = constant = k_C. Note that $\Sigma_i k_i = n_C$ (the total number of data points obtained for the calibration) where this summation is over i = 1 to m_C; if $k_i = k_C$ = constant, then $k_C.m_C = n_C$.

In the general case for the fitting function given as Equation [8.53], the following equations apply (Schwartz 1977):

$$\bar{x}_w = \Sigma_i(W_i.x_i)/\Sigma_i(W_i); (\bar{x}^2)_w = \Sigma_i(W_i.x_i^2)/\Sigma_i(W_i);$$
$$\bar{Y}_w = \Sigma_i(W_i.Y_i)/\Sigma_i(W_i) \qquad [8.54]$$

$$S'_{xx} = \Sigma_i(W_i.x_i^2) - (\Sigma_i W_i).(\bar{x}_w)^2 \qquad [8.55a]$$

$$S'_{xY} = \Sigma_i(W_i.x_i.Y_i) - (\Sigma_i W_i).(\bar{x}_w.\bar{Y}_w) \qquad [8.55b]$$

$$S'_{x2x2} = \Sigma_i(W_i.x_i^4) - (\Sigma_i W_i).[(\bar{x}^2)_w]^2 \qquad [8.55c]$$

$$S'_{x2x} = \Sigma_i(W_i.x_i^3) - (\Sigma_i W_i).\bar{x}_w.(\bar{x}^2)_w \qquad [8.55d]$$

$$S'_{x2Y} = \Sigma_i(W_i.x_i^2.Y_i) - (\Sigma_i W_i).[(\bar{x}^2)_w].\bar{Y} \qquad [8.55e]$$

where all summations are over $(i = 1$ to $m_C)$, $(\bar{x}_w)^2$ is the square of the (weighted) mean value of x, and $(\bar{x}^2)_w$ is the (weighted) mean value of the square of x. Then:

$$B' = [S'_{x2x2}.S'_{xY} - S'_{x2x}.S'_{x2Y}]/$$
$$[S'_{xx}.S'_{x2x2} - (S'_{x2x})^2] \qquad [8.56a]$$

$$C' = [S'_{xx}.S'_{x2Y} - S'_{x2x}.S'_{xY}]/$$
$$[S'_{xx}.S'_{x2x2} - (S'_{x2x})^2] \qquad [8.56b]$$

$$A' = \bar{Y}_w - B'.\bar{x}_w - C'.(\bar{x}_w)^2 \qquad [8.56c]$$

$$V'_{res} = \Sigma_i[W_i.(Y_i - Y_{i,pred})^2]/dof'_{res} \text{ with } dof'_{res}$$
$$= (n_C - 3) \qquad [8.57]$$

$$V_{B'} = V'_{res}.S'_{x2x2}/[S'_{xx}.S'_{x2x2} - (S'_{x2x})^2] \qquad [8.58]$$

$$V_{C'} = V'_{res}.S'_{xx}/[S'_{xx}.S'_{x2x2} - (S'_{x2x})^2] \qquad [8.59]$$

$$V_{A'} = (V'_{res}/n) + V'_{res}\{(\bar{x}_w^2).S'_{x2x2} + [(\bar{x}^2)_w]^2.S'_{xx} -$$
$$(2.\bar{x}_w).[(\bar{x}^2)_w].S'_{x2x}\}/[S'_{xx}.S'_{x2x2} - (S'_{x2x})^2]$$
$$\qquad [8.60]$$

Inversion of Equation [8.53] to give a value of x* based on a measured value of Y* is still reasonably straightforward since this is a simple quadratic equation in x; one of the two roots of this quadratic will be physically meaningless (e.g., negative value) leaving the other root as the unambiguous desired result. However, the procedures required to calculate confidence limits $CL(Y_{i,pred})$ and $CL(x^*)$, corresponding to Equations [8.29] and [8.32], are now much more complicated and usually employ an appropriate iterative numerical method implemented on a computer. Indeed, the entire evaluation of Equations [8.54–8.60] plus the confidence limits for x and Y are generally too extensive and complex to be attempted by hand; this feature is made even worse by the requirement that a large number of data points, distributed evenly across the entire range, must be used in order to properly define the coefficients in Equation [8.53].

8.3.7 *Summarized Procedure to Determine Best-Fit Calibration Curve*

The several possibilities described in Section 8.3 may seem confusing but in practice the problem is not usually too complicated for analytical methods based on mass spectrometric detection, as a result of the excellent degree of linearity observed for mass spectrometric response over ranges of several orders of magnitude in the concentration (amount) of analyte introduced. The 'best' way to proceed will vary with the details of the required analysis and the

purpose for which it is to be conducted; criteria that are 'fit for purpose' for one application need not be suitable for others. However, the following series of steps can provide a useful starting point.

(1) Determine an initial calibration curve covering the desired range of concentration (amount), with replicate determinations (absolute minimum three, better five to seven) at each of the chosen concentrations spaced evenly across the range.

(2) Apply simple (unweighted) linear regression (Equations [8.19] and [8.23]); calculate and plot residuals $(Y_i - Y_{i,pred})$ vs x_i to search for evidence of heteroscedacity and/or nonlinearity of the data (Figure 8.9). At this point determine whether or not the calibration obtained is 'fit for purpose'; the statistical test for goodness of fit (Equations [8.33–8.38]) will be helpful here, as will the calculated confidence limits for x* (Equation [8.32]).

(3) If the tests in (2) are deemed to have failed as a result of significant heteroscedacity, assign statistical weights w_i based on the experimental $s(Y_i)$, and try a weighted linear regression. Plot the residuals to check that the weighting has accounted satisfactorily for the variation in $s(Y_i)$ and examine for indications of 'significant' (i.e. unfit for purpose) nonlinearity; again the *F*-test for goodness of fit and an evaluation of $CL(x^*)$ can add a quantitative dimension to this decision.

(4) If the tests in (2) indicate nonlinearity but no significant deviation from homoscedacity, perform an unweighted quadratic regression (Section 8.3.6). A plot of residuals can give a qualitative assurance that the situation has improved but the quantitative aspect is more complex and generally requires access to a suitable computer-based statistical package.

(5) If the weighted linear regression in (3) is judged to be inadequate for the purpose of the analysis, perform a weighted regression on a quadratic fitting function. As in (4) a plot of residuals can still give a qualitative assurance that the situation has improved but the quantitative aspect is again more complex.

(6) If the calibration is still deemed to be unfit for purpose, the analytical method itself needs to be examined with a view to improvement and/or an unusual fitting function or a linearizing data transformation will have to be applied. The two latter possibilities can introduce significant problems with respect to inversion of the calibration function and determination of the confidence limits.

NonLinear Least-Squares Regression

By 'linear least-squares regression' we mean fitting experimental data (x_i, Y_i) to a theoretically justified function $Y_{i,pred} = f(x_i, A, B, l...)$, that is such that the partial derivatives (e.g., $\partial(\Sigma_i r_i^2)/\partial A$) with respect to each of the parameters $A, B, C. . .$, are all linear with respect to the parameters. Examples of functions $f(x_i, A, B, C. . .)$ that can be fitted in this way are polynomial functions like Equations [8.19a–c] (see Equations [8.21] for the partial derivatives for Equation [8.19a]), although when more than two parameters are involved the calculation of the least-squares parameter estimates is best done using matrix algebra.

An example of a more complex function that is 'linear' in this sense is $Y_{i,pred} = A + B.\exp(x_i)$, but $Y_{i,pred} = A + B.\exp(C.x_i)$ can *not* be fitted by linear least-squares since $\partial(\Sigma_i r_i^2)/\partial C$ is not linear in C; in such cases there is no simple algebraic solution for the parameters as there is for the linear case (e.g., Equations [8.21]–[8.24]), so we must use nonlinear regression techniques that involve successive *iterations* to the final best result.

The steps involved in a nonlinear regression are:

(1) Provide initial estimated values for each parameter A_{nl}, B_{nl}, C_{nl}, etc. in the nonlinear fitting function.

(2) Calculate the curve defined by these initial values and then $(\Sigma_i r_i^2)$ as before.

(3) Adjust the parameters so as to reduce $(\Sigma_i r_i^2)$. There are several algorithms used for adjusting the parameters to achieve this, the most common being the Levenberg–Marquardt method that is a combination of the methods of Steepest Descent (for early iterations) and Gauss–Newton (for later iterations). These algorithms exploit the partial derivatives $\partial(\Sigma_i r_i^2)/\partial A_{nl}$) etc. Excellent accounts of these iteration methods and all other aspects of least-squares regression (Motulsky 2004) are also found at http://curvefit.com.

(4) Adjust the parameters again and again so that $(\Sigma_i r_i^2)$ becomes even smaller, until the adjustments make no significant difference in $(\Sigma_i r_i^2)$.

(5) Examine the best-fit results. Ensure that the parameter values are physically meaningful relative to what you know about the problem, and also test for 'goodness of fit' e.g., by plotting values of residuals:

$$r_i = (Y_i - Y_{i,pred}) \text{ vs } x_i.$$

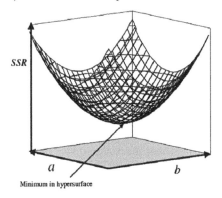

SSR

a   *b*

Minimum in hypersurface

Also ensure that the results obtained do *not* correspond to a local (rather than a global) minimum for $(\Sigma_i r_i^2)$ (see graphics for two-parameter and one-parameter functions); there is no foolproof way of deciding this. Poor performance of the residuals plot could indicate that a local minimum was found and/or an inappropriate model function. A good idea is to fit the same function several times using different initial estimates for the parameters; consistency among the resulting final parameter estimates indicates that the global minimum has indeed been identified.

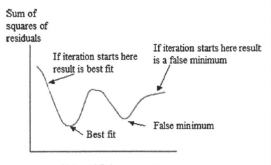

Sum of squares of residuals

If iteration starts here result is best fit

If iteration starts here result is a false minimum

False minimum

Best fit

Value of fitting parameter

8.3.8 Nonlinear Least-Squares Regression

This Section deals briefly with least-squares regression on a fitting function $Y = f(x, A_{nl}, B_{nl}, C_{nl}, \ldots)$ that is not linear with respect to $(A_{nl}, B_{nl}, C_{nl}, \ldots)$, the parameters of the function, e.g., Equation [8.19d] is not linear with respect to the parameter C. Section 8.3.6 discussed a function that is not linear (in fact quadratic) with respect to the measured quantities (x_i, Y_i), but is linear in the parameters. The importance of linearity with respect to the parameters can be understood qualitatively by recalling that the least-squares criterion involves minimizing the sum of squares of residuals $\Sigma_i(r_i^2)$ with respect to each and every parameter in the fitting function, and in turn this involves application of the condition $\partial[\Sigma_i(r_i^2)]/\partial A_{nl} = $ zero, and similarly for all parameters; Equation [8.21] is an explicit example for the simple linear function given as Equation [8.19a]. (Strictly the second derivatives should also be shown to be positive to ensure that we are locating a minimum rather than a maximum, but this is taken for granted here.) Since $r_i = [Y_i - f(x_i, A_{nl}, B_{nl}, C_{nl}, \ldots)]$, if the fitting function is not linear in e.g. the parameter A_{nl}, the partial derivative $\partial[\Sigma_i(r_i^2)]/\partial A_{nl}$ will be a complex function of the parameters and the resulting simultaneous equations will not yield a straightforward algebraic solution for the parameters. For example, in the case of Equation [8.19d]:

$$Y = A + B.\exp(C.x) \text{ so } \partial f/\partial C = B.x.\exp(C.x).$$

In such nonlinear cases a numerical rather than an algebraic approach must be used to determine values of the parameters that minimize $\Sigma_i(r_i^2)$. Several approaches have been devised to accomplish this (see the accompanying text box), but all require initial estimates for the values of the parameters Φ, followed by calculation of the corresponding first estimate of $\Sigma_i(r_i^2)$. An iterative process is then used to adjust the parameter values in the direction that corresponds to a negative value for $\partial[\Sigma_i(r_i^2)]/\partial\Phi$ and similarly for the other parameters, i.e., we are heading towards a minimum for $\Sigma_i(r_i^2)$; it is at this stage that different strategies have been employed (see text box). The procedure is then repeated using the new adjusted values of the parameters; after several iterations the absolute values of $\partial[\Sigma_i(r_i^2)]/\partial\Phi$ become progressively smaller until they approach zero, indicating that a minimum in $\Sigma_i(r_i^2)$ has been found. However, it is possible that the minimum thus discovered is not the true 'global' minimum but a 'local' minimum (see text box), resulting from poor initial estimates for the parameters. Accordingly it is advisable to repeat the procedure using significantly different initial estimates for the parameters to determine whether or not the iteration leads

to the same minimum; in addition, the best-fit parameters should be examined critically to ensure that they are physically reasonable within the bounds of the data and the theoretical justification for choice of the fitting function.

Obviously such iterative procedures are realistic only if performed using a computer-based algorithm. An excellent account of the iteration methods, and indeed of all other aspects of least-squares regression (Motulsky 2004), can also be found at http://curvefit.com. Depending on the details of the nonlinear fitting function, inversion to give a value of x^* from a measured value of Y^* can be difficult and itself require an iterative procedure. As always the confidence limits of the best-fit curve are equally important as the parameter values. However, the confidence limits from linear regression are calculated using straightforward unambiguous mathematical methods (e.g., Equations [8.28] and [8.29]), and if the assumptions of linear regression are valid for a set of experimental data the confidence limits of slope and intercept, calculated for a chosen confidence level, can be interpreted quite rigorously. In contrast it is not straightforward to calculate the CLs of the parameters, or of the values of x^* predicted from measured Y^*, from nonlinear regression, and indeed mathematical shortcuts are required (Motulsky 2004); such 'shortcut CLs' (supplied by most nonlinear regression algorithms) are sometimes referred to as 'asymptotic CLs', and in some cases these CLs can be too narrow (i.e., too optimistic) and should not be interpreted too rigorously. If time, availability of standards and finances permit, a better feel for the true CLs is obtained by repeating the same experiment (calibration) several times. Section 8.5.2c describes a common example where such nonlinear regression is required for construction of calibration curves, as a result of cross-contributions of analyte and internal standard to signal intensities at the *m/z* values used to monitor these compounds, e.g., a natural-abundance isotopolog of the analyte might contribute significantly at the *m/z* value used to measure the isotope-labeled internal standard. If at all possible it is best to avoid circumstances in which the calibration function is nonlinear in the parameters.

8.4 Limits of Detection and Quantitation

Sections 8.1–8.3 presented an introduction to some basic principles of statistical theory with emphasis on some applications of importance for analytical chemists, particularly instrument calibration and its use in determination of unknown concentrations. However, in many cases the critical questions to be answered by analysis are: 'Is the target analyte really present if it is barely detectable; what degree of confidence is fit for purpose in the circumstances?' and/or 'If the presence of the target analyte is confirmed to within a low degree of uncertainty, how high must the concentration be in order that the concentration can be determined with a degree of precision that is fit for purpose?'. Such questions can arise in several situations, e.g., in forensic tests for drug abuse or in environmental concerns about harmful pollutants. These questions are implicit in the concepts of limit of detection (LOD) and lower limit of quantitation (LLOQ); historically these concepts have led to considerable debate and confusion. We shall deal initially with the LOD since definitions of LLOQ tend to track the former in some cases.

8.4.1 Limit of Detection

An important distinction to be kept in mind is that between the instrumental detection limit (IDL) and the method detection limit (MDL); the IDL refers to an LOD determined using clean solutions of the analytical standard injected into the analytical instrument (e.g., LC–MS, GC–MS/MS) operated under stated conditions, while the MDL involves blank matrix samples spiked with known amounts of the calibration standard and taken through the entire extraction–clean-up–analysis procedure. This distinction is irrelevant to the question of how either of these LOD quantities is to be defined and measured (see below).

The simplest and most intuitively obvious definition of LOD is the lowest concentration (or amount) of analyte for which the observed signal/noise ratio (S/N) = Z, where Z is (somewhat arbitrarily) chosen to be in the range

3–6. Figure 4.27 showed an example in which $Z = 2$ was used as the criterion but also where the noise was defined as the maximum peak-to-peak noise (background in this case) on either side of the peak rather than some suitably defined average of this background signal. This ambiguity (definition of the noise measurement) is not always resolved as clearly as it might be. A somewhat more explicit form of such a definition (IUPAC 1977) defines the LOD in terms of the smallest response signal Y_{min} that can be detected with 'reasonable certainty':

$$Y_{min} = \bar{Y}_{blank} + Z.s_{blank} \qquad [8.61a]$$

$$LOD = (Y_{min} - \bar{Y}_{blank})/S = Z.s_{blank}/S \qquad [8.61b]$$

where $\bar{Y}_{blank}$ is the mean of the signals observed for several blank samples and s_{blank} is its standard deviation; s_{blank} is thus dominated by the baseline noise and to a first approximation (when $\bar{Y}_{blank} = 0$) is measured as the square root of the mean of the squares of the noise signals, i.e., $\{[(Y_{noise})^2]_{mean}\}^{\frac{1}{2}}$ (compare Equation [8.2c]). The sensitivity (S) of the analytical method is defined as the slope of the calibration curve, so $S = B$ (Equation [8.19a]) if the calibration is linear in the region around the LOD (the role of the sensitivity in determining the LOD can be appreciated from Figure 8.10). This approach assumes that a signal more than $Z.s_{blank}$ above the sample blank value with $Z \geq 3$ could only have arisen from the blank much less than 1 % of the time (Figure 8.6) and therefore is likely to have arisen from something else, i.e., the analyte.

However, even this more meaningful definition of the LOD is subject to uncertainties (Williams 1991), since it

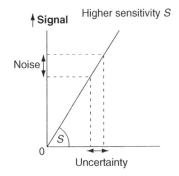
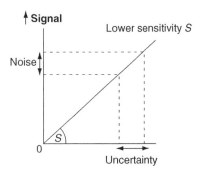

Figure 8.10 A simple example (linear calibration function with intercept $A = 0$) demonstrating the effects of both noise (or background) and sensitivity (S, Equation [8.61]) on the error propagation when predicting the content (concentration or amount of analyte) from the noisy signal by projecting the Y-axis onto the content axis (x-axis) through the calibration line. A low noise level combined with a high sensitivity S (slope of calibration line) will thus allow small changes in concentration (or amount) to be detectable, and by extension lead to a lower LOD (Equation [8.61]); this is the origin of the more colloquial use of the term 'sensitivity' to indicate lower LOD (and possibly LLOQ) values.

assumes that the noise (or background) levels in the signal and in the background are the same, and also requires that the number of measurements used to determine $\bar{x}_{blank}$ and s_{blank} is sufficiently high that uncertainties in the values of these two parameters are sufficiently small. Moreover (Meier 2000), it does not take heteroscedacity into account. Relative to the hypotheses [H_o: the analyte is not present] and [H_1: the analyte is present], provided that Z is sufficiently large (and thus that α is sufficiently small) this general simple approach to the LOD does tend to protect against false positive conclusions (i.e., the conclusion that the analyte is present when in fact it is not).

However, a false negative result with respect to H_1, *i.e.*, deducing that analyte is not present when in fact additional experiments confirm it is, can also occur. The latter condition can be clarified by considering a case in which a large number of replicate measurements is made on a sample containing the analyte at the LOD level. Then the measured reponses (Y) will be distributed symmetrically around x = LOD with a Gaussian (normal) distribution (if the number of measurements is sufficiently high) with a standard deviation that can be assumed to be given to a good approximation by s_{blank}. Then exactly 50 % of the experimental values of Y determined for the LOD sample will be below the mean value $\bar{Y}_{LOD}$. If we take as our criterion that any sample that gives an analytical signal that is less than $\bar{Y}_{LOD}$ shall be interpreted as consistent with H_o (i.e., analyte is not present), the probability of a false negative conclusion will always be 50 % regardless of any other considerations. This is obviously not a helpful situation and it is for this reason that a critical level is set to define a decision limit for the response Y, so that reasonable probabilities can be set for both false positives and false negatives. This concept is discussed more fully later in this section but for now it is mentioned only that, e.g., if the LOD is defined with $Z = 3$ (Equation [8.61]) then choice of a decision limit of $\bar{Y}_{blank} + 1.5 s_{blank}$ will result in probabilities for both false positives and false negatives of 6.7 % (resulting from the appropriate integrated areas of the assumed normal distributions). Of course the choice of these parameters will be made based on the requirements of the particular test including the relative importance of the implications of false positives and negatives.

A rather different approach is to construct the calibration curve using a sufficient number of replicate determinations that reliable values of $s(Y_i)$ can be determined, then calculate relative standard deviations $[s(Y_i)/\bar{Y}_i]$, and interpolate or extrapolate these values to $[s(Y_i)/\bar{Y}_i] = 0.15$ (or other value selected in the context of fitness for purpose). The corresponding value of x_i is assigned as the LOD. This approach appears to be somewhat related to the US FDA rules for method validation (FDA 2001),

that specify the definition of the LLOQ (*not* the LOD), discussed later. Unfortunately that document gives only the following definition for the LOD: 'The lowest concentration of an analyte that the bioanalytical procedure can reliably differentiate from background noise'. Such a lack of a more specific definition of the LOD (what is the operational meaning of 'reliably'?) is understandable for a government agency that is concerned with quantitation rather than mere detection, and this tendency is made explicit in the European Guide to Quantifying Uncertainty (EuraChem–CITAC 2000): 'It is widely accepted that the most important use of the 'limit of detection' is to show where method performance becomes insufficient for acceptable quantitation, so that improvements can be made'. An exhaustive discussion (IUPAC 1995) of theoretical considerations of the LOD (including recommended changes in nomenclature) emphasizes the somewhat nebulous nature of the concept if viewed as a robust statistical quantity.

For present purposes more operational approaches will suffice. In many applications, e.g., sports doping and environmental contamination, the ability to have a scientifically and statistically defensible method for demonstrating the presence of the smallest possible amount of a substance present in a sample becomes essential. For example, when monitoring hazardous waste facilities, the detection of a single volatile organic compound in a groundwater sample can be taken as an indication that the facility may have affected environmental quality. The analytical results are thus crucial in determining whether the facility is in or out of compliance with environmental regulations, with potentially costly implications for the facility (possibly unjustified if the analytical finding of a detectable concentration can not be scientifically justified, i.e., a false positive); on the other side of the question, if the analytical result turns out to be a false negative this could be detrimental to public health. Thus it is not surprising that environmental agencies concerned with development of scientifically defensible definitions of a LOD have concentrated on the method detection limit (MDL) with only limited interest in instrumental detection limits as contributors to the overall MDL.

The MDL is a statistical concept based on the ability of a specified measurement method to determine an analyte in a specified sample matrix, together with data interpretation using a specified statistical method. The aim is to provide an unambiguous approach to determining a scientifically defensible value for the MDL, that enables different laboratories to be consistent in how they determine these values. The MDL will of course vary from substance to substance and from matrix to matrix and, perhaps less obviously, with the measurement procedure used. Because it is also a statistical concept the probabilities of false positive and false negative conclusions must be taken into consideration; these will depend on the confidence levels specified in the statistical evaluation.

All definitions of an MDL involve determinations of the standard deviation (Equation [8.2c]) of measurements on matrix blanks that have been spiked with very low concentrations of the target analyte (or sometimes on nonspiked blanks). Most of the MDL definitions that have been proposed can be classified into two groups: single concentration designs that assume the measurement variability at a specified spike concentration to be equal to the variability at the true MDL, and calibration designs that take into account the concentration dependence of the measurement variability.

The best known of the single concentration definitions of MDL (Glaser 1981) is that adopted by the US Environmental Protection Agency (EPA 1997). The statistical theory incorporated into this methodology for determining an MDL was fully discussed in the original work (Glaser 1981) and only a brief review is given here. As defined by the EPA, the MDL is the minimum concentration (amount) of a substance that can be measured and reported with 99 % confidence that the analyte concentration is greater than zero; this MDL protects against incorrectly reporting the presence of a compound at low concentrations in cases when noise (background) and analyte signal may be indistinguishable (false positive), but is correspondingly subject to reporting false negatives. (The MDL concentration intrinsically does not carry any implications regarding the accuracy of the quantitative measurement.)

The EPA MDL determination specifies a minimum of seven ($k = n = 7$) replicate spikes prepared at a single 'appropriately' low concentration (generally one to five times the expected MDL). The spiked matrix typically is reagent water, or clean sand or sodium sulfate for sediment samples, or a well characterized natural material that does not contain the substance for tissue samples; the point here is to avoid matrix interferences, a condition that will apply to real world analytical samples only if appropriate clean-up procedures are employed. These spiked blank matrices are processed through the entire analytical method, preferably completed within a short period (a few days). It is assumed that the calibration function is linear in both concentration–response and in the fitting parameters (i.e., Equation [8.19a] is valid) at least in the low concentration region, and that the frequency distribution of measured concentrations in the low concentration spiked blank has a normal distribution (Figure 8.11(a)). Another important assumption of the EPA MDL calculation is that the variance $V_{Y,i}$

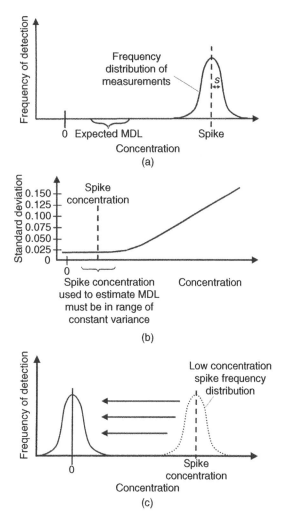

Figure 8.11 Graphical representations of the definition and implications of the EPA definition of an MDL. (a). *Assumed* normal frequency distribution of measured concentrations of MDL test samples spiked at one to five times the expected MDL concentration, showing the standard deviation s. (b) *Assumed* standard deviation as a function of analyte concentration, with a region of constant standard deviation at low concentrations. (c) The frequency distribution of the low concentration spike measurements is *assumed* to be the same as that for replicate blank measurements (analyte not present). (d) The MDL is set at a concentration to provide a false positive rate of no more than 1 % (t = Student's t value at the 99 % confidence level). (e) Probability of a false negative when a sample contains the analyte at the EPA MDL concentration. Reproduced with permission from *New Reporting Procedures Based on Long-Term Method Detection Levels and Some Considerations for Interpretations of Water-Quality Data Provided by the US Geological Survey NationalWater Quality Laboratory* (1999), Open-File Report 99–193.

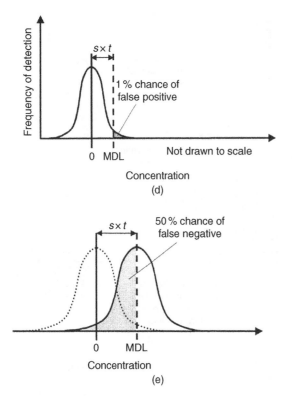

Figure 8.11   (Continued)

of measured responses Y_i at a selected known concentration (Equation [8.27]) will become constant at some low concentration and will remain constant for sequentially lower concentration replicate spikes down to zero concentration (Figure 8.11(b)). Note that this assumption of homoscedacity at low concentrations implies that the standard deviation s^*_x (Equation [8.31]) of the concentration value x^* deduced by inverting the linear calibration in this low concentration range, also becomes constant; the quantity s^*_x appears explicitly in the algebraic definition of the EPA MDL, Equation [8.62].

While it is true that the experimental standard deviations observed for very low concentrations become similar because of the inability to adequately measure small differences in the diminishing signal, this assumption has been criticized; note, however, that the EPA does recommend that an iterative process should be used in which the spike concentration x_{spike} is reduced to successively lower values to help ensure that the region of practically constant standard deviation near the MDL has been reached. (An *F*-test, Section 8.2.5, is performed on the two data sets to ensure that the difference between

the variances is statistically insignificant.) Under this assumption of constant standard deviation $s(Y_i)$ from the low concentration spike concentration down to zero, the frequency distribution determined experimentally for the low concentration spikes will thus also represent the distribution expected when measuring signals from instrumental noise or unspiked blank matrix (or both) in a series of blank samples (Figure 8.11(c)). Of course it is impractical to meaningfully measure noise in repetitive blank samples, so these hypothetical blank measurements are then used to calculate the concentration at which no more than 1 % of the blank measurements will result in the reporting of a false positive (Figure 8.11(d)), i.e., analyses resulting in claimed detection at concentrations $\geq$ the EPA MDL concentration should be true detections 99 % of the time. It is worthwhile to repeat the assumptions made: minimal matrix interferences, linear calibration line (at least at low concentrations) and normal distributions for the response Y with standard deviations that become independent of concentration at low concentrations.

In algebraic terms, the EPA MDL is calculated as follows:

$$MDL = s^*_x.t[(k-1)]; \quad \text{evaluated for } (1-\alpha) = 0.99,$$
$$\text{i.e., } p = 0.01 \quad [8.62]$$

where s^*_x is the standard deviation of concentration x^* in the spiked blank determined by analyzing k replicate samples of the blank matrix spiked to a known selected concentration x_{spike} (that should turn out to be one to five times the expected MDL), α is the level of significance (chosen by EPA to be 0.01) and t is the one sided Student's-t value at (k–1) degrees of freedom at a $100(1-\alpha)$ confidence level (chosen to be 99 %, i.e., $p = 0.01$ in Table 8.1). Note that for k = 7 (the EPA recommendation) and $\alpha = p = 0.01$, $t = 3.143$ (Table 8.1), which is consistent with the rule of thumb estimate of LOD based on a S/N ratio of three (see above). A one sided t-test is employed because we are interested in the probability of only a positive deviation of the signal from zero.

Limitations of the EPA MDL procedure (USGS 1999) arise mainly from the assumptions of normal distribution and constant standard deviation over the low concentration range (from x_{spike} down to zero). Generally, as one measures increasingly higher concentrations, larger variabilities are associated with these measurements but the EPA MDL approach performs no extended calibration to account for such trends. Therefore, in principle the measured MDL will vary in parallel with the spike concentration used. The only verification performed in the EPA procedure is a two sample variance test in which the variance of the first set of spiked samples is compared against that of another set of samples spiked at a 'slightly' higher concentration that is not defined, so that it is possible to perform this check with only a small concentration change to ensure measurement of variances that are 'not significantly different'. Thus, because of the uniformity assumption, the lower the spiked concentration the lower the measured MDL, and indeed it is possible (USGS 1999) in this way to report MDLs that are not statistically achievable!

In addition to the preceding criticism, the EPA MDL typically is determined by using a fairly small number of spike replicates ($k \geq 7$) measured over a short period, in an attempt to reach a compromise between statistical respectability and requirements of cost and time. Such an approach provides only a narrow estimate of the overall method variation and thus of the standard deviation, particularly for production laboratories in which there are multiple sources of method variation including multiple instruments, instrument calibrations, instrument operators and, similarly, multiple sample preparation protocols and staff. Consequently, in such circumstances the MDL needs to be determined by using routine conditions and procedures, i.e., by conducting the determination over an extended period by using several brands of instrumentation and as many experimental variables as feasible to obtain a more accurate and realistic measurement of the standard deviation near the MDL (USGS 1999). A more general critique of the EPA MDL has been published by the EPA itself (EPA 2004); this critique also provides an exhaustive but instructive discussion and comparison of the many variants of MDL definition that have appeared over the years.

All definitions of the MDL, including that of the EPA, focus on minimizing the risk of reporting a false positive. However, at the EPA MDL concentration there is a significant risk of a false negative, e.g., it can be shown (Keith 1992) that a sample with a true concentration equal to the EPA MDL has a 50 % chance of not being detected. In other words, up to 50 % of the measurements made on a sample with a true concentration equal to the MDL would be predicted to be false negatives, i.e., be measured as less than the MDL (the shaded region in Figure 8.11(e) where the frequency distribution drawn with the solid curve is centered on the calculated MDL). Such high probabilities for false negatives are generally regarded as unacceptable for purposes of environmental monitoring (USGS 1999). Therefore some agencies and laboratories, recognizing the inadequacies of the MDL as

a reporting level, set minimum reporting levels at concentrations such that quantitative determinations can be justified (appreciably greater than the determined MDLs). For example, reporting levels might be set at a so-called 'practical quantitation limit' that is five or ten times the MDL (EPA 1985) or at a lower limit of quantitation (LLOQ) defined as a concentration higher than the average blank response by 10 times the standard deviation. Similarly, the EPA has suggested (EPA 1993) the use of a minimum level (ML) that is 3.18 times the MDL for $k = 7$ (note that $3.18(t[(k-1) = 6; (1-\alpha) = 0.99]) = 10$, see Equation [8.62]). At present, within the context of single concentration definitions of MDL, there does not seem to be any generally accepted definition for a reporting level that sets acceptable rates of both false positives and false negatives, e.g., at no more than 1 %, although some agencies have set their own working definitions (USGS 1999).

While the original EPA definition of the MDL (Glaser 1981) did strive to achieve a useful compromise between scientific and statistical defensibility on the one hand, and cost in terms of time, money and effort on the other, there appears to be a growing acceptance (EPA 2004) that a more rigorous approach is required. Because there is always some variability in the calibration data, the precision of a measurement of x^* for an unknown sample will obviously be poorer than the precision estimated from replicate measurements of response Y^* for the same sample. Thus, in estimating any parameter such as the LOD, LLOQ etc., one must take into account the variability in both the calibration and the analysis of the unknown sample.

The EPA MDL is an example of a single concentration definition. Evaluations of LOD that involve calibration designs, taking into account all the information available from a full calibration via the confidence limit function (Equations [8.27–8.32]), have been described by several authors (Hubaux 1970; Oppenheimer 1983; Schwartz 1983; Zorn 1997, 1999; Wilson 2004). The earliest of such theories assumed a linear calibration function with homoscedastic data but later developments extended the approach to weighted least squares regression and nonlinear functions. All of these treatments assume that the calibration was obtained using matrix matched standards (i.e., these are intrinsically method detection limits though not the single point type exemplified by the EPA MDL) and that no systematic errors are present. They all build on a prior discussion of the general problem (Currie 1968) that proposed that three specific quantities should be defined before any discussion of LOD can be initiated:

(1) a *decision limit* L_C, a value of *concentration* x defined via the calibration line by a *critical level* R_C, the value of the *instrument signal* level (Y) above which an observed signal may be recognized with sufficient confidence (defined later) to be detected (thus R_C is the signal level at which one may decide whether or not the analytical result indicates detection);

(2) a *detection limit* R_D, the 'true' signal level that *a priori* can be expected to lead to detection (this is the limiting value of Y at which a given analytical procedure may be relied upon to lead to detection);

(3) a *determination limit* R_Q, the signal level at which the measurement precision will be adequate for quantitative determination.

These concepts are illustrated graphically in Figure 8.12. These critical signal levels (R_C, R_D and R_Q) have corresponding values of analyte concentration (or amount), e.g., R_D and R_Q are values of Y with corresponding values of x (L_D and L_Q, Figure 8.12, determined from R_D and R_Q via the inverted calibration function) that we refer to as the LOD and LLOQ. The calibration design approach determines all these parameters by considering the confidence limits of the calibration curve obtained using matrix matched standards (spiked blanks), thus making explicit their dependence on calibration data whose variability includes that arising from the entire analytical method.

Figure 8.12(b) shows the calibration line flanked by the two confidence level curves; these are curves because of the dependence of $CL(Y_{i,pred})$ on $V_{Y,i}$ that in turn varies with $(x_i - \bar{x})^2$ (e.g., Equations [8.27] and [8.29]), so that $CL(Y_{i,pred})$ increases with distance from the average point of the calibration ($\bar{x}$, $\bar{Y}$). Of course these confidence limits also vary with the user-selected level of confidence determined by the *p*-value (Figure 8.12(a)). The calibration line and its two confidence limits represent a synthesis of our knowledge about the relationship between concentration (amount) x and signal Y.

The confidence band is also used in reverse; for a measured value of the signal (Y*) for an unknown, the range of the corresponding concentration (or amount) of analyte x* can be predicted; this is shown graphically in Figure 8.10, and in the case of Figure 8.12(b) is equivalent to transforming $CL(Y)$ into $CL(x^*)$ as in Equations [8.29] and [8.32]. In particular, for a measured signal equal to R_C (the intercept of the upper confidence level with the Y-axis), the corresponding lower limit of x* is zero (see Figure 8.12(b)). For signals $Y* \leq R_C$ there is a non-negligible probability $100(\alpha + \beta)\%$ that they arise from a sample with a null concentration ($x^* = 0$) and hence

we cannot state within our chosen level of confidence whether or not the target analyte is really present.

It is important to note that the critical limit L_C (also known as decision limit) is often confused with the limit of detection L_D. However, the two limits happen to coincide only when 50 % false negatives are deemed to be acceptable; such a high value for β is not often useful in practice, i.e., the concept of 'detection capability' is not credible if the analyte is missed in 50 % of cases! A related misconception is that it is impossible to detect the analyte when the actual level x is below the limit of detection. However, Figure 8.12 makes it clear that the analyte will be detected (by definition) as long as the experimental signal exceeds the critical level R_C. However, the probability of a false negative (β) increases with decreasing (actual) analyte concentration (amount), but the risk of a false positive (α) remains smaller than the pre-selected value (0.05 in Figure 8.12) as long as the result exceeds the critical level. The crucial point is that this calibration based definition of LOD refers to detection with quantifiable probabilities for false positives and false negatives, not simply detection with some minimal specified value of signal/noise ratio.

In Europe a different notation is used instead of L_C and L_D (Figure 8.12(c)). In the 2002/657/EC decision (European Community 2002) the decision limit was designated CCα and was defined as 'the limit at and above which it can be concluded with an error probability of α that a sample is non-compliant', and the detection capability was designated CCβ and defined as 'the smallest content of the substance that may be detected, identified and/or quantified in a sample with an error probability of β'. As emphasized previously (Antignac 2003) these concepts had already been introduced in a general sense (i.e., not specifically formulated for analytical chemistry) by the International Standards Organization (ISO 1997) as a method of clearly defining a limit relative to which a system can be declared different from its 'basic state'. In our case the 'system' is a LC–MS chromatogram (usually SIM or MRM, Chapter 6) chosen to be selective for the analyte and the 'basic state' corresponds to such a chromatogram for a blank sample (in the case of forbidden substances such as illegal drugs in urine) or, in the case of regulated compounds such as pesticides in food, for a sample containing the analyte at or below the maximum regulatory limit (MRL) concentration. These concepts assume that the qualitative identification of the analyte (Section 10.5.3) is satisfactory (see Chapter 9) but do provide a statistically defensible definition of the lowest concentration at which the analyte can be said to have been detected.

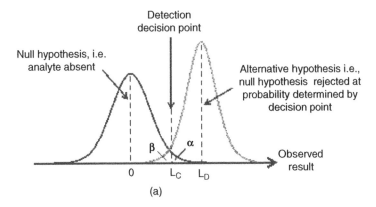

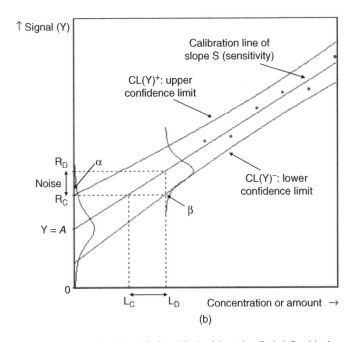

Figure 8.12 (a) Relationship between critical level (L_C) and limit of detection (L_D) defined in the concentration (amount) domain, with probabilities of false positives (α) and false negatives (β) given as the areas of the indicated small portions of the distribution curves. If L_D is chosen to be L_C, the curve for the alternative hypothesis moves to the left resulting in a probability of a false negative (β) of 50 %. (b) Relationship between concentration (or amount) x, and signal (S $\equiv$ Y), for the blank (x=0 and Y=B), at the critical level (x = L_C and Y = S_C), and at the limit of detection (x = L_D and Y = S_D). True x values of 0 and L_D give rise to distributions of observed signal that overlap to some extent (see also Figure 8.12(a)). The detection *decision* is taken at the Y = S_C level, i.e. in the signal domain but usually reported at the corresponding x level, i.e., in the concentration (amount) domain. As a result of uncertainties in the signal of the test sample and the estimated calibration model, a (true) zero content can give rise to a false positive result with probability α. Similarly, a false negative (nondetect) result is predicted with probability β when the analyte is present at the L_D level. The requirement that the target analyte can be reliably detected implies specifying sufficiently small values for α and β (this Figure assumed $\alpha = 0.05 = \beta$). (c) Similar to (b) but using the CCα – CCβ notation for the critical and detectability limits recommended in the ISO/11843–1 document (ISO 1997). Reproduced from Antignac, *Anal. Chim. Acta* **483**, 325 copyright (2003), with permission from Springer–Verlag.

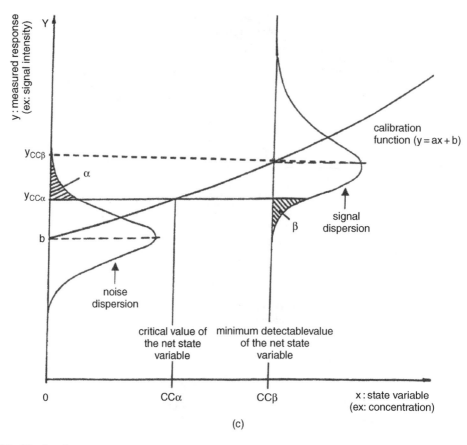

Figure 8.12   (Continued)

In very general terms, the procedure for determination of an LOD by such a calibration design approach is as follows:

(1) Obtain a full calibration with sufficient replicates at each concentration (amount) of the calibration standard that reliable values of $s(Y_i)$ are obtained. Determine the least-squares calibration function (Equations [8.22–8.24] for the linear homoscedastic case), together with the confidence limits CL(Y) (Equations [8.25–8.29]) using a chosen values of α. It may be helpful to plot these values as in Figure 8.12(b).

(2) Determine the value of $R_C = CL(Y)^+$ from Equation [8.29] with $V_{Y,i}$ evaluated from Equation [8.27] with $x_i = 0$ and the positive sign in Equation [8.29]; graphically this corresponds to the Y-axis intercept of the upper confidence level curve, i.e., $CL(Y)^+$ vs x. Figure 8.12(b) is drawn for a positive value of the intercept A at $x = 0$ but, if A

turns out to be negative, identify the value of $x = x'$ at which the calibration line crosses the x-axis, and take R_C to be the corresponding value of $CL(Y)^+$ at $x = x'$.

(3) Determine the decision (critical) limit L_C by inserting the value of $Y = R_C$ into the least-squares regression calibration function (Equation [8.19a] in the linear case); graphically this is equivalent to drawing a horizontal line through $Y = R_C$ in Figure 8.12(b) until it intersects the calibration line and then dropping a perpendicular to the x-axis.

(4) Graphically the detection limit is now determined by extending the horizontal line through $Y = R_C$ until it intersects with the lower confidence limit $CL(Y)^-$; algebraically this corresponds to inserting the value $Y = R_C$ into the equation for $CL(Y)^-$ setting $x_i = L_D$, and solving for L_D, probably best done numerically with a computer-based algorithm.

It is worthwhile to caution that, as for single concentration definitions of LOD such as the EPA MDL, even the calibration design definitions do not yield unique values. The most obvious variable is the selection of values for α and β according to the levels of risk of false conclusions that the analyst is prepared to accept, i.e., that have been judged fit for purpose. However, even for fixed values of α and β, the value obtained for the LOD is only a best estimate of the 'true' value, with uncertainties guaranteed to be present by the limited number of replicates used in constructing the calibration curve and its confidence levels; when a second series of standards, identical to the first, is made up and analyzed, we would obtain signals differing randomly from the 'true' signals and also from the first data set. In other words, L_C and L_D are only best estimates, i.e., random variables, but with statistically defensible confidence limits.

It is emphasized that the present discussion of critical and detection limits is concerned with statistical evaluation of S/N or S/B values, where the signal monitored has been chosen ahead of time to be sufficiently selective for the target analyte (Section 9.6). The two concepts complement one another for the purpose of qualitative and quantitative analysis.

The additional complications in defining detection limits for transient signals, including those encountered in chromatography, have been briefly addressed (Williams 1988); this contribution discussed quantitatively the qualitatively obvious point that, by not taking into consideration the additional information provided by knowledge of the time window in which the transient signal (e.g., chromatographic peak) is expected to appear, the detection limit could be significantly overestimated.

8.4.2 Limits of Quantitation

A related problem concerns a scientifically defensible definition of the lower limit of quantitation (LLOQ). This is sometimes referred to simply as 'the LOQ' but in fact the upper limit of quantitation (ULOQ) can also be an important parameter for validated analytical methods (see Chapter 9). A very important definition of LLOQ is that of the US FDA with regard to bioanalytical data used in human clinical pharmacology and in bioavailability and bioequivalence studies requiring pharmacokinetic evaluation (FDA 2001); this guidance also applies to bioanalytical methods used for non-human pharmacology–toxicology studies and preclinical studies. According to this protocol, the LLOQ is accepted as the lowest standard on the calibration curve provided that (a) the analyte response for this concentration (amount)

is at least five times the blank response, and (b) the analyte peak (response) should be identifiable, discrete and reproducible with a precision of 20 % and accuracy of 80–120 %. In this context, the *accuracy* of an analytical method, i.e., the closeness of mean test results obtained by the method to the 'true value' (concentration or amount) of the analyte, is determined by replicate analyses of samples containing known amounts of the analyte (usually spiked blanks), measured using a minimum of five determinations per concentration. The mean value should be within 15 % of the 'true' value except at the LLOQ where it should not deviate by more than 20 %. That same document (FDA 2001) also defines the *precision* of an analytical method, i.e., the closeness of individual measures of an analyte when the procedure is applied repeatedly to multiple aliquots of a single homogeneous volume of matrix containing the analyte; this definition of precision requires a minimum of five determinations per concentration at a recommended minimum of three concentrations in the range of expected concentrations. The required precision is specified similarly (15 % except at the LLOQ, where it should not exceed 20 %). Precision is further subdivided into within-run, intra-batch precision or repeatability (which assesses precision during a single analytical run) and between-run, inter-batch precision or repeatability (which measures precision with time and may involve different analysts, equipment, reagents, and laboratories).

Such a definition of LLOQ is obviously directed at ensuring that quantitative analytical methods developed for bioanalytical purposes are properly validated, rather than at describing a definition of LLOQ as a statistically defensible estimate of the minimum concentration that can be measured 'reliably'. In practice bioanalytical laboratories are therefore not necessarily interested in the ultimate performance of their quantitative methods, but rather in ensuring that the method meets the regulatory requirements over the concentration range pertinent to their study; in other words, provided that the method satisfies the 20 % demands on accuracy and precision at the lowest concentration relevant to that particular study, that concentration is taken to be the LLOQ. The FDA allows no interpolation or extrapolation of data on accuracy and precision to concentration values other than those at which the experimental determinations were made.

Other less stringent but scientifically defensible definitions of an LLOQ, which can be fit for purpose in some cases, are extensions of specific approaches to calibration design definitions of the LOD (see above). For example, one such approach (Zorn 1997) defines the LLOQ as the concentration (amount) corresponding to a signal R_Q representing an increase over the intercept A that is 10 times the standard deviation of the signal at the critical point (Figure 8.12):

$$R_Q = A + 10.s_{R,C} \qquad [8.63a]$$

Then the LLOQ is the quantitation level L_Q in concentration units, given (Zorn 1997) by:

$$LLOQ \equiv L_Q = (R_Q - A)/B \qquad [8.63b]$$

This, and other calibration design definitions of LLOQ (and LOD), have been critically evaluated (Zorn 1999).

8.5 Calibration and Measurement: Systematic and Random Errors

Thus far the discussion in this chapter has concentrated on the statistical treatment of random errors in calibration data and calibration equations, with only passing mention of the implications of the practicalities involved in the acquisition of these data. For example, no mention has been made of what kind of calibration data are (or can be) acquired in common analytical practice, under which circumstances one approach is used rather than another and (importantly) what are the theoretical equations to which the experimental calibration data should be fitted by least-squares regression for different circumstances. Moreover, it is important to address the question of analytical accuracy to complement the discussion of precision that we have been mainly concerned with thus far; the meanings of accuracy and precision in the present context are discussed in Section 8.1. The present section represents an attempt to express in algebra the calibration functions that apply in different circumstances, while exposing the potential sources of systematic uncertainty in each case.

Thus the present section does not refer extensively to the statistical considerations summarized in Sections 8.2 and 8.3; rather we are dealing here with the realities of different situations in which the analyst can find himself/herself when confronted with situations that can be appreciably less than ideal, e.g., how to calibrate the measuring instrument when no analytical standard or internal standard or blank matrix is available. It is understood that the statistical considerations of Sections 8.1–8.3 would then be applied to the data thus obtained. Also, note that none of the statistical evaluations of random errors, discussed in this chapter, would reveal any systematic errors. These aspects are addressed in this section and build upon the preceding discussion of analytical standards (Section 2.2), preparation of calibration solutions (Section 2.5) and the brief simplified introduction (Section 2.6) to calibration and measurement.

The working expressions derived below for the various real-world circumstances necessarily involve a complex algebraic notation, so a list of the symbols used is added in Appendix 8.7. Note that the following treatment is based upon an earlier discussion (Boyd 1993) but that some of the symbols used are different to avoid confusion with other notations used in this book.

The best modern practice in trace analytical chemistry uses a fully validated method (FDA 2001) employing a surrogate internal standard (Chapter 2, preferably an isotope-labeled version of the target analyte) together with matrix matched calibrators. However, it is not always possible to meet this level of rigor. If no blank matrix is available, the Method of Standard Additions (see later) is a useful possibility provided that the analytical sample is available in sufficient quantity; otherwise the only option is external calibration using clean solutions of analytical standard, an approach that is subject to several sources of uncertainty as discussed in subsequent sections. Similar remarks apply to situations in which no suitable internal standard is available. Because such situations do arise in real life, they will be discussed with a view to exposing the sources of systematic errors. We first address cases in which no surrogate internal standard is available, although it is emphasized that use of a less than perfect surrogate IS (Table 2.1) is better than using none at all.

8.5.1 *Analyses Without Internal Standards*

Lack of any internal standard leads to several sources of uncertainty and, with the exception of the Method of Standard Additions (Section 8.5.1b), it is emphasized that in general analyses without use of an appropriate internal standard are inherently less reliable and are fit for purpose only in less demanding applications. For convenience of discussion these approaches have been grouped into subcategories.

8.5.1a *External Linear Calibration With a Zero Intercept*

If the intercept of a linear calibration line ($R_a'' = A + B.C_a''$) can be shown to be statistically indistinguishable from zero, the resulting expressions describing inversion of the calibration function are simplified. The confidence limits for the intercept are given by ($A \pm t.V_A^{\frac{1}{2}}$), where the value of t is chosen at the desired confidence level (e.g., 0.05, two tailed values since we are concerned with both positive and negative deviations from zero), with (n–2) degrees of freedom, and the variance (V_A) is given by Equation [8.26]. If these confidence limits include zero then the intercept is accepted as zero to within the chosen confidence level. (See also the discussion in the final paragraph of Section 8.3.2.)

Within this sub-category probably the worst possible scenario is that where neither a surrogate internal standard nor a blank matrix is available; this requires external calibration using calibrators consisting of clean solutions of analytical standard. We shall work through this example in some detail in order to establish the notation, defined in the list of symbols in Appendix 8.7; note that single primes (') denote quanties pertinent to experiments on extracts of analytical samples ready to be injected into the analytical instrument, while double primes ('') denote those pertinent to calibration solutions of analytical standard in clean solvent. The desired quantity is the concentration of analyte in the analytical sample, $C_a = Q_a/W_S$ where Q_a is the amount (preferably in moles, sometimes in mass units) of analyte in a mass W_S of analytical sample; however, the measured quantity is R_a', the analytical response observed upon injection of a volume v' of extract solution (total volume V'). The intermediate quantity linking the desired and measured quantities is q_a', the amount of analyte in the volume v' of extract that is injected into the chromatograph; this is related to the desired quantity C_a via:

$$q_a' = Q_a'.(v'/V') = (Q_a.F_a').(v'/V')$$
$$= (C_a.W_S.F_a').(v'/V') \qquad [8.64a]$$

where F_a' is the fractional recovery of analyte from the original sample into the final extract (original volume V') that is analyzed. Now q_a' is also related to the observed detector response R_a' via the detector sensitivity corrected for both proportional and fixed losses of analyte:

$$R_a'' = S_a.[q_a''.f_a - l_a] = S_a.[C_a''.v''.f_a - l_a] \qquad [8.64b]$$

where S_a is the intrinsic sensitivity of the detector (e.g., mass spectrometer), i.e., the response per unit amount of analyte that actually reaches the detector from the chromatograph or other input device, f_a is the fractional transmission efficiency of analyte through the chromatograph to the detector, and l_a is any constant loss of analyte after injection into the chromatograph (arising from e.g., adsorption on some fixed number of active adsorption sites in the chromatographic train). In this section the $(C_a'' = 0)$ intercept of the calibration curve $(A = -S_a.l_a)$ is assumed to be zero to within statistical error, indicating negligible bias error $(l_a = 0)$. In this case the calibration line is obtained as a linear least squares regression of the observed response R_a'' vs C_a'', the concentration of analyte in the standard solutions:

$$R_a'' = (S_a.f_a).q_a'' = (S_a.f_a).(v''.C_a'')$$
$$= (S_a.f_a).(v''/V'').w''_a \qquad [8.64c]$$

where w''_a is the mass of calibration standard weighed out and dissolved in volume V'' to prepare that particular calibration solution. The experimental least-squares slope B'' is the derivative of R_a'' with respect to C_a'', so differentiation of Equation [8.64c] gives:

$$B'' = (dR_a''/dC_a'') = S_a.f_a.v'' \qquad [8.64d]$$

Now for the unknown sample extract, the response is $R_a' = q'_a.(S_a.f_a)$, so:

$$q'_a = R_a'/(S_a.f_a) = R_a'.v''/B'' \qquad [8.64e]$$

where we used Equation [8.64d] to introduce the experimentally derived slope B''. Eliminating q_a' from Equations [8.64a–e] gives:

$$C_a = (R_a'/S_a.f_a).(V'/v')/(F_a'.W_S)$$
$$= (R_a'.v''/B'').(V'/v')/(F_a'.W_S) \qquad [8.64f]$$

Equation [8.64f] is the final working expression for the desired concentration of analyte in the original sample for the present case of calibration without an internal standard and where the intercept of the calibration curve is statistically indistinguishable from zero. The disadvantages of such an approach are now apparent. Volumes appear explicitly so potential systematic uncertainties in $C_a''(= w_a''/V'')$ and V' due to solvent evaporation, as well as random errors in the injection volume v' (not well controlled in GC, good precision but poor accuracy in HPLC using injection loops, Section 2.4.3), directly affect the final result. The peak area (or height) R_a' for the analytical extract, and its calibration parameter B'', must be measured in separate chromatographic experiments with no correction possible for potential systematic error due to instrumental drift between the two sets of experiments. The fractional recovery F_a' is not measurable from such experiments alone, and the requirement for an assumed value of unity for F_a' in Equation [8.64f] in order to obtain a value for C_a introduces a potential proportional systematic error into the values of Q_a, thus yielding values for C_a that are lower limits to the true value. Note also that uncertainties in the injection volumes v' imply that the ordinate (independent variable) in the calibration plot is not strictly free of experimental uncertainties as strictly required for linear regression, although generally these are much less than those in the instrument signals.

It is important to note that it is tacitly assumed in Section 8.5 that the intrinsic detector sensitivity S_a is a constant. This assumption amounts to saying that there are no significant ionization suppression effects (Section 5.3.6a) operating for sample extracts and not for clean solutions of calibration standard. A slightly better scenario is that of a linear calibration with zero intercept and no internal standard but using spiked matrix blanks to prepare calibrators. A matrix blank is taken to mean a sample that is identical in every way to the sample to be analyzed except that it contains an undetectable quantity of the analyte. Triple primes (''') denote quantities relevant to experiments involving spiked matrix blanks. The degree to which a matrix blank can be said to exist will vary strongly with the situation but usually it is relatively easy to achieve in a bioanalysis context, for example. This approach involves many of the same assumptions as those described in the derivation of Equation [8.64f] but now the calibration curve is obtained by analyzing different aliquots of blank matrix spiked with varying amounts P_a''' of the analytical standard and then taken through the complete analytical procedure. The most reliable method of determining the quantities P_a''' is by direct weighing of the pure standard but in practice the spiking is usually done (Section 9.5.4) by dispensing volumes V_a''' of a stock solution of concentration C_a''' of analytical standard, i.e., $P_a''' = w_a'''.(v'''/V''')$; this method is more convenient but carries the risk of introducing both systematic errors (e.g., from evaporation of solvent and inaccuracy of the volumetric equipment) and random errors (via imprecision in the volumes dispensed). Now the calibration is obtained as a linear regression of R_a''' vs P_a''' that takes the form (again assuming zero intercept, i.e., $A = 0$):

$$R_a''' = (S_a.f_a).q_a''' = (S_a.f_a).(v'''/V''').Q_a'''$$
$$= (S_a.f_a).(v'''/V''').(P_a'''.F_a''') \qquad [8.65a]$$

In this case the calibration is likely to be obtained as a plot of R_a''' vs P_a''' (the amount of calibration standard spiked into a fixed mass of blank matrix prior to extraction, cleanup etc.). Then the experimental slope B''' corresponds to:

$$B''' = (dR_a'''/dP_a''') = (S_a.f_a).(v'''/V''').F_a''' \qquad [8.65b]$$

Now for the analysis of the extract of the analytical sample:

$$R_a' = (S_a.f_a).q_a' = (S_a.f_a).(v'/V').Q_a'$$
$$= B'''.(V'''/v''').(1/F_a''')(v'/V').(F_a'.Q_a)$$

where we substituted for $(S_a.f_a)$ from Equation [8.65b] and $(F_a'.Q_a)$ for Q_a. Simple rearrangement and writing $Q_a = C_a.W_s$ gives:

$$C_a = (R_a'/B''').(v'''/v').(V/V''').(F_a''/F_a')/W_S \qquad [8.65c]$$

This final working equation is a better situation than that summarized in Equation [8.64f] since, although the various volumes still appear explicitly, the ratio of injection volumes (v'''/v') is likely unity with little or no uncertainty if an HPLC loop injector is used. Moreover, we now must assume only $F_a' = F_a'''$ rather than $F_a' = 1$; the only reason why F_a''' might not be exactly equal to F_a' is that the extraction efficiency F_a' for the native analyte in the analytical sample might be less that for the spiked-in standard analyte (F_a''') as a result of occlusion effects of some kind, as can happen with some samples (Boyd 1996, Ke 2000).

The complete or partial cancellation of the recovery factors in Equation [8.65c] is a major advantage but in itself does not provide an estimate of the recovery from the unknown sample (F_a'). However, access to blank matrix does permit straightforward measurement of F_a''' (and thus a reasonable estimate of F_a') by comparing responses R_a''' for a spiked matrix blank with the corresponding response for the extract of a matrix blank that is spiked with the same amount of standard after the extraction and clean-up. Naturally the extract and injection volumes must be the same in the two cases, preferably the same mass of matrix should be extracted, and a sufficient number of replicates should be analyzed to provide reasonable estimates of statistical uncertainties.

8.5.1b Method of Standard Additions

The two cases discussed thus far are obviously not 'best practice' examples but such circumstances may arise on occasion; also, the preceding discussion allowed demonstration of the benefits of matrix matched calibrators over clean solutions of the analytical standard. In situations where neither a surrogate internal standard nor suitable blank matrix are available, the Method of Standard Additions (MSA) is a preferred analytical method; it is included in this sub-category for convenience, although strictly speaking it involves a non-zero intercept for the experimental line that serves both as calibration and measurement. In this method the analytical sample itself is used as a kind of 'blank' into which the calibration standard is spiked. As will become apparent, linearity of response is a prerequisite for accurate results and this implies that multiple experiments must be performed at

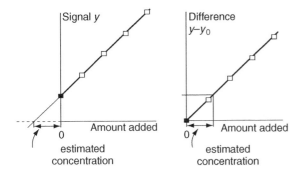

Figure 8.13 Idealized plots according to the Method of Standard Additions. Each point plotted is assumed to be the mean of several replicate determinations. The traditional method (left panel) simply plots the observed analytical signal Y vs the amount of calibration standard added x (the black square corresponds to the nonspiked sample, $Y = Y_0$), and estimates the value of x by extrapolation of a least-squares regression line to $Y = 0$ (see text); however, this procedure implies that the confidence interval at this point (not shown, compare Figure 8.12) has widened considerably. By using a simple transformation from Y to $(Y-Y_0)$ the extrapolation procedure is replaced by one of interpolation, thus improving the precision (more narrow confidence interval). Reproduced from Meier and Zünd, *Statistical Methods in Analytical Chemistry*, 2nd *Edition* (2000), with permission of John Wiley & Sons Inc.

different spiking levels, using fresh aliquots of analytical sample each time; if the amount of sample is limited, thus restricting the number of spiked aliquots and data points, the accuracy and precision of the method will be correspondingly limited. A typical plot of MSA data is shown in Figure 8.13. The traditional data treatment performs a linear least-squares regression on the data (R_{msa} vs Q_{msa}, where Q_{msa} is the amount of calibration standard spiked into the fixed mass W_S of untreated analytical sample), to give:

$$R_{msa} = A_{msa} + B_{msa} \cdot Q_{msa}$$

Now when $R_{msa} = 0$, the intercept on the Q_{msa} axis is:

$$Q_{msa,0} = -A_{msa}/B_{msa} \qquad [8.66a]$$

i.e., the ratio of the intercept to the slope of the linear least-squares regression line for the raw data. To relate $Q_{msa,0}$ to the desired quantity Q_a, we must examine the nature of A_{msa} and B_{msa} in the same way as was done for Equations [8.64–8.65] assuming that $l_a = 0$:

$$R_{msa} = (S_a \cdot f_a) \cdot (v'''/V''') \cdot [F'_a \cdot Q_a + F'_{msa} \cdot Q_{msa}] \qquad [8.66b]$$

where the recoveries of the native (F'_a) and spiked-in (F'_{msa}) analyte are not necessarily the same as discussed above for Equation [8.65]. For $R_{msa} = 0$, $Q_{msa} = Q_{msa,0}$ as before, so:

$$Q_a = -Q_{msa,0} \cdot (F'_{msa}/F'_a) \qquad [8.66c]$$

Combining Equations [8.66a–c] gives:

$$C_a = Q_a/W_S = (A_{msa}/B_{msa}) \cdot (F'_{msa}/F'_a)/W_S \qquad [8.66d]$$

This is a remarkably simple result but a few cautionary remarks are necessary. The MSA method has a great advantage in that, since the recovered spike and analyte from the sample are contained in the same sample extract solution, the volumes v and V are the same for both and thus do not appear explicitly in the final expression for Q_a. An assumption that was glossed over in the derivation of Equation [8.66d] was that the reponse level to which the extrapolation should be done was indeed $R_{msa} = $ zero; this is valid in the absence of any bias systematic errors (such as can arise if e.g., $l_a \neq 0$). However, if any such error is present, extrapolation to a value of S_{msa} other than zero will be observed; this possibility is considered later (see discussion leading to Equation [8.75]).

Again F'_a can not be measured since C_a is not known *a priori*, and in most cases it is assumed that $F'_a = F'_{msa}$ in Equation [8.66d], thus leading to a potential proportional systematic error in the measured value of C_a since generally $F'_{msa} \leq F'_a$ as a result of occlusion effects. However, an estimate of F'_{msa} can be obtained by noting from Equation [8.66b] that:

$$B_{msa} = (S_a \cdot f_a) \cdot (v'''/V''') \cdot F'_{msa}$$

which permits an estimate of F'_{msa} provided that $(S_a \cdot f_a)$ can be measured; an estimate of $(S_a \cdot f_a)$ can be obtained by measuring a calibration curve for pure solutions of known concentrations C_a injected through the chromatograph (recall that f_a represents the fractional losses through the chromatograph and S_a the sensitivity of the mass spectrometer or other detector for direct injections of the analyte). Then $S_a \cdot f_a = B''/v''$ (Equation [8.64d]).

Extrapolation to well outside the range of the acquired data in order to estimate $Q_{msa,0}$ (Figure 8.13) implies that the confidence limit band will be widened considerably (see discussion of Figure 8.12), thus increasing the statistical uncertainty (precision) in the estimate of $Q_{msa,0}$. This aspect can be alleviated by a simple trick (Meier 2000) in which R_{msa} is replaced by $R'_{msa} = (R_{msa} - Y_0)$, where Y_0 is the value of R_{msa} with zero spike ($Q_{msa} = 0$), i.e the intercept on the response axis in the left panel of

Figure 8.13. Then linear least-squares regression gives a value of A'_{msa} that is identically zero (the line necessarily includes the origin) and with slope $B'_{msa} = B_{msa} = [(S_a \cdot f_a) \cdot (v''/V'') \cdot F'_{msa}]$. When $Q_{msa} = Q_a$, $R_{msa} = 2Y_0$, so $R'_{msa} = Y_0$; thus the value of Q_a is determined by interpolating the value of Q_{msa} when $R'_{msa} = Y_0$ (right panel of Figure 8.13), where the confidence interval is less wide as it is closer to the mean point of the dataset (see discussion of Figure 8.12). Equation [8.32] gives the best estimate in the resulting uncertainty in the interpolated value of $x = x^*$ (i.e., of $Q_{msa} = Q_a$ in this case).

Several investigations have examined the optimization of precision and accuracy in MSA analyses (Ratzlaff 1979; Gardner 1986, 1988; Franke 1978; Renman 1997); some of these compare the performance of MSA with that of conventional external calibration for situations in which both approaches are possible. There appears to be general agreement that the precision achievable using MSA is generally less than that for conventional calibration, although the use of MSA has the advantage that each determination is independently calibrated thus minimizing the effect of within-batch calibration drift. An important disadvantage of MSA is the reduction of analytical range (even under optimum conditions) over which acceptable precision is likely to be obtained. The circumstances most favourable to the use of MSA rather than external calibration include use of an analytical technique that is subject to changes in sensitivity or calibration drift within a batch of analyses, application to analyses that are performed relatively infrequently and in small batches (i.e., where the economy of effort that favors external calibration is not so important), and sample concentrations that are spread over a relatively narrow range and are near to the lower end of the range of the measurement technique. The effect of the size of the incremental additions and their spacing on the precision depends on the number of additions and on the relation between precision of measurement and concentration (i.e., the variation of the confidence level). For the MSA technique in general, optimum precision and accuracy are obtained by using a smaller number of standard additions and performing a larger number of replicate determinations on each of these; a similar conclusion holds for the external calibration approach applied to samples with very low analyte concentrations. A weighted regression is preferable when multiple additions are made (i.e., multiple data points in plots like Figure 8.13); the benefit of multiple additions is that they provide a form of internal quality control, i.e., a check on the linearity of response on which the MSA depends. It has been shown (Bader 1980) that there are no advantages to a converse MSA method, i.e.,

one in which the amount of analyte spike Q_{msa}' is held constant while the amount of analytical sample W_S is varied. However, as discussed in Section 8.5.1d, a variant of this approach (Youden 1947, 1960) permits detection and measurement of bias systematic errors and leads to what is probably the most reliable estimate of the true concentration available from the Method of Standard Conditions.

8.5.1c External Linear Calibration With a Nonzero Intercept

We now consider the complications that arise when a zero value for the intercept A of a linear calibration curve can not be statistically justified. If observed for solutions of pure standards, such a finding usually indicates a fundamental problem in the analytical method. There are two broad classes of such problems, corresponding to positive and negative values for the Y-intercept A. A positive value for A usually corresponds to a non-zero analytical signal for a solution known to contain none of the analyte, i.e., a lack of analytical specificity arising from chemical interferences. It would be highly unusual if the instrumental calibration curve, obtained using pure standard in clean solutions (rather than calibrators prepared from spiked blank matrix), were to show a postive value for A; one would suspect presence of an impurity in the solvent in such a case. An impurity in the standard sample of the analyte would show up as a proportional error in the external calibration curve, i.e., an erroneous sensitivity (slope B) rather than as a constant bias. Where possible, it is good practice to attempt to remove or minimize problems resulting in positive A values by first identifying their causes.

The second class of non-zero intercepts corresponds to negative values for A; their physical meaning is best approached via the implied positive value for the x-intercept ($x_0 = -A/B$). Such a circumstance implies a threshold value for x (concentration or amount of analyte injected) below which no signal is observed. Such observations, particularly for calibration experiments using solutions of the pure standard, are usually interpreted in terms of irreversible losses of analyte on 'active sites' on the column (Grob 1978, Self 1979) or on the injector or other components of the chromatographic train (Fehlhaber 1978), or even in the ion source of the mass spectrometer (Wegmann 1978); such effects ($l_a \neq 0$) are more important for GC/MS analyses and tend to be irreproducible and thus not susceptible to being taken into account for accurate calibration. Thus, it is preferable to investigate the source of such effects and if possible to eliminate them to the point where the calibration curve passes through the origin to within the experimental precision.

Such non-zero intercepts in calibration curves are examples of bias errors. The detection and characterization of bias errors, often appearing together with simultaneous proportional systematic errors (e.g., those associated with values of $F'_a < 1$), is a major thrust of the approach (Section 8.5.1d) promoted by Youden and Cardone (Youden 1947, 1960; Cardone 1983,1986; Ferrus 1987).

Again a worst scenario corresponds to a linear external calibration with a non-zero intercept, no internal standard and no matrix blank available. Such circumstances lead to a modified form of Equation [8.64a] by following the same reasoning:

$$R_a'' = A'' + (S_a.f_a).q_a'' = A'' + (S_a.f_a).(v''.C_a'')$$
$$= A'' + (S_a.f_a).(v''/V'').w''_a \qquad [8.67a]$$

so that $B'' = (dR_a''/dC_a'') = (S_a.f_a).v''$, as for the case discussed in Section 8.5.1a (Equation [8.64d]). Then the relationship analogous to the zero intercept case (Equation [8.64e]) is:

$$q'_a = (R_a' - A'')/(S_a.f_a) = (R_a' - A'').v''/B'' \qquad [8.67b]$$

However, Equation [8.64a] that relates q_a' to the concentration of analyte in the unknown sample still applies, so elimination of q'_a from Equations [8.64a] and [8.67b] as before, gives:

$$C_a = [(R_a' - A'')/(S_a.f_a)].(V'/v')/(F_a'.W_S)$$
$$= [(R_a' - A'')v''/B''].(V'/v')/(F_a'.W_S) \qquad [8.67c]$$

The derivation of Equation [8.67c] as an extension of Equation [8.64f] is thus algebraically trivial but its implications for the propagation of experimental error are not. The values obtained for the calibration slope B'' and intercept A'' will not necessarily apply to the sample extracts. For example if A'' is found to be positive, often interpreted in terms of coeluting interferences, the amount of such interfering substances in the sample extract could be significantly very different from that in the calibration solution. If such interfering substances were derived from the solvent, the effective value for A'' would depend upon the total volumes of solvent employed in extracting and dissolving the sample as opposed to dissolving the analytical standard. On the other hand, if A'' were negative, corresponding to a positive x-intercept often interpreted in terms of a constant loss of analyte on active adsorption sites in the chromatographic train, this amount of

lost analyte could vary depending on the quantity of co-extractives from the sample that could compete for these active sites.

This discussion emphasizes that non-zero values for A'' usually indicate potential uncertainties which are best avoided, if possible, by diagnosing the cause and taking remedial action, e.g., if $A'' > 0$ a more selective analytical method such as different cleanup and chromatography and/or increased mass spectrometer resolution or, if $A'' < 0$, changing columns and/or silylation to deactivate appropriate portions of the chromatographic train etc. However, if none of these remedies improve the situation, assumption of the applicability of the calibration parameters B'' and A'' to analysis of sample extracts gives the working relation Equation [8.67].

As for the zero intercept case, a somewhat better situation arises for a linear calibration with non-zero intercept and no internal standard but using spiked matrix blanks to prepare calibrators. In such cases a value $A'' > 0$ again usually implies that a co-eluting impurity is contributing to the analytical response R_a'' but now the direct interference (i.e. not a matrix suppression effect that would affect the sensitivity S_a) could arise from the blank matrix as well as from the solvent. Similarly, if $A'' < 0$ the implied positive value for the x-intercept could now reflect loss of a fixed quantity of analyte on 'active sites' during the extraction and/or clean-up procedures, as well as in the chromatographic train, as discussed above. The same comments concerning the advisability of diagnosing and removing the causes of nonzero intercepts also apply here. However, if this cannot be done, the working relationship is readily deduced to be analogous to Equation [8.65c]:

$$C_a = [(R_a' - A''')/B'''].(v'''/V''').(V'/v').$$
$$(F_a'''/F_a')/W_S \qquad [8.68]$$

Since in general $F_a''' \le F_a'$ (reflecting the possibility of occluded analyte in the analytical sample but not in the spiked matrix blank), the assumption that the recoveries are equal will result in a proportional error. This reinforces the recommendation that, if possible, the cause of a calibration curve not passing through the origin should be sought and rectified. This is a good example of the statement in the first paragraph of Section 8.5.1 to the effect that, in general, analyses without use of an appropriate internal standard are inherently less reliable and are fit for purpose only in less demanding applications; accordingly, it becomes a matter of judgement for the analyst as to the extent to which systematic errors should be investigated and rectified in any given case.

8.5.1d Systematic Errors in the Method of Standard Additions: Youden Plots

Systematic errors in the Method of Standard Additions can be investigated by methods (Youden 1947, 1960; Cardone 1983, 1986; Ferrus 1987) that, in their most developed form, provide a powerful protocol of great generality for the detection of both bias and proportional systematic errors in an MSA procedure. To explain the principles involved in a reasonable space, a particular model for the analytical procedure will be discussed here:

(i) the chromatography detection system used is sufficiently selective that no co-extractives interfere directly in the analysis (though indirect interferences via ionization suppression that affects the sensitivity S must still be considered, Section 5.3.5a);

(ii) the detector response is directly proportional to the quantity of analyte reaching it, i.e.:

$$R_a' = S_a.(f_a.q_a' - l_a) \qquad [8.69]$$

where l_a represents the (assumed fixed) quantity of analyte lost to 'active sites' located between the injector and detector, and f_a is the fraction of injected analyte that reaches the detector if l_a is zero;

(iii) extraction of analyte from the sample and the associated clean-up procedures are subject to losses of both the proportional and bias types, where the proportional loss may vary according to whether the analyte is that originally present or spiked into the sample (i.e., $F_a' \neq F_{msa}'$), but the fixed loss (L_a', possibly on 'active sites') is assumed to be the same for both, i.e.:

$$Q_a' = (F_a'.Q_a) + (F_{msa}'.Q_{msa}) - L_a' \qquad [8.70]$$

Although the model specified in (i)–(iii) is not completely general, it does cover the majority of circumstances that can lead to significant error in quantitative trace analysis by MSA.

A general approach (Youden 1947, 1960) to determining whether or not bias errors are present in such an MSA analysis involves acquisition of a response curve in which the only variation in quantity of analyte Q_a is due to controlled variations in sample size W_s:

$$Q_a' = F_a'.(Q_a/W_S).W_S - L_a' = (F_a'.C_a).W_S - L_a' \qquad [8.71]$$

Note that Equation [8.71] is a special case of Equation [8.70] with $Q_{msa} = 0$ (no spiking) and rewritten so

that W_s becomes the independent variable. The implicit assumption that $C_a = (Q_a/W_S)$ is constant amounts to ensuring that the sample is homogeneous over the range of values of W_S to be used (for analytical samples like soil or sediments this effectively places a lower limit on the values of W_S, see Section 8.6). Then, the functional form of the Youden sample response curve can be derived from Equations [8.69–8.71] as follows:

$$R_a' = S_a.(f_a.q_a' - l_a) = S_a.[f_a.(v'/V').Q_a' - l_a]$$
$$= [(S_a f_a.).(v'/V').F_a'.C_a].W_S - S_a.[l_a + f_a.(v'/V').L_a']$$
$$\qquad [8.72]$$
$$= B_Y.W_S + A_Y$$

where $A_Y = -S_a.[l_a + f_a.(v'/V').L_a']$ and $B_Y = [(S_a f_a.).(v'/V').F_a'.C_a]$ are the intercept and slope of the Youden plot. Note that, for the present model (assumptions (i)–(iii)), only negative or zero values are predicted for A_Y (Figure 8.14). If A_Y turns out to be positive, the most probable cause is that the bias errors are dominated by co-eluting interferences as discussed previously. While a non-zero value for A_Y that is statistically significant (via a *t*-test) indicates the presence of bias errors in the analytical procedure, this alone cannot determine whether these errors arose during the extraction

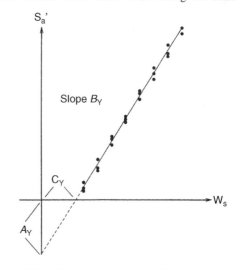

Figure 8.14 Illustration of a typical Youden plot (fictitious data) in which the independent variable is the quantity of (unspiked) analytical sample taken through the entire analytical method, keeping the volumes v' and V' constant. The 'threshold' value for W_s is $c_Y = -A_Y/B_Y$, and reflects the magnitude of bias errors arising from a fixed analyte loss on 'active sites' during either or both of the extraction/clean-up stage (L_a') and in the final analytical train (l_a). Modified from Boyd, *Rapid Commun. Mass Spectrom.* **7**, 257 (1993), with permission of John Wiley & Sons, Ltd.

and clean-up steps ($L_a' > 0$) and/or in the chromatographic train ($l_a > 0$). Under the present assumptions (i)–(iii), this question could be approached by determining the dependence of A_Y on (v'/V') since (see Equation [8.72]) the contribution of l_a to A_Y is independent of the fractional volume injected but this is not true of L_a'. Note that a value for C_a can be obtained from the Youden slope B_Y if it is assumed that the proportional error factors F_a' and f_a are both unity and if S_a can be measured. However, this is not generally feasible and the Youden method is usually applied only as a diagnostic approach to detection of bias errors (but see Equation [8.73] below).

The simple idealized version of the Method of Standard Additions, which assumed no significant bias errors as described in the discussion leading to Equation [8.66], can be modified to the more realistic model defined by assumptions (i)–(iii). In such a procedure Equation [8.70] applies with the restriction that W_S and thus Q_a are fixed, and with Q_{msa} as the independent variable as before. Then, combining Equations [8.69–8.70] gives:

$$R_a' = S_a.(f_a.q_a' - l_a) = S_a.f_a.(v'/V').Q_a' - S_a.l_a$$

$$= [S_a.f_a.(v'/V').F_{msa}'].Q_{msa} + \{S_a.f_a.[(v'/V').F_a'.$$

$$C_a.W_S - L_a] - S_a.l_a\} \qquad [8.73]$$

$$= B_{msa}.Q_{msa} + A_{msa}$$

Comparison of Equation [8.72] (summarizing the Youden plot) with Equation [8.73] (the MSA taking bias errors into account, i.e., L_a' and l_a not necessarily zero) gives:

$$B_{msa} = B_Y.(F_{msa}'/F_a')/C_a \qquad [8.74a]$$

$$A_{msa} = A_Y + S_a.f_a.(v'/V').F_a'.C_a.W_S \qquad [8.74b]$$

The most reliable value of $C_a = Q_a/W_S$ as determined by the Method of Standard Additions is probably that given by Equation [8.75] obtained by combining Equations [8.73–8.74]:

$$C_a = Q_a/W_S = [(A_{msa} - A_Y)/B_{msa}].$$

$$(F_{msa}'/F_a')/W_S \qquad [8.75]$$

where A_{msa} and B_{msa} are determined exactly as for the idealized case (Figure 8.13). The use of the Youden intercept A_Y as the appropriate correction for A_{msa} to take account of systematic bias errors (Equation [8.74b]), rather than e.g., the intercept A'' of the calibration curve determined using clean solutions of pure analytical standard (see discussion of Equations [8.67–8.68]), is consistent with the more general conclusions (Cardone 1986)

of the less restrictive model rather than that described by assumptions (i)–(iii), above. As before, the ratio of analyte recovery efficiencies (F_{msa}'/F_a') for spiked and original analyte is difficult to determine experimentally and is always a source of uncertainty; the claim (Cardone 1986) that the Youden and MSA experimental approaches may be regarded as simple extensions of one another is thus valid only if $F_{msa}' = F_a'$. This condition is readily fulfilled if both recoveries are close to 100 %, as was probably the case for the relatively simple matrices (pharmaceutical preparations) that were the primary concern of the original work (Cardone 1986), but is less likely to be valid for e.g., an aged sediment sample contaminated with an anthropogenic pollutant.

Such a combination of Youden experiments (e.g. Figure 8.14) and conventional MSA determinations involves a great deal of labor, and presumes a supply of analytical sample that is both plentiful and also of sufficient homogeneity that the Youden plot experiments can cover a suitably large range of values of W_S. Again this is generally not a problem for pharmaceutical preparations, but in most other applications different analytical approaches must be used.

8.5.1e Strategies When Analytical Standards are not Available

Unfortunately situations can arise in which not only is a suitable internal standard not available, neither is an analytical standard. A common occurrence is in environmental analyses for groups of closely related compounds, e.g., the polychlorinated biphenyls (PCBs), where a laboratory may not have access to standards (analytical or internal) for all 210 congeners; in such multi-residue analyses a limited number of standards are used and quantitation of a congener for which no standard is available uses data for one that is believed to have chemical and physical properties as closely similar as possible. Of course this does introduce possible systematic errors into the determination.

Another example where this problem can arise involves drug discovery requiring bioanalytical quantitation of drug metabolites in plasma when reference standards are not available. Two innovative approaches to this fairly common problem were proposed recently. The first of these approaches (Valaskovic 2006) exploits very low flow rate nanospray ionization that exhibited a distinct trend towards equimolar responses of similar compounds (in this case a set of well known drugs) when the flow rate was reduced from 25 to less than 10 nL.min^{-1} (see also Section 5.3.5b). A more uniform response between a parent drug and its metabolites (arising from hydroxylation, dealkylation, hydrolysis, and glucuronidation) was obtained at flow rates $\leq 10\,nL.min^{-1}$. Based on this observation, nanospray was used (Valaskovic 2006) to provide a calibration for conventional LC–ESI–MS/MS analysis. As an example of application of this principle, normalization factors were obtained in this way and applied to quantitation of an acylglucuronide metabolite of a proprietary compound in rat plasma; the standard curve obtained with the nanospray calibration method yielded quantitative results for the drug metabolites within $\pm 20\,\%$ of that obtained with reference standards and conventional ESI. In general it was found (Valaskovic 2006) that nanospray ($\leq 10\,nL.min^{-1}$) behaved as a detector with almost equimolar responses for samples prepared from a complex plasma matrix, with the largest discrepancy within 50 %. The success of this approach is presumably related in some way to the demonstration that low flow rate nanospray ceases to behave like a concentration dependent detector (with respect to its variation with flow rate) when precautions are taken (Schneider 2006) to ensure complete 'total solvent consumption' (Section 5.3.5b). This strategy is discussed further in Section 11.5.1.

A rather different approach to the same problem (Yu 2007) uses LC–MS in combination with radiometric detection. Biologically-derived ^{14}C-labeled metabolites from preclinical *in vitro* and *in vivo* matrices were used as metabolite standards, with concentrations in these matrices calculated based on the measured levels of radioactivity. The amount of drug-related metabolites in plasma samples was estimated (Yu 2007) by determining relative MS responses for the metabolites between the plasma extracts and the ^{14}C-metabolite standards, using the calculated concentrations of metabolite standards as calibrants (in effect as rather special surrogate internal standards). As emphasized in the original work (Yu 2007), use of a LC method with adequate separation efficiency and a simple matrix mixing step prior to LC–MS analysis minimized the consequences of inconsistent matrix effects when measuring the MS responses of metabolites in plasma and reference samples. Further, use of the same LC–MS method developed for preclinical metabolite profiling can unambiguously bridge the data obtained for human metabolites back to the preclinical species, thus preventing possible misinterpretation of these analytical results particularly when the drug is extensively metabolized in human and lab animals, e.g., if multiple mono-oxygenated metabolites are present in both, it is critical to verify that the major mono-oxygenated metabolites in humans are the same as those observed in preclinical species used in toxicity studies. Another advantage of this

method (Yu 2007) is the simple sample preparation step that does not require isolation and purification of metabolites for use as analytical standards in more conventional quantitation methods.

However, in general it is obvious that analyses in which no analytical standard is available can be classed as 'heroic' and the reliability of data thus obtained can not match that of data acquired using calibrations with standards.

8.5.2 Analyses Using Internal Standards

Internal standards are used in quantitative chromatography for several reasons, including alleviation of problems arising from variations in sensitivity S_a arising from suppression or enhancement of ionization efficiency (Section 5.3.5a), as well as removal or reduction of the dependence of the final result on the values of volumes of solutions (v', V' etc.) and estimation of the fractional recovery of analyte from the sample into the extract (F_a'). A compound used only to address uncertainties arising from liquid volumes is referred to as a volumetric internal standard (VIS) and is usually added to the extract solution; a surrogate internal standard (SIS, or sometimes simply IS), on the other hand, is added in known quantities to the sample at as early a stage as possible (Figure 1.2). The fractional recovery of the SIS is measurable and thus provides some information on the recovery of the native analyte. The principles upon which these two objectives can be achieved, and their limitations, are discussed below. However, it will be convenient here to repeat the desirable properties for any internal standard (Section 2.2.2): (a) it should be completely resolved in the chromatogram from all other known and unknown substances in the sample extract, though for mass spectrometric detection the retention time of the internal standard need not be well separated from those of other components, provided that it is characterized by at least one unique m/z value within the appropriate retention time window and that ionization suppression effects are negligible; (b) the IS should elute as closely as possible to the target analyte, in order to minimize effects of instrumental drift; (c) the IS must be chemically stable under the conditions of the analytical procedure; (d) best accuracy and precision are obtainable if the peak height (area) for the internal standard is as close as possible to that for the target analyte; (e) the IS must be wholly absent from the analytical sample; (f) in the case of a VIS, lack of volatility is an important consideration especially when volatile solvents are used; (g) a SIS must be as chemically similar to the analyte as possible (where possible a stable-isotope-labelled version of the analyte is used, see Chapter 2, implying that mass spectrometric detection must be used); (h) as a footnote to requirements (e) and (g), for an isotopically labeled SIS (an isotopolog of the analyte) the degree of isotopic substitution should be sufficiently large that the probability of observing such an isotopolog in the natural material (with natural isotopic abundances) is negligible; this requirement can be difficult to satisfy in practice for analytes containing large numbers of carbon atoms (e.g., peptides, see Section 11.6.3) and if it is not satisfied special data processing methods (nonlinear regression) must be used (see Section 8.5.2c).

8.5.2a Volumetric Internal Standards for GC Analyses

The general intent of a VIS is to minimize uncertainties associated with volumes such as v' and V'. The most common of such uncertainties arise from relatively low precision in GC injection volumes v' and v'' and, to some extent, from potential uncontrolled systematic errors in total solution volumes V' and V'' when volatile solvents must be used. These volume parameters are replaced by quantities Q_{VIS} of VIS, that are ultimately defined by weighing procedures although in practice these quantities are most often dispensed as volumes of a solution of VIS. This apparent contradiction is resolved by noting that all that is required of the volume dispensing in this context is that it be reproducible, in the sense that the same volumes of VIS solution must be added to the sample extract solution to be analyzed and to the external standard solutions used to obtain the calibration curve; the absolute values of the dispensed volumes and of the concentration of VIS solution are not required. Modern digital pipets can achieve a high degree of precision in the delivered volumes (Section 2.4.2). In order that the amount of VIS added to each solution to be analyzed is the same for all, it is important that all such additions be made at the same time to all solutions particularly if the solvent is volatile. (Similar remarks apply also to applications of surrogate internal standards, discussed below.)

To understand the effectiveness and limitations of a VIS in eliminating uncertainties in dispensed volumes, we shall consider in some detail the example of a standard calibration curve using clean solutions of analytical standard together with a VIS, analogous to the derivation of Equation [8.64]. As for any procedure incorporating a calibration curve its characteristics must be determined, particularly the question of whether or not it encompasses the origin, and also the extent of the linear dynamic range.

In the present context involving use of a VIS, a suitable calibration procedure could involve the following steps:

(a) using accurate weighings and appropriate dilutions (best not to use serial dilutions of a single stock solution, see Section 9.5.4), prepare a series of standard solutions of the analyte but, before filling to the calibration mark on each standard volumetric flask, add a fixed volume V_{VIS} of the solution of internal standard of concentration C_{VIS}. If a volatile solvent is used, it is important to perform all of these additions at the same time, so that each solution contains the same quantity $Q_{VIS,a}$ of volumetric internal standard;

(b) using weighings and dilutions, prepare a series of standard solutions of the VIS alone;

(c) perform the quantitative analyses on all solutions prepared in (a) and (b), including as many replicate injections of each solution as are feasible.

The present treatment assumes the validity of a linear response curve, possibly with a non-zero intercept (this assumption must apply to both the analyte and the VIS). Then, as before, for the analyses of the calibration solutions but for now ignoring the data obtained for the spiked-in VIS:

$$R_a'' = S_a \cdot (f_a \cdot q_a'' - 1_a) = [S_a \cdot f_a \cdot (v''/V'')] \cdot$$
$$Q_a'' - S_a \cdot 1_a \qquad [8.76a]$$

where the intercept $A_a'' = -R_a \cdot 1_a$ is measured as the intercept on the R_a'' axis (at $Q_a'' = 0$) of the least-squares regression line. Similarly for the analysis of the solutions of only VIS:

$$R_{VIS}'' = S_{VIS} \cdot (f_{VIS} \cdot q_{VIS}'' - 1_{VIS}) = [S_{VIS} \cdot f_{VIS} \cdot (v''/V'')] \cdot$$
$$Q_{VIS}'' - S_{VIS} \cdot 1_{VIS} \qquad [8.76b]$$

where $A_{VIS}'' = -R_{VIS} \cdot 1_{VIS}$ is again measured as the intercept of the least-squares regression line. The main purpose thus far has been to evaluate the intercepts, but as before it is always preferable to investigate and alleviate the causes of nonzero values if feasible. However, in the general case, the data obtained in analyses of the VIS-spiked calibration solutions are processed somewhat differently. The values of R_a'' and of $R_{VIS,a}''$ are both measured for each individual spiked calibration solution; the values of v'' are thus identical for the analyte and VIS since only one injection of a mixed solution is involved, and similarly for V'', so the cancellation of volumes for analyte and VIS is exact for each individual injection, regardless of random inter-injection variations. Then combination of the response equations for analyte and VIS gives:

$$(R_a{}'' - A_a{}'')/(R_{VIS,a}{}'' - A_{VIS,a}{}'') =$$

$$\{[(S_a \cdot f_a)/(S_{VIS} \cdot f_{VIS})]/Q_{VIS,a}\} \cdot Q_a{}'' \qquad [8.77]$$

The intercept $A_{VIS,a}{}'''$ pertains to the analyses of the VIS in the VIS-spiked calibration solutions and can not be determined since these solutions all contained the same quantity $Q_{VIS,a}$. Instead, it is assumed that this parameter is equal to $A_{VIS}{}''$ determined from analysis of the series of solutions of only VIS (Equation [8.76b]). (Of course it is always preferable to ensure that intercepts of calibration curves are statistically zero, see Section 8.3.2.) The quantity $\{[(S_a \cdot f_a)/(S_{VIS} \cdot f_{VIS})]/Q_{VIS,a}{}''\}$ is determined as the slope of the least-squares regression of the values of the left side of Equation [8.77] (i.e., the ratio of signals for analyte to VIS) against $Q_a{}'$; this regression line should have a zero intercept (to within the confidence limits determined as before) if the two intercepts were measured properly. This completes the calibration procedure.

Application of such a calibration to quantitative analysis of sample extracts requires that these extracts be spiked with the identical quantity $Q_{VIS,a}$ of volumetric internal standard as used in the calibration experiments. (It is emphasized that all of these spiking procedures, i.e. calibration solutions plus sample extracts, should be done at the same time, to minimize any drift in the value of C_{VIS} as a result of uncontrolled evaporation.) Then it is straightforward to show that:

$$Q_a{}' = [(R_a{}' - A_a{}')/(R_{VIS}{}' - A_{VIS}{}')]/ \qquad [8.78]$$

$$\{[(S_a \cdot f_a)/(S_{VIS} \cdot f_{VIS})]/Q_{VIS,a}{}''\}$$

Evaluation of Equation [8.78] uses the value of the slope of the calibration line (Equation [8.77]) and also requires the assumption that the values of the two intercepts can be taken to be those determined in the calibration experiments (Equation [8.76]). It is not possible to generalize about the validity or otherwise of this assumption and this emphasizes yet again the importance of investigating the causes of any non-zero intercepts with a view to remedial action to reduce them to low values (zero if possible).

The way in which a VIS corrects for random errors associated with irreproducible injection volumes and/or with uncontrolled evaporation of volatile solvent can now be appreciated to arise from use of the signals from known amounts of VIS as the normalizing parameter, rather than solution volumes. It is thus important that the VIS itself should be involatile and that it be added in precisely equal quantities to the external calibration solutions and to the sample extracts. The effect of a VIS on the overall analytical precision is clear-cut if a significant contribution to the random error is due to uncertainties in the injection volume v′, as is commonly the case for GC analyses. However, modern sample loop injectors for HPLC can achieve a precision of injection of $\leq 0.1\%$ for complete loop filling (Section 2.4.3); it seems inherently unlikely that use of a VIS could improve on this and indeed the question arises as to whether it might actually impair the overall precision. This question has been considered in detail (Haefelfinger 1981) in an interesting analysis of the propagation of error (Section 8.2.2) in such experiments; the statistical analysis was illustrated using by real-world examples, including at least one in which the random errors introduced by the manipulations of the VIS were large enough to significantly decrease the overall precision relative to that achieved using only the simple external standard procedure (Equations [8.64–8.65]). Clearly, an informed assessment is required in each individual case. Nonetheless, one advantage of use of a VIS if a surrogate internal standard is not available, not mentioned thus far, is that it can always provide a correction for instrumental drift.

A VIS can also be used to advantage for GC analyses when no surrogate internal standard is available but when either a blank matrix is available or when there is sufficient sample available that the Method of Standard Additions can be deployed. Here only a brief indication of the appropriate relationships (Boyd 1993) is given. For the case of a calibration using spiked matrix blanks together with a VIS, it is assumed here that analyte losses associated with 'active sites' in the chromatography train have been reduced to negligible values, i.e., $l_a \approx 0$; this assumption is made only to keep the algebra reasonably uncomplicated although there is no intrinsic difficulty in extending to the full treatment. However, the corresponding assumption is not made for the extraction–clean-up procedure (i.e., $L_a{}'$ is not necessarily zero). In this method a fixed quantity $Q_{VIS,a}$ of VIS is used to spike all final extracts from both the analytical samples and from the blank matrix samples, which were themselves spiked with variable amounts P_a of analyte standard before extraction–clean-up. Again the ratio of signals for analyte to VIS is used as the dependent variable in the calibration as a function of P_a. The empirical calibration equation takes the form:

$$R_a{}'''/R_{VIS}{}''' = A''' + B'''.P_a \qquad [8.79a]$$

and the interpolation equation for determination of unknown samples (Q_a replaces P_a) is:

$$C_a = Q_a/W_S = [(R_a'/R_{VIS}') - A''']/(B'''.W_S) \quad [8.79b]$$

The least-squares-fit parameters A''' and B''', determined from the calibration experiments, are assumed to be directly applicable to analyses of real sample extracts, also spiked with the fixed amount of VIS. For Equation [8.79b] to be valid, the full treatment (not shown) shows that the following conditions must also be satisfied:

$$F_a' = F_a''' \text{ and } L_a' = L_a'''$$

i.e., the proportional and fixed losses of analyte must be the same for the analytical samples and for the spiked matrix blanks.

It is also possible to use a VIS in conjunction with the Method of Standard Additions. Again, to keep the algebra tractable it is assumed here that l_a and l_{VIS} are zero to within experimental uncertainty but the possibility of bias errors L_a' and L_{VIS}', arising during the extraction and clean-up steps, is not excluded. The same development as that leading to Equations [8.73–8.75] for MSA combined with Youden analysis applies here, but with $l_a = 0$; the volume ratio (v'/V') again cancels exactly despite random variations in v' (GC injections), and is replaced by Q_{VIS}', the quantity of VIS added to all sample extracts (not the samples before extraction and cleanup, in contrast with surrogate internal standards as discussed later). The theoretical expression for the Youden sample response curve (compare Equation [8.72]) now becomes:

$$R_a'/R_{VIS}' = [(S_a.f_a/S_{VIS}.f_{VIS}).(F_a'/Q_{VIS}').C_a].W_S -$$
$$[(S_a.f_a/S_{VIS}.f_{VIS}).(L_a'/Q_{VIS}')]$$
$$= B_{Y,VIS}.W_S + A_{Y,VIS} \quad [8.80]$$

As discussed for MSA without a VIS (Section 8.5.1d), the main purpose of the Youden procedure is to evaluate any bias errors via the value obtained for $A_{Y,VIS}$. If random errors in v' are significant, use of the VIS generally (but not invariably) improves the precision (Haefelfinger 1981) and the degree of confidence in evaluating a nonzero value for $A_{Y,VIS}$ will correspondingly increase. In the analyses corresponding to the MSA itself the extract solutions are spiked with the same fixed quantity Q_{VIS}' and we obtain a relationship independent of (v'/V'):

$$R_a'/R_{VIS}' = [(S_a.f_a/S_{VIS}.f_{VIS}).(F_{msa}'/Q_{VIS}')].P_a$$
$$+ [(S_a.f_a/S_{VIS}.f_{VIS}).(F_a'.Q_a - L_a')]/Q_{VIS}'$$
$$= B_{msa,VIS}.P_a + A_{msa,VIS} \quad [8.81]$$

As for the case of MSA without a VIS, the most reliable value of C_a is obtained by combining with the results of a Youden analysis:

$$C_a = Q_a/W_S = [(A_{msa,VIS} - A_{Y,VIS})/B_{msa,VIS}].$$
$$(F_{msa}'/F_a')/W_S \quad [8.82]$$

again with the necessary assumption that $F_e' = F_a'$. Further, if the Youden experiment is not performed, i.e., the intercept $A_{Y,VIS}$ is not known, it is necessary to assume it is zero, i.e., $L_a' = 0$. The same comments concerning the chance that use of the VIS will improve or impair the overall precision (Haefelfinger 1981) apply here also. However, if volatile solvents are used, incorporation of a non-volatile VIS will always provide insurance against bias errors introduced by uncontrolled solvent evaporation from the final sample extract. Even if a VIS is included for this latter reason, there is of course no requirement to evaluate the data using Equations [8.80–8.82]; the VIS can be ignored if it is judged appropriate to do so (Haefelfinger 1981), and Equations [8.72–8.75] are then used instead.

8.5.2b Use of Surrogate Internal Standards

In general the most reliable analytical methods in the present context are those employing a suitable surrogate internal standard (SIS); the possible choices of compounds as SIS were described in Chapter 2 (see discussion of Table 2.1). The ideal SIS is an isotope-labelled version (an isotopolog) of the target analyte, preferably with sufficient isotope labels in each molecule that the natural abundance of this species in the natural sample is negligible, so that the mass spectrometric responses corresponding to analyte and SIS do not interfere with one another; this simplification is assumed to be valid in the first examples discussed below (but see Section 8.5.2c). It is worth noting again that a high level of deuteration usually results in a significant shift of retention time relative to that of the nondeuterated analyte, thus providing further separation of the two responses (see the text box describing physico-chemical effects of isotopic substitution in Chapter 2). However, the analyte and SIS will often co-elute and the (usually higher) m/z value monitored for the SIS surrogate will in principle contain interfering contributions from naturally occurring isotopologs (mostly natural abundance ^{13}C) of the native analyte. In such cases it is necessary to deconvolute the two signals from one another, a complicated procedure especially if MS/MS is used to monitor fragment ions; with MS/MS one may also have to worry about interferences of the SIS isotopolog on the m/z value monitored as characteristic of the unlabelled native analyte.

This problem has been extensively reviewed (De Leenheer 1985). In some cases it may even be preferable to use a closely related compound (e.g., a methyl homolog) as SIS rather than an isotopolog if clean separation of the mass spectrometric signals turns out to be more critical than increased assurance that the recovery efficiency F_{SIS}' provides a good estimate for F_a' (see below). Note, however, that if a chemically similar compound (e.g., a simple homolog) is used as SIS the relationship $F_{SIS}' \geq F_a'$ is not necessarily valid; the idea that F_{SIS}' provides an upper limit to F_a' is based on the combined assumptions that the SIS is chemically identical to the analyte with respect to extraction and clean-up procedures, and that the native analyte may be more difficult to extract (occlusion effects) than the SIS added externally. Despite all these reservations, however, when a SIS is available it provides a means to achieve an analytical result with a high degree of reliability when a limited quantity of sample is available. (In contrast the Method of Standard Additions requires, in principle, multiple sample aliquots that must be of a size sufficient that sample homogeneity is assured, Section 8.6.1.)

In general, a fixed quantity Q_{SIS} of the surrogate internal standard is added to a measured quantity W_S of the analytical sample, rather than to the cleaned-up extract as for a VIS (see Figure 1.2). The spiked sample is then taken through the extraction and clean-up steps and analyzed for both native analyte and the SIS in the same chromatography–MS experiment. In the following treatment it is assumed that no bias errors exist in the chromatographic quantitation of either the analyte or SIS, i.e., that l_a and l_{SIS} are both zero. If this is found experimentally to not be the case, it is recommended that the causes of this effect should be identified and remedied, as otherwise the algebraic expressions become intractable and the inversion of the calibration curve to provide estimates of concentration less reliable.

The simplest application of an SIS is to spike known amounts of the analytical sample (preferably weighed to give values of W_S) with known amounts Q_{SIS} of the internal standard, without the simultaneous use of an analytical standard. In practice Q_{SIS} is usually defined by a measured volume v_{SIS} of a solution of concentration C_{SIS}, since in trace analysis it is not realistic to weigh out the very small quantities of the SIS that are comparable with the unknown amount of analyte Q_a. The resulting uncertainties in v_{SIS} and C_{SIS} represent a disadvantage of this simple spiking strategy (Section 9.5.4). However, if time is a crucial factor in the purpose for which the analysis is to be conducted, this procedure may be fit for purpose provided that the caveats evident in the following development are taken into account.

Under the assumption that l_a and l_{SIS} are both zero, the theoretical expressions for the analytical signals for analyte and SIS in a cleaned-up extract of a spiked sample are:

$$R_a' = S_a.f_a.q_a' = (S_a.f_a).(v'/V').F_a'.Q_a$$
$$- (S_a.f_a).(v'/V').L_a \qquad [8.83a]$$
$$R_{SIS}' = (S_{SIS}.f_{SIS}).(v'/V').F_{SIS}'.Q_{SIS}$$
$$- (S_{SIS}.f_{SIS}).(v'/V').L_{SIS} \qquad [8.83b]$$

On dividing these two equations, (v'/V') cancels exactly for each individual analytical sample since analyte and SIS are present in the same extract solution:

$$(R_a'/R_{SIS}') = [(S_a.f_a)/(S_{SIS}.f_{SIS})].(F_a'/F_{SIS}').$$
$$[(Q_a - L_a'/F_a')/(Q_{SIS} - L_{SIS}'/F_{SIS}')]$$
$$[8.83c]$$

To obtain a value of Q_a from the measured signal ratio via Equation [8.83c], further information and/or assumptions are required. The ratio of response factors $[(S_a.f_a)/(S_{SIS}.f_{SIS})]$ for chromatography–MS analysis can be measured using clean solutions containing known quantities of both analyte and SIS although applicability of this value to sample extract analysis assumes that suppression effects on S are the same for both analyte and SIS. If the SIS is an isotopolog of the analyte, it is frequently assumed that these response factors are equal for mass spectrometric detection. While this assumption is valid for most cases if the ions that are monitored are the intact ionized molecules, it can break down due to kinetic isotope effects (see Chapter 2) if fragment ions are monitored, or even if the ionized molecules are monitored but are subject to extensive fragmentation; a dramatic example of this effect has been published (Takahashi 1987). It is often possible to establish whether or not such an effect is operating by comparing mass (or MS/MS) spectra of the analyte and its isotopolog as a function of ion source parameters (or MS/MS collision energy). However, in general such complicating effects are significant only if the isotope labelling involves H/D replacement and are negligible for $^{13}C/^{12}C$ or other heavier atom substitution (see text box in Chapter 2).

A value for Q_a from Equation [8.83c] also requires knowledge of, or assumptions regarding the recovery parameters F_a', F_{SIS}', L_a' and L_{SIS}'. Some experimental information can be obtained if sufficient sample is available that several aliquots can be analyzed using a range of values of Q_{SIS}, since Equation [8.83b] predicts that a plot of R'_{SIS} vs Q_{SIS} provides a value of (L_{SIS}'/F_{SIS}') from the ratio of the intercept to the slope; values for F_{SIS}' and

L_{SIS}' separately can be obtained from the same experimental data if independent information on (S_{SIS}/f_{SIS}) and (v'/V') is available. The most usual procedure, however, is to assume that $F_a' = F_{SIS}'$ (in most cases $F_{SIS}' \geq F_a'$ if the SIS is an isotopolog, reflecting possible occlusion effects) and that the constant loss parameters L_a' and L_{SIS}' are both zero. (Note that it is possible that nonzero values of L_a' and L_{SIS}' may interact with one another, due to competition for the 'active sites' in the clean-up procedure.) Finally, the best accuracy and precision are achieved when Q_{SIS} is determined by direct weighing of an SIS of known purity (both chemical and isotopic) but it is common practice to dispense trace level quantities Q_{SIS} as measured volumes v_{SIS} of a solution of concentration C_{SIS}. Then $Q_{SIS} = v_{SIS}.C_{SIS}$, so that resulting errors and uncertainties in Q_{SIS} are reflected directly in the values deduced for Q_a.

In summary, the simple method employing a SIS spiked into the analytical sample without use of a calibration standard, together with the various assumptions and approximations described above, gives a value for C_a given by:

$$C_a = (Q_a/W_S) = (R_a'/R_{SIS}').(C_{SIS}.v_{SIS})/W_S \quad [8.84]$$

A more useful application of an SIS is in conjunction with an external calibration using solutions of analytical standard. This represents an extension of the approach summarized as Equation [8.67] and eliminates much of the uncertainty inherent in the latter method. In this method the SIS is used to spike both the raw sample and a series of external standard solutions of concentrations C_a'', resulting in measurement of Q_a relative to a multi-point calibration line rather than to a single value of $(C_{SIS}.V_{SIS})$ as in Equation [8.84]. The SIS plays a multiple role of correcting (at least partially) for extraction efficiency and for effects (via sensitivity S_a) of suppression of ionization efficiency by co-eluting matrix components (Section 5.3.5a), as well as providing the advantages of a VIS, as will become apparent in the following development.

For a clean standard solution of analytical standard spiked with SIS, the analytical signals can be described by:

$$R_a'' = (S_a.f_a).q_a'' - RS_a.1_a = (S_a.f_a).(v''.C_a'') - S_a.1_a \quad [8.85a]$$

$$R_{SIS}'' = (S_{SIS}.f_{SIS}).q_{SIS}'' - S_{SIS}.1_{SIS}$$
$$= (S_{SIS}.f_{SIS}).(v''.C_{SIS}'') - S_{SIS}.1_{SIS} \quad [8.85b]$$

where $C_{SIS}'' = Q_{SIS}/V''$ is the concentration of SIS in the spiked calibration solution of total volume V''; note that $Q_{SIS} = C_{SIS}.v_{SIS} = (w_{SIS}/V_{SIS}).v_{SIS}$ where w_{SIS} is the mass

of SIS weighed out to make the stock solution of SIS (original volume V_{SIS}) used to spike the analytical standard solutions (and the raw sample before extraction) with a dispensed volume v_{SIS} of the solution of concentration $C_{SIS} = (w_{SIS}/V_{SIS})$, so:

$$C_{SIS}'' = C_{SIS}.(v_{SIS}/V'') = Q_{SIS}/V''$$

The problems introduced by non-zero values of the constant (bias) losses of analyte and SIS within the chromatographic train are such that it is always best to eliminate them (these are more commonly observed in GC than in HPLC); then under this restriction, i.e. 1_a and 1_{SIS} are both zero, division of Equations [8.85a–8.85b] gives:

$$(R_a''/R_{SIS}'') = \{[(S_a.f_a)/ $$
$$(S_{SIS}.f_{SIS})]/[C_{SIS}.(v_{SIS}/V'')]\}.C_a'' \quad [8.85c]$$

Equation [8.85c] represents the calibration equation; a least-squares regression of n measured values of (R_a''/R_{SIS}'') vs C_a'' should give an intercept A'' that is not statistically distinguishable from zero as determined using a t-test; as before the confidence limits for the intercept A'' are given by $(A \pm t.V_A^{\frac{1}{2}})$ where the value of t is chosen at the desired confidence level, (e.g. 0.05, two tailed values) with $(n-2)$ degrees of freedom, and the variance V_A is given by Equation [8.26]. If these confidence limits ($\pm$) include zero then the intercept is taken to be zero to within the chosen confidence level, but if the test fails the possibility of nonzero values for 1_a and/or 1_{SIS} should be examined. The calibration slope is given by:

$$B_{SIS}'' = \{[(S_a.f_a)/(S_{SIS}.f_{SIS})]/[C_{SIS}.(v_{SIS}/V'')]\} \quad [8.85d]$$

with confidence levels given by Equation [8.26] plus a user-selected confidence level to define the t-value.

Analogous considerations apply to analysis of the solutions of the extracts of the analytical samples spiked (pre-extraction) with SIS. Now, however, the problem of constant (bias) losses L_a' and L_{SIS}' occurring on 'active sites' during extraction and clean-up, must be faced. As for the corresponding chromatographic losses (1_a and 1_{SIS}), L_a' and L_{SIS}' may interact with one another in a manner related to the so-called carrier effect (Millard 1977; Haskins 1978; Self 1979). The best hope for an accurate and precise analysis is that these bias losses (unlike the proportional losses F_a' and F_{SIS}') are reduced to zero by appropriate experimental precautions. Under these conditions the volume ratios (v'/V') for analyte and

SIS again cancel exactly for each individual injection of a sample extract, to give:

$$(R_a'/R_{SIS}') = [(S_a.f_a)/(S_{SIS}.f_{SIS})].(F_a'/F_{SIS}').(Q_a/Q_{SIS})$$

$$= \{[(S_a.f_a)/(S_{SIS}.f_{SIS})]/[C_{SIS}.(v_{SIS}/V')]\}.$$

$$(F_a'/F_{SIS}').Q_a \qquad [8.86a]$$

$$= B_{SIS}''.(F_a'/F_{SIS}').Q_a$$

so that:

$$C_a = Q_a/W_S = (R_a'/R_{SIS}').(F_{SIS}'/F_a')/(B_{SIS}'.W_S)$$
$$[8.86b]$$

where B_{SIS}'' from Equation [8.85d] can be directly applied to interpolate a value of Q_a (or C_a) from the measured signal ratio (R_a'/R_{SIS}') provided that relative matrix effects (Section 5.3.5a) are negligible, i.e., $[(S_a.f_a)/(S_{SIS}.f_{SIS})]$ is the same for extracts of the analytical sample and of the spiked blank matrix used to prepare the matrix matched calibrator; this implies that suppression/enhancement effects are the same for the two extract types. Use of B_{SIS}'' in Equation [8.86b] also requires that the same spiking volume v_{SIS} is used for both calibration solutions and sample extracts, and that the total final volumes V'' and V' are also equal; if this is not the case appropriate volume ratio corrections must be made.

An important point is that for Equation [8.86a] to be useful the absolute values of C_{SIS}, v_{SIS}, and $V'(= V'')$ need not be known; the only condition required is that they be constant and reproducible and the same for both calibration solutions and sample extracts. This lack of dependence on the absolute value of C_{SIS} is particularly important when a volatile solvent must be used so that C_{SIS} might change if the solution is stored for an extended period. It also implies that the chemical and isotopic purities of the SIS (Section 2.2.3) need not be known with the highest possible levels of accuracy and precision. Since surrogate internal standards are usually scarce and expensive, they are usually not available in quantities sufficient that a quantity Q_{SIS} can be weighed out accurately and precisely each time (a minimum of several milligrams for most analytical balances). The same restriction of limited quantities does not usually apply to the analytical standard, so sizable quantities can be weighed out with good accuracy and precision, and external standard solutions made up fresh as required.

For Equation [8.86] to provide a value for C_a, either the recovery efficiencies F_a' and F_{SIS}' must be known or must be assumed to be equal; of course it is never possible to know F_a' since C_a is unknown so equality must be assumed, an

intrinsic limitation of the method. However, the SIS modification of external calibration does permit an estimate of F_{SIS}' to be made. The best procedure is to compare the SIS analytical signal obtained for an extract of an extract of an analytical sample spiked before extraction and clean-up (as above, for the determination of C_a) with the signal for a nonspiked sample but for which the SIS spike is added after extraction and clean-up. Provided that $Q_{SIS}''(= v_{SIS}.C_{SIS})$, v' and V' are all the same for both types of extract solution, the ratio of the two signals gives F_{SIS}' directly. If insufficient analytical sample is available to permit these additional experiments, a slightly less reliable estimate can be obtained by using one of the SIS-spiked calibration solutions in place of the sample extract spiked with SIS post-extraction. Such an estimate of F_{SIS}' is subject to propagation of uncertainties in the volumes and in any event provides only an upper limit to F_a' if the SIS is an isotopolog of the analyte. However, the assumed equality of F_a' and F_{SIS}' is most likely to be valid when F_{SIS}' is close to unity, so an experimental estimate of its value provides a check on the internal consistency of this procedure.

It is probably true to say that the gold standard for trace analysis, in the context of analytical samples that are not available in sufficiently large quantities for MSA, is provided by a method employing an isotope-labeled SIS together with a blank matrix that permits calibration using matrix matched standards. In other words, the calibrator solutions are prepared by spiking aliquots (of fixed mass W_{BM}) of blank matrix with varying amounts P_a of analytical standard plus a fixed amount Q_{SIS} of the internal standard. Again under the assumption that the bias losses of analyte and SIS in both the extraction–clean-up and chromatography steps are all considered to be negligible, the calibration relationships are:

$$R_a''' = (S_a.f_a).F_a'''.P_a.(v'''/V''') \qquad [8.87a]$$

$$R_{SIS}''' = (S_{SIS}.f_{SIS}).F_{SIS}'''.(v_{SIS}.C_{SIS}).(v'''/V''') \qquad [8.87b]$$

$$(R_a'''/R_{SIS}''') = \{[(S_a.f_a)/(S_{SIS}.f_{SIS})].(F_a'''/F_{SIS}''')/$$

$$(v_{SIS}.C_{SIS})\}.P_a \qquad [8.87c]$$

$$= B_{a,SIS}'''.P_a$$

Similar remarks apply as to those above for Equation [8.85]; however, note that now F_a''', the recovery efficiency for analyte spiked into the blank matrix, appears in the expression for the calibration slope parameter. The relationship for analysis of an analytical sample spiked with SIS is similarly analogous to that in Equation [8.85] but with 1_{SIS} assumed zero:

$$(R_a'/R_{SIS}') = \{[(S_a \cdot f_a)/(S_{SIS} \cdot f_{SIS})] \cdot (F_a'/F_{SIS}')/$$

$$(v_{SIS} \cdot C_{SIS})\} \cdot Q_a \qquad [8.87d]$$

$$= B_{a,SIS}''' \cdot (F_{SIS}'''/F_{SIS}') \cdot (F_a'/F_a''') \cdot Q_a$$

which allows interpolation of values of Q_a (and thus C_a) if $F_a' = F_a'''$, i.e., if the recovery efficiency of native analyte from the analytical sample is assumed equal to that for the analyte standard spiked into the blank matrix. Note that use of the calibration slope $B_{a,SIS}'''$ to interpolate the data obtained for the analytical extract via Equation [8.87d] assumes that S_a and S_{SIS} are the same for both matrix matched calibrators and the analytical extract, thus presuming that any relative matrix effect has been detected and accounted for (Sections 5.3.6a and 9.11). This is a general comment applicable to all examples in which matrix matched calibrators are used. The additional assumption that the recovery efficiency for the spiked-in SIS is the same for the blank matrix as for the analytical sample, $F_{SIS}''' = F_{SIS}'$, is less questionable. The condition $F_a' = F_a'''$ is generally more likely to be valid than the corresponding condition for validity of Equation [8.86] that pertains to external calibration using clean solutions of analytical standard plus SIS; however, this is not always a valid assumption (Section 4.2), e.g., if the native analyte is somehow occluded in the sample matrix (Boyd 1996) or strongly bound to another component in the analytical sample (Ke 2000).

Use of an SIS that is not an isotopolog of the analyte can be a realistic option. Again an estimate of F_{SIS}''' can be obtained by comparing signals for the SIS in extracts of blank matrix spiked with SIS both before and after extraction and clean-up; estimates of confidence limits for F_{SIS}''' are obtained from several replicate measurements (best practice calls for at least three such independent measurements at each of three spiking levels covering the desired dynamic range).

Use of matrix matched calibrators has advantages other than those described above, including a degree of protection against errors arising from suppression or enhancement of ionization efficiencies (thus S_a affecting and S_{SIS}), although this is not a fail-safe assumption in view of potential differences between the analytical matrix and the 'blank matrix' used to prepare the calibrators (relative matrix effects, Section 5.3.5a). However, in most cases if the blank matrix is sufficiently similar to the sample matrix, such effects arising from co-eluting interferences will be similar for both and may cancel at least partially (i.e., $S_a' = S_a$ even if neither is equal to S_a measured for the pure analyte), and similarly for the SIS. However, such effects are inherently irreproducible and it is best to eliminate them as far as possible by improving clean-up

and/or chromatographic methods; discovery of retention time windows, in which suppression effects are negligible, is described in Section 5.3.5a.

It is appropriate to add here some caveats about the application of linear least-squares regression to calibration data obtained using an SIS (including isotopologs of the analyte) in conjunction with external calibration using an analytical standard. The potential for significant uncertainties arising from simple propagation of error (Haefelfinger 1981) has already been mentioned. A fundamental assumption in all of the regression methods described thus far, including those that take account of heteroscedacity and/or that use a polynomial extension of the simple linear relationship between mole ratio Q_a''/Q_{SIS}'' and the observed ratio of responses R_a''/R_{SIS}'', is that the random uncertainties in the former (independent variable) are negligible compared with those in the response ratio. However, with the use of high precision mass spectrometric measurements or in the presence of an isotope effect (most likely for a deuterated SIS) that might vary as a result of small changes in the analytical procedure, the precision with which the mole ratios are established may be comparable to or worse than that of the R_a''/R_{SIS}'' measurement (Schoeller 1976). The variance (V_S, or standard deviation $\sigma_S = V_S^{\frac{1}{2}}$, Equation [8.2]) introduced by the mass spectrometric determination of the R_a''/R_{SIS}'' values can be obtained from replicate measurements on a single standard mixed solution. The variance ($V_P = \sigma_P^2$), introduced during the preparation of the mixtures of analytical standard and SIS cannot be obtained directly because any ratio determination will also be subject to measurement uncertainty, so instead the total variance (V_T) is calculated for the R_a''/R_{SIS}'' values measured for a set of standard mixtures prepared at the same nominal concentrations of analyte and SIS. The variance introduced by the solution preparation is then taken as the difference between the total variance and that in measurement of R_a''/R_{SIS}'':

$$\sigma_P^2 = \sigma_T^2 - \sigma_S^2$$

If σ_P^2 is small it may be difficult to determine it in this way; only when $\sigma_P^2 > \sigma_S^2$ will the difference ($\sigma_T^2 - \sigma_S^2$) become significant for a small set of replicate determinations (Schoeller 1976). The F-test (see discussion of Equation [8.17]) is used to test for significant differences between σ_T^2 and σ_S^2. If σ_P^2 is found to be significantly greater than σ_S^2, then a regression of mole ratios (Q_a''/Q_{SIS}'') as the dependent variable on the R_a''/R_{SIS}'' ratios as the independent variable should be performed. Under these conditions replicate observations of R_a''/R_{SIS}'' for a single standard mixture do not increase the precision of the calibration; instead, it is necessary to analyze replicate preparations of the nominally identical relative concentration Q_a''/Q_{SIS}'' to obtain greater precision, a laborious task and one that requires access to appreciable quantities of SIS. If σ_P^2 and σ_S^2 differ by a factor of less than three, neither can be said to be much more precise than the other and more general regression techniques (York 1966) must be used. Performing the regression the 'wrong way round', i.e., with R_a''/R_{SIS}'' treated as the independent variable and Q_a''/Q_{SIS}'' as the dependent variable, will not substantially change the regression line for well-correlated data (correlation coefficient $R \sim$ unity, Equation [8.18]), but even a small change can cause large relative errors at the extremes of the calibration curve (Schoeller 1976).

It is also the case that calibration data obtained as R_a''/R_{SIS}'' as a function of Q_a''/Q_{SIS}'' are intrinsically heteroscedastic in cases where the concentrations are so low that statistical variations in ion counts (shot noise, Section 7.1.1) dominate the experimental variance. This can be demonstrated for a model (Schoeller 1976) that assumes that a fixed total amount of material is analyzed for each Q_a''/Q_{SIS}'' ratio, and that the ion currents for the two species (analyte and SIS) are integrated for the same length of time; then it is straightforward to show (not done here) that σ_S^2, the variance of the measurement of R_a''/R_{SIS}'', will increase linearly with R_a''/R_{SIS}'' for values < 0.1, and as $(R_a''/R_{SIS}'')^3$ for $R_a''/R_{SIS}'' > 10$. For $0.1 < R_a''/R_{SIS}'' < 10$, σ_S^2 increases as $[(R_a''/R_{SIS}'') + (R_a''/R_{SIS}'')^3]/Q_{SIS}$. In practice other experimental uncertainties will contribute to the total variance, but it is clear that data of this type can never be exactly homoscedastic.

It seems worthwhile at this point to emphasize again that, for all the calibration and determination (interpolation) equations, described here and below, appropriate statistical analysis to obtain appropriate least-squares regression, including determination of the accompanying confidence levels for the measured signals and for the subsequent interpolated values of concentration/amount, should be performed. Additionally, it can not be emphasized too strongly that many of the analytical strategies discussed here have intrinsic shortcomings that should be taken into account in a considered judgement based on a fitness for purpose approach.

8.5.2c *Cross-Contributions Between Analyte and Internal Standard – a Need for Nonlinear Regression*

It is important to note that all of the calibration and measurement procedures involving a VIS and SIS discussed thus far have assumed that there is no cross-talk between the mass spectrometric signals for analyte and internal standard, e.g., the response labeled R_{SIS}' can be attributed to only the SIS with no contributions from the analyte, and vice versa. This represents one of the major advantages of mass spectrometry as a chromatographic detector, since it is possible to find an SIS (e.g., a ^{13}C-labelled isotopolog of the analyte) that behaves essentially identically to the analyte in extraction, clean-up and chromatography stages, but is distinguishable by mass spectrometry *via* the m/z values. However, in view of significant natural abundances of heavier isotopes (especially ^{13}C) it is possible that a naturally occurring isotopolog of the analyte will contribute intensity to the m/z value used to monitor the SIS, so that R_{SIS} for both calibrator solutions and sample extracts can contain contributions from the analyte. The converse situation is much less common if the SIS is an isotopolog of the analyte labeled with a heavy isotope, but can arise if the isotope-labeled SIS is of low isotopic purity (and thus contains some natural abundance analyte as an 'isotopic impurity'), or if the SIS is a compound of related structure but of slightly lower molecular mass so that a natural isotopolog of the SIS can contribute to the signal used to quantitate the analyte; if the SIS is a similar compound but of higher molecular mass, a fragment ion might similarly interfere.

Fortunately, from knowledge of natural isotopic relative abundances of the elements (Table 1.4) it is possible to calculate the relative abundances of naturally occurring isotopologs of any analyte of known molecular formula (see accompanying text box). A simple graphical extension (Meija 2006) of the Pascal's Triangle approach to determining the binomial coefficients (see text box) facilitates calculation of isotope distributions for unit mass spectra of compounds containing more than one multi-isotopic element. The restriction to unit mass spectra implies that small mass differences between e.g. $^{12}C_2H_5{}^{18}OH$ and $^{13}C_2H_5{}^{16}OH$ are not resolved so that a sample of ^{18}O-labeled ethanol will yield a mass spectrum in which the signal at m/z 34 contains contributions from the intended $^{12}C_2H_5{}^{18}OH^{+\bullet}$ and also from naturally occurring $^{13}C_2H_5{}^{16}OH^{+\bullet}$ present as unlabelled impurity. A high resolution instrument would resolve these two signals but for a unit mass spectrum the combined signals would be registered and this is an underlying assumption of the calculations of isotopic distributions discussed in the accompanying text box. A method of determining isotopic patterns of ions, appropriate to any chosen instrumental resolving power (Rockwood 1995, 1995a), provides a powerful tool for cases where high resolution is used to increase the analytical specificity. A more subtle problem is that of deducing isotopic patterns for fragment ions, produced by dissociation of either the full isotopic distribution of the precursor ions (Rockwood 1997) or from

Natural Abundance Isotopic Distribution in a Molecule or Ion

Most elements on planet Earth exist naturally as several isotopes (see Table 1.4) as a more-or-less homogeneous mixture. Therefore when a molecule is 'assembled' from its constituent atoms, the likelihood that one isotope (e.g., ^{12}C) will be incorporated rather than another (^{13}C) is a matter of random probability controlled by the relative abundances of the two isotopes. (The situation is exactly analogous to a typical problem of elementary statistics in which a very large stockpile containing different percentages of black and white balls is sampled to give several collections each containing n balls.)

Consider a molecule that contains n atoms of an element with two isotopes with percentage abundances a and x (where $a + x = 100$). For example, if the element is carbon, $a = 98.93$ for ^{12}C and $x = 1.07$ for ^{13}C. Then the relative probabilities that molecules will contain k ^{12}C atoms plus $(n-k)$ ^{13}C atoms are given by the terms in the Binomial Expansion of $(a+x)^n = \Sigma[n!/k!(n-k)!]a^{n-k}.x^k$, where the summation is over $k = 0$ to n so that there are $(n+1)$ terms. This is particularly useful in cases where $x << a$ (as is often the case for relative abundances of isotopes of an element, the main exceptions being chlorine and bromine see Table 1.4), since in such cases the first few terms provide a good approximation to the complete expression.

The coefficients $[n!/k!(n-k)!]$ can be calculated directly, or given by the famous Pascal Triangle (named after the 17th century French mathematician Blaise Pascal). In a Pascal Triangle the first line gives the coefficients for $n = 1$, the second for $n = 2$, and so on; note that each value (other than unity at each end of each line) is the sum of the two values in the previous line that lie on either side. Thus the first few coefficients are readily generated even for large values of n.

As a simple example, a molecule containing four carbon atoms will occur naturally (i.e., other than as a synthetic SIS) as a mixture of 5 isotopologs $^{12}C_4$, $^{12}C_3^{13}C_1$, $^{12}C_2^{13}C_2$, $^{12}C_1^{13}C_3$ and $^{13}C_4$, with relative probabilities given by the terms in the expansion of:

1 1
1 2 1
1 3 3 1
1 4 6 4 1
1 5 10 10 5 1
1 6 15 20 15 6 1
etc.
Pascal's Triangle

$$(98.93 + 1.07)^4 = 98.93^4 + 4(98.93^3)(1.07) + 6(98.93^2)(1.07^2) + 4(98.93)(1.07^3) + (1.07^4)$$

For mass spectrometry it is customary to set a as the value for the most abundant isotope and to normalize the expansion so that the first term $a^n = 100$. For our simple example the relative probabilities for the five isotopologs are 100, 4.326, 0.070, 5×10^{-4}, and 1×10^{-8}, and this is the predicted mass spectrometric pattern for an ionized C_4 molecule in which no other element makes significant contributions to the isotopic pattern.

If the element of interest exists as more than two isotopes (e.g., oxygen, Table 1.4, with three isotopes) the relative probabilities for the isotopomers are still given by an expansion of $(a+x+y)^n$ but now its evaluation becomes laborious since it must be done step-wise:

$$(a+x+y)^n = [a+(x+y)]^n = [a+z]^n$$

where $z = (x+y)$, and each power m of z in the expansion of $(a+z)^n$ must be itself expanded out as $(x+y)^m$.

Things become even more complicated when the molecule contains more than one multi-isotopic element, since the relative abundances for ions appearing at successive *integral mass numbers* (NOT exact m/z values) are calculated from: $(a_1 + x_1 + y_1 + \ldots)^{n_1}. (a_2 + x_2 + y_2 + \ldots)^{n_2}. (a_3 + x_3 + y_3 + \ldots)^{n_3}\ldots$.

Fortunately all modern mass spectrometer data systems include algorithms that calculate isotope patterns for ions with any combination of elements, at both unit mass resolution (as above) and at high resolution; a brief review (Meija 2006) lists the mathematical approaches that have been applied to devising efficient algorithms for this purpose. A simple example of application of these principles (determination of the 'cross-over value') was given in Section 1.5.

precursor ions m/z selected at unit mass resolution (Rockwood 2003) and thus potentially containing contributions from several species with slightly different precise m/z values. Again this represents a significant advance in interpretation of tandem mass spectra but, from the viewpoint of use of an isotopolog of the analyte as an SIS, the possibility of overlap of signals in MS/MS spectra is usually approached experimentally by examining the spectra.

Thus we can predict how many isotopic labels (e.g., ^{2}H, ^{13}C, ^{15}N) need to be incorporated into an isotopolog SIS to ensure that the molecular mass of the SIS is sufficiently large that contributions to R_{SIS} from higher natural isotopologs of the analyte are negligible if low resolution MS is used. Unfortunately this is not always a practical possibility, especially for larger analytes (such as peptides, see Section 11.6) for which the natural isotopic distribution extends to higher m/z values. The general problem of cross-contributions to signals at m/z values selected as characteristic of analyte and SIS is well established in elemental analysis (Heumann 1992; Rodríguez-González 2005) and has been the subject of several papers (Schoeller 1976; Pickup 1976; Trager 1978; Min 1978; Patel 1978; Colby 1979, 1981; Howald 1980; Bush 1981; Barbalas 1991; Whiting 2001; Chang 2001) in the context of analysis of organic compounds by chromatography coupled to mass spectrometry. These authors use different detailed approaches and algebraic notations but the following discussion adapts the notation used thus far in this book to this purpose.

The quantities Q_a, Q_{SIS}, q_a and q_{SIS} are for convenience taken here to refer to numbers of moles (not the raw weighed masses). R_a and R_{SIS} refer to the total responses measured at the m/z values chosen as characteristic of the analyte and the SIS, respectively; in principle each of these can contain contributions from both of the analyte and SIS. Then $R_a{}^a$ and $R_{SIS}{}^{SIS}$ refer to the contributions to R_a and R_{SIS} from the respective expected compound (i.e., in all of the foregoing discussions that assumed cross-contributions to be zero, $R_a = R_a{}^a$ and $R_{SIS} = R_{SIS}{}^{SIS}$). However, $R_a{}^{SIS}$ is the cross-contribution to R_a by the SIS, and $R_{SIS}{}^a$ is that to R_{SIS} by the analyte. A similar notation is used for the intrinsic mass spectrometric sensitivities (i.e., those for the pure compounds introduced directly into the mass spectrometer, not via the chromatograph); thus $S_X{}^Y$ represents the sensitivity (response factor) for compound Y at the m/z value assigned as characteristic of compound X, e.g., $S_{SIS}{}^a$ is the sensitivity for the analyte at the m/z value assigned to monitor the SIS. The proportional systematic errors f_a and f_{SIS} refer to the transmission efficiencies of the compounds between chromatographic injector and ion source and are not affected by the mass spectrometric cross-contributions. We shall also keep the algebra tractable by assuming that the systematic bias errors L_a, l_a etc. are zero or have been reduced to insignificant values by careful experimentation.

Then the signals $R_a{}''$ and $R_{SIS}{}''$, observed at the respective m/z values for an injected volume of a calibration solution containing $q_a{}''$ and $q_{SIS}{}''$ moles of the two compounds, can be written:

$$R_a{}'' = (R_a{}^a)'' + (R_a{}^{SIS})'' = S_a{}^a \cdot f_a \cdot q_a{}'' + S_a{}^{SIS} \cdot f_{SIS} \cdot q_{SIS}{}''$$
[8.88a]

$$R_{SIS}{}'' = (R_{SIS}{}^{SIS})'' + (R_{SIS}{}^a)'' = S_{SIS}{}^{SIS} \cdot f_{SIS} \cdot q_{SIS}{}''$$
$$+ S_{SIS}{}^a \cdot f_a \cdot q_a{}''$$
[8.88b]

The measured quantities used as the dependent and independent variables in the calibration are the response ratio $R_a{}''/R_{SIS}{}''$ and the ratio $Q_a{}''/Q_{SIS}{}''$ of weighed amounts of analyte and SIS used to make the calibrator solutions; the latter $= q_a{}''/q_{SIS}{}''$ since the volumes V'' and v'' are the same for both compounds, present in the same solution. Then dividing Equation [8.88a] by Equation [8.88b] and dividing numerator and denominator by $(S_{SIS}{}^{SIS} \cdot f_{SIS} \cdot q_{SIS}{}'')$ gives:

$$R_a{}''/R_{SIS}{}'' = [(S_a{}^a/S_{SIS}{}^{SIS}) \cdot (f_a/f_{SIS}) \cdot (Q_a{}''/Q_{SIS}{}'') +$$
$$(S_a{}^{SIS}/S_{SIS}{}^{SIS})]/[(S_{SIS}{}^a/S_{SIS}{}^{SIS}) \cdot (f_a/f_{SIS}) \cdot (Q_a{}''/Q_{SIS}{}'')$$
$$+ 1] = [\Phi \cdot (Q_a{}''/Q_{SIS}{}'') + \Theta]/[\Psi \cdot (Q_a{}''/Q_{SIS}{}'') + 1]$$
[8.89]

Equation [8.89] is linear with respect to parameters Φ and Θ, but nonlinear with respect to Ψ, and therefore the data must be fitted to this calibration function using nonlinear least-squares regression (Section 8.3.8); it is emphasized that it is very important to ensure that the initial estimates for the unkown parameters should be reasonably close to the final best estimates (see the text box dealing with nonlinear regression). In the present example (Equation [8.89]) excellent initial estimates can be obtained experimentally (see below) but if this is not possible tricks can be employed to obtain reasonable first estimates. One way is to plot the experimental data for $R_a{}''/R_{SIS}{}''$ vs $Q_a{}''/Q_{SIS}{}''$ and draw an approximate curve though the points by hand. Experimental data expected to be well represented by Equation [8.89] should extrapolate to a value of $(R_a{}''/R_{SIS}{}'') = \Theta$ as $(Q_a{}''/Q_{SIS}{}'') \rightarrow$ zero, and to $(R_a{}''/R_{SIS}{}'') = (\Phi/\Psi)$ as $(Q_a{}''/Q_{SIS}{}'')$ becomes large, or alternatively as a separate plot of $(R_a{}''/R_{SIS}{}'')$ vs $[1/(Q_a{}''/Q_{SIS}{}'')] \rightarrow$ zero. Together with data from a point on the hand-drawn curve near the centre of the experimental range, initial estimates for Φ and Ψ separately can then be obtained.

However, in our case of a calibration curve described by Equation [8.89], good initial estimates can be obtained by some simple additional experiments. For example, $\Phi = [(S_a{}^a/S_{SIS}{}^{SIS}).(f_a/f_{SIS})]$ is the ratio of response factors for chromatography–MS analysis of pure analyte and SIS, respectively, each monitored at its own characteristic m/z value in the absence of cross-contributions, and can be determined experimentally, although use of an additional VIS is recommended (Chang 2001) since the SIS and analyte are no longer analyzed together in the same solution. The parameter $\Theta = (S_a{}^{SIS}/S_{SIS}{}^{SIS})$ represents the ratio of intrinisic sensitivities (direct introduction without chromatography) for the SIS at the m/z values characteristic of the analyte to that for itself and can be determined directly from the mass spectrum (MS^1 or MS^2, whichever is being used) of the SIS. Similarly, $\Psi = [(S_{SIS}{}^a/S_{SIS}{}^{SIS}).(f_a/f_{SIS})]$ is the ratio of sensitivities at the m/z value used for $R_{SIS}{}''$ for chromatography–MS analysis of pure analyte and SIS separately, i.e., Ψ is the cross-contribution of the analyte to the response $R_{SIS}{}'''$ at the m/z value monitored as characteristic of the SIS, and can be determined experimentally in a similar fashion to that for Φ (Chang 2001).

Note that $\Theta = (S_a{}^{SIS}/S_{SIS}{}^{SIS}) = 0$ if the SIS is of higher molecular mass, either a heavy-isotope-labelled version of the analyte or a higher mass homolog or analog of the analyte, and provided (a) that there is no SIS fragment ion with the m/z value characteristic of the analyte used to measure $R_a{}''$ and (b) that the SIS does not contain any unlabelled analyte as an impurity. If the SIS has a lower molecular mass than the analyte (e.g., a lower mass analog or homolg) $\Theta = 0$ if the natural abundance isotopologs of the SIS do not contribute significantly at the m/z value used to measure $R_a{}''$. However, if the SIS is indeed of higher molecular mass than the analyte (usually a heavy isotope labeled version of the analyte) and if natural isotopologs of the analyte contribute to $R_{SIS}{}''$, then $\Psi > 0$. This is the most common source of nonlinearity if the isotope-labelled SIS contains insufficient heavy isotope labels; similar comments apply if the SIS is a higher-mass homolog or analog of the analyte. If the SIS is a lower mass analog or homolog, $\Psi = 0$ provided (a) that no fragment ion of the analyte contributes at the m/z value used for $R_{SIS}{}''$ and (b) that there is no analyte present as an impurity in the SIS. It is unusual (though possible in principle) for cross-contributions between analyte and SIS to be present in both directions, i.e., for both Θ and Ψ to be significantly different from zero; the most common nonlinear circumstance is for $\Theta = 0$ and $\Psi > 0$. Of course if $\Theta = 0 = \Psi$, i.e., no cross-contribution in either direction, Equation [8.89] reduces to the simple linear form of Equation [8.85c] (but with all bias systematic errors (L_a,

l_a, etc.) assumed to be negligible, note that $Q_{SIS}{}''$ in Equation [8.89] corresponds to $[C_{SIS}.v_{SIS}]$ in Equation [8.85c]).

Examples of such nonlinear calibrations are shown in Figure 8.15; these data are for GC–EIMS analysis of a derivatized barbiturate using a D_5-substituted version as SIS (Whiting 2001). The EI mass spectra are quite complex so the probability of cross-contributions in both directions between analyte and SIS is fairly high. In fact, initial estimates (Whiting 2001) of the GC/MS sensitivity ratios ($S_a{}^{SIS}.f_{SIS}/S_a{}^a.f_a$) and ($S_{SIS}{}^a.f_a/S_{SIS}.f_{SIS}$), i.e., respectively the cross-contributions by the SIS to the signal at the m/z value characteristic of the analyte and vice versa, were (0.93/99.7) and (0.01/99.9) for the m/z pair 196 and 201 (for analyte and SIS respectively), and (4.47/95.3) and (7.64/92.36) for the m/z pair 138 and 143. Thus the calibration data for the second pair of m/z values are expected to show a much greater degree of nonlinearity than the first, and indeed this is clearly the case (Figure 8.15). The functional form used for the nonlinear regression was equivalent to Equation [8.89]:

$$R_a{}''/R_{SIS}{}'' = [Q_a{}'' + (S_a{}^{SIS}.f_{SIS}/S_a{}^a.f_a).Q_{SIS}{}'']/$$
$$[(S_{SIS}{}^a/S_a{}^a).Q_a{}'' + (S_{SIS}{}^{SIS}.f_{SIS}/S_a{}^a.f_a).Q_{SIS}{}'']$$

where $Q_{SIS}{}''$ was held constant for all the calibration solutions. Unfortunately, insufficient spectral information was provided to permit full interpretation of the best-fit parameters (Whiting 2001) in terms of the theoretical expressions, but the relative degrees of nonlinearity (Figure 8.15) are clearly qualitatively consistent with relative values of the cross-contributions predicted from the GC/EIMS spectra.

The example discussed here (Whiting 2001) involved a small molecule analyte. However, the most common example in which cross-contribution effects are significant involves absolute determinations of protein levels determined via specific proteolytic peptides as surrogate analytes and using isotopologs of these peptides as SIS. The cross-contribution effects can be large in such cases as a result of the generally higher masses of the peptide analytes compared to typical 'small molecules', and the consequent wider distribution of naturally occurring isotopologs of the analyte resulting from the larger number of carbon atoms. This example is discussed further in Section 11.6.

The cross-contribution phenomenon discussed above, with its implications for the need to adopt nonlinear regression in some form, can be a genuine difficulty for the analyst in real-world situations. Approaches used to mitigate the problem without having to resort to nonlinear regression are described in Section 9.4.5b.

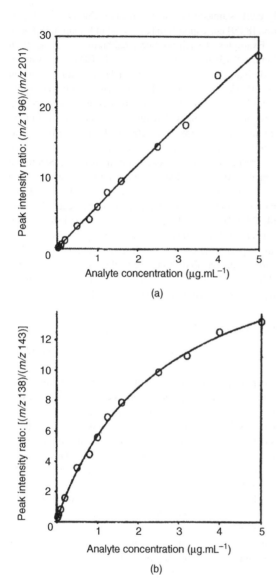

(a)

(b)

Figure 8.15 Examples of GC/EIMS calibration curves that are nonlinear as a result of cross-contributions between the analyte (butalbital) and SIS (butalbital-D_5), both as the methylated derivatives. The SIS was present in all calibration solutions at a concentration of 200 ng mL^{-1}. The m/z values (all for fragment ions) monitored were: (a) analyte 196, SIS 201; (b) analyte 138, SIS 143. See main text for data on cross-contributions. In both cases the curves represent nonlinear least-squares regression fits of the data to Equation [8.89], accounting for cross-contributions in both directions. Reproduced from Whiting, *J. Anal. Toxicol.* **25**, 179 (2001), with permission of Preston Publications, A Division of Preston Industries, Inc.

As discussed in Section 8.3.8, the confidence limits (CLs) obtained from nonlinear least-squares regression are not very reliable. If time, availability of standards and finances permit, a better feel for the true CLs is obtained by repeating the same experiment (calibration) several times. Similar comments apply to CLs of interpolated values of $(Q_a'/Q_{SIS}')^*$ from measured values of (R_a'/R_{SIS}') in an actual analysis using the nonlinear regression parameters evaluated from the calibration experiments (compare Equation [8.32] for the linear regression case). Of course, as before, the calibrators can in principle be made up either as mixed solutions of pure analytical and internal standards or as extracts of spiked matrix blanks.

8.6 Statistics of Sampling of Heterogeneous Matrices

Measurement of analyte concentration in an analytical sample requires, by definition, the acquisition of the analytical sample. This truism is (like all truisms) true, in this case because it is impossible to measure every part of what has been called the 'sampling target' (Ramsey 2004). The inherent heterogeneity of sampling targets such as contaminated soils or sediments, and even biological tissues, presents challenges to personnel responsible for the 'primary sampling' of hazardous waste sites etc. because, of course, the entire site can not be analyzed. This heterogeneity must be addressed by statisticians and chemometricians in development of sampling strategies, and also affects the manner in which analytical chemists sub-sample in the laboratory (see later).

The way in which heterogeneous sampling targets are sampled has a direct effect on the quality of the final analytical result; the uncertainty generated by the sampling process is propagated into this final result and affects the interpretation of the analytical data and the decisions made on the basis of these data about actions to be taken, e.g., to remediate contamination at a polluted site. However, most analytical chemists currently estimate the uncertainty of only the analytical measurement result (as the confidence limits within which the true analyte concentration is claimed to lie to within a stated confidence level). Such estimates of uncertainty do not include contributions from the process of primary sampling or possibly not even of secondary sub-sampling in the laboratory (see later). It has been stated (Ramsey 2004) that, while an analytical measurement for a heterogeneous matrix may be quoted as having confidence limits of ±5 % at a 95 % confidence level (calculated without taking into account uncertainty arising from the primary sampling), if the entire measurement procedure including sampling is repeated from the start the results may differ by as much as 50 %!

All samples are heterogeneous to some degree depending on the scale on which the degree of heterogeneity is to be defined; at one extreme even a well mixed liquid solution of several components is heterogeneous at the molecular scale. However, we are concerned here with heterogeneity on a macroscopic scale, e.g., all the way from differences in average pollutant concentrations between different locations within a contaminated site to concentration differences among soil or sediment particles of different sizes. Complete homogeneity of such particulate samples, even within one narrowly defined location within the site, is impossible to achieve because many factors, including both gravity (smaller particles tend to find their way to lower levels as the more easily pass through spaces among the larger particles) and discrimination with respect to particle size introduced by the nature of the primary sampling tool, work against it. However, the extent of heterogeneity, and its effect on sampling and thus on the final analytical result, can be minimized.

The aim of the sampling process is to ensure that the analytical measurement of the chemical composition of samples taken from a sampling target is an unbiased indicator of the composition of the entire sampling target. This is seldom possible using a single step sampling operation and it is therefore useful (Heydorn 2004) to divide the process into primary and secondary sampling (ISO 2002), as discussed later. The problems arising from heterogeneity in chemical assay of geological samples were addressed (Ingamells 1973, 1985) by introduction of the concept of a sampling constant K_S that is useful for the characterization of secondary samples that are often used in environmental analysis; this concept is briefly described below, based on a published discussion (Heydorn 2004). A comprehensive theory of sampling, that is in principle applicable to both primary and secondary sampling, has been developed (Gy 1992, 1998, 2004; Pitard 1993); since primary sampling from the field is a specialized discipline not normally conducted by laboratory analytical chemists it will be discussed here only briefly, and the main emphasis placed on the secondary sampling procedures used for sub-sampling of heterogeneous primary samples in the laboratory.

Gy's theory identifies, and proposes techniques for minimization of, seven major categories of sampling error, covering differences within samples. Other problems exist, including errors involving sampling over space (e.g., soil samples that provide a poor representation of a polluted site) and over time. The internal sample error categories are (Gy nomenclature):

Fundamental Error: this category concerns loss of precision arising from inherent properties of the sample such as variations in chemical and physical composition, including particle size distribution effects; it can be reduced by decreasing the diameter of the largest particles selected for analysis or by increasing the total mass of the primary sample.

Grouping and Segregation Error: this error arises from nonrandom distribution of particle sizes, usually a result of gravitational effects; it can be minimized by selecting samples for analysis from many randomly selected primary samples or by careful homogenization and splitting of the sample.

Long Range Heterogeneity (or Point Selection) Error: errors of this type are nonrandom and arise from existence of spatially localized 'hot spots' in the sampling target; such errors can be identified and reduced by taking many sampling increments to form the final sample.

Periodic Heterogeneity (or Point Selection) Error: this error is the periodic counterpart of the long range heterogeneity error, corresponding to variations in the chemical and/or physical properties of the sampling target that vary in a periodic fashion with respect to space (e.g. depth at which a soil sample is taken) and/or time; the recommendations for minimization are similar to those for long range heterogeneity.

Increment Delimitation Errors: these are errors arising from inappropriate sampling design and the wrong choice of sampling equipment.

Increment Extraction Error: error of this kind arises when the mechanics used to actually take a sample from the sampling target fail to precisely extract the intended sample; properly designed sampling equipment recommended by Gy, e.g., scoops and spatulas that are flat (not spoon-shaped) and have parallel sides to avoid the preferential sampling of larger particles, and good protocols, are essential.

Preparation Error: as its name suggests, this category of errors arises from (possibly selective) loss, and/or contamination and alteration of, a previously acquired sample or subsample; well designed protocols for both field and laboratory techniques address this problem.

The foregoing are very brief indications of the problems addressed by Gy's comprehensive theory (see also Figure 8.16); detailed discussions (Gy 2004) of these concepts provide complete explanations. An interesting discussion (Minkinnen 2004) describes use of these principles in optimization of real-world applications in the context of a fitness for purpose approach. The following

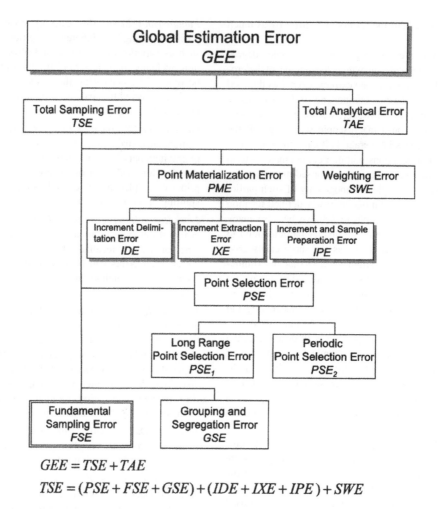

$$GEE = TSE + TAE$$

$$TSE = (PSE + FSE + GSE) + (IDE + IXE + IPE) + SWE$$

Figure 8.16 Classification of sampling errors as developed by Gy according to their origin; sampling errors can be classified into two main groups, those that originate from incorrect design or operation of the sampling equipment (the materialization errors $IPE + IXE + IPE$) and those from statistical errors ($PSE + FSE + GSE$). Weighting errors represent an additional category that arise if the primary sample consists of sub-samples of different sizes that are analyzed separately and a simple arithmetical mean is calculated without weighting for sub-sample size. The 'Total Analytical Error' is that arising in the analytical laboratory, as discussed in the present book. Since all these errors can be considered to be independent, they can be combined to give the Global Estimation Error according to the rules of propagation of error. Reproduced from Minkinnen, *Chemometrics and Intelligent Laboratory Systems* **74**, 85, copyright (2004), with permission from Elsevier.

discussion emphasizes those aspects that most directly involve the analytical chemist in the laboratory to whom the primary sample has been sent. Gy theory addresses representative sampling of systems that are effectively infinite and nonuniform, but finite systems (such as field samples sent for analysis) can be made more uniform by mechanical treatment and mixing. Errors in sub-sampling such well mixed finite systems do not contain many of the error types (e.g., the point selection errors) accounted for in the complete Gy theory (Figure 8.16), and the heterogeneity of such systems can be characterized by a sampling constant K_S (expressed in units of mass) for each specific analyte being determined. The immediate practical importance of K_S lies in its ability to predict the minimum size of sub-sample of a heterogeneous analytical sample (e.g., soil or sediment particles) that must be taken to minimize random errors arising from the heterogeneity. The following discussion summarizes the definition of this characteristic (Ingamells 1973) as developed later (Heydorn 2004).

Consider a sampling target composed of a population of N_S unspecified particles, each with its own mass m_i and value y_i of the chemical concentration to be determined, thus contributing to the heterogeneity of the system:

$$M_S = \Sigma_i(m_i) \text{ and } Y = \Sigma_i(m_i.y_i)/M_S \text{ for } i = 1 \text{ to } N_S \qquad [8.90]$$

where M_S is the total mass and Y the representative average value of analyte concentration (the 'measurand') for the entire system. Define the contribution h_i of particle i to the overall heterogeneity with respect to the analyte concentration y_i as:

$$h_i = [(y_i - Y)/Y].(m_i/M_S).N_S \qquad [8.91]$$

If all N_S particles are equally likely to be selected in acquisition of the primary sample, the particles are referred to (Heydorn 2004) as 'sampling units' and then h_i has a mean value of zero (by definition) and a weighted variance V_{SU} given by:

$$V_{SU} = \sigma_{SU}^2 = \Sigma_i[w_i.(y_i - Y)^2]/[Y^2.\Sigma_i(w_i)] \qquad [8.92]$$

where the statistical weight $w_i = m_i.(N_S/M_S)$. If the sampling units are chosen to be analyte molecules, σ_{SU}^2 becomes an intrinsic property of the system independent of all subsequent processing and is referred to as its constitutional heterogeneity (CH).

The preceding concepts can be applied to primary sampling of the sampling target, with the intention of providing a gross sample that is truly representative of the target, i.e., all N_S particles have the same probability of being selected. In the special case that the sampling units are not individual analyte molecules but rather groups of molecules that are spatially close to one another, the variance σ_{SU}^2 reflects the degree of grouping (segregation) in the sampling target and is referred to as the distributional heterogeneity (DH); an objective of representative sampling is to obtain a gross primary sample (i.e., the combined primary sub-samples) with the same DH value as the overall sampling target itself. If this gross primary sample contains n_S of these sampling units (spatially confined groups of analyte molecules), their contribution to the relative uncertainty $(u_{Y,SU}/Y)$ in the final value Y of the measurand is:

$$(u_{Y,SU}/Y)^2 = \sigma_{SU}^2/n_S \qquad [8.93]$$

and the relationship between CH and DH can be written (Heydorn 2004):

$$DH = CH.(1 + G.Z)/(1 + G) \qquad [8.94]$$

where G is a grouping parameter ≥ 0 such that $G = 0$ represents no localized grouping (DH = CH) and Z is a segregation parameter such that $0 \leq Z \leq 1$ with $Z = 1$ indicating complete segregation of analyte molecules (and also DH = CH). Generally DH < CH, and mixing the material to increase uniformity (smaller Z) further decreases its value.

A secondary sample (or test sample), that is to be subjected to chemical analysis, differs from the primary gross sample in that it is made uniform by grinding, mixing etc.; these procedures do not affect the CH value but reduce the Z value to as close to zero as possible thus reducing DH (Equation [8.94]). The sampling contribution to the uncertainty in the final result of the combined sampling–analysis procedure (Equation [8.93]) varies as n_S^{-1}; for a uniform test sample an appropriate measure of n_S is simply the mass of the sample. The sampling constant K_S is thus defined as the mass of uniform test sample for which (u_Y/Y) is 1 % (an arbitrary choice) and the test sampling contribution to the uncertainty in the final result becomes (Heydorn 2004):

$$(u_Y/Y)^2 = K_S/m_S \qquad [8.95]$$

where m_S is the mass of test sample that is actually analyzed in a single analysis. The sampling constant K_S (Ingamells 1973) is analogous (Heydorn 2004) to the 'heterogeneity invariant (HI)' in the Gy theory, such that:

$$HI = CH.M_S/N_S = 10^{-4}.K_S \qquad [8.96]$$

where the factor $10^{-4} = (1 \%)^2$. The practical importance to the analyst of K_S (or HI) is that it is a single parameter that can specify the minimum size (m_S) of uniform test sample that should be chosen for analysis in order to obtain a final result Y in which the sampling contribution to the overall uncertainty is a specified (low) amount. The preceding concepts were developed for particulate solid materials of interest in geological and geochemical problems (Ingamells 1973, 1985). The assumptions underlying the sampling constant K_S may not be fulfilled for materials that do not consist of solid particles (e.g. biological tissues), and should be verified for such cases; an example is discussed briefly below.

Since sampling errors and analytical errors are essentially independent of one another they can be combined using a simple application of the rules of propagation of error:

$$u^2{}_{total} = u^2{}_{sampling} + u^2{}_{analysis}$$

where $u^2{}_{sampling}$ in turn can be expressed in terms of contributions from primary and secondary sampling:

$$u^2{}_{sampling} = u^2{}_{prim} + u^2{}_{sec}$$

where $u^2{}_{prim}$ is given as $u_{Y,SU}$ in Equation [8.93] and can be reduced to insignificant values if the gross primary sample contains a sufficiently large number n_S of sampling units. The contribution from the uniform secondary (test) sample is given as u_Y in Equation [8.95], so the combined sampling uncertainty under the described conditions is given by:

$$u^2{}_{sampling} = Y^2 . [(u_{Y,SU})^2/n_S + 10^{-4} . (K_S/m_S)] \quad [8.97]$$

Modern mass spectrometers and ancillary analytical instrumentation require ever-decreasing amounts of test samples to provide analytical data with acceptably low values of $u_{analysis}$, so that knowledge of K_S is becoming of increasing importance in order to avoid appreciable errors resulting from sampling uncertainties ($u_{sampling}$) by choosing m_S values that are too small. Evaluation of K_S in any specific case can be achieved (Heydorn 2004) either by statistical analysis of experimental data exploiting Equation [8.97] or by using *a priori* information regarding the gross primary sample, e.g., particle size distributions. The former strategy is preferred, exemplified by instructive work (Chatt 1990) on experimental determination (by neutron activation analysis for selenium) of K_S values for some certified reference materials. Analysis of six replicate test samples at each of several values of sample weight m_S showed that the observed variation was statistically significantly greater than that expected from the analytical measurements alone, indicating a significant contribution from sampling uncertainty; the value of K_S was determined (Chatt 1990) as the slope of the weighted linear regression of the sampling variance (determined as the difference between observed variance and that for the analytical determinations) vs $m_S{}^{-1}$ (Equation [8.95]). A somewhat different re-evaluation (Heydorn 2004) of the same data (Chatt 1990) calculated the sampling constant K_S and its uncertainty for each of the m_S values used, and compared the values by a *t*-test. Lack of statistically significant differences among the values obtained for different values of m_S would indicate that the material is uniform in the present sense, and K_S can then be

calculated as the weighted average of all individual determinations. In fact it was found (Heydorn 2004) that the K_S values were indeed internally consistent when tested at $t = 0.81$ at five ($= 6$-1) degrees of freedom and a confidence level $p > 95\%$, and the final weighted mean value of K_S was in good agreement with the value reported by the original researchers (Chatt 1990). This approach (Heydorn 2004) accounted for the uncertainty in K_S while confirming the invariance of the values relative to the analytical sample size (m_S).

The second (*a priori*) approach is exemplified by determination of K_S for vanadium concentration in a certified reference material consisting of orchard leaves (Ingamells 1974). Application of an analogous approach to calculation of K_S for determination of iron in blood (Esbensen 2004) showed that, under a plausible set of assumptions, K_S is about 5×10^{-6} g; therefore, even with the smallest samples of 5–10 mg used for determination of iron in blood, the contribution from sampling (Equation [8.97]) to the overall uncertainty of the result is negligible relative to the analytical uncertainty of iron determination ($\sim 2 - 3\%$).

This specific result is encouraging for continued reliable analyses for all kinds of analytes in small blood samples, although the numerical evaluation will be different for different analytes. However, this conclusion is not valid for many other types of biological material (Esbensen 2004). For example, diagnostic investigations of human organs are conducted using analyses of small samples (a few mg) taken *in vivo* by means of biopsy needles. The most frequently sampled organs are liver and kidney, and each of these has a complex internal structure that does not allow a simple theoretical calculation of the expected contribution from the sampling process to the uncertainty of a final result. As an example (Esbensen 2004), the K_S values for iron and copper in parenchymatous (functional) tissue were estimated on the basis of available experimental data estimated to be about 200 g, corresponding (Equation [8.97]) to a sampling uncertainty exceeding 100% for a 10 mg total sample. This uncertainty is obviously at least one order of magnitude larger than the analytical uncertainty and thus becomes decisive for the reliability of conclusions made on the basis of analytical results for small biopsies.

Determinations of K_S are laborious, as is clear from the preceding discussion. Determination of this (or an equivalent) parameter are desirable in the case of a certified reference material (CRM), that could serve as an indicator for minimum sample size of an analytical sample of general type similar to that of the CRM. But in general there is not enough time or sample to perform the extensive experiments required (e.g., Chatt 1990) and the analyst must use judgment plus whatever information is available (including that for analogous CRMs) to estimate the minimum sub-sample size for analysis.

8.7 Summary of Key Concepts

Statistical methods provide an approach that yields quantitative estimates of the random uncertainties in the raw data measurements themselves and also in the conclusions drawn from them. Statistical methods do not detect systematic errors (e.g. bias) present in an assay nor do they give a clear-cut answer to the question as to whether or not a particular experimental result is 'acceptable.' An **acceptability criterion** must be chosen *a priori* based on the underlying assumption that the data follow a **Gaussian (normal) distribution**. A common acceptability criterion is the 95 % confidence level, corresponding to a *p*-value of 0.05. Because work is with small data sets in trace quantitative analyses, as opposed to the infinitely large data sets required for idealized statistical theory, use is mode of tools and tests based on the T_f (*t*) distribution (Sudent's t distribution) developed specifically for the statistical analysis of small data sets.

Statistical methods provide tools for assessing **univariate data** (replicate measurements of a single parameter) resulting in measurements of: i) **accuracy**, defined as the difference between an experimental value and the 'true' value; the latter is generally not known for a real-world analytical sample, so that accuracy must be estimated using a surrogate sample e.g., a blank matrix spiked with a known amount of analytical standard); ii) **precision**, such as the relative standard deviation (RSD, also known as the coefficient of variance, COV or CV); iii) methods for calculation of **propagation of experimental error** in calculations.

Statistical methods for assessing **bivariate data** (two dimensional data where one parameter is measured as a function of another) are used to perform **regression analysis of calibration data** and to determine the **goodness of fit** of the calibration curve. **A simple, linear equation is desirable** when fitting quantitative calibration data for several reasons: i) only two fitting parameters ($A = $ intercept and $B = $ slope) need to be calculated; ii) it is straightforward to invert the equation so as to calculate an unknown value of x^* (e.g., analyte concentration) from a measured value of Y^* (e.g. mass spectrometer response), i.e. $x_i^* = (Y_i^* - A)/B$; and iii) relatively few experimental measurements (x_i, Y_i) are required to establish reliable values of A and B in the calibration equation

A linear fit is determined using a **simple linear least-squares regression** that is based on two assumptions: i) the experimental uncertainty in x_i is much smaller than

that in Y_i,; and ii) the data are homoscedastic, i.e. the variances in the Y-values are essential identical at each x. Under these assumptions the regression line includes the average point, resulting in narrower confidence limits (CLs) towards the center of the curve. When calculating a linear regression, **it is advised not to constrain the calibration curve to pass through the origin (zero intercept)** as the imposition of this constraint implies that the point ($x = 0 = Y$) is both accurate and has zero variance, a condition rarely encounted in trace quantitative analyses.

Once the predicted regression line is calculated, a **plot of the residuals** ($Y_i - Y_{i,pred}$) vs x_i, can be conducted to indicate the reason(s) why the data might not conform well to a linear fit, i.e. the residuals can point out the reason(s) for a poor **goodness of fit**, such as reproducibility concerns, nonlinearity and heteroscedacity. If reproducibility concerns turn out to be the major cause of poor goodness of fit, they must be addressed by revisiting the method.

If the residuals plot indicates heteroscedasticity, then a **weighted linear regression** is usually required. A weighted regression essentially 'evens out' differences in the variance of the Y data (responses) at different x-values (amounts or concentrations), thus making the weighted data 'homoscedastic.' If the **data are nonlinear**, one approach to regression analysis is to transform the Y data (and possibly also the x data) in such a way that the transformed data $(x_{tr,i}, Y_{tr,i})$ display a linear dependence; a **common transformation is the logarithmic function**. While such a transformation can be useful in *exhibiting* the full range of data on a single graph, simple linear regression of the transformed data $(x_{tr,i}, Y_{tr,i})$ followed by back-transformation to the original variables (x_i, Y_i) can lead to undesirable consequences. **If the nonlinear data** have no significant deviation from homoscedacity, the nonlinearity can be handled by using a **simple quadratic relationship** instead of a linear relationship. i.e. $Y = A' + B'.x + C'.x^2$. Inversion of this equation to give a value of x^* based on a measured value of Y^* is still reasonably straightforward since this is a simple quadratic equation in x. However, **since a mass spectrometer is intrinsically a linear detector**, it is always best to determine the reasons for the nonlinearity and correct them. One such cause of nonlinearity when an SIS is used is possible **isotopic overlap** between the isotopologs of the target analyte and those of the SIS.

Goodness of fit (correlation) for both unweighted and weighted linear least-squares regression can be measured using: i) correlation coefficient (R) and/or the coefficient of determination (R^2); and ii) *F*-test of the residuals calculated for the fit line. The *F*-test is preferred over R or R^2, since modern analytical instruments commonly generate

$R^2 > 0.999$, leaving only the 4th digit after the decimal point as the only meaningful test statistic to determine correlation.

Inversion of the calibration equation is used to determine the unknown amount of the analyte in the sample, x^*, and is simplest for calibration equations that give Y as a linear function of x, although a quadratic function is only slightly less convenient. **Confidence limits (CLs)** for x^* can be determined from those measured for Y^* to establish the main sources of uncertainly in x^*. CLs can be regarded as the bivariate equivalents of the RSD and other measures of uncertainty for univariate data.

Once a best fit calibration curve has been determined, the following **statistical figures of merit** that characterize the calibration curve can be determined for the assay (method):

Limit of Detection (LOD): The simplest definition of LOD is the lowest concentration (or amount) of analyte for which the observed signal/noise ratio (S/N) = Z, where Z is (somewhat arbitrarily) chosen to be in the range three to six. Other, more rigorous definitions are also used, but the rule of thumb estimate of LOD is based on a S/N = three.

Instrumental detection limit (IDL): Refers to an LOD determined using clean solutions of the analytical standard injected into the analytical instrument operated under stated conditions.

Method detection limit (MDL): Involves measuring and LOD using blank matrix samples spiked with known amounts of the calibration standard and taken through the entire extraction–clean-up–analysis procedure. MDL measures the ability of a specified measurement method to determine an analyte in a sample matrix, with data interpretation using a specified statistical method. The term 'MDL' is often taken to mean the particular single point estimate determined using the prescription of the US EPA.

Lower limit of quantitation (LLOQ, or LOQ): Is defined in various ways depending on the end purpose. For some applications the LLOQ is defined as the concentration of analyte that yields a signal higher than the average blank response by 10 times the standard deviation of the latter. It is sometimes also defined as the lowest standard on the calibration curve, e.g., according to US FDA protocol, the LLOQ is accepted as the lowest standard on the calibration curve provided that (a) the analyte response for this concentration (amount) is at least five times the blank response, and (b) the analyte peak (response) should be identifiable,

discrete, and reproducible with a precision of 20 % and accuracy of 80–120 %.

Upper limit of quantitation (ULOQ): Is generally taken to be the highest point on the calibration curve.

Range of reliable response (often called the **linear dynamic range**): the range of analyte amounts (concentration) for which a given method gives rise to levels of accuracy, precision and linearity that have been decided by the analyst to be fit for purpose.

The **form of the calibration equation**, to which the calibration data are to be fitted by least-squares regression, varies with circumstances. The **gold standard** in this context involves calibration using **matrix matched calibrators**, i.e., extracts of aliquots of a **blank (or control) matrix** (several different lots of the latter to check for *relative* matrix effects) that have been spiked with known amounts of both the analytical standard and a **surrogate internal standard (SIS)** that is preferably a heavy atom (e.g. ^{13}C rather than ^2H) isotopolog of the analyte; it is also preferable to ensure that the best-fit calibration curve has a zero intercept to within the associated confidence limits, though it is possible to work with such a curve if it is deemed fit for purpose. For the most demanding work **it is not advisable to force the curve through the origin by specifying zero intercept in the calibration equation.** Instead the reasons for a nonzero intercept should be investigated and changes made to the overall method to minimize it. Such a gold standard approach is not always possible, e.g., a truly blank matrix that provides a good match to the matrix of the analytical sample, and/or an isotope-labeled SIS, may not be available. In such cases significant compromises with respect to accuracy and precision of the final result must be accepted. If the m/z values monitored for the analyte and SIS are not sufficiently separated, **cross-contributions** between the two can arise as a result of, e.g., natural isotopologs of the analyte contributing to the signal used to measure the amount/concentration of SIS; this leads to a requirement for **nonlinear regression**, since the calibration equations are then not linear with respect to the fitting parameters.

The **Method of Standard Additions** uses the analytical sample itself as the 'blank matrix' and does not require an SIS. The best results are obtained when a conventional MSA analysis is combined with a **Youden plot** analysis. The MSA is realistic only when sufficiently large quantities of the analytical sample are available but does minimize many of the uncertainties that can arise when a true blank matrix is not available.

The problem of acquiring **representative analytical samples of a matrix that is intrinsically heterogeneous** is one that still presents problems for the analyst. **Primary**

sampling of, e.g., a contaminated environmental site, is a specialized activity in which the analyst is not usually involved. **Secondary sampling (sub-sampling)** of the primary samples is, however, a task for the analytical laboratory, since usually the primary samples are sufficiently large that several sub-samples can be obtained from it for analysis to provide data suitable for statistical evaluation of the results. The main concern in this context is to determine the **minimum size of sub-sample** that must be processed to minimize uncertainties arising from the intrinsic heterogeneity, e.g., distribution of particle sizes for soil samples. While appropriate theory and methods do exist to determine this critical parameter, they require large amounts of sample, time and effort. In practice the minimum sample size is determined by some other means, e.g., comparison with a certified reference material (CRM) similar to the analytical sample.

Appendix 8.1

A Brief Statistics Glossary

This is a much abbreviated version of a very complete glossary (that includes helpful examples) made available by the Department of Statistics, University of Glasgow, at: http://www.stats.gla.ac.uk/steps/glossary/

Alternative Hypothesis: The alternative hypothesis H_1 is formulated in contrast with the *null hypothesis*, and usually represents a more 'radical' interpretation than the 'safer' H_0. For example, if the experiment is designed to determine whether results from two laboratories for the same sample are in agreement, H_0 would state that indeed they do agree and H_1 that the two sets of data are different within some agreed *confidence limits* at some agreed *confidence level*.

Analysis of Variance (ANOVA): In general this term refers to statistical tests applied to more than two data sets. *One way analysis of variance* is a method of comparing several data sets, all of which are independent but possibly with a different mean value for each data set; a very important example is a test to determine whether or not all the means are equal. An example would be replicate analyses of a homogeneous test analytical sample by three independent analytical techniques, yielding three independent data sets each providing independently obtained mean values. The 'one way' designation refers to the fact that the only difference between the data sets is the analytical method used. A *two way analysis of variance* is a method of studying the effects of two potential predictor variables separately (their 'main effects', i.e., the effect of each predictor variable alone averaged across the levels of the other predictor variables), and sometimes together

(their 'interaction effect'). For example, the three different analytical methods referred to above may have been used in three different laboratories, thus introducing a second potential predictor variable.

Confidence Interval: A confidence interval gives an estimated range of values, calculated from a given set of sample data, that is likely to include an unknown population parameter (e.g., mean, standard deviation). The width of the confidence interval gives us some idea about how uncertain we are about the unknown parameter (see precision). Confidence intervals are more informative than the simple results of *hypothesis tests* that decide 'reject Ho' or 'do not reject H_0' since they provide a range of plausible values for the unknown parameter.

Confidence Interval for the Difference Between Two Means: A confidence interval for the difference between two means specifies a range of values within which the difference between the means of the two populations may lie. The confidence interval for the difference between two means contains all the values of $(\mu_1 - \mu_2)$ that would *not* be rejected in the *two sided hypothesis test* of: [H_0 : $\mu_1 = \mu_2$] vs [H_1 : $\mu_1 \neq \mu_2$], or alternatively [H_0 : $(\mu_1 - \mu_2) = 0$] vs [H_1 : $(\mu_1 - \mu_2) \neq 0$]. In the latter case if the *confidence interval* includes 0 it is concluded that there is no significant difference between the means of the two populations, at the stated *confidence level*.

Confidence Level: The confidence level (CL) is the probability value $(1 - p)$ associated with a confidence interval. It is often expressed as a percentage, e.g., if $p = 0.05$, then $CL = (1 - 0.05) \times 100 = 95\%$. A *confidence interval* for a mean of a distribution, calculated at e.g. a 95 % CL, means that we are 95 % confident that the stated interval (limits) contains the true population mean or, in other words, that 95 % of all confidence intervals formed in this manner (from different samples of the population) will include the true population mean.

Confidence Limits: Confidence limits are the lower and upper boundaries (values) of a confidence interval, i.e., the values defining the range of a *confidence interval*, e.g., the upper and lower bounds of a '95 % confidence interval' are the '95 % confidence limits', and similarly for other *confidence levels*.

Correlation Coefficient: A correlation coefficient is a number between −1 and +1 which measures the degree to which two variables are *linearly* related. If there is perfect linear relationship with positive slope between the two variables, the correlation coefficient is +1, but if the perfect linear relationship has negative slope the correlation coefficient is −1; a correlation coefficient of 0 means that there is no linear relationship between the variables. The so-called *multiple regression correlation coefficient*, often referred to simply as 'the correlation coefficient R^2, is a measure of the proportion of variability accounted for by the *regression* (linear relationship) in a sample of paired data (x_i, Y_i). A very high value of R^2 can arise even though the relationship between the two variables is nonlinear, and the goodness of fit of data to a linear model should never be judged from only the R^2 value.

Critical Region: The critical region (or rejection region) is a set of values of the *test statistic* for which the *null hypothesis* is rejected in a *hypothesis test*. The range of values for the *test statistic* is partitioned into two regions; if the experimental value of the *test statistic* falls inside the critical region, the *null hypothesis* H_0 is rejected, but if it falls outside the critical region then 'do not reject H_0' is the result of the *hypothesis test'*.

Critical Value(s): The critical value for a *hypothesis test* is a threshold to which the value of the *test statistic* (e.g., Student's *t*, or Fisher's *F* parameter) for a data set is compared to determine whether or not the *null hypothesis* is rejected. The critical value for any *hypothesis test* depends on the *significance level* at which the test is carried out, and whether the test is *one sided or two sided*. The critical value(s) determine the limits of the *critical region*.

Hypothesis Test: To formulate a statistical test, usually some theory has been proposed. The question of interest is simplified into two competing hypotheses (claims), the *null hypothesis* H_0 and the *alternative hypothesis* H_1, that are not treated on an equal basis since special consideration is given to the *null hypothesis*. Thus, the outcome of a hypothesis test is either 'reject H_0 in favour of H_1' or 'do not reject H_0'; the result of a statistical hypothesis test is never 'reject H_1' and in particular is never 'accept H_1'. The result 'do not reject H_0' does not necessarily mean that H_0 is true, it only suggests that there is not sufficient evidence against H_0 in favour of H_1; the result 'reject H_0' does, however, suggest that the alternative hypothesis H_1 *may* be true. The hypotheses are often statements about population parameters like expected *mean value* and *variance*.

Least-Squares: The method of least-squares is a criterion for fitting a specified model to observed data, in which the parameters of the model are adjusted so as to minimize the sum of the squares of the *residuals*. For example, it is the most commonly used method of defining a straight line through a set of data points on a scatterplot.

Linear Regression: A linear regression equation is an important special case of a *regression* equation; the defining criterion is that the function relating in dependent variable Y to the independent variable x must be linear with respect to the *parameters* of the function (not necessarily linear with respect to x). The simplest case is usually written:

$$Y = A + B.x + e$$

where Y is the 'dependent (or response) variable', *A* is the 'intercept', *B* is the 'slope' or 'regression coefficient', x is the 'independent (or predictor) variable' and e is the error term. This example happens to also be linear in the independent variable. The regression equation is often represented on a scatter plot by a 'regression line'. *Simple Linear Regression* is a method that evaluates a possible linear relationship between a response variable and one predictor variable by the method of *least-squares*. *Multiple linear regression* attempts to find a linear relationship between a response variable and several possible predictor variables.

Matched (Paired) Sampling: This term refers to circumstances in which the data values within the data set to be statistically evaluated occur in clearly defined pairs. There are two common circumstances in which this condition can arise. (a) Data sets in which the members are deliberately paired by the investigator as a result of the nature of the experiment, e.g., analytical determinations for several identical analytical samples using two independent analytical techniques for each analytical sample. (b) Those data sets in which the same attribute is measured twice on each subject under different circumstances (commonly called repeated measures), e.g., analytical data obtained by the same experimentalist on several sub-samples of a certified reference material, repeated on a later occasion.

Nonlinear Regression: Nonlinear regression aims to describe the relationship between a response variable and one or more predictor variables via a function that is not linear in all the *parameters* of the function (compare *linear regression*).

Null Hypothesis: A null hypothesis is usually formulated in one of two ways depending on the context of the theory to be tested. The experimental test may have been performed in an attempt to reject a particular pre-existing hypothesis, and in such cases this hypothesis is assigned as the *null hypothesis* so it cannot be rejected unless the evidence against it is sufficiently strong. Alternatively, if the experiment was conducted in a truly 'agnostic' context and two competing hypotheses have been proposed as a result of the experimental observations, H_0 is usually chosen as the simpler hypothesis so that a more complex theory is not adopted unless there is sufficient evidence against the simpler one.

One Sided and Two Sided Tests: A one sided test is a statistical *hypothesis test* in which the values for which the *null hypothesis* H_0 can be rejected are located entirely in one tail of the probability distribution. Thus the *critical region* for a one sided test is the set of values less than (or greater than, depending on the nature of the phenomenon under test) the *critical value* of the *test statistic*. The choice between a one sided and a two sided test is determined by the purpose of the investigation. For example, a supplier claims that a box of pipet tips contains 100 tips on average, so the following hypotheses could be set up: $[H_0: \mu = 100]$ vs $[H_1: \mu < 100]$ or $[H_1: \mu > 100]$. Either of these two alternatives would lead to a one sided test but it would be more useful to know the probability of fewer than 100 tips in a box since no one complains if there are the correct number (or more!). Yet another alternative hypothesis could be tested against the same H_0, namely, $[H_0: \mu = 100]$ vs $[H_1: \mu \neq 100]$; this would require a two sided test and its conclusion would say only that, if H_0 can be rejected, the average number of tips in a box is likely to differ from 100.

p-**Value:** The probability value (*p*-value) of a statistical *hypothesis test* is the threshold value that defines the probability (significance level α) of observing an experimental value of the *test statistic* as extreme as, or more extreme than, the value predicted by chance alone if the *null hypothesis* H_0 were true; it is thus the probability of wrongly rejecting H_0 if it is in fact true, and is equal to the *significance level* of the test for which H_0 would be only just rejected. In practice the p-value is compared with the *significance level* and, if it is smaller, the result is considered to be significant, i.e., if the investigator had decided that H_0 should be rejected at $\alpha = 0.05$, this would be reported as '$p < 0.05$'. Note that small *p*-values suggest that H_0 is unlikely to be true; the smaller the *p*-value, the more convincing is the rejection of H_0. Thus this approach indicates the strength of evidence for e.g., rejecting H_0, in contrast with a simple binary choice between 'reject H_0' and 'do not reject H_0'.

Regression Equation: A regression equation expresses the relationship between two (or more) variables algebraically and indicates the extent to which some variables

can be predicted by knowing others, or the extent to which some are associated with others. *Linear regression* is an important special case.

Sensitivity *(S)*: strictly the slope of the calibration line but often used in a more colloquial sense to indicate low values of LOD (the two are related in Figure 8.10).

Residual: The residual represents 'unexplained' (or residual) variation after fitting (e.g., by *least squares*) experimental data to a regression model. It is the difference between the observed value of the dependent variable and the value suggested by the regression model for each experimental value of the independent (predictor) variable.

Significance Level: The significance level α of a statistical *hypothesis test* is a user-selected probability of wrongly rejecting the *null hypothesis* H_0 (i.e., if it can be shown that H_0 is in fact true). It is thus the probability of a *type I error* and is set by the investigator in the context of the consequences of such an error, i.e., the value of α is chosen to be as small as possible in order to protect the *null hypothesis* and to prevent the investigator from inadvertently making false claims (as far as possible). Usually, the significance level is chosen to be $\alpha = 0.05 = 5\%$. The *p*-value is the threshold level defining the limit of the integrated probability α.

Test Statistic: A test statistic is a quantity calculated from the experimental data set, that is used to decide whether or not the *null hypothesis* should be rejected in a *hypothesis test*. The choice of a test statistic depends on the assumed probability model and the hypotheses under question; common choices are the Student's-*t* and Fisher's *F* parameters.

Time Series: A time series is a sequence of observations that are ordered in time (or possibly space). A relevant example would be the analytical data obtained for a quality control (QC) sample or a certified reference material (CRM) on a weekly basis. If such measurements are made as a function of time, it is sensible to display the data in the order in which they were obtained, best in a scatter plot in which the measured value x is plotted as a function of time as the independent variable (in this case, however, something over which we have little control!). There are two kinds of time series data, *continuous* in which data are recorded at every instant of time (e.g., temperature and/or humidity in a climate-controlled laboratory), and *discrete* in which data are recorded at (usually regularly) spaced intervals, e.g, the QC analyses mentioned above.

Type I and Type II Errors: In a *hypothesis test*, a type I error occurs when the *null hypothesis* is rejected when it is in fact true, i.e., H_0 is wrongly rejected. For example, a test designed to compare data from two different laboratories would set up [H_0: the two data sets are statistically in agreement] and [H_1: the two data sets are statistically different] given a pre-selected value for the *confidence level*. A type I error would arise if H_0 were rejected when in fact further investigation shows that there is no difference between them at the stated *confidence level*. Such an error is sometimes referred to as a *false positive* (i.e., a premature conclusion that H_1 may be valid). A *type II error* arises when H_0 is *not* rejected when in fact further evidence shows that significant differences do exist, and is sometimes referred to as a *false negative* (a premature conclusion that H_0 is valid and thus that H_1 has no validity). For any given set of data, type I and type II errors are inversely related, i.e., the smaller the risk of one, the higher the risk of the other, and these two risks must be carefully balanced against one another. A type I error is often considered to be more serious, and therefore more important to avoid, than a type II error; the hypothesis test procedure is therefore adjusted so that there is a guaranteed 'low' probability of wrongly rejecting the *null hypothesis* (of course this probability is never zero); this is done by judicious choice of the *probability of a type I error* given by *P(type-I error) = significance level* = α. The exact probability of a type II error is generally unknown; a type II error is frequently due to sample sizes being too small.

Appendix 8.2

Symbols Used in Discussion of Calibration Methods

Superscripts (primes):
Nonprimed symbols refer to the analytical sample prior to extraction, clean-up, etc. Symbols marked with a single prime (′) refer to the sample extract solution (possibly spiked with internal standard). Symbols marked with a double prime (″) refer to a solution of pure analytical standard used for calibration, possibly spiked with an internal standard. Symbols marked with three primes (‴) denote a calibrator prepared by spiking a matrix blank with a known amount of calibration standard, possibly spiked with an internal standard, and taken through the entire analytical procedure.

Subscripts are used as follows:
a = target analyte; VIS or SIS = internal standard (volumetric or surrogate); Y = pertaining to the Youden Sample Response Curve; msa = pertaining to the Method of Standard Additions.

Major symbols used (in alphabetical order):
A = Y-intercept of experimental calibration curve, related to $-S_a.l_a$;

B = slope of calibration line, related to the sensitivity ($S_a.f_a$) of the chromatograph–detector combination;

C_a = concentration of the target analyte in the analytical sample (the desired end result = Q_a/W_s);

C_{IS} = concentration of the stock solution of internal standard used to spike the sample, sample extract, external standard solution etc.;

C_a'' = concentration of analyte in a calibration solution of pure standard in clean solvent;

F_Z' = fractional recovery of substance Z (= analyte, or SIS etc.) from analytical sample into final extract solution to be analyzed;

f_Z = fractional transmission efficiency of compound Z from chromatographic injector to the mass spectrometer detector (ideally unity);

L_Z' = constant loss of substance Z during extraction and clean-up (losses on active sites etc.) leading to a bias error (ideally zero);

l_Z = fixed loss of compound Z after injection into chromatograph, leading to a bias error (ideally zero);

P_a = quantity (amount of substance) of analytical standard spiked into raw sample as for Method of Standard Additions;

P_a''' = quantity (amount of substance) of analytical standard spiked into a matrix blank to prepare a matrix matched calibrator;

Q_Z = total quantity (amount of substance, i.e., number of moles, although the mass can be used as an equivalent in some circumstances) of compound Z in an analytical sample, sample extract, calibrator solution etc.;

q_Z = quantity (amount of substance) of substance Z contained in the volume v of any solution injected into the analytical chromatograph;

R_z = signal response (usually chromatographic peak area or height relative to the background baseline) observed in analysis of a quantity q_Z of substance Z *injected into a detector via a chromatograph* (the number of primes indicates the type of solution injected);

S_Z = intrinsic sensitivity of a detector (e.g., mass spectrometer) for compound Z, i.e. response R_Z per unit amount of Z reaching the detector, i.e., the slope of a plot of response vs analyte amount if the latter were introduced *directly* into the detector (mass spectrometer), not via a chromatograph; note that for introduction of analyte via a chromatograph, $R_Z = S_Z.(f_Z.q_Z - 1_Z)$, i.e., S_Z does *not* include corrections for proportional or bias losses in the chromatographic introduction, nor does it account for suppression or enhancement of ionization efficiency (Section 5.3.5a);

V = total volume of a solution *prior* to removal of the first injection volume v;

v = chromatographic injection volume of a solution;

v_{IS} = volume of solution of internal standard (concentration C_{IS}) used to spike the sample, sample extract or external standard solution;

w_Z = mass of compound Z (pure calibration standard or internal standard) weighed out to prepare a solution for calibration and/or spiking;

W_S = mass of analytical sample represented in the sample extract solution to be analyzed (sometimes replaced by V_S in the case of liquid analytical samples).

9

Method Development and Fitness for Purpose

9.1 Introduction

Other than experiments designed to develop or improve new methodologies, the vast majority of chemical analyses are performed to meet a specified purpose and the results are likely to be scrutinized by a person or group from outside the laboratory. Such evaluations may be done by people who have some expertise, e.g. a regulatory agency or a client such as a chemical company, or possibly by members of the public who have little or no experience with the scientific approach and statistical evaluation in general, quite apart from specialized knowledge of the techniques of analytical chemistry.

As emphasized by R Baldwin in a 1996 Workshop organized by the American Society for Mass Spectrometry (Baldwin 1997), analytical chemists should not become too engrossed in the array of often ingenious technologies that are available but should also take a larger view to ensure the credibility of their work, not only to their immediate peer group (our first professional responsibility) but also to a wider scientific community as well as to the general public, elected officials etc. Thus it is important for analytical chemists to appreciate the interests of this wide range of consumers of the data that they produce. Sections 9.2 and 9.3 provide an introduction to some of these communication issues. The remainder of the chapter is devoted to formulating a strategy for the development of the analytical method and to the underlying principles and analytical techniques that should be considered prior to validation.

9.2 Fitness for Purpose and Managing Uncertainty

This message was pursued by M. Kaiser and D.G. Gibian in the 1996 ASMS workshop (Baldwin 1997) in the context of their advice to analysts called as expert witnesses in legal proceedings. It is important to remember that judge and jury will hear possibly conflicting opinions from expert witnesses appearing for the two sides in the dispute, so it is important to be able to defend one's decisions concerning the chosen analytical method and the evaluation of the data thus obtained. It is here that the 'fitness for purpose' concept, introduced briefly in the Preface (see especially Figure P.2) and referred to periodically in the preceding chapters, plays a crucial role.

The concept of fitness for purpose is a concept that is certainly not limited to analytical chemistry and is considered here in more detail. The first formal definition of the concept in the analytical chemistry literature appears to be that of Thompson and Ramsey (Thompson 1995) in a study of sampling protocols: 'Fitness for Purpose is the property of data produced by a measurement process that enables the user of the data to make technically correct decisions for a stated purpose.'

The refinement of the original definition (Thompson 1995) proposed by Kaiser and Gibian (Baldwin 1997) explicitly takes into account the inevitable uncertainties in any quantitative measurements: 'Fitness for purpose refers to the magnitude of the uncertainty associated with a measurement in relation to the needs of the application area.' This definition takes what would otherwise be a nebulous or subjective idea and makes it quantifiable.

Since then, the concept of fitness for purpose has become common currency in analytical chemistry; this is well exemplified by a very extensive discussion (Eurachem 1998) of the issue that also emphasizes the close relationship of this concept with that of validation. This relationship can be appreciated by comparing the preceding definitions with that of validation applied to all measurements (ISO 1994) and quoted in the fitness for purpose document (Eurachem 1998): 'Validation: Confirmation by examination and provision of objective evidence that the particular requirements for a specified intended use are fulfilled.' The application of this very general definition of validation to analytical methods establishes the link with fitness for purpose. Indeed, demonstration of fitness for purpose is now an essential ingredient in many validation schemes (e.g. van der Voet 1999; Bethem 2003).

The document that is most directly applicable to concerns of this book (Bethem 2003) proposed a definition of fitness for purpose that was given in the Preface together with some of the other recommendations extracted from the Executive Summary provided by these authors. It is worthwhile to repeat here, together with the remaining recommendations.

(1) Ultimately it is the responsibility of the analyst to make choices, provide supporting data, and interpret results according to scientific principles and qualified judgment.

(2) Analysts should use methods that are fit for purpose. Analysts should be able to show that their methods are fit for purpose.

(3) Fitness for purpose means that the uncertainty inherent in a given method is tolerable given the needs of the application area.

(4) Targets for measurement uncertainty describe how accurate and precise the measurements need to be. Targets for identification confidence describe how certain one needs to be that the correct analyte has been identified.

(5) Establishing method fitness consists of showing that the targets for measurement uncertainty and identification confidence have been met.

(6) There are two divergent approaches to assessing uncertainty. These two approaches are applicable to both quantitative methods and qualitative methods. Empirical or top-down approaches work with data acquired within the method's working range and use a holistic view of method performance. Statistical or bottom-up approaches look to differentiate signals from background and consider method performance as a combination of individual steps.

(7) A method's limit is the point where the targets for acceptable measurement uncertainty or identification confidence can no longer be met. Given this definition, the empirical or top-down approach is preferable for describing method limits when

situations where the analyst's data and its interpretation must be defended.

(8) All choices that are inherent in the chosen methodology should be documented.

(9) Identification criteria are neither absolute nor arbitrary: they arise from the level of identification confidence that is considered acceptable for a given application.

(10) Method limits may be determined according to either qualitative or quantitative considerations; the point where identification confidence is unacceptable could be different from the point where precision and accuracy are unacceptable. For qualitative methods, the desired balance between the acceptable rates of false positives and false negatives should be described.

Note that point (3) is a definition of fitness for purpose that closely resembles that proposed by Kaiser and Gibian (Baldwin 1997), in that it explicitly ties the concept to that of the uncertainty inherent in quantitative measurements. In deciding on uncertainty tolerances the fitness for purpose criterion becomes crucially important. For example, if when developing and validating a bioanalytical method the achievable accuracy and precision lead the analyst to conclude that the current method is not suitable relative to the original assay requirements, the method in its current form might be discarded altogether, used as the basis for further development of a method that does satisfy the specific criteria and/or retained as a useful method that can be applied to studies conducted in a less demanding environment.

Some of these 10 recommendations (Bethem 2003) are essentially self-explanatory but others (especially (6)–(10)) require some amplification, given later in this chapter. However, it is worthwhile to first make some comments concerning how the 'purpose' is to be established in discussions with a client and how the colloquial language in which such a purpose is often expressed can be translated into technical terms in a manner that can be understood by the client. Planning should start from the first contact between analyst and client; it involves work outside the laboratory such as literature review, setting uncertainty targets, establishing the availability of suitable analytical and internal standards etc. This important first aspect is discussed in Section 9.3 and the subsequent stages of method development in the rest of the chapter. The issues of method validation and sample analysis in such an environment are discussed in Chapter 10.

9.3 Issues Between Analyst and Client: Examining What's at Stake

It is emphasized (Bethem 2003) that evaluation of fitness for purpose should begin with a clear understanding of the purpose and it is here that communication with the 'customer' for whom the analyses are to be undertaken is key. The authors (Bethem 2003) make some interesting distinctions among types of 'customer' with whom an analyst is likely to work, each of which involves a different kind of relationship.

9.3.1 Define the Analytical Problem

Quite apart from the degree of scientific knowledge and sophistication that the customer brings to the table, the purpose of the proposed analyses can cover a wide range, e.g. a requirement for analytical information as a basis for a decision to be made within the customer's organization, or analyses undertaken in order to meet some regulatory requirement of a government agency, or participation in an accreditation exercise, or for a customer embroiled in a potential legal action etc. It is always best to discuss with the customer at the earliest stage the purpose for which he/she wishes the analyses to be performed, in order to clearly delineate the nature of the problem that in turn can be used to define the performance criteria appropriate for the customer's circumstances. If this is done in a frank and open manner, in which the analyst explains the strengths and limitations of possible analytical strategies, it can help avoid conflicts arising from unrealized expectations on the part of the customer after the analyses have been performed.

The fitness for purpose principle is essential here. The illustrative example described by Kaiser and Gibian (Baldwin 1997) was an expansion of an earlier one (Thompson 1995). This example concerns the determination of gold concentration in various materials for different purposes. Measurement of the gold content in scrap metal returned for credit, for example, calls for the highest possible accuracy and precision in the assay and the time required to complete the analysis is generally not a critical factor. In contrast, in survey assays conducted to aid geochemical prospecting for promising ore bodies, high accuracy and precision are not required and indeed would be too expensive and time consuming for the purposes of the geologists in the field. (In passing it is interesting to compare either of these circumstances with the legendary task assigned to Archimedes with respect to the gold content of his king's crown, see the text box in Chapter 2.)

Once the analytical purpose has been agreed upon with the customer, the analyst should then formulate the implications of this purpose for the corresponding analytical problem, including both qualitative and quantitative aspects, acceptable levels of uncertainty (including false positive/negative identifications (Section 8.2.4) where this is an issue etc. This process requires not only the specialized professional skills of the analyst but also consideration of the potential consequences for the customer (including cost of analysis vs potential costs if the analyses are not fit for purpose). It may be necessary to consult with other organizations (regulatory agencies etc.) and possibly with legal advisers to come to the optimal decisions at this stage. In the end, a careful evaluation of the perceived risk associated with the analytical task and the consequences associated with the final result must be thoroughly evaluated.

9.3.2 Consider the Needs of all Interested Parties

It is usually the case that the client and analyst are not the only interested parties. An example would be a situation where the client requests that a method be developed and validated to provide evidence that a compound is, or is not, present in a given matrix such that the data could very likely be used for regulatory and/or litigation purposes. In these circumstances, all analysts understand the importance of ensuring the scientific integrity of the analysis by following the acceptance criteria established as a result of the validation process. However, extra precautions should be taken to be sure that the method objectives suit the interests of all other parties (regulatory bodies, legal advisers etc.) as well (see Section 9.3.5 concerning the role of the analyst as an expert witness). If it is in the best interests of the client to show that the analyte is not present, then setting criteria that only considers the impact of false positives at the expense of incurring a greater rate of false negatives (Section 8.2.4) would be inappropriate. After these considerations have been taken into account, a discussion of the method objectives and assumptions made during development and validation should be documented in the final validation report and method prior to the analysis of any samples. A more detailed discussion of the requirements of the documentation requirements for validation and reporting of sample results is found in Chapter 10.

9.3.2a Client's Perspective vs Analyst's Perspective

From the analyst's viewpoint, the more demanding the acceptance criteria, the higher the risk of run failure; the severity of these problems increases with the number of compounds to be analyzed by a given method, since failure to meet the established run acceptance criteria for any one analyte triggers a repeat analysis of the samples in order to generate reportable data for that analyte (Dadgar 1995). For clients unfamiliar with the concepts and language of trace analysis, it may be necessary to emphasize some fundamentals at the onset of the discussions. In particular, it may be important to emphasize the uncertainties that are present in all measurements. An example could be the impossibility of reporting that it is certain that a specific compound is absent and that the best that can be done is to report it is not present at a concentration above the limit of confirmation (identification) to within a clearly defined confidence level (see Section 9.3.3b for details).

With regard to uncertainty limits, it is essential to be as clear as possible when establishing the criteria in consultation with the client before the analyses are performed, so that the analysts can design the analytical method accordingly. This should be done bearing in mind that, at the end of the day, some hard-nosed questions will be raised, e.g. were the decision criteria satisfied, was the analytical method fit for the agreed purpose of meeting specified confidence levels for identification and/or uncertainty in quantitative measurement, and is the analyst prepared to sign off on conclusions that the target analyte is indeed present at or above some stated concentration to within statistically defensible confidence limits. To address these concerns, the analyst will be required to balance the expected concentration of all analytes being measured with the qualitative and quantitative requirements of the assay, and be able to explain the nature of the compromise with the client. It must also be borne in mind that any assay will only be able to determine if the analytes are present above the defined limit of confirmation and quantitation and that it is advisable to avoid client demands for a simple YES/NO result.

An earlier publication (Bethem 2003) offered advice concerning the careful choice of words when communicating the analyst's decision about qualitative identification of a target analyte to the client who may not be well-versed in the statistical language of science; Figure 9.1 presents suggested terminology for this purpose, although at present no numerical values of confidence limits and levels can be assigned to these descriptors. Finally, for those assays that are intended to detect and quantify multiple analytes, the criteria for establishing criteria and reporting of results should be established for each analyte and guidelines for how data are to be reported for the individual analytes must be established.

These generalizations are easy to understand for a trained analyst but are often difficult for a client whose main concern is to ensure that he/she is not in breach of

Prose	Numeric
	(Highest selectivity)
Identified with utmost certainty	
Confirmed	
Identified with confidence	
Identified	
Indicated	
Tentative identification	
Suspected	
Presumptive	
Non-negative	
	(Lowest selectivity)

Figure 9.1   Terminology suggested for communication of the degree of confidence concerning analyte identification to a client. At present no quantitative confidence limits and levels can be assigned to these descriptors. However, in this regard it is advisable to avoid the use of 'absolute' terms such as 'detected', 'positive identification' and 'absent'. Reproduced from Bethem, *J. Amer. Soc. Mass Spectrom.* **14**, 508 (2003), copyright (2003), with permission from Elsevier.

some regulations. The differences between the perspectives of client and analyst, with respect to interpretation of the 'Zone of Uncertainty' that inevitably arises as result of limited experimental accuracy and precision, are illustrated in Figure 9.2 both for analyses whose primary objective is qualitative identification of an analyte and for those whose objective is quantitation. If the client is unable to provide guidance as to tolerances for confirmation criteria (and for accuracy and precision if quantitation is required), it is essential that the analyst documents all assumptions built into the analytical method (it is probably a good rule to always do so anyway). A more detailed discussion of setting uncertainty targets for identification and quantitation, in the context of regulatory requirements or guidelines and possible changes to standard operating procedures (SOPs) in the laboratory, will be found in Section 9.3.3 and with respect to method validation in Section 10.2.1. It is also necessary for the analyst to balance his/her natural desire for thoroughness against available time and resources (including cost, see Section 9.3.4), and to make it clear to the client that all of these considerations must be taken into account when deciding on the analytical procedures to be used.

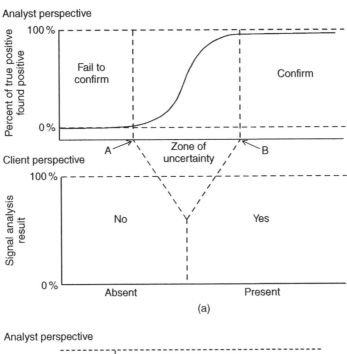

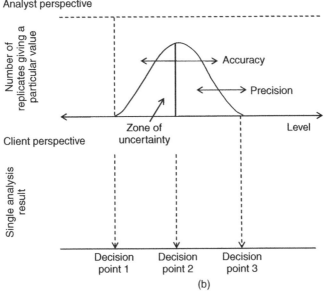

Figure 9.2 Illustration of differences between the perspectives of the analyst and client with respect to the statistical 'Zone of Uncertainty' for (a) analyses principally directed at qualitative identification and (b) quantitation. Reproduced with permission from material prepared by Dr David N Heller, FDA Center for Veterinary Medicine.

9.3.3 Define Uncertainty Tolerance

The analyst must always be concerned with the different viewpoints of expert assessors in the regulatory agencies or other bodies to whom the analytical data will eventually be sent, and of clients many of whom may be unfamiliar and even uncomfortable with the notion of experimental uncertainty and its implications (Chapter 8). The contrast between Figures 9.2(a) and 9.2(b) is a good example of this dichotomy. The analyst must therefore both pay close attention to his/her professional obligations to produce and report data with well defined uncertainties that can be defended when necessary in litigation proceedings in an adversarial environment, and also act as 'interpreter' of the specialist concepts and language used to express them for the client.

Perhaps surprisingly at first sight, it is more straightforward to define uncertainty tolerances for quantitative measurements than those for qualitative confirmation of analyte identity. This is the result of decades of development of statistical methods for evaluation of quantitative measurements of all kinds (Chapter 8). In contrast, chemical structure is intrinsically a much more complex matter (see, e.g., Section 1.4) and it is correspondingly more difficult to define and express degrees of uncertainty in confirmation of analyte identity. Since the principles of defining uncertainty in quantitative data are covered in some detail in Chapter 8, while practical approaches to determining identification uncertainty are still under development, the discussion for qualitative uncertainty (Section 9.3.3b) is much longer than that for quantitative uncertainty (Section 9.3.3a).

9.3.3a Setting Targets for Quantitation

In deciding on uncertainty tolerances the fitness for purpose criterion becomes crucially important. For example, when developing and validating a bioanalytical method, if the achievable accuracy and precision lead the analyst to conclude that the current method is not suitable relative to the original assay requirements, the method in its current form might be discarded altogether, used as the basis for further development of a method that does satisfy the specific criteria, and/or retained as a useful method that can be applied to studies conducted in a less demanding environment.

While experienced analysts will often be able to estimate *a priori* the likelihood that a particular LLOQ or reporting limit will be readily achieved, it is often necessary to first obtain the reference standard and run some initial feasibility experiments; the need to run these initial experiments will increase as the required LLOQ decreases. In many instances a simple sensitivity check, performed by injecting a standard of known concentration using routine chromatography conditions, may be adequate for initial estimates. For this type of experiment, assumptions are made as to what the sample size, extraction technique and final volume of extract might be in order to determine an appropriate concentration for the solution used for the sensitivity check. Even in cases where this quick test using clean standard solutions is done, however, it is best to follow up using an extract of the analyte from a spiked control matrix using a first guess sample preparation method followed by analysis using the initial chromatography conditions developed when running the analyte in solvent. Quick estimates for analyte recovery can also be made at this point and are used to establish a scheme for further development of sample preparation and chromatography. Obviously the complexity of this process and the time required to develop the final method increase significantly with the number of analytes to be measured in a single method.

Related to concerns about whether or not a particular LLOQ will be achievable is the expected range of concentration of the analyte(s) in the samples to be assayed, and the impact of that on the range of reliable response that will be required. Ideally, the validated range of reliable response (validated linear dynamic range) will be adequate for quantification of all analytes in all samples using the same sample aliquot, but in many instances additional dilution of the sample with control matrix will be required to quantify one or more of the analytes within the range of the validated calibration curve; such cases are often referred to as over-range samples. To minimize the need for re-assays with diluted sample (Section 10.2.9d), the analyst should strive to validate a method that encompasses the expected concentration of analyte in the unknown samples without sacrificing method linearity, precision and accuracy.

Even after extensive method development has been conducted, a compromise may be required with respect to what was originally requested. In the end, the analyst and client must balance expectations with respect to the acceptable level of qualitative and quantitative uncertainty with the results obtained during method development. The determination of the final quantitative method is ultimately based upon a reiterative process of method development followed by method validation that, when considered together, determine whether additional method development is required or if a modification to the original method objectives must be considered. Throughout this entire process it is crucial that the results obtained and the decisions based upon those results are well documented. Any modifications to the original method objectives along with a justification for any changes should also be documented in support of the final method.

9.3.3b Setting Targets for Identification

It is important to start this section by discussing the relationship between confirmation of analyte identity and detection selectivity (Section 6.2.1). These are not identical concepts, although in a sense confirmation of identity could be viewed as the ultimate case of detection selectivity. However, in practical terms in the context of trace analysis, selectivity refers to the degree to which one can be confident that the analytical signal used as the basis for quantitation arose only from the analyte that passed all the identification criteria determined for the analytical standard, e.g. chromatographic retention time plus relative responses at different *m/z* values in SIM or MRM (Section 6.2.2) that fall within pre-set limits of acceptability. In contrast, more demanding criteria are required for enforcement or similar methods that could be subject to litigation in an adversarial environment, e.g. the confidence level at which an additive found in a foodstuff can be identified as a banned substance. The lowest concentration at which the specified identification criteria can be satisfied is referred to as the limit of confirmation (sometimes 'identification' is used rather than 'confirmation'). This contrast is discussed further below.

The question of identification of an analyte can take several forms depending on the predicted chance of making wrong identification decisions (either false positive or false negative results). For applications where it is known that the compound of interest was applied or dosed at a given concentration and it is wished to follow the fate of that compound over time, e.g. pharmacokinetic and soil dissipation studies, minimizing the risk associated with false positives is usually only critical insofar as it ensures the integrity of the limit of quantitation by demonstrating method selectivity. In other words, it is known that the compound should be there and the goal is to quantify it with acceptable accuracy and precision while minimizing the risk of false negative results. For applications of this type, selectivity may be demonstrated via the analysis of method and matrix blanks from one or several sources of control matrix.

For bioanalytical methods, screening multiple lots of the biological matrix (e.g. plasma, serum, urine etc.) is usually sufficient (see Section 10.2.2) for demonstrating that the method is adequately selective to minimize false positives. Similarly, if the customer is studying the environmental fate of a herbicide applied to a specific location, then soil from an adjacent control plot where the chemical has never been applied and/or water from a control well upstream from the water flow gradient could serve as adequate controls for demonstrating method selectivity. Such analyses are often performed using chromatography with mass spectrometric detection in SIM or MRM modes rather than using full spectral acquisition. This choice is made in order to increase the duty cycle of a scanning mass spectrometer such as a triple quadrupole instrument, and thus lower the LOD and/or LLOQ; scanning instruments in SIM or MRM mode are currently the instruments of choice if optimal accuracy and precision are desired at the lowest levels (Section 6.2). But such a choice implies that spectral information that could further confirm the analyte identity is discarded.

In contrast, if the identification test is for purposes of 'enforcement' of regulations, e.g., presence of performance-enhancing drugs in the urine or blood of an athlete, or of forbidden pesticides or growth promoters in food, it is extremely important that the probability of false positives be controlled as otherwise an innocent athlete or farmer could be unjustly accused (see Section 10.4.3 for further discussion). ('Risk assessment' methods in general are expected to view as acceptable an appreciable probability of false positives, e.g., 5 %, but 'enforcement methods' must necessarily be designed to provide a much lower probability of false positives, Figure 9.3.) Of course,

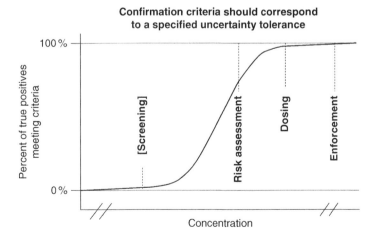

Figure 9.3 Illustration of the principle that statistical criteria for confirmation of analyte identity, that are fit for purpose, vary with the nature of that purpose (see text). Reproduced with permission from material prepared by Dr David N Heller, FDA Center for Veterinary Medicine.

not all circumstances can be said to fall into these rigid categories, e.g., biomedical analyses for metabolites of a drug candidate etc.

So the question arises as to what type of identification criteria or compromise should be adopted between the desire to lower the detection and quantitation limits one the one hand, and retain confidence that the measured signals really do arise from the target analyte and only that analyte on the other. More specifically, how are reliable (and defensible) estimates of probabilities of false positives and negatives obtained? These questions

have aroused considerable controversy and it must be said that they are not completely settled at present. The following discussion is largely based on a much more complete one published previously (Bethem 2003) which views the problem from a fitness for purpose perspective (Figure 9.4). Different applications will likely require different compromises between probabilities of false positive and false negative identifications, but evaluating these probabilities is not always straightforward.

The problem is less severe in developing defensible criteria for claimed rates of false negative than for false

What's at stake
- Set targets for identification confidence
- Set targets for false positives/false negatives
- Set target for decision level (minimum level of full performance)

Define the method
- Define **terms** to be used in reporting
- Define the identification criteria to be used in the method
- Define the diagnostic power of the identification criteria
  - Ions, ion ratios, mass losses, transitions, isotopes, patterns
  - Mass accuracy--**exact mass** sufficient to exclude alternatives
- **Validate** in working range
  - Analyze control tissues--exclude endogenous interferences
  - Interference testing--exclude similar compounds
- Characterize lower limit of performance
  - Extrapolate from working range, fortify through lower limit range
  - Lower limit is where false negative / false positive target is exceeded

Assess uncertainty
- Show that empirically-determined lower limit meets target for decision level
- Show that validation data support claim for diagnostic power of method identification criteria, rates of false outcomes
  - Harmonize with **core confirmation criteria**
  - Compare diagnostic spectrum with relevant **library of spectra**
  - Estimate **selectivity index** of total method
  - Report identification confidence in **numeric or prose terms**

Report results
- YES/NO outcome based on whether criteria met, identification confidence achieved.
- YES/NO outcome based on interpretation by **qualified analyst**
- ACCEPTABLE based on review by **qualified expert**
- Suitable methodology based on **precedent, references**

Figure 9.4 A fitness for purpose approach to determining criteria required for qualitative identification of an analyte. The underlined items are considered to require further clarification. Reproduced from Bethem, *J. Amer. Soc. Mass Spectrom.* **14**, 528 (2003), copyright (2003), with permission from Elsevier.

positive identifications. Assuming that there is adequate evidence for selectivity of the method, replicate analyses of standards of target analytes can yield means and standard deviations of relative abundances of ions considered diagnostic for that analyte. For electron ionization (EI) mass spectra the reproducibility achievable for different instruments under well specified conditions (Section 5.2.1) is sufficient that universal spectral libraries are feasible (Stein 2004, Halket 2005). For soft ionization techniques that usually produce only ionized intact molecules in the MS^1 spectra, MS^n spectra typically exploiting CID of the ionized molecules (most often with n = 2) are required to provide sufficient diagnostic fragment ions and, with care, spectral libraries can provide useful indications of analyte identity (Baumann 2000, Bristow 2004, Halket 2005, Milman 2005, Mueller 2005). However, library-base identifications should be treated with some reserve (see below).

For purposes of identification of analytes with statistically defensible estimates of rates of false negatives, it is best to use contemporary spectral data for analytical standards, obtained using the same instrument and experimental conditions as those used in analysis of the analytical samples. It is then possible to define a matching range for the relative abundances of structurally diagnostic ions (see Section 9.5.3 for a discussion of what is meant by this phrase) that can be translated into a specified confidence level, e.g., a 95 % confirmation rate is achieved with an abundance matching range ± 2 standard deviations about each abundance mean, thus providing a false negative rate of 5 %. A very complete example of this approach (van der Voet 1998; de Boer 1999) takes account of the fact that the precision in relative ion abundances ratios decreases at low analyte concentrations, as expressed in the relative standard deviation (equivalent to the standard deviation after a logarithmic transformation of the relative abundances). A robust measure of the variability of the log (relative abundances) was determined in terms of the median absolute deviation (Section 8.2.7) and the nonlinear relationship between this measure of variability and the analyte level (concentration) was incorporated into the procedure for reaching a decision as to whether samples were classified as either positive (analyte identified) or negative. This test does rely on the details of the analytical method and this dependence is represented in the statistical approach by a 'tolerance parameter ε', the value of which must be determined empirically. In this case (de Boer 1999) this was done by performing the statistical test on the data obtained for blank samples using increasing test values of ε; the highest value of ε giving no false positives for the blanks was taken as the value to be applied to the other data.

As emphasized previously (Bethem 2003), as an alternative to careful evaluation of the probabilities of false positive and/or false negative identifications, it is also possible to specify uncertainty in identification using numerical scores such as 'Purity' and 'Reverse Fit' for the comparison of the spectrum of a putative analyte with standard spectra in a library. To be meaningful such numerical scores require a sufficiently large library of comparable mass spectra because uniqueness values can only be calculated in relation to such a library. For electron ionization (EI), libraries of sufficient size exist but only a small sub-set of compounds of interest are amenable to analysis by e.g. GC–EIMS. Libraries of standard MS/MS spectra or product ion spectra acquired by LC–MS/MS with 'soft' atmospheric pressure ionization techniques (Baumann 2000, Bristow 2004, Halket 2005, Milman 2005, Mueller 2005) can not yet be said to provide the same degree of confidence for identification by spectral matching, although use of contemporary standard spectra (rather than universal libraries) in the matching algorithms can improve the situation. Accordingly, analyses that must use LC–MS/MS rather than GC–EIMS must involve a much more extensive evaluation of identification uncertainty if cited in adversarial situations.

Current attempts to derive such a scheme are discussed below, but the ultimate aim would be to develop such criteria to the point where the appropriate Null Hypothesis (see Section 8.2.4), i.e., [H_o: the experimental data for compound identification in the analytical sample and for the standard are indistinguishable] can be affirmed with a defensible confidence level. As was emphasized previously (Bethem 2003), all that science can do is show that the alternative hypothesis [H_1: the two sets of data are statistically different] is highly improbable. In other words, the scientific method does not claim to be able to prove absolutely that one or other of the opposing hypotheses (appropriately expressed, by no means a trivial task in this case) is valid. If H_1 is accepted on the basis of the currently available data, this does not mean that the suspected analyte has been confirmed as not present, only that it can not be proved to be present at a pre-agreed confidence level based on the available data. There is always some probability (however low) that a negative result can be changed to positive as a result of additional data acquired in the future, or that a positive identification might have arisen as a result of the presence of some hitherto unknown compound or some unsuspected scientific or technical phenomenon.

It might be emphasized here that the discussion (Section 8.4) of critical and detection limits L_C and L_D (or CCα and CCβ, Figure 8.12) is concerned with questions of signal:noise and signal:background ratios. These concepts assume that the signals being monitored have already been demonstrated to be sufficiently selective for the target analyte and are different from the Limit of Confirmation that would be determined by, e.g., experiments of the type illustrated in Figure 9.3.

Currently there are two general defensible approaches to the problem of analyte identification with stated probabilities of false positives, though it is fair to say that both require further work to establish a clear connection to statistical definitions of confidence limits and confidence levels. These approaches were previously (Bethem 2003) conveniently distinguished as 'top-down' and 'bottom-up'. Chronologically the first example of the 'top-down' approach (Sphon 1978) was concerned with GC/MS analysis using EI. Confirmation criteria were based on the analysis of a reference compound contemporaneously with unknowns and required acquisition of signals for three diagnostic ions (i.e. three m/z values, the so-called 'three-ion criterion') at roughly unit mass resolution, with relative abundance matching tolerances in selected ion monitoring (SIM) measurements; it was tacitly assumed that the GC retention times also matched, although no matching tolerance was specified. This work (Sphon 1978) used diethylstilbestrol (DES) as a model test analyte and an EI–GC–MS database containing ∼30 000 spectra. Monitoring the relative abundances of three ions with sufficiently tight tolerances was sufficient to select DES as the only candidate in a computer-assisted library search. A later re-evaluation (Baldwin 1997) used a database containing ∼270 000 spectra and, with even tighter tolerance windows for relative abundances, namely, 90–100 %, 50–70 % and 45–65 % for m/z 268, 239 and 145, respectively, DES still showed up as the only matching candidate in the library. More recently (Stein 2006) it was shown that the identification confidence of an SIM method improves by about one order of magnitude for each additional ion monitored; full scan spectra still represent the best choice if instrument sensitivity is adequate.

This approach (Sphon 1978) is valuable and scientifically valid in that it uses an approach of exclusion of possibilities without claiming to achieve a positive identification. The 'three-ion criterion' was also extended (Li 1996) to LC–ESI–MS/MS methods; at least two and preferably three precursor–product ion pairs were used to replace the three-ion MS^1 (SIM) criterion appropriate for GC–EI–MS, with relative abundance criteria for the monitored ion abundances of $\pm 20 \%$. Moreover, in this case (Li 1996) the LC retention time was explicitly included as an identification criterion, with an acceptance window of $\pm 2 \%$. Quite apart from its contribution to the issue of qualitative identification, this work (Li

1996) is exemplary in its approach to quantitative trace analysis (in this case for sulfonylurea herbicides in soil).

A more general 'top-down' approach to acceptable identification uses aliquots of blank matrix spiked with analytical standard at a series of concentrations from well below an estimated identification limit to well above it. The fraction of such spiked samples found to meet the preselected identification criteria (e.g. Sphon 1978; Li 1996) at each concentration level is plotted against concentration level (Figure 9.3); the value of the latter at which this fraction of 'true positives' exceeds the chosen criterion, e.g., 95 % or 99 % etc. (depending on the objectives of the analysis), is taken as the concentration level at which the rate of false positives is less than 5 % or 1 % etc. If this empirically determined critical concentration level is less than that specified in the objectives for the analysis, then the analytical method has been demonstrated to be fit for purpose with at least some connection to quantitative statistical confidence levels. The specified acceptance criteria will of course vary with the purpose and therefore so will the concentration levels at which the criteria will be satisfied for a given analytical method (Figure 9.3). For example, the US Food and Drug Administration (FDA) criteria for confirmation of the presence of drug residues in edible tissues of treated farm animals (an example of analyses conducted for enforcement purposes, Figure 9.3) include demonstration of a zero false positive rate and of a false negative rate $\leq 10 \%$ at or above the regulatory tolerance level (FDA 2003); in addition very complete requirements on chromatographic and mass spectrometric figures of merit are specified in detail (see Section 10.4.3a). When feasible this general 'top-down' approach is a powerful and generally applicable method (Bethem 2003). However, it does require the availability of large quantities of a blank matrix that is truly representative of the matrices involved in the actual analytical samples, of analytical standard and also of time and financial commitment.

The 'bottom-up' approach to analyte identification (Bethem 2003) is conceptually rather different. One promising approach is that of selectivity scoring (de Ruig 1989; Van Ginkel 1993; De Brabander 2000; André 2001; European Commission 2002; de Zeeuw 2004). In this approach an overall confidence level of an analytical method is estimated by combining measures of the estimated selectivities of all steps including extraction, clean-up, chromatography and mass spectrometric analyses of varying resolving power and number (n) of stages in MS^n analysis. Some steps are widely recognized as major contributors to high selectivity, such as monitoring multiple diagnostic ions in SIM or MRM particularly with higher mass spectrometric resolution, applying acceptance limits on ion abundance ratios and good chromatography that cleanly separates matrix components from target analytes. Table 9.1 summarizes the 'identification points (IPs)' to be awarded (European Commission 2002) to mass spectrometry used in SIM or MRM mode as chromatographic detector for trace analysis provided that the relative abundance ratios agree with those determined contemporaneously for the analytical standard to within specified limits (Table 9.2).

Table 9.1 Identification points (IPs) assigned (European Commission 2002) for different SIM/MRM monitoring techniques for chromatography, subject to satisfaction of relative abundance criteria (Table 9.2)

Mass Spectrometric Technique	Identification Points per Ion Monitored
Low Resolution Mass Spectrometry (LRMS)	1.0
LRMS^n precursor ion	1.0
LRMS^n product ion	1.5
High Resolution Mass Spectrometry (HRMS)	2.0
HRMS^n precursor ion	2.0
HRMS^n product ion	2.5

Note that each ion monitored may be counted only once, e.g., in MS^3 the first generation fragment ion used as precursor ion for the second generation fragments may only be counted once. High resolution is defined as 10^4 at 10 % valley across the entire m/z range monitored (i.e. $\sim 2 \times 10^4$ for a FWHM definition). Also if e.g., GC–EI–MS and GC–CI–MS are both used, these are regarded as independent monitoring techniques and the IPs awarded for each separately should be added. Also if derivatization of analyte is used, analyses of different derivatives are considered to be independent only if the derivatization chemistries are clearly different (e.g., trimethylsilyl and heptafluorobutyryl, see Section 5.2.1b).

Table 9.2 Maximum permitted tolerances (European Commission 2002) for relative ion abundances in mass spectrometric identification of analytes, for varying abundances of each ion relative to the base peak among the m/z values monitored, and for several chromatography–mass spectrometry techniques

Abundance of monitored ion (%) relative to the base peak	GC-EI-MS (% tolerance)	GC–CI–MS GC–MS^n LC–MS LC–MS^n (% tolerance)
> 50	± 10	± 20
> 20 and ≤ 50	± 5	± 25
> 10 and ≤ 20	± 20	± 30
< 10	± 50	± 50

According to this bottom-up approach, the (somewhat arbitrary) minimum number of IPs that must be obtained for the confirmation of regulated substances is proposed (European Commission 2002) to be three, but for positive identification of banned compounds the number is increased to four in order to decrease the risk of false positives; examples of IPs for various combinations are shown in Table 9.3. For banned substances at least some IPs earned by mass spectrometry are mandatory but, since some analytes do not yield a sufficient number of ions, other techniques can be used to contribute a maximum of one IP each, including LC coupled with full scan diode array spectrophotometer (DAD) or with fluorescence detection or immunochemical detection, or two-dimensional thin-layer chromatography (TLC) coupled with spectrometric detection.

Table 9.3 Examples of IPs proposed (André 2001) for several chromatography–MS techniques (see Table 9.2) and some combinations

Technique(s)	Number and Type of Ions Monitored	IPs
GC–LRMS (EI *or* CI)	n	n
GC–LRMS (EI *plus* CI)	2 for EI plus 2 for CI	4
GC–LRMS (EI *or* CI) for 2 derivatives	2 for each derivative (must be different derivatization chemistries, see Table 10.1)	4
LC–LRMS (ESI *or* APCI)	n	n
GC–LRMS/MS *or* LC–LRMS/MS	1 precursor ion plus 2 product ions (i.e., 2 MS/MS transitions)	4
GC–LRMS/MS *or* LC–LRMS/MS	2 precursor ions (2 IPs) each with 1 product ion (3 IPs)	5
LC–MS^3	1 precursor ion (1 IP), 1 first-generation product ion, and 2 second-generation ions (4.5 IPs)	5.5
GC–HRMS	n	$2n$
GC–LRMS *plus* LC–LRMS	2 each	4
GC–LRMS *plus* GC–HRMS	Same 2 ions for each (2 IPs each for the HRMS, no additional IPs for the LRMS, same 2 ions).	4

Such a selectivity scoring approach clearly corresponds in some way to a real situation but the rationale for the particular IP scheme (Tables 9.1–9.3) is not clear and, more important, it has not as yet been shown how any such scheme can be linked to a scientifically defensible estimate of the rate of false positives (Stephany 2000). Nonetheless, this approach does offer some promise (Bethem 2003) and hopefully will be pursued on an international basis and extended to include IPs to be awarded for aspects such as the selectivity of extraction and clean-up procedures, the selectivity and resolving power of the chromatography (Sections 3.4.3 and 3.4.6) etc. In the meantime, this bottom-up approach does at least provide a useful guideline when there is not sufficient time and/or availability of blank matrix and analytical standard to conduct the top-down determination. Less stringent bottom-up criteria (e.g. lower values of the minimum required number of identification points) may be considered fit for purpose in some cases, e.g. when it is already known that a certain class of compounds is present.

A special case of problematic identification of analytes is that of proteins in biological extracts, in which proteolytic peptides are used as surrogate analytes for quantitation (a reflection of practical limits on molecular mass in trace level quantitation, see Section 1.5). In the case of 'shotgun proteomics' (Section 11.6) the global protein extract is digested by a specific enzyme (typically trypsin) before LC–MS analysis, so that all connectivity between the target analyte (protein) and surrogate analyte (peptide) is lost. This can lead to serious uncertainties in analyte identification for protein quantitation that call for strict criteria (Section 11.6) over and above those applied for the small molecule case.

Whatever approach (top-down or bottom-up) is used, it is essential to clearly define and document the intended identification criteria before starting the analyses. These agreed criteria should be tested as much as possible through validation procedures. If the criteria can not be validated with acceptable reliability, then new criteria will have to be considered and the process of method development and validation will have to be repeated. As the final choice, and the underlying rationale and process for choosing the final identification criteria, could be aggressively challenged in an adversarial environment the entire planning and validation process should be thoroughly documented.

9.3.4 Balancing Thoroughness vs Time and Resources

Some important parameters that must be taken into account in balancing the degree of analytical thoroughness vs available time and resources (financial and

otherwise) include the analytical objectives and requirements (Sections 9.3.1–9.3.3) and what is at stake if these are not met, the nature of the analysis as one that is likely to be repeated for additional samples in the future or is simply a one-time request, the number of samples to be analyzed and the required timelines, the time required vs that available for method development and validation, the applicability of high throughput automated methodologies etc.

Approaches to minimizing costs for a pre-determined level of risk of quantitative uncertainty have been described (Ramsey 2001; Fearn 2002). In these schemes the optimal strategy is taken to be the one that minimizes the expected total cost, which includes the analytical cost plus the losses anticipated as a consequence of using the analytical results for their intended purpose. Lower cost analytical procedures usually have larger probable errors, which in turn typically lead to larger expected losses arising from correction of the actions taken, so a trade-off between the two parts of the total cost is required. At the planning stage the uncertainty ranges in the candidate analytical procedures are themselves uncertain, as are those in the predicted financial losses arising from use of the analytical results. Therefore, the problem amounts to making an optimal decision based on input information that is subject to degrees of uncertainty which are themselves not well defined. Both of the cited approaches (Ramsey 2001; Fearn 2002) are developed for situations in which the intended use of the analytical result is to make a decision based upon some limiting concentration of analyte, e.g., a regulatory upper limit for some pollutant in a foodstuff, or a lower limit for the amount of the active principle in a pharmaceutical formulation. The consequences of both false positive and false negative analytical results must be estimated with respect to the likely subsequent costs, e.g., in the example of a foodstuff a false positive could result in unnecessary rejection of a production batch, while a false negative could lead to legal proceedings over undetected contamination and/or loss of reputation of the food company. As emphasized by the original authors, it is unlikely that such approaches can be developed to the stage where an analytical method can be chosen so as to minimize the probable total cost via an exact mathematical calculation; however, when appropriate even crude economic estimates of this kind can provide a valuable starting point.

From a practical perspective, several key factors will need to be considered in order to determine how much time should be invested in the development of the method. Assuming that the method criteria and reporting requirements are the same, the number of samples to be analyzed will have a direct impact on the method objectives and development strategy in terms of whether or not time and resources should be dedicated to developing a high sample throughput and/or automated method. For example, if the total number of samples to be analyzed is small, then a method capable of analyzing as few as 10–20 samples per day may be perfectly adequate. On the other hand, if thousands of samples need to be analyzed over a relatively short amount of time, as is the case for many bioanalytical methods supporting drug development studies, then high throughput methods capable of analyzing hundreds of samples per day may be required. The time that a laboratory requires from the receipt of sample to reporting of final results is typically referred to as the turnaround time and, even in instances where there are relatively few samples to be analyzed, a requirement for fast turnaround time may, in turn, require that a method with a relatively fast analysis time be developed.

In addition to the number of samples to be analyzed, another factor that must be taken into account is whether or not the method will be used subsequent to its initial application over a long period, or if there will be a requirement to transfer the method to another laboratory after validation. The parameter describing method performance in this regard is method ruggedness, a measure of the degree of reproducibility (% RSD) of the results obtained under a variety of operating conditions, including different laboratories, instruments, instrument operators etc. (This terminology is replaced by 'intermediate precision' in the appropriate guideline (ICH 2005) of the International Conference on Harmonization of Technical Requirements for Registration of Pharmaceuticals for Human Use.) 'Robustness' and 'ruggedness' are sometimes used interchangeably, but method robustness is generally taken to indicate the demonstrated capacity of a method to remain unaffected by small deliberate variations in method parameters. For example, for an HPLC method if performance remains fit for purpose when organic content in the mobile phase is adjusted by 2 %, the mobile phase pH by 0.1 pH units and column temperature by +5 °C, it could be declared suitably robust. Another example (Heller 2007) discusses the role of matrix effects in restricting the degree of ruggedness (level of success in transferring between instruments or laboratories) of LC/MS assays employing API methods. Further discussion of robustness and ruggedness can be found in Section 9.8.4.

The compromise between a higher precision (and thus higher cost) analytical method and lower financial risk can sometimes be significantly shifted by improving the quality of the chemical analysis (using the same optimum method) as it is actually performed in the laboratory. This can potentially be achieved by using standardized methods

supported by laboratory accreditation with strict internal quality control and proficiency testing requirements. For example, many jurisdictions require that analytical data submitted to regulatory bodies to support claims of environmental regulatory compliance must be obtained by fully accredited laboratories (see Section 9.4.2b).

However, even if successful, this accreditation approach does not necessarily guarantee a match of the laboratory performance to the particular requirements for all situations. Generally the intuitive judgment of an experienced investigator is relied on to select the most appropriate analytical method (from sampling protocol to final analysis and data processing), based on the objectives of the investigation and the financial resources available. This approach is obviously dependent on the experience of the investigator but may not provide a transparent logic that can be followed by a less experienced investigator, or a method to objectively justify the expenditure (Ramsey 2001). Of course the intuitive approach is the appropriate one for urgent tasks when cost optimization may not be an over-riding concern. Here, as in all situations in which analyses are conducted for a client, it is essential that the analyst fully documents all aspects of the planning and the reasons for the various compromises adopted.

If, at this stage, the analyst feels that it is not possible to perform the analyses of the quality required to meet the customer's purpose within the latter's budget and/or within the time available, it is always best to be frank with the customer about the dilemma and help him/her to decide what to do about the situation. Any compromises that could lead to increased uncertainty in the method should be documented prior to sample analysis, along with how results will be reported in general. If an agreement (documented!) is reached to proceed, development of the analytical method in accord with the agreed analytical purpose can begin.

An example in which the compromise between analytical speed and accuracy is crucial is provided by biomedical analyses (Section 11.5); in some cases there is a possibility of co-elution of the drug analyte with certain metabolites as a result of compromising on-column efficiency to achieve the desired sample throughput. Some common metabolites (e.g. glucuronides) are highly susceptible to collision induced decomposition (CID) in an API interface (in-source CID, Section 6.5), which produces ions indistinguishable from the ionized drug molecules. If this occurs the result determined for the concentration of the surviving drug will be too high (Jemal 1999; Weng 1999; Tong 2001, Liu 2002). In other words, there are intrinsic limits to the extent to which sample throughput can be increased by using higher speed chromatography.

9.3.5 *The Analyst as Expert Witness*

In legal cases, analytical chemists can find themselves questioned by legal experts who possess as little scientific training as the chemist possesses legal training. It is important for analysts to realize that they may be called to testify as an expert witness, e.g. to explain and defend their analytical results that have been introduced as evidence in the case. In a 1996 ASMS workshop an informative presentation by M Kaiser and DG Gibian emphasized the culture shock that can be experienced in such circumstances; a summary (Baldwin 1997) of this presentation is worth reading and a few of the more important points are reproduced here. As emphasized by Kaiser and Gibian (Baldwin 1997), the differences between 'scientific peer review' and review by a 'jury of one's peers' are enormous. Probably most important, the role of legal proceedings is not to establish the degree of scientific validity to be ascribed to a set of measurements and their interpretation, but to determine winners and losers in the dispute. In the courtroom, lawyers are not primarily interested in scientific validity; rather their professional duty lies in doing their utmost to win the case for their clients in the confrontational and hierarchical milieu of a courtroom. Within this milieu the analytical chemist called as an expert witness can expect no respect merely as a result of the years of training, study and experience that were necessary for appointment to his or her present position. The judge, who occupies the top position in the court's hierarchy, can disallow an expert's testimony for a variety of reasons (Baldwin 1997), e.g. the analytical method used had not been subjected to appropriate validation or testing and thus had not been evaluated with respect to its reliability (what we would refer to as the confidence limits for a stated confidence level, see Chapter 8). There is also an ongoing legal dispute in the United States of America concerning the admissibility of 'technical' vs 'scientific' expert testimony (Borenstein 2001) that illustrates the culture gap.

Most scientists are unprepared for these cultural differences between the scientific community, with its search for the most valid consensus interpretation of the available observations, and the courtroom's 'win or lose' attitude. This is an important reason for the advice to document the reasons for choosing a particular analytical method (and for any subsequent changes made in the light of experience gained in testing the method) and for development and enforcement of extensive standard operating procedures (SOPs) in the laboratory before producing the final

results. Some insight into the perspective of lawyers on expert testimony from scientists and analytical chemists in particular, can be found at a website with the somewhat misleading URL www.scientific.org. In fact, this is an online journal (published by faculty and students of the Department of Criminology, Law and Society at the University of California, Irvine) for lawyers concerned with scientific testimony; not unexpectedly this online journal has a strong emphasis on DNA testing, but it also includes tutorials for lawyers on 'Expert Qualifications and Testimony', 'Investigating Forensic Evidence', as well as one on 'GC/MS'. While the last of these is unlikely to teach readers of this book very much about GC or MS, it does provide insight into the communications and culture gap that an analyst must face if called as an expert witness. The same is true in a more general sense of the other tutorials.

9.4 Formulating a Strategy

9.4.1 *Defining the Scope and Method Requirements*

The previous sections of this chapter discussed the need to assess the requirements of an analytical method from the perspectives of the client and analyst based on a fitness for purpose approach taking into account the acceptable tolerance for uncertainty and the end use of the data. This section considers a general overview of basic practicalities that should be considered at the onset of method development in order to meet those objectives, and builds on material discussed earlier in this chapter and in Chapter 2. A more detailed consideration of some of these practicalities is given in Section 9.5. The method scope will also impact the required validation procedures described in Chapter 10. In general, the topics in the remainder of this chapter and the next will be approached from the perspective of developing and validating a novel method for a single analyte, but will also address the added complexity for development of multi-analyte methods when these issues arise.

No matter how urgent a new analytical task might be, it is always best to expend some time in formulating a strategy. Of course if the task will involve a long series of analyses such planning becomes more critical, and will demand proportionately more time and effort than for an urgent 'one-off' task. As discussed at some length earlier in this chapter, a fitness for purpose perspective that includes a consideration of who the end user of the data will be, and how that data will be used for decision making purposes, needs to be clearly understood and evaluated in conjunction with the other technical factors that will impact the final method design. Finally, as much as the analyst would like to believe that there is no problem that could not solved or some type of method that could not be developed in the laboratories, the analyst should also face the tough question that relates to whether or not his/her laboratory is equipped with the appropriate equipment and laboratory infrastructure to address the challenge at hand. An obvious example might be a case where the client requests an assay for several nonvolatile polar compounds in a single run, and the laboratory has only GC/MS capabilities. If there were no straightforward means to derivatize the analytes to make them amenable for analysis by GC/MS, then another laboratory that has LC/MS capabilities would have to be identified. Even if derivatization were possible, a consideration of factors such as analysis time, costs, method ruggedness and limits of quantitation would have to be carefully evaluated versus what might be possible with an LC/MS method.

Beyond technology, the ability to meet the client's needs for a particular project may be limited by the laboratory infrastructure. For instance, if the method required that the laboratory had SOPs and systems in place to support analysis under GLP guidelines (Section 9.4.2a), and the laboratory being considered was not established in this area, the obvious answer to this request would be that the laboratary could not take on the project. A more subtle example along these lines would be one where, even though the project was not being run in accordance with GLP standards, the requirements for method and laboratory facility documentation were just as demanding as those found under GLP. Once again, the responsible analyst should be ready to face the possibility that, even though the scientific knowledge and required equipment are in place, another laboratory with better defined documentation and operational procedures in place might be better suited for the task at hand.

The first technical considerations that the analyst must evaluate are the number of analytes that are to be detected and quantified, their structures and molecular masses, and the specific requirements for limits of detection and quantitation, including the dynamic range of the assay (Section 8.4). Although these may seem to be obvious points at first glance, the answers to these questions will have a profound impact on the method design and feasibility. The next question has to do with the availability of standards and particularly whether or not a Certificate of Analysis (Sections 9.4.4c and 2.2.1) is available. In many instances the client will be able to provide standards of known purity and stability, or they may be available through commercial sources, but in other cases it may be necessary to have reference substances synthesized.

In any event, some type of documentation that establishes the purity, source and lot number and expiration date of each reference standard must be provided or generated through appropriate measures. A more detailed discussion of the minimum requirements for establishing and documenting the source and purity of the reference standard is given in Section 9.4.4. In addition to identifying a source for obtaining the reference standards for the analytes of interest, it is necessary to consider the likelihood that the samples will contain potential interferences, e.g. metabolites in the case of biomedical analyses, breakdown products in the case of environmental samples etc., and if analytical standards are also available for these compounds. Even if the metabolites or degradates are not to be quantified, it is often necessary to demonstrate that their presence in the sample will not interfere and thus bias the quantitative measurement of the analytes of interest (Section 9.6).

The importance of the surrogate internal standard (SIS) has been discussed throughout this book (Sections 2.2.3, 2.6.3 and 8.5.2b) and the type and availability of the SIS will have a profound impact on method design and reliability. In addition to what has been discussed previously, several practical considerations with respect to the SIS are covered in Section 9.4.5.

Of course, none of the issues introduced above can be considered without knowledge of the analytical sample. As well as the nature and properties of the sample itself, the amount of sample available will affect the method design in terms of the required limit of quantification and the size of the sample aliquot that will be used. The practical considerations around the sample as it affects the method design are discussed in Section 9.4.6 along with requirements for sample receipt, storage and use.

Related to the analytical sample is the control matrix that will be used for method development and for preparing QCs and matrix matched calibrators (Section 9.4.7). In addition, the amount of available control matrix will also impact the design of method development and validation studies, especially in instances where a suitable control is not available or in the case where a similar control matrix is very rare and/or extremely costly.

Finally, before any method design or development studies are initiated, the analyst should perform a thorough search of the available literature for published methods of the same analyte or similar compounds (Section 9.4.8). In addition to any published material that may be available, the client should also be encouraged to search for any other method or background information about the analyte. In addition, the analyst should review any existing laboratory SOPs and determine whether they will be sufficient for supporting laboratory procedures and meeting any regulatory or other fitness for purpose requirements.

9.4.2 Considering Prevailing Guidelines or Regulatory Requirements

No analysis is performed in a total absence of any context, even if no reference to previous methods can be found in either the peer-reviewed or 'grey' literature. Depending on the general purpose of the proposed analysis, e.g. early phase drug discovery, drug development, regulatory requirements including enforcement etc., one or more government bodies may have some oversight responsibilities and authority. It is essential for the analyst to be fully aware of the restrictions and requirements imposed by any and all such government authorities when planning an analytical method.

Different jurisdictions (too numerous to attempt to list here but including local, state or province, national, regional like the European Union etc.) can have independent regulations for essentially the same analysis. However, some common underlying principles are generated by the ISO (www.iso.org), a nongovernmental organization that is a network of the national standards institutes of 155 countries on the basis of one member per country, with a Central Secretariat in Geneva, Switzerland, that coordinates the system. An example of the underlying contribution provided by the ISO is a glossary of terminology (ISO 2004a) used in international metrology (including chemical metrology).

9.4.2a Good Laboratory Practice

An essential concept in any laboratory undertaking analytical work intended to meet regulatory requirements, or that may form part of evidence in legal proceedings, is good laboratory practice (GLP). GLP is regulation and, while different jurisdictions may have different definitions and requirements, a general definition for GLP that would probably be common to all might be the following:

> GLP is a quality system of management controls for laboratories and research organisations involved in nonclinical studies conducted for the assessment of the safety of chemicals to humans, animals and the environment; it is designed to ensure the consistency and reliability of results, and specifies the manner in which studies are to be planned, performed, monitored, recorded, reported and archived.

Note that GLP regulations do not attempt to specify every operation conducted in a laboratory. However, they

do have a significant impact on management and operation of the analytical laboratory as well as how procedures and results are documented during method validation and sample analysis. GLP regulations provide a broad overview of how a laboratory should be managed in order that the data produced by that laboratory can be reviewed from the perspective of who conducted the work, how that work was done and the overall traceability (Section 2.2.1) of the final results. Also, if there were any problems during the conduct of the experiments and data production, a description of what was done to investigate and correct those problems, along with an assessment of how the final results may have been impacted, is required. It is also important to emphasize that the GLP regulations expect that it should be possible to audit all aspects of study sample analysis for several years after the work was performed. Essentially the principles of GLP specify commonsense procedures that allow analytical samples, actions taken, results achieved etc. to be fully documented and be accessible in the future. The main difference between GLP and non-GLP work is the level of documentation that is generally required and maintained. The regulations are extensive and cover so many aspects of laboratory and quality management that it is impractical to attempt to summarize all of them here. However, this section will discuss some of the general requirements common to all GLP regulations with an emphasis on how they pertain to equipment and mass spectrometry instrument systems in particular (Boyd 1996a).

The basic components, or chapters, that are in common to all GLP regulations include Facility Management and Personnel, Quality Assurance Unit, Equipment and Computer Systems, Standard Operating Procedures, Reagents and Solutions (generally referred to as solvents throughout this book), Test and Control Articles, Conduct of Study, (record keeping, retention and retrieval) and raw data including any electronic data that may be generated.

Facility Management and Personnel: The regulations do not necessarily specify any minimum qualifications for personnel, but rather that there should be an adequate number of personnel at the laboratory who are sufficiently trained and qualified to perform the work. The requirements that pertain to facility management include not only components related to the actual building and rooms where work is conducted, but also a requirement for a *Study Director* who is the individual responsible for the overall conduct of the study. In many cases a *Principal Investigator* is an individual who, for a multi-site study, acts on behalf of the Study Director and has defined responsibility for only some delegated phases of the study. Management has many responsibilities with respect to ensuring that adequate facilities and qualified personnel are available for the conduct of the study and to assure that a *Quality Assurance Unit (QAU)* is available. The Quality Assurance Unit serves in the capacity of an internal control unit; it is responsible for monitoring each study to ensure that the management of the facilities and all aspects of laboratory operations and data generation are conducted according prevailing laboratory Standard Operating Procedures (SOPs) and applicable GLP regulations.

GLP regulations may also include requirements that address the *Equipment and Computer Systems*. The key components of this section address instrument design, maintenance, calibration, qualification, maintenance and record keeping (Bansal 2004).

Standard Operating Procedures (SOPs) are documented procedures that describe how commonly used tests or activities not specified in detail in study plans should be performed, e.g. weighing analytical standards using a semi-microbalance, maintaining a laboratory notebook, archiving data etc. Additional discussions with respect to SOPs and documenting data supporting sample analysis are provided throughout this chapter and in Chapter 10. Once again, it is beyond the scope of this book to discuss or even list all of the critical SOPs and documentation requirements in support of GLP regulations, but it is interesting to note that three of the major deviations cited by the FDA relate to SOPs that are not available, are not adequate, or are not followed.

GLP regulations also address the preparation, labeling and maintenance of all *Reagents and Solutions* that are used in the laboratory. A record for how all reagents and solutions were prepared (including traceability of all components used in their preparation, Section 2.2.1) and stored must be retained either in the study or facility records. All storage containers should be labeled with the identity, concentration, storage requirements and expiry date. Expired reagents should not be used and should be discarded immediately. Laboratory SOPs are used to establish how the expiry dates for reagents are established, but laboratory management and scientific personnel must be prepared to justify how this was decided.

Fundamental to all GLPs are the regulations concerning the *Reference Standards* that are also referred to as *Reference Substances* or *Test and Control Articles*. Establishing the identity, purity and stability of the reference standard is fundamental to all quantitative analysis and a discussion of this topic can be found in Sections 2.2.1 and 9.4.4.

Finally, the GLP regulations have extensive requirements that address the overall *Conduct Of The Study*, record keeping, raw data (including electronic raw data) and retention and retrieval of records including data, archival requirements. A *study plan or protocol* is a document that defines the objectives and experimental design for the conduct of a particular project (the 'study'), which may include aspects other than analytical chemistry, and includes any amendments made in the light of experience with the original version. The study plan is a detailed prescription for the particular project for which it was designed and, at a minimum, should include the purpose of the study, test facility, study director, identification of the analytical sample and reference standards, sponsoring organization, description of the experimental design, description of the analytical procedures to be used and records to be maintained. Additional components of the analytical portion of the study that should be recorded to support the conduct of the study include the name of the laboratory, analytical objectives, statistical methods, reference standards and internal standards used, description of the sample and method, and finally a description of all calculations and data reporting procedures (text and tables).

The general issue of data quality also has an important international dimension in that, if regulatory authorities in one country can rely on test data developed abroad, exchange of information is facilitated, duplicate testing can be avoided and costs saved, and nontariff barriers to trade can be avoided while still contributing to the protection of human health and the environment. A good introduction to the GLP concept, including some detail, is given in a document developed by the Organization for Economic Co-operation and Development (OECD 1998); the URL given as the source for this freely available document also provides access to several other OECD documents related to GLP and quality management. Other good sources of information on GLP include guidance documents from the US FDA (www.fda.gov/opacom/morechoices/industry/guidedc.htm) and the FDA Preamble to the GLP regulations (FDA 1978).

As described above, the GLP requirements are commonsense precautions to ensure the *traceability* of laboratory data and to document the study with sufficient detail that it can be reconstructed from the archived version. Even if a laboratory is not doing work under GLP guidelines and regulations, a basic understanding of the underlying principles with respect to systems and procedures used to ensure that the data produced is traceable and will stand up to rigorous data audit and scrutiny is invaluable for those conducting quantitative analyses.

9.4.2b Laboratory Accreditation

Laboratory accreditation addresses and certifies the technical competence of a laboratory to conduct specified measurements on a continuing basis in conformity with defined standards. There is an international network of accrediting institutions which mutually recognize one another's accreditation certificates, e.g., the International Laboratory Accreditation Cooperation (ILAC, www.ilac.org), the Asia-Pacific Laboratory Accreditation Cooperation (APLAC, www.aplac.org) and the Canadian Association of Environmental Analytical Laboratories (CAEAL, www.caeal.ca). All of these organizations conform with the ISO/IEC 17011 standard for accreditation organizations and apply the tests for accreditation of laboratories described in the ISO/IEC 17025 standard and its later (2005) alignment with ISO 9000:2000 that specifies the general requirements for the competence to conduct tests and/or calibrations (including sampling, see Section 8.6). The key document here is Section 5.9.1 of the ISO/IEC 17025 standard. (For the ISO/IEC 17025 standard and its later (2005) alignment with ISO 9000:2000, see www.caeal.ca/17025_changes_explained.pdf.) The ISO document is intended for use by laboratories which are developing their management system for quality, administrative and technical operations, and may be of help for laboratory customers, regulatory authorities and accreditation bodies in confirming or recognizing the competence of the laboratory, but are not intended as the last word with respect to accreditation.

Note that GLP compliance (traceability of data and method documentation) may be part of, but not identical with, accreditation that a laboratory is technically competent to conduct the experimental tests (e.g. chemical analyses) that it claims to be able to perform (OECD 1994). A laboratory accredited for specific determinations (e.g., analytical chemistry procedures) according to the ISO requirements (ISO 2005, 2006) or to equivalent standards can indeed be considered to have satisfied many of the GLP requirements. However, certain requirements of the GLP principles are not covered by laboratory accreditation if defined as compliance with the ISO or equivalent standards, e.g. the use of study plans and the Study Director. Some other requirements, although called for under ISO laboratory accreditation, are much more stringent under GLP; these are related to recording and reporting of data, management of data retained in archives to allow complete reconstruction of a study and a program of independent quality assurance (QA) including internal audits of every study. Therefore, data generated by a laboratory accredited solely under the ISO or equivalent standards are unlikely to be accepted by regulatory authorities for purposes addressed directly by GLP regulations.

9.4.2c Method Development and Validation Without Established Guidance

Sometimes a client will approach an analyst with a request that has no connection to any regulatory protocols or similar guidance as to the requirements for method development and validation. If the client can not provide any guidance as to what is required, it is essential for the analyst to explicitly apply all fitness for purpose principles and to meticulously document the rationale for all stages of method development, validation and the manner in which the method will be applied to the client's samples. Often in such situations there are strong demands on turnaround time for large numbers of samples, so any initial screening procedure should be carefully designed to minimize the number of false negatives, leaving detection of the false positives that will inevitably result from such a strategy to a more selective second stage method. This is just one example of the kind of situation that can arise but clearly any such strategy and the reasons for it should be clearly documented in view of the possible need to defend it in court (see Section 9.3.5) or in other situations where the strategy might be challenged.

9.4.3 The Analyte

9.4.3a Structures and Chemical Properties – MS Considerations

Although this book is concerned with applications of mass spectrometry to trace level quantitation, it has already become apparent that structural identification of analytes also plays an inevitable role (Section 9.3.3b). Although nuclear magnetic resonance spectroscopy (NMR) is the gold standard for this task, it requires the availability of milligram amounts of analyte and is difficult to use on-line with chromatographic separations. Mass spectrometry is unique in its ability to provide (admittedly considerably less) structural information for trace amounts emerging from a chromatographic column, via natural abundance isotopolog distributions and (for higher resolving power instruments) accurate and precise measurements of molecular mass (can confirm the molecular formula), as well as library matching (preferably using contemporaneous spectra obtained from reference standards) and/or *a priori* structural interpretation (Section 5.2.1) of full scan MS^n spectra (usually $n = 1$ and 2).

Some physico-chemical properties of the purified analyte (the analytical standard, Section 2.2.1) will determine some broad features of the analytical method. Analytes that are volatile and stable at typical GC temperatures are usually best analyzed by GC/MS; such analytes are generally of lower molecular mass (up to ∼500Da) and also of low polarity or can be made so by suitable derivatization (Section 5.2.1b) though at the expense of an extra chemical step added to the integrated analytical method. Mass spectrometric detection can use positive ions (EI or CI) or, if the analyte is electronegative (possibly as a result of derivatization), using negative ion CI (Section 5.2.1). A low polarity also has implications for details of extraction and clean-up procedures. For analytes with higher molecular mass and/or low volatility and thermal stability, HPLC with either APCI, APPI or ESI (Sections 5.3.3–5.3.6) is currently indicated for the final analytical measurement although HPLC with off-line MALDI–MS (Section 5.2.2) may become viable for some applications. Generally APCI works well for analytes of moderate polarity since the ionization step occurs in the gas phase (the higher effective volatility from fine nebulized droplets subjected to rapid heating was discussed in a text box in Chapter 5). For larger analytes that are acidic or basic, or highly polar for other reasons, either ESI or MALDI are directly applicable; APCI could be used if there is a willingness to accept an additional chemical derivatization step.

9.4.3b Structures and Chemical Properties – Analytical Considerations

It should not be forgotten that the safety implications of the analyte (and/or the matrix) for the analyst is an important consideration. When indicated by available information on this aspect (e.g. a Materials Safety Data Sheet (MSDS) sheet or a published LD_{50}), any proposed analytical method must take into account the safety of laboratory personnel. In extreme cases specialized equipment (fume hoods, glove boxes etc.) may be necessary.

Selection of sample preparation methodologies (Section 4.3) must obviously take into account the chemical nature and properties of analyte and matrix, but also the possibility of related compounds (metabolites, breakdown products etc.) and unrelated matrix constituents that could lead to matrix effects in the analytical step (Sections 5.3.6a and 9.6). These considerations affect the extent of necessary clean-up procedures following extraction, but of course the situation is complicated by the role of the analytical chromatography in possibly resolving

the analyte(s) away from potential interferences and of the susceptibility of the ionization technique to matrix effects. At this stage of initial planning of the analytical strategy without practical testing in the laboratory, the best that can be done with respect to this complex problem will be based on background knowledge and experience of the analyst. The nature of the chromatography to be used will also depend largely on the chemical properties, the most important choices being GC (ideal for volatile and thermally stable analytes) vs HPLC and, in the latter case, normal phase (analytes with a greater affinity for organic than aqueous solvents) or reversed phase (vice versa) or possibly HILIC (suitable for extremely polar or ionic analytes, Section 4.4.2c). In turn, the choice of chromatography strongly affects that of ionization technique (usually EI for GC analyses, and APCI and ESI for HPLC and CE). It is also important to remember that solvent compatibility can be a concern, e.g. if the solvent used for the cleaned up extract is a stronger solvent than the proposed HPLC mobile phase, a solvent exchange step must be included to avoid significant loss of column efficiency, resolution etc.

At this first planning stage it may be advisable to consider the likely benefits of initial development of two analytical methods rather than just one, as an 'insurance policy' against eventual failure to meet fitness for purpose criteria. Of course the price paid for this insurance involves increased initial costs and laboratory effort.

9.4.4 The Reference Standard

9.4.4a Availability and Source

In many instances, the analytical reference standard will be readily available from one or more commercial sources but this is not always the case. Certified reference materials (Section 2.2.2, not natural matrix CRMs but calibration solutions that are solutions of a reference standard at a certified concentration in a clean solvent) must be prepared and characterized in accordance with the strict guidelines in ISO Guide 34-2000 (ILAC 2000). The reference standard may also be supplied by the client and this is often the case when working with proprietary compounds in development. If neither of these two options exist, the standard must be synthesized or in some instances extracted from a natural product or commercial formulation and then purified. The requirements for obtaining a certificate of analysis (Section 9.4.4c) or other documentation that establishes the identity and purity of the analyte are discussed in Section 2.2.2 and also below in 9.4.4c, but the amount of standard available is also an important factor in terms of formulating a strategy for the development of a method which will ultimately have to be validated and used for routine sample analysis.

When using standards that are available in only very limited quantities, the analyst must consider how this might impact the method design and indeed the ability to develop, validate and run the anticipated number of study samples. In addition to the obvious need to prepare stock solutions of sufficient volume and concentration (typically 0.1 mg.mL^{-1} or greater), limited quantities of reference standard might also impact other aspects of the method, such as the type of sub-stocks and spiking solutions that might be used for the preparation of QCs and the standard curve. A more likely problem, which might be encountered in instances where very limited amount of standard is available, has to do with establishing stock stability. If stability of the reference standard in solution is not known, stock stability studies will have to be initiated as early as possible in support of the method validation (Section 10.2.7). Stock stability studies require fresh weighings at various time intervals, so in cases where the amount of standard is very limited the stock stability investigation will have to be carefully planned to balance the need to establish stability for the longest time interval possible with minimum consumption of the standard.

In some instances, especially when working with metabolites of compounds in the early development stage, the analyte will be provided in a solution at specified concentration. Obviously a Certificate of Analysis will not be provided in these circumstances but some supporting documentation that identifies the identity, source, concentration and, at a minimum, the preparation date if no expiration date is available, should be provided. When using standards of this type careful consideration should be given to the original method objectives and client requirements from a fitness for purpose perspective. In addition to the normal documentation requirements described in Section 9.4.4b, the fact that this type of standard was used for validation and sample analysis should be highlighted in the associated validation and sample analysis reports, along with some comments relating to how the accuracy of the final result might be compromised.

The expiration date is also an important consideration in terms of source and availability of the reference substance. Based on the study design, expiration date and amount of reference standard available, the analyst must determine if a new lot of reference standard will be required at some future date to support on-going study sample analyis. If it appears that the standard will be consumed or will expire prior to the completion of the study, the client should be contacted immediately to discuss the need to identify a source for additional material. In some instances, a new

expiration date may be provided and the standard will be 're-certified'.

9.4.4b Receipt and Documentation

A thorough audit of the final data requires a significant review of all records relating to the authenticity and integrity of the reference standard. Documentation of receipt, identity and storage is therefore vital in terms of defending the accuracy of all stock solutions prepared and, ultimately, the calibration solutions that are susequently prepared from them and used for quantitation. Failure to show proper documentation to support the authenticity and integrity of the reference standard could invalidate all of the subsequent methodology and results. The guidelines and material that are provided in this section describe the general requirements for reference standard documentation, but it is ultimately the responsibility of the analyst and management to ensure that the specific requirements in their laboratory are followed.

Upon receipt, the reference standard should be checked for any physical damage due to inadequate packing and protection during shipment, loss of standard due to inadequate/improper sealing and possible contamination due to inadequate separation of multiple standards packaged in one shipment. Also, the recipient should check that any special shipping requirements required to preserve the integrity of the standard were followed. If the reference standard is to be stored frozen, it should be arrive frozen at the laboratory. If not, the temperature upon receipt should be recorded and retained with the rest of the records.

Reference standard documentation (chain of custody, certificate of analysis) should correspond to the information given on the container labels. If labeling is unclear, inadequate or inconsistent with paperwork or nonexistent, the supplier should be contacted for clarification. If discrepancies cannot be resolved and proper identification cannot be made, the reference standard should not be used.

Each container of reference standard should be labeled by name or Chemical Abstracts (CA) number, log or batch number (as provided by the supplier), expiration or reassay date (if any) and storage conditions necessary to maintain the identity, strength, purity and composition of the reference standard. Observations related to receipt of shipped materials and any unresolved discrepancies should be documented. All shipment papers such as bills of lading and express shipment forms should be retained and stored with the rest of the project or study data.

All reference substances are to be stored according to the instructions submitted by the supplier or per the container label instructions. If storage conditions are not provided, the analyst should contact the supplier and request that information.

The temperatures of the refrigerators, freezers and ambient storage areas containing reference standards should be recorded on a regular basis and this information should be readily retrievable upon audit or inspection. For hygroscopic compounds or in other instances when storage in a dessicator is specified, precautions must be taken to ensure that desiccant is not saturated (typically desiccants with color indicators are used) or, in the case where electronic desiccators are used, procedures to ensure that they are functioning properly should be developed and maintained.

9.4.4c Certificate of Analysis

Ultimately all quantitative analyses are traced back to a comparison with a known amount of the analytical standard (Section 2.2.1), most commonly a weighed amount. It is essential to ensure that the analytical standard is indeed the correct compound, is of a high and known degree of chemical purity, and whether or not it contains water of crystallization (in which case weighed quantities must be corrected for the water content) or is highly hygroscopic (i.e. highly variable amounts of water). Confirmation of identity is best achieved using spectroscopic techniques such as nuclear magnetic resonance (NMR) and infrared spectroscopy together with information from mass spectrometry (accurate mass measurement and natural isotopolog abundances to support the assigned molecular formula), while purity analysis by chromatography with mass spectrometric and/or UV (diode array) detection is usually the best choice. In the case of highly hygroscopic material the weighed amount can not be corrected for the indeterminate water content; a strategy employing NMR (Burton 2005) to determine the amount of standard in a solution was discussed in Section 2.3.2.

The requirements for a Certificate of Analysis (COA) were described in Section 2.2.2, with emphasis on matrix CRMs rather than analytical standards, so some of these requirements are not applicable in this case. The crucial parameters for an analytical standard, apart from names and addresses of the certifying body and its certifying officers, material code and batch numbers, hazard warnings, date of certification and expiry date etc., are purity (both chemical and chiral if relevant, Section 2.2.1) and moisture (and possibly inorganic salt or residual solvent) content, and stability information.

In some circumstances an analytical standard that does not meet all of the criteria required for a COA might be fit for purpose. For compounds that are available only in very limited quantities, e.g. metabolites of drugs in early development, it may not be possible to generate the type of purity and moisture data that is typically required for a COA but, depending on the application and experimental objectives, this may not necessarily be required. In any event, some type of documentation that states the estimated purity, source and lot number should be considered to be the minimum requirement for any reference standard used for any quantitative analysis.

In other instances, a lot of material without a COA might be available for use in early method development with the anticipation that a certified lot will be available prior to method validation and sample analysis. In these circumstances it is acceptable practice to begin method development studies, but the analyst should try to obtain any information about the purity that might be available in order to assess early sensitivity and the ability to achieve the required limit of quantitation based on the method being developed. When working with substances of this type it is imperative that all stock and spiking solutions that were made from the initial batch of standard material received be destroyed after the new lot with an accompanying COA is received, in order to prevent inadvertent use of the older solutions.

9.4.4d Assigned Purity

The Assigned Purity, sometimes referred to as the potency or strength, is derived from a correction factor that is applied to neat (as received) materials (requiring weighing) for use in the preparation of stock solutions and is calculated after correcting for purity, moisture content, residual solvents and salt correction factors if applicable. The Salt Correction Factor is used for used for salt forms of an analyte, e.g. a hydrochloride of an amine or sodium salt of an acid, and is equal to the molecular mass of the free acid or base divided by the molecular mass of the salt form for use in determining the assigned purity:

$$\text{Assigned Purity} = [100\,\% - (\%\ \text{moisture content}$$
$$+\ \%\ \text{residual solvents})]$$
$$\times [\text{purity}/100] \times [\text{salt correction factor}]$$

If a % counter ion (or equivalent parameter) is provided:

$$\text{Assigned Purity} = [100\,\% - (\%\ \text{moisture content}$$
$$+\ \%\ \text{residual solvents}$$
$$+\ \%\ \text{counter ion})] \times [\text{purity}/100].$$

Other equations may be used for calculating assigned purity and documented in the raw data, depending on information provided by the supplier or the client directive. If an assigned purity value is provided by the supplier, then a calculation for assigned purity need not be made and only the assigned purity needs to be documented.

9.4.4e Chiral Purity

About 56 % of the drugs currently in use are chiral compounds, and 88 % of these chiral synthetic drugs are used therapeutically as racemates (Rentsch 2002). Only a few of these display significant efficacy differences between enantiomers but enantiomer specific analyses are increasingly important. This raises the question of requirements on an analytical standard for a specified chiral purity; this was discussed in Section 2.2.1 and also in Section 4.4.1d with regard to chromatographic separation of enantiomers. If the analytical standard was synthesized as the racemate the relative proportions of the two enantiomers is known to be exactly 50:50, and enantiomer specific quantitation then requires only the development of a chiral chromatographic separation plus knowledge of the elution order of the enantiomers; the latter generally requires a sample of the analyte that is at least partially enriched in one enantiomer, which can be subjected to, e.g., optical rotatory dispersion analysis.

Note, however, that ambiguity can sometimes arise in all chiral analyses as the result of chiral conversion. Analytes containing chiral functional groups of the type $R_1R_2R_3CH$, in which the hydrogen atom is labile as a result of the chemical nature of the R-groups, can be epimerized and thus partially or completely racemized (Testa 1993). Similarly, functional groups that can adopt E or Z configurations can undergo isomerization (Xia 1999). Such transformations are perhaps not too surprising under the influence of natural enzymes, e.g. removal of glucuronic acid from a metabolite by glucuronidase (Hazai 2004). However, other reports (e.g. Vakili 1996) suggest that chiral conversion can occur without enzymatic intervention. The cited example involved derivatization of Tiaprofenic acid, a nonsteroidal anti-inflammatory drug (NSAID) that is chemically a chiral 2-arylpropionic acid marketed as a racemate. Reaction of the enantiomers with 2,2,2-trichloroethyl chloroformate in the presence of a base, followed by reaction of the resulting mixed anhydride with L-leucinamide in the presence of triethylamine, resulted in partial chiral conversion that was attributed to use of the strong base in the second step; in fact, replacement of triethylamine with the weaker base *N*-methylmorpholine significantly reduced the observed conversion (Hazai 2004).

9.4.4f Storage and Stability

A crucial property of the analyte (and thus of the analytical standard) is its stability, both in the purified form and in solution. Relevant information pertaining to the stability of the pure compound, though not necessarily of the compound in solution in the chosen solvent, may be found on the Certificate of Analysis (Sections 2.2.2 and 9.4.4c). In many instances, instead of an expiration date a re-assay date will be provided. An extreme form of instability is the propensity of the analyte to explode or be set afire; such properties should be available in the MSDS information and the Certificate of Analysis that should specify the appropriate storage conditions under which the stated expiration or re-assay date is valid. In any event, the reference standard should be stored under the same conditions that were used to establish stability and should not be used past the expiration or re-assay date unless other documentation is provided to substantiate the stability of the analyte under the storage conditions used.

9.4.5 The Surrogate Internal Standard (SIS)

9.4.5a General Considerations

The need for a surrogate internal standard (SIS) and the ways in which it is used in practice were discussed in Sections 2.6.3 and 8.5.2. Questions concerning choice and availability of an SIS were introduced in Section 2.2.3 and Table 2.1, and are discussed further below. All of the concerns about chemical and chiral purity, applicable to the analytical standard (Section 9.4.4), apply also to the SIS; however, the absolute chemical purity is not as crucial for the SIS in cases where both calibration and measurement are performed using response ratios of analyte:SIS where the absolute quantity of SIS detected does not enter into the final determination (Section 8.5.2b, essentially the signal from the SIS is used as a normalizing factor for the analyte responses). In the few instances where a volumetric internal standard is required (Sections 2.2.4 and 8.5.2a) the only concern is that the VIS is stable and provides no interferences (direct or indirect) that co-elute with the analyte.

9.4.5b Stable Isotope Internal Standards

An isotopolog of the analyte that has been labeled with 'heavy' element isotopes (usually ^{13}C, ^{15}N or ^{16}O) is the preferred choice as SIS since kinetic isotope effects (text box in Chapter 2) and back exchange of the isotopic labels with natural abundance atoms in the solvent are negligible, and co-elution of analyte and SIS is essentially guaranteed. Labeling by substitution of ^1H with ^2H (D) is often fit for purpose but can be susceptible to both isotope effects and back exchange, and can lead to appreciable chromatographic separation of analyte and SIS with resulting concerns as to whether or not matrix effects are the same for both (Section 9.6). If no stable isotope-labeled SIS is available a close chemical homolog or analog can be used, although this inevitably introduces some uncertainty (Table 2.1) especially with respect to recovery efficiency (F_a' and F_{SIS}' can no longer be assumed to be equal, Section 8.5.2b) and significant uncertainty as to the equivalence (cancellation) of matrix effects since analyte and SIS no longer co-elute.

It is also important to ensure that the degree of labeling (number of atoms in the analyte that have been replaced by their heavier isotopes) is sufficient that the highest mass isotopolog of the native (natural isotopic abundance) analyte does not contribute significantly to the ion abundance at the m/z value corresponding to the lowest mass isotopolog of the labeled SIS (see later in this section). If significant cross-contributions of this kind do exist (also referred to as isotopic overlap) the calibration equation is intrinsically nonlinear with respect to the fitting parameters (Section 8.5.2c).

Finally, for an isotope-labeled SIS it is desirable to have a measured isotopic purity (Section 2.2.3) as high as possible (and particularly to ensure that the SIS contains negligible amounts of unlabeled analyte), although similar remarks concerning the degree of isotopic purity required apply here as for chemical purity of the SIS (see above) when the response ratios of analyte:SIS are used in both the calibration and the measurement itself. In practical terms, if a new lot of isotope labeled SIS is introduced, it is important to compare the degrees of labeling and acquire a new calibration curve if necessary.

Since the cross-contribution between the analyte and SIS needs to be avoided if at all possible, it is worth taking some time to discuss ways that minimize it and also check for its significance during the method development process. Fortunately, software for calculating the relative isotopolog abundances for a given molecular formula is now provided as part of the standard software package for most modern mass spectrometer systems. By simply entering the molecular formula, the predicted relative ion abundances, which can be generated in table form, can be used to predict what degree of labeling is required for a stable isotope SIS.

After the reference standard and SIS are received, a simple and rapid two compound chromatography–MS method (analyte and SIS) can be developed for screening these two compounds in solution. The two solutions can then be analyzed independently at relatively high concentration (S/B >100 for compound injected) as an initial indication for any potential cross-contribution from the

analyte→SIS or SIS→analyte. Ultimately, the impact of either type of cross-contribution will depend on the details of the final method and the relative concentration of the analyte and SIS. Once the final method is established, an extracted sample with only the analyte spiked at the upper limit of quantitation (ULOQ/Blank), and an extracted blank with only the SIS (Blank/SIS), can be analyzed to check for cross-contributions. A further discussion of the use of blanks during method development, validation and analysis can be found in Section 9.5.6b, but for present purposes only the ULOQ/Blank and Blank/IS will be considered.

Obviously, a SIS with a sufficient degree of labeling and sufficient purity to eliminate the potential for cross-contribution would be the first choice for the analyst (Section 8.5.2c), but this ideal situation is not always what is found in real-world method development. However, even if there are cross-contributions between the analyte and SIS, it may be possible to circumvent the problem by making small modifications to the method. For instance, if a small cross-contribution is found from the analyte→IS due to insufficient isotope labeling, then the amount of SIS in the method can be increased and/or the ULOQ can be decreased. If, on the other hand, a small cross-contribution from the SIS→analyte is detected, the amount of SIS used in the method can be decreased.

Such a strategy implies a question as to the acceptable level of such cross-contributions that can be tolerated within a fitness for purpose context in terms of ability to use a simple linear regression on the linear calibration curve equation given as Equation [8.19a]. In general, however, for analyte→SIS cross-contributions (the most common case when an isotope-labeled SIS is used) a maximum contribution at the ULOQ of 5 % of the SIS response at the concentration used in the analytical method is considered an acceptable guideline; if this guideline is exceeded the effect can be mitigated by first increasing the concentration of the SIS used in the method. If that does not produce an acceptably low cross-contribution, the only remaining general option is to drop the ULOQ and thus the maximum possible contribution from the analyte. However, if each of the analyte and SIS contains just one chlorine or bromine atom in its structure, and if the SIS contains at least two isotopic labels (e.g. ^2H or ^{13}C), the SIS isotopolog ion containing ^{37}Cl or ^{81}Br can be used to avoid significant overlap with the isotopic distribution of the analyte. If more than one Cl or Br atom is present, correspondingly more isotopic labels are required. For SIS→analyte cross-contributions (this usually occurs for an isotope-labeled SIS of poor isotopic purity or for a homolog or analog SIS of molecular mass lower than the analyte) a general industry standard for

FDA work will tolerate an SIS cross-contribution of up to 20 % of the response of the analyte being quantified at the LLOQ concentration. Note that these fitness for purpose guidelines are based largely on practical experience without (thus far) any statistical justification. Ultimately this question should be settled by visual examination of the experimental calibration curve together with careful evaluation of the accuracy and precision over the entire range of analyte concentration for the specified SIS concentration used to generate the calibration. In any event, the cross-contributions (if any) must be carefully monitored during all phases of method validation and sample analysis and also must be fully discussed in the method description and final report.

9.4.5c Analog or Homolog Internal Standards

When an isotope-labeled SIS is not available, an analog or homolog of the analyte can be used as the SIS. While this can introduce appreciable bias (systematic error) into the quantitative results (see discussion of Table 2.1), in most cases it is better to use a less-than-perfect SIS than no SIS at all (Section 8.5.1). In addition to uncertainties in relative extraction efficiencies of analyte and SIS, the analyte and SIS will no longer co-elute in the analytical chromatography, opening the possibility of unequal degrees of ionization suppression/enhancement (matrix effects) that therefore do not cancel. Also, matrix effects may vary throughout the course of the analytical run due to late-eluting compounds that are not fully rinsed off the column after each injection.

Poor precision due to matrix effect variability within a multi-sample analytical run, that is not accounted for by the SIS, is an example of what is often referred to as 'curve divergence'. Curve divergence can result from response drift or from variable matrix effects on the analyte and/or SIS. A simple test to determine whether the divergence is associated with the SIS is to perform linear regression on the raw data ignoring the SIS (i.e. simply analyte response vs concentration rather than ratio of analyte/SIS responses vs concentration). If the precision and accuracy for the calibration curve improve, then modifications to the method and/or finding another SIS will be required.

If the particular analog happens to be a structural isomer (Table 2.1), chromatographic resolution is necessary to distinguish the analytical signals for analyte and SIS (identical molecular masses) unless MS/MS is used and the isomers yield different product ions. Since the molecular mass of a homologous structure differs from that of the analyte by at least 14 Da, there is seldom any problem of cross-contributions arising from overlap of the isotopolog distributions (Section 8.5.2c).

A different potential problem with use of an analog or homolog of the analyte as SIS arises in the case of biomedical analyses in which metabolites can have a molecular mass indistinguishable (at unit mass resolution) from that of the SIS (Matuszewski 1998; Jemal 2002). This problem is more fully discussed in Sections 9.4.7b and 10.2.9c in the context of analysis of incurred samples during method validation or to assess a previously validated method.

9.4.5d SIS for Multi-analyte Assays

An additional dilemma with respect to choice of SIS arises when several analytes in the same matrix are to be quantitated simultaneously. From the point of view of using an SIS to correct for (or at least alleviate) uncertainties arising from extraction efficiencies, matrix effects etc., it is clearly preferable to use the best available SIS for each analyte. The same SIS may be used for several different analytes but in these instances the frequency of run failures for one or more analytes may be higher than what can be tolerated. In addition to poor precision for the assay in general, if the sample needs to be re-assayed more than once the amount of available sample may be depleted, resulting in the inability to report results for that sample. As discussed in Section 10.5.3b, failure to meet acceptance criteria for any one analyte triggers a repeat analysis of the samples in order to generate reportable data for that analyte (Dadgar 1995). To address these concerns, additional method development may be required in order to generate a cleaner extract and/or demonstrate method reliability with ruggedness testing prior to and during method validation (Section 9.8.4).

On the other hand, if a unique SIS is used for each analyte the number of SIM or MRM channels that must be monitored in the course of a single chromatographic run can become so large that it is not possible to obtain an adequate number of data points to properly define a chromatographic peak (Sections 4.4.8 and 7.1). Certainly this duty cycle problem can be greatly alleviated through use of time scheduling of the SIM/MRM channels, so that appropriate sub-sets are monitored only over chromatographic time windows in which the corresponding sub-set of analytes and SIS elute. Nonetheless, this is an intrinsic dilemma that becomes particularly pressing when the number of analytes becomes large, e.g. as in so-called multi-residue analyses for pesticides in foodstuffs.

9.4.6 The Analytical Sample

The word 'sample' is used in the analytical chemistry literature (including this book) in a variety of ways, both as a noun and as a verb. The term 'analytical sample' is taken to refer to the raw untreated piece of matter that is presented to the analytical laboratory and comprises both the matrix (the vast majority of the analytical sample that is of little current interest, e.g., the water, protein and salt content of blood plasma, or the silicate, carbonate and other inorganic constituents of soil etc.) and the analyte(s) of interest.

Although the matrix is, by definition, not of primary interest for the analytical task at hand, its physical and chemical nature are extremely important in determining the details of all steps of the analytical method (Figure 1.2), particularly the earlier steps. Thus, acquisition of a representative primary sample to be delivered to the laboratory is relatively easy for homogeneous matrices (usually gases and liquids), but can be challenging for solids including particulates like soils, sediments etc. (Section 8.6). The extraction step involves selective removal of analyte(s) from other matrix constituents of the analytical sample, and clearly the degree of selectivity provided by a particular extraction technique will depend on the chemical nature of both analyte and matrix. Similar remarks apply to the clean-up step, and indeed to all subsequent steps (Figure 1.2) via the matrix effects that have preoccupied us in this book. Examples of matrix components that can be of particular concern include lipids and proteins in biological samples, organic matter in soils etc.

9.4.6a Sample Availability vs Method Requirements

An important practical concern that is easily overlooked is the amount (size) of the analytical sample that is available. If this is too small given the expected concentrations of analyte(s) and the minimum size of the aliquots required per analysis in order to avoid statistical limitations on precision (see discussion of Equations [8.95–8.97, Section 8.6) it will limit the strategic options available to the analyst, e.g., number of replicate analyses, applicability of the Method of Standard Additions etc.

The amount of available sample needs to be considered both from the perspective of how this impacts the method and LLOQ and also from the standpoint of having enough sample for re-analysis in the event of run failure or other sample re-assay directives. Even in cases where an adequate amount of sample is available for re-assay, repeated run failures could lead to depletion of the sample. This is particularly problematic with multi-analyte assays where the probability of run failure due to one or more of the analytes in the assay increases as a function of the number of analytes in the method (Dadgar 1995).

Another example where the amount of available sample is a critical consideration with respect to method design

and procedures is the case where the entire sample is consumed in the first analysis and there is no archive or back-up sample available for re-analysis, if needed. This is commonly encountered with air or incinerator monitoring studies where only one air sampling trap is used for the experiment. Another example would involve the analysis of difficult to obtain samples such as ocular tissue or cerebral spinal fluid. In this type of situation, run failure due to poor method ruggedness or human error can result in extremely costly delays and expense due to the need to repeat the study. In addition to having a limited amount of sample for analysis, complications with respect to method development and validation due to limited or no available control matrix for method development, validation or for the preparation of calibrators and QCs, must be considered (Section 9.4.7c).

The issue of having very limited sample, and the associated need to protect against run failure, can be addressed in a number of ways, including more stringent validation and ruggedness and robustness testing of the method. Additional procedures designed to reduce human error in the laboratory can be implemented as well. For instance, a common practice for critical work such as this is to have other chemists acting as 'witnesses' during critical steps in the method, such as spiking QCs or the internal standard. Other SOPs can be established to reduce laboratory contamination or, in extreme cases, by designating segregated laboratories and equipment for a particular type of analysis.

9.4.6b Documentation of Sample Receipt, Storage and Use

Similar to the need for documenting the receipt and storage of the reference standard (Section 9.4.4b), detailed records for the receipt and storage of the sample are required in order to verify sample authenticity and integrity. The exact requirements for documenting sample receipt and storage may vary depending on what regulations or SOPs may be in effect, but the basic requirements for sample documentation and for ensuring the integrity of the sample that are common to most circumstances are outlined below.

Information with respect to the receipt of the sample shipment should be recorded and maintained with the study or project file. If a Chain of Custody is sent along with the sample shipment, the instructions as to how to complete the form must be reviewed and completed as prescribed. At a minimum, all shipping documentation should be retained, and the condition of the sample shipping container and the condition of the samples themselves should be recorded. For samples that are shipped frozen, a common practice is to record the temperature inside the shipping container immediately upon receipt. Also, the identity of the person who received the sample shipment should be recorded in the sample receipt records. If the samples are not logged in immediately upon receipt, the samples should be stored at the appropriate temperature until they are logged into the laboratory's sample receipt database or records. In this situation it is common practice to record the number of samples received prior to placing them into temporary storage.

After the required procedures (laboratory SOPs) are completed for sample receipt and documentation, the samples must be logged into the laboratory's Laboratory Information System (LIMS) database, or other method that is described in the applicable SOPs. This step is also commonly referred to as 'sample accessioning'. If sample log-in cannot be completed upon receipt or within the time prescribed by the laboratory SOPs, an explanation for the delay should be documented in the sample receipt records. All samples should arrive with sufficient labeling for unique identification (sample ID), including any duplicates received. In many instances, the laboratory will add another label with additional information, such as a unique laboratory ID.

On a comments page or similar form it should be noted whether the paperwork enclosed with the sample shipment contains any of the following: cross-out, write-over, changes without dated initials, or any pencil or whiteout marks. Anomalies should be documented. This includes but is not limited to the date sampled, time points, initials, accession, subject and/or numbers on the client sample label that do not match the applicable sample documentation or protocol. If there is a difference in information between a chain of custody and client sample label, which source was used for the log-in process should be documented. The anomalies must be documented at the time the observation occurs, either during sample log-in or reconciliation of paperwork.

After the samples have been checked and logged into the laboratory's database or similar program or system, they should be stored in a secure location under conditions described in the sample paperwork or applicable study protocol or plan. Depending on the applicable SOPs or regulations, access to the samples may be restricted to specified personnel in order to substantiate the integrity of the sample and final results.

9.4.7 The Control Matrix

As emphasized in Section 8.5, use of a control ('blank') matrix in the preparation of matrix matched calibrator solutions provides several significant advantages over use

of clean calibration solutions of analytical standard in solvent. These advantages include considerable alleviation of uncertainties in extraction efficiencies ('recoveries') and also of matrix effects (Section 5.3.6a); both of these aspects can introduce very serious doubts as to the reliability of the results if not properly addressed. It is highly desirable where possible to test different lots or sources of control matrix to check for relative matrix effects arising from differences between control matrix and that constituting the analytical sample. Other practical aspects are discussed in the following Sections.

9.4.7a Obtaining a Suitable Control Matrix

The availability of a true control (blank) matrix, particularly of multiple lots and/or sources, varies widely and this represents a major consideration in method development, validation and sample analysis. This concern is even greater when there is a possibility of significant matrix effects (suppression/enhancement of ionization efficiency), in which case *relative matrix effects* (Sections 5.3.6a and 9.6) must be tested for by comparing nominally similar control matrices from different lots and/or suppliers.

In some cases a suitable control matrix may not be available, e.g., soils from different sources around the world (and thus widely variable). In some cases it may be necessary to resort to use of a matrix containing known analyte concentrations that lie well below the LLOQ required for the particular analysis at hand, or analyte-stripped or surrogate matrices (Section 9.4.7c), but all of these approaches are problematic and may require additional validation procedures. Unavailability of a truly blank matrix was until very recently a major intrinsic problem in quantitation of endogenous analytes but novel new procedures have been recently developed for this particular application (Section 11.7).

A control matrix, once obtained, is just as important as the analytical or internal standard and should be stored under appropriate conditions to maintain stability, such that any effects on extraction efficiency or ionization efficiency of the analyte do not change with time.

9.4.7b Using the Control Matrix to Assess Selectivity

Selectivity, and its often necessary compromise with sensitivity, has previously been discussed mainly with respect to mass spectrometric detection (Section 6.2.1, also discussion of enhanced resolution quadrupoles in Section 6.4.2 especially Figure 6.12). However, for an integrated analytical method this is just one of the factors that determines the overall selectivity. The use of a control matrix to test the overall selectivity with respect to matrix interferences (direct contributions from matrix components to the analytical signals assigned to the analyte or internal standard) and matrix effects (ionization suppression and enhancement, Sections 5.3.6a and 9.6) are discussed in this section.

It has been emphasized several times in this book thus far that use of a control matrix affords great advantages with respect to providing better estimates of recoveries (extraction efficiencies F_a', Section 8.5), monitoring and correcting for matrix effects (both direct interferences, i.e. a selectivity test, and ionization suppression/enhancement), and quality control (QC samples). However, the analyst should be aware of potential limitations. It was mentioned (Section 5.3.6a) that different lots and/or suppliers of control matrix should be investigated whenever possible to identify any possibility of relative matrix effects, whereby significant differences might exist between the control matrix used to prepare calibrators and the matrix of the analytical sample. This check is most often done in biomedical analyses, in which the matrix is a body fluid like plasma or urine, but would be more complicated for, e.g., pesticide residues in soil where the variation of soil types is much greater. The recommended sequence for screening a new control matrix is to set up a run that begins with a reagent blank (solvent only), LLOQ (analyte extracted from matrix at the LLOQ concentration), extracts of individual controls, extracts of the pooled controls that are intended to be used for subsequent studies and, finally, another LLOQ .

In the biomedical context another check can involve comparison of LC–MS and/or LC–MS/MS chromatograms obtained for control matrix spiked with analytical standard with those obtained for incurred samples, i.e. post-dose samples from the same individual (human or lab animal) as that from which the control matrix was obtained. In this way the presence of metabolites and/or of matrix components that are peculiar to the test subject (individual) can sometimes be detected (see below). When several different control matrices from different sources are available, it can be useful to compare chromatograms from the individual control matrices with that obtained for a pool (mixture) of all of them, since this comparison can highlight regions in the chromatogram that are potentially problematic as a consequence of variations in matrix composition. If undetected, such variations in matrix composition can give rise to false positive identifications or to quantitative results that are significantly in error.

In view of the potential problems summarized in the preceding paragraph, if it becomes necessary to obtain additional control matrix part way through a study (a series of analyses of samples from a complete study,

e.g. a pharmacokinetic evaluation or a study of pollutant levels in environmental samples collected in different places or at different times) an extensive screening of the new batch of control matrix (as described above) will be required.

The benefits of further validation of a bioanalytical method by judicious use of incurred samples have been described (Jemal 2002), with particular emphasis on detection of contributions of formerly undetected co-eluting metabolites to signal intensity in the SIM or MRM channels used to monitor the target analyte and/or the SIS; Table 9.4 summarizes accumulated experience permitting prediction of metabolite types resulting from specific functional groups in the analyte. Such spurious contributions (direct interferences, not matrix effects) will be absent from the samples prepared from control matrix spiked with analytical standard and used for the validation procedures and as QC samples. The following sequence of checks was recommended (Jemal 2002):

(1) Investigate the stability of the incurred sample both before and after sample preparation. This step is intended to check for the possibility that a metabolite can undergo degradation or even reversion back to the original analyte during sample preparation and before LC–MS analysis.

(2) Re-analyze the incurred samples using the same validated method, except that now the LC–MS/MS method is extended to include MRM channels for the most likely metabolites predicted on the basis of the analyte structure (Table 9.4). Observation of signals in these additional channels will alert the analyst to the possibility of direct interference by metabolites in the analyte's MRM channel if the metabolite can undergo CID in the API interface to yield ions indistinguishable from those of the analyte (Jemal 1999; Weng 1999) and which can thus interfere if the metabolites are not chromatographically resolved from the analyte (see Section 9.3.4).

(3) Repeat step (2) but using a different chromatographic method designed to enhance separation of the analyte and its metabolites. This step can be particularly informative for methods designed for high throughput in which such separation can be incomplete or even zero. Indeed this step can succeed in detection of a metabolite that is a simple isomer of the analyte, e.g., $E - Z$ isomers, that are likely to yield responses in the same MRM channel and are thus not detectable in steps (1) and (2).

Another potential complication (Matuszewski 1998; Jemal 2002) arising from metabolites in incurred samples

Table 9.4 Putative metabolites arising from various functional groups in an analyte molecule and their potential for direct spurious contributions to the analytical signal if not chromatographically resolved and subjected to CID in an API interface followed by CID in a collision cell

Functional group in analyte structure	Metabolite and mass shift	CID reactions of metabolite that can lead to spurious signals in analyte MRM channel	Potential for metabolite to lead to spurious signals for SIS in SIM (and possibly MRM) channel
Carboxylic acid	Acylglucuronide (+176 Da)	$[A + 176 + H]^+ \rightarrow$ $[A + H]^+ \rightarrow P^+$	–
Methyl ester	Demethylated (−14 Da)	–	Significant potential for SIS that is a lower homolog
γ- or δ-hydroxy-carboxylic acid	Lactone (−18 Da)	$[A − 18 + H]^+ \rightarrow$ $[A + H]^+ \rightarrow P^+$	Significant potential for SIS that is an anhydro analog
Lactone	Hydroxy acid (+18 Da)	$[A + 18 + H]^+ \rightarrow$ $[A + H]^+ \rightarrow P^+$	Some potential for SIS that is a +18 Da analog
Alcohol or phenol	O-glucuronide (+176 Da)	$[A + 176 + H]^+ \rightarrow$ $[A + H]^+ \rightarrow P^+$	–
Alcohol or phenol	O-sulfate (+80 Da)	$[A + 80 + H]^+ \rightarrow$ $[A + H]^+ \rightarrow P^+$	–
Amine	N-glucuronide (+175 Da)	$[A + A + 176 + H]^+ \rightarrow$ $[A + H]^+ \rightarrow P^+$	–
Amine	N-oxide (+16 Da)	$[A + 16 + H]^+ \rightarrow$ $[A + H]^+ \rightarrow P^+$	Significant potential for SIS that is an oxidative analog
Thiol	Disulfide (−1 Da per thiol group)	$[A + A − 2 + H]^+ \rightarrow$ $[A + H]^+ \rightarrow P^+$	–
Thiol	Methylated (+14 Da)	–	Significant potential for SIS that is a higher homolog
Sulfide	S-oxide	$[A + 16 + H]^+ \rightarrow$ $[A + H]^+ \rightarrow P^+$	Significant potential for SIS that is an oxidative analog
CH₃, CH₂ and CH groups	Oxidation to hydroxyls (+16 Da)	–	Significant potential for SIS that is an oxidative analog
CH₃ and CH₂ groups	Oxidation to aldehyde and ketone groups (+14 Da)	–	Significant potential for SIS that is a higher homolog

In all cases considered here it is assumed that the $[A + H]^+$ ion of the analyte is selected as the precursor ion and that a characteristic product ion P^+ is monitored in the MRM channel. In the case of interference in the analytical signal for the SIS the potential is greatest if SIM monitoring is used and potentially less for the more selective MRM. This table is modified from one shown previously (Jemal 2002).

that will not be detected in conventional method development and validation procedures is that of direct interference in the MRM channel used to monitor the internal standard (SIS) if the latter is a homolog or analog of the analyte (Table 2.1). Functional groups within the analyte structure that give rise to metabolites that can potentially interfere in this way include:

(a) *functional groups yielding metabolites with molecular mass differing from that of the analyte by ± 14 Da (potential interferences with an SIS that is a homolog)*: methoxy, methyl ester and

N–oxide groups can undergo enzymatic demethylation; thiol and catchol groups that can undergo methylation; some methyl (CH₃) and methylene (CH₂) groups can be enzymatically oxidized to produce aldehyde and ketone groups, respectively (−2H + O = +14 Da).

(b) *functional groups yielding metabolites with molecular mass differing from that of the analyte by +16 Da (potential interferences with an SIS that is an analog)*: methyl, methylene and methyne (CH) groups that can be enzymatically oxidized to yield hydroxy groups; amine and sulfide groups that can be oxidized to yield N– and S–oxides.

Of course, coincidence of molecular masses will always result in direct interferences in an SIM channel used for the SIS and thus directly affect the quantitation of the analyte, but not necessarily in an MRM channel for the SIS depending on the nature of the product ions (a good example of the increase in detection selectivity provided by MRM). Another approach to facilitating detection of metabolites is via a similar comparison for pre-dose and post-dose samples from the same test subject. It is also emphasized that all of the preceding discussion refers to direct interferences with the analytical signals for analyte and SIS, not to indirect interferences arising from suppression or enhancement of ionization efficiency. These matrix effects are almost always the result of endogenous components of the matrix rather than of metabolites of the analyte.

9.4.7c Surrogate Matrices

Sometimes it is simply not possible to find a control matrix, e.g., when working with very rare matrices such as ocular tissues. In such circumstances the analyst is forced to use alternatives. A stripped matrix is one that originally closely matched the matrix of the analytical sample but has had all detectable amounts of the target analyte(s) removed, e.g. using an immunoaffinity column specific for the analyte(s) or, less desirably, activated charcoal as a highly nonspecific adsorbent. A surrogate matrix is one that either bears little chemical resemblance to the analytical sample or has been chemically synthesized to resemble it, e.g. synthetic urine (MacLellan 1988; Yan 2007). Use of either of these will require considerable additional effort to demonstrate that the differences between the natural sample matrix and the stripped/surrogate type used as a control do not affect the validity of the overall analytical method when used for sample analysis, i.e., the method must be shown to be robust relative to such variations in matrix.

The use of surrogate matrices was a major concern in biomedical analyses in the context of quantitation of *endogenous analytes*, but this particular problem was solved recently by an ingenious approach (Li 2003a; Jemal 2003) discussed in Section 11.7.

9.4.8 Evaluate Analytical Options

This section is concerned with consideration of the various tools of the trade discussed in previous chapters (particularly Chapters 4–6) in the context of a new analytical task. Considering the nature of the analyte (structure and molecular mass) and the matrix together with the likely concentration range, it is necessary to evaluate the existing instrumentation and capabilities of the laboratory in relevant techniques of sample preparation (Section 4.3), analytical chromatography (Section 4.4), ionization technique (Chapter 5) and mass spectrometer (Chapter 6, particularly the triple quadrupole instruments that are almost always the best choice for a laboratory in view of their unique combination of figures of merit, Section 6.2.3). It is important to remember that, while it is useful to first consider these various tools separately, it must also be taken into account that the chosen 'ideal' techniques must be mutually compatible; for example, while it is possible to use an API source together with a high resolution double focusing magnetic sector instrument, this combination is impractical for any kind of trace analysis if there are any restrictions on availability of time and/or sample availability. Some of these considerations are reviewed in the following section.

It is also important at an early stage to recognize the possibility that the laboratory does not possess the necessary equipment and/or capabilities. The client will need to know about this one way or another as early as possible.

9.4.8a Reviewing The Literature and Previous Methodologies

It is an old adage in scientific investigations that a morning spent in the library can avoid months of wasted effort in the laboratory, an important modification of the 'learning by doing' principle to include 'learning from what others have done'. In this case the search should be for any available literature on the background to the situation that underlies the requirement for the proposed analyses as well as for any reviewed publications or reports describing previous method development and/or validation of related analytical problems. Information on sample preparation and chromatographic separations, as well as stability data for the analyte(s), is of particular interest. It can also be useful during a literature search to keep in mind the possibility of collaboration with other scientists who have experience with this or closely related analyses. In this respect, the client should be encouraged to participate by checking with any background records that might be in their possession or readily obtained. This simple request can often save weeks of method development time but, by the same token, not obtaining data that was previously generated can lead to significant frustration after learning that a problem that has been worked on of for days, or even weeks, was studied previously by another analyst.

9.4.8b Review Existing Laboratory SOPs

Laboratory Standard Operating Procedures (SOPs) are used to describe the procedures used in the laboratory as well as the requirements for documentation of those procedures once carried out. The analytical method can be viewed as one type of SOP but rarely will these methods describe all of the procedures conducted at the laboratory in support of generating and defending a final result. To adequately defend the data derived from an analytical method, all factors that could potentially impact final analytical results must be documented and be verifiable upon final inspection or data audit, which could occur months and, in some cases, several years after the data were originally generated. A valuable perspective that can be used to determine whether or not data will stand up to a rigorous data audit is to imagine starting with the final result and then go backwards to determine if every step or other factor that could contribute to the generation of the final calculated amount could be reviewed and verified upon inspection. It is certainly possible to validate a method and run samples for quantitative analysis without detailed laboratory SOPs, but the final result will only be as good as the supporting laboratory records. Using SOPs that describe the conduct of laboratory procedures along with the documentation requirements for those procedures is the most effective and efficient way to maintain a quality operation that consistently generates accurate and reliable results. This is essentially a discussion of documentation or, in the classic sense, 'the laboratory notebook', and the need to defend data at all levels.

The documentation of critical procedures, data and results is discussed throughout this chapter and also in Section 10.6 but, beyond the requirements that are highlighted in this book, additional record keeping and documentation procedures may be required for adequate support and defense of the final data generated. In addition to verifying that the analytical method was run properly, critical factors such as sample storage and integrity (Section 9.4.6), preparation of standards and solvents (Section 9.5.4), equipment calibration (Section 9.5.1) and software used for final calculations must be taken into account (to name just a few). Many regulatory agencies have specific requirements for data documentation but one of the more widely recognized sets of regulations is that issued by the United States Food and Drug Administration (FDA 1978, see Part 10.7 entitled *Good Laboratory Practices for Nonclinical Laboratory Studies*). Although the applicability of these regulations for quantitative analysis is rather narrow in scope, they can serve as an excellent reference for determining if the necessary SOPs and data documentation procedures that are required to support the final results are in place in the analyst's laboratory. They are also helpful when reviewing the analyst's existing SOPs relative to fitness for purpose when considering if the laboratory infrastructure is adequate for the task at hand in terms of generating and defending the final data.

9.5 Method Development

9.5.1 Instrument Qualification, Calibration and Maintenance

It is often said that 'it is a poor craftsman that blames his tools', but along with that is the understanding that professional craftsmen take very good care of their tools and that they see this as integral to doing a quality job. Quantitative analysis by mass spectrometry is no different in this respect and instrument operation, maintenance and calibration procedures are at the core of this particular concept. The challenges around maintaining tools are more demanding if they are being shared, and these challenges increase with the number of individuals using them. In the analytical laboratory this can be addressed by developing SOPs that describe how the instruments should be run and maintained, as well as by specifying the type of training that is required before someone is permitted to use the instrument. However, even with such thorough SOPs there are certain perspectives and potential issues that should be considered during the normal course of development, validation and sample analysis. For the purposes of this section the focus is on instrument qualification, operation and maintenance for the mass spectrometer, but the basic principles and procedures described here can also be applied to any type of equipment in the laboratory or to other equipment (e.g. a chromatograph) that may be connected to the mass spectrometer. Finally, many of the recommended procedures described below draw significantly from GLP regulations; although these may not be applicable in the laboratory performing the work, they serve as an excellent guideline for the minimum requirements around the proper installation, maintenance, calibration and operation of laboratory equipment.

Obviously, only trained and qualified analysts should be allowed to operate the mass spectrometer. However, in addition to the practical considerations around this topic, an up-to-date record of the qualifications of each analyst should be maintained in the laboratory's facility records, which can be cited as verification that the analyst who performed the analyses in support of the generation of data used to support validation and sample analysis was indeed qualified to do so.

Prior to the initiation of any work it is essential that the analyst is confident that the instrument is properly maintained, calibrated and running at maximum sensitivity; in this regard strategic decisions based on initial optimization and sensitivity checks need to be made. In addition to the instrument maintenance and calibration, the analyst should also consider what types of methods and samples may have been run previously on the same instrument. For instance, if an ion pairing reagent was used by another analyst just prior to developing a new method, the analyst would have to be aware of the possibility of ionization suppression that could result from the residual effects of using that mobile phase. Another example that is commonly encountered is when extremely 'dirty' samples have been run just prior to method development, resulting in loss of instrument sensitivity and/or ionization suppression. To mitigate the impact of problems of this kind it is wise to have available some type of general system suitability check (Section 9.8.3), or other quick way to assess instrument performance, prior to initiating any new type of method development or sample analysis on a particular instrument. It is essential that the mass spectrometer be kept in a state of optimum sensitivity at all times to ensure good data reliability and method performance. Even if the method does not require the best sensitivity that could be achieved, allowing instrument performance to degrade never pays off in the long run.

A list of common terms and definitions that are used with respect to instrument qualification, calibration and maintenance is provided below. More details are provided in cited sections or later in this section.

Calibration: An operational check that generally involves the use of traceable standard materials or test instruments. In the present context this term applies to both calibration of a particular apparatus, e.g. an analytical balance (Section 2.3.1) or the m/z scale of a mass spectrometer (Sections 6.4.1 and 9.5.1b), and to calibration of the response vs analyte concentration (or amount) for the analytical instrument (e.g. GC/MS or LC/MS) or alternatively for the entire analytical method (Sections 2.6 and 8.5).

Equipment (also referred to as instrument or system): Analytical measurement or processing hardware including firmware (Section 6.1). In some cases, a reference to equipment may actually refer to an integrated system that is comprised of two or more instruments. In a computerized system the equipment may be controlled by the computer system (instrument work station). The instrument work station may also collect measurement data from the equipment and be included as part of the overall system definition.

Equipment Qualification (sometimes referred to as equipment validation): Documented verification that equipment is properly installed, performs as specified throughout operating ranges, and reliably executes desired functions.

Installation Qualification (IQ): An early phase of equipment qualification that establishes that the instrument is received as designed and specified, that it is properly installed in the selected environment, and that this environment is suitable for the operation of the instrument.

Nonroutine Maintenance: Any repair or instrument maintenance that is not planned or scheduled.

Operational Qualification (OQ): Following IQ, OQ provides documented verification that the equipment-related system or modular sub-system performs as specified throughout representative or anticipated operating ranges.

Performance Qualification (PQ): Documented verification that the equipment or integrated system is reliably executing the desired function for which the system was procured. This qualification is to be performed on an ongoing basis by the user.

Routine Maintenance (preventative maintenance): Any scheduled maintenance that is performed on a routine basis at defined time and/or frequency intervals.

Validation: Establishing documented evidence that provides a high degree of assurance that a specific process will consistently produce a product or result meeting its pre-determined specifications and quality attributes.

9.5.1a Equipment Qualification

An interesting and useful classification of steps used to assure quality of analytical data (Bansal 2004) has drawn a clear distinction between qualification of apparatus used to obtain the data and validation of the methods developed to use the apparatus to obtain the data pertinent to a particular analytical problem. Overlaid on this distinction is another, that between tests that must be completed satisfactorily before acquisition of the desired analytical data can begin (instrument qualification and method validation) on the one hand, and those that are conducted immediately before or simultaneously with data acquisition (system suitability and quality control checks) on the other. The original paper (Bansal 2004) represented the outcome of a workshop intended to fill a need for a more detailed scientific approach to what was termed 'Analytical Instrument Qualification (AIQ)', particularly in the context of applications in the pharmaceutical industry. Note in particular that qualification of both hardware and software plays an important role in method validation.

AIQ clearly falls into the categories of quality procedures concerned with the apparatus and those that must be undertaken before proceeding to any acquisition of analytical data (including data acquired in the method validation!). The workshop considered some key objectives, including proposal of effective procedures for instrument qualification that focus on outcomes and not only on generating documentation; these procedures should establish the essential parameters for AIQ and be risk-based and founded on competent science. An essential element is to clearly define the roles and responsibilities of all those involved in the various stages of AIQ. The following account of the extensive discussion in the original paper attempts to summarize the more important points.

Four phases of AIQ were identified, namely, design of the instrument or apparatus, its installation in the user's laboratory, operational qualification and performance qualification, each with its own objectives and assignments of responsibility. These phases are not intended to be watertight compartments and some of the required activities overlap two or more phases. Design qualification (DQ) is most suitably performed by the instrument developer/manufacturer, but users/purchasers should ensure that the apparatus is suitable for their intended applications and that the manufacturer has adopted a quality system for developing, manufacturing and testing. Users should also establish that manufacturers and vendors adequately support installation, service and training via vendor audits or vendor-supplied documentation. In addition, informal personal communications and networking with peers at conferences or user group meetings are often highly informative about the suitability of specific instrument designs for various applications and the quality of vendor support services. Informal site visits to laboratories of other users, and more formal visits to facilities of competing vendors, to obtain data on representative samples using the specified instruments are usually considered essential components of the user's responsibility with respect to design qualification.

Installation qualification (IQ) is a more formal procedure that should be documented and applies to an instrument that is new, pre-owned or exists on-site but has not previously been qualified. The documentation should include a description of the apparatus, including its manufacturer, model, serial number, software version etc. On delivery of a new instrument to the user's laboratory it should be carefully checked and documented that the instrument itself and all associated software, manuals, supplies, accessories etc. are complete as specified in the purchase order and that they are undamaged. For a pre-owned or existing instrument, manuals and documentation should be obtained as completely as possible. The installation site should be checked to ensure that it satisfactorily meets vendor-specified requirements; this includes any data storage capabilities and/or network connections. While relatively simple apparatus (see below) can be assembled and installed by users with in-house experts (e.g. IT professionals), for complex instruments these tasks are best done by the vendor or specialized installation engineers. Acceptance of a new instrument should be based on criteria established in advance and specified in the purchase order; for complex instruments vendor-established installation tests and guides provide a valuable baseline reference for such criteria. Any abnormal event observed during assembly and installation should be documented. If a pre-owned or unqualified existing instrument requires assembly and installation, the preceding tasks should be performed by the user or by a qualified consultant engineer. Once acceptable results have been obtained on initial diagnostics and tests, and confirmed by both the installing engineer and the user, this should be documented before proceeding to the next phase of the AIQ.

Operational qualification (OQ) is best defined in terms of an example; that described in the original work (Bansal 2004) is that of a stand-alone HPLC system equipped with a UV detector, and the operational parameters that would be appropriately checked in this AIQ phase include pump flow rate, gradient linearity, detector wavelength accuracy, detector linearity, column oven temperature and precision of peak area and retention time. Test parameters for a mass spectrometer in this phase would include attainable vacuum under operating conditions, resolution–sensitivity characteristics, stability of m/z calibration, response linearity etc. Secure data storage, back-up and archiving should also be tested at the user site according to specified documented procedures. The extent of such operational qualification testing depends to some extent on its proposed applications, i.e., it should be fit for purpose and can be modular or holistic. Modular testing of individual components of a system (e.g. the chromatograph in GC/MS and LC/MS systems) may facilitate interchange of such components without re-qualification and should be done whenever possible. In contrast, holistic tests involve the entire system. Note that operational qualification tests are intended to initially verify the operation of the instrument compared with documented specifications within the user's environment, and are more demanding than routine analytical tests (system suitability tests, Section 9.8.3). When the instrument undergoes major repairs or modifications, relevant operational qualification tests should be repeated to check whether the instrument continues to operate satisfactorily.

The AIQ phases described thus far are intended to check that the instrument is capable of general performance that meets expectations. The performance qualification (PQ) phase checks on the continued suitability of the instrument for the kinds of application (e.g. bioanalytical, food safety, environmental etc.) to be conducted. Several kinds of procedures fall under this category. Performance checks are set up to verify an acceptable performance of the instrument for its intended use, i.e., are based on typical on-site applications and are performed routinely on a working instrument, not just on a new instrument at installation. Therefore, performance checks can be slightly less rigorous than those applied for operational qualification (see above), but should test for trouble-free instrument operation vis-à-vis the intended applications. Again, these tests can be modular or holistic. Testing frequency depends on the proven degree of ruggedness of the instrument and on the critical nature of the instrument's performance for the applications to be conducted. The same performance tests should be repeated each time so that it is possible to track the performance history of the instrument. Some system suitability tests or quality control checks (i.e., checks that run concurrently with the test samples) also imply that the instrument is performing suitably, but are best regarded as supplements to periodic performance checks. The performance qualification tests should be repeated after maintenance or repair to ensure that the instrument remains qualified. It is recommended that standard operating procedures (SOPs) should be established to maintain and calibrate the instrument, including appropriate record keeping of such activities.

When changes to the instrument and software are recommended when manufacturers add new features or correct known defects, users should carefully assess these changes and adopt only the changes they deem useful or necessary. The laboratory should establish a Change Control process to be followed in such circumstances. Operational and performance criteria should be carefully re-checked and documented after any such changes.

Responsibility for the instrument operations and data quality rests with the "users group", which includes analysts, their supervisors and the organizational management. Users should be adequately trained in the instrument's use (and their training records should be maintained as required by the regulations) and should be responsible for qualifying their instruments. Their training and expertise make them the best-qualified group to design the instrument test(s) and specification(s) necessary for successful AIQ. The users must also maintain the instrument in a qualified state by routinely performing required maintenance procedures and performance quality tests. The role of the quality assurance group in AIQ is the same as that in any other regulated study, i.e., they should acquire an understanding of the instrument qualification process by working with the users, and should review the AIQ process to determine whether it meets regulatory requirements and that the users attest to its scientific validity. The instrument manufacturer is responsible for design qualification and also for validating relevant processes for manufacturing and assembly of the hardware and for validating software associated with the instrument, as well as the stand-alone software used in analytical work. The manufacturer should also notify all known users about hardware or software defects discovered after release of a product, offer user training and installation support, and invite user audits as necessary.

It is a major theme of this book that every piece of apparatus used in an analytical method is as important as any other in assuring the quality of the analytical data. However, it is self-evident that qualification of, e.g., a volumetric standard flask is less arduous than that of a tandem mass spectrometer. It was proposed (Bansal 2004) that apparatus could be classified with respect to the level of effort required for its qualification, although to some extent the assignment to these categories will vary somewhat with the particular needs of a laboratory. Group A instruments, e.g. light microscopes, magnetic stirrers, mortars and pestles, nitrogen evaporators, ovens, spatulas and vortex mixers, require no independent qualification other than visual inspection. Group B instruments require relatively simple qualification and causes of their failure are readily discernable by simple observations. Examples include incubators, infrared spectrometers, melting point apparatus, muffle furnaces, pH meters, refrigerator–freezers, thermocouples and thermometers, vacuum ovens etc. Group C instruments require qualification procedures that are highly method specific, and these should be performed using a full AIQ process as outlined above. Examples of such instruments include diode-array detectors, gas chromatographs, HPLC instruments, mass spectrometers etc.

For integrated systems that are comprised of different pieces and types of equipment, the configuration of the system at any given time should be documented in the equipment maintenance logbook for the main system component. This procedure allows individual pieces of equipment to be exchanged as needed but preserves the ability for a reviewer to determine how the system was configured on any given date. This procedure also allows the laboratory to keep a complete maintenance and qualification record for any given piece of equipment in its own maintenance logbook and facilitates the review of historical maintenance and qualification procedures.

9.5.1b Mass Spectrometer Calibration

The importance of careful calibration of the m/z scale of a mass spectrometer was introduced in Section 6.4.1, together with cautionary comments concerning automated calibration routines that rely on peak-finding algorithms. An example where an error could easily occur concerns a mass peak for an organic calibrant near the crossover point ($\sim$1350–1400 Da) where the all-^{12}C ion ('monoisotopic ion') no longer corresponds to the dominant peak in the natural abundance isotopic cluster; this problem was described in Section 1.5. It is good practice to conduct regular checks on instrument calibration and to re-calibrate if necessary. The instrument does not necessarily need to be calibrated every day and, in fact, some instruments will maintain their calibration for several weeks, if not months. The frequency for required calibration should be based on the type of instrument, historical data and any specific requirements that are established in the laboratory SOPs. In the case of magnetic sector analyzers operated at high resolving power, in which magnet hysteresis (Section 6.4.4) plays a significant role in limiting the precision and accuracy with which an SIM channel can be re-set, it is necessary to ensure that a suitable lock mass ion is always present so that the m/z scale can be continuously adjusted to reflect the contemporary field settings and ensure that the SIM does indeed monitor the desired peak maximum.

For tandem quadrupole instruments it is recommended, but not required, that both Q_1 and Q_3 be calibrated separately but at the same time. For some instruments, a separate calibration is required for the positive ionization and negative ionization modes. Solutions of polypropylene glycol (PPGs) are often used for LC–MS calibration but any solution containing a mixture of compounds of known mass and stability that covers the mass calibration range (see Section 6. 4.1) is sufficient provided that the solutions used are traceable and that the final calibration results are documented. At a minimum, the difference between the found and expected mass based on the previous calibration should be recorded along with the solution ID and/or lot number for the solution that was used to calibrate the instrument. A record of all calibrations should be kept in either the equipment maintenance logbook, in another notebook (or equivalent) that is stored near the instrument or archive (see above discussion regarding the instrument logbook).

9.5.1c Maintenance

Appropriate maintenance of a mass spectrometer makes sense from a scientific viewpoint and also for financial reasons, since the mass spectrometer is by far the most expensive single apparatus in the laboratory. The mass spectrometer and peripheral systems such as GC, LC pump, autoinjectors etc. need to be well maintained with the objective of keeping them at a performance level at or near to what it was on the first day of installation. In addition, procedures need to be in place to ensure that the instrument performance does not deteriorate or drift over time. Any routine maintenance procedures that are suggested by the instrument manufacturer should be followed in addition to any routine maintenance procedures that are established at the laboratory based upon how the instruments are used, what types of samples are typically run and the number of users that have access to the instrument, along with a consideration for their level of training and expertise.

Maintenance procedures should be distinguished as routine or nonroutine. In the event of a nonroutine operation, the record should indicate the nature of the failure, how and when it was discovered and any remedial action taken in response to the failure. All records must contain the date of maintenance and the initials of the person performing the procedure. A record of all routine and nonroutine maintenance should be maintained for each piece of equipment or system, e.g. LC–MS/MS with autoinjector, in the laboratory. An instrument maintenance logbook should be maintained for each mass spectrometer system and be readily available. In addition to any requirements in the laboratory's SOPs, the following information should be included in the equipment logbook: the equipment identification, model number, serial number, the date the equipment was put into service, current location, reference(s) to any relevant SOP(s) and contact information for the equipment manufacturer and/or service organization.

9.5.1d Validation of Computerized Laboratory Instruments and Equipment

In modern practice, which is so heavily dependent on instrumentation controlled by computers, validation and change control of the computer equipment and software are essential for demonstrating that data are processed accurately and consistently. When reviewing data it should be possible to determine from the documentation which computer equipment and software were used to process the data and generate the final results. The computer and software should be validated prior to use and documentation of all validation tests conducted should be readily available. Instrument manufacturers maintain internal change control and validation of programs in accordance with ISO and other good manufacturing standards, but ultimately it is the responsibility of the end user to validate the system in the operating environment in which it is used.

An excellent overview of the problem, particularly as it affects bioanalytical data and information submitted to the US FDA, was published by the Society for Quality Assurance (SQA 2000). The relevant regulatory documents on electronic documents and signatures are found as Part 11 of the US Code of Federal Regulations (CFR21 2003) and an FDA Guidance document is available (FDA 2003a). (Note that the FDA interprets the word 'equipment' to include hardware and software as well as instrumentation.) The review of qualification of analytical instrumentation (Bansal 2004) includes a section on software validation and the same phases (DQ, IQ, OQ and PQ, see Section 9.5.1a) apply.

While complexity of automated equipment can vary widely, the basic concepts of validation are the same although the extent of testing required will vary based on its functional requirements. For example, single task nonconfigurable instruments, such as a pH meter, will have minimal validation requirements that can be described in a single document. In contrast, more complex instruments such as a mass spectrometer or HPLC that are capable of many functions, such as analyzing a number of samples while maintaining associations with pre-loaded sample identification tags, storing data for reprocessing and possibly transferring information to a central laboratory information management system (LIMS), require a much more demanding validation procedure with usually separate documentation of aspects like functional requirements, validation plan, installation and qualification testing. If an instrument provides some functions that are not required by the user, these should be clearly identified as exclusions in the test plan and should be disabled and/or clearly identified as such in user manuals and written procedures to ensure that they are not inadvertently used.

All regulatory agencies adopt what amounts to a 'life cycle' approach to developmental and operational control of computerized equipment, with emphasis on documentation of software development and quality management; especially in GLP environments. The 'life cycle' for analytical instruments from the user's perspective includes the purchasing phase (including system/vendor qualification), the IQ, OQ and PQ phases of AIQ (Section 9.5.1a), the maintenance phase and, finally, the retirement phase, which probably occurs more frequently with respect to computers and software than the analytical instruments

themselves. Regulatory requirements for data and records retention and retrieval make the data storage and retrieval aspects of instrument retirement a crucial issue that should be considered at the time of purchase relative to the requirements of the appropriate regulatory body.

Somewhat different software validation approaches are appropriate depending on whether the component is firmware, instrument control software, or data processing software. Firmware (low level software that controls an integrated chip in an instrument of some kind that is not accessible to the user) is considered a component of the instrument itself, and since qualification of the hardware (Section 9.5.1a) is not possible without its firmware, when the hardware (analytical instrument) is qualified the integrated firmware is automatically validated also. Any changes made to firmware versions should be tracked through the Change Control process for the instrument (Section 9.5.1a).

As emphasized previously (Bansal 2004), software for instrument control, data acquisition and processing is nowadays usually all loaded on a computer connected to the instrument. Thus the functions of both hardware and software in providing analytical data are inextricably intertwined and the DQ should be performed by the manufacturer, who should also validate this software and provide users with a summary of the validation. However, in the installation phase, qualification of the entire instrument and software system is more efficient and indeed meaningful than modular validation of the software and hardware separately. Thus, the user qualifies the instrument control, data acquisition and processing software by testing the instrument according to the AIQ process described in Section 9.5.1a.

Many analytical laboratories also use a laboratory information management system (LIMS), an example of stand-alone software. Validation of such software is administered by the software developer but user-site testing is also an essential part of the software development and validation cycle. The FDA requirements for these systems are described in a guidance document (FDA 2002).

Change Control for changes to the firmware and the instrument control, data acquisition and processing software follows the DQ/IQ/OQ/PQ classification process for the affected instrument but for stand-alone software requires user-site testing of the changed functionality. While changes to the instrument and software are common as manufacturers add new features and correct known defects, implementation of all changes may not always benefit specific users who should therefore implement only the changes they consider to be useful or necessary.

The Change Control process enables them to do this in a controlled and documented manner.

9.5.2 Instrument Optimization

Modern practice of trace level quantitative analysis is possible only because of the sensitivity of modern mass spectrometers, so it is important to ensure that they are tuned to optimum performance. Note, however, that the limiting factor in determining limits of detection, for example, is usually the chemical background in the chromatogram (see discussion of ion beam shot noise vs chemical background in Section 7.1.1). Tuning of a mass spectrometer is related to calibration of the m/z scale (Section 9.5.1b) but refers instead to adjustment of instrumental parameters in order to optimize performance characteristics such as peak shape (and thus resolving power) and ion transmission efficiency (and thus instrument sensitivity). Most tuning involves adjustment of electrical potentials on lenses and other components of the ion optics (Section 6.3.2). Adjustment of the voltage on the electron multiplier (V_{SEM}, Section 7.3) to set the gain is not usually included in this category of tuning parameters but is a separate issue, although some autotune algorithms do include V_{SEM} adjustment. Other nonelectrical tuning parameters can include choice of CI reagent gas (Section 5.2.1), distance from an ESI needle to the mass spectrometer entrance orifice or flow rate and temperature of nebulizing gas, or the potential applied to a collision cell (and thus the collision energy) and the pressure of collision gas. It is useful to distinguish between tuning parameters that are compound independent (these are generally parameters involved in ion transmission through the mass spectrometer) and those that are compound dependent (mostly those that control efficiency of ionization and transmission of ions from the API region into the mass spectrometer, also ion fragmentation in MS/MS mode). However, there are no iron-clad rules that apply to this distinction and to some extent an experienced analyst will develop an understanding of which parameters are strongly compound dependent and which are not. Note that in GC/MS most tuning parameters are compound independent. However, in environmental analyses in which EI–GC/MS with quadrupole mass filters (or even ion traps) are used, a tuning compound such as DTTPP (decafluorotriphenylphosphine) may be used to tune for specified ion abundance ratios across the m/z range to provide consistency when using library matches for compound identification. (These EI libraries were mostly assembled using data obtained from magnetic sector instruments for which the mass discrimination characteristics can be significantly different from those for the quadrupole analyzers.)

Experience with a particular instrument will also develop an understanding for which tuning parameters are most likely to lead to the greatest gains in sensitivity if carefully optimized for a particular method. In other words, it is frequently the case that some tuning parameters can be adjusted even slightly to provide large sensitivity gains for a given analyte, but other parameters might provide only find 5–10 % increase, in which case the experienced analyst will not waste time on those unless desperate to find more absolute signal.

Most modern mass spectrometers nowadays are provided with an 'autotune' algorithm of some kind, analogous to the automated calibration routines mentioned in Section 9.5.1b, using some user-selected tuning compound or mixture. These autotune facilities are useful but the analyst should not rely on them to provide the best possible tuning under all circumstances, e.g. the fit for purpose compromise between sensitivity and peak shape (and thus resolution). Just as for the m/z scale calibration, it is wise to review the results of an autotune to ensure that they make some sense relative to the known behavior of the particular instrument. (It is possible for an autotune routine to find a false optimum, similar to the problem of false minima in fitting experimental data to a specified algebraic relation by minimizing the squares of deviations, see the text box on nonlinear regression in Chapter 8.) However, there are advantages to the autotune routines, e.g. they provide a means for electronic storage of the optimized parameters that provide a sensible starting point for later work. Similarly for a given analytical method it is possible to optimize several MS/MS transitions and store the relevant parameters for each in an electronic file; an example of optimization of collision energy for four MRM channels for the same precursor ion is shown in Figure 9.5. If the MRM channel that originally provided the best signal was later found to have problems, e.g. unexpected levels of 'chemical noise (background)', or lack of robustness (Section 9.8.4) with respect to drifts in the electrical potentials that control the collision energy (exemplified by the 4th channel in Figure 9.5 that exhibits a fairly sharp maximum), the other optimized transition(s) may be quickly retrieved and evaluated to determine if better selectivity, robustness etc. can be achieved relative to what was originally chosen (Section 9.5.3b). As always, the judgment of the analyst in a fitness for purpose context should be the final arbiter in deciding the extent to which automated routines of any kind are useful.

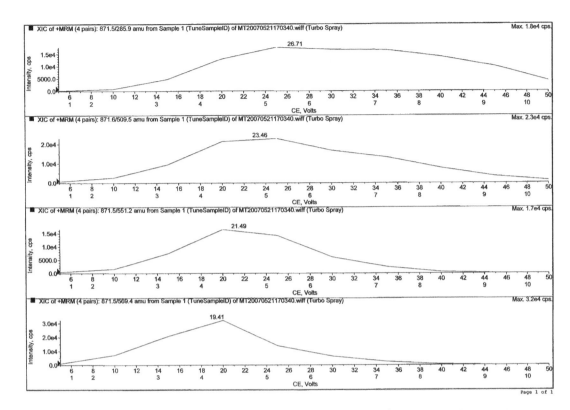

Figure 9.5 Illustration of optimization of collision energy for CID of m/z 871.5 ([M+H]$^+$ ion of paclitaxel) for four MRM channels; m/z values of product ions from top to bottom are 285.9, 509.5, 551.2 and 569.4. The optimum collision energy in each case is noted at each maximum. Note that the MRM channel providing the highest signal of 3.2×10^4 cps (m/z 569.4) also displays a rather sharp maximum in the signal vs collision energy graph and is less likely to be robust with respect to drifts in collision energy than the others that have broad flat maxima.

9.5.3 LC–MS/MS Optimization for MRM Analysis Using QqQ

In view of the dominant role currently played by triple quadrupole instruments in trace quantitative analyses (Section 6.4.3), it is appropriate to devote special attention to optimization (including instrument tuning) of LC–MS/MS performance for this special case. The present section is concerned with providing some general guidelines for optimizing the mass spectrometer (including the ion source) for LC–MS/MS analysis; chromatography development is discussed in Section 9.5.5. Additional information and training support for a specific instrument platform should be obtained from the instrument manufacturer.

Prior to beginning, the monoisotopic molecular mass and compound structure of the analytical standard should be checked again; even if this information is provided by the supplier, and in almost all cases it is, nonetheless it is good practice to calculate the molecular mass independently to verify that the information provided is accurate. Note that both ESI and APCI can produce adduct ions of various kinds (Section 5.3.3 and Table 5.2) and this could be misleading in this check (see also Figure 9.6).

The molecular structure should be evaluated for any structural information that may be useful during the optimization process. For instance, even though both positive and negative ionization modes should be checked routinely in almost all instances, an understanding for whether the compound is neutral, basic or acidic will be useful in terms of predicting which ionization mode can be expected to provide the best sensitivity and selectivity. For some analytes, it is difficult to predict whether or not the best response will be in the positive or negative ionization modes, but if there is no response in the ionization mode that is predicted based on structure then factors such as whether or not the correct reference standard was used for preparing the tuning solution, or if the reference standard could possibly be unstable in the solvents used in its preparation, should be investigated. The compound structure should also be evaluated to determine whether it contains any elements such as chlorine, bromine or boron, which would result in a unique isotope pattern (see the text box on isotope patterns in Section 8.5.2c). If the predicted isotope pattern is not observed (remember to include the adducting ion (Table 5.2) in the formula used to calculate the isotopic distribution), then the analyst would have to expect that either the incorrect reference standard was used or that the response from the compound was weak and the observed isotopic pattern was distorted as the result of background interference. The analyte structure also can provide a good basis for the initial choice of ESI or APCI as the ionization method likely to provide the stronger response in LC/MS (Section 9.5.3a). In general, ESI is favored with respect to sensitivity for analytes with any likelihood of thermal lability and responds well if the structure includes strongly acidic or basic functional groups. APCI is often usually more sensitive if the most polar functional groups are of moderate polarity (hydroxyl groups are a common example) and if the molecule seems likely to possess some degree of thermal stability. APCI is also less susceptible to matrix effects and this can be an important criterion in some cases (e.g. with respect to robustness and ruggedness).

It is also important to have a sense for whether or not the instrument is performing at the level of sensitivity that would normally be expected. If only one person has been running the instrument he/she will probably have good idea of its current level of performance. However, if other analysts had been running other methods and samples it might be difficult to assess poor response, in terms of whether sensitivity loss reflected instrumental problems or was related to the compound or tuning solution being used. To address this dilemma many laboratories use a tuning check solution containing a compound(s) of known stability and response under a defined set of operating conditions. If this approach is used, the check solution can be infused just prior to optimization of a new compound to verify that the mass spectrometer is performing as expected.

The basic steps performed for optimizing a new compound for analysis include: optimization of the API interface and Q_1, generating product ion scans at different collision energies and selecting product ion candidates, optimization of collision cell conditions (gas pressure and collision energy) in SRM mode, optimization of Q_3 (if

necessary) and final optimization of any flow dependent parameters.

9.5.3a API Interface and Q_1 Optimization

The initial optimization of the API interface and Q_1 parameters is usually performed by infusing a tuning solution of the analyte. Typically, a syringe pump is used to deliver a constant flow between 10–50 μL/min; note that this is appreciably lower than flow rates commonly used in LC–MS analysis and that absolute ionization yields can be flow rate dependent (Section 5.3.6b). Since the chromatography mobile phase is generally not known at this stage, a tuning solution containing approximately 50 % acetonitrile or methanol in water with 5 mM of ammonium acetate can be used. Other solvent compositions can be used providing that they contain low concentrations of a volatile buffer based on acetic or formic acid plus ammonia to provide pH conditions that will enhance ionization efficiency for ESI (Section 5.3.6). Stock or sub-stock solutions (Section 9.5.4) can be used for preparing tuning solutions. The concentration of the tuning solution should be high enough to ensure a strong and stable signal but not so high that it would result in detector saturation. Use of tuning solutions that are too concentrated can also result in a lingering elevated baseline for several hours after optimization and this prevents the analyst from gaining information about sensitivity under the initial chromatography conditions (Section 9.5.5) until the background returns to normal. If a stable isotope SIS is available in sufficient quantity it is good practice to first optimize using a solution of that compound and, after this is completed, optimize using the analyte at a lower concentration that is determined based on the response of the SIS.

In some instances prediction of the ionization mode based on analyte structure is relatively straightforward. For example, a compound with one or more basic amine or amide functional groups will yield a strong signal in the positive ionization mode, but in other instances it may be difficult to predict whether positive or negative ionization will be appropriate. Other compounds may yield similar response in both ionization modes so it may not be possible to determine which will result in the best selectivity and sensitivity until the ionization mode can be assessed by testing the analyte in matrix. For this reason, it is wise to optimize in both positive and negative ionization modes and retain both sets of conditions for potential future use. Since acid solutions may suppress negative ionization for some compounds, solutions at low pH should be avoided when optimizing new compounds that could potentially be analyzed under negative ionization conditions.

As discussed in Section 9.5.2, an experienced analyst familiar with a particular instrument, or with similar models from the same manufacturer, will gain an understanding for which tuning parameters are compound dependent, which parameters will have the greatest impact on method sensitivity, and also typical values for specific tuning parameters. Using this gained knowledge, a new compound can be manually optimized relatively quickly and with confidence that the final values obtained are reasonable for that instrument. When using autotune software this experience can still be applied to the evaluation of the final tuning result to determine whether or not the final values for each parameter are reasonable. In addition to parameters that are compound dependent, the analyst should be aware of any parameters that could be flow or temperature dependent. To optimize these it will be necessary to tune at LC flow and mobile phase conditions similar to those anticipated to be close to those used with the final method (Section 9.5.5).

In some cases it is observed that, under the experimental conditions used (mobile phase composition, ionization and API interface parameters), more than one ionized form of the intact analyte molecule is observed, i.e. adduct ions of various kinds (see Section 5.3.3 and Table 5.2). An example is shown in Figure 9.6, in which a well known anticancer drug (paclitaxel, Figure 9.6(a)) was analyzed by positive ion ESI–MS (infusion of a 'clean' solution). The first spectrum (Figure 9.6(b)) shows four different adducts (with H^+, NH_4^+, Na^+ and K^+). Adjustment of the cone (skimmer) potential (Section 5.3.3a), to lower values in this case, enabled production of the ammonium ion adduct to dominate the MS^1 spectrum (Figure 9.6(c)) in a robust fashion, and this ion yielded a useful product ion spectrum (that appeared to proceed via a first loss of ammonia to give the protonated molecule) which was exploited to develop an MRM method that was successfully validated and used. It is advisable to avoid use of analyte adducts with alkali metal ions (commonly Na^+ and to some extent K^+) since, when subjected to collisional activation, these adducts frequently yield the metal ion as the dominant product ion with only a few low abundance product ions derived from the analyte molecule. However, when feasible, both the ammonium adduct and protonated molecule should be investigated as potential precursor ions at least until it becomes clear that one will provide superior performance (sensitivity/selectivity compromise) than the other.

If the API interface conditions, particularly the cone potential (Section 5.3.3a), are too extreme, the dominant analyte ion transmitted to Q_1 might correspond to a fragment ion formed by in-source CID if the analyte happens to undergo a facile neutral loss. The most common of these is water loss, so if the predominant ion is observed at a m/z value that is 18 Th lower than expected it is advisable to check whether lowering the cone potential results in observation of the expected ion. Of course, such a fragment ion might still be entirely acceptable for use as a precursor ion in an MRM assay, in the same way as ammonium adduct ions, for example.

In addition to the general instrument tuning procedures described above, the resolution at the precursor m/z value(s) being used for compound optimization should be evaluated for an optimum compromise (Section 6.2.1) between selectivity (via higher resolving power) and sensitivity (response via lower resolving power). In most triple quadrupole instruments the Q_1 mass filter is designed to provide unit mass resolution over an m/z range up to 3000–4000. By degrading the resolving power (often referred to as 'opening up the resolution') enormous gains in transmission efficiency can be obtained (see Figure 6.6), but eventually any resulting gains in signal intensity are more than offset by simultaneous increases in chemical background in the SIM or MRM chromatograms resulting from the accompanying loss of selectivity, i.e. the signal:background ratio is degraded. Three spectra obtained using the same instrument in Q_1 mode but at different resolving powers are illustrated in Figure 9.7; at unit mass resolution (peak FWHM = 0.7 Th) the isotopologs of the test ion are almost baseline resolved and this represents a case where the compromise between transmission efficiency and selectivity is that optimized for general use by the instrument manufacturer. The slightly under-resolved spectrum (valleys between isotopolog peaks are 30–50 % of peak heights) may improve signal level somewhat via the higher transmission efficiency, especially at higher m/z values. However, very low resolution (and thus very high transmission efficiency) results in an unresolved envelope of the isotopolog ions (Figure 9.7c) and caution is advised in resorting to this extreme as the loss of selectivity in the mass spectrometer can result in poorer signal:background in the resulting chromatograms. Sometimes it is advantageous with respect to signal:background to set Q_1 at unit mass resolution (or better, see next paragraph) and degrade the resolution in Q_3; this can often greatly improve the absolute signal level with only a relatively minor accompanying increase in background, reflecting the much 'cleaner' MS^2 spectra compared with the corresponding MS^1 spectra.

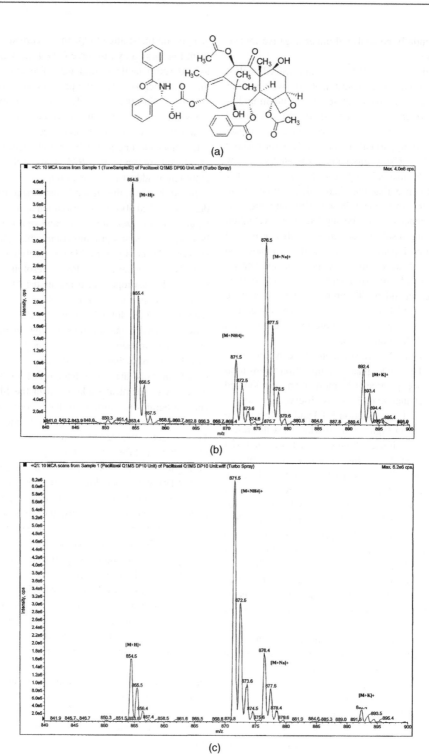

Figure 9.6 (a) Structure of paclitaxel. (b) and (c) ESI mass spectra (MS[1]) of paclitaxel using different solvent and API interface conditions, demonstrating the ability to favor formation of one or other of the possible adduct ions.

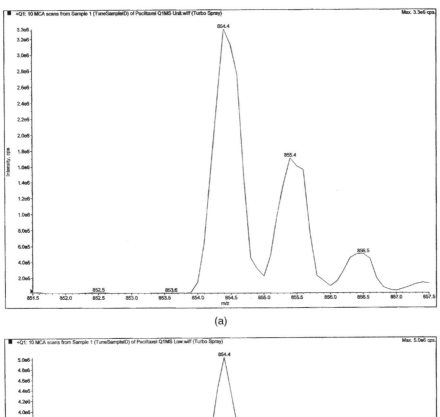

(a)

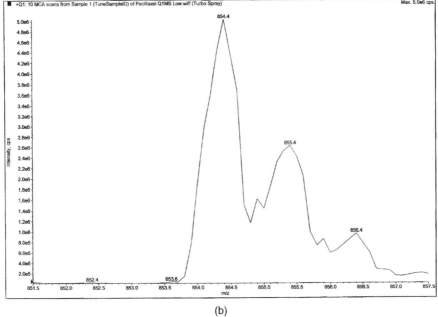

(b)

Figure 9.7 ESI mass spectra (MS^1) showing the $[M+H]^+$ isotopologs of paclitaxel obtained at different resolving power settings. (a) FWHM ~0.5 Th; (b) FWHM ~0.7 Th; (c) FWHM ~1.0 Th. Note that the maximum signal strengths (10^6 cps) are (a) 3.3, (b) 5.0 and (c) 5.8.

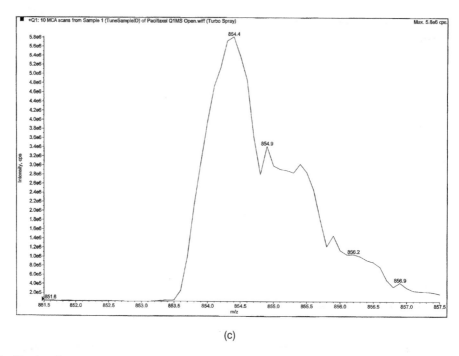

(c)

Figure 9.7 (Continued)

By judicious choice of instrumental parameters for a quadrupole mass filter (see the discussion of Equations [6.19–6.20] in Section 6.4.2) it is possible to trade off the usable m/z range for much higher resolving power (FWHM $\sim$0.1–0.2 Th) while maintaining transmission efficiency comparable with that obtained at 0.7 Th by more conventional designs (see Section 6.4.2). This is certainly a significant advance in the range of options open to the analyst and some impressive results have been reported (exemplified by Figure 6.12). However, it should be noted that when operated at the enhanced resolution settings, these instruments are reported (Jemal 2003a) to be subject to variability, arising at least in part from drift of the m/z axis with small temperature variations of the analyzer; this resulted in a requirement to continuously monitor the m/z calibration throughout a batch run. For more conventional quadrupole instruments it is good practice to perform quick checks on the m/z calibration before starting a batch run.

9.5.3b Selecting Product Ions and CID Conditions

The ideal product ion for use in MRM is one that is both structurally characteristic of the analyte (good selectivity) and can be observed at good relative abundance in the spectrum by adjusting collision conditions (the collision energy is generally the most important, although pressure and type of gas can also have an effect). It is somewhat advantageous if the chosen product ion can be assigned a structure that makes sense in terms of chemical intuition, but this is not necessary. Choice of product ions that are close in m/z to the precursor implies that the neutral loss fragment is of low molecular mass. Water, ammonia, carbon monoxide and carbon dioxide are common examples.

For both chemical and physical reasons, product ions of 'low mass' can be problematic for MRM detection. The practical interpretation of 'low mass' in this context varies with circumstances. Thus, the chemical disadvantage is mainly concerned with the observation (Aebi 1996) that 'chemical noise' (background) from API sources is considerably more intense at lower m/z values. While this observation directly affects SIM rather than MRM detection, it can also apply to MRM to some extent. The physical problem with low mass product ions for MRM concerns the low mass cut off that exists for all RF-quadrupolar field devices (Section 6.4.2). This problem is extremely important for 3D ion traps, as expressed in the so-called 'one third rule' whereby the low mass (really low m/z) cut off for product ions in MS/MS is $\sim$30 %

of that of the precursor ion (Section 6.4.5). However, for the linear quadrupoles that are of major interest here, the importance of the low mass cut off arises mainly with respect to the RF-only quadrupole collision cells, for which the minimum value of m/z with a stable trajectory corresponds to the maximum value of the stability parameter $q = 0.908$. This cut off is instrument dependent and is given by Equation [6.15] (Section 6.4.2) as:

$$(m/z)_{min} = (2e/m_u)(V_0/\omega^2 r_0^2)/(q_{max} = 0.908)$$

In practice, for quadrupole collision cells operated with typical values of r_0 and V_0, the low mass cut off problem is not often a major concern. In such cells that are operated with a sufficiently high pressure of collision gas to be in the collisional focusing regime (Section 6.4.3), the loss to the quadrupole rods of ions with $q > 0.908$ and $m/z < (m/z)_{min}$ (predicted for collision free conditions) is countered to a large degree by the collisional focusing of ions on the main axis of the device.

Just as for the analyte, the choice of product ion for MRM detection of the SIS requires the same considerations. In addition, usually the m/z value of the SIS precursor ion is different from that of the analyte, so there is no problem if the best choice of product ion turns out to be the same for both. This conclusion does not apply if there is appreciable overlap between the isotopic distributions of analyte and SIS, i.e. cross-contributions between the precursor ions (Section 8.5.2c), since the latter circumstance requires uniquely different product ions for there to be any hope of distinguishing the two. It is good practice to make a choice of product ion as the primary candidate for the MRM method, but also to choose one or two others as back-up candidates that should be independently optimized (collision energy etc.). Then, during subsequent stages of the overall method development, continue to use all of these candidate ions until the final optimization, as an insurance against unforeseen selectivity problems with the primary candidate for the real-world samples.

For multi-analyte method development, it is possible to use one tuning solution containing several different compounds for MRM optimization. However, there is always a possibility that impurities in any of the reference standards being used could result in the optimization being done for some solute other than the analyte of interest. For this reason it is safer to use independent tuning solutions for the initial optimization of Q_1 and especially for product ion optimization.

If an MRM method is taken from the literature, as the specified precursor and product icons used in the method should still be evaluated using the instrument that is to be used for the analytical samples. This advice relates to the

difficulties (Section 6.2) in preparing libraries for product ion spectra compared with conventional EI MS[1] spectra.

9.5.3c Final MRM Optimization

After the precursor and product ions have been selected and optimized by infusion experiments, the final step is to optimize any API interface parameters that are known to be flow dependent. The specific parameters that need to be considered depend on the instrument model and interface design but, in general, additional optimization using the mobile phase and flow conditions that will likely be used for chromatography include interface gas flow, temperatures and, possibly, the cone voltage. For methods being developed for the first time, approximately 50 % acetonitrile or methanol in water with 5 mM of ammonium acetate can be used for the initial optimization. However, if the mobile phase composition or concentration of the organic modifier is significantly different from that used during this process, it may be necessary to repeat this step at a later time to ensure that final conditions used are appropriate. (Note that if gradient elution is used it is the mobile phase composition at elution that affects the ionization conditions.) This is particularly important if it is determined that the analyte elutes at high aqueous concentration, or in the case of a HILIC method, high organic.

The quickest way to optimize at the typical mobile phase flows that are anticipated (usually in the range 0.1–1.0 mL/min), based on a sense for what type of chromatography development will be initially attempted, is to inject the analyte repeatedly every 5–10 seconds without a column in-line while making incremental adjustments to the parameter being optimized. This technique is commonly referred to as optimization by loop injection. The actual time between injections will depend on the flow being used, but it should be set such that there is an adequate amount of time between injections to allow the signal to return to baseline and/or equilibrate the parameter being adjusted, e.g., desolvation gas temperature (electrical potentials re-adjust almost instantaneously). For this procedure, the peaks will be only a few seconds wide at best, so the dwell times will have to be adjusted such that an adequate number of data points (SIM/MRM acquisitions, or scans) are acquired across the peak (typically $\sim$100 ms). For this purpose the analyte should be prepared in the mobile phase at a concentration of approximately 10–100 ng/mL; the exact concentration will depend on the sensitivity achieved for that analyte, but it is advisable to start with lower concentrations in order to reduce the risk of contaminating the interface, resulting in elevated baselines until the interface is cleaned. This is particularly important if the analyst intends to begin chromatography development with the goal of evaluating sensitivity in solution using the initial method. If the solution used for loop injections is too concentrated, the baselines will likely be elevated and little information will be gained with respect to the signal:background until the latter dissipates (can be up to several hours).

The experienced analyst should have a sense for the typical settings that are used for most compounds on a particular instrument, and essentially should evaluate the compound currently being optimized relative to that historical information. Particular attention should be paid to any settings that are significantly different from those typically used; if these arise the instrument should be checked to ensure that it is running properly. Prior to beginning the loop injections, if ESI is the chosen technique the analyst should check to be sure that the ESI needle is not partially clogged and that a good spray is observed. Temperature optimization should begin at a low temperature relative to what is known to be typical for that instrument, and then increase incrementally to detect any indication that the compound my be thermally labile. If APCI is chosen for evaluation, similar loop injection optimization procedures can be used.

9.5.4 Preparation of Stock and Spiking Solutions

Once the analytical standard has been acquired and its purity established (Section 9.4.4), the next step is to prepare solutions of known concentration that can be further diluted as required. As discussed in Section 2.2.1, the amounts of analytical standard required in any calibrator solution or QC sample are far too small to be weighed out accurately, so a strategy of weighing out an amount large enough that a good semi-micro or microbalance (Section 2.3) can provide negligible uncertainty in the mass followed by solution to form a stock solution of accurately and precisely known concentration is adopted. By subsequent aliquoting and dilution of this primary stock solution using Class A volumetric equipment (Section 2.4), the small amounts of analytical standard required can be dispensed with degrees of accuracy and precision that make negligible contributions to the overall uncertainty in the final analytical results. Details of how this creation and dilution of the primary stock solution is done in practice are described below.

9.5.4a Primary Stock Solutions

An obvious question to be settled before beginning any preparation of solutions of standards concerns the range

of concentrations required for the final calibration solutions (or matrix matched calibrators). Parameters involved in this decision include a rough estimate of the concentration in the unknown sample (obtained from the client or by a quick experimental estimate) and the resulting concentration in the analytical extract prepared by the proposed sample preparation scheme taking into consideration the sample aliquot size and final volume of the extract. Regulatory limits on permitted concentration or targeted concentrations to be measured will also impact these decisions. Essentially, the analyst must consider the range of reliable response that needs to be validated and the method that will be used to develop a scheme for making stock solutions and any subsequent sub-stocks or spiking solutions that will be required. Once this has been established it is possible to estimate the quantity of analytical standard that should be weighed out to make the initial stock solution with acceptable accuracy and precision and dissolved in a suitable volume that, together with an appropriate sequence of volumetric dilutions, can provide standard solutions covering the required concentration range.

Stock and spiking solutions (see below) are eventually used to make the calibration solutions and matrix matched calibrators that are the basis for the comparative measurement of analyte concentration in the unknown analytical sample (Sections 2.6 and 8.5), so the reliability of their concentrations is directly related to that of the measured concentrations. Accordingly, meticulous care and monitoring of these primary solutions is every bit as important as that expended in weighing the analytical standard in the first place (Sections 2.3).

The considerations summarized in this section are largely associated with basic chemistry. Obviously the solvent used to dissolve the weighed aliquot of analytical standard or SIS must be such that it is capable of dissolving these compounds up to concentrations that are appreciably larger than that used for stock solutions. Also, the solvent used for the stock solutions and/or any substocks or spiking solutions subsequently prepared must also be compatible with other requirements of the analytical method, e.g. must be miscible with the solvent used to extract the analyte from the matrix (see discussion of solvent miscibility in Section 4.4.2a). Most solvents of interest here (for analytes amenable to reversed phase HPLC) are likely to be volatile organics (e.g. methanol or acetonitrile, boiling points 65 and 81 °C), or mixed organic–aqueous, so precautions (suitable containers and storage conditions) are necessary to prevent evaporation and thus uncontrolled increase of concentration in the stock solution. A more detailed discussion of these principles and guidelines for preparing stock solutions, spiking solutions and calibrators can be found below, but first some additional comments relative to the process of weighing a substance, and what is required to substantiate the accuracy of that weighing, need to be addressed.

Primary stock solutions are those made by dissolving a weighed aliquot of analytical standard in a suitable solvent in a Class A volumetric flask (Section 2.4.1) fitted with a ground glass stopper to prevent solvent evaporation, and are used to produce sub-stocks, spiking solutions etc. by accurate dilutions (see below). Stock solutions should be prepared from reference standards of known purity. For analyses demanding the most extreme rigor, it is good practice to prepare more than one primary stock solution *using independent weighings* as a precautionary check on weighing errors. A common practice is to prepare quality control (QC) samples (Section 9.8.2) from stock and spiking solutions made with weighed aliquots of analytical standard independently of those used to make calibration solutions and matrix matched calibrators. These precautions can alert the analyst to the possibility of weighing errors or deterioration of stock solutions through, e.g., adsorption or contamination, if analyses of the QC samples indicate a lack of accuracy (measurement bias); generally any such errors <5 % are considered fit for purpose in view of other uncertainties in the overall method that can increase the overall uncertainty to 10–20 %. However, if independently prepared stock solutions have been shown to be equivalent and adequate stability has been established (upon storage and use at room temperature and upon multiple use), a single stock solution containing the analyte(s) of interest is sometimes used to prepare both QC samples and calibration standards.

Another important concern is related to the chemical stability in solution of the analyte or SIS itself. Special consideration should also be given to the dangers inherent in use of protic solvents (water, alcohols etc.), e.g., hydrolysis or transesterification of ester analytes, D/H back exchange of a D-labeled SIS.

The stability of the reference substance in a stock solution made from a particular solvent and stored under a given set of conditions should be established prior to use (Section 10.2.7). Method development studies can commence prior to establishing known stability but the analyst must be aware that the analyte could be degrading in solution (or during all phases of sample processing) and take that into consideration especially when unusual or unexpected results are obtained during development. In no instance should solutions without established analyte stability be used for sample analysis. Also, additional stability data should be generated for any other solutions that are prepared in a different solvent and/or stored under

different conditions. In some cases this information may be available from the supplier or client and can be very useful in terms of selecting the appropriate solvents and storage conditions at the onset of work, but unless the supporting data to substantiate the stability are readily available, it is good practice to instigate a protocol for regular testing of stability and concentrations of stock and spiking solutions from the day that they are made up, and indeed this may also be required in support of the method validation process; this is discussed in greater detail in Sections 10.2.7 and 10.4.1h.

For compounds that are light sensitive, additional precautions may be required during the preparation and storage of the stock and spiking solutions. Rooms equipped with yellow lights may be required during the weighing and handling of the reference standard, and solutions should be protected from light during storage by using amber colored containers, wrapping storage containers with aluminum foil or equivalent techniques that will minimize exposure to light. In extreme cases, the entire sample preparation and analysis procedure may need to be conducted in a laboratory specifically designed for the analysis of light sensitive compounds. Once again, information about the stability of the compound, in particular if it is sensitive to normal light conditions in the laboratory, can often be obtained from the supplier or client prior the initiation of work.

In addition to the impact from analyte instability in a given solvent and storage conditions, adsorption effects due to different types of container that are used to store solutions can also be assessed and monitored though the careful planning and execution of the stock stability studies. This special case of poor storage stability is especially pronounced with very polar compounds and/or when solutions are prepared at very low concentrations. To circumvent this relatively common problem, care must be taken when selecting the storage vessel. In many instances, silanized glass or plastic containers are used to address this problem.

Once stability is established, storage containers should be clearly labeled with the solution name or ID, appropriate storage condition and expiration date. Any solution that is found to be past the established expiration date should be removed from the refrigerator or freezer and destroyed.

Similarly, stock and spiking solutions of an SIS should be made entirely independently of those for the analytical standard. Note that the SIS is usually available in much lower quantities than the standard analyte (particularly if it is a stable isotope labeled SIS, see Section 9.4.5), so the strategy used to prepare spiking solutions from stocks can be quite different.

9.5.4b Sub-Stocks and Spiking Solutions

Sub-stocks and spiking solutions are used for early method development studies and for the preparation of instrument tuning solutions, of matrix matched calibrators and QCs (Sections 9.8.1 and 9.8.2) after the final method has been established. The weighing of milligram amounts of analytical standard is a meticulous and time-consuming procedure, often involving several replicate weighings at each stage using different analytical balances to check on accuracy and precision. In trace quantitative analyses it is not realistic to use this procedure every time a different spiking solution etc. is required, so the compromise adopted is to use a smaller number of independent weighings to make up independent primary stock solutions and to create the required sub-stocks and spiking solutions by a series of volumetric dilutions. Low-tech devices like standard flasks and pipets are every bit as important as the newest chromatographs and mass spectrometers in this regard. To perform the necessary volumetric dilutions with adequate precision and accuracy, correct use of properly calibrated pipets (Section 2.4.2) together with Class A pipets and volumetric flasks is essential.

Sub-stocks are also used for multi-analyte assays where a mixture of analytes in a single solution is prepared for subsequent preparation of QCs, calibration solutions or spiking solutions used for the preparation of matrix matched calibrators. In this case, individual stock solutions are combined to make one or more sub-stock solutions containing all of the analytes in one solution. This practice will reduce the number of steps that would be required versus making an individual spiking solution for each analyte. However, for methods that required a different LLOQ for each analyte, the sub-stocks and spiking solutions need to be prepared at the corresponding appropriate concentration for each analyte.

At the onset of method development the appropriate concentrations of the sub-stock and spiking solutions may not be known, but estimates can be based upon the initial method development scheme and by considering what types of spiking solutions will be required for early sensitivity and recovery assessments. Also, to determine the required concentrations for these solutions, consideration must be given to the type of solvent and the volume of solution that will be used for any subsequent dilutions or for spiking the analyte into the control matrix (Section 9.5.6c).

9.5.4c Miscellaneous Considerations

When using reusable glassware in the laboratory, cross-contamination (Section 9.7.1) is always a concern but taking adequate precautions to prevent this from happening is especially critical during the preparation of solutions that will eventually be used for calibration and for the preparation of QCs. Extra care must also be taken when using pipets and standard flasks that may have been used previously for dilution of stock solutions that are often prepared at high concentrations relative to sub-stocks and spiking solutions. Before using glassware such as pipets and volumetric flasks, always check visually that they are sufficiently clean by determining whether its surface is uniformly wetted by reagent water or distilled water. If grease is present it will be observed that, after draining for ~1 min, water droplets will adhere to the walls of the glassware instead of the unbroken film of water that indicates cleanliness. (If the interior walls are not uniformly wetted, drainage of liquid from a pipet will be uneven resulting in loss of accuracy and precision.) To clean glassware a cleaning solution is used. Solutions of laboratory detergents (available from any chemical supply company) are often sufficient, but for more difficult cases a solution of sodium or potassium dichromate in concentrated sulfuric acid is commonly used. Other cleaning solutions can also be used, e.g. a solution of one volume of 3 M hydrochloric acid added to one volume of reagent grade methyl alcohol. These cleaning solutions are best stored and used in polyethylene containers. (Appropriate safety and handling precautions should be taken when using chemicals of this type.)

The glassware to be cleaned should first be rinsed with water and then filled with the cleaning solution for a few minutes (be sure to use a rubber bulb to fill a pipet, Section 2.4.2). Then empty out the cleaning solution (some laboratories re-cycle these solutions) and thoroughly rinse the glassware inside and out with reagent water or distilled water (tap water can be used for the first few rinses if desired to save the expensive purified water that must, however, be used for the final rinses). The glassware can then be dried under a gentle stream of clean air or nitrogen. (Note that laboratory compressed air is often compressed using apparatus that uses oil and contamination from oil vapor or even droplets can result; compressed gases of suitable quality supplied in cylinders are recommended.) If it is necessary to speed up the drying process further, a final rinse with a volatile organic solvent (HPLC grade or better) can be used. Never expose Class A pipets or flasks to extreme conditions such as placing them in an oven.

Once again, the documentation of all procedures conducted and equipment used is critical in terms of supporting the final data. SOPs specific to the laboratory where the solutions were prepared will dictate the exact documentation requirements but, at a minimum, the following information should be recorded: solution ID(s), expiration date, storage location, analyst who prepared the solution, date prepared, pipets used, concentration of standard solution(s) used and amount used, final volume and solvent, storage container, and finally, the final concentration of the resulting solution.

9.5.5 Chromatography Method Development

Considerable background information on chromatography and mass spectrometry, for the purposes discussed in this book, is provided in Chapters 3–7. As emphasized previously it is not a trivial task to sort through all the numerous combinations of available technologies (Hopfgartner 2003a) to decide on an optimum method for the purpose at hand. The present section describes some practical approaches to development of a chromatographic method that is fit for purpose.

In this context it is helpful to bear in mind three major objectives: (i) the chromatographic method should be sufficiently fast to meet requirements for sample throughput (determined by number of samples and any deadlines set by the client); (ii) if a value of LLOQ has been specified, during development of a chromatographic method it is advisable to aim for a signal:background ratio ≥ 10 at this LLOQ concentration whenever possible; (iii) the method should be rugged and reliable in the presence of matrix, i.e. analyte in extracts of control matrix.

9.5.5a General Considerations

Most of the following discussion is devoted to LC/MS analysis. However, it is always a good idea to first consider if the analyte properties (volatility and thermal stability) are compatible with analysis by GC/MS, since chromatographic development for GC is generally much simpler (only one realistic choice of mobile phase, no significant problems with matching conditions with those of the mass spectrometer). While the majority of analytes are not suitable (or may require a derivatization step, Section 5.2.1b), there are many advantages if GC/MS can be used.

If HPLC is the only option, a start can be made based on chromatographic methods that have been described previously for the present analyte(s) or those that are structurally similar; start with an initial method using standard(s) in solution. Optimization at this stage will not only provide a good start towards establishing the best chromatographic method for analytes in matrix, but also will be used to make estimates regarding sensitivity and its impact on sample preparation requirements (minimum size of sample aliquot and final extract volume). If no

good starting point can be found in the literature or elsewhere, the analyst must start with the analyte structure and look for chromatographic columns whose stated application areas appear to match the present case; manufacturer's information about columns will usually suggest the general type of mobile phase to be used. In such circumstances analytical experience is undoubtedly a great advantage. If at all possible, reversed phase methods are preferred for LC/MS assays as the API methods are most compatible with such mobile phases.

It is also a good idea at this initial stage to look for signs of potential trouble with respect to carryover (Section 9.7.2). Although this will need to be fully evaluated later for analyte in matrix, these exploratory experiments will at least provide an indication of any truly 'sticky' compounds among the analytes. Other pitfalls that can be avoided at this stage concern the robustness and ruggedness (Section 9.8.4) of the chromatographic method with respect to column changes in the sense of use of nominally identical columns from a different batch from the manufacturer, or of a new column vs one that had been used previously for a different assay. Injection of very 'dirty' extracts can possibly provide advantages in some cases where sample throughput is of paramount importance, but will lead to problems with method ruggedness, column life and requirements for cleaning and maintenance of the mass spectrometer as well as applicability for future uses. If in doubt it will often be more cost-efficient in the long run to start method development with a new column rather than to try to adapt a much-used column for this purpose. Some laboratories have standard procedures for documenting past uses of each column and criteria for their disposal.

The chromatographic method is a central component of the overall analytical method and its optimization overlaps with that of the sample preparation and mass spectrometric analysis steps. Thus, if signal:background at the LLOQ turns out to be problematic for a proposed chromatographic method, it may be possible to avoid starting a search for a new method if the sample preparation can be adjusted to provide a more concentrated extract. Similarly, the mobile phase should be compatible with the mass spectrometer (e.g. no involatile buffers, preferably low water content if reversed phase) and permit good yields of the preferred precursor ion (Section 9.5.3a).

9.5.5b Sample Throughput and Selectivity

An important consideration in the development of any analytical method is the compromise necessary between speed of LC–MS analysis (sample throughput) and the need to provide reliable data. Particularly in bioanalytical analyses undertaken for the pharmaceutical industry, there is a demand to reduce the time per sample by minimizing the effort put into sample preparation, reducing chromatographic elution times and restoring selectivity by using elaborate MS^n techniques. One potential reason for loss of selectivity in LC–MS/MS bioanalyses is interference from a co-eluting metabolite that can undergo in-source conversion to an ionized molecule identical to that of the drug molecule itself. The most commonly observed mechanism of conversion is collision induced dissociation (CID) activated by collisions occurring in the cluster dissociation region of an API interface (Section 5.3.3) (Jemal 1999; Weng 1999), although the conversion may be due to thermal decomposition during sample ionization as observed in an APCI source (Tong 2001). As a result, the MRM transition(s) monitored for the analysis of the surviving drug will respond not only to the drug, but also to the metabolite. Thus, if adequate chromatography is not provided to separate the drug from its metabolite, the LC–MS/MS method will give measured signals that are not specific to the drug but also contain a contribution from the metabolite; an analogous effect could be observed if the pharmaceutical compound under test were actually a pro-drug rather than the drug itself. Examples described in the literature (Jemal 2000, 2000a; Weng 1999, 2000; Ayrton 1999; Ramanathan 2000; Romanyshyn 2000; Yan 2003) include drugs with lactone, carboxylic acid and thiol functional groups that produce hydroxy acid, acylglucuronide and disulfide metabolites, respectively, that can interfere in this way; other examples of such interferences arising from in-source CID of a metabolite or a pro-drug include a variety of compounds in the presence of their *O*- and *N*-glucuronide metabolites or of the *N*-oxide metabolites, and analysis of fosinoprilat in the presence of its pro-drug fosinopril and analysis of simvastatin acid in the presence of its pro-drug simvastatin. Clearly this kind of interference is sufficiently common that it should always be a concern in bioanalytical analyses and sufficient attention paid to adequate chromatographic separation. An example (Jemal 1999) illustrating the effect is shown in Figure 9.8. It is important to note that this phenomenon can not in principle be detected in conventional method validation procedures using control matrix spiked with analytical standard since these do not contain any metabolites. Generally analytical standards for metabolites are not available, so re-analysis of incurred samples (discussed in Sections 9.4.7b and 10.2.9c) is the only viable approach to investigate the degree to which such problems may exist. While this problem of interferences by metabolites has received most attention for biomedical analyses, in principle it can also apply to other applications, e.g. analysis of environmental

samples for pesticides that are subject to formation of oxidation and other breakdown products (the analogs of metabolites). The development of ultra-small particles for HPLC stationary phases (Section 3.5.7), with its resulting intrinsic improvements in separation efficiency and selectivity, does provide a significant extension of the range of acceptable compromises between chromatographic analysis time and selectivity despite reservations (Butchart 2007) about the extent to which such advantages can be realized in practice (see Section 3.4.7).

With respect to optimizing a proposed method for sample throughput, it is not a trivial task to sort through the numerous combinations of available technologies (Hopfgartner 2003a); experience acquired through 'learning by doing' plays a key role here and an article (Lagerwerf 2000) has described the results of extensive experience in the area of bioanalytical analysis. Standard approaches (Jemal 1997; Køppen 1998) to develop LC–MS methods have used either trial-and-error or intentional variation of experimental parameters ('one factor at a time'); the latter method involves variation of one experimental factor with the others held constant and thus provides an estimate of the effect of that particular factor at selected fixed conditions of the others. However, such estimates assume that the effect of that single experimental parameter is the same over the entire range of all the others and thus fail to account for interactions among them, so that the final result may not be optimal for the combination. However, it is possible to use chemometrics approaches that use multi-variate designs, in which experimental factors are varied simultaneously over selected levels and thus take into account interactions between factors. Excellent examples of this approach describe a multi-variate approach to developing a generic ion pair LC–MS method for the analysis of acidic compounds (Seto 2002) and for analysis of iron-chelating siderophores (Moberg 2006).

9.5.5c Multi-dimensional Separations

'Multi-dimensional' separations, in which more than one mechanism of separation is applied to a sample with each mechanism considered as an independent separation dimension, have a long history as described in a recent review (Dixon 2006). It is characteristic of human nature that, while modern developments in chromatography have greatly increased separation power, this achievement has moved the objective to separations of ever more complex mixtures. However, multi-dimensional separation techniques are not only employed in analyses of samples containing overwhelming numbers of analytes, but are also of use when a critical separation of two or more

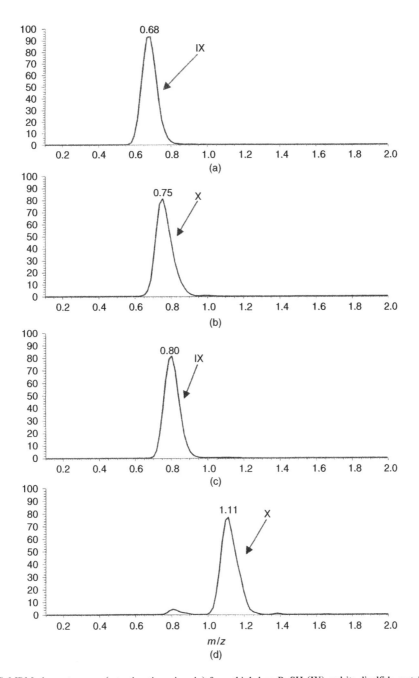

Figure 9.8 LC–MRM chromatograms (retention times in min) for a thiol drug R–SH (IX) and its disulfide metabolite R–S–S–R (X). (a) and (b) show the MRM chromatograms of IX and X, obtained using the (M–H)$^-$ ions of IX (*m/z* 407) and X (*m/z* 813) as precursor ions and the MRM transition *m/z* 407 → 280, with an isocratic mobile phase of 45 % aqueous ammonium acetate (5 mM, pH 5.5) and 55 % acetonitrile at a flow rate of 0.8 mL min^{-1}; clearly IX and X were not resolved under these LC conditions, with k′ values of 0.43 and 0.56, respectively. In contrast, (c) and (d) show that IX and X were well resolved by using an isocratic mobile phase of 55 % aqueous ammonium acetate (5 mM, pH 5.5) and 45 % acetonitrile, with k′ values of 0.68 and 1.3 and a resolution of 1.7. The small interference from IX in chromatogram (d) for analysis of X presumably corresponds to a low level amount of IX in this sample. Reproduced from Jemal, *Rapid Commun. Mass Spectrom.* **13**, 97 (1999), with permission of John Wiley & Sons, Ltd.

analytes cannot be achieved using one particular column or phase type; in such cases two sequential separations using two very different phase columns (e.g., ion exchange and reversed phase HPLC) may be required.

The general term 'multi-dimensional chromatography' includes both the so-called heart-cutting methods and also the more ambitious comprehensive multi-dimensional chromatography. In the former case, selected fractions of the eluant from the first column are collected and stored separately, and are subsequently analyzed by conventional injection onto the second column that provides increased resolution for these 'heart-cut' fractions. The chromatographic equipment required is essentially the same as for conventional one-dimensional chromatography, with a fraction collector and facilities (on-line or off-line) for preparing the heart-cut fractions to be suitable for injection onto the second column. In contrast, the comprehensive approach aims to subject each and every component in the sample to the full combined separatory power of the two (or more, in principle) chromatographic dimensions. By convention, comprehensive strategies are denoted using a 'x' notation, e.g., LCxLC represents comprehensive 2D LC analysis (Dixon 2006, Shellie 2006), GCxGC the corresponding GC case (Marriott 2002; Dimandja 2004), while LCxGC and LCxCE would represent comprehensive separations by LC with CE and by LC with GC, respectively. Equipment required for comprehensive multi-dimensional chromatography is specialized and considerably more complex; a key component is always the interface between the two columns.

With respect to the problems addressed in this book, i.e., those involved in quantitative analysis of trace level small molecule target analytes, comprehensive multidimensional chromatography is not often appropriate since there is usually no interest in complete separation of each and every component of the mixture; however, this may not be the case for environmental analyses that must accommodate multi-residue methods in which comprehensive GCxGC–MS methods are becoming accepted (Panić 2006). Heart-cutting multi-dimensional separations as such are not much used in target analyses; however, any hard and fast meaningful distinction between elaborate clean-up methods followed by one-dimensional chromatography and heart-cutting multi-dimensional chromatography in its usual sense (Dixon 2006; Marriott 2002) becomes difficult to sustain.

9.5.5d Miscellaneous Components

It is a truism (but therefore true!) to say that the strength of a chain is that of its weakest link and this certainly applies to integrated systems for trace analysis. In this section some of the 'minor' components of a chromatography/MS system that are a concern for overall performance (Kazakevich 1996) are briefly mentioned.

Chromatography columns, and tubing used to connect the different components, must be constructed of inert materials; this is particularly important for GC columns that are subjected to high temperatures. Packed GC columns are therefore usually constructed of a chemically inert metal that can be formed into a coil to fit inside a GC oven, while wall-coated capillary columns are invariably made of fused silica with enhanced mechanical strength provided by an external coating of a polyimide polymer. Tubing used to connect the carrier gas supply to the GC column is usually of copper. Packed capillary columns for HPLC are also usually made of fused silica, while larger columns ($\geq$ 1 mm i.d.) are usually high grade stainless steel; it is common for manufacturers to supply such columns with various internal diameters but with a fixed outer diameter of 0.25 inches or equivalent, in order to facilitate switching between columns without having to replace the fittings and connectors.

Tubing used to connect various HPLC components together must be inert, have as low a volume as possible and be mechanically strong and flexible. A polyimide-coated fused silica capillary is a common choice, though tubing made of an inert organic polymer such as PEEK is also used. (PEEK: 'PolyEtherEtherKetone', oxy-1,4-phenyleneoxy-1,4-phenylene-carbonyl-1,4-phenylene; it can be used continuously at up to 250 °C and is chemically highly inert). Connector tubing should have low internal diameter to maintain separated components in narrow bands as they are transferred from column-to-column (for two-dimensional LC) or from column-to-detector. Tubing with an internal diameter > 0.5 mm can lead to band broadening and even band mixing, and it is advisable to use tubing with an internal diameter less than ~150 μm to minimize risk of any such effects.

It is also essential to pay attention to end fittings for columns and connectors of various kinds, since sudden changes of internal diameter, corners and other potentially 'stagnant' regions can ruin what is otherwise a good separation system. Several manufacturers supply suitable 'zero dead volume' connectors, T-unions etc. made of suitably inert materials (e.g., PEEK). Many reducing unions and column end fittings incorporate special frits designed to hold packings in place and prevent them from being transported into the detector. The frits are usually disks made of porous stainless steel or other inert material that are fitted tightly into the ends of a column or in its end fittings (connectors). It is important that the frit be a tight fit to avoid stagnant spaces at the edge where the disk is in contact with the wall.

Tight temperature control of a chromatographic column is essential for the most precise and accurate work. The importance of the GC oven was emphasized in Section 4.4.3b but it is also preferable to maintain an HPLC column to within ~0.2 °C.

Guard columns are sometimes used to protect the analytical column from contamination from sample extracts that have been subjected to only minimal clean-up. The stationary phase used in a guard column should ideally be matched to that the analytical column, both in the chemical nature of the phase and also in particle size. Connection between the two columns should be as short as possible to avoid dead volumes. The guard column should be short enough that no discernable effect on the chromatographic performance of the analytical column is observed; a 10 mm guard column is a common choice. Naturally, since the purpose of a guard column is to protect the expensive analytical column from contamination, the former should be changed much more frequently than the latter; increase in back pressure or deterioration of chromatographic quality are useful indicators that a guard column should be changed.

9.5.6 Sample Preparation – Extraction and Clean-Up

Depending on circumstances, different compromises between extensive clean-up on the one hand, and high selectivity provided by the final analysis by chromatography–MS on the other, will be appropriate. Extensive clean-up will tend to increase analyte losses (lower F_a' and higher L_a, Section 8.5) and thus reduce overall sensitivity, but this is compensated for by generally better performance of the final analysis and increased column lifetime and decreased contamination of the mass spectrometer. Less extensive clean-up saves time and will generally result in lower losses, but often results in lower precision and accuracy in the analytical measurements as a result of increased matrix effects etc. and introduces contaminants into the column and the mass spectrometer. This is an example where the judgment of the analyst must be exercised in order to determine which approach is the most fit for purpose in the given circumstances.

The technologies currently available for sample preparation were discussed in Section 4.3; a recent review (Hopfgartner 2003a) provides additional information particularly concerning the robots currently available to assist in high throughput analyses. The present section describes a practical approach to appropriate deployment of these technologies to create a functioning analytical method. It is important to choose an extraction method that does not destroy the analyte, e.g., avoid high temperatures for thermally labile analytes. Bear in mind that the recoveries F_a' and F_{SIS}' (Section 8.5.2) are not necessarily equal in view of possible occlusion of analyte within the analytical sample (e.g., in small internal pores of soils or sediments); similar remarks apply to F_a' and F_e', the recoveries for the native and spiked-in analyte relevant to the Method of Standard Additions.

The following discussion presents some recommendations for aspects that should be borne in mind from the earliest stages of sample preparation development to the final optimization. Obviously this can not start until a suitable control matrix (Section 9.4.7) has been acquired. It is important to always remember the sensitivity requirement of the assay (particularly the LLOQ); in turn this implies initial and continuing assessments of recovery and of extract cleanliness (freedom from matrix interferences and matrix effects). Quick estimates of recovery are made by comparing the LC/MS response for an extract of control matrix that was spiked with analyte before sample preparation with that for a similar extract spiked after extraction and clean-up. It is also important to ensure that the final extract has a solvent strength compatible with that of the HPLC mobile phase. Throughout development the analyst should also check for signs of instability of the analyte during the sample preparation procedure and in the final extract; in this respect a good rule is to always use the least aggressive extraction and clean-up conditions (moderate temperature, no highly reactive solvents or additives) consistent with good recovery and extract cleanliness. When feasible it is advisable to evaluate more than one approach to sample preparation (e.g. LLE and SPE) until it becomes clear that the preferred approach does not suffer from serious problems with regard to sensitivity (signal:background at the LLOQ) and selectivity (i.e. no problems from matrix interferences and/or matrix effects).

When evaluating SPE (Section 4.3.3c) it is a good idea to run elution profiles to determine the lowest organic concentration that provides good recovery; this minimizes co-extractives and also makes it less likely that the final extract will have a solvent strength greater than that of the chromatographic mobile phase, thus avoiding the need for a solvent switching step. If LLE (Section 4.3.1) is investigated the pH is an obvious parameter in optimization; if the sample matrix is plasma, hexane or ether vs ethyl acetate should be considered as the organic solvent to extract the drug. If simple protein precipitation (Section 4.3.1f) is used as sample preparation for plasma samples, a 3:1 ratio of acetonitrile:plasma followed by a

1:10 dilution of the resulting supernatant is suggested as a good starting point for optimization.

Other aspects that should be checked throughout development include matrix interferences (Section 9.6.3), contamination (either general laboratory sources or arising within the sample preparation process itself) and carryover (Section 9.7); all of these can be evaluated by appropriate use of extracts of analyte-free control matrix.

9.5.6a Evaluating Options

To proceed with a logical development of a sample preparation method for a new analysis, the analyst should first assemble all available information known about the analyte, i.e. molecular mass, structure, stability and solubility characteristics, any previous analytical methods in the reviewed literature or elsewhere, and similar information about the matrix. Of course it is often the case that only partial information is available, in which case the initial development phase will necessarily be longer but, based on the molecular structrure and whatever additional information is available, it is necessary to seriously consider which of the more common sample preparation techniques (Section 4.3) is likely to be most appropriate. The most likely candidates are liquid–liquid extraction (LLE, Section 4.3.1), simple protein precipitation (PPT, Section 4.3.1f) for bioanalytical analyses involving plasma as matrix, some form of liquid extraction from a solid matrix (Section 4.3.2), or some variant of solid phase extraction (SPE, Section 4.3.3) for extraction of analyte from liquid or gaseous matrices. In addition to such considerations based on chemical knowledge, it is essential from the outset to also take into consideration the information derived from any initial sensitivity checks that may have been run using the analytical standard in clean solution (Section 9.5.5), together with considerations of the total amount of analytical sample available (Section 9.4.6a). This will help establish the volume and analyte concentration of the final analytical extract that should result from the sample preparation method.

Sample preparation must also be compatible with the chromatographic separation and detection method. The analyte concentration in the extract is one obvious example of this since it must be sufficiently large that the analytical measurement can provide sufficient signal to permit acceptable levels of accuracy and precision. Demands on signal:background ratio also arise here since the extract must be sufficiently free of matrix components that might cause matrix interferences and/or matrix effects such as ionization suppression, but this aspect can only be tested later in the development once some tests have been run on analyte in matrix (Section 9.5.7). Another compatibility issue that can be dealt with in the early stages is that of solvent strength (Section 4.4.2a), since for HPLC analysis the solvent strength of the analytical extract must be no greater than that of the chromatographic mobile phase into which the extract is to be injected; this might require inclusion of a solvent switching step into the sample preparation method.

Stability of the analyte during the sample preparation procedure, as well as in the final extract, is another concern that should be considered from the earliest stages of method development. Again, a proper investigation of this aspect must await testing of the method with analyte in matrix but some indications can be obtained using clean solutions of analytical standard, e.g., thermal stability, chemical stability in acidic or basic solutions etc., and such information can be used to exclude some possible candidates for the sample preparation technique.

Special problems can arise in cases requiring measurement of multiple analytes that differ appreciably in their chemical properties and thus may not be amenable to the same extraction and/or clean-up technique. Well known examples of this situation are provided by some of the analytical methods approved by the US EPA in which different procedures are specified for acidic and basic analytes. A similar situation might arise in bioanalytical applications, e.g. if both a pro-drug and drug are to be measured but have different chemical functional groups. In such cases if all else fails the analyst might have to resort to two complementary methods, but it is desirable to avoid this choice if at all possible. Possible approaches to avoiding more than one sample preparation method include the use of mixed mode SPE cartridges (Section 4.3.3c) with an appropriate back extraction scheme. If both acidic and basic analytes can somehow be combined in the same final extract, the two types will likely yield appreciably better signal:background ratios in one or other of positive and negative ionization modes. This last problem can often be accommodated to some extent in modern mass spectrometers that provide facilities for switching between the two polarities in a time-scheduled program, although the pH and other properties of the mobile phase may still favor one polarity over the other.

The preceding discussion has tried to emphasize the importance of considering some important parameters at the earliest stages of development of a sample preparation method. These parameters are important not only for the sample preparation step itself but for its compatibility with the other steps of the overall analytical method.

9.5.6b Use of Blanks During Method Development and Analysis

The different types of blanks that can be prepared and how they are used are discussed in this section. Although the focus of this chapter is on method development, it should be stated that blanks are also a fundamental component of method validation and sample analysis, and in fact are critical in terms of supporting the validity of the final analysis.

The two general types of blanks that are used in method development are solvent blanks and extracted blanks. Solvent blanks may be as simple as pure solvent or mobile phase that is injected to demonstrate the lack of background or carryover in the absence of matrix. Extracted blanks are generated by extracting a given matrix or control using the analytical method being tested and are generally much more useful than solvent blanks for method development, validation and sample analysis studies. The most obvious example of an extracted blank would be a blank sample that is generated by extracting a representative aliquot of the control matrix being used in the study. This type of blank is often referred to as an extracted matrix blank or sample blank. Another type of extracted blank that is commonly used is a method blank, which is designed to investigate the exogenous background or interference that is a result of the sample preparation procedure itself. For aqueous samples, purified water is commonly used for the preparation of a method blank, whereas sand is often used for a soils analysis.

A matrix blank with no other analytes or internal standard spiked into it prior to extraction is also referred to as a double blank. This terminology distinguishes this type of blank from that resulting from the extraction of control matrix that is spiked with SIS only, which will be referred to throughout this book as a Bl/IS. Another type of blank, which is generally only used to check for cross-contribution of the higher isotopologs of the analyte to the m/z value used to monitor the SIS (Section 9.4.5b), is one that is generated by extracting the control matrix spiked with only the analyte of interest at the upper limit of quantitation, and is typically referred to as a ULOQ/Bl. A variation of the extracted blank is one that is spiked with either one or both of the analyte and SIS after extraction and just prior to analysis. This is referred to as spiked extract matrix blank and is commonly use to evaluate the effect of sample matrix effects and to determine the actual recovery of the sample extraction process, as compared to the combined effect of losses during sample extraction plus ion suppression or other matrix effects (Section 9.6.1). How the various types of blanks are used

in the method development and analytical quality control processes can now be considered.

As previously discussed (Section 9.4.7b), extracted control matrix is used to assess the selectivity for the method under development. The selectivity of the method is a function of the sample preparation (extraction and clean-up), chromatography and mass spectrometry conditions that are used for the method. Assuming that the control matrix is representative of the sample matrix to be analyzed, and that method blanks have been used to demonstrate that the method is free of exogenous interferences due to solvents, or to containers or other apparatus (a source of interferences that is often overlooked in the method development process), an extracted blank is used to demonstrate that the method has sufficient selectivity for the intended analytical purpose. When interferences at the expected retention time of the analyte being quantified are detected, modifications to the sample preparation and chromatography (and sometimes the ions monitored by the method) can be made to improve the selectivity of the method. Recall (Section 9.4.7b) that only re-analysis of incurred samples can reveal interferences resulting from metabolites or degradates of the analyte with either or both of the analyte and SIS.

Blanks are a vital component of the laboratory quality control process. They are incorporated into all method development, validation and sample analysis schemes to monitor and mitigate laboratory contamination (Section 9.7.1). Also, with LC–MS calibration curves often covering up to three orders of magnitude, carryover due to insufficient rinsing of the syringe needle or autoinjector is common and blanks (extracted blanks in particular) are used to assess and quantify the impact of the carryover on the final method (Section 9.7.2).

9.5.6c Using Spikes and 'Recovery Samples'

After the initial instrument and chromatography conditions have been established using analyte in solvent, the next step is to establish a basic analytical method that can then be used as a basis for further optimization of sample extraction and chromatography strategies. A practical approach is to use matrix spiked with analyte and internal standard (referred to here as 'spikes'), together with additional extracted blanks to assess recovery, ($F_a{}'$ in Section 8.5), selectivity, sensitivity, carryover (Section 9.7.2) and matrix effects (Section 9.6). In addition to the obvious goals of obtaining high extraction efficiencies (recovery) and suitable chromatography conditions that are optimized for selectivity and speed, the initial method will also be used to assess analyte stability in solution and matrix during the early stages of development. This latter requirement is essential when working with compounds that are known, or suspected, to be unstable in solution and/or matrix. Clearly, a plentiful supply of appropriate control matrix (Section 9.4.7) is needed, preferably from more than one source, for these early exploratory experiments.

In many instances the compound structure will lead the experienced analyst to a likely approach, e.g. liquid/liquid extraction or SPE (Section 4.3), but even then systematic development and optimization of the extraction method will be required to arrive at the best option. For example, when evaluating SPE, experiments designed to evaluate recoveries and matrix effects should be planned in a manner such that several different columns and conditions can be tested within a single experimental set. Even if the analyst is convinced that one mode of extraction will be superior to another based on compound structure, it is wise to evaluate other options at this stage of development in case the primary method is eventually shown to be subject to selectivity or matrix effects. For instance, if the primary method was SPE, but upon further investigation an interfering peak was detected in the chromatogram near the LLOQ, the analyst could quickly generate some extracts using the conditions established for the back-up method to determine if better selectivity could be achieved.

One possible approach is to prepare spikes in duplicate at approximately 10 times the concentration of the LLOQ (or higher) along with three extracted blanks. Two of the extracted blanks are then spiked with analyte after extraction in order to permit calculation of the extraction efficiency and to test for matrix effects (Section 9.6). Under this scenario several variations of solvent, pH or other critical extraction parameters can be tested with minimal effort and use of instrument time. The size of the sample aliquot and the final volume of the extract will depend on the estimated sensitivity obtained during early chromatography development in solution and the amount of sample available. For instance, a client might require an LLOQ of 0.1 ng/mL (in the original sample), and it is determined that the amount of sample available for the analysis is 500 µL; taking into consideration that it is desirable to retain some sample for re-assay if needed, and that the final extract volume will be 200 µL for reasons of practicality, the final extract concentration would be 0.25 ng/mL assuming a recovery of 100 %. Given that the analyst should expect losses during extraction and be cautious with respect to the limitations imposed by matrix effects, a reasonable target for the method would be that upon injection of a 0.1 ng/mL solution (equivalent to a 40 % recovery) of analyte in clean solvent, the signal:baseline should be 10:1 or greater (Figure 9.9). At this point it is worth emphasizing this objective of a signal: baseline

ratio of at least 10:1, or more whenever possible, for this exploratory experiment using analyte in clean solvent. There are many examples of methods that are validated with a S/B of less than 10:1 at the LLOQ but, in general, methods with higher sensitivity at the LLOQ tend to be more rugged reliable with improved precision. However, depending on the desired calibration range and concentration of the analyte at the ULOQ, care must be taken to not have a LLOQ standard that is too concentrated because problems such as carryover (Section 9.7.2) and saturation of the detector (Section 7.3) could then come into play. In the end, the analyst will have to balance the instrumental sensitivity with other factors such as the LLOQ, method recovery, matrix effects, carryover and stability, in order to arrive at the most appropriate approach; the use of spikes during early development play an important role in this process. Obviously, adjustments could eventually be made to the final volume of the extract or injection volume, but this type of approach reveals what level of instrumental sensitivity is required, and establishes the basis for determining the amount of analyte to be added to the control matrix when preparing spikes for method development. In this example, spikes at a concentration of approximately 2.5 ng/mL would be appropriate even if the initial recoveries were low and/or the matrix effect was severe.

Now that the sample size and concentration of method development spikes are known (2.5 ng/mL in the present example), the concentrations of the analyte and SIS spiking solutions to be used in preparing calibrators and QC samples can be calculated based on the volume of each solution to be added. If the analyst desires to use a spiking volume of 100 µL, then a spiking solution prepared at 12.5 ng/mL would be required (remember that the sample volume in this example is 0.5 mL).

At this stage it is also important to investigate the question of an appropriate and available surrogate internal standard (Section 9.4.5). If an isotope-labeled SIS is not available an analog SIS that is as structurally similar as possible must be selected (Section 9.4.5c). However, early development to assess recovery, sensitivity, stability and matrix effects can be initiated with what might be considered to be a poor internal standard, while looking for a more suitable analog or synthesizing a stable labeled compound. If the full benefit of an SIS is to be achieved, it should be added to the sample (spiked control matrix in these exploratory experiments) before any other sample preparation steps (see Figure 1.2), at a concentration of at least 50 times the LLOQ to ensure a strong response while the method is being optimized. However, there are some pitfalls that should be avoided, e.g. in the case of bioanalytical methods for analyte in plasma it is important to

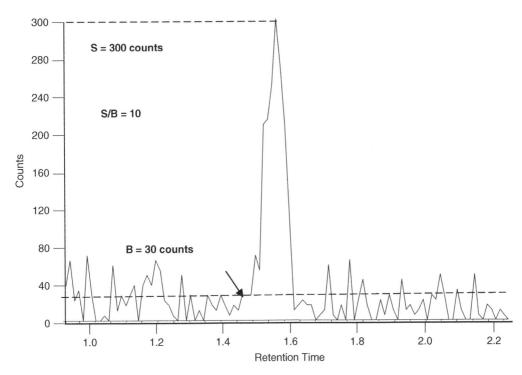

Figure 9.9 A chromatogram obtained from the analysis of a plasma extract showing a signal:background ratio (S/B) of approximately 10:1.

not add the SIS in large volumes of organic solvent since that will precipitate the protein before the SIS has time to equilibrate in the same way that the analyte already has done, i.e. even for an isotope-labeled SIS the extraction efficiencies F_a' and F_{SIS}' (Section 8.5) could be very different and their ratio irreproducible. Similar consideration should also be given to the preparation of the analyte spiking solution. In a more general sense it should be borne in mind that the experimental conditions adopted in one step of the analytical method can have a strong influence on later steps; the solution strength of solvents is a good example of this, e.g. the solvent used to reconstitute the final analytical extract, after extraction and taking to dryness, should not be of higher solvent strength (Section 4.4.2a) than the HPLC mobile phase into which it is injected.

As method development progresses, modifications to the method and chromatography will generally lead to higher recoveries and better sensitivity, resulting in improved signal:baseline at the LLOQ. In many instances the size of the sample aliquot can be reduced to improve efficiencies and conserve sample. This approach has the added benefit of generating a more dilute extract, which

in many instances reduces matrix effects with improved precision and method ruggedness.

9.5.7 Evaluating Sensitivity and Linearity in Matrix

Once a basic method (including sample preparation and instrumental analysis) has been developed, and initial estimates of sensitivity, selectivity and recovery have been made using clean solutions of analytical standard (Section 9.5.6), the next step is to further investigate the calibration curve using matrix matched calibrators for sensitivity (slope of curve, S_a in Section 8.5), linearity, robustness etc. If there are problems in this regard it is important to discover them and take remedial measures as early as possible. The desired characteristics of the calibration curve must be determined first, including LLOQ and ULOQ (see comments in Section 9.5.6), the number of concentration values over the desired range and the number of replicates at each concentration (see Section 2.6.1). At this stage the concentrations of the spiking solutions and the volumes that will be used for spiking the matrix have already been determined (Section 9.5.6). Practical details of how to prepare matrix matched calibrators are described in Section 9.8.1. It is important to also include extracts of matrix blanks in this initial investigation of analyte-in-matrix in order to check the selectivity.

It is always preferable to work with a calibration equation (Section 8.5) that is linear with respect to both analyte concentration and the parameters of the equation (i.e. equation [8.19a], Section 8.3.2), although an equation that is quadratic in concentration (Equation [8.19b]) can be practically usable. Nonlinearity of a calibration curve (Figure 9.10) with respect to concentration can arise for a variety of reasons, some of them intrinsic to the method (e.g. nonlinearity with respect to equation parameters as a result of overlap of isotopolog distributions of analyte and SIS, Section 9.4.5b) and some of them instrumental (i.e. in that portion of the entire analytical method illustrated in Figure 1.3). Some of the more common instrumental reasons for non-linearity at both the low and high concentration ends are listed in Table 9.5. In a general sense, negative deviations from linearity at the high concentration end can be attributed to saturation effects of some kind, e.g. saturation of an ion detector operating in ion counting mode (see discussion of detector dead time in Section 7.3), or of the ionization capacity of the ion source (most common for ESI sources, see discussion of Figure 5.32 in Section 5.3.6). At the low end, deviations can arise from losses of analyte on adsorption sites

(accounted for by the parameters f_a and l_a in Section 8.5) or from losses of ions as a result of the contamination of metal surfaces in the ion optics. Another possibility is simply a failure to notice that the concentration is close to the LOD and the data system is attempting to integrate peaks whose signal:background ratios are too low for accurate and precise data to be obtained; this is a good example of the importance of visual inspection of chromatograms to check that the automated peak-finding and integration routines are performing properly.

A possible approach to addressing nonlinearity that is not the result of cross-contributions between analyte and SIS (Section 9.4.5b) is to simply limit the range of reliable response to the linear region. While this is usually defensible with respect to negative deviations at high concentrations where the reasons are more clear-cut and reproducible, it could be problematic at the low end depending on the causes of the nonlinearity there since some of these will be intrinsically irreproducible. If the low end nonlinearity sets in at concentrations close to the desired LLOQ, it is advisable to undertake additional method development work to discover the cause and eliminate it.

Among the experimental variables that might be involved, matrix effects (Section 5.3.6a) are always a possibility. The ruggedness with respect to injection of large sets, and robustness of the calibration curve towards

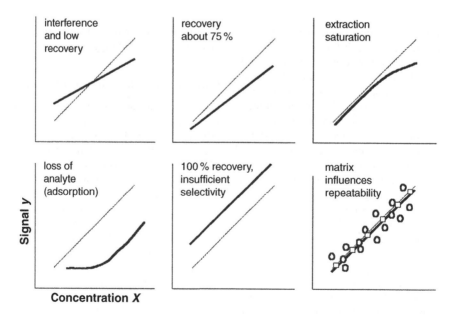

Figure 9.10 Schematic representation of some distortions of the calibration curve caused by various experimental effects. Dotted line = reference (ideal) method; solid line = test method. Modified from Meier and Zünd, *Statistical Methods in Analytical Chemistry*, 2nd *Edition* (2000), with permission of John Wiley & Sons Ltd.

Table 9.5   Summary of possible instrumental reasons for deviations from linearity in the calibration curve

Instrument component	Low end of curve	High end of curve
Ion Source		
EI/CI:	Reached LOD.	Running too concentrated (ion detector saturated).
ESI/APCI:	Reached LOD. Not running in optimum ESI solvent.	Matrix effects leading to poor release of analyte from droplets (ESI). Running too concentrated (ion detector saturated)
Ion transport optics		
EI/CI:	Dirty optics (not at maximum ion transmission efficiency).	
ESI/APCI:	Dirty optics (not at maximum ion transmission efficiency).	
Mass Analyzer		
Q or QqQ in SIM/MRM:	Reached LOD.	Running too concentrated (ion detector saturated).
3D and 2D ion trap in full scan MS/MS:	Reached LOD. Reached maximum ion fill time.	Running too concentrated (ion detector saturated). Experiencing space charge.

changes in control matrix used to prepare the calibrators, are obvious possibilities for investigation of any such nonlinearity. Other robustness factors to be checked in this regard include the mobile phase composition that might not be suitable for the ionization conditions.

Having thus established a first attempt at a calibration curve, it can be advisable at this point to start initial testing of robustness by preparing some QC samples in bulk (Section 9.8.2) and injecting some large sample sets. Based on the results obtained, it may be necessary to undertake some systematic robustness testing (Section 9.8.4) and make adjustments to the method if necessary. Once the resulting second-generation method has been optimized to this stage, it is ready to be finalized as described in Section 9.8 and eventually taken on to validation.

9.6 Matrix Effects

There has been a great deal of discussion of matrix effects on ionization efficiency in previous chapters, particularly Section 5.3.6a. (Not being discussed here is '*matrix interference*', i.e. direct contributions from matrix components to the analytical signals in the SIM or MRM channels used to monitor the analyte and SIS; these interferences are easily detected in analyses of extracts of control matrix or incurred sample). Suppression of ionization efficiency by co-eluting matrix components is the more common, but enhancement is also observed (Section 5.3.6a). While it is true that operating under conditions of 'total solvent consumption', e.g. by using very low flow rates for ESI combined with additional heating, can reduce matrix

effects to low levels or even eliminate them (Schneider 2006), this is not a practical solution in the context of the great majority of real analytical tasks. In this section a practical approach to evaluating and dealing with matrix effects in development of real-world analytical methods is described.

A practically useful distinction is that between endogenous and exogenous matrix effects. The former correspond to the effects of components that were co-extracted from the sample matrix together with the analyte and SIS, and also co-elute in the analytical chromatography step; these effects are initially evaluated using an appropriate control matrix (Section 9.4.7) and ultimately, by reanalysis of incurred samples (Section 10.2.9c). In contrast, exogenous matrix effects are, in one sense, an extreme case of the phenomenon of 'chemical background' that can arise from various sources not connected to the analytical sample (Aebi 1996). In the present context, however, exogenous matrix effects are best regarded as a special case of chemical background of significantly greater magnitude and which affect the analytical signal indirectly via suppression or enhancement of ionization efficiency. They arise from components extracted from sample containers or connecting tubing (Mei 2003; Xia 2005), or that possibly were present as impurities in extraction solvents and survived the solvent switching procedure that was required to ensure that the analytical extract injected into the HPLC had elution strength no higher than that of the mobile phase. Other sources of exogenous matrix effects are additives commonly used to stabilize samples during storage, e.g., Li-heparin commonly used as an anti-coagulant for blood samples

(Mei 2003). A special case is that of the dosing vehicles typically employed in high concentrations to dissolve the drug candidates in dose formulations in drug discovery investigations (Shou 2003). The practical usefulness of the distinction lies in the different methods available to remove or alleviate the two kinds of matrix effect, since endogenous effects are intrinsic to the matrix of the analytical sample and can thus only be dealt with by improving sample clean-up and/or analytical chromatography, while exogenous effects can also be addressed by changing storage containers, solvents etc.

A related problem in biomedical analyses (possibly with analogies in other applications) arises from the presence of metabolites of the analyte in incurred analytical samples, but not in the control matrix spiked with analytical and internal standards used in method development and validation. Problems arising from this distinction are most likely to be important when rate of sample throughput is emphasized at the expense of chromatographic efficiency (Section 9.3.4) and can be addressed by re-analysis of incurred samples as described in Section 9.4.7b. The discussion in the present section is concerned with matrix effects during method development, but these must also be addressed in the context of method validation (Section 10.4.1d).

9.6.1 Evaluating Ion Suppression in Method Development

Methods for detecting and evaluating ionization suppression (or occasionally enhancement) resulting from matrix effects in LC–MS methods were described in Section 5.3.6a. The most reliable of these methods are those that incorporate post-column infusion of a clean solution of analyte during the HPLC separation of an extract of a blank control) matrix (Bonfiglio 1999). This approach (Section 5.3.6a) permits determination of elution time windows in which the analyte and internal standard will be subject to minimal (hopefully zero) suppression and/or indicate how the chromatographic conditions (or possibly those for sample preparation) should be altered to provide such windows. Such a procedure is essential for methods in which calibration curves are obtained using clean solutions of analytical standard (Section 8.5) rather than matrix matched calibrators, but is highly preferable even in the latter case as it minimizes any uncertainties arising from variability of the level of matrix effects. Once elution conditions have been optimized with respect to matrix effects, the level of the residual suppression or enhancement can be determined by a procedure (Matuszewski 2003) that separates out the closely related parameters ME (matrix effect) and recovery efficiency

(RE, equivalent to F_a' in Section 8.5) that together determine the overall process efficiency (Section 5.3.6a). It is worth noting again that the results obtained using the Method of Standard Additions (Section 8.5.1b) are not subject to uncertainties arising from ionization suppression (although the latter may well be occurring), since here the sample matrix acts as its own control matrix. As discussed in Section 5.3.6a, it was shown recently (Matuszewski 2006) that relative matrix effects can also be detected via the precision of the slopes of calibration curves (CV %, Equation [8.4]) determined using LC–ESI–MS and different lots of a blank matrix (biofluid in this example). The empirically determined acceptance limits were CV $\leq 2.4\%$ for methods that incorporate an SIS and ≤ 3–4% for those that do not use an internal standard.

Extracted blanks that are spiked with the analyte and SIS after extraction but before analysis are commonly used to distinguish between loss due to the sample extraction procedure (recovery, RE) and loss due to ion suppression effects (Section 5.3.6a). By comparing the results from samples spiked prior to sample extraction with those spiked after extraction but before analysis, the contribution to recovery due to loss during extraction (F_a' in Section 8.5) can be distinguished from the total apparent recovery (loss due to sample extraction plus the effect of ion suppression). When responses of extracts of blanks that are spiked post-extraction are compared to similarly prepared matrix free blanks (i.e., clean solvent at the same concentration) the degree of ion suppression resulting from the control matrix extract can be easily estimated. These approaches are simplified versions of the general procedure (Matuszewski 2003) that separates out the parameters ME (matrix effect) and recovery efficiency (RE, equivalent to F_a' in Section 8.5) that together determine the overall process efficiency (PE, see Section 5.3.6a).

Because of the pervasive and pernicious occurrence of matrix effects, it is usually advisable to build a routine check on the extent of these effects into any method that has been shown to be subject to them, e.g. the ME/RE procedure (Matuszewski 2003) described in Section 5.3.6a. A particularly deceptive cause of ionization suppression that is not really a matrix effect is the mutual interference of an analyte and its co-eluting SIS (Liang 2003; Sojo 2003), discussed in some detail in Section 5.3.6a. While any level of suppression (or enhancement) of ionization efficiency is undesirable, the mutual suppression of analyte and an isotope-labeled SIS appear to be equal, with minimal effect on the validity of the quantitative analysis, although it may adversely affect limits of detection and quantification as a result of the depressed signal levels. Recall that all suppression effects can be minimized by reducing ESI flow rates and this is true also of the analyte–SIS mutual suppression problem.

A somewhat different origin for co-eluting compounds that can interfere either directly in the SIM or MRM channel for the analyte, or indirectly via suppression or enhancement of ionization efficiency, is that of late eluting compounds (presumably derived from the matrix) from the previous analysis that remain on the column at the injection time of the next sample. In one sense this phenomenon is another example of the negative consequences of achieving high sample throughput at the expense of other considerations. Such interferences are liable to appear at any time during subsequent analyses. The characteristic of an unexpected peak observed in a chromatogram that indicates its origin as a late eluting compound is its width, generally much greater than that of expected peaks observed under the experimental conditions (Dolan 2001d). Their potential to cause trouble in practice can be assessed by analyzing an extract of blank (control) matrix using the same chromatographic conditions as for the analytical samples but extending the run time well beyond that planned.

This problem is more common with isocratic HPLC methods because gradient methods usually end with conditions of high solvent strength (Section 4.4.2a) for the mobile phase that will remove strongly retained compounds from the column. For isocratic methods the late eluter problem can be removed by extending the chromatographic run time if acceptable from a throughput viewpoint, or by introducing column flushes using a strong solvent between injections (this requires re-equilibration of the column as for gradient methods). Alternatively, the compound(s) that are strongly retained relative to the target analytes could, by the same token, likely be removed by a more rigorous sample clean-up, e.g. SPE. Another approach that can sometimes be helpful is to use a monolithic column (Section 3.5.8) that permits much higher flow rates and shorter analysis times while maintaining chromatographic separatory power, thus flushing out the strongly retained compounds without time penalty relative to using conventional particle-packed columns. Note that the late eluter problem is different from that of carryover (see Section 9.7.2) and requires different remedial actions.

It is good practice in the context of checking for matrix effects to conduct regular checks on the response ($R_a{'}$ and $R_{SIS}{'}$ in Section 8.5) during the course of method development (and also validation, and analysis in a multi-sample run). This can be done in a number of ways but one of the best and most convenient is to simply plot the IS response from beginning to end (recall that it is essentially universal practice to add the SIS to a fixed constant concentration). It may not be practical to develop universal criteria to determine acceptability limits for observed variations, e.g. the criterion for a stable isotope SIS would probably differ from that for an analog SIS, as would that for an LLOQ defined at a signal:background ratio of 10:1 vs 50:1. Nonetheless, it is highly desirable to have an analytical method with a reasonably consistent response for both SIS and analyte from the beginning to end of method development in the interest of contributing to acceptable degrees of both robustness and ruggedness (Section 9.8.4). For example, during method development a check of this kind might be conducted by programming an autosampler to repetitively analyze the same sample set overnight to monitor SIS response over a period that would be significantly longer than what would be typically of routine sample analysis.

9.6.2 Addressing Ion Suppression/Enhancement

Discovery of significant suppression (or enhancement) of ionization efficiency for the analyte and/or SIS as a result of co-eluting matrix components, while serious, need not constitute a roadblock to use of the analytical method. There are several ways in which the effect can be mitigated, often to an extent sufficient for the purposes of the analysis. Use of an appropriate isotope-labeled SIS that co-elutes with the analyte, in conjunction with matrix matched calibrant standards, usually provides reliable results if all other appropriate precautions are taken (Section 8.5.2) since co-elution of SIS and analyte should expose both equally to the same matrix effects. A potential complication to this strategy, which has arisen in some cases, is the mutual suppression of SIS and analyte even at trace level concentrations (Liang 2003; Sojo 2003), discussed in Section 5.3.6a. It appears that this phenomenon has minimal effect on the validity of the quantitative analysis provided that analyte and SIS do genuinely co-elute, although it may adversely affect limits of detection and quantification as a result of the depressed signal levels. Where possible, this is probably the best available approach to dealing with matrix effects. However, a recent publication (Wang 2007) described an example in which a deuterium-labeled SIS was sufficiently separated from the analyte that the matrix effects were significantly different in that the analyte to internal standard peak area ratio changed with two specific lots of commercially supplied human plasma (a relative matrix effect). Such differences in elution times of SIS and analyte are limited to deuterium-labeled SIS and are not observed in the case of heavy atom labeling (^{13}C, ^{15}N etc.

When a heavy atom isotope-labeled SIS is not available, other approaches to dealing with matrix effects must be taken. In general (but with exceptions, see Section 5.3.6a), APCI is less subject to matrix effects than ESI, so if the analyte responds well in APCI conditions it is worth trying a change of ionization technique. Otherwise, a more rigorous clean-up step in sample preparation may be necessary, with consequent decrease in sample throughput and possibly lower overall recovery. An example of how introduction of SPE clean-up improved the variability of matrix effects in the course of a multi-sample run is shown in Figure 9.11. If it is suspected that the problem is one of exogenous matrix effects (Mei 2003; Xia 2005), discussed in the introductory comments in Section 9.6, changes to materials used in storage containers, connecting tubing etc. may solve the problem.

Modifications to the existing chromatographic method, guided by the post-column addition method (Bonfiglio 1999) to find elution time windows relatively free of matrix effects (Section 5.3.6a), are another valid approach. If this does not result in satisfactory performance, possibly a complete change of chromatographic technique is called for, e.g., use of HILIC (Section 4.4.2c) for highly polar analytes or even normal phase chromatography for less polar compounds. Of course such a dramatic change also implies a need for corresponding changes to the ionization conditions in the mass spectrometer and probably also to the solvent chosen for the final analytical extract that must match the mobile phase in eluotropic strength. The mass spectrometer itself might offer the answer via the very high sensitivity available in modern instruments, so that dilution of sample extracts might be possible while still meeting targets for LOD and LLOQ. The relative importance of matrix effects is generally found to decrease with overall dilution, presumably reflecting the same fundamental phenomena that lead to the same result when decreasing flow rates and/or ensuring complete evaporation of droplets in the ion source (Section 5.3.6a). This dilution strategy is sometimes used for analyses of blood plasma that use simple protein precipitation (Section 4.3.1f) as the only sample preparation method. Cleaning the API interface (build up of salts and other nonvolatile extract components) in the API interface will also help to control ion suppression in some instances but diluting the extracts will help to keep the interface clean for longer periods, similar to what would be expected with cleaner sample extracts in general.

Other approaches to alleviating matrix effects that are possible but not often realistic in real-world practical situations include the use of nano-electrospray (Section 5.3.6a) with packed capillary columns, post-column splitting so as to deliver flow rates compatible with nanospray, post-column addition of signal enhancing agents, use of pre-dose plasma samples from the same test animal to be dosed as the 'blank matrix' used to prepare matrix matched calibrators and QC samples.

9.6.3 Interferences

Matrix interferences (direct contributions to the analytical signals used to monitor the analyte and/or SIS) as distinct from matrix effects (indirect influence on analytical signals as a result of suppression or enhancement of ionization efficiency) are discussed briefly here. Thus, it is essentially the detection selectivity of the method that is being dealt with. Direct interferences are much more readily detected (through analyses of control matrix extracts and solvent blanks). As for matrix effects, as discussed above, it is useful to consider whether the interferences might be endogenous or exogenous in nature, since such a determination (comparison of the chromatograms obtained for the matrix and solvent blanks) indicates what kind of mitigating approach is appropriate (see the corresponding discussion for matrix effects in Section 9.6.2).

Another aspect that is common to matrix interferences (direct contributions of matrix components to the signal measured for analyte and/or SIS) and matrix effects (suppression or enhancement of ionization) is that of the consequences of the presence of metabolites or other types of degradates when analyzing incurred analytical samples. Such interferences are in principle absent from the control matrix used for matrix matched calibrators, QC samples etc. Thus use of re-analysis of incurred samples to evaluate and consequent matrix effects was discussed in Section 9.4.7b, and applies equally to matrix interferences arising from presence of metabolites. Variations of metabolite levels among samples (e.g. from different time points in a pharmacokinetic study), which can lead to parallel variations in the extent of both matrix effects and matrix interferences, are an example of how some problems can arise unexpectedly despite prior precautions.

As for matrix effects (ionization suppression), improvement of chromatography conditions (e.g. use of ultra-small particle stationary phases to improve chromatographic separation without significant increase in analysis time, Section 3.5.7) can remove or mitigate matrix interferences, including those arising from presence of metabolites or degradates in incurred samples. More extensive clean-up procedures can achieve the same objective although, if it is required to retain the metabolites or degradates in the final extracts, some care is required to ensure that only unwanted matrix components are removed.

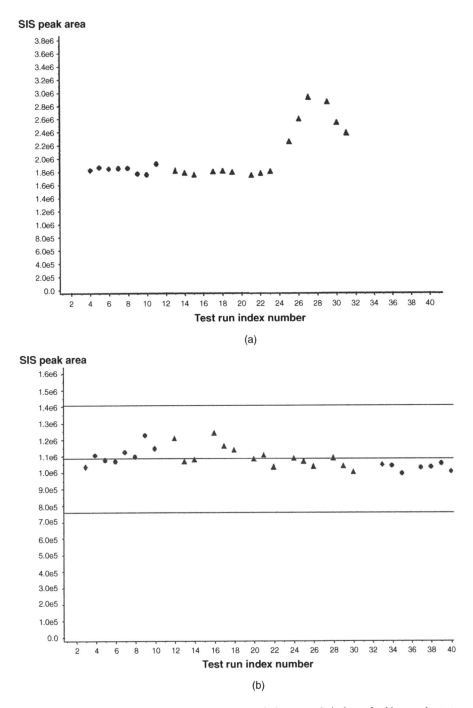

Figure 9.11 Example of how the extent of matrix effects can vary during an analytical run. In this case the test samples were QC samples (extracts of control matrix that had been spiked with an analog internal standard). In (a) the extraction was done using liquid–liquid extraction and in (b) a SPE extraction/clean-up was performed. In (b) the horizontal lines indicate the mean ± twice the standard deviation.

9.7 Contamination and Carryover

9.7.1 Laboratory Contamination

9.7.1a Common Causes and Impact

An on-going threat in trace analysis is that of contamination. A common example involves inadvertent transfer of analyte and/or internal standard from a location or piece of apparatus used to handle large quantities (e.g., synthesis or handling of pure standards) to one used for trace level samples and extracts. It is best to define entirely different work areas for low, medium and high level quantities, and to apply strict rules about not transporting materials or apparatus (e.g., syringes, standard volumetric flasks etc.) directly from the high level to low level areas. Another common example is the case when a sample containing an unusually high and unexpected concentration of the analyte is analyzed along with other samples with analyte concentrations several orders of magnitude lower. This type of situation cannot be predicted in advance, so adequate quality control procedures including the use of blanks must be employed to detect and mitigate the contamination as soon as possible before too many samples are affected (see below). In any event it is not acceptable practice to segregate apparatus, such as rotary evaporators and pipets for the generation of quality control blanks, from similar apparatus used to process the study samples.

Sloppy analytical technique can also lead to widespread laboratory contamination, so it is imperative that any analyst working at trace levels has a demonstrated expertise when it comes to general sample handling and analytical techniques, including the operation and maintenance of apparatus used in the laboratory. This is especially true for methods that use automated liquid handling devices that could be prone to dripping or other phenomena that cause cross-contamination. Extraction schemes and formats in which extracts are arranged in close proximity to one another, e.g. 96-well or other similar formats, can be especially problematic in this respect (an example is described in Section 9.7.1b).

Disposable glassware and plastic containers should be used whenever possible but, if glassware is to be re-used, which is often the case when using rotary evaporators, Class A pipets, volumetric flasks, reflux condensers, round bottom flasks or other expensive apparatus, it is important to ensure it is thoroughly cleaned, e.g. with a strong oxidizing cleaner or detergent, followed by thorough rinsing with water and then clean organic solvent, with final drying in an oven. A very effective way to eliminate contamination on glassware that is not used for critical measurements, e.g. round bottom flasks, can be to bake in a furnace for approximately four hours at 400 °C.

9.7.1b Monitoring Contamination in the Laboratory

The only direct technique for monitoring contamination in the laboratory is to use various types of blanks (Section 9.5.6b) at strategic locations or steps in the method. Depending on the nature of the analyte and matrix, laboratory contamination may only be detected in extracted matrix blanks, and therefore at least one extracted matrix blank should be included with every method development, validation and sample analysis batch; note that the preparation of extracted blanks in duplicate is often considered to be the minimum requirement for many methods. Solvent blanks and method blanks can also be invaluable with respect to detecting contamination in the laboratory, but these should always be used in conjunction with extracted blanks unless specific experiments or historical data show that they are equivalent to extracted blanks in this respect for the method in question.

In some instances laboratory contamination may occur but not be detected in the blanks that are run concurrently with analytical samples. This situation is often discovered via unusually high concentrations of analyte detected in the standards or quality control samples at the lower end of the curve. This relatively common indicator is often overlooked until the problem is so bad that the incidence of run failures increases significantly.

When contamination is detected in the laboratory, an investigation and corrective action strategy should be developed to determine the source of the contamination and what procedures will be necessary to control it. No samples should be analyzed until it is shown that the problem has been corrected and that blanks and curves can be generated successfully on a routine basis. Additional procedures or modifications to the analytical method may be required to prevent the reoccurrence of the contamination. If it was determined that the blanks used concurrently did not adequately indicate that contamination existed, an additional investigation should be conducted to determine why, and what additional steps or types of blanks will need to be generated for the method.

Even if laboratory contamination is not suspected it may be deemed necessary to run one or more control sets with blanks only, prior to running samples. This is most common with extremely valuable samples where there is no archive and the analyst has essentially only one chance to generate reliable results.

There is no one procedure to determine which types of blanks would be best to use. Nonetheless, a method depending on the analyte, matrix, method and LLOQ or LOD required will have to be developed to ensure that laboratory contamination is detected as early as possible.

An interesting case (Chang 2006) in which a regimen was developed to detect laboratory contamination arising from cross-contamination among samples prepared in adjacent wells of a 96-well plate (Section 4.3.3c) is discussed here as an example. Such contamination cannot be detected by analysis of calibration standards and QC samples interspersed with analytical samples. As noted above, in more general current practice cross-contamination is discovered only when analytes are detected in blanks of various kinds (Section 9.5.6b), or in placebo or pre-dose samples in bioanalytical applications, or at abnormally high levels in low concentration standards or QCs. Judicious insertion of appropriate blanks and low concentration QCs in a sample stream provides some protection on this basis. However, it was emphasized (Chang 2006) that in general there is no approach that allows discovery of cross-contamination among other samples of higher concentration, or among adjacent wells of a high density device, with the result that erroneous results can be reported even though a batch analysis meets all the usual acceptance criteria.

The strategy used to monitor inter-well contamination (Chang 2006) involved addition of two different marker compounds added in an appropriate pattern to the 96 wells. In the original example of analysis of plasma samples for lopinavir and ritonavir (the active ingredients of a widely prescribed human immunodeficiency virus (HIV) protease inhibitor), two related compounds (amprenavir as marker A and saquinavir as marker B) were selected to monitor the contamination of individual wells. These markers were selected because they co-extracted with the analytes but did not co-elute with either the analytes or the internal standard (a proprietary structural analog of ritonavir), thus avoiding the possibility of ionization suppression. Solutions of the marker compounds were added to the 96-well plate in the checkerboard pattern shown in Figure 9.12. The plate was then evaporated to dryness; these additions and the drying process must be done with care to avoid cross contamination at this stage, a precaution that was aided in the original work (Chang 2006) by the fact that the marker solutions were mostly aqueous (high viscosity and surface tension) and the solutions were deposited at the bottoms of the wells. The additions of plasma and internal standard were then performed following the validated procedures developed for this assay. When the plate was subjected to analysis in the usual way, any cross-contamination into well 5D, for example, would be indicated by detection of marker compound B from any or all of wells 5C, 4D, 6D or 5E. The distances between well 5D and neighbouring wells also containing marker A, i.e. wells 4C, 6C, 4E and 6E, are appreciably larger than those from its neighbours

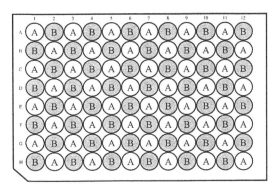

Figure 9.12 The pattern of applying contamination markers A and B to a 96-well plate to determine the extent of inter-well contamination. Note that 'A' and 'B' are also used (in smaller font size) to label the first two rows of wells. Reproduced from Chang, *Rapid Commun. Mass Spectrom.* **20**, 2190 (2006), with permission of John Wiley & Sons Ltd.

containing marker B so the risk of cross-contamination from these neighbouring wells is significantly lower (and would not be detectable using this two marker scheme).

The original paper (Chang 2006) gives many more details of the approach, and tests of its effectiveness indicated a detection power of close to 100 % compared with 3–50 % when relying on the conventional practice of detection of analyte in blanks or of high levels in low concentration calibrators or QC samples. It was also emphasized that the marker technique should not be regarded as a replacement for careful design of the sample preparation process and sample sequence within a batch. Rather, it provides a means to monitor the performance of assays incorporating 96-well technology and to detect any cross-contamination.

9.7.2 Carryover

Carryover can be regarded as a special case of contamination but it is useful to distinguish between the two; the term is usually applied to the chromatography–MS step and/or on-line sample preparation followed by chromatography–MS and refers to a detector response from a previously injected sample that occurs when a subsequent injection is made. Chromatographic peaks attributed to the previously analyzed sample may be observed in the subsequent chromatogram for the blank and these spurious peaks may co-elute or interfere with the target analytes when analyzing samples. Carryover can also sometimes be observed via the mass spectrometric signature of the chemical background in the vicinity of the analyte retention time; in such cases some of the analyte is probably trapped in some way at some site(s) between injection valve and mass spectrometer and is gradually leached away. One reason for the increasing concern about carryover is the ever-increasing sensitivity of analytical mass spectrometers with the consequent lowering of LOD and LLOQ values.

Carryover is sometimes confused with the problem of late eluting peaks (Dolan 2001d) that also manifests itself through observation of unexpected peaks in a chromatogram obtained on injection of a blank of some kind following a previous injection of an analytical extract. However, the 'late eluter' problem is more a result of being in too much of a hurry to start the next analysis in a batch, before elution of all components in the previous injection have had a chance to elute, and is discussed in Section 9.6.1. In contrast to the late eluter problem, carryover results from a failure to completely eliminate residual compounds (not only target analytes) from a previous sample injection and can reflect inadequate flushing of some components of the analytical train after the previous injection or temporary trapping of compounds in a small dead volume, or some particularly 'sticky' compounds becoming adsorbed to a surface somewhere in the analytical train (Dolan 2001, 2001c; LEAP Technologies 2007). In general the degree of carryover (see below) is compound specific, reflecting the compound 'stickiness' contribution to the problem.

9.7.2a Evaluating Carryover During Method Development

In general, carryover (and also contamination) is not a point-source problem, i.e. 'sticky compounds' will stick everywhere they can, and the analyst should not expect to be able to fix such problems by eliminating it at only one location in the analytical train. The main exception to this rule arises when the problem is the result of dead volumes. For example, fittings and ferrules made of the inert polymer PEEK (Section 9.5.5d) are popular because of the convenience of finger-tightening and behave well at back pressure, up to ~4000 psi, but can slip and distort at higher pressures conditions creating dead volumes that can temporarily 'hide' some of the injected sample. Another source of dead volumes are grooves worn in the polymer seal in the rotor seal of the loop injector valve (Section 4.4.4a). Examples of adsorption sites for 'sticky' compounds include stainless steel or other metal surfaces (ionic interactions with ionic or basic compounds) and organic polymers (hydrophobic interactions with lipophilic compounds). Insufficient solubility of the analyte can result in adsorption of some analyte on internal walls of almost any of the components of

the analytical train; if the solvent can not be changed for some reason, extensive rinsing of all components of the injector between injections is necessary to reduce carryover to acceptable levels. Since the liquid flows are usually laminar, i.e. much slower near the walls, a large amount of solvent is required to properly flush out adsorbed analyte and other compounds in the sample extract (including the SIS).

For convenience in determining appropriate approaches to elimination or mitigation, carryover is sometimes classified as 'classic' or 'constant'. If the carryover peaks regularly decrease in size as blanks are injected consecutively, the phenomenon is categorized as classic, and is often associated with the existence of dead volumes. If this is suspected it is advisable to carefully check and possibly replace all fittings, ferrules etc., with the objective of eliminating dead volumes as the source of the problem. If the problem persists, the analyst must proceed on the basis that adsorption sites for 'sticky' compounds are the cause of the carryover. In contrast, if consecutive injections of blanks produce spurious peaks of approximately the same size, the phenomenon is termed 'constant carryover' and the most likely source of this problem is contamination in the sense of Section 9.7.1, e.g. in the solvent used to prepare the analytical extract, or from improperly cleaned laboratory glassware. Of course in real life intermediate situations can arise, e.g. if the sizes of the carryover peaks do decrease for consecutive injections, but not to a great extent. Another way in which carryover can be detected (Dolan 2001) is by comparing two calibration curves, one in which the calibrators are injected in order from low concentration to high and the other in the reverse order. Effects of carryover can be observed in the latter case at the low concentration end if a statistically significant positive intercept (a signal for zero concentration injected) is obtained.

Before discussing steps to address carryover, it is appropriate to mention how the extent of the problem is evaluated relative to the requirements of the analytical method. For a quantitative assay the problem is obviously going to be most important with respect to percentage error for a sample at the LLOQ that is injected immediately after an injection of a sample with a concentration at the ULOQ (or greater). With this in mind, one possible experimental measure of carryover would be $(PA_{bl1}/PA_{ULOQ}) \times 100\%$, where 'PA' denotes peak area and the subscript 'bl1' denotes the peak observed on injection of the first blank sample (could be a solvent blank or a matrix blank) injected immediately following injection of a sample (e.g. a calibrator or a QC sample) at the ULOQ. Then the contribution of such a carryover to a sample at the LLOQ can be estimated as $([ULOQ]/[LLOQ]).(PA_{col}/PA_{ULOQ}) \times 100\%$, where e.g. [ULOQ] represents the ULOQ concentration. If the specified accuracy limit for an assay at the LLOQ is 20%, say, and $([ULOQ]/[LLOQ]) = 10^3$, this implies an upper acceptability limit of 0.02% for the observed ratio $(PA_{cbl1}/PA_{ULOQ}) \times 100\%$. As always, carryover should be evaluated and dealt with on a fitness for purpose basis, i.e. the problem should be assessed and controlled to within acceptable limits rather than making strenuous efforts to eliminate it altogether.

Carryover is one of the major problems faced by trace quantitative analysts and should be checked regularly throughout the entire process of method development; it must also be evaluated and assessed during method validation (Section 10.2.9b) and sample analysis (see Section 10.5.5, where a more detailed method of evaluation of the effect of carryover on the accuracy of an assay (Zeng 2006) is described).

9.7.2b Addressing Carryover in the Method

Detailed protocols for addressing carryover, once detected and evaluated as significant relative to the needs of the assay, have been described (Dolan 2001c), and an approach particularly relevant attention to such problems arising within an autosampler (LEAP Technologies 2005) is complemented by an instructive account of a case study in which the source of the problem within an autosampler was tracked down (Vallano 2005). The following abbreviated account is drawn from all these sources.

A suggested procedure for assessing and addressing carryover might be as follows:

(1) Check all compression fittings and tube connections from those on the injection valve and everything else downstream to the detector (mass spectrometer in our case). When connecting tubing to a component or an injection loop to the injector valve, ensure that the ends are cleanly cut and perpendicular to the length of the tubing/loop, and that the tubing ends are fully inserted into the fitting; the tubing should be pushed in as the fitting is tightened to eliminate gaps between the end of the tubing and the bottom of the fitting. Some authors recommend foregoing the convenience of finger-tightened PEEK connectors and ferrules for the higher reliability of stainless steel fittings that can be checked by slight tightening using a wrench. It is unwise to try to tighten PEEK fittings when the system is still pressurized since the tubing can slip in the fitting when the nut is rotated; rather shut off the pump, loosen each fitting, push the tubing firmly to the bottom and retighten the fitting.

Other dead volume possibilities include scratches on the rotor of the injection valve. Simple matters, such as arranging that the tubing that carries excess sample solution from a loop injector operated in the 'overfill' mode (Sections 2.4.3 and 4.4.4a) to the waste container is positioned with its exit below the waste port of the injection valve, can avoid one source of carryover (back drainage of the excess solution if the exit tube is not arranged with a continuous downhill slope). Other mechanical possibilities include incorrect adjustment of the autosampler syringe with respect to the injector port of the injection valve, or a damaged seal in the rotor valve. Once all fittings and other suspect components have been checked and/or tightened (or even replaced), the carryover test should be repeated using a ULOQ sample followed by several blanks. If this has not resolved the problem, proceed with the following steps.

(2) Classify the nature of the carryover problem as classic or constant (Section 9.7.2a) by injecting a ULOQ sample followed by several blank injections (one or more of mobile phase or solvent used for the analytical extract or extracted matrix blanks).

(3) If the carryover peak is of approximately constant size contamination should be suspected; this can be tested by replacing the blank with a fresh sample believed to be free of contamination, e.g. mobile phase or solvent acquired from a different laboratory. If the carryover peaks persist the source may be another reagent or possibly contaminated glassware. To test the hypothesis that general contamination is the source of the problem, try adjusting the blank injection volume by a factor of at least two; if the carryover peak intensity increases or decreases in proportion to injection volume, it is highly likely that the blank is contaminated.

(4) At some point it can be helpful to determine whether the carryover occurs only for the analyte(s) of interest or also for other compounds. If indeed the carryover is a more general phenomenon it is likely to be a result of physical problems with the autosampler hardware or system plumbing. On the other hand, if the carryover is compound specific, chemical approaches to the problem are indicated.

(5) If the constant carryover problem persists, or if the observations match the classic carryover symptoms, it is likely that the problem is related to adsorption of 'sticky' compounds on one or more components. A good strategy at this point is to attempt to narrow down the location of the adsorption site(s) by systematically removing components from the

system. Autosamplers are a frequent source of carryover problems (Dolan 2001; Vallano 2005), so a first attempt might be to disconnect this component and replace it with a manual injector that has not been exposed to the analyte or samples of interest. If the carryover problem disappears or is reduced to insignificant levels, the task is now to identify the source of the problem within the autoinjector. Frequently a change of the solvent used in the autosampler's wash cycle, from one that is equivalent in solvent strength to the mobile phase (no involatile buffers!) to one containing more organic component, or at a different pH, can solve the problem; alternatively the volume of wash solvent may be inadequate. Otherwise the problem may have arisen from the detailed electromechanical mechanisms of the autosampler, e.g. the method used to wash the outside surface of the autoinjector needle. Such problems are likely to be specific to the type and make of the autosampler so only some general suggestions can be made. The stator of the injection valve can be cleaned (e.g. with appropriate solvent(s) with ultrasonic assistance) and the rotor can be replaced perhaps with one made of a different material; the autosampler syringe may be incorrectly adjusted with respect to the injector port of the injection valve; the seal in the rotor valve may be damaged; the sample loop can be replaced possibly with one made of a different material. The autosampler syringe should be checked for wear and discoloration and for tightness of fit of the plunger in the syringe barrel.

(6) The loop filling technique (overfill vs partial fill, Section 2.4.3) can have a significant effect on the extent of carryover. Partial loop fill injections isolate the sample solution with a volume of mobile phase that was left in the loop from the previous injection, and this can minimize the sample solution coming into contact with valve rotor grooves, the waste port outlet and other fittings; if this approach is to be used to mitigate carryover it is important to arrange that the valve is plumbed such that the sample leaves in a 'last in first out' configuration, thus ensuring that contact of the sample solution with these problematic regions is minimized. On the other hand there are internal volumes of an injector valve that are not flushed with rinsing solvent from the autoinjector between injections, and this can give rise to carryover problems when partial fill injections are used (Vallano 2005). In contrast, in full loop injection mode, after the initial flush step

the sample loop is filled by injecting two or more times the loop volume of sample solution, so that analyte adsorbed on the problematic regions of the stator surface can be flushed to waste during the overfill step (Vallano 2005).

A third alternative filling technique that can alleviate carryover problems is known as the 'sandwich technique'; the autosampler can be programmed so that a solvent plug is picked up before the sample solution in order to isolate the syringe plunger tip from contact with analytes. This same solvent plug will also come to rest in the injection port inlet and adjacent rotor groove after the sample has been injected so that in a series of such injections the sample solution is sandwiched between two such plugs. Note that this procedure is successful only for partial loop fill injections. It can also be done with or without an air gap between the solvent and sample, and minimizes the contact made between the sample solution and the crossover port on the injector valve. Essentially, there is no universal solution to carryover problems associated with the injector.

If all such attempts to reduce carryover to acceptable levels fail, it may be necessary to develop a new method including one or both of sample preparation and chromatography. For example, if gradient elution is used in the method, carryover can sometimes arise either from adsorption of analyte on the entrance frit of the column that is released upon the next injection, or from analyte leaching from the injector during re-equilibration of the column between injections. In such cases the problem can be removed or significantly alleviated by increasing the organic concentration at the initial equilibration state, or by switching to an isocratic method. For bioanalytical analysis, carryover is expressed as described above as $(PA_{bll}/PA_{ULOQ}) \times 100\%$ where in this case the blank is an extracted blank (not a solvent blank); a blank response of $< 20\%$ of that of the LLOQ is often considered to be acceptable as it is consistent with the FDA Guideline (FDA 2001) that an interfering peak is defined as any signal that exceeds 20 % of the analytical signal at the LLOQ. (Note, however, that this approach can not estimate the effect of carryover on accuracy for unknown samples with concentrations higher than the ULOQ). If the carryover test fails this criterion, the LLOQ may be increased or the ULOQ may be reduced (or even both). Although this type of measure might be considered to be 'last resort' by many, it may be necessary if the carryover cannot be reduced to a level were it does not bias the results to a significant extent. More detailed accounts of carryover mitigation for methods using autosamplers with LC/MS are given elsewhere (Dolan 2001c; LEAP Technologies 2005). A novel approach to dealing with carryover (Dethy 2003; Yang 2004) in bioanalytical work uses direct infusion nanospray ionization MS/MS with a chip device (Section 4.4.7) without any HPLC or CE separation; serial sample introduction was done with a robotic interface and individual disposable pipet tips. This approach reported reduction of carryover to unobservable levels.

9.8 Establishing the Final Method

This section deals with final decisions about details of the new method discussed previously, ensuring that the separate steps (appropriate size of analytical sample aliquot, sample preparation, analysis) are mutually compatible and can deliver the required selectivity, sensitivity (LLOQ), linear range (ULOQ) etc. Practical issues concerned with some of the key aspects are also included here.

9.8.1 Curve Preparation

A brief introduction to the concept of a calibration curve in quantitative analysis was provided in Section 2.6, while Section 8.5 represented an attempt to express in algebra the calibration functions that apply in different circumstances to expose the sources of systematic uncertainty in each case. In this section some practical issues that arise in preparation of the calibration curve are discussed. For example, how many concentration values should be included (often covering a dynamic range of 10^3) and how many replicates at each concentration should be acquired, in order to define the calibration curve to within confidence limits that are fit for purpose? This question was considered in Section 2.6.1. For example, the FDA Guideline (FDA 2001) specifies that a calibration curve should consist of a blank sample (control matrix processed without internal standard), a zero sample (control sample processed with internal standard only) and six to eight nonzero matrix matched calibrators covering the expected range including LLOQ, but does not specify the number of replicates at each concentration. For a linear range as large as 10^3 a total of nine data points in addition to the blanks is recommended; if the goodness of fit tests (Section 8.3.3) indicate significant nonlinearity and a quadratic fitting function must be used (Section 8.3.6), a larger number of concentration values should be included to better define the function parameters. With respect to the number of replicates, the first decision must be on the precise meaning of 'replicate' appropriate to the purpose of the analysis (Section 2.6.1), e.g. replicate injections of the same set of nine calibrators or replicate sets of calibrators all prepared from first principles. For the most rigorous statistical tests a minimum of three replicates

at each concentration is required to provide a reasonable estimate of the standard deviation of the response. In bioanalytical work the industry standard considered fit for purpose is to use two replicates at each concentration, all prepared from first principles (i.e. extracts of control matrix that had been fortified with analyte and SIS).

Acquisition of a working calibration often precedes final acceptance of the complete integrated analytical method, so it is important to keep in mind that the solvent used to prepare the calibrator solutions must match the requirements of the overall method, e.g. solvent strength should be no greater than that of the HPLC mobile phase.

9.8.1a Calibration Solutions

For the preparation of calibration curves in solvent free of matrix, the stock solution (or sub-stock) of analytical standard must be diluted by appropriate factors to produce the desired concentration range; such dilutions are performed using calibrated pipets (Section 2.4.2) and Class A volumetric flasks. For the most accurate work it is not recommended that this be done by serial dilution, e.g., using a dilution factor of 10 to prepare calibration solution I, then diluting an aliquot of calibration solution I to form calibration solution II, and so on. Such a procedure can create unacceptably large uncertainties in the concentration of the most dilute solution by propagation of error. Rather it is best to devise a scheme (ahead of time!) that involves, at most, two dilutions of the stock solution in preparation of any particular calibration solution. Recall (Section 2.4.2) that Class A volumetric flasks in the range of interest here can produce liquids with certified volumes to within 1–2 %; for well-calibrated pipets in the range of a few microlitres the delivered volumes at or near the calibrated value are accurate to within 1–2 % with precision <1 %. Thus by using at most two dilution steps the propagated uncertainties (Section 8.2.2) are limited to ~3 %. One possible strategy would be to first prepare three solutions A–C, each by a single dilution of the stock solution. Standard solutions of intermediate concentrations would then be produced by appropriate second dilutions of one of solutions A–C. An example of such a dilution scheme is shown in Table 9.6.

If an SIS is to be used, a calibrated pipet would be used to deliver the same fixed amount of SIS to each volumetric flask before making the total volume up to the mark. Since all of the calibration procedures involving an SIS that are discussed in Sections 8.5.2b–c assume a fixed amount of SIS added to all calibrators and analytical extracts, an obvious question concerns the quantity of SIS that should be used. It turns out that this is a complex

Table 9.6 An example of a preparation scheme for a calibration curve in solution using three sub-stocks each of which was prepared by direct dilution from a stock solution

Solution ID	Spike Solution	Concentration of Spike Solution (ng/mL)	Volume of Spike Solution (mL)	Final Volume (mL)	Analyte Concentration (ng/mL)
Sub Stock A	Stock	100,000	5.0	50	10,000
Sub-Stock B	Stock	100,000	0.5	50	1,000
Sub-Stock C	Stock	100,000	0.2	100	200
STD1000	Sub-Stock A	10,000	1.0	10	1,000
STD800	Sub-Stock A	10,000	2.0	25	800
STD500	Sub-Stock B	1,000	5.0	10	500
STD250	Sub-Stock B	1,000	2.0	10	200
STD100	Sub-Stock B	1,000	1.0	10	100
STD50	Sub-Stock B	1,000	0.5	10	50
STD 10	Sub-Stock C	200	0.5	10	10
STD4	Sub-Stock C	200	0.5	25	4
STD2	Sub-Stock C	200	0.5	50	2

and important issue. A useful hint, whenever possible, is to determine the approximate range of analyte concentrations in the sample extracts; it is then good practice to use an SIS spiking level that results in a concentration of SIS in the final extracts (calibrators and samples) that lies somewhere in the lower third of the likely calibration range for the analyte, but sufficiently higher than the estimated lower level of quantitation that the SIS always gives a response that can be reliably measured with good precision and accuracy. Restriction to the lower third of the working concentration range minimizes the risk that the SIS will suppress (or enhance) the ionization efficiency of the analyte (Liang 2003; Sojo 2003) in cases where the two co-elute (see the discussion of Figure 5.36 in Section 5.3.6a). In addition, the best precision and accuracy are obtained when the responses of analyte and SIS are approximately equal and use of an SIS concentration in the lower third will apply this advantage to this more difficult range near the LLOQ).

9.8.1b Matrix Matched Calibrators

Matrix matched calibrators are the best choice whenever possible since matrix effects (Sections 5.3.6a and 9.6) will tend to cancel (Section 8.5.2b), but even so caution is required in view of the distinction (Matuszewski 2003) between an absolute matrix effect and a relative matrix effect that refers to comparison of such effects between different sources (batches) of blank matrix, e.g., biofluids such as plasma or urine for bioanalysis, or environmental matrices such as soil or water. The calibration schemes discussed in Section 8.5 did not take relative matrix effects into account (e.g. the calibration slope $B_{a,SIS}'''$ from Equation [8.87c] was assumed to be applicable

to the evaluation of the unknown concentration in the analytical extract via Equation [8.87d]. The importance of checking for relative matrix effects was discussed in Section 9.6.

For the preparation of matrix matched calibrators, the same general principles used for the preparation of calibration solutions apply except that a fixed known volume of one or more spiking solutions is added to each aliquot of blank (control) matrix (fixed amount, either weighed or possibly dispensed by volume for liquid matrices such as plasma, urine etc.), followed by the fixed volume of SIS solution.

When adding standard or SIS spiking solutions to the matrix, care must be taken with the selection of the solvent used to prepare the spiking solution and the volume added to prevent the sample from being altered prior to extraction. It may be necessary to use a spiking solvent that is different from that used in the primary stock solutions or sub-stocks when preparing matrix matched calibrators. For instance, when adding standard or SIS spiking solutions to plasma, it is desirable to keep the final organic concentration of the fortified control matrix to <10 % to prevent precipitation of the protein from the plasma prior to the sample preparation step.

Matrix matched calibrators are often prepared fresh on the day of analysis using spiking solutions of known stability. However, if the stability of the analyte in matrix has been established (Section 10.2.8), calibrators may be prepared in bulk, divided into smaller aliquots and then thawed and used on the day of analysis.

For reasons such as cost and availability, it may be necessary to use a matrix with a composition differnt from that of the samples which are to be analyzed. The choice of an appropriate alternate matrix may be as simple as using the same matrix from another similar source. For example, the high cost and limited availability of a biological matrix that is difficult to obtain in human may be circumvented by preparing a calibration curve and QCs (Section 9.8.2) in the same matrix from an animal species. For tissue analysis a lower cost tissue may be a suitable substitute for another more expensive but similar tissue from the same species. When employing such schemes, additional validation experiments designed to demonstrate equivalency between the surrogate and sample matrix must be designed.

Depending on the analytical range desired, one possible approach to limited matrix availability would be to employ a dilution method whereby the matrix or a homogenate of the control matrix is diluted in a manner consistent with the dilution used for sample preparation. Calibration curve and QC spiking solutions are then added to the diluted matrix. In this fashion, the rare matrix is conserved and the matrix effects will be equivalent. This type of methodology has the added benefit of allowing repeat sample analysis where typically the entire sample would be consumed upon extraction.

It is important to note that extraction of analytes from rare matrices presents a unique set of challenges that may not always permit the use of a substitute matrix. In any event, the use of a stable isotope labeled internal standard is often the key to successfully validating methods when applying these alternative techniques.

The discussion thus far has assumed that the control matrix is completely or relatively homogeneous, e.g., water, plasma. For many types of assay, e.g. soil or tissue analysis, the control matrix (and sample) must be homogenized as much as possible prior to using it for the preparation of calibrators (see Section 8.6 for a summary of the theory of sampling and sub-sampling heterogeneous matrices). In practice 'homogenizing' can imply any one or a combination of blending, mixing, disrupting, emulsifying, dispersing, stirring etc., depending on the matrix. Numerous techniques and types of equipment are used and can be classified into three main categories, namely ultrasonic, pressure and mechanical. A highly specialized discussion (Schumacher 1990) describes techniques used for homogenization of soil samples. The following discussion is an abbreviated version of a more detailed description (Yacko 2007).

If the matrix is a biological tissue for which it is desired to disrupt the cells and homogenize the resulting lysate, ultrasonic disruption is widely used, exploiting the cavitation phenomenon discussed in Section 4.3.2c. That earlier discussion was mainly concerned with extracting analytes from slurries of small particles and by implication referred mainly to thermally stable analytes. When applied to biological tissues when the analytes are likely to be thermally fragile, the temperature of the cell suspension should be as low as possible. In addition to addressing the concerns about temperature lability of analytes, low temperatures also promote high-intensity shock front propagation and thus more efficient disruption; ideally the temperature of the suspension should be kept just above its freezing point. For microorganisms (especially spores and yeast cells) the addition of glass beads in the 0.05 to 0.5 mm size range enhances cell disruption by focusing energy released by the bubble implosions and by physical crushing. Tough tissues such as skin and muscle should be macerated first in a mechanical homogenizer (see below) before ultrasonication. Recall that the localized extreme conditions produced by cavitation (Section 4.3.2c) can produce highly reactive oxidative free radicals that could possibly create artifactual modifications of analytes; this can be minimized by including radical scavengers (e.g. cysteine, dithiothreitol or other thiol compounds) and/or by saturating the sample with a protective atmosphere of helium or hydrogen gas to prevent oxidation.

The second broad category of homogenizers (Yacko 2007) are the high pressure homogenizers that have been found to be suitable for a variety of bacteria, yeast and mycelia. Homogenizers of this type work by forcing cell suspensions through a very narrow channel or orifice under pressure; the most effective type then causes the cells to impinge at high velocity on a surface or against another high velocity stream of cells coming from the opposite direction. A familiar laboratory scale high pressure homogenizer is the 'French press' that uses a motor-driven piston inside a steel cylinder to develop pressures up to 40 000 psi. Because the process generates heat, the sample, piston and cylinder are usually precooled. Most high pressure homogenizers require several passes of the cell suspension to achieve high levels of cell disruption.

The third broad category is that of mechanical homogenizers, which can be sub-divided (Yacko 2007) into rotor–stator and blade-type homogenizers, that are widely used to homogenize biological tissues. In the rotor–stator type the cellular material is drawn up into the apparatus by a rapidly rotating rotor (blade) positioned within a static head or tube (stator) containing slots or holes, where the material is centrifugally thrown outward exit through the slots or holes. Because the rotor (blade) turns at a very high speed (the crucial operating parameter is not the number of revolutions per minute (rpm) but the speed of the blade tip), the tissue material is rapidly reduced in size by a combination of extreme turbulence, cavitation and mechanical shearing occurring within the narrow gap between the rotor and the stator. Depending on the toughness of the tissue sample and the degree of homogenization required, the process is usually complete within 15–120 seconds. Cell disruption with the rotor–stator homogenizer involves hydraulic and mechanical shear, as well as cavitation (see above comments about concerns with cavitation). There is some degree of safety concern with rotor–stator homogenizers with respect to contamination of the air in the laboratory as a result of foaming and aerosol formation, which can be minimized by using appropriately covered vessels. On the positive side, rotor–stator homogenizers generate minimal heat during operation, readily dissipated by cooling the homogenization vessel in iced water.

The other type of mechanical homogenizer is represented by the blade homogenizers ('blenders') commonly used to produce extracts from plant and animal tissues. The cutting blades rotate at speeds of 6000–50 000 rpm. Blenders are not suitable for disruption of microorganisms unless glass beads or other abrasives are added to the media, but this leads to the same problems of foaming and aerosol formation as mentioned above for rotor–stator homogenizers (Yacko 2007).

Another type of mechanical homogenizer for biological samples that is mentioned but not discussed in the otherwise excellent review (Yacko 2007) is the bead mill. Agitation of cell suspensions in the presence of small steel or silica beads (usually 0.2–1.0 mm diameter) leads to cell lysis by the high liquid shear gradients and collision with the beads. Any type of biomass can be disrupted by bead milling but, in general, larger sized cells will be broken more readily than small bacteria. A large number of small beads will be more effective than a smaller number of larger beads because of the increased likelihood of collisions between beads and cells. Heat production is a problem in the use of bead mills, particularly on a larger scale (e.g. 20 L). Smaller vessels may be cooled adequately through cooling jackets around the bead chamber. High throughput bead mills, based on shaking 96-well plates, are now available.

It must be mentioned that sample to sample contamination can become a major concern unless the homogenization equipment is scrupulously maintained.

9.8.2 Preparation of QCs

Quality Control samples (QCs) can be generally defined as samples that are prepared from control matrix fortified with the analyte(s) of interest and are used to demonstrate that a method is performing according to the established uncertainty targets within a particular run. They are also used to assess intra-laboratory method performance over a longer period and to demonstrate that the method is performing within uncertainty limits (see Section 2.5.3). This important aspect of quality control, together with many other relevant aspects of quality of chemical measurements, are described in detail in an excellent monograph (Taylor 1987). For method validations, QCs are used to establish a variety of method components such as precision and accuracy, LLOQ, recovery, stability and method robustness and ruggedness (Section 10.2). In addition to their use in validation experiments or sample analysis, QC samples are also used during method development to assess the final method prior to validation. Experimental runs that use QCs for this purpose are often referred to as assay pre-qualifications or pre-study assay evaluations (PSAE).

QCs are prepared in solvent or in a suitable control matrix that is the same or similar to that used for the calibration standards. Specific guidelines for the preparation of QCs may vary according to analytical method, matrix being used and laboratory SOPs or any prevailing regulatory guidelines. Stock or sub-stock solutions to be used in fortifying the control matrix for QC samples should be prepared at an appropriate concentration and volume based on the amount of QC sample being prepared and the control matrix used (Section 9.5.4), and should be prepared independently of the stock solution used to prepare the calibrators (to check on weighing errors etc.). Typically, QCs are prepared in large batches (in bulk), aliquoted into smaller samples, and then stored for subsequent use at the same conditions that are used to store the samples. A QC preparation scheme should be prepared prior to making the QCs. An example is shown in Table 9.7.

For some studies, instead of preparing QCs in bulk, individual samples are fortified with a known amount of analyte and analyzed fresh; although this may be acceptable practice for some applications, it is less common and adds additional variability to any procedures that may be used to monitor the method performance over time (QC plots, Section 2.5.3b).

Complete documentation for the material and procedures use to prepare the QCs should be maintained in the study records, and it is advisable to create SOPs that specify the minimum documentation requirements. At a minimum, the following information should be recorded: QC ID, analyte(s), description of the matrix and ID, preparation date, expiration date, solutions and amounts used, procedures, storage location and temperature, and the person who prepared the QC.

Certified Reference Materials (CRMs, Section 2.2.2) are generally too expensive and valuable to be used as

Table 9.7 An example of a Quality Control sample spiking scheme for the preparation of liquid quality control samples prepared in bulk at three concentrations (low, mid, high)

QC Type	QC Spiking Solution ID	Spiking Solution Concentration (ng/mL)	Spike Volume (mL)	Final Vol. (mL)	Final Concentration (ng/ml)	Aliquot Size (mL)	Lot ID	# QCs Stored
High	Spike ID-1	10, 000	4.00	50	800	1.00	Lot ID-H	50
Mid	Spike ID-1	10, 000	2.5	50	500	1.00	Lot ID-M	50
Low	Spike ID-2	100	1.5	50	3	1.00	Lot ID-L	50

QC samples in method development or sample analysis. However, when a suitable CRM in a matrix that approximates that of the analytical sample is available, it can provide a check on the validity of the method and also provide an additional degree of traceability (Section 2.2.1).

For those cases where the availability of suitable control matrix is very limited or expensive, surrogate matrices may be used for the preparation of the QCs provided that data to support equivalency with the study matrix are available. The use of surrogate control matrix was discussed in Sections 9.4.7c and 9.8.1b with respect to the preparation of matrix matched calibrators. For the preparation of QCs a surrogate matrix, which may or may not be from a similar origin, should be as analogous as possible to the analytical sample in order to mimic the complexity of any type of matrix effect. In all cases, the appropriate precision and accuracy of QCs in surrogate matrices should be confirmed prior to sample analysis. To demonstrate the equivalency of the surrogate sample matrix, a surrogate matrix calibration curve, blanks and a set of QCs at multiple levels should be evaluated for precision and accuracy alongside a set of QCs prepared in the sample matrix whenever possible.

For method validation studies, QCs are often prepared and 'analyzed fresh' on the same day, but when QCs are being prepared and then stored for subsequent use during sample analysis it is advisable to run some type of QC check of the prepared batch of QCs prior to use. Generally, QC replicates are prepared and assayed along with a standard curve and a set of blanks. The number of replicates used will depend of the availability of control matrix (and thus the number of available QCs), laboratory SOPs and/or historical information about the method with respect to previous batches of QCs made.

The stability of the QCs should be established prior to use for sample analysis and is established during the method validation process. Expiration dates of validation QC samples may also be based on known established stability if available. If no stability has been established, the laboratory SOPs or similar documentation should be

used to establish describes how the QCs may be used until stability is established. In addition to storage stability, stability at room temperature and during the sample preparation process and in the sample extract must also be established (Section 10.2.8). Extension of the expiration date is acceptable if long term stability is established after the original QC preparation date and expiration date assignment.

When selecting flasks and containers for the preparation and storage of QC samples and calibrators, two potential problems should be borne in mind. The first of these concerns the possibility of adsorption of analyte on the walls of the container. This problem tends to be highly compound-specific, but in general can be minimized in glassware by silanizing the surface to eliminate the most active adsorption sites. Use of labware made from some type of plastic can be a good solution for some analytes and bad for others. The storage stability of QC samples, calibrators, stock solutions etc. is an important aspect of method development and validation (Sections 10.2.7 and 10.2.8). The second potential problem with storage containers is that of extraction of compounds from the container material, especially plastics. This is essentially the problem of exogenous matrix effects and interferences discussed in Section 9.6.

9.8.3 System Suitability

This concept has been defined as follows (ICH 2005): 'System suitability testing is an integral part of many analytical procedures. The tests are based on the concept that the equipment, electronics, analytical operations and samples to be analyzed constitute an integral system that can be evaluated as such. System suitability test parameters to be established for a particular procedure depend on the type of procedure being validated'. In practice, system suitability tests are applied to analytical methods to confirm the continuing suitability of the method for use before obtaining valid data on different occasions following the initial method validation; usually such tests are performed prior to an analytical run (FDA 2001) but a contrary viewpoint is described below. The concept should also be addressed at the method development stage where the acceptable performance limits can be established with the help of results from a method robustness investigation (Hund 2002).

The concept of system suitability testing appears to have been first introduced (King 1974) to analytical chromatography for the pharmaceutical industry. Early schemes proposed only two performance criteria (reproducibility of replicate injections and resolution (Section 3.4.6)), though occasionally peak asymmetry (Section 3.5.12) and chromatographic efficiency (Section 3.4.5) were also suggested for inclusion. Later, in the same context it was proposed (Wahlich 1990) that system suitability tests should be considered for each of the performance criteria investigated for validation of the method, although these authors concluded that peak asymmetry and column efficiency were likely to be of secondary importance in this regard. For example, the following suitability tests were proposed (Wahlich 1990) for selected validation parameters: accuracy (QC samples, try re-extraction of the sample and calculate a mass balance); precision (RSD of data from replicate injections and from replicate sample preparations); selectivity (resolution check, remember that this work considered only HPLC with UV detection); stability of extract solutions and of the chromatographic system (compare analyses of standard samples and QCs at the beginning and end of a series of analyses); linearity (check using calibration solutions or calibrators); LOD/LLOQ (check signal:background ratio near these limits); general acceptability (visual comparison of a test chromatogram with a reference chromatogram).

These early approaches to system suitability testing laid the groundwork for modern practice. Thus the FDA guidelines for bioanalytical method validation (FDA 2001) specify that, based on the analyte and technique, a specific SOP should be identified to ensure optimum operation of the system used with respect to, e.g., sensitivity and chromatographic retention by analysis of a reference standard prior to running the analytical batch. Most such SOPs in analytical laboratories satisfy this requirement by analyzing a few reference standards covering a range of analyte concentration.

Very recently (Briscoe 2007) the question of system suitability testing for LC–MS/MS applications in biomedical analysis was addressed in detail, and a generic test that monitors instrument performance throughout an analytical run was described. This procedure includes tests for signal stability, carryover and instrument response in a fashion that is integrated throughout an analytical run of multiple samples. The importance of this work is illustrated by two examples in which QC samples and calibration standards met all acceptance criteria based on SOPs and the FDA Guideline for bioanalytical method validation (FDA 2001), but failed the proposed system suitability test that had been integrated with the analytical run. In these two examples (Briscoe 2007) the concentrations of > 35 % of clinical samples were in this way flagged as unacceptable, resulting in changes of > 15 % when the samples were reanalyzed. It was clear in these examples that poor performance of the LC–MS/MS system adversely affected the calculated concentrations of unknown samples even though the results for QC samples appeared to meet the usual acceptance criteria for bioanalytical methods (FDA 2001)!

The original work (Briscoe 2007) provides a full account of the proposed procedure and only a brief outline is given here. As mentioned previously the three critical performance criteria that are monitored are signal stability, carryover (Section 9.7.2) and instrument response as determined via the signal:background ratio observed for a system suitability test sample with concentration near the LLOQ. A typical injection sequence (injection numbers are given in sequence), used to analyze a batch of clinical samples (CS) using system suitability test samples (SSTS) in an integrated fashion, is as follows: #1–5, high-concentration SSTS; #6, solvent blank; #7, low concentration SSTS; #8, 9, control matrix spiked with SIS only; #10–18, calibrators; #19, control matrix; #20–25, CS; #26, QC; #27–33, CS; #34, QC; #35–40, CS; #41, QC; #42–43, high concentration SSTS; #44–48, CS; #49, QC; #50–54, CS; #55, QC; #56–60, CS; #61, QC; #62–67, CS; #68, QC; #69–72, CS; #73, QC; #74, control matrix; #76–76, high concentration SSTS; #77, solvent blank; #78, low concentration SSTS. Thus a total of 78 injections is used to analyze just 44 actual analytical samples while providing ongoing checks of various kinds.

In accord with usual practice, QC samples and calibrators are inserted at regular intervals in the sample stream. Note however that the first five injections are high concentration SSTS (that must yield data with a coefficient of variance RSD% < 6%, thus checking system stability), injection #6 is a solvent blank to check for carryover, and #7 is a low concentration SSTS to check the signal:background ratio near the LLOQ. Injections #8 and 9 are extracts of matrix blanks spiked with SIS only to check the stability of the SIS, #10–18 are the nine calibrators covering the validated range to check on linearity and sensitivity (slope of the calibration curve) and #19 is an extract of control matrix to check for matrix interferences. It is not till injection #20 that the first analytical sample is analyzed! It is considered good practice (Briscoe 2007) to analyze high concentration SSTS every 40 injections or so (see #42–43 and 76–76 in the above example) to provide on-going checks on signal stability. Less frequent checks on carryover (#77) and system response (#78) are found to be adequate in practice. But the main point is that the system suitability checks are conducted regularly throughout the analytical run, not only at the beginning.

Note that the system suitability test samples (SSTS), while somewhat similar to QC samples, are prepared in a different fashion and the data evaluated entirely separately from the results of the various calibrators and QC samples, reflecting the fact that the purposes are quite different. Calibrators are used to calibrate the linearity and sensitivity of the method, and QCs are used to mimic the treatment of an unknown sample thus providing an indication of any problems with sample storage, thawing, taking an aliquot of the sample, extraction, analysis, data integration etc. QC sample extracts are prepared within the sample stream together with the unknowns. In contrast, the sole purpose of the system suitability test is to check that the instrument continues to perform up to a standard that can produce accurate and reproducible data. It is thus important that all high concentration SSTS used within a given run should be exactly the same, so a large amount of extract of spiked control matrix is made up ahead of time and used in appropriate aliquots; the procedure is similar for the low concentration SSTS, a marked difference from how QC samples are inserted into the sample stream.

Thus the SSTS analyses in principle can not check for the various sample preparation effects that the QCs monitor. On the other hand, variations in the quantitative data obtained by the QCs will reflect all the steps of the analytical method combined, so that instrument performance can not be unambiguously determined from the combined effects that lead to variability in the QC data. Indeed, it is possible that some of these effects can partially cancel one another out, e.g., a poor extraction efficiency might be cancelled by an upward drift in the instrument sensitivity.

In both of the examples described (Briscoe 2007) as examples where something of this sort must have occurred, the use of QC samples followed the FDA guidelines, i.e. a minimum of three concentrations (one within $3\times$ of the LLOQ, one in the midrange and one approaching the high end of the range), with a minimum number of QC samples that should be at least 5 % of the number of unknown samples or six in total, whichever is greater. Moreover, the data obtained for the QCs met or exceeded the FDA acceptance criteria, i.e. at least 67 % ($\frac{2}{3}$) of the QC samples should be within 15 % of their respective nominal (theoretical) values (the 33 % of the QC samples that can be outside the ± 15 % of

the nominal value must not be all replicates at the same concentration). Thus the data for the unknown samples were acceptable with respect to the usual QC criteria (FDA 2001). However, when the data for the SSTS were examined (Briscoe 2007) for the two cited cases, they failed the system suitability test for signal stability, indicating significant instrumental drift. When the instrumental problem had been rectified and the analytical run repeated, in one case 33 of the 112 unknown samples, and in the second example 16 out of 40, had measured concentrations that changed by more than 15 %.

The take-home message from this work (Briscoe 2007) appears to arise from two aspects that, when combined, can potentially cause problems, namely the role of QC samples in monitoring the integrated variability of the entire analytical method and the potential for instrumental drift in response of LC–MS/MS instruments for a variety of reasons (e.g. accumulation of contamination). Thus system suitability checks conducted only before an analytical run is started may not suffice to guarantee validity of data obtained and should be designed to monitor instrument performance regularly in the course of an analytical run.

However, as always, the extent to which such measures are necessary in any given case should be evaluated on a fitness for purpose basis. In most cases system suitability tests appreciably less rigorous than those described (Briscoe 2007) are fit for purpose, but this work does draw attention to the possibility of occasional problems that could possibly arise. A system suitability test for an LC–MS/MS system that is fit for purpose in most circumstances is described in Section 10.5.2b.

9.8.4 Testing Method Robustness and Ruggedness

The two terms 'robustness' and 'ruggedness' are sometimes used interchangeably but are usually accepted as referring to two distinct though related concepts. Method robustness is generally taken to indicate the demonstrated capacity of a method to remain unaffected by small deliberate variations in method parameters; such tests are undertaken with the objective to determine that the method can still provide valid data even if such small variations in conditions occur unnoticed in the course of a long run involving multiple samples, QCs, calibration check samples etc. Method ruggedness is concerned with transferability of a method between operators, instrumental platforms or laboratories; a commonly used measure of ruggedness is the degree of reproducibility (% RSD) of the results obtained for aliquots of the same test sample by the different operators, laboratories etc. Note that the term 'ruggedness' is replaced by 'intermediate precision' in an important international guideline on analytical validation (ICH 2005).

Evaluation of robustness should be undertaken during method development and, if the results are found to be too susceptible to variations in some of the analytical conditions, these conditions should be specified in the method description as requiring special control. Before such evaluations are started, a plan should be formulated in which the analytical conditions to be varied are specified together with the ranges of variation; the performance criteria to be used to judge whether or not the method is suitably invariant to the small changes in the analytical conditions should be similarly specified. Many of these performance criteria for method robustness used in method development will coincide with parameters included in a system suitability test used in preparation for using the method to analyze real samples (Section 9.8.3). Examples of analytical conditions that could be varied include the composition of solvent used to extract the analyte from a solid or liquid matrix, the extraction time, the composition, pH and flow rate of an HPLC mobile phase, the temperature program in GC and tuning parameters (both compound-dependent and independent, Section 6.3.2) of a mass spectrometer. Corresponding examples of performance criteria might be the extraction efficiency (recovery), chromatographic peak shape, resolution and retention times, and mass resolution and instrumental sensitivity (in the narrowly defined sense of slope of a calibration curve, Section 6.2.1) of the mass spectrometer.

Of course, other parameters and criteria could be added to this list but the main point is that evaluation of method robustness is intrinsically a multi-variable problem. An excellent review of the principles involved, together with an account of a real-world test case study (Vander Heyden 2001), describes the setting up of a multi-factorial design procedure to test the method robustness systematically. The type of multi-factorial design normally used in method robustness tests is the so-called Plackett–Burman design (Plackett 1946), which is easier to construct than fractional factorial designs but gives results in which two-parameter interaction effects are mixed with the main effects; however, the two-parameter interactions in a robustness test of the kind considered here are generally negligible (Vander Heyden 1995). It is not appropriate to attempt a summary of the procedure here in view of the clear and complete description given previously (Vander Heyden 2001), so only a few comments are included.

For a given number of analytical parameters (called 'factors' in the original paper), there are two general options, one involving minimal designs, i.e. designs with the absolute minimal number of experiments for the number of parameters considered, and the second that requires more experiments but permits a more extensive statistical interpretation of the effects. In the examples discussed previously (Vander Heyden 2001) the smallest number of analytical parameters to be examined in an experimental design to evaluate robustness was considered to be three. Moreover, for statistical reasons concerning interpretation of the observed effects, designs with fewer than eight experiments are not used while those with more than 24 experiments are considered impractical. A computer program designed to assist the setting up and interpretation of method robustness tests has been described (Questier 1998). A recent example of application of these principles (van Nederkassel 2003) describes tests for robustness of several analytical methods for a range of analytes following replacement of conventional silica-based HPLC columns by monolithic silica columns of five and ten centimeters in length to reduce analysis times to below three minutes; for those cases considered successful with respect to chromatographic efficiency and selectivity for the sets of test analytes (the performance criteria), the new methods were robust with respect to variation of analytical conditions including flow rates of $1-9 \, \mathrm{mL.min^{-1}}$, changes in organic content of the mobile phase with pH values varying from 3.5 to 7.

Such systematic investigations of method robustness (Vander Heyden 2001) represent the best and most complete approach within reasonable bounds of time and overall effort. However, other less demanding approaches might be fit for purpose in other circumstances, e.g. a simple one-at-a-time evaluation of the effects of a limited set of analytical parameters. An example (Danaher 2000) is that of a new alumina-based SPE clean-up method for HPLC analysis (fluorescence detection) of macrocyclic lactone drugs (avermectins and moxidectin) used for the treatment of parasitic infections in food-producing animals; the matrix was animal liver. The robustness of the new SPE method was tested for several analytical conditions, i.e. volumes and composition of wash and elution solvents, volume of sample extract applied, and source and amount of alumina in the SPE cartridge; the only performance criterion was the recovery. Variation in the volumes and composition of the wash and elution solvents were tested using a laboratory-prepared alumina cartridge; no significant differences were observed, with recovery for all drugs under all conditions $> 80 \, \%$. Similarly, no significant differences were found (Danaher 2000) for variation in quantity of alumina or for addition of sample extract as $3 \times 15 \, \mathrm{mL}$ or as $1 \times 45 \, \mathrm{mL}$ to the cartridge. Analyte recovery was most influenced by the source of alumina used for laboratory-prepared

SPE cartridges and the type of pre-packed commercial alumina cartridges used. Although the robustness evaluation was not planned using a multi-factorial method, the results were evaluated using simple analysis of variance (ANOVA, Section 8.2.6).

In contrast, the definition of method ruggedness or 'intermediate precision' (ICH 2005) used here corresponds to that of the US Pharmacopeia: 'The ruggedness of an analytical method is the degree of reproducibility of test results obtained by the analysis of the same sample under a variety of normal test conditions, such as different laboratories, different analysts, different instruments, different lots of reagents, different elapsed assay times, different assay temperatures, different days etc. Ruggedness is a measure of reproducibility of test results under normal, expected operational conditions from laboratory to laboratory and from analyst to analyst' (US Pharmacopeia 1990). In other words, ruggedness tests are tests for transferability of an analytical method in which the method is executed in accordance with the instructions in its operating procedure, but on different days, by different analysts, on different instruments etc. The resulting variance in the results obtained reflects differences in conditions that were not investigated in the original method development robustness tests (e.g. laboratory temperature, instrumental differences etc.).

The traditional approach to ruggedness testing involves collaborative trials between several laboratories, which is expensive and time consuming. A recent investigation (Thompson 2002a) examined the suggestion that a relatively inexpensive, properly designed ruggedness test conducted in a single laboratory, could provide information similar in quality to that offered by the much more expensive inter-laboratory studies. Unfortunately, in a comparison of the reproducibility obtained using such a carefully designed single laboratory evaluation with that obtained in a traditional collaborative inter-laboratory trial, it was concluded that the single laboratory variability was at least a factor of two lower. At least in part this disappointing result could be attributed to calibration bias that is believed (Thompson 2002a) to be a major contributor to inter-laboratory variation, and could have been simulated in the intra-laboratory approach only if each repeat analysis of the test sample was done in a separate run together with its own *ab initio* calibration. This would require a set of calibrators made from separate batches of the pure analyte by different analysts for each analytical repeat, a laborious task that would obviate the original intention to develop a fast and cheap replacement for the inter-laboratory study. At present there does not appear to be any alternative to actual inter-laboratory testing to estimate method ruggedness. A strategy designed to systematize ruggedness testing (but mainly for variations in the kinds of analytical conditions appropriate to method robustness) has been described (Vander Heyden 1998).

9.8.5 Establishing Final Stepwise Procedure

At this point the most essential aspects of the analytical method have been established, but a few practical details may need to be finalized. For example, appropriate timing of operation of divert valves between chromatograph and mass spectrometer can be crucial in protecting the mass spectrometer from contamination, thus permitting long reliable operating times. Similarly, an optimized procedure for rinsing an HPLC column between injections can achieve the same result. Other parameters that can require adjustment include delay times between injection and initiation of mass spectrometric data acquisition (to minimize storing useless data early in the chromatogram), and SIM or MRM dwell times (to optimize the compromise between acquiring sufficient ion statistics, see Section 7.1.1, and recording at least 10 data points for each SIM/MRM channel across a chromatographic peak). Time scheduling of SIM/MRM channels across the total chromatographic time can also assist in achieving this compromise.

The document describing the final method should include a cover page that includes a descriptive title and method identification, along with the name and address of the laboratory where the work was conducted. In many instances, an approval page that includes the names of the principal scientists and any other individuals who may have reviewed and/or approved the final method is provided along with their signatures and date that they signed the method.

Every method should include an introductory section that describes the scope and applicability of the method. The scope should include (at a minimum) a brief description of the method including the analyte and matrix, LLOQ and range of reliable response. For methods that have been successfully validated, a reference to the validation report number or ID should be provided. Any revisions to the original method (if previously validated) should be listed along with a brief description of the revisions made and reference to any other validation studies that were conducted as a result of those revision.

A written stepwise analytical procedure is a document that provides a detailed description of the extraction and analysis techniques used for validation (Chapter 10). If the method has been previously validated, the stepwise procedure will be a component of the final method or SOP. Many laboratories will in fact create a draft method or SOP that includes the stepwise procedure, and then finalize that document with other information after the method validation has been completed. In any event, a detailed stepwise procedure should be documented prior to the validation process, and the procedures that are described in this document should be followed exactly as written throughout the entire validation process; any deviation(s) from the method during the conduct of validation or sample analysis must be documented by the analyst or any other scientists involved in the validation study. The stepwise procedure (or SOP) should be reviewed, signed and dated by the responsible scientist or management prior to the initiation of work. Guidelines for preparing a stepwise procedure suitable for most applications are given here but, as pointed out throughout this Chapter, the SOPs at the laboratory where the work is conducted along with any other regulatory requirements will prevail over the specific recommendations given here.

The stepwise procedure should include a description of the sample preparation procedures including a list of materials required. Any specialized or critical equipment or apparatus must be described; however, standard laboratory equipment such as common glassware and pipets may not necessarily need to be included. A description of all chemicals, solvents and solutions used should be included along with examples for the preparation of any buffers, reagents or any other solutions used. Examples for the preparation of stocks sub-stocks and spiking solutions should be provided along with detail descriptions and examples for the preparation of the calibration curve and QC samples. The stepwise extraction procedures should be described in detail and any special instructions or critical steps that the analyst needs to be aware of should be highlighted along with any critical details associated with the analyte and/or matrix, such as matrix instability or light sensitivity, or special techniques involving extraction.

The chromatography method, including the manufacturer and model for the LC system and autoinjector, if applicable, should be provided. Ideally, the exact same equipment that was used for method development should be used to validate the method. After the method has been validated, other types of equivalent equipment may be used but, since factors such as system dead volume and carryover may impact the method, additional validation experiments may be required to support these changes. (Section 10.2.10). Similar to what was described above for the preparation of solutions and reagents, a description of all chemicals and solvents used for the preparation of mobile phases should be provided along with examples of how these are prepared and stored when

not in use. A description of the chromatography conditions should be provided in sufficient detail to ensure that the method conditions are reproduced and run exactly as intended. The manufacturer, column type, and preferably part number for the analytical column should also be provided, along with any special instructions for the use, rinsing or storage of the column. This is particularly important with columns that are used for chiral separations or any others that require special handling and storage procedures. A discussion of any potential for carryover, along with any established procedures for monitoring and preventing it, should also be discussed.

A description of the make and model of mass spectrometer used should be provided along with a description of the operating parameters used for method development. Some of the the settings for the mass analyzer or interface will depend on the optimized conditions for a particular instrument, e.g. gas flows, interface potentials, collision cell conditions, but the conditions (or range) that were used during method development and validation should be provided. Any parameter settings that are considered to be critical, such as interface temperatures, should be described in the procedure.

All critical steps, method parameters or restrictions for any aspect of the stepwise procedure that may have been established during method robustness testing (Section 9.8.4) should also be described in the appropriate section of the procedure. If any special test injections or system suitability procedures have been established (Section 9.8.3) a detailed description of these procedures, along with instructions and examples for the preparation of any required test solutions, should be described.

At least one representative chromatogram, preferably one that shows an injection of an extracted LLOQ with SIS, should also be provided. Although this is not really a procedure or requirement for reproducing the method as written, the example chromatogram is invaluable with respect to confirming that the method has been run as written and that the chromatography and sensitivity obtained are typical of the validated method. Finally, many laboratories will use a standard form for recording critical instrument parameters when the method is used, and these may be included in the method to ensure proper documentation of all relevant chromatography and/or mass spectrometer conditions.

9.8.6 Final Method Qualification/Assessment

Method development and validation (Chapter 10) are connected in the sense that an analyst will not know if the procedures that were developed will be fit for purpose until after the validation has been conducted. If, as a result of the validation, it is determined that the method as written is not suitable for sample analysis, additional method development will be required followed by another validation. Method validation is designed to test the method as written and no modifications to the method should be made during this process. Also, repeated run failures during validation would result in a failed validation. For these reasons, the final stepwise procedure should be tested thoroughly prior to validation and any modifications that may be required should be made before it is used to validate the method.

The method should be assessed using the same or similar calibration standards, QCs and blanks that will be used for validation. In addition to a final assessment of sensitivity, linearity, accuracy and precision, additional experiments designed to assess analyte stability during the sample preparation and analysis processes, method robustness and ruggedness, carryover, and potential for laboratory contamination, should be designed and included. If carryover is present, procedures that will be required to assess the extent of carryover during validation and sample analysis should be developed and tested (Section 9.7.2). Method ruggedness and robustness (Section 9.8.4) should also be evaluated prior to validation and, if possible, these experiments should be conducted with multiple lots of control matrix. In addition to checking precision and accuracy, particular attention should be focused on determining whether there is any indication of sensitivity or SIS response drift throughout the runs. If, as a result of these tests, additional method development is warranted, the process of development and testing should be repeated until there is reasonable confidence that the method will pass the criteria set for validation successfully. If any modifications are made to the method, the final stepwise procedure should be updated before method validation is started.

10

Method Validation and Sample Analysis in a Controlled Laboratory Environment

10.1 Introduction

There are many organizations in different jurisdictions that issue documents concerning the validation of analytical methods and these can seem bewildering at first. However, these documents are all devoted to the same final result no matter how they may differ in details of definitions or even terminology i.e., that a given method has been validated to ensure that the final results generated during sample analysis with a given matrix are accurate and precise to within acceptable uncertainty requirements. The present chapter is not intended to be an all-inclusive cookbook that attempts to synthesize all variants of 'validation' but rather an overview with examples of how the fitness for purpose principle has been applied to address 'what's at stake'. Previous chapters will be cross-referenced for more in-depth coverage of topics when applicable.

Definitions of the basic concept of 'validation' can vary from single line sentences to much longer descriptions. Here one of the longer descriptions is quoted (Thompson 2002) since it appears that it better addresses the complexity of the real situation at the expense of some measure of perceived 'elegance'.

'Method validation makes use of a set of tests that both test any assumptions on which the analytical method is based and establish and document the performance characteristics of a method, thereby demonstrating whether the method is fit for a particular analytical purpose. Typical performance characteristics of analytical methods are: applicability, selectivity, calibration, trueness, precision, recovery, operating range, limit of quantification, limit of detection, sensitivity,

and ruggedness. To these can be added measurement uncertainty and fitness for purpose.'

'Strictly speaking, validation should refer to an 'analytical system' rather than an 'analytical method', the analytical system comprising a defined method protocol, a defined concentration range for the analyte, and a specified type of test material.'

'Method validation is regarded as distinct from ongoing activities such as internal quality control QC or proficiency testing. Method validation is carried out once, or at relatively infrequent intervals during the working lifetime of a method; it tells us what performance we can expect the method to provide in the future'.

The foregoing definition was prepared (Thompson 2002) on behalf of the International Union of Pure and Applied Chemistry (IUPAC) but many other documents have been published by other organizations representing national bodies and/or for more limited applications. For example Eurachem (a network of organizations from most European countries with the objective of establishing a system for the international traceability of chemical measurements and the promotion of good quality practices) has published at least two detailed guides (Eurachem 1998; CITAC 2002) dealing with this subject, sometimes jointly with the Cooperation on International Traceability in Analytical Chemistry (CITAC), a similar organization but with a wider international reach. Similarly the Association of Official Analytical Chemists International (AOAC International) has published a guide for single laboratory validation of methods for dietary supplements (AOAC International 2002) that mirrors the other more general documents.

As far as chemical analysis of a final pharmaceutical product is concerned, the International Conference on Harmonization of Technical Requirements for Registration of Pharmaceuticals for Human Use (which brings together the regulatory authorities of Europe, Japan and the United States of America with experts from the pharmaceutical industry to recommend ways to achieve greater harmonization with respect to the requirements for product registration) has published a guide for validation of analytical procedures (ICH 2005). There are also some other documents on related matters, e.g., traceability in chemical analyses that helps achieve comparable results (Eurachem 2003), quality assurance for research activities and non-routine analyses (Eurachem 1998a) and a guide for the use and interpretation of proficiency testing schemes (Eurachem 2000).

Elucidation of how the general principles underlying the concept of validation should be expressed in practice is an evolving process, as exemplified by the ongoing evolution of validation requirements for bioanalytical assays in the pharmaceutical industry (Shah 1992, 2000; FDA 2001; Viswanathan 2007). The complementary principle of fitness for purpose (Section 9.2) applies not only to the assay method but also to the validation process itself. Procedures that are considered to be fit for purpose in validation of an analytical method to be used in drug development, for example, need not necessarily apply to, e.g., methods used to screen pesticide residues in foodstuffs. As noted in Section 9.2, this point of view appears to be consistent with the definition of validation applied to all measurements (ISO 1994): 'Validation: Confirmation by examination and provision of objective evidence that the particular requirements for a specified intended use are fulfilled.' Of course, some basic principles are common to all validation schemes.

However, whatever the details, method validation is a critical prequisite to ensuring that the reported results are reliable and defensible. A flowchart summarizing the overall process involved in a new analysis, from initial discussions with the client to conducting the analyses, is shown in Figure 10.1. (Note that the important aspects of within-run validation of sample analyses, scientific review of results, and preparation of reports, are not included in this chart.)

10.2 Method Validation

10.2.1 Figures of Merit for Full Validation

'Doing a thorough method validation can be tedious, but the consequences of not doing it right are wasted time, money, and resources (Green 1996).'

Full validation is a general term to describe the process for validating a method for the first time. As discussed above, the specific types of studies and procedures that are conducted for a full validation may vary according to the analytical objectives or guidelines established by other governing bodies or regulatory agencies. However, the basic components of a full validation include investigations of: (1) selectivity, (2) sensitivity, (3) range of reliable response, (4) accuracy, (5) precision and (6) reproducibility. In addition to these fundamental components of a validation, other types of validation studies, such as (7) analyte recovery (extraction efficiency) and (8) stability, should be considered, or may be in fact be required, for a method to be considered fully validated. Using the example of bioanalytical methods to be validated according to FDA guidelines (FDA 2001; Viswanathan 2007) the laboratory should also consider studies designed to evaluate (9) the ability to dilute samples originally above the upper limit of the standard curve, (10) carryover and contamination, (11) matrix effects and (12) incurred sample reproducibility (which may be demonstrated during study sample analysis). The remainder of this section addresses the major figures of merit underlying the validation of an analytical method as well as other types of validation studies that may be conducted in special circumstances. Guidelines for the overall conduct of the validation and documentation of the results are discussed in Section 10.3. A general description for several types of methods used for routine quantitative analysis can be found in Section 10.4. The full validation of bioanalytical assays according to the US Food and Drug Administration (FDA), and generally accepted industry guidelines, is an excellent example for fit for purpose validation design and a discussion of the fundamental components for this particular application can be found in Section 10.4.1. The key figures of merit that are addressed in validation of an analytical method are listed in Table 10.1

10.2.2 Selectivity

As discussed previously in Section 6.2.1, the selectivity of an analytical method can be summarized as the ability of the method to measure and differentiate the analytes in the presence of components that may be expected to be present, e.g. metabolites, impurities, degradants or matrix components. The terminology 'specificity' is sometimes used in the literature as a synonym for selectivity but more properly describes the property of an analytical method for which only the target analyte gives any response whatsoever, i.e. the extreme case of selectivity. (It seems unlikely that a truly specific method exists for any analyte.) One of

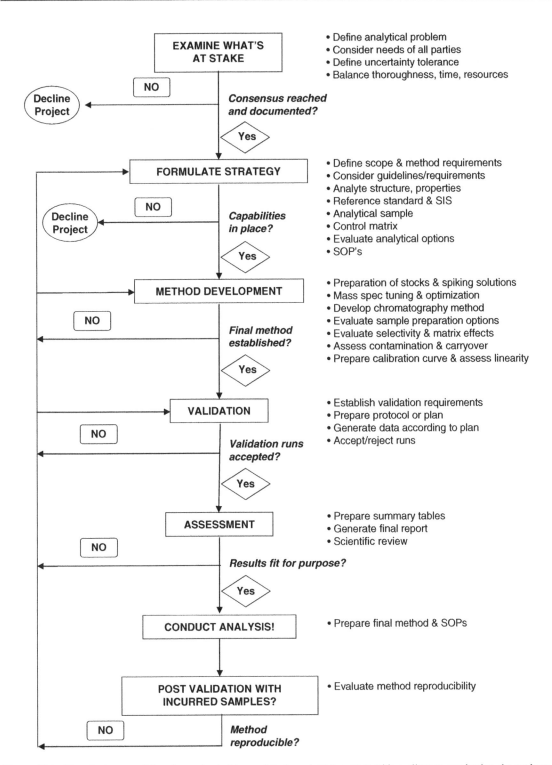

Figure 10.1 Flowchart summarizing the main decision points from initial contact with a client to conducting the analyses, emphasizing the key role of validation in the process.

Table 10.1 Key figures of merit that are addressed in validation of an analytical method

Full Validation	Figure of Merit
Basic Components	Selectivity
	Sensitivity
	Range of reliable response
	Accuracy
	Precision
	Reproducibility
Additional Components	Analyte recovery
	(extraction efficiency)
	Stock stability
	Matrix stability
	Post process stability (extract)
	Integrity of dilution
	Carryover and contamination
	Matrix effects
	Incurred sample reproducibility
	Incurred sample stability

the major advantages of mass spectrometry as a chromatographic detector is the significant improvement in selectivity compared with single parameter detectors such as flame ionization detectors for GC and single wavelength UV detectors for LC (*m/z* is far more selective than UV wavelength). Selectivity can be increased in several ways, including a more rigorous clean-up step, improved chromatography and use of MS/MS detection (Section 6.2.2).

The best assessment for selectivity of a method is achieved with the use of matrix extracted blanks that are prepared from control matrix that is representative of the sample matrix. This allows checking that ions with *m/z* values characteristic of analyte or SIS are not observed at the crucial retention times in chromatograms for the blank. A detailed account of how blanks of this type are used in method development to ensure selectivity is given in Section 9.4.7b.

Depending on the application, one or more source of control matrix may be available for selectivity screening. For instance, when testing for herbicides in a soil dissipation study the control matrix may come from the actual field where the study is to take place prior to application of the herbicide or from an adjacent control plot. For bioanalytical analysis several lots of control matrix may be screened independently and as a pool (Section 10.4.1c), and additional selectivity data may be obtained from actual subjects prior to dosing (pre-dose samples). When a representative control matrix is not available or if additional confidence is required for analyte identification additional measures such as the inclusion of multiple

ion confirmation criteria may be required (see discussion of top-down vs bottom-up identification criteria in Section 9.3.3b).

In any event, a strategy for demonstrating selectivity should be developed and included as a part of every method validation and the results should be presented and discussed in the final validation report (Section 10.3.3). The amount of tolerable interference will depend on the analytical objectives and on the needs of the end user of the data. For a bioanalytical assay, background interference of less than 20 % of the response at the LLOQ is often considered to be acceptable; this criterion is related to the recommendation (Viswanathan 2007) that the analyte response at the LLOQ should be at least five times the response due to blank matrix. Laboratory SOPs will dictate how this is measured in practice. If just a single measurement was made for each, the ratio of area (or height) of the LLOQ vs that of the blank is used, but if replicate measurements were made for the calibration curve at each concentration (as in bioanalytical practice), the lowest of the responses at the LLOQ would most likely be used for this purpose.

Whether or not background interference is observed, the impact of potential interference during sample analysis and the criteria used to measure and set maximum amount tolerable should be established and documented prior to sample analysis.

10.2.3 Sensitivity, Range of Reliable Response and Linearity

Sensitivity strictly refers to the slope of the calibration curve for the method but is widely used (including in this book) in a more general sense of a low value for the limit of detection (LOD, the lowest concentration of an analyte that the bioanalytical procedure can reliably differentiate from background and noise, see discussion in Section 8.4.1). 'Sensitivity' can also refer to the lower limit of quantitation (LLOQ, the lowest amount of an analyte in a sample that can be quantitatively determined with acceptable precision and accuracy, Section 8.4.2). While it is true that a higher sensitivity in the strict sense can lead to lower values of LOD and LLOQ, the latter parameters also vary with the level of chemical background and statistical noise (see discussion of Figure 8.10 in Section 8.4). While the LLOQ is usually the more important parameter in quantitative analyses, the LOD can be significant for some applications (e.g. dioxin analysis, Section 11.4.1).

The range of reliable response (also known as the calibration range or quantitation range) is the range of concentration that can be reliably and reproducibly quantified with specified accuracy and precision through the use of a concentration–response relationship. The latter is referred to as the standard curve (or calibration curve), the relationship between the experimental response value and the analytical concentration (or amount of analyte). It is always preferable if this functional relationship can be satisfactorily represented by a function that is linear with respect to both the analyte concentration and the parameters of the function (see Section 8.3). The preparation of QCs is discussed in Section 9.8.2 and further discussions relating to the use of these QCs to demonstrate adequate precision and accuracy at the LLOQ (and other concentrations within the range of reliable response) can be found in Section 10.2.4. The assessment of the LLOQ with respect to the original validation objectives and guidelines for calculating and reporting data can be found in Section 10.3.3.

10.2.4 Accuracy and Precision

Accuracy refers to the degree of closeness of the determined value to the nominal or known 'true' value under prescribed conditions and is sometimes termed 'trueness'. A deviation from trueness can be constant (bias or constant systematic error) or vary with the size of the sample and/or the analyte concentration (proportional systematic error, Section 8.1). Precision, in contrast, describes the closeness of agreement (degree of scatter) between a series of measurements obtained from multiple sampling of the same homogenous sample under the prescribed conditions. For a more detailed discussion see Section 8.2.

Analytical QC samples, analyzed in replicate at different concentrations representative of the calibration range, are used for assessing the precision and accuracy of the method during validation. It is important to note that the replicates are generated from distinct aliquots of the QCs prepared in bulk (Section 9.8.2), as opposed to simple replicate injections of the same QC extract. The specific validation requirements in terms of the concentration and number of QCs to be analyzed will depend on any prevailing guidelines or regulations for a particular application. However, it is helpful to cite the example of bioanalytical methods to be validated according to FDA guidelines (FDA 2001; Viswanathan 2007), in which accuracy and precision are determined by analyzing QC samples at a minimum of three concentrations with a minimum of five replicate assays at each concentration. The concentration of the low QC should in general be near the LLOQ and for bioanalytical validations no more than three times the LLOQ concentration. The mid QC should be, as the name implies, somewhere near the middle of the calibration range or at approximately the geometric mean of the low and high QC. The high QC should be near the ULOQ and for bioanalytical validations it is recommended that this concentration should be in the upper quartile of the calibration range. Additional QCs should be analyzed at the LLOQ itself to confirm the lowest amount of an analyte in a sample that can be quantitatively determined with acceptable precision and accuracy (Section 10.2.3). A discussion of the conduct of the validation and how precision and accuracy are evaluated relative to the original 'uncertainty targets' (Section 9.3.3), and how validation data should be calculated and reported, can be found in Section 10.3.

10.2.5 Reproducibility

Reproducibility can be defined (Bansal 2007) as the ability of the method to yield similar concentration values for a sample when measured on different occasions, and thus is intimately connected to the concepts of accuracy and precision. A useful distinction that is often made is that between intra-day and inter-day measures of accuracy and precision. This distinction is related to the concepts of robustness and ruggedness (Section 9.8.4) in that it is less demanding to maintain a prescribed level of precision for experiments conducted all on the same day (intra-day) than for a set of experiments conducted over several days (inter-day) possibly with other tasks intervening between the days.

Some writers (though not all) use the term repeatability in the present context as the precision among determinations made by the same operator using the same apparatus within a short period, often a single day (in which case it is referred to as intra-day repeatability or precision). In this nomenclature convention, use of the term reproducibility is limited to the precision of the data obtained under varying conditions, e.g. using the method in the same laboratory under the same operating conditions but after a period (inter-day precision), or the precision of data obtained by the method performed by two or more laboratories (and thus related to the ruggedness concept, Section 9.8.4). The context will make it clear in which sense the term 'reproducibility' is being used.

In validation studies, inter-run reproducibility (one or both of intra- and inter-day precision and accuracy) are determined in a given laboratory by analysis of the QC samples. However, there are other important instances where reproducibility is important (Bansal 2007). Thus, some apparatus used in a method may have to be changed (e.g. a chromatography column will typically have to be replaced after a certain number of injections). To validate the robustness of the method (Section 9.8.4) relative to such inevitable changes, it is recommended (Bansal 2007) that the reproducibility of the method on an alternative column (or other components of the analytical apparatus) instrument should be tested using appropriate QC samples on one of the days of validation. This is a good practice but is not required for all validations. Another instance where special reproducibility testing is recommended arises when there is reason to believe that incurred samples, i.e. post-dose bioanalytical samples from individuals (human or laboratory animal), may contain components (e.g. metabolites) that in principle can not be present in the control matrix used to prepare calibrators and QC samples. (This difference can also arise in other applications such as soil analysis where the matrix contains biochemically active microorganisms.) This reproducibility test can be performed during validation if incurred samples are available but can be postponed and performed during sample analysis where it is more important to prove the reproducibility of analyses of the incurred samples that are collected for a given study. A more complete discussion is given in Section 10.4.11.

10.2.6 Recovery

Recovery is a term often used to describe the extraction efficiency viewed as the percentage of the amount of analyte carried through the sample extraction and processing steps of the method into the sample extract (see the discussions of F'_a in several portions of Section 8.4). The best measurements of F'_a are made by comparing the response observed for the extract of a spiked QC sample to that for an extract of the same control matrix used to prepare the QCs that is spiked post-extraction with the same quantity of analyte. Clean solutions of pure analytical standard can be used instead of the post-extraction-spiked extract of control matrix, but such a procedure may confound the desired extraction efficiency with matrix effects (see Section 5.3.6a for a discussion of the inter-relationships (Matuszewski 2003) among extraction efficiency, matrix effects and process efficiency). Of course, if the control matrix is in short supply a clean solution of analytical standard must be used for this purpose.

Although the method recovery should be evaluated during the method development process (Section 9.5.6c), demonstrating recovery may or may not be a requirement of the validation process in every given case. If recovery is included in the validation plan, it may only be necessary to conduct the appropriate experiments and report the average recovery in the validation report and to demonstrate that it is consistent and reproducible. In such cases the distinction between extraction and process efficiencies (Matuszewski 2003) is irrelevant to the purpose for which the recovery is measured. For other applications, demonstrating that the recovery is within specified limits, e.g. > 70 % and < 120 %, may be required. Similar considerations for demonstrating recovery during validation should also be given to the SIS, especially if it is an analog or homolog.

The ratio of the extraction efficiencies of the analyte and SIS provides a so-called SIS-normalized extraction efficiency.

The recovery of analyte in matrix should be evaluated at a minimum of two concentrations using a sufficient number of replicates (minimum of three). In practice the extraction efficiency is often determined by comparing the mean area response from processed QC samples that are used to determine intra-assay accuracy and precision to the mean area response from recovery samples at each QC concentration.

10.2.7 Stability in Solution

The chemical stability of an analyte in a given solution that is stored under specific conditions for given time intervals is an important validation parameter that was discussed in general terms in Section 9.4.4f. An extreme form of instability is the propensity of the analyte to explode or be set afire, and such properties should be available in the MSDS information or, if the analyte is a new compound, the chemists associated with its synthesis or isolation will have discovered such properties before the analyst! A crucial special case is the stability of the analyte (and thus of the analytical standard) in solution, as in stock, sub-stock and spiking solutions. The stability of analyte or of an analog internal standard in solution should be determined by comparing stored stock solution(s) to freshly made-up stock solution(s). For a stable isotope-labeled internal standard, stability data for the corresponding analyte are often used to establish stability for such an SIS.

Fresh stock solutions should be used immediately after preparation if possible and certainly within 24 hours (one day) of preparation when used for comparison to stored stock solutions that are kept at the anticipated storage conditions. The number of replicate injections of the freshly prepared and stored stock solutions (or dilutions thereof) should be sufficient to achieve the precision required to make such comparisons statistically meaningful (see Section 8.2). The solutions should be analyzed at relevant intervals spanning the maximum anticipated storage duration in order to expose any trends in the stability with time. The storage stability of an analyte stock solution is defined by the period over which the mean peak area or peak area ratio of the stored solution deviates from the mean peak area or peak area ratio of the fresh solution by no more than the acceptable limit established in advance (10 % is a value that is often chosen). If a stock solution falls outside of acceptance criteria, an additional comparison can be made to verify instability. If the verification indicates that the analyte in solution was in fact stable (e.g. indicating a prior preparation error rather than instability), then at least one additional assessment or interval should be run to demonstrate that stability has been established.

Stability should be clearly established for the anticipated duration of storage and use by comparing peak areas for stored solutions against those for solutions made from fresh weighings of analytical standard at selected time intervals such as two weeks, one month, three months etc. As a practical consideration, it is often desirable to establish a full year of stock stability under appropriate storage conditions. If sub-stocks or spiking solutions are used and made in different solvents, then similar stability studies would be required for these but, for reasons of practicality, probably for shorter periods.

Additional discussion of the various types of stability data that are typically generated in support of validation studies is given here, and also for the special case of bioanalytical applications in Section 10.4.1.

10.2.8 Stability in Matrix

While this is an important validation parameter in general since it addresses the requirements for sample storage conditions, it is particularly so for bioanalytical samples in which the analyte is plasma or other biofluid containing natural enzymes that can potentially metabolize the analyte, and also for soil samples where metabolism (due to active microorganisms) and/or degradation is common. For the bioanalytical case it is useful to be aware of structural groups that are susceptible to metabolic modifications and the differences in molecular mass that arise. Small molecule drugs are usually lipophilic substances and the corresponding drug metabolism is a process of introducing hydrophilic functional groups. The most common phase I reactions are oxidative processes involving cytochrome P450 enzymes that catalyze aromatic hydroxylation, aliphatic hydroxylation, N-, O-, and S-dealkylation, N-hydroxylation and N-oxidation, as well as sulfoxidation, deamination and dehalogenation. CP450 enzymes can also catalyze some reduction reactions under oxygen-deficient conditions. Hydrolysis is also commonly observed for a variety of drugs, catalyzed by esterases, amidases and proteases and yielding hydroxyl or amine groups suitable for phase II conjugation reactions that introduce hydrophilic functionalities, such as glucuronic acid, sulfate, glycine or acetyl groups, onto the drug or drug metabolite molecules. Glucuronic acid is essentially a glucose molecule oxidized to a carboxylic acid and is extremely hydrophilic. Some common functional groups susceptible to metabolism are: aliphatic/aromatic carbons (hydroxylation), methoxyl/methylamine groups (demethylation), amine (N-oxidation or deamination), sulfur-containing groups (S-oxidation), phenol/alcohol

(glucuronidation/sulfation) and esters/amides (hydrolysis). The shifts in molecular mass introduced by such metabolic reactions are: glucuronidation, +176 Da; sulfation, +80 Da; oxidation (N-, S-), +16 Da; hydroxylation (aliphatic, aromatic), +16 Da per site; dealkylation, (loss of the alkyl group mass); hydrolysis, (loss of the mass of $(R_2–H)$ for hydrolysis of an ester $R_2–CO–OR_1$ to the acid).

For stability in matrix studies, the replicate 'stability QCs' are generally prepared at $N = 2$, $N = 3$ or $N = 5$, depending on the particular validation requirement or specification. Typically, the concentration determined for the stability QC will be calculated against 'nominal' (the theoretical concentration) but in some instances stability will be evaluated by comparison to the average value of the QCs that were analyzed on what is referred to as 'time zero'. The problem with using time zero values is that this value may have a bias due to imprecision of the assay, and this could impact the interpretation of the stability study.

Analyte stability in matrix is generally assessed by analysis of stored QC samples (minimum of low and high QC samples, appropriate N value and separate storage tube for each replicate) with comparison to nominal concentration. The mean accuracy for a set of stability QC samples should be within ± 15 % of the nominal concentration values. If the run (set of calibrators, QC and blanks) that was used to analyze the stability samples fails due to failure to pass the acceptance criteria for calibrator and/or analytical QCs, then the stability experiment(s) within that run must be rejected. If a set of stability QC samples does not meet acceptance criteria at the specified interval and the failure does not appear to be a result of clear analyte instability (e.g. single replicate outlier), then the test at that interval can be repeated. If this is done the number of replicates should be increased, e.g. to $N \geq 6$, to establish stability. Alternatively, two additional later time/cycle intervals may be run, but both intervals must meet acceptance criteria to establish stability.

It may not be necessary to conduct some or all of the stability experiments described below if stability data have been generated previously by the analytical laboratory or by the client. Other stability experiments may be added to those described here.

10.2.8a In-Process Stability (Room Temperature Stability)

This terminology refers to the stability of the analyte, particularly while still in matrix, while undergoing the extraction and clean-up procedures in the laboratory. The stability of the analyte(s) in matrix at room temperature (benchtop stability) should be determined by extracting and analyzing a set of QC samples removed from storage and left at room temperature for some time (a minimum of four hours) prior to extraction. It is considered good practice to establish about 24 hours of room temperature stability after removal from storage to address the circumstance when a sample is inadvertently left on the bench for an extended period of time. Finally, if a sample is re-assayed several times as in freeze–thaw stability tests (Section 10.2.8b), it might be appropriate to consider the total accumulated time that the sample was exposed to room temperature on multiple occassions.

If an analyte is found to be unstable in matrix at room temperature during a validation, or if it is known prior to initiating the validation that room temperature matrix instability is problematic, the analytical method may be designed to make use of a stabilization technique (e.g. all operations conducted at 4 °C or addition of a chemical stabilizer). Analyte stability should then be assessed under the conditions of the assay using the stabilization technique.

10.2.8b Freeze–Thaw Stability

A special aspect of stability concerns subjection of the sample to multiple freeze-thaw cycles which could occur if the sample needs to be re-assayed one or more times. While storage in a freezer may seem to be an obvious precaution, it does have its drawbacks since the freezing process can segregate the solvent from the solutes to some extent as described by multi component phase diagrams governed by Gibbs' Phase Rule (see Section 4.3.2e and any physical chemistry textbook). Thus, on thawing the solid does not melt uniformly all at once and the analyte can find itself temporarily in a small pocket of liquid that might also contain high concentrations of other components (e.g. added acid or base) that react with the analyte. This principle is illustrated in Figure 10.2. Generally, the first solid to separate on cooling the liquid is richer in the solvent, and this process proceeds until the final portion of liquid solution to freeze is relatively concentrated in solutes (both analyte and any other components, e.g. acids or bases). When the entire frozen mass is thawed the process is reversed, i.e. the first liquid to form can be much more concentrated in solutes than the overall composition, and it is at this stage that a labile analyte is susceptible to decomposition and/or chemical reaction.

10.2.8c Long Term Storage Stability (Sample Stability)

Long term storage stability of analytical samples can be a major issue in several applications, e.g. analysis of pesticide residues in environmental samples and bioanalytical samples. For some validation purposes (e.g. bioanalytical

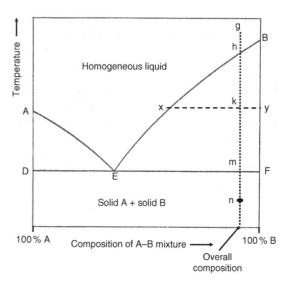

Figure 10.2   A simple example of a solid–liquid phase diagram for two components ($C = 2$ in the Phase Rule, Chapter 4) that are completely miscible in the liquid phase (the area above the curves ACB) but completely immiscible in the solid state (below the line DCE). The pressure is assumed to be fixed, thus removing one degree of freedom so $F = (C + 1 – P) = (3 – P)$, where P is the number of phases present at equilibrium. The area ACDA represents liquid in equilibrium with solid A ($P = 2$ so $F = 1$, i.e. fixing the temperature determines everything about the system including the compositions of the two phases), and the area BCEB represents liquid in equilibrium with solid B. Above the curves ACB there is just one phase (liquid) so $F = 2$, i.e. to completely describe the system in this condition both temperature and composition must be specified. Below DCE there are two immiscible solid phases so again $F = 1$ (temperature, the phase compositions are fixed as pure A and B in this simple example). For a mixture with overall composition given by the line *sqpl*, if the completely frozen mixture (solid A + solid B) is warmed up starting at the temperature corresponding to point *s*, liquid first begins to form when the temperature reaches the value indicated by the line DCE, i.e., at the point *q*, and this liquid has composition indicated by the point C; here $P = 3$ (solids A and B plus liquid C) so $F = $ zero, i.e. the temperature can not rise further until one of the solid phases (A in this case since B is in excess) has been transformed into liquid C. The important point is that liquid C is much more concentrated in A than is the overall system (e.g. the final liquid l), and this will be true no matter how close the composition line sqpl is to 100 % B, i.e. even for very dilute solutions of A in B; this temporary concentration effect can lead to unwanted reactions of A when thawing frozen solutions.

methods) the stability of the analyte in matrix may be determined at room temperature and at each of −20 °C and −70 °C (the approximate temperatures of conventional freezers and ultra-freezers). It has been claimed, on the basis of physico-chemical considerations of the Arrhenius description of the temperature dependence of chemical rate constants, that demonstration of analyte stability in bioanalytical samples at, e.g., $-20\,^{\circ}$C automatically guarantees stability at the lower $-70\,^{\circ}$C. Objections to this view have been based (Viswinathan 2007) on considerations of matrix degradation rather than chemical stability of the analyte. Thus, it was suggested that the lower temperature could cause additional denaturation or precipitation of matrix proteins and that this could affect protein binding and the ability to extract the drug from the matrix. While no firm recommendation was made concerning this question (Viswinathan 2007), it was suggested that validation of stability in this sense should be undertaken on a fitness for purpose basis. However, in early development of stages of development when no stability has been established, $-70\,^{\circ}$C is often the default storage temperature if that capability exists. Similar precautions may be required for other matrix types.

The stability of the analyte(s) in matrix during extended freezer storage should be determined by analyzing stored stability QC samples at appropriately selected time intervals and at temperatures that reflect the intended storage conditions and anticipated storage periods for study samples. For many applications, the compound is determined to be stable as long as the calculated concentration of the analyte is within 15 % of the nominal concentration, or the concentration that was established for the same batch of QCs when analyzed immediately after preparation (time zero). Freshly prepared QC samples, or QC samples prepared and stored within an established period of stability, should be used as analytical QCs in the same set that the stability QCs are analyzed to confirm run acceptance. If the analytical QC samples do not meet assay acceptance criteria, the run should be rejected and the stability interval should be repeated. Incurred samples or sample pools may also be analyzed for assessment of stability in a similar manner.

Alternatively, stability can be assessed over at least three intervals (including the initial analysis as one interval) to determine if any instability trends are present. Determination of the statistical significance of any differences in concentration observed at the different time intervals can in the first instance use standard statistical tests for differences between means and variances (Section 8.2.5). It is more difficult to determine whether or not any such differences represent a trend. In principle the analyst could determine the slope of the plot of measured concentration vs time and apply statistical tests to the variance of the slope to determine whether the slope differs significantly from zero (Section 8.3); three points on such a plot are the bare minimum to establish a slope, and in any event for any statistical test of significance, appropriate confidence limits and confidence levels must first be decided on. Thus, while

statistical testing will provide an invaluable tool in this regard, informed judgment of the analyst will be a major factor in deciding whether or not to discard a QC batch (or indeed a stock solution for which the same dilemma arises). For these reasons, the assesment of stability for regulatory applications is often based upon the evaluation of stored stability QCs relative to the nominal or time zero concentration as described above.

10.2.9 Other Special Validation Requirements

The following special requirements do not necessarily constitute either an exhaustive or required list for all circumstances. Ultimately a fitness for purpose approach should be adopted with respect to these special requirements.

10.2.9a Extract and Re-Injection Stability

Once the extraction–clean-up steps have been completed the stability of the analyte in the analytical extract must be validated, both with respect to long term storage if the final analytical step can not be conducted immediately and also with regard to stability while the extract is sitting for some time in a vial in an autosampler waiting to be analyzed as part of a multi-sample run.

Analyte stability in processed extracts for re-injection is assessed by analysis of stored QC sample extracts (minimum of low, mid and high QC samples, $N \geq 3$) and comparison of these results to nominal concentration. The mean accuracy value for a set of stability QC extract samples should be within ± 15 % of the nominal concentration values. The stability of the analyte in reconstituted extracts should be determined by analyzing a set of QC sample extracts that have been stored at room temperature (or under other conditions consistent with storage at the autosampler). Extract stability is measured by at least one of the following four procedures:

1. QC samples are prepared and injected together with calibrators to establish a calibration curve; the QC set is then re-injected. The re-injected QC samples are quantified against the same calibration curve used for the initial injections ('Original Curve'). The period of re-injection is calculated from the time the last calibration standard was analyzed in the original analysis set to the time of the first re-injection of the QC samples. This experiment assesses the ability to re-inject samples after completion of an analytical run without the need for re-injecting an entire set of calibration curve and QC samples.

2. QC samples are prepared and injected together with calibrators to establish a calibration curve as in 1.;

the calibration curve and QC samples are then re-injected. The re-injected QC samples are quantified against the re-injected calibration curve. The period of re-injection is calculated from the time the last calibration standard was analyzed in the original analysis set to the time of the re-injection of the first calibration standard. This experiment assesses the ability to re-inject samples after completion of an analytical run with the requirement for re-injecting an entire set of calibration curve and QC samples.

3. QC samples are extracted, stored at the appropriate temperature, and analyzed against a freshly prepared calibration curve with analytical QC samples. (Stored QC samples from a previously analyzed set may also be used for assessing extract stability.) The storage time is calculated from the time the final extracts were prepared to the time the first of these stability QC samples was injected. If extract stability is established with this experiment alone, then the requirement for re-injecting an entire set of calibration curve and QC samples applies.

4. To allow for prolonged storage of reconstituted sample extracts, the stability of the analyte in reconstituted extracts may be determined by analyzing a set of QC extracts that have been stored in a refrigerator (4 °C). Stored extracts should be quantified against a freshly prepared calibration curve. Alternatively, the QC samples can be analyzed against a calibration curve obtained using calibrators extracted and stored in the refrigerator together with the stability QC samples. The assessed storage time is calculated from the time the extracts were placed in the refrigerator to the time the extracts were removed from the refrigerator.

10.2.9b Assessing Carryover and Potential for Laboratory Contamination

The concepts of carryover within the analytical train and of general laboratory contamination were discussed in Section 9.7. Recommendations for how to address these ubiquitous potential problems in bioanalytical method validation are given elsewhere (Viswinathan 2007). Important aspects are discussed here and in Section 10.5.5.

A consideration that can be important more generally for carryover problems is that of positional differences (Bansal 2007). During a typical LC–MS analysis, samples are injected in sequence over several hours so it is important to determine whether or not sample position in the sample queue has an influence on the observed response (e.g. carryover from previous samples, or possibly instrumental drift

over the course of the run). An evaluation of this possibility should be done during the validation of the method and monitored during sample analysis. A procedure that can be used to assess positional differences during validation (Bansal 2007) can be summarized as follows. Calibration standards and QC samples should be analyzed in a predefined order: (i) One set of calibration standards, preferably in ascending order of concentration, at the beginning of the run (defining the 'front curve'); (ii) QC and any other validation samples in the middle of the run, distributed randomly or in a sequence designed to help assess positional differences, e.g. place a blank matrix sample or zero standard after the high concentration QC to help assess carryover; (iii) One set of calibration standards, preferably in descending order of concentration, at the end of the run (defining the 'back curve').

After the analysis is complete, the combined sets of calibration standards are then used to construct a calibration curve that is used to determine the intra- and inter-run accuracy and precision. In addition, construct individual 'front' and 'back' calibration curves and determine the concentrations of the QC samples using these curves separately; these two sets of QC concentrations will show a bias if there are positional differences.

The peak response in the blank matrix sample or zero standard placed after the high concentration sample should be reviewed. It is accepted practice that the analyte response in this blank sample should be less than 20 % of that of the LLOQ sample. This aspect is discussed in more detail in Section 10.5.5, together with some practical considerations (Zeng 2006) that can help determine optimum conditions (including sample ordering) that minimize the quantitative effects of carryover if the latter can not be avoided. If injection carryover exceeds 20 % of the response of the LLOQ calibration standards and/or QC samples, and continues to do so even after additional or more rigorous rinsing techniques in the injector or autosampler (Section 9.7.2) are employed in subsequent runs of the validation, then appropriate changes to the method, either prior to or during sample analysis, must be made to prevent injection carryover from having deleterious effects on quantification of study samples.

The general problem of laboratory contamination was discussed in the context of method development (Section 9.7.1). This problem might have to be addressed also during validation through appropriate use of blank samples but the same principles apply.

10.2.9c Incurred Sample Re-Analysis and Stability

The terminology 'incurred sample' is used to distinguish the true analytical sample from a control matrix that is fortified with known amounts of the analyte for purposes of creating a matrix matched calibrator or a QC sample. 'Incurred stability' is essentially the same thing as long term stability in matrix (Section 10.2.8c) but is assessed using incurred samples. Incurred reproducibility demonstrates good repeatability upon re-analysis of incurred samples that might be subject to matrix interferences and matrix effects not necessarily present in calibrator and QC samples made by spiking a control matrix. Any such differences could reflect problems with the method but more likely are the result of metabolites or other interferences that are in the incurred samples but not in the control matrix (Section 9.4.7b). The implications for validation of bioanalytical methods are described in Section 10.4.11 but the problem is more general, e.g. soil samples usually contain metabolically active microorganisms that are not present in the control matrices normally used for such analyses. In lieu of any regulatory requirements or other similar directives, the decision as to whether or not additional validation experiments designed to assess stability and reproducibility using incurred samples should be based upon the nature of the analyte, matrix and the analytical method being used.

10.2.9d Integrity of Dilution

On occasion, the validated ULOQ for an assay is not sufficiently high to accommodate high concentration study samples that are anticipated to arise. It is possible to analyze such samples if it can be demonstrated that dilution of the high concentration samples does not invalidate the method. To assess whether such a a sample can be diluted with matrix for accurate quantitation within the range of the calibration curve, additional validation experiments are required. For this purpose QC samples, spiked at above the ULOQ concentration (the so-called 'above quantitation limit' (AQL) QCs), are diluted with control matrix before sample preparation using the following procedure.

Typically, a minimum of five individual dilutions of the AQL QC with control matrix are prepared using at least one dilution scheme. Dilutions may be made outside of the extraction tube/plate in separate tubes and aliquots assayed at the volume specified for the assay. Alternatively, aliquots smaller than that required by the assay can be transferred using a calibrated pipet into the extraction tube/plate and then diluted with control matrix to the volume required for the assay. The mean accuracy value of diluted AQL samples should be within ±15 % of the nominal concentration value for each dilution scheme. The precision around the mean (% CV) should not exceed 15 % for each dilution scheme. If a given dilution scheme fails to meet acceptance criteria with no assignable cause for the failure, then the experiment may be repeated but it is recommended to use a

greater number of replicates in this case to establish dilution integrity for that dilution scheme.

In practice, if it is necessary to dilute samples during analysis, additional AQL QCs are added to the sample set (using the same dilutions as those used for the samples on that day) and the same type of acceptance criteria are used as for the other analytical QCs. If the AQL QCs fail, the entire sample set is not failed but the data from the diluted samples are rejected.

10.2.10 Abbreviated and Partial Validation

Before a newly developed method can be used for the first time, it must undergo a full validation as described above and later in this chapter. A full validation of the revised assay is also important if additional target analytes, such as metabolites of the original analyte, are added to an existing assay. However, for various reasons a full validation may not be possible or even necessary. For example, control matrix may be extremely scarce so that only an abbreviated validation is possible. For rapid screening methods that are designed only to protect against false negatives and are not primarily concerned with either false positives (because confirmation procedures will follow) or accuracy (because more rigorous quantitative methods will be used after the screening) full validation may not be required. Other examples where abbreviated validations may be appropriate include applications where the end user of the data is gathering preliminary information of the fate of the analyte and does not require the level of precision, accuracy or other parameters that are typically part of a full validation. An example of this latter type of application would be bioanalytical methods used in support of late discovery or lead optimization studies, where the results are used for rapid screening of large compound libraries for bioavailability in rat plasma, to aid decisions as to whether or not a particular compound should be investigated further as a potential drug candidate (see Section 11.5.1).

The FDA Guidance document (FDA 2001) describes a circumstance somewhat related to abbreviated validations but one that is applicable to modifications of existing fully validated methods. The validation procedures required for such circumstances are called partial validations. In general, the required components of a partial validation will depend on what type of modification was made and the validation parameters that may have been affected; the requirements can range from as little as one determination of intra-assay accuracy and precision to a nearly full validation. Examples of the modifications to existing methods that require some form of partial validation (FDA 2001) include: method transfers from one laboratory to another (but it has been argued that this circumstance is more properly described as a cross-validation if the method was not modified, Section 10.2.11); change in species within a given matrix, e.g. rat plasma to mouse plasma; change in matrix within species, e.g. human plasma to human urine; change in sample processing procedures (e.g. change in sample preparation such as modifications of the validated extraction method, or in the type of chromatographic column); change in relevant concentration range; limited sample volume, e.g. as in a pediatric study; selectivity issues for an analyte in the presence of concomitant medications or of metabolites not present in control matrix (this example corresponds to the incurred sample issue, Section 10.2.9c); change in instrument platform (e.g. model of mass spectrometer); or change in the SIS.

The nature and extent of partial validations must therefore be decided by the analyst on a fitness for purpose basis. The additional validation might be as little as one-day precision and accuracy checks, or even the same over three days depending on the change and prevailing SOPs. For example, for a change in methodology, matrix and stock stability and possibly other parameters would not have to be re-validated because matrix and stock stability have nothing to do with the actual method. However, selectivity would have to be re-validated if the changes involved sample preparation, details of the chromatography or of SIM or MRM detection parameters or a change in the LLOQ. A change in the mass spectrometer between manufacturers or different models from the same manufacturer would certainly require a partial validation to assess precision and accuracy (especially at the LLOQ). Testing the ability to run a method on different instruments of the same manufacturer and model is less critical, although it is advisable to run a partial validation since subtle changes in, e.g. the API interface, might affect the original validation. If an SIS from a previously validated analytical method is changed from one structural analog to another, a minimum of a two-day partial validation or more may be required. If the change is made from a structural analog to an isotope-labeled SIS a minimum of a one-day partial validation should be sufficient. When only minor modifications are made to a method (e.g. slight adjustments in the mobile phase to improve peak shape or resolution, column temperature changes or changes in the final reconstitution volume) re-validation may not be necessary but at a minimum, the analyst must properly document any such minor modifications and update the procedure accordingly.

Depending on the governing body, regulatory guidelines or the SOPs in place at the laboratory where the validation is taking place, the particular requirements for partial validation requirements may vary, but ultimately a thorough evaluation of the impact for any method modification or application must be conducted and the extent of the partial validation must be justified. In many instances laboratories chose to conduct full or nearly full validations for all but the most minor modifications to the method as a means to ensure that the integrity of the originally validated method is adequately investigated and demonstrated.

10.2.11 Cross Validations

The term 'cross validation' is often used incorrectly when describing validations that would typically fall under the category of an abbreviated validation (Section 10.2.10), and vice versa. A more appropriate use of the term cross validation is for describing validations designed to compare different kinds of analytical data generated in the same laboratory to support the same study (e.g. a GC/MS method with one using LC/MS), or by different laboratories using similar or different methodologies but used to support the same study. It is important to ensure that the interpretation of the results is not biased by the fact that a change was made within or between laboratories; in such situations not only should the QC data from within a particular laboratory be scrutinized, but QC and incurred results between laboratories should also be compared.

According to an earlier comprehensive review (Muirhead 2000) the term 'cross validation' is generally used to cover three different circumstances, including that mentioned above in which two or more laboratories running the same analytical procedure are compared. A second example for which these authors consider the term to be appropriate was also mentioned above, i.e. a comparison of two different methods for the same assay (e.g. LC–MS/MS vs radioimmunassay and/or LC–UV). The third circumstance considered refers to validation of a new operator, or of a small change to the method (e.g. altering an extraction time), or of a switch from a common to a rare matrix; this example was said (Muirhead 2000) to be a misuse of the term 'cross validation' and instead should be referred to as an abbreviated (or partial) validation (Section 10.2.10).

However, for the agreed cases of inter-laboratory and inter-method comparisons, this review (Muirhead 2000) did propose a study design, together with statistical evaluation criteria to assess equivalence, that accounted for inter-batch variability and variability within both laboratories or assay procedures. This approach proposed the use of incurred samples, but unfortunately no follow-up publication describing the application of this study design appears to have been published.

In many instances when performing cross-validations between laboratories, full statistical analyses are not performed; rather, criteria based on acceptable precision and accuracy, with at least two-thirds of the samples passing, are used in practice. Using this scenario, the first laboratory sends a set of its analytical QCs as well as incurred samples to the other laboratory. The data obtained for analytical QCs and incurred samples that were run at both laboratories are then evaluated.

When two or more analytical methods are used to generate data within the same study, a cross validation study should incorporate calibration standards and a set of QC samples ($N \geq 5$ per QC concentration) and/or incurred samples run by both analytical methods. The mean values of a set of QC samples and/or incurred samples from one method to the other must not deviate by more than the pre-determined acceptable bias for the method. Other statistical tests can be used to assess comparability between the methods, e.g. cross validation data evaluated by the client may follow acceptance criteria based upon their own internal SOPs.

10.3 Conduct of the Validaton

The basic requirements for the conduct of a validation and reporting of results are presented in this section. However, it must be emphasized that it is the responsibility of the analytical site to ensure that the necessary SOPs and other required documents are in place before the initiation of work. In addition to the procedural practices which are described in the SOPs, a documented plan or protocol that describe the objectives, conduct and data reporting requirement should be prepared and approved by the responsible scientist, and preferably also by laboratory management.

10.3.1 Validation Plan or Protocol

A detailed validation plan should be prepared prior to beginning the validation. This document should provide a detailed description of the various experiments that will be run, the methodology that will be used to run the validation samples and the requirements for processing, reporting and interpreting the final results. In laboratories that have detailed SOPs to describe the validation and documentation procedures that are followed on a routine basis, an approved document that summarizes specific validation parameters to be tested with references to the method and all prevailing SOPs may be suitable. In other instances, depending on the prevailing regulatory or laboratory SOP requirements, a formal protocol is often prepared.

A formal protocol, in addition to describing the validation requirements and procedures, will specify the person, often referred to as the Principal Investigator or Principal Scientist, who has the responsibility for the overall conduct of the study. In many cases the protocol will be relatively brief and will simply refer to existing laboratory SOPs or documents to describe the procedures that need to be followed. If detailed descriptions of validation requirements and procedures are not available, the protocol can serve as the primary

document for both planning and establishing the laboratory procedures and data interpretation requirements. In a comprehensive book on quality assurance as it relates to chemical measurements in the laboratory (Taylor 1987), the author presents the general requirements for *Protocols For Specific Purposes* (PSP). This will be used here as a basis for establishing guidelines for a validation protocol and ensuring that the minimum requirements for the conduct of a validation are established prior to the initiation of work.

As described by the author, the Principal Investigator (PI) is assigned the overall responsibility for implementing the work. Other persons may be designated by the PI to perform certain tasks but ultimately it is the responsibility of the PI to ensure that the work was conducted according to the specifications and requirements of the protocol. For the purposes of a validation, the PI would be responsible for determining the exact validation parameters that need to be tested and for reviewing the final validation report to make sure that the work was conducted according to plan and that the final results are reported and interpreted properly.

The Problem (validation requirements in this case) should be clearly defined with highly specific definitions for the scope and objectives of the work to be undertaken. If part of a larger program, the protocol may include an overview of all other activities relating to the work, e.g. cross validation activities at another laboratory, but more likely a validation protocol will be focused on the analytical portion of the study and the procedures to be conducted in a particular laboratory.

The Model includes the hypotheses to be tested (validation parameters), as well as the general data requirements including the precision and accuracy targets that must be established for each experiment.

The Specification of Samples includes details relating to the types and numbers of samples to be analyzed. In the context of a validation this will essentially be a description of the QCs and how they will be used to validate specific parameters. A reference to the SOPs that describe the procedures and documentation requirements for the preparation and storage of QCs should also be provided. If these do not exist, they should be created or included in the validation protocol.

A detailed Specification of the Methodology is a critical component of the protocol so that there are no ambiguities with respect to how the assay should be performed. In addition to the analytical method, any other procedures that are required but not described in the method should described in detail in the protocol, or references to the prevailing SOPs should be provided.

The Specification of Controls section of a protocol describes any activities that will be undertaken to assure the quality of the data. This also includes any preliminary activities that may be required and could include a description of extra precautions or system checks based on the particular analyte or data requirement. For validation this information should be available in the method and supporting laboratory SOPS. For sample analysis, the specification of controls would also include a description of any control charts that should be maintained, plus guidelines for the interpretation of results and any investigations that may be required based on those data.

In addition to an assessment of the validation parameters described in Section 10.2, an important goal for any validation should be to establish the guidelines that will ultimately be used in sample analysis for the Release of Data. For the validation, these guidelines will simply be the data acceptance and reporting requirements used for interpreting and reporting the validation results including procedures for calculations, rules for rounding and use of significant figures, and will also establish the minimum requirements for data acceptance. However, for sample analysis, additional guidelines based on the validation results may need to be established for situations where data fall outside the accepted quantitative or qualitative limits of uncertainty. For a quantitative analysis this might include specific reporting procedures for any detections less than the LLOQ but greater than the LOD. From a qualitative perspective this would include procedures for reporting data that met some but not all of the confirmation requirements described in the method. In this respect, the validation protocol needs to consider fitness for purpose in general for all of the protocol requirements described in this section but, in particular, with respect to embracing uncertainty and the guidelines based on the validation results and established for releasing the final data to any outside agency or user. At this point it becomes apparent that the issues addressed in Section 9.3 ('Examining What's at Stake') before any other method development or validation has begun, will be crucially important since these will form the basis of the criteria embodied in the protocol for releasing the final data.

10.3.2 Validation Sample Analysis

A validation plan or protocol (Section 10.3.1) that specifies the experiments to be run on each day should be prepared prior to validation sample analysis. This document should specify how the validation should be conducted and, for validations that are conducted over multiple days, the plan should specify what experiments should be run on each day. Prior to the initiation of work, the protocol should be reviewed. Any questions or discrepancies should be resolved and, if necessary, an amendment should be prepared to resolve these issues.

The analytical method should also be thoroughly reviewed for completeness and accuracy. All paperwork and documentation for the reference standard, stock and spiking solutions, and control matrix that will be used to make the QCs should be reviewed. All supplies, chemicals and solvents required for the method should be reviewed and the general purpose laboratory items should be checked to verify that adequate supplies are in stock prior to commencement of work.

If another analyst is going to be performing the work, a meeting between the scientist who developed the method and the analyst should be scheduled to review each step of the method and to ensure that there are no ambiguities or questions about the stepwise procedure or other method requirements. All critical steps and procedures should be identified and discussed with the analyst.

The analytical run list should be constructed in such a way that it meets any prevailing method or SOP requirements; in particular, consideration should be given to the need to include a sufficient number of sample blanks in order to detect the potential for carryover. Positional differences (Section 10.2.9b) should also be considered when constructing the run list.

It is essential that the analysis is conducted following the exact procedures described in the method with no modifications; if any deviations do occur, this must be documented in the laboratory notebook and validation report. The control matrix should be screened for interferences prior to use and a single stock solution of the internal standard (and subsequent dilutions) must be used within a single validation run.

10.3.2a Data Review

A general description of the review process for an analytical run with respect to chromatography and the integration of peaks can be found in Section 10.5.3. The same general principles apply to validation but, if any of the problems such as drift, loss of sensitivity, curve divergence etc. are observed during validation, the responsible scientist may determine that the validation should be abandoned and that further method development will be required. Automatic integration of the chromatographic peak using the same integration parameters for the entire run is highly recommended, but for purposes of assessing interfering peaks with low signal intensity, manual integration of peaks is acceptable when automatic integration parameters fail to provide adequate integration of peak areas. When present as broad peaks with regions that are outside the retention time limits for the analyte peak(s), or if direct matrix interferences prevent adequate integration, peak height may be a more appropriate measurement parameter for quantifying the impact of matrix interference on a method. However, even if the validation runs pass the original qualitative and quantitative uncertainty targets, it may be determined that additional method refinement to address this type of situation in order to improve method ruggedness or robustness will be necessary.

In addition to reviewing the analytical data generated from the validation runs, a detailed review of all notebooks, sample preparation records and documentation for all other supporting data should be completed; it should be possible to verify that the method was run exactly as written. A detailed description of the supporting data that is required for a validation is provided in Section 10.3.3 and these data must also be carefully reviewed for completeness and accuracy.

10.3.2b Addressing Validation Run Failures

Validation runs may not be re-injected simply on the basis of failure to meet the method targets for precision and accuracy. Data for an entire validation run may be rejected only if there is an assignable cause. Rejection of data due to assignable cause includes documented sample processing errors or for other documented reasons, e.g., the data were lost or corrupted during acquisition or data processing or there was a documented hardware failure. Complete documentation of the failed run and the reasons for the failure should be maintained in the study file.

If a validation run fails for any other reason, the data should still be processed and included in the validation report (Section 10.3.4). If the validation fails repeatedly, the method notebooks and procedures should be reviewed to ensure that it is being run correctly. If there is no apparent reason for the run failures, it will be necessary to perform additional method development experiments, modify the method and re-start the validation. Under no circumstances should the method be modified between validation runs.

10.3.3 Documentation of Supporting Data

Thorough documentation of all data that support the conduct of a validation is necessary to reconstruct the study and validity of the results upon inspection. The analytical site should have established SOPs that describe the data documentation requirements for the laboratory based on regulatory and quality program requirements. Specific data documentation requirements have been recently published by the FDA and other industry collaborators (Viswanathan 2007). Although these were written specifically for laboratories generating bioanalytical data for regulatory applications, the requirements can be readily adapted to other types of work and applications. In this book the authors distinguish the documentation requirements for: the analytical

site, the validation report and amendments, and the sample analytical report and amendments. The requirements for the analytical site are very similar for the validation and sample analysis, and are summarized below. Specific requirements for the validation report and sample analytical report are summarized in Sections 10.3.4 and 10.6.2, respectively.

10.3.3a Assay Procedure

The description of the assay, including a detailed stepwise procedure that was established during method development should be followed exactly as written during the validation and ultimately documented in a final analytical method or SOP. A description of the specific documentation requirements for the method can be found in Section 9.8.5. The method may also include references to other laboratory SOPs that describe routine procedures (e.g. equipment calibration).

10.3.3b Standards and Stock Solutions

Documentation of preparation of stock and sub-stock solutions must start with receipt of the reference standard, its Certificate of Analysis, assigned purity (and possibly chiral purity) and its history of storage and use after receipt (Section 9.4.4). Procedures for preparation and subsequent dilution of stock solutions are described in Section 9.5.4 and stability testing for these solutions in Section 10.4.1h. Some of the relevant documentation might be included in general laboratory SOPs but full documentation of all study-specific procedures and data regarding preparation, storage and validation of stock solutions is required.

10.3.3c Preparation of Calibrators and QCs

General principles for preparation of calibration solutions and matrix matched calibrators are described in Section 9.8.1 and similarly for QC samples in Section 9.8.2. Records to support the preparation, storage conditions (location and temperature) and use of these samples must be maintained at the analytical site, together with SOPs and information for all relevant apparatus including freezers and refrigerators, analytical balances, volumetric equipment etc.

10.3.3d Acceptance Criteria

The validation process is used to support acceptance criteria used for study sample analysis. The targeted acceptance criteria to be used for sample analysis will be dictated by the analytical requirement, sponsor, or regulatory guidelines but ultimately, the final criteria that will be used for study sample analysis are based upon the method validation results. *In any event, the exact procedures for evaluating the run for acceptance must be established prior to sample analysis and specified in the analytical method, an approved SOP, or equivalent.*

With regard to the calibration standards, the chromatographic response-concentration data should initially be subjected to the regression method established during method development but it may be appropriate to evaluate the results using other regressions as well after all of the validation days have been completed. For the validation of bioanalytical methods (Section 10.4.1a), the current FDA Guidelines states 'The simplest model that adequately describes the concentration–response relationship should be used. Selection of weighting and use of a complex regression equation should be justified.' When the sample set contains multiple analytes, the data for each analyte are evaluated independently.

The calibration curve should be inspected to verify that all of the calibration standards are within the acceptance criteria specified in the method. If any point on the calibration curve is considered to be an outlier, that standard should not be used in the calculation of the regression function for the curve. The method describing evaluation of calibrators should also describe the minimum number of standards that should be included in the analytical run. (For bioanalytical applications, acceptable analytical runs should yield calibration curves in which at least 75% of the individual standards are used in the regression.) In many instances, specific guidelines and procedures for identifying calibration curve outliers, as well as criteria for how many can be excluded from the regression, are established prior to the validation process. However, even if these requirements are met, an excessive number of curve outliers throughout the validation process may indicate the need to reconsider the type of regression used, the dynamic range for the method (range of reliable response) or the analytical method itself.

For study sample analysis the calibration curve and QC samples are evaluated separately, and run acceptance is based upon criteria established for both curves and QCs. For validation runs, however, only the standard curve and other factors such as carryover are considered for run acceptance and all of the results for the various types of validation QCs, e.g., precision and accuracy, stability, etc., are reported and used for statistical analysis. It is important at this time to emphasize the distinction between a 'failed' and 'rejected' validation run. Runs may be rejected for specific assignable cause such as documented evidence that the method was run incorrectly or hardware failure (Section 10.5.2c). Data from failed runs on the other hand, such as those where an excessive number of calibrators are considered to be outliers or QCs used to assess precision and accuracy do not meet the method objectives, must be retained in the validation study file and addressed in the validation report.

Reporting requirements for validation runs are addressed in Section 10.3.4. For runs that fail due to an excessive number of calibrator outliers, a reference to the failed run should be provided in the report along with a description of underlying cause for the failure. For validation QCs, accuracy and precision calculations should include all data although additional tables with outliers excluded can be provided also. Poor precision and accuracy for a particular experiment e.g. stability, may indicate the need to repeat that study in a subsequent validation run. If poor results are observed for the QCs used for intra-day and inter-day precision and accuracy, it may be necessary to repeat the entire validation run. In these instances, a reference to the failed run must be included in the final validation report.

Validations with one or more failed days or experiments may still be valid as long as the potential impact of those failures relative to the final method is addressed in the validation report. For instance, if there is only one run failure due to unassignable cause and additional validation days were successfully completed, the validation may be considered to be a success and used for subsequent sample analysis. Alternatively, it may be determined that the results of the validation do not support the original targets for precision and accuracy but the method can still be used for sample analysis using other acceptance criteria, e.g. 20% versus 15%. However, if there are repeated run failures during the course of validation it will probably be necessary to conduct more method development and re-start the entire validation process. Stated differently, it is not acceptable practice to simply keep repeating validation runs until the required number of acceptable validation days are successfully completed. A more detailed discussion of how run acceptance criteria are applied during validation can be found in Section 10.5.3b.

10.3.3e Sample Analysis

Background information that should be fully documented for validation sample analysis includes dates of extraction and analysis, identification of instruments and operators, identify of calibrators, QCs and samples. With respect to the analytical instrument itself (GC/MS, LC–MS/MS etc.), in addition to the maintenance and calibration records, the chromatography conditions, mass spectrometer and interface operating parameters, injection sequence and quantitation parameters are often documented automatically in the electronic raw data file that is associated with each injection, but critical parameters that are not captured electronically must be recorded manually; alternatively, all of this information is written on laboratory forms or in notebooks. Examples of supporting information that may need to be recorded manually in the laboratory or study notebook include; column type and serial number, mobile phase (identification and expiration dates), and any other flow, temperature or interface alignment conditions that are critical to intended method performance but not captured by the data system. For parameters that are automatically captured as electronic raw data, hard copy representations of these data may be used to assist the review process.

10.3.3f Run Summary Sheets

Run summary sheets are paper records that are generated at the instrument, such as results tables, run lists etc. These should be maintained with the study record as essential documentation.

10.3.3g Chromatograms

All chromatograms generated in the study, including original and re-integrated chromatograms for accepted runs, along with the reason for changing integration parameters across a run or for individual samples within a run, must be stored so as to be identifiable and retrievable.

10.3.3h Communications

Accurate records for any communications that result in a decision or directive that may impact the conduct of the study must be maintained as part of the documentation. For instance, if during the validation process it becomes clear that the original targets for quantitation or identification will not be achieved, it may be necessary to do more method development and re-initiate the validation with a new method. Alternatively, the validation may be completed with the current method with the understanding that the final acceptance criteria used for actual study sample analysis would have to be modified. An example would be where the original objective was to validate a method with a target accuracy and precision of 15% but based on the validation results, it was determined that this would likely result in a high rate of run failures unless the acceptance criterion was increased to 20%. Another important example that could be especially critical in an adversarial environment, would be a decision to use the existing method, but re-start the validation with a higher LLOQ because the reproducibility at the original target level was not supported by the validation results. In any event, detailed documentation of any communications that impact the validation processor interpretation of the results must be maintained in order to reconstruct the study and any critical decisions made during that process.

10.3.4 The Validation Report and Scientific Review

A validation report that describes overall conduct of the validation, and includes all of the relevant tables needed to support the interpretation of results and final conclusions, should be written prior to initiating sample analysis. The report should be assigned a unique identification number that can be used as a reference in the sample analytical report. A cover page should contain, at a minimum, the name and address of the laboratory and the report and amendment number (if applicable). A signature page that includes, at a minimum, the author(s), reviewer(s) and laboratory management should also be included.

Prior to approving the validation and associated report, the responsible scientist must review all aspects relating to the conduct of the validation and to the final results to confirm that the method meets the established targets of qualitative and quantitative uncertainty and is fit for purpose (Section 9.2). In addition to making sure that the correct validation experiments were run and documented properly, each experiment designed to validate a particular parameter should be carefully reviewed to ensure that the results support the original objectives. Any anomalies, run failures or other observations should be addressed in the body of the validation report and discussed. If additional method validation experiments are required, they should be performed prior to finalizing the report. In instances where the additional experiments are scheduled to occur over a longer period, e.g., for stability, the report can be finalized and the additional data can be added by means of an amendment.

For methods which require an additional qualitative identification component, such as an FDA method for the analysis of pesticide residues in food (Section 10.4.3a), the analysis of the validation data should be reviewed from the perspective of whether or not the results support the need to demonstrate an extremely low risk for false positives. In this respect, a 'zero rate of false positives' refers to method validation, under controlled circumstances with known samples. If there is one false positive for a confirmatory method, the validation should be failed.

For other applications that require reporting of data below the LLOQ, the report should address the 'region of uncertainty' discussed previously in Section 9.3.2, in terms of how the data will be interpreted and how the final results will be reported for identification and quantitation.

The remainder of this section addresses some of the specific requirements of the validation report. These requirements are drawn from a recently published paper published by the US FDA and other industry collaborators (Viswanathan 2007) and are an extension of the requirements described in Section 10.3.3 which were also derived from the same document. Once again, although these requirement were written specifically for laboratories generating bioanalytical data for regulatory applications, they provide an excellent summary of what is needed in support of any type of method validation and can be readily adapted to other applications.

10.3.4a Reference Standard and Solutions

At a minimum the validation report should include the batch or lot number for the reference standard, purity and source as well as the stability at the time of use. A record for all stock solutions used for the preparation of calibration standards and QCs may be included but is not necessarily required if the IDs for these samples are traceable to their source and preparation.

10.3.4b QCs and Calibrators

Typically, fresh calibration curves in control matrix are prepared for each day of analysis. Alternatively, calibration standards may be prepared in bulk, sub-divided and stored under conditions typical for sample storage for as long as storage stability has been established. Quality control samples are typically prepared in bulk. After preparation the QC pool is sub-divided and stored under conditions typical for study sample storage. As a result of stability concerns or unusual matrix limitations, QC samples may be prepared fresh from spiking solutions (individually or in bulk). In any event the identity, preparation dates and storage location and conditions used for calibration standards and QCs should be included in the report.

10.3.4c Assay Procedures

The report shall include reference to the analytical method (Section 10.3.3a) and a brief description of the method of extraction and analysis.

10.3.4d Run Acceptance Criteria

A brief description of the criteria used to assess run acceptance (Section 10.3.3d) should be included in the validation report. For method validation all of the data generated from the QC samples and used for statistical analysis are generally reported, and run acceptance is based primarily on the calibration standard run acceptance criteria. This is distinct from the procedure for sample analysis where QCs are also used to establish run acceptance or failure based on the criteria established during validation.

10.3.4e Sample Analysis Tables

A table of all runs that includes the instrument IDs and analysis dates should be provided. In addition to the runs that

were used for the validation statistics, a table that includes the identity of any failed runs, along with a reason for the failure, should also be included. Summary tables for all other validation experiments are also required. With the exception of calibration curve samples, analytical results may be excluded only with scientific justification (e.g. poor chromatography, unusual internal standard recovery, sample preparation error). Accuracy and precision calculations should include all data, although additional tables can be included with any outliers excluded.

The validation protocol should include a description of all statistical calculations that are used to generate the validation summary tables, including rules for how the final values will be reported with respect to the use of significant figures or other precision based techniques. Alternatively, the protocol can reference a laboratory SOP that establishes the routine procedures that are used in the laboratory. Calculated concentration/amount values should be determined using peak areas, peak heights, dilution or aliquot factors and regression coefficients with no intermediate rounding. For tabulation of data, the reported values described above should be used to calculate statistics such as the mean and standard deviation (SD, Equation[8.2]). Calculations such as percentage coefficient of variation (% CV, Equation[8.4]), should be performed without intermediate rounding. (i.e. do not use a value for the mean or SD which has been previously rounded). A detailed discussion for the use of significant figures and propagation of error can be found in Section 8.2.2.

Calibration curve data should be assessed to determine the appropriate mathematical regression that describes the instrument's response over the range of the calibration curve (Section 8.5). The report should include the back-calculated concentration values, accuracies, slopes, y-intercepts and correlation coefficients (R) and the coefficients of determination (R^2) (Equation[8.18] in Section 8.3.1) for all curves used in the validation. The R^2 value should be ≥ 0.98 for each calibration curve. The R value (if used) must be ≥ 0.99 for each calibration curve. An example table used to summarize the calibration curve statistics for each run used for method validation is shown as Table 10.2.

Table 10.2   Example of a report table for linear calibration curve parameters (fitted to Equation[8.19a]) determined on three separate days of analysis

Run #	Analysis Date	Slope	Intercept	R^2
12345_001	05-Dec-06	0.00248271	0.000320958	0.9986
12345_002	07-Dec-06	0.00246787	0.000109029	0.9995
12345_003	08-Dec-06	0.00240585	0.000434735	0.9983

Tables of all within-run (intra-day) and between-run (inter-day) QC results for demonstrating accuracy and precision should also be provided. An example of a table that can be used for recording intra-day and inter-day precision and accuracy statistics is shown Table 10.3, although additional tables showing the results for all individual QC results should also be provided.

Table 10.3 Example of a table used to record between-run precision and accuracy in analysis of QC samples during validation

Run #		Quality Control Concentrations (ng/mL)		
		6.00	400	750
Date 1	Mean	5.87	401	747
	SD	0.200	3.35	7.36
	% CV	3.41	0.834	0.986
	% Acc.	97.8	100	99.6
Date 2	Mean	5.92	397	745
	SD	0.0815	5.49	10.3
	% CV	1.38	1.38	1.38
	% Acc.	98.7	99.3	99.4
Date 3	Mean	6.07	391	742
	SD	0.240	5.22	20.6
	% CV	3.95	1.33	2.78
	% Acc.	101	97.7	98.9
Date4	Mean	5.90	384	722
	SD	0.293	4.22	13.8
	% CV	4.96	1.10	1.91
	% Acc.	98.3	96.1	96.3
Between-	Mean	5.94	394	739
Batch	SD	0.210	7.76	16.2
	% CV	3.54	1.97	2.19
	% Acc.	99.0	98.4	98.6

Tables summarizing results for all stability experiments conducted in the validation should be provided. These may include bench-top, freeze–thaw, long term, stock and post-preparative extract stability. Example tables that can be used for freeze–thaw stability and stability in matrix at room temperature are shown in Tables 10.4 and 10.5.

Tables for analyte and SIS recovery (extraction efficiency) should also be provided to demonstrate that the recovery is consistent, precise and reproducible. An example of a table that can be used for the recovery of the analyte is shown in Table 10.6.

If matrix effects are evaluated during validation, a table that summarizes the response data and the calculated matrix effect (ME, Section 9.6.1) should also be provided. An example table for demonstration of ME data is shown in Table 10.7.

Table 10.4 Example of a report table for cycles 2 and 3 of a freeze–thaw stability test of analyte in matrix, conducted in method validation

Run #	Quality Control Concentrations (ng/mL)					
	6.00		400		750	
	Amount Found (%)	Acc. (%)	Amount Found (ng/mL)	Acc. (%)	Amount Found (ng/mL)	Acc. (%)
Cycle-2	5.77	96.2	379	94.8	708	94.4
	5.98	99.7	386	96.5	732	97.6
	5.72	95.3	377	94.3	712	94.9
Mean	5.82		381		717	
SD	0.138		4.73		12.9	
% CV	2.37		1.24		1.79	
% Acc.	97.1		95.2		95.6	
Cycle-3	5.83	97.2	388	97.0	718	95.7
	5.44	90.7	375	93.8	725	96.7
	5.69	94.8	382	95.5	723	96.4
Mean	5.65		382		722	
SD	0.198		6.51		3.61	
% CV	3.49		1.70		0.499	
% Acc.	94.2		95.4		96.3	

Table 10.5 Example of a report table for a short term stability test for analyte stability in matrix left at room temperature for 26 hours

Run # (Time Stored)	Quality Control Concentrations (ng/mL)					
	6.00		400		750	
	Amount Found (%)	Acc. (%)	Amount Found (ng/mL)	Acc. (%)	Amount Found (ng/mL)	Acc. (%)
(26 hours)	5.86	97.7	382	95.5	719	95.9
	5.76	96.0	384	96.0	712	94.9
	5.45	90.8	391	97.8	726	96.8
Mean	5.69		386		719	
SD	0.214		4.73		7.00	
% CV	3.76		1.23		0.974	
% Acc.	94.8		96.4		95.9	

Additional tables that summarize any other experiments conducted during method validation should also be included in the final report.

10.3.4f Failed Runs

The data from all failed validation runs should be maintained with the rest of the validation data and reported in

Table 10.6   Example of a report table for results of extraction efficiency at three different concentrations, obtained during validation

Run #	Peak Area for Analyte1					
	6.00 ng/mL		400 ng/mL		750 ng/mL	
	Processed	Unprocessed	Processed	Unprocessed	Processed	Unprocessed
Run ID	11588	11480	640080	718537	1299658	1242810
	10089	10512	610240	658468	1111883	1241377
	10992	10278	611379	657735	1064294	1216074
	11117	10105	664878	762477	1286234	1285230
	10810	10431	566192	746920	1194208	1259968
Mean	10919	10561	618554	708827	1191255	1249092
SD	546	537	36985	48913	103923	
% CV	5.00	5.08	5.98	6.90	8.72	
% Recovery	103		87.3		95.4	

$$\% \text{Recovery} = \frac{\text{Mean peak area (processed)}}{\text{Mean peak area (unprocessed)}} \times 100$$

Here 'processed' and 'unprocessed' data refer to control matrix spiked with analytical standard before and after extraction, respectively. Note that the ratios of the two peak areas represent a combination of extraction efficiency plus differences in matrix effects (if any).

Table 10.7   Example of a report table for results of tests for matrix effects, conducted during validation

Run #	Peak Area of Analyte1			
	6.00 ng/mL		1600 ng/mL	
	A	D	A	D
Run ID	52826	51689	10314216	10204342
	62219	56039	10795471	10452173
	63483	63017	10442402	10308786
Mean	59509	56915	10517363	10321767
SD	5822	5715	249231	124424
% CV	9.78	10.0	2.37	1.21
Matrix Effect Ratio	1.05		1.02	

A = Analyte (post process), Internal Standard (post process)
D = Analyte (solution only), Internal Standard (solution only)

$$\text{Matrix Effect Ratio} = \frac{\text{Mean Peak Area (A)}}{\text{Mean Peak Area (D)}}$$

the validation report. A table that identifies all failed runs, and the assay date with a supporting discussion in the report that describes the suspected reason for the failure, should also be included in the report. It is not acceptable practice to simply repeat the validation runs until the minimum number of acceptable runs is achieved. (Section 10.3.3d). An excessive number of validation run failures generally indicates a need to return to method development and begin a new validation with a modified method.

10.3.4g Deviations From Method or SOPs

Any significant deviations from the prevailing SOPs, validation plan or protocol must be noted in the report with a statement that addresses the impact, if any, the deviation may have had on the study or results. Significant deviations or an excessive number of deviations may require repeating some of the validation runs or repeating the entire validation in accordance with the specific procedures described in the analytical method.

10.3.4h Chromatograms

The report should contain relevant chromatograms. At a minimum, representative chromatograms for at least one LLOQ and for each type of matrix blank used in the validation should be included.

10.3.4i Amendments

Any additions, corrections or changes to an existing validation report should be written as an amendment of the original report. An amendment report will contain amendment notes that clearly identify the part(s) of the report added to, corrected or changed, including appropriate justification(s) if applicable. The amendment shall be issued in either of two forms: either the entire report may be reissued containing the amendment(s); or the amendment(s) may be issued independently from the original report. The original signature copy or exact copy of the validation report and any subsequent amendments should be maintained in the laboratory archives with the associated validation data, validation plan or protocol (if applicable).

10.4 Examples of Methods and Validations Fit for Purpose

The intent of this section is to provide the reader with several examples of different types of methods and validations, aimed at addressing a variety of applications and data requirements. Depending on the needs of the client, regulatory requirements etc., there is more than one way to validate an analytical method. All of the examples given in this section are considered to fall into the general category of target compound analysis where it is known in advance what analytes are to be detected and quantified, and where properly characterized reference substances are used for characterization and calibration. However, what distinguishes one type of quantitative analysis from another depends on the purpose of the analysis and on how the quantitation range and subsequent reporting of results are determined and implemented in a particular laboratory.

With respect to determining the quantitation range used for reporting results, including the LLOQ, ULOQ and LOD if applicable, methods generally fall into one of the following two categories: novel methods that are developed in a particular laboratory for a specific application or study; or previously validated methods that are transferred to and applied in the laboratory according to predetermined method transfer guidelines, SOPs or published regulations.

With respect to data reporting requirements, most quantitative methods fall into one of following four general categories: reporting of all results that fall within a specified calibration range where the LLOQ and calibration range are based on the results obtained during method development and verified through validation; reporting of results using a reporting limit (RL) that is generally established as part of a risk assessment requirement; reporting of results that are above a specific enforcement level; reporting of all quantified results even if the values fall below the validated limit of quantitation. In the fourth example all values are reported with no filtering of data using criteria of validated LLOQ etc. In this regard the analyst should use appropriate numbers of significant figures (Section 8.2.2) decided on the basis of overall precision and confidence limits (Section 8.2.4) and possibly also an analysis of propagation of error (i.e. an uncertainty budget).

In the following examples these particular applications are covered to varying degrees. It is worth emphasizing at this point, however, that the intent is not to provide a 'how to manual', but rather to use these examples to illustrate how validations are designed based on fitness for purpose principles and to demonstrate some of the parameters that must be considered when validating novel methods.

10.4.1 Bioanalytical Method Validation

Quantitative methods are often developed and validated with the purpose of reporting data to a validated Lower Limit of Quantitation (LLOQ) or pre-defined reporting limit (RL). Bioanalytical methods used for the quantitative analysis of drugs and metabolites in biological tissues and fluids provide an important example of methods used to report results down to a validated LLOQ according to industry and regulatory guidelines. A thorough discussion covering not only the specific current requirements for bioanalytical validations, but also how these guidelines have developed and evolved over time, provides an excellent example of how method validations are designed under the principle of fitness for purpose.

The specifications given in the current FDA guideline (FDA 2001) and the subsequent Workshop Report (Viswinathan 2007) for the appropriate figures of merit (discussed later in this section) are designed for bioanalytical data used to support drug safety and pharmacokinetic studies in support of new drug investigations and for other types of human clinical pharmacology, bioavailability and bioequivalence studies requiring pharmacokinetic evaluation. However, with appropriate changes they can also be taken as a guide in other application areas requiring the best possible quality of data, such as environmental science, toxicology etc. Not all analyses require the same highest level of rigor in validation; indeed a fitness for purpose approach has been advocated (Antignac 2003) with respect to European regulations on validation (European Commission 2002).

Beginning with a workshop co-sponsored by the FDA and the American Association of Pharmaceutical Scientists in 1990 and subsequent publication of the Workshop Report (Shah 1992), there has been an on-going collaboration between regulators and industry scientists aimed at establishing guidelines for developing and validating bioanalytical methods in support of bioequivalence, pharmacokinetic (PK) and toxicokinetic studies. A draft FDA Guidance on Bioanalytical Methods Validation was produced in January of 1999 which led to a 2nd AAPS/FDA workshop in January 2000; this resulted in a second Workshop Report (Shah 2000). These workshops, publications and subsequent discussions between industry and FDA representatives led to the publication of the FDA Guidance on Bioanalytical Methods Validation in May 2001 (FDA 2001). While this guidance document provided industry with a foundation for development and validation of bioanalytical methods, many questions still remained with respect to specific interpretations and application of the guidance document. These questions, combined with the ever-evolving technology and techniques and regulatory experience, led to the 3rd AAPS/FDA Bioanalytical Workshop in May 2006 that also resulted in a Workshop Report (Viswanathan 2007).

In the Introduction of the most recent report (Viswanathan 2007) the authors make the following statement: 'Bioanalysis, employed for the quantitative determination of drugs and their metabolites in biological fluids, plays a significant role in the evaluation and interpretation of bioequivalence, pharmacokinetic (PK) and toxicokinetic studies. The quality of these studies, which are often used to support regulatory filings, is directly related to the quality of the underlying bioanalytical data. It is therefore important that guiding principles for the validation of these analytical methods be established and disseminated to the pharmaceutical community'.

It is important to recognize at this juncture that it is not the intent of the FDA or authors of these documents to prescribe rigid 'rules' for the validation of bioanalytical methods. In fact, given the broad scope of applications and evolving technologies, combined with the need for each laboratory to develop and maintain their own SOPs according to independent policies and operational constraints, any attempt to develop a single set of rules to cover all validation and application contingencies would be foolish if not impossible. The guidance documents are just that, guidance for how to generally approach the validation of bioanalytical methods. Even so, the published guidance documents and associate workshop reports still leave many questions with respect to how methods should be fully validated and how different aspects of the data should be interpreted. Many of these questions will eventually be addressed through similar workshops and guidance documents. However, even with the limitations and questions that remain unanswered, the workshop reports and FDA guidance have provided analytical scientists in the pharmaceutical community with a common foundation for developing methods that are fit for purpose for the needs of those interpreting the data or using the data to make regulatory decisions.

The fundamental components of a bioanalytical validation (and all validations in general) are: selectivity; sensitivity; accuracy; precision; range of reliable response and linearity; and reproducibility. Other characteristics of the method that should be addressed are: stability; carryover and control of lab contamination; and matrix effects, including interferences from metabolites in incurred samples or other co-extracted compounds, as well as ionization suppression.

10.4.1a Sensitivity and the Calibration/Standard Curve

With respect to sensitivity, the FDA specifications do not define this parameter as such, although it is stated in the key document (FDA 2001) to be one of the crucial parameters. Rather, the term appears to be used in its more colloquial sense covering the LLOQ (and LOD although the FDA is not directly concerned with the latter), rather than as the more narrow definition as the slope of the response (calibration) curve. The concept of sensitivity in the colloquial sense is covered by the criteria required (FDA 2001) for the calibration curve and the LLOQ: 'A calibration curve should be prepared in the same biological matrix as the samples in the intended study by spiking the matrix with known concentrations of the analyte. The number of standards used in constructing a calibration curve will be a function of the anticipated range of analytical values and the nature of the analyte/response relationship. Concentrations of standards should be chosen on the basis of the concentration range expected in a particular study. A calibration curve should consist of a blank sample (matrix processed without internal standard), a zero sample (matrix sample processed with internal standard), and six to eight nonzero samples covering the expected range, including LLOQ'.

The guidance document thus chose a strictly operational definition of LLOQ, rather than one based on statistical theories applied to responses near the signal:background limit (Section 8.4); the latter are of more relevance with respect to detection limits rather than LLOQ. (Note that the FDA Guidance document (FDA 2001) refers to 'signal:noise' but in this book the limiting feature of a chromatogram is referred to as chemical background, Section 7.1.1.) The specification of matrix matched standards confers all the advantages described in Section 8.5, but again this might be difficult to achieve in other analytical applications.

The guidelines (FDA 2001) continue with a discussion of the LLOQ and state: 'The analyte response at the LLOQ should be at least five times the response compared to a blank response', and 'Analyte peak (response) should be identifiable, discrete and reproducible with a precision of 20 % and accuracy of 80–120 %'. With respect to what types of regressions are acceptable and how residual outliers should be handled in practical application, the guidelines state: 'The simplest model that adequately describes the concentration–response relationship should be used. Selection of weighting and use of a complex regression equation should be justified. The following condition should be met in developing a calibration curve: 20 % deviation of the LLOQ from nominal concentration and 15 % deviation of standards other than the LLOQ from nominal concentration. At least four out of six nonzero standards should meet the above criteria, including the LLOQ and the calibration at the highest concentration. Excluding the standards should not change the model used'.

In the following what these guidelines do and do not tell us, and how they are typically applied in the bioanalytical laboratory are discussed. In practice, even though the guidelines require only six nonzero calibration points to define a curve (Section 9.8.1), they also provide a definition for how residual outliers can be excluded from the regression. The guidelines allow for the user to exclude standards from the regression as long as at least 75 % of the standards prepared and analyzed pass within the 15–20 % criteria. So, in general, the minimum number of standard concentrations that most laboratories would use is eight, which allows for two to be excluded and still pass the run acceptance criteria. A caveat to this acceptance criterion is that all runs must begin and end with the injection of a standard. The remaining standard calibrators may then be distributed evenly among all other sample injections. In practice, however, it is common to have all standards prepared in duplicate. In this scenario the run typically begins with the blank and zero samples, followed by one injection for each concentration of the calibration curve. After these injections all samples from the run are injected with QCs interspersed, and finally the run ends with the injection of the remaining standards at the end of the run. Comparison of these two calibration curves permits an additional check on the validity of the run by checking for curve divergence (see discussion of Figure 10.8 in Section 10.5.3b). Once again, the guidelines allow the user to exclude standards from the regression as long as at least 75 % of the standards prepared and analyzed pass within the 15–20 % criteria. For example, if a curve is prepared with eight concentrations in duplicate, then 12 must be within ± 15 % of the nominal concentration (± 20 % at the LLOQ) for the calibration run to be accepted. Once again, the calibration curve must be defined by including at least one standard at the LLOQ and one at the ULOQ.

10.4.1b Accuracy and Precision

The FDA requirement on accuracy states (FDA 2001): 'Accuracy is determined by replicate analysis of samples containing known amounts of the analyte. Accuracy should be measured using a minimum of five determinations per concentration. A minimum of three concentrations in the range of expected concentrations is recommended. The mean value should be within 15 % of the actual value except at LLOQ, where it should not deviate by more than 20 %. The deviation of the mean from the true value serves as the measure of accuracy'. As discussed in Section 8.4.2, such a definition implies that proportional systematic errors

(L_a', L_a''') are negligible, although in principle the specification of three different spiking levels should provide a check. Also, estimating accuracy by such measurements of F_a''' in experiments on blank matrix spiked with analytical standard presumes that this parameter is equal, within experimental uncertainty, to F_a' for the native analyte in the analytical sample. In the case of pharmaceuticals in body fluids this assumption is probably valid since occlusion effects are likely to be negligible although similar effects resulting from strong adsorption of analyte on endogenous proteins have been reported (Ke 2000). It is important to note the distinction between intra-day and inter-day assessments of accuracy and precision (Section 10.2.5).

For assessment of accuracy and precision the replicate samples referred to in the guidelines (see above) are QC samples (Sections 2.5.3 and 9.8.2). In practice it is common to prepare the QCs for this purpose in bulk, in an amount sufficient for storing a number of aliquots sufficient for the entire validation and subsequent long term stability studies. Also, it is wise to prepare and store additional aliquots to address unplanned contingencies such as failed runs or repeating one or more components of the validation run. Many laboratories will use freshly prepared QCs for the first validation run and then use QC aliquots that were stored in the freezer for all subsequent validation runs. If a sufficient number of QC aliquots were prepared and stored, they can also be used for monitoring subsequent sample analysis providing that supporting stability data are available.

In terms of reporting and calculating both accuracy and precision, QCs may be excluded during sample analysis according to outlier criteria used by that laboratory and described in their SOPs. However, if this approach is used for validation studies, a table should be prepared showing precision and accuracy before and after the rejection of an outlier, and this should be included in the final report. It is generally agreed that observation of some outliers through the course of an entire validation is acceptable but, if it appears that there is a general tendency towards 'QC failure', some type of investigation should be conducted, and a decision made as to whether or not to move forward with the validation and subsequent sample analysis versus doing more method development and starting the validation over again with new conditions.

With respect to the precision of an analytical method, the FDA requirement states that this figure of merit 'describes the closeness of individual measures of an analyte when the procedure is applied repeatedly to multiple aliquots of a single homogeneous volume of biological matrix. Precision should be measured using a minimum of five determinations per concentration. A minimum of three concentrations in the range of expected concentrations is recommended. The precision determined at each concentration level should not exceed 15 % of the coefficient of variation (CV) except for the LLOQ, where it should not exceed 20 % of the CV. Precision is further subdivided into within-run, intra-batch precision or repeatability, which assesses precision during a single analytical run, and between-run, inter-batch precision or repeatability, which measures precision with time, and may involve different analysts, equipment, reagents, and laboratories' (FDA 2001). Recall that the coefficient of variation (CV) is also referred to as the relative standard deviation (RSD, Equation[8.4] in Section 8.2.1).

For validation studies, precision calculations are based on the same intra-day and inter-day validation QCs and runs that are used for accuracy assessment. Once again, the same rules apply in terms of the reporting and interpretation of outliers in the final report.

10.4.1c Selectivity

The FDA specification of selectivity (the ability of an analytical method to differentiate and quantify the analyte in the presence of other components in the sample), requires that 'analyses of blank samples of the appropriate biological matrix (plasma, urine or other matrix) should be obtained from at least six sources. Each blank sample should be tested for interference, and selectivity should be ensured at the lower limit of quantification (LLOQ). Potential interfering substances in a biological matrix include endogenous matrix components, metabolites, decomposition products, and in the actual study, concomitant medication and other exogenous xenobiotics. If the method is intended to quantify more than one analyte, each analyte should be tested to ensure that there is no interference' (FDA 2001). Here the definition of 'interference' refers to direct contributions by co-eluting components to the signal attributed to the analyte (e.g. SIM or MRM channel of a triple quadrupole mass spectrometer). The specification of checking for interferences using at least six different sources of blank matrix is a wise precaution in the applications of concern to the FDA, but may not be possible and/or applicable for all other applications in which suitable blank control matrix may not be readily available.

10.4.1d Matrix Effects

With the increased awareness of matrix effects, i.e., suppression or enhancement of the ionization efficiency R_a' (Section 8.4) of the analyte (see Section 5.3.5a for mass spectrometry based assays), new guidelines have been included in the Conference Workshop Report (Viswanathan 2007) for assessment as part of the validation. As defined in the conference report: 'Matrix effect is the suppression or enhancement of ionization of analytes by the presence of matrix components in the biological samples'. The report goes on to define the term 'Matrix Factor (MF)' as follows: 'a ratio of the analyte peak response in the presence of matrix ions to the analyte peak response in the absence of matrix, ie, Matrix Factor = Peak response in presence of matrix ions/Peak response in absence of matrix ions'. By this definition, 'An MF of one signifies no matrix effects. A value of less than one suggests ionization suppression. An MF of greater than one may be due to ionization enhancement and can also be caused by analyte loss in the absence of matrix during analysis'. Note that the parameter MF (Viswanathan 2007) appears to correspond to the ME parameter described previously (Matuszewski 2003) and discussed in Section 5.3.6a.

The Workshop Report (Viswanathan 2007) continues with a brief discussion of the advantages of using stable isotope internal standards to compensate for ion suppression ('IS-normalized MF') since the matrix effect observed for the internal standard will be similar to that for the matching analyte. It also acknowledges that analog internal standards may also be effective in compensating for matrix effects but concludes that stable isotope internal standards 'should be used whenever possible'. It is also stated (Viswanathan 2007) that, while it is not necessary to demonstrate a MF approximately equal to one, it is necessary to determine the variability of the MF since this will directly affect the reproducibility of the method. The recommendation is written as follows: 'To predict the variability of matrix effect in samples from individual subjects, determine the MF (or IS-normalized MF) for six individual lots of matrix. The variability in matrix factors, as measured by the coefficient of variation, should be less than 15 %'. The report recognizes that in some instances, such as when using rare matrix, six individual lots may not be readily available. In these instances, the absolute requirement for six individual lots may be waived. Finally, the report also recognizes the value of using stable isotope internal standards and states that in these instances, '. . . it is not necessary to determine the IS-normalized MF in 6 different lots.'

The opinions written in the workshop conference report (Viswanathan 2007) represent a practical approach to assessing matrix effects with certain assumptions, i.e. using multiple individual lots of matrix spiked with known amounts of analyte will provide a good indication for matrix effects for the method in general, and stable isotope internal standards will provided significant compensation for matrix effects that will result in lower variability of response and better method reproducibility. A further discussion of how to assess and minimize matrix effects as part of the development process can be found in Chapter 9, and the guidelines discussed above can be incorporated into the method development process for mass spectrometry method develop in general. However, these experiments and validation guidelines rely on control matrix as a representative sample, but do not address the effects of metabolites or other co-eluting endogenous compounds that may be present when analyzing incurred samples. For this reason, additional post-validation studies should be considered for assessing the matrix effects using incurred samples (see Section 10.4.11 below).

10.4.1e Recovery

The recovery of an analyte in an assay is defined by the FDA in a strictly operational way as 'the detector response obtained from an amount of the analyte added to and extracted from the biological matrix, compared to the detector response obtained for the true concentration of the pure authentic standard. Recovery pertains to the extraction efficiency of an analytical method within the limits of variability. Recovery of the analyte need not be 100 %, but the extent of recovery of an analyte and of the internal standard should be consistent, precise, and reproducible. Recovery experiments should be performed by comparing the analytical results for extracted samples at three concentrations (low, medium, and high) with unextracted standards that represent 100 % recovery' (FDA 2001). In terms of the symbols used in Section 8.4, the recovery is thus defined as the ratio (R'_a/R''_a), and is equivalent to determination of F'_a provided that no suppression or enhancement effects give rise to differences between R'_a and R''_a and that the proportional systematic errors L'_a and l'_a are negligible. The FDA definition of recovery also corresponds to that of the *PE* ('process efficiency') parameter (Matuszewski 2003) discussed in Section 5.3.6a, since the former (FDA 2001) measures a combination of extraction efficiency and matrix effects (if any).

For certain types of regulatory or forensic assays there may be specific requirements for the absolute recovery required for the assay, e.g. 70–120 %, but the bioanalytical guidelines do not impose any specific requirements for recovery other than that it should be shown to be consistent and precise to ensure method reproducibility. Experiments designed to investigate and optimize the absolute recovery in conjunction with the cleanliness of the extract should be a part of any method development activities, and then tested during validation with the objectives described above.

10.4.1f Reproducibility

The FDA guideline (FDA 2001) does not enlarge further upon the meaning of reproducibility other than the definition given in Section 10.2.2; however, repeatability was defined within the prescription for precision. As a result of the latest bioanalytical workshop (Viswanathan 2007) there is now a much more heightened awareness of the need to evaluate method reproducibility using incurred samples, reflecting the fact that standards and QCs prepared in control matrix may not adequately mimic method accuracy and/or reproducibility during the analysis of samples from dosed subjects (Section 9.4.7b). Further discussion of this topic in the context of bioanalytical validation can be found in Section 10.4.11.

10.4.1g Stability – Overview

The last of the figures of merit singled out in the document is stability and for pharmaceutical drugs in body fluids and tissues this is a key parameter. Accordingly, stability testing is accorded a great deal of attention by FDA: 'Drug stability in a biological fluid is a function of the storage conditions, the chemical properties of the drug, the matrix, and the container system. The stability of an analyte in a particular matrix and container system is relevant only to that matrix and container system and should not be extrapolated to other matrices and container systems. Stability procedures should evaluate the stability of the analytes during sample collection and handling, after long term (frozen at the intended storage temperature) and short term (bench top, room temperature) storage, and after going through freeze and thaw cycles and the analytical process. Conditions used in stability experiments should reflect situations likely to be encountered during actual sample handling and analysis. The procedure should also include an evaluation of analyte stability in stock solution. All stability determinations should use a set of samples prepared from a freshly made stock solution of the analyte in the appropriate analyte-free, interference-free biological matrix. Stock solutions of the analyte for stability evaluation should be prepared in an appropriate solvent at known concentrations' (FDA 2001).

10.4.1h Stock Solution Stability

Generalized requirements for validation of the stability of stock and sub-stock solutions were described in Section 10.2.7. The FDA guidance document (FDA 2001) does not, however, recommend any exact procedures about how this should be done: 'Conditions used in stability experiments should reflect situations likely to be encountered during actual sample handling and analysis. The procedure should also include an evaluation of analyte stability in stock solution'. Also the following: 'The stability of stock solutions of drug and the internal standard should be evaluated at room temperature for at least six hours. If the stock solutions are refrigerated or frozen for the relevant period, the stability should be documented. After completion of the desired storage time, the stability should be tested by comparing the instrument response with that of freshly prepared solutions'. The most recent Workshop Report (Viswanathan 2007) adds the following: 'The need to characterize the stability of stock solutions was emphasized throughout the meeting and accepted as a core validation experiment. However, there was no agreement on the degree of degradation that defines acceptable stability. The consensus was that lower degradation in the standard acceptable ranges is desirable since these stock solutions are used for making other solutions and this error may be propagated in the concentrations reported for biological samples'. Clearly the last sentence of this recommendation is the most important. The same report (Viswanathan 2007) also adds the following recommendation: 'In general, newer stock solutions within their established stability period (e.g. a solution with established 60-day stability used on Day 55) should not be used to measure stability of an older solution (e.g., 120 days old). Although the newer stock may meet stability criteria for bioanalytical purposes, the chance of misinterpreting the stability of the older solution is high. The suggestion is to make a solution fresh from powder when determining the stability of any older stock'.

To understand the meaning of 'standard acceptable ranges' in this context, it is necessary to return to the FDA guideline that is mainly concerned with stability of analyte in matrix: 'All stability determinations should use a set of samples prepared from a freshly made stock solution of the analyte in the appropriate analyte-free, interference-free biological matrix. Stock solutions of the analyte for stability evaluation should be prepared in an appropriate solvent at known concentrations' (FDA 2001). The following Sections all refer to such stability tests for analyte in matrix samples. In practice, results for QCs must fall within ±15% of the nominal values but the laboratory should also have some means for detecting any downward trends with time within this acceptability range. Most laboratories currently handle the trend detection problem within a scientific review process. In any event, no samples should be analyzed outside of the established storage stability time that is established during validation and then for selected intervals after that (Section 10.4.1k).

10.4.1i Freeze and Thaw Stability

The FDA guidelines specify the following procedure: 'Analyte stability should be determined after three freeze and thaw cycles. At least three aliquots at each of the low and high concentrations should be stored at the intended storage temperature for 24 hours and thawed unassisted at room temperature. When completely thawed, the samples should be refrozen for 12 to 24 hours under the same conditions. The freeze–thaw cycle should be repeated two more times, then analyzed on the third cycle. If an analyte is unstable at the intended storage temperature, the stability sample should be frozen at −70 °C during the three freeze and thaw cycles' (FDA 2001). Some laboratories take the extra precaution of analyzing after each of the three freeze–thaw cycles, not only the last one. The most recent Workshop Report (Viswanathan 2007) adds the following: ' the freezing and thawing of stability samples must mimic the intended sample handling conditions to be used during sample analysis. ... If during the sample analysis for a study, a sample was thawed through greater than three cycles or if storage conditions changed and/or exceeded the sample storage conditions evaluated during method validation, stability must be established under these new conditions in order to demonstrate that the concentration values from these study samples are valid'.

10.4.1j Short Term Temperature Stability

The FDA guideline (FDA 2001) states: 'Three aliquots of each of the low and high concentrations should be thawed at room temperature and kept at this temperature from 4 to 24 hours (based on the expected duration that samples will be maintained at room temperature in the intended study) and analyzed', and also:' If study samples are to be stored on wet ice, for thawed periods greater than four hours, then these conditions should be evaluated during validation as well' (Viswanathan 2007). In practice, most laboratories will attempt to establish at least 24 hours of short term matrix stability at room temperature to address situations where the analyst forgets to return the unused sample to the freezer, or for situations where the sample is re-assayed, to capture the total cumulative amount of time during which the sample is left at room temperature.

10.4.1k Long Term Stability

With the exception of quantitative specifications as to what are 'acceptable' stability limits, the FDA guidelines for long term stability are quite explicit: 'The storage time in a long term stability evaluation should exceed the time between the date of first sample collection and the date of last sample analysis. Long term stability should be determined by storing at least three aliquots of each of the low and high concentrations under the same conditions as the study samples. The volume of samples should be sufficient for analysis on three separate occasions. The concentrations of all the stability samples should be compared to the mean of back-calculated values for the standards at the appropriate concentrations from the first day of long term stability testing' (FDA 2001). The recent Workshop Report adds the following recommendations: '... long term measurements are initiated during method validation, possibly evaluating analyte stability for a period of a few weeks, with the remaining long term storage time points evaluated post-method validation. These post-validation data can then be added to the original validation data in the form of a validation report addendum or as a stand-alone stability report. Long term stability should be evaluated at the expected storage conditions, including expected satellite storage temperature and duration (e.g., prior to shipment to the analytical laboratory). In consideration of this, there may be the need to include both –70 and –20 °C evaluations (e.g. when samples are stored under different conditions at the various study locations)' (Viswanathan 2007). The concerns about the assumption that validated stability at –20 °C can be taken to also imply stability at –70 °C were described in Section 10.2.8c.

10.4.1l Incurred Sample Re-Analysis

Validation of analytical methods is now accepted good practice, and of course is a requirement in many cases. Current practice accepted within e.g., the pharmaceutical industry and its regulatory bodies (e.g., FDA 2001) specifies validation of a bioanalytical LC–MS/MS method using analytical standards and QC samples prepared by spiking the appropriate blank matrix (e.g. human plasma). If the specified procedures (FDA 2001) are followed carefully, the method is accepted as adequately validated for analyzing incurred biological samples (e.g. plasma from human subjects who have been dosed with the drug). However, unlike QC samples, incurred samples may contain an isomer of the drug, e.g. an epimer or a Z or E isomer, and such metabolites will interfere with the LC–MRM transition(s) used for quantitation of the drug. Moreover, incurred samples may also contain metabolites with molecular mass different from that of the drug (see Section 9.3.3 for common examples of drug metabolites) that nevertheless can still contribute to the MRM transition used for quantitation of the drug as a result of insufficient chromatographic resolution plus in-source CID of the metabolite ion to the drug ion (Jemal 1999). Further, a metabolite can potentially revert back to the drug during the multiple steps of sample preparation that precede introduction of the processed sample into the LC–MS/MS system. This general problem applies also to applications other than bioanalytical, e.g. soil samples contain active microorganisms that can metabolize agricultural chemical analytes.

The FDA guideline (FDA 2001) mentions incurred samples only briefly and then only with respect to microbiological and ligand binding assays. It is interesting (Gallicano 2006) that a Health Canada 1992 Guideline required that 15 % of incurred samples be randomly selected and re-assayed, and the results reported separately from those obtained using the conventional validation spiked samples; however, this requirement was revoked in 2003! Shortly before this, a publication dealing with this problem (Jemal 2002) proposed that a set of procedures should be used with incurred samples, as soon as such samples are available, in order to further validate an analytical method based on LC–MS/MS. The recommendations, proposed (Jemal 2002) to investigate the various possible complications in incurred samples, were discussed in Section 9.4.7b.

The most recent bioanalytical Workshop Report (Viswanathan 2007) devotes considerable space to this topic and some recommendations not discussed previously (Section 9.4.7b) are included below. 'There should be some assessment of both reproducibility and accuracy of the reported concentration. Sufficient data should be generated to demonstrate that the current matrix (i.e. the incurred sample matrix) produces results similar to those previously validated. It is recognized that accuracy of the result generated from incurred samples can be more difficult to assess. It requires evaluation of any additional factors besides reproducibility upon storage, which could perturb the reported concentration. These could include metabolites converted to parent during sample preparation or LC–MS/MS analysis, matrix effects from high concentrations of metabolites, or variable recovery between analyte and internal standard' (Viswanathan 2007). Most of these phenomena are those described previously (Jemal 2002) and discussed in Section 9.4.7b.

'If a lack of accuracy is not due to assay performance (i.e., analyte instability or interconversion) then the reason for the lack of accuracy should be investigated and its impact on the study assessed. The extent and nature of these experiments is dependent on the specific sample being addressed and should provide sufficient confidence that the concentration being reported is accurate. The results of incurred sample reanalysis studies may be documented in the final bioanalytical or clinical report for the study, and/or as an addendum to the method validation report'.

'In selecting samples to be re-assayed, it is encouraged that issues such as concentration, patient population and special populations (e.g., renally impaired) be considered, depending on what is known about the drug, its metabolism and its clearance. First-in-human, proof-of-concept in patients, special population and bioequivalence studies are examples of studies that should be considered for incurred sample concentration verification. The study sample results obtained for establishing incurred sample reproducibility may be used for comparison purposes, and do not necessarily have to be used in calculating reported sample concentrations' (Viswanathan 2007). The preceding paragraph emphasizes the complexity of the phenomena that can affect an apparently simple LC–MS/MS bioanalytical assay and the need for the judgment of the informed analyst and his/her colleagues to determine the nature and extent of re-analysis of incurred samples.

Further implications of differences between incurred bioanalytical samples and calibrators and QC samples prepared in accord with accepted practice (FDA 2001) have been discussed (Gallicano 2006). It was emphasized that if the imprecision of data obtained for incurred samples differs from that of spiked samples used in the original validation, this will lead to questioning the estimates of accuracy, LLOQ and stability for the incurred samples. If the incurred sample data are of low precision it will be necessary to analyze more samples to investigate these parameters. It was proposed (Gallicano 2006) that '. . . an LLOQ higher than the lowest standard concentration (LSC) should be considered if the incurred samples show more imprecision than spiked samples, because the main criteria for selection of the LSC is its precision on the basis of spiked samples'.

This presentation (Gallicano 2006) describes alternative experimental approaches and corresponding statistical evaluations of differences between data obtained for incurred samples and validation data for spiked control matrix samples, mainly concentrating on precision as the test parameter; these details are not reproduced here. It was emphasized that an assay with consistent imprecision of $>15\%$ ($>20\%$ at the LLOQ) is not necessarily a 'bad' assay, depending on its intended purpose; such a method may be acceptable for, e.g., comparative PK studies with high ($>30\%$) CV values (within-subject for crossover designs, or between-subject for parallel designs) so that the contribution of analytical imprecision to overall study variability is therefore small. However, imprecision of $>15\%$ may not be acceptable for Phase I pharmacokinetic (PK) studies in which the data are used, e.g., to predict multiple dose from single dose PK data.

It has also been reported (Gallicano 2007) that matrix effects on a structural analog SIS standard have been observed in incurred samples that were not apparent in spiked samples. It was suggested that the %CV of SIS peak response should be reported for each batch of study samples. If a variable response is observed, the question arises as to whether this is the result of random variability, or the result of spiked and incurred sets of samples

having consistent responses within each set but different responses between sets ('sample-by-set interaction'). If the SIS response is different in incurred samples from that in spiked samples, then it is not possible to sustain the assumption that analyte response changes proportionally to that of the internal standard in the incurred samples (the basis of all calibration approaches employing an SIS, Section 8.5.2). This observation suggests that the effect of matrix needs to be determined for both analyte and SIS for incurred samples, e.g. see discussion of Figure 10.6 in Section 10.5.3a.

The question as to when repeat analyses of incurred samples should be investigated with respect to reproducibility was also addressed (Gallicano 2006). It was recommended that ideally these should be conducted during method validation, particularly for 'analytically high risk' drugs, i.e. those present at 'low' concentrations and/or that indicate potential for metabolite interference of one kind or another. Another indicator that repeat analyses of incurred samples should be done during validation is fore-knowledge that the test subjects are likely to involve a significantly different matrix (indicated by e.g. obesity or other serious disease) from that used for the conventional validation spiked samples, thus potentially leading to different matrix interferences and matrix effects. For 'analytically low risk' drugs the incurred sample re-analysis can be concurrent with analysis of study samples, although this is potentially risky if unexpectedly high and inconsistent imprecision is observed. In the latter case the analyst is faced with the question as to what should be done with the data since such an observation may compromise the original results. To address this dilemma, strategies for creating individual pools of sample that are representative of the sampling time profile can be developed or the re-assay data can be excluded from the final report.

These and other strategies have been proposed and are being used in different laboratories. However, no matter which approach is used, the requirement to describe the procedures used for incurred sample re-analysis and date reporting must be described in an approved laboratory SOP. If the incurred sample re-analysis indicates that there is a potential problem with method reproducibility, the analyst could stop the study analyses and investigate the causes of the imprecision. Sample analysis should not resume until the required corrective action has been implemented, including the potential need to develop a new assay, re-validate and then re-assay all of the samples in the study. These are not easy questions and their resolution demands the best possible judgment on the part of the analyst. It can be added that the recent emergence of the incurred sample issue is a result of the greatly increased sensitivity of modern analytical instrumentation and the methods developed to exploit it.

10.4.2 Risk Assessment Methods

As their name suggests, risk assessment methods are quantitative methods designed to evaluate risks to, e.g., public health on a longer term basis, rather than situations that could pose immediate potential harm; an example (an enforcement method in this case) is described in Section 10.4.3a. Typical examples of quantitative risk assessment methods can be found among US Environmental Protection Agency (EPA) environmental methods that require analysis down to specified reporting levels. Laboratories using these methods are required to demonstrate method proficiency based on the requirements and procedures described in these methods. In a sense, the independent laboratory validates a pre-established method in their operating environment prior to the analysis of actual samples that may be submitted from a variety of sources, e.g. the US EPA itself, chemical manufacturers or engineering companies under contract for environmental reclamation projects. For methods of this kind, the validation requirement at the performing laboratory may be as simple as running a batch of reference samples or certified reference materials (Section 2.2.2) to demonstrate proficiency in a method that includes the quantitative and qualitative requirements for reporting values down to the established reporting limit. (In the EPA examples, the qualitative confirmation requirements may include full scan mass spectra with accompanying searches of a library that is generated on the instrument being used.) On-going calibration checks, quality control samples and established method criteria are used to verify individual run acceptance and reliability of the data.

A special case that falls under the present definition of risk assessment methods is provided by the methods for analysis of 'dioxins' and related compounds included in the calculation of 'toxic equivalent quantity' (see Section 11.4.1).

10.4.3 Enforcement Methods

A good example of situations that fall under this general heading is that of regulations covering maximum permitted concentration limits for residues of pesticides or veterinary drugs in foodstuffs. At first sight this question would appear to be susceptible to a simple yes or no answer, i.e. the concentration is below the regulatory limit (and is thus acceptable) or it is not. However, this view does not take into account the experimental uncertainties in any quantitative measurement, especially those made for amounts of a substance present at trace levels in a complex matrix. The problem in terms of four generalized possibilities is illustrated in Figure 10.3. The simple 'yes–no' answer applied to the mean values of several replicates in each case would indicate that results (i) and (ii) are not in compliance with the regulation, while (iii) and (iv) are compliant. However, it is understood from the discussion of Section 8.2.4 that such an interpretation of result (ii) would involve significant probability of a false positives relative to the null hypothesis [H_o: the true concentration is below the regulatory limit]; recall that a false positive refers to acceptance of the alternative hypothesis H_1 when it can be assumed or shown later (e.g. additional data) that in fact H_o is valid. Similarly the naïve interpretation of result (iii) ignores a significant probability of a false negative.

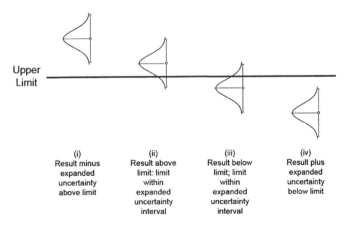

(i)	(ii)	(iii)	(iv)
Result minus expanded uncertainty above limit	Result above limit: limit within expanded uncertainty interval	Result below limit; limit within expanded uncertainty interval	Result plus expanded uncertainty below limit

Figure 10.3 Typical scenarios arising when measurements of the concentration of analyte are used to assess compliance with an upper specification limit. The vertical lines show the expanded uncertainty $\pm U$ on each result (see main text) and the associated curve indicates the inferred probability density function for the value of the measured concentration, emphasizing that there is a larger probability of the value of the measured concentration lying near the centre of the expanded uncertainty interval than near the ends. For a full discussion see Eurachem–CITAC Guide (2007) www.eurachem.ul.pt

How the decision criteria are set depends on the relative importance of such outcomes, e.g. in our example of a pesticide or drug residue in a foodstuff, in some cases it may be important to guard strongly against false negatives if public health risks are considered to be sufficiently serious as to justify facing the financial and other consequences of a false positive decision. In other cases where the health risks are deemed to be minor or highly uncertain and the regulatory limit is precautionary, decision criteria may be set to minimize the probability of false positives. The first parameter to be settled in this regard is the regulatory limit itself, and this is derived from information on food safety issues or other scientific information; usually such limits are set so as to minimize risk to public health while still recognizing the legitimacy of e.g. using drugs to treat sick animals. The other parameter shown in Figure 10.3 is the expanded uncertainty (U) that is derived from the experimental standard deviation s_x (Equation[8.2], Section 8.2.1) by multiplying by a coverage factor (k) chosen on the basis of the desired level of confidence to be associated with the interval defined by $U = k.s_x$. If a normal (Gaussian) distribution for the experimental measurements is assumed (Figure 10.3), a choice $k = 2$ (i.e. $U = 2s_x$) defines an interval having a confidence level of ∼95 %, while $k = 3$ ($U = 3s_x$) defines an interval with a confidence level > 99 % (Figure 8.6). The value chosen for k is thus important in the choice of decision criteria. (The coverage factor k is related to the similar factor Z introduced in Equation[8.61] although in that case Z is a factor used to multiply s_{blank}.) A detailed discussion of this procedure (Eurachem–CITAC 2007) emphasizes the importance of stating such decision rules concerning regulatory limits as clearly and unambiguously as possible, and provides the following example:

1. The result of an analysis shall be considered noncompliant if the decision limit of the confirmatory method for the analyte is exceeded.
2. If a permitted limit has been established for a substance, the decision limit is the concentration above which it can be decided with a statistical certainty of $(1-\alpha)$ that the permitted limit has been truly exceeded.
3. If no permitted limit has been established for a substance, the decision limit is the lowest concentration level at which a method can discriminate with a statistical certainty of $(1-\alpha)$ that the particular analyte is present.
4. The value of α is specified as $\leq 1\%$ for compounds of particular concern (this implies $k \geq 2.33$), and $\leq 5\%$ ($k \geq 1.64$) for other substances. Note that the quantity $CC\alpha$ introduced in Section 8.4.1 is the lowest measured concentration at which it can be certain to

within probability $(1-\alpha)$ that the true concentration is above the regulatory limit, i.e. $CC\alpha$ corresponds to the decision limit L_D (Section 8.4.1) with risk α that the true value is in fact below the limit, i.e. the probability of a false positive.

Rather than continue in this generalized statistical vein, for purposes of this book it is more appropriate to discuss a more specific example with particular reference to trace quantitation by mass spectrometry as the key technology.

10.4.3a FDA Enforcement Methods for Drug Residues in Animal Food Products

Farm animals raised for food do fall ill and are treated with veterinary medicines. This poses no problems in principle for consumers of animal-based foodstuffs, but in practice the residues of these drugs in the foodstuffs could pose a threat to public health if allowed to be present at levels that are too high. Accordingly, regulatory agencies set maximum allowed levels for each approved veterinary drug (and sometimes zero permitted levels for others), so that foodstuffs that do not satisfy this regulation are not approved for human consumption. But of course, realities of measurement uncertainty, as noted above with reference to Figure 10.3, must be taken into account. The following discussion is an abbreviated summary of a 'Guidance for Industry' document from the Center for Veterinary Medicine of the US FDA (FDA 2003); this document is a particularly relevant reference for this book in view of its explicit direction towards 'technical professionals familiar with mass spectrometry', and the clarity and completeness of its exposition.

The document is also explicit about a commitment to a fitness for purpose approach to development, validation and application of analytical methods intended for confirmation of drug residues above regulatory limits in edible animal tissues. The detailed elements of the Guide are described under four main headings, intended to assist laboratories wishing to submit their analytical methods for approval by the FDA. It is not appropriate to reproduce these details here. Instead some particular points that appear to relate most closely to the fundamental themes of this book are highlighted below. The Guide is applicable only to analytes for which a reference standard (analytical standard) is available.

I. Validation package from the originating laboratory. All analytical data submitted should represent five replicates. These should include analyses of five controls (blank matrices, preferably from different sources to check for relative matrix interferences and possibly matrix effects, Section 5.3.6a), analyses of 'fortified controls' (spiked blank matrix) at the regulatory limit specified for the drug that is the target analyte, and examples of analyses of real-world incurred samples. These data should show zero false positives in the analyses of the control matrix samples, and $\leq 10\%$ false negatives for the controls fortified at the regulatory limit. In this case, which addresses residues of approved veterinary drugs for which a regulatory level has been already set so as to minimize risk to public health as its guiding criterion, it is appropriate to weight the false positive/false negative compromise (see discussion of Figure 10.3) in confirmatory tests so as to guard against false positives while accepting a reasonable risk of false negatives. It must be demonstrated that these data can be replicated on different days (i.e. a test of method ruggedness, Section 9.8.4). Selectivity testing should include not only a concern for direct matrix interferences (via testing of control matrices) but also for interferences from other veterinary drugs that might reasonably be expected to be present in the particular class of tissues under test.

II. Method description (SOP). In addition to the expected unambiguous, stepwise description of all reagents, apparatus and operational steps including the mass spectrometric details, the SOP should also include a structure and full mass spectrum (or possibly MS^2 spectrum) of the target analyte (the 'marker residue' that might be the original drug molecule or possibly a metabolite). For purposes of confirmation of analyte identity at least three structurally characteristic ions should be identified, or more than three if the selectivity of some of the available fragment ions is low. These ions may include the intact ionized molecule (referred to as the 'parent ion' in the Guide), but use of isotopolog ions is discouraged as are fragment ions resulting from water loss (low selectivity, see Section 9.5.3b). These confirmatory ions can be monitored in SIM or MRM mode, or in selected scan ranges, depending on the mass spectrometric details, but their choice should be justified with respect to the degree of selectivity thus obtained. The SOP should also include the analyte confirmation criteria (see item *III* below). (The Guide uses the term 'specificity' rather than 'selectivity' that is preferred in this book, as explained in Sections 6.2.1 and 10.2.2). System suitability parameters (Section 9.8.3) should also be included.

III. Confirmation criteria. The importance of confirming the identity of the compound whose quantitative concentration is being compared to the regulatory limit is emphasized by the amount of detail in this section of the Guide (FDA 2003). The specifications in the Guide are entirely consistent with the extended discussion of this topic in Section 9.4.3b which in turn is a summary of an earlier publication (Bethem 2003). Confirmation of identity is ultimately based on comparisons with the analytical standard

and all analyses of the latter should be done contemporaneously with the incurred samples; explicit details of this comparison should be included in the SOP. If matrix interferences and/or matrix effects alter the MS or chromatographic behaviour of the analytical standard, an extract of a spiked control matrix, together with an extract of the nonspiked control matrix, can be substituted for the pure standard. The chromatographic peak to be identified as that of the target analyte must exhibit a signal:background ratio of $< 3:1$, where the definition of 'noise' (chromatographic background) and its measurement must be specified (Figure 4.27). An acceptability criterion for matching retention times should not exceed 2% for GC/MS and 5% for LC/MS and should be specified in the SOP. Relative abundances of the ≥ 3 structure specific ions (see item *II* above) should fall within specified limits depending on the nature of the mass spectrometric method, e.g. for three ions, $\pm 120\%$ for MS1 full scan spectra, $\pm 10\%$ for SIM etc., where the acceptability range is specified by addition and subtraction, so that for a monitored ion observed at 50% relative abundance the acceptability range is 40–60% not 45–55%. An interesting comment on the use of spectral libraries and library search algorithms as identification tools in the present context (FDA 2003) echoes that made elsewhere (Bethem 2003) and in Section 9.4.3b, namely that library match scores can be useful as indications of analyte identity but can not be used as statistically defensible criteria in view of their intrinsically incomplete nature and possible lack of selectivity and discrimination among similar compounds. It is for this reason that contemporaneous comparisons with an analytical standard, using well defined criteria for both chromatographic and mass spectrometric characteristics, are required (FDA 2003).

The Guide lists many details of the identity confirmation procedure that are not reproduced here, but essentially these details are designed to ensure compliance with the general principles outlined above.

IV. Quality control. Under this heading the Guide lists several requirements in common with similar QC protocols (Section 10.5.3b). For example, system suitability (Section 9.8.3) should be established before valid data can be obtained. At least one control and one fortified control must be analysed each analysis day (and the results obtained should meet the appropriate acceptance criteria!). Sufficient solvent blanks and/or control extracts should be analyzed immediately after standards or fortified control extracts to ensure that carryover (Section 9.7.2) does not cause a false positive result.

Other examples of documents describing development and validation of enforcement methods could have been chosen, but the example selected (FDA 2003) for discussion seemed particularly appropriate for a book devoted to applications of mass spectrometry. It is also commendable in that, while still attending to the potential need to defend the analytical data in litigation proceedings, it also maintains a scientific outlook. An example of the latter aspect is the inclusion of requirements for *ad hoc* confirmatory methods, which may be needed in unanticipated situations where fully validated procedures are not available and time does not permit a new method to be developed and validated. In such cases any regulatory agency will favor methods submitted by laboratories that can demonstrate good quality assurance, good training and expertise, and general adherence to the principles of GLP (Section 9.4.2a).

10.5 Validated Sample Analysis

The previous sections in this chapter addressed the process of validating the method and documenting the results in a final validation report. When reviewing the analytical data for the study samples the validation report and associated tables and supporting documentation must be carefully reviewed for accuracy and scientific content, to ensure that the necessary experiments have been run and that the data supports the original uncertainty tolerances (Section 9.3). In addition to the validation report, the final method must be reviewed and approved for sample analysis. Specific acceptance criteria that will be used during sample analysis should be documented in the approved method or equivalent SOP.

10.5.1 Sample Batch Preparation and Extraction

The apparently mundane task of collecting together the analytical samples to be analyzed in a sample batch run, together with the appropriate blanks, calibrators and QCs, is a crucial component of a successful analysis. A batch list that describes of all the components needed for an analytical run should be prepared prior to the initiation of any work. In addition, a list of all samples that will be analyzed in the run, a description of all standards, QCs and blanks that need to be prepared is often included as part of the batch list. Depending on the application, the types and number of blanks (extracted and/or solvent) that will be used may vary but in many instances at least one matrix blank with no internal standard (double blank) and one blank with internal standard should be included. The control matrix that is used for preparation of the extracted blanks should be screened prior to use (Section 9.4.7), to ensure that no appreciable interfering peaks that would impede the ability to meet acceptance criteria elute at the retention time(s) of the analyte(s) of interest.

For bioanalytical analyses, it is recommended that the number of QC samples in a run must be at least 5 % of

that of the number of study samples, and it is good practice (that should be included in the laboratory SOPs) to include QCs at a minimum of three concentrations covering the validated range. (Note that bioanalytical regulations require a minimum of six QC samples (N=2 at three concentrations) for each sample set.) The SOPs should also specify the number of QC sample analyses that can fail to meet the established accuracy criteria before the entire run is declared to have failed. The FDA Guidance (FDA 2001) specifies that, for validation, at least two-thirds of QC samples should be within 15 % of their respective nominal values and that each QC concentration should be represented by at least 50 % of the QCs prepared at that level. For cases where matrix dilution is required, for each dilution scheme used for study samples, one or more additional QC samples at concentrations several times higher than the upper limit of the calibration curve should be prepared, covering the maximum expected dilution (Bansal 2007).

In cases where study sample concentrations are likely to differ significantly from the established analytical QC concentrations specified in an SOP, or by client directive, additional QC samples may be prepared and analyzed to ensure that the QC samples are representative of the sample concentrations anticipated in the study. The concentrations of these additional QCs should be determined through agreement with the client and should be fully documented with all supporting raw data. These additional QC samples are to be analyzed as a number of replicates that should be specified in the appropriate SOP. Acceptance criteria for the additional QCs should follow the criteria used for the other analytical QCs in the batch.

It is equally important that the sample preparation (extraction and clean-up) of the batch must be done carefully in accordance with the procedures established by the validated method and any prevailing SOPs. The analysts conducting these tasks must be trained on the applicable SOPs and this training must be documented in their training records. Study samples must be analyzed within known analyte stability constraints for the matrix and the processed extracts. When sufficient stability has not been established at the time of sample reporting (e.g. long term freezer stability may be on-going at the time of analysis), this must be clearly documented within the study report.

10.5.2 Sample Analysis

10.5.2a Analytical Run Lists

Prior to sample analysis, a run list that describes the samples, calibration standards, blanks and QC to be injected, in addition to the injection sequence that is necessary to minimize effects of potential carryover (Sections 9.7.2 and 10.2.9b) and to provide the best monitoring of intra-run variability, must be generated. At a minimum at least one calibration standard should be injected at the beginning and end of each run along with a representative blank. The QC samples should be interspersed randomly in the sample injection stream.

The choice of the chronological sequence should be chosen such that the impact due to positional differences due to carryover. In general, the need to include additional carryover blanks in the run should be assessed during validation and specified in the analytical method. The run list is also used to generate an injection batch sequence, including rack and vial number, for the mass spectrometer data system and autoinjector.

10.5.2b Instrument Set-Up and System Suitability

System suitability checks should be performed before the first injection of the analytical batch and, for some methods, it is advisable to repeat this check when the analyses are completed. A detailed discussion of system suitability is given in Section 9.8.3, but the following description of a practical check procedure for LC–MS/MS systems has been found to be fit for purpose in many situations.

To demonstrate that the LC–MS/MS system is performing with adequate sensitivity, that peaks are eluting at approximately the expected retention time(s) and that there is method appropriate instrument background, system suitability test injections followed by at least one blank injection should precede the sample run. The system suitability test and blank injections should be evaluated and documented as acceptable in the instrument notebook prior to injection of the run(s). In cases where several runs are injected in sequence, the system suitability may be used for all runs in the sequence. Any significant disruptions between runs (e.g. system failure) would require a new system suitability assessment. The chromatograms for the system suitability and blank injections are generally not used to assess run acceptance but should be included with the sample run data.

Examples of typical solutions to be used for system suitability checks are a standard solution (in clean solvent) containing the analyte(s) and internal standard(s) approximating the concentrations in a LLOQ or low QC sample, or an extracted calibration standard or QC at or near the LLOQ. Additional system suitability standards at concentrations at or near the ULOQ can be used to test for carryover at the beginning of the run.

In some instances it may be necessary to inject one or more conditioning samples prior to the start of a chromatographic run for purposes of equilibrating the LC–MS/MS system. The conditioning samples may be derived from the run or may be separately extracted samples. Study samples should not be used as conditioning samples. The conditioning samples are not used in curve regression or evaluation of the run in regards to run acceptance. If a method requires the injection of an excessive number of conditioning samples, e.g. > 10, or if matrix injections are required to maintain chromatographic integrity throughout the run, then the analytical method must specify this requirement as part of the exact procedures that are described in the method and must be followed.

10.5.2c Failed Runs and Re-Analysis

The review of data and evaluation of the run for acceptance based on the criteria established during method validation can be found in Section 10.5.3, but in this section the instance considered is that when the analytical run fails prior to completion of the injection sequence or upon inspection of the data and it is apparent that there was a hardware or software failure of some kind. For events of this kind the system will need to be checked for performance to determine the nature of the failure, and repaired if necessary prior to resumption of any additional analyses. Examples for reasons why a run could fail prior to completion include hardware or computer failure, intermittent software failures, clogged injectors or broken syringes.

For situations where it appears that the system is running properly but there is a brief interruption in the run, it may be possible to resume the injection sequence immediately provided that some procedure (preferably defined in a sample analysis SOP) is followed to assure that the system is equilibrated and running properly. For other circumstances where it is discovered that the system has been idle for an extended period but appeared to be running properly prior to stopping, a system re-start suitability test may be used to establish that the instrument is equilibrated prior to re-starting the run at the point where it stopped. Solutions that are used for this purpose can be a calibration curve standard or a QC sample. The back-calculated concentration of the standard or QC used for re-start suitability must meet the acceptance criteria stated in the method or SOP. The result obtained from a re-start suitability injection should not be used in the calibration curve or reported with the sample set QC's. In any event, the fact that there was an interruption in the run and the reason for a system re-start must be documented in the raw data and possibly flagged in the analytical results table. If the system re-start criteria that are specified in the laboratory SOPs are not met, then the analytical run should be rejected (Section 10.5.3c) and the entire run should be re-injected provided that adequate extract stability has been established. Alternatively, the samples may be re-assayed (extraction and analysis of a new sample aliquot).

However, it is important to emphasize that, based on the validation results of analytical restrictions that may be in force at a particular laboratory or for a given method, the restart of analytical runs may not be permitted and the course of action may be restricted to either re-injecting the entire batch, or re-assaying the sample set.

10.5.3 Data Review

10.5.3a Evaluating Chromatography and Instrument Response

Most modern mass spectrometer data systems today have post-acquisition data processing software capable of automatically detecting chromatographic peaks and generating response data used for the calculation of final results. Although quite reliable in most circumstances, the quality of the final results is dependent upon the proper use and optimization of this software; even so, a thorough and careful review of the integrated chromatograms and instrument response throughout the entire run is imperative. This section provides a discussion of the procedures commonly used for chromatographic peak integration, manual and automatic, as well as other important considerations that relate to the review of the response data and analytical run. Procedures for evaluating the curve and QCs for run acceptance are discussed in Section 10.5.3b, and additional procedures for rejection of data are described in 10.5.3c.

Before discussion of peak integration, it is important to emphasize that in no circumstances must peak integration parameters for any particular sample be modified by the analyst for the sole purpose of meeting the acceptance criteria requirements of the method. Although this comment may seem obvious to most readers, the demand for rapid turnaround times and the costs associated with run failures can be very stressful in the high-paced modern laboratory, and the temptation to make 'minor adjustments' to calibration standards or QCs could potentially lead the data reviewer to make poor decisions and manipulate the final results. Ideally, the review of the chromatographic data and peak integrations should be done without knowledge of the final back-calculated results and accuracies but this is often not practical in many settings. Alternatively, the analytical method can essentially prohibit, in almost all circumstances, any type of integration of calibration standards or quality control samples that is not consistent with what is done in the rest of the run. When unusual integration procedures must be used, extra documentation and, in some instances, management approval, will be required.

Variability in extract cleanliness, instrument sensitivity and baseline background may require daily optimization of parameters that affect peak detection and integration. This daily evaluation and optimization process assists in ensuring the appropriateness of the software-generated integrations for all samples within a given run. Instrument software has advanced to the point where algorithms exist that can assess a chromatogram and determine the optimum integration parameters for each run based on a representative sample. These parameters can then be applied to integration of peaks for the analysis of all samples in a run. However, instances do occur in which peak detection or integration parameters using the software-generated parameters are not appropriate for a particular sample, due to irregularities in the baseline that may result from other closely eluting compounds, matrix interferences or poor chromatography. In these cases, it may be necessary to use a different set of parameters or to manually integrate the peak in order to generate the most appropriate integration.

Different software packages use different algorithms and integration parameters for processing post-acquisition data but, in general, the process of peak integration with automated software systems involves a two-step process: peak detection (see discussion of Figure 7.11 in Section 7.4.3) and peak integration. In general, the same peak-detect parameters are used for all samples in the set; however, peak-detect parameters must be evaluated on a run-by-run basis to ensure that that all peaks at or above the LLOQ will be detected and subsequently integrated in a consistent manner for all injections in the run. In some cases it may be necessary to modify peak-detect parameters on a per sample basis to ensure that all peaks within the validated calibration range are detected, i.e. no false negatives.

Best practice dictates that the same set of integration parameters be used for all peaks in a run, and optimizing the integration parameters for a particular run to ensure consistent and accurate integrations is generally accepted practice. However, it is also recognized that automated algorithms do not always produce the best integrations for all peaks consistently throughout the entire run. This is especially true for chromatographic peaks falling at, near or below the LLOQ, for peaks that have shoulders due to background, for tailing peaks and for peaks that are not completely resolved. In any event, every attempt should be made to evaluate smoothing, bunching, baseline noise and other parameters that affect integration before resorting to manual integration.

If it is not possible to apply the same integration parameters for all peaks without producing inconsistent results for all of the integrations, modification of the integration parameters using a different integration method (set of parameters) or manual integration of individual chromatographic peaks (i.e. redrawing the baseline) will be required. Depending on the chromatographic resolution from other peaks, the preferred method for drawing a baseline will be baseline to baseline (b/b), baseline to valley (b/v) or valley to valley (v/v). Every attempt should be made to draw the manual integration baseline in a manner consistent with the automatically integrated peaks. When manual integration is required, the laboratory should have established SOPs that clearly define the required procedures and documentation requirements. At a minimum, an entry must be made in the laboratory notebook, or equivalent, indicating which peaks were manually integrated and why. It should be possible, using hard copy or electronic records, to reproduce the original integration and results before the manual integration was performed. In addition to having unambiguous SOPs to describe this process, many laboratories require additional approvals from the laboratory manager or principal scientist prior to accepting the data for the generation of final sample results. Exceptions to this procedure may include instances when a manual integration is required for the assessment of

carryover or selectivity, when the detected peaks are significantly lower than the LLOQ and the automated integration parameters are not appropriate.

After the peaks have been integrated it is important to review the run for consistent sensitivity and response from the beginning to the end of the run. This is an important attribute of the data that should be reviewed for all runs, but is particularly critical for methods exhibiting low S/B ratios at the LLOQ, as exemplified in Figure 10.4.

The system suitability (Section 9.8.3) should have been previously reviewed before beginning the sample run but, for cases when multiple analytical sets are run back to back overnight, additional suitability test samples may be included between sets. The chromatography should be inspected to ensure that the peak shape quality is consistent

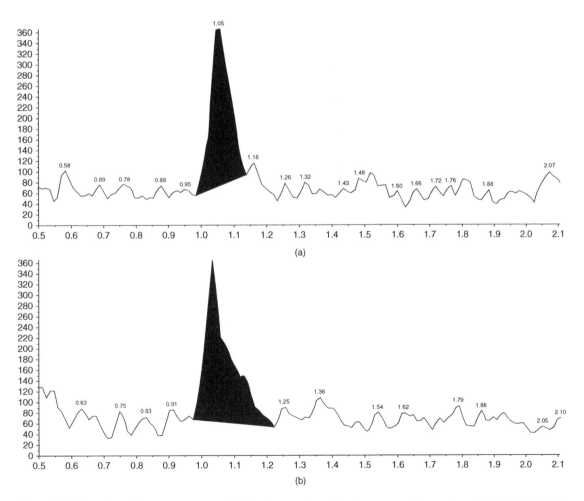

Figure 10.4 Examples of chromatograms generated from two samples within the same run with very low signal:background ratios leading to large uncertainties in peak integration.

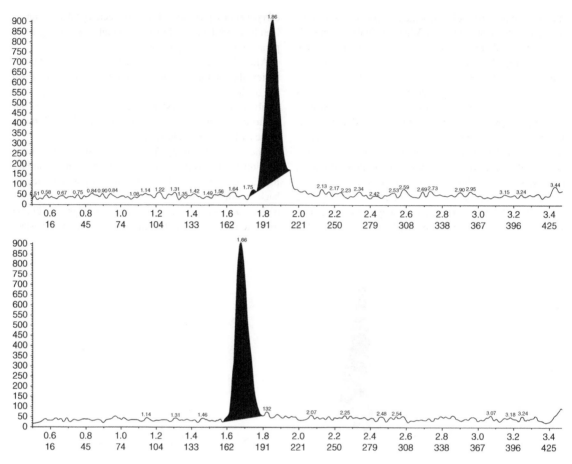

Figure 10.5 An example of the effect of retention time drift on the uncertainty in peak integration for the analyte at the LLOQ. The top chromatogram shows the peak observed after the retention time had drifted to a point that significant background contributed to a shoulder on the trailing edge of the peak resulting in uncertainty in how to draw the baseline for integration. This is compared with the peak observed when the system was working properly (bottom chromatogram).

with what was obtained during validation and in previous runs. The retention times for the analyte and SIS should be inspected to verify that there was no significant drift within the run (see discussion of Figure 4.27 in Section 4.4.8 for a description of baseline drift within a single chromatography elution). Many data systems can generate a tabular list of response data and retention times in chronological order of injection, and these can be used to rapidly inspect absolute and relative retention times. If retention time drift is detected, extra care should be taken to ensure the correct peaks are integrated properly and the potential reasons for the drift should be considered and investigated. Possible reasons include pump cavitations, clogging, dirty extracts, column failure and inconsistent extracts, to name just a few.

An example of the effect that retention time drift can have on the uncertainty in peak integration is shown in Figure 10.5.

The response from the SIS should also be evaluated to determine whether there is any evidence for response drift, matrix effects or inconsistent preparation of the extracts. Analytical runs with particularly dirty samples can negatively impact the chromatography and sensitivity. For some methods just one or two samples can have a negative impact on the samples injected immediately following or, in some cases, for the remainder of the run. Two internal standard response plots from separate runs using the same method for study sample analysis are shown in Figure 10.6.

In Figure 10.6 (a) a run is shown where the internal standard response exhibited large variation throughout the run and was rejected. The samples were re-extracted and upon re-analysis the internal relative standard deviation was found to be typical for the method based on validation and other study sample results (Figure 10.6 (b)). See Section 10.5.6 for an additional discussion of investigation and corrective action procedures.

Another potential anomaly that should be investigated is that arising from the presence of metabolites in incurred samples that in principle will not be present in the QC samples prepared using control matrix. This problem and a suggested strategy to test for it were described in Sections 9.4.7b and 10.2.9c. In Figure 10.7 an example is shown of a comparison between QC sample made in control human plasma matrix and a patient sample with an interfering glucuronide metabolite. In this case, the chromatography conditions were modified to resolve the analyte from the glucuronide and the method was re-validated.

Following the review procedures described above, each chromatogram for analyte and SIS should be inspected

for consistent integration and peak detection as described above. The temptation to rely only on run acceptance criteria, with a cursory review of standards, blanks and QCs, should be resisted (see also Section 10.5.6 for a discussion of the importance of scientific review in general). Experience and skill will assist with the process of optimizing the most appropriate automated integration parameters but, in the end, the final scientific review relies on the eyes of a trained scientist in the laboratory.

10.5.3b Evaluating the Curve and QCs

The acceptance criteria used to accept or reject the run based on the curve and QCs are established during method validation (Section 10.3.3d) and must be documented in the method or appropriate laboratory SOP. As previously discussed, the targeted acceptance criteria to be used for

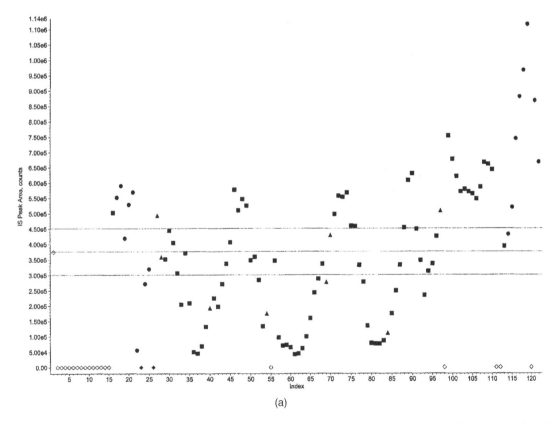

(a)

Figure 10.6 Two internal standard response plots from separate analytical runs using the same method for study sample analysis. (a) Data from a run in which the internal standard response exhibited large variation throughout the run, that as a result was rejected (the limit bars correspond to ±20 %). (b) Plot of the internal standard response generated from the same set of samples that were re-extracted and analyzed in a subsequent run.

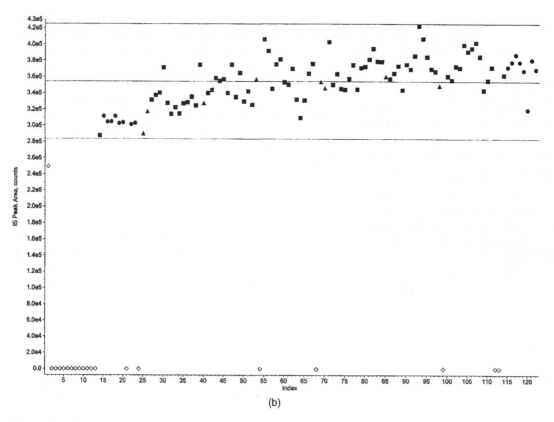

(b)

Figure 10.6   (Continued)

sample analysis will be dictated by the analytical requirement, sponsor or regulatory guidelines. Procedures for identifying calibration curve outliers, and criteria for how many can be dropped before a run is rejected, must also be documented prior to sample analysis.

In some instances during sample analysis, if the LLOQ calibrator fails the data can be reported with a raised LLOQ (perform the regression with the next highest standard as the revised LLOQ). Samples with concentrations below that of the raised LLOQ must then be re-assayed in another run if sufficient sample volume remains.

Run failures due to failed curves or QCs can be due to a large number of reasons that include failure due to assignable cause, e.g., documented human error or system failure, or as a result of a method that is not rugged and reliable for the given application. When using analog or homolog surrogate internal standards, the relative responses between the analyte and SIS can drift during the course of the run. This situation is often referred to as curve divergence and was discussed in Section 9.4.5c. An example is illustrated in Figure 10.8. Curve divergence due to matrix

effects is often not observed during method validation but can potentially have a significant impact during analysis of incurred samples. If the problem persists on a regular basis, sample analysis should be halted until the problem is resolved. In many instances, this may mean doing more method development and, if significant changes are made to the method, re-validation of the new method and repeating the study sample analysis.

10.5.3c Rejection of Data

Data for individual calibration, QC and study samples may be rejected based on assignable cause. Rejection of data due to assignable cause includes documented sample processing errors or when there was a suspected hardware failure (e.g. autoinjector, chromatography, interface or mass spectrometer problem). Assuming that criteria have been established prior to the analytical run, data may also be rejected for other observed conditions, such as when unusual internal standard response is observed, e.g. samples with double the expected internal standard area or samples with no internal standard

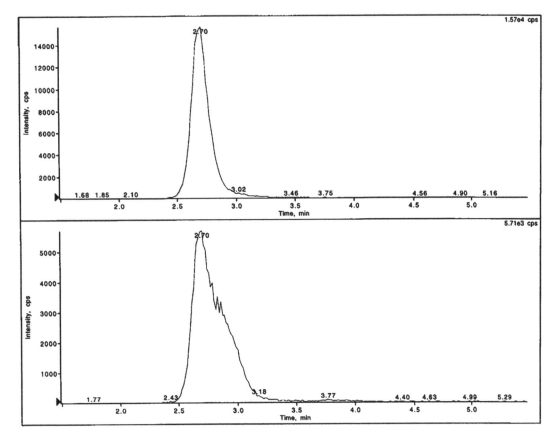

Figure 10.7 Comparison of a chromatogram obtained for the analyte in a QC sample made from control matrix with that for an incurred sample with a co-eluting metabolite (glucuronide). In this case the chromatography conditions were modified and the method was revalidated.

area. Rejection of samples for other reasons must be scientifically justified. All rejected data must be documented with the reason for the rejection; if adequate extract stability has been established the samples may be re-injected according to procedures established in the laboratory SOPs.

Analytical runs may not be rejected simply on the basis of failure to meet acceptance criteria. In these instances, the data should be retained and a reference to the failed set should be included in the study sample report. However, if sufficient re-injection extract stability has been established, analytical runs may be rejected and then re-injected for the following analytical reasons: instrument hardware failure; poor chromatography throughout the run; curve divergence of replicate calibration curve samples from the beginning to the end of the analytical run; loss of sensitivity through the run, which impacts quantitation; excessive injection carry-over; erratic internal standard response throughout the run.

Rejection of runs for other reasons must be scientifically justified; all rejected data must be documented together with the reason for the rejection.

10.5.4 Sample Re-Assays

Repeat analysis (sample re-assay) is defined as the re-extraction and analysis of a sample with a new calibration curve and QC samples. The procedures to be followed for determining how the re-assay should be conducted and final results reported for sample re-assay data should be described in an SOP. An example of a sample re-assay flow chart is shown in Figure 10.9 although other similar schemes can be employed. In any event, the procedures used for re-assay of samples and reporting of results must be established prior to study sample analysis!

If a result is above the upper limit of the calibration curve (OCR), or if a result of a diluted sample is below the

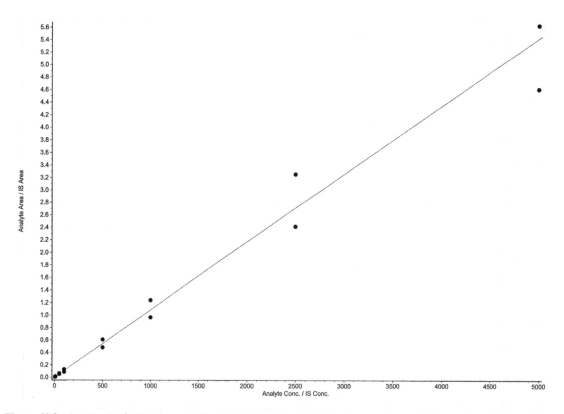

Figure 10.8   An example of calibration curve divergence between two identical calibrator sets run at the beginning and at the end of an analytical run. The points represent the ratios of peak area of analyte to that of the SIS vs a corresponding concentration ratios, and the consistently increasing discrepancies between the two determinations indicate that the relative sensitivities (S_a and S_{SIS} in Section 8.5) have changed over the course of the run.

lower limit of the calibration curve (DCR), or if the data are rejected and there is inadequate re-injection stability data and/or insufficient sample for re-injection, a single repeat analysis may be performed. In these instances only the repeat analysis value should be reported. In either of the cases where a sample was diluted and re-assayed because the original result was over the calibration range (OCR) or because it was diluted below the calibration range (DCR), if the result from the re-assay appears to be incongruous with the original predicted concentration based on the extrapolated result, a re-analysis in duplicate should be performed.

In general, samples may not be re-assayed simply because the result appears to be incongruous with the results from other samples in the set. However, if a repeat analysis is performed based on other established outlier criteria, the samples should be re-assayed in duplicate and reported according to a pre-established re-assay flow chart. Documentation of this activity and the results for each of the sample assays must be maintained in the study file. Samples from control groups, pre-dose or time

zero samples with detectable concentrations of analyte should be re-assayed in duplicate (at a minimum) to confirm the positive result, if sufficient sample exists. If a sample is inadvertently re-assayed, the result of the repeat analysis will be treated as a single repeat analysis.

A detailed list identifying the location of all analytical data associated with the repeated samples must be generated and kept as part of the study file. A table of repeat analysis results should be included in the final report, together with the reason(s) for the repeat analyses, the original, repeat and reported results, and the justification for selecting the reported results.

10.5.5 Addressing Carryover During Sample Analysis

A general discussion of carryover and of methods to address it can be found in Section 9.7.2 and some implications for validation are described in Section 10.2.9b. Here an approach is described that facilitates formulation of a

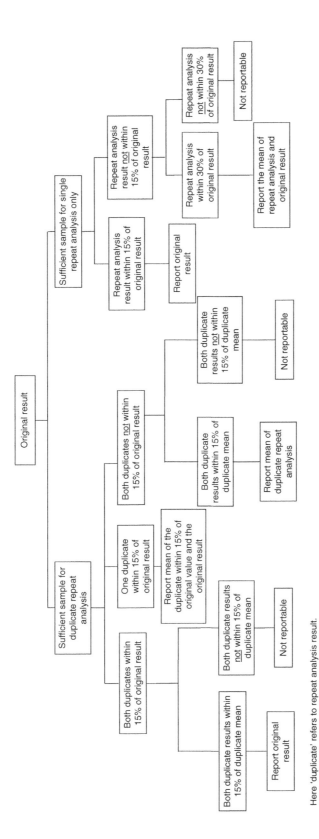

Here 'duplicate' refers to repeat analysis result.

Figure 10.9 A flowchart summarizing a typical strategy for determining how to perform a sample re-assay and report final results.

strategy to minimize effects of carryover during sample analysis once some basic characteristics of its extent for the analytical method have been determined. This approach is an example of the importance of positional differences discussed in Section 10.2.9b, and is based on an investigation of these characteristics for a specific method for a specific analyte (Zeng 2006) but should clearly be of general applicability. It should be noted first that the authors established that random errors made a negligible contribution to observed deviations of measured from nominal concentration values under the conditions used.

The basis of this approach was the experimental demonstration of the constancy over a wide range of concentration of the relative carryover *(RC)* defined as:

RC = ratio of (peak area for a blank sample) to (peak area for a preceding sample with nonzero concentration)

Several different experiments were conducted (Zeng 2006) to confirm the constancy of *RC* to within a few percent; these are not summarized here but this result should be re-confirmed for any different method (of course the value of *RC* need not be the same as the value of 0.05 % = 5×10^{-4} found for the method in the original work). Based on this result, the approach can be based on evaluation of the estimated carryover influence *(ECI)* on a chromatographic peak area as:

$$ECI\,(\%) = RC \times CR \times 100$$

where *CR* is the concentration ratio for the sample injected immediately preceding sample of interest to the concentration of the sample of interest. This relationship was also tested experimentally (Zeng 2006) by comparing *ECI* values with experimentally estimated systematic errors. Note that since an SIS is always present at the same concentration in bioanalytical samples for a given method, $CR = 1$ so $ECI = RC$ which is usually negligibly small.

For the original method (Zeng 2006) it was found that the *ECI* gave a reasonable estimate of the relative error in peak area (a measure of systematic bias error and thus of accuracy) for a wide range of concentrations for both the sample of interest and the preceding sample, and covering *CR* values ranging from 20 to 2000. For $RC = 5 \times 10^{-4}$ and $CR \leqslant 100$, the average accuracy of the measurements ranged from 96.9–105 % and the influence of carryover on the quantitation of the second sample was negligible, but for *CR* values between 400 and 1000 the accuracy was in the range 116–157 %, i.e. the influence of carryover on the quantitation of the second sample becomes significant and cannot be neglected. For $RC = 0.05\%$ and $CR = 200$ the

average accuracy was between 109–112 % and thus meets the requirements of the FDA guidelines (FDA 2001), but the probability that the accuracy of a particular individual measurement > 115 % is still non-negligible.

In general the total uncertainty in a measured value can be considered to be the sum of the systematic error(s) plus the random error (often expressed as the relative standard deviation, RSD). The systematic error can be estimated as the *ECI* plus any other systematic errors in the system.

The *ECI* concept has two useful applications (Zeng 2006). Firstly, the contribution of carryover to the total error of a measurement can be estimated, so the sum of the RSD and the *ECI* can be used as the criterion to evaluate the accuracy of a measurement. Thus if the RSD of a validated bioanalytical method is less than 10 %, a repeat analysis is not necessary if the $ECI < 5\%$. For example, when $RC \sim 0.05\%$, the second sample of the two consecutive samples does not need re-analysis if $CR < 100$ since the corresponding $ECI < 5\%$. This simple guideline is practical and easy to apply because the *CR* value for two consecutive samples can be readily estimated based on their measured concentrations. The second application follows from the fact that the *ECI* on the second sample of a consecutive pair is proportional to the *CR* value. In the case of application of an analytical method to a pharmacokinetic study, when the sequence of sample injection for each subject is in chronological order by collection (PK) time, the influence of carryover on the quantitation of each unknown sample is significantly decreased because the *CR* of the two such consecutive samples is rarely greater than 50. By also arranging that QC samples and calibrators are appropriately interspersed in the sample stream, it is possible in principle to eliminate carryover as a significant concern in PK studies. For cases where the concentration of a sample is higher than the ULOQ, the calculated *ECI* can provide a quantitative estimate of the influence of carryover on the analysis of the following sample.

The general conclusions of this work (Zeng 2006) are that it is possible in at least some cases to 'manage' sample arrangement so as to mitigate the effects of carryover, without having to take more drastic measures such as limiting the dynamic range of the assay (Section 9.7.2). Generally, if the RSD of a validated bioanalytical method < 10 % and the $ECI < 5\%$, the accuracy of the concentration measurement meets the FDA guideline requirement (FDA 2001).

10.5.6 Investigation and Corrective Action

If the analytical method goes out of control, e.g. as indicated by the control chart maintained for the QCs (Section 2.5.3) repeated run failures without cause, or loss of selectivity due to metabolites or other interferences, it is imperative that any further sample analysis is stopped, the events are documented and a determination is made as to what type of investigation and corrective action will be required prior to commencing with any additional sample analysis.

All of the hardware, software, clever methodologies and validation go for nothing if analysts are unable to assure the end user of the data that the method was in control throughout sample analysis and that the results can be trusted within the stated uncertainty limits. In addition to within-run acceptance criteria based on analysis of QCs and calibrators within a single batch run, control charts are also often used as a means for assessing whether or not the method is in control. Use of only the former as quality control indicators can often lead to missing problems that affect even the most fully validated methods.

Corrective action investigations are used to identify, investigate and evaluate the correction of observations within a procedure, process or system. These investigations provide a means to determine root cause and assess procedures for preventing recurrence of problems. The precision and accuracy established during validation together with the established acceptance criteria for the method are essential, but many other on-going influences that could bias sample analysis, e.g., the impact of matrix effects and direct interferences with incurred samples will require that 100 % of the data and associated chromatograms are critically reviewed on a routine basis.

It is all too easy to forget that trace level quantitation of analytes, present at trace levels ($\sim 10^{-9}$) in complex matrices, is at root a scientific investigation, one that pushes a series of technologies (Chapters 3–7) to their limits (see discussion in Section 1.4). When such a method has been shown to be out of control, the pressure to generate results with rapid turnaround times must be set aside temporarily to allow for a fundamental scientific investigation of the problem. Any nonconformities, unexpected events or quality system deficiencies that are observed by the analyst or other laboratory should be brought to the attention of the principal investigator and management.

A recent FDA guidance for industry (FDA 2006) provides the Agency's current thinking on how to evaluate out-of-specification (OOS) test results. For purposes of this document, the term 'OOS results' includes all test results that fall outside the specifications or acceptance criteria established in drug applications, drug master files (DMFs), official compendia or by the manufacturer. The term also applies to all in-process laboratory tests that are outside of established specifications. This document is intended for GMP studies, but it can be generalized for all applications. The following discussion represents an attempt to do so, as an adaptation of seven key recommendations (FDA 2006).

1. The Principal Investigator and possibly laboratory management should discuss the situation with the analyst(s) responsible for acquiring the suspect data, and confirm that the analysts fully understand and have correctly performed all steps specified in the method and in laboratory SOPs.

2. Examine the raw data obtained in the analysis, including visual inspection of chromatograms, and identify anomalous or suspect information.

3. Verify that the calculations used to convert raw data values into final results are scientifically sound, appropriate and correctly performed; in particular, determine whether unauthorized or unvalidated changes have been made to automated calculation methods.

4. Confirm the performance of the instruments, e.g. chromatographic performance, *m/z* calibration of mass spectrometers, other system suitability checks.

5. Determine that appropriate reference standards, solvents, reagents and other solutions were used, and that they met quality control specifications (e.g. had been stored appropriately, were not out of date etc.).

6. Evaluate the performance of the method to ensure that it is performing according to the standard expected based on method validation data and historical data. This could include a check on the overall calibration of the method by comparison of the within-run calibrators with freshly prepared calibrators, and similarly for the QC samples.

7. Fully document and preserve records of this laboratory assessment.

These seven steps represent a systematic means to attempt to narrow down the constituent part(s) of the method and/or its performance that led to its going out of control. Discovery of the specific cause must now rely on the scientific abilities of the Principal Investigator and his/her colleagues, and this will be greatly facilitated if the retained study samples, calibrators and QCs are examined promptly. This part of the procedure is an example of the scientific method, i.e. framing hypotheses and designing experimental tests of their validity.

Laboratory error should be relatively rare (FDA 2006). Frequent errors suggest a problem that might be due to inadequate training of analysts, poorly maintained or improperly calibrated equipment, or careless work. Whenever laboratory error is identified, the laboratory should determine the source of that error, take corrective action to prevent recurrence and maintain adequate documentation of the incident and the corrective action.

Situations where re-testing is indicated (FDA 2006) include identified instrument malfunctions or sample handling problems, e.g. a dilution error. A re-testing plan should be in place (e.g. as an SOP), but all decisions to re-test should be based on the objectives of the testing and scientific judgment. The maximum number of re-tests to be performed should be specified in this SOP; the number may vary depending on the variability of the particular method employed but should be based on scientifically sound principles. The number of re-tests should not be adjusted depending on the results obtained! In the case of a clearly identified laboratory error, the re-test results would substitute for the original results. All original data should be retained, and an explanation of the error recorded and signed.

Once the cause(s) of the problem is (are) identified, it is important to determine also whether samples analyzed previous to detection that the method was out of control are still valid and whether or not the entire study could have been negatively impacted or is suspect. Retrospective investigations of this kind are even more difficult and it is here that the maintenance of QC control charts, for various types of QCs (e.g. blanks, analyte-only and SIS-only QCs), shows up as an essential feature of assurance of the validity of the final results reported to the user(s).

For sample analysis in a regulated environment, each laboratory should have an SOP that describes the procedural and documentation requirements for how to evaluate out-of-specification (OOS) test results and any corrective action performed. If, as a result of this investigation changes are made to the method, additional validation may be required; the various circumstances are discussed as examples of full vs partial validation in Section 10.2.10 and need not be repeated here.

10.6 Documentation

The documentation requirements at the analytical site for sample analysis are essentially the same as those described for the original validation (Section 10.3.3). Additional requirements for the analytical samples and the final report are discussed briefly below in this section.

10.6.1 Sample Tracking

The requirements for tracking study samples, from receipt at the laboratory through storage and analysis to data reporting, are described in Section 9.4.6b. The exact requirements for chain of custody handling and documentation ('sample accessioning') may vary depending on what regulations or SOPs may be in effect. Depending on the applicable SOPs or regulations, access to the samples may be restricted to specified personnel in order to substantiate the integrity of the sample and final results. The overall objective is to be able to prove that the reported data refer to samples that were received, stored and analyzed with an unambiguous identification trail plus a record of what was

Table 10.8 Minimum requirements of the sample analytical report

Reference Standard	Batch/Lot, Purity and Source, Stability at time of use.
Calibration Standards and QCs	The identity and storage conditions for all calibration standards (Section 9.8.1) and QCs (Section 9.8.2).
Method and Run Acceptance Criteria	A description of the method. Run acceptance criteria (Section 9.8.5 and 10.3.3d).
Sample Tracking Information	Date of receipt of shipments and contents. Sample condition at time of receipt. Storage location and condition (Section 9.4.6)
Results Tables	Reference to calculations, rounding etc. (SOP). Tables of all runs (final results) and analysis dates.
	Tables of calibrator results from all runs with mean and CVs.
	Table of QC results for passed runs with accuracy and precision.
Chromatograms	Representative number based on reporting requirement (5–20%)
Failed Run	Identify runs, assay dates and reason for failure (Section 10.5.2c)
Deviations	Description of deviations and impact on results.
	Description of any supporting investigations and data (Section 10.5.6)
Re-assay Tables	Table of sample IDs. Reason for re-assay. Re-assay values. Run IDs. Re-assay SOPs (Section 10.5.4)

A more complete listing of reportable parameters can be found elsewhere (Viswanathan 2007).

done to each sample, by whom and when. It is here that a computerized laboratory information management system (LIMS) is invaluable.

10.6.2 Sample Analytical Report

The FDA Guidance document (FDA 2001) provides a good model for analytical reports in general. However, since the publication of that document additional details are commonly addressed (Viswanathan 2007). Table 10.8 was drawn from these publications and although the documents specifically address reporting requirements for bioanalytical analysis in support of drug development studies, they provide an excellent overview of the requirements for a analytical report in general.

10.7 Traceability

Thus far analytical method validation has been discussed mainly from the viewpoint of agencies regulating bioanalytical measurements, taking the US FDA regulations as a widely quoted example (FDA 2001). There are issues that are understandably not a primary concern of agencies like the FDA but are nonetheless background issues; one of these is traceability (Eurachem 2003). (This concept was briefly discussed with respect to the analytical standard in Section 2.2.1.) Accurate and reliable measurements of all kinds that can be compared in different jurisdictions and at different times are necessary so that the concept of 'tested once, accepted everywhere' can be widely accepted and technical (nontariff) barriers to trade can be minimized. In addition to the criteria for GLP and validation already discussed, all the individual measurement results must be related to some common, stable reference or measurement standard (see the discussion of the Système Internationale in Chapter 1); this strategy of linking results to an internationally agreed reference is termed traceability, formally defined by the ISO as the 'property of the result of a measurement or the value of a standard whereby it can be related to stated references, usually national or international standards, through an unbroken chain of comparisons all having stated uncertainties' (ISO 2004).

A very complete account of traceability in the context of chemical measurements (Eurachem 2003) emphasizes that laboratories are under ever increasing pressure to demonstrate that their use of measurement and reference standards is appropriate and sufficient. Many of the physical quantities used in routine chemical measurement (e.g., mass and volume, Chapter 2) are underpinned by extensive and effective calibration and traceability systems, making the establishment of traceability for these quantities relatively straightforward. However, the values of chemical quantities

involved are typically drawn from a wide range of reference materials and data, reflecting the much greater degree of complexity intrinsic to chemistry. Such references typically come with varying pedigree and provenance (not all are fully certified CRMs, Section 2.2.1), so that considerable care and judgment is required in selection of references (Eurachem 2003). The general problem faced by chemists in this regard is further compounded by the fact that chemical analyses frequently involve measurement of a species in a complex matrix, and of course this is the subject of this book. Chemical measurements also typically require confirmation of identity as well as measurement of amount, and this aspect is discussed later in this chapter.

The relationship between validation and traceability can be illustrated by the following discussion, adapted from one given previously (Eurachem 2003). Firstly assume that the measurement of a quantity Y (in this case, typically the concentration of a target analyte in a matrix) can be accomplished by combining measurements of other quantities via a specified functional relationship:

$$Y = f(x_1, x_2, x_3, \ldots, x_m) \text{ under experimental}$$
$$\text{conditions } x_{(m+1)}, x_{(m+2)}, \ldots x_n$$

$$[10.1]$$

Here x_1–x_m represent measured quantities including physical ones (e.g., mass of analytical standard weighed out, volumes of standard flasks etc.) and chemical properties (e.g., chemical purity of analytical standard, isotopic purity of an internal standard, ratios of analytical signals for analyte and internal standard etc.). Specific examples of such functions are those discussed in Section 8.4. The experimental variables ($x_{(m+1)}$–x_n) that do not appear directly in the functional relationship could include temperature of various components of all forms of apparatus used, mobile phase composition and flow rate, operating parameters of the mass spectrometer, etc.

The process of validation checks, using appropriate tests (see above), that the functional relation (Equation[10.1]) above is adequate under a stated range of conditions ($x_{(m+1)}$–x_n). If this validation check is found to be obeyed to within an acceptable level of uncertainty, Y is then said to be traceable to ($x_1 \ldots x_n$). Then, to demonstrate complete traceability for Y, it is necessary to show that all the values (x_1–x_n) are themselves either traceable to reference values (and via these, ultimately to the SI standards), or are defined values (i.e., unit conversion factors, mathematical constants like π, or the values of constants used to relate some SI units to fundamental physical constants). An example would involve calibration of a semi-microbalance against a set of 'weights' that have been certified relative to a national mass standard that has in turn been calibrated against the international prototype of the kilogram (Chapter 2). The overall uncertainty in Y is then derived from uncertainties in (x_1-x_m) via appropriate application of the principles of propagation of error (Section 8.2.2). Extensive discussion (Eurachem 2003) of realistic chemical examples clarify the relationship between validation and traceability (note that the former is a prerequisite for the latter).

The discussion thus far has been limited to what might be called 'traceability theory', which is readily applicable to purely physical measurements of quantities that are not affected by chemical composition. The extent to which the 'theory' can or should be applicable to measurements of amount-of-substance at the trace level in complex matrices is still a topic for discussion (Antignac 2003; Quevauviller 2005). In cases where the analytical problem is reasonably well defined, e.g., in analyses of plasma or urine for a pharmaceutical drug, conducted in accord with criteria specified by an organization like the US FDA (FDA 2001), traceability (including a reliable estimate of the overall uncertainty in the measurement, both accuracy and precision) is relatively straightforward to demonstrate. However, in cases that are chemically more complex, e.g., analyses of environmental pollutants in matrices that are neither well defined nor reproducible, the 'unbroken chain of comparison' that theoretically connects the actual measurements to SI definitions implies that no loss of information (from, e.g., sampling uncertainties, recovery efficiency that is not well defined due to potential occlusion effects) should occur during the entire analytical procedure.

This question assumes increased importance in view of the ISO/IEC 17025:2005 laboratory accreditation standard (ISO 2005), which requires explicit attention to a number of activities including client interactions, method validation and traceability (including measurement uncertainty).

It has been argued (King 2001) that a 'fitness for purpose' approach should be taken with respect to the degree of rigor to be applied with respect to formal traceability in any particular case. This recommendation (King 2001) seems to be particularly appropriate for environmental analyses (Quevauviller 2005), in which the 'traceability chain' will be more or less strong from case to case. In general, traceability to well defined and accepted references and accuracy can become an almost meaningless ideal in some cases, and extremely difficult to demonstrate in others. Comparability of data in environmental analyses can be established by proficiency testing but this does not necessarily mean that the data thus produced are accurate, i.e. close to the 'true' value. Traceability to pure analytical standards is unsatisfactory since the complications of sampling and extraction are not addressed properly if at all, while CRMs or 'documented standards' often correspond to 'consensus' values and not 'true' values. Thus, traceability compromises that are 'fit for purpose' must be sought while still adhering to validation procedures and using the best state-of-the-art analysis techniques. An example of a traceability chain that is intrinsically subject to some uncertainty is shown in Figure 2.1.

11

Examples from the Literature

11.1 Introduction

The task of selecting examples from the reviewed literature to illustrate the general principles outlined in the
preceding chapters is paradoxically both easy and difficult for the same reason – there is a wealth of exemplary
papers available. The choices made largely reflect the
idiosyncratic interests of the present authors but many
other completely different sets of examples would have
served equally well. Most of the papers discussed here
were chosen because they combined general best practices with innovative twists that represent advances in
the field. In most cases the emphasis is upon method
development, although some examples that incorporate
full method validation are included. With respect to use
of internal standards, the chosen examples reflect the full
range from a heavy atom isotopically-labeled SIS (best
practice) to use of no SIS at all (as a result of difficulties in
finding a suitable compound). In the following summaries
an attempt has been made to point out in each case the
particular difficulties that were faced by the analysts, and
how these difficulties were approached in terms of the
chosen combination of extraction/clean-up methods with
chromatographic and mass spectrometric technologies.

It is appropriate to add here a few general comments
concerning details of mass spectrometric parameters given
in published papers. Some of these details that should
always be included are unfortunately not always available.
For example, it is not uncommon to read that MS/MS
detection was used but, while the m/z values of precursor
and fragment ions are specified, the effective resolution
of both stages of m/z filtering (e.g., the 'isolation width'
for the precursor) are not. This is important because,
in trace analysis, it is always necessary to find a 'fit
for purpose' compromise between selectivity (increases
with resolution, i.e., with decreasing peak widths) and
response ('sensitivity' in the colloquial sense) for which

the trend with resolution is in the opposite direction as
exemplified for a linear quadrupole in Figure 6.9. On the
other hand, some details (e.g., precise values of optimum
settings of ion source parameters and of collision energies and potentials on ion optical lenses) are dependent on
the instrument used (and on its state of cleanliness etc.!)
and are best regarded as recommended starting points for
independent optimization if the analytical method is to
be reproduced. However, the general point being made
here is that development and publication of an analytical
method is a branch of experimental science, and in the
true scientific spirit we owe it to our colleagues to provide
sufficient detail that they can form judgements as to the
reliability of our quantitative measurements and can, if
desired, test this reliability by repeating our experiments.
A good example of this important feature of the scientific
approach is described below in Section 11.2.1.

It is hoped that the present choice of examples provides
a broad view perspective of quantitative mass spectrometry and successfully emphasizes that this is truly an
extremely powerful and versatile interdisciplinary technology. Two books relevant to this chapter deal with trace
analysis techniques applied to environmental samples
(Loconto 2005) and the use of mass spectrometry in drug
metabolism studies (Korfmacher 2004).

11.2 Food Contaminants

The foodstuffs we consume contain large numbers of
compounds at trace levels, some natural and some anthropogenic. All chemical compounds are 'toxic', depending
on the dose relative to the size of the organism subjected
to the compound and the period over which the dose(s)
is/are administered (acute vs chronic toxicity). A great
deal of effort has been devoted to determining toxicological dose–effect relationships for man-made chemicals

Trace Quantitative Analysis by Mass Spectrometry   Robert K. Boyd, Cecilia Basic, Robert A. Bethem
© 2008 John Wiley & Sons, Ltd

used in the production of foodstuffs, and to analytical methods for determining their concentrations in a wide variety of food matrices. Probably the compounds that have attracted most attention and regulatory control are the pesticides used to protect crops against attacks by insects, viruses, moulds etc. Analytical methods for one or a few target compounds, as well as multi residue methods, continue to be developed (Niessen 2006; Alder 2006) for these man-made pesticides.

However, nature is not delinquent in providing plants with chemical protection, the only kind of defence available to organisms that are rooted to the ground and can not flee their attackers. Tens of thousands of natural secondary metabolites of plants have been identified and it has been estimated that hundreds of thousands of these compounds exist, many at trace levels; it is believed that many if not most of these compounds are involved in the defense of the plant from microbial pests and insects. Plants also produce many compounds that are poisonous to mammals, e.g. strychnine is used as a commercial rodenticide, while warfarin is a chronic poison that is also used as a rodenticide but nowadays mainly as a life-saving drug in carefully controlled sub-lethal doses as a blood anti-coagulant in treatment of thrombosis and embolism.

Many plant metabolites marketed as 'natural pesticides' are in fact more toxic than their synthesized competitors; for example, rotenone (extracted from the roots of certain members of the bean plant family) has been used as a crop insecticide since the mid-19th century to control leaf-eating caterpillars, but is six times more toxic to mammals on a strictly comparable basis than carbaryl, a synthetic chemical also effective for caterpillar control. Nicotine sulfate is extracted from tobacco by steam distillation or solvent extraction and has been used as a pesticide since the early 20th century; it is six times more toxic than diazinon, a widely available synthetic insecticide sold for control of many of the same pests. The best known work in this area (Ames 1990, 1990a, 1997) used the Ames test (Ames 1973, 1973a) to compare potencies of natural and synthetic pesticide compounds with respect to mutagenicity in special bacterial strains. While some of the conclusions of this work are controversial (Tomatis 2001), it does at least emphasize the importance of development of analytical methods for natural as well as synthetic compounds in foodstuffs. In this section an example of each is considered.

11.2.1 Acrylamide

Acrylamide (2-propenamide, $CH_2=CH-CO-NH_2$) is manufactured for use as the monomer feedstock for poly-acrylamides that are used as water-soluble thickeners,

e.g., in wastewater treatment, paper making, manufacture of permanent-press fabrics, and also in the laboratory as the stationary phase in gel electrophoresis of biopolymers (SDS–PAGE, see Section 11.6). There is no question that acrylamide (AA) poses a considerable health risk as a potent neurotoxin; it consistently produces various types of cancer in laboratory experimental mice and rats but studies in human populations have not yielded consistent results (Rice 2005). This could possibly be due to difficulties inherent in such studies on humans, e.g. it might be difficult to isolate the effects of acrylamide in epidemiological studies because it is so ubiquitous in most diets in the industrialized world that it is difficult to find a true experimental control (zero concentration). Current evaluations (Hagmar 2003; Mucci 2003, 2003a, 2005, 2005a; Taeymans 2004) suggest that cancer risks derived from consumption of foodstuffs containing traces of acrylamide are likely to be low. Nonetheless, the known dangers associated with acrylamide in general dictate continued investigation of the issues associated with its presence in food, and in turn this will require continued development of analytical methods for acrylamide and its metabolites in various matrices. The challenges to the analyst arise mainly from the combination of low molecular mass (71 Da) and high polarity and reactivity.

The recent level of interest in acrylamide in food was sparked by a press release in April 2002 from the Swedish National Food Administration and the University of Stockholm (Swedish National Food Administration 2002), reporting their observation and measurements of acrylamide in a range of cooked foods, especially in starchy and carbohydrate-rich foods that had been fried or oven baked, e.g. potato crisps (known as chips in North America) and chips (French fries), biscuits, crispbreads and breakfast cereals. These observations were the result of some meticulous analytical chemistry (Tareke 2002); the reported levels in laboratory-heated foods were temperature dependent, ranging from 5–50 $\mu g.kg^{-1}$ in heated protein-rich foods and 150–4000 $\mu g.kg^{-1}$ in carbohydrate-rich foods such as potato and also in some heated commercial potato products and crispbread. Acrylamide was not detected (method detection limit was 5 $\mu g.kg^{-1}$) in unheated control or boiled foods. To put these findings in perspective, at the time of the original press release the World Health Organization had set a drinking water limit of 0.5 $\mu g.kg^{-1}$ (parts per 10^9) for many countries, and in Europe the recommended limit was about to be lowered to 0.1 $\mu g.kg^{-1}$. With respect to acrylamide in foodstuffs the then existing European Union limits on chemical migration from plastics used as food containers or wrappings required that migration levels into the foods should be below 10 $\mu g.kg^{-1}$. These findings

(Tareke 2002) created considerable concern and were quickly confirmed in an independent study (Ahn 2002).

This major public health issue was initiated by a team of analytical chemists (Tareke 2002) and was motivated by findings of mass spectrometric studies of adducts of the N-terminal valine residue (see Section 11.6) in hemoglobin (Hb) with several reactive and potentially mutagenic and carcinogenic compounds in humans; this aspect of the work is not described in detail here. Such protein adducts are commonly used as biomarkers for such exposures (Fennell 1992; Ogawa 2006) since they are not subject to *in vivo* repair mechanisms, as are the corresponding DNA adducts, and can thus provide a measure of the integrated exposure over the average *in vivo* lifetime (for Hb in blood $\sim$120 days, the same as red blood cells). In the case of the Swedish work a ubiquitous 'background signal', which appeared to correspond to the acrylamide adduct to the N-terminal valine residue of Hb, had been consistently observed in control persons who had not been occupationally exposed to acrylamide (details of this work are not given here); the measured levels of this adduct suggested an average daily intake approaching $100\,\mu g$.

In a major effort to confirm that the source of this ubiquitous Hb adduct is indeed acrylamide and, if so, to determine acrylamide sources and mechanisms of formation, two independent analytical methods (Tareke 2002) were developed. With respect to sources, pre-existing evidence of acrylamide in tobacco smoke, which was reflected in increased levels of the Hb adduct of interest, suggested that acrylamide could be formed during incomplete combustion or heating of organic matter. Also, lower background levels of this Hb adduct were observed in wild animals than in humans and laboratory animals, hypothesized to be due to consumption of unheated food by wild animals. Previous laboratory animal experiments (Tareke 2000) had provided some confirmation that acrylamide is formed in fried rat food by verification of the identity of the marker Hb adduct acrylamide using MS/MS analysis, and by demonstration that the amounts of acrylamide determined in fried feed were correlated with the observed increases in adduct levels.

To provide additional confidence in these analytical results, two independent methods were developed for acrylamide itself rather than the Hb adduct (Tareke 2002). As mentioned above, experimental difficulties arose from the fact that acrylamide is an unusually small (71 Da), highly polar and highly reactive compound present in a wide variety of food matrices and arising from a variety of known and unknown sources and chemistries (see below). Despite the presence of the polar amide group the compound is amenable to high

resolution separation by capillary GC but, to move the detected m/z to higher values well away from the ubiquitous low mass background, acrylamide was converted to the dibromocompound $CH_2Br–CHBr–CO–NH_2$ (2,3-dibromopropionamide) that behaved well in low resolution (single quadrupole) EIMS. In this case the selectivity of the final analytical step relied on a combination of the high resolution chromatography with that provided by low resolution ('unit mass') EIMS, as described in more detail below. To confirm these findings an LC–ESI–MS/MS method (triple quadrupole) was also developed (Tareke 2002), as described below; a challenge here was to find an HPLC column that would provide sufficient retention ($k' > 1$) for the small polar analyte. The chromatographic resolution is not as high in HPLC as in GC but additional selectivity was provided by MS/MS analysis (again using 'unit mass' resolution for both precursor and fragment ions).

In the method based on GC/MS analysis of brominated acrylamide, as described previously (Castle 1993), extraction and clean-up of analytical samples involved mixing (with a food blender) 10 g of sample with 100 mL water, followed by filtration and clean-up on a graphitized carbon black column. Originally the internal standard was N,N-dimethylacrylamide, but at an early stage $^{13}C_3$-acrylamide became available and was added before the mixing–extraction step. The samples were derivatized at 4 °C overnight using brominated water containing potassium bromide (KBr) and hydrogen bromide (HBr) (acidification to pH 1–3); the excess bromine was decomposed by adding sodium thiosulfate solution dropwise until the yellow color disappeared. The solution was extracted with ethyl acetate/hexane (1:4, 2 × 20 mL) and the pooled extracts were dried to $\sim$200 μL. Two microlitres of each extract were injected (splitless injector) into a GC–EIMS instrument (70 eV electrons) for analysis using SIM. Several columns (30 m × 0.25 mm i.d. × 0.25 μm film thickness) with a range of stationary phases were used to ensure that the results were not dependent on this parameter. The ions monitored for the brominated analyte, i.e. 2,3-dibromopropionamide, were $C_3H_5{}^{81}BrNO^+$ (m/z 152, 100 %), $C_3H_5{}^{79}BrNO^+$ (m/z 150, 100 %), and $C_2H_3{}^{79}Br^+$ (m/z 106, 65–70 %), using m/z 150 for quantitation and the others for confirmation of analyte identity. The ions monitored for the $^{13}C_3$-internal standard were $^{13}C_3H_5{}^{81}BrNO^+$ (m/z 155, 100 %) and $^{13}C_2H_3{}^{81}Br^+$ (m/z 110, 65–70 %) using m/z 155 for quantitation; a variation of $\pm$10 % in the abundance ratios between fragment ions was allowed for identification, consistent with accepted practice in EIMS (Section 9.3.3b).

In view of the narrow GC peaks (see Figure 11.1 for examples) the SIM dwell times were correspondingly

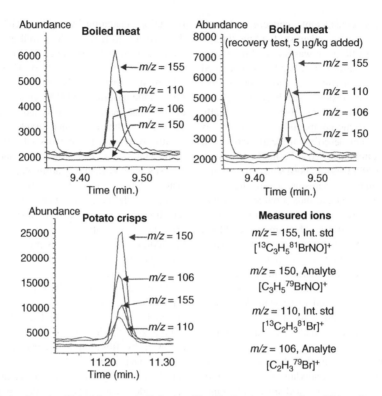

Figure 11.1 Analysis of acrylamide in laboratory-boiled minced beef both prior to and after addition of acrylamide standard, and of potato crisps ('chips' in North America) with a high content of acrylamide. The food samples were extracted and brominated (see main text) and analyzed by GC–EIMS in SIM mode. SIM chromatograms are shown for two selected ions for each of the brominated analyte and the $^{13}C_3$-acrylamide internal standard. Slight variations in retention times are due to use of different GC columns. The SIM chromatograms for m/z 152 are omitted because of an overlapping contribution from the internal standard; m/z 150 and 155 were used for quantitation, the others for confirmation of identity. Reproduced from Tareke, *J. Agric. Food Sci.* **50**, 4998 (2002), copyright (2002) with permission from the American Chemical Society.

short ($\sim$0.04 s per SIM channel with 0.01 s switching time) to provide sufficient data points across each GC peak. Quantitation was achieved using a calibration curve of peak area ratios vs acrylamide concentration in water in the linear range 0.5–$50\,\mu g.L^{-1}$, corresponding to 5–$500\,\mu g.kg^{-1}$ in the original food sample. (Note that in this GC-EIMS approach, matrix effects giving rise to ionization suppression are expected to be negligible, so the fact that clean calibration solutions were used rather than matrix matched calibrators should be an acceptable procedure. In any case the ubiquitous presence of the analyte in foodstuffs would have made it difficult to find a suitable blank matrix.) Recoveries (98 %, RSD 7.5 %) were estimated as the ratio of concentrations of extracts of each foodstuff before and after the addition of known amounts of additional acrylamide (an approach imposed by the unavailability of true blank matrices). The reproducibility

of the overall method was ±5 % when using the $^{13}C_3$-internal standard and the limit of detection for the method was stated (Tareke 2002) to be $5\,\mu g.kg^{-1}$, although the precise definition of this parameter (Section 8.4.1) was not specified. Generally, however, it is clear that this GC/MS method was well designed and the results obtained can be regarded with a high degree of confidence; examples of some of the raw data are shown in Figure 11.1.

Nonetheless, in view of the potential importance of the experimental results and of the limited selectivity of 'unit mass resolution' SIM detection, an independent LC–MS/MS method using ESI was also developed (Tareke 2002) to confirm the GC–EIMS data. In this case the samples were homogenized and 100 mL of water and 1 mL of the $^{13}C_3$-internal solution ($1\,\mu g.mL^{-1}$ in water) were added to 10.0 g of the homogenate. After centrifugation the supernatant was cleaned up on an Isolute Multi Mode SPE column (International Sorbent

Technology Ltd.); this column is based on mixed C_{18}, SAX and SCX phases that retain interferences through nonpolar, anion and cation exchange retention mechanisms. 500 μL of the cleaned up extract were ultrafiltered until 200 μL had passed through. (Ultrafiltration or nanofiltration is a pressure driven separation process through an organic semi-permeable membrane. Because of its selectivity, one or several components of a dissolved mixture are retained by the membrane despite the pressure driving force, while water and substances with a molecular mass less than ~200 Da, depending on the nature of the membrane, are able to permeate the membrane which also can exhibit selectivity for the charge of the dissolved components, e.g., singly charged ions will pass through the membrane while more highly charged ions will be rejected.) The resulting solution was analyzed by LC–MS/MS in positive mode using a Hypercarb column, a porous graphitic carbon phase with retention characteristics rather different from those of a conventional C_{18} reverse phase column; Hypercarb exhibits excellent retention of extremely polar compounds that elute close to the void volume on conventional C_{18} reverse phase columns and are thus likely to co-elute with acrylamide. Deionized water was used as isocratic mobile phase; the injection volume was 20 μL into a mobile phase flow rate of 200 μL.min^{-1}, and after elution the column was carefully washed with 80 % aqueous acetonitrile followed by reconditioning with water.

MRM analysis using a triple quadrupole analyzer monitored the acrylamide $(M+H)^+$ ion (m/z 72) together with its product ions $H_2C=CH-C=NH^+$ (m/z 54, loss of water (H_2O)) and $H_2C=CH-C=O^+$ (m/z 55, loss of ammonia (NH_3)); the abundance ratio of product ions was 1:35 ($\pm$20 % accepted) and the major m/z 55 ion was used for quantitation together with the corresponding $^{13}C_3H_3O^+$ (m/z 58) product ion of the $(M+H)^+$ precursor (m/z 75) of the internal standard. The ESI source parameters (e.g. 'cone voltage') and MS/MS conditions (especially collision energy) were optimized for maximum response using continuous infusion of standard solutions of analyte and internal standard. The much wider HPLC peaks (Figure 11.2) could be adequately defined for peak integration using correspondingly longer MRM dwell times thus ensuring acquisition of sufficient ion statistics (Section 6.5). For further confidence in analyte identification, complete product ion spectra of acrylamide standard and analyte were compared at various collision energies. Quantitation used a linear calibration over the range 1–500 μg.L^{-1} in water, corresponding to 10–5000 μg.kg^{-1} in the original food samples. Recoveries were determined to be 99.5 % (RSD 6.5 %), the reproducibility as 2–9 %, and the LOD as ~10 μg.kg^{-1} for

the samples investigated (Tareke 2002) though again the precise definition of the latter was not described (probably $S/N \geq 3$); some typical results are shown in Figure 11.2.

The most obvious deficiency in this method is its lack of concern about ionization suppression caused by matrix interferences, since the calibration curves were obtained using clean solutions of analytical standard in water rather than spiked blank matrix extracts (that were of course unavailable, see above). However, use of a co-eluting ^{13}C-labeled surrogate internal standard certainly mitigates this concern though the possibility of mutual suppression of analyte and internal standard remains (Section 5.3.6a). Also, the excellent agreement obtained (Tareke 2002) with the results obtained using the GC–EIMS method indicates that any such concerns about the LC–MS/MS method are negligible, probably as a result of the good clean-up procedure plus the unusually selective HPLC column used. Essentially the same two analytical methods were used later (Ahn 2002) in a meticulous confirmation of the results of the original work (Tareke 2002). This confirmatory work used different internal standards (e.g., d_3-acrylamide) and additional MS/MS transitions, namely m/z 72 → 44 (loss of CO) and 27 (loss of NH_3 + CO), to confirm analyte identity, although the relative abundances of the latter fragment ions were sufficiently low that they could be measured with sufficient precision only for acrylamide levels > 1 ppm in the foodstuff.

However, in the flurry of activity that followed the original findings, later experience did not always result in such satisfactory agreement between results of GC–EIMS and LC–ESI–MS/MS methods. In a 2003 review (Wenzl 2003) it was concluded that the extraction step was the main culprit in this regard, a view subsequently verified in a very thorough investigation (Petersson 2006); in particular, coffee and cocoa powder were identified as particularly difficult matrices. In general, an assessment of several round-robin exercises concluded (Petersson 2006) that both incomplete extraction and *in situ* formation of acrylamide during the extraction process itself were possible pitfalls for many matrices. For example, several authors proposed *in situ* formation in Soxhlet extraction (Section 4.3.2a) of potato crisps ('chips' in North America) at 60 °C in methanol, and a group which used accelerated solvent extraction (Section 4.3.2b) with ethyl acetate at elevated temperature and pressure found incomplete extraction of acrylamide from cocao and milk powder compared to the results obtained using water as extraction solvent. Potential sources of analytical error other than the extraction step include formation of the analyte from precursors in the heated GC injector, pick-up of extraneous acrylamide from apparatus such as filters

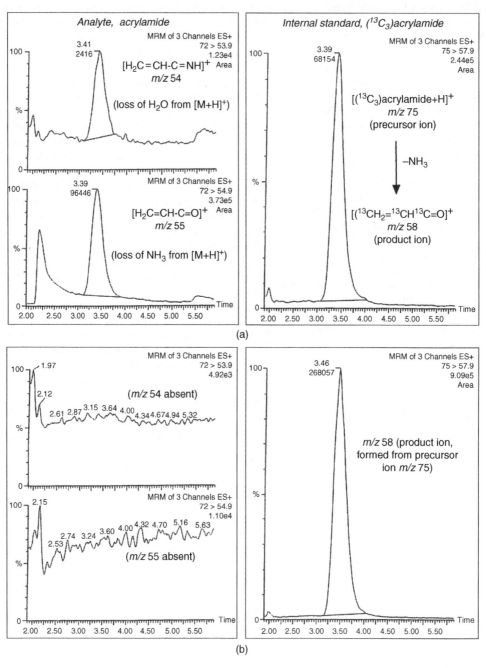

Figure 11.2 LC–ESI–MS/MS analysis of acrylamide in foodstuffs. (a) French-fried potatoes; MRM chromatograms for product ions m/z 55 and 54 from the $(M+H)^+$ precursor ion m/z 72 quantified relative to the $^{13}C_3$-internal standard (precursor ion m/z 75, product ion m/z 58). (b) Analysis of a blank sample. Reproduced from Tareke, *J. Agric. Food Sci.* **50**, 4998 (2002), copyright (2002) with permission from the American Chemical Society.

and thermal degradation (rather than formation) of acrylamide (Petersson 2006).

In view of these uncertainties an extensive investigation of extraction of acrylamide from several foodstuff matrices was undertaken (Peterssen 2006). The test extracts were analyzed using a two step SPE clean-up with different stationary phases, together with an LC–ESI–MS/MS approach that essentially followed the original method (Tareke 2002). It was shown that acrylamide can be efficiently extracted from various food matrices (coffee, crispbread, mashed potatoes, milk chocolate and potato crisps) using plain water; the use of aqueous alcohol, or removal of fat by an organic solvent, gave negative or nonsignificant effects on the extraction yield. Disintegration of the samples to small particles prior to extraction was effective, with no need for homogenisation and vigorous mixing; quantitative extraction was achieved by gentle mixing at room temperature, and varying extraction times between 30 minutes and 17 hours yielded the same results. These optimization experiments (Peterssen 2006) indicated that incomplete extraction, rather than accumulation of additional acrylamide from the procedures themselves, is the most likely cause of erroneous results in the extraction step, and can arise if the food matrix is not divided sufficiently finely prior to extraction, or when using organic solvent or a short extraction time or a low extraction temperature, particularly when several of these parameters are used together. In contrast, destruction of acrylamide during extraction and clean-up was found (Peterssen 2006) to be small or undetectable for the investigated foodstuffs. Based on this work, an optimised extraction procedure was devised, suitable for a wide range of food matrices, consisting of disintegration of the food samples (particle size < 1 mm) together with use of water as the extraction solvent with shaking of the sample at 25 °C for 45 minutes. The authors (Peterssen 2006) emphasize that, in order to fully benefit from this optimised extraction procedure, it should be combined with optimized cleanup and chromatography/detection-steps.

This is a telling example of the importance of the relatively 'unglamorous' extraction and clean-up steps for the success of any analytical method. Modern advances in high resolution separation science and mass spectrometry have made these latter components of the overall analytical method much less subject to errors as long as well known problems (e.g., sufficient chromatographic resolution, ionization suppression by matrix interferences etc.) are recognized and dealt with.

Many analytical methods for acrylamide in foodstuffs have been described in the literature and have been summarized in a review (Zhang 2005). Instructive examples are consider to be the methods developed for such analyses in the 'difficult' matrices, coffee and cocoa powder. A method designed for analysis of acrylamide in ground and instant coffees, both solids and brewed liquids (Andrzejewski 2004), used mild extraction of the solids with water at room temperature, as later shown (Peterssen 2006) to correspond to an optimized procedure. This extraction was followed by clean-up using two SPE steps, the first an Oasis HLB cartridge (a mixed hydrophobic–hydrophilic polymeric phase) and the second a Bond Elut–Accucat cartridge (a mixed SAX–SCX phase). Brewed coffee samples were subjected directly to the SPE clean-up. As before, $^{13}C_3$-acrylamide was used as surrogate internal standard. Analysis was by LC–ESI–MS/MS, using a Synergi Hydro RP 80A column. This is a C_{18} reversed phase column with 4 μm particle size that is end capped not with trimethylsilyl or similar hydrophobic residues, but with polar groups, and is thus suitable for separation of both nonpolar and highly polar analytes; it also has the advantage of being stable in highly aqueous mobile phases (Section 4.4.1a). The mobile phase was 0.5 % methanol in water (an equilibration time of ~1.5 hours was needed before any analyses could begin, to avoid variations in acrylamide retention times). The HPLC effluent was supplemented with 0.1 % acetic acid in 2-propanol via a Tee-connector, prior to introduction into the ESI source of the mass spectrometer, in order to modify the mass spectrometric background at low m/z values so that the $(M+H)^+$ ions of the labeled and unlabeled acrylamide could be detected without direct interference from intense mobile phase ions especially at m/z 74 and 76. The isolation widths for the precursor ions, corresponding to 'unit mass' resolution, were such that intense background ions at neighbouring m/z values could lead to significant interferences in the MRM signals; the precursor isolation widths could not be greatly reduced because of the resulting drop in ion transmission efficiency (compare Figure 6.7) and thus overall sensitivity. The triple quadrupole analyzer was operated in MRM mode, monitoring the $(M+H)^+$ ions of analyte (m/z 72) and internal standard (m/z 75) as well as their MS/MS fragment ions at m/z 55 and 27 for analyte, and m/z 58 and 29 for the $^{13}C_3$-internal standard. The acrylamide stock solutions used to prepare calibration solutions were found to be stable for periods of several months when stored in red glass at 7 °C.

The careful method development in this case (Andrzejewski 2004) also provides good lessons with respect to unexpected pitfalls that must often be faced. Thus, sample sizes had to be changed from the intended one gram size to accommodate acrylamide contents that were higher (instant coffee crystals) or lower (brewed coffees) than anticipated. Another problem arose when it was observed that, after repeated injections of coffee extracts onto the LC column, no signal was observed for either labeled or unlabeled analyte; this problem was solved by increasing the column oven temperature to 35 °C. A related problem of limited useful life of the HPLC column due to loss of chromatographic resolution was addressed by thorough washing of the column (50:50 methanol:water for two hours followed by 50:50 methanol:acetonitrile for up to 12 hours) following analysis of a complete set of acrylamide standards and coffee extracts.

Problems also arose (Andrzejewski 2004) with the mass spectrometric detection of this low mass analyte in a complex matrix like coffee. Despite two SPE clean-up steps, a selective LC separation and the additional selectivity of MRM detection, several LC peaks were observed close to that corresponding to the analyte (Figure 11.3), and it was feared that the variability of HPLC resolution might lead to direct interferences in the MRM channel used for quantitation; to check for such direct interferences, relative responses for the three MRM transitions for acrylamide were compared to those for the labeled standard, and agreement to within ±10 % was taken to indicate no direct interferences. This degree of acceptable variability for analyte identification is stringent (Section 9.3.3b) and qualifies for the full three identification points (Tables 9.1–9.2). Relative abundance values outside this window would indicate a contribution from a compound other than acrylamide. However, since the internal standard co-eluted with the analyte, this check would not detect indirect interference via ionization suppression (Section 5.3.6a). In fact, use of only the first SPE clean-up step (Oasis HLB cartridge) was found to result in direct interferences (contributions of co-eluting components to at least one of the MRM transitions for the analyte), that led to inclusion of the second SPE step.

Even this intense search for sources of error was insufficient, as demonstrated (Andrzejewski 2004) by the effect of HPLC run time and its impact upon successive injections. A response in the MRM channel used to quantitate acrylamide (m/z 72 → 55) was observed at ~23 minutes for all of the extracts; this can affect the corresponding peak used for quantitation of acrylamide with repetitive injections, e.g. if the analyses are run at 15 minute intervals instead of 10 minutes, the tailing edge of the acrylamide response overlaps with the 23 minute response for the (m/z 72 → 55) channel from the previous injection. By adjusting the inter-injection interval to 10 minutes this late eluter problem (Section 9.7.2) was eliminated.

This long list of pitfalls that were identified and dealt with (Andrzejewski 2004) provides an exemplary lesson for all analysts. As a result of this meticulous investigation the final method provided excellent results, including a recovery of 92 ± 6 %, and an inter-day precision (RSD) of 5.1 %. Confirmation of acrylamide in the coffee extracts required four criteria to be satisfied, including the 10 % limit on relative intensities of MRM signals already mentioned, plus coincidence of retention times for analyte and internal standard, clear separation of these LC peaks from those arising from co-extractives, and a minimum 10:1 S/B value for the (m/z 72 → 55) MRM signal used for quantitation. This last criterion corresponds to a rather crude estimate of both the LOD and LLOQ (Section 8.4) of 10 μg.kg^{-1} for coffee solids and 1 μg.L^{-1} for coffee brews; the detailed significance of this criterion is difficult to evaluate since the signal:background ratios were determined automatically by an algorithm supplied in the mass spectrometer software that was not described. As for the original work on acrylamide in heated foodstuffs (Tareke 2002; Ahn 2002), this method for coffee did not explicitly explore the possibility of ionization suppression, but use of a co-eluting internal standard would minimize the consequences of any such effects such as those observed (Granby 2004) for coffee extracts. The levels of acrylamide thus determined (Andrzejewski 2004) ranged from 45 to 374 ng.g^{-1} in unbrewed coffee grounds, from 172 to 539 ng.g^{-1} in instant coffee crystals, and from 6 to 16 ng.mL^{-1} in brewed coffee. Acrylamide proved to be quite stable in brewed coffee as indicated by the observation that no significant decrease in concentrations was observed after five hours of heating. However, the acrylamide levels measured in ground coffees stored at room temperature for ~6 months after the containers were first opened and analyzed were 40–65 % lower than those measured the first time, whereas samples stored for the same period in a freezer at −40 °C showed no such decrease, which may be good news for those who store their coffee at room temperature!

A later study (Aguas 2006) adopted a somewhat different approach to the analysis of acrylamide and extended it to the case of cocoa, which is at least as complex a matrix as coffee. The initial extraction of acrylamide and its $^{13}C_3$-internal standard with water was essentially unchanged, but the aqueous extract was then 'defatted' by liquid–liquid extraction with dichloromethane. The subsequent clean-up strategy was very different from the earlier methods (Tareke 2002; Andrzejewski 2004) in that a normal phase SPE sorbent was used to selectively elute the amide from more polar impurities. The SPE phase chosen for this purpose was of the aminopropyl type that has weak anion exchange properties in addition to the characteristic normal phase adsorption properties. However, such an approach requires that the cocoa extract must be in a relatively nonpolar solvent so that acrylamide is retained on

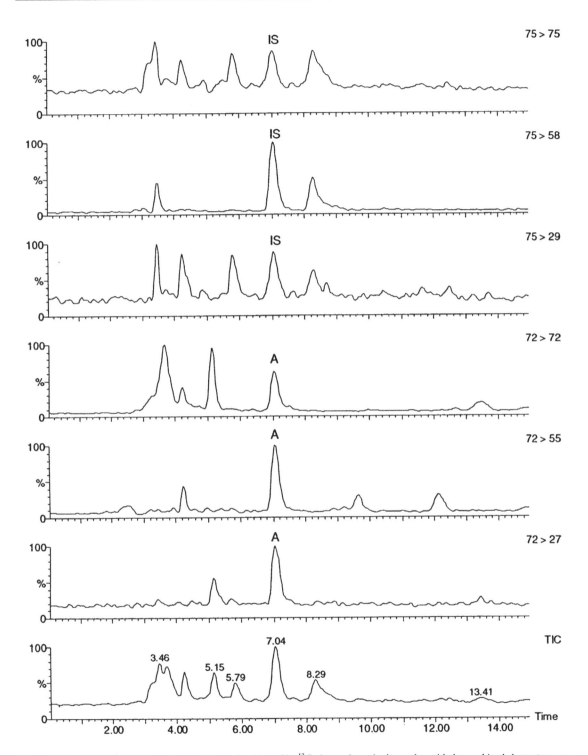

Figure 11.3 HPLC–MRM chromatograms for acrylamide and its $^{13}C_3$-internal standard, together with the combined chromatograms (TIC chromatogram), in a typical coffee extract. Reproduced from Andrzejewski, *J. Agric. Food Chem.* **52**, 1996 (2004), copyright (2004) with permission from the American Chemical Society.

the aminopropyl cartridge. This required a fairly complex series of liquid–liquid extractions to transfer the acrylamide from water solution into a mixture of acetonitrile and ethyl acetate (that had the added benefit of efficient precipitation of proteins), and finally into a solvent mainly composed of ethyl acetate and *cyclo*hexane. This solution was then applied to an aminopropyl SPE cartridge that was first eluted with pentane; this was followed by one mL of a 1:2 acetone:pentane mixture that was discarded, and then by a further five mL of the same solvent that eluted the acrylamide off the cartridge. The solvent was carefully evaporated to dryness under a gentle stream of nitrogen with mild heating (40 °C) and the resulting colorless residue was dissolved in 0.5 mL HPLC grade water to give a final extract in a solvent compatible with reversed phase HPLC.

The LC–MS/MS analysis of the final extract (Aguas 2006) was also different from that used previously (Tareke 2002; Andrzejewski 2004) in that APCI was used instead of ESI; this choice was made on the basis that APCI usually performs better that ESI at higher flow rates ($> 0.2\,mL.min^{-1}$) unless additional heat is supplied to the ESI source, and that APCI appears to be somewhat less susceptible to ionization suppression by co-eluting matrix components (Section 5.3.6a). (The mobile phase flow rates used in this and the earlier work were $0.2–0.3\,mL.min^{-1}$.) This new method (Aguas 2006) was not explicitly tested for ionization suppression effects, but did provide encouraging results. Derminations of acrylamide levels in coffee and cocoa samples showed precision (RSD, $n = 5$) of 2.2–2.5 % in the range $12.3–139\,\mu g.kg^{-1}$, and 0.6 % at $465\,\mu g.kg^{-1}$. The accuracy of the method was tested by applying it to a coffee CRM with a certified acrylamide content of $174\,\mu g.kg^{-1}$; the value obtained (Aguas 2006) was $179 \pm 9\,\mu g.kg^{-1}$.

In addition to many other published analytical methods for quantitation of acrylamide, related studies have also appeared. For example, the origins of this compound in heated foodstuffs is a continuing subject of investigation. A study (Granvogl 2004) investigated 3-aminopropionamide (3-APA, $NH_2–CH_2–CH_2–CO–NH_2$) as a possible transient intermediate in acrylamide formation during thermal degradation of the amino acid asparagine (Asn, $NH_2–CO–CH_2–CH(–NH_2)–COOH$) initiated by reaction with reducing carbohydrates or aldehydes (the so-called Maillard reaction responsible for the colors and flavors of partially burned foodstuffs such as toasted bread). A highly simplified description of the Maillard reaction is given in Equations [11.1–11.3], in which the reducing sugars are represented as their open chain structures:

$$R\text{-CH(OH)-CHO} + NH_2\text{-R'} \rightarrow R\text{-CH(OH)-CH(OH)}$$
$$\text{-NHR' (a glycosylamine)} \qquad [11.1]$$

$$R\text{-CH(OH)-CH(OH)-NHR'}$$
$$\leftrightarrow R\text{-CH(OH)-CH=NR'} + H_2O \qquad [11.2]$$

$$R\text{-CH(OH)-CH=NR'} \leftrightarrow R\text{-CO-CH}_2\text{-NHR'}$$
$$\text{(a 1-amino-1-deoxyketose)} \qquad [11.3]$$

The 1-amino-1-deoxyketose (or 'ketosamine') can dehydrate further to produce caramel-like compounds, or undergo hydrolysis to shorter chain products. If the original amine NH_2-R' happens to be asparagine, one of these products can be 3-APA.

A different route to 3-APA could be enzymatic decarboxylation of asparagine without the required participation of reducing carbohydrates (Figure 11.4). An LC–MS/MS analytical method for quantitation of 3-APA was used to analyze several potato cultivars and demonstrated that 3-APA is formed during storage of intact potatoes at moderate temperatures (20–35 °C) or after crushing the potato cells. Heating 3-APA under either aqueous or low water conditions at temperatures of 100–180 °C in model systems always generated more acrylamide than asparagines under the same conditions; the highest yields measured in the presence of carbohydrates (170 °C, aqueous buffer) were about 28 mol %, but in the absence of carbohydrates 3-APA was converted to acrylamide by about 63 mol % under the same conditions. Taken together, the results obtained (Granvogl 2004) suggested a reaction pathway to acrylamide other than Maillard reactions of asparagine, involving a combination of enzymatic conversion of asparagine to 3-APA that then generates acrylamide in high yields on heating. Although 3-APA is present in many foodstuffs at lower concentrations compared to free asparagine, the higher efficiency of conversion of 3-APA suggest its potential role as an additional precursor. Although in potatoes (the subject of this study, Granvogl 2004) 3-APA was not indicated to be a more effective precursor of acrylamide than the asparagine–carbohydrate Maillard reaction, once 3-APA is formed acrylamide can be effectively generated under aqueous conditions without proceeding via the Maillard reaction; this reaction might account for the observation of acrylamide formed from raw (unheated) foodstuffs with low asparagine content.

The other obvious question concerns the metabolism of acrylamide (AA) and its mechanism of action as a neurotoxin and possible carcinogen. Glycidamide (GA), the epoxide of acrylamide (Figure 11.5(a)), is believed to be the metabolite mainly responsible for its toxic

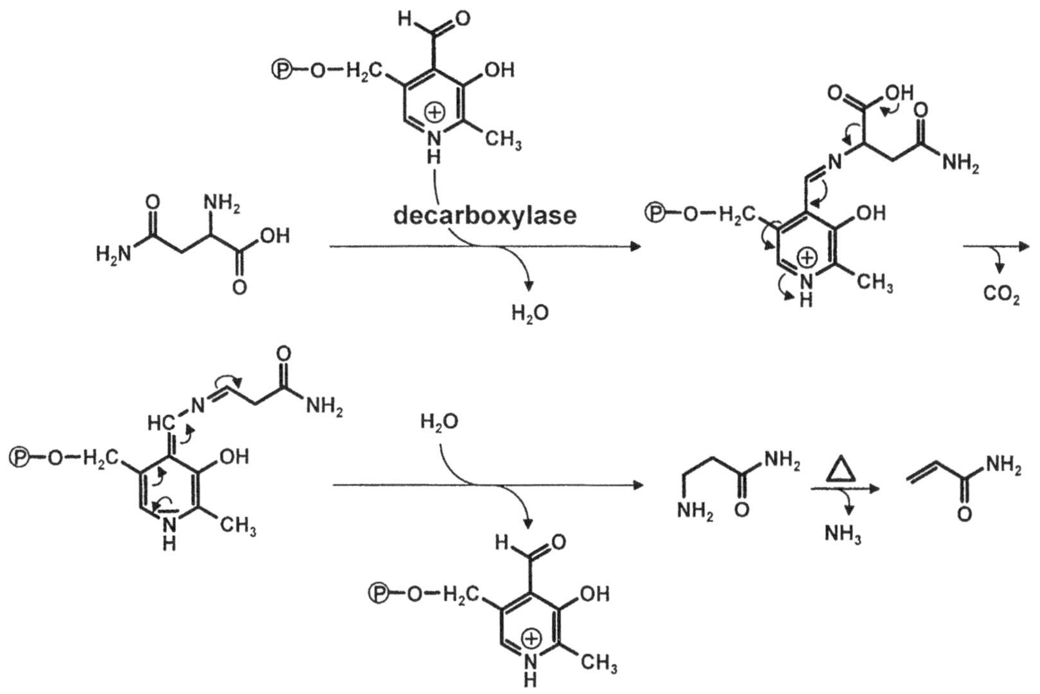

Figure 11.4 Proposed enzymatic pathway (decarboxylase with pyridoxal phosphate as co-factor) for conversion of asparagine into 3-aminopropionamide that is then thermally de-aminated to give acrylamide. Reproduced from Granvogl, *J. Ag. Food Chem.* **52**, 4751 (2004), copyright (2004) with permission from the American Chemical Society.

action, so measurement of *in vivo* levels of GA is essential for assessment of cancer risk arising from exposure to acrylamide. It was addressed (Paulsson 2003) by analyzing the concentrations in blood of the hemoglobin (Hb) adduct of GA, formed by reaction with the Hb N-terminal valine (Val). This was done by exploiting a modification of the Edman degradation reaction used in the traditional wet chemistry method for sequencing the amino acid sequences in proteins; using this well known chemistry but with pentafluorophenylisothiocyanate (PFPITC) reagent, the GA–Val adduct was detached from the Hb as a pentafluorophenylthiohydantoin (PFPTH, Figure 11.5(b)) and measured by GC–MS/MS using negative ion chemical ionization (NICI) as a high sensitivity method taking advantage of the high electron affinity conferred by the perfluorophenyl ring. However, the highly polar character of the GA–Val–PFPTH derivative made it necessary to modify the method via further derivatization by acetonization of the adjacent –OH and –NH$_2$ groups in the adduct (Figure 11.5(c)). Quantitation was possible through availability of N-(2-carbamoyl-2-hydroxyethyl)valine (GA–Val, the analytical standard) and N-(2-carbamoyl-2-hydroxyethyl)(d$_7$)valine–PFPTH

(i.e., GA–d$_7$Val–PFPTH, the internal standard). The sensitivity obtained was sufficient for studies of background adduct levels of GA in animals and humans. For example, GA–Val and AA–Val adducts in humans ($n = 5$) were measured (Paulsson 2003) as 26 ± 6 and 27 ± 6 pg per g Hb, respectively.

More recently, an LC–MS/MS method was reported (Boettcher 2005) for analysis of the mercapturic acids of both acrylamide and glycidamide in urine (Figure 11.6). Mercapturic acids are conjugates of both xenobiotic and toxic endogenous molecules with *N*-acetyl cysteine, formed in the liver and excreted through the kidneys into the urine. These metabolites are initially formed as conjugates with glutathione, a tripeptide (γ-glutamylcysteinylglycine, Figure 11.6(a)) with many important physiological functions including that of safe removal of toxic substances from the body. This method used negative mode ESI with MS/MS detection; a particularly complete description of the mass spectrometric parameters is provided (Boettcher 2005) but it must be borne in mind that the final optimized ESI–MS/MS conditions are always instrument dependent to some extent and the published conditions should be taken as a useful starting point for the optimization. The [M–H]$^-$ ions of the

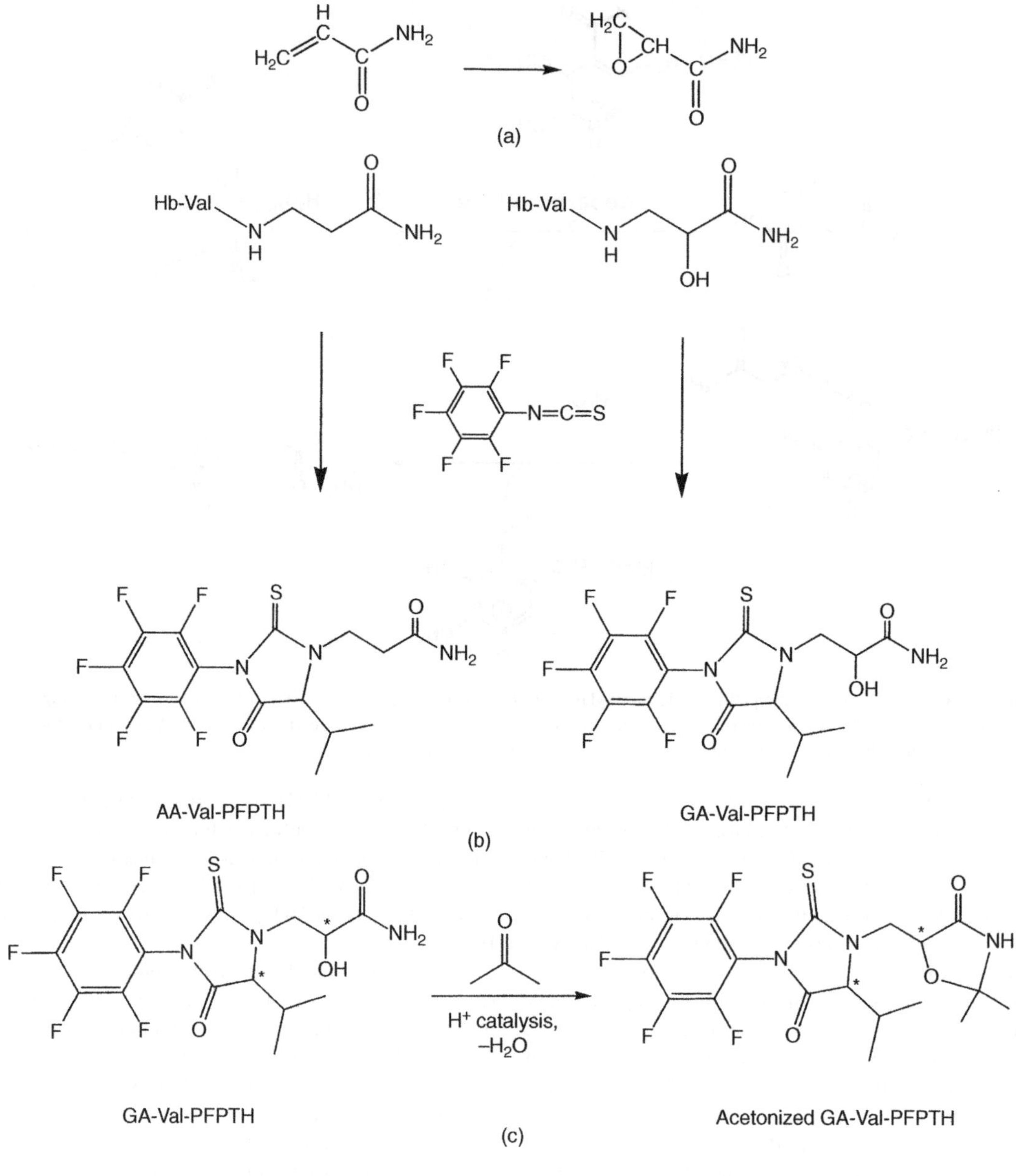

Figure 11.5 (a) *In vivo* metabolism of acrylamide (AA) to glycidamide (GA). (b) Adducts of AA or GA to the N-terminal valine in Hb followed by derivatization with pentafluorophenylisothiocyanate (PFPITC) and detachment of AA–Val–PFPTH or GA–Val–PFPTH. (c) Derivatization of GA–Val–PFPTH with acetone to form a less polar derivative suitable for GC, N-(2,2-dimethyl-4-oxazolidinonylmethyl)valine–PFPTH (acetonized GA–Val–PFPTH). Asymmetric carbon atoms are denoted with *. Reproduced from Paulsson, *Rapid Commun. Mass Spectrom.* **17**, 1859 (2003), with permission of John Wiley & Sons Ltd.

mercapturic acids (AAMA and GAMA, Figure 11.6(b)) yielded dominant fragment ions under the collision conditions corresponding to $NH_2–CO–C_2H_4–S^-$ (m/z 104) and to $NH_2–CO–CH(OH)–CH_2–S^-$, respectively, that were ideal for MS/MS monitoring; the d_3-isotopologs were used as internal standards and yielded the corresponding d_3 fragment ions. The detection limit, defined as the concentration yielding a S/B ratio of three for each quantifier MRM chromatographic peak, was estimated to be $1.5\,\mu g.L^{-1}$ for both analytes, and the limits of

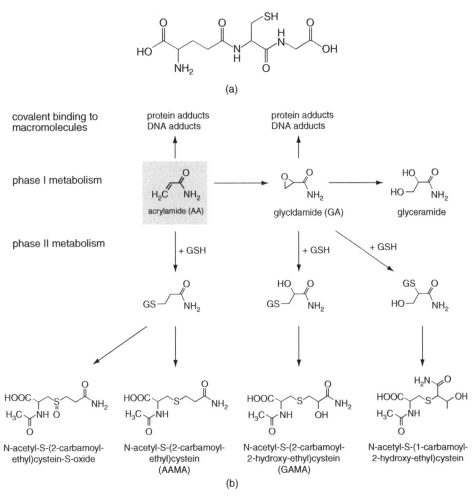

Figure 11.6 (a) Structure of glutathione (GSH). (b) Presumed scheme for metabolism of acrylamide (AA), and conversion of AA and its phase I epoxide metabolite glycidamide (GA) to the corresponding mercapturic acids (AAMA and GAMA) and other metabolites excreted into the urine; the enzymes responsible for the conversions are not shown. Reproduced from Fuhr, *Cancer Epidemiol. Biomarkers Prev.* **15**, 266 (2006), with permission of the American Association for Cancer Research.

quantitation, defined relative to a S/B value of 10, were thus estimated to be 5.0 μg.L^{-1}, with reproducibility between 2 % and 6 % for both analytes (intra- and inter-assay). It was important to be able to measure the mercapturic acid of GA, believed to be the ultimate carcinogenic agent derived from AA.

This method (Boettcher 2005) is suitable for determinations of these acrylamide-derived metabolites over a wide concentration range and is thus applicable for monitoring levels in the general population and also in people who are occupationally exposed to acrylamide. For example, results obtained for a small group of 29 persons out of the general population ranged from 5 to

338 μg.L^{-1} in urine. This analytical method was used subsequently (Boettcher 2005a, 2006; Fuhr 2006) in an extensive investigation of the toxicokinetics of acrylamide in humans. The most complete of these studies (Fuhr 2006) used a 0.94 mg dose of acrylamide ingested with food by human volunteers, and found that most of the dose is absorbed in humans to produce the mercapturic acid of acrylamide (AAMA, Figure 11.6) as the major metabolite. This observation suggests that detoxification by glutathione, together with elimination of unchanged acrylamide, is more efficient than formation of the more reactive epoxy metabolite glycidamide. Formation of the glycidamide mercapturic acid (GAMA) was estimated to

---

Shellfish Toxins

It is believed that the dangers of shellfish toxins were known thousands of years ago, as one of the seven plagues of Egypt: '— and all the water that was in the Nile turned to blood. And the fish in the Nile died and the Nile became foul, so that the Egyptians could not drink water from the Nile and there was blood throughout all the land of Egypt.' (Exodus **7**, 20–21). The phenomenon was also known to First-Nation North Americans, who considered it dangerous to eat shellfish when the water was colored red (whence the colloquial name 'red tide toxins' although the color can also be green or brown). Unfortunately, this native knowledge came too late for the crew of Captain George Vancouver, who in 1793 landed at a place still known as Poison Cove in the Fiordland area of British Columbia (a favorite side-trip location for cruises to Alaska via the inside passage). The crew ate a meal of fresh mussels and soon fell extremely sick; one unfortunate sailor, James Carter, died as a result and was buried in nearby Carter Bay.

Although many different shellfish toxins are now known (Figure 11.7), Vancouver's description of the symptoms makes it likely that this incident was caused by some of the paralytic shellfish poisoning (PSP) toxins. Incubation periods are short (5–30 min), with symptoms including numbness, tingling and burning sensations, beginning with lips, tongue and fingertips spreading to arms and legs, accompanied by dizziness and drowsiness with possible incoherence and general weakness; respiratory paralysis can ensue for heavy doses. The first 12 hours are critical for the outcome; complete recovery from lower doses usually takes a few days. The estimated acute lethal dose is 0.3–1 mg, depending on the characteristics of the person and the types of PSP toxins present in the filter-feeding shellfish that accumulate the toxins from the plankton on which they feed; a single mussel can contain up to ∼30 mg of PSP toxins. PSP toxins are potent, reversible blockers of voltage-activated sodium channels on excitable cells; no antidote is known, but emetics to induce vomiting together with respiratory support can increase chances of survival. Other kinds of shellfish toxins include those responsible for amnesic shellfish poisoning that induces memory loss and even coma and death, diarrhetic shellfish poisoning for which the symptoms are obvious from the name, and many others (Figure 11.7). Even today over 2000 cases of shellfish poisoning per year are reported worldwide, with ∼15 % proving fatal. In technologically advanced countries such occurrences are extremely rare in commercially grown or harvested shellfish thanks to careful monitoring by regulatory authorities, but cases do arise in which people disregard public warnings and collect and consume wild mussels, often from officially quarantined areas!

The traditional method of monitoring shellfish for PSP toxins was the 'mouse bioassay' prescribed by the Association of Official Analytical Chemists (AOAC International). Although this method has the advantage of being nonselective and integrative, and therefore well suited for protection of public health, the bioassay suffers from considerable variability and gives little information on toxin composition; there is also pressure from animal rights groups to discontinue such tests in which mice suffer and die, and some countries have banned the use of this test. As a result there is a strong effort to develop alternative analytical methods that do not involve sacrificing laboratory animals. Among the thousands of plankton species in the ocean (they constitute the first link in the food chain), only a few produce the secondary metabolites ('phycotoxins') that we know as shellfish toxins. Dangerous incidents can occur when heavy 'blooms' of these species are formed near shellfish-producing areas. Unfortunately, the frequency of toxic plankton blooms and their geographical range, as well as the content of the phycotoxins, appear to be increasing. While plankton blooms are generally completely natural events, their intensification has been attributed to fertilizer from agricultural run-off into coastal waters, worldwide distribution of previously localized plankton species in ships' ballasts, and uncontrolled international trade in shellfish.

---

be lower than in rats by a factor of at least two, and by a factor of at least four than in mice, at the same doses relative to body weight. These results of the LC–MS/MS analyses suggest that extrapolations of human cancer risk caused by oral acrylamide exposure, from similar studies in rodents, should be corrected by these factors since glycidamide is believed to be the agent largely responsible for the carcinogenicity of acrylamide.

11.2.2 Paralytic Shellfish Poisons

It is believed that paralytic shellfish poisoning (PSP) was known thousands of years ago as one of the seven plagues of Egypt (see accompanying text box on shellfish toxins). PSP is a wholly natural phenomenon, caused by the ingestion of filter-feeding shellfish that have fed on a bloom of particular species of plankton (mostly belonging to the genus *Alexandrium*) which produces a

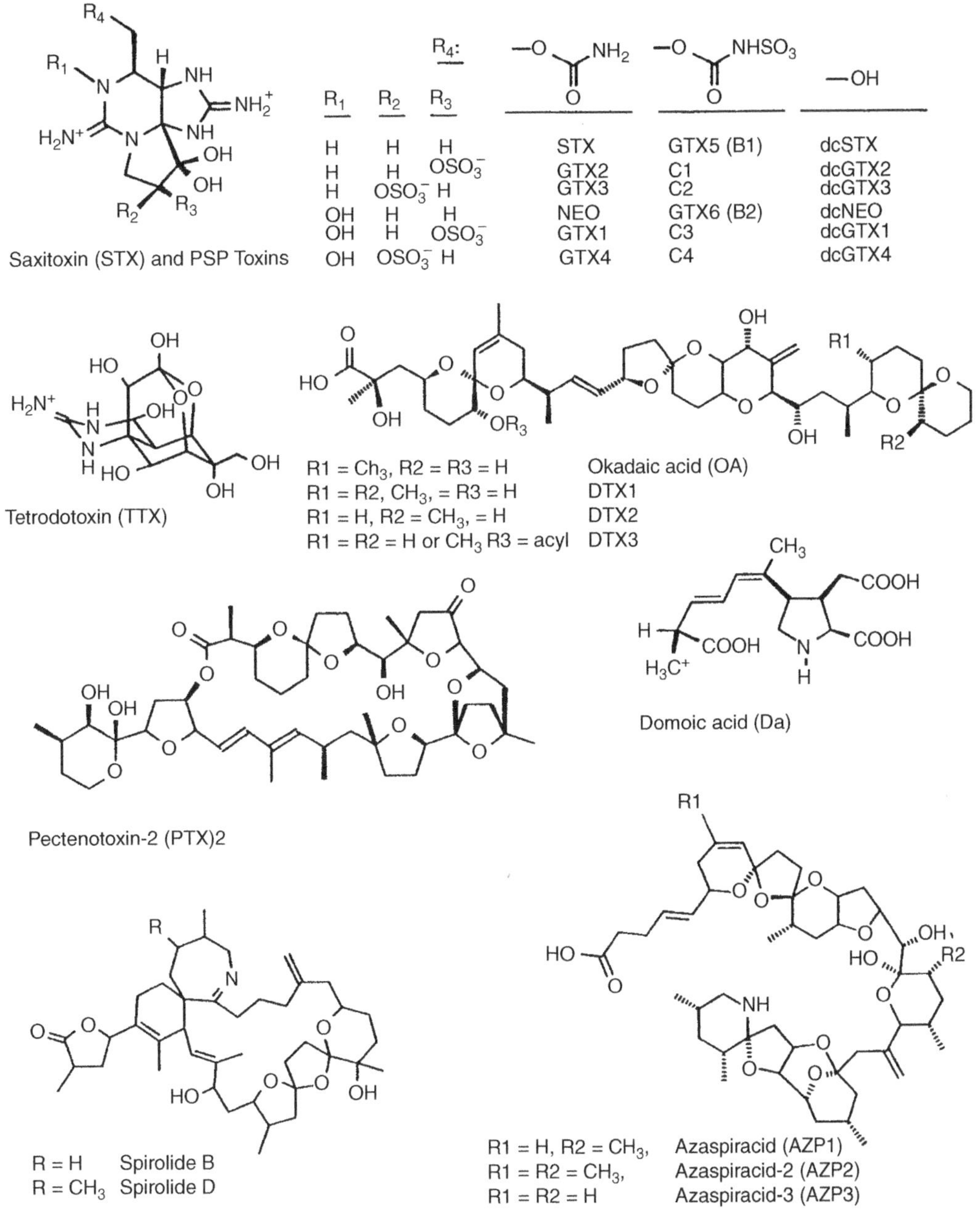

The structures of some natural shellfish toxins are shown.

Saxitoxin (STX) and PSP Toxins

R_1	R_2	R_3	$-O\overset{O}{\underset{}{}}NH_2$	$-O\overset{O}{\underset{}{}}NHSO_3$	$-OH$
H	H	H	STX	GTX5 (B1)	dcSTX
H	H	OSO_3^-	GTX2	C1	dcGTX2
H	OSO_3^-	H	GTX3	C2	dcGTX3
OH	H	H	NEO	GTX6 (B2)	dcNEO
OH	H	OSO_3^-	GTX1	C3	dcGTX1
OH	OSO_3^-	H	GTX4	C4	dcGTX4

Tetrodotoxin (TTX)

R1 = Ch₃, R2 = R3 = H   Okadaic acid (OA)
R1 = R2, CH₃, = R3 = H   DTX1
R1 = H, R2 = CH₃, = H   DTX2
R1 = R2 = H or CH₃ R3 = acyl   DTX3

Domoic acid (Da)

Pectenotoxin-2 (PTX)2

R = H   Spirolide B
R = CH₃   Spirolide D

R1 = H, R2 = CH₃,   Azaspiracid (AZP1)
R1 = R2 = CH₃,   Azaspiracid-2 (AZP2)
R1 = R2 = H   Azaspiracid-3 (AZP3)

Figure 11.7 Structures of some natural shellfish toxins. Reproduced from Quilliam, *J. Chromatogr. A* **1000**, 527 (2003), copyright (2003) with permission from Elsevier.

suite of secondary metabolites with structures related to that of saxitoxin, the best known member of the family (Figure 11.7). In fact, the PSP toxins are only one of several families of natural toxins produced by various plankton species and bio-accumulated in filter-feeding bivalves; these compounds (Figure 11.7) are structurally extremely diverse and present a major challenge to the analyst (Quilliam 2003). Probably the most difficult challenge is presented by the PSP toxins, in view of the highly polar (basic) functionality and extreme hygroscopic nature of these compounds that cause difficulties in obtaining accurate standard calibration standards (not weighable in view of the hygroscopic nature, Section 2.3). Moreover, the high polarity of these compounds definitively rules out GC separation and also makes reverse phase HPLC impossible without the use of ion pairing reagents that in turn present problems of greatly reduced sensitivity in ESI. In fact, development of suitable chromatographic separations for these compounds continues to be a problem and, as discussed below, ion exchange chromatography (Section 4.4.1c) and HILIC (Section 4.4.2c) have recently been employed as alternatives to strategies incorporating pre- or post-column chemical modification. On the other hand, the highly basic functional groups of the PSP toxins confer high responses in positive mode ESI and low resolution MS/MS (as with a triple quadrupole instrument) provides adequate selectivity. Accordingly the major problems faced in devising quantitative methods for PSP toxins did not involve mass spectrometry, but rather the difficulties in preparing calibration solutions and in devising efficient procedures for extraction and HPLC separation compatible with ESI–MS. It should be added that lack of any suitable SIS has thus far required that all quantitative analyses have been conducted using external calibration with clean solutions of standards.

The traditional method of analysis is a mouse bioassay (AOAC International 2000) in which an acidic extract of suspect shellfish tissue is injected intraperitoneally into a mouse of specified body weight; the read-out parameters are the resulting distress symptoms and the time before death that is usually inversely related to the toxin concentration. This method was developed (Stephenson 1955) as a 'blind' screening procedure before anything was known about the chemical nature of the toxins and is still invaluable in bioassay-directed fractionations that are a key step in identification of previously unknown toxins (Quilliam 2003). However, apart from the ethical considerations in subjecting animals to such treatment, this bioassay has a large intrinsic uncertainty ($\pm 20\%$) and insufficient sensitivity for most of the shellfish toxins other than those associated with PSP. Accordingly, chemical analytical methods have been sought for some time.

Until relatively recently a major problem in developing chemical analytical methods for PSP toxins was the unavailability of analytical standards. These compounds (Figure 11.7) are extremely hygroscopic and thus can not be weighed reliably even if the purity can be established somehow; it is impractical to synthesize them in useful quantities and they are found in nature as mixtures of closely related structures (Figure 11.7) that are not easy to separate. This is nowadays a major problem for regulatory laboratories that require certified analytical standards and reference materials in order to conduct accurate

analyses. Some commercial sources of these compounds have now been shown to be unsuitable for such applications and are really marketed for biological applications. However, a suite of 10 certified standard solutions including most of the PSP toxins is now available from the National Research Council of Canada (http://imb-ibm.nrc-cnrc.gc.ca/crmp/shellfish/psp_e.php). Preparation of these calibration standards required access to highly contaminated shellfish tissues and/or large scale laboratory cultures of the appropriate phytoplankton, together with meticulous separation and purification methods. Careful attention to degree of purity and stability of the toxins benefited greatly from the introduction of LC–MS methodology, which exploits the suitability of ESI for these highly polar compounds (Quilliam 1989). Accurate quantitation (Quilliam 2003) of these analytical standards involves cross-comparisons of results from several independent procedures, including gravimetry, nuclear magnetic resonance (Burton 2005) and separation methods particularly HPLC and capillary electrophoresis coupled with ESI–MS (Pleasance 1992, 1992a), but also UV–visible and fluorescence detection.

Unfortunately the family of highly polar PSP toxins can not be separated readily using conventional reversed phase HPLC since they are not retained and elute at the solvent front. It was possible (Quilliam 1998) to achieve reversed phase separation by adding a volatile ion pairing reagent (heptafluorobutyric acid) that formed overall-neutral ion pairs with the highly basic PSP toxins, thus permitting retention on the non polar reversed phase sorbent but at the cost of a large reduction in ESI sensitivity; this problem is further discussed below. The problems associated with the inability to accurately weigh these highly hygroscopic materials were solved by devising a proton-counting approach to determination of molar quantities of a compound specified by its high field NMR spectrum (Burton 2005) and was discussed previously (Section 2.3.1). The availability of these analytical standards has greatly improved the quality of PSP toxin analyses; unfortunately it has not yet been possible to certify a shellfish tissue certified reference material (CRM, Section 2.2.2) for the PSP toxins, although these are available for some of the less polar toxins (e.g., http://imb-ibm.nrc-cnrc.gc.ca/crmp/shellfish/crm/ASP-Mus-c_e.php).

The most commonly used chemical method for analysis of PSP toxins in shellfish is the combination of LC with on-line post-column oxidation by periodate under alkaline conditions, and with fluorescence detection (Sullivan 1987). In this method the PSP toxins are successfully separated by reversed phase HPLC by using an alkaline mobile phase (thus deprotonating the basic functionalities in the PSP toxins), consistent with conditions required for the post-column oxidation. A special reversed phase column was required that was effective and stable under these conditions, and a PRP-1 column based on a poly(styrene-divinylbenzene) cross-linked polymer was found to give adequate resolution. On oxidation the heterocyclic rings of the PSP toxins are converted to purine ring systems that give good fluorescence yields. This method (Sullivan 1987) provides sufficient selectivity and sensitivity that are adequate for routine protection of human health, but the system including the post-column oxidation reactor is difficult to set up and requires considerable daily maintenance.

Therefore, an alternative method, still based on periodate oxidation to fluorescent derivatives but now precolumn rather than on-line post-column, was devised (Lawrence 1991, 1991a; Janaček 1993). Since the oxidation products are appreciably less polar than the parent toxins they are more amenable to separation on a reversed phase column; indeed gradient elution on a C_{18}-silica phase, using a mobile phase containing aqueous ammonium formate (pH 6) with a gradient from $0-5\%$ acetonitrile, was found (Lawrence 1991a) to provide good peak shape and reproducibility, better than that provided by the PRP-1 column for the parent toxins (Lawrence 1991a). Moreover, because the periodate oxidation was conducted off-line before HPLC separation, the conditions could be optimized to improve yields of the fluorescent products, and a SPE clean-up was incorporated to reduce interference from co-extracted components of the shellfish tissue or plankton biomass. Later (Janaček 1993) further improvements were made including much improved sensitivity, and the entire oxidation–HPLC procedure was automated. However, a major problem with this pre-column oxidation procedure is that some PSP toxins yield identical oxidation products and others give multiple products, so that it is impossible to provide reliable toxin profiles. The identities of these oxidation products were elucidated (Quilliam 1993) using LC–ESI–MS but unfortunately the ESI–MS sensitivity was much poorer for the oxidation products than for the parent PSP toxins, so this approach in not competitive for trace analysis of these compounds. Nevertheless, despite its failure to distinguish between some members of the PSP toxin family, the pre-column oxidation method with fluorescence detection (Lawrence 1991, 1991a; Janaček 1993) does provide a sensitive, reproducible screening method that can be implemented and maintained with relative ease. Of course the post-column oxidation approach (Sullivan 1987) is capable of distinguishing among all the PSP toxins via their retention times.

The usual advantages of mass spectrometric detection, i.e. a unique combination of high sensitivity and selectivity, were for some time considered to be inapplicable to trace analysis of PSP toxins because of the difficulty of simultaneously satisfying the requirements of efficient chromatographic separation with those of good ESI efficiency. Problems with HPLC ion pairing agents (Quilliam 1998) were mentioned above and, while capillary electrophoresis (Section 3.7) with mass spectrometric detection (Section 4.4.6) is ideally suited to the analysis of the highly charged PSP toxins (Locke 1994), it is not possible to analyze all the toxins in a single analysis due to their different charge states. Moreover, the technique is susceptible to interference from co-extracted salts in real-world samples, and the extremely small injection volumes (a few nanolitres) result in LOD and LLOQ concentration values about an order of magnitude higher for CE–MS compared to those for LC with fluorescence detection. More recently two different approaches to the efficient coupling of HPLC with ESI–MS for PSP toxins were described.

The first of these methods (Jaime 2001) uses ion exchange (Section 4.4.1c) chromatographic separation of the toxins followed by electrochemical post-column oxidation, with subsequent detection by fluorescence and ESI–MS. A simple extraction procedure involved homogenization of algal or shellfish tissue with acetic acid, followed by centrifugation and filtration. A combination of an anion exchange column followed by a cation exchange column was used to separate toxins from co-extractives and from one another. Gradient elution was used, with both mobile phases consisting of aqueous ammonium acetate at pH 6.9 (at which the highly basic PSP toxins will be largely protonated) but at different concentrations (20 and 450 mM). The electrochemical oxidation (Jaime 2001) alleviates many of the experimental difficulties associated with the post-column chemical reactor (Sullivan 1987) and permits use of volatile buffers that provide good HPLC efficiency and are compatible with ESI–MS. A single quadrupole analyzer was used for ESI–MS detection and the evaporation of the entirely aqueous charged droplets (no organic solvent content) required assistance from a hot-air gun in view of the relatively low vapor pressure of water at room temperature.

Good linearity was observed using clean solutions of analytical standards, over ranges corresponding to about an order of magnitude for each of the seven PSP toxins investigated; unfortunately there is some ambiguity in the published work (Jaime 2001) in that the linear ranges are listed as corresponding to concentrations but the units given are nanograms (i.e. mass). The same is true of the LOD values (approximated via extrapolation of measured S/B values to a value of 3), that for fluorescence detection are quoted in the range 0.01–0.8 'ng' for the seven tested toxins and 0.5–2 'ng' for ESI–MS. Injection volumes were $100 \mu L$ into a mobile phase flow rate of $800 \mu L.min^{-1}$, so it seems likely that the quoted 'concentrations' in 'ng' are really the amounts injected on-column, since this would then correspond to concentrations in the low nM range, which is the range typically achieved by modern implementations of the post-column oxidation–fluorescence method (Oshima 1995). Whatever the correct interpretation might be, the much lower sensitivity values achieved by ESI–MS were likely due in part to the fact that the splitting of the LC eluent to the mass spectrometer was done after the electrochemical oxidation, i.e. it was the oxidation products of the PSP toxins that were detected by ESI–MS, and it is known (Quilliam 1993) that this greatly reduces the ESI–MS sensitivity. Figure 11.8 shows the combined SIM chromatograms obtained for a mussel tissue extract compared with that obtained (Jaime 2001) using fluorescence detection in parallel in the same experiment.

The other recent approach to use of LC–MS for analysis of PSP toxins (Dell'Aversano 2005) exploits the unique properties of hydrophilic interaction chromatography (HILIC, Section 4.4.2c) that are ideally suited to these highly polar analytes. Moreover, the high organic content of the mobile phase leads to ready evaporation of the charged electrosprayed droplets and thus high ionization efficiency.

Extraction of PSP toxins from plankton samples again used a simple acid extraction followed by centifugation and filtration (Dell'Aversano 2005), but the procedure used for contaminated shellfish tissue was slightly more complicated, involving three extractions of a homogenized sample with acidic acetonitrile:water followed by centrifugation and a SPE clean-up step. No experiments to estimate extraction–clean-up efficiencies were reported (Dell'Aversano 2005), but a large effort was expended in optimizing both the HILIC and ESI–MS efficiencies. For the first time the ESI–MS and ESI–MS/MS characteristics of all of the 20 most important PSP toxins (Figure 11.7) were reported. The various families of the PSP toxins exhibited different MS/MS fragmentation behaviour that is described in detail in the original work (Dell'Aversano 2005); interestingly, STX, NEO, dcSTX and dcNEO gave $[M+H]^+$ ions as the base peak in their ESI–MS[1] spectra, with no $[M+2H]^{2+}$ ions despite the fact that they are known to exist as dications in aqueous solution.

Two HILIC columns known to provide good retention for guanine (which contains the same highly basic functional group as the PSP toxins) were tested and eventually a TSK-gel Amide-80® column ($5 \mu m$ spherical silica particles covalently bonded with carbamoyl groups) was selected. The mobile phase had to be simultaneously optimized for both chromatographic and ESI efficiencies; the final choice used isocratic elution with a mixture of 35 % water plus 65 % of acetonitrile containing 5 % water, each containing ammonium formate and formic acid at concentrations 2.0 and 3.6 mM, respectively, at pH 3.5 where all PSP toxins are fully protonated. This apparently peculiar choice of mobile phase (i.e. why not simply specify the water:acetonitrile ratio) arose because of the need for extensive testing of HPLC conditions, including the possibility that gradient elution might be required while maintaining a constant buffer concentration.

Figure 11.9 shows the final results thus obtained for HILIC–MS separation of all 20 PSP toxin standards investigated and Figure 11.10 illustrates application of this analytical method to extracts of toxic plankton and contaminated wild mussel tissue; note that these two samples were collected from the same location on the same occasion, leading to the reasonable supposition that it was these plankton that had been responsible for contaminating the mussels (Dell'Aversano 2005). In this regard it is interesting that unknown compounds, labeled M1, M2 and M3 in Figure 11.10, were observed in the mussel sample but not in the plankton, and are thus reasonably assigned as metabolites and/or degradation products formed in the mussel. The M2 peak had an exact match of retention time and product ion MS/MS spectrum for 11-hydroxy-STX, but the structures of M1 and M3 were unknown at the time of publication.

The LC–MS/MS instrumental detection limits were roughly estimated by extrapolation of S/B values measured for the clean standard solutions to S/B = 3, and comparisons of the results obtained using different instruments nicely illustrates the advances in ESI–MS/MS technology over the last 20 years or so. The LOD values for the 20 toxins obtained using optimized MRM on the first commercial LC–MS/MS instrument (MDS–SCIEX API III, manufactured 1988) were in the range 40–7000 nM for an injection volume of $5 \mu L$, but when a modern version (API 4000) was used under identical conditions the LOD range was 5–30 nM, that compares favorably with values 5–100 nM obtained (Oshima 1995) for the LC–oxidation–fluorescence method.

Despite the meticulous attention to detail, this method (Dell'Aversano 2005) could not be finally validated at that time because of some residual problems. Thus, a slight shift of retention times was observed for sample extracts

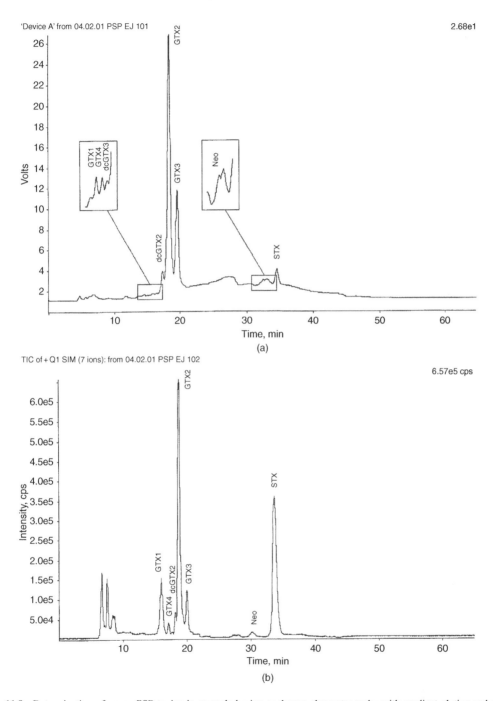

Figure 11.8 Determination of seven PSP toxins in mussels by ion exchange chromatography with gradient elution and parallel fluorescence and ESI–MS (100 μL of extract injected into a mobile phase flow rate of 800 μL.min⁻¹. (a) Fluorescence chromatogram; (b) TIC chromatogram (combination of individual SIM chromatograms). Striking differences in relative response factors are evident when comparing the two chromatograms. Reproduced from Jaime, *J. Chromatogr. A* **929**, 43 (2001), copyright (2001), with permission from Elsevier.

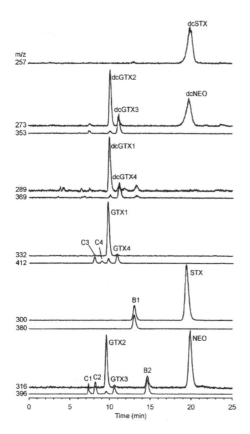

Figure 11.9 HILIC–MS separation of a standard mixture containing 20 PSP toxins, monitoring either protonated molecules or abundant fragment ions in SIM mode (Chromatographic conditions are described in the main text). Reproduced from Dell'Aversano, *J. Chromatogr. A*, **1081**, 190 (2005), copyright (2005) with permission from Elsevier.

compared with those for the standards and attributed to a matrix effect that worsened as more concentrated crude extracts were analyzed. Moreover, in analyses of sample extracts, observation of a 'hump' near the end of the chromatogram appeared to correlate with earlier elution with sharper peaks of the last eluting compounds, i.e. saxitoxin (STX) and neosaxitoxin (NEO); this made it difficult to match retention times of sample peaks with those of standards. The nature of the extracted material that causes this effect is currently unknown; the effect can be mitigated by dilution of the extract but at the cost of a corresponding reduction in sensitivity. Another problem is caused by high levels of salts in some crude extracts that can partially suppress ionization of the C-toxins (Figure 11.7). Development and validation of a quantitative method will follow development of clean-up procedures that will eliminate both the salt suppression effect

and the retention time shift effects (Dell'Aversano 2005). This method should then provide a gold standard reference method for PSP toxins as a result of its sensitivity combined with the high selectivity of HILIC–MS/MS; however, the cost of the apparatus required together with the skills required to maintain and operate it suggest that lower cost screening methods, such as LC–oxidation–fluorescence (Sullivan 1987; Oshima 1995; Jaime 2001), will still be important for routine public health protection.

An even quicker and cheaper screening test involving none of mass spectrometry, HPLC or mouse bioassay, suitable for use by operators untrained in analytical chemistry or indeed in any scientific discipline, would be of great benefit on shellfish farms or on-board ships at sea (scallop draggers). To be of practical use such a test would have to be shown to detect all of the different toxin analogs in the PSP toxin family in complex matrices like the acid extracts specified in the mouse bioassay method (AOAC International 2000), from a variety of shellfish tissue types; furthermore, it should be able to detect the majority of samples containing half of the regulatory PSP toxic level. Such a rapid screening test (MIST Alert[TM]) based on an immunoassay method (Laycock 2000), has been shown (Jellett 2002) to detect 100 % (no false negatives) of the toxic extracts, i.e. those containing at least $80\,\mu g$ saxitoxin equivalents (STX equiv.) in 100 g of shellfish tissue as measured by the mouse bioassay, in over 2100 regulatory samples. Only one potentially toxic sample, measured to contain 78 and 86 mg STX equiv./100 g tissue in two different mouse bioassays, was recorded as negative in one replicate MIST Alert[TM] assay; the test also detected the majority of extracts containing toxin levels greater than $32\,\mu g$ STX equiv./100 g, which is the mouse bioassay detection limit. False positive results, evaluated relative to the mouse bioassay, were an average of about 14 %; analysis (Jellett 2002) of these false positive extracts by the LC–oxidation–fluorescence method (Sullivan 1987; Oshima 1995) showed that many contained PSP toxicity in the range of 20–$40\,\mu g$ STX equiv./100 g, i.e. below the detection limit of the mouse bioassay.

This immunoassay (Jellett 2002) detects each of the PSP toxins with different efficiencies in an integrated manner and therefore it is impossible to specify an exact detection limit given the sometimes wildly different toxin profiles found in different parts of the world, but was stated (Jellett 2002) to be about $40\,\mu g$ STX equiv./100 g for the 'average' PSP toxin profile. These results were largely confirmed (Mackintosh 2002) in field tests, some of which were conducted by trained analysts and some by non-scientific personnel who were able to conduct and interpret the observations. This suggests that the MIST Alert[TM] test may be suitable as an initial screen for PSP

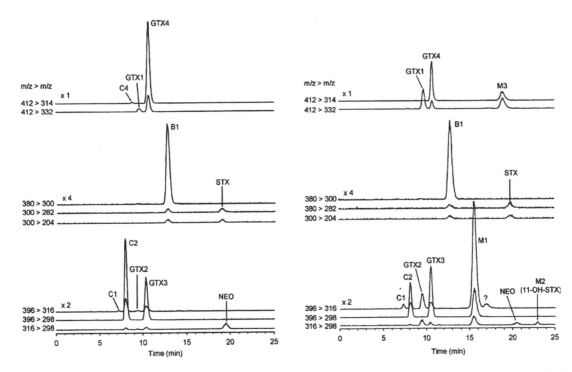

Figure 11.10 HILIC–MS/MS analyses (MRM mode) of an extract of (left side) a plankton sample (*Alexandrium tamarense*) and (right side) of tissue from contaminated mussels (*Mytilus edulis*), each containing several PSP toxins. HILIC conditions were the same as those used with analytical standards to obtain Figure 11.9. Some SRM chromatograms are plotted with expanded scales as indicated. Reproduced from Dell'Aversano, *J. Chromatogr. A*, **1081**, 190 (2005), copyright (2005) with permission from Elsevier.

toxins as part of routine monitoring programs, thereby greatly reducing the number of mouse bioassays and the overall cost of shellfish monitoring. Of course any such 'yes–no' screening method must be fully validated by comparison with more definitive measurements, up to and including the gold standard methods based on mass spectrometric detection.

11.3 Anthropogenic Pollutants in Water

The benefits of the rise of the chemical and pharmaceutical industries for human health and well being are clearly evident when examining the variation of life expectancy with time. Estimates of the life expectancy at birth in ancient Greece and Rome give ~28 years, rising to ~33 years in medieval Britain, to an average ~37 in Western Europe in the late 19th century, the latter increasing to >80 in the early 21st century, with a world-wide average ~66. There are also strong differences on the basis of geography (see the world distribution in 2003 at http://www.worldpolicy.org/globalrights/econrights/maps-life.html). Of course, other factors also enter the equation,

including gender, average income within a given country at a given period, infant mortality rate etc., but the trend in the life expectancy estimates for Western European countries over the 100 years or so in which these industries developed and grew (see above), and between Swaziland (~33 years) and Japan (~81 years) in the later 20th century, do seem to indicate a major effect. However, such progress does not come without prices of various kinds. One of these prices concerns the concentrations of anthropogenic pollutants in the water supply, and this section is concerned with this problem and the analytical challenges it presents. A recent comprehensive review (Richardson 2005) covers many kinds of pollutants found in waters, including drinking water. Environmental pollutants, and the application of mass spectrometry in detecting and quantitating them, are the subject of comprehensive biennial reviews (Richardson 2004, 2006).

11.3.1 Disinfection By-Products

One important class of anthropogenic pollutants in drinking water comprises the compounds produced

---

The Search for Safe Drinking Water

Safe drinking water is crucial to human survival and, unfortunately, even at the start of the 21st century literally billions of people do not have routine access to this basic need. The realization that drinking water can be a source of deadly disease can be traced to very early Sanskrit and Egyptian records. A fascinating summary of this history up till modern times (Jesperson 1996) can be found at the website of the US National Drinking Water Clearinghouse (www.nesc.wvu.edu/ndwc/ndwc_dwhistory.htm). The ancient Greeks and Romans are famous for their elaborate engineering schemes to transport clean water from less inhabited regions into cities. The Romans even introduced aeration of this water and constructed long supply routes including aqueducts over valleys and rough terrain to facilitate this transport. After the fall of the Roman Empire the tradition was continued by alchemists in the new Islamic society, who introduced stills for purifying water that used wicks made of a fibrous cord that could siphon water from one vessel to another. Unfortunately over the next 1000 years these engineering contributions to public health were forgotten and diseases like cholera, typhus, malaria and yellow fever took a terrible toll. In Western Europe one of the first steps forward was made in 1685 by Lucas Antonius Portius, an Italian physician who suggested a multiple sand filtration method. This approach was not exploited at the time, although a version was proposed by La Hire in 1703 to the French Academy of Sciences.

The first modern municipal water treatment plant was not installed (in Paisley, Scotland) until 1804; it consisted of concentric sand and gravel filters, and its distribution system consisted of a horse and cart! In 1807 Glasgow was the first city to pipe filtered water to consumers. By 1827 slow sand filters were put into use at Greenock (Scotland) and similar systems were completed in London in 1829. Subsequent developments of filtration technology (Jesperson 1996) improved matters still further but did produce notable decreases in deaths from cholera etc. A major step forward was the result of application of the scientific method by the English doctor, John Snow, who traced a local concentration of cholera cases in London to a water pump that became contaminated by an overflowing cesspool. This famous event is said to mark the founding of the science of epidemiology and is celebrated at a website (www.ph.ucla.edu/epi/snow.html) maintained by the Epidemiology Department at UCLA, that provides references to this major advance in public health (e.g. Brody 2000).

The microbial origins of waterborne diseases were identified by the work of Louis Pasteur and Robert Koch, a German scientist who in 1884 identified *Vibrio cholerea* as the causative agent of cholera. While proper filtration and installation of sanitary sewer systems greatly improved the situation with respect to waterborne diseases, it did not provide a perfect solution. The next major advance was the introduction of chemical disinfection of drinking water, made possible by the rise of the chemical industry in the 19th century. The disinfectant qualities of chlorine were first used in 1846 in the maternity ward of the Vienna General Hospital in Austria to prevent 'child bed fever.' In 1905 chlorine was added to London's water supply and a raging typhoid epidemic suddenly ceased. Chlorination of drinking water has saved untold numbers of lives and is still widely applied even when other approaches (ozone or UV irradiation) are the main strategies used for disinfection since chlorine has a usefully long lifetime in the distribution system, unlike ozone. But this great advance in public health comes with a price in the form of harmful compounds resulting from the reaction of chlorine with organic compounds present (often naturally) in the water. Identification and measurement of the resulting trace level chlorinated compounds continues to present a considerable challenge for the analytical chemist.

---

by chemical disinfection, particularly chlorination (Richardson 2003). These procedures represent one of the major advances in preventing water-borne diseases that for thousands of years had devastated the human race (see accompanying text box), but the byproducts arising from chemical reactions of the disinfection agents with organic compounds (many of natural origin) can be detrimental to health. A very complete discussion (WHO 2004) of the nature, origins and health effects of these compounds is supplemented by guidelines (WHO 2006) as to the maximum safe levels (Table 11.1). These levels are under constant revision, as are the similar (but not necessarily exactly equivalent) regulatory levels set for the most common disinfection byproducts (DPBs) by bodies such as the US EPA (http://www.epa.gov/safewater/mdbp/dbp1.html) and the European Union (http://www.epa.gov/safewater/mdbp/dbp1.html).

Despite this progress it is sobering to realize (Richardson 2002, 2003) that, although ~500 DBPs are currently known, few have been investigated for their

Table 11.1 WHO Guideline maximum concentrations (WHO 2006) for some disinfectants (first two entries) and disinfection byproducts in drinking water

Disinfectant or Disinfection By-product	Guideline Concentration
Chlorine (hypochlorous acid and hypochlorite)[a]	$5\,mg.L^{-1}$
Monochlororamine	$3\,mg.L^{-1}$
Bromodichloromethane[b][c]	$60\,\mu g.L^{-1}$
Dibromochloromethane[c]	$100\,\mu g.L^{-1}$
Bromoform[c]	$100\,\mu g.L^{-1}$
Chloroform[c]	$300\,\mu g.L^{-1}$
Dichloroacetic acid[b]	$50\,\mu g.L^{-1}$ (P)
Trichloroacetic acid	$200\,\mu g.L^{-1}$
Dibromoacetonitrile	$70\,\mu g.L^{-1}$ (P)
Dichloroacetonitrile	$20\,\mu g.L^{-1}$ (P)
Bromate[b]	$10\,\mu g.L^{-1}$ (P)
Chlorite	$70\,\mu g.L^{-1}$
Cyanogen chloride (as cyanide)	$70\,\mu g.L^{-1}$
2,4,6-Trichlorophenol[b]	$200\,\mu g.L^{-1}$

(a) For effective disinfection, there should be a residual concentration of free chlorine of $\geq 0.5\,mg.L^{-1}$ after at least 30 minutes contact time at pH < 8.0.

(b) For substances considered to be carcinogenic, the guideline value is the concentration in drinking water associated with an upper-bound excess lifetime cancer risk of 10^{-5} (one additional cancer per 100 000 of the population ingesting drinking water containing the substance at the guideline value for 70 years).

(c) For these trihalomethanes the sum of the ratios of the concentration of each to its respective guideline value should not exceed 1.

(d) For cyanide as total cyanogenic compounds.
Note that some byproducts included on earlier WHO guideline lists (e.g., chloral and chloral hydrate, formaldehyde, trichloroacetonitrile) are not included as they are now considered to be present in drinking water at concentrations far below those considered to present a health risk.

quantitative occurrence and health effects (DBPs that have been quantified in drinking water are generally present at ng–$\mu g.L^{-1}$ levels and, until recently, studies of health effects have been directed primarily toward linking chronic DPB exposure with cancer initiation or mutagenicity). Further, currently only ~50 % of total chlorinated organics formed during chlorination of drinking water have been identified, and the same is true of ~60 % of the 'assimilable organic carbon' compounds in ozonated drinking water (Richardson 2003). The initial (pre-disinfection) nature of the water, especially pH and concentrations of bromide, iodide and natural organic matter (surface water tends to contain a lower dissolved organic load than groundwater), can significantly affect the levels and types of DBP species formed, e.g. ozonization can significantly reduce or eliminate trihalomethanes and halogenated acetic acids (Table 11.1)

but can produce bromate (a potent carcinogen) in source water with high levels of bromide (Richardson 2003).

11.3.1a MX [3-chloro-4-(dichloromethyl)-5-hydroxy-2(5H)-furanone]

This compound (Figure 11.11) is one of the most potentially dangerous DBPs found in chlorinated drinking water. It was first discovered and shown to be highly mutagenic in effluents from paper mills that use the Kraft

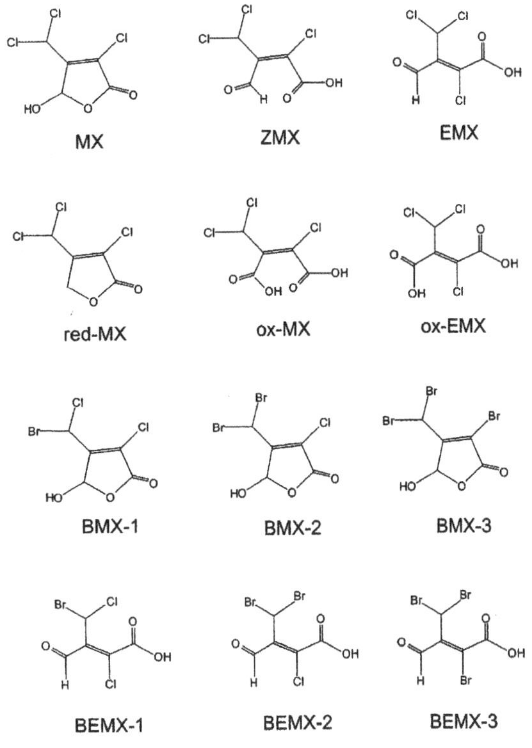

Figure 11.11 Structures of MX, i.e., 3-chloro-4-(dichloromethyl)-5-hydroxy-2(5H)-furanone, and related compounds observed as disinfection byproducts. Chemical names are: *Z/E*-2-chloro-3-(dichloromethyl)-4-oxo-butenoic acid (ZMX/EMX), 3-chloro-4-(dichloromethyl)-2-(5H)-furanone (red-MX), (E)-2-chloro-3-(dichloromethyl)-butenedioic acid (ox-MX), (E)-2-chloro-3-(dichloromethyl)-4-oxobutenoic acid (EMX), 2,3-dichloro-4-oxobutenoic acid (mucochloric acid), 3-chloro-4-(bromochloromethyl)-5-hydroxy-2(5H)-furanone (BMX-1), 3-chloro-4-(dibromomethyl)-5-hydroxy-2(5H)-furanone (BMX-2), 3-bromo-4-(dibromomethyl)-5-hydroxy-2(5H)-furanone (BMX-3), (E)-2-chloro-3-(bromochloromethyl)-4-oxobutenoic acid (BEMX-1), (E)-2-chloro-3-(dibromomethyl)-4-oxobutenoic acid (BEMX-2), (E)-2-bromo-3-(dibromomethyl)-4-oxobutenoic acid (BEMX-3). Reproduced from Richardson, *Trends Anal. Chem.* **22**, 666 (2003), copyright (2003) with permission from Elsevier.

chlorination process (Holmbom 1981,1984) and later was shown (Hemming 1986; Backlund 1988; Kronberg 1988) to be present also in chlorinated drinking water and in the products of various chlorinating agents with humic waters. Humic substances are major constituents of soil organic matter that contribute to the chemical and physical quality of soil; they are complex colloidal supramolecular mixtures that have never been separated into pure components. The fraction of humic substances that is soluble in water under all pH conditions is called fulvic acids, the fraction that is not soluble in water under acidic conditions (pH < 2) but is soluble at higher pH values is called humic acids and is the major component of extractable humic substances, and the fraction that is not soluble in water at any pH value are the humins. Since then MX has been shown to account for up to 50 % of the mutagenic activity measured in chlorinated drinking water (Kronberg 1988a) and has also been shown to be a multi-site carcinogen in laboratory rat strains (Komulainen 1997) with an estimated cancer potency 170 times greater than that of chloroform (Wright 2002). The mutagenic activity of MX is affected by pH and temperature, which together control the equilibrium between the closed hydroxyfuranone form of MX and its open chain isomeric form EMX (Figure 11.11). Mutagenicity is usually evaluated using some form of the Ames Test (Ames 1973, 1973a; Maron 1983), which uses several strains of the bacterium *Salmonella typhimuium* that carry mutations in genes involved in histidine synthesis so that they require histidine for growth. The ability of the mutagen to cause a genetic reversion to the native state in which histidine synthesis is restored is tested via observed growth on a histidine-free medium. The procedure is not a test for carcinogenicity *per se* but is used as a fast and cheap screen for mutagenicity as an indicator for possible carcinogenic activity.

Analytical methods for MX are complicated from the outset by the fact that a large number of structurally similar halogenated compounds, many not yet identified, are present at different concentrations in source and drinking waters. Current best practice, described below, uses a well developed clean-up followed by derivatization (methylation) for GC–HRMS using a double focusing magnetic sector instrument (Section 6.4.4) with EI and SIM. This may seem like a cumbersome approach and it might be asked why some form of LC–MS/MS could not be applied thus avoiding the need for derivatization. The answer to this question is partly historical (the MX problem was first addressed in 1986 before LC–MS/MS instruments became commonly available) and partly because MX and its structural relatives (Figure 11.11) in

complex real-world samples are not well suited to efficient separation by HPLC compared with the high resolution separations available with capillary GC. Having made the decision to stay with derivatization and GC–EIMS, however, further complications arise as a result of multiple methylation sites yielding derivatized extracts that are particularly complex. As a result, high resolution SIM is necessary to provide sufficient specificity, as opposed to MS/MS, because one of the critical ions monitored for MX is in fact a mass doublet of two fragment ion species whose m/z values differ by only 0.0186 Th. The crucial point is that the relative abundances of these two ions can vary significantly with experimental conditions, so that any low resolution MS or MS/MS method that measures the sum of the unresolved abundances of these two ions will be inherently of low accuracy and precision. This aspect is discussed in some detail later. An additional difficulty is the current unavailability of an isotopolog SIS of MX, so that less satisfactory internal standards must be used; this is also discussed below.

Analytical methods for MX and related compounds have evolved over the years (Hemming 1986; Charles 1991, 1992; Kanniganti 1992; Suzuki 1995; Smeds 1997; Zwiener 2001; Wright 2002) but in general have used some variant of an early procedure (Hemming 1986) in which the water samples to be analyzed are acidified to pH < 2 to stabilize the MX (conversion to the open chain forms is complete at pH ~5.3) and extracted using an XAD-4 or XAD-8 resin; the latter, a macropous resin composed of an acrylic ester polymer of intermediate polarity and large internal surface area, is generally preferred. After a simple derivatization MX is sufficiently volatile and stable to be amenable to GC separation and ionization by electron ionization (EI) and, as discussed below, provides an exemplary case in which low resolution SIM is unable to provide unambiguous quantitative results and high mass resolution is essential.

Analytical methods for MX at present are still limited, as in the earliest work (Hemming 1986), by the unavailability of an isotope-labeled SIS and various analogous compounds have been used instead (Section 2.2.3). An acceptable analog internal standard, 2,3-dibromo-4-oxo-2-butenoic acid (HOOC–CBr=CBr–CHO, mucobromic acid, MBA), was originally used (Hemming 1986), and this or its chloro-analog have been the most popular choices. In a modern variant (Smeds 1997) of the original procedure, a 7 litre sample of the water to be analyzed was adjusted to pH 2 with concentrated hydrochloric acid and passed through a column filled with 50 mL of XAD-8 resin at a flow rate of 0.3 bed volumes.min^{-1}. The XAD-8 resin had been pre-washed with 1 M sodium hydroxide (NaOH), methanol (MeOH) and ethyl acetate (EtOAc) and stored in methanol; before use the resin was conditioned with distilled water. The adsorbed organics were eluted with six bed volumes of EtOAc, and the combined EtAc extracts were concentrated to a final volume adjusted to 300 μL per L of original water (2.1 mL for a 7 L sample). At this stage the MBA internal standard was added at a concentration corresponding to 40 ng/L of original water (Smeds 1997), rather than before extraction by the XAD resin, for reasons not made clear although later reports (Wright 2002) have also added the internal standard after the extraction. The original work (Hemming 1986) spiked the internal standard into the water sample before extraction and stated that extraction of MX by the XAD resin was quantitative, as confirmed later (Wright 2002). Presumably this is the justification for adding the internal standard after extraction from water but it is still unclear why it could not be added initially as a check on the claim of 100 % extraction efficiency.

Because of the free hydroxyl group on the MX structure (Figure 11.11) the compound is not amenable to GC analysis without derivatization. All methods appear to use methylation under acidic conditions as described (Hemming 1986), following spiking with internal standard, evaporation of the EtOAc solution to dryness and re-dissolution in acidified methanol to provide a final extract solution suitable for injection into a GC. The internal standard thus corrects for losses during derivatization (with the extraction efficiency assumed to be 100 % as discussed) and also plays the role of a volumetric internal standard (Section 8.5.2a) in the subsequent analysis by GC/MS.

Note that MBA and mucochloric acid (its chlorinated analog) were investigated as internal standard previously, but were shown (Charles 1992) to give two products on methylation. One of these was the expected methyl ester, but the other contained three methoxy groups and was attributed to a structure in which the carboxyl group is methylated as expected, but the aldehyde group has been converted to a dimethoxy moiety $-CH(OCH_3)_2$. Moreover, mucochloric acid can be formed on chlorination of humic waters and so can not be guaranteed to be absent from the water samples to be analyzed. Nonetheless, later work (Smeds 1997; Zwiener 2001; Wright 2002) used MBA as in the original work (Hemming 1986). This formation of multiple derivatization products from the internal standard is not necessarily a concern as long as it is established that the relative yields of the mono- and tri-methoxy derivatives of MBA in the derivatization reaction are reproducible and known. In HRMS–SIM the EI response of MX relative to that of the internal standard was determined (Hemming 1986) using a standard solution containing both MX and MBA. However, as pointed

out previously (Charles 1992), this method is only semi-quantitative because it relies on the response of a single standard of MX to that of MBA, i.e. only a single point calibration. For this reason a six point calibration curve was generated (Charles 1992) using standard solutions of methylated MX in the range pg.μL^{-1} plus, in each case, 50 pg.μL^{-1} of methylated $^{13}C_6$-benzoic acid used mainly as a volumetric internal standard (VIS, Section 8.5.2a). This external calibration approach (i.e. not matrix matched calibrators and without a true SIS) implies that the recoveries (F'_a in Section 8.5) are not known but at least matrix effects are not a concern in the case of GC–EIMS.

As for the 'dioxin' analyses (Section 11.4.1), the complexity of the extracts of real-world samples containing MX is such that a high degree of selectivity in the mass spectrometric detection is required. The most popular choice has been high resolution ($\sim10^4$ at 10 % valley) SIM using a double focusing instrument (Section 6.4.4) although some preliminary experiments (Zwiener 2001) have explored the possibility of screening with a 3D ion trap in MS^2 mode. Choice of HRMS–SIM (rather than LRMS) for mass spectrometric detection was justified (Charles 1992) on the basis that one of the ions monitored, that at m/z 201, is in fact a mass doublet (Charles 1991, Figure 11.12); one component corresponds to the expected $(M–CH_3O)^+$ fragment ion $(C_5H_2O_2{}^{35}Cl_2{}^{37}Cl)^+$, m/z 200.9091), and the other to a $(M–CHO)^+$ ion $(C_5H_4O_2{}^{35}Cl_3$, m/z 200.9277). This is an exemplary case where the use of high resolution SIM possesses an essential advantage over low resolution, since the relative abundances of the $(M–CH_3O)^+$ and $(M–CHO)^+$ ions were shown to change with ion source conditions, so that a calibration obtained using m/z 201 at low resolution might not be applicable when the water sample extracts are analyzed. Furthermore, use of HRMS allows use of criteria for compound identification that set a limit on the deviation between theoretical and measured abundance ratios of isotopolog ions. The abundance ratio of the ions at m/z 199 and 201 is 62:100 at low resolution (Figure 11.12(a)) but, by using high resolution and monitoring the appropriate member of the doublet at m/z 201, i.e., $C_5H_2O_2{}^{35}Cl_2{}^{37}Cl^+$ at m/z 200.9091, the ratio is observed (Figure 11.12(b)) to be 100:98 which matches the theoretical ratio for a Cl_3 species. Confirmation of analyte identity can thus be established using HRMS–SIM on the basis of the theoretical abundance ratio of the ions monitored together with a retention time criterion, but this would not be possible with LRMS since the ratio would deviate from the theoretical value in a manner dependent on the $(M–CH_3O)^+/(M–CHO)^+$ abundance ratio. In fact,

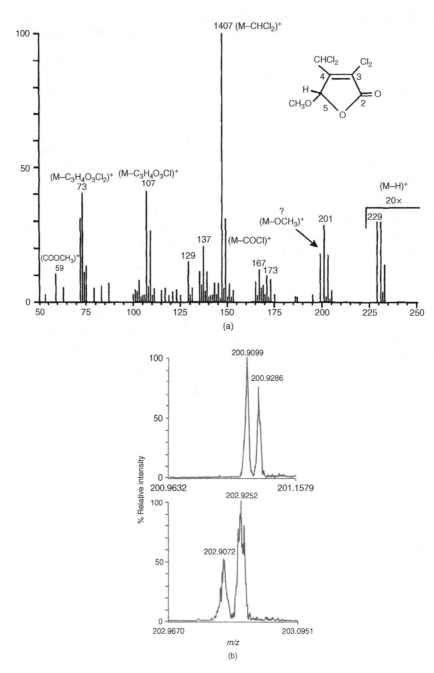

Figure 11.12   70 eV EI mass spectrum of the methyl derivative of 3-chloro-4-(dichloromethyl)-5-hydroxy-2(5H)-furanone (MX). (a) Low resolution spectrum; the fragment ion at m/z 199 has been attributed to the loss of (OCH)$^\bullet$ from the $^{35}Cl_3$-molecular ion at m/z 230, and if so the Cl_3 isotopic pattern should be similar to that observed in the m/z 229 fragment ion (M–H)$^+$. This is clearly not the case since the intensities of the peaks at m/z 221 and 203 are much too high. The discrepancy arises because of an overlapping isotopic distribution for (M–OCH$_3$)$^+$ fragment ions. (b) Partial mass spectra obtained at resolving power 20 000, demonstrating the existence of mass doublets at m/z 201 and 203, arising from overlap of the (M–CHO)$^+$ and (M–CH$_3$O)$^+$ isotopic distributions. Reproduced from Charles, *Biol. Mass Spectrom.* **20**, 529 (1991), with permission of John Wiley & Sons Ltd.

use of LRMS can actually miss the presence of MX (i.e. a false negative result, as indicated by the highly selective HRMS analysis) as a result of intense interferences at *m/z* values falling within the LRMS window at the MX retention time (Figure 11.13).

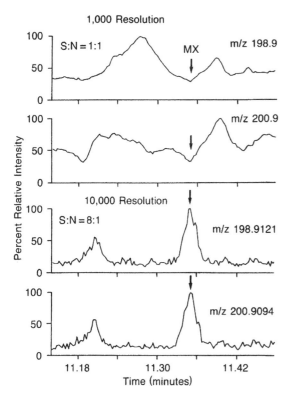

Figure 11.13 GC–MS chromatograms acquired in analysis for MX of a derivatized extract of a chlorinated water sample from a municipal facility. The use of LRMS–SIM did not detect the MX at all (S/N $\leq$ 1), but HRMS permitted confirmation of identity via the abundance ratio of the isotopomeric ions monitored, and also led to quantitation of the MX. Reproduced from Charles, *Env. Sci. Technol.* **26**, 1030 (1992), copyright (1992) with permission from the American Chemical Society.

Initially a large variability in the relative response factors of the methyl derivative of MX to that of isotopically labeled benzoic acid was observed as a function of relative concentrations, indicating a nonlinear response of the analyte relative to that of the VIS. This nonlinear behavior was shown (Charles 1992) to not arise from competition between MX and the benzoic acid standard in the derivatization reaction (the methylating agent was in very large excess), but rather from a peculiarity of the SIM

procedure when a double focusing magnetic sector instrument is used. As a consequence of the impossibility of switching a magnetic field rapidly so as to reproduce the transmitted *m/z* values consistent with a resolving power of 10^4 (10 % valley definition), it is common practice to leave the magnetic field B fixed and instead switch the accelerating voltage V_{acc} as the latter is not subject to hysteresis effects (Section 6.4.4):

$$m/z = [(e/m_u).(r_M^2/2)].(B^2/V_{acc}) \qquad [6.25]$$

Thus, with B fixed the transmitted *m/z* value is inversely proportional to V_{acc}. In the case of methylated MX (*m/z* 199/201 monitored) using $^{13}C_6$-methyl benzoate (*m/z* 128) as VIS, the *m/z* ratio for the molecular ions is ~1.56 and so this is also the ratio for the values of V_{acc} required to successively transmit analyte and standard ions. This is a very large ratio in this context, since such a large change in V_{acc} strongly affects the electric fields within the ion source itself and also in the extraction region where the ions emerge and are focused into the analyzer. In turn this results in very large effects on the transmission efficiency of ions from source to analyzer, and it was this that caused the observed nonlinearity (Charles 1992). The problem was overcome by monitoring the benzoic acid ions with one setting of the magnetic field B, and the MX ions with another so that only small variations in V_{acc} were required; this was easy to arrange since the retention times were well separated, but is more complicated in the dioxin case (Section 11.4.1).

The analytical method for MX was partially validated (Charles 1992). Recoveries ~100 % were obtained by analyzing water samples spiked with 12.6 ng.L^{-1} of MX. An 8 % RSD was observed in triplicate injections of a derivatized extract, demonstrating instrument stability and reproducibility, but in analysis of replicate aliquots of a sample of chlorinated water from a municipal supply, spiked at 12.6 mg.L^{-1} of MX the RSD was 32 %, demonstrating much poorer overall method reproducibility. Identification of the methyl derivative of MX in the sample extracts was based on the relative retention times (varied <1 %) of analyte and internal standard and the abundance ratio (variations from the theoretical value <10 % for HRMS–SIM) of the ions monitored for MX. To measure the instrumental detection limit, a 42 pg.μL^{-1} standard solution of methylated MX was repeatedly analyzed over three days, yielding S/B values ranging from 5:1 to 9:1; if these values are extrapolated linearly back to the origin, S/B values of 3:1 provide estimates of the instrumental detection limit of ~14–25 pg.μL^{-1} of MX in the derivatized extract solution (Charles 1992). The method detection limit was estimated by analyzing a chlorinated water sample for MX, giving a value of 38.6 ng.L^{-1} of MX in the original water sample; again assuming that this single data point can provide the basis for a linear extrapolation back to the origin, a method detection limit of $\sim 0.6\,\text{ng.L}^{-1}$ of MX in the water sample can be estimated on the basis of a S/B value of 3:1. This estimate implies that method detection limits of around one part in 10^{12} can be obtained by extracting a 2L water sample and concentrating the extract to $100\,\mu\text{L}$. In fact, MX concentrations in the range $2\text{--}40\,\text{ngL}^{-1}$ were measured in the water samples (Charles 1992), but no statistically defensible estimates of LOD or LLOQ were reported.

Some variations of this general approach (Hemming 1986; Charles 1992) have been described. Methanol with sulfuric acid does not methylate the di-acidic MX analogs (ox-MX and ox-EMX, Figure 11.11), diazomethane does not methylate MX, but boron trifluoride in methanol was found to successfully derivatize all these compounds (Kanniganti 1992). The initial exploration of a 3D trap/MS (Zwiener 2001) as a low-cost replacement for, or complement to, HRMS–SIM with a double focusing analyzer appeared to provide similar detection limits and selectivities although this preliminary evaluation did not involve validation in anything like the depth of the HRMS work (Charles 1992). It must be added that, although this investigation (Charles 1992) was exemplary in many ways, the quality of the validation was not up to the best modern standards. The LOD values were crude estimates without any attempt at deducing confidence limits at a specified confidence level (Section 8.4), no stability studies (including freeze–thaw studies) were reported and the entire problem was (and still is) hampered by lack of a fully satisfactory internal standard. On the other hand, the question of a regulatory limit is presently unclear. The World Health Organization (WHO 2006, p.414) states that a health-based maximum allowable value of $1.8\,\mu\text{g.L}^{-1}$ can be estimated for MX based on cancer studies on rats; this is significantly larger than the concentrations typically found in drinking water (usually $\leq 60\,\text{ng.L}^{-1}$). For this reason the WHO considered it unnecessary to propose a formal guideline value for MX in drinking water, thus implying that fully validated methods should not be necessary. However, as emphasized previously (Richardson 2003), MX levels as high as $80\,\text{ng.L}^{-1}$ were found (Wright 2002) in drinking waters from Massachusetts and a US Nationwide DBP Occurrence Study (Weinberg 2002) found levels of MX frequently $>100\,\text{ng.L}^{-1}$ and as high as $310\,\text{ng.L}^{-1}$) in drinking waters. Further, BMX and BEMX (Figure 11.11) were also found at levels as high as 170 and $200\,\text{ng.L}^{-1}$, respectively. As a result it is possible that regulatory limits for MX-like compounds might be set in the future.

11.3.1b N-nitrosodialkylamines

These compounds, frequently referred to as 'nitrosamines', have the structure $R_2N\text{--}N = O$. The simplest member of this family is *N*-nitrosodimethylamine (NDMA), another low mass high polarity analyte that has been known for many years to be carcinogenic to many laboratory animal species (Barnes 1954; Magee 1956), and has been shown (Magee 1962) to alkylate rat liver nucleic acids *in vivo*. NDMA is a yellow liquid that is moderately soluble in water and is stable in the absence of light, strong oxidizing and reducing agents, and strong bases. The MSDS entry for this compound emphasizes its extreme toxicity (see the public health statement for NDMA at http://www.atsdr.cdc.gov/toxprofiles/phs141.html). NDMA is very harmful to the liver of animals and humans; people who were poisoned on one or several occasions with unknown levels of NDMA in beverage or food died of severe liver damage accompanied by internal bleeding. Animals that ate food, drank water, or breathed air containing high levels of NDMA over a period of days or several weeks also developed serious, non-cancerous, liver disease, and when animals ate food, drank water or breathed air containing lower levels of NDMA for periods of more than several weeks, liver cancer and lung cancer as well as non-cancerous liver damage occurred. Although there are no confirmed reports of NDMA causing cancer in humans, it is reasonable to assume that exposure to NDMA could do so.

Nitrosamines are produced in foodstuffs from reactions of nitrites (added as preservatives to bacon and other processed meat products) with amines in (or derived from) the foodstuff, particularly during cooking, e.g., frying; anti-oxidants such as ascorbic acid (vitamin C) are added to such foodstuffs to inhibit formation of nitrosamines. A source of nitrosamines that may be of interest to some analytical chemists is beer, in this case attributed (Sen 1983) to reaction of nitrogen oxides with alkaloids (usually present in germinated malt) during the drying process. NMDA can also be formed inadvertently in a number of industrial processes.

Because of the very high toxicity of NDMA, sensitive analytical methods (suitable for levels in parts per 10^{12} range) have been developed, all involving GC separation. Various mass spectrometric detection strategies have been employed to overcome the problem of possible direct interferences at the low *m/z* values employed (molecular mass of NMDA is 74 Da), as described below. Earlier methods (e.g. Billedeau 1987; Tomkins 1995) used GC with detection by the so-called 'thermal energy analyzer', which is in fact a GC detector based on a chemiluminescent reaction of ozone with nitric oxide (NO) produced by on-line pyrolysis of nitrosamines (and also of nitro-compounds when pyrolyzed at higher temperatures); this approach does possess some degree of selectivity but not enough to satisfy modern requirements for analyte identification. One of the first reports of GC–MS quantitation of nitrosamines (Stephany 1976) used glass capillary columns (this was before the introduction of fused silica columns, Section 4.4.3) with detection by EI–MS in SIM mode at a resolving power ~4000 (10 % valley definition), but without the benefit of an internal standard. It was also shown (Gaffield 1976) that use of chemical ionization (CI) in positive ion mode significantly increased the sensitivity of MS detection of nitrosamines.

A more recent example of this GC–CI–MS approach (Longo 1995) described a quantitative method for NDMA in beer samples. The analyte was isolated by distillation and subsequent extraction from the distillate using dichloromethane, and quantitation was achieved using $[d_6]$-NDMA as internal standard. The analysis monitored the $(M + H)^+$ ions of analyte (*m/z* 75) and internal standard (*m/z* 81) at unit mass resolution using a single quadrupole analyzer; this low resolution detection had low selectivity, as indicated by the SIM chromatogram for *m/z*

75 that contained a number of peaks. Fortuitously there did not appear to be any close co-elutions with the NDMA analyte in this case (Figure 11.14), but for more complex matrices additional detection selectivity (via either high resolution SIM or MS/MS) is essential (see below). An instrumental (not method) detection limit, defined (Longo 1995) as the lowest analyte amount yielding a signal equal to the blank signal plus three standard deviations of the blank, was about $0.04 \mu g.kg^{-1}$ (8 pg injected), i.e. about 40 parts in 10^{12}. The concentration of NDMA in one particular beer sample was $0.862 \mu g.kg^{-1}$ with an RSD ($n = 5$) of 0.73 %, but no other indication of LLOQ was reported.

The occurrence of NDMA as a drinking water contaminant (Mitch 2003; Andrzejewski 2005) includes some dramatic instances in California in 1998 that were traced to drinking water wells near a testing facility for rocket engines that used unsymmetrical dimethylhydrazine $((CH_3)_2N–NH_2)$ as fuel. Groundwater concentrations of NDMA as high as $400\,000\,ng.L^{-1}$ on-site, and $20\,000\,ng.L^{-1}$ off-site, were detected (Mitch 2003; California Department of Health Services 2006). On a

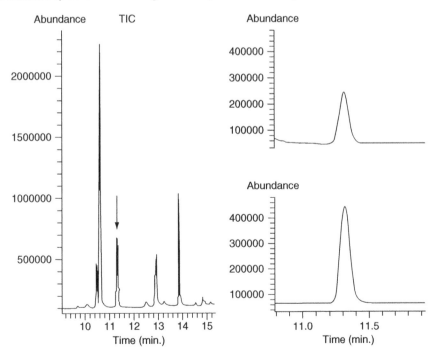

Figure 11.14 Analysis of the extract of a beer sample by GC–EI–SIM at unit mass resolution for NMDA (*m/z* 75) and its d_6-labeled internal standard (*m/z* 81). The total ion current (TIC) chromatogram is the sum of the SIM chromatograms for *m/z* 75 (top) and *m/z* 81 (lower); the extra peaks in the TIC trace are almost all due to *m/z* 75 but no co-elutions with NDMA are apparent. The concentration measured for NDMA in this beer sample was $1.36 \mu g.kg^{-1}$ (1.36 parts in 10^9). Reproduced from Longo, *J. Chromatogr. A* **708**, 303 (1995), copyright (1995) with permission from Elsevier.

less dramatic note, NDMA was first discovered as a disinfection by-product in chlorinated drinking water in Ontario (Jobb 1992, 1994; Taguchi 1994, 1994a; Graham 1995; Jenkins 1995) as a result of follow-up investigations after a discovery in 1989 of NDMA concentrations of 80000–10000 ng.L^{-1} resulting from industrial pollution (Taguchi 2007). These investigations required development (Taguchi 1994; Jenkins 1995) of sensitive and selective analytical methods for NDMA present in water down to the low ng.L^{-1} level (a few parts in 10^{12}), with sufficiently high throughput to meet the demands of the Ontario monitoring program on raw and treated drinking water as well as wastewater streams.

Liquid–liquid extraction was found to be inefficient, so a SPE extraction using a carbonaceous Ambersorb 572 resin was used in conjunction with d$_6$-labeled NDMA as surrogate internal standard (SIS). In this method it is important to treat the adsorbent gently; gentle rolling with no shaking with the water sample was specified in order to avoid pulverization to powder of the Ambersorb 572 granules. It is believed (Taguchi 2007) that this was the reason for the disappointing performance of this extraction method reported in an independent investigation (Tomkins 1996). The analytes are adsorbed onto the surface and into the pores of the granules and then desorbed into dichloromethane in the autosampler vial (Taguchi 1994a; Jenkins 1995); no further clean-up is required if these precautions are taken. Use of EI with a single quadrupole mass filter at unit mass resolving power gave GC chromatograms for *m/z* 74 that contained several extraneous peaks, some of which co-eluted with NDMA even when different GC columns were tried (Taguchi 1994a). These interferences, attributed to common pollutants like chlorobenzene, ethylbenzene and toluene, were resolved when a double focusing magnetic instrument was used at a resolving power of 2300 (10 % valley definition). However, the ion at *m/z* 74 arising from EI of methyl esters requires a resolving power of 6600 to be distinguished from NDMA, so a value of 7000 was used as a precaution (Taguchi 1994, 1994a).

Despite all these (and other) precautions, significant nonlinearity was observed in the response curves (GC peak area vs. pg injected), initially attributed to adsorption on active sites in the liner of the splitless injector and on the GC column. In the original work (Taguchi 1994, 1994a) use of the peak area ratio for analyte:SIS with a multi-point calibration permitted good quality data to be obtained in the range 5–200 ng.L^{-1}, with a MDL (EPA definition, Section 8.4) of 1 ng.L^{-1} determined from analysis of QC samples spiked at 10 ng.L^{-1}. Difficulties were reported (Taguchi 1994a) in obtaining a true blank, since HPLC-grade water usually contained 1–2 ng.L^{-1} of

NDMA; later attempts (Jenkins 1995) used the Ambersorb 572 resin SPE procedure and high purity water obtained using a reverse osmosis–ion exchange–UV exposure system, that gave much lower NDMA blanks. More recently, use of a more modern double focusing instrument with greatly improved resolution–transmission characteristics has permitted use of resolution of 10 000 (10 % valley definition). The calibration curves are now linear over the desired range with a calculated method detection limit in the 0.4–0.6 ng.L^{-1} range and a reporting detection limit of 0.9 ng.L^{-1} (Taguchi 2007).

Since the early work many improvements in analytical methodology for NDMA in water have been reported (Mitch 2003; Richardson 2003, 2004, 2006), mainly associated with improved SPE media and different mass spectrometric approaches to increase detection selectivity. Chemical ionization has come to be preferred over EI, e.g. a method using positive ion ammonia CI with low resolution SIM, in conjunction with a SPE procedure with two adsorbent beds in series (the carbonaceous Ambersorb 572 plus LiChrolutEN, polymeric (styrene-divinylbenzene), was reported (Charrois 2004) to give good results for several alkylnitrosamines in water. However, the increased selectivity afforded by CI in combination with MS/MS detection appears to be preferred, e.g., in an investigation of precursors of NDMA in chlorinated waters, methanol CI was used together with GC–MS/MS in an ion trap (Mitch 2003a). More recently (Munch 2006) a method using SPE on activated coconut charcoal (80 to 120 mesh size) with dichloromethane as the eluent to concentrate 0.50 L water samples to 1 mL was combined with GC–CI–MS/MS using a PTV injector (Section 4.4.4b) for large volume (20 μL) injection and methanol or acetonitrile as CI reagent; d$_6$-NDMA was again used as surrogate internal standard. This method (Munch 2006) was used as the basis for the US EPA method 521 for seven nitrosamines (including NDMA) in drinking water (EPA 2004a). The tandem mass spectrometer is specified as either a triple quadrupole or an ion trap, but must be capable of scanning at a minimum from *m/z* 40 to 160 with a complete scan cycle time (including MS/MS scan overhead) of ≤ 0.5 seconds such that a minimum of five scans is recorded across each analyte chromatographic peak (though 7–10 scans are recommended). The detection limit for this NDMA method (EPA 2004a) was determined using the EPA MDL method (Glaser 1981; EPA 1997) on laboratory blanks (purified water, $n = 8$) fortified at 1.0 ng.L^{-1}, to be 0.28 ng.L^{-1}, and the values for the other six nitrosamines were all <0.7 ng.L^{-1}. The lowest concentration minimum reporting levels (LCMRLs) were also determined for the seven target nitrosamines to be in the range 1.2–2.1 ng.L^{-1}. (In EPA terminology the method reporting limit (MRL) is the lowest analyte concentration that demonstrates a selected quantitative quality. The LCMRL (EPA 2004b) is the single laboratory determination of the lowest true concentration for which future recoveries are predicted with high confidence (99 %) to be measured between 50 and 150 %.)

A very complete review of the mechanisms of formation of NDMA, its occurrence and means for mitigating its formation (Mitch 2003), emphasizes the complexity of the situation. Nonetheless, some clear-cut conclusions have been possible. One route to NDMA involves nitrosation, via formation of nitrosyl cation NO$^+$ or other reactive nitrogen containing species such as dinitrogen trioxide (N$_2$O$_3$) as a result of acidification of nitrite:

$$HNO_2 + H^+ \leftrightarrow H_2O + NO^+ \qquad [11.4]$$

The nitrosyl cation then reacts with an amine such as dimethylamine to form NDMA:

$$NO^+ + (CH_3)_2NH \rightarrow (CH_3)_2N\text{-}N = O + H^+ \qquad [11.5]$$

This route to NDMA occurs most efficiently at pH ~3.4, a compromise between increased protonation of nitrite (pK$_a$ of HNO$_2$ = 3.35) and a decreasing fraction of dimethylamine in the reactive deprotonated form, with decreasing pH (pK$_a$ of H$_2$N(CH$_3$)$_2$$^+$ = 10.7). Formation of NDMA in fish, and especially in meat products treated with nitrite to prevent growth of dangerous bacteria on storage, is believed to occur via the nitrosation mechanism summarized in Equations [11.4 and 11.5]. Of course, consumption of nitrite as a component of cured meats (that provide the proteins as raw materials for dimethylamines) can thus lead to formation of NDMA in the acidic environment of the stomach. For this reason many food regulatory agencies limit the amount of nitrite that can be added to processed meats, and some manufacturers add ascorbic acid (vitamin C) to provide a reducing agent to minimize NDMA formation in the stomach. This mechanism can also contribute to NDMA formation in the environment at pH values close to neutral (Mitch 2003), but does not appear to be a major contributor in this case. Recently it was demonstrated (Mitch 2002; Choi 2002, 2002a) that a major route to NDMA from dimethylamines involved unsymmetrical dimethylhydrazine (UDMH) as an intermediate (not as a pollutant as in the rocket fuel case) in reaction with chloramines that can arise either from chlorination of amines in drinking water or as the chlorinating agent itself. The mechanism proposed by these workers is shown as Figure 11.15; consistent with this mechanism, removal of ammonia prior to chlorination

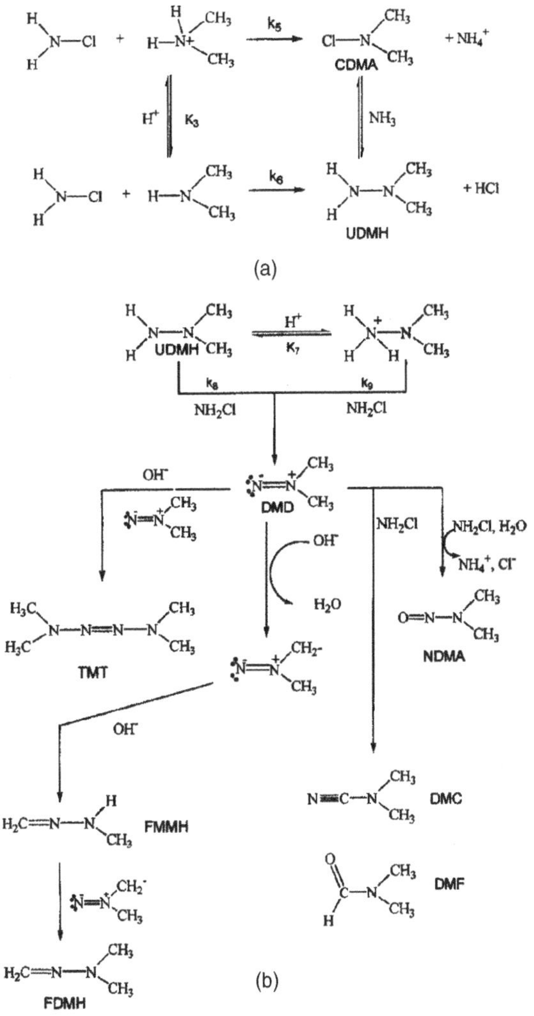

Figure 11.15 (a) First mechanism proposed for formation of NDMA by chlorination of water via reaction of chloramines with a dimethylamine to form chlorodimethylamine (CDMA) and thence unsymmetrical dimethylhydrazine (UDMH). (b) Further reactions of UDMH with chloramines to form NDMA plus other identifiable reaction products dimethyldiazene (DMD), tetramethyltetrazene (TMT), dimethylcyanamide (DMC), dimethylformamide (DMF), formaldehyde dimethylhydrazone (FDMH), formaldehyde monomethylhydrazone (FMMH). Reproduced from Mitch, *Env. Sci. Technol.* **36**, 588 (2002), copyright (2002) with permission of the American Chemical Society.

using hypochlorite (ClO$^-$), thus preventing formation of chloramines, reduces NDMA formation by an order of magnitude (Mitch 2003).

As a result of its chemical structure dimethylamine is expected to be the most effective organic

nitrogen precursor for formation of NDMA by both
the nitrosation (Equations [11.4 and 11.5]) and UDMH
(Figure 11.15) pathways, and indeed this has been demonstrated (Mitch 2003). Tertiary amines containing the
dimethylamine functional group (e.g., trimethylamine and
dimethylethanolamine) also can produce NDMA but at
lower yields, whereas monomethylamine, the quaternary
tetramethylammonium ion, and amino acids or proteins,
do not form significant concentrations of NDMA in
reaction with chloramine. Dimethylamine can arise from
various sources including industrial activity, and its
concentrations in wastewater influent have been found
(Mitch 2003) to be in the range 20–80 $\mu g/L^{-1}$. However,
it is readily degraded by bacteria and, as a result, concentrations in wastewater effluents are generally low (average
$\sim 4 \mu g/L^{-1}$) so that it could account for only $\sim 10\%$ of
the NDMA formed when wastewater effluent was chlo-
raminated. Clearly other precursors of NDMA are important. Recently (Schreiber 2006) a re-examination of the
mechanism of formation of NDMA in chlorinated water
has indicated critical roles for dichloramine ($NHCl_2$) and
dissolved oxygen. In this revised pathway for formation of
nitrosamines, dichloramine reacts with secondary amines
to form chlorinated unsymmetrical dialkylhydrazine intermediates that can be oxidized by dissolved oxygen to
form nitrosamines in competition with oxidation by chloramines (Figure 11.16). Even when pre-formed monochloramine was added nearly all NDMA formed could be
attributed to this new mechanism from traces of dichloramine formed via disproprtionation of monochloramine:

$$2\,NH_2Cl \rightarrow NHCl_2 + NH_3 \qquad [11.6]$$

Such mechanistic studies are important as they provide
information crucial to development of water treatment
technologies that minimize formation of nitrosamines
while still providing safe disinfection of drinking water.
Clearly still more needs to be done on this issue, and
development of analytical methods for NDMA precursors and reaction intermediates will continue to play a
key role.

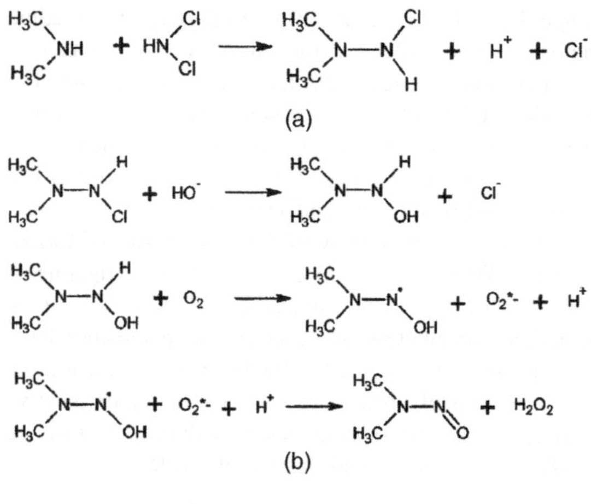

11.3.2 Multi-residue Methods for Pharmaceutical Residues

Multi-residue analytical methods are those designed for
circumstances where several analytes are to be measured
in the same procedure. These multiple analytes could be
closely related (e.g., antimicrobial compounds of a given
structural type), or be used for a common purpose (e.g.,
antimicrobials of all kinds), or an even broader class of
compounds (e.g., pharmaceutical compounds in general).

Figure 11.16 Revised mechanism for formation of NDMA
from dimethylamine during water chlorination by chloramine or
by other chlorinating agents with ammonia present. The first
step in disproportionation of chloramines to dichloramine (Equation [11.3]). (a) The dichloramine then reacts with dimethylamine in a fashion analogous to the reaction of chloramine in
the original mechanism (Figure 11.15) to form chloro-NDMA.
(b) Subsequent oxidation of the substituted hydrazine by oxygen
competes strongly with oxidation by chloramine. Reproduced
from Schreiber, *Env. Sci. Technol.* **40**, 6007 (2006), copyright
(2006) with permission of the American Chemical Society.

Clearly, different compromises must be applied compared
with those for analytical methods that target only one or
two specific compounds, with respect to analysis time,
degree and level of validation criteria including accuracy
and precision, recovery, confirmation of identity etc. The
broader the class of compounds to be analyzed, the lower
the level of validation that is possible, unless it is accepted
that large amounts of time and resources are available.
Usually broad class multi-residue analyses are designed as
semi-quantitative screening methods usually followed by
identity confirmation. This section describes some examples of such methods developed for pharmaceutical and
pesticide residues in water.

Pharmaceutical drugs contribute significantly to the rise
in life expectancy particularly in industrialized countries
and also in the general quality of life. These are, by
definition, highly bioactive substances that may express
negative side effects in some patients as a result of
intrinsic variations in human biochemistry (or in that of
animals in the case of veterinary drugs). Another safety
issue concerns the drug's therapeutic index (therapeutic
ratio), the ratio of the dose required to produce a toxic
effect divided by the therapeutic dose. A commonly used
measure of therapeutic index is the lethal dose of a drug

for 50 % of the population (LD_{50}) divided by the effective dose for 50 % of the population (ED_{50}).

Less well known is the potential for either the active pharmaceuticals or their metabolites to enter and accumulate in the environment; data to support environmental risk assessments are required to support registration of pharmaceutical products in North America, Europe and other countries. For example, in the United States of America formal assessments are required to be submitted to the FDA for any new drug with projected use that could result in a surface water concentration above one part in 10^{12} ($1\,ng.L^{-1}$). A recent assessment (Schwab 2005) of health risks associated with concentrations (measured or estimated, Anderson 2004) of 26 representative pharmaceuticals in surface waters in the USA concluded that the concentration levels in question are well below the no-effect values. This conclusion was based on a strict statistical analysis of the very complete toxicology data from laboratory tests and the information from human clinical trials that are required by regulation for drug registration. However, the conclusion could also be rationalized (Schwab 2005) on the basis that safe exposure levels for active pharmaceuticals are normally directly related to the therapeutic dose. Thus, the preferred safety profile for pharmaceuticals is that the therapeutic effect should be the first effect observed, i.e., at the lowest dose, and indeed the therapeutic dose for the most sensitive patients is usually used as the starting point for estimation of the acceptable daily intake and thence the predicted no-effect concentration (PNEC) for each pharmaceutical. On this basis there is also a positive direct correlation between the PNEC and the total amount of the drug entering the environment. For a given use rate by the population (and thus a given rate of excretion of the drug and its metabolites) only low production volumes are needed for potent pharmaceuticals with a low therapeutic dose; for the same population use rate, a low potency drug with a high therapeutic dose requires higher production volumes. That is to say, the total amount of a pharmaceutical drug entering the environment is generally inversely correlated to its potency, and the two effects tend to cancel one another.

Despite these encouraging conclusions for a limited set of test compounds (Schwab 2005), concerns remain about the long term effects of low concentrations of the wide range of pharmaceutical drugs and their metabolites that can end up in drinking water, e.g. the potential effect of antimicrobial residues on the rise in antibiotic resistance among bacteria, and of oral contraceptive residues on fertility of humans and other species. Accordingly, considerable effort is devoted to development of analytical techniques capable of quantitating pharmaceutical residues in water at the $ng.L^{-1}$ level. The following discussion starts

with an example of a fully validated method for two specific antibiotics, to be contrasted with the lower levels of validation possible for true multi-residue methods discussed later. In general, mass spectrometry is the only detection technology possessing a suitable combination of applicability, sensitivity, and selectivity, for multi-residue analyses other than those for members of a narrowly defined class of closely related compounds.

11.3.2a Validated Method for Macrolide Antibiotics

A major question with regard to pharmaceutical residues in water concerns the effectiveness of their removal in the treatment of wastewater. A recent study (Yang 2006) of the concentrations of two widely used macrolide antibiotics, erythromycin (ETM) and tylosin (TLS) (Figure 11.17(a)) in the influent and effluent streams of two wastewater treatment plants, is an excellent example of development of a validated analytical method for such applications. These two compounds are the most important macrolide antibiotics approved for use in human and veterinary medicine in North America but, while they have been widely used in the prevention and treatment of disease, they have also been widely used as feed additives to promote growth in animal feeding operations and this is likely a major source of such residues in waters. In general macrolides are metabolized to only a minor extent except in the case of erythromycin (ETM) whose main metabolite is an antibacterially inactive product corresponding to loss of water ($ETM–H_2O$). This dehydration reaction also occurs readily under the acidic conditions used to extract and analyze these compounds (Volmer 1998), so in this work (Yang 2006) ETM was analyzed only as ($ETM–H_2O$), i.e. no information was available concerning the relative amounts of the two forms in the original sample. Spiromycin I is a macrolide antibiotic structurally related to the present analytes but is not approved for use in either human or veterinary medicine in the USA, which is where the water treatment plants of interest were located (Yang 2006). As a very similar compound not found in the analytical samples of interest, it was thus suitable for use as internal standard in this work.

Extraction of the analytes used an SPE procedure, and these water soluble compounds were obvious candidadtes for analysis by LC–ESI–MS/MS (positive ion mode was used). Considerable pains were taken (Yang 2006) to optimize these steps and these details are not repeated here. However it is worth pointing out that a 3D ion trap was used as the MS/MS detector, certainly advantageous from the viewpoint of capital cost but with the penalty of a much lower duty cycle (Section 6.4.5a) than a triple

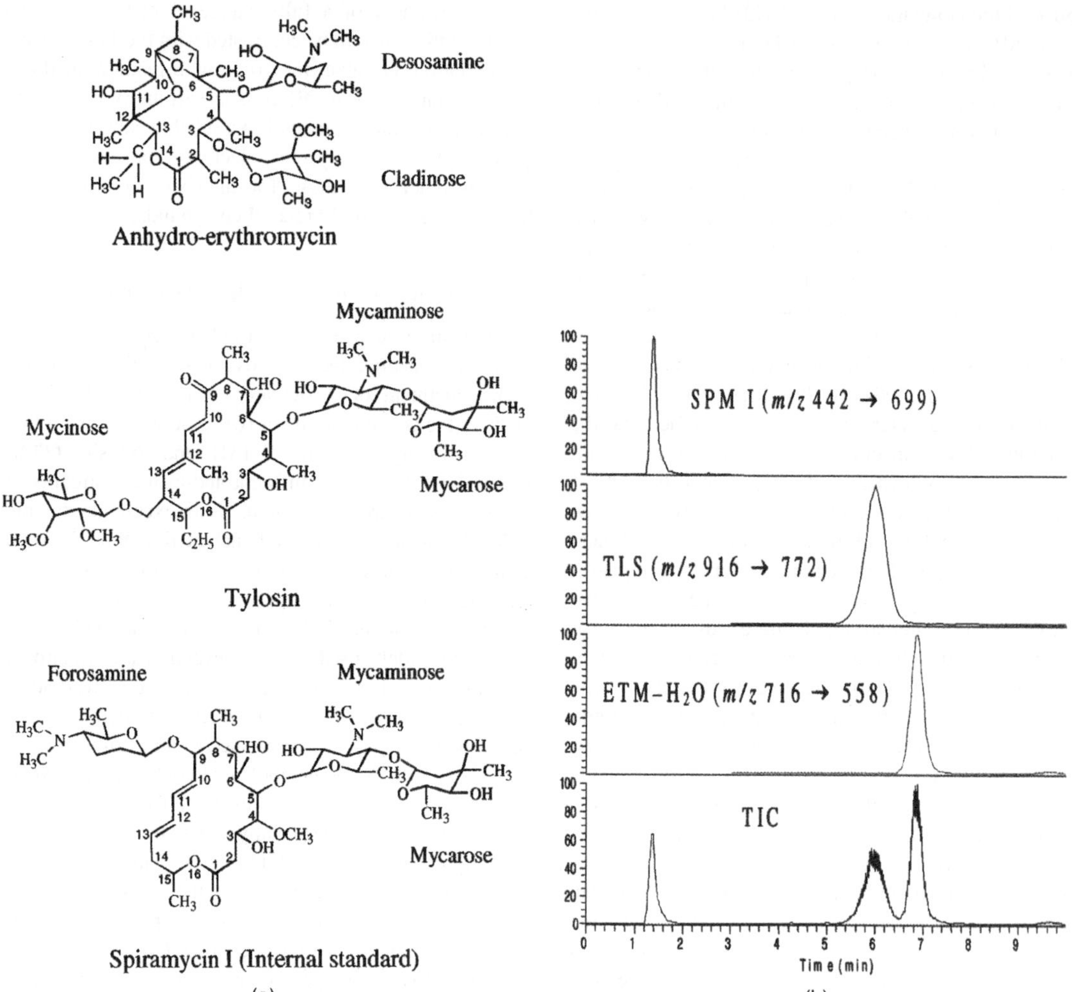

Figure 11.17 (a) Chemical structures of macrolide antibiotics (note that erythromycin, ETM, was analyzed as its anhydro form shown, (ETM–H$_2$O)), but tylosin (TLS) was analyzed as the intact protonated molecule. Spiromycin (SPM) was used as the internal standard. (b) Multiple reaction monitoring (MRM) chromatograms of macrolides, each spiked at a concentration of 300 ng.L^{-1} before extraction from 100 mL of influent wastewater from a water reclamation plant. The MRM transitions used for quantitation are shown in each case (the precursor ion for Spiromycin I was the (M+2H)$^{2+}$ ion, the others were (M+H)$^+$ ions), and the total ion current (TIC) chromatogram is simply the sum of the three MRM traces. Reproduced from Yang, *Anal. Bioanal. Chem.* **385**, 623 (2006), with permission from Springer Science and Business Media.

quadrupole in MRM mode and thus also higher LLOQ as well as limited accuracy and precision. The isolation widths of the precursor ions for MS/MS were quite large (3–4 Th), presumably to maximize sensitivity, but the selectivity of the method did not appear to be adversely affected in this case. The method detection limit was estimated by analyzing seven influent and seven effluent extracts from each of two wastewater treatment plants, each spiked at 200 ng.L^{-1}, using the EPA definition of

MDL (Glaser 1981; EPA 1997) that is subject to some criticism (Section 8.4); the MDL values for these two compounds were found to be in the range 10–60 ng.L^{-1}. Linear calibration curves were obtained over the range 100–2000 ng.L^{-1} using matrix matched calibrators taken through the entire analytical procedure, thus compensating for matrix effects on ionization efficiency that were observed (Yang 2006) to be larger for the more complex influent samples, as expected. Accuracy, within-run precision and between-run precision (day-to-day repeatability) were estimated from relative standard deviations (RSDs) obtained for QC samples spiked at three concentration levels (100, 500 and 1000 ng.L^{-1}). Aliquots of six spiked samples each of influent and effluent were extracted to obtain six independent replicates of each QC, and these six replicates were all analyzed on each of three different days. The accuracy ranges in the influent and effluent water matrices were respectively −10.8 to +12.0 % and −6.5 to +10.4 % for the two analytes. The within-run RSDs for the influent and effluent ranged from 6.3 to 14.1 % and from 4.3 to 10.7 %, well within the pre-determined (Yang 2006) acceptance criteria (≤ 15 %) at each concentration level. Furthermore, the analytical data for the 18 replicates (six per day for three days) of each influent and effluent QC sample led to estimates for the between-run precision (repeatability) that were all between 5.8–13.7 % for both analytes in both matrices, again within the pre-set acceptance criteria, with an overall accuracy always within 7 %. Thus the effective lower limit of quantitation (LLOQ) was taken as the lowest concentration on the calibration range that was also the lowest QC sample concentration in the validation, i.e. 100 ng.L^{-1}. Recoveries of the two analytes (ETM–H$_2$O and TLS) were estimated as the ratios of the measured concentrations of analytes for extracts spiked before extraction to the values for extracts spiked after extraction, for influent, effluent and distilled water matrices; average recoveries for the two compounds ranged from 89.2 ± 9.7 % for the more complex raw influent to 93.7 ± 6.9 % for effluent. Essentially the only validation parameter that was not assessed in some fashion was the analyte stability under various conditions (Sections 10.2.7 and 10.2.8).

A notable feature of this investigation (Yang 2006) is the degree of attention paid to estimates of all sources of uncertainty, i.e. repeatability, recovery and calibration, in accord with the European standard (Eurachem–CITAC Guide CG4 2000, Chapter 10). The uncertainty component arising from sampling was considered to be negligible in view of the high degree of homogeneity of the water samples and the care taken to ensure that the samples used for analysis provided a true representation of the waters over a period of 12 months. Sampling was conducted using a flow proportioned automatic sampler, so that the 24-hour composite samples of the final effluent were collected in duplicate at the same time as the influent (Yang 2006), twice per month for each treatment plant investigated. The results of this error analysis are summarized in Figure 11.18; note that the contribution of calibration to the overall uncertainty is negligible except for the lowest

concentration QC sample, that the recovery uncertainty increases somewhat at lower concentrations, and that the general uncertainty levels are higher for the more complex influent matrix than for the effluent, all in accord with expectations.

Another positive aspect was the explicit concern for confirmation of analyte identity that used the 'identification points (IPs)' approach (European Commission 2002, see discussion of Tables 9.1–9.3 in Section 9.3.3b). For low resolution MS/MS detection using an ion trap as was done here, one IP is awarded for the *m/z* selected precursor ion and 1.5 IPs for each MRM transition monitored, with three IPs required (European Commission 2002) for acceptable confirmation of analyte identity. To qualify for the IPs, at least one ion abundance ratio for two MRM product ions must be measured and fall within tolerance intervals of contemporary values measured for an authentic analytical standard, defined (European Commission 2002) as ±20 % for a relative abundance of >50 %, to ±25 % for 10–50 % and to ±50 % for <10 %, although the US FDA recommends (FDA 2003) that these tolerances should be within ±10 % regardless of the absolute values of the relative abundances. The present work (Yang 2006) satisfied both tolerance criteria over the calibration range, although it was considered unlikely that they could be satisfied at the MDL.

It is of interest to summarize the findings made possible by this carefully designed and executed analytical method. Both treatment plants mainly treat domestic sewage from 125 000 inhabitant-equivalents. In both treatment plants, mechanically pre-treated wastewater first passes through conventional biological (activated sludge) treatment, but in Plant I chlorine was used for disinfection while UV was used in Plant II. The influent and effluent water concentrations measured for (ETM–H$_2$O) for Plant I (chlorination) over the 12 month sampling period (48 samples total) were 350 and 120 ng.L^{-1}, respectively, for a removal efficiency of 66 %; the corresponding values for Plant II (UV disinfection) were 90 and 40 ng.L^{-1} for a removal efficiency of 56 %. In the case of TLS, the Plant I influent and effluent concentrations were 180 and 50 ng.L^{-1} for a removal efficiency of 72 %, while for Plant II the influent value was 60 ng.L^{-1} and the effluent concentration was below the MDL concentration for this analyte in this matrix (30 ng.L^{-1}) for an estimated removal efficiency >50 %. Note that many of these quoted concentrations are below the lowest concentration (100 ng.L^{-1}) within the validated calibration range and should thus be taken as estimates of uncertain accuracy, but are well within the range of concentrations for these analytes determined at many other treatment plants in North America and Europe (Yang 2006).

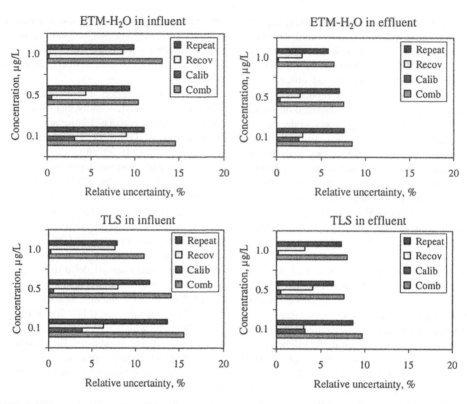

Figure 11.18   Relative uncertainty values (%) determined at three concentration levels (0.1, 0.5 and 1.0 µg.L^{-1}) for anhydro-erythromycin (ETM–H$_2$O) and tylosin (TLS) in spiked wastewater QC samples both before (influent) and after (effluent) treatment. The combined uncertainty was calculated from the separate values for repeatability, recovery and calibration using Equation [8.6] (see Section 8.2.2). Reproduced from Yang, *Anal. Bioanal. Chem.* **385**, 623 (2006), with permission from Springer Science and Business Media.

11.3.2b Extraction/Clean-up for Multi-residue Analysis of Compounds of Widely Different Polarity etc.

The preceding example of a method developed for just two closely related analytes has been discussed at some length as it sets a benchmark for multi-residue methods that must necessarily settle for less stringent validation criteria. There is a growing realization that multi-residue analytical methods are essential tools in acquiring wider knowledge about the nature and extent of contamination in various waters in the environment. Application of such methods in general permits acquisition of a large amount of data after a single sample preparation step and in a single analytical run (usually LC–MS/MS with ESI or APCI in the present context of pharmaceutical compounds since most of the ~3000 such compounds currently in use are polar and thus amenable to ionization by these two techniques). However, simultaneous analysis

of compounds with often very different physicochemical characteristics (especially polarity) requires significant compromises in the selection of experimental conditions that in many cases are far from optimum for all the analytes studied.

While chromatographic separations and efficient ionization in the mass spectrometer do present difficulties for mixtures of widely different compound types, the most difficult problems in development of multi-residue methods usually concern the need for simultaneous extraction and clean-up of analytes with widely different polarities. With respect to multi-residue extraction and clean-up, SPE approaches (Section 4.3.3c) are the most popular; these can use either a single SPE medium in a single extraction step or a combination of two SPE materials used either in series or in parallel, thus delivering the analytes as two or more groups defined by their physico-chemical properties relevant to the SPE process.

A recent review of multi-residue analysis for pharmaceutical compounds (Gros 2006) draws interesting comparisons among several SPE media including Oasis–HLB SPE cartridges (Waters Inc.), C_{18}-silica media, and Strata-X cartridges (Phenomenex Inc.) with respect to their suitability for extraction from aqueous matrices of compounds with a wide range of physico-chemical properties. A theoretical investigation (Dias 2002), using a semi-empirical solvation parameter model, discusses some of these SPE media among others. The Oasis–HLB media are based on a porous copolymer of the lipophilic divinylbenzene and the hydrophilic N-vinylpyrrolidone, and can extract acidic, neutral and basic compounds from water and water–methanol mixtures over a wide range of pH values including neutral. Increasing solute size and electron lone pair interactions favor retention from aqueous solutions, but Oasis–HLB is not competitive with water for dipole type and hydrogen bond interactions, resulting in lower analyte retention. Strata-X is a polymeric sorbent with properties somewhat similar to those of Oasis–HLB and can retain a wide range of analytes via hydrophilic, hydrophobic and π–π interactions. C_{18}-silica, especially with high carbon loading, is also capable of extracting a wide range of compound types but, depending on the nature of the compounds included, pH adjustment prior to extraction is generally required. Thus for acidic compounds a lower pH is used to ensure that the acidic components are in their neutral (nondissociated) forms, since the lipophilic C_{18} material does not retain the ionized forms, and for neutral and basic compounds samples are adjusted to neutral or basic pH values. Oasis–HLB was preferred in this evaluation (Dias 2002) for the extraction of compounds of low molecular mass and higher polarity, and the C_{18}-silica was the best compromise for extraction of compounds covering a wide polarity range provided that several pH adjustment steps are acceptable. The latter approach generally results in two or more fractions containing analytes with different ranges of physico-chemical characteristics.

11.3.2c Screening Method for Multiclass Antibiotics

A recent review (Díaz-Cruz 2006) has summarized recent work on determinations of antibacterial residues in aquatic matrices (and gives useful information on structures and MS/MS fragmentations characteristic of each compound class), but with only brief mention of multi-residue methods covering more than one class of such compounds. However, an example of a multi-residue analytical method (Hao 2006) was developed for determination of 27 target compounds, including 18 commonly used veterinary and nine prescribed human pharmaceuticals that are representative of use in Ontario, Canada. This method was developed to monitor Ontario's source waters and applied to analysis of waters collected from tributaries and downstream sites of the Grand River watershed, affected by both agricultural and urban activities. Of the 27 chosen analytes, 18 were antibiotics and the following discussion will concentrate on these. The antibiotic types included several tetracyclines, several sulfonamides, macrolides (including erythromycin and tylosin, Figure 11.17), a polyether ionophore, a cyclic peptide and two small molecule drugs (trimethoprim, a pyrimidine-based antibacterial, and chloramphenicol), but no representatives of the β-lactams or the quinolones.

Samples were collected between April and November 2003 using one litre brown glass bottles at sites from midstream to avoid the collection of sediment, and stored at ∼4 °C without preservatives. Field QC samples, spiked with analytical standards, were used to monitor their stability during transport and storage; under these conditions it was shown that the compounds studied (Hao 2006) are stable for at least 35 days. Sample extraction was done in batches of 12, which included three QC samples (method blank, method spike sample and spiked field sample) and nine regular field samples. QC samples were prepared by spiking 400 mL of deionized water with a standard solution of the analytical standards with addition of two grams of disodium EDTA and the method surrogate solution ($^{13}C_6$-sulfamethazine); the pH was adjusted to an experimentally optimized value of 8.2 using either sulfuric acid (H_2SO_4) or sodium hydroxide (NaOH) solution. Extraction was by SPE with 60 mg Oasis–HLB cartridges (Section 11.3.2b) preconditioned with 5 mL of methanol followed by 5 mL of water. Following extraction the cartridges were rinsed with 5 % methanol in deionized water, dried, and the analytes were eluted with 5 mL of methanol.

In preparation for LC–MS analysis, 1 mL of a sample extract and 1 mL of each of the calibration standard solutions were taken to dryness at ambient temperature using a flow of nitrogen, and were reconstituted in 100 µL of deionized water, representing concentration of the 400 mL water sample by a factor $> 10^3$. Gradient elution HPLC was used, but in view of the wide range of structural types represented by the analytes two separate methods with different mobile phases were used for compounds that responded in positive and negative ion ESI–MS, i.e. in this case of multiresidue analysis a single extraction step was used but two analytical steps optimized separately for different groups of compounds. It was not possible to provide surrogate internal standards for all analytes (a common circumstance in multi-residue analysis), so a single method labeled standard was used to provide some information on variations in extraction efficiency, mass spectrometer sensitivity etc. For this semiquantitative method 3-point calibration curves, each covering a 50-fold concentration range, were generated for all analytes using matrix matched calibrants taken through the extraction procedure. The HPLC run time was sufficient to achieve an optimized chromatographic separation; about 90 minutes were required for the analysis of one sample in both positive and negative detection modes, including the time to recondition the columns.

A QqQ_{trap} mass spectrometer (Section 6.4.6) was used in conventional triple quadrupole MRM mode for routine quantitation, but in some experiments this was combined using the information dependent acquisition (IDA) software facility function with the ion trap function for identification of unknowns and/or confirmation of target analytes. Only one MRM transition was monitored for each analyte, thus gaining only 2.5 identification points rather than the specified three (European Commission 2002, see Table 9.1); this was undoubtedly a consequence of the large number of analytes monitored in this multiresidue analysis. A retention time within 5 % of the standard value determined for pure analytes in distilled water was also a required criterion (Hua 2006) for confirmation of identity.

Recoveries were estimated in the range 50–130 % (with most RSDs <20 %) at pH 8.2 by comparing signals for extracts spiked at the same nominal concentrations pre- and post-extraction; the tetracycline derivatives accounted for most of the high recoveries ~130 %. Average concentrations of the analytes varied over 3 orders of magnitude. These differences are naturally reflected in the method detection limits determined using the US EPA approach (Glaser 1981; EPA 1997) with eight replicate analyses of blank matrix spiked with analytes each at its own selected concentration, (unfortunately not specified, Hao 2006). The MDLs thus determined for the antimicrobial analytes ranged from 30 ng.L^{-1} for trimethoprim to 600 ng.L^{-1} for chlortetracycline.

As emphasized previously (Section 8.4), while the EPA MDL does have the attribute of transparency and some level of connection to confidence limits at a stated confidence level, it has several disadvantages including its dependence on the value of the arbitrarily chosen concentration used to determine its value. This problem is also evident in this work (Hao 2006), which also estimated quantitation limits (LLOQs) for the target analytes as follows. Serial dilutions of the extract solution used in the MDL determination were analyzed until the S/B value for one or more analytes fell to 5:1. The LLOQs of the

analytes were then calculated as 10 times the standard deviation of the quantitative results of ten replicate analyses of this diluted sample. The resulting LLOQ values were all significantly lower than the corresponding MDLs, by factors in the range 4–170. It does not seem to be meaningful to have a detection limit that is much higher than a quantitation limit. This is not, of course, a criticism of the experimental work (Hua 2006), but rather of the use of a single point estimate of a detection limit; different results would surely have been obtained if appreciably lower test concentrations of the analytes had been used in the procedure for the EPA MDL. Indeed, analyses of the real-world source water samples gave concentrations that were mostly below the EPA MDL values reported, but many were well above the corresponding LLOQ values determined as described above and were reported on that basis (Hua 2006).

Apart from this MDL ambiguity, which did not arise from the details of the analytical method itself but rather from the procedure for determining the MDL, it is clear from the preceding summary description that this multi-residue method for antibacterials is much less rigorous than that developed (Yang 2006) for just two structurally similar macrolide antibiotics. This is obviously a consequence of the different sets of compromises that must be reached under the two sets of circumstances. The appropriate assessment of a multi-residue method does not involve its degree of rigor and validation, but rather concerns its fitness for purpose. In fact, the stated purpose (Hua 2006) was development of a broad-range quantitative screening method to monitor pharmaceuticals in surface waters, and indeed the method did permit confirmation and semi-quantitation of a wide range of multiple target antibacterials (plus some other pharmaceuticals not mentioned here). These data support exposure assessment investigations for a variety of pharmaceuticals present at low levels in the environment. Antibacterial residues detected and quantitated at the ng.L^{-1} level in tributaries receiving predominantly agricultural inputs included monensin, trimethoprim, sulfamethazine and incomycin, while those receiving both agricultural and urban inputs still showed trace levels of agriculturally related compounds together with some commonly used human prescription pharmaceuticals, e.g. sulfachloropyridazine.

As a result of the design of multi-residue methods for such purposes, they are intrinsically prone to provide false positive results. A good example is provided by work designed to monitor residues of some macrolide antibiotics as well as some steroid hormones (Schlüsener 2005) in unfiltered influents and effluents of sewage treatment plants. These target analytes represent a less chemically diverse range than those in the preceding study (Hua 2006) and the method developed was correspondingly more rigorous. Thus, the analytes were extracted by SPE followed by clean-up using size exclusion chromatography, isotope-labeled internal standards were available for some of the steroid analytes, and APCI was used in place of ESI when the matrix effects in the latter were excessive. In addition, two MRM transitions per analyte were used to attain three identification points (European Commission 2002, Table 9.1) and this minimized the probability of false positive identifications. Figure 11.19 shows an example that would have resulted in a false positive identification of ethinylestradiol (the estrogen that is an active component in almost all modern oral contraceptive pills and is one of the most commonly used pharmaceuticals) if only the more intense MRM transition (used for quantitation) had been monitored (Schlüsener 2005). The probability of such false positives will increase as the number of residues monitored and the diversity of their physico-chemical properties also increases, as a result of the compromises that must be used to provide the required screen. This requirement to reach appropriate compromises is the key to assessment of the fitness for purpose of any proposed multi-residue analytical method.

An interesting recent example (Castiglioni 2006) describes an LC–ESI–MS/MS method for the determination of the illicit drugs cocaine, amphetamines, morphine, cannabinoids, methadone and some of their metabolites in wastewater (both treated and untreated). Quantitation was achieved in the low ng.L^{-1} range, with overall variability $<10\%$. Cocaine and metabolites, amphetamines, morphine and metabolites, methadone and its main metabolite, and 11-nor-9-carboxy-Δ9-tetrahydrocannabinol (the main metabolite of tetrahydrocannabinol), were all measured in significant amounts in influents and effluents of two wastewater treatment plants, indicating possible environmental risks to public health.

11.4 GC–MS Analyses of Persistent Environmental Pollutants

Chronologically this class of analytical problems was the first to be extensively addressed using GC/MS methods that pre-date reliable LC/MS technology by a considerable time, and is still very important. Because target analytes amenable to GC methods must necessarily possess considerable thermal stability, they can also be extracted using methods that expose them to higher temperatures such as those described in Section 4.3.2, and this will become apparent in the following discussion.

11.4.1 'Dioxin-like' Compounds

One of the most demanding analytical tasks is still that of quantitative analysis of individual congeners of polychlorodibenzo-*p*-dioxins ('PCDDs', or 'dioxin') and related 'look-alike' compounds (Figure 11.20) at levels of parts per 10^{12} in a range of environmental, industrial product and biological matrices. (Note that 10^{12} is referred to as a 'trillion' in some countries including the USA, but as a 'billion' in many other countries although this usage appears to be decreasing in countries in which English is the official language; in view of this ambiguity it is best to specify values numerically, see Section 1.1.) Determinations at these ultra-low levels is a challenge in any event, but are made more difficult for these specific chlorinated aromatic compounds in view of the many other related compounds found in environmental samples that can directly interfere in GC–MS analyses even when high performance GC is used. As an example, some potential interferences with the ions used to monitor tetrachlorodibenzo-*p*-dioxins (TeCDDs) are shown in Table 11.2. The currently accepted 'gold standard' approach to avoiding such interferences is to employ meticulous matrix-specific clean-up procedures plus high resolution GC with selected ion monitoring (SIM) using a double focusing magnetic sector instrument at a resolving power of at least 10^4 (10 % valley definition). Less vigorous but faster and cheaper screening approaches are described below. It must also be emphasized, however, that it is the availability of ^{13}C-labeled internal standards of most of the 29 dioxin-like compounds (Table 11.3) in the 1990s, as well as a selection of CRMs, that has permitted implementation of high quality methods for precise, accurate, isomer-specific determination of all these compounds at ultra-trace levels (down to parts per $10^{12} - 10^{15}$) in a wide range of matrices.

These very extensive clean-up procedures and selective analytical methods required for analysis of 'dioxin' and its 'look-alike' compounds (see below) are necessary to produce data of the highest quality possible that can be supported and defended in a court of law. The reason for all the effort lies in the extremely low levels of these compounds that can produce severe health effects. With respect to acute toxicity, a one-time dose equivalent to one microgram of the most toxic compound (2,3,7,8-tetrachlorodibenzo-*p*-dioxin) per kilogram of body weight is lethal to a laboratory strain of guinea pigs. However, there is an ongoing debate about some features of the effects of this and related compounds on other animal species, particularly mammals (especially humans). In fact, it appears that guinea pigs are remarkably sensitive to 'dioxin', much more so that other mammals including humans. With respect to longer term chronic toxicity,

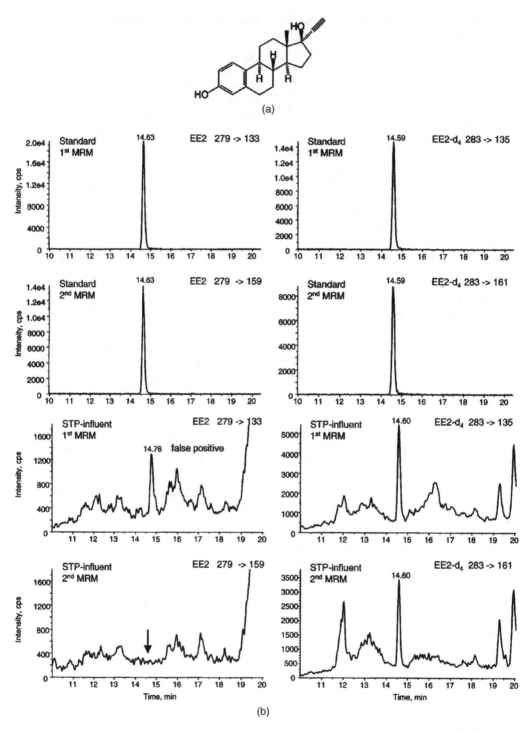

(a)

(b)

Figure 11.19   (a) Structure of ethinylestradiol. (b) LC–MS/MS chromatograms for ethinylestradiol (left-side panels) and its internal standard ethinylestradiol-d₄ (right-hand panels) in the extract of an influent sample from a sewage treatment plant. A false positive determination is indicated using only retention time and the first MRM transition (top chromatograms) but is detected by the second MRM chromatogram (lower panels). Reproduced from Schlüsener, *Rapid Commun. Mass Spectrom.* **19**, 3269 (2005), with permission of John Wiley & Sons Ltd.

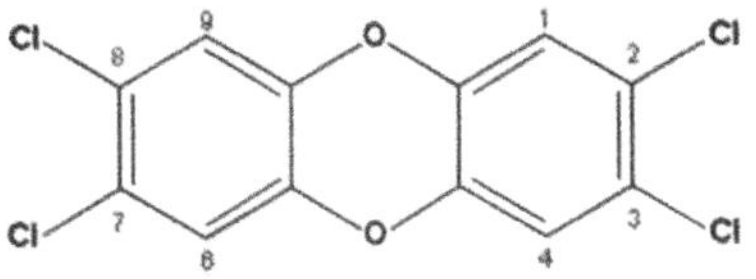

2,3,7,8-Tetrachlorodibenzo-p-dioxin

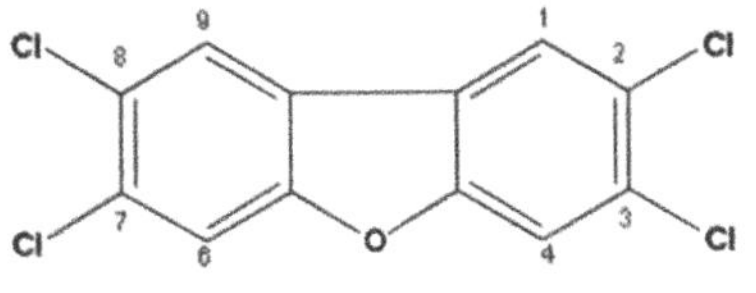

2,3,7,8-Tetrachlorodibenzofuran

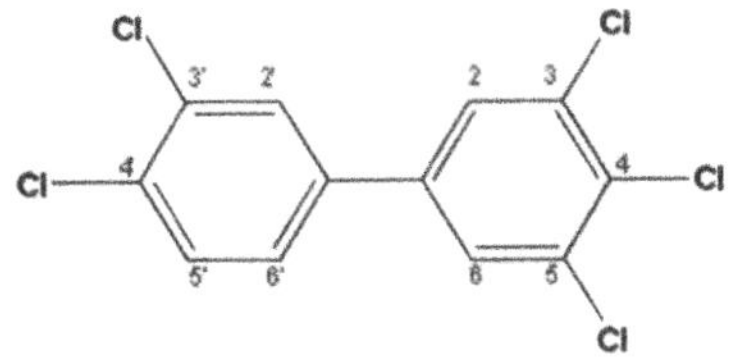

3,3',4,4',5-Pentachlorobiphenyl (PCB126)

Figure 11.20 Typical structures of the polychlorinated dibenzo-*p*-dioxins (PCDDs, i.e., 'dioxins'), polychlorinated dibenzofurans (PCDFs, i.e., 'furans') and polychlorinated biphenyls (PCBs). The PCDD and PCDF congeners considered significantly toxic are those with chlorine substitution at all of the 2,3,7 and 8 positions; only seven PCDD congeners (out of a total of 75) and 10 PCDF congeners (out of 135) have 2,3,7,8-substitution. There are 209 possible PCB congeners (mono- through decachloro substitution); the 12 dioxin-like PCBs (DLPCBs) are those with no (or only one) chlorine substitution *ortho* to the biphenyl linkage (and are thus approximately planar like the PCDDs and PCDFs as a result of minimal steric hindrance around the biphenyl linkage), that are also substituted in combinations of the 3,3',4,4',5,5' positions so that the spatial distribution of the chlorine substituents 'looks like' that in the 2,3,7,8-substituted PCDDs and PCDFs. The total of 29 'toxic' congeners is listed in Table 11.1. Reproduced from Reiner, *Anal. Bioanal. Chem.* **386**, 791 (2006), with permission from Springer Science and Business Media.

while there is no doubt that levels in the μg .kg^{-1} range can cause severe health effects including liver necrosis, immune system impairment, teratogenesis and the severe skin disorder known as chloracne, there is still controversy concerned with the question of whether or not there is a threshold dose below which these effects are not expressed (and if so what is it), whether or not these compounds are in any sense direct carcinogens etc. There are many references concerning these controversies, particularly regarding some well known incidents involving 'dioxin' exposure (Boffey 1971; Anderson

1985; CDC 1988; Geusau 1999, 2001; Kim 2003; Hays 2003; Starr 2003; Ngo 2006; NAS 2006; see also www.roche.com/com_his_sev-e.pdf).

It is advisable at this point to describe the shorthand notation used in the following discussion. The 'P' in PCB, PCDD and PCDF is taken to indicate 'poly', as in 'poly-chlorinated biphenyl', or 'polychloro-dibenzo-*p*-dioxin' or 'polychloro-dibenzofuran'. When the specific degree of chlorination is specified M, Di, Tr, Te, Pe, Hx, Hp, O and De, indicate substitution by 1–10 chlorine atoms (denoting mono, di, tri, tetra, penta, hexa, hepta, octa, nona, deca), respectively, e.g. TeCDD indicates a tetrachloro-*p*-dibenzodioxin, HxCDF a hexachloro-dibenzofuran and DiCB a dichlorobiphenyl. Some authors use subscript numbers instead of lower case letters to distinguish e.g., hexa from hepta-chloro compounds (thus H_6CDD instead of HeCDD, H_7CDF instead of HpCDF etc.). The notation TCDD is often used instead of TeCDD.

There is no doubt that the literature on dioxin toxicity can be very confusing, and so it helps to be aware of potential sources of confusion (Tuomisto 1999). Different measured quantities are used for different purposes. Thus, the amount of PCBs may be given as a sum of all PCB congeners (Figure 11.20) in the sample, and the amount of PCDDs plus PCDFs are sometimes given as a sum of the 17 most toxic congeners (Table 11.3, those that include a 2,3,7,8-chlorine substitution pattern, Figure 11.20). The total toxicity risk is frequently expressed as a toxic equivalent quantity (TEQ), a concept first introduced by the Ontario Ministry of Environment (OME 1984) for 'dioxins plus furans'; extension to PCBs was first proposed a few years later (Safe 1990). The scientific basis for this concept can be traced to our understanding of the mechanism of action of dioxin-like chemicals on living organisms. There is a considerable body of evidence (Reiner 2006) supporting the hypothesis that all major toxic effects of these chemicals are expressed via their binding to an intracellular protein, the aryl hydrocarbon receptor (AHR), whose normal function is to regulate transcription of multiple genes that are important in development, physiological function and (important in the present context) responses to xenobiotic chemicals. It seems probable that these dioxin-like chemicals (Table 11.3) exert their toxicity by disrupting expression of genes that are under control of the AHR. If, indeed, this hypothesis of a common mode of action for these dioxin-like compounds is correct, it is reasonable to propose that the combined effects of a mixture of them can be expressed as a sum of their independent contributions:

$$\text{TEQ} = \Sigma_i \{[\text{PCDD}_i].(\text{TEF}_i)\} + \Sigma_j \{[\text{PCDF}_j].(\text{TEF}_j)\}$$
$$+ \Sigma_k \{[\text{DLPCB}_k].(\text{TEF}_k)\} \qquad [11.7]$$

Table 11.2 Sources of potential interferences for molecular ions of tetrachlorodibenzo-*p*-dioxins (TeCDDs) at *m/z* 319.8966 and 321.8936, if not chromatographically resolved. Adapted from Clement, *Mass Spectrom. Revs.* **7**, 593 (1988), with permission.

Potential Interfering Compound	*m/z* of Interfering Ion	Required Resolving Power (10 % valley)
Heptachlorobiphenyls (HpCBs)	321.8678	12 500
Nonachlorobiphenyls (NCBs)	319.8521	7200
	321.8491	7300
Tetrachloromethoxybiphenyl	319.9329	8900
	321.9299	8900
Tetrachlorobenzylphenyl ether	319.9329	8900
	321.9300	8900
Pentachlorobenzylphenyl ether	319.9143	18 100
	321.9114	18 200
Tetrachloroxanthene	319.9143	18 100
	321.9114	18 200
Hydroxytetrachlorodibenzofuran	319.8966	Cannot resolve
	321.8936	Cannot resolve
Tetrachlorophenylbenzoquinone	319.8966	Cannot resolve
	321.8936	Cannot resolve

where [X] represents the mass concentration of compound X and the toxic equivalent factors (TEFs) reflect the species-specific toxicities relative to that of the most toxic compound (2,3,7,8-TeCDD). This simple linear combination of independent contributions from the $(i + j + k = 29)$ dioxin-like compounds (including the dioxin-like PCBs, Table 11.3) is found to work reasonably well despite ignoring potential synergistic and antagonistic effects, although discrepancies by as much as a factor of two can arise occasionally. The values assigned to the TEFs are under constant review, e.g. the values recommended by the World Health Organization in 1998 (Van den Berg 1998) were revised in 2005 (Van den Berg 2006) on the basis of a re-evaluation of the toxicological database (Haws 2006); both sets of values for human are shown in Table 11.3. Note that only the dioxin-like PCBs (DLPCBs) have the same mode of action as the chlorinated 'dioxins' and 'furans'. Those PCBs with more than one chlorine substituent *ortho* to the biphenyl linkage adopt a configuration that is far from planar resulting from steric interactions between the bulky chlorine atoms, and appear to have a different mode of toxicological interaction; for this reason they are not included in the TEQ/TEF approach. The TEQ value is useful for purposes of estimating toxicological risk, but the individual concentrations should also be reported since these can often indicate the source of the contamination.

Another potential source of confusion in the 'dioxin' literature (Tuomisto 1999) is that different units for compound concentration [X] are used for different purposes. Sometimes concentrations are given as 'ng/kg b.w.' (nanograms per kilogram body weight), but more often as 'pg/g fat' (picograms per gram fat); the difference between these two units can be as much as a factor of 10 since the human body contains 10–15 % of fat tissue. In addition to this source of ambiguity, use of the nonstandard units 'ppm' (parts per million, i.e. $\mu g.g^{-1}$ or $mg.kg^{-1}$), 'ppb' (parts per 'USA' billion = 10^9, i.e. $\mu g.kg^{-1}$), and 'ppt' (parts per 'USA' trillion = 10^{12}, i.e., $ng.kg^{-1}$) further confuses the issue; note that 'ppt' is also used in different contexts to denote 'parts per thousand'! Even worse, different measures are used for different matrices. In foodstuffs, concentrations of 'dioxins' and PCBs are often expressed 'per wet weight' ('fresh weight', really mass of course) to facilitate the calculation of human intake of these compounds via food. In contrast, in contaminated soils or sediments they are usually expressed 'per dry weight', and differences between these two measures may be substantial. An unambiguous report of an analysis of dioxin-like compounds would specify the quantity used to normalize the concentrations, i.e. mass ('weight') of total matrix or of fat content in matrix, whether this mass is 'wet weight' or 'dry weight', and unambiguous definition of the units used (e.g. $ng.kg^{-1}$); if 'ppb' and/or 'ppt' are used, the

Table 11.3 Toxic equivalent factors (TEFs) for 29 dioxin-like compounds for human, produced for the World Health Organization (WHO) in 1998 (Van den Berg 1998) and revised in 2005 (Van den Berg 2006), normalized to a value of unity for 2,3,7,8-TeCDD

Compound	WHO 1998 TEF	WHO 2005 TEF
Chlorinated dibenzo-*p*-dioxins (PCDDs)		
2,3,7,8-TeCDD	1	1
1,2,3,7,8-PeCDD	1	1
1,2,3,4,7,8-HxCDD	0.1	0.1
1,2,3,6,7,8-HxCDD	0.1	0.1
1,2,3,7,8,9-HxCDD	0.1	0.1
1,2,3,4,6,7,8-HpCDD	0.01	0.01
Octachloro-DD (OCDD)	0.0001	**0.0003**
Chlorinated dibenzofurans		
2,3,7,8-TeCDF	0.1	0.1
1,2,3,7,8-PeCDF	0.05	**0.03**
2,3,4,7,8-PeCDF	0.5	**0.3**
1,2,3,4,7,8-HxCDF	0.1	0.1
1,2,3,6,7,8-HxCDF	0.1	0.1
1,2,3,7,8,9-HxCDF	0.1	0.1
2,3,4,6,7,8-HxCDF	0.1	0.1
1,2,3,4,6,7,8-HpCDF	0.01	0.01
1,2,3,4,7,8,9-HpCDF	0.01	0.01
Octachloro-DF (OCDF)	0.0001	**0.0003**
Non-*ortho*-substituted PCBs		
$3,3',4,4'$-TeCB (PCB 77)	0.0001	0.0001
$3,4,4',5$-TeCB (PCB 81)	0.0001	**0.0003**
$3,3',4,4',5$-PeCB (PCB 126)	0.1	0.1
$3,3',4,4',5,5'$-HxCB (PCB 169)	0.01	**0.03**
Mono-*ortho*-substituted PCBs		
$2,3,3',4,4'$-PeCB (PCB 105)	0.0001	**0.00003**
$2,3,4,4',5$-PeCB (PCB 114)	0.0005	**0.00003**
$2,3',4,4',5$-PeCB (PCB 118)	0.0001	**0.00003**
$2',3,4,4',5$-PeCB (PCB 123)	0.0001	**0.00003**
$2,3,3',4,4',5$-HxCB (PCB156)	0.0005	**0.00003**
$2,3,3',4,4',5'$-HxCB (PCB 157)	0.0005	**0.00003**
$2,3',4,4',5,5'$-HxCB (PCB 167)	0.00001	**0.00003**
$2,3,3',4,4',5,5'$-HpCB (PCB189)	0.0001	**0.00003**

Values in bold font indicate values changed in 2005 from 1998 values. PCB numbering is consistent with both the original version of Ballschmiter and Zell (1980) and with the later consensus numbering approved by IUPAC, see: http://www.epa.gov/toxteam/pcbid/defs.htm

meanings of these units must be clearly defined in terms of scientific notation (e.g. 'ppt' refers to parts per 10^{12}). If a TEQ is reported the source of the TEF values used should be specified and the concentrations of the individual dioxin-like compounds should also be reported as mentioned above.

The actual body burden of dioxin-like compounds at a given point in time is the important quantity in defining the toxicological risk to that individual; a single acute dose and an average daily dose mean very different things. Approximately similar body burdens of 'dioxins' could

be achieved either by a single dose of 5000 pg or as the steady-state concentration established by a lifelong intake of 1 pg/day (Tuomisto 1999). Therefore, it is important to be careful in comparing, e.g., the amounts in the body and the amounts in the food.

The actual analysis of trace amounts of these dioxin-like compounds has evolved over the years. A remarkable early paper (Crummett 1973) used GC/MS with a packed 3 mm i.d. GC column and a single focusing magnetic sector instrument (Section 6.4.4) operated at RP = 600 (10 % valley definition), with a pen recorder to record the total ion current (TIC) chromatogram and a UV oscillograph to record the mass spectrometric peaks for three m/z values characteristic of TeCDD (isotopologs of the molecular ion). An instrumental detection limit of 6 pg of 2,3,7,8-TeCDD that still yielded correct abundance ratios for the monitored ions (calculated based on chlorine isotope abundance ratios was achieved. Admittedly the analytical samples analyzed (Crummett 1973) were not complex by modern standards, and it is likely in view of the limited resolution in both the chromatographic and mass spectrometric steps that the GC–SIM peaks measured may have contained contributions from compounds other than 2,3,7,8-TeCDD. Nonetheless this was a remarkable achievement for its time (e.g. fused quartz capillaries were not available till some seven years later, Section 4.4.3c).

Since that time, the performance of analytical methods for these compounds, at ever-decreasing required LODs and LLOQs in a wide variety of matrices, has steadily improved. Every detailed step in the analytical procedure must be scrutinized in order to meet the ever more demanding regulatory requirements. Representative sampling (Section 8.6) is always problematic for heterogeneous samples but becomes particularly so when the target analytes are present at levels of the order of a few parts in 10^{12} (or even 10^{15} in some cases). Extraction of the compounds from the sample matrix and transfer to an organic solvent appropriate for subsequent steps can be far from trivial, although the thermal and chemical stabilities of these persistent pollutants do permit use of technologies not suitable for more labile analytes. Clean-up is a particularly important phase in this case, in view of the ubiquitous presence of a wide range of other chlorinated aromatic compounds, often at levels orders of magnitude larger than those of the dioxin-like compounds. These potential interferences often can not be resolved from the target analytes even by modern high resolution GC (HRGC) with SIM monitoring at resolving power $\sim 10^4$ (10 % valley definition), and so must be selectively removed in the clean-up process that can involve several distinct steps. Methodologies developed through the 1980s have been fully described (Clement 1988) and updated recently (Reiner 2006). As mentioned above, the use of HRGC/HRMS in the final analytical step, combined with extensive meticulous clean-up procedures (silica, alumina and carbon-based columns), is now accepted as the 'gold standard'.

Some comments about the use of internal standards in this case are appropriate. In the approved method used by the Ontario Ministry of the Environment (Reiner 2007), 15 of the 17 toxic dioxins/furans, as well as all of the dioxin-like PCBs (Table 11.3), are analyzed by isotope dilution

mass spectrometry (IDMS) using ^{13}C-labeled isotopologs as surrogate internal standards (Section 8.5.2b); the remaining two compounds are analyzed by internal standard quantification. In IDMS the system is set up so that the native and SIS concentrations fall into the working range; for example, in the US EPA method EPA1613 (EPA 1994) the native compound concentration should lie in the range 0.5–$200 \, pg.\mu L^{-1}$ for the tetrachlorodioxins/furans, while the SIS is fixed at $100 \, pg.\mu L^{-1}$. Although ideally the EIMS responses of each native and ^{13}C-labeled SIS should be equal, in the real world the relative response factor (ideally unity) is determined experimentally and applied as a correction factor to the peak area ratio (Reiner 2007). Some additional labeled internal standards are used for quality control purposes, e.g. $^{13}C_{12}$-labeled 1,2,3,4-TeCDD is used as an injection (volumetric, or 'syringe') standard for tetra- and penta-chloro compounds. Finally, $^{37}Cl_4$-labeled 2,3,7,8-TeCDD is used as a 'clean-up standard'; this compound is typically added after extraction and its recovery monitors the losses during the extensive sample clean-up. The entire integrated procedure is extremely demanding. It will be difficult to further decrease the LODs and LLOQs because of ubiquitous background levels of potentially interfering compounds that are largely removed by the meticulous clean-up presently employed; even more extensive clean-up runs the risk of significant losses of analytes.

Analytical methods of this kind are now specified by regulatory agencies worldwide; typical examples (EPA 1994; European Commission 2000a, 2000b) are designed to provide data that will withstand scrutiny in a court of law (Section 9.3.5). Such analyses are very expensive in both money and resources as well as time (a complete analytical procedure can require up to 20 hours depending on the nature of the matrix etc.), as exemplified by the long detailed prescriptions for the various steps (EPA 1994). There is not much point in trying to summarize these here. Instead it is noted that, while these 'gold standard' methods are fit for purpose in demanding cases, there is a need for faster, less expensive screening procedures, e.g. LRMS methods for some environmental applications where high levels are expected. There have been serious cases of foodstuffs contaminated at significant levels. A brief review (Focant 2004) of the known incidents dating from the 1960s reveals that, surprisingly, some of these incidents could be traced to incorporation of natural inorganic materials, such as a clay material used as an anticaking agent in the production of animal feed. Geothermal processes were implicated (Ferrario 2002) as the source of the dioxin-like compounds at levels as high as 460 ng TEQ.kg^{-1} in this clay material. Other incidents of food contamination (Focant 2004), however, were the result of anthropogenic production of dioxin-like compounds.

As a result of such events, analytical procedures based on different technologies are currently being assessed as cost-effective screening methods for monitoring dioxin levels in food and animal feed, complementary to the GC–HRMS approach regarded as the ultimate confirmatory method. These approaches are described below as examples of less stringent approaches that nonetheless are designed to be fit for their intended purpose. Clean-up procedures are crucial for any successful 'dioxin' analysis and considerable effort has been devoted to development of faster approaches for the 'gold standard' method, as well as for proposed screening methods for a range of matrices (Focant 2004; Reiner 2006). Particular attention has been devoted to automation and integration of the extraction and clean-up steps to facilitate higher throughput for biological sample preparation (Focant 2004). Pressurized liquid extraction (PLE, Section 4.3.2b) has received considerable attention in this regard since it lends itself to automation such that up to 24 samples can be extracted in unattended operation. Microwave assisted extraction (MAE, Section 4.3.2d) and supercritical fluid extraction (SFE, Section 4.3.2e) have also been investigated in this context. However, none of SFE, MAE and PLE can readily accommodate large volumes of liquid samples with low lipid content required for 'dioxin' analyses in water samples (the aqueous solubilities of these hydrophobic compounds are extremely low), and more complex (and thus slower) methods must be considered, e.g. C$_{18}$-based SPE (Section 4.3.3.c).

Clean-up procedures can not be allowed to become less effective in order to achieve higher throughput for these demanding analyses at the expense of less reliable results, so this presents a particularly difficult challenge. Classical column chromatography using stationary phases such as silica, alumina and Florisil (a proprietary magnesium silicate with basic properties) is well established in this analytical application. Sulfuric acid digestion followed by clean-up on silica gel and alumina has been widely used for 'dioxins' in biological matrices; sulfuric acid–silica columns were introduced to remove the bulk of the lipids and other oxidizable components, and basic alumina columns were then used to separate dioxins from common environmental halogenated compounds, e.g. pesticides. Activated carbon sorbent provides a complementary fractionation tool to alumina due to its affinity for planar aromatic systems, especially those with adjacent aromatic rings and electronegative substituents; as a result it can fractionate the planar dioxins, furans and PCBs from other classes of aromatic compounds.

Some of these approaches have been incorporated into a system as interchangeable cartridges (silica, alumina, carbon) that can be used in different configurations (Turner 1992). A commercially available system (Power-Prep system, FMS Inc., Waltham, MA, USA, Figure 11.21(a)) can clean up 10 extracted samples each containing up to one gram of lipids in less than two hours (Eljarrat 2001). A further refinement (Pirard 2002) permitted adjustment of the system to isolate fractions containing PCDD plus PCDFs, coplanar PCBs (i.e. no chlorine substitution *ortho* to the biphenyl linkage), mono *ortho*-PCBs, and the seven WHO marker PCBs, from a single sample clean-up. This approach achieved analysis for all congeners assigned a TEF value and thus a total TEQ for each of the processed samples. Recoveries, repeatability, reproducibility and robustness were evaluated for commonly processed foodstuffs, and a validated procedure involving PLE, automated clean-up and HRGC/HRMS was reported. Compared to manual sample preparation, a significant reduction in sample handling and glassware use was achieved resulting in the lowering of blank levels, a crucial point for analysis of ultra-trace 'dioxins' especially when disposable glassware is not used. This advantage was pursued by interfacing the automated clean-up system on-line with an automated PLE system (Figure 11.21(a)); the flowchart of the integrated procedure is shown in Figure 11.21(b). Many experimental difficulties had to be overcome (Focant 2002, 2004) and these problems were different for different matrices, e.g., solids vs aqueous liquids. Nonetheless, the authors (Focant 2002) were able to validate a method using the fully integrated apparatus for 'dioxins' in bovine and human milk as well as human serum, by comparison with a method incorporating Soxhlet extraction and manual SPE. As hoped, a further reduction in the blank levels and thus in the LLOQs was achieved; this was attributed to the disposable nature of all columns used. The total sample preparation time between reception of the sample and final concentration for GC/HRMS injection was decreased from several days to less than one-half of a working day (Focant 2002). This remarkable achievement emphasizes yet again the importance of the extraction and clean-up steps for mass spectrometric analyses at the trace and ultra-trace levels.

A somewhat different approach was adopted for a specialized analysis for dioxin-like compounds in human serum at the sub-picogram level (Focant 2006). This sample preparation procedure is based on manual small

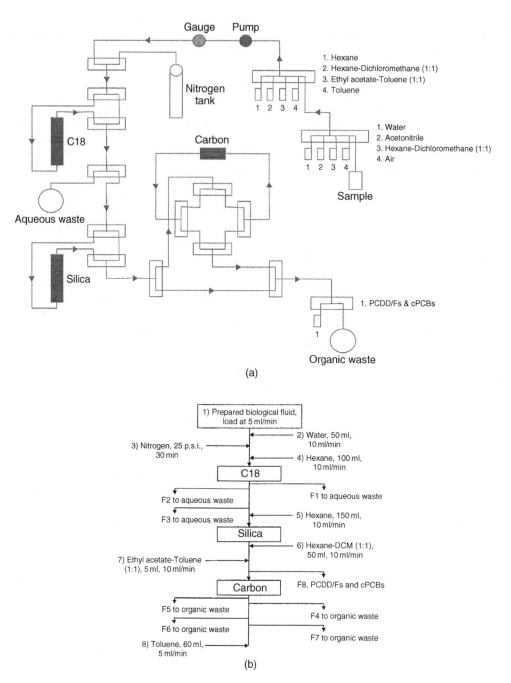

Figure 11.21 (a) Modified diagram of the integrated extraction and clean-up system for analysis of dioxin-like compounds. The set of solvents shown here was used for preparation of milk samples; if serum samples are analyzed, acetonitrile is replaced by methanol. PC-controlled valves are used for solvent selection and flow direction within the system. (b) Flow chart of the events occurring during automated extraction and clean-up; each step occurs sequentially from sample load (step 1) to final back-flush elution of the carbon column using toluene (step 8), that produces the PCDD/PCDFF plus cPCB fraction (F8). The solvation and conditioning steps are not shown for clarity. Reproduced from Focant, *J. Chromatogr. B* **776**, 199 (2002) copyright (2002) with permission from Elsevier.

sized SPE cartridges (to replace the PLE method shown in Figure 11.21) followed by the automated clean-up and fractionation discussed above; all SPE cartridges and clean-up columns are disposable. The manual extraction procedure was chosen in order to be able to analyze marker-PCBs in addition to the dioxin-like compounds that contribute to the TEQ. Samples were processed in batches of 20 that included one blank control sample and one QC sample, and were analyzed by GC–HRMS using ^{13}C-labeled internal standards. The sample throughput, once the system had been set up and optimized, corresponded to one set of 20 samples in one day, from sample reception to data quality cross-checking and reporting; four analysts were required to ensure acceptable performance of the overall procedure, which was validated with respect to sensitivity, LODs, LLOQs, accuracy (trueness), precision and inter-laboratory studies for 'dioxin' analysis. The procedure was also tested using more than 1500 unknown samples during various epidemiological studies.

The gold standard 'dioxin analysis' still requires GC with HRMS; no other analytical technique provides the same combination of selectivity and sensitivity (Reiner 2006). However, the capital and maintenance costs of a double focusing magnetic instrument are high, and some effort has been expended in finding alternative technologies for screening methodologies. In particular, tandem mass spectrometry in various formats has been extensively investigated since, in general, low resolution MS/MS detection is more selective for 'dioxins' and 'furans' than HRMS since the facile neutral loss of COCl (molecular mass 63 and 65 for the two Cl isotopes) has not been observed for any other halogenated organic compound (Reiner 2006). However, this selectivity advantage of MS/MS (e.g. Charles 1989) is not universal; it was shown (Fraisse 1989) that, in analyses of fly ash extracts, more interferences were observed in the GC–MS/MS chromatograms than in those obtained using GC–HRMS. Thus at present GC–HRMS retains its status as the ultimate confirmatory method despite its (generally) lower selectivity for 'dioxins' and 'furans' because it reliably provides good selectivity with very good sensitivity, and also because there is no unique neutral loss that is characteristic and selective for the dioxin-like PCBs (facile loss of Cl_2 is observed for essentially all polychlorinated aromatics) so MS/MS exhibits poor selectivity for the PCBs.

For GC–HRMS, EI with a nominal electron energy $\sim$30 eV is generally used for 'dioxin' analysis instead of the otherwise almost universal 70 eV. Although such a decrease of electron energy results in a decrease in the total ionization yield, this is more than compensated by the simultaneous decrease in extent of fragmentation, so the end result is an increase in the yield of the molecular ions monitored by HRMS. Another more subtle effect (Green 1986; Maggio 1990) concerns a limitation on EI source efficiency arising from the space charge created by the high levels of ions from the helium carrier gas. Because of the uniquely high ionization energy of helium ($\sim$24.5 eV) this space charge suppression effect can be greatly decreased, while maintaining reasonable ionization efficiency for the analyte (ionization energies $\leq$ 10 eV), by operating at an appropriately lower electron energy. Note, however, that this space charge effect appears (Waddell 1987) to be appreciably less important for quadrupole analyzers, presumably reflecting the less demanding ion optics compared with double focusing magnetic instruments.

Some attempts have been made to retrieve the situation with respect to the sensitivity of low cost MS/MS instruments, particularly 3D ion traps, by investigating alternative ionization techniques. In particular, electron capture negative ion CI (NICI, Section 5.2.1) has drawn considerable attention in view of the high electronegativity of PCDDs, PCDFs and PCBs (Chernestsova 2002) that seemed likely to provide increased sensitivity, but this turns out to not be the case. In general the ionization efficiencies are equal to, or lower than, those for EI although the NICI efficiencies did tend to increase with increasing degrees of chlorination as expected from the increase in electronegativity. A notable exception to the general trend was observed (Koester 1992) for the key compound 2,3,7,8-TeCDD, for which the NICI efficiency was almost two orders of magnitude lower than for EI. The intrinsic irreproducibility of NICI (Section 5.2.1) is evident for these compounds also via their variation with minor changes in operating conditions. Thus, in the presence of traces of oxygen or water $(M − Cl + O)^−$ ions are observed (Hass 1979; Fung 1985) and for PCDDs ether oxidative cleavage of the two aromatic rings also occurs (Hunt 1975; Fung 1985). At lower source temperatures, with measures adopted to minimize oxygen and water, PCDFs yielded $(M − 34)^−$ ions with isotope patterns indicating replacement of one chlorine atom by a hydrogen atom, somewhat similar to behavior reported previously (Deinzer 1982) for alkoxy-PCDFs. With respect to MS/MS detection, the selectivity advantage via COCl loss observed in EI–MS/MS is lost in NICI since the predominant fragmentation reaction for almost all chlorinated aromatic compounds leads to formation of chloride ion. This fact can be exploited (Curtis 1997) in use of a GC/MS instrument as a chlorine-specific detector in seeking unknown components in complex biological and environmental extracts

(Milley 1997), but confers zero specificity on MS/MS detection for 'dioxins' and 'furans'.

One way to speed up the overall process is by manipulating the GC separation. Comprehensive two-dimensional GC (GC × GC, Section 9.5.5c) offers the possibility of avoiding co-eluting interferences and thus perhaps requiring a less stringent clean-up step in analyses for persistent pollutants including 'dioxins' (Panić 2006). Because of the nature of comprehensive GC × GC analysis, the second GC stage can be characterized as 'fast GC' requiring correspondingly fast MS response. In one example (Korytár 2005) NICI was used with a novel fast scanning single quadrupole analyzer (Section 6.4.2) in quantitation of several chlorinated pollutants including 'dioxins' and 'furans'. Promising results were obtained, but the remarkably low sensitivity for 2,3,7,8-TeCDD in NICI (see above) was again observed such that this particular approach will not be applicable to trace analysis of these compounds.

3D ion traps can also provide full spectral acquisition for both MS and MS/MS at modest capital cost. An interesting approach to exploitation of this property as a screening method by GC–EI–MS/MS (Eppe 2004) addressed the duty cycle limitation on sensitivity (Section 6.4.5) by increasing the volume of cleaned up sample extract injected into the GC. MS/MS rather than GC × GC was used to increase selectivity in this case. The use of ion trap MS/MS for analysis of PCDDs and PCDFs (but not PCBs) had been thoroughly investigated previously (Plomley 2000; March 2000). The alternative to increasing the injection volume was to increase the size of the analytical sample taken through the extraction–clean-up steps, but this would have increased the effort required in these steps thus negating the purpose of a screening method. The large injection volume of $10\,\mu L$ was accommodated by use of a large volume programmable temperature vaporizer (PTV–LV) injector (Section 4.4.4b), and was found (Eppe 2004) to be the best compromise between sensitivity requirements and the robustness required for a high throughput screening method.

Examples of the chromatograms obtained using the PTV–LV–GC/MS/MS method are shown in Figure 11.22. A comparison of the data obtained using this new method with those from splitless GC–HRMS involved analyses of five different matrices, namely, beef fat, yolk eggs, milk powder, animal feed and bovine serum, covering a concentration range of two orders of magnitude in TEQ (i.e. $0.2–25\,ng.kg^{-1}$). An analysis of variance (ANOVA, Section 8.2.6) was performed on the data for the two techniques (the cleaned up extracts analyzed were the same for both). In general, excellent agreement was obtained, although a systematic bias was observed for 2,3,7,8-TeCDF and proportional systematic errors (i.e. discrepancy increasing with increasing levels) were noted for 1,2,3,4,7,8-HxCDD and 1,2,3,4,6,7,8-HpCDD (Eppe 2004). It was concluded that the PTV–LV–GC/MS/MS method can be used as a cost-effective complementary method to splitless GC–HRMS for monitoring food and animal feed for PCDD and PCDF levels (but not the coplanar PCBs). However, the original purpose of the development as a high throughput technique, dealing with hundreds or thousands of samples per year, was not assessed (Eppe 2004). It was also emphasized that, although adequate sensitivity was obtained, this was achieved not only through use of the large volume injector but also by modifying the ion trap, e.g. by increasing the pressure of helium damping gas by a factor of $\sim$5 from the pre-set default value. As a result of this high pressure, the motion of the precursor ions was damped more than usual and a greater collisional excitation energy (RF voltage of 5–6V between the end caps) was required to provide sufficient CID efficiency. These modifications, together with continuous optimization of trap conditions, resulted in lowering the instrumental detection limit for 2,3,7,8-TeCDD by a factor of $\sim$3. It was emphasized (Eppe 2004) that such modifications required maintenance of the mass spectrometer at its maximum performance in order to detect levels in the low parts per 10^{12} range, and such a requirement is a key problem for any high throughput screening method.

A different approach to the same problem seeks to exploit the fast spectral acquisition rate of TOFMS as a detector for GC × GC. This approach has been compared to the GC–HRMS method in analyses for dioxin-like compounds in ash, sediment, vegetation and fish samples (Focant 2004a), and in fish, pork and cow's milk (Focant 2005a). It was also investigated for selected polybrominated diphenylethers, polybrominated and polychlorinated biphenyls, and organochlorine pesticides, in human serum and milk (Focant 2004b). A schematic illustrating the experimental arrangement of the GCxGC-TOFMS apparatus and how the data are handled and processed is shown in Figure 11.23. Note that, in order to provide a low cost instrument the TOFMS used was a single stage (no reflectron) analyzer providing full spectral acquisition rates up to 500 spectra s^{-1}, at effectively unit mass resolving power. The GC × GC separation enhances the sensitivity of the TOFMS analyzer by compression of the chromatographic peaks, and specificity is improved by using a dioxin-specific first stationary phase that separates congeners from one another and a second phase that isolates the target compounds from sample matrix interferences (Focant 2004b). The instrumental

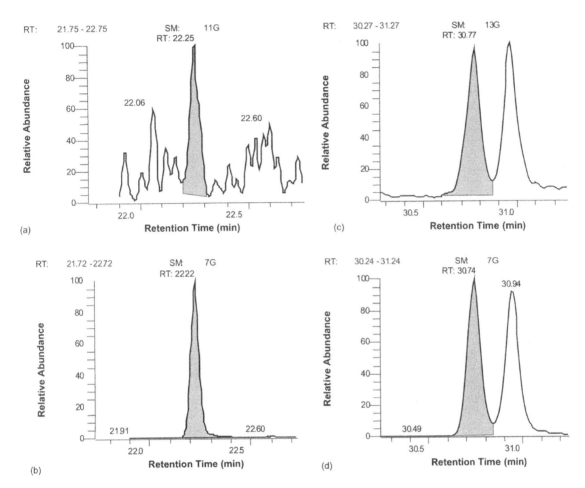

Figure 11.22 Typical PTV–LV–GC/MS/MS chromatograms for 2,3,7,8-TeCDD (a and b) and for 1,2,3,4,7,8-HxCDF (c and d) from a bovine serum QC sample. Chromatograms (a) and (c) correspond to the merged (summed) ion currents of the product ions (m/z $257 + 259$) and (m/z $309 + 311$) monitored for the native 2,3,7,8-TeCDD and 1,2,3,4,7,8-HxCDF, respectively. Chromatograms (b) and (d) are for the corresponding internal standards $^{13}C_{12}$-2,3,7,8-TeCDD and $^{13}C_{12}$-1,2,3,4,7,8-HxCDF, monitoring product ions (m/z $268 + 270$) and (m/z $320 + 322$), respectively. The large peaks eluting just after those for 1,2,3,4,7,8-HxCDF in (c) and (d) are for other congeners of HxCDF. The concentrations in the serum sample were 0.02 and 0.1 ng.kg^{-1} for 2,3,7,8-TeCDD and 1,2,3,4,7,8-HxCDF, respectively. Reproduced from Eppe, *Talanta* **63**, 1135 (2004), copyright (2004) with permission from Elsevier.

LOD, defined here as the lowest quantity of analyte injected on-column to be detected with respect to peak identification criteria of S/B > 3, plus verification of the isotope abundance ratio for at least two isotopologs of the molecular ion and retention time consistency in both GC dimensions, was 0.5 pg for 2,3,7,8-TeCDD (Figure 11.24). This represents an improvement by a factor of 5–10 compared with conventional GC–TOFMS but is an order of magnitude higher than for GC–HRMS instruments (typically 0.04 pg for 2,3,7,8-TeCDD).

The sensitivity of the GC × GC–TOFMS instrument is acceptable for measurements of dioxin-like compounds

in environmental matrices using the same sample size as for GC–HRMS, but is on the borderline of accept-ability for biological samples. Larger sample sizes are recommended (Focant 2004a) for accurate measurements. This presumably reflects at least in part the limited duty cycle (Section 6.2.3e) of a TOF analyzer compared with the high duty cycle of a magnetic sector instrument used in SIM mode. The stability was similar for the GC × GC–TOFMS and GC–HRMS instruments, but the tuning and everyday use of the former did not require such a high skill level as the latter, and the instru-ment cost is about half. The poor mass resolution of

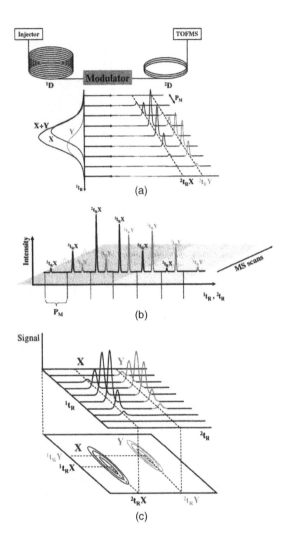

(a)

(b)

(c)

Figure 11.23 Schematic of the column coupling in the GC ×
GC–TOFMS apparatus and how the data are handled (not to
scale). (a) The modulator allows rapid sampling of the analytes
eluting out of the first-dimension GC (1D) and re-injection
into 2D. The modulation process is illustrated for two overlapping compounds (X and Y) eluting from 1D at a defined first-
dimension retention time ^1t_R. As the modulation process occurs
during a defined modulation period P_M, narrow bands of sampled
analytes enter 2D and appear to have different second-dimension
retention times $2tR(X)$ and $^2t_R(Y)$. (b) Raw data signal as
recorded by the TOFMS through the entire two-dimensional separation process. (c) Construction of the two-dimensional contour
plot from the high-speed secondary chromatograms obtained in
(b), in which similar signal intensities are connected by contour
lines. Reproduced with permission from Focant, *Talanta* **63**, 1231
(2004), copyright (2004) Elsevier.

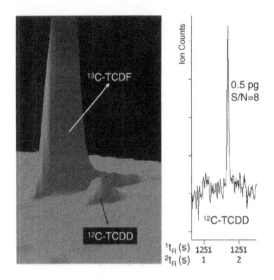

Figure 11.24   GC × GC analysis corresponding to injection of
1 μL of a standard solution containing native and labeled TeCDD
and TeCDF at concentrations of 0.5 pg. μL^{-1} and 100 pg. μL^{-1},
respectively. The deconvoluted ion current chromatogram was
reconstructed for *m/z* 322, the most abundant isotopolog of
the molecular ion of TeCDD. The GC × GC surface plot (left)
shows the small signal corresponding to native TeCDD; the
intense peak for labeled TeCDF arose from the unit mass resolution of the TOFMS instrument that cannot distinguish between
^{12}C-TeCDD (*m/z* 321.8936) and ^{13}C-TeCDF (*m/z* 321.9330).
Only one modulated peak was obtained indicating the low
concentration of the TeCDD analyte (right). Reproduced from
Focant, *Talanta* **63**, 1231, copyright (2004) with permission from
Elsevier.

the TOFMS analyzer is partly compensated for by the
GC × GC separation efficiency, which reduces the risk
of co-elution with interfering analytes of similar molecular masses and provides two retention time values for
enhanced selectivity. However, GC × GC–TOFMS full
spectral data files are appreciably larger and require
extended storage capacity (Focant 2004b). However, they
enabled identification of other environmentally significant compounds. Apart from the sensitivity issue, it will
be necessary to reduce data handling and processing
time requirements for GC × GC–TOFMS to be established as a true alternative to GC–HRMS (as opposed
to a screening method) for routine ultra-trace measurement of dioxin-like compounds in challenging foodstuff matrices (Focant 2005a). Although the cost per
sample for GC × GC–TOFMS is currently similar to
that for GC–HRMS, reduction of human effort in data
processing should eventually reduce the cost significantly
and make this single injection multiresidue method an

effective approach to obtaining a quick and complete picture of all the dioxin-like compounds (Focant 2005a).

A critical assessment (Focant 2005) of mass spectrometry-based analytical techniques for analysis of dioxin-like compounds compared the 'gold standard' GC–HRMS–SIM method with large volume injection GC with ion trap MS/MS (PTV–LV–GC/IT–MS/MS), fast GC coupled directly to a TOFMS (FGC–TOFMS, not discussed here) and GC × GC–TOFMS, all using EI and ^{13}C-labeled internal standards. None of the newer methods rivals the sensitivity (as indicated by LOD and LLOQ) usually provided by GC–HRMS–SIM; at least in part this must reflect the very high duty cycle available with SIM operation compared with limitations of TOF and ion trap analyzers in this regard (Chapter 6). It thus seems surprising that, e.g., GC × GC with a triple quadrupole operating in MRM and/or SIM mode has not been investigated as a medium cost alternative in view of the promising results obtained (Reiner 1991) using an early-model instrument in GC–MS/MS mode. The linear dynamic range was greater than three orders of magnitude and the repeatability was demonstrated by the RSD (<10 %) of the 2,3,7,8-TeCDD concentrations of five fish extracts rerun after three months without optimization of the tuning. Other advantages of this GC–MS/MS method (Reiner 1991) included selectivity for selected samples, infrequent tuning and reduced routine maintenance compared with the HRMS instrument. The sensitivity corresponded to a LOD~1 pg for all congeners of PCDD and PCDF, rather poor by GC–HRMS standards but surely capable of improvement by using a more modern triple quadrupole design. However, some laboratories have observed distortion of the chlorine isotope ratios with such MS/MS methods (a critical confirmation dimension for the analysis).

Nonetheless, the comparison of the three newer analytical methods with GC–HRMS (Focant 2005) provides some interesting information as a guide to choice of technique based on a fitness for purpose criterion. Figure 11.25 shows a qualitative comparison of the four compared techniques with respect to sensitivity (particularly LLOQ), selectivity and speed (sample throughput). The PTV–LV–GC–MS/MS (3D trap) and GC × GC–TOFMS methods both represent reasonably good compromises among these three figures of merit and have similar instrumental LOD values (0.2 pg), but the GC × GC–TOFMS method seems better able to handle more matrix interferences than the ion trap (reflecting the higher chromatographic resolution available) thus potentially reducing the clean-up requirements, and can also indicate the presence of other unsuspected compounds in the extract. More detailed information on the comparison among the four techniques

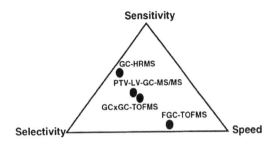

Figure 11.25 Comparison of rival techniques for GC/MS analysis of dioxin-like compounds with respect to three key figures of merit. Reproduced from Focant, *J. Chromatogr. A* **1067**, 265, copyright (2005) with permission from Elsevier.

is provided in Table 11.4 (Focant 2005). Clearly, although some of these methods require further investigation, they do offer (together with innovations in extraction–clean-up technologies, see above) a range of possibilities to suit different purposes (e.g., fast screening vs confirmatory analyses) in this extremely demanding analytical field.

As mentioned previously, the cost of analysis for dioxin-like compounds is much higher than that for most other analytical problems as a result of the challenging sample preparation and selectivity requirements to achieve the necessary ultra-low LODs and LLOQs. The lower cost MS-based methods discussed above represent one type of response to the challenge to produce effective screening methods with a low rate of false negative (false compliant) results and reasonably low rates of false positives (false non-compliant data). Another approach to the same problem exploits immunoassay and bioassay methods that can reduce costs by a factor ≥2 and provide fast results. The relative potencies for bioassays and cross-reactivities for immunoassay methods roughly mimic TEQs determined by mass spectrometric methods (see above). The main disadvantages of such methods (Reiner 2006) are their inability to use labeled internal standards to correct for recovery (the Method of Standard Additions, Sections 2.6.2 and 8.5.1b, can avoid this problem at the expense of many more analyses), low selectivity, and lack of information on congener profiles that can indicate the source of contamination.

A cell-based assay that has attracted a lot of attention in this regard is Dioxin-Responsive Chemically Activated LUciferine gene eXpression (DR-CALUX), a technique that detects all aryl hydrocarbon receptor (AHR) agonists (Murk 1996). This approach uses genetically modified

Table 11.4 Comparison among four analytical methods for dioxin-like compounds with respect to eight figures of merit (all use EI and ^{13}C-labeled surrogate internal standards)

Figure of Merit	GC–HRMS	PTV–LV– GC–IT– MS/MS	FGC–TOFMS	GC × GC– TOFMS
Capital investment cost Euros (USD)	~350 000 (~460 000)	~140 000 (~180 000)	~170 000 (~220 000)	~240 000 (~315 000)
Operating cost	+ + +	+ +	+	+
Sample throughput	+	+	+ + +	+
Number of analytes measured per unit time	+	+	+ + +	+ + +
Instrumental LOD	+ + +	+ +	+	+ +
Suitability for PCDD/PCDF ultra-trace analysis	+ + +	+ +	−	+ +
Suitability for PCB analysis	+ + +	+ + +	+ + +	+ + +
Ability to detect and measure unknown compounds	−	−	+ +	+ + +

The relative instrumental LOD values (iLODs) are presented as a measure of the intrinsic instrumental sensitivity; method detection limits will in general be significantly higher but should follow the same trend. The methods employing SIM or MS/MS detection are designed for target (known) analytes and thus can not provide information on other unknown compounds present in the extract; this same characteristic also implies that they are 'blind' to the presence of other compounds of potential interest. Instrumental LOD is a major factor in determining suitability for ultra-trace analysis. Capital costs are estimates based on 2006 information. Modified from Focant, *J. Chromatogr. A* **1067**, 265 (2005).

mammalian hepatoma cells that contain a transfected AHR-responsive luciferase reporter gene that responds to dioxin-like compounds via the induction of luciferase gene expression in a time-, dose- and compound-specific manner. The technique has undergone several comparisons with GC–HRMS and other validation procedures (Van Wouwe 2004, 2004a; Scippo 2006). One difficulty is that the relative potencies of dioxin-like congeners in the DR-CALUX assay are not additive, unlike the TEQ contributions in GC–MS methods (Equation [11.7]). Also, co-extracted compounds displaying either antagonistic effects or agonistic effects can modulate the DR-CALUX response. For example, the requirement to ensure, by additional clean-up steps if necessary, that brominated dibenzo-*p*- dioxins, furans and biphenyls are absent or removed for DR-CALUX to provide a TEQ value comparable with that obtained using GC–MS methods was demonstrated (Van Wouwe 2004). However, it was shown (Scippo 2006) that the DR-CALUX screening led to less than 10 % false positive identifications (false non-compliant) and no false compliant (false negative) results, and that a good correlation between GC–HRMS and DR-CALUX data was obtained. If these preliminary results can be confirmed on a larger number of samples the method should be capable of providing a screening assay for these compounds in food and animal feed.

11.4.2 Other Persistent Pollutants

Some recent reviews of analytical methods for persistent pollutants other than dioxin-like compounds are mentioned briefly here. Again, mass spectrometry is the key enabling technology that permits any progress in analysis of these various compound classes, though as always the sampling, extraction, clean-up and chromatographic steps must also be optimized. A review of methods for organochlorine pesticides and PCBs in biological samples (fish, aquatic and terrestrial mammals and birds, blood and milk), as well as in soils and sediments (Muir 2006), summarizes best-practice methods for these analytes but also emphasizes the need for laboratories in developing countries to develop some capability to perform analyses for these compounds. Examples of 'legacy' chlorinated pollutants that require specialized methods are the polychlorinated naphthalenes and toxaphene (Kucklick 2006). The latter is a pesticide produced by chlorination of bornane, consisting mostly of chlorobornanes with lesser amounts of chlorinated camphenes, dihydrocamphenes, bornenes and bornadienes. For the chlorobornanes only, the potential number of individual congeners containing from one to 18 chlorine atoms is 32 768 of which 32 256 are enantiomers (i.e. 16 128 pairs of enantiomers). However, this number of compounds is not found in technical toxaphene, and the actual number is probably closer to 1000, still a major challenge for the analyst.

Chlorinated paraffins are commercial mixtures of poly-chlorinated *n*-alkanes with carbon chain lengths from 10–30 and a degree of chlorination of 30–70 % by weight, obtained by direct chlorination of n-alkane fractions derived from petroleum distillation. The resulting formulations are complex mixtures containing thousands of different isomers, diastereomers and enantiomers that present yet another extremely challenging problem for the analyst (Santos 2006). Different formulations are used in a wide range of applications, including as additives in cutting fluids and high temperature lubricants for metal-working, plasticizers in paints, coatings and sealants, and as flame retardants in rubbers and textiles. Short chain chlorinated alkanes have been the most extensively used and are frequently detected in environmental samples. Such formulations are of particular interest because they exhibit the highest toxicity of all such products; they are very toxic to aquatic organisms, and studies in rats have demonstrated that they are carcinogenic. They are persistent, not hydrolyzed in water and are not biodegradable, and have a high potential for bioaccumulation (Santos 2006).

Polybrominated diphenyl ethers (PBDEs) are another class of persistent pollutant that has attracted increasing attention. PBDEs are used as flame retardants, incorporated into a number of polymers and resins and are used in many consumer products. Until recently almost 70 000 tons were produced every year, half in products sold in the USA and Canada. In 2004 the European Union phased out the use of two of the three commercial PBDE mixtures (PentaBDE and OctaBDE) and, shortly afterwards (January 2005), the only manufacturer in the USA voluntarily agreed to stop manufacturing these two PBDE commercial mixtures. Again these compounds present a difficult analytical challenge, whether present in environmental matrices (Stapleton 2006) or in food (Gómara 2006). Although analytical methods are available for quantifying tribrominated through heptabrominated congeners found in the commercial PentaBDE and OctaBDE mixtures, the analysis of congeners with eight or more bromines has proven to be difficult. For example, an interlaboratory comparison on PBDE measurements in 1999–2000 gave acceptable agreement for most of the lower brominated congeners, but reported values for the fully-brominated congener (BDE 209) varied over almost two orders of magnitude (Stapleton 2006).

Paper mill effluents contain a large range of compounds and possess high toxicity overall, yet another analytical challenge (Lacorte 2003). Classes of compounds include natural products such as resin and fatty acids (wood extractives), additives used during paper making such as biocides, surfactants and phenolic compounds, and by-products generated during bleaching including PCDDs and PCDFs. Obviously such a wide range of compounds can not all be accommodated within a single analytical method. Several extraction methods including LLE and SPE are used, in combination with GC/MS or LC–MS with APCI or ESI (Lacorte 2003).

Finally, polycyclic aromatic hydrocarbons (PAHs) have long been known to be dangerous pollutants. Under the right conditions PAHs can persist in the environment and accumulate, e.g. in anaerobic sediments, to such an extent that the potential for adverse health effects is high. Some PAHs are relatively potent carcinogens and biological and mutagenic effects are well documented. GC (rather than LC) with mass spectrometric detection is usually the preferred analytical method for extracts containing PAHs because GC generally provides better selectivity, resolution and sensitivity than LC (Poster 2006). Enhanced GC methods, e.g. large volume injection (PTV–LV–GC), fast GC, and coupling of GC to LC, have all been used.

11.5 Bioanalytical Applications

This category includes one of the most active areas in trace analysis using mass spectrometry, namely, activities conducted to support the discovery, development and eventual regulatory approval of new pharmaceutical drugs. Indeed, development of reliable LC–MS/MS instrumentation can fairly be said to have revolutionized this important activity since about 1990. Because life processes occur in aqueous media, most of the chemical compounds involved (including pharmaceutical drugs and their metabolites) are polar and well suited to analysis by LC–MS/MS. Several papers in this area have already been cited in previous chapters to illustrate some point of importance more generally in trace analysis. Recently an excellent book (Korfmacher 2004) containing chapters by expert authors has described the practices currently used in this demanding area; some recent reviews (Hopfgartner 2003a, 2005; Xu 2005; Zhou 2005a) are also sources of valuable information and overviews. The somewhat similar term 'biomedical analysis' is usually understood to refer to applications such as those undertaken in support of clinical diagnoses by physicians; indeed 'clinical chemistry' is a well recognized field in which mass spectrometry is playing an increasingly important role, but is not discussed here for reasons of space.

The demands on the analyst in the pharmaceutical environment are somewhat different in the drug discovery and development phases. In drug discovery the data generally are not required to be submitted to regulatory authorities and it is often possible to develop generic quantitative analytical methods for large libraries of potential drug compounds, 'new chemical entities' (NCEs). These methods need not be completely validated and are intended to provide pharmacokinetic data of reasonable quality (accuracy and precision ~30% compared with 15% for drug development) necessary to determine if the compound should be selected for further drug development studies. In contrast, data acquired during the drug development process must be validated to satisfy regulatory requirements (e.g., FDA 2001). In the development phase the current emphasis is on leveraging the selectivity and sensitivity afforded by LC–MS/MS to increase the speed of development and validation of the analytical method for compounds selected for further development, and also on reducing times required per analysis (higher sample throughput) that can be important when the drug candidate is at the stage of clinical trials. Mass spectrometry, particularly LC–MS technology in combination with developments in extraction–clean-up and chromatography, has revolutionized both of these aspects and examples from both have been chosen for discussion. While the recent emphasis has been on analytical speed, it is pertinent to ask not only whether the method is fast enough, but also whether it is too fast. In the latter case it is possible that interference from a co-eluting metabolite that can undergo in-source conversion to an ionized molecule (e.g., $[M + H]^+$), identical to that of the drug molecule itself (Section 9.3.4), will occur (Jemal 1999; Naidong 1999; Tong 2001).

11.5.1 Drug Discovery Methods

An informative introduction to the concepts involved in drug discovery (Korfmacher 2005) also emphasizes the role played by LC–MS techniques. As discussed in that review, the discovery chain involves at least the following identifiable steps:

1. chemistry (e.g. systematic modifications of suspected pharmacophores by combinatorial syntheses);
2. biological high throughput screening of the mixtures of compounds produced in step 1., usually by exposing the mixture to a protein selected as a potential target for pharmaceutical intervention, and using a fluorescence technique (e.g., fluorescence is observed only when a compound binds to the protein) to find those compounds in the mixture that exhibit this desired *in vitro* activity (LC–MS is then used to identify these lead compounds);

3. an *in vitro* passive permeability test performed early in the discovery process to rapidly screen (using LC–MS) the ability of these lead compounds to cross the intestinal barrier, a crucial step in the absorption of an oral dose of the drug candidate into the blood (the most common of these tests is that based on a human colon adrenocarcinoma cell line Caco-2);
4. a screen to determine the potential of a candidate compound for drug–drug interactions by measuring the inhibition or induction of metabolism of a standard set of compounds by the cytochrome P450 (CYP450) enzymes (this assay is unique in that it does not measure the concentration of the new drug candidate, but rather those of the suite of test compounds, so that it is not necessary to develop a new analytical method for each new candidate);
5. a high throughput *in vitro* assessment of the likely *in vivo* clearance rate of the drug candidate, controlled partly by enzymatic conversion to readily excreted metabolites in the liver, usually performed by incubation of the compound with human hepatocytes (whole liver cells) or their microsomes (sub-organelles that contain the CYP450 family of enzymes responsible for metabolic activity), with tracking of the metabolism kinetics using LC–MS assays;
6. an *in vivo* version of step 5. using laboratory animals (often a rat model) to indicate more closely the likely pharmacokinetics of the drug candidate (kinetics of absorption into and clearance from the blood and possibly also important body organs, i.e. absorption–distribution–metabolism–excretion (ADME) studies) in humans;
7. identification of metabolites and measurement of their rate of appearance and yield in excreta;
8. safety checks and initial estimates (again using animal models) of likely effective pharmaceutical dose and pharmaceutical ratio (ratio of the dose at which adverse effects start to appear to the effective pharmaceutical dose).

This multistep procedure, used to discover likely drug candidates for treatment of a specific disorder identified via the nature of the target protein(s) used in step 2., will reduce the number of compounds in the chemists' original mixture from thousands to just a few (or possibly zero!).

Only after this meticulous series of screens and tests has been completed can a drug candidate proceed to the development phase using clinical trials on human patients. These trials are conducted in four phases; Phase 1 trials, typically conducted on a few (10–30) healthy volunteers, are designed to indicate the effective safe dose via dose escalation and pharmacokinetics (PK) studies. Phase II–IV trials use increasingly large patient cohorts and post - marketing studies to better define the ADME–PK parameters, side effects, dose response characteristics and comparisons with existing treatments (drug-drug interaction studies).

At all of these phases it is important for the LC–MS laboratory to provide fast sample turnaround of PK and related data in order to provide the clinicians conducting the trials with timely information concerning the individual patients. Thus analytical speed and efficiency are important. However, unlike in drug discovery, these data must be obtained using validated methods with high levels of accuracy, precision and reproducibility (Chapter 10), since they form part of the submission to regulatory authorities for approval for use. A recent book (Wells 2003) provides an extensive discussion of sample preparation methods (extraction and clean-up) for high throughput bioanalyses emphasizing protein precipitation, liquid–liquid extraction and SPE approaches. An interesting account (Murphy 2002) of the effect of HPLC flow rate on quantitative analysis of plasma samples with only protein precipitation (Section 4.3.1f) as extraction–clean-up demonstrated yet again that ESI sensitivity decreased with flow rate (Section 5.3.5b) in the range often used for high throughput analyses in a drug discovery context. A major finding of this work was that the price paid for increasing flow rate (an approximately proportional decrease in analyte response) was not fully recompensed by a proportional decrease in cycle time. An excellent critical review (Tiller 2003) discusses approaches to fast HPLC separations that also meet mimimal criteria for acceptable chromatographic parameters, i.e. for isocratic methods the analyte(s) must elute with $k' > 1$ (Section 3.3.4), while for gradient elution methods anaytes that elute before two column volumes (i.e. still within the solvent front) and after the end of the gradient were considered unacceptable. Another requirement for a fast HPLC method to be deemed acceptable was appropriate interfacing to the mass spectrometer used. An indication of the rate of innovation in this area is provided by the mention (Tiller 2003) of monolithic columns (Section 3.5.8) and ultra-small particle packings (Section 3.5.7) as emerging technologies that are now well established (only four years later).

It is striking that, despite all the technological progress in achieving effective high throughput analytical methods for drug discovery screens, of the very small fraction of drug candidates that survive the screening procedures to enter Phase 1 clinical trials a significant fraction fail at this early stage, and only about one in five will finally receive regulatory approval (Nolan 2007). Furthermore, even approved drugs can fail later when prescribed to a large cohort of the general population since some toxic side effects can be discovered only in the light of experience with much larger numbers of patients than are feasible in any clinical trial. It has been suggested (Nolan 2007) that the reason for this disappointingly large failure rate does not lie within the analytical chemistry involved but can be attributed to inadequate initial biological screening (step 2 in the above summary of the screening procedures). Essentially, it is argued that most current biological screens monitor interactions with a single purified protein or sometimes with some specialized cell line, and that such screens are too simplistic to account for the complex networks of interactions that can follow administration of this compound to a human being. If the quality of the drug candidates that enter Phase 1 trials could be improved by devising biological screens that take better account of this complexity, the resulting savings in time and money from reducing the number of failed clinical trials should more than compensate for the extra effort expended in the biological screening step (Nolan 2007).

The preceding paragraph is intended to put into perspective the role of the analytical chemist in the overall drug discovery process; this is not to say that further improvements are not possible or desirable in the analytical methods currently practised. The examples chosen to illustrate some of these activities in the present book can not claim to provide anything like complete coverage, and were chosen mainly on the basis of idiosyncratic interests of the authors to illustrate some innovative development in application of mass spectrometry. The cited introduction/review (Korfmacher 2005) discusses good examples illustrating all the steps involved in the drug discovery process.

11.5.1a Passive Permeability Test (Caco-2 Assays)

Caco-2 cell assays (Figure 11.26 shows a schematic of the general principle) are nowadays automated so that permeability screening has been speeded up. Simultaneously, the number of drug candidates emerging from the biological high throughput screening has significantly increased and this combination places a strong demand on the LC–MS analyst to keep up with the demand. A major concern for MS analysis of the solutions sampled from the Caco-2 cell assay (Figure 11.26) is the concentration of nonvolatile buffers, such as Hanks balanced salt solution (HBSS), normally used in these assays. These buffers must either be removed prior to introduction of samples into the ESI–MS source or, if LC–MS is used, diverted away from the source as they elute with the void volume

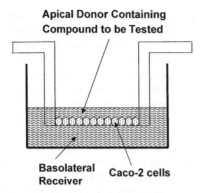

Apical Donor Containing Compound to be Tested

Basolateral Receiver   **Caco-2 cells**

Figure 11.26 A schematic diagram of an apparatus used to obtain estimates of the passive permeability of a drug candidate across the intestinal mucosa using Caco-2 cells. A monolayer of Caco-2 cells is grown on a porous polyethylene terephthalate (PET) membrane (a so-called 'confluent' monolayer of cells that grows only in two dimensions on such a substrate from an initial small inoculation). In the experiment the cells are submerged in Hanks's Balanced Salt Solution (HBSS) buffer (contains Na^+, K^+, Cl^-, phosphate, glucose, and in some formulations also Ca^{2+}, Mg^{2+} and SO_4^{2-}); the Caco-2 cell layer provides the only connection between an apical (donor) reservoir, into which the drug candidate is dosed, and a basolateral (receiver) reservoir. For the assay, aliquots are removed for analysis from the apical reservoir at 0 min, and from both reservoirs at 120 min. Reproduced from Van Pelt, *Rapid Commun. Mass Spectrom.* **17**, 1573 (2003), with permission of John Wiley & Sons Ltd.

of the column before the target analytes elute. Since these quantitative assays are not required to be validated in any regulatory sense, and since the analytical samples represent relatively simple matrices, it has been possible to devise LC–MS/MS methods that combine minimal sample clean-up with short LC retention times (using columns only <50 mm long) and the high selectivity of MS/MS detection, that are fit for purpose in this context. However, when gradient elution is required the time per analysis (including column re-equilibration) can be as high as five minutes. Accordingly, there are on-going efforts to devise ways of increasing the throughput of these analyses.

In one such approach (Van Pelt 2003) off-line protein precipitation (using acetonitrile, a commonly used simple clean-up in drug discovery experiments) and a desalting step (using a C_{18} ZipTip, Millipore), are followed by spiking with an internal standard and analysis by nanoESI–MS/MS using a liquid-handling robot (Nano-Mate™ 100, Advion Biosciences) and a chip containing a 10×10 array of nanosprayer nozzles (ESI Chip™). The mass spectrometer used (Van Pelt 2003) was not a triple quadrupole (the instrument of choice for validated analyses) but rather a 3D ion trap that is fit for purpose

in this case. This approach offered reduced method development time and the automated nanoESI–MS/MS analysis did not exhibit any carryover (Section 9.7.2) between samples; moreover, the approach required only low sample consumption and provided increased ease-of-use as compared with conventional pulled-capillary nanoESI. Comparison of the permeability values obtained using this automated infusion–ESI–MS/MS approach (no HPLC separation) with those obtained using a conventional HPLC–MS/MS approach (Van Pelt 2003) for drug candidate compounds indicated quantitative agreement in the range 5–15%. The rate-limiting step in the chip-based approach was the cycle time (40 seconds) of the liquid-handling robot, so in principle this device could analyze samples in a 96-well plate in under 75 minutes, probably ~9 times faster than the conventional HPLC–MS/MS approach. However, no further reference to this approach to Caco-2 screening assays appears to be available in the literature at this time.

A different concept devoted to increasing the throughput of Caco-2 assays involves testing several compounds simultaneously in the same assay ('sample pooling' or 'cassette dosing'). An early investigation of the validity of this approach (Bu 2000) was motivated by concerns that some of the tested compounds might interfere with one another. These concerns are founded on details of the membrane transport process itself. Transport of orally-dosed drugs across the intestinal membrane (for which the Caco-2 cell layer provides an experimental surrogate) is a complex dynamic process that involves mechanisms for both influx (from the intestine into the blood) and efflux. The latter limits the bio-availability of such compounds, and in particular P-glycoprotein (P-gp) is a well known mediator of efflux for several widely used drugs. In the context of simultaneous multicompound screening it was of concern that possible drug–drug interactions (e.g. inhibition of P-gp-mediated transport of one drug by another and/or competition of the drugs for transport pathways) might give rise to inconsistent results compared with those obtained using a traditional single drug dosing approach. However, the apparent permeability coefficients of up to five test drugs measured by the sample pooling or cassette dosing approach were in good agreement with the data obtained by single drug dosing followed by discrete sample analysis (Bu 2000). Nonetheless, this reservation concerning multi-compound screening must be borne in mind, as confirmed in a later elaboration (Laitinen 2003) of the earlier investigation.

These early investigations of compound pooling in the Caco-2 assay (Bu 2000; Laitinen 2003) used conventional gradient LC–MS/MS analytical methods with diversion of the first one to two minutes of the eluent to waste to avoid fouling of the ESI source by the buffer salts; the LC–MS/MS cycle times were several minutes. A more recent development (Smalley 2006) achieved sample clean-up and de-salting using on-line extraction turbulent flow chromatography (Sections 3.5.9 and 4.3.3d) coupled to a conventional C_{18} reversed phase column and a triple quadrupole mass spectrometer equipped with a hot air gun to assist vaporization of the high aqueous eluent. The LC–MS/MS cycle time was 6.5 minutes, but this was effectively reduced to 3.5 minutes through the use of a dual column arrangement that allows the mass spectrometer to acquire data from one column while the other is re-equilibrating (Smalley 2006). An entirely different approach to increasing sample throughput (Fung 2003), that did however develop further the multiplexing of HPLC columns (Figure 11.27), showed that it was not

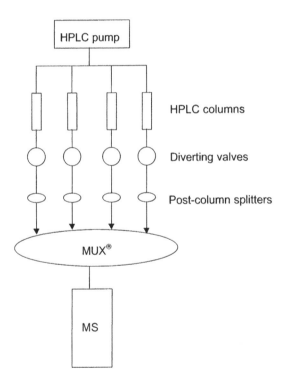

Figure 11.27 Schematic diagram of a multiplexed HPLC–MS system employing a single mobile phase pumping system (gradient elution) and just one shared triple quadrupole mass spectrometer. Four samples using the same HPLC gradient are sequentially introduced into the mass spectrometer in a sampling cycle that is fast relative to the LC peak width. The multiplexed ion source (MUX1™) consisted of four electrospray needles attached to a rotating dish that contains an orifice, allowing independent sampling from each sprayer. The re-equilibration time of each column was used to generate data from the next one. Reproduced from Fung, *Rapid Commun. Mass Spectrom.* **17**, 2147 (2003), with permission of John Wiley & Sons Ltd.

necessary to obtain and use a calibration curve for each new compound in determination of Caco-2 permeabilities, since the crucial measured quantity is the concentration ratio between apical and basolateral compartments (Figure 11.26); this realization permitted considerable time savings and thus higher throughput.

11.5.1b Drug–Drug Interaction Studies (CYP450 Assays)

Evaluation of potential drug–drug interactions is an important consideration in drug discovery and the main technique used to screen for this activity involves the inhibition or induction of metabolism of drug candidates by the cytochrome P450 (CYP450) enzymes. This superfamily of oxidases are structurally related, but the differences confer different kinds of substrate specificity to individual CYP450 isoenzymes that are responsible for oxidative metabolism of the majority of drugs and chemicals to which the population is exposed. Three of the families (CYP1, 2 and 3) are primarily responsible for the majority of metabolism of xenobiotics and other families are mainly involved in metabolism of endogenous compounds in the body. About five to six isozymes within the first three families (CYP3A4, CYP2D6, CYP2C9, CYP2C19, CYP2E1 and CYP1A2) have been found to metabolize > 95 % of pharmaceutical drugs currently prescribed and are therefore the primary targets for screening drug candidates for potential inhibitory and/or induction activities. For example, inhibition of one or more of these enzymes by a drug could result in concentration of a co-administered drug in body tissues, which may in turn cause adverse side effects due to levels above the toxicity threshold.

In practice, these screens for potential drug–drug interactions use a cocktail of five to six small molecule compounds, each of which has been determined to match closely the metabolic specificity of one of the CYP450 enzymes of interest. Several different cocktails of probe compounds have been used in the past (Scott 1999) and variations continue to appear (Peng 2003). One such cocktail is shown in Figure 11.28, together with examples of the MS and MS/MS spectra used to choose MRM transitions used to monitor the LC–MS/MS analyses of the metabolites, as well as an example of the MRM chromatograms obtained for an incubation of a candidate drug plus the probe cocktail with human microsomes containing the CYP450 enzymes (Peng 2003). A generic LC–MS/MS method was used, as first described earlier (Scott 1999), that permits analysis of all six probe metabolites in a single injection rather than the laborious one-by-one analyses that had been used previously. In this newer method, however, very fast HPLC separations (< 0.5 minutes, Figure 11.28) were achieved (Peng 2003) through use of a fast gradient and a short monolithic column (50 mm × 4.6 mm i.d.) that permitted (Section 3.5.8) high flow rates (5 mL.min^{-1}), together with fast MRM switching of a triple quadrupole instrument.

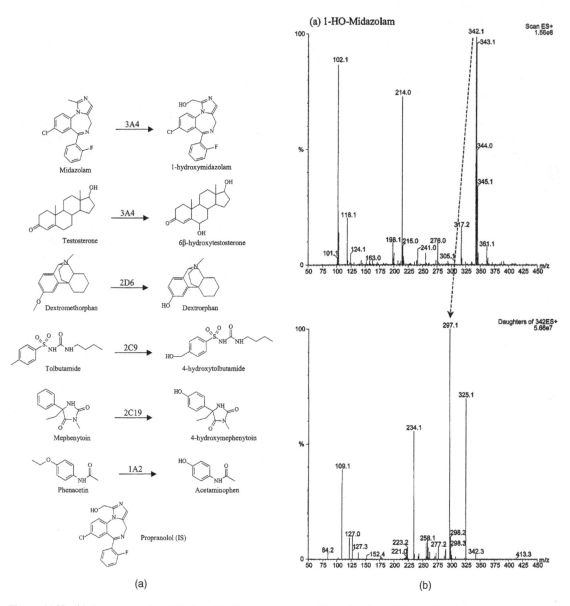

(a)   (b)

Figure 11.28 (a) Six compounds and their metabolites chosen as specific probes for activity of six CYP450 isozymes mainly responsible for metabolism of xenobiotic compounds. (b) An example of the ESI–MS and ESI–MS/MS spectra used to select MRM transitions used in the LC–MS/MS assays for inhibition/induction of a specific isozyme. (c) HPLC–MRM chromatograms for the internal standard (propronanol) and for the metabolites of the six probe compounds obtained for a microsomal incubation. Reproduced from Peng, *Rapid Commun. Mass Spectrom.* **17**, 509 (2003), with permission of John Wiley & Sons Ltd.

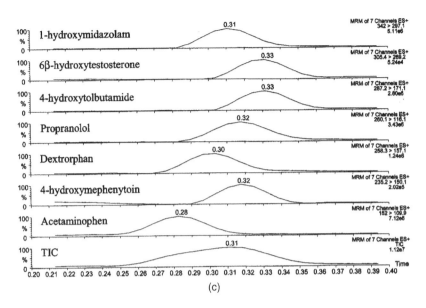

Figure 11.28 (Continued)

Although CYP450 inhibition screens are not required in the submission to regulatory bodies for drug approval, some degree of validation is advisable to ensure sufficient accuracy and precision for decision making. In this case (Peng 2003) the intra-assay precision of peak area ratios for metabolite:internal standard were always <10 %, while the inter-assay precision was surprisingly better at <8 %; the corresponding accuracies (percentage relative error) were −10 to +11 % and −6.5 to +7.7 %. Positive QC samples (incubation of the probe cocktail with a known inhibitor for a specific enzyme, but with no drug candidate present) and negative ones (neither inhibitor nor candidate compound present) were included in each assay to ensure the integrity of the microsomal incubation system. For each drug candidate a series of assays were performed at different concentrations and the IC_{50} value was obtained by nonlinear regression analysis of the curve describing the relative enzymatic activity (expressed as a percentage of the data obtained for the negative QC sample) as a function of concentration of the drug candidate (Peng 2003). (The IC_{50}, the half maximal inhibitory concentration of a specific compound with respect to a specific biological process, is a measure of how much of the compound is needed to inhibit the process by 50 %). This detailed method was made feasible by the very short analysis time that allowed multiple determinations per day but, as for any fast LC–MS method, the possibility of suppression of ionization efficiency by co-eluting compounds (Section 5.3.5a) must still be taken into consideration.

11.5.1c Clearance Rate Tests (Metabolic Stability)

The next *in vitro* high throughput screen to which drug candidates are usually subjected during drug discovery is the metabolic stability test that provides some information on the likely clearance rate of the compound *in vivo*. Again the CYP450 enzymes are the agents mainly responsible but, unlike the Caco-2 permeability screens in which the drug candidates themselves are not analyzed, in this case they are indeed the target analytes. The state of the art up until about 2003 is the subject of an extensive review (Ansede 2004) and is well exemplified by development (Xu 2002) of an eight-channel parallel LC–MS system (Figure 11.29) combined with an incubation procedure based on 96-well plates and robotic liquid handling systems. This highly automated parallel LC–MS system was capable of analyzing up to 240 samples per hour and thus permitted complete profiling of two 96-well plates per day (i.e. 176 drug candidates plus 12 controls, Figure 11.29). Column-to-column variation and reproducibility over a 17 hour period (approximately 500 injections per column) indicated (Xu 2002) that variations in retention time and peak area were < 2 and 10 %, respectively, in both tests; the results of time-course microsomal incubations at four time points (each analyzed in triplicate, Figure 11.29) allowed estimation of the half-life ($t_{\frac{1}{2}}$) to characterize the microsomal stability of each candidate compound. Any such highly automated high throughput LC–MS assay requires considerable software development to control the data acquisition and,

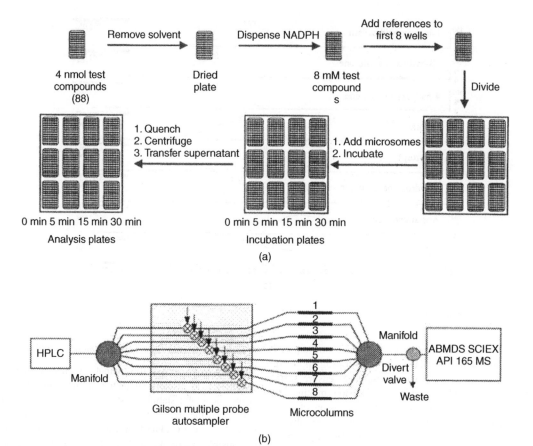

Figure 11.29 (a) High throughput microsomal stability procedure conducted in 96-well plate format. Each original plate containing 88 drug candidates was dried, and nicotinamide adenine dinucleotide phosphate-oxidase (NADPH) is added (the essential coenzyme for the CYP450 enzymes that generates the highly reactive free radical superoxide O_2^- by transferring electrons from inside the cell across the membrane and coupling these to O_2). Eight reference compounds with known behaviour under the enzymatic conditions are then added to the empty eight wells to provide a check on the potency of the enzymes in the liver microsomal incubation. The contents of each of the original 96 wells is then subdivided into 12 sub-samples in low volume 96-well plates, microsomes are added and incubation intitiated. Samples are taken for analysis from all wells of the first set of three low volume plates at time zero, thus providing t = 0 analyses in triplicate, and similarly for the remaining sets of three plates at 5, 15 and 30 min. Thus each of the 96 wells in the original sample plate yields 12 analytical data points (4 time-points each in triplicate). (b) High throughput 8-channel parallel LC/MS system used to analyze samples for the microsomal incubation. The microbore columns ($10\,mm \times 1\,mm$ i.d.) were packed with $3\,\mu m$ C_{18} particles, full loop injections ($20\,\mu L$) into a gradient flow rate of $250\,\mu L.min^{-1}$ for each of the 8 columns ($2\,mL.min^{-1}$ total). The flow was diverted to waste by a switching valve in the early part of the gradient to protect the mass spectrometer from contamination with salts etc. Reproduced from Xu, *J. Amer. Soc. Mass Spectrom.* **13**, 155, copyright (2002), with permission from Elsevier.

equally important, to process the data into the desired final results while monitoring the quality of the data. This software aspect is emphasized in later works (Fung 2004; Yan 2005).

A different approach to the analysis of the large numbers of samples generated in metabolic stability screens (Zhang 2000a; O'Connor 2006) used a combination of HPLC (in one case with an ultra-small particle stationary phase, UPLC, Section 3.5.7) with TOFMS analysis. The latter is currently not fit for purpose with respect to the fully validated pharmacokinetic analyses required in the drug development stage, but can provide data of adequate quality for a drug discovery screen and indeed offers two advantages over triple quadrupole MRM detection in this context (Figure 11.30). Development of different MRM conditions for each analyte is not required, since now

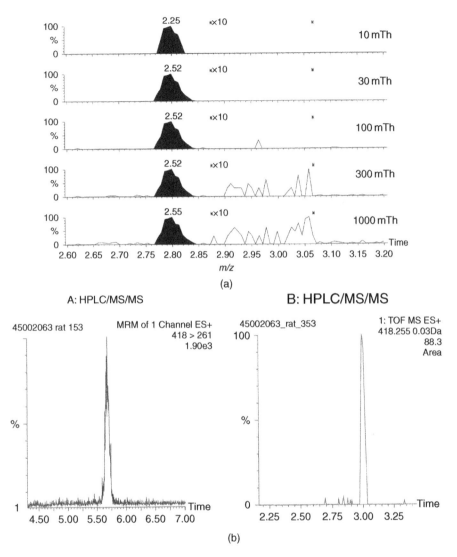

(a)

(b)

Figure 11.30 (a) Demonstration of the effect of using more narrow *m/z* windows centered on the observed *m/z* value (significant to within a few mTh) of the [M+H]$^+$ ion when extracting SIM chromatograms from the full spectral UPLC–TOFMS dataset for a metabolic stability sample; a portion of the baseline is magnified ×10 in each chromatogram. The filled peaks are those for the analyte and are annotated with the peak areas arbitrary units); note that on reducing the *m/z* window to 30 mTh the UPLC peak is significantly more narrow and its peak area is ~10% less than that obtained using wider windows. (b) Comparison of analyses of the same metabolic stability sample using LC–MS/MS (triple quadrupole, MRM) and UPLC–TOFMS (SIM chromatogram extracted using a 30 mTh window). This sample contained the lowest analyte concentration (97% metabolic conversion of drug candidate) in the study. Note the better S/N ratio for UPLC–TOFMS despite the much lower absolute signal (peak areas 88 and 1900, top right of each panel). (c) Extracted (25 mTh window) SIM chromatograms for (A) (M+O+H)$^+$ and (B) (M–C$_2$HF$_3$+H)$^+$ ions corresponding to metabolites of a drug candidate, obtained by UPLC–TOFMS analysis of a 1 μM microsomal incubation in which the compound turnover was 70%. Spectra obtained for the corresponding UPLC peaks are shown in each case; the expected (calculated) *m/z* values are 407.136 and 309.139. Reproduced from O'Connor, *Rapid Commun. Mass Spectrom.* **20**, 851 (2006), with permission of John Wiley & Sons Ltd.

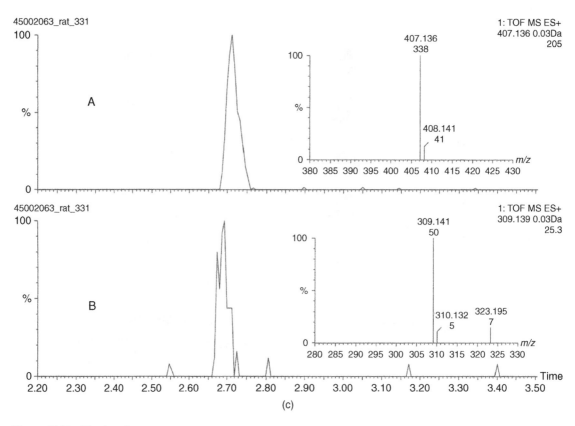

Figure 11.30　(Continued)

detection specificity for each analyte is obtained from the post-acquisition extraction of appropriate selected ion chromatograms defined by a narrow *m/z* window from the medium resolution full spectral LC–TOFMS data. The second advantage, important for metabolic stability applications, is that data providing preliminary information on metabolic routes are collected simultaneously with no penalty with respect to sensitivity for the drug candidate compound itself or to method development time. The robustness and reproducibility of the UPLC system (O'Connor 2006) were equivalent to the performance reported for other fast HPLC systems, and the combined UPLC–TOFMS system allowed generic analytical conditions, with run times as short as 2.5 minutes, to be used to provide data on both parent compound stability and metabolic routes. The only potential disadvantage arose from the absolute sensitivity that can probably be traced to the duty cycle of the TOFMS (Section 6.4.7), which is much lower than that of a triple quadrupole in MRM mode (Section 6.4.3), although the S/B values were better (Figure 11.30(b)) in this particular application.

11.5.1d Pharmacokinetics and ADME Studies

An important feature of drug development is the estimation of pharmacokinetic parameters in animal models. Pharmacokinetics is the study of the time dependence and mechanism of absorption of a compound dosed into the body, its distribution throughout the fluids and other body tissues, the sequential metabolic transformations of the compound and its first-generation metabolites, and the elimination of the original compound and its metabolites (whence the common abbreviation 'ADME' studies). The usual experimental raw data consist of concentrations of the test compound (and sometimes of its metabolites) in body tissues and body fluids (blood plasma, urine) as a function of time following a single dose. Extraction of quantitative parameters characterizing this behavior is determined by the theoretical model used to interpret the data. For example, if the dose is administered intravenously and the compound concentrations are measured in the blood, there will be an immediate drop of compound concentration with time as the compound is re-distributed, metabolized and excreted, but if an oral dose is used (as is the most common) the compound must first enter the gastrointestinal tract, dissolve and be absorbed via the gut wall (as estimated previously by the Caco-2 permeability assay); after passing through the portal vein and liver, it enters the systemic blood circulation. This process of absorption may take some time, resulting in a lag-time before the concentration in the blood plasma starts to rise, but once absorption starts the drug compound is simultaneously being removed from the plasma via distribution to other tissues and elimination, either unchanged via the liver and/or kidneys, or by metabolism mainly in the liver.

At some point the rates of these competing processes (appearance in and disappearance from the plasma) will be equal and the maximum plasma concentration C_{max} is reached at time t_{max} (an idealized pharmacokinetic curve is shown in Figure 11.31). The subsequent fall-off of

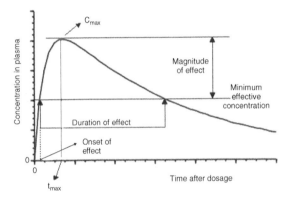

Figure 11.31 An idealized pharmacokinetic curve illustrating the rise and fall of the test compound concentration following an oral dose. The maximum concentration C_{max} corresponds to the point at which the rate of absorption from the gastrointestinal tract is just balanced by the removal of the compound by distribution to other body tissues, excretion unchanged and metabolism. The desired physiological effect extends only for the time over which the plasma concentration is greater than the minimum effective concentration. The total area under the curve is a measure of the amount of the dosed compound that effectively reaches the plasma and other tissues (the 'bioavailability' is the ratio of this quantity to the actual oral dose).

plasma concentration is not necessarily a simple exponential decay; if the distribution rate is larger than that of elimination, the curve is generally a combination of at least two distinguishable exponentials. The time required for the concentration of the drug in the plasma to decrease by 50 % from some initial value (e.g. C_{max}) is called the half life ($t_{\frac{1}{2}}$), and more than one such parameter can be defined and measured for the case where the fall-off curve is clearly composed of more than one exponential portion. In such cases the initial, more rapid decline after t_{max} is

characterized by the distribution half life, and the later slower decline that reflects various elimination processes is characterized by the terminal half life. In practice, in the drug discovery pharmacokinetic screens often no distinction is made between the several half lives that can be distinguished, and instead an overall value is assigned.

An instructive introduction (Mehvar 2004) to the definition and determination of pharmacokinetic parameters (an exercise in least-squares regression of data on a theoretical equation that is nonlinear with respect to the fitting parameters, Section 8.3.8) provides a more extensive description than is appropriate here. Here the main concern is with the experimental methods used to generate the raw data. In the drug discovery phase high throughput is a major concern but pharmacokinetic investigations are time consuming by their nature, since plasma concentrations at many time points are required to define the curve (Figure 11.31). An extensive review (Hopfgartner 2003a) of the many approaches that have been proposed to increase throughput of this crucial screen covers all aspects of sample preparation (extraction and clean-up) and analysis (separation and detection). Many of the principles involved have been described in the preceding chapters and the following discussion will focus on the concept of sample pooling, in which several test compounds are analyzed in the same LC–MS run.

Apart from the usual requirements for multiresidue method development (Section 11.3.2), the crucial decision concerns the point in the process at which the drug candidates are pooled. The earliest descriptions of sample pooling in this context (Berman 1997; McLaoghlin 1997; Olah 1997; Beaudry 1998) used administration of several compounds (up to 12) to the same test animal (the so-called 'N-in-one' approach). The analysis of plasma samples taken from this single animal at several time points thus provided pharmacokinetic data for all of the test compounds simultaneously, permitting a considerable increase in throughput. The data obtained using this then-new approach were compared (Olah 1997) with results from the traditional one-in-one approach and found to be acceptable provided that the accuracy, precision and LLOQ of the method were monitored using appropriate QC samples. While this work represented a real breakthrough in this area, later experience (White 2001) showed that complications can arise as the result of drug–drug interactions in the test animal (leading to both false positives and false negatives), and that doses must be minimized to avoid saturation of the physiological receptors. As a result, the low dose requirement of the N-in-one approach implies that the LC–MS methods must be extremely sensitive. An interesting survey of the pharmaceutical industry (Ackermann 2004) confirmed that the potential problems with the N-in-one approach, particularly the possibility of drug–drug interactions, have led to this approach being adopted as routine by only a small fraction (~8 %) of the companies surveyed.

The alternative pooling strategy (Kuo 1998) is to dose one compound per test animal, but then pool the plasma samples from several animals at each time point for simultaneous LC–MS analysis. This approach, like the N-in-one method, is a multiresidue method and care must be taken in selecting the compounds that are to be pooled in the combined plasma samples, particularly compounds with significantly different physico-chemical characteristics that can affect the efficiency of extraction and cleanup, as well as the suitability for chromatographic separation and the sometimes marked variations in sensitivity in mass spectrometric detection. However, this approach (Kuo 1998) does avoid the problems associated with the N-in-one method although at the cost of using proportionately more laboratory animals. A more recent example of application of the concept (Korfmacher 2001) optimized the approach in a strategy that was named the 'cassette-accelerated rapid rat screen (CARRS)'. In this method the drug candidates ('new chemical entities, NCEs') are dosed individually (n = 2, i.e., two rats/compound), in batches of six compounds per set. The six plasma samples (one per compound) are pooled (i.e., two such batches for each of the six time points) and a semi-automated protein precipitation procedure in a 96-well plate format was found to be fit for purpose in pharmacokinetic screening at the drug discovery stage. Throughput was further enhanced by resorting to calibration curves containing just three points per compound. Method development (Korfmacher 2001) was performed for each cassette of six compounds, not for each compound separately, and a single internal standard was used for all six (Figure 11.32).

11.5.1e Metabolite Identification and Quantitation

The final screen in the drug discovery process is the identification and quantitative estimation of metabolites; a description of the metabolite identification strategies available (Clarke 2001) is still valid although subsequent developments in mass spectrometric technology may have changed the emphasis somewhat. There is now considerable accumulated experience in the nature of metabolites likely to be observed for candidate pharmaceuticals in mammals; this information is summarized in Table 9.4.

Although the very low duty cycle of a triple quadrupole instrument in full spectral scan mode severely limits the sensitivity, this design is in principle ideal for finding

metabolites via a combination of tandem mass spectrometry scans (Section 6.2.3h); in addition to the conventional MS^1 mass spectra, the hardware implementation of precursor ion and constant neutral loss scan functions are unique to triple quadrupoles. The precursor ion scan can find metabolites if the chosen target fragment ion from the original (non-metabolized) compound is likely to be conserved in the common metabolism processes (Table 9.4). In contrast, the constant neutral loss scan searches for precursor ions in the analytical sample that fragment via expulsion of neutral fragments with particular molecular masses that are characteristic of metabolites of common functional groups. Once a mass peak in the MS^1 spectrum has been thus flagged as a metabolite, a simple fragment ion spectrum can then provide some structural information concerning the location of the metabolic modification in the molecule. Use of an ion trap in MS^n mode can be invaluable in this regard. Additional confirmation can be obtained from more precise mass measurements of the metabolite and its fragment ions to provide information on the molecular formula of the ions (not necessarily uniquely defined but rather a limited set of possibilities), e.g. using LC–QToF (Zhang 2000a; O'Connor 2006). Several manufacturers of MS/MS instruments provide software tools to assist the discovery and characterization of metabolites.

Several attempts have been made to integrate the discovery stage pharmacokinetic analyses of the parent drug with metabolite identification in order to further speed up the overall process. An early attempt (Poon 1999) used a triple quadrupole instrument but this was only partly successful as a result of the poor duty cycle and to some extent inadequate software for data acquisition and processing; it was necessary to acquire the quantitative and qualitative data in separate HPLC runs. A different approach using HPLC–TOF (Zhang 2000a; O'Connor 2006) exploited the parallel detection and higher precision mass measurement capabilities of the TOF analyzer in analyses of drug candidates in rat plasma, and was thus able to identify predicted metabolites via correspondence of measured m/z values with those predicted for common metabolic pathways; however, no MS/MS data were obtained (Zhang 2000a; O'Connor 2006) even though a QTOF instrument was used in the latter case. A 3D ion trap was used (Tiller 2002) to quantitatively monitor both parent drug and predicted metabolites via HPLC–MS/MS analyses in a metabolic stability screen but further characterization of the metabolites was not possible as a consequence of the poor duty cycle in MS/MS mode. A different compromise using an ion trap (Kantharaj 2003) obtained quantitative data in a metabolic stability assay for the parent drug

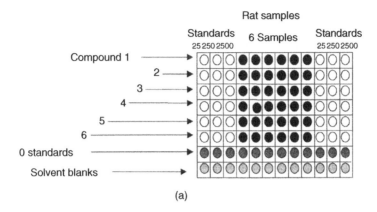

(a)

Results of the rapid oral pharmacokinetic screen
in the rat (10 mg/kg; n = 2 rats; Pooled each hour)

SCH #	Est. AUC (ng/ml.hr)	Estimated concentration (ng/ml)					
		0.5-hr	1-hr	2-hr	3-hr	4-hr	6-hr
SCH 123	4270	981	1060	1130	706	515	376

Samples below the limit of quantitation (100 ng/ml) are reported as zero

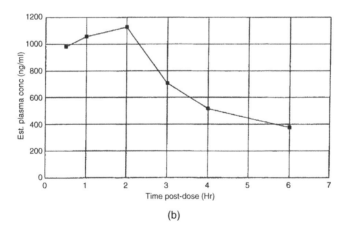

(b)

Figure 11.32 An example of a report (Excel® format) for pharmacokinetic data obtained in an oral dose experiment using a pooled plasma strategy (six compounds analyzed per pooled plasma sample). The sample concentrations, the calculated AUC (area under the curve, a measure of bioavailability) and the time vs concentration plot for the dosed compound, are shown. Reproduced from Korfmacher, *Rapid Commun. Mass Spectrom.* **15**, 335 (2001), with permission of John Wiley & Sons Ltd.

together with MS^2 spectra of metabolites within the same HPLC run.

However, metabolite identification has more recently been significantly aided by introduction of the new QqQ_{trap} instrument (Section 6.4.6) that combines the capabilities of a triple quadrupole with those of a linear ion trap with axial ejection (Hager 2002). The advantages of this new instrument in finding and characterizing metabolites (Hopfgartner 2003) emphasized the speed and efficiency that is now possible, particularly in combination with information dependent data acquisition (IDA) that permits appropriate MS/MS experiments to be performed 'on-the-fly' on ions selected on the basis of user-specified criteria during HPLC analyses. Compared with 3D ion traps, the QqQ_{trap} instrument did not suffer from a low mass cutoff (Section 6.4.5a), and full scan quadrupole CID spectra could be obtained with ion trap sensitivity; MS^3 spectra obtained using the new instrument provided further information but it was found (Hopfgartner 2003) that, unlike the 3-D traps in which the very low energy CID limits the extent of fragmentation at any particular stage in an MS^n cascade, there was no need for MS^n (n > 3) capability since extensive fragmentation could be obtained in the MS^2 spectra created in the quadrupole collision cell and these fragment ions could be further examined by CID in the third quadrupole (Q3) operated as a linear trap (the 'enhanced product ion scan' mode). A subsequent application of the QqQ_{trap} technology to characterization of metabolites in plasma (Li 2007) used information dependent acquisition in which the enhanced mass spectrum scan (Q_1 and q_2 in RF-only funnel mode, Q_3 in linear trap scan mode) was used as the survey scan to trigger multiple enhanced product ion scans (precursor ion selected by Q_1 and Q_3 in linear trap MS/MS mode). For a known pharmaceutical drug (nefazodone) as test compound, 22 metabolites (including seven not previously reported) were detected and characterized in a single LC–MS/MS run.

Exploitation of the unique capabilities of this instrument in acquisition of quantitative information for a drug candidate in discovery stage pharmacokinetic studies as well as full scan fragment ion spectra of metabolites has been described (King 2003), but separate HPLC runs were required for the two different experiments although full scan MS^1 spectra data were recorded in the same run as the MRM quantitative data. More recently (Li 2005), it was demonstrated that the instrument could provide quantitation of the parent drug candidate for discovery stage pharmacokinetic studies and full scan MS^2 metabolite identification information, all within the same HPLC run. To test the quality of the MRM data obtained in the 'survey' experiments, in which the parent drug MRM data acquisition was combined with information dependent acquisition used to select the metabolite precursor ions for full scan MS^2 spectra, they were compared with conventional MRM data acquired in a different HPLC run with the QqQ_{trap} operating in only conventional triple quadrupole MRM mode; as shown in Figure 11.33(a), the agreement was excellent. Further, the MS^2 fragment ion spectra (Figure 11.34(b)), whose acquisition was triggered by detection of significant signal levels in the metabolite MRM channels (predicted on the basis of the common metabolic pathways, Section 9.3.3), were

used post-acquisition to construct extracted MRM chromatograms for the metabolites (Figure 11.34(a)) that led to construction of peak area vs time profiles for these metabolites (Figure 11.33(b)). The latter represent relative quantitation of these identified metabolites since analytical standards were not available to determine response factors.

The problem of inability to quantitate metabolites at the drug discovery stage because of a lack of analytical standards was addressed (Valaskovic 2006) using an ingenious application of the findings (Section 5.3.6b) concerning equimolar response factors in ESI when the solution flow rate is reduced to very low values ($\sim$10 nL.min^{-1}, Schmidt 2003). The concept (discussed briefly in Sections 5.3.6b and 8.5.1e) involves analyzing a plasma extract containing drug and metabolites using nanospray flow rates and obtaining the relative signal intensity for each metabolite compared with that for the parent drug. If the equimolar response hypothesis is valid, these ratios of mass spectrometric intensities are also those of the molar concentrations. The absolute concentration of the parent drug (for which an analytical standard is available) in the same extract can be determined by conventional LC–ESI–MS. Then, by combining the parent drug concentration with the concentration ratios, it is possible to evaluate the response factors for all the metabolites since their concentrations are now known relative to that of the parent compound. This is essentially a single point calibration strategy, (assumed to be linear between this single point and the origin although it is possible to extend the strategy to a multi-point version), but one that can be used when no calibration standards are available for the metabolites.

This investigation (Valaskovic 2006) selected some existing pharmaceutical drugs (codeine, dextromethorphan, tolbutamide, phenobarbital, cocaine and morphine) whose metabolites (formed via hydroxylation, dealkylation, hydrolysis and glucuronidation) are well known (and available) as test cases. The first step confirmed the trend towards equimolar response at flow rates $\sim$10 nL.min^{-1} observed previously (Schmidt 2003) for these test compounds and their major metabolites. This was found to be the case in general although in one example (cocaine and its metabolite benzoylecgonine) for unknown reasons the metabolite response reached only 50 % of that of the parent compound at the lowest flow rate tested (Valaskovic 2006). However, this was only the first step in devising a usable protocol. There is an additional difficulty arising from possible differences in extraction efficiencies of the parent drug and its metabolite(s). If the same extraction method could be used to prepare the same extract for both the nanospray estimation

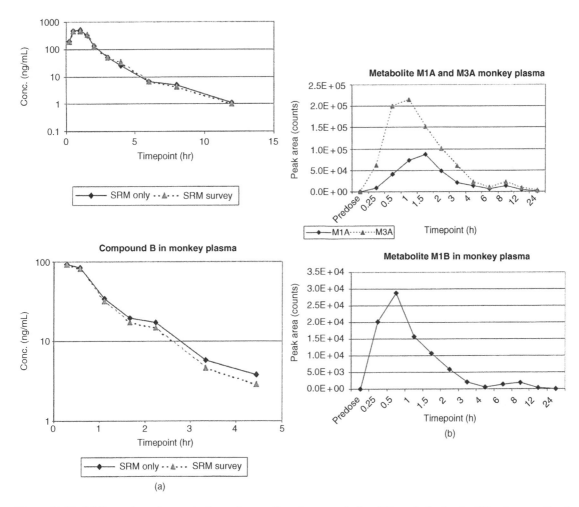

Figure 11.33 (a) Comparison of concentration vs time profiles of compounds A and B acquired using the QTrap in conventional triple quadrupole MRM mode ('SRM Only') with those acquired in 'survey' mode in which the conventional MRM acquisition for the parent compound was combined with MRM-triggered information dependent acquisition of 'enhanced' full scan fragment ion spectra (Q3 operated in linear trap mode) for metabolites predicted to be present on the basis of known metabolic pathways. (b) Peak area vs time profiles of metabolites of compounds A and B based on data acquired in the 'survey' experiments (see Figure 11.34). Reproduced from Li, *Rapid Commun. Mass Spectrom.* **19**, 1943 (2005), with permission of John Wiley & Sons Ltd.

of relative concentrations (via the equimolar responses) and the conventional LC–ESI–MS quantitative assays, any such differences would be automatically accounted for in the peak height ratio obtained in the nanospray spectrum of that extract. However, the inability of nanospray ionization to provide good quality data for plasma extracts prepared using a nonselective extraction method like protein precipitation without some additional clean-up step (and with no chromatographic pre-separation) implies that one of two strategies must be used; either a selective extraction–clean-up procedure must be used for both at the cost of appreciably lower throughput and with the added

uncertainty arising from likely differences in extraction efficiency, or different procedures must be used in preparation of extracts for the nanospray calibration experiment and the LC–ESI–MS production assays.

The strategy finally chosen (Valaskovic 2006) is illustrated in Figure 11.35. The approach was tested by application to the quantitation of an acylglucuronide metabolite of a proprietary compound in rat plasma; the metabolite was available as a pure standard, so the nanospray calibration method could be directly compared with a conventional quantitation via a calibration curve for LC–ESI–MS/MS. The nanospray calibration method yielded quantitative data within 20 % of those generated by the conventional quantitative assay. While this approach is unlikely to be capable of full validation of an analytical method to be used in drug development assays, it does appear to be fit for purpose in providing otherwise inaccessible information in discovery stage pharmacokinetics studies.

More recently (Ramanathan 2007) this general equimolar nanospray response approach was developed into a more usable approach to quantitation of metabolites in plasma and other tissues when analytical standards are not available. Chromatographic separation of extracts is required in any case for several reasons; these include avoidance or minimization of serious matrix effects (Section 5.3.6b) that can not be accounted for in the absence of an isotope-labeled SIS, and separation of isomeric metabolites from one another and from the parent drug itself in cases where ions of fragile metabolites

can be transformed into those of the drug by in-source CID. However, the equi-molar responses that are essential for this approach to be valid can not be achieved when gradient elution is used, since ESI responses vary with the solvent composition except under the 'total solvent consumption' conditions (Schneider 2006) discussed in Section 5.3.6b. Such conditions are unrealistic for a high throughput analytical method. The problem was solved in an elegant fashion (Ramanathan 2007) by post-column addition of a flow of mobile phase whose composition variation is the exact converse of that used in the HPLC separation itself. In this way the solvent composition presented to the ESI source is constant and thus any variations in nanospray molar responses among structurally related compounds are intrinsic and not a function of solvent composition. Of course this approach results in dilution of the analyte by a factor of two, but in the low

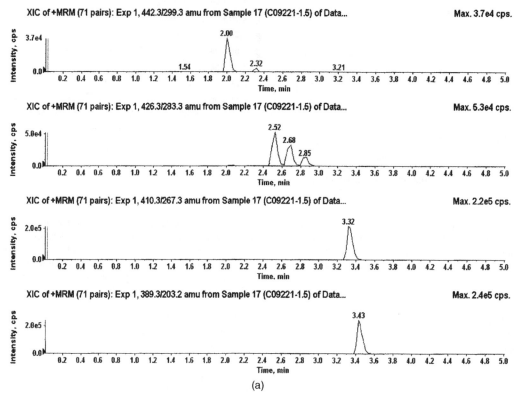

(a)

Figure 11.34   (a) Chromatograms of extracted MRM channels of metabolites (top two panels) and of conventional MRM analyses of compound A and the internal standard (bottom panel), acquired in the same HPLC run. Retention times (min): M1A 2.00; M2A 2.32; M3A 2.52; M4A 2.68; M5A 2.85; compound A 3.32 and internal standard 3.43. (b) 'Enhanced' product ion spectra (Q3 in linear trap mode) of compound A and of five metabolites (the extracted MRM chromatograms in Figure 11.34(a) were obtained from these 'enhanced' MS² spectra). Reproduced from Li, *Rapid Commun. Mass Spectrom.* **19**, 1943 (2005), with permission of John Wiley & Sons Ltd.

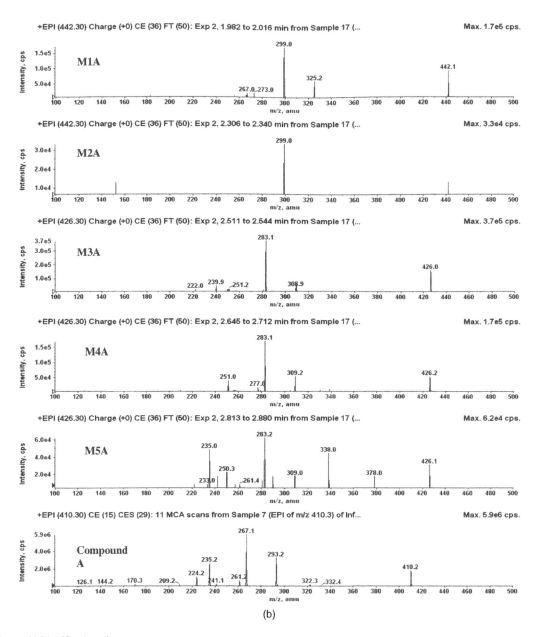

Figure 11.34 (Continued)

flow rate regime compatible with nanospray the concentration dependence of response is minimal.

As discussed in Section 8.5.1e, an alternative solution to the problem of quantitation of metabolites without analytical standards in discovery phase (pre-clinical) screening is based on the use of radioactive (^{14}C) metabolites (Yu 2007). This method was used to test the quantitative validity of the

gradient-LC–nanospray-MS approach (Ramanathan 2007) to quantitation of metabolites. The advantage of the latter approach is that it provides a means to acquire quantitative information, with a useful degree of accuracy, during metabolite structural characterization in early stage exploratory drug discovery studies without the dangers inherent in use of radio-labeled drugs.

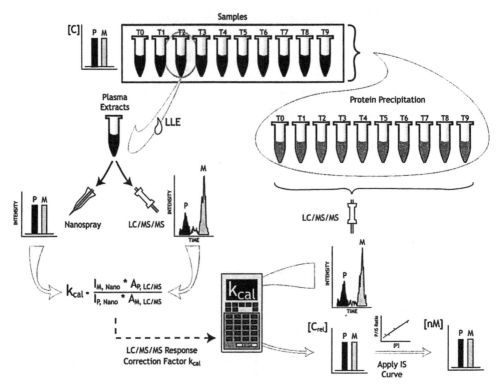

Figure 11.35 Diagram illustrating a method for providing absolute quantitation of metabolites in incurred serum samples when no analytical standard is available. The method exploits the trend towards equimolar responses for a drug candidate and its metabolite at low flow rates ($\sim 10\,\text{nL.min}^{-1}$) that permits measurement of the concentration ratio of metabolite to parent compound via the observed ratio of signals for the two using nanospray ionization ($I_{M,nano}/I_{P,nano}$). However, use of a fast nonselective extraction procedure like protein precipitation for the LC–ESI–MS/MS assays is not suitable for the nanospray experiment since the extracts are too complex, so a more selective procedure, liquid–liquid extraction (LLE) in this case, must be used to prepare these extracts. The connection between the two procedures is achieved by analyzing the LLE extracts by both nanospray and the LC–ESI–MS method used for the assays of the incurred samples, yielding an LC–ESI–MS peak area ratio ($A_{M,LC/MS}/A_{P,LC/MS}$) for comparison with ($I_{M,nano}/I_{P,nano}$) that is interpreted as the concentration ratio. This comparison yields a calibration factor $k_{cal} = (I_{M,nano}/I_{P,nano})/(A_{M,LC/MS}/A_{P,LC/MS})$ that enables quantitation of the metabolite *relative* to that of the parent compound; subsequently, absolute quantitation of the latter by conventional LC-MS using an analytical standard to prepare a calibration curve permits absolute quantitation of the metabolite also. Only a limited number of nanospray analyses (one in the example shown) is required to derive a k_{cal} value applicable to a complete set of study samples. Reproduced from Valaskovic, *Rapid Commun. Mass Spectrom.* **20**, 1087 (2006), with permission of John Wiley & Sons Ltd.

11.5.2 Drug Development Phase Validated Methods

Of the thousands of chemical compounds produced by the medicinal chemists that were screened, only one or a few will be considered worthwhile subjecting to animal studies under GLP conditions followed by Phase 1 clinical trials. High throughput analytical methods that were fit for purpose in the drug discovery phase may be no longer adequate to meet the regulatory requirements (e.g. FDA 2001). For example, 3D ion trap mass spectrometers (2D yet to be proven) are generally not now fit for purpose as a result of several factors, including a

poor duty cycle (particularly in MS/MS mode) that limits the number of data points that can be acquired over a chromatographic peak, and the space charge limitation on the total number of ions (not just those derived from the analyte) that can be accommodated (selective ejection methods require considerable time overhead and thus worsen the duty cycle problem, Section 6.4.5a). Development and full validation to the required level of an analytical method (Chapter 10) is a task requiring meticulous attention to detail. Many such methods have been fully described in the literature, and the illustrative examples summarized below were chosen on the basis that they should involve some additional level of challenge in addition to that imposed by the validation requirements.

11.5.2a Enantiomer-Specific Analyses

It is nowadays a truism (but therefore also true) to say that life is a chiral process. The implication for the subject of this book is that those pharmaceutical compounds that contain one or more chiral centers could well show differences between enantiomers with respect to their efficacy. Since most pharmaceutical drugs are administered as racemates, it is sometimes important in drug development to be able to track the enantiomers separately, and this presents a significant challenge to the analyst in addition to the usual demands of validation and GLP etc. Since mass spectrometry can not in general distinguish between enantiomers, the chiral selectivity must be provided by chromatographic separation (Section 4.4.1d).

The example discussed below concerns the development and validation of an LC–MS method for simultaneous quantitation of (R)- and (S)-propranolol in plasma (Xia 2003a). Method development involving chiral separation tends to be a lengthy process relative to that for achiral separations, since it is intrinsically a highly empirical process as it is difficult to predict the enantiomeric selectivity of any particular chiral stationary phase for some new target analyte. The chiral stationary phase used in this work was based on the natural product teicoplanin, a glycopeptide with antibiotic properties whose use as a stationary phase (when covalently bound to 5 μm silica particles) was originally described (Berthod 1996) for application to analysis of amino acids and peptides. In this case (Xia 2003a) the column was commercially available, a Chirobiotic T CSP column (5 μm particle size, 4.6 × 100 mm; Advanced Separation Technologies, Whippany, NJ, USA). These columns work best with sample extracts that have been subjected to extensive clean-up, thus potentially reducing the throughput significantly. This aspect was addressed by adopting a strategy based on on-line very high flow chromatography (Section 4.3.3d), in which the plasma samples were injected directly onto an extraction column packed with large particles that retained the analytes in the particle pores while allowing the proteins to be washed away by a fast flow rate of ∼4 mL.min^{-1}. To maintain an acceptable throughput, sample preparation and injection onto this 'turbulent flow' system, followed by transfer to the analytical column, were achieved using 96-well plate technology plus robotic autosamplers and liquid handling as described previously (Xia 2001) for an achiral analytical method. This was an important component of the overall analytical method; a schematic diagram of the apparatus is shown in Figure 11.36. In summary,

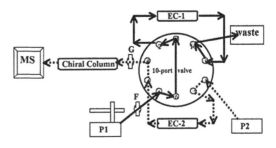

Configuration A

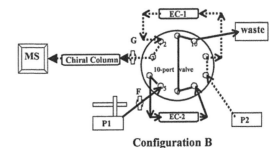

Configuration B

Loading LC gradient program used by P1

Time (min)	Flow rate (ml/min)	A(%)	B(%)	10-Port valve	Event
0.0	4.0	100	0		Loading
0.5	4.0	100	0	Open	
0.6	4.0	0	100		Elution and
2.6	4.0	0	100		equilibration
2.7	0.1	100	0		
9.0	0.1	100	0		
9.1	4.0	100	0		
10	4.0	100	0		

A, 10 mM ammonium acctate in water; B, 100 % methanol.

Figure 11.36 Schematic diagram of the dual-column extraction system on-line via a Valco 10-port switching valve with the chiral LC–MS/MS system. The components were as follows: P1, a Leap Technologies autosampler with two HPLC pumps delivering mobile phases A and B to the extraction column EC-1 or EC-2 (Waters Oasis HLB 25 μm, 1 × 50 mm); P2, HPLC pump system delivering isocratic elution mobile phase through the extraction column to the chiral analytical column; full bold arrows, pathway for mobile phase A (Table) used to load plasma sample onto extraction column; dashed arrows, pathway for mobile phase (neither A nor B) used to elute analytes from extraction column to the chiral analytical column; F, in-line filter; G in-line guard column; MS, a triple-quadrupole instrument in MRM mode. Reproduced from Xia, *J. Chromatogr. B* **788**, 317, copyright (2003) with permission from Elsevier.

the total analysis time (loading, eluting and equilibration) per sample was 10 minutes of which 0.5 minutes was for loading and 9.5 minutes for elution/analysis plus simultaneous equilibration of the extraction column.

To increase throughput two high flow extraction columns were used so that one could be used while the other was still equilibrating. The procedure for odd-numbered analyses is illustrated in Figure 11.36. The 0.5 minute loading process involves loading the $10\,\mu L$ plasma sample onto extraction column EC-1 using the high flow mobile phase described in the table shown in Figure 11.36. During this time the analyte (propranolol) is retained on EC-1 while the proteins and salts are washed away to waste and, simultaneously, pumping system P2 is delivering the isocratic mobile phase (ammonium trifluoroacetate in methanol) for the analytical elution $(1.5\,mL.min^{-1})$ through the other extraction column EC-2 to the chiral column (Configuration A, Figure 11.36). After 0.5 minutes the switching valve is switched to Configuration B in which EC-1 is now supplied by pumping system P2 that back-elutes the analytes on EC-1 through the chiral column for LC–MS/MS analysis. Meanwhile EC-2 is equilibrated by P1. For the next (even-numbered) injection the roles of EC-1 and EC-2 (Figure 11.36) are reversed. In this way the total analytical cycle time was kept low (10 minutes per sample of which only 0.5 minutes was required for the turbulent flow loading and clean-up) without compromising the extraction efficiency (Xia 2003).

A problem with all chiral HPLC separations is that there is no way of knowing which of the enantiomer peaks should be attributed to which enantiomer. Fortunately in this case standard samples of both (R)- and (S)-propranolol were available, and also racemic propranolol and d_7-labeled propranolol (as an internal standard). MRM chromatograms for each enantiomer injected separately, and for racemic propranolol and d_7-propranolol, are shown in Figure 11.37(a). Although the peaks are quite wide ($\sim30\,s$ at half height but with long tails) the separation of the enantiomers is very good; the estimated values of the selectivity factor (Section 3.4.3) and resolution (Section 3.4.6) for the two enantiomers are 1.16 and 1.83, and the retention (capacity) factors k' (Section 3.4.4) are 2.57 and 2.97 for the (S)- and (R)- enantiomers, respectively (i.e., good retention, well removed from the solvent front). The APCI–CID fragment ion spectra used to select the m/z values for the MRM transitions are shown in Figure 11.38.

Analyte recoveries were evaluated (Xia 2003a) by comparing the response for a spiked plasma extracted and analyzed on-line as described above, with that obtained by spiking a blank plasma extract at the same nominal

concentration. The control plasma extract was obtained by injecting blank rat plasma samples into an extraction column of the apparatus as usual, discarding this eluent and then, instead of eluting the retained components from the extraction column onto the analytical column, they were back-eluted (still using the same elution mobile phase) into a container and evaporated to dryness under a nitrogen stream. This dried residue of endogenous plasma components was then reconstituted in elution mobile phase spiked with an appropriate quantity of propranolol standard, and analyzed by (chiral)LC–APCI–MS/MS for comparison with the extract of blank plasma that had been spiked at the same nominal concentration prior to extraction. This was done using six different lots of blank plasma in order to check for variability in this parameter. In this way the recoveries at $2\,ng.mL^{-1}$ were measured to be 85–90 %, and at $2000\,ng.mL^{-1}$ as 105–106 % for the two enantiomers; for the deuterated internal standard at $250\,ng.mL^{-1}$ the recoveries were 103–105 %.

The specificity of this (chiral) LC–APCI–MS/MS method was examined by analyzing extracts of blank plasma. As shown in Figure 11.37(b), no signals were observed in the MRM channels for analyte and internal standard. The possibility of indirect interferences in the MRM channels, resulting from suppression of APCI efficiency by co-eluting matrix components, was investigated by comparing the response of an extract of a spiked plasma extract with that of a solution of analytical standard (racemate) at the same nominal concentration in methanol. The resulting difference represents the sum of the loss during recovery for the extract of the spiked plasma sample plus the matrix interference effect. It was found (Xia 2003a) to correspond to a total reduction in the range 9–17 % considering both enantiomers tested at the LLOQ and ULOQ and the internal standard at $250\,ng.mL^{-1}$. After correcting for the recovery losses, ionization suppression by 1.5–15 % was estimated. These estimated values of absolute matrix suppression are not large, as expected for APCI; note that these tests (recovery and matrix effects) were all conducted using six different batches of blank plasma, thus checking also on possible relative matrix effects (Section 5.3.6a).

Matrix matched calibrators were prepared in blank rat plasma for each enantiomer separately and also for racemic propranolol. Prior to online extraction–analysis each $25\,\mu L$ plasma sample was spiked with an equal volume of a solution $(250\,ng.mL^{-1})$ of the internal standard (d_7-propranolol) in 20 % methanol:80 % 0.5 M formic acid in water, so the internal standard concentration in all analytical samples was $125\,ng.mL^{-1}$. To monitor the accuracy and precision of the assay, four different sets

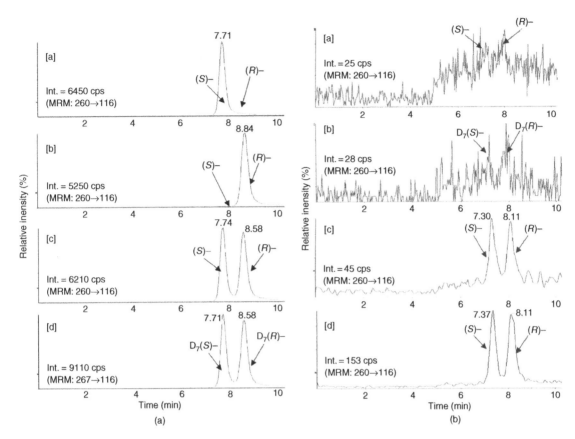

Figure 11.37 (A) MRM chromatograms of QC samples for propranolol obtained using the automated dual extraction column system with chiral LC–MS/MS system: [a], (S)-propranolol; [b] (R)-propranolol; [c]racemic propranolol; [d] racemic d_7-propranolol. Concentrations in rat plasma were 100 ng.mL^{-1} for [a]–[c], and 250 ng.mL^{-1} for [d]. (b) MRM chromatograms for propranolol obtained using the automated dual extraction column system with chiral LC–MS/MS system: [a] MRM channel for propranolol in a blank rat plasma extract; [b] MRM channel for d_7-propranolol in a blank rat plasma extract; [c] MRM channel for propranolol in an extract of rat plasma spiked with racemic propranolol at 0.5 ng.mL^{-1}; [d] MRM channel for propranolol in an extract of rat plasma spiked with racemic propranolol at 2 ng.mL^{-1}. Reproduced from Xia, *J. Chromatogr. B* **788**, 317, copyright (2003) with permission from Elsevier.

of QC samples were prepared independently of the calibrators (thus checking on possible errors in weighing the analytical standard when preparing the spiking solutions for calibrators). Each of the first three sets contained only (R)-, or (S)-, or racemic propranolol, respectively, and each included rat plasma spiked at six different concentrations covering the range 2–5000 ng.mL^{-1} (the last of these was included as a dilution QC sample to check on the dilution procedure used to bring concentrated samples to below the upper limit of quantitation), while the fourth set contained the two enantiomers in ratios of 10:1 and 1:10.

Already it is evident that the requirements placed upon validation of a chiral LC–MS method are appreciably more demanding than those for an achiral assay, for which only one set of QC samples covering the analytical concentration range would be required. The calibration range was linear over the range 2–2000 ng.mL^{-1} in plasma and the calibration curves for each enantiomer was constructed using a 1/x-weighted least-squares linear regression (see comments in Section 8.3.5). All four sets of QC samples covered this concentration range and the intra- and inter-day precision (RSD %, $n = 5$) for the (R)-enantiomer were within 7.6 and 3.6 %, respectively, while the corresponding values for (S)-propranolol were within 6.9 and 4.5 %. Accordingly, the LLOQ and ULOQ were taken to be the two limits of the linear calibration range. The detection limit (LOD) was not considered to be a crucial parameter (Xia 2003a) since regulatory

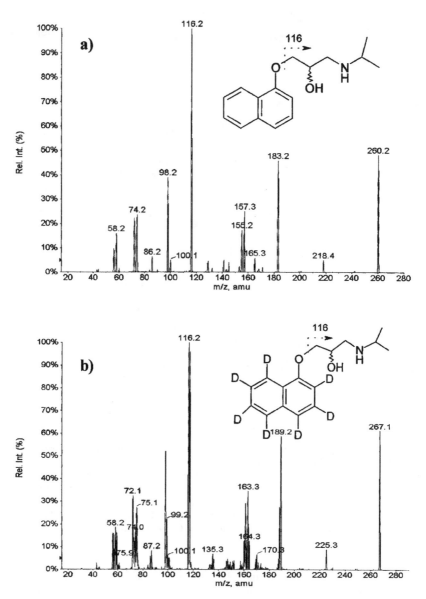

Figure 11.38 Fragment ion CID spectra (triple quadrupole) for protonated propranolol (m/z 260) and d_7-propranolol (m/z 267); the chiral center is the carbon atom carrying the hydroxyl group. The intense signal at m/z 116 is attributed to the ion $CH_2 = C(OH) - CH_2 - H_2N^+ - CH(CH_3)_2$, used as the selected fragment ion for the MRM channels in both cases. Reproduced from Xia, *J. Chromatogr. B* **788**, 317, copyright (2003) with permission from Elsevier.

concentration limits are not a concern in such cases, but was estimated to be 0.5 ng.mL^{-1} in plasma on the basis of the S/B value in the MRM chromatogram shown in Figure 11.37(b)[c].

The automated on-line turbulent flow extraction with analysis by (chiral)LC–APCI–MS/MS was demonstrated to exhibit excellent reproducibility and ruggedness for the propranolol enantiomers. The relative standard deviations of the retention times for the two enantiomers (measured for the deuterated internal standard) were within 1.6 % in over 130 injections and the RSDs for the corresponding peak areas were both within 8.3 % although a small number (~5) of determinations were demonstrated to be outliers.

Finally, stability of propranolol in rat plasma was investigated using QC samples at four concentrations in the range 6–1600 ng.mL^{-1}, for each of the individual enantiomers plus the racemate. Different storage conditions were examined (tests done in triplicate). The enantiomeric analytes were found to be stable at room temperature for at least four hours and in the autosampler (after mixing with internal standard solution) maintained at 10 °C for at least 48 hours. It was stated (Xia 2003a) that propranolol in rat plasma was stable at −70 °C for at least 30 days but no details of number of freeze–thaw cycles were given. Propranolol in clean methanol was stable for at least 30 days at 4 °C.

The preceding account of this work is quite long but actually represents a summary of the complete detailed account (Xia 2003a). The point here is that not only did this development and validation of a chiral analytical method involve a large quantity of work, it all had to be performed and evaluated with meticulous care. The only partial criticisms that might be made of this otherwise exemplary work concern the lack of information on stability in freeze–thaw cycles, and the investigation of absolute matrix suppression effects that were somewhat uncertain as a result of the difficulty in distinguishing these effects from those arising from imperfect recoveries (see above). Even here, however, the effects were not large and would be at least partly accounted for by the use of matrix matched calibrators (no relative matrix effects were detected) plus a co-eluting surrogate internal standard. An example of application of the method to the pharmacokinetics of the individual enantiomers in the rat model (Figure 11.39) confirmed previous findings (Vermeulen 1992) that the combined distribution-metabolism-excretion rate is higher for the (S)-enantiomer (in this experiment the dose was administered intravenously so absorption of the compound was not involved).

In an interesting more recent development of chiral analysis using propranolol as test compound (Chen 2006), supercritical fluid chromatography (SFC, see Section 4.3.2e) was used for pharmacokinetic studies in a drug development environment. Although this was developed as a high throughput method rather than as a fully validated method for clinical studies, as was the case for the method just discussed (Xia 2003a), it is interesting to briefly compare the two. For the SFC method a carbohydrate-type chiral stationary phase was used. A triple quadrupole mass spectrometer was again used in MRM mode, with APCI as the ionization method but in this case (Chen 2006) the APCI plasma was dominated by ions derived from methanol added as a polar modifier to the supercritical carbon dioxide. As befits a high throughput method, sample extraction used a simple protein precipitation procedure. The most striking difference involved the chromatographic run time, 10 minutes (Figure 11.37) for the HPLC method and three minutes for SFC. Despite this much shorter chromatography time, separation of the enantiomers was much improved in the SFC case (Figure 11.40). However, it is difficult to directly compare the two methods with respect to overall throughput since the 10 minute analysis time in the HPLC case includes the time required for the on-line extraction and elution, while extraction was done off-line in the SFC method. The linear calibration range was 2–2000 ng.mL^{-1} for the HPLC method (Xia 2003a) and 5–5000 ng.mL^{-1} for SFC (Chen 2006). The two methods provided comparable levels of accuracy and precision. However, as was appropriate, the method intended for drug development

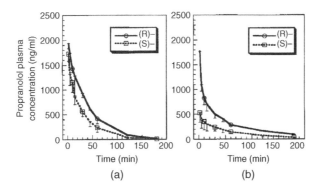

Figure 11.39 Profiles of rat plasma concentrations of (*R*)- and (*S*)-propranolol; 5 mg.kg^{-1} of the free base was administered intravenously (accounting for absence of the characteristic rise to a maximum for oral dosing, compare Figure 11.32). Results are expressed as mean ± SD (*n* = 3). Enantiomers administered (a) separately; (b) as racemate. Reproduced from Xia, *J. Chromatogr. B* **788**, 317, copyright (2003) with permission from Elsevier.

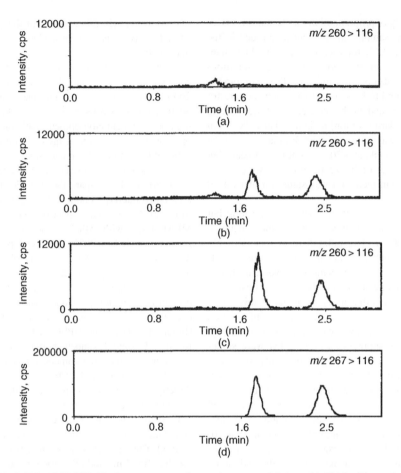

Figure 11.40 Chiral SFC–APCI–MS/MS chromatograms of racemic propranolol from (a) blank, (b) standard at $25 \, \text{ng.mL}^{-1}$, (c) study mouse blood samples, and (d) d_7-racemic propranolol from study mouse blood samples. Reproduced from Chen, *Anal. Chem.* **78**, 1212, copyright (2006) with permission from the American Chemical Society.

assays (SFC) did not require the same level of validation (e.g., no extensive analyte stability studies, no checking for relative matrix effects using different batches of plasma etc).

11.5.2b Methods for Two Concentration Ranges

The second example of development and validation of an analytical method for pharmacokinetic studies of a candidate compound in drug development (clinical studies) concerns an additional complication rather different from the enantiomeric distinction discussed above. In this case (Sennbro 2006) the drug candidate (laquinimod) was a synthetic compound intended for oral treatment of multiple sclerosis. In pre-clinical studies the pharmacokinetics in animal models had been determined using

a fast LC–UV method, but its LOD and LLOQ were 400 and $750 \, \text{nmol.L}^{-1}$, respectively. Since Phase I trials involving human subjects always start with low doses to avoid any toxicity risks, a much more sensitive and selective method was required leading to development of an LC-MS/MS method with LLOQ in the sub-nmol.L^{-1} range. However, in later Phase I and also Phase II trials a method with a much wider validated concentration range and which could handle larger sample throughput (reflecting the larger numbers of subjects in Phase II trials) was required. To this end method 1 (Sennbro 2006) was developed primarily to determine low levels of the drug, with throughput a secondary consideration. An efficient clean-up procedure was necessary in this regard and this used automated off-line SPE cartridges. As usual (Section 4.3.3c) the solvent used to elute the cartridges (methanol) was a stronger solvent than the isocratic mobile phase used for reversed phase HPLC (a 55:10:35 mixture of methanol:0.1 % aqueous TFA:water, so the SPE eluent had to be evaporated to dryness and reconstituted in a 25:75 mixture of methanol:0.1 % aqueous TFA. (Note the unusual use in this method of TFA, which is not popular among mass spectrometrists as it is difficult to flush from the system, and even trace amounts can swamp negative ion mode; however, note also that only 0.1 % TFA was used, an order of magnitude lower that the amount required for, e.g., formic or acetic acid.) The HPLC–MS/MS cycle time was 2.5 minutes, and during this time the autosampler washed the needle and injection valve several times with a 50:50 mixture of 1 % aqueous potassium hydroxide:methanol followed by 0.2 % aqueous TFA.

Method 2 was developed with method 1 as starting point but with a different compromise between throughput and a low quantitation limit, intended for use at a later stage in the clinical development when the dose levels and sample load were both increased. The sample preparation step was simplified to a protein precipitation procedure (acetonitrile) using 96-well plate technology with vortex mixing and centrifugation. The LC–ESI–MS/MS analytical method used a fast gradient (total cycle time 3.5 minutes) unlike the isocratic elution used for method 1. Although the analytical step was slower for method 2, the time savings in the extraction–clean-up step more than made up for this.

The two methods were validated in accordance with regulatory guidelines (FDA 2001) and it is not necessary to repeat all details here. However, this paper (Sennbro 2006) provides an excellent example of validation of a biomedical method with full concern for aspects emphasized in Chapter 10, particularly robustness, carryover and matrix effects. Typical MRM chromatograms for processed plasma samples are compared for the two methods in Figure 11.41. The method of estimating the LOD was not described (Sennbro 2006) but presumably was an approximate approach involving a S/B of three (Section 8.4.1), since statistically defensible LOD values are not relevant in drug development assays; for method 1 the LOD was $0.1 \, nmol.L^{-1}$ and for method 2 it was $0.2 \, nmol.L^{-1}$. More to the point, method 1 gave a linear dynamic range of $0.4–100 \, nmol.L^{-1}$, satisfying requirements of accuracy and precision, i.e., according to the regulatory protocol use (FDA 2001) the LLOQ was $0.4 \, nmol.L^{-1}$. For method 2, the dynamic range was $0.75–750 \, nmol.L^{-1}$ for an injection volume of $10 \, \mu L$ but if the injection volume was lowered to $0.5 \, \mu L$ a dynamic range of $15–15\,000 \, nmol.L^{-1}$ was achieved

within the accuracy and precision requirements. This flexibility of method 2 is an advantage but is somewhat offset by the fact that a quadratic calibration curve resulted (Section 8.3.6). The reasons for the departure from linearity were not described (Sennbro 2006) but, in view of the unusually high ULOQ, saturation of the ionizing power of the ESI source seems a likely explanation.

Plasma concentration profiles for individual healthy volunteers following a low single oral dose (0.05 mg), and those for patients with multiple sclerosis undergoing a continuous regimen of 1.2 mg/day, are also shown in Figure 11.41; method 1 was used for the low dose control group and method 2 for the patients. The intra- and inter-day precision for both methods were 1.6–3.5% and 2.1–5.7%, respectively (Sennbro 2006). These individual profiles for different subjects illustrate an important point in such work; the analytical uncertainty is generally appreciably smaller than the biological differences among test subjects, but of course detection of such differences using adequate analytical methods can sometimes provide useful information to the clinician.

As mentioned above, many other examples of development of analytical methods, fit for purpose in either the drug discovery or drug development phases, could have been chosen. Some of these are discussed in earlier chapters in the context of general principles of trace quantitative analysis.

11.6 Quantitative Proteomics

Mass spectrometry is not very useful for direct quantitation of intact large proteins at the trace levels required by biologists. In particular, the efficiencies of ionization and detection of intact proteins are not sufficient and reliable identification of such analytes would be problematic in view of the large absolute uncertainties in molecular mass measurement in the range up to 100 kDa. Also limited information is available from fragment ion spectra of such large ions, which are diffcult to activate sufficiently to give rich product ion spectra. As a result, quantitative analysis of proteins present at trace levels in biological samples at some point invariably involves specific enzymatic hydrolysis of the protein (usually using trypsin as the enzyme) to form fragment peptides of which one or a few are quantitated as surrogates for the parent protein (see the text box describing controlled hydrolysis of proteins). When incorporated into the so-called 'shotgun strategy' for protein identification, enzymatic hydrolysis can lead to uncertainties with respect to analyte (protein) identification, as discussed further below, since in a complex proteome mixture there is not necessarily a unique correspondence between a single tryptic peptide

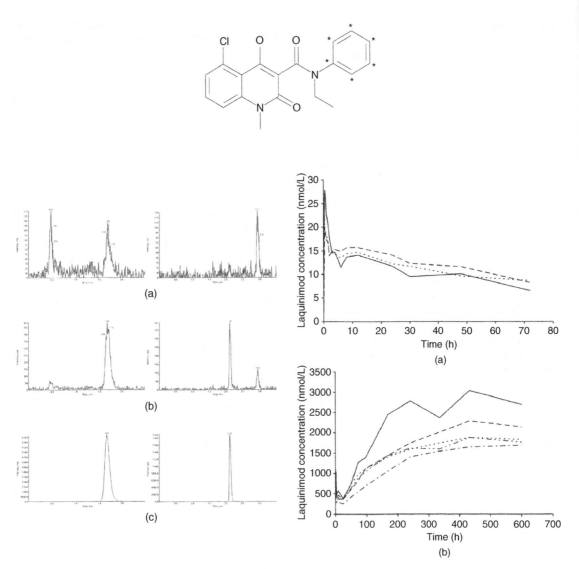

Figure 11.41 Structure of laquinimod (the $^{13}C_6$-internal standard is labeled as shown). The LC-MS/MS chromatograms (MRM for m/z 357.1 → 236.1) obtained using method 1 (left) and method 2 (right) after injection of extracts of (a) blank plasma, (b) plasma spiked with laquinimod at the limit of quantitation, and (c) plasma with 30 nmol.L^{-1} of laquinimod obtained from a patient suffering from multiple sclerosis. Retention times are about 1.7 min for method 1 and 2.1 min for method 2 (the other peaks are co-extractives). The plasma concentration profiles (determined using method 1) are shown for three healthy volunteers (top) after a single oral dose of 0.05 mg, and for five patients suffering from multiple sclerosis (bottom) after repeated oral dose administration of 1.2 mg per day (method 2 was used in this case). Reproduced from Sennbro, *Rapid Commun. Mass Spectrom.* **20**, 3313 (2006), with permission of John Wiley & Sons Ltd.

and a parent protein. Of course quantitative proteomics also shares all the problems faced in quantitation of small molecules discussed in previous chapters, e.g. the need for effective clean-up and chromatographic separation. The molecular masses of the proteolytic peptides can range from a few hundred Da (well within the range considered optimum for quantitation by LC–MS/MS) to larger fragments of a few thousand Da that are of limited value in this context. Many of them fall in the range 1000–1500 Da that is manageable as far as LC–MS/MS technology is

Selective Digestion of Proteins to Peptides

The most common agent used to digest (hydrolyze) a protein into constituent peptides in a controlled manner is the enzyme trypsin (such an enzyme is called a protease). Trypsin cleaves the peptide (amide) bond on the C-terminal side of the highly basic amino acid residues of lysine (Lys, K) and arginine (Arg, R), thus producing 'tryptic peptides' that all have a basic residue at the C-terminus (except for the original C-terminus of the protein in the great majority of cases). Tryptic peptides are usually of a convenient length for mass spectrometric (MS/MS) determination of amino acid sequence; they readily form doubly charged ions $[M+2H]^{2+}$ in electrospray ionization, and these in turn generally yield fragment ion spectra that are readily interpretable in terms of sequence. Tryptic digestion also yields at least a few peptides that are of a size suitable for quantitative analysis by LC/MS. An over-simplified picture of a protein or peptide containing n amino acd residues is as follows:

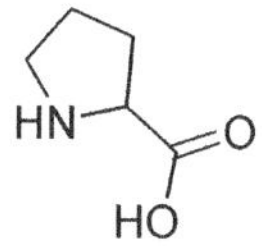

$$H_2N-CHX_1-CO-NH-CHX_2-CO-NH—CO-NH-CHX_{(n-1)}-CO-NH-CHX_n-COOH$$

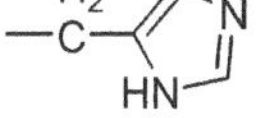

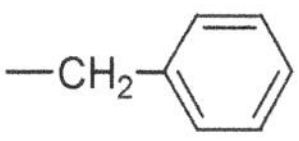

where the X_i groups are the amino acid sidechains shown below.

Glycine, G	$-H$	Aspartate, D	$-CH_2C(=O)OH$
Alanine, A	$-CH_3$	Glutamine, Q	$-CH_2CH_2C(=O)NH_2$
Serine, S	$-CH_2OH$	Glutamate, E	$-CH_2CH_2C(=O)OH$
Proline, P		Lysine, K	$-CH_2CH_2CH_2CH_2NH_2$
Valine, V	$-CH(CH_3)_2$	Methionine, M	$-CH_2CH_2SCH_3$
Threonine, T	$-\overset{OH}{C}H_2CH_3$	Histidine, H	
Cysteine, C	$-CH_2SH$	Phenylalanine	
Leucine, L	$-CH_2CH(CH_3)_2$	Arginine, R	$-CH_2CH_2CH_2NHC(=NH)NH_2$
Isoleucine, I	$-\overset{CH_3}{C}HCH_2CH_3$	Tyrosine, Y	

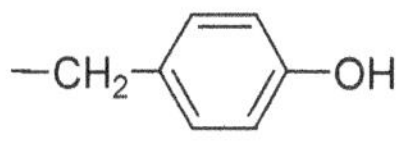

(Continued)

Asparagine, N $-CH_2\overset{\displaystyle O}{\overset{\displaystyle \|}{C}}NH_2$ Typtophan, W

Structures of sidechains X of the 20 common amino
acids NH_2-CHX-COOH (but note unusual case of proline).

However, trypsin can not hydrolyze the lysine–proline bond (probably a result of steric hindrance), and if a protein sequence contains adjacent basic residues (KK, KR, RK or RR) the resulting digestion products include single amino acids (K or R) that are difficult to detect. As a check on this and other possibilities, other proteases are used in addition to trypsin, e.g., Lys-C, Arg-C and Glu-C are proteases that hydrolyze peptide bonds on the C-terminal side of only lysine, arginine and glutamic acid, respectively. Cyanogen bromide chemically hydrolyses peptide bonds on the C-terminal side of methionine to form peptides with a homoserine residue (X $= -CH_2-CH_2-OH$) at the C-terminus.

---

concerned but do present a significant problem arising from potential cross-contributions between MRM channels for analyte and isotope-labeled internal standard, in a more extreme fashion than that encountered for small molecule analytes (Section 8.5.2c); this aspect is also discussed later.

There is now a large amount of literature on 'quantitative proteomics', sufficiently large that it requires a book dedicated to this subject alone. Publications on this subject do not always make it clear in titles and abstracts as to whether they are concerned with absolute quantitation (determination of the amount of substance or an equivalent concentration) of a specified protein in a given analytical sample, or with relative quantitation (comparison of protein amounts or concentrations in two or more analytical samples, e.g., one tissue sample from an organism in a 'normal' healthy state and one from a similar organism in a disease state). The great majority of publications on quantitative proteomics thus far have described investigations of relative concentrations, whereas the focus of the present book is on measurement of amount of substance. However, the following discussion starts with an overview of relative quantitation of proteins since some of the concerns for absolute quantitation were first recognized and addressed in the comparative investigations.

The body tissue that is most often sampled for proteomics analyses is the blood, partly because of its relatively easy access and partly because, during its circulation through all other organs, it picks up 'leakage' proteins and thus reflects to some extent all the physiological processes occurring within the body. Of course the downside of this aspect is the consequent complexity

of the protein complement. The red color of blood is due to the abundant protein hemoglobin that is localized in the red blood cells. The homogeneous solution part of blood in which the blood cells are suspended is called the plasma, a pale yellow liquid that contributes about two-thirds of total blood volume and contains all the extracellular proteins of blood. Serum is plasma collected from blood that has first been allowed to 'clot' naturally, thus removing clotting factor proteins (fibrinogen etc.). The total protein concentration in serum is 60–80 mg.mL^{-1} and the 12 major protein types (albumin, immunoglobulins, transferrin, haptoglobin, and lipoproteins) contribute $\sim$90 % of the total. The range of individual protein concentrations in plasma covers >10 orders of magnitude (Anderson 2002) from $\sim$1 pg.mL^{-1} for the cytokines (signaling proteins) to $>10^{10}$ pg.mL^{-1} for the albumins. It is thus not surprising that a considerable simplification of a plasma sample, about to be analyzed for a small molecule drug candidate and/or its metabolites (Section 11.5), can be obtained by simply precipitating out the vast majority of the complex protein mixture by adding an organic solvent (usually acetonitrile, Section 4.3.1f).

11.6.1 Identification of Proteins by Mass Spectrometry

The number of different protein species in plasma is in the hundreds of thousands with some estimates as high as 500 000; all of these are ultimately derived from the 'genetic blueprint' carried in the DNA of all cells in the individual organism, but in humans the number of identifiable genes is only $\sim$30 000. One important reason for the disparity between number of genes and number of proteins derived from them is the phenomenon of alternative splicing in eukaryotes (organisms whose DNA is enclosed within a cell nucleus, i.e. excluding the prokaryotes, bacteria and archea). It is now known that the genes of eukaryotes are 'split', i.e., present in several distinct segments ('exons') along the DNA chain. Those regions of the gene that do code for proteins are separated by noncoding stretches of DNA ('introns') that do not contribute to protein 'blueprints'. The DNA blueprint is expressed via the messenger RNA (*m*RNA) that transcribes the permanent DNA blueprint into a working blueprint used in the actual assembly of the protein.

Before the *m*RNA proceeds to the sub-cellular structure (the 'ribosome') that assembles individual amino acids into the peptide chains (see the protein hydrolysis text box for names, structures and one letter codes for the 20 principal amino acids), the noncoding regions of the *m*RNA are excised and the coding regions are joined up, or 'spliced'; thus the split-gene structure results in shorter portions of *m*RNA (each corresponding to a different exon coding region) that float around within the cell and can thus re-assemble in more than one way ('alternative splicing'), giving rise to several proteins that are different from all of those that would have arisen if the exons were all strung together with no noncoding DNA separating them. It must be emphasized that even this apparently complicated description is actually a considerably simplified version of the detailed process, but it will suffice to explain some of the complications that are intrinsic to proteomic experiments based on analyses of the proteolytic fragment peptides used as surrogate analytes for the proteins.

Another process adding to the complexity of the situation is the transformation of the 'bare' protein (the string of amino acids emerging from the ribosome) into the functional protein. This transformation involves appropriate folding into the functional 3D shape and also the post-translational chemical modifications (PTMs) required for the protein to carry out its functions within the cell. The most important of these PTMs are phosphorylation and glycosylation of specific amino acid residues within the protein. Each of these processes can give rise to several modified forms of the 'bare' protein that could itself be a member of a family of proteins arising from alternative splicing! If the one or few peptides formed by specific enzymatic cleavage that are selected as analytical surrogates for a protein do not contain any region that contains a PTM, the analytical procedure will be 'blind' to the existence of these modified forms and could give rise to erroneous identifications of the protein that is the true

target analyte for quantitation. Other protein molecules found in plasma can be assigned as shortened versions of original proteins formed by enzymatic cleavage occurring naturally within the blood itself.

Because correct identification of the analyte being quantitated is a crucial aspect (Section 9.3.3b), it is worthwhile to consider further the ambiguities that can arise from the alternative splicing phenomenon. This is the subject of an extensive tutorial review (Nesvizhskii 2005) and only a brief summary is given here. First, it is necessary to compare two competing strategies used for characterization of a proteome, viewed as the complete set of proteins present in cells of a given differentiated type (e.g., human skin cells or human liver cells) that are in a specified state (e.g., 'normal' or 'diseased') at the time of sampling. The traditional method involves first separating the proteins from one another as far as possible followed by characterization of each separated species, e.g., by specific enzymatic digestion followed by HPLC–MS/MS analysis. The other approach (termed 'shotgun proteomics') dispenses with pre-separation of the proteins and proceeds directly to enzymatic hydrolysis followed by HPLC–MS/MS. The key to the success of the shotgun approach involves extensive computer-based comparisons of MS/MS spectra with spectra predicted theoretically based on known fragmentation reactions and predicted amino acid sequences of the proteolytic peptides from databases of known DNA sequences (and thus amino acid sequences in the corresponding proteins) of the organism in question. Even in the context of this very broad description of the two approaches, it is clear that the shotgun approach loses all experimental connectivity between the characterized peptides and the proteins that are the real objective of the analysis.

In order that a protein can be unambiguously recognized via a surrogate peptide that has been detected and characterized (hopefully via an unambiguous amino acid sequence but often via a less than perfect match between experimental and theoretical MS/MS spectra), it is necessary that the peptide sequence is unique to that particular protein. This is a difficult condition to satisfy, particularly for higher organisms in view of the alternative splicing mechanism and PTM formation described briefly above (Nesvizhskii 2005), and sometimes it is impossible to decide on the basis of shotgun information whether a particular protein is or is not present. Unfortunately, any given peptide, especially a small peptide (<10 amino acids) may be a part of the sequences of several different proteins, e.g., as a result of the alternative splicing phenomenon, complicating the problem of deducing which proteins are present in the sample based only on the sequences of identified peptides. Balancing this disadvantage, the success of the shotgun proteomics approach lies in the realtively high throughput that can be achieved, although at the expense of increased difficulty in computational interpretation and statistical evaluation of the large volumes of LC–MS/MS data that are acquired. In general, protein identification is less complex and ambiguous if the proteins are first separated from one another as far as possible, thus retaining the protein–peptide connectivity (Nesvizhskii 2005).

The traditional method of pre-separation of proteins, and still the most widely used, is electrophoresis (Section 3.7) on a thin layer of polyacrylamide gel supported on a glass plate. (It is important to remove all unpolymerized acrylamide monomer, not only for health reasons as discussed in Section 11.2.1, but also because free acrylamide can readily derivatize the proteins applied to the gel.) Proteins are generally first separated on the basis of their pI values. All proteins contain some amino acids with acidic side chains (e.g., aspartic and glutamic acids) plus some with basic side chains (e.g., arginine and lysine), so the nett charge on a protein molecule depends strongly on the pH of the medium that determines the number of negatively charged carboxylate groups vs the number of positively charged nitrogen atoms. At one special pH value the numbers of positive and negative charges will be equal so that the overall charge is zero, and this is the pI of that particular protein (that can be used as an identifying characteristic).

The way in which this property is exploited to separate proteins is to use a gel that has been permeated with ampholytes (small molecules that also contain both acidic and basic groups, i.e., have their own pI values). By inserting different ampholytes with gradually increasing pI values along the length of a gel strip, a pH gradient can be established along the strip. There are many proprietary ampholyte series available commercially but a published example (Righetti 1975) describes the use of linear polyamines in which the amino groups are separated from one another by two methylene groups. When a mixture of proteins is applied to such a strip, and an electric field is applied in the direction such that high pH corresponds to the negative terminal of the applied voltage, those proteins that are positively charged at the local pH along the strip will migrate towards the negative terminal until they reach a (higher) pH value corresponding to their own characteristic pI where they are overall uncharged and will thus stop migrating. Similarly, proteins that find themselves at a pH at which they bear a nett negative charge will migrate towards the positive

terminal, i.e., in the direction of decreasing pH, so that the nett charge will become progressively less negative, until the pI is reached. This simple device thus achieves a one-dimensional separation of proteins according to the pI values.

A second dimension of separation can then be achieved in a direction perpendicular to that used for the pI separation by connecting the original strip to a plate that is free of ampholytes in the polyacrylamide gel layer. Application of an appropriate field can now achieve a second separation on the basis of molecular mass provided that certain conditions are satisfied. The electrophoretic mobility, and thus the speed of motion in this second dimension, depends not only on the nett charge but also on the hydrodynamic radius (basically determined by the overall size and 3D shape) of the protein molecule (Section 3.7).

To achieve a separation that reflects only molecular mass it is necessary to account for differences in nett charge in the new electrophoretic medium (recall that the proteins are initially distributed along the first strip so that all have zero nett charge) and in the original shapes (degrees of protein folding). This is achieved through addition of excess detergent, almost invariably sodium dodecylsulfate (SDS), that denatures all the protein molecules so that to a first approximation all assume long rod-like shapes. It also turns out that the number of dodecylsulfate anions that attach themselves to the denatured protein molecules are proportional to the size of the latter, i.e., roughly proportional to the molecular mass, so that to within this approximation all the protein molecules now have the same general shape and the same mass-to-charge ratio. As a result the electrophoretic mobility now depends only on the hydrodynamic radius (Section 3.7), i.e., on molecular size and thus mass, so by applying the external field for a fixed time the proteins will move along their separate lanes on the polyacrylamide gel plate towards the positive terminal (each lane pre-defined by the pI value) by a distance determined largely (approximately) by its molecular mass.

Since its original introduction (Laemmli 1970) this combination of polyacrylamide gel electrophoresis (PAGE) with denaturation by SDS has become well established but further improvements to the SDS–PAGE method are still being developed (e.g., Schägger 1987). Note that getting rid of the SDS detergent is essential before LC–MS/MS analysis is attempted, since the dodecylsulfate anions form ion pairs with the protonated peptides and severely reduce the electrospray response of the latter.

In shotgun proteomics a two-dimensional separation of the proteolytic peptides is also generally required before MS/MS analysis. The most common combination is an LC × LC approach (Section 9.5.5c) in which the first dimension uses a strong cation exchange mechanism and the second involves an essentially conventional reversed phase approach. Interestingly, it has recently been suggested (Cargile 2005) that the first gel separation (by pI value) can be successfully used as the first step in a two-step separation strategy for shotgun proteomics, replacing the commonly used strong cation exchange (SCX) chromatography (Section 4.4.1c). Since pI is an intrinsic property of a protein, unlike its retention characteristics on an SCX column that vary with the nature of both stationary and mobile phases, it can be used as an additional identification filter in the computerized database matching.

The question of protein identification, especially in shotgun proteomics (qualitative and quantitative), by matching mass spectrometric data for the surrogate peptides with theoretical predictions based on DNA sequence databases, has been a contentious issue. It is fair to say that some early investigations did not fully appreciate the need to account for the potential ambiguities arising from the loss of protein–peptide connectivity intrinsic to any shotgun approach. A succinct account of what are now considered to be the minimal requirements for publication of such claimed identifications (Taylor 2005) lists 11 points. It is not appropriate to reproduce the complete list here but details are given of the requirements in reporting how the primary MS and/or MS/MS data for the peptides were acquired, processed and used to search the databases, as well as access to all information concerning the peptide–protein matches thus obtained. Further, it is emphasized (Taylor 2005) that the false positive identification rate should be estimated, e.g. by searching the experimental data against a reverse sequence database (Peng 2003a), and that a false positive rate > 10 % is unacceptable (for mamalian organisms, identifications based on a single peptide sequence are generally too risky in this context). Indeed, it is strongly recommended (Taylor 2005) that, wherever possible, protein identifications should be made using a software algorithm that provides an objective statistical evaluation of the proposed identification.

The preceding discussion may seem like a long preamble concerning analyte identification before discussing the quantitation methods that are the true subject of interest here but, in the case of quantitative proteomics, identification can be a major issue (Nesvizhskii 2005).

11.6.2 Relative (Comparative) Quantitation

As for all quantitative measurements, proteomic analyses are intrinsically comparisons (Chapter 1) and, as for most of the chemical analyses discussed in this book, such comparisons are facilitated through use of isotopologs of the target analyte(s) as comparison standards. The use of stable isotope labeling is also important for the widely used methods of relative quantitative proteomics, and these are now briefly described. All of them involve tagging the proteins or peptides from one biological sample (e.g., cells in a normal healthy state) with a functional group containing one specific isotopic composition (e.g., all-^{1}H, ^{12}C, ^{14}N), and those from another biological sample (e.g., otherwise identical cells but now in some specific disease state) with a different isotopic composition (e.g., all-^{1}H, ^{14}N, but ^{13}C$_n$ and all the rest ^{12}C, where n is preferably sufficiently large to keep the separation between the isotopic distributions for the two samples sufficiently large that there is no significant overlap, Section 8.5.2c). In such a case the physico-chemical properties of the two tagged samples are identical with respect to all steps in the analytical procedure up to and including ionization (e.g., ESI), but the two samples are distinguishable on the basis of MS^1 and MS^2 spectra.

A detailed introduction to the principles involved (MacCoss 2005) emphasizes the relationship of these relative quantitation methods for proteomics to those established for quantitation of small molecules using isotope-lableled internal standards, and also summarizes the major approaches in current use. Another extensive review (Schneider 2005) emphasizes the advantages and disadvantages of each isotope-labeling technique with respect to biomarker discovery, target validation, efficacy and toxicology screening and clinical diagnostic applications. An earlier review (Hamdan 2002) of this rapidly developing field also emphasizes potential pitfalls in the all-important sample preparation steps (including solubilization denaturation and, possibly, reduction and alkylation of the thiol groups in cysteine residues) that can result in undesirable artifacts that can have serious consequences for the final outcome of protein separations and subsequent analyses.

Most present methods attempt to address a major potential flaw (which also affects non-quantitative comparisons to some extent) arising from lack of adequate reproducibility of 1D and 2D gel electrophoresis, or of other separation techniques for proteins based on chromatography of some kind. It is very difficult to make two 2D gel plates, or two chromatography columns, behave in exactly the same way. So when, e.g., cells in a normal healthy state are compared with those in a disease state, it is best to avoid analyzing the two protein extracts on two different gels (or columns, or at different times on the same column). This can be achieved by first derivatizing each sample separately using different isotopically labeled tags, and then analyzing them simultaneously on the same gel or column by exploiting the unique capabilities of the mass spectrometer to distinguish between them. Most methods in current use ensure that this different labeling occurs at as early a stage as possible in the total integrated procedure, so that the protein extracts from two (or more) cell states are separated on the same gel or column at the same time, thus avoiding all ambiguities arising from variability in the separation procedures. In the case of culture-based labeling (Section 11.6.2e) the proteins are actually created in the cells with the isotopic labels already in place, so that this 'ideal' situation is automatically satisfied. In other methods (see below) the protein extracts to be compared are mixed together after the protein extraction and isotopic labeling steps. Both of these steps must be developed to be essentially 100 % efficient so that no distortion of relative abundance can arise.

Because some of the proteolytic peptides used as surrogate analytes for the target protein can be appreciably larger than typical small molecule analytes, the potential problem (Section 8.5.2c) of cross-contributions between isotope-labeled internal standard and native (natural isotopic abundance) analyte at the m/z values chosen to monitor the two becomes more serious. Unless a large number of heavy isotope labels is used in the internal standard, sufficiently large that only a negligible contribution is made by higher isotopologs of the native analyte peptide at the lowest m/z value for the internal standard, it may be necessary to adopt a calibration function that is non-linear in the parameters of the function. This applies to both relative quantitation (in which the biological sample that includes the isotopic label(s) plays the role of the internal standard) and to absolute quantitation as discussed in Section 8.5.2c. As a separate but related issue, many measurements of relative abundance of a protein in two (or more) samples derive this quantity as the simple ratio of LC–MS peak areas for the surrogate peptide and its isotope-labeled analog; this is equivalent to a single point calibration that assumes a linear calibration curve that includes the origin. While it is true that any uncertainties that might arise from this assumption may not be significant relative to those from the many other causes of uncertainty in these difficult experiments, it is a limitation that should be borne in mind.

At present there are three isotopic labeling strategies in common use for such relative quantitation experiments in proteomics. For convenience these can be referred to

as proteolytic, culture-based and chemical, and have been mainly concerned with applications within the shotgun proteomics context although there is no intrinsic reason why they should not be applicable also to experiments involving some pre-separation of proteins; a recent review (Righetti 2003) describes a wide range of pre-separation techniques that have been used for this purpose. An important classification of all these labeling methods concerns the point in the overall procedure at which the isotopic labeling is performed. Since the object of the exercise is to compare the amounts or concentrations of specific proteins (via chosen proteolytic peptides as surrogates) in two different biological samples, it is important that the two samples should be treated in exactly the same way as far as possible; for this reason, those methods in which the labeling is performed at as early a stage as possible (preferably on the proteins themselves), so that the two samples can then be mixed and thereafter treated in the identical manipulations (enzymatic proteolysis, separation etc.), have an intrinsic advantage.

Since the chemical methods are the most widely used and also those most directly related to methods employed in absolute quantitation, this is the main focus of the following summary. Chemical labeling methods can be further classified as those that target specific residues (e.g., cysteine or lysine) and those that are 'global' in the sense that they label functional groups that are present in all proteins and peptides (usually the N-terminal amino or C-terminal carboxyl groups). Unfortunately, as emphasized previously (Zhang 2004), several difficulties have combined to make approaches exploiting the peptide termini less suitable than alternative methods. Firstly, the labeling can only be done after the enzymatic proteolysis, so that the two samples to be compared are mixed only after the digestion and labeling steps, leading to uncertainties arising from possible differences in the efficiencies of these steps as applied to the two samples (one of which is labeled with light and one with heavy isotopologs). Secondly, as emphasized previously in an extensive review (Zhang 2004), it is difficult to arrange that the derivatization reaction used in the labeling can distinguish the N-terminal amino and C-terminal carboxyl groups from their counterparts in amino acid side chains; thus multiple labeling of varying degrees can occur. Thirdly, the termini (particularly the N-termini) of a significant fraction of proteins are modified and therefore unreactive in the derivatization reactions, so that the proteolytic peptides that contain one or other of such termini are excluded from consideration. However, methods for labeling C-termini (Goodlett 2001) and N-termini (Gevaert 2003; Huang 2006; Leitner 2006) have been described. In the latter case labeling of amino groups on lysine side chains was avoided by acetylation of all amino groups (including the protein N-terminus) before enzymatic hydrolysis of the proteins, so that only the N-termini of the newly formed peptides were available for the derivatization with isotopic labeling. This avoided the multiple-labeling problem but at the cost of restricting the sites of proteolysis by trypsin to the arginine residues (a problem for all labeling methods involving prior derivatization of lysine sidechains).

11.6.2a Isotope-Coded Affinity Tag (ICAT)

The best known method that targets specific residues for labeling is the ICAT (isotope-coded affinity tag) method (Gygi 1999). In this case that targeted residue is cysteine, $[-NH-CH(CH_2SH)-CO-]$. The ICAT reagents (Figure 11.42(a)) include a reactive group (alkyl iodide) with specificity toward thiol (i.e., $-SH$) groups (specifically cysteine residues for proteins), a linker that can incorporate stable isotopes (8 deuterium atoms in the original version) and an affinity tag (biotin, sometimes known as vitamin H or vitamin B_7) used to isolate ICAT-labeled peptides via immunoaffinity chromatography (Section 4.4.1e) exploiting the very strong affinity of biotin for avidin (a glycoprotein); the dissociation constant for the avidin–biotin complex is $\sim 10^{-15}$ M, the lowest known for any ligand–protein combination. This immunochromatographic purification is thus highly specific for peptides containing cysteine residues and this greatly simplifies the subsequent LC–MS/MS analysis. A workflow plan for comparative quantitation of proteins in cells grown under different conditions is shown in Figure 11.42(b).

An improved version of the ICAT reagents (Li 2003) uses ^{13}C in the isotope-labeled linker group instead of deuterium (thus ensuring HPLC co-elution of the 'light' and 'heavy' labeled peptides), and also included a hydrolyzable linkage between the linker group and the biotin moiety. The latter feature (Li 2003) allows removal of the biotin after immunoaffinity purification of the derivatized peptides, beneficial in widening the retention time window over which the derivatized peptides elute (i.e., increasing the separation efficiency). Another modification of the original ICAT method (Zhou 2002) introduces a solid phase approach, in which a small isotope-labeled tag is attached to cysteine residues following the initial capture and then release of the peptides containing this residue. The ICAT method has been very successful and widely used; its main drawback concerns the relatively low cysteine content in most proteins. This implies that only a very few tryptic peptides are captured and characterized (qualitatively and quantitatively), so

that many partial sequence strings (including those with post-translational modifications) will be missed and the proteins thus prone to mis-identification.

11.6.2b Isotope Tagging for Relative and Absolute Quantitation (ITRAQ)

Another labeling strategy (Ross 2004) that also achieves global labeling of digested proteins has come to be known as ITRAQ (isotope tagging for relative and absolute quantitation). This method uses a set of reagents that derivatize free amino groups at the N-termini and lysine side chains of the peptides in a digest mixture. These isotopically labeled reagents (Figure 11.43) are all isobaric, have the same molecular mass to within the small differences (a few mDa) among different combinations of ^{13}C, ^{15}N and ^{18}O used to synthesize the different reagents, i.e. well within m/z windows normally used to isolate precursor ions in MS/MS. In this way all derivatized peptides of a given sequence (but derived from different biological samples that are to be compared) are isobaric and co-elute. However, these reagents are constructed (Figure 11.43) so that, upon CID in MS/MS experiments, they yield signature (reporter) ions that differ in m/z value among the reagents and can be used to track, and provide relative quantitation for, proteins from individual biological samples. Absolute quantitation of targeted proteins can also be achieved using synthetic peptides tagged with one of the reagents to be used as a surrogate internal standard in the usual fashion (Sections 2.6.3 and 8.5.2d).

The ITRAQ method benefits from a more complete coverage of the original protein sequence than does ICAT simply because every proteolytic peptide is tagged at the N-terminus (i.e. tagging is not limited to relatively uncommon cysteine residues), thus permitting increased confidence in identification and a reliable estimate of confidence limits with respect to the relative quantitation through comparisons of data acquired via different peptides. The ITRAQ method also has an intrinsic advantage with respect to MS/MS duty cycle in the following sense; since all (up to four) precursor ions, originating from the different biological samples to be compared, are isobaric (m/z values all agree to within typical widths of the precursor ion isolation window), there is no need to spend precious LC–MS/MS time in sequentially selecting the differently-labeled precursors over the time course of the HPLC peak in order to obtain the MS/MS spectra for identification purposes. The same MS/MS spectrum gives information on peptide sequence regardless of which label is present, and also the data required for relative quantitation via the relative signal intensities of the reporter fragment ions.

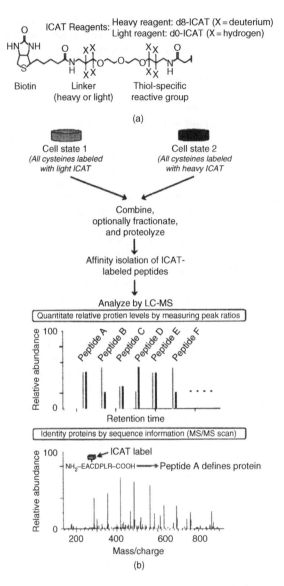

Figure 11.42 (a) Structure of the original ICAT reagent, consisting of three structural elements: an affinity tag (biotin), which is used to isolate ICAT-labeled peptides via immunoaffinity chromatography exploiting the very strong affinity of biotin for avidin (a glycoprotein; the dissociation constant for the avidin–biotin complex is $\sim 10^{-15}\,\mathrm{M}^{-1}$, the lowest known for any ligand–protein combination); a linker that can incorporate stable isotopes (eight deuterium atoms in the original version); and a reactive group (alkyl iodide) with specificity toward thiol (i.e., –SH) groups (specifically cysteine residues for proteins). The 'light' ICAT reagent has X=H, and the 'heavy' version has X=D. (b) Workflow for a differential protein expression (relative quantitation) analysis. Following separate ICAT derivatizations of proteins from cells in the two states (e.g., 'normal' and 'diseased') using the isotopically distinct ICAT reagents, the two derivatized protein extracts are combined, and the mixture of differentially-labeled proteins is the fractionated (e.g., by SDS–PAGE) such that the two forms of a given protein are co-located (or co-eluted if the fractionation is conducted chromatographically) since the H/D difference between the two forms is too small to lead to separation of the two forms. All the following steps in the procedure are thus applied to the two labeled forms of any given protein in the same solution, i.e., they are treated exactly the same. Note that a new version of the ICAT reagent (Li, *Molec. Cell Proteomics* **2**, 1198 (2003)) uses ^{13}C to replace deuterium as the isotopic label (thus avoiding potential complications from back-exchange in solution and noncoelution of differently-labeled tryptic peptides), and also permitted cleavage of the bulky biotin moiety before analysis by LC–MS/MS. Reproduced with permission from MacMillan Publishers Ltd: Gygi, *Nature Biotech.* **17**, 1994 (1999), copyright (1999).

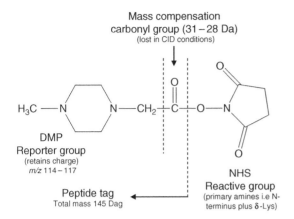

Mass compensation
carbonyl group (31 – 28 Da)
(lost in CID conditions)

DMP
Reporter group
(retains charge)
m/z 114 – 117

Peptide tag
Total mass 145 Dag

NHS
Reactive group
(primary amines i.e N-
terminus plus δ-Lys)

Figure 11.43 Diagram showing the components of the ITRAQ multiplexed isobaric tagging chemistry. The complete molecule consists of a reporter group based on *N, N*-dimethylpiperazine (DMP), a mass balance carbonyl group, and a peptide-reactive group (ester of N-hydroxysuccinimide, NHS). The overall mass of reporter plus balance components is kept constant using differential isotopic tagging using combinations of ^{13}C, ^{15}N, and ^{18}O (no problems with chromatographic separation arising from deuterium substitution); the *m/z* value of the reporter group (DMP) ranges from 114.1 to 117.1, while the balance group (CO) mass is 28–31 Da such that the combined mass remains constant (145.1 Da) for the four reagents. When reacted with a peptide the tag forms an amide linkage to any peptide amine (N-terminal or lysine amino group). When subjected to CID these amide linkages fragment in the same way as backbone peptide bonds, but in this fragmentation the mass balancing carbonyl group is lost while the charge is retained by the reporter group fragment.

The disadvantages include possible errors in the quantitation arising from differences in the efficiencies of enzymatic digestion and in any peptide pre-fractionation step among the different samples (the tagging is performed only after all these steps, unlike the case with the ICAT approach), the cost of the reagents and difficulties arising from the low mass cutoff (*m/z* values are low for the reporter ions relative to those for the precursors) if a 3D ion trap is to be used (Section 6.4.5a). Indeed, the linear ion trap with axial ejection (Section 6.4.5b) is currently the only ion trap instrument that can be used to provide quantitative data with appropriate sensitivity using the ITRAQ approach.

11.6.2c Other Chemical Labeling Methods

Other methods of chemical labeling of reactive groups in proteins have been described. Lysine residues [–NH–CH(C$_4$H$_8$–NH$_2$)–CO–] are much more plentiful in proteins than cysteines (typically 3-8 % of all residues)

and offer an attractive target for isotopic labeling, since the presence of multiple representative peptides from each protein can provide a higher degree of assurance in protein identification and higher quantitative accuracy and precision. However, as mentioned above, derivatization of the amino group in lysine makes it inactive with respect to digestion by trypsin, leaving only arginine residues as tryptic cleavage sites and thus tending to yield fewer and thus larger peptides.

A different, essentially global approach involves dimethylation of peptide amino groups of pre-digested protein mixtures via reductive amination with isotopically coded formaldehydes (CH$_2$O and CD$_2$O) in the presence of the reducing agent sodium cyanoborohydride (CNBH$_3$); a recent example (Melanson 2006) also describes earlier examples of this approach. This produces a mass difference of 4 Da [–N(CH$_3$)$_2$ vs –N(CHD$_2$)$_2$] per labeling site (lysine or N-terminus) for tryptic peptides. Since all peptides possess an N-terminal labeling site (except for the N-terminus peptide from N-terminally blocked proteins), all peptides are labeled at least once (similar to the ITRAQ approach). The reaction is found to proceed to completion (no unlabeled or monomethylated peptides observed, probably a result of the large excess (100-fold) and small size of the derivatizing reagents (no steric hindrance). However, since this reductive amination approach performs the derivatization after enzymatic digestion and just before LC–MS analysis, it is subject to potential errors arising from variability of the efficiency of the digestion and other procedures as discussed above as an advantage of the ICAT method. On the other hand, the fact that all tryptic peptides are labeled and thus potentially detected in this method means that it is possible to use multiple peptide pairs to both identify and quantitate (in the relative sense) the proteins with an appreciably higher statistical confidence than is possible with methods such as ICAT. The latter usually have only one or two peptide pairs available per protein as a result of the scarcity of cysteine residues. An example is shown in Figure 11.44.

An unusual approach to chemical tagging for purposes of identification and quantitation of proteins has been described (Ornatsky 2006) in proof-of-principle experiments. Very briefly, this approach uses chemical tags that contain one of a series of metal atoms (gold, europium, samarium or terbium) that are then detected and quantitated using inductively-coupled-plasma (ICP)–MS, a standard analytical technique for elemental metal analysis (Nelms 2005) that can provide very high sensitivity and precision. At present this approach is very much in its initial stages of development, but the sensitivity is such that single cell proteomics may not be out of reach.

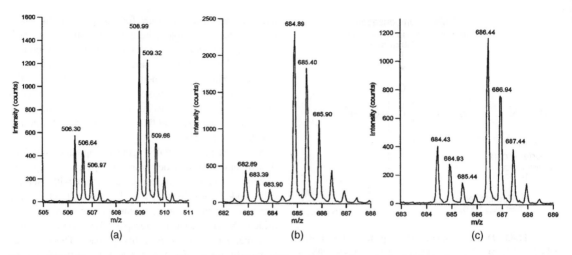

Figure 11.44   Three examples of peptide-pair spectra (out of 11 peptide pairs isotopically labeled using the reductive amination method and used for relative quantitation in this case) for myelin basic peptide; the peptide sequences are (a) *TQDENPVVHFF*K, (b) *YLATASTMDHAR, and (c) *HRDTGILDSIGR, where * denotes labeling sites (N-terminus on the left in all cases plus the C-terminal lysine (K) for (a), the C-terminal arginines (R) do not contain an amino group and are not derivatized). Note that spectrum (a) is for triply charged peptides while the other two are doubly charged. The higher abundance peptides ('heavy' labels) were derived from brain tissue from a patient who had suffered from Alzheimer's Disease, and the 'light' labeled peptides to a normal healthy brain. The average abundance ratio over all 11 peptide pairs was 3.0 ± 1. After tryptic digestion and labeling ($100\,\mu$g of each brain tissue separately) the two samples were mixed and then fractionated by strong cation exchange into 30 fractions; each fraction was then analyzed by reversed phase HPLC–MS using a QqTOF instrument. The acquisition method consisted of a 1 s TOF–MS survey acquisition (Q1 in RF-only funnel mode, examples shown above) followed by three 1 s QqTOF–MS/MS spectral acquisitions for the three largest precursor peaks detected in each survey scan (the latter for protein identification). Reproduced from Melanson, *Proteomics* **6**, 4466 (2006), copyright (2006) with permission from Wiley–VCH.

A method has been described (Weckwerth 2000) for relative quantitation of phosphoproteins to simultaneously compare the phosphorylation status of proteins under two different conditions. This is important because reversible protein phosphorylation, at the hydroxyl groups of serine (Ser), threonine (Thr) or tyrosine (Tyr) residues, is one of the major regulatory mechanisms for controlling intracellular protein functionality. In this case (Weckwerth 2000) relative quantitation was achieved by β-elimination of phosphate from phospho-Ser/Thr to give dehydroalanine and dehydroamino-2-butyric acid, respectively, followed by Michael addition of ethanethiol or ethane-d_5-thiol across the vinyl moieties resulting from elimination of the phosphate. This approach can give good data but is susceptible to unwanted side reactions (Wolschin 2003). There are also methods for relative quantitation that involve no labeling but rely instead on parameters such as counts of identifiable peptides (Wiener 2004; Higgs 2005; Old 2005; Silva 2006), but since these approaches are not applicable to absolute quantitation they are not discussed here.

11.6.2d Proteolytic Labeling Methods

The proteolytic approach to isotopic labeling involves conducting the enzymatic digestion ('proteolysis') of the protein in water with different oxygen isotopic compositions. An early work (Bender 1957) described $^{16}O - ^{18}O$ exchange on hydrolysis of small peptides with chymotrypsin (a proteolytic enzyme that specifically hydrolyzes peptide bonds on the C-terminus side of tyrosine, tryptophan and phenylalanine residues). This phenomenon also is observed for other enzymes including trypsin and has been exploited for differential labeling of peptides in the course of enzymatic digestion in $H_2^{16}O$ or $H_2^{18}O$ (Stewart 2001; Yao 2001; Reynolds 2002); this application has been reviewed (Miyagi 2007). The mechanism of, e.g., trypsin-catalyzed cleavage involves formation of a peptide–enzyme ester (acyl-enzyme intermediate) at the C-terminus of the newly formed peptide; this ester is then hydrolyzed to form the free peptide and regenerate the enzyme, and at this stage the oxygen atom of the water molecule is incorporated into the terminal carboxyl group of the peptide. This offers a simple method of isotopic labeling in the course of enzymatic digestion without the need for exotic reagents and also provides global labeling (apart from the C-terminus carboxyl of the original protein), and thus multiple peptide pairs for protein identification and relative quantitation.

However, there are drawbacks to this approach, particularly the variability in the extent of ^{18}O incorporation in the 'heavy'-labeled peptides. This arises because of the possibility of back-exchange (^{16}O replaces ^{18}O) when the peptide–trypsin complex is re-formed and subsequently hydrolyzed again. For this and other reasons this approach is not often used, since many precautions have to be taken in the acquisition and interpretation of the data (Stewart 2001). However, such labeling has been exploited (Shevchenko 1997) in experiments designed to assist in the interpretation of MS/MS spectra of peptides by distinguishing between primary fragments that contain the C-terminus (exhibit an appropriate mass shift) from N-terminal fragments (no such mass shift).

11.6.2e Culture-Based Labeling Methods

The culture-based approach to differential isotopic labeling has many advantages; this approach was first applied (Oda 1999) in the present context by growing wild-type yeast cells in a growth medium containing nitrogen atoms in their natural relative abundance (99.6 % ^{14}N, 0.4 % ^{15}N) and comparing these with mutant-line cells grown in a uniformly ^{15}N-substituted medium (^{15}N content > 96 %) that was otherwise identical. The two cultures can thus be first mixed together and only then subjected to cell lysis to liberate the proteins, enzymatically digested, and analyzed by LC–MS, with no possibility of variability in the sample processing for the two cultures (even the ICAT approach must subject the different cell cultures to lysis separately, Figure 11.42). However, this approach has turned out to be somewhat limited in its range of applications, for several reasons. Thus the degree of incorporation is not necessarily 100 %, and since the numbers of nitrogen atoms vary in the different amino acids it has proved difficult to devise algorithms for computer-based interpretation of the spectra. Further, it is difficult to conduct any culture-based experiments for mammalian cells, so the experiments that have successfully used this approach have generally been limited to those microorganisms that can be cultured in this way.

A different approach to the same general objective was first suggested (Veenstra 2000) in a somewhat different context, that of assisting interpretation of LC–MS data for intact proteins (no digestion involved) to identify the proteins using a combination of accurate molecular mass measurements (by FT–ICR MS, Section 6.4.8), partial knowledge of amino acid content, and searches of genomic databases. The test case was an auxotrophic strain of the bacterium *Escherichia coli* (*E. coli*) (an auxotrophic strain is a mutant form that is unable to synthesize one or more chemical compounds necessary for its growth); in this case the mutant strain was unable to produce the essential amino acid leucine (Leu). Proteins extracted from the organism grown in a natural isotopic abundance medium containing Leu, and those from another culture to which d_{10}-Leu was added, were mixed and analyzed by ESI–FTICR–MS to give molecular masses with uncertainties generally $\leq 1\,Da$. The difference between the molecular masses of the natural isotopic abundance and corresponding d_{10}-Leu-labeled protein was used (Veenstra 2000) to determine the number of Leu residues present in that particular protein, and combined knowledge of the molecular mass and number of Leu residues permitted unambiguous identification of the intact protein from the *E. coli* genomic database.

This labeled-Leu approach was adopted and applied to relative quantitation of proteomes practically simultaneously by two groups. One (Jiang 2002) demonstrated the new method using d_{10}-Leu incorporated by an auxotrophic strain of yeast (*S. cerevisiae*) and the other (Ong 2002) used d_3-Leu with specialized mammalian (mouse) cells. (Note that leucine is one of the 'essential amino acids' in mammals, i.e., mammals are unable to produce it from other compounds so it must be obtained from external sources in food; thus isotope-labeled Leu is suitable for incorporation into proteins in mammalian cells and has the further advantage of being one of the most abundant amino acids in proteins.) Both of these investigations successfully applied the new method and one of them (Ong 2002) coined the acronym 'SILAC' (Stable Isotope Labeling by Amino acids in Cell culture) for the approach. An example of the power of the method is provided by the use of $^{13}C_6$-lysine (another essential amino acid) in a SILAC strategy to investigate prostate cancer cells (Everley 2004).

Thus far all of these culture-based methods of isotopic labeling for relative quantitation have involved cell cultures. More recently (Wu 2004) the general approach was extended to live mammals (laboratory rats in this case) fed on a diet consisting of a protein-free rodent diet supplemented with algal cells containing either only natural abundance isotopes (control rat) or algal cells enriched with >99 atomic % excess ^{15}N. The long term (six weeks) metabolic labeling of rats with this ^{15}N-enriched diet resulted in no observed adverse health consequences. The ^{15}N-enriched rat liver (92.2 atomic %)

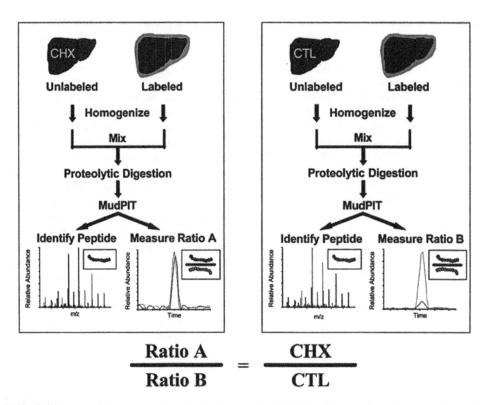

$$\frac{\text{Ratio A}}{\text{Ratio B}} = \frac{\text{CHX}}{\text{CTL}}$$

Figure 11.45 Relative quantitative proteomic analysis using metabolically labeled mammalian tissue as an internal standard. The postnuclear supernantant (PNS, a fraction isolated from centrifugation of homogenized tissue) from untreated (CTL) and cycloheximide-treated (CHX) liver and ^{15}N-enriched liver were mixed early during the sample preparation to account for protein losses during the digestion and measurement including identification using tandem mass spectrometry and measurement of ratio using mass chromatograms. (MudPIT, i.e., Multidimensional Protein Identification Technology (Washburn 2001) is an LC × LC–MS/MS technique for separation and identification of complex peptide mixtures that uses a strong cation exchange (SCX) stationary phase in series with a reversed phase material inside a fused silica capillary). Changes in protein level were estimated by measuring the ratio of two ratios using the following: (1) ratio A (left-hand panel), the abundance (peak area) ratio of a tryptic peptide from cycloheximide-treated rat vs that from the ^{15}N-enriched liver; and (2) ratio B (right-hand panel), the corresponding ratio for the untreated rat vs the ^{15}N-enriched liver. By taking the ratio of ratio A to ratio B, the abundance of the ^{15}N-labeled internal standard cancels, giving the relative abundance of the peptide derived from treated to untreated rats, and thus of the protein for which the chosen peptide is used as a surrogate. Reproduced from Wu, *Anal. Chem.* **76**, 4951 (2004), copyright (2004) with permission from the American Chemical Society.

was used (Figure 11.45) as internal standard for relative quantitation of proteins on comparing the proteomes of rats treated (or not) with doses of cycloheximide (a compound widely used in biomedical research to inhibit protein biosynthesis in eukaryotic organisms, generally in culture-based *in vitro* experiments because of its significant toxic effects). In this way, statistically significant changes in individual protein levels in response to cycloheximide treatment were measured for 310 proteins and revealed 127 proteins with altered protein levels ($p < 0.05$). As emphasized in the original work (Wu 2004), this approach has many potential applications in biomedical science. For example, diseased and healthy tissues can be compared directly in mammalian model organisms (as opposed to cell cultures for example) for identification of novel drug targets, with the advantage that the new method can monitor clinically important drug interactions that can occur when one drug modulates the level of cellular metabolic enzymes thus affecting the activity of co-administered drugs.

Finally in this section, it must be emphasized that relative quantitative proteomics measurements currently use a range of methodologies other than those based on mass spectrometry. Recently (Unwin 2006) this range of techniques, including those summarized above plus protein arrays and flow cytometry, has been reviewed and compared. This review also covers the related issue of absolute quantitation of concentrations of specific proteins, and this subject is discussed in the following section.

11.6.3 Absolute Quantitation

Sometimes relative quantitation of proteins, to compare protein concentrations between two or more cell cultures (or whole multicellular organisms), is not sufficient to answer the biological questions being addressed. In such cases absolute quantitation is required to provide measurements of, e.g., average copy numbers of specific protein molecules per cell. Unfortunately, single cell proteomics analyses, qualitative or quantitative, are currently unattainable in view of the substantial gap between analytical limits of detection and quantitation and the very low sample sizes typically available, although ICP–MS methods (Ornatsky 2006) might possibly help to bridge the gap. The most important missing link is a method of multiplication of protein molecules such as that provided for DNA analysis by the polymerase chain reaction, PCR. As emphasized previously (MacCoss 2005), absolute quantitation of specific proteins in complex mixtures relies on all the same principles as those established for small molecule analytes plus the additional problems arising from the use of proteolytic peptides as surrogate analytes, including difficulties in confirming analyte (protein) identity particularly in shotgun proteomics approaches (Section 11.6.1). The other potential problem shared with relative quantitation measurements, concerning the possible cross-contributions from higher isotopologs of the native analyte peptide (Section 8.5.2c), can be avoided by using isotope-labeled internal standards with a sufficient number of heavy isotope labels.

In practice, development of methods for absolute quantitation of proteins by mass spectrometry methods has been delayed by difficulties in obtaining the necessary isotope-labeled peptides for use as surrogate internal standards. An early report (Barr 1996) not only synthesized two isotopically labeled peptides (and their unlabeled analogs) characteristic of the protein analyte (apolipoprotein-A1), one as the surrogate analyte used for quantitation and the other to provide internal validation of the method as a confirmation standard, but also actually succeeded in using an LC–MS method based on a flow-FAB interface (Section 5.3.1) to a double focusing magnetic mass spectrometer (Section 6.4.4). To those who have experience with this instrument combination, this work (that involved 96 independent measurements) is seen as heroic even though the sample analyzed was

not a complex proteomics mixture but a purified protein sample supplied as a CRM (certified on the basis of amino acid analysis). Calibration curves were obtained using calibration solutions of the unlabeled peptide standards plus the isotopically labeled peptides (containing six to nine heavy isotopic labels to avoid cross-contributions, Section 8.5.2c). Thanks to the use of more than one internal standard a sophisticated statistical analysis was possible (Barr 1996) that led to independent confirmation of the certified content of the apo-lipoprotein A1 in the CRM.

Both the ICAT (Jenkins 2006) and ITRAQ methodologies (Ross 2004) can be adapted to enable absolute rather than relative quantitation of proteins, essentially by tagging a (natural isotope) synthetic version of the peptide chosen as the surrogate analyte with one of the labeled reagents. There is no question that such approaches can work well but they suffer from the same disadvantages as when used for relative quantitation and, moreover, are more laborious than some more recent methods described below.

In early work dating from the 1980s that is somewhat related to absolute proteomics quantitation, Desiderio and colleagues used isotope-labeled internal standards to measure absolute quantities of small neuropeptides in tissues such as the pituitary gland and cerebrospinal fluid (Desiderio 1998, 1999). This work is important in its own right and also as a guide for the more complex problem of using proteolytic peptides as surrogate analytes for target proteins in complex proteomic extracts. The use of FTICR to quantitate biological molecules ranging from cyclic peptides of molecular mass $\sim$1200 Da (Gordon 1999) to nondigested medium sized ($\sim$12 kDa) proteins (Gordon 1999a) was extensively investigated using close analogs as internal standards rather than isotopically labeled analytes; this work was mainly devoted to the conditions necessary for FTICR to be used in quantitation of biomolecules rather than applications to complex proteomics samples.

11.6.3a AQUA

Applications of the isotope dilution approach to absolute quantitation of target proteins in complex proteomics samples began to appear in 2003–2004 (Barnidge 2003, 2004; Gerber 2003; Kuhn 2004); one of these (Gerber 2003) was given the acronym AQUA and was later reviewed by its originators (Kirkpatrick 2005). An example of the use of this approach to determine absolute quantity of a phosphoprotein and its extent of phosphorylation at a serine (Ser) residue is shown in Figure 11.46. All of these developments used classical

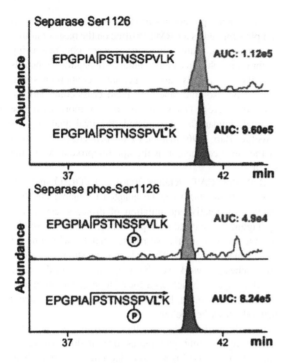

Figure 11.46 HPLC–MRM quantitative analysis of the phosphorylation state of human separase protein at Ser-1126 using the surrogate tryptic (AQUA) peptide EPGPIAPSTNSSPVLK LC with Leu-1129 isotopically labeled ($^{13}C_6 - ^{15}N_1$) in the internal standard versions (both phosphorylated and nonphosphorylated as shown). The MRM transitions monitored are indicated on the peptide sequences given in the Figure; 'AUC' denotes area under the curve (LC peak). Quantitation of both phosphorylated and nonphosphorylated forms provides information on absolute concentrations and the fraction of this protein that is phosphorylated (34 % in this case). These data were acquired from the equivalent of only 16 μg of starting material and 10 fmol of each AQUA peptide using an enhanced-resolution triple quadrupole mass spectrometer. The calculated percent of phosphorylation was 34 %. Reproduced from Gerber, *Proc. Natl. Acad. Sci. USA* **100**, 6940 (2003), copyright (2003) with permission from the National Academy of Sciences, USA.

solid-supported peptide synthesis to prepare the isotope-labeled peptide internal standards. In the AQUA method (Gerber 2003) Leu was again chosen as the labeled amino acid ($^{13}C_6 - ^{15}N_1$) to be incorporated into the internal standard, and in some cases a serine or threonine residue was phosphorylated to provide measurements of the extent of phosphorylation. The N-terminal tryptic peptide of prostate-specific antigen (PSA) was synthesized to incorporate labeled glycine ($^{13}C_2 - ^{15}N_1$), used as internal standard in the LC–MS/MS determination of this biomarker (Barnidge 2004). Although good analytical performance

was obtained despite using minimal sample preparation (denaturing the serum with urea) to permit high throughput, the levels of PSA that could be determined by this method were $\sim 10^3$ times higher than those detected and measured by standard immunological tests in serum from patients. At least part of this disappointing aspect could be attributed to significant ionization suppression by co-eluting serum components that could, in principle, be alleviated by a more complicated sample preparation procedure and/or optimized HPLC separation, but at the cost of considerably decreased throughput.

This same problem was encountered (Kuhn 2004) in development of a method to determine C-reactive protein (CRP, molecular mass ~ 25 kDa), a diagnostic marker of rheumatoid arthritis, in serum. In this case the ionization suppression problem was mitigated by enriching small volumes of patients' serum for low abundance proteins through selective removal of the most abundant proteins (human serum albumin (HSA), immunoglobulin G (IgG) and haptoglobin). The complexity of the protein mixture was further simplified using size exclusion chromatography (SEC, Section 4.4.1) to fractionate denatured proteins into discrete molecular mass ranges. After trypsin proteolysis and spiking with four synthetic isotope-labeled peptide standards derived from C-reactive protein, nanoflow chromatography–MS/MS using MRM on a triple quadrupole mass spectrometer with up to five MRM transitions per peptide yielded absolute quantitative data in good agreement with an immunoassay method.

It was emphasized (Kuhn 2004) that, in selecting peptides for use as internal standards, their hydrophobicity, number of amino acid residues, peptide location within the protein sequence and lack of sites that could be subject to potential post-translational modification should be considered in addition to a high degree of uniqueness in the protein–peptide connectivity (Section 11.6.1). Although multiple MRM transitions were not required (Kuhn 2004) to obtain data of good accuracy and precision for more abundant target proteins, use of relative intensities of more than one (up to five) MRM signals increased the specificity of detection for peptides derived from low abundant proteins in complex proteomics mixtures and thus also increased the confidence of protein identification.

11.6.3b QCAT Peptides

As mentioned several times already, access to appropriate quantities of synthetic isotope-labeled peptides is a significant barrier to more widespread use of isotope dilution mass spectrometry in absolute quantitation in proteomics. This is especially true when several proteins are targeted for quantitation simultaneously, as in screening for biomarkers (Pan 2005). Fortunately a recent innovation (Beynon 2005) has improved this situation considerably. This ingenious approach allows nature to perform the job of assembling the amino acids into the required peptide sequences. Several tryptic peptides to be used as analytical standards are produced as a concatemer of peptides (QCAT peptides) generated by gene design *de novo*. Essentially the concept exploits the relative ease (compared with peptide synthesis) with which stretches of DNA can be synthesized from the component nucleotides with high fidelity, and can then be amplified using the polymerase chain reaction. By thus creating an artificial gene that codes for several QCAT peptides strung together and expressing this gene in *E. coli* culture using well established techniques of molecular biology, the concatenated peptides are produced in good yield (measurable in molar terms by determination of the purified QCAT protein) and in strict 1:1 stoichiometric ratios to one another after digestion of the QCAT protein with trypsin. The intact QCAT protein is added to the protein mixture before enzymatic digestion so that some degree of correction for absolute efficiency of digestion is possible.

Labeled QCAT peptides can be readily produced by conducting the *E. coli* fermentation in a medium containing $^{15}NH_4Cl$ as the only nitrogen source to produce uniformly ^{15}N-labeled peptides (Beynon 2005), although it was emphasized that in some cases it might be preferable to select Q-peptides that all contain the same isotope-labeled amino acid by supplying it to the *E. coli* expression system. For example, incorporation of $^{13}C_6$-lysine and $^{13}C_6$-arginine (the N-terminal residues in tryptic peptides) would ensure that the Q-peptides thus created would be singly labeled and thus display a mass offset between 'heavy' and 'light' peptides by a constant 6 Da, which would facilitate peptide recognition in LC–MS spectra. Full details of the procedure are given in the supplementary material for the paper (Beynon 2005).

Figure 11.47 shows an example of application of a QCAT protein containing 20 tryptic peptides, of which 13 acted as surrogate analytes for proteins present in a preparation of soluble skeletal muscle proteins from chick. The absolute quantitation results are compared with those obtained by densitometry following two-dimensional SDS–PAGE separation, with reasonable agreement considering that the latter method probably detects all isoforms (including post-translationally modified forms) that are not measured by the QCAT–LC–MS method as originally applied. For purposes of monitoring

post-translationally modified proteins, subsequent chemical modification of the QCAT protein or its constituent peptide standards would be required and this would imply a need for a great deal more work to achieve the goal of facile multiplexed quantitation of proteins (Beynon 2005). On the other hand, as emphasized by the authors, once a QCAT gene has been synthesized it can readily be stored, amplified by PCR if required and re-expressed to give a fresh supply of the QCAT peptides when required; in contrast, peptides produced by traditional chemical synthesis are a finite resource.

11.6.3c Stable Isotope Standards and Capture by Anti-Peptide Antibodies (SISCAPA)

More recently (Anderson 2006) a strategy similar to QCAT was described, but in this case incorporation of $^{13}C_6 - ^{15}N_2$-labeled lysine was used to create the internal standard peptides. This method was developed to help meet the demand for internal standards in a strategy described earlier (Anderson 2004a) for quantitation of lower abundance proteins in complex tryptic digests such as those derived from plasma. This strategy (Figure 11.48(a)), denoted SISCAPA, uses anti-peptide antibodies immobilized on 100 nL nanoaffinity columns to enrich specific peptides along with spiked stable-isotope-labeled internal standards of the same sequence. In the original work (Anderson 2004a) these labeled standards were synthesized by traditional wet chemistry methods and this step was a limiting one for the extent of multiplexing that could be achieved.

Peptides in the digest that are not bound by the immobilized antibodies are not retained and the bound peptides can then be eluted by changing the composition of the mobile phase and analyzed by LC–MS. An important component of this strategy is the ability to obtain antibodies that specifically bind the tryptic peptides selected as surrogate analytes for the target proteins. In the SISCAPA work (Anderson 2004a) the antibodies used were polyclonal, i.e., were produced by injecting mammals (rabbits in this case) with the 'antigen'. The selected peptide is attached to a suitable carrier moiety, and this induces the B-lymphocytes (a class of white blood cells responsible for adaptive immune response in mammals) to produce IgG immunoglobulins that are specific for the antigen and can be purified from the serum of the immunized mammal. (In contrast, monoclonal antibodies are derived from a single cell line, not a living mammal.) Clearly this key step involves specialized immunology skills not commonly found among analytical chemists! The peptide-specific IgG immunoglobulins were then immobilized on a solid support and used in

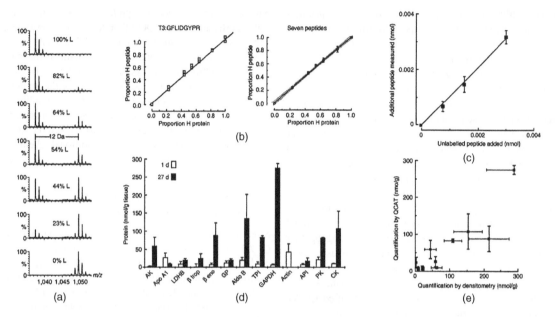

Figure 11.47 Preliminary results exemplifying the use of a QCAT protein and its 20 constituent tryptic peptides in quantitation of proteins via their surrogate tryptic peptides; 'L' and 'H' denote QCAT peptides that are unlabeled and uniformly ^{15}N-labeled, respectively. All mass spectrometric measurements were by MALDI–TOFMS. (a) Partial mass spectra for mixtures of L- and H-forms of one of the QCAT peptides (T3 = GFLIDGYPR, contains 12 N-atoms); the percentage of 'L'-peptide in the mixture is indicated in each spectrum. (b) Graphs indicating the relationship between the fraction of H-QCAT protein in a mixture and the corresponding ratio proportion of H-peptide (measured by MALDI–TOFMS) after tryptic digestion of the QCAT protein; data are shown for the T3 peptide only and for a group of seven QCAT peptides (n = 3). The plots are linear with a slope of unity (1.008 ± 0.008, n = 7), as expected for a clean and quantitative proteolysis. (c) Plot of data from a Method of Standard Additions experiment (Section 8.5.1b) in which known amounts of unlabeled (L–QCAT) protein were added to different aliquots of a preparation of soluble skeletal muscle proteins (from chick) prior to tryptic digestion. Data were obtained for seven different peptides (n = 3); error bars represent standard error of the mean (Equation [8.5], Section 8.2.1). The objective of this experiment was to demonstrate the linearity of response of the MALDI–TOFMS assay applied to analysis of the complex protein mixture. (d) Absolute quantitative levels (ng protein per g *chick* tissue) measured for 13 proteins in muscle protein preparations (1Day-old and 27-day-old chicks) by mixing with known amounts of the U[15]N-labeled QCAT protein, before tryptic digestion and MALDI–TOFMS analysis (n = 3). (e) Correlation between the QCAT data for seven of these proteins with those obtained previously (Doherty 2004) by 2D gel electrophoresis and densitometric measurements of amounts of protein on each spot; the correlation coefficient is 0.82 (p < 0.001). Reproduced by permission from MacMillan Publishers Ltd: Beynon, *Nature Methods* **2**, 587 (2005), copyright (2005).

the nanoaffinity columns to provide an average 120-fold enrichment of the antigen peptide relative to others as measured by LC–MS. (Such antibodies are not 100 % specific for the antigen peptide, i.e., there is appreciable cross-reactivity with peptides of similar structures.) The entire SISCAPA strategy was devised as an approach to increased sensitivity of MS-based assays for targeted proteins. Figure 11.48(b) illustrates the degree of specificity of one of the immunoaffinity columns for one isotope-labeled peptide relative to three others.

The isotope-labeled standard peptides with labeled lysine residues, made by expression of the artifical genes coding for the desired amino acid sequences (Anderson 2006), were used in a demonstration of quantitative

multiplexed LC–MS/MS assays (in MRM mode) for 53 high and medium abundance proteins in human plasma (this work did not include immunoaffinity enrichment). Exhaustive testing of the precision obtainable showed that data for 47 of the 53 target proteins exhibited CV values (coefficients of variation, Section 8.2.1) of 2–22 % for repeat LC–MS analyses (n = 10) of the same digest. 78 % of these assays had CV < 10 % and some peptides gave CVs in the range 2–7 % in five entirely different experiments (10 replicate LC–MS analyses each) while continuously measuring 137 MRM channels. In this way, proteins present in the complex digest could be quantitated within these precision limits at concentrations down to 0.67 μg.mL[−1], with minimal sample preparation. Protein

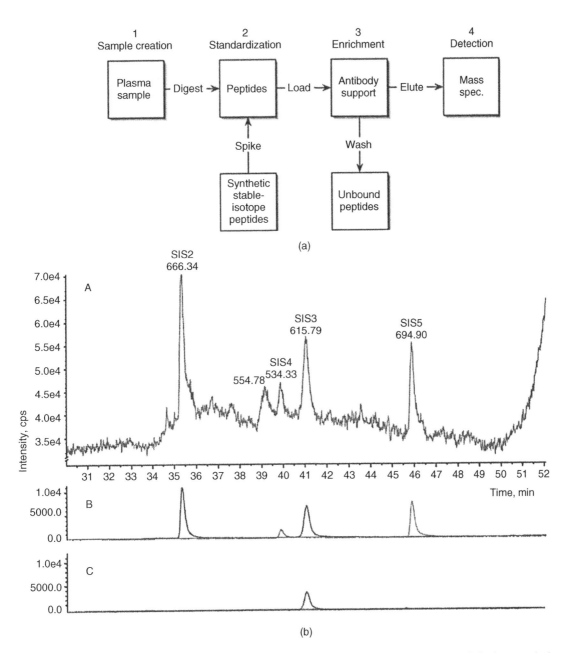

(a)

(b)

Figure 11.48 (a) Schematic diagram showing the major steps in the SISCAPA strategy for enrichment and absolute quantitation of tryptic peptides selected as surrogate analytes for low abundance proteins in complex mixtures. (b) LC–MS analysis of a mixture of four synthetic stable-isotope-labeled peptides (SIS1–4, amounts range from 23.5–48 fmol). Chromatogram A shows the total ion current observed during a positive control run (peptides transferred directly to the LC–MS instrument without exposure to the immunoaffinity column that was specific for SIS 3. Chromatogram B is the same as A (same LC–MS data set) except that the extracted ion chromatograms (0.25 Th windows) for the $(M+2H)^{2+}$ ions of the four SIS peptides are combined. Chromatogram C is the same as B but for a different experiment in which the mixture of SIS peptides was first subjected to the immunoaffinity column (specific for SIS3) before LC-MS analysis, demonstrating the specificity of the immunoaffinity column for SIS3 compared with the other three. Reproduced from Anderson, *J. Proteome Res.* **3**, 235, copyright (2004) with permission from the American Chemical Society.

concentrations covering a dynamic range of 4.5 orders of magnitude were measurable in a single experiment. This demonstration (Anderson 2006) is a notable achievement but corresponds to only the lower limit of the more common plasma proteins (Anderson 2003). As emphasized by the authors, this lower limit can be improved to cover some of the less abundant proteins by incorporating the SISCAPA strategy (Anderson 2004a). It might also be possible to apply the new LC–MALDI–MRM technologies (Section 5.2.2) to further decrease the limits of quantitation; this was recently demonstrated (Melanson 2006a) in the context of relative quantitation of a biomarker that could not be reliably analyzed using LC–ESI–MS/MS.

The rapid rate of development of technologies designed to facilitate quantitative proteomics measurements, an extremely difficult field of study, is evident in the present section and provides a prime example of the temporary validity of any discussion such as this. The ingenuity of our colleagues in incorporating a wide range of scientific and technical skills other than those described in the first few chapters, and including protein chemistry, molecular biology, immunology etc., will rapidly render this section out of date. It is for this reason that the present book has in general sought to emphasize basic scientific principles in addition to describing the current state of the art.

11.7 Analysis of Endogenous Analytes

The phrase 'endogenous analytes' refers to situations in which all matrices of interest contain observable amounts of the target analyte(s); acrylamide in foodstuffs (Section 11.2.1) is an example. The most common example is that of analysis of biological tissues for compounds that are essential for the life of the organism from which the tissue was derived. One of the earliest applications of mass spectrometry to such problems (Aubourg 1985; Stellaard 1990; Vallance 1994) involved application of GC/MS to quantitative analysis of long chain fatty acids ($C_{22} - C_{26}$) in blood plasma as a diagnostic tool for peroxisomal disorders. (Peroxisomes are sub-cellular organelles with several functions, a major one being the breakdown of fatty acid molecules two carbons at a time by so-called β-oxidation. Malfunction of this mechanism can lead to a range of neurological disorders that can be diagnosed via the consequent elevated levels of the long chain fatty acids that thus act as biomarkers for these disorders). Another early example of biomarker analysis (Millington 1990, 1991, 1994; Chace 1993, 1997; Rashed 1995; Van Hove 2000) was that of acylcarnitines for disorders of mitochondrial fatty acid oxidation (organic acidurias). This work is interesting from the viewpoint of mass spectrometry technological development since these thermally fragile and polar molecules were first analyzed using FAB (Section 5.3.1) with MS/MS (no chromatography), then by ESI–MS/MS, and most recently by LC–ESI–MS/MS (Vernez 2003).

A problem faced in this early work on quantitative analysis of endogenous biomarkers was lack of a control (blank) matrix containing no observable amounts of the analytes. This meant that it was not possible to conduct experiments to determine extraction efficiencies ($F_a{}'$ in Section 8.5) by spiking known amounts of the analytical standard into aliquots of the blank matrix, and it was necessary to assume that $F_a{}' = 100\%$. This problem, and its impact on the very concept of LLOQ for the method, is intrinsic to all quantitative analyses of endogenous analytes as defined above and was not solved until recently (see below). In principle, the situation could be greatly alleviated by using the Method of Standard Additions (Sections 2.6.2 and 8.5.1b) in which (in a sense) the analytical sample is used as its own control matrix, but this time consuming method requires a significant amount of the unknown analytical sample that is not available in the context of blood samples (a few millilitres) available for clinical diagnosis.

A comprehensive review of the role of mass spectrometry in the discovery and quantitative analysis of endogenous biomarkers (Ackermann 2006) covers both large molecules (mainly proteins) and small molecules (in the sense used in this book, Section 1.5). Fitness for purpose considerations determine the degree of analytical rigor applied, as in any quantitative analysis (see the Preface and Section 9.2), but additional concerns for biomarkers (Ackermann 2006) include the availability of reference substances, the intrinsic instability of endogenous compounds in the presence of the natural enzymes whose functions include control of their concentrations, provision of adequate selectivity for biomarkers that belong to sizable families of closely related compounds, and the need to overcome the lack of a control matrix. The last of these concerns is important for the most rigorous assays for absolute concentrations and is the main subject of the following discussion.

As emphasized throughout this book, the importance of a suitable control matrix lies both in the sample preparation stage (so that the best possible estimate of extraction efficiency can be obtained by spiking experiments, Section 8.5), and in the final analysis by mass spectrometry (so that matrix effects (i.e. suppression or enhancement of ionization efficiency, Section 5.3.6a) can be adequately accounted for. Further, determination of meaningful estimates of the LOD and LLOQ (Section 8.4) of the integrated analytical method requires access to a control matrix. Strategies that have been used to work around the problem (see Section 9.5.6) include use of a control matrix containing levels of the endogenous analyte that are sufficiently small to lie well below the desired LLOQ, use of a natural matrix that has been 'stripped' of the analyte (e.g., using an immunoaffinity column specific for the analyte or, less desirably, activated charcoal as a highly nonspecific adsorbent), or use of a surrogate matrix (e.g. synthetic urine (MacLellan 1988; Yan 2007)). All of these approaches may be fit for purpose in specific cases, but all are subject to errors and uncertainties even if the appropriate materials are available.

The problem was essentially solved by a mass spectrometric approach that is remarkably simple (once someone else has thought of it!), introduced practically simultaneously by two independent groups (Li 2003a; Jemal 2003). The basis of this approach is to use a stable-isotope labelled version as a 'surrogate analyte', spiked into a natural matrix containing the endogenous analyte at natural levels, in order to obtain the calibration curve. As emphasized in Section 2.2.3, it is best to choose a heavy atom isotope label rather than hydrogen/deuterium

and to ensure that the higher isotopologs of the natural analyte do not overlap with those of the isotope-labelled surrogate (Section 8.5.2c). Moreover, since the labelled surrogate analyte is now to be used as the standard against which the concentration of the endogenous analyte is to be compared via the calibration curve, the chemical and isotopic purities of the surrogate must be high and known. This requirement contrasts with the situation in which an isotope-labelled version of the analyte is used as a surrogate internal standard (SIS) in more conventional analyses for compounds that are not ubiquitous in the matrices of interest. In the latter case neither of these purity characteristics is crucial (see Sections 2.2.3 and 8.5.2b, especially Equation [8.86]).

This simple but clever idea (Li 2003a; Jemal 2003) avoids all the uncertainties associated with lack of a proper control matrix free of analyte. It also lends itself to the use of standards and QC samples prepared in the biological matrix of interest at concentrations below the endogenous levels of the analyte, thus facilitating routine monitoring of the performance of the analytical method at the LLOQ level. Note, however, that it is still subject to uncertainties arising from possible differences between extraction efficiencies for the native and spiked-in analytes, reflecting possible occlusion or adsorption effects (Section 8.5) for the endogenous analyte (Boyd 1996; Ke 2000).

Epilog

'New directions in science are launched by new tools much more often than by new concepts. The effect of a concept-driven revolution is to explain old things in new ways. The effect of a tool-driven revolution is to discover new things that have to be explained.'

Freeman Dyson, 'Imagined Worlds', Harvard University Press, 1997.

This quotation from a Physics Nobel Prize Winner is amply confirmed by the progress since about 1980 in the multitude of techniques involved in trace level chemical analysis. New types of sample handling robots, chromatographic phases, ionization methods, mass analyzers etc. continue to appear. As a result, significant aspects of any book such as this one must inevitably become out of date relatively rapidly. It is for this reason that a serious attempt was made while writing the book to ensure that fundamental principles are described at an accessible level, intended to be sufficient to provide a basic understanding of the principles of physics and physical chemistry that are exploited so ingeniously by our colleagues in devising new and improved apparatus and experimental methods. Of course, in a book devoted to the entire range of technologies involved in trace analysis, it is not possible (or desirable) to attempt a rigorous treatment of every topic. Readers of this book will not emerge as experts in chromatography theory, or ion optics, or mathematical statistics etc. but, hopefully, will feel comfortable while evaluating new developments in the context of the particular applications with which they are involved.

As emphasized in the Preface, quantitative chemical analysis at trace levels is a highly demanding *experimental* discipline, that can be truly learned only by actually performing experiments in the laboratory, i.e., through *learning by doing*. One can not truly understand the intricacies involved in devising a complete analytical procedure in which all the steps (Figure 1.2) are compatible with one another and in which carryover and similar problems are minimized, or in developing a chromatographic method that achieves a suitable compromise between speed (throughput) and separation efficiency, or in evaluating experimental uncertainties, statistically defensible LOD values etc., without having actually performed all such procedures. Any book on this subject, including this one, can be judged successful only to the extent that it helps the reader to perform well in the laboratory. In this spirit we close the book by repeating the ancient wisdom that was already quoted in the Preface:

'Those who are good at archery learnt from the bow and not from Yi the Archer. Those who know how to manage boats learnt from boats and not from Wo (the legendary boatman). Those who can think learned for themselves and not from the Sages.' Kuan Yin Tze, 8th century.

Trace Quantitative Analysis by Mass Spectrometry   Robert K. Boyd, Cecilia Basic, Robert A. Bethem
© 2008 John Wiley & Sons, Ltd

References

A

Abian J (1999), *J. Mass Spectrom.* **34**, 157.

Ackermann BL (2004), *J. Amer. Soc. Mass Spectrom.* **15**, 1374.

Ackermann BL, Hale JE, Duffin KL (2006), *Current Drug Metab.* **7**, 525.

Adlard ER (2006), *LCGC North America* **24**, 1102.

Adriaens A, Van Vaeck L, Adams F (1999), *Mass Spectrom. Revs.* **18**, 1.

Aebi B, Henion JD (1996), *Rapid Commun. Mass Spectrom.* **10**, 947.

Agah M, Potkay JA, Lambertus G *et al.* (2005)*, J. Microelectromechanical Systems* **14**, 1039.

Aguas PC, Fitzhenry MJ, Giannikopoulos G, Varelis P (2006), *Anal. Bioanal. Chem.* **385**, 1526.

Ahmadi F, Assadi Y, Hosseini SMRM, Rezaee M (2006), *J. Chromatogr. A* **1101**, 307.

Ahn JS, Castle L, Clarke DB *et al.* (2002), *Food Additives and Contaminants* **19**, 1116.

Alder L, Lüderitz S, Lindtner K, Stan H-J (2004), *J. Chromatogr. A* **1058**, 67.

Alder L, Greulich K, Kempe G, Vieth B (2006), *Mass Spectrom. Revs.* **25**, 838.

Aleksandrov ML, Gall LN, Krasnov VN *et al.* (1984), *Dokl. Akad. Nauk. SSSR* **277**, 379; English translation (1985), *Doklady Phys. Chem.* **277**, 572.

Aleksandrov ML, Gall LN, Krasnov VN *et al.* (1984a), *Bioorg. Kim.* **10**, 710.

Aleksandrov ML, Barama GI, Gall LN *et al.* (1985), *Bioorg. Kim.* **11**, 700, 705.

Aleksandrov ML, Kondratsev VM, Kusner Yu.S *et al.* (1988), *Bioorg. Kim.* **14**, 852.

Alikanov SG (1957), Sov. Phys. JETP 4, 452.

Alm RS, Williams RJP, Tiselius A (1952), Acta Chem. Scand. **6**, 826.

Alpert AJ (1990), *J. Chromatogr.* **499**, 177.

Altria KD, Rogan MM (1998), *Introduction to Quantitative Applications of Capillary Electrophoresis in Pharmaceutical Analysis* [available at: http://www.beckmancoulter.com/ and at http://www.ceandcec.com/538703%20-%20CE%20Primer%206.pdf].

Amad MH, Cech NB, Jackson GS, Enke CG (2000), *J. Mass Spectrom.* **35**, 784.

Ames BN, Gurney EG, Miller JA, Bartsch H (1973), *Proc. Natl. Acad. Sci USA* **69**, 3128.

Ames BN, Durston WE, Yamasaki E, Lee FD (1973a), *Proc. Natl. Acad. Sci USA* **70**, 2281.

Ames BN, Profet M, Gold LS (1990), *Proc. Natl. Acad. Sci. USA* **87**, 7777.

Ames BN, Profet M, Gold LS (1990a), *Proc. Natl. Acad. Sci. USA* **87**, 7782.

Ames BN, Gold LS (1997), *FASEB J.* **11**, 1041.

Amster IJ (1996), *J. Mass Spectrom.* **31**, 1325.

Anacleto JF, Ramaley L, Boyd RK *et al.* (1991), *Rapid Commun. Mass Spectrom.* **5**, 149.

Anacleto JF, Perreault H, Boyd RK *et al.* (1992), *Rapid Commun. Mass Spectrom.* **6**, 214.

Anacleto JF, Ramaley L, Benoit FM *et al.* (1995), *Anal. Chem.* **67**, 4145.

Analytical Methods Committee (1994), *Analyst* **119**, 2363.

Anderson A (1985), *Nature* **315**, 90.

Anderson NL, Anderson NG (2002), *Mol. Cell Proteomics* **1**, 845; see also *Erratum* (2003), *Mol. Cell Proteomics* **2**, 50.

Anderson PD, D'Aco VJ, Shanahan P *et al.* (2004), *Env. Sci. Technol.* **38**, 838.

Anderson NL, Anderson NG, Haines LR *et al.* (2004a), *J. Proteome Res.* **3**, 235.

Anderson L, Hunter CL (2006), *Molec. Cell. Proteomics* **5**, 573.

André F, De Wasch KKG, De Brabander HF *et al.* (2001), *Trends Anal. Chem.* **20**, 435.

Andreev VP, Rejtar T, Chen H-S *et al.* (2003), *Anal. Chem.* **75**, 6314.

Andrzejewski D, Roach JAG, Gay ML, Musser SM (2004), *J. Agric. Food Chem.* **52**, 1996.

Andrzejewski P, Kasprzyk-Hordern B, Nawrocki J (2005), *Desalination* **176**, 37.

Ansede JH, Thakker DR (2004), *J. Pharm. Sci.* **93**, 239.

Antignac J-P, Le Bizec B, Monteau F, Andre F (2003), *Anal. Chim. Acta* **483**, 325.

AOAC International (Association of Official Analytical Chemists International) (2000), *Official Methods of Analysis of AOAC International, 17th Edn* (Horwitz W, Ed.), Gaithersburg, MD, USA.

AOAC International (Association of Official Analytical Chemists International) (2002), *Guidelines for Single Laboratory Validation of Chemical Methods for Dietary Supplements and Botanicals* [available at: http://www.aoac.org/dietsupp6/Dietary-Supplement-web-site/ ValidProc.html].

Ardrey RE (2003), *Liquid Chromatography Mass Spectrometry: an Introduction*, John Wiley & Sons Ltd, Chichester, UK.

Armstrong DW (1984), *J. Liq. Chromatogr.* **7**, 353.

Armstrong DW, DeMond W, Czech BP (1985), *Anal. Chem.* **57**, 481.

Armstrong DW, Ward TJ, Armstrong RD, Beesley TE (1986), *Science* **232**, 1132.

Armstrong DW, Tang YB, Chen SS *et al.* (1994), *Anal. Chem.* **66**, 1473.

Arpino P (1990), *Mass Spectrom.Rev* **9**, 631.

Arpino P (1992), *Mass Spectrom.Rev.* **11**, 3.

Arshady R, Mosbach K (1981), *Macromol. Chem. Phys.* **182**, 687.

Arthur C, Pawliszyn J (1990), *Anal. Chem.* **62**, 2145.

Ashworth WJ (2004), *Science* **306**, 1314.

Asperger A, Efer J, Koal T, Engewald W (2002), *J. Chromatogr. A* **960**, 109.

Aston FW (1919), *Phil. Mag.* **38**, 707.

Aubourg P, Bougnères PF, Rocchiccioli F (1985), *J. Lipid Res.* **26**, 263.

Audunsson GA (1986), *Anal. Chem.* **58**, 2714.

Auroux P-A, Iossifidis D, Reyes DR, Manz A (2002), *Anal. Chem.* **74**, 2637.

Ayrton, J, Dear GJ, Leavens WJ *et al.* (1997), *Rapid Commun. Mass Spectrom.* **11**, 1953.

Ayrton, J, Dear GJ, Leavens WJ *et al.* (1998), *J. Chromatogr. A* **828**, 199.

Ayrton J, Clare RA, Dear GJ *et al.* (1999), *Rapid Commun. Mass Spectrom.* **13**, 1657.

Azenha MA, Nogueira PJ, Fernando Silva A (2006), *Anal. Chem.* **78**, 2071.

B

Backlund P, Kronberg L, Tikkanen L (1988), *Chemosphere* **17**, 1329.

Bader M (1980), *J. Chem. Educ.* **57**, 703.

Bainbridge KT (1932), *Phys. Rev.* **42**, 1.

Bainbridge KT, Jordan EB (1936), *Phys. Rev.* **50**, 282.

Bakalyar SR, Henry RA (1976), *J. Chromatography* **126**, 327.

Baldwin R, Bethem RA, Boyd RK *et al.* (1997), *J. Amer. Soc. Mass Spectrom.* **8**, 1180.

Ballschmiter K, Zell M (1980), *Fresenius Z. Anal. Chem.* **302**, 20.

Baltussen E, Sandra P, David F, Cramers CA (1999), *J. Microcol. Sep.* **11**, 737.

Bansal SK, Layloff T, Bush ED *et al.* (2004), *AAPS PharmSciTech* **5 (1)**, Article 22.

Bansal SK, DeStefano A (2007), *AAPS PharmSciTech* **9 (1)**, Article 11.

Barbalas MP, Garland WA (1991), *J. Pharm. Sci.* **80**, 922.

Barber M, Bordoli RS, Sedgwick RD, Tyler AN (1981), *J. Chem. Soc. Chem. Commun.* 325.

Barnes JH IV, Sperline R, Denton MB *et al.* (2002), *Anal. Chem.* **74**, 5327.

Barnes JH IV, Hieftje GM, Denton MB *et al.* (2003). *Amer. Lab.* Oct. 2003, 15.

Barnes JM, Magee PN (1954), *Brit. J. Industr. Med.* **11**, 167.

Barnes JH IV, Schilling GD *et al.* (2004), *J. Amer. Soc. Mass Spectrom.* **15**, 769.

Barnett, D A, Guevremont R, Purves RW, (1999), *Appl. Spectrosc.* **53**, 1367.

Barnidge DR, Dratz EA, Martin T *et al.* (2003), *Anal. Chem.* **75**, 445.

Barnidge DR, Goodmanson MK, Klee GG, Muddiman DC (2004), *J. Proteome Res.* **3**, 644.

Barr JR, Maggio VL, Patterson DG Jr *et al.* (1996), *Clin. Chem.* **42**, 1676.

Barron LD (2007), *Nature* **446**, 505.

Bartlett MS (1937), *Proc. Royal Statistical Society A* **160**, 268.

Barton GW Jr, Gibson LE, Tolman LF (1960), *Anal. Chem.* **32**, 1599.

Basso E, Marotta E, Seraglia R *et al.* (2003), *J. Mass Spectrom.* **38**, 1113.

Bateman RH, Green MR, Scott G, Clayton E (1995), *Rapid Commun. Mass Spectrom.* **9**, 1227.

Baumann C, Cintora MA, Eichler M *et al.* (2000), *Rapid Commun. Mass Spectrom.* **14**, 349.

Baumbach JI, Eiceman GA, Klockow D *et al.* (1997), *Int. J. Environ. Anal. Chem.* **66**, 225.

Beaudry F, LeBlanc JYC, Coutu M, Brown N (1998), *Rapid Commun. Mass Spectrom.* **12**, 1216.

Beavis R (1992), *Org. Mass Spectrom.* **27**, 653.

Becker EW, Bier K, Burghoff H, Zigan F (1957), *Z. Naturforsch. Teil. A* **12**, 609.

Beckey HD (1977), *Principles of Field Desorption and Field Ionization Mass Spectrometry*, Pergamon Press, Oxford, UK.

Bedair MF, Oleschuk RD (2006), *Anal. Chem.* **78**, 1130.

Beesley TE, Lee JT (2004), *LC/GC, LC Column Technology Supplement (June)* **22**, 30.

Bélanger JMR, Paré JRJ (2006), *Anal. Bioanal. Chem.* **386**, 1049.

Belardi RG, Pawliszyn J (1989), *Water Pollut. Res. J. Can.* **24**, 179.

Bender ML, Kemp KC (1957), *J. Am. Chem. Soc.* **79**, 116.

Beneke K (2003), *Friedrich Feigl und die Gescichte der Chromatographie und der Tüpfelanalyse* [available at: http://www.uni-kiel.de/anorg/lagaly/group/klausSchiver/Feigl. pdf].

Benetton S, Kameoka J, Tan A *et al.* (2003), *Anal. Chem.* **75**, 6430.

Berli U, Bönzli M (1991), Electronic Weighing Principles, Article #4 in the series Weighing Uncertainty [available on the website of Mettler-Toledo Inc: http://us.mt.commt/ filters/products-applications_autochem/Autochem_0x0000100 83f6f05f140006007.jsp].

Berman J, Halm K, Adkison K, Shaffer, J (1997), *J. Med. Chem.* **40**, 827.

Berthod A, Liu Y, Bagwill C, Armstrong DW (1996), *J. Chromatogr. A* **731**, 123.

Bethem R, Boison J, Gale J *et al.* (2003), *J. Amer. Soc. Mass Spectrom.* **14**, 528.

Beuhler RJ, Flanigan E, Greene LJ, Friedman L (1974), *J. Am. Chem. Soc.* **96**, 3990.

Beuhler RJ, Friedman L (1977), *Int. J. Mass Spectrom. Ion Phys.* **23**, 81.

Beynon JH (1960), *Mass Spectrometry and its Applications to Organic Chemistry*, Elsevier, Amsterdam, The Netherlands.

Beynon RJ, Doherty MK, Pratt JM, Gaskell SJ (2005), *Nature Methods* **2**, 587.

Bharadwaj R, Santiago JG (2005), *J. Fluid Mech.* **543**, 57.

Bicchi C, Brunelli C, Cordero C *et al.* (2004), *J. Chromatogr. A* **1024**, 195.

Bill JC, Green BN, Lewis IAS (1983), *Int. J. Mass Spectrom. Ion Phys.* **46**, 147.

Billedeau SM, Thompson HC Jr (1987), *J. Chromatogr.* **393**, 367.

Birkemeyer C, Kolasa A, Kopka J (2003), *J. Chromatogr. A* **993**, 89.

Björksten M, Almby B, Jansson ES (1994), *Appl. Ergonomics* **25**, 1.

Blakley CR, McAdams MJ, Vestal ML (1978), *J. Chromatogr.* **158**, 261.

Blakley CR, Carmody JJ, Vestal ML (1980), *Anal. Chem.* **52**, 1636.

Blakley CR, Carmody JJ, Vestal ML (1980a), *J. Am. Chem. Soc.* **102**, 5931.

Blakley CR, Vestal ML (1983), *Anal. Chem.* **55**, 750.

Blau K, Halket JM, Eds. (1993), *Handbook of Derivatives for Chromatography, 2nd Edn*, Wiley-Interscience, Chichester, UK.

Blewett JP (1952), *Phys. Rev.* **88**, 1197.

Bligh EG, Dyer WJ (1959), *Can. J. Physiol.* **37**, 911.

Blom KF (2001), *Anal. Chem.* **73**, 715.

Blues J, Bayliss D, Buckley M (2004), *The use and calibration of piston-operated volumetric pipettes*, ISSN 1368-6550 [available at: http://www.npl.co.uk/mass/guidance/balancetypes.html].

Boettcher MI, Angerer J (2005), *J. Chromatogr. B* **824**, 283.

Boettcher MI, Schettgen T, Kutting B *et al.* (2005a), *Mutat. Res.* **580**, 167.

Boettcher MI, Bolt HM, Drexler H, Angerer J (2006), *Arch Toxicol.* **80**, 55.

Boffey PM (1971), *Science* **171**, 43.

Bogan MJ, Agnes GR (2002), *Anal. Chem.* **74**, 489.

Bogan MJ, Agnes GR (2004), *J. Amer. Soc. Mass Spectrom.* **15**, 486.

Bogdanov B, Smith RD (2005), *Mass Spectrom. Revs.* **24**, 168.

Bojarski J, Aboul-Enein HY, Ghanem A (2005), *Current Analytical Chemistry* **1**, 59.

Bonfiglio R, King RC, Olah TV, Merkle K (1999), *Rapid Commun. Mass Spectrom.* **13**, 1175.

Bonner RF, March RE, Durup J (1976), *Int. J. Mass Spectrom. Ion Phys.* **22**, 17.

Bonner RF (1977), *Int. J. Mass Spectrom. Ion Phys.* **23**, 249.

Borenstein J (2001), *Professional Ethics Report* XIV (**3**), 1 [available at http://www.aaas.org/spp/sfrl/per/per26.pdf].

Borra C, Soon MH, Novotny M (1987), *J. Chromatogr.* **385**, 75.

Bortolini O, Spalluto G, Traldi P (1994), *Org. Mass Spectrom.* **29**, 269.

Bos SJ, van Leeuwen SM, Karst U (2006), *Anal. Bioanal. Chem.* **384**, 85.

Boswell CE (1999), *15th Annual Waste Testing & Quality Assurance Symposium* [available at: www.clu-in.org/download/char/dataquality/EBoswell.pdf].

Boyd RK, Kingston EE, Brenton AG, Beynon JH (1984), *Proc. Roy. Soc. Lond. A* **392**, 59.

Boyd RK (1993), *Rapid Commun. Mass Spectrom.* **7**, 257.

Boyd RK (1994), *Mass Spectrom. Revs.* **13**, 359.

Boyd RK, Crain SM, Curtis JM *et al.* (1996), *Polycyc. Arom. Compnds., Proc. 15th. Int. Conf. Polycyc. Aromatic Compnds.*, Belgirte, Italy (Sept. 1995) **9**, 217.

Boyd RK, Henion JD, Alexander M *et al.* (1996a), *J. Amer. Soc. Mass Spectrom.* **7**, 211.

Brauman JL (1966), *Anal. Chem.* **38**, 607.

Brenton AG, Krastev T, Rousell DJ *et al.* (2007), *Rapid Commun. Mass Spectrom.* **21**, 3093.

Brereton RG (2000), *Chemometrics: Data Analysis for the Laboratory and Chemical Plant*, John Wiley & Sons Ltd, Chichester, UK.

Brinkmann U (1972), *Int. J. Mass Spectrom. Ion Phys.* **9**, 161.

Briscoe CJ, Stiles MR, Hage DS (2007), *J. Pharm. Biomed. Anal.* **44**, 484.

Bristow PA, Knox JH (1977), *Chromatographia* **10**, 279.

Bristow AWT, Webb KS, Lubben AT, John Halket J (2004), *Rapid Commun. Mass Spectrom.* **18**, 1447.

Brody H, Rip MR, Vinten-Johansen P *et al.* (2000), *Lancet* **356**, 64.

Brown L, Koerner T, Horton JH, Oleschuk RD (2006), *Lab Chip* **6**, 66.

Bruins AP, Covey TR, Henion JD (1987), *Anal. Chem.* **59**, 2642.

Bruins AP (1991), *Mass Spectrom. Revs.* **10**, 53.

Bruins CHP, Jeronimus-Stratingh CM, Ensing K *et al.* (1999), *J. Chromatogr. A* **863**, 115.

Brunner W (2002), *Weighing the Right Way* [available from Mettler-Toledo Company at: http://us. mt.com/mt/brochures/Weighing_the_right_way_0x000246700 005761700059b6e.jsp].

Bu H-Z, Poglod M, Micetich RG, Khan JK (2000), *Rapid Commun. Mass Spectrom.* **14**, 523.

Buckley JA, French JB, Reid NM (1974), *Can. Aeronaut. Space J.* **20**, 231.

Buckley JA (1974a), *A Hypersensitive Real Time Trace Atmospheric Gas Analyzer*, PhD Thesis, University of Toronto, Canada.

Bucknall M, Fung KYC, Duncan MW (2002), *J. Am. Soc. Mass Spectrom.* **13**, 1015.

Buhrman D, Price PI, Rudewicz PJ (1996), *J. Am. Soc. Mass Spectrom.* **7**, 1099.

Burton IW, Quilliam MA, Walter JA (2005), *Anal. Chem.* **77**: 3123.

Buryakov I, Krylov E, Nazarov E, Rasulev U (1993), *Int. J. Mass Spectrom. Ion Proc.* **128**, 143.

Bush ED, Trager WF (1981), *Biomed. Mass Spectrom.* **8**, 211.

Butchart K, Potter T, Wright A, Brien A (2007), *Intl. Labmate* **32**, Technical Article (August 2007) [available at http://www.internationallabmate.com/006_007_008_ILM_AUG_07.pdf].

Buttrill SE Jr, Shaffer B, Karnicky J, Arnold JT (1992), *Proceedings of the 40th ASMS Conference on Mass Spectrometry and Allied Topics*, Washington, DC, USA p. 1015.

Bzik TJ, Henderson PB, Hobbs JP (1998), *Anal. Chem.* **70**, 58.

C

Cabrera K, Lubda D, Eggenweiler H-M *et al.* (2000), *J. High Resol. Chromatogr.* **23**, 93.

Cai Y, Kingery D, McConnell O, Bach AC II (2005), *Rapid Commun. Mass Spectrom.* **19**, 1717.

California Department of Health Services (2006), *California Drinking Water: NDMA-Related Activities* [http://www.dhs.ca.gov/ps/ddwem/chemicals/ndma/default.htm].

Cameron AE, Eggers DF (1948), *Rev. Sci. Instrum.* **19**, 605.

Campana JE (1980), *Int. J. Mass Spectrom. Ion Phys.* **33**, 101.

Campana JE, Jurs PC (1980a), *Int. J. Mass Spectrom. Ion Phys.* **33**, 119.

Campargue R (1984), *J. Phys. Chem.* **88**, 4466.

Campbell JM, Collings BA, Douglas DJ (1998), *Rapid Commun. Mass Spectrom.* **12**, 1463.

Caprioli RM, Fan T, Cotrell JS (1986), *Anal . Chem.* **58**, 2949.

Caprioli RM (Ed., 1990), *Continuous Flow Fast Atom Bombardment Mass Spectrometry*, John Wiley & Sons Ltd, Chichester, UK.

Cardone MJ (1983), *J. Assoc. Off. Anal. Chemists* **66**, 1257, 1283.

Cardone MJ (1986), *Anal. Chem.* **58**, 433, 438.

Cardone MJ, Willavize SA, Lacy ME (1990), *Pharm. Res.* **7**, 154.

Cargile BJ, Sevinsky JR, Essader AS *et al.* (2005), *J. Biomolec. Techniques*, **16**, 181.

Carroll DI, Dzidic I, Stillwell RN *et al.* (1975), *Anal. Chem.* **47**, 2369.

Cassiano NM, Lima VV, Oliveira RV *et al.* (2006), *Anal. Bioanal. Chem.* **384**, 1462.

Castiglioni S, Zuccato E, Crisci E *et al.* (2006), *Anal. Chem.* **78**, 8421.

Castle L, Campos MJ, Gilbert J (1993), *J. Sci. Food Agric.* **54**, 549.

CDC (US Center for Disease Control, 1988), *J. Amer. Med. Assocn.* **260**, 1249.

Cech NB, Enke CG (2001), *Mass Spectrom. Revs.* **20**, 362.

CFR21 (US Code of Federal Regulations, 2003) *Title 21, Food and Drugs, Part 11* [available at: http://www.access.gpo.gov/nara/cfr/waisidx_04/21cfrv1_04.html].

Cha B, Blades M, Douglas DJ (2000), *Anal. Chem.* **72**, 5647.

Chace DH, Millington DS, Teradan N *et al.* (1993), *Clin. Chem.* **39**, 66.

Chace DH, Hillman SL, Van Hove JLK, Naylor EW (1997), *Clin. Chem.* **43**, 2106.

Chang W-T, Lin D-L, Liu RH (2001), *Forensic Sci. Intl.* **121**, 174.

Chang MS, Kim EJ, El-Shourbagy TA (2006), *Rapid Commun. Mass Spectrom.* **20**, 2190.

Charles MJ, Green B, Tondeur Y, Hass JR (1989), *Env. Sci. Technol.* **19**, 51.

Charles MJ, Marbury GD, Chen G (1991), *Biol. Mass Spectrom.* **20**, 529.

Charles MJ, Chen G, Kanniganti R, Marbury GD (1992), *Envir. Sci. Technol.* **26**, 1030.

Charrois JWA, Arend MW, Froese KL, Hrudey SE (2004), *Environ. Sci. Technol.* **38**, 4835.

Chatt A, Rao RR, Jayawickreme CK, McDowell LS (1990), *Fresenius J. Anal. Chem.* **338**, 399.

Chen L, Wang TL, Ricca TL, Marshall AG (1987), *Anal. Chem.* **59**, 449.

Chen J, Korfmacher WA, Hsieh Y (2005), *J. Chromatogr. B* **820**, 1.

Chen J, Hsieh Y, Cook J *et al.* (2006), *Anal. Chem.* **78**, 1212.

Chernestsova ES, Revelsky AI, Revelsky IA *et al.* (2002), *Mass Spectrom. Revs.* **21**, 373.

Chernushevich IV (2000), *Eur. J. Mass Spectrom.* **6**, 471.

Chernushevich IV, Loboda AV, Thomson BA (2001), *J. Mass Spectrom.* **36**, 849.

Chia K-J, Huang S-D (2006), *Rapid Commun. Mass Spectrom.* **20**, 118.

Chirica GS, Remcho VT (20000), *Anal. Chem.* **72**, 3605.

Choi BK, Hercules DM, Zhang T, Gusev A (2001), *LC/GC* **19**, 514.

Choi JH, Valentine RL (2002), *Water Sci. Technol..* **36**, 65.

Choi JH, Valentine RL (2002a), *Water Res.* **36**, 817.

CITAC–Eurachem Guide (2002), *Guide to Quality in Analytical Chemistry: an Aid to Accreditation'* [available at http://www.citac.cc or http://www.eurachem.org].

Clarke NJ, Rindgen D, Korfmacher WA, Cox KA (2001), *Anal. Chem.* **73**, 430A.

Clauwaert KM, Van Bocxlaer JF, Major HJ *et al.* (1999), *Rapid Commun. Mass Spectrom.* **13**, 1540.

Clegg GA, Dole MJ (1971), *Biopolymers* **10**, 821.

Clement RE, Tosine HM (1988), *Mass Spectrom. Revs.* **7**, 593.

Clerk-Maxwell J (1870), Address to the Mathematical and Physical Sections of the British Association, Liverpool (September 15, 1870); from the British Association Report, Vol. XL, reproduced in *The Scientific Papers of James Clerk Maxwell*, Niven WD (Ed), Vol. 2 Cambridge University Press, Cambridge (1890), p. 225.

Cody RB, Laramée JA, Durst HD (2005), *Anal. Chem.* **77**, 2297.

Cohen LH, Gusev AI (2002), *Anal. Bioanal. Chem.* **373**, 571.

Colby BN, McCaman MW (1979), *Biomed. Mass Spectrom.* **6**, 225.

Colby BN, Rosencrance AE, Colby ME (1981), *Anal. Chem.* **53**, 1907.

Cole RB (2000), *J. Mass Spectrom.* **35**, 763.

Collings BA, Campbell JM, Mao D, Douglas DJ (2001), *Rapid Commun. Mass Spectrom.* **15**, 1777.

Collings BA, Stott WR, Londry FA (2003), *J. Am. Soc. Mass Spectrom.* **14**, 622.

Comisarow MB, Marshall AG (1974), *Can. J. Chem.* **52**, 1997.

Comisarow MB, Marshall AG (1974a), *Chem. Phys. Lett.* **25**, 282.

Comisarow MB, Marshall AG (1974b), *Chem. Phys. Lett.* **26**, 489.

Constapel M, Schellenträger M, Schmitz OJ *et al.* (2005), *Rapid Commun. Mass Spectrom.* **19**, 326.

Cooks RG, Beynon JH, Caprioli RM, Lester GR (1973), *Metastable Ions*, Elsevier, Amsterdam, The Netherlands

Cooks RG, Ast T, Mabud MDA (1990), *Int. J. Mass Spectrom. Ion Proc.* **100**, 209.

Cooks RG, Rockwood AL (1991), Rapid Commun. Mass Spectrom. **5**, 93.

Cooks RG (1995), *J. Mass Spectrom.* **30**, 1215.

Cornish T J, Cotter RJ (1993), *Anal. Chem.* **65**, 1043.

Corr JJ, Kovarik P, Schneider BB *et al.* Hendrikse J, Loboda A, Covey TR (2006), *J. Am. Soc. Mass Spectrom.* **17**, 1129.

Cotte-Rodríguez I, Mulligan CC, Cooks RG (2007), *Anal. Chem.* **79**, 7069.

Courant ED, Livingston MS, Snyder HS (1952), *Phys. Rev.* **88**, 1190.

Covey TR, Lee ED, Henion JD (1986), *Anal. Chem.* **58**, 2453.

Covey TR, Bonner RF, Shushan BI, Henion JD (1988), *Rapid Commun. Mass Spectrom.* **2**, 249.

Covey TR, Anacleto JF (1995), *US Patent 5,412,208.*

Crummett WB, Stehl RH (1973), *Env. Health Perspectives* **5**, 15 [freely available at: http://www.ehponline.org/members/1973/005/05003.PDF].

Cui W, Thompson MS, Reilly JP (2005), *J. Am. Soc. Mass Spectrom.* **16**, 1384.

Curcuruto O, Fontana S, Traldi P, Celon E (1992), *Rapid Commun. Mass Spectrom.* **6**, 322.

Currie LA (1968), Anal. Chem. **40**, 586.

Curtis JM, Boyd RK (1997), *Int. J. Mass Spectrom. Ion Proc.* **165/166**, 623.

D

Dadgar D, Burnett PE, Choc MG *et al.* (1995), *J. Pharm. Biomed. Anal.* **13**, 89.

Daly NR (1960), *Rev. Sci. Instrum.* **31**, 264.

Danaher M, O'Keeffe M, Glennon JD (2000), *Analyst* **125**, 1741.

Dandeneau RD, Zerenner EH (1979), *J. High Resolut. Chromatogr.* **2**, 351.

Dandeneau RD, Zerenner EH (1990), *LC/GC* **8** (12), 908.

Däppen R, Arm H, Meyer VR (1986), *J. Chromatogr.* **373**, 1.

Darling S, Santos M (2005), *Operational Tips for Improving the Robustness of Capillary Electrophoresis Methodology in Your Laboratory* [available at: www.beckmancoulter.com].

Daves GD Jr (1979), *Acc. Chem. Res.* **12**, 359.

David F, Tienpont B, Sandra P (2003), *LC/GC* **21**, 108.

Davidson S, Perkin M, Buckley M (2004), *Measurement Good Practice Guide No. 71: The Measurement of Mass and Weight*, ISSN 1368-6550 [available from the UK National Physical Laboratory at: http://www.npl.co.uk/mass/guidance/balancetypes.html].

Davies PL (1988), *Fresenius Z. Anal. Chem.* **331**, 513.

Davis JM, Giddings JC (1983), *Anal. Chem.* **55**, 418.

Davis RS (1992), *Metrologia* **29**, 67.

Dawson PH, Whetten NR (1968), *Research/Development, February 1968*, 46.

Dawson PH, Hedman JW, Whetten NR (1969), *Rev. Sci. Instrum.* **40**, 144.

Dawson PH, Whetten NR (1970), US Patent 3,527,939.

Dawson PH (Ed., 1976), *Quadrupole Mass Spectrometry and its Applications*, Elsevier, 1976; re-issued (1995) by the American Institute of Physics Press, New York, NY.

Dawson PH, French JB, Buckley JA *et al.* (1982), *Org. Mass Spectrom.* **17**, 205.

Dawson PH (1986), *Mass Spectrom. Revs.* **5**, 1.

Dawson JHJ, Guilhaus M (1989), *Rapid Commun. Mass Spectrom.* **3**, 155.

Dear G, Plumb R, Mallett D (2001), *Rapid Commun. Mass Spectrom.* **15**, 152.

De Boer WJ, van den Voet H, de Ruid WG, *et al.* (1999), *Analyst* **124**, 109.

De Brabander HF, De Wisch K, Van Ginkel L *et al.* (2000), *Proceeding of the EuroResidue IV Conference on Residues of Veterinary Drugs in Food*, p.248 [freely available at: http://www.euroresidue.nl/ER_IV/Contributions%20A-H/idx.htm].

Dehmelt H (1990), *Revs. Modern Phys.* **62**, 525.

de Hoffmann, E (1996), *J. Mass Spectrom.* **31**, 129.

Deinzer M, Griffin D, Miller T *et al.* (1982), *Biomed. Mass Spectrom.* **9**, 85.

Delaunay-Bertoncinni N, Hennion M-C (2004), *J. Pharm. Biomed. Anal.* **34**, 717.

De Leenheer AP, Lefevere MF, Lambert WE, Colinet ES (1985), *Advan. Clin. Chem.* **24**, 111.

Dell'Aversano C, Hess P, Quilliam MA (2005), *J. Chromatogr. A* **1081**, 190.

de Mello A (2002), *Lab Chip* **2**, 48N.

Dempster AJ (1921), *Phys. Rev.* **18**, 415.

Dempster AJ (1921a), *Proc. Natl. Acad. Sci. US* **7**, 45.

De Nardi C, Bonelli F (2006), *Rapid Commun. Mass Spectrom.* **20**, 2709.

Deng Y, Henion J, Li J *et al.* (2001), *Anal. Chem.* **73**, 639.

Deng Y, Zhang H, Henion J (2001a), *Anal. Chem.* **73**, 1432.

De Ruig WG, Dijkstra G, Stephany RW (1989), *Anal. Chim. Acta* **223**, 277.

Desiderio DM, Zhu X (1998), *J. Chromatogr. A* **794**, 85.

Desiderio DM (1999), *J. Chromatogr. B* **731**, 3.

de Silva AP, James MR, McKinney BOF *et al.* (2006), *Nature Materials* **5** 787.

Dethy J-M, Ackerman BL, Delatour C *et al.* (2003), *Anal. Chem.* **75**, 805.

Devakumar A, Thompson MS, Reilly JP (2005), *Rapid Commun. Mass Spectrom.* **19**, 2313.

Dévé J, Bartlen T (2007), *Nature Methods* **4**, June 2007 (advertising feature).

de Zeeuw RA (2004), *J. Chromatogr. B* **811**, 3.

Dias NC, Poole CF (2002), *Chromatographia* **56**, 269.

Díaz-Cruz MS, Barceló D (2006), Anal. Bioanal. Chem. **386**, 973.

Diddams SA, Bergquist JC, Jefferts SR, Oates CW (2004), *Science* **306**, 1318.

Dimandja J-M (2004), *Anal. Chem.* **76**, 167A.

Dionex Corporation (1998), Technical Note 208, *Methods Optimization in Accelerated Solvent Extraction (ASE®)* [available at: http://www1.dionex.com/en-us/webdocs/4736_Tn208.pdf].

Dittrich PS, Tachikawa K, Manz A (2006), *Anal. Chem.* **78**, 3887.

Dixon SP, Pitfield ID, Perrett D (2006), *Biomed. Chromatogr.* **20**, 508.

Dodonov AF, Chernushevich IV, Laiko VV (1991), *Proceedings of the 12th International Mass Spectrometry Conference*, Amsterdam, The Netherlands, p. 153.

Dodonov AF, Kozlovski VI, Soulimenkov IV *et al.* (2000), *Eur. J. Mass Spectrom.* **6**, 481.

Doerffel K, Herfurth G, Liebich V, Wendlandt E (1991), *Fresenius J. Anal. Chem.* **341**, 519.

Doherty MK, McLean L, Hayter JR *et al.* (2004), *Proteomics* **4**, 2082.

Dolan JW, Gant JR, Snyder LR (1979), *J. Chromatogr.* **165**, 31.

Dolan JW (2001), *LC/GC North America* **19**, 164.

Dolan JW (2001a), *LC/GC North America* **19**, 386.

Dolan JW (2001b), *LC/GC North America* **19**, 478.

Dolan JW (2001c), *LC/GC North America* **19**, 1050.

Dolan JW (2001d), *LC/GC North America* **19**, 32.

Dolan JW (2002), *LC/GC North America* **20**, 430.

Dole M, Mack LL, Hines RL *et al.* (1968), *J. Chem. Phys.* **49**, 2240.

Doroshenko VM, Cotter RJ (1993), *Rapid Commun. Mass Spectrom.* **7**, 822.

Doroshenko VM, Cotter RJ (1994), *Rapid Commun. Mass Spectrom.* **8**, 766.

Doroshenko VM, Cotter RJ (1996), *Rapid Commun. Mass Spectrom.* **10**, 65.

Doroshenko VM, Cotter RJ (1996a), *Anal. Chem.* **68**, 463.

Doroshenko VM, Cotter RJ (1997), *J. Mass Spectrom.* **32**, 602.

Doroshenko VM, Cotter RJ (1999), *J. Am. Soc. Mass Spectrom.* **10**, 992.

Douglas DJ, French JB (1992), *J. Am. Soc. Mass Spectrom.* **3**, 398.

Douglas DJ (1993), US Patent 5,179,278.

Douglas DJ, Frank AJ, Mao D (2005), *Mass Spectrom. Revs.* **24**, 1.

Droste S, Schellenträger M, Constapel M *et al.* (2005), *Electrophoresis* **26**, 4098.

Drozd J (1981), *Chemical Derivatization in Gas Chromatography*, Elsevier, Amsterdam, The Netherlands.

Duckworth HE, Ghoshal SN (1963), in *Mass Spectrometry* (McDowell CA Ed.,) McGraw-Hill, New York, USA

Duncan DB (1955), *Biometrics* **11**, 1.

Duncan MW, Gale PJ, Yergey AL (2006), *Quantitative Mass Spectrometry*, Rockpool Productions LLC, Denver, CO, USA.

E

Eiden GC, Cisper ME, Alexander ML, Hemberger PH (1993), *J. Am. Soc. Mass Spectrom.* **4**, 706.

Eiden GC, Garrett AW, Cisper ME *et al.* (1994), *Int. J. Mass Spectrom. Ion Proc.* **136**, 119.

Eiden G C, Barinaga CJ, Koppenaal DW (1997), *Rapid Commun. Mass Spectrom.* **11**, 37.

Eksigent LLC (2005), *How to achieve repeatable, high throughput HPLC 6 times faster than your current method* [available at: http://71.136.6.99/data/presentation/HPLC6times faster.pdf].

Eksigent LLC (2006), *Capillary LC Beliefs and Realities* [available at http:71.136.6.99/data/presentation/hplc2006.pdf].

Eljarrat E, Sauló J, Monjonell A *et al.* (2001), *Fresenius J. Anal. Chem.* **371**, 983.

Ellerton RRW (1980), *Anal. Chem.* **52**, 151.

El-Faramawy A, Siu KWM, Thomson BA (2005), *J. Am. Soc. Mass Spectrom.* **16**, 1702.

Ellison SLR, Rosslein M, Williams A (2000), Eurachem –CITAC Guide CG4, *Quantifying Uncertainty in Analytical Measurement*, 2nd edn, [freely available at: http://www.eurachem.org].

Ells B, Barnett DA, Purves RW, Guevremont R (2000), *J. Environ. Monitoring* **2**, 393.

Engelhardt H (2004), *J. Chrom. B* **800**, 3.

Ensing K, Berggren C, Majors RE (2001), *LC/GC* **9**, 942.

Ensing K, de Boer T (1999), *Trends Anal. Chem.* **18**, 138.

EPA (US Environmental Protection Agency) (1985), *U.S. Code of Federal Regulations, Title 50*, 46906.

EPA (US Environmental Protection Agency) (1993), *Guidance on evaluation, resolution, and documentation of analytical problems associated with compliance monitoring*, USEPA 821-B-93-001.

EPA (US Environmental Protection Agency) (1994), *Method 1613, Revision B: Tetra- through Octa-Chlorinated Dioxins and Furans by Isotope Dilution HRGC/HRMS* [freely available at: http://www.epa.gov/waterscience/methods/1613.pdf].

EPA (US Environmental Protection Agency) (revised July 1, 1997), *Guidelines establishing test procedures for the analysis of pollutants (App. B, Part 136, Definition and procedures for the determination of the method detection limit): U.S. Code of Federal Regulations, Title 40*.

EPA (US Environmental Protection Agency) (2003), *Purge-and-Trap for Aqueous Samples: Method 5030C, Revised May 2003*.

EPA (US Environmental Protection Agency) (2004), *Revised Assessment of Detection and Quantitation Approaches*, EPA-821-B-04-005.

EPA (US Environmental Protection Agency) (2004a), *Determination of Nitrosamines in Drinking Water by Slid Phase Extraction and Capillairy Column Gas Chromatography with Large Volume Injection and Chemical Ionization Tandem Mass Spectrometry,Version 1.0*, EPA/600/R-05/054.

EPA (US Environmental Protection Agency) (2004b), *Statistical Protocol for the Determination of the Single-Laboratory Lowest Concentration Minimum Reporting Level (LCMRL) and Validation of Laboratory Performance at or Below the Minimum Reporting Level (MRL)*, EPA-815-R-05-006.

Eppe G, Focant J-F, Pirard C, De Pauw E (2004), *Talanta* **63**, 1135.

Erickson D, Li D (2004), *Anal. Chim. Acta* **507**, 11.

Esbensen KH, Heydorn K (2004), *Chemometrics Intell. Lab. Systems* **74**, 115.

Ettre LS (1987), *J. High Resol. Chromatogr.* **10**, 221.

Ettre LS (2001), *LC/GC* **19**, 48.

Eurachem Guide (1998), *The Fitness for Purpose of Analytical Methods: a Laboratory Guide to Method Validation and Related Topics* [available at http://www.eurachem.org].

Eurachem–CITAC Guide (1998a), *Quality Assurance for Research and Development and Non-routine Analysis* [available at: http://www.eurachem.org].

Eurachem–CITAC Guide CG4 (2000) *Quantifying Uncertainty in Analytical Measurement (2nd Edition)* [available at http://www.eurachem.org].

Eurachem Guide (2000), *Selection, Use and Interpretation of Proficiency Testing Schemes by Laboratories* [available at: http://www.eurachem.org].

Eurachem–CITAC Guide (2003), *Traceability in Chemical Measurement* [available at: http://www.eurachem.org]

Eurachem–CITAC Guide (2007), *Use of Uncertainty Information in Compliance Assessment* [available at: http://www.eurachem.org].

European Commission (2002), *Implementing Council Directive 96/23/EC concerning the performance of analytical methods and the interpretation of results*, 2002/657/EC Decision 12 August 2002.

European Commission (2002a), *Commission Directive 2002/69/EC of 26 July 2002 laying down the sampling methods and the methods of analysis for the official control of dioxins and the determination of dioxin-like PCBs in foodstuffs* [available at: http://europa.eu.int/eur-lex/en/archive/2002/l_20920020806en.html].

European Commission (2002b), *Commission Directive 2002/70/EC of 26 July 2002 establishing requirements for the determination of levels of dioxins and dioxin-like PCBs in feedingstuffs* [available at: http://europa.eu.int/eur-lex/en/archive/2002/l_20920020806en.html].

Everley PA, Krijgsveld J, Zetter BR, Gygi SP (2004), *Molec. Cell Proteomics* **3**, 729.

F

Fafet A, Bonnard J, Prigent F (1999), *Oil & Gas Science and Technology – Rev. IFP*, **54**, 439–452.

Fanali S (1995), *An Introduction to Chiral Analysis by Capillary Electrophoresis* [available at: http://www.bio-rad.com/LifeScience/pdf/Bulletin_1973.pdf].

Farmer JB (1963), in *Mass Spectrometry* (Ed. McDowell CA), McGraw-Hill, New York, USA.

FDA (US Food and Drug Administration) (1978), *Nonclinical Laboratory Studies; Good Laboratory Practice Regulations* [freely available at: http://www.fda.gov/ora/compliance_ref/bimo/glp/78fr-glpfinalrule.pdf].

FDA (US Food and Drug Administration) (2001), *Guidance for Industry, Bioanalytical Method Validation (Docket No. 98D-1195)* [available at: www.fda.gov/cder/guidance/4252fnl.htm].

FDA (US Food and Drug Administration) (2002), *General Principles of Software Validation; Final Guidance for Industry and FDA Staff* [available at: http://www.fda.gov/cdrh/comp/guidance/938.pdf].

FDA (US Food and Drug Administration) (2003), *Mass Spectrometry for Confirmation of the Identity of Animal Drug Residues* [freely available at: http://www.fda.gov/cvm/Guidance/guide118.pdf].

FDA (US Food and Drug Administration) (2003a), *Guidance for Industry Part 11, Electronic Records; Electronic Signatures – Scope and Application* [freely available at: http://www.fda.gov/cder/guidance/5667fnl.pdf].

FDA (US Food and Drug Administration) (2005), *FDA's Policy Statement for the Development of New Stereoisomeric Drugs*, available at: http://www.fda.gov/cder/guidance/stereo.htm.

FDA (US Food and Drug Administration) (2006), *Guidance for Industry, Investigating Out-of-Specification (OOS) Test Results for Pharmaceutical Production* [freely available at: htt://www.fda.gov/cder/guidance/3634fnl.pdf].

Fearn T, Fisher SA, Thompson M, Ellison SLR (2002), *Analyst* **127**, 818.

Fehlhaber H-W, Metternich K, Tripler D, Uihlein M (1978), *Biomed. Mass Spectrom.* **5**, 188.

Fenn JB, Mann M, Meng CK *et al.* (1989), *Science* **246**, 64.

Fenn JB, Mann M, Meng CK, Wong SF (1990), *Mass Spectrom. Revs.* **9**, 37.

Fennell TR, Sumner SCJ, Walker VE (1992), *Cancer Epidemiology Biomarkers Prevention* **1**, 213.

Fernandez de la Mora J (1992), *Mass Spectrom. Revs.* **11**, 431.

Ferrario F, Byrne C (2002), *Chemosphere* **46**, 1297.

Ferrus R, Cardone MJ (1987), *Anal. Chem.* **59**, 2816.

Feser K, Kögler W (1979), *J. Chromatogr. Sci.* **17**, 57.

Fetterolf DD, Yost RA (1984), *Int. J. Mass Spectrom. Ion Proc.* **62**, 33.

Fialkov AB, Steiner U, Jones L, Amirav A (2006), *Int. J. Mass Spectrom.* **251**, 47.

Fies WJ Jr. (1988), *Int. J. Mass Spectrom.* **82**, 111.

Fischer E (1959), *Z. Physik* **156**, 1; *ibid.* 26.

Fisher RA (1925), *Statistical Methods for Research Workers*, Hafner Publishing Co., New York, NY, USA, re-issued (1990) as *Statistical Methods, Experimental Design, and Scientific Inference: A Re-issue of Statistical Methods for Research Workers, The Design of Experiments, and Statistical Methods and Scientific Inference* (Ed. Bennett JH), Oxford University Press, Oxford, UK.

Flanagan RJ, Morgan PE, Spencer P, Whelpton R (2006), *Biomed. Chromatogr.* **20**, 530.

Fleming CM, Kowalswi BR, Apffel A, Hancock WS (1999), *J. Chromatogr. A* **849**, 71.

Flowers J (2004), *Science* **306**, 1324.

Focant J-F, De Pauw E (2002), *J. Chromatogr. B* **776**, 199.

Focant J-F, Pirard C, De Pauw E (2004), *Talanta* **63**, 1101.

Focant J-F, Reiner EJ, MacPherson K *et al.* (2004a), *Talanta* **63**, 1231.

Focant J-F, Sjödin A, Turner WE, Patterson DG Jr (2004b), *Anal. Chem.* **76**, 6313.

Focant J-F, Pirard C, Eppe G, De Pauw E (2005), *J. Chromatogr. A* **1067**, 265.

Focant J-F, Eppe G, Scippo M-L *et al.* (2005a), *J. Chromatogr. A* **1086**, 45.

Focant J-F, Eppe G, Massart A-C *et al.* (2006), *J. Chromatogr. A* **1130**, 97.

Folch J, Lees M, Stanley GHS (1957), *J. Biol. Chem.* **226**, 497.

Fornstedt T, Zhong G, Guiochon G (1996), *J. Chromatogr. A* 742, 55.

Fortier M-H, Bonneil E, Goodley P, Thibault P (2005), *Anal. Chem.* **77**, 1631.

Franco P, Senzo A, Oliveros L, Minguillón C (2001), *J. Chromatogr. A* **906**, 155.

Fraisse D, Gonnord MF, Becchi M (1989), *Rapid Commun. Mass Spectrom.* **3**, 79.

Franks F (1981), *Polywater*, The MIT Press, Cambridge, MA, USA.

Franke JP, de Zeeuw RA, Hakkert R (1978), *Anal. Chem.* **50**, 1374.

French JB, Reid NM, Poon CC, Buckley JA (1977), *Proceedings of the 25th ASMS Conference on Spectrometry and Allied Topics.*

Fries HE, Aiello M, Evans CA (2007), *Rapid Commun. Mass Spectrom.* **21**, 369.

Fuhr U, Boettcher MI, Kinzig-Schippers M *et al.* (2006), *Cancer Epidemiol. Biomarkers Prev.* **15**, 266.

Fulford J, March RE (1978), *Int. J. Mass Spectrom. Ion Phys.* **26**,155.

Fung D, Boyd RK, Safe S, Chittim BG (1985), *Biomed. Envir. Mass Spectrom.* **12**, 247.

Fung EN, Chu I, Li C *et al.* (2003), *Rapid Commun. Mass Spectrom.* **17**, 2147.

Fung EN, Chu I, Nomeir AA (2004), *Rapid Commun. Mass Spectrom.* **18**, 2046.

G

Gaffield W, Fish RH, Holmstead RL *et al.* (1976), *IARC Sci. Publ.,* **14**, 11.

Gale B (2001), *Micro Total Analysis Systems* [available at: http://www.latech.edu/tech/engr/bme/gale_classes/mems/lecture23.pdf].

Gallicano K (2006),*Random repeat analysis (reaffirmation) of incurred samples,* presentation at the Applied Pharmaceutical Analysis Conference (September 11–12, 2006, Boston, MA. USA).

Gallicano K (2007), *private communication.*

Gamero-Castaño M, Fernandez de la Mora J (2000), *J. Mass Spectrom.* **35**, 790.

Gangl ET, Annan M, Spooner N, Vouros P (2001), *Anal. Chem.* **73**, 5635.

Gao S, Zhang Z-P, Karnes HT (2005), *J. Chromatogr. B* **825**, 98.

Gardner MJ, Gunn AM (1986), *Fresenius. J. Anal. Chem.* **325**, 263.

Gardner MJ, Gunn AM (1988), *Fresenius. J. Anal. Chem.* **330**, 103.

Garden JS (1980), *Anal. Chem.* **52**, 2310.

Garrett AW, Cisper ME, Nogar NS, Hemberger PH (1994), *Rapid Commun. Mass Spectrom.* **8**, **174**.

Gauthier GL (2006), *LC/GC, The Peak Online Edition,* May 2006.

Geear M, Syms RRA, Wright S, Holmes AS (2005), *J. Microelectromechanical Systems* **14**, 1156.

Gelpi E (2002), *J. Mass Spectrom.* **37**, 241.

George C (2001), *LC/GC* **19**, 578.

Gerber SA, Rush J, Stemman O *et al.* (2003), *Proc. Natl. Acad. Sci. USA* **100**, 6940.

Gernand W, Steckenreuter K, Wieland G (1989), *Fresenius Z. Anal. Chem.* **334**, 534.

Geromanos S, Freckleton G, Tempst P (2000), *Anal. Chem.* **72**, 777.

Geusau A, Tschachler E, Meixner M *et al.* (1999), *Lancet* **354**, 1266.

Geusau A, Abraham K, Geissler K *et al.* (2001), *Environ. Health Perspect.* **109**, 865 [freely available at: http://www.ehponline.org/members/2001/109p865-869geusau/geusau.pdf].

Gevaert K, Goethals M, Martens L *et al.* (2003), *Nature Biotechnol.* **21**, 566.

Gibson DK, Reid ID (1984), *J. Phys. E* **17**, 443.

Giddings JC (1961), *J. Chromatogr.* **5**, 46.

Gieniec J, Mack LL, Nakamae K *et al.* (1984), *Biomed. Mass Spectrom.* **11**, 259.

Gioti EM, Skalkos DC, Fiamegos YC, Stalikas CD (2005), *J. Chromatogr. A* **1093**, 1.

Glaser JA, Foerst DL, McKee GD *et al.* (1981), *Env. Sci. Technol.* **15**, 1426.

Gobey J, Cole M, Janiszewski J *et al.* (2005), *Anal. Chem.* **77**, 5643.

Goeringer DE, Asano KG, McLuckey SA *et al.* (1994), *Anal. Chem.* **66**, 313.

Golay MJE (1957), *Anal. Chem.* **29**, 928.

Golay MJE (1958), in *Gas Chromatography 1958 (Ed. Desty DH)*, Butterworths, London, UK, 36.

Gómara B, Herrero L, González MJ (2006), *Env. Sci. Technol.* **40**, 7541.

Gomez A, Tang K (1994), *Phys. Fluids* **6**, 604.

Gong M, Wehmeyer KR, Limbach PA *et al.* (2006), *Anal. Chem.* **78**, 3730.

Good ML (1953), Report No. 4146, University of California Radiation Laboratory, Berkeley, CA, USA.

Goodlett DR, Keller A, Watts JA *et al.* (2001), *Rapid Commun. Mass Spectrom.* **15**, 1214.

Gordon EF, Muddiman DC (1999), *Rapid Commun. Mass Spectrom.* **13**, 164.

Gordon EF, Mansoori BA, Carroll CF, Muddiman DC (1999a), *J. Mass Spectrom.* **34**, 1055.

Gorin G (1994), *J. Chem. Ed.* **71**, 114–116.

Gosset WS (1908), *Biometrika* **6**, 1.

Graham JE, Andrews SA, Farquhar GJ, Meresz O (1995), *Proceedings of the American Water Works Association Water Quality Technology Conference*, New Orleans, LA, USA, p. 757.

Granby K, Fagt S (2004), *Anal. Chim. Acta* **520**, 177.

Grange AH, Winnik W, Ferguson PL *et al.* (2005), *Rapid Commun. Mass Spectrom.* **19**, 2699.

Granvogl M, Jezussek M, Koehler P, Schieberle P (2004), *J. Ag. Food Chem.* **52**, 4751.

Grayson MA (Ed., 2002), *Measuring Mass: From Positive Rays to Proteins*, Chemical Heritage Foundation, Philadelphia: Chapter 3.

Green BN, Gray BW, Guyan SA, Krolik ST (1986), *Proceedings of the 34th ASMS Conference on Mass Spectrometry and Allied Topics.*

Green CE (1996), *Anal. Chem.* **68**, 305A.

Grill V, Shen J, Evans C, Cooks RG (2001), *Rev. Sci. Instrum.* **72**, 3149.

Griffiths IW (1997), *Rapid Commun. Mass Spectrom.* **11**, 2.

Grob K Jr., Grob G, Grob K (1978), *J. Chromatogr.* **156**, 1.

Gros M, Petrović M, Barceló D (2006), *Anal. Bioanal. Chem.* **386**, 941.

Gross ML, Giordani AB, Price PC (1994), *J. Am. Soc. Mass Spectrom.* **5**, 57.

Grumbach ES, Wagrowski-Diehl DM, Mazzeo JR *et al.* (2004), *LC/GC* **22**, 1010.

Gu M, Wang Y, Zhao X-G, Gu Z-M (2006), *Rapid Commun. Mass Spectrom.* **20**, 764.

Guan S, Marshall AG (1993), *Anal. Chem.* **65**, 1288.

Guevremont R, Purves RW (1999), *Rev. Sci. Instrum.* **70**, 1370.

Guevremont R (2004), *J. Chromatogr. A* **1058**, 2.

Guilhaus M (1995), *J. Mass Spectrom.* **30**, 1519.

Guilhaus M, Selby D, Mlynski V (2000), *Mass Spectrom. Revs.* **19**, 65.

Guo X, Bruins AP, Covey TR (2006), *Rapid Commun. Mass Spectrom.* **20**, 3145.

Gy P (1992), *Sampling of Heterogeneous and Dynamic Material Systems*, Elsevier, Amsterdam, The Netherlands.

Gy P (1998), *Sampling for Analytical Purposes*, John Wiley and Sons Ltd, Chichester, UK.

Gy P (2004), *Chemometrics Intell. Lab. Systems* **74**, 7, 25, 39, 49,

Gygi SP, Rist B, Gerber SA *et al.* (1999), *Nature Biotech.* **17**, 1994.

H

Haefelfinger P (1981), *J. Chromatogr.* **218**, 73.

Hager JW (1999), *Rapid Commun. Mass Spectrom.* **13**, 740.

Hager JW (2002), *Rapid Commun. Mass Spectrom.* **16**, 512.

Hager JW, Le Blanc JCY (2003), *Rapid Commun. Mass Spectrom.* **17**, 1056.

Hager JW, Le Blanc JCY (2003a), *J. Chromatogr. A* **1020**, 3.

Haginaka J (2001), *J. Chromatogr. A* **906**, 253.

Hagmar L, Törnqvist M (2003), *Br. J. Cancer* **89**, 774.

Halász I (1964), *Anal. Chem.* **36**, 1428.

Halász I, Endele R, Asshauer J (1975), *J.Chromatogr.* **112**, 37.

Halket JM, Zaikin VG (2003), *Eur. J. Mass Spectrom.* **9**, 1.

Halket JM, Waterman D, Pryzborowska AM *et al.* (2004), *J. Exptl. Botany* **56**, 219.

Halket JM, Zaikin VG (2004a), *Eur. J. Mass Spectrom.* **10**, 1.

Halket JM, Zaikin VG (2005), *Eur. J. Mass Spectrom.* **11**, 127.

Halket JM, Zaikin VG (2006), *Eur. J. Mass Spectrom.* **12**, 1.

Hamdan M, Curcuruto O (1991), *Int. J. Mass Spectrom.* **108**, 93.

Hamdan M, Righetti PG (2002), *Mass Spectrom. Revs.* **21**, 287.

Hamilton Company (2006), *Hamilton Precision Syringes: Care and Use* [available at: http://www.hamiltoncomp.com/newdev/syringes/pdfs/L20058_care&use_guide.pdf].

Handy R, Barnett DA, Purves RW *et al.* (2000), *J. Anal. At. Spectrom.* **15**, 907.

Hanold KA, Fischer SM, Cormia PH *et al.* (2004), *Anal. Chem.* **76**, 2842.

Hao C, Lissemore L, Nguyen B *et al.* (2006), *Anal. Bioanal. Chem.* **384**, 505.

Harris FM, Trott GW, Morgan TG *et al.* (1984), *Mass Spectrom. Revs.* **3**, 209.

Harrison AG (1992), *Chemical Ionization Mass Spectrometry, 2nd Edn*, CRC Press, Boca Raton, FL, USA

Hass JR, Friesen MD, Hoffman MK (1979), *Org. Mass Spectrom.* **14**, 9.

Hatsis P, Brombacher S, Corr J *et al.* (2003), *Rapid Commun. Mass Spectrom.* 17, 2303.

Hawkridge AM, Zhou L, Lee ML, Muddiman DC (2004), *Anal. Chem.* **76**, 4118.

Haws LC, Su SH, Harris M *et al.* (2006), *Toxicol. Sci.* **89**, 4.

Hays SM, Aylward LL (2003), *Regul. Toxicol. Pharmacol.* **37**, 202.

Hazai E, Gagne PV, Kupfer D (2004), *Drug Metabolism and Disposition* **32**, 742.

Heck AJR, de Koning LJ, Nibbering NMM (1991), *J. Am. Soc. Mass Spectrom.* **2**, 454.

Heeren RMA, Kleinnijenhuis AJ, McDonnell LA, Mize TH (2004), *Anal. Bioanal. Chem.* **378**, 1048.

Heller DN (2007), *Rapid Commun. Mass Spectrom.* 21, **644.**

Hemming J, Holmbom B, Reunanen M, Kronberg L (1986), *Chemosphere* **15**, 549.

Hemmings I (2004), *Ten Tips to Improve Your Pipetting Technique* [available at the website of PipetteDoctor Inc.: www.pipettedoctor.com/downloads.asp].

Hemmings I (2004a), *Artel PCS2 Explained* [available at the website of PipetteDoctor Inc.: www.pipettedoctor.com/downloads.asp].

Henderson SC, Valentine SJ, Counterman AE, Clemmer DE (1999), *Anal. Chem.* **71**, 291.

Henion JD, Thomson BA, Dawson PH (1982), *Anal. Chem.* **54**, 451.

Henry MP, Ratnoyake CK (2005), *J. Chromatogr.* **1079**, 69.

Heumann KG (1992), *Mass Spectrom. Revs.* **11**, 41.

Heydorn K, Esbenson K (2004), *Accred. Qual. Assur.* **9**, 391.

Higgs RE, Knierman MD, Gelfanova V *et al.* (2005), *J. Prot. Res.* **4**, 1442.

Him T, Tolmachev AV, Harkewicz R *et al.* (2000), *Anal. Chem.* **72**, 2247.

Hines RL (1966), *J. Appl. Phys.* **37**, 2730.

Hinshaw JV (2000), *LC/GC* **18**, 1142.

Hinshaw JV (2002), *LC/GC* **20**, 948.

Hinshaw JV (2002a), *LC/GC* **20**, 1034.

Hinshaw JV (2002b), *LC/GC* **20**, 276.

Hinshaw JV (2003), *LC/GC* **21**, 1056.

Hinshaw JV (2004), *LC/GC* **22**, 1160.

Hinshaw JV (2004a), *LC/GC* **22**, 624.

Hinshaw JV (2005), *LC/GC* **23**, 36.

Hipple JA, Condon EU (1945), *Phys. Rev.* **68**, 54.

Hoaglund CS, Valentine SJ, Sporleder CR *et al.* (1998), *Anal. Chem.* **70**, 2236.

Holle A, Haase A, Kayser M, Höhndorf J (2006), *J. Mass Spectrom.* **41**, 705.

Holmbom BR, Voss RH, Mortimer RD, Wong A (1981), *Tappi J.* **64**, 172.

Holmbom BR, Voss RH, Mortimer RD, Wong A (1984), *Env. Sci. Technol.* **18**, 333.

Hopfgartner G, Bean K, Henion J, Henry R (1993), *J. Chromatogr.* **647**, 51.

Hopfgartner G, Husser C, Zell M (2003), *J. Mass Spectrom.* **38**, 138.

Hopfgartner G, Bourgogne E (2003a), *Mass Spectrom. Revs.* **22**, 195.

Hopfgartner G, Varesio E, Tschäppät V *et al.* (2004), *J. Mass Spectrom.* **39**, 845.

Hopfgartner G, Varesio E (2005), *Trends Anal. Chem.* **24**, 583.

Horning EC, Carroll DI, Dzidic DI *et al.* (1974), *J. Chromatogr. Sci.* **12**, 725.

Horning EC, Carroll DI, Dzidic DI *et al.* (1974a), *J. Chromatogr.* **99**, 13.

Horváth CG, Preiss BA, Lipsky SR (1967), *Anal. Chem.* **39**, 1422.

Horváth CG, Lipsky SR (1969), *Anal. Chem.* 41, 1227.

Howald WN, Bush ED, Trager WF *et al.* (1980), *Biomed Mass Spectrom.* **7**, 35.

Hsieh Y, Wang G, Wang Y *et al.* (2002), *Rapid Commun. Mass Spectrom.* **16**, 944.

Hsieh Y, Merkle K, Wang G *et al.* (2003), *Anal. Chem.* **75**, 3122.

Hu Q, Noll RJ, Li H *et al.* (2005), *J. Mass Spectrom.* **40**, 430.

Huang EC, Wachs T, Conboy JJ, Henion JD (1990), *Anal. Chem.* **62**, 713A.

Huang SY, Tsai ML, Wu CJ *et al.* (2006), *Proteomics* **6** 1722.

Hubaux A, Vos G (1970), *Anal. Chem.* **42**, 849.

Huber JFK, Hulsman JARJ, Meijers CAM (1972), *J. Chromatogr.* **62**, 79.

Huertas ML, Fortan J (1975), *Atmospheric Environ.* **9**, 1018.

Hund E, Vander Heyden Y, Massart DL, Smeyers-Verbeke J (2002), *J. Pharm. Biomed. Anal.* **30**, 1197.

Hunt DF, Harvey TM, Russell JW (1975), *J. Chem. Soc. Chem. Commun.* 151.

Hyötyläinen T, Tuutijärvi T, Kuosmanen K, Riekkola M-L (2002), *Anal. Bioanal. Chem.* **372**, 732.

Hyötyläinen T, Lüthje K, Rautianen-Rämä M, Riekkola M-L (2004), *J. Chrom. A* **1056**, 267.

I

ICH (International Conference on Harmonization of Technical Requirements for Registration of Pharmaceuticals for Human Use) (2005), *Validation of Analytical Procedures: Text and Methodology – Q2(R1)* [available at: http://www.ich.org/LOB/media/MEDIA417.pdf].

Ihle HR, Neubert A (1971), *Int. J. Mass Spectrom. Ion Phys.* **7**, 189.

Ikegami T, Tanaka N (2004), *Current Opinion Chem. Biol.* **8**, 527.

Ingamells CO, Switzer P (1973), *Talanta* **20**, 547.

Ingamells CO (1974), *Talanta* **21**, 141.

Ingamells CO, Pitard FF (1985), *Applied geochemical analysis*: Monographs on analytical chemistry and its applications, Vol 88, John Wiley & Sons Inc., New York, USA.

ILAC (International Laboratory Accreditation Cooperation) (2000), *ILAC-G12: 200: Guidelines for the Requirements for the Competence of Reference Materials Producers*, [available at: http://www.ilac.org/guidanceseries.html].

Iribarne JV, Thomson BA (1976), *J. Chem. Phys.* **64**, 2287.

Iribarne JV, Dziedzic PJ, Thomson BA (1983), *Int. J. Mass Spectrom. Ion Phys.* **50**, 331.

Ishii D, Asai K *et al.* (1977), *J. Chromatogr.* **144**, 157.

Ishii D, Hibi K, Asai K, Jonokuchi M (1978), *J. Chromatogr.* **151**, 147.

Ishii D, Hibi K, Asai K, Nagaya J (1978a), *J. Chromatogr.* **152**, 341.

Ishii D, Hirose A, Hibi K *et al.* (1978b), *J. Chromatogr.* **157**, 43.

Ishizuka N, Minakuchi H, Nakanishi K *et al.* (1998), *J. High Resol. Chromatogr.* **21**, 477.

ISO (International Standards Organization) (1994), *Terminology concerning assessment criteria for a software module, package or product; document ISO 8402* [freely available at:http://www.issco.unige.ch/ewg95/node69.html#SECTION00630000 000000000000].

ISO (International Standards Organization) (1997), *ISO/11843–1, Capability of Detection (Part 1): Terms and Definitions.*

ISO (International Standards Organization) (2002), *ISO 8655, Piston-operated volumetric apparatus – Part 2: Piston pipettes* [available *for purchase* at: http://www.iso.org/iso/en/CatalogueDetailPage.CatalogueDetail?CSNUMBER=297267].

ISO (International Standards Organization) (2002), *Statistical aspects of sampling from bulk materials. Part 1: General principles*, ISO/FDIS (2002) 11648–1.

ISO (International Standards Organization) (2004), *International vocabulary of basic and general terms in metrology* (draft for discussion, update of the 1993 document) [available at: http://www.abnt.org.br/ISO_DGuide_99999_(E).PDF].

ISO (International Standards Organization) (2005), *ISO/IEC 17025:2005: General requirements for the competence of testing and calibration laboratories* [available *for purchase* at:http://www.iso.ch/iso/en/CatalogueDetailPage.Catalogue Detail?CSNUMBER=39883].

ISO (International Standards Organization) (2006), *ISO/IEC 17025:2005/Cor1:2006, Amendments to ISO/IEC 17025:2005* [available for purchase at: http://www.iso.org/iso/iso_catalogue/catalogue_detail.htm?csnumber=44644].

Ito Y, Takeuchi T, Ishii D, Goto M (1985), *J. Chromatogr.* **346**, 161.

IUPAC Analytical Chemistry Division (1977), *Spectrochim. Acta B* **33**, 242.

IUPAC Analytical Chemistry Division (1995), *Pure Appl. Chaem.* **67**, 1699.

J

Jaime E, Hummert C, Hess P, Luckas B (2001), *J. Chromatogr. A* **929**, 43.

Jakeway SC, de Mello AJ, Russell EL (2000), *Fresenius J. Anal. Chem.* **366**, 525.

James AT, Martin AJP (1952), *Biochem. J.* **50**, 679.

James AT (1979, Aug.13), *Current Contents* **33**, 45 [available at: http://www.garfield.library.upenn.edu/classics1979/A1979HE73400001.pdf].

Janeček M, Quilliam MA, Lawrence JF (1993), *J. Chromatogr.* **644**, 321.

Janasek D, Franzke J, Manz A (2006), *Nature* **442**, 374.

Jeannot MA, Cantwell FF (1996), *Anal. Chem.* **68**, 2236.

Jeannot MA, Cantwell FF (1997), *Anal. Chem.* **69**, 235.

Jeanville PM, Woods JH, Baird TJ III, Estapé ES (2000), *J. Pharm. Biomed. Anal.* **23**, 897.

Jellett J, Roberts RL, Laycock M *et al.* (2002), *Toxicon* **40**, 1407.

Jemal M, Almond R, Ouyang Z, Teitz D (1997) *J. Chromatogr. B* **703**, 167.

Jemal M, Qing Y, Whelan DB (1998), *Rapid Commun. Mass Spectrom.* **12**, 1389.

Jemal M, Xia Y-Q (1999), *Rapid Commun. Mass Spectrom.* **13**, 97.

Jemal M, Huang M, Mao Y *et al.* (2000), *Rapid Commun. Mass Spectrom.* **14**, 1023.

Jemal M, Ouyang Z, Powell M (2000a), *J. Pharm. Biomed. Anal.* **23**, 323.

Jemal M, Ouyang Z, Powell M (2002), *Rapid Commun. Mass Spectrom.* **16**, 1538.

Jemal M, Schuster A, Whigan DB (2003), *Rapid Commun. Mass Spectrom.* **17**, 1723.

Jemal M, Ouyang Z (2003a), *Rapid Commun. Mass Spectrom.* **17**, 24.

Jenkins SWD, Koester CJ, Taguchi VY *et al.* (1995), *Environ. Sci. Pollut. Res.* **2**, 207.

Jenkins RE, Kitteringham NR, Hunter CL *et al.* (2006), *Proteomics* **6**, 1934.

Jennings ME II, Matthews DE (2005), *Anal. Chem.* **77**, 6435.

Jensen J (2004), *Technical Note:Eksigent Microfluidic Flow Control* [available at: www.eksigent.com].

Jesperson K (1996), *Search for Clean Water Continues*; *Contaminated Water Makes a Deadly Drink*; *Safe Water Should Always Be On Tap* [available at: www.nesc.wvu.edu/ndwc/ndwc_dwhistory.htm].

Jessome LL, Volmer DA (2006), *The PEAK. LCGC* Electronic Edition (May 2006) [available at: http://www.lcgcmag.com/lcgc/article/articleDetail.jsp?id=327354].

Jiang H, English AM (2002), *J. Proteome Res.* **1**, 345.

Jiang X, Lee HK (2004), *Anal. Chem.* **76**, 5591.

Jobb DB, Hunsinger RB, Meresz O, Taguchi VY (1992), *Proceedings of the American Water Works Association Water Quality Technology Conference*, Toronto, ON, Canada p. 103.

Jobb DB, Hunsinger RB, Meresz O, Taguchi VY (1994), *Removal of N-Nitrosodimethylamine From the Ohsweken (Six Nations)Water Supply*; Final Report, Ontario Ministry of Environment and Energy, Toronto, ON, Canada.

Johnson JV, Yost RA, Kelley PE, Bradford DC (1990), *Anal. Chem.* **62**, 2162.

Jönsson JÅ, Mathiasson L (2001), *J. Sep. Sci.* **24**, 495.

Jönsson JÅ, Mathiasson L (2003), *LC/GC North America* **21**, 424.

Jönsson JÅ, Lövkvist P, Audunsson G, Nilvé J (1993), *Anal. Chim. Acta* **227**, 9.

Jorgenson J, Lukacs KD (1981), *J. Chromatogr.* **218**, 209.

Julian RK, Cooks RG (1993), *Anal. Chem.* **65**, 1827.

Juraschek R, Dülcks T, Karas M (1999), *J. Amer. Soc. Mass Spectrom.* **10**, 300.

K

Kaiser RE Jr, Cooks RG, Moss J, Hemberger PH (1989), *Rapid Commun. Masc Spectrom.* **3**, 50.

Kaiser RE Jr, Cooks RG, Stafford GC Jr *et al.* (1991), *Int . J . Mass Spectrom . Ion Proc.* **106**, 79.

Kambara H, Kanomata I (1976), *Mass Spectroscopy (Japan)* **24**, 229.

Kambara H, Kanomata I (1976a), *Mass Spectroscopy (Japan)* **24**, 271.

Kameoka J, Craighead HG, Zhang H, Henion J (2001), *Anal. Chem.* **73**, 1935.

Kandimalla VB, Hunagxian J (2004), *Anal. Bioanal. Chem.* **380**, 587.

Kanniganti R, Johnson JD, Ball LM, Charles MJ (1992), *Envir. Sci. Technol.* **26**, 1998.

Kantharaj E, Tuytelaars A, Proost PEA *et al.* (2003), *Rapid Commun. Mass Spectrom.* **17**, 2661.

Kapron JT, Jemal M, Duncan G *et al.* (2005), *Rapid Commun. Mass Spectrom.* **19**, 1979.

Kapron J (2006), *The PEAK LCGC* Electronic Edition (Nov 2006) [available at: http://chromatographyonline.findpharma.com/lcgc/article/articledetail.Jsp?id=3906778pageID=/&sk=&date=].

Kapron J, Wu J, Mauriala T *et al.* (2006a), *Rapid Commun. Mass Spectrom.* **20**, 1504.

Karas M, Bachmann D, Bahr U, Hillenkamp F (1987), *Int. J. Mass Spectrom. Ion Proc.* **78**, 53.

Karas M, Hillenkamp F (1988), *Anal. Chem.* **60**, 2299.

Karas M, Krüger R (2003), *Chem. Revs.* **103**, 427.

Karger BL, Martin M, Guiochon G (1974), *Anal. Chem.* **46**, 1640.

Karlberg B, Thelander S (1978), *Anal. Chim. Acta* **98**, 1.

Karlsson K-E, Novotny M (1988), *Anal. Chem.* **60**, 1662.

Katz E, Ogan KL and Scott RPW (1984), *J.Chromatogr.* **289**, 65.

Kauppila TJ, Kuuranne T, Meurer EC *et al.* (2002), *Anal. Chem.* **74**, 5470.

Kauppila TJ, Östman P, Martilla S *et al.* (2004), *Anal. Chem.* **76**, 6797.

Kauppila TJ, Kostiainen R, Bruins AP (2004a), *Rapid Commun. Mass Spectrom.* **18**, 808.

Kazakevich Y, McNair H (1996), *Basic Liquid Chromatography* [available at: http://hplc.chem.shu.edu/NEW/HPLC_Book/index.html].

Ke J, Yancey M, Zhang S *et al.* (2000), *J. Chromatogr. B* **742**, 369.

Kebarle P, Tang L (1993), *Anal. Chem.* **64**, 972A.

Kebarle P (2000), *J. Mass Spectrom.* **35**, 804.

Keith, LH (1992), *Environmental sampling and analysis: A practical guide*, Lewis Publishers, Chelsea, MI, USA, pp. 93–119.

Kelley PE (1992), US Patent 5134286.

Kelvin, Lord (1904), *Baltimore Lectures*, Clay, London.

Kenndler E (2004), *Gas Chromatography* [available at: http://www.univie.ac.at/anchem/publikation/Gas_Chromatography_in_Capillaries.pdf].

Kennedy GJ, Knox JH (1972), *J. Chromatogr. Sci.* **10**, 549.

Kennedy RT, Jorgenson JW (1989), *Anal. Chem.* **61**, 1128.

Kerwin L (1963), in *Mass Spectrometry*, (Ed. McDowell CA), McGraw-Hill, New York, USA.

Ketkar SN, Dulak JG, Dheandhanoo S, Fite WL (1991), *Anal. Chim. Acta.* **245**, 267.

Keulemans AI (1959), *Gas Chromatography*, Van Nostrand-Reinhold, New York, USA.

Kim JS, Lim HS, Cho SI *et al.* (2003), *Ind. Health* **41**, 149.

Kim T-Y, Thompson MS, Reilly JP (2005), *Rapid Commun. Mass Spectrom.* **19**, 1657.

Kimball J (2006), *Kimball's Biology Pages* [available at: http://users.rcn.com/jkimball.ma.ultranet/BiologyPages/].

King R, Bonfiglio R, Fernández-Metzler C, Miller-Stein C *et al.* (2000), *J. Am Soc. Mass Spectrom.* **11**, 942.

King RH, Grady LT, Reamer JJ (1974), *J. Pharm. Sci.* **63**, 1591.

King B (2001), *Fresenius J. Anal. Chem.* **371**, 714.

King RC, Gundersdorf R, Fernández-Metzler CL (2003), *Rapid Commun. Mass Spectrom.* **17**, 2413.

Kipiniak W (1981), *J. Chromatogr. Sci.* **19**, 332.

Kirkpatrick DS, Gerber SA, Gygi SP (2005), *Methods* **35**, 265.

Kłodziska E, Moravcova D, Jandera P, Buszewski B (2006), *J. Chromatogr. A* **1109**, 51.

Knapp DR (1979), *Handbook of Analytical Derivatization Reactions*, John Wiley & Sons Ltd, Chichester, UK.

Knight AK, Sperline RP, Hieftje GM *et al.* (2002), *Int. J. Mass Spectrom.* **215**, 151.

Knochenmuss R, Dubois F, Dale MJ, Zenobi R (1996), *Rapid Commun. Mass Spectrom.* **10**, 871.

Knochenmuss R, Karbach V, Wiesli U *et al.* (1998), *Rapid Commun. Mass Spectrom.* **12**, 529.

Knochenmuss R, Stortelder A, Breuker K, Zenobi R (2000), *J. Mass Spectrom.* **35**, 1237.

Knochenmuss R, Zenobi R (2003), *Chem. Revs.* **103**, 441.

Knox JH (1980), *J. Chromatogr. Sci.* **18**, 453.

Knox JH, Grant IH (1987), *Chromatographia* **24**, 135.

Knox JH, Grant IH (1991), *Chromatographia* **32**, 317.

Koerner T, Brown L, Xie R, Oleschuk RD (2005), *Sensors and Actuators B* **107**, 632.

Koerner T, Oleschuk RD (2005a), *Rapid Commun. Mass Spectrom.* **19**, 3279.

Koester CJ, Harless RL, Hites RA (1992), *Chemosphere* **24**, 421.

Kohli BM, Eng JK, Nitsch RM, Konietzko U (2005), *Rapid Commun. Mass Spectrom.* **19**, 589.

Komulainen H, Kosma V-M, Vaittinen S-L *et al.* (1997), *J. Natl. Cancer Inst.* **89**, 848.

Kopp MU, Crabtree HJ, Manz A (1997), *Current Opinion Chem. Biol.* **1**, 410.

Køppen B, Spliid NH (1998), *J. Chromatogr. A* **803**,157.

Koppenaal DW, Barinaga CJ, Denton MB *et al.*, Sperline RP, Hieftje GM, Schilling GD, Andrade FJ, Barnes JH IV (2005), *Anal. Chem.* **77**, 418A.

Korfmacher WA, Cox KA, Ng KJ *et al.* (2001), *Rapid Commun. Mass Spectrom.* **15**, 335.

Korfmacher WA (Ed., 2004), *Using Mass Spectrometry for Drug Metabolism Studies*, CRC Press, Boca Raton, FL, USA.

Korfmacher WA (2005), *Drug Discovery Today* **10**, 1357.

Korytár P, Parerac J, Leonards PEG *et al.* (2005), *J. Chromatogr. A* **1067**, 255.

Kosevich MV, Shelkovsky VS, Boryak OA, Orlov VV (2003), *Rapid Commun. Mass Spectrom.* **17**, 1781.

Kosevich MV, Chagovets VV, Shelkovsky VS *et al.* (2007), *Rapid Commun. Mass Spectrom.* **21**, 466.

Kostianinen R, Tuominen J, Luukkanen L *et al.* (1997), *Rapid Commun. Mass Spectrom.* **11**, 283.

Kovarik P, Grivet C, Bourgogne E, Hopfgartner G (2007), *Rapid Commun. Mass Spectrom.* **21**, 911.

Kronberg L, Holmbom B, Reunanen M, Tikkanen L (1988), *Envir. Sci. Technol.* **22**, 1097.

Kronberg L, Vartiainen T (1988a), *Mutation Res.* **206**, 177.

Krutchinsky AN, Loboda AV, Spicer VL *et al.* (1998), *Rapid Commun. Mass Spectrom.* **12**, 508.

Kucera P (1980), *J. Chromatogr.* **198**, 93.

Kucklick JR, Helm PA (2006), *Anal. Bioanal. Chem.* **386**, 819.

Kuhlmann FE, Apffel A, Fisher SM *et al.* (1995), *J. Am. Soc. Mass Spectrom.* **6**, 1221.

Kuhn E, Wu J, Karl J *et al.* (2004), *Proteomics* **4**, 1175.

Kuo B-S, Van Noord T, Feng MR, Wright DS (1998), *J. Pharm. Biomed. Anal.* **16**, 837.

Kurz EA (1979), *Amer. Lab.*, March 1979, 67.

L

Lacorte S, Latorre S, Barceló D *et al.* (2003), *Trends Anal. Chem.* **22**, 725.

Laemmli UK (1970), *Nature* **227**, 680.

Lagerwerf FM, van Dongen WD, Steenvoorden RJJM *et al.* (2000), *Trends Anal. Chem.* **19**, 418.

Laiko VV, Baldwin MA, Burlingame AL (2000), *Anal. Chem.* **72**, 652.

Laitinen L, Kangas H, Kaukonen AM *et al.* (2003), *Pharm. Res.* **20**, 187.

Lam RB, Isenhour TL (1980), *Anal. Chem.* **52**, 1158.

Lane DA, Sakuma T, Quan ESK (1980), in *Polynuclear Aromatic Hydrocarbons: Chemistry and Biological Effects* (eds Bjorseth A, Dennis AJ), Battelle Press, Columbus, OH, USA, p. 199.

Langmuir I (1916), *J. Am. Chem. Soc.* **38**, 2221.

Langmuir RV (1967), US Patent 3,334,225.

Lavagnini I, Magno F, Seraglia R, Traldi P (2006), *Quantitative Applications of Mass Spectrometry*, John Wiley & Sons Ltd, Chichester, UK.

Lavagnini I, Magno F (2007), *Mass Spectrom. Revs.* **26**, 1.

Lawrence JF, Menard C, Charbonneau CF, Hall S (1991), *J. Assoc. Off. Anal. Chem.*, **74**, 404.

Lawrence JF, Menard C (1991a), *J. Assoc. Off. Anal. Chem.*, **74**, 1006.

Laycock MV, Jellett JF, Belland ER *et al.* (2000), in *Harmful Algal Blooms*, (Eds Hallegraeff GA, Blackburn SI, Bolch CJ, Lewis RJ), *Proceedings of the 9th International Conference on Harmful Algal Blooms*, Hobart, Tasmania, Australia, February 2000 Intergov. Oceanogr. Comm., Paris.

Lazar IM, Grym J, Foret F (2006), *Mass Spectrom. Revs.* **25**, 573.

LEAP Technologies (2007), Steps to Assess and Remedy Carryover [available at:http://www.leaptec.com/chromatogra phy/carry-over-prevention-and-abatement.php].

Le Blanc JCY, Bloomfield N (2004), *Proceedings of the 52nd ASMS Conference on Mass Spectrometry and Allied Topics*, Nashville, Tennessee, USA.

Leclercq PA, Cramers CA (1998), *Mass Spectrom. Revs.* **17**, 37.

Lee SJ, Lee SY (2004), *Appl. Microbiol. Biotechnol.* **64**, 289.

Lee SSH, Douma M, Koerner T, Oleschuk RD (2005), *Rapid Commun. Mass Spectrom.* **19**, 2671.

Leitner A, Linder W (2006), *Proteomics* **6**, 5418.

Lesney MS (1998), *Today's Chemist at Work*, **7**, 67–72.

Lesney MS (2001), *Today's Chemist at Work*, Special Supplement 'Chromatography: Creating a Central Science', 8 [available at: http://pubs.acs.org/journals/chromatography/ chap1.html (regularly updated)].

Letellier M, Budzinski H (1999), *Analusis* **27**, 259.

Levin S (2002), *Solid Phase Extraction* [available at: www.forumsci.co.il/HPLC/SPE_site.pdf].

Li LYT, Campbell DA, Bennett PK, Henion J (1996), *Anal. Chem.* **68**, 3397.

Li J, Steen H, Gygi SP (2003), *Molec. Cell Proteomics* **2**, 1198.

Li W, Cohen LH (2003a), *Anal. Chem.* **75**, 5854.

Li AC, Alton D, Bryant MS, Shou WZ (2005), *Rapid Commun. Mass Spectrom.* **19**, 1943.

Li AC, Gohdes MA, Shou WZ (2007), *Rapid Commun. Mass Spectrom.* **21**, 1421.

Liang HR, Foltz RL, Meng M, Bennett P (2003), *Rapid Commun. Mass Spectrom.* **17**, 2815.

Lichtenberg J, de Rooij NF, Verpoorte E (2002), *Talanta* **56**, 233.

Little JL (1999), *J. Chromatogr. A* **844**, 1.

Liu S, Dasgupta PK (1995), *Anal. Chem.* **67**, 2042.

Liu DQ, Pereira T (2002), *Rapid Commun. Mass Spectrom.* **16**, 142.

Loboda AV, Krutchinsky AN, Bromirski M *et al.* (2000), *Rapid Commun. Mass Spectrom.* **14**, 1047.

Loboda A (2006), *J. Amer. Soc. Mass Spectrom.* **17**, 691.

Lock CM, Dyer EW (1999), *Rapid Commun. Mass Spectrom.* **13**, 422.

Lock CM, Dyer E (1999a), *Rapid Commun. Mass Spectrom.* **13**, 432.

Locke SJ, Thibault P (1994), *Anal. Chem.* **66**, 3436.

Loconto P (2005), *Trace Environmental Quantitative Analysis*, 2nd Edn, CRC Press, Boca Raton, FL, USA.

Loeb LB (1958), *Static Electrification*, Springer-Verlag, Berlin, Germany

Londry FA, Morrison RJS, March RE (1995), *Proceedings of the 43rd ASMS Conference on Mass Spectrometry and Allied Topics*, Atlanta, GA, USA, 1124.

Londry FA, Hager JW (2003), *J. Am. Soc. Mass Spectrom.* **14**, 1130.

Longo M, Lionetti C, Cavallaro A (1995), *J. Chromatogr. A* **708**, 303.

Loo JA, Udseth HR, Smith RD (1988), *Rapid Commun. Mass Spectrom.* **2**, 207.

Loss RD (2003), *Pure Appl. Chem.* **75**, 1107.

Louris JN, Brodbelt-Lustig J S, Cooks RG *et al.* (1990), *Int. J. Mass Spectrom. Ion Proc.* **96**, 117.

Luo Q, Shen Y, Hixson KK *et al.* (2005), *Anal. Chem.* **77**, 5028.

M

MacCoss MJ, Matthews DE (2005), *Anal. Chem.* **77**, 295A.

Mack LL, Kralik P, Rheude A, Dole MJ (1970), *J. Chem. Phys.* **52**, 4977.

Mackintosh FH, Gallacher S, Shanks AM, Smith EA (2002), *J. AOAC Intl.* **85**, 632.

MacLellan JA, Traub RJ, Fisher DR (1988), *Performance Testing of Radiobioassay Laboratories: In Vitro Measurements (Urinalysis). Final Report*. National Technical Information Service, Springfield, VA, USA.

Magee PN, Barnes JM (1956), *Brit. J. Cancer* **10**, 114.

Magee PN, Farber E (1962), *Biochem. J.* **83**, 114.

Maggio VL, Alexander LR, Green VE *et al.* (1990), *Toxicol. Env. Chem.* **28**, 143.

Majors RE (1991), *LC/GC* **9**, 16.

Majors RE (2002), *LC/GC* **20**, 1098.

Majors RE (2004), *LC/GC* **22**, 1062.

Majors RE, Shukla A (2005), *The PEAK, LCGC* Electronic Edition (July 2005) [available at:http://chromatographyon-line.findpharma.com/lcgc/article/articleDetail.jsp?id=171149].

Majors RE (2006), *The PEAK, LCGC* Electronic Edition (February 2006) [available at :http://chromatographyonline.findpharma.com/lcgc/article/ articleDetail.jsp?id=302241].

Makarov A (2000), *Anal. Chem.* **72**, 1156.

Mallet CR, Lum Z, Mazzeo JR (2004), *Rapid Commun. Mass Spectrom.* **18**, 49.

Malz F, Jancke H (2005), *J. Pharm. Biomed. Anal.* **38**, 813.

Mamyrin BA, Karataev DV, Shmikk DV, Zagulin VA (1973), *Sov. Phys. JETP* **37**, 45.

Mansoori BA, Dyer EW, Lock CM *et al.* (1998), *J. Am. Soc. Mass Spectrom.* **9**, 775.

Manz A, Graber N, Widmer HM (1990), *Sensors and Actuators B*, **1**, 244.

Manz A, Miyahara Y, Miura Y *et al.* (1990a), *Sensors and Actuators B*, **1**, 249.

Manz A, Harrison DJ, Verpoorte EMJ *et al.* (1992), *J. Chromatogr.* **593**, 253.

March RE (1992), *Int. J. Mass Spectrom. Ion Proc.* **118/119**, 71.

March RE, Todd JFJ (Eds, 1995), *Practical Aspects of Ion TrapMass Spectrometry*, Modern Mass Spectrometry Series, Vols 1–3, CRC Press, Boca Raton, FL, USA.

March RE (1997), *J. Mass Spectrom.* **32**, 351.

March RE (1998), *Rapid Commun. Mass Spectrom.* **12**, 1543.

March RE, Splendore M, Reiner EJ *et al.* (2000), *Int. J. Mass Spectrom.* **197**, 283.

March RE, Todd JF (2005), *Quadrupole Ion Trap Mass Spectrometry*, 2nd Edn, John Wiley & Sons Ltd, Chichester, UK.

Maron DM, Ames BN (1983), *Mutation Res.* **113**, 173.

Marotta E, Seraglia R, Fabris F, Traldi P (2003), *Int. . Mass Spectrom.* **228**, 841.

Marriott P, Shellie R (2002), *Trends Anal. Chem.* **21**, 573.

Marshall AG, Wang TC, Ricca TL (1985), *J. Am. Chem. Soc.* **107**, 7893.

Marshall AG, Schweikhard L (1992), *J. Mass Spectrom.* **118/119**, 37.

Marshall AG, Hendrickson CL, Jackson GS (1998), *Mass Spectrom. Revs.* **17**, 1.

Marshall AG, Hendrickson CL (2002), *Int. J. Mass Spectrom.* **215**, 59.

Marshall AG, Rodgers, RP (2004), *Acc. Chem. Res.* 37, 53–59.

Martin AJP, Synge RLM (1941), *Biochem. J.*, **35**, 1358.

Martin K (2002), *Pipetting, Ergonomics and You* [available at the website of Rainin Instrument LLC: www.rainin.com/lit_ergopaper.asp].

Martin M, Herman DP, Guiochon G (1986), *Anal. Chem.* **58**, 2200.

Matuszewski BK, Chavez-Eng CM, Constanzer ML (1998), *J. Chromatogr. B* **716**, 195.

Matuszewski BK, Constanzer ML, Chavez-Eng CM (2003), *Anal. Chem.* **75**, 3019.

Matuszewski BK (2006), *J. Chromatogr. B* **830**, 293.

Maynard BJ (2000), *Chemistry Summer 2000*, 17 [available at: http://www.scs.uiuc.edu/suslick/pdf/chemistry.summer00.sonochem.pdf].

McClellan JE, Murphy JP III, Mulholland JJ, Yost RA (2002), *Anal. Chem.* **74**, 402.

McCooeye MA, Ells B, Barnett DA *et al.* (2001), *J. Anal. Toxicology* 25, 81.

McCooeye MA, Mester Z, Ells B *et al.* (2002), *Anal. Chem.* **74**, 3071.

McCooeye MA, Ding L, Gardner GJ *et al.* (2003), *Anal. Chem.* **75**, 2538.

McCoy RW, Aiken RL, Pauls RE *et al.* (1984), *J. Chromatogr. Sci.* 22, 425.

McEwen CN, McKay RG, Larsen BS (2005), *Anal. Chem.* **77**, 7826.

McFadden WH, Schwartz HL, Evans S (1976), *J. Chromatogr.* **122**, 389.

McGilvery DC, Morrison JD (1978), *Int. J. Mass Spectrom. Ion Phys.* 25, 81.

McGlashan ML (1970), *Pure Appl. Chem.* 21, 1.

McGuffin VL, Novotny M (1983), *J. Chromatogr.* **255**, 381.

McLafferty FW, Turecek F (1993), *Interpretation of Mass Spectra*, 4th Edn, University Science Books, Mill Valley, CA, USA.

McLean J, Russell W, Russel D (2003), *Anal. Chem.* **75**, 648.

McLoughlin D A, Olah TV, Gilbert J D (1997), *J. Pharm. Biomed. Anal.* **15**, 1893.

McLuckey SA, Van Berkel GJ, Goeringer DE, Glish GL (1994), *Anal. Chem.* **66**, 689A.

McLuckey SA, Goeringer DE (1997), *J. Mass Spectrom.* **32**, 461.

McNeff CV, Yan B, Stoll DR, Henry RA (2007), *J. Sep. Sci.* **30**, 1672.

Medzihradszky KF, Campbell JM, Baldwin MA *et al.* (2000), *Anal. Chem.* **72**, 552.

Mehvar R (2004), *Am. J. Pharm. Educ.* **68 (2)**, Article 36 [freely available at http://www.ajpe.org/aj6802/aj680236/aj680236.pdf].

Mei H, Hsieh Y, Nardo C *et al.* (2003), *Rapid. Commun. Mass Spectrom.* **17**, 97.

Mei H (2004), in *Using Mass Spectrometry for Drug Metabolism Studies* (Ed. Korfmacher WA), CRC Press, Boca Raton, FL, USA

Meier PC, Zünd RE (2000), *Statistical Methods in Analytical Chemistry*, 2nd edn, John Wiley & Sons Ltd, Chichester, UK.

Meija J (2006), *Anal. Bioanal. Chem.* **385**, 486.

Melanson JE, Avery SL, Pinto DM (2006), *Proteomics* 6, 4466.

Melanson JE, Chisholm KA, Pinto DM (2006a), *Rapid Commun. Mass Spectrom.* **20**, 904.

Melwanki MB, Huang S-D (2005), *Anal. Chim. Acta* **555**, 139.

Meng CK, Mann M, Fenn JB (1988), *Z. Phys. D* **10**, 361.

Merenbloom SI, Koeniger SL, Valentine SJ *et al.* (2006), *Anal. Chem.* **78**, 2802.

Millard BJ (1977), *Quantitative Mass Spectrometry*, John Wiley & Sons Ltd, Chichester, UK.

Miller JC, Miller JN (1988), *Analyst* **113**, 1351.

Miller JN (1991), *Analyst* **116**, 3.

Miller JN (1993), *Analyst* **118**, 455.

Milley JE, Boyd RK, Curtis JM *et al.* (1997), *Envir. Sci. Technol.* **31**, 535.

Milman BL (2005), *Rapid Commun. Mass Spectrom.* **19**, 2833.

Millington DS, Kodo N, Norwood DL, Roe CR (1990), *J. Inher. Metab. Dis.* **13**, 321.

Millington DS, Kodo N, Terada N *et al.* (1991), *Int.J. Mass Spectrom. Ion Phys.* **111**, 211.

Millington DS, Chace DH, Hillman SL *et al.* (1994), in *Biological Mass Spectrometry: Present and Future* (eds Matsuo T, Caprioli RM, GrossML, Seyama Y), John Wiley & Sons Ltd, Chichester, UK, p. 559.

Min BH, Garland WA, Khoo K-C, Torres GS (1978), *Biomed. Mass Spectrom.* **5**, 692.

Minakuchi H, Nakanishi K, Soga N *et al.* (1996), *Anal. Chem.* 68, 3498.

Minakuchi H, Nakanishi K, Soga N *et al.* (1997), *J. Chromatogr.* **762**, 135.

Minkinnen P (2004), *Chemometrics Intell. Lab. Systems* 74, 85.

Mitch WA, Sedlak DL (2002), *Environ. Sci. Technol.* **36**, 588.

Mitch WA, Sharp JO, Trussell RR *et al.* (2003), *Env. Eng. Sci.* **20**, 389.

Mitch WA, Gerecke AC, Sedlak DL (2003a), *Water Res.* **37**, 3733.

Miyagi M, Rao KCS (2007), *Mass Spectrom. Revs.* **26**, 121.

Moberg M, Bergquist J, Bylund D (2006), *J. Mass Spectrom.* 41, 1334.

Moir RW (1989), *J. Fusion Research* 8, 93.

Moon JH, Yoon SH, Kim MS (2005), *Rapid Commun. Mass Spectrom.* 19, 3248.

Moore WJ (1972), *Physical Chemistry*, 4th Edn, Prentice–Hall, Englewood Cliffs, NJ, USA, p. 481.

Mordehai AV, Henion JD (1993), *Rapid Commun. Mass Spectrom.* 7, 1131.

Morris M, Thibault P, Boyd RK (1994), *J. Amer. Soc. Mass Spectrom.* 5, 1042.

Morris HR, Paxton T, Dell A *et al.* (1996), *Rapid Commun. Mass Spectrom.* 10, 889.

Motulsky H, Christopoulos A (2004), *Fitting Models to Biological Data Using Linear and Nonlinear Regression: A Practical Guide to Curve Fitting*, Oxford University Press, Oxford, UK.

Moyer SC, Cotter RJ (2002), *Anal. Chem.* 74, 469A.

Mucci LA, Dickman PW, Steineck G *et al.* (2003), *Br. J. Cancer* 88, 84.

Mucci LA, Dickman PW, Steineck G *et al.* (2003a), *Br. J. Cancer* 89, 774.

Mucci LA, Sandin S, Bälter K *et al.* (2005), *J. Amer. Med. Assocn.* 293, 1326.

Mucci LA, Adami H-O, Wolk A (2005a), *Int. J. Cancer* 118, 169.

Muddiman DC, Rockwood AL, Gao Q *et al.* (1995), *Anal. Chem.* 67, 4371.

Mueller CA, Weinmann W, Dresen S *et al.* (2005), *Rapid Commun. Mass Spectrom.* 19, 1332.

Muir D, Sverko E (2006), *Anal. Bioanal. Chem.* 386, 789.

Muirhead DC, Smart TS (2000), *Chromatographia* 52 (Suppl.), S72.

Munch JW, Bassett MV (2006), *J. AOAC Intl.* 89, 486.

Munson MSB, Field FH (1966), *J. Am. Chem. Soc.* 88, 2621.

Murk AJ, Legler J, Denison M *et al.* (1996), Fundament. Appl. Toxicol. 33, 149.

Murphy AT, Berna MJ, Holsapple JL, Ackermann BL (2002), *Rapid Commun. Mass Spectrom.* 16, 537.

Murray KK (1997), *Mass Spectrom. Revs.* 16, 283.

N

Nakanishi K, Soga N (1992), *J. Non-Cryst. Solids* 139, 1 and 14.

Nakanishi K, Minakuchi H, Soga N, Tanaka N (1997), *J. Sol-Gel Sci. Technol.* 8, 547.

Nakano T (2001), *J. Chromatogr. A* 906, 205.

Nappi M, Weil C, Cleven CD *et al.* (1997), *Int. J. Mass Spectrom. Ion Proc.* 161, 77.

NAS (US National Academies of Science, 2006), *EPA Assessment of Dioxin Understates Uncertainty About Health Risks and May Overstate Human Cancer Risk* [press release freely available at: http://www8.nationalacademies.org/onpinews/newsitem.aspx?RecordID=11688; public summary freely available at: http://dels.nas.edu/dels/rpt_briefs/dioxin_brief_final.pdf; full report available for sale at: http://books.nap.edu/catalog/11688.html].

Nelms, SM (2005), *Inductively Coupled Plasma Mass Spectrometry Handbook*, Blackwell Publishing Ltd, Oxford, UK (distributed in USA and Canada by CRC Press, Boca Raton, FL, USA).

Nesvizhskii AI, Aebersold R (2005), *Mol. Cell Proteomics* 4, 1419.

Ngo AD, Taylor R, Roberts CL, Nguyen TV (2006), *Int.J. Epidemiology* 35, 1220.

Nier AO (1940), *Rev. Sci. Instr.* 11, 212.

Nier AO (1947), *Rev. Sci. Instr.* 18, 398.

Niessen WMA, Tinke AP (1995), *J. Chromatogr. A*, 703, 37.

Niessen WMA (1998), *J. Chromatogr. A* 794, 407.

Niessen WMA (Ed., 2001), *Current Practice of Gas Chromatography–Mass Spectrometry*, Marcel Dekker, New York, USA.

Niessen WMA, Manini P, Andreoli R (2006), *Mass Spectrom. Revs.* 25, 881.

Nolan GP (2007), *Nature Chem. Biol.* 3, 187.

Novotny M (1981), *Anal. Chem.* 53, 1294A.

Novotny M (1988), *Anal. Chem.* 60, 502A.

NPL (National Physical Laboratory, UK) (2002), *Buoyancy Correction and Air Density Measurement* [available at: http://www.npl.co.uk/mass/guidance/balancetypes.html].

O

O'Connor D, Mortishire-Smith R, Morrison D *et al.* (2006), *Rapid Commun. Mass Spectrom.* 20, 851.

Oda Y, Huang K, Cross FR *et al.* (1999), *Proc. Natl. Acad. Sci. USA* 96, 6591.

OECD (Organisation for Economic Co-operation and Development) (1994), *The Use of Laboratory Accreditation with Reference to GLP Compliance Monitoring: Position of the OECD Panel on Good Laboratory Practice* [available at: http://www.oecd.org/document/63/0,2340,en_2649_34381_2346175_1_1_1_37465,00.html].

OECD (Organisation for Economic Co-operation and Development) (1998), *OECD Principles of Good Laboratory Practice (revised 1997)* [available at: http://www.oecd.org/document/63/0,2340,en_2649_34381_2346175_1_1_1_37465,00.html].

Ogawa M, Oyama T, Isse T *et al.* (2006), *J. Occup. Health* 48, 314.

Oh JY, Moon JH, Kim MS (2004), *J. Am. Soc. Mass Spectrom.* 15, 1248.

Oh, JY, Moon JH, Kim MS (2005), *Rapid Commun. Mass Spectrom.* 18, 2706.

Ohnesorge J, Neusüss C, Wätzig H (2005), *Electrophoresis* 26, 3973.

Olah TV, McLoughlin DA, Gilbert JD (1997), *Rapid Commun. Mass Spectrom.* 11, 17.

Old WM, Meyer-Arendt K, Aveline-Wolf L *et al.* (2005), *Molec. Cell. Proteomics* 4, 1487.

Oleschuk RD, Harrison DJ (2000), *Trends Anal. Chem.* 19, 379.

OME (Ontario Ministry of the Environment, 1984), *Scientific criteria document for standard development. No. 4–84. Polychlorinated Dibenzo-p-dioxins (PCDDs) and polychlorinated dibenzofurans (PCDFs)*.

Ong S-E, Blagoev B, Kratchmarova I *et al.* (2002), *Molec. Cell. Proteomics* **1**, 376.

Oppenheimer L, Capizzi TP, Weppelman RM, Mehta H (1983), *Anal. Chem.* **55**, 638.

O'Reilly J, Wang Q, Setkova L *et al.* (2005), *J. Sep. Science* **28**, 2010.

Ornatsky O, Baranov VI, Bandura DR *et al.* (2006), *J. Immun. Methods* **308**, 68.

Oshima Y (1995), *J. AOAC Intl.* **2**, 528.

Östman P, Luosojärvi L, Haapala M *et al.* (2006), *Anal. Chem.* **78**, 3027.

Ouyang G, Pawliszyn J (2006), *Anal. Bioanal. Chem.* **386**, 1059.

Owens PK, Karlsson L, Lutz ESM, Andersson LI (1999), *Trends Anal. Chem.* **18**, 146.

P

Pan S, Zhang H, Rush J *et al.* (2005), *Molec. Cell Proteomics* **4**, 182.

Panić O, Górecki T (2006), *Anal. Bioanal. Chem.* **386**, 1013.

Papantonakis MR, Kim J, Hess WP, Haglund RF Jr (2002), *J. Mass Spectrom.* **37**, 639.

Paradisi C, Todd JFJ, Traldi P, Vettori U (1992), *Org. Mass Spectrom.* **27**, 251.

Paradisi C, Todd JFJ, Vettori U (1992a), *Org. Mass Spectrom.* **27**, 1210.

Patel IH, Levy RH, Trager WF (1978), *J. Pharmacol. Exp. Therapeutics* **206**, 607.

Paul W, Steinwegen HS (1953), *Z. Naturforsch.* **8a**, 448.

Paul W, Steinwedel H (1953a), U.S. Patent Application, granted in 1960 (Patent No. 2,939,952).

Paul W, Raether M (1955), *Z. Physik.* **140**, 262.

Paul W, Reinhard HP, Van Zahn U (1958), *Z. Physik.* **152**, 143.

Paul W, Steinwedel (1960), US Patent 2,939,952.

Paul W (1990), *Revs. Modern Phys.* **62**, 531.

Paul G, Winnik W, Hughes N *et al.* (2003), *Rapid Commun. Mass Spectrom.* **17**, 561.

Paulsson B, Athanassiadis I, Rydberg P, Törnqvist M (2003), *Rapid Commun. Mass Spectrom.* **17**, 1859.

Pedersen-Bjergaard S, Rasmussen KE (1999), *Anal. Chem.* **71**, 2650.

Peng Y, Plass WR, Cooks RG (2002), *J. Am. Soc. Mass Spectrom.* **13**, 623.

Peng SX, Barbone AG, Ritchie DM (2003), *Rapid Commun. Mass Spectrom.* **17**, 509.

Peng J, Elias JE, Thoreen CC, Licklider LJ, Gygi SP (2003a), *J. Proteome Res.* **2**, 43.

Petersson EV, Rosén J, Turner C *et al.* (2006), *Anal. Chim. Acta* **557**, 287.

Phillips GR, Eyring EM (1983), *Anal. Chem.* **55**, 1134.

Picart D, Jacolot F, Berthou F, Floch HH (1980), *Biol Mass Spectrom.* **7**, 464.

Pickup JF, McPherson K (1976), *Anal. Chem.* **48**, 1885.

Pirard C, Focant J-F, DePauw E (2002), *Anal. Bioanal. Chem.* **372**, 373.

Pirkle WH, How DW, Finn JM (1980) *J. Chromatogr.* **192**, 143.

Pirkle WH, Hyun MH, Bank,B (1984), *J. Chromatogr.* **316**, 586.

Pirkle WH, Pochapky TC, Burke JA, Deming KC (1988), in *Chiral Separations* (Eds Stevenson D, Wilson ID), Plenum Press, New York, USA, p. 29.

Pirkle WH, Welch CJ, Lamm B (1992), *J. Org. Chem.* **57**, 3854.

Pitard F (1993), *Pierre Gy's Sampling Theory and Sampling Practice*, CRC Press, Boca Raton, FL, USA.

Plackett RL, Burman JP (1946), *Biometrika* **33**, 305.

Pleasance S, Ayer SW, Laycock MV, Thibault P (1992), *Rapid Commun. Mass Spectrom.* **6**, 14.

Pleasance S, Thibault P, Kelly J (1992a), *J. Chromatogr.* **591**, 325.

Plomley JB, Mila Lauševic M, March RE (2000), *Mass Spectrom. Revs.* **19**, 305.

Plumb R, Castro-Perez J, Granger J *et al.* (2004), *Rapid Commun. Mass Spectrom.* **18**, 2331.

Pohl C (2004), *LC/GC, LC Column Technology Supplement* (June), 22.

Pongpun N, Mlynski V, Crisp PT, Guilhaus M (2000), *J. Mass Spectrom.* **35**, 1105.

Poole CF (2003), *Trends Anal. Chem.* **22**, 362.

Poon GK, Kwei G, Wang R *et al.* (1999), *Rapid Commun. Mass Spectrom.* **1**, 1943.

Poster DL, Schantz MM, Sander LC, Wise SA (2006), *Anal. Bioanal. Chem.* **386**, 859.

Pramann A, Rademann K (2001), *Rev. Sci. Instrum.* **72**, 3475.

Preisler J, Foret F, Karger BL (1998), *Anal. Chem.* **70**, 5278.

Preisler J, Hu P, Rejtar T, Karger BL (2000), *Anal. Chem.* **72**, 4785.

Pretorius V, Smuts T (1966), *Anal. Chem.* **38**, 274.

Pretorius V, Hopkins BJ, Schieke JD (1974), *J. Chromatogr.* **99**, 23.

Pringle SD, Giles K, Wildgoose JL *et al.* (2005), *Int. J. Mass Spectrom.* **261**, 1.

Prokai L (1990), *Field Desorption Mass Spectrometry*, Marcel Dekker, New York, USA.

Przybyciel M (2004), *LC/GC, LC Column Technical Supplement* (June), 26.

Psillakis E, Kalogerakis N (2003), *Trends Anal. Chem.* **22**, 565.

Ptolemy A, Britz-McKibbin P (2006), *J. Chromatogr.* **1106**, 7.

Pursch M, Sander LC (2000), *J. Chromatogr. A* **887**, 313.

Purves RW, Guevremont R, Day S *et al.* (1998), *Rev. Sci. Instrum.* **69**, 4094.

Q

Qin J, Chait BT (1996), *Anal. Chem.* **68**, 2102.

Queiroz MEC, Lanças FM, Majors RE (2004), *LC/GC* **22**, 970.

Questier F, Vander Heyden Y, Massart DL (1998), *J. Pharm. Biomed. Anal.* **18**, 287.

Quevauviller P (2005), *Anal. Bioanal. Chem.* **382**, 1800.

Quilliam MA, Thomson BA, Scott GJ, Siu KWM (1989), *Rapid Commun. Mass Spectrom.* **3**, 145.

Quilliam MA, Janaček M, Lawrence JF (1993), *Rapid Commun. Mass Spectrom.* **7**, 482.

Quilliam MA (1998) in *Harmful Algae* (Eds Reguera B, Blanco J, Fernandez ML, T. Wyatt T), Xunta de Galicia and IOC/UNESCO, Vigo, Spain, p. 509.

Quilliam MA (2003), *J. Chromatogr. A* **1000**, 527.

Quinn HM, Takarewski JJ (1997), International Patent No. WO 95 22555 and US Patent No. 5795469.

R

Raffaelli A, Saba A (2003), *Mass Spectrom. Revs.* **22**, 318.

Rainville S, Thompson JK, Pritchard DE (2004), *Science* **303**, 334.

Rainin Instrument LLC (2005), Procedure for Evaluating Accuracy and Precision of Rainin Pipettes [available at: www.rainin.com/pdf/ab15.pdf].

Ramanathan R, Su A-D, Alvares N *et al.* (2000), *Anal. Chem.* **72**, 1352.

Ramanathan R, Zhong R, Blumenkrantz N *et al.* (2007), *J. Am. Soc. Mass Spectrom.* **18**, 1891.

Ramsey RS, Goeringer DE, McLuckey SA (1993), *Anal. Chem.* **65**, 3521.

Ramsey MH, Lyn J, Wood R (2001), *Analyst* **126**, 1777.

Ramsey MH (2004), *Accred. Qual. Assur.* **9**, 727.

Ramström O (1997), *Molecular Imprinting Technology* [available at: http://www.smi.tu-berlin.de/story/MIT.htm].

Rashed MS, Ozand PT, Bucknall MP, Little D (1995), *Pediatr. Res.* **38**, 342.

Ratcliffe LV, Rutten FJM, Barrett DA *et al.* (2007), *Anal. Chem.* **79**, 6094.

Ratzlaff KL (1979), *Anal. Chem.* 51, 232.

Rayleigh Lord (1882), *Phil. Mag.* 14, 184.

Reedy V (2006), *Optimizing Split/Splitless Flows in Capillary Gas Chromatography* [available at: http://www.uhl.uiowa.edu/newsroom/hotline/1997/1997_07/labcert.xml].

Reichmuth, A (2001a), *Adverse Influences and their Prevention in Weighing*, Article #1 in the series *Weighing Uncertainty* [available on the website of Mettler-Toledo Inc: http://us.mt.com/mt/library/Weigh_Uncertain_Number1_0x0003d6750003db6700091746.jsp].

Reichmuth, A (2001b), *Weighing Accuracy: Estimating Measurement Bias and Uncertainty of a Weighing"*, Article #3 in the series *Weighing Uncertainty* [available on the website of Mettler-Toledo Inc: http://us.mt.com/mt/library/Weigh_Uncertain_Number3_0x0003d6750003db6700091740.jsp].

Reiner EJ, Schellenberg DH, Taguchi VY (1991), *Env. Sci. Technol.* **25**, 110.

Reiner EJ, Clement RE, Okey AB, Marvin CH (2006), *Anal. Bioanal. Chem.* **386**, 791.

Reiner EJ (2007), private communication.

Renman L, Jagner D (1997), *Anal. Chim. Acta* **357**, 157.

Rentsch KM (2002), *J. Biochem. Biophys. Methods* **54**, 1.

Revel'skii IA, Yashin YS, Kurochkin VK, Kostyanovskii RG (1991), *Chem. Physicochem. Methods Anal.*, 243–248 (translated from *Zavodskaya Laboratoriya* **57** (3), 1–4).

Reyes DR, Iossifidis D, Auroux P-A, Manz A (2002), *Anal. Chem.* **74**, 2623.

Reynolds J, Yao X, Fenselau C (2002), *J. Proteome Res.* **1**, 27.

Rezaee M, Assadi Y, Milani Hosseini M-R *et al.* (2006), *J. Chromatogr. A* **1116**, 1.

Rheodyne LLC (2001), *Achieving Accuracy and Precision with Rheodyne Manual Sample Injectors* [available as Technical Note #5 at: www.rheodyne.com/support/product/technotes].

Rice JM (2005), *Mutation Res./Genetic Toxicol. and Env. Mutagenesis* **580**, 3.

Richardson SD, Simmons JE, Rice G (2002), *Environ. Sci. Technol.* **36**, A198.

Richardson SD (2003), *Trends Anal. Chem.* **22**, 666.

Richardson SD (2004), *Anal. Chem.* **76**, 3337.

Richardson SD, Ternes TA (2005), *Anal. Chem.* **77**, 3807.

Richardson SD (2006), *Anal. Chem.* **78**, 4021.

Righetti PG, Pagani M, Gianazza E (1975), *J. Chromatogr.* **109**, 341.

Righetti PG, Castagna A, Herbert B, Reymond F, Rossier JS (2003), *Proteomics* **3**, 1397.

Rimmer CA, Piraino SM, Dorsey JG (2000), *J. Chromatogr. A* **887**, 115.

Rinn K, Müller A, Eichenauer H, Salzborn E (1982), *Rev. Sci. Instrum.* **53**, 829.

Rizzo V, Pinciroli V (2005), *J. Pharm. Biomed. Anal.* **38**, 851.

Robb DB, Covey TR, Bruins AP (2000), *Anal. Chem.* **72**, 3653.

Rockwood AL (1995), *Rapid Commun. Mass Spectrom.* **9**, 103.

Rockwood AL, Van Orden SL, Smith RD (1995a), *Anal. Chem.* **67**, 2699.

Rockwood AL (1997), *Rapid Commun. Mass Spectrom.* **11**, 241.

Rockwood AL, Kushnir MM, Nelson GJ (2003), *J. Am. Soc. Mass Spectrom.* **14**, 311.

Rodríguez-González P, Marchante-Gayón JM, García-Alonso JI, Sanz-Medel A (2005), *Spectrochim. Acta B* **60**, 151.

Rolčík J, Lenobel R, Siglerová V, Strnad M (2002), *J. Chromatogr. B* **775**, 9.

Romanyshyn L, Tiller PR, Hop CECA (2000), *Rapid Commun. Mass Spectrom.* **14**, 1662.

Rosenstock HM, Wallenstein MB, Wahraftig AL, Eyring H (1952), *Proc. Nat. Acad. Sci.* **38**, 667.

Rosenthal D (1982), *Anal. Chem.* **54**, 63.

Rosman KJR, Taylor PDP (1998), *Pure Appl. Chem.* **70**, 217.

Ross PL, Huang YN, Marchese JN *et al.* (2004), *Molec. Cell Proteomics* **3**, 1154.

Roussis SG (1999), *Rapid Commun. Mass Spectrom.* **13**, 1031–1051.

Rubey WA (1991), *J. High Res. Chromatogr.* **14**, 452.

Rubey WA (1992), *J. High Res. Chromatogr.* **15**, 795.

Rubin Y (2000), *Flash Chromatography* [available at: http://siggy.chem.ucla.edu/VOH/136/Flash_Chromatography.pdf].

Ryhage R (1964), *Anal. Chem.* **36**, 759.

S

Said AS (1956), *J. Am. Soc. Chem. Eng.* **2**, 477.

Safe S (1990), *Crit. Rev. Toxicol.* **21**, 51.

Sandahl M, Mathiasson L, Jönsson JÅ (2000), *J. Chrom. A* **893**, 123.

Sangster T, Spence M, Sinclair P *et al.* (2004), *Rapid Commun. Mass Spectrom.* **18**, 1361.

Santesson S, Barinaga-Rementaria Ramirez I, Viberg P *et al.* (2004), *Anal. Chem.* **76**, 3030.

Santesson S, Nilsson S (2004a), *Anal. Bioanal. Chem.* **378**, 1704.

Santos FJ, Parera J, Galceran MT (2006), *Anal. Bioanal. Chem.* **386**, 837.

Savitzky A, Golay MJE (1964), *Anal. Chem.* **36**, 1627.

Schägger H, von Jagow G (1987), *Anal. Biochem.* **166**, 368.

Schantz MM (2006), *Anal. Bioanal. Chem.* **386**, 1043.

Schlüsener MP, Bester K (2005), *Rapid. Commun. Mass Spectrom.* **19**, 3269.

Schmidt A, Karas M, Dülcks T (2003), *J. Am. Soc. Mass Spectrom.* **14**, 492.

Schneider BB, Douglas DJ, Chen DDY (2002), *J. Am. Soc. Mass Spectrom.* **13**, 906.

Schneider LV, Hall MP (2005), *Drug Discovery Today* **10**, 353.

Schneider BB, Javaheri H, Covey TR (2006), *Rapid Commun. Mass Spectrom.* **20**, 1538.

Schoeller DA (1976), *Biomed. Mass Spectrom.* **3**, 265.

Schoen AE, Dunyach J-J, Schweingruber H *et al.* (2001), *Proceedings of the 49th ASMS Conference on Mass Spectrometry and Allied Topics*, Chicago, IL, USA, Abstract A011079.

Schoonover RM, Jones FE (1981), *Anal. Chem.* **53**, 900.

Schrader W, Klein H-W (2004), *Anal. Bioanal. Chem. Anal. Bioanal. Chem.* **379**, 1013.

Schreiber IM, Mitch WA (2006), *Env. Sci. Technol.* **40**, 6007.

Schumacher BA, Shines KC, Burton JV, Papp ML (1990), *A Comparison of Soil Sample Homogenization Techniques*, EPA600//X-90/043 (US Environmental Protection Agency) [available at http://www.epa.gov/nerlesd1/cmb/research/papers/bs120.pdf].

Schurig V (2001), *J. Chromatogr. A* **906**, 275.

Schwab BW, Hayes EP, Fiori JM *et al.* (2005), *Reg. Toxicol. Pharmacol.* **42**, 296.

Schwartz LM (1977), *Anal. Chem.* **49**, 2062.

Schwartz LM (1979), *Anal. Chem.* **51**, 723.

Schwartz LM (1983), *Anal. Chem.* **55**, 1424.

Schwartz LM (1986), *Anal. Chem.* **58**, 246.

Schwartz LM (1989), *Anal. Chem.* **61**, 1080.

Schwartz JC, Wade AP, Enke CG, Cooks RG (1990), *Anal. Chem.* **62**, 1809.

Schwartz JC, Senko MW, Syka JEP (2002), *J. Am. Soc. Mass Spectrom.* **13**, 659.

Schwartz JC, Senko MW, Syka JEP (2002a), *Proceedings of the 50th ASMS Conference on Mass Spectrometry and Allied Topics*, Orlando, Florida, USA, paper AO21187.

Scippo M-L, Rybertt S, Eppe G *et al.* (2006), *Accred. Qual. Assur.* **11**, 38.

Scorer T, Perkin M, Buckley M (2004), *Weighing in the Pharmaceutical Industry*, ISSN 1368-6550 [available at: http://www.npl.co.uk/mass/guidance/balancetypes.html].

Scott RPW, Scott GC, Munroe M, Hess J Jr (1974), *J. Chromatogr.* **99**, 395.

Scott RPW, Kucera P (1979), *J. Chromatogr.* **169**, 51.

Scott RPW, Kucera P, Munroe M (1979a), *J. Chromatogr.* **186**, 475.

Scott RPW (http://www.chromatography-online.org/), Chrom-Ed Books [free online textbooks].

Scott RJ, Palmer J, Lewis IAS, Pleasance S (1999), *Rapid Commun. Mass Spectrom.* **13**, 2305.

Self R (1979), *Biomed. Mass Spectrom.* **6**, 315.

Sellegren B (2001), *J. Chromatogr. A* **906**, 227.

Sen NP, Tessier L, Seaman SW (1983), *J. Agric. Food Chem.* **31**, 1033.

Senko MW, Schwartz JC (2002), *Proceedings of the 50th ASMS Conference on Mass Spectrometry and Allied Topics*, Orlando, FL, USA, Paper AO21179.

Sennbro CJ, Olin M, Edman K, Hansson G, Gunnarsson PO, Svensson LD (2006), *Rapid Commun. Mass Spectrom.* **20**, 3313.

Seto C, Bateman KP, Gunter B (2002), *J. Amer. Soc. Mass Spectrom.* **13**, 2.

Shaffer SA, Tang KQ, Anderson GA, Prior DC, Udseth HR, Smith RD (1997), *Rapid Commun. Mass Spectrom.* **11**, 1813.

Shah VP, Midha KK, Dighe SV, McGilveray IJ, Skelly JP, Yacobi A, Layloff T, Viswinathan CT, Cook CE, McDowall RD, Pittman KA, Spector S (1992), *Pharm Res.* **9**, 588.

Shah VP, Midha KK, Findlay JWA, Hill HM, Hulse JD, McGilveray IJ, McKay G, Miller KJ, Patnaik RN, Powell ML, Tonelli A, Viswanathan CT, Yacobi A (2000), *Pharm. Res.* **17**, 1551.

Shah S, Richter RC, Kingston HM (2002), *LC/GC* **20**, 280.

Shakespeare W (1601), *The Tragedy of Hamlet, Prince of Denmark*, Act IV, Scene 4, line 47.

Shapiro M, Luo R, Zang X, Song N, Chen T-K, Bozigian H (2005), *Proceedings of the 53rd ASMS Conference on Mass Spectrometry and Allied Topics*, San Antonio, TX, Abstract TP046.

Shellie RA, Haddad PR (2006), *Anal. Bioanal. Chem.*, **386**, 405.

Shevchenko A, Chernushevich I, Ens W *et al.* (1997), *Rapid Commun. Mass Spectrom.* **11**, 1015.

Shewchuck S (1953), *Summary of the Research Progress Meetings of February 12, 19 and 26, 1953, Contract No. W-7405-eng-48, Document UCRL-2209,* University of California Radiation Laboratory, Berkeley, CA, pp. 5–7.

Shou WZ, Chen Y-L, Eerkes A *et al.* (2002), *Rapid Commun. Mass Spectrom.* **16**, 1613.

Shou WZ, Weng N (2003), *Rapid Commun. Mass Spectrom.* (2003), **17**, 589.

Shushan B, Douglas DJ, Davidson WR, Nacson S (1983), *Int. J. Mass Spectrom. Ion Phys.* **46**, 71.

Siegel MW, Fite WL (1976), *J. Phys. Chem.* **80**, 2871.

Silva JC, Denny R, Dorschel C *et al.* (2006), *Molec. Cell Proteomics* **5**, 589.

Skelton R, Dubois F, Zenobi R (2000), *Anal. Chem.* **72**, 1707.

Sleno L, Volmer DA (2004), *J. Mass Spectrom.* **39**, 1091.

Sleno L, Volmer DA (2005), *Rapid Commun. Mass Spectrom.* **19**, 1928.

Sleno L, Volmer DA (2005a), *Anal. Chem.* **77**, 1509.

Sleno L, Volmer DA, Marshall AG (2005b), *J. Am. Soc. Mass Spectrom.* **16**, 183.

Sleno L, Volmer DA (2006), *Rapid Commun. Mass Spectrom.* **20**, 1517.

Smalley J, Kadiyala P, Xin B *et al.* (2006), *J. Chromatogr. B* **830**, 270.

Smeds A, Vartiainen T, Mäki-Paakkanen J, Kronberg L (1997), *Envir. Sci. Technol.* **31**, 1033.

Smith A (1763), in *Lectures on Justice, Police, Revenue and Arms; Delivered in the University of Glasgow and reported by a Student in 1763* (Ed. Cannan E), Oxford University Press, Oxford, UK, 1896, p. 183.

Smith RD, Loo JA, Orgazalek RR *et al.* (1991), *Mass Spectrom Revs.* **10**, 359.

Smith RD, Loo JA, Orgazalek RR *et al.* (1992), *Mass Spectrom Revs.* **11**, 434.

Smith N (2001), *Capillary Electrochromatography* [available at http://www.beckmancoulter.com/ and at http://www.ceandcec.com/AP-8508A%20-%20CEC%20Primer.pdf].

Snyder LR (1964), *J. Chromatogr.* **13**, 415.

Snyder LR (1974), *J. Chromatogr.* **92**, 223.

Snyder LR, Dolan JW, Gant JR (1979), *J. Chromatogr.* **165**, 3.

Snyder LR, Kirkland JJ (1979a), in *Introduction to Modern Liquid Chromatography*, 2nd Edn, John Wiley & Sons Ltd, Chichester, UK.

Snyder LR (1980), in *High-Performance Liquid Chromatography, Vol. 1* (Ed. Horváth CG), Academic Press, New York, USA, p. 207.

Snyder LR, Stadallus MA, Quarry MA (1983), *Anal. Chem.* **55**, 1412A.

Sojo LE, Lum G, Chee P (2003), *Analyst* **128**, 51.

Soxhlet F (1879), *Polytechnisches J. (Dingler's)*, **232**, 461 [available at: http://www.cyberlipid.org/extract/soxhlet.PDF].

Spégel P. Schweitz L, Nilsson S (2002), *Anal. Bioanal. Chem.* **372**, 37.

Spengler B, Kirsch D, Kaufmann R *et al.* (1990), *Rapid Commun. Mass Spectrom.* **4**, 301.

Sphon JA (1978), *J. Assoc. Official Anal. Chemists* **61**, 1247.

SQA (Society for Quality Assurance, 2000), *Validation of Computerized Laboratory Instruments and Equipment*, SQA Newsletter, Fall 2000.

Stafford GC Jr. (1980), *Environ. Health Perpectives* **36**, 85.

Stafford GC Jr, Kelley PE, Syka JEP *et al.* (1984), *Int. J. Mass Spectrom. Ion Proc.* **60**, 85.

Stafford GC Jr (2002), *J. Am. Soc. Mass Spectrom.* **13**, 589.

Stanton HE, Chupka WA, Inghram MG (1956), *Rev. Sci. Instrum.* **27**, 109.

Stapleton HM (2006), *Anal. Bioanal. Chem.* **386**, 807.

Starr TB (2003), *Environ. Health Perspect.* **111**, 1443 [freely available at: http://www.ehponline.org/members/2003/6219/6219.pdf].

Stein S (2004), *How we handle mass spectra* [available at: http://math.nist.gov/mcsd/Seminars/2004/2004-10-26-stein-presentation.pdf].

Stein SE, Heller DN (2006), *J. Am. Soc. Mass Spectrom.* **17**, 823.

Steiner F, Scherer B (2000), *J. Chromatogr. A* **887**, 55.

Stellaard F, Ten Brink JJ, Kok RM *et al.* (1990), *Clin. Chim. Acta* **192**, 133.

Stephany RW, Freudenthal J, Egmond E *et al.* (1976), *J. Agric. Food Chem.*, **24**, 536.

Stephany RW (2000), *Proc. EuroResidue IV Conf. on Residues of Veterinary Drugs in Food*, p. 137 [freely available at: http://www.euroresidue.nl/ER_IV/Key%20lectures/idx.htm].

Stephenson NR, Edwards HI, MacDonald BF, Pugsley LI (1955), *Can. J. Biochem. Pharmacol.* **33**, 849.

Stevenson D (1999), *Trends Anal. Chem.* **18**, 154.

Stewart II, Thomson T, Figeys D (2001), *Rapid Commun. Mass Spectrom.* **15**, 2456.

Still WC, Kahn M, Mitra A (1978), **43**, 2923.

Stöckl D, Budzikiewicz H (1982), *Org. Mass Spectrom.* **17**, 470.

Stolker AAM, Niesing W, Hogendoorn EA *et al.* (2004), *Anal. Bioanal. Chem.* **378**, 955.

Stolker AAM, Niesing W, Fuchs R *et al.* (2004a), *Anal. Bioanal. Chem.* **378**, 1754.

Strife RJ, Kelley PE, Weber-Grabau M (1988), *Rapid Commun. Mass Spectrom.* **2**, 105.

Strong FCS (1979), *Anal. Chem.* **51**, 298.

Strong FCS (1980), *Anal. Chem.* **52**, 1152.

Stüber M, Reemtsma T (2004), *Anal. Bioanal. Chem.* **378**, 910.

Sullivan JJ, Wekell MM (1987), in *Seafood Quality Determination* (Eds Kramer DE, Liston J), Elsevier/North-Holland, New York, p. 357.

Suslick RS (1994), *The Yearbook of Science & the Future 1994*; Encyclopaedia Britannica, Chicago, pp. 138–155 [available at: http://www.scs.uiuc.edu/suslick/britannica.html].

Suzuki N, Nakanishi J (1995), *Chemosphere* **30**, 1557.

Swedish National Food Administration (2002) [freely available at: http://www.konsumentverket.se/html-sidor/livsmedelsverket/engakrylpressmeddelande.htm].

Syage JA, Evans MD (2001), *Spectroscopy* **16**, 14.

Syka JEP (1995), in *Practical Aspects of Ion Trap Mass Spectrometry, Volume 1: Fundamentals of Ion Trap Mass Spectrometry* (Eds March RE, Todd J FJ), CRC Press, Boca Raton, FL, USA p. 169.

Szejtli J (1988), *Cyclodextrin Technology, Vol. 1*, Springer, New York.

T

Tachibana K, Ohnishi A (2001), *J. Chromatogr. A* **906**, 127.

Taeymans D, Wood J, Ashby P *et al.* (2004), *Crit. Revs. Food Sci. Nutrition* **44**, 323.

Taguchi V (1994), *The Determination of N-Nitrosodimethylamine (NDMA) in Water by Gas Chromatography-High Resolution Mass Spectrometry (GC–HRMS)*, Environment Ontario (Canada), Laboratory Services Branch, Quality Management Office, Ontario; Method catalogue code NDMA-E3291A (originally approved February 19, 1993; revised and approved April 7, 1994).

Taguchi VY, Jenkins SWD, Wang DT *et al.* (1994a), *Can. J. Appl. Spectrosc.* **39**, 87.

Taguchi VY (2007), private communication.

Takeuchi T, Hirata Y, Okumura Y (1978), *Anal. Chem.* **50**, 659.

Takahashi S (1987), *Biomed. Mass Spectrom.* **14**, 257.

Takats Z, Wiseman JM, Gologan B, Cooks RG (2004), *Science* **306**, 471.

Takats Z, Cotte-Rodriguez I, Talaty N *et al.* (2005), *Chem. Commun.* (2005), 1950.

Tal'roze VL, Lubinova AK (1952), *Rept. Soviet Acad. Sci.* **LXXXVI**, N5. Reprinted as *J. Mass Spectrom.* (1998) **33**, 502.

Tan A, Benetton S, Henion JD (2003), *Anal. Chem.* **75**, 5504.

Tanaka K, Ido Y, Akita S, (1987), *Proceedings of the 2nd. Japan–China Joint Symposium on Mass Spectrometry*, Osaka, Japan, p. 185.

Tanaka K, Waki H, Ido Y, Akita S, Yoshida Y, Yoshida T (1988), *Rapid Commun. Mass Spectrom.* **2**, 151.

Tang L, Kebarle P (1993), *Anal. Chem.* **65**, 3654.

Tanner SD, Baranov VI (1999), *J. Am. Soc. Mass Spectrom.* **10**, 1083.

Tareke E, Rydberg P, Karlsson P, Eriksson S, Törnqvist M (2000), *Chem. Res. Toxicol.* **13**, 517.

Tareke E, Rydberg P, Karlsson P, Eriksson S, Törnqvist M (2002), *J. Agric. Food Chem.* **50**, 4998.

Taylor GI (1964), *Proc. Roy. Soc. Lond. Ser. A* **280**, 383.

Taylor JR (1982), *An Introduction to Error Analysis*, University Science Books, Mill Valley, CA, USA.

Taylor JK (1987), *Quality Assurance of Chemical Measurements*, Lewis Publishers Inc.

Taylor, BN (1995), *Guide for the Use of the International System of Units*, NIST Special Publication 811 [available at: http://physics.nist.gov/cuu/Units/bibliography.html].

Taylor BN (2001), *The International System of Units*, NIST Special Publication 330 [available at: http://physics.nist.gov/cuu/Units/bibliography.html].

Taylor S, Tindall RF, Syms RRA (2001a), *J. Vac. Sci. Technol. B* **19**, 557.

Taylor GK, Goodlett DR (2005), *Rapid Commun. Mass Spectrom.* **19**, 3420.

Terry SC, Jerman JH, Angell JB (1979), *IEEE Trans. Electron. Devices* **ED-26**, 1880.

Testa B, Carrupt PA, Gal J (1993), *Chirality* **5**, 105.

Thompson M (1989), *Anal. Chem.* **61**, 1942.

Thompson M, Ramsey M (1995), *Analyst* **120**, 261.

Thompson M, Ellison SLR, Wood R (2002), *J. Pure Appl. Chem.* **74**, 835.

Thompson M, Guffogg S, Stangroom S *et al.* (2002a), *Analyst* **127**, 1669.

Thomson W (1871), *Phil. Mag.* **42**, 448.

Thomson JJ (1913), *Rays of Positive Electricity and Their Application to Chemical Analysis*, 2nd Edn (1921), Longmans, London, UK.

Thomson BA, Iribarne JV (1975), *Proceedings of the 5th International Conference on Atmospheric Electricity* (Eds Reiter R, Dolezalek), Steinkopff, Darmstadt, Germany.

Thomson BA, Iribarne JV (1979), *J. Chem. Phys.* **71**, 4451.

Thomson BA, Iribarne JV, Dziedzic PJ (1982), *Anal. Chem.* **54**, 2219.

Thomson BA, Danelewych-May L (1983), *Proceedings of the 31st ASMS Conference on Mass Spectrometry and Allied Topics*, Boston MA, USA, 852.

Thomson BA, Chernushevich IV (1998), *Rapid Commun. Mass Spectrom.* **12**, 1323.

Tiller PR, Romanyshyn LA (2002), *Rapid Commun. Mass Spectrom.* **16**, 1225.

Tiller PR, Romanyshyn LA (2002a), *Rapid Commun. Mass Spectrom.* **16**, 92.

Tiller PR, Romanyshyn LA, Neue UD (2003), *Anal. Bioanal. Chem.* **377**, 788.

Todd JFJ, Waldren RM, Mather RE (1980), *Int. J. Mass Spectrom. Ion Phys.* **34**, 325.

Todd JFJ (1991), *Mass Spectrom. Revs.* **10**, 3.

Tomatis L, Melnick RL, Haseman J *et al.* (2001), *FASEB J.* **15**, 195.

Tomkins BA, Griest WH, Higgins CE (1995), *Anal. Chem.* **67**, 4387.

Tomkins BA, Griest WH (1996), *Anal. Chem.* **68**, 2533.

Tong W, Chowdhury SK, Chen J-C *et al.* (2001), *Rapid Commun. Mass Spectrom.* **15**, 2085.

Torto N, Mwatseteza J, Laurell T (2001), *LC/GC* **19**, 462.

Touchstone JC (1993), *J. Liq. Chromatogr.* **16**, 1647.

Trager WF, Levy RH, Patel IH, Neal JN (1978), *Anal. Lett.* **11**, 119.

Traldi P, Curcuruto O, Bortolini O (1992), *Rapid Commun. Mass Spectrom.* **6**, 410.

Traldi P, Favretto D, Catinella S, Bortolini O (1993), *Org. Mass Spectrom.* **28**, 745.

Tran CD, Grishko VI (1994), *Anal. Biochem.* **218**, 197.

Tran CD, Grishko VI, Baptista MS (1994a), *Appl. Spectrosc.* **48**, 833.

Tubaro M, Marotta E, Seraglia R, Traldi P (2003), *Rapid Commun. Mass Spectrom.* **17**, 2423.

Tucker DB, Hameister CH, Bradshaw SC *et al.* (1988), *Proceedings of the 36th ASMS Conference on Mass Spectrometry and Allied Topics*, San Francisco, CA, USA, pp. 628–629.

Tuomisto J, Vartiainen T, Tuomisto JT (1999), *Synopsis on Dioxins and PCBs*, KTL (National Public Health Institute, Finland), Division of Environmental Health, Kuopio, Finland [freely available at: http://www.ktl.fi/dioxin/index.html].

Turiel E, Martin-Esteban A (2004), *Anal. Bioanal. Chem.* **378**, 1976.

Turner WE, Isaacs SG, Patterson DG Jr (1992), *Chemosphere* **25**, 805.

Tyler AN, Clayton E, Green BN. (1996), *Anal. Chem.* **68**, 3561.

U

Ulberth F (2006), *Anal. Bioanal. Chem.* **386**, 1121.

Unwin RD, Evans CA, Whetton AD (2006), *Trends Biochem. Sci.* **31**, 473.

USGS (US Geological Survey) (1999), *New Reporting Procedures Based on Long-Term Method Detection Levels and Some Considerations for Interpretations of Water-Quality Data Provided by the U.S. Geological Survey National Water Quality Laboratory*, Open-File Report 99-193 [(http://water.usgs.gov/owq/OFR_99-193/)].

US Pharmacopeia XXII, The National Formulary XVII (1990), *United States Pharmacopeial Convention* (Rockville, MD), p. 1712.

V

Vakali M, Jamali F (1996), *J. Pharm. Sci.* **85**, 638.

Valaskovic GA, Utley L, Lee MS, Wu J-T (2006), *Rapid Commun. Mass Spectrom.* **20**, 1087.

Vallano PT, Shugarts SB, Woolf EJ, Matuszewski BK (2005), *J. Pharm. Biomed. Anal.* **36**, 1073.

Van Berkel GJ (2000), *J. Mass Spectrom.* **35**, 773.

Van Berkel GJ (2000a), *J. Am. Soc. Mass Spectrom.* **11**, 951.

van Deemter JJ (1952), *Phys. Rev.* **85**, 1049.

van Deemter JJ (1953), *Ind. Eng. Chem.* **45**, 1227.

van Deemter JJ (1954), *Ind. Eng. Chem.* **46**, 2300.

van Deemter JJ, Zuiderweg FJ, Klinkenberg A (1956), *Chem. Eng. Sci.* **5**, 271.

van Deemter JJ (1967), *Proc. Int. Symposium on Fluidization*, Netherlands University Press, Amsterdam, The Netherlands.

van Deemter JJ (1981),*Citation Classics* **3**, 245 [available at http://www.garfield.library.upenn.edu/classics1981/A1981KX 02700001.pdf].

van de Merbel NC (1999), *J. Chromatogr. A* **856**, 55.

Van den Berg M, Birnbaum L, Bosveld AT *et al.* (1998), *Environ. Health Perspect.* **106**, 775.

Van den Berg M, Birnbaum LS, Denison M *et al.* (2006), *Toxicol. Sci.* **93**, 223.

Van der Greef J, Niessen WMA (1992), *Int. J. Mass Spectrom. Ion Proc.* **118/119**, 857.

Vander Heyden Y, Luypaert K, Hartmann C *et al.* (1995), *Anal. Chim. Acta* **312**, 245.

Vander Heyden Y, Questier F, Massart DL (1998), *J. Pharm. Biomed. Anal.* **17**, 153.

Vander Heyden Y, Nijhuis A, Smeyers-Verbeke J *et al.* (2001), *J. Pharm. Biomed. Anal.* **24**, 723.

van der Voet H, de Boer WJ, de Ruig WG, Van Rhijn JA (1998), *J. Chemometr.* **12**, 279.

van der Voet H, van Rhijn JA, van de Wiel HJ (1999), *Anal. Chim. Acta* **391**, 159.

Van Eeckhaut A, Michotte Y (2006), *Electrophoresis* **27**, 2880.

Van Ginkel LA, Stephany RW (1993), *Proceedings of the EuroResidue II Conference on Residues of Veterinary Drugs in Food* (Ed. Haagsma N), Veldhoven, The Netherlands, p. 303.

Van Hout MWJ, Niederländer HAG, de Zeeuw RA, de Jong GJ (2003), *Rapid Commun. Mass Spectrom.* **17**, 245.

Van Hove JLK, Kahler SG, Feezor MD *et al.* (2000), *J. Inherit. Metab. Dis.* **23**, 571.

van Nederkassel AM, Aerts A, Dierick A *et al.* (2003), *J. Pharm. Biomed. Anal.* **32**, 233.

Van Pelt C, Zhang S, Fung E, Chu I, Liu T, Li C, Korfmacher WA, Henion J (2003), *Rapid Commun. Mass Spectrom.* **17**, 1573.

Van Vaeck L, Adriaens A (1999), *Mass Spectrom. Revs.* **18**, 48.

Van Wouwe N, Windal I, Vanderperren H *et al.* (2004), *Talanta* **63**, 1157.

Van Wouwe W, Windal I, Vanderperren H *et al.* (2004a), *Talanta* **63**, 1269.

Vas G, Vékey K (2004), *J. Mass Spectrom.* **39**, 233.

Veenstra TD, Martinovic S, Anderson GA *et al.* (2000), *J. Am. Soc. Mass Spectrom.* **11**, 78.

Venn R, Merson J, Cole S, Macrae P (2005), *J. Chromatogr.* B **817**, 77.

Venter A, Cooks RG (2007), *Anal. Chem.* **79**, 6398.

Vernez L, Hopfgartner G, Wenka M, Krähenbühl S (2003), *J. Chromatogr. A* **984**, 203.

Vermeulen AM, Belpaire FM, Moerman E *et al.* (1992), *Chirality* **4**, 73.

Vessman, J, Stefan RI, Van Staden JF *et al.* (2001), Poster presented at the IUPAC Congress/General Assembly, July 2001 [available at: http://www.iupac.org/projects/posters01/vessman01.pdf].

Vestal M, Wahraftig AL, Johnston WH (1962), *J. Chem. Phys.* **37**, 1276.

Vestal M L, Futrell JH (1974), *Chem. Phys. Lett.* **28**, 559.

Vestal M (1983), in *Ion Formation from Organic Solids* (Ed. Benninghoven A), Springer Series in Chemical Physics **25**, 246, Springer–Verlag, New York, USA.

Vestal ML, Juhasz P, Martin SA (1995), *Rapid Commun. Mass Spectrom.* **9**, 1044.

Vilkner T, Janasek D, Manz A (2004), *Anal. Chem.* **76**, 3373.

Vissers JPC, Claessens HA, Cramers, CA (1997), *J. Chromatogr. A* **779**, 1.

Vissers JPC (1999), *J. Chromatogr., A*, **856**, 117.

Viswanathan CT, Bansal S, Booth B *et al.* (2007), *Pharm. Res.* **24**, 1962.

Valaskovic GA, Utley L, Lee MS, Wu J-T (2006), *Rapid Commun. Mass Spectrom.* **20**, 1087.

Vallance H, Applegarth D (1994), *Clin. Biochem.* **27**, 183.

Venter A, Cooks RG (2007), *Anal. Chem.* **79**, 6398.

Volmer DA, Hui JPM (1998), *Rapid Commun. Mass Spectrom.* **12**, 123.

Volmer DA, Niedziella S (2000), *Rapid Commun. Mass Spectrom.* **14**, 2143.

Volmer DA, Brombacher S, Whitehead B (2002), *Rapid Commun. Mass Spectrom.* **16**, 2298.

Von Busch F, Paul W (1961), *Z. Physik.* **164**, 581; *ibid.* 588.

von Lieshout M, von Deursen M, Derks R, Janssen H-G (1999), *J. Microcol. Sep.* **11**, 155.

Vonnegut K (1963), *Cat's Cradle*, Bantam Doubleday Dell Publishing Group, Inc., New York, USA.

W

Wadell DS, McKinnon HS, Chittim BG *et al.* (1987), *Biomed. Env. Mass Spectrom.* **14**, 457.

Wahlich JC, Carr GP (1990), *J. Pharm. Biomed. Anal.* **8**, 619.

Wang GH, Aberth W, Falick AM (1986), *Int. J. Mass Spectrom. Ion Proc.* **69**, 233.

Wang C, Oleschuk R, Ouchen F *et al.* (2000), *Rapid Commun. Mass Spectrom.* **14**, 1377.

Wang H, Liu W, Guan Y, Majors RE (2004), *LC/GC* **22**, 16.

Wang G, Hsieh Y, Korfmacher WA (2005), *Anal. Chem.* **77**, 541.

Wang S, Cyronak M, Yang E (2007), *J. Pharm. Biomed. Anal.* **43**, 701.

Ward TJ, Farris AB III (2001), *J. Chromatogr. A* **906**, 73.

Washburn MP, Wolters D,Yates JR III (2001), *Nature Biotechnol.* **19**, 242.

Watt AP, Morrison D, Locker KL, Evans DC (2000), *Anal. Chem.* **72**, 979.

Weber-Grabau M, Kelley PE, Bradshaw SC, Hoekman DJ (1988), *Proceedings of the 36th ASMS Conference on Mass Spectrometry and Allied Topics*, San Francisco, CA, USA, pp. 1106–1107.

Weckwerth W, Willmitzer L, Fiehn O (2000), *Rapid Commun. Mass Spectrom.* **14**, 1677.

Wegmann A (1978), *Anal. Chem.* **50**, 830.

Weickhardt C, Moritz F, Grotemeyer J (1996), *Mass Spectrom. Revs.* **15**, 139.

Weinberg HS, Krasner SW, Richardson SD, Thruston AD Jr (2002), *The Occurrence of Disinfection By-Products (DBPs) of Health Concern in Drinking Water: Results of a Nationwide DBP Occurrence Study*, EPA/600/R02/068 (US Environmental Protection Agency) [freely available at: http://www.epa.gov/athens/publications/reports/EPA_600_R02_068.pdf].

Weinkauf R, Walter K, Weickhardt C *et al.* (1989), *Z. Naturforsch. A* **44**, 1219.

Wells DA (Ed., 2003), *High Throughput Bioanalytical Sample Preparation*, Vol. 5 of *Progress in Pharmaceutical and Biomedical Analysis*, Elsevier, Amsterdam, The Netherlands.

Wells RJ, Cheung J, Hook JM (2004), *Accred. Qual. Assur.* **9**, 450.

Welter E, Neidhart B (1997), *Fresenius J. Anal. Chem.* **357**, 345.

Weng N, Lee JW, Jiang X *et al.* (1999), *J. Chromatogr. B* **735**, 255.

Weng N, Jiang X, Newland K *et al.* (2000), *J. Pharm. Biomed. Anal.* **23**, 697.

Weng N, Shou WZ, Addison T *et al.* (2002), *Rapid Commun. Mass Spectrom.* **16**, 1965.

Weng N, Zhou W, Song Q, Zhou S (2004), *Rapid Commun. Mass Spectrom.* **18**, 2963.

Wenzl T, de la Calle MB, Anklam E (2003), *Food Addit. Contam.* **20**, 885.

White RE, Manitpisitkul P (2001), *Drug Metab. Dispos.* **29**, 957.

Whitehouse CM, Dreyer RN, Yamashita M, Fenn JB (1985), *Anal. Chem.* **57**, 675.

Whitesides GM (2006), *Nature* **442**, 368.

Whiting TC, Liu RH, Chanf W-T, Bodapati MR (2001), *J. Anal. Toxicol.* **25**, 179.

WHO (World Health Organization, 2004), *Disinfectants and Disinfectant Byproducts*, Environmental Health Criteria 216 (Web Version) [freely available at: http://www.who.int/ipcs/publications/ehc/ehc_216/en/].

WHO (World Health Organization, 2006), *Guidelines for drinking-water quality incorporating first addendum. Vol. 1, Recommendations, 3rd Edn (Electronic version for the Web)* [freely available at: http://www.who.int/water_sanitation_health/dwq/gdwq3rev/en/index.html].

Wien W (1898), *Verhanal. Phys. Ges.* **17**.

Wiener MC, Sachs JR, Deyanova EG, Yates NA (2004), *Anal. Chem.* **76**, 6085.

Wiley WC, McLaren IH (1955), *Rev. Sci. Instrum.* **26**, 1150.

Williams TW, Salin ED (1988), *Anal. Chem.* **60**, 727.

Williams RR (1991), *Anal. Chem.* **63**, 1638.

Williams JP, Patel VJ, Holland R, Scrivens JH (2006), *Rapid Commun. Mass Spectrom.* **20**, 1447.

Willoughby RC, Browner RF (1984), *Anal. Chem.* **56**, 2626.

Wilm M, Mann M (1994), *Int. J. Mass Spectrom. Ion Proc.* **136**, 167.

Wilm M, Mann M (1996), *Anal. Chem.* **68**, 1.

Wilson MD, Rocke DM, Durbin B, Kahn HD (2004), *Anal. Chim. Acta* **509**, 197.

Windig W, Phalp JM, Payne AW (1996), *Anal. Chem.* **68**, 3602.

Winkler PC, Perkins DD, Williams DK, Browner RF (1988), *Anal. Chem.* **60**, 489.

Wise SA, Poster PL, Kucklick JR, *et al.* (2006), *Anal. Bioanal. Chem.* **386**, 1153.

Wiza JL (1979), *Nucl. Instrum. Methods* **162**, 587.

Wolf C, Pirkle WH (2002), *Tetrahedron* **58**, 3597.

Wolff J-C, Fuentes TR, Taylor J (2003), *Rapid Commun. Mass Spectrom.* **17**, 1216.

Wolschin F, Lehmann U, Glinski M, Weckwerth W (2003), *Rapid Commun. Mass Spectrom.* **19**, 3626.

Wong SF, Meng CK, Fenn JB (1988), *J. Phys. Chem.* **92**, 546.

Wood DC, Miller JM, Christ I (2004), *LC/GC* **22**, 516.

Wright JM, Schwartz J, Vartiainen T *et al.* (2002), *Environ. Health Perspect.* **110**, 157.

Wu JT, Zeng H, Deng Y, Unger SE (2001), *Rapid Commun. Mass Spectrom.* **15**, 1113.

Wu CC, MacCoss MJ, Howell KE *et al.* (2004), *Anal. Chem.* **76**, 4951.

Wuerker RF, Shelton H, Langmuir RV (1959), *J. Appl. Phys.* **30**, 342.

Wulff G, Sarhan A (1972), *Angew. Chem.* **84**, 364

X

Xia YQ, Whigan DB, Jemal M (1999), *Rapid Commun. Mass Spectrom.* **13**, 1611.

Xia Y-Q, Hop CECA, Liu DQ *et al.* (2001), *Rapid Commun. Mass Spectrom.* **15**, 2135.

Xia Y-Q, Miller JD, Bakhtiar R *et al.* (2003), *Rapid Commun. Mass Spectrom.* **17**, 1137.

Xia Y-Q, Bakhtiar R, Franklin RB (2003a), *J. Chromatogr. B* **788**, 317.

Xia Y-Q, Patel S, Bakhtiar R *et al.* (2005), *J. Am. Soc. Mass Spectrom.* **16**, 417.

Xia Y-Q, Jemal M, Zheng N, Shen X (2006), *Rapid Commun. Mass Spectrom* **20**, 1831.

Xie R, Oleschuk R (2005), *Electrophoresis* **26**, 4225.

Xu Y (1996), *Chem. Educator* **1**, 1. Available at: http://www.rit.edu/~pac8612/Biochemistry/505(705)/pdf/CE_tutorial.pdf

Xu R, Nemes C, Jenkins KM *et al.* (2002), *J. Am. Soc. Mass Spectrom.* **13**, 155.

Xu X, Lan J, Korfmacher WA (2005), *Anal. Chem.* **77**, 389A.

Y

Yacko RM (2007), *The Field of Homogenizing* [freely available at the website of PRO Scientific Inc.: http://www.proscientific.com/Homogenizing.shtml].

Yamashita M, Fenn JB (1984), *J. Phys. Chem.* **88**, 4451.

Yan Z, Caldwell GW, Jones WJ, Masucci JA (2003), *Rapid Commun. Mass Spectrom.* **17** 1433.

Yan Z, Lu C, Wu J-T *et al.* (2005), *Rapid Commun. Mass Spectrom.* **19**, 1191.

Yan W, Byrd GD, Ogden MW (2007), *J. Lipid Res.* **48**, 1607.

Yang FJ (1982), *J. Chromatogr.* **236**, 265.

Yang L, Amad M, Winnik WM *et al.* (2002), *Rapid Commun. Mass Spectrom.* **16**, 2060.

Yang Y, Kameoka J, Wachs T *et al.* (2004), *Anal. Chem.* 76, 2568.

Yang S, Cha J, Carlson K (2006), *Anal. Bioanal. Chem.* **385**, 623.

Yashima E (2001), *J. Chromatogr. A* **906**, 105.

Yao X, Freas A, Ramirez J *et al.* (2001), *Anal. Chem.* **73**, 2836.

Ye L, Mosbach K (2000), *Proceedings of the Euro-Residue IV Conference on Residues of Veterinary Drugs in Food*, p. 162 [freely available at: http://www.euroresidue. nl/ER_IV/ Key%20lectures/idx.htm].

Yergey AL, Edmonds CG, Lewis IAS, Vestal ML (1990), *Liquid Cromatography/Mass Spectrometry*, Plenum Press, New York, USA.

Yin H, Killeen K, Brennen R *et al.* (2005), *Anal. Chem.* **77**, 527.

York D (1966), *Can. J. Phys.* **44**, 1079.

Yost RA, Enke CG (1978), *J. Am. Chem. Soc.* **100**, 2274.

Yost RA, Enke CG, McGilvery DC *et al.* (1979), *Int. J. Mass Spectrom. Ion Phys.* **30**, 127.

Yost RA, Murphy JP III (2000), *Rapid Commun. Mass Spectrom.* **14**, 270.

Youden WJ (1947), *Anal. Chem.* **19**, 946.

Youden WJ (1960), *Anal. Chem.* **32**, 23A.

Yu K, Little D, Plumb R, Smith B (2006), *Rapid Commun. Mass Spectrom.* **20**, 544.

Yu C-P, Chen CL, Gorycki FL, Neiss TG (2007), *Rapid Commun. Mass Spectrom.* **21**, 497.

Z

Zaikin VG, Halket JM (2003), *Eur. J. Mass Spectrom.* **9**, 421.

Zaikin VG, Halket JM (2004), *Eur. J. Mass Spectrom.* **10**, 555.

Zaikin VG, Halket JM (2005), *Eur. J. Mass Spectrom.* **11**, 611.

Zeng W, Musson DG, Fisher AL, Wang AQ (2006), *Rapid Commun. Mass Spectrom.* **20**, 635.

Zhang H, Heinig K, Henion J (2000), *J. Mass Spectrom.* **35**, 423.

Zhang N, Fountain ST, Bi H, Rossi DT (2000a), *Anal. Chem.* **72**, 800.

Zhang H, Henion J (2001), *J. Chromatogr. B* **757**, 151.

Zhang H, Yan W, Aebersold R (2004), *Current Opinion Chem. Biol.* **8**, 66.

Zhang Yu, Zhang G, Zhang Y (2005), *J. Chromatogr. A* **1075**, 1.

Zhou H, Ranish JA, Watts JD, Aebersold R (2002), *Nature Biotech.* **20**, 512.

Zhou L, Yue B, Dearden DV *et al.* (2003), *Anal. Chem.* **75**, 5978.

Zhou S, Zhou H, Larson M *et al.* (2005), *Rapid Commun. Mass Spectrom.* **19**, 2144.

Zhou S, Song Q, Tang Y, Weng N (2005a), *Curr. Pharm. Anal.* **1**, 3.

Zhu Y, Wong PHS, Cregor M *et al.* (2000), *Rapid Commun. Mass Spectrom.* **14**, 1695.

Zlotorzynski A (1995), *Critical Rev. Anal. Chem.* **25**, 43.

Zorn ME, Gibbons RD, Sonzogni WC (1997), *Anal. Chem.* **69**, 3069.

Zorn ME, Gibbons RD, Sonzogni WC (1999), *Environ. Sci. Technol.* **33**, 2291.

Zubarev RA, Kelleher NL, McLafferty FW (1998), *J. Am. Soc. Chem.* **120**, 3265.

Zubarev RA, Haselmann KF, Budnik B *et al.* (2002), *Eur. J. Mass Spectrom.* **8**, 337.

Zubarev RA (2003), *Mass Spectrom. Rev.* **22**, 57.

Zwiener C, Kronberg L (2001), *Fresenius J. Anal. Chem.* **371**, 591.

Zworkin VK, Morton GA, Malter L (1936), *Proc. IRE* **24**, 351.

Zworkin VK, Rajchman JA (1939), *Proc. IRE* **27**, 559.

Index

Printed and bound by CPI Group (UK) Ltd, Croydon, CR0 4YY

16/04/2025

14658561-0004